International Scientific Conference on Radiation-Thermal Effects and Processes in Inorganic Materials 2014

IOP Conference Series: Materials Science and Engineering
Volume 81

Tomsk, Russia
3 - 8 November 2014

ISBN: 978-1-5108-0295-7
ISSN: 1757-8981

Printed from e-media with permission by:

Curran Associates, Inc.
57 Morehouse Lane
Red Hook, NY 12571

Some format issues inherent in the e-media version may also appear in this print version.

Printed by Curran Associates, Inc. (2015)

For permission requests, please contact the Institute of Physics
at the address below.

Institute of Physics
Dirac House, Temple Back
Bristol BS1 6BE UK

Phone: 44 1 17 929 7481
Fax: 44 1 17 920 0979

techtracking@iop.org

Additional copies of this publication are available from:

Curran Associates, Inc.
57 Morehouse Lane
Red Hook, NY 12571 USA
Phone: 845-758-0400
Fax: 845-758-2634
Email: curran@proceedings.com
Web: www.proceedings.com

TABLE OF CONTENTS

IOP Conf. Series: Materials Science and Engineering **81** (2015) 011001 doi:10.1088/1757-899X/81/1/011001

INTRODUCTION

The International Scientific Conference on "Radiation-Thermal Effects and Processes in Inorganic Materials" is a traditional representative forum devoted to the discussion of fundamental problems of radiation physics and its technical applications. The first nine conferences were held four times in Tomsk, then in Ulan-Ude (Russia), Bishkek (Kyrgyzstan), Tashkent (Uzbekistan), Sharm El Sheikh (Egypt), and the island of Cyprus. The tenth conference was held in Tomsk, Russia.

The program of the Conference covers a wide range of technical areas and modern aspects of radiation physics, its applications and related matters. Topics of interest include, but are not limited to:

- Physical and chemical phenomena in inorganic materials in radiation, electrical and thermal fields;
- Research methods and equipment modification states and properties of materials;
- Technologies and equipment for their implementation;
- The use of radiation-thermal processes in nanotechnology;
- Adjacent to the main theme of the conference issues

The conference was attended by leading scientists from countries near and far abroad who work in the field of radiation physics of solid state and of radiation material science. The School-Conference of Young Scientists was held during the conference. The event was held with the financial support of the Russian Foundation for Basic Research, projects № 14-38-10210 and № 14-02-20376.

THE ORGANIZING COMMITTEE OF THE CONFERENCE RTEP-2014:

A. Surzhikov, Chairman of Scientific Program Committee, Deputy Director on Research, Professor, Tomsk, Russia
G. Remnev, co-chair of Scientific Program Committee, TPU, Professor, Tomsk, Russia
V. Krivobokov, co-chair of Scientific Program Committee, TPU, Professor, Tomsk, Russia

S. Gyngazov, Academic Secretary of the Conference, TPU, Tomsk, Russia

E. Lysenko, Secretary of the Conference, TPU, Tomsk, Russia

V. Lisitcin, Section Chairman, Professor, TPU, Tomsk, Russia

N. Koval, Section Chairman, Professor, TPU, Tomsk, Russia

Yu. Yakovlev, Section Chairman, Professor, TPU, Tomsk, Russia

Yu. Ivanov, Section Chairman, Professor, TSU, Tomsk, Russia

Yu. Tyurin, Section Chairman, Professor, TPU, Tomsk, Russia

A. Potekaev, Section Chairman, Professor, TSU, Tomsk, Russia

E. Naiden, Section Chairman, Professor, TSU, Tomsk, Russia

B. Aduev, Section Chairman, Professor, KSU, Kemerovo Tomsk, Russia

PROGRAM ORGANIZING COMMITTEE

K. Baktybekov, Professor, ENU, Astana, Kazakhstan
R. Bayazitov, Professor, KPhTI, Kazan, Russia
G. Bondarenko, Professor, MIEM, Moscow, Russia
V. Brudnyi, Professor, TSU, Tomsk, Russia
B. Goshchitskii, Professor, corresponding member of RAS, IPhM, Ekaterinburg, Russia
E. Gorkunov, Professor, academician of RAS, Institute of Engineering, Ekaterinburg, Russia
B. Kalin, Professor, MEPHI, Moscow, Russia
M. Kidibaev, Professor, Corresponding Member of the Kyrgyz Academy of Sciences, IPhTPMS, Bishkek, Kyrgyzstan
N. Kolin, Professor, IAE, Obninsk Russia
V. Kortov, Professor, UFU, Ekaterinburg, Russia
A. Kupchishin, Professor, KNU, Almaty, Kazakhstan
Yu. Gulyaev, Professor, academician of RAS, KIREE, Moscow, Russia
E. Martynovych, Professor, ISU, Irkutsk, Russia
V. Ovchinnikov, Professor, IEP, Ekaterinburg, Russia
N. Ratakhin, Professor, academician of RAS, IHCE, Tomsk, Russia
A. Sigov, Professor, academician of RAS, MSUREA, Moscow, Russia
P. Selishchev, Professor, KNU, Kiev, Ukraine

IOP Conf. Series: Materials Science and Engineering **81** (2015) 011003 doi:10.1088/1757-899X/81/1/011003

List of delegates of the conference RTEP-2014

1. E Krasikov E-mail: ekrasikov@mail.ru
2. V A Zhuravlev E-mail: ptica@mail.tsu.ru
3. A I Kupchishin E-mail: ankupchishin@mail.ru
4. F V Konusov E-mail: konusov@hvd.tpu.ru
5. A V Gradoboev E-mail: gradoboev1@mail.ru
6. A. I. Potekaev E-mail: potekaev@mail.ru
7. A V Malyshev E-mail: malyshev@tpu.ru
8. Yu F Ivanov E-mail: yufi55@mail.ru
9. Yu. Ya. Gafner E-mail: ygafner@khsu.ru
10. S V Plotnikov Email: SPlotnikov@ektu.kz
11. I. V. Chepkasov E-mail: ilya_chepkasov@mail.ru
12. S A Ghyngazov E-mail: ghyngazov@tpu.ru
13. N A Shaburova E-mail: shaburovana@susu.ac.ru
14. E B Kashkarov E-mail: egor_kashkarov@mail.ru
15. N N Nikitenkov E-mail: nikitenkov@tpu.ru
16. O. B. Okhrimenko E-mail: olga@isp.kiev.ua
17. V Lisitsyn E-mail: lisitsyn@tpu.ru
18. A D Pogrebnyak E-mail: alexp@i.ua
19. F. F Umarov E-mail: umarov@mail.ru
20. A V Khaneft E-mail: avkhaneft@mail.ru
21. L A Lisitsyna E-mail: lisitsyna@mail.ru
22. T V Fursa E-mail: fursa@tpu.ru
23. T S Frangulyan E-mail: fts@tpu.ru
24. G F Karlova karlova@mail.ru
25. A A Demikhova demikhova@tpu.ru
26. S Kh Djanibekova E-mail: nanocentre@kerc.msk.ru
27. S L Gafner E-mail: ygafner@khsu.ru
28. A I Blesman E-mail: blesm@mail.ru
29. G A Melnikov E-mail: melnikovga@mail.ru
30. Z I Popov E-mail: zipopov@mail.ru
31. V P Poltavtseva E-mail: poltavtseva@inp.kz
32. O Kazmina E-mail: kazmina@tpu.ru
33. I V Khromushin E-mail: khromushin@mail.ru
34. B P Aduev E-mail: lesinko-iuxm@yandex.ru
35. V Perminov E-mail: perminov@tpu.ru
36. V I Jordan E-mail: jordan@phys.asu.ru
37. L N Orlikov E-mail: oln4@yandex.ru
38. N Kh Ibrayev E-mail: niazibraev@mail.ru.
39. V V Shagaev E-mail: shagaev_vv@rambler.ru
40. B R Ilyassov E-mail: baurdinho@mail.ru
41. G A Raspopova E-mail: raspopova@imp.uran.ru

42. S S Perepelkin E-mail: sperepelkin@kipt.kharkov.ua
43. Ye Khuanbay E-mail: esen_d@mail.ru
44. S E Sakipova E-mail: sesaule@mail.ru
45. S N Yanin E-mail: yanin@tpu.ru
46. V I Torganchuk E-mail: torganv@mail.ru
47. V A Meshcheriakov E-mail: mva@mail.tsu.ru
48. P I Fedotov E-mail: pif@tpu.ru
49. I V Lopatin E-mail: lopatin@opee.hcei.tsc.ru
50. S P Zhuravkov E-mail: zhursp@yandex.ru
51. K M Mukashev E-mail: mukash_kanat@mail.ru
52. V P Surzhikov E-mail: iva1954@ mail.ru
53. S Bordunov E-mail: bvv@academ.tsc.ru
54. M A Fedotov E-mail: mikle_fed@mail.ru;
55. O A Gorbunova E-mail: olg-gorbunova@yandex.ru
56. O A Golovanova E-mail: golovanoa2000@mail.ru
57. V V Denisov E-mail: volodyadenisov@yandex.ru
58. A S Akimov E-mail: zerobox70@mail.ru
59. A D Aleutdinov E-mail: a_alet@mail.ru
60. I Plotnikova E-mail: inna@tpu.ru
61. A I Filkov E-mail: aifilkov@gmail.com
62. O B Nazarenko E-mail: obnaz@mail.ru
63. I Bolotina E-mail: bolotina@tpu.ru
64. A Lider E-mail: lider@tpu.ru
65. D Sednev E-mail: sednev@tpu.ru
66. M E Fedorov E-mail: evgeny_fyodorov@list.ru
67. K I Machekhina E-mail: mauthksu@yandex.ru
68. L N Shiyan E-mail: lshiyan@rambler.ru
69. E N Gryaznova E-mail: t_elena@mail2000.ru
70. M Korovkin E-mail: mvk@tpu.ru
71. V M Le ya808@ya.ru
72. Y A Salchak E-mail: yasalchak@gmail.com
73. E N Lysenko E-mail: lysenkoen@tpu.ru
74. S P Osipov E-mail: osip1809@rambler.ru
75. V V Redko E-mail: nadushasns@tpu.ru
76. V P Vavilov vavilov@tpu.ru
77. Yu V. Borodin E-mail: uryborodin@tpu.ru
78. Y. V Shefer E-mail: yuri@shefer.ru
79. V A Vlasov E-mail: vlvitan75@mail.ru
80. A V Chernyavskii E-mail: chernyavski@tpu.ru
81. A N Vtorushina E-mail: anl@tpu.ru
82. I F Nam E-mail: irenanam@gmail.com
83. E V Larionova E-mail: evl@tpu.ru
84. A P Leonov E-mail: leonov_ap@tpu.ru, ultratone@rambler.ru
85. O B Nazarenko E-mail: obnaz@mail.ru

86. T G Nesterenko E-mail: ntg@tpu.ru
87. A Okhorzina E-mail: ohra.avit@gmail.com
88. V A Perminov E-mail: p_valer@mail.ru
89. T S Rein E-mail: tsrein@mail.ru
90. A O Chulkov E-mail: chulkovao@tpu.ru
91. D V Shashev E-mail: dshashev@mail.ru
92. A P Shestak E-mail: shestak@tpu.ru
93. Yu V Shulgina E-mail: shulgina@tpu.ru
94. E A Vasendina E-mail: vasendina@tpu.ru
95. A A Bespalko E-mail: besko48@tpu.ru
96. A A Bombizov E-mail: unclelab@gmail.com
97. S V Gorbachev E-mail: spp03@sibmail.com
98. E Gromakov E-mail: Gromakov@tpu.ru
99. V N Izvekov E-mail: izvekovvn@tpu.ru
100. O S Kyrmakova E-mail: olia_917@mail.ru
101. F V Savrasov E-mail: savrasov@tpu.ru
102. D P Starikov E-mail: dstarikov@me.com
103. A V Yurchenko E-mail: slawa_v@mail.ru
104. Yu ANikolaev E-mail: barton.kz@mail.ru
105. F A Gubarev E-mail: gubarevfa@tpu.ru
106. A I Soldatov E-mail: asoldatof@mail.ru
107. D A Afanasyev E-mail: a_d_afanasyev@mail.ru
108. A Maslovskaya E-mail: maslovskayaag@mail.ru
109. I A Fedorov E-mail: ifedorov@kemsu.ru
110. T M Serikov E-mail: serikov-timur@mail.ru
111. U V Ancharova E-mail: ancharova@gmail.com
112. M V Korobeinikov E-mail: mkor@inp.nsk.su

Tributes to colleagues

Editorial board of the collection (journal) thanks the organizers of the **International Scientific Conference "Radiation-Thermal Effects and Processes in Inorganic Materials"**:

- A. Surzhikov, 1th Section Chairman, Professor, TSU, Tomsk, Russi
- B. Aduev, 2th Section co-Chairman, KSU, Professor, Kemerovo, Russia
- E. Gorkunov, 3th Section Chairman, academician of RAS Institute of Engineering, Ekaterinburg, Russia
- V. Ovchinnikov, 4th Section Chairman, Professor, IEP, Ekaterinburg, Russia
- A. Potekaev, 5th Section Chairman, Professor, TSU, Tomsk, Russia
- Yu. Tyurin, 6th Section Chairman, Professor,TPU, Tomsk, Russia
- S. Gyngazov, Academic Secretary of the Conference, TPU, Tomsk, Russia
- E. Lysenko, Secretary of the Conference, TPU, Tomsk, Russia
- I. Sharkov, Department for development of publication activities, TPU, Tomsk, Russia

Conference photos

Peer review statement

All papers published in this volume of *IOP Conference Series: Materials Science and Engineering* have been peer reviewed through processes administered by the proceedings Editors. Reviews were conducted by expert referees to the professional and scientific standards expected of a proceedings journal published by IOP Publishing.

Change Spectrum Characteristics Modification of Films Deposited by Magnetron Sputtering with the Assistance of Argon Ions Beam

S Umnov[1] **and O Asainov**[1]
[1] Physical Technical Institute,
Tomsk Polytechnic University,
634050, Lenina 2, bdg. 11, Tomsk, Russia

E-mail: usp@tpu.ru

Abstract. Thin aluminum films were prepared using the method of magnetron sputtering with and without argon ion beam assistance. The influence of argon ion beam on the reflectivity in the UV range and the structure of aluminum films was studied. The structure of the films was studied by transmission electron microscopy (TEM), X-ray diffractometry (XRD) and atomic-force microscope (AFM). The study has shown that the films deposed with the assistance of the argon ion beam have more significant microstresses associated with an increase of crystallites microstructure defects as compared to the films deposed without ion assistance. Comparison of the measured reflectivity of aluminum films deposed without and with the assistance of the ion beam has shown that the films characterized by a higher level of microstructure def ects have increased reflectivity in the UV range. The studies suggest that the defects of thin aluminum films crystal structure influence its optical properties.

1. Introduction

High-reflectance Al thin films have been widely used in optics and optical applications. Optical characteristics of these coatings have been studied quite extensively. The effects of such parameters as film thickness and deposition conditions (including pressure, deposition rate, substrate temperature, and residual gasses) have been previously examined in several papers [1-4]. Additionally, as described in the studies [5, 8], optical properties of the films can vary depending on the deposition method (magnetron sputtering, electron-beam evaporation, thermal deposition, laser deposition, etc.), which is connected with the modification of their structure. In turn, the structure of a film coating is highly determined by the initial stage of the crystallites' nuclei formation and growth processes [8, 9]. Upon formation the crystallite structure may change as a result of plastic deformation, exposure to neutral or charged particle radiation, or electromagnetic radiation. Thus, the application of an assist ion beam when depositing film coatings leads to their structure modification, which in turn results in the modification of their optical properties [6, 7, 10,11].

The results obtained in the above mentioned studies demonstrate the potential of using charged particle beams when depositing film coatings in order to modify their structure with a view to

[1] E-mail address: usp@tpu.ru .

changing their optical characteristics. It is also worth noting that the relationship between the structural changes and optical parameters has been relatively understudied due to the variety of the accompanying processes.

This paper investigates the effect of the concomitant Ar ion irradiation of the Al films deposited by magnetron sputtering on the modification of their reflectance in the UV region.

The relationship between the reflectance and the crystalline structure is not always obvious. Structure modification induced by any reason results in the modification of the solid-state plasma parameters. These modifications change the nature of the electromagnetic field interaction with matter. In our study, the films' reflectance was measured by a photometer. The TEM, XRD, and AFM methods were used to examine the Al films' structure and topology. The reflection factors of the Al films produced with and without Ar ion-beam assist were compared.

2. Experimental details

Al thin films were deposited by the magnetron sputtering process with a flat target made of high-purity aluminium (99.99 wt.%). The films were deposited on glass substrates previously cleaned with a chalk suspension in isopropyl alcohol. Argon with 99.996 wt.% was used as the sputtering gas. The gas flow was controlled by an electronic flow meter; the gas pressure was measured by a thermocouple-ionization vacuum gage. Prior to the deposition process, the aluminium target was cleaned by scattering during 4 min. All the samples were prepared at a pressure in the vacuum chamber of $8x10^{-4}$ Pa before sputtering and $1.4x10^{-1}$ Pa during sputtering.

On one part of the samples the Al films were deposited by using only magnetron sputtering while on the other part of the samples the films were deposited with the use of the Ar ion-beam assist at the initial stage of the deposition process. The ion beam was perpendicular to the sample surface. The energy characteristics of the beam are given in Figure. 1.

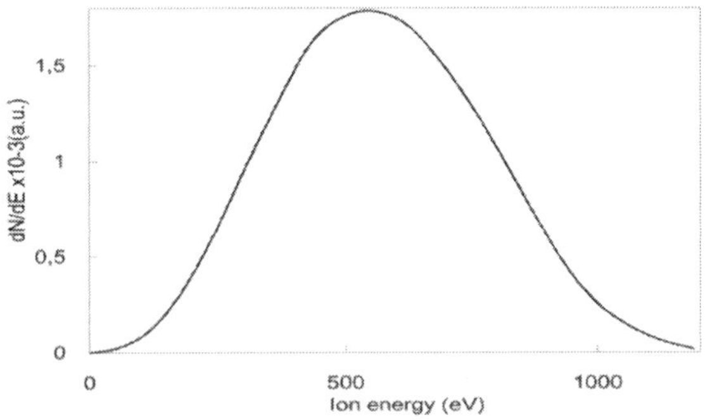

Figure 1. Ar ions distribution in the beam with regard to the energies

Upon assisting the ion current was 320 mA. In both cases the thickness of the deposited Al films was approximately 85 nm; the deposition rate was 7.5 nm/s.

The reflectance spectra of the grown films were measured by the SF-256 spectrophotometer (LOMO, Russia) with the use of a specular reflectance attachment. Five samples of each type were prepared for the investigation; each sample was prepared in a separate sputtering process under similar deposition conditions.

The TEM, XRD and AFM measurements were carried out by using the CM 12 Transmission Electron Microscope (PHILIPS, Netherlands), XRD6000 X-ray Diffractometer (SHIMADZU, Japan), and Solver HV Atomic Force Microscope (NT-MDT, Russia) correspondently.

3. Results and discussion

The samples were held in air for more than 120 hours. Then, the Al films optical characteristics were measured. The results of the Al films' reflection factor measurements in the 200-240 nm spectrum range for the samples obtained with and without ion assist are given in Figure. 2. It is seen that the reflection factor of the Al films produced by ion-assisted deposition is 4 – 5% higher than that of the films obtained without ion assist.

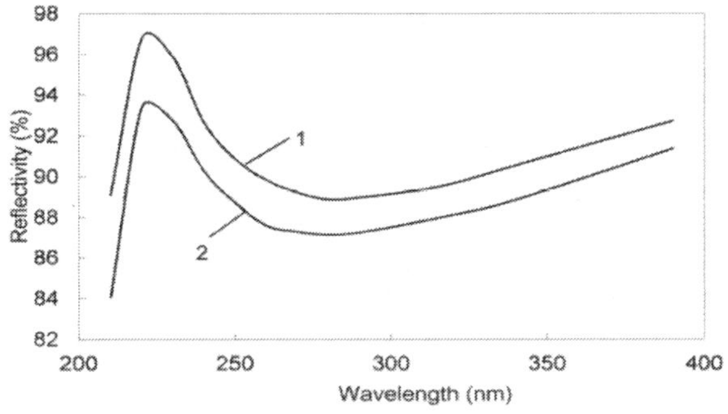

Figure 2. Al films reflectivity: 1 – with assistance; 2 – without assistance

The reflection factor of a coating is known to depend on a number of factors: topography, crystallite size, and the level of the crystal lattice perfection. To determine the reason that accounted for the increase in the reflection factor we performed the TEM, AFM, and XDR analyses of the obtained films.

The conducted TEM analysis showed that the Al films had similar nanocrystallite structure. The size of the Al crystallites was 15 – 470 nm for the films deposited without ion assist and 70 – 300 nm for the films deposited with ion assist.

In the case of ion assist, the change in the size of the crystallites at upper and lower borders is caused by the impact of the Ar ions on the processes that occur upon the condensation of aluminium on the deposited surface area. Under the impact of an ion beam the formed islands get partially or fully sputtered on the deposited surface, which leads to an increase in the adsorbed aluminium atoms' concentration. This process is schematically shown in Figure. 3.

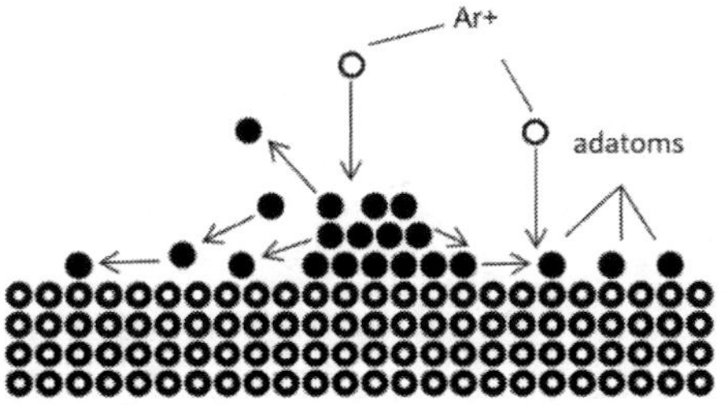

Figure 3. Scheme of Ar ion interaction with the surface

Large islands are sputtered only partially while s mall islands of a critical size get fully sputtered. It is also worth noting that the ion-assisted processes are influenced not only by the ions' energy but also by the density of the ion current [7]. The additional formation of the adsorbed atoms contributes to the growth of the small, unsputtered islands. Thus, as a result of the competing processes, islands with a less size spread are formed on the surface. The result of the ion assist application is the increased crystallite size at the lower border and reduced crystallite size at the upper border. It coincides with the experimental results described in [7].

To determine the microrelief of the deposited film surface we have performed the AFM studies. Figure.4 shows the AFM images of the Al films deposited with and without ion assist. As seen in this figure, the surface roughness of the films produced by the ion-assisted deposition slightly increases. Indeed, the root-mean-square surface roughness of the films deposited without ion assist is 2.289 nm, with ion assist – 2.397 nm.

The conducted TEM and AFM examination of the Al films produced by ion-assisted deposition has shown that the resultant changes in the crystallite size and roughness are insignificant. Hence, the geometrical factor cannot be the reason for the reflectance growth.

Another reason for the reflectance modification can be the imperfection level of the crystalline structure since real crystals are always characterized by the presence of defects. The defect concentration depends on the conditions of the crystalline structure formation. Under the impact

IOP Conf. Series: Materials Science and Engineering **81** (2015) 012001 doi:10.1088/1757-899X/81/1/012001

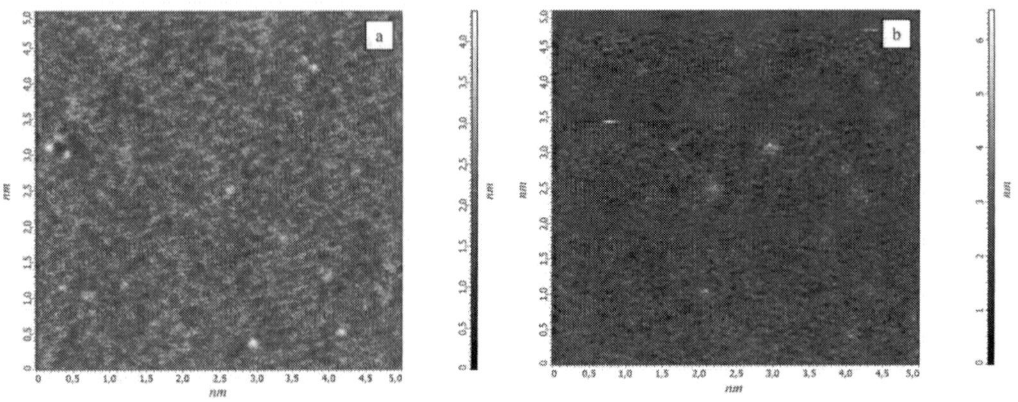

Figure 4. AFM images: 1 – without assistance; 2 – with assistance

of the assist ions of a relatively high energy, point defects (vacancies and dislocated atoms) are formed in the growing film. The defects of given type have a considerable effect almost on all the properties of solid bodies including optical characteristics. Point defects in crystals provide a slight lattice distortion, which, in turn, results in the appearance of internal microstress. The microstress value is known to be proportional to the density of the defects. Consequently, the density of the defects in a film can be estimated by the microstress value.

We have conducted the XRD studies of the films obtained by various deposition methods. The diffraction patterns have been analysed by using the Williamson-Hall method. The measurement results are given in Figure 5 and Table 1. As seen from the table, the microstress value is more than 5 times higher in the films produced by the ion-assisted deposition. Hence, there is an increase in the concentration of the crystalline structure defects of the film coating.

Table 1. XDR measurement results

	scaling lattice, a, $x10^{-10}$, m	coherent scattering region (CSR), x 10^{-9} m	relative strain, x 10^{-3}	preferred orientation, March-Dollase (111)
without assistance	4.0550	15	1.5	0.5350
with assistance	4.0161	62	8.5	0.7517

A point defect interacts with all the conduction electrons. The electrons located in the defect's neighborhood are "bound" by it and do not take part in conductivity, i.e. the N_e electron concentration decreases. This leads to the decrease in the plasma frequency ω_p, electron collision rate Y, the parameter $Q = v_e w_p / c$ that characterize the influence on the absorption of the spatial dispersion effects of the conductivity, etc. Thus, the parameters of the solid-state plasma undergo changes. Since the reflection factor R(w) and the absorption factor A(w) are determined by these parameters, their modification leads to the change in the nature of the electromagnetic wave interaction with matter. In the UV region, the role of the forced oscillations of the crystal lattice bound electrons that intensely absorb light in the area close to the resonance frequency is increased. The reflection and absorption factors in this area are highly dependent on the frequency.

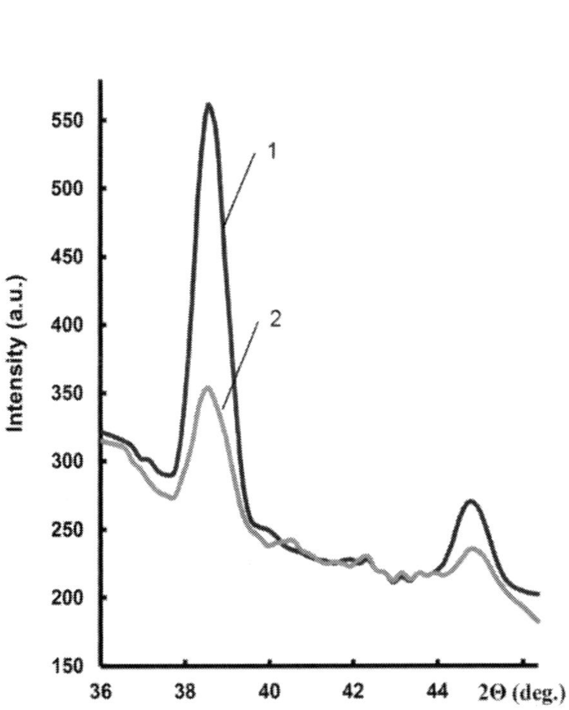

Fig.5 XRD diagram: 1 – without assistance; 2 – with assistance

The estimation values in our case are: w = 8.2x10^{15}, w_p = 5.7x10^{16}, Q = 2x10^{14} c-1, i.e. $w_p > w > Q$. Hence, the absorption factor is $A(w) \sim (w_p / w)^2$. It means that with such a relation between the incident wave frequency and the metal (aluminium) plasma parameters the absorption decreases.

Thus, an increase in the reflectance is related to a decrease in the losses connected with the absorption of electromagnetic radiation.

4. Conclusions

In this paper, we have investigated the consequences of using an assist Ar ion beam in the magnetron sputtering deposition of the Al films. The photometric studies have shown that the films deposited with ion assist are characterized by a 4-5% higher reflection factor in the 200-240 nm range compared to the films deposited without ion assist. The conducted TEM and AFM analyses of the films have revealed that the crystallite size and roughness change insignificantly. It means that the geometrical factor cannot be considered the reason for the enhanced reflectance. The X-ray diffraction (XRD) studies have shown that in the films produced by the ion-assisted deposition the value of microstress is 5 times higher than in the films deposited without ion assist. The microstress value is proportional to the concentration of the crystallites' point defects. In other words, the application of ion assist leads to an increase in the point defect concentration. Thus, an increase in the concentration of point defects in the deposited films results in their enhanced reflectance. It is connected with the change in the conditions of the electromagnetic wave interaction with matter.

References

[1] Hass G and Waylonis J E 1961 *JOSA* **51** pp 719-722
[2] Nguyen H V, An I and Collins R W 1992 *Phys. Rev. Lett.* **68** pp 994-997
[3] Monard H and Sabary F 1997 *Thin Solid Films* **310** pp 265-273
[4] Du H, Xiao J Q, Zou Y S *et al* 2006 *Opt. Mater.* **28** pp 944-949
[5] Her S C and Wang Y H 2010 *Appl. mechanics and materias* **121-126** pp 3930-3940
[6] Kim S H and Hwangbo C K 2005 *Thin Solid Films* **475** pp 155-159
[7] Ensinger W 1997 *Nuc. Instrum. and Methods Phys. Res. Sect.* B **127-128** pp 796-808
[8] Novotny M, Bulir J, Lancok J *et al* 2011 *Journal of Nanophotonics* **5** 051503
[9] Nguyen H V, An I and Collins R W 1993 *Phys. Rev.* B. **47** 3947
[10] Martin P, Netterfield R, Kinder T et al 1992 *Appl. Opt.* **31** pp 6734-6740
[11] Asainov O and Umnov S 2012 *Int. Forum on Strategic Technology (Tomsk: Russia)* vol 1 p 292

Radiation intensification of the reactor pressure vessels recovery by low temperature heat treatment (wet annealing)

E Krasikov
Laboratory Head, National Research Centre «Kurchatov Institute»,
1, Kurchatov Sq., Moscow, 123182, Russia.

E-mail: ekrasikov@mail.ru

Abstract. As a main barrier against radioactivity outlet reactor pressure vessel (RPV) is a key component in terms of NPP safety. Therefore present-day demands in RPV reliability enhance have to be met by all possible actions for RPV in-service embrittlement mitigation. Annealing treatment is known to be the effective measure to restore the RPV metal properties deteriorated by neutron irradiation.
There are two approaches to annealing. The first one is so-called «dry» high temperature (~475°C) annealing. It allows obtaining practically complete recovery, but requires the removal of the reactor core and internals. External heat source (furnace) is required to carry out RPV heat treatment.
The alternative approach is to anneal RPV at a maximum coolant temperature which can be obtained using the reactor core or primary circuit pumps while operating within the RPV design limits. This low temperature «wet» annealing, although it cannot be expected to produce complete recovery, is more attractive from the practical point of view especially in cases when the removal of the internals is impossible.

1. Introduction

Modern nuclear power engineering is based on LWR type plant reactors. As a main barrier against radioactivity outlet reactor pressure vessel is a key component in terms of NPP safety. Therefore present-day demands in RPV reliability enhance have to be met by all possible actions for RPV in-service embrittlement mitigation. Annealing treatment is known to be the effective measure

There are two approaches to annealing. The first one is so-called «dry» high temperature (~475°C) annealing. It allows obtaining practically complete recovery, but requires the removal of the reactor core and internals. External heat source (furnace) is required to carry out RPV heat treatment.

The alternative approach is to anneal RPV at a maximum coolant temperature which can be obtained using the reactor core or primary circuit pumps while operating within the RPV design limits. This low temperature «wet» annealing, although it cannot be expected to produce complete recovery, is more attractive from the practical point of view especially in cases when the removal of the internals is impossible.

As a rule there is no recovery effect up to annealing and irradiation temperature difference of 70°C. It is known, however, that along with radiation embrittlement neutron irradiation may mitigate the radiation damage in metals. Therefore we have tried to test the possibility to use the effect of radiation-induced ductilization in «wet» annealing technology by means of nuclear heat utilization as heat and neutron irradiation sources at once.

2. Prerequisites to the substantiation of RPV embrittlement controlling through wet annealing

Irradiated metal condition after annealing can be expressed in terms of recovery index (RI) by the following dependence: $RI=(TT_{irr}- TT_{ann})/(TT_{irr}- TT_{init})$,
where TT_{irr}, TT_{ann}, and TT_{init} are ductile-to-brittle transition temperatures (DBTT) of the metal in the irradiated, annealed and initial conditions respectively.

RI variation as a function of furnace annealing and irradiation temperatures (T_{ann} and T_{irr}=270°C) respectively) difference at a holding duration of 100 h for Russian 15Cr2MoV type RPV steel irradiated at the fast neutron (E≥0,5 MeV) fluxes from $5\times10^{10}sm^{-2}s^{-1}$ to $3\times10^{12}sm^{-2}s^{-1}$ in NPP reactors is presented in [1]: there are no recovery effect up to annealing and irradiation temperature difference of 70°C while T_{ann}=340°C (T_{irr}+70°C) is practically maximum accessible value for wet annealing technology.

The first RPVs «wet» annealing were done using nuclear heat (US Army SM-1A) or primary pump heat (Belgian BR-3). The annealing temperature in the former case was 293-300°C (72-79°C above the service temperature). The degree of recovery in this case was about 70%. In the BR-3 reactor the service temperature was 260°C and the vessel was annealed at 343°C. The recovery was estimated to beat least 50%. The planned annealing of the Yankee Rowe vessel at 343°C (83°C above the service temperature) was estimated to give a 45-55% recovery [2].

The "wet" annealing technique is easy to implement because there is no need to remove fuel from the RPV, but it can be effectively utilized only in reactors which have a relatively low service temperature. Due to limited recovery, «wet» annealing with primary water is not a practical solution for power reactors or it needs to be frequently repeated.

It is known, however, that along with radiation embrittlement neutron irradiation may mitigate radiation damages in metals [3]. Therefore we have tried to study the possibility to use this effect in «wet» annealing technology by means of nuclear heat utilization as heat and neutron irradiation sources at once.

3. Experimental procedures and test results

In support of the above-mentioned conception the 3-year duration reactor experiment on RPV steel was carried out.

Material. Metal used for the studies described here is RPV steel of the 15Cr3NiMoV-A type. Table 1 provides the chemical composition of the metal.

Table 1. RPV steel chemical composition (wt. %%).

C	P	Cu	Cr	Mo	Ni
0,14	0,009	0,08	2,00	0,90	1,15

Irradiation. Preliminary (primary) irradiation was carried out in the commercial power plant. Cylindrical billets of the steel were placed into the perforated stainless steel capsules and therefore were irradiated in direct contact with running primary water. Full size and sub size Charpy type specimens were manufactured from irradiated billets by means of electro discharge machining (EDM).

To examine «wet» annealing potentiality, experiment on effectiveness of the irradiated RPV steel extra irradiation was conducted. Pre-irradiated to $9\times10^{19}sm^{-2}$ (E>0,5MeV) at a nominal temperature of 270°C metal was additionally extra irradiated in test reactor IR-8 at 330°C at neutron flux level of $3\times10^{11}sm^{-2}s^{-1}$ (E≥0,5MeV) during 86,7 hours (fluence ~$10^{17}sm^{-2}$). Figure 1 represents result obtained.

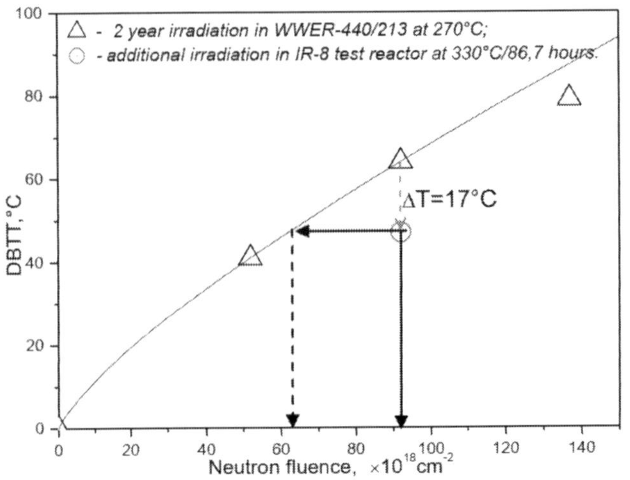

Figure 1. Result of the experiment on potential effectiveness of the RPV «wet» annealing.

One can see that despite the presence of irradiation during annealing treatment reduction of the transition temperature (DBTT) took place. In fact, embrittlement was partly suppressed up to value equivalent to 1,5 fold neutron fluence decrease.

4. Discussion

In connection with data received question of enhanced annealing effectiveness at the presence of the neutron field arise. One can imagine that during heat treatment at the presence of the neutron field annealing competes with damage process and exceeds it that result in recovery. Possibly neutron bombardment during heat treatment reduces the hardening efficiency of the existing precipitates. Figure 2 represents dependence of the precipitate hardening efficiency on dimension of the features [4] that act as obstacles to dislocation glide.

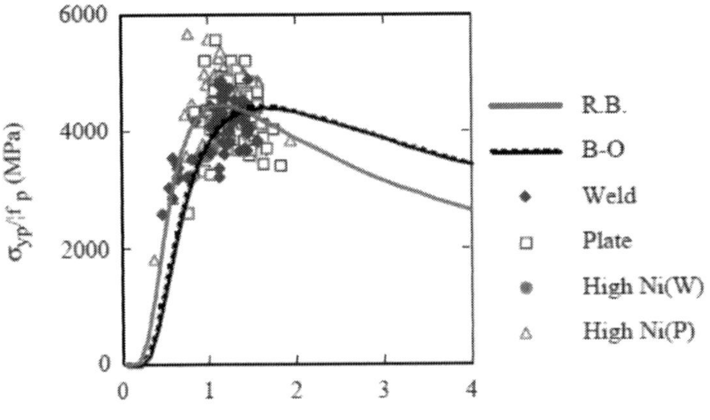

Figure 2. The precipitate hardening efficiency versus feature dimension (radius) showing both the Russell-Brown (RB) model along with the Bacon-Osetsky (BO) model.

It is seen that turning point are at the area of the 1,5 nm and that left branch is very sharp. Hypothetically neutron bombardment leads to partial precipitates dissolution that in turn results in their hardening efficiency reduction.

In this context one can remember that there are data on beneficial effect on mild steel properties of low dose neutron irradiation at 80°C up to fast neutron fluencies of $3,9\times10^{16}$sm^{-2}; $2,8\times10^{17}$sm^{-2}; $2,0\times10^{18}$sm^{-2} and $1,4\times10^{19}$sm^{-2} (E≥1MэB) [4]. Composition of this steel is given in table 2.

Table 2. Chemical composition of the steel examined (%% mass.).

C	Si	Mn	S	Cu	Cr	Mo	Ni	N	O
0,05	0,001	0,39	0,012	0,09	0,04	0,16	0,03	0,004	0,012

Original load-elongation curves of mild steel observed at 100°C are reproduced in Figure 3.

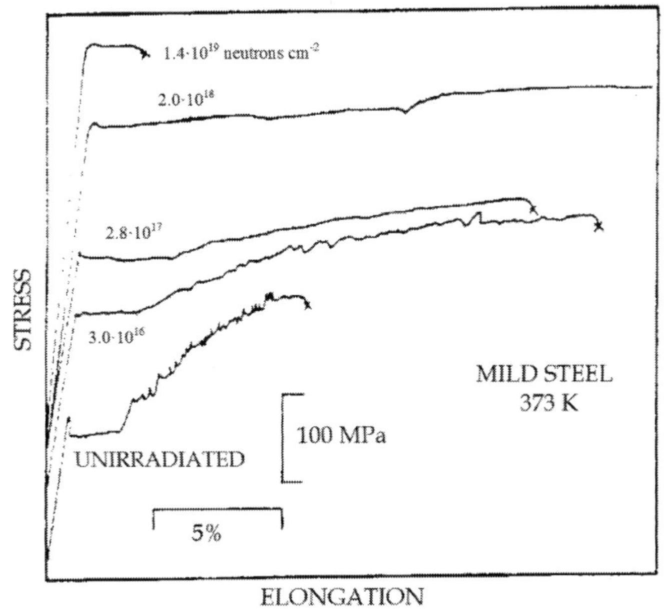

Figure 3. Load-elongation curves of mild steel observed at 100°C.

One can see that irradiation up to neutron fluence of $2,0\times10^{18}$sm^{-2} improved both the mechanical strength and ductility. In our case steel was extra irradiated in test reactor IR-8 up to fast neutron fluence ~10^{17}sm^{-2} for E≥0,5MeV and roughly 6×10^{17}sm^{-2} for E≥1MeV, that is in good conformity with data depicted in Figure 3.

Similar effect of the radiation-induced ductilization was recorded for Ti47Al alloy [5].

Thus it is possible to suppose that just neutron field by means of cascade dissolution of the existing radiation defects at low doses can bring to beneficial modification of the steel mechanical properties.

References

[1] Alekseenko N, Amayev A, Gorynin I, Nikolaev V 1997 *Radiation damage of nuclear power plant pressure vessel steels* (American Nuclear Society, La Grande Park, Illinois p 97)

[2] Potapovs U, Knighton G, Denton A 1968 Critique of in-place annealing of SM-1A nuclear reactor pressure vessel *Nuclear Engineering and Design* **8** № 1 pp 39-57

[3] Murty K 1984 Is neutron exposure always detrimental to metal (steels)? *Nature* **308** pp 51-52.E

[4] Eason E, Odette G, Nanstad R 2007 *Physically based correlation of irradiation-induced transition temperature shifts for RPV steels* ORNL/TM-2006/530 (Oak Ridge, Tennessee)

[5] Hishinuma A 1996 Ductilization of TiAl intermetallic alloys by neutron irradiation *Intermetallics* **4** p 179

Radiation-thermal synthesis of W-type hexaferrites[1]

V A Zhuravlev*, E P Naiden*, R V Minin, V I Itin**, V I Suslyaev* and E Yu Korovin***

*Tomsk State University, Lenina av., 36, Tomsk, 634050 Russia
**Tomsk Scientific Center SB RAS, pr. Academicheskii 10/4, Tomsk, 634021 Russia

E-mail: ptica@mail.tsu.ru

Abstract. The results of investigations of the phase composition, structural parameters, static and dynamic magnetic properties of $BaCo_{0.7}Zn_{1.3}Fe_{16}O_{27}$ hexaferrites obtained by the method of self-propagation high-temperature synthesis in combination with mechanochemical activation and radiation-thermal post-sintering are presented. The prospects of the proposed energy-saving approach for the production of ferrite ceramics with a hexagonal structure is shown.

1. Introduction

Oxide ferromagnetic materials (ferrites) are widely used in various fields of modern technology and medicine. Industrial production of ferrites is based on the conventional multi-operational ceramic technology which involves a number of long-running operations. This technology has been used in the world practice for many decades, and its development is associated with optimization by varying the composition and processing method of the charge [1].

In the last decades, new methods of ferrite synthesis have been proposed based on self-propagating high-temperature synthesis (SHS) in the regime of filtration combustion of the mixture of powders comprising iron and oxide of some elements in the atmosphere of reacting gases: oxygen or air [2]. This method utilizes the internal chemical energy of the initial reagents, and is organized most favorably from the thermal viewpoint. In the process of obtaining ferrites, it substitutes the most important operation – ferritization – which in the conventional method of sintering is very long and proceeds at high temperatures.

The SHS method allows various cubic ferrites and hexagonal ferrites with M-phase composition to be synthesized [3]. However, the yield of the end product of synthesis of complex hexagonal barium and strontium ferrites with W-, Y-, and Z-compositions is low. In this regard, new resource-saving methods of synthesis of magnetic materials based on complex hexaferrites with W-composition are suggested with the SHS, mechanochemical pre- or post-activation, and final ferritization in electric furnaces [3, 4] that in comparison with the conventional methods allow the energy and material expenses for the production process to be reduced significantly. It is of interest to substitute the final ferritization operation in an electric furnace by radiation-thermal sintering (RTS) upon exposure to an electron beam. This process sharply accelerates ion diffusion due to joint thermal and radiation influence on the material and as a result, improves magnetic properties of the synthesized materials [5].

[1] Work carried out in the framework of the increase of competitiveness of Tomsk State University

The present work is aimed at investigation of the influence of radiation-thermal sintering on the phase composition, structural parameters, and magnetic properties of the multiphase product based on barium hexaferrite with W-composition obtained by the method of self-propagating high-temperature synthesis.

2. Preparation of samples

To synthesize barium hexaferrite with W-composition, we used the chemical reaction

$$BaO_2 + 5Fe_2O_3 + 0.7CoO + 1.3ZnO + 6Fe + 4O_2 = BaCo_{0.7}Zn_{1.3}Fe_{16}O_{27}\,(Co_{0.7}\,Zn_{1.3}W).$$

Barium hexaferrite with W-type composition was synthesized in following regimes:

- No 1: SHS,
- No 2: SHS with mechanical activation of the initial charge for 3 min and the grinding body mass to the powder mass in the ratio 20:1 that corresponded to a milling process power of 60g (MA3+SHS),
- No 3: SHS with mechanical post-activation of the product for 40 min for the grinding body mass to the powder mass in the ratio 10:1 that corresponded to a milling process power of 40g (SHS+MA40).

Radiation-thermal sintering of samples was carried out at the Institute of Nuclear Physics of the Siberian Branch of the Russian Academy of Sciences (Novosibirsk) using an ILU-6 pulsed electron accelerator [6]. Sample preparation technique is described in detail in [7].

3. The results of x-ray researches

The x-ray analysis was performed using a SHIMADZU XRD-6000 polycrystalline diffractometer. For a qualitative analysis of the phase composition, the computer database of x-ray powder diffractometry or PDF4+ of the International Center of Diffraction Data (ICDD, Denver, USA) was used. The quantitative analysis of the phase composition and the refinement of the structural parameters of phases being detected were performed using the program Powder Cell 2.4 for full profile analysis. Results of x-ray structural analysis of ferrite ceramic samples after RTS are presented in table 1.

Table 1. Influence of RTS regimes of the SHS product on the phase composition and structural parameters of barium hexaferrite.

Sample	Phase composition of the product, vol. %			Mean W-phase crystallite size, nm	$\Delta d/d\cdot 10^3$
	W-phase	Fe_3O_4	Fe_2O_3		
No 1 + RTS at 1050°C	45	50	5	67	0.6
No 1 + RTS at 1100°C	54	40	6	46	1.3
No 1 + RTS at 1200°C	93	7	–	170	1.9
No 1 + RTS at 1250°C	98.4	1.4	0.2	>500	0.2
No 2 + RTS at 1050°C	39	54	6	100	0.9
No 2 + RTS at 1100°C	53	46	1	140	0.7
No 2 + RTS at 1200°C	89	11	–	>300	0.2
No 2 + RTS at 1250°C	54	40	6	400	0.2
No 3 + RTS at 1050°C	64	28	8	110	0.3
No 3 + RTS at 1100°C	53	41	6	135	0.7
No 3 + RTS at 1200°C	88	12	0	140	0.4
No 3 + RTS at 1250°C	93	7	–	>300	0.2

From table 1 it follows that the yield of the target (W) phase at RTS temperatures of 1200–1250°C is about 90–98 vol.%. Moreover, sufficiently high internal elastic microstresses proportional to the relative change in interplanar distances $\Delta d/d$ remain for all compositions. The mean microcrystallite size estimated on the size of coherent scattering regions (CSR) increases with RTS temperature.

4. The results of scanning electron microscopy

The morphological characteristics of hexagonal ferrimagnetic materials synthesized by a combination of SHS and RTS methods were analyzed using a Phylips 515 electron microscope with different magnifications. Results of the scanning microscopy for some samples are shown in figure 1.

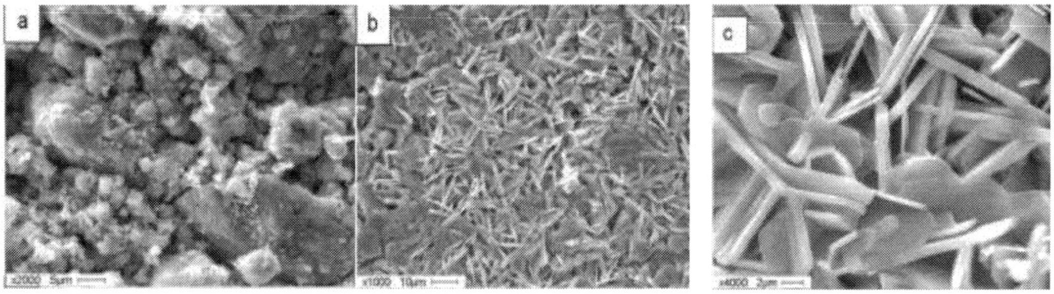

Figure 1. Scanning microscopy: *a*) No 1 + RTS at 1100°C; *b*) and *c*) No 1 + RTS at 1250°C with indicated magnifications.

The morphology of samples obtained at RTS temperatures of 1100–1200°C was characterized by the presence of grains having quasi-spherical shapes characteristic for cubic ferrospinels along with particles shaped as hexagonal plates whose linear size and thickness were in the aspect ratio $a/c \approx 10$ typical for ferrites with hexagonal composition. An increase in the RTS temperature to 1250°C caused a significant change of the morphology of ferrite ceramics with hexagonal structure. Quasi-spherical particles practically disappeared, and the aspect ratio a/c of hexaferrite crystallites considerably increased (to ≈ 20).

5. Results of the study of the magnetization curves

The static magnetic characteristics of ferrite ceramic samples, including the specific saturation magnetization and effective magnetic anisotropy field, were investigated using an automated complex intended for investigation of the magnetic characteristics in pulsed magnetic fields. The measurement procedure was described in detail in [8]. The main magnetic characteristics: the saturation magnetization M_S and the effective anisotropy fields $H_{a\ eff}$ for samples with maximum content of the $Co_{0.7}Zn_{1.3}W$ target phase are presented in table 2.

Table 2. Results of static measurements of the magnetic parameters of samples.

Sample	M_S, Gs	$H_{a\ eff}$, kOe
No 1 + RTS at 1200°C	493	2.03
No 2 + RTS at 1200°C	487	2.23
No 3 + RTS at 1200°C	482	2.41
No 1 + RTS at 1250°C	484	2.22
No 2 + RTS at 1250°C	487	2.43
No 3 + RTS at 1250°C	447	1.89

6. The study of the ferromagnetic resonance spectra

The ferromagnetic resonance (FMR) spectra were measured by the standard waveguide transmission technique in the frequency range 26–37 GHz using an automated radio spectroscope built on the base of a R2-65 analog scalar circuit analyzer. To investigate the FMR, hexaferrite powders were charged into quartz tubes with inner diameter of 0.7 mm and length of 10 mm. The density of powder samples was identical and equal to ≈ 2.9 g/cm^3. The spectra were processed using the procedure described in [9, 10].

Investigations of the ferromagnetic resonance spectra of powder and polycrystalline oxide ferrimagnetic materials with hexagonal crystal structure allowed a number of magnetic parameters of these materials important for practical applications to be determined [9]. Such as: magnitude and sign of magnetocrystalline anisotropy fields (H_{al}); saturation magnetization (M_S); value of the gyromagnetic ratio $\gamma = ge/2mc$, where g is the effective g-factor of the examined material, e is the electron charge, m is the electron mass, and c is the velocity of light.

Table 3 presents the gyromagnetic ratios, anisotropy fields, and saturation magnetizations of the synthesized materials. The parameters $\gamma/2\pi$ and H_{a1} were estimated from a comparison of the positions of maxima for theoretical and experimental curves of ferromagnetic resonance at different frequencies.

Table 3. Magnetic parameters of materials measured by the FMR method

Sample	$\gamma/2\pi$, GHz/kOe	H_{a1}, kOe	M_S, Gs
No 1 + RTS at 1200°C	2.56	–0.8	440
No 2 + RTS at 1200°C	2.62	–1.5	520
No 3 + RTS at 1200°C	2.62	–1.5	525
No 1 + RTS at 1250°C	2.58	–1.0	490
No 2 + RTS at 1250°C	2.58	–1.0	480
No 3 + RTS at 1250°C	2.58	–1.0	450

Values of the saturation magnetization M_S given in table 3 are close to M_S values estimated from the measured magnetization curves (see table 2). The difference between M_S values measured by different methods did not exceed 10%. We note that the negative sign of the anisotropy field H_{a1} demonstrates that the anisotropy of easy magnetization type is observed for these materials. The measured fields are close to those obtained from analysis of magnetization curves in pulsed fields and to the literature data for these materials synthesized by the conventional ceramic technology [10].

In the first series of samples (RTS at a temperature of 1200°C), the mechanical activation leads to an increase $\gamma/2\pi$ and H_{a1} values estimated from the FMR experiment. Processing of the data of experiments for samples of the second series demonstrated that $\gamma/2\pi$ and H_{a1} values for these materials coincide within the limits of the experimental errors.

7. Spectra of permittivity and permeability of composites based on a sample No 3 + RTS at 1250°C

This section presents the results of measurements of the complex permittivity ($\varepsilon = \varepsilon' - i\varepsilon''$), permeability ($\mu = \mu' - i\mu''$) spectra of composites made on the basis of the sample No 3 + RTS at 1250°C. Investigations were carried out on the equipment Center of collective use "Center of radio physical measurements, diagnostics and research parameters of natural and artificial materials» [11].

To prepare composite materials samples of No 3 + RTS at 1250°C were ground in a ball mill to a particle size less than 100 microns. As the matrix in the manufacture of composite samples used urethane alkyd varnish UNICA SUPER. Mass fraction of ferrite in the composites was 50 % (sample H-50) and 79.5% (sample H-79.5).

Since all methods of measurement of spectra complex permittivity and permeability are not free of systematic and random error of measurement the verification of the experimental spectra $\mu = \mu' - i\mu''$ using the Cramers-Kronig dispersion relations was carried out. The method was developed by us in [12]. During the verification of the spectra is necessary to know the magnitude of the initial magnetic permeability μ_0. It was determined with using a precision LCR meter Agilent E4980A on the same toroidal samples, which are used in the waveguide method measurements of complex permittivity and permeability. The measured values of μ_0 samples are shown in table 4.

Table 4. The initial magnetic permeability μ_0 and the permittivity of composites

Sample	μ_0, rel. units	ε', rel. units	ε'', rel. units
H-50	2.3 ± 0.1	7.3 ± 0.1	0.9 ± 0.1
H-79.5	3.2 ± 0.1	9.1 ± 0.1	2.5 ± 0.1

Measurements permittivity spectra of these materials in the frequency range 0.01 ÷ 12 GHz shown that an appreciable frequency dependence of the $\varepsilon = \varepsilon' - i\varepsilon''$ in this range is absent. Table 4 shows the

average value over the frequency range of the real and imaginary parts of the permittivity of the composite materials. According to table 4, the increase in concentration of the magnetic fraction in the composite from 50 to 79.5 wt. % most significantly affected on the value of ε''. It has been increased by 64 %, whereas μ_0 at 28 % and ε' at 20 %.
The spectra of the permeability of composite materials investigated are presented in figure 2.

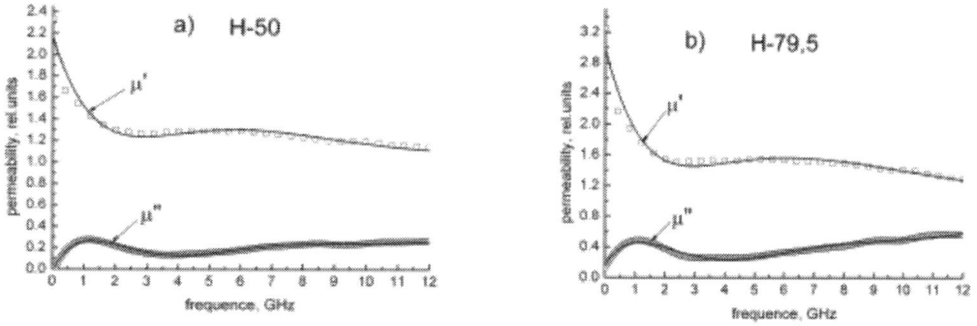

Figure 2. Spectra of permeability composites: *a*) sample H-50, *b*) sample H-79.5. Points - experiment, the line - the results of processing of the spectra by the method of [12].

It is seen that the spectra of the imaginary part of permeability of both materials present a maximum near the frequency of 1 GHz, due to oscillations of the domain boundaries. With further increase of frequency we observed growth of losses due to natural ferromagnetic resonance (NFMR) in the presence of a domain structure. Similar behavior of the permeability spectrum was observed in [13] for polycrystalline hexaferrites Co_{2-x} Zn_xW system with close to x = 1.3 composition. The essential differences spectrum of $\mu''(f)$ powder composite materials from polycrystalline is higher bandwidth of NFMR region. With increasing concentration the magnetic losses in NFMR region significantly increased.

8. The study of radar absorbing properties of the composite H-79.5

Theoretical analysis and experimental study of radar absorbing properties of materials are usually carried out for the simplest case of normal incidence of a plane electromagnetic wave (EMW) on a magnetodielectrics (MD) layer. Permeability (μ) and permittivity (ε) of the layer are considered as complex scalar quantities. For reduction of radar signature a model of MD layer disposed on the metal surface is used. Expression for the reflection coefficient (R) for located on the metal plate layer of MD has the form [14]:

$$R = [\rho - \exp(-2i\gamma d)]/[1 - \rho\exp(-2i\gamma d)] \quad (1)$$

Here: $\rho = (Z-1)/(Z+1)$ – reflection coefficient from the front edge of MD. $Z = \sqrt{\mu/\varepsilon}$ – wave impedance. $\gamma = k_0\sqrt{\varepsilon\mu}$ – EMW propagation constant in MD. $k_0 = \omega/c$ – the wave number of free space. $\omega = 2\pi f$ – circular frequency of the EMW. c – speed of light. d – thickness of the layer. $i = \sqrt{-1}$.

The study of radar absorbing properties have been conducted on the composite H-79.5 with different thicknesses of MD layer. The results of experimental studies and calculations of the reflection coefficient $|R|^2$ from the measured permeability and permittivity spectra are shown in figure 3. Points – experimental date, solid lines - calculation by formula (1).

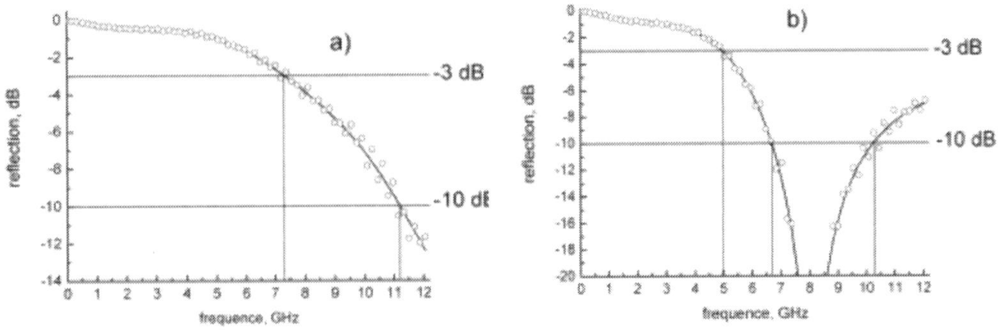

Figure 3. The frequency dependence of the reflection coefficients of of the absorber layers: *a*) – composite H-79.5, $d = 1.5$ mm thick, *b*) – composite H-79.5, $d = 2.5$ mm thick. Points – experiment, line – calculation.

From the figure 3a it is seen that for a coating thickness of $d = 1.5$ mm reflectance smaller –3 dB is observed at frequencies above 7.3 GHz and lower then –10 dB at frequencies more than 11.2 GHz. The increase in coating thickness up to $d = 2.5$ mm (figure 3b) leads to a shift in the reflection minimum to lower frequency region. Reflectance less than –3 dB, begins with a frequency of 5 GHz and $|R|^2 < -10$ dB in the frequency band 6.7 ÷ 10.3 GHz. Results in figure 3 show that there is good agreement between measured and calculated from the spectra of μ and ε reflection coefficients.

9. Conclusion

Thus, the technology combining SHS, mechanical activation, and radiation-thermal sintering allows complex multicomponent ferromagnetic materials with hexagonal crystal structure to be synthesized and magnetic characteristics no worse than and in some cases even exceeding the corresponding characteristics of samples synthesized by the conventional ceramic technology. An advantage of the suggested technology is a significant decrease of the time of synthesis and energy consumption.

Acknowledgments

The authors thank the staff of INP SB RAS Korobeinikov M.V., Bryazgin A.A., Tolochko B.P. for help in radiation-thermal sintering experiment. The staff of Tomsk Polytechnic University Gyngazova S.A., Lysenko E.N. for their valuable advice.

10. References

[1] Letyuk L M , Nauman G, Gonchar A B, *et al.* 2006 in: Materials of the 4th Russian–Japanese seminar *"Perspective Technologies and Equipment for Materials Science and Micro- and Nanoelectronics,"* Part 2 [in Russian], Moscow; Astrakhan' 12

[2] Merzhanov A G and Nersesyan M D 1990 *Zh. Vsesoyuzn. Khim. Obshch. Imen D. I. Mendeleyeva* **35** 700

[3] Minin R V 2008 *Technology of synthesis of hexagonal oxide ferrimagnetic materials with W-, M-, and Z-compositions by the method of self-propagating high-temperature synthesis*, Cand. Tech. Sci. Dissert., Tomsk Polytechnic University, Tomsk)

[4] Itin V I, Kirdyashkin A I, Minin R V, *et al.* 2006 *Izv. Vyssh. Uchebn. Zaved. Tsvetn. Metallurg..* **5** 83

[5] Vereshchagin V I, Pletnev P M, Surzhikov A P, *et al.* 2004 *Functional Ceramics*, Vereshchagin V I, ed. [in Russian], Publishing House of the Institute of Inorganic Chemistry of the Siberian Branch of the Russian Academy of Sciences, Novosibirsk

[6] Auslender V L, Bryazgin A A, Voronin L A, *et al.* 2003 *Nauka – Proizv.* **7** 11
[7] Naiden E P, Zhuravlev V A, Itin V I, *et al.*, *Russ. Phys. J.* 2013 **56** 674
[8] Kreslin V Yu and Naiden E P 2002 *Prib. Tekh. Eksp.* **1** 63
[9] Naiden E P, Zhuravlev V A, Itin V I, *et al*, 2013 *Russ. Phys. J.* **55** 869
[10] Zhuravlev V A 1999 *Fiz. Tverd. Tela,* **41** 1050
[11] The electronic resource: http://www.ckp.tsu.ru/ckp_3/
[12] Zhuravlev V A and Suslyaev V I 2006 *Russ. Phys. J,* **49** 840
[13] Zhuravlev V A and Suslyaev V I 2006 *Russ. Phys. J,* **49** 1032
[14] Baker-Jarvis J, Janezic M D, Grosvenor J H Jr, Geyer R G 1993 Transmission/Reflection and Short-Circuit Line Methods for Measuring Permittivity and Permeability *NIST Technical Note 1355 (revised)*

Development of technology and investigation of structure and properties of composites on the basis of shungit and polymer resins

A I Kupchishin[1*], M K Nauryzbaev[2], A A Kupchishin[1], F F Komarov[3], G E Remnev[4], N A Voronova[1], M Kaykanov[4]
[1] Abai Kazakh National Pedagogical University, Almaty, Kazakhstan
[2] Al-Farabi Kazakh National University, Almaty, Kazakhstan
[3] Belarusian State University, Minsk, Belarus
[4] Research Tomsk Polytechnic University, Tomsk, Rassia

E-mail: ankupchishin@mail.ru

Abstract. The technology for manufacturing of composites based on polymer (epoxy) resins and shungites was developed by using of the electron irradiation and experienced samples with different percentages of the components a) 90 and 10 %, b) 70 and 30 %, c) 50 and 50 %, were manufactured respectively. It was carried out the series of experimental studies of the structure and properties of the obtained materials, including the atomic force microscope images were obtained and Raman spectra of these composites were measured and their analysis was performed.

Introduction

It is known from the literature that carbon materials have a high mechanical strength. In the case of mixing them with various polymeric materials, it is possible to obtain sufficiently thin plate with high strength and resistance to external influences, particularly to radiation. Such materials can be used in conventional and in the aviation industry, medicine, agriculture, engineering, space objects; in home appliances, car industry, etc. In this regard, it is evident that carrying out such works is very important as the creation of new materials with sufficiently stringent requirements (high mechanical stresses, large changes in temperature, aggressive media, acids, alkalis, organic solvents, etc., the ambient humidity, etc.) is caused by the their practical application in various industries, including Kazakhstan, Russia, CIS and non-CIS countries. The use of high-energy electron fluxes gives undoubtedly a substantial industrial and economic effect. Effect of ionizing radiation on solids for such materials is specific and has a number of significant features associated with the properties of the solid state: by a regularity of the structure (crystallinity and defectiveness), by the electronic interactions between the components of the solid particles, by shortness of diffusion of large particles on the surface and volume, etc. Taken together these factors determine the kinetics of chemical reactions generated by ionizing radiation, as well as characterize the output of stabilized intermediate and final products of transformation [1 – 4]. In contrast to the photochemical reactions at radiative initiation of transformation occurs not only on the surface and in the subsurface layers, but also in the whole volume of the material and first of all in the fields of micro-, macro- and nanodefects.

The experimental technique

We have developed the technology for manufacturing of composites based on polymer (epoxy) resins and shungites, which includes the following steps: 1. The choice of powder on the basis of shungit. 2. The qualitative analysis of the granulometric composition of the powder and study of its physical and mechanical properties. Additional grinding of the material. 3. The analysis and selection of the type of binder (resin, etc. to improve the parameters by pressing and sintering). 4. A detailed analysis and sub-

IOP Conf. Series: Materials Science and Engineering **81** (2015) 012004 doi:10.1088/1757-899X/81/1/012004

sequent selection of the parameters of electron irradiation (energy of high-energy electrons, beam intensity and dose of electron irradiation). 5. Fabrication the target holder and adaptations. 6. The analysis of beam parameters and the development of a technique of irradiation components of composite material with high-energy electrons. 7. Development of a technique of cooling material under the beam of high-energy electrons. 8. Radiation processing of components of the composite material. 9. The analysis of parameters and selection of the type of a pressing apparatus for producing composite materials with given parameters.

The analysis and modification of the individual blocks of the installation relation to specific technological modes of pressing and sintering of the experimental samples. 10. Fabrication of the press - form, working in a given range of pressure, temperature and time. 11. The oil lubrication of the press-form and filling components of the future composite material into the press- form.12. The analysis and verification of the parameters of the press installation in different intervals of temperature, time and mechanical loads. 13. Development of measurement techniques of the temperatury.14. The analysis and choice of intervals of the mechanical loadings (stresses). 15. The analysis and choice of time intervals of pressing to achieve the necessary homogeneous structures of the composite by volume of the material. 16. The trial test works, pressing and agglomeration of powder. The recess samples from the press-form and their drying in air or in a gas atmosphere at a specific temperature for a given time (from several hours to several days). 17. The mechanical treatment and manufacturing of the experimental samples of the composite of a given shape (cutting, grinding, polishing, and others.). 18. Studying the physical and mechanical properties of obtained composite and revealing the features of the material deviations from a given characteristics. 18. The semi-industrial tests and correction techniques and modes of manufacturing of the composite material (temperature, size and mechanical load). 19. Establishment of the parameters and making an experimental batch. 20. The analysis and giving specific advice. In view of developed technology the experimental samples with the following percentages of the components a) 90 and 10 %, b) 70 and 30 %, c) 50 and 50 %, were made respectively.

1 – a transformer T1; 2 – generator housing; 3 – compressor; 4 – cover protection

Figure 1. The appearance of pulsed high-current electron accelerator ASTRA-M SP

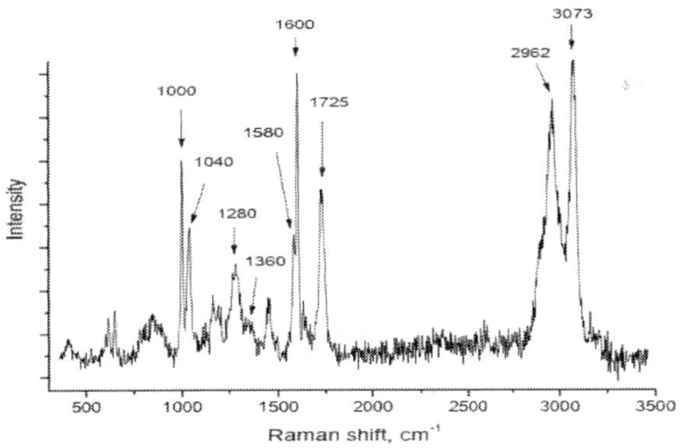

Figure 2. The Raman spectrum of the unirradiated composite on the basis of epoxy resin and shungite (90 and 10 %)

The irradiation was performed on a pulsed electron accelerator ASTRA-M (Figure 1) [5], which has the following parameters: 1. The amplitude of the accelerating voltage is 500 kV. 2. The duration the voltage pulse (on base) is 450 ns. 3. The extracted electron current is up to 1 kA. 4. The duration of the extracted electron current (on base) is 110 ns. 5. The pulse repetition frequency is from 1 to 50 pulses/sec. 6. The energy of the extracted electron beams is 20 J.

The experimental results

Studying of composites on the basis of shungite and epoxy resins was allowed us to conclude that the increase in the concentration of shungite in a composite leads to a slight darkening of the samples, while the irradiation with electrons to different integral doses (103 – 105 Gy) does not cause substantial discoloration. The series of experimental studies were carried out, including Raman spectra were measured (Figures 2, 3) on obtained composites. In the composite 1 (90 and 10 %) were found peaks at frequencies of 1000; 1040; 1280; 1360; 1580; 1600; 1725; 2962 and 3073 cm^{-1}. For the composite 2 (70 and 30 %) on the one hand there are the same bands 1000; 1040; and 1365, 1600; 1725; 2950, 3070 (with a slight shift) cm^{-1}. On the other hand the bands in 1280; 1580 cm^{-1} were completely disappeared. These two bands were also disappeared for the sample number 3 (50 and 50 %). It was found decrease in band intensity of 1000 cm^{-1} with increasing concentrations of the second component (several times). AFM images of the non-irradiated and irradiated (by electrons with an energy of 2 MeV) composites to various integral doses (Figure 4) were obtained. It is seen a substantial effect of the concentration of components and dose of electron irradiation on the structure and some properties of the material.

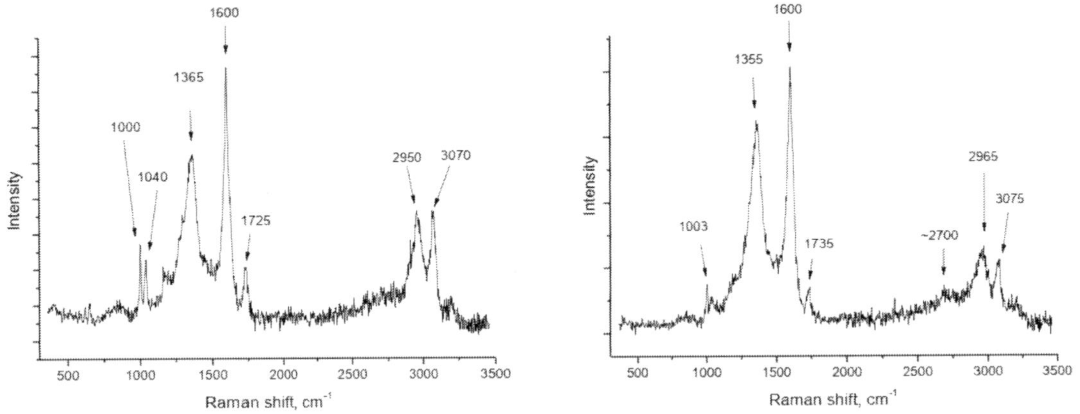

Figure 3. A Raman spectrum of the unirradiated composite based on epoxy resins and shungite (70, 30, 50 and 50 %)

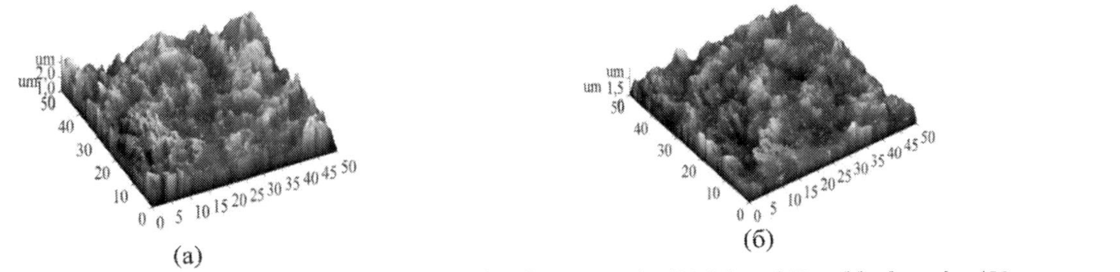

Figure 4. AFM image of the composite on the basis of epoxy resin (50 %) and Kazakh shungite (50 %), irradiated by electrons with an energy of 2 MeV up to integral doses 5 (a) and 10 (b) kGy

Findings

1. The technology for manufacturing of composites based on epoxy resins and shungites with the use of electron irradiation was developed. The experimental samples with different percentages of the components were made.
2. The experimental studies of the properties and structure of composites based on epoxy resins and shungites were carried out. Raman spectra of these composites were measured. The peaks at 1,000;

1040; 1280; 1360; 1580; 1600; 1725; 2962 and 3073 cm^{-1} were found. For the composites (70 and 30 %) and (50 and 50 %) on the one hand there are the same bands 1000; 1040; 1365 (there is a slight shift), on the other hand the bands in 1280; 1580 cm^{-1} were completely disappeared. It was found decrease in intensity of band 1000 cm^{-1} and others with increasing concentrations of the second component (in several times).
3. Based on the analysis of carried out data it was concluded that the main contribution to the intensity of the bands gives the carbon OH, CH CH_2 bonds. With increasing irradiation dose for various concentrations of the second component Raman spectra behave differently - there is also decrease in and increase in the intensities that is connected with competing processes of rupture and cross linking of polymer bonds.
4. AFM images of the non-irradiated and irradiated (by electrons with an energy of 2 MeV) composites up to various integral doses were obtained. It was revealed a substantial effect of the concentration of components and of electron irradiation on the surface structure of the material.

References

[1] Dyukkiev E F, Kalinin Yu K, Sokolov V A 1981 Shungite rocks of Karelia Petrozavodsk. Karelia 182
[2] Huczko A 2002 Synthesis of aligned carbon nanotubes *Appl. Phys. Lett.* A 74 387 - 91
[3] Li Z et al 2004 Synthesis of 4 Å single-walled carbon nanotubes in catalytic Si-substituted $AlPO_4$-5 molecular sieves *Appl. Phys. Lett.* 85 7 1253 - 55
[4] Golubev E A, Glebashev S G 2006 Supramolecular Structure anthraxolite Bakyrchik deposit, Eastern Kazakhstan *Bulletin* 4 4 - 7
[5] Remnev G E, Stepanov A V, Kaykanov M I, Sazonov R V, Kolokolov D Y, Merinova L R, Egorov I S 2013 Pulsed-beam processing of industrial and domestic wastewater *Proceedings of the higher educational institutions. Physics* **56** 7 / 2 42 - 48

Accumulation of free radicals in irradiated Polytetrafluoroethylene and study of its properties

K B Tlebaev [1], A A Kupchishin[1] and A I Kupchishin[2*]

[1]Institute of mathematics, physics and informatics, Abai Kazakh National Pedagogical University, 86 Tole bi Street, Almaty 050010, RK
[2]Al-Farabi Kazakh National University, Almaty, Kazakhstan

E-mail: ankupchishin@mail.ru

Abstract. The accumulation of free radicals in polytetrafluoroethylene (PTFE) is experimentally studied and its properties are investigated by ESR, Raman spectroscopy and optical microscopy.

1. Introduction

At present, the question of the effect of ionizing radiation on the dielectric systems seriously arises in connection with their wide use in space technology, nuclear industry and nuclear power.

The necessity of such studies is due to clarify the behavior of polymeric materials working in the fields of different streams of particles. Data in the literature give a large spread in the results, since the radiation resistance of dielectric solids can depend strongly on the specific conditions of their production and testing (type of radiation, exposure mode - continuous or pulsed, short or long, in air or vacuum, at different dose rates in during the tests, etc.). In particular, the effects of radiation oxidation in the air may result in greater damages even at low dose rates with fixed integral doses.

When electron and gamma-irradiation (electromagnetic impact) in the polymers the rupture of the bonds and cross-linking of the polymer occurs. In practically all polymers and composites with increasing radiation dose occurs deterioration of mechanical strength (predominates radiation degradation), except for polyimide (in which at not very high doses the predominance of cross-linking over degradation is observed).

Most of the reactions and effects in this case are irreversible [1 – 2]. In the interaction of electrons with polymers the following processes occur: 1. The scattering of incident particles by electrons of atoms. At the same time there goes or excitation of an atom or its ionization. In this process, about 99% of primary energy is lost by a charged particle. These so-called ionization losses. In the polymer when ionization the polymer chains are scissioned, from which free radicals, volatile substances (degradation) are appeared. A significant part of scissioned bonds of atoms reacts with the nearest neighbors, forming stable compounds (cross-linking). The charged particles depending on the energy can produce from a few dozen to several thousand collisions. Ultimately, the particle is stopped, forming a profile of defects, of scissioned bonds and radiation damages. In this case, almost all the physico-chemical properties are substantially changed. 2. Scattering of the primary particles by atoms. In this case knocking of the atom as a whole occurs and the so-called primary- knocked atom (PKA) is formed. PKA generates secondary, tertiary knocked atoms, etc. and atom-atomic cascade arises. The cascading areas are formed from which the vacancy clusters and clusters of interstitial atoms are finally formed. The mechanism of defects generation during gamma irradiation is about the same as in the electron irradiation. However, the effectiveness of gamma-irradiation (per unit volume of material) is approximately 100 times smaller than for electron impact. When considering of the single trajectories, the process of passage of electrons can be considered as chaotic and for a finding of any physical quantity it is necessary to gain a large number of measurements. After a lot of trials and averaging over the angles, energies and the depths of a physical quantity (streams of particles, the concentration of defects, etc.) are reduced to a minimum scatters and errors.

The difficulties arising when interpreting experimental data are primarily associated with the development of an adequate model, by the complexity of the mathematical description of the phenomena, incomplete informativeness of methods and lack in practice of the net materials. The presence of defects or accelerates or slows down the processes (especially diffusional) in the tens or hundreds of times. Naturally, this is affected and on the end result – the formation of different compounds and structures and on the physical properties of the material. [1 – 3].

Experimental technique

The industrial Teflon – 4 was used. The irradiation of material samples was conducted by electrons in air at room temperature on a accelerator of quasi-continuous action ELU–6 under the following beam parameters: 1. Energy – 2 MeV. 2. The current density j = 0, 5 mA. 3. The sample temperature was 273K. The irradiation by gamma rays was conducted on the installation Co^{60}. The samples were represented themselves as a packing from the Teflon film and 2 cylinders with a diameter of 2.5 mm and a length of 5 mm and a mass of ~ 62 mg. The irradiation was conducted in a thermal bath at temperatures 273 K, after which the samples were stored in liquid nitrogen. The annealing was held at room temperature (T = 293 K). Isotopes in the form of cylinders were located along the azimuth around the target in order to obtain uniform treatment.

ESR spectra were measured with a spectrometer ESP 300E, which is designed to study paramagnetic centers and their identifications. The essence of ESR lies in resonance absorption of radiofrequency power with redundant number of unpaired electrons in the samples that are in a constant magnetic field. Raman spectrometer of the system NTMD INNTEGRA SPECTRA was used for the study of the Raman spectra. The modular structure of the spectrometer, a wide range of lasers and microscopes, the use of automatic three-position elements for operations with three lasers, the availability of switching or simultaneous use of three lasers with automatic tuning of optical and mechanical elements for obtaining maximum efficiency of the work makes the system flexible and allows us to solve a large number of tasks. The main technical characteristics of the Raman spectrometer: 1. The working wavelength range is 244 – 1050 nm. 2. Diffraction gratings depend on the configuration. 3. The spectral resolution is 0.1 A. 4. Focal length is 520 mm.

The study of the surface structure was carried out at the optical microscope Leica DM 6000M, which is a system with a high-resolution digital cameras and software for analyzing and storing images. The device has the following features and options: 1. Automatic mode of the incident light (reflection) for the methods of light and dark field. 2. Automated mode of the transmitted light (on clearance). 3. Automated dimmers, changes of the aperture and aperture of the illuminator. 4. Motorized drive focus and a motorized objective little table responsible for x, y and z movements. 5. x / y panel motorized stepping motor, the removable objective little table with external dimensions: 234 x157 mm, the traverse range: 76 x50 mm, the minimum step: 0.3 μm, z panel corresponds to removable plate, the traverse range – 25 mm, the minimum step – 0.015μm, the maximum speed –5 mm / s, the minimum speed – 1mm / s, the maximum load – 4 kg. 6. Motorized, encoded turret head with 6 lens with magnification 5x, 10x, 20x, 50x, 100x, 150 x times. 7. The unique memory function for simultaneous switching of the lens and the contrast method. Color selection of image (for detection of small or subtle elements of the sample in contrast).

Experimental results

The intensity of the ESR signal in Polytetrafluoroethylene on the dose of γ-irradiation ^{60}Co at 273 K is shown in Figure 1. As can be seen from the figure, with an increase in irradiation dose the signal intensity increases and gradually reaches saturation, indicating an increase in the number of free radicals.

Figure 2 is shown the dependence of the accumulation curve of the concentration of radicals on the absorbed dose of radiation at a temperature of 333 K. As shown in Figure 2, the radical concentration initially increases in proportion to the dose, reaches a maximum and then decreases exponentially.

The curve of the intensity depending on the irradiation dose can be explained. Such move of depending is explained by the competing processes of accumulation and radical loss, the mobility of which increases with increasing temperature.

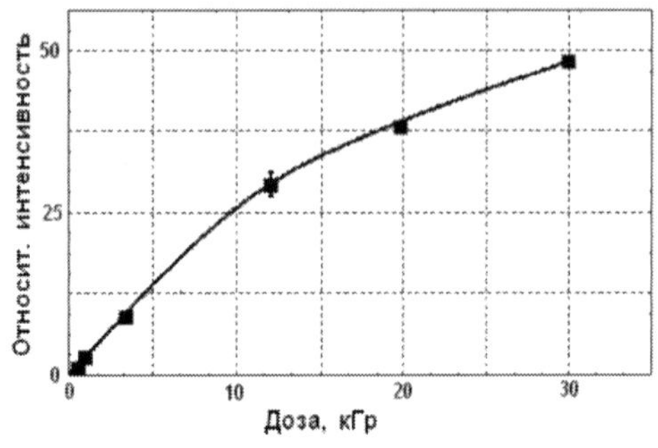

Figure 1. The dependence of the intensity of the ESR signal in Polytetrafluoroethylene on the irradiation dose at 273K

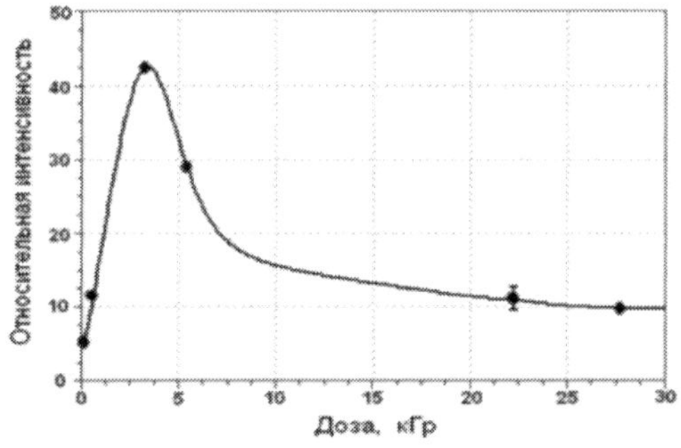

Figure 2. The dependence of the intensity of the EPR signal in PTFE on the dose of γ-irradiation ^{60}Co at 333 K

Further Raman spectra of unirradiated and irradiated PTFE were measured. Figure 3 shows the Raman spectrum of the irradiated PTFE with a dose of 13.5 kGy (film thickness was d = 40 micrometers). There are seven precisely expressed peaks. For irradiated materials in comparison with unirradiated the frequencies are slightly changed.

The intensity of the peaks in the spectrum after irradiation of Polytetrafluoroethylene films are significantly increased (more than 2 times), which is associated with changes in the structure of the material. This is confirmed by optical studies. The changes of bands in the spectrum are associated with a very quick and active reaction of transition of primary fluoroalkyl radical ~ CF_2 - $\dot{C}F_2$ - CF_2 ~ into the peroxygen radical ~ CF_2 - CF ($OO^{\cdot}$) - CF_2 ~ which is due to oxygen is always present in the polymer matrix.

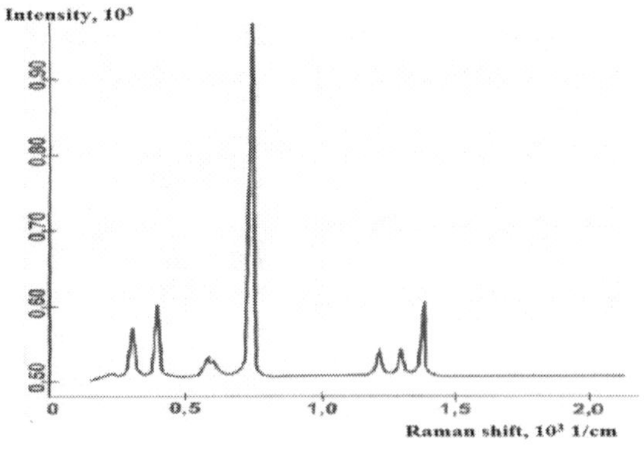

Figure 3. Raman spectrum of the irradiated PTFE with a dose of 13.5 kGy

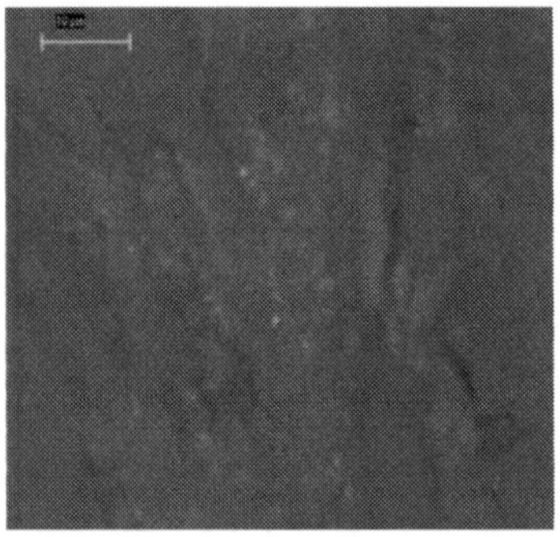

Figure 4. The image of fluoroplastic film surface after irradiation with a dose of 13.5 kGy. Field of view is 44 x 33 mm

The study surfaces of unirradiated PTFE films when viewing specimens on a gleam in an optical microscope shows that most part of surface of the samples are homogeneous, smooth surface and looks as matte blue. After irradiation with a dose of D = 13.5 kGy it can be seen that the materials have completely inhomogeneous surface with varying degrees of transmission. In addition, dark and light spots of different densities ranging in size from one to several micrometers with clear boundaries for their localization (Figure 4) are seen.

Findings

1. The experimental studies of ESR spectra in Polytetrafluoroethylene irradiated by gamma rays at 273 K are carried out. It was found that the introduction of radiation defects significantly increases the intensity of the signal amplitude. With increasing irradiation dose the signal intensity increases and gradually reaches saturation, indicating an increase in the number of free radicals. At a temperature of 333 K on the curve has a pronounced maximum. Such move of depending is explained by competing processes of accumulation and radical loss, the mobility increases with increasing temperature.
2. The experiments show that Raman spectra of the irradiated (and unirradiated) PTFE have seven precisely expressed peaks. Their intensity after irradiation are significantly increased (more than 2 times) which is associated with changes in the structure of the material. The bands changes in the spectrum are associated with a very quick and active reaction of transition of primary fluoroalkyl radical ~ CF_2 - $\dot{C}F_2$ - CF_2 ~ into the peroxygen radical ~ CF_2 - CF ($OO^{\cdot}$) - CF_2 ~ which is also due to oxygen is always present in the polymer matrix.
3. The surface of the unirradiated PTFE films when viewing specimens on a gleam in the optical microscope shows that most of it is a homogeneous, smooth surface. The irradiation leads to inhomogeneity of the surface with varying degrees of transmission. The dark and light spots of different densities ranging from one to several microns with clear boundaries of the regions of their localization are observed.

References

[1] Bondarenko V N, Goncharov A V, Suhostavets VI 2007 Definition sections of radiative-insulating polymer depletion of the individual elements of the data analysis with a beam of accelerated ions *Nuclear Energetics fizika that* 1(19) 23-34
[2] Slutsker AI, Gilyarov VL, Polikarpov YI 2000 The influence of mechanical loading on the kinetics of the electrical breakdown of polymers *ZHTF* **78** iss.11 60 - 63
[3] Vorobev SA, Zabaev VN, Caplin V. Kalinin BN and others 1991 The observation of X-ray transition radiation of 900-MeV electrons in a layered target *Letters in Zh* **53** iss. 7 332-36

Electrophysical properties of polycrystalline diamond films deposited from an abnormal glow discharge

A V Gaydaychuk, A V Kabyshev, F V Konusov, S A Linnik and G E Remnev
National Research Tomsk Polytechnic University, 30, Lenin Avenue, 634050, Tomsk, Russia

E-mail: konusov@hvd.tpu.ru

Abstract. Electrophysical properties of polycrystalline diamond films (PDF), deposited from the abnormal glow discharge, were obtained. The energetic and kinetic characteristics of surface dark and photoconductivity of PDF and their temperature, field and spectral dependences were investigated. Dominant carrier transport mechanisms, their type and the energetic spectrum of localized defect states were established. Current–voltage characteristics, photosensitivity and activation energy are determined by conductivity of PDF, which varies from 10^{-14} to 10^{-4} S depending on deposition conditions. Vacuum annealing of films up to 600–800 K stabilizes the electrical characteristics. Thermal stability of PDF properties higher than semiconductor films, deposited under nonequilibrium conditions by pulsed laser and ion ablation. PDF films not inferior on properties to films, obtained by alternative CVD methods. Electrical characteristics, mechanism of charge transport of PDF were caused by defects of different nature, the energetic levels of which are continuously distributed on energy in the band gap. Dominant n–type of activation conduction is complemented by hopping mechanism through the localized states distributed near the Fermi level with density $5.6\cdot10^{17}$–$2.1\cdot10^{21}$ eV^{-1} cm^{-3}. Trapping and recombination centers are heterogeneously distributed on grain boundaries.

1. Introduction

Unique chemical, mechanical, electrical and thermal, optical and photoelectrical properties of diamond contribute to its wide application in high–frequency, high–temperature electronics and semiconductor manufacturing of a special purpose [1–5]. Due to the limited possibility of using diamond single crystals and epitaxial diamond films, because of their high cost in the most appropriate instruments are the polycrystalline diamond films (PDF), which are obtained by developed methods of the chemical vapor deposition (CVD) [1–9]. PDF are successfully used for the production of stable radiation-resistant detectors of UV and ionizing radiation, as well as for laser and photodiode structures [1, 3]. Dielectric PDF are used for these purposes with conduction $\sigma=10^{-15}$–10^{-11} S·cm^{-1}. Depending on the characteristics of polycrystalline structure of PDF, the content of alloying impurity atoms and defects of crystal structure, electrical characteristics, as well as the mechanism of the transport and type of the charge carriers vary widely in range $\sigma=10^{-6}$–10^{3} S·cm^{-1} [1, 3–15]. Investigations of peculiarities of the PDF electrical properties and reasons determining property changes and deposition conditions effect on characteristics are very important. Among the various PDF deposition methods, there are microwave plasma, hot filament, arc–jet and glow discharge CVD to be marked out. But each of them has a number of disadvantages: has a high cost and limited maximal deposition area; the growth rate is low and constant replacement of filaments is necessary, the disadvantages of arc–jet CVD method are small deposition area and high gas consumption. Among the others, glow discharge CVD is considered to be an effective diamond film deposition method because of its simplicity and high growth rate.

The aims of the work are to determine the electrical and photoelectrical characteristics and dominant transport charge mechanism also, to examine the energetic spectrum and nature of localized defect levels in PDF films, deposited from plasma abnormal glow discharge.

2. Experimental

Methods of films deposition from abnormal glow discharge and the study of their structural features are described in detail in [2, 3]. We submit a new discharge system for diamond film deposition at AC abnormal glow discharge. In this process the discharge in form of a plasma line operates between two small electrodes, and the inter electrode spacing reaches 20 cm. The electrodes are moved away from the substrate and do not contaminate the film [2]. Dark surface conductivity σ and photoconductivity $\Delta\sigma_{ph}=\sigma_{ph}-\sigma$ (σ_{ph} is conductivity at lighting), photosensitivity $K(U, T, h\nu)=\Delta\sigma_{ph}/\sigma$ were measured at a constant voltage between the electrodes U=0.01–300 V, at temperature T=300–700 K and photon energy $h\nu$=1.5–4.0 eV. The electrodes were deposited on the surface by rubbing the soft graphite. Temperature dependences of the σ, $\Delta\sigma_{ph}(T)$ were approximated by equation for the activation mechanism

$$\sigma_a(T)=\sigma_0\times e^{(-\varepsilon_\sigma/k\cdot T)}, \qquad (1)$$

where σ_0 – pre–exponential factor, ε_σ – activation energy, k – Boltzmann constant and equation for the hopping mechanism of transport between the localized states (LS) near the Fermi level E_F in frame model of Mott

$$\sigma_p(T)=\sigma_0'\times e^{(-(T_0/T)^{0.25})}, \qquad (2)$$

where σ'_0 – pre-exponential factor, T_0–activation energy [16–18]. The density of states $N(E_F)$ was calculated from values T_0 according to [16–18]. The sign of the dominant charge carriers was determined from amplitude of the photo–stimulated currents $I_{PhTSC}(T, h\nu, U=0)$ [16].

3. Result and discussion

Current voltage characteristic (CVC) I(U), field dependence of K(U) and characteristics of carrier transport ε_σ, σ_0, T_0, σ'_0, $N(E_F)$, I_{PhTSC} vary in thickness and deposition plane, due to the inhomogeneity of PDF properties (table 1). Quantitative differences in the parameters σ and $\Delta\sigma_{ph}$ were recorded from the growth smooth and opposite rough sides (figure 1). Conductivity from the growth side (samples 12, 42, 62) is higher than from the side of the substrate (samples 11, 31, 41, 51) and the ratio between K is opposite (table 1, figure 1). In vacuum σ in 1.1–1.3 times lower than in the air, due to the weak influence of the absorption currents. Only physical adsorption of molecules of atmospheric gases on the surface of PDF affects the change of $\sigma(T)$. So, stable chemical bindings are not formed. CVC coincidence at U>0 and U<0 indicates a weak influence of volume space charge on σ and $\Delta\sigma_{ph}$. The annealing of the films before T_{ann}<600 K stabilizes the properties. CVC of the PDF, which having

Table 1. Electrical and photoelectrical characteristics of deposited PDF.

Sample number	σ (S)	ε_σ (eV)	$N(E_F)$ ($eV^{-1}\cdot cm^{-3}$)	R (nm)	E_g (eV)	$\alpha\cdot 10^4$ (cm^{-1})	I_{PhTSC} (A)
11	$6\cdot 10^{-11}$	0.21	$5.6\cdot 10^{17}$	8.1	1.48	4.8	$-4\cdot 10^{-11}$
12	$1.1\cdot 10^{-4}$	0.007	$1.5\cdot 10^{21}$	1.1	1.11	2.6	$-1\cdot 10^{-8}$
31	$4\cdot 10^{-10}$	0.12	$1.5\cdot 10^{18}$	6.4	1.28	3.3	$-1\cdot 10^{-11}$
41	$7\cdot 10^{-10}$	0.15	$5.6\cdot 10^{17}$	8.1	1.28	3.2	$-6\cdot 10^{-11}$
42	$4\cdot 10^{-9}$	0.08	$4.2\cdot 10^{18}$	4.9	1.55	2.7	
51	$2\cdot 10^{-13}$	0.09	$5\cdot 10^{18}$	4.7		3.0	$-2\cdot 10^{-13}$
62	$8\cdot 10^{-13}$	0.12	$1.6\cdot 10^{18}$	6.2		2.8	$-5\cdot 10^{-13}$

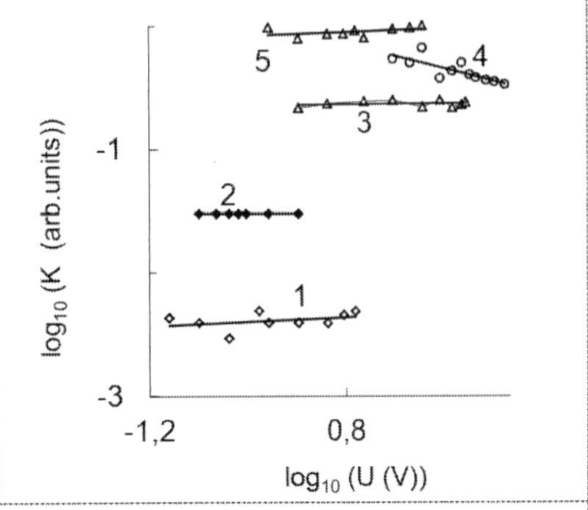

Figure 1. Field dependence of photosensitivity K(U) of PDF from the growth side (curves 1 and 2) and from the substrate side (curves 3–5) in the centre (curves 1,3) and on edge deposition (curves 2,4 and 5).

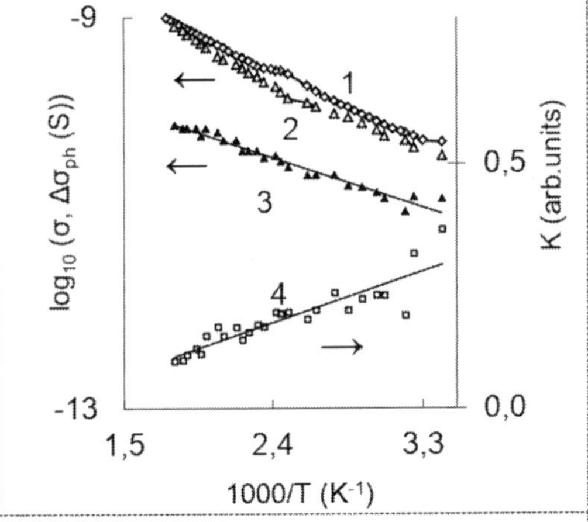

Figure 2. Temperature dependences of conductivity σ at the sequential heating (curves 1 and 2), photoconductivity $\Delta\sigma_{ph}$ (curve 3) and photosensitivity K (curve 4) of PDF from the growth side in the centre of deposition.

$\sigma=10^{-10}$–10^{-4} S is ohmic. These films have a low (K=0.005–0.1), but stable photosensitivity to heat and change U, as compared with dielectric PDF with K=0.1–1.0 (figure 1). In PDF with $\sigma=10^{-14}$–10^{-11} S deviation from the linear law is fixed and CVC obeys the equation $I\propto\exp(\beta U)$ (β=0.02–0.05) which indicates the effect from currents, limited by volume space charge, as in [8]. In CVC nonlinearity makes the spatial distribution of the energy barrier that exists on the border with electrodes and grain boundaries [6–8]. Variation in PDF electrical parameters may be caused by inhomogeneity in the distribution of crystallites on size and growth defects in their crystallites and on the boundaries or by the influence of hydrogen impurity in the surface layers [4, 11–14].

Dependences σ, $\Delta\sigma_{ph}(T)$ in interval T=300–600 K are determined by thermally stimulated electron exchange between the shallow donor levels with energy ε_σ=0.007–0.21 eV and the conduction band (CB) (figure 2). N–type σ and $\Delta\sigma_{ph}$ are dominated in PDF layers as shown I_{PhTSC} (T, hν) (table 1). At illumination the value ε_σ is lower at 0.03–0.05 eV than for σ by reducing the population of LS upon excitation by 5–10 times. There is usual thermal quenching of $\Delta\sigma_{ph}(T)$ at T≥340 K (figure 2, curve 4). It cannot be excluded the influence on characteristics the exchange of holes between LS defects and the valence band (VB) [13]. Differences in deposition methods reflect on the specific character of dependence $\sigma(\varepsilon_\sigma)$ of PDF films deposited from the abnormal glow discharge (figure 3). The electrical parameters indicate the effect on transport from the hopping conductivity, which is confirmed by approximation $\sigma(T)$ by equation (2) in interval ΔT=300–600 K. Density of LS calculated from σ, $\Delta\sigma_{ph}(T)$ is $N(E_F)=5.6\cdot10^{17}$–$1.5\cdot10^{21}$ $eV^{-1}cm^{-3}$ in PDF with ε_σ=0.007–0.15 eV (table 1). In other PDF films a density of LS, exponentially distributed near E_F, consist $N(E_F)=(4$–$8)\cdot10^{19}$ [17, 18] and 10^{15}–10^{18} $eV^{-1}cm^{-3}$ [19]. The magnitude of the most probable jump length is calculated as in [16–18] for values $N(E_F)$ is R=1.1–8.1 nm and under lighting is reduced to 1.0–6.0 nm accordingly with the increase of the LS density in 3–5 times for PDF with $\sigma<10^{-10}$ S.

Dependences between the electrical parameters allow establishing the predominance of one transport charge mechanisms. Interrelation between σ, σ_0 and ε_σ indicates the predominance of activation transport (table 1, figure 3). This confirmed by interrelation $N(E_F)(\varepsilon_\sigma)$ which is typical for

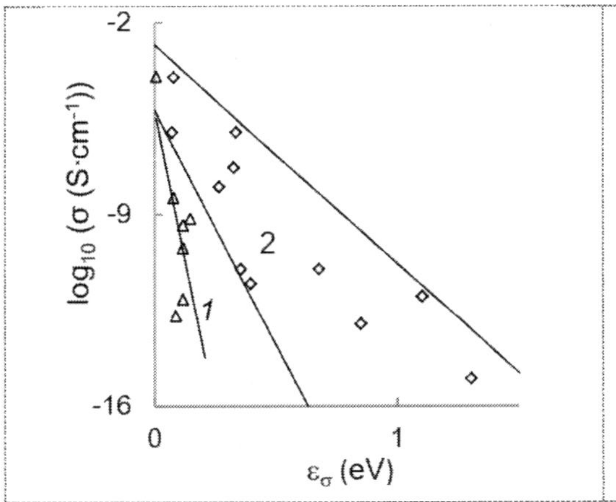

Figure 3. Interrelation between conductivity σ and activation energy ε_σ in PDF on our data (curve 1) and on literature data [4–7, 11] (curve 2).

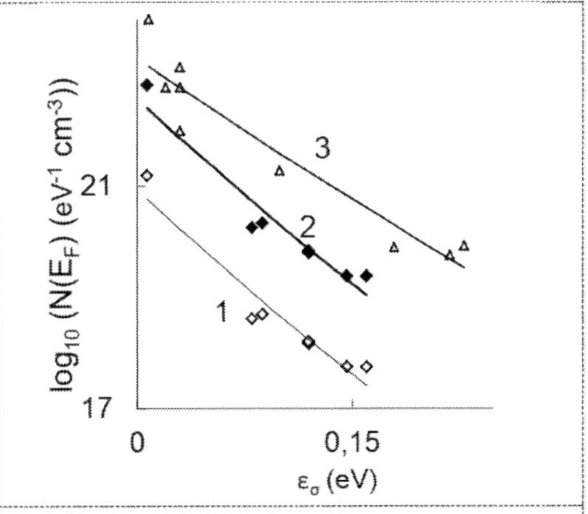

Figure 4. Interrelation between $N(E_F)$ and ε_σ in PDF on our data (curves 1 and 2) and on literature data [17] (curve 3). Decay constant $\alpha=5.7\cdot10^6$ (curve 1) and $2\cdot10^7$ cm^{-1} (curve 2).

the uncrystalline semiconductors having a high content of defects (figure 4) [16–19]. By analogy with [6–13, 16–19], we can conclude that PDF hopping conductivity on LS near the Fermi level dominates over the activation in interval T=300–400 K. At T=400–600 K the characteristics of PDF is determined by the electrical conductivity activation $\sigma_a(T)$ with LS defects distributed on grain boundaries and localized in band gap (BG) in area of "tails" of the allowed bands. Charge carriers are excited through the mobility gap in the area of delocalized states. The relationship between the parameters σ'_0 and $T_0^{0.25}$, obtained in accordance with the equation (2), confirms the applicability of hopping transport through LS distributed near E_F, and demonstrates the differences in the structural

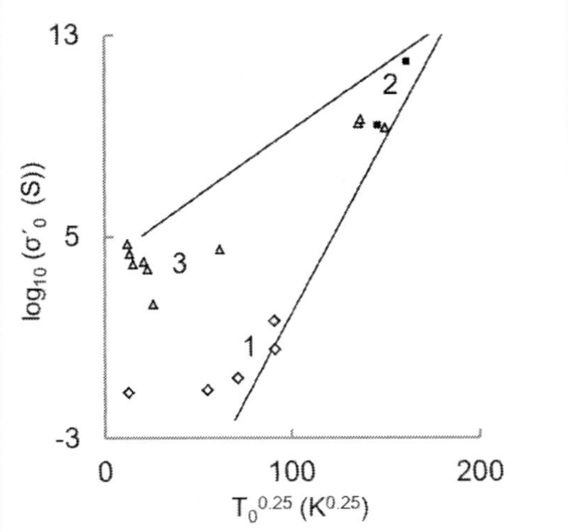

Figure 5. Interrelation between pre-exponential factor σ'_0 and activation energy $T_0^{0.25}$ (equation (2)) in PDF on our data (curve 1) and literature data [18] (points 2) and [17] (points 3).

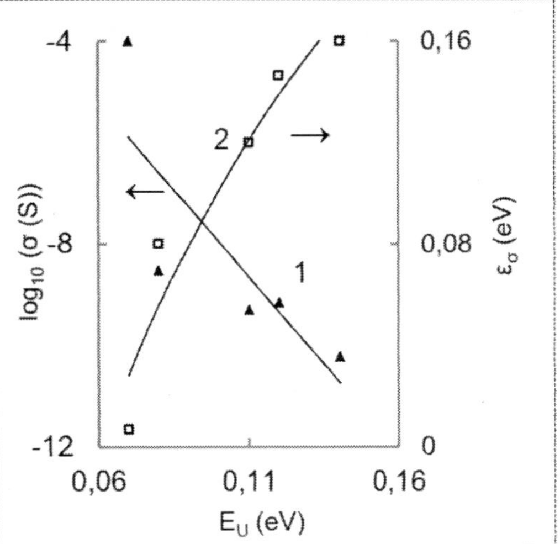

Figure 6. Interrelation between conductivity σ and Urbach energy E_U (curve 1) and activation energy ε_σ and E_U (curve 2) in PDF. Values E_U were calculated in the energetic interval $h\nu$=1.15–1.32 eV.

disorder (figure 5) [17, 18]. At T>600 K, the influence of the barrier between the crystallites is increases [6, 8]. The largest contribution to $\Delta\sigma_{ph}$ gives the activation component, as confirmed by K(T, U) (figures 1 and 2). The relationships between σ, ε_σ and direct BG width E_g, extracted from absorption spectra $\alpha(h\nu)$ in interval $h\nu$=1.5–4.2 eV, are consistent with the concept of carrier transport in amorphous or highly defective materials by exchange of carriers between CB and LS in its "tails" through the mobility gap (table 1) [1, 16]. Spectrum K(hν) is similar to $\alpha(h\nu)$ of PDF [1, 4]. A same spectra is proper to other PDF in intervals hν=1.8–2.9 eV and is typical for most structures in detectors of UV irradiation [4]. In work [12] a spectrum K(hν) was determined by electronic transitions between the VB and LS defects in thin oxidized layer after hydrogenation. Such a spectrum in nano–crystalline films was induced by transitions in LS of "tails" and levels belonging to graphite–like phases [1]. At that, oxygen arriving from the air and adsorbing on structural defects may have a passivation effect on the surface photosensitivity.

Accumulation of concentration of the optical active levels was connected with decreasing of LS density near the Fermi level which are participated in hopping transport, as it is shown by its reduction from $N(E_F)=10^{19}$ to $5\cdot10^{17}$ $eV^{-1}\cdot cm^{-3}$ an increase of the absorption coefficient from $\alpha=2\cdot10^4$ to $5\cdot10^4$ cm^{-1} (table 1). Reduction of conduction and LS density responsible for hopping transport are correlated also with an increase in ε_σ and in Urbach energy E_U of the optical active levels which determine $\alpha(h\nu)$ in intervals hν=1.15–1.3 eV and 1.6–3.0 eV (table 1, figure 6). Growth of its value from E_U=0.07 to 0.15 eV with the decreasing of conductivity from $\sigma=10^{-4}$ to 10^{-11} S and the increasing of the activation energy from ε_σ=0.007 to 0.16 eV is determined by influence on properties from the disorder induced with defects in the crystalline phase material due to the presence of disordered graphite–like phases in films. As a result, the exponential distribution of LS on energy is formed in BG of material [8, 17, 19].

4. Conclusion

Films deposited by abnormal glow discharge not inferior to films obtained by alternative CVD methods. Vacuum annealing of films up to 600–800 K stabilizes the electrical characteristics. Electrical and photoelectric characteristics, mechanism of transport charge in the deposited PDF are caused by the defects, the levels of which are continuously distributed on energy in the BG. Trapping and recombination centers of carriers are heterogeneously distributed in grain boundaries. Activation component of dark and photoconductivity of n–type is realized by exchange of the charge carriers between the conduction band and shallow donor levels with activation energy $\varepsilon\leq0.21$ eV. Thermal quenching of photoconductivity is appeared. Activation conductivity was complemented by a hopping mechanism, involving localized states distributed in materials band gap near a Fermi level. Density of states, where the hopping conduction occurs, is significant $N(E_F)=5.6\cdot10^{17}$–$2.1\cdot10^{21}$ eV^{-1} cm^{-3}. Their variation over the surface and the film thickness is determined by content of the growth defects, heterogeneously distributed in the grain boundaries. Electrical and spectral parameters are subjected by strong effects from disorder induced by defects in crystallites and presence of disordered graphite–like phase in films.

Acknowledgments
The paper has been completed within megaproject of National Research Tomsk Polytechnic University «Materials for Extreme Conditions».

References

[1] Williams O A 2011 *Diamond and Relat. Mater.* **20** 621
[2] Linnik S A and Gaydaychuk A V 2013 *Diamond and Relat. Mater.* **32** 43
[3] Linnik S A and Gaydaychuk A V 2012 *Tech. Phys. Lett.* **38** 258
[4] Polyakov V I, Rukovishnikov A I, Avdeeva L A, Kun'kova Z E, Varnin V P, Teremetskaya I G and Ralchenko V G 2006 *Diamond and Relat. Mater.* **15** 1972
[5] Chiquito A J, Berengue O M, Diagonel E, Galzerani J C 2007 *J. Appl. Phys.* **101** 033714

[6] Kopylov P G, Lotonov A M, Apolonsky I A and Obraztzov A N 2009 *Vestn. of MSU. ser. 3(In Russian)* 2 54

[7] Trucchi D M, Cappelli E, Conte G, Mattei G, Gramaccioni C and Ascarelli P 2005 *Diamond and Relat. Mater.* **14** 575

[8] Conte G, Rossi MC, Spaziani F and Arcangeli R 2005 *Diamond and Relat. Mater.* **14** 570

[9] Hikavyy A, Clauws P, Maes J, Moshchalkov V V, Butler J E, Feygelson T, Williams O A, Daenen M and Haenen K 2006 *Phys. status solidi* **A 203** 3021

[10] Muret P and Saby Ch 2004 *Semicond. Sci. and Technol.* **19**1

[11] Hubik P, Mares J J, Kozak H and Kromka A 2012 *Diamond and Relat. Mater.* **34** 63

[12] Stallhofer M, Seifert M, Hauf M V, Abstreiter G, Stutzmann M, Garrido J A and Holleitner A W 2010 *Appl. Phys. Lett.* **97** 111107

[13] Yutaka Itoh, Yu Sumikawa, Hitoshi Umezawa and Hiroshi Kawarada 2006 *Appl. Phys. Lett.* **89** 203503

[14] Chaudhary A, Welch J O and Jackman R B 2010 *Appl. Phys. Lett.* **96** 242903

[15] Ri Sung-Gi, Takeuchi Daisuke, Kato Hiromitsu, Ogura Masahiko, Makino Toshiharu, Yamasaki Satoshi, Okushi H, Rezek B and Nebel Ch E 2005 *Appl. Phys. Lett.* **87** 262107

[16] Kabyshev A V, Konusov F V, Remnev G E and Pavlov S K 2013 *Proc. of the universities. Physics (In Russian)* **56** 30

[17] Achatz P, Williams O A, Bruno P, Gruen D M, Garrido A and Stutzmann M 2006 *Phys. Rev.* **B 74** 155429

[18] Gan L, Bolker A, Saguy C, Kalish R, Tan D L, Tay B K, Gruen D and Bruno P 2009 *Diamond and Relat. Mater.* **18** 1118

[19] Nath S and Wilson J I B 1996 *Diamond and Relat. Mater.* **5** 65

IOP Conf. Series: Materials Science and Engineering **81** (2015) 012007 doi:10.1088/1757-899X/81/1/012007

Research on the radiation exposure “memory effects” in AlGaAs heterostructures

A V Gradoboev, V V Sednev
Tomsk polytechnic university, Tomsk, Russia

E-mail: gradoboev1@mail.ru

Abstract. Radiation exposure and long running time cause degradation of semiconductors' structures as well as semiconductors based on these structures. Besides, long running time can be the reason of partial radiation defects annealing. The purpose of the research work is to study the “memory effect” that happens during fast neuron radiation in AlGaAs heterostructures. Objects of the research are Infrared Light Emitting Electrodes (IRED) based on doubled AlGaAs heterostructures. During the experimental research LEDs were preliminarily radiated with fast neutrons, and radiation defects were annealed within the condition of current training with high temperatures, then emission power was measured. The research proved the existence of the “memory effect” that results in radiation stability enhancement with subsequent radiation. Possible mechanisms of the “memory effect” occurrence are under review.

1. Introduction

It is necessary to give a brief observance of microelectronic articles utilization. Microelectronic articles are utilized in different environments. More specifically, they serve in outer space, in upper atmosphere and at nuclear energy centers [1, 2] being subjected to different kinds of ionizing radiation. As ionizing radiation takes place, active layers of microelectronic articles get different radiation defects, which change articles specifications and finally cause their break-down.

Similar effect occurs as a result of a longtime service when electric and thermal fields cause defects of articles' active layers.

Consequently, ionizing radiation and longtime exploitation lead to defects in microelectronic articles' active layers and to degradation of their output characteristics. It is worth noticing that defects caused by ionizing radiation can differ in their structure and properties from defects caused by long operation time.

However, it is well known that longtime service factors might result in full or partial annealing of defects that occur as a consequence of ionizing radiation (e.g., [3]). In this case if we preliminarily subject microelectronic articles to ionizing radiation and then to longtime service, we shall obtain full or partial recovery of articles specifications, which can be observed in actual practice.

Behavior of prepared articles newly subjected to radiation is expected to be different from that of primary ones. Whence, if ionizing radiation and/or annealing provoke defect migration to the stock region, where it might interact with other defects, there is a little hope it will move back.

Corresponding author: Gradoboev Alexander Vasilievich, professor, Doctor of Technical Science,
+7 913 866 84 05, e-mail: gradoboev1@mail.ru,
634050, Tomsk polytechnic university, 30 Lenina av., Tomsk, Russia.

Consequently, difference in primary and prepared articles behavior being newly radiated can be called the "memory effect". Similar effects were observed previously in metal alloys [4].
The purpose of this research work is to study the "memory effect" in AlGaAs doubled heterostructures meant to be used to produce infrared light emitting diodes (LED) radiated with fast neutrons.
In the first instance it is necessary to prove the existence of this effect. The "memory effect" investigations (conditions that cause its occurrence, its impact on radiated articles criteria parameters changes, its stability over time, etc.) will hereafter allow us to forecast microelectronic articles behavior when subjected to complex (simultaneous) or combined (spread out over a period of time) ionizing radiated emission or to longtime service factors.

2. Research objects and methods

Commercial LEDs based on doubled AlGaAs heterostructures with 1μm thickness of an active layer, obtained through liquid-phase epitaxi with monocrystall n-GaAs as a base-member, were taken as research objects.
LEDs were fabricated through the standard sandwich technology where deposition techniques and layer metallization to form ohmic contacts are applied; photolithography and chemical etching to form crystals (chips) as well as scribing to divide wafers into separate chips were used. LEDs had corpses and lens elements to form necessary diagram of light flux directivity from optical compound. According to the preliminary research data optical compound used to fabricate lenses does not change its optical properties even at $F_n = 2 \cdot 10^{14}$ cm^{-2} fast neutron radiation. From there all changes of LEDs′ optical properties resulted from radiation within the mentioned above fast neutrons transfer might be considered as conditioned by changes of diodes active region optical properties.
When battery mode is continuous the intensity of forward operating current of LEDs under review makes I_{op} = 100 mA, for this purpose voltage supply does not exceed U_{op} = 3 V. Maximum wave mode is within 0,82 .... 0,90 μm range.
Watt – ampere specification of each LED before and after radiation were evaluated in a blob. Data obtained were analyzed by mathematical statistics technique. Each lot of LEDs was classified by average measured values. Dispersion in the radiation power in primary LEDs did not exceed ±10%, but it increased after fast neutron radiation up to ±15%.
It is worth noticing that equipment used to measure vat-ampere specification in a blob provided each individual LED with radiation measurement repeatability with less than ± 2,5% deflection for (1-200) mA operating current range.
Fast neutron radiation was delivered in passive battery mode without operating current, while fast neutron impact level was characterized by F_n (cm^{-2}) particle transfer.
Each lot of LEDs was provided with a sequential number of fast neutron transfers with each definite pitch. Results from the analysis of different lots of LEDs and vat-ampere specification measurements make it possible to prevent any radiation defects annealing during measurement. Therefore an observable change of radiant power is determined by fast neutrons impact only.
Three lots of LEDs were investigated during the research work. Each lot specifications are shown in table 1. Experiments allowed obtaining and describing corresponding relations between radiation power decrease and fast neutrons transfer.
Table 1 shows that the "memory effect" research was carried out by means of comparative study of relation between LEDs′ radiant power decrease when radiated with fast neutrons (for primary LED-0) and LEDs′ radiation power decrease when radiated with fast neutron (other two lots of LED-1 and LED-2) which were obtain under following conditions:
LEDs were preliminarily subjected to radiation at different fast neutrons transfer rates that resulted in radiant power decrease (ground for neutron transfer rate choice will be given in the following chapter);
Then LEDs were subjected to current training with I_{op}= 75 mA operating current during 24 hours with + 65 ^{0}C environment temperature, which led to radiant power recovery;
Further LEDs′ radiant power decrease when radiated with fast neutrons was investigated. During the experiments a sequential increase in effect level was applied.

IOP Conf. Series: Materials Science and Engineering **81** (2015) 012007 doi:10.1088/1757-899X/81/1/012007

Table 1. Investigated lots of LEDs specifications

	Preconditioning	Primary	Radiation:	Annealing:
LED-0		LED	non	non
	Radiant power P, mW	34,82	-	-
	Forward voltage U_{op}, V	1,63	-	-
	Preconditioning	Primary	Radiation:	Annealing:
		LED	$F_n = 6{,}08 \cdot 10^{11}$ cm^{-2}	$t = 24$ h, $T = +65$ ^{0}C,
LED-1				$I_{op} = 75$ mA
	Radiant power P, mW	36,00	27,18	36,02
	Forward voltage U_{op}, V	1,55	1,54	1,56
	Preconditioning	Primary	Radiation:	Annealing:
		LED	$F_n = 3{,}57 \cdot 10^{13}$cm^{-2}	$t = 24$ h, $T = +65$ ^{0}C,
LED-2				$I_{op} = 75$ mA
	Radiant power P, mW	32,67	10,08	24,89
	Forward voltage U_{op}, V	1,45	1,45	1,48

Practically observable LEDs′ specification changes (radiant power and forward voltage at predetermined operating current) for each lot subjected to different types of preconditioning are presented in table 1. The table contains half-values for each lot of LEDs.

3. Experimental findings and discussion

In the first instance it is necessary to study the findings obtained during the investigation of LED-0 lot's radiant power decrease when radiated with fast neutrons, presented in table 1. Radiant power measured after radiation P_F was normalized to P_0 in case with primary LEDs, thereby power level measured when applying I_{op} = 100 mA operating current was used. Similar normalization of LEDs′ radiant power will be employed further on using for starting value its value before it obtained these relations.

Radiation model [5] for LEDs based on AlGaInP heterostructures with multiple quantum wells is the most relevant to describe experimental findings. Data presented in figure 1 show that process of power decrease can be described through three periods.

During the first period radiant power decrease occurs as a result of radiation-stimulated

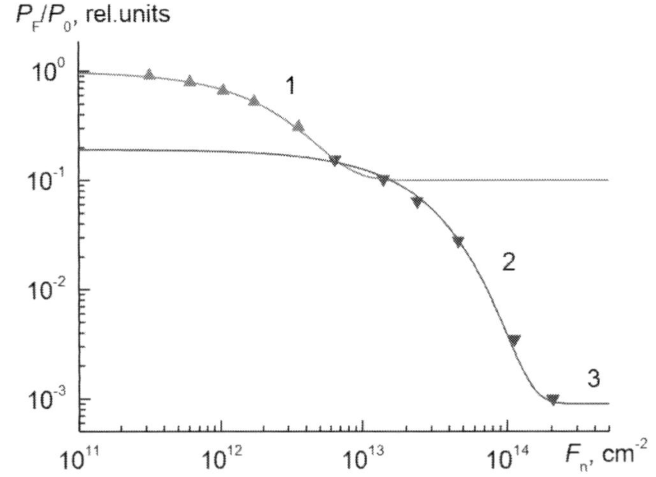

Figure 1. LED emissive power lowering in passive power mode under fast neutron irradiation.

reconfiguration of present defect structure what allows explaining the period finiteness (radiant power decrease process saturation). In this case radiant power decrease is described with the following equation:

$$\frac{P_F}{P_0} = 1 - A \cdot \left[1 - \exp\left(-C_1 \cdot F_n\right)\right] \quad (1)$$

where P_F, P_0 – radiant power with I_{op} = 100 mA, measured before and after radiation consequently; A – coefficient of proportionality that characterizes the first period input into the overall radiant power decrease when radiated with fast neutrons; C_1 – damage coefficient that specifies the speed of radiant power decrease during the first period at neutron transfer growth [cm^2].

The second period is characterized by LED's radiant power change with neutrons transfer growth that occurs due to the radiation defects and is described with the equation

$$\frac{P_F}{P_0} = \frac{P_{min}}{P_0} + B \cdot \exp\left(-C_2 \cdot F_n\right) \quad (2)$$

where P_{min} – LED radiant power with low injection of electrons into its active region without any dependence from current flow; B – coefficient of proportionality that characterizes the second period input into the overall radiant power decrease when radiated with fast neutrons; C_2 – damage coefficient that specifies the speed of radiant power decrease during the second period at neutron transfer growth [cm^2].

The third period corresponds to LED fast neutron radiation limiting value. LED occurs in a mode of low injection of electrons into its active region. Radiant power is described in the equation

$$P_F = P_{min} \quad (3)$$

During the third period of radiant power decrease, the majority of LEDs catastrophically break down.

More precise comparison of 1-3 formulae and results, presented in table 1, makes some discordance observable. The second period starts with some higher level of LED's radiant power, than it comes from formula (1).

In particular, from formulae (1,2) goes the following formula

$$A + B = 1 \quad (4)$$

Actually, results from table 1 and equations we have

$$\delta\left({}^{P_F}\!/\!_{P_0}\right) = (A + B) - 1 \geq 0 \quad (5)$$

Earlier investigations [6] in AlGaInP heterostructured LEDs' radiant power decrease when radiated with gamma-ray quantum proved that at the edge of the first and the second periods of power decrease some relaxation processes occur. These processes result in partial recovery of radiant power. At this LEDs' radiant power at the edge of the first and the second periods were observable in an explicit form.

In this research radiant power growth $\delta\left({}^{P_F}\!/\!_{P_0}\right)$ on the edge of the first and the second periods can also be explained by alike relaxation processes with partial recovery of radiant power which are not observable in an explicit form.

Summing up, LEDs under consideration being subjected to fast neutron radiation get through the period when radiant power conditioned by radiation-stimulated reconfiguration of present defect structure decreases. This fact ground the decision to take these LEDs as "memory effect" research objects. This is a serious possibility to assume the "memory effect" at this period.

Research findings submitted in the paper allow explaining the choice of fast neutrons transfer for preliminary radiation (figure 1). The first transfer of neutrons (LED-1) is chosen to the effect that radiant power decrease occurs within the first period and the second transfer of neutrons (LED-2) corresponds to the second period of radiant power while radiated.

Further the influence of a preliminary fast neutron radiation mode in the identical conditions of annealing on radiant power decrease with further fast neutron radiation (LFD-1, LED-2) is described.

Figure 2 shows relative changes of radiant power for given LEDs with further fast neutron radiation. In this case LED s′ radiant power normalization was up to its value after annealing (see table 1). Here are the experimental findings obtained for lot LED-0 to make comparison possible.
Results in figure 2 prove that radiant power changes after radiation are explained by earlier presented equations (1-3). Table 2 contains proportionality and damage coefficients for 1-3 equations, which were obtained during experimental findings analysis, presented in figure 2.

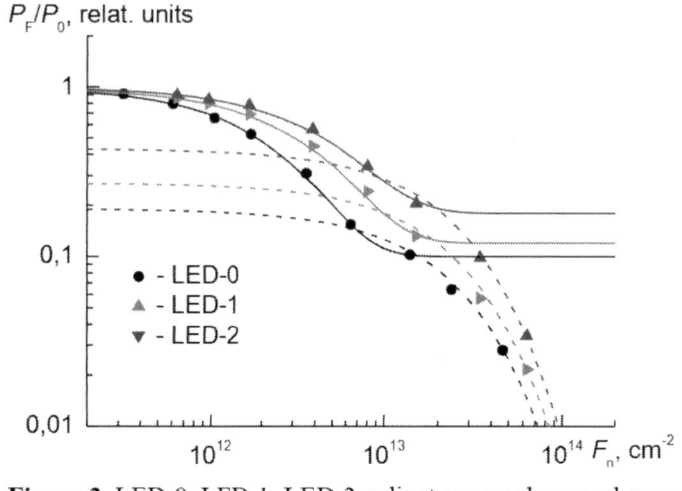

Figure 2. LED-0, LED-1, LED-2 radiant power change when radiated with neutrons: symbols – experimental data; lines – determined relations

Table 2. Proportionality and damage coefficients for different lots of LEDs

LED lot	A	C_1, cm^2	$\delta\left(P_F/P_0\right)$	B	C_2, cm^2
	First period			Second period	
LED-0	0.90	$4{,}35\cdot10^{-13}$	0,09	0,19	$4{,}17\cdot10^{-14}$
LED-1	0.88	$2{,}70\cdot10^{-13}$	0,15	0,27	$4{,}17\cdot10^{-14}$
LED-2	0.82	$2{,}12\cdot10^{-13}$	0,25	0,43	$4{,}17\cdot10^{-14}$

Obtained data show that the "memory effect" brings into change A, B, C_1,proportionality coefficients value of which depends on the LED's history, whereas in contrast C_2 coefficient is identical to all presented lots of LED. The fact that C_2 coefficient does not depend on the history of LED additionally proves that during the second period fast neutron radiated LED s′ radiant power decrease is the result of radiating defects, at that, defect structure of LED s′ primary active layer is absolutely prevented. Besides, the "memory effect" exceeds recovery degree of radiant power at the edge of the first and the second periods of LED radiant power decrease when radiated with fast neutrons.
Thereby, "memory effect" occurrence when AlGaAs heterostructures are radiated with fast neutrons has been proved. Experiment findings presented in figure 2 demonstrate a possibility to increase LED s′ stability to fast neutron radiation by means of the "memory effect". Of course, it is possible to object that preconditioning result in radiant power decrease in some cases, in spite of its partial (sometimes full) recovery during annealing. It is worth pointing out that any microelectronic article with high ionizing radiation resistance does not possess peak specifications as its basic property is parameter permanence. If parameters change, they change in some limited boundaries being subjected to maximal value of ionizing radiation.

Moreover, taking into consideration that the purpose of the research work is to prove experimentally the anticipation that the ionizing radiation "memory effect" exists; chosen conditions cannot be admitted as optimum. In future radiation and annealing modes optimization might make it possible to develop some practical techniques of the "memory effect" application in order to improve exploitation specifications of different microelectronic articles. At this, ionizing radiation stability might grow without any significant decrease of primary microelectronic articles' parameters.

Fast neutrons radiation stability growth can be demonstrated with the help relations presented in figure 3. Here are relations of the maximum allowable transfer of fast neutrons that allows to keep radiant power decrease at the maximum permissible value $(P_F/P_0)_{limit}$ for the lots of LEDs under consideration.

It is obvious that the "memory effect" causes fast neutron influence stability growth not only within the first period but also within the second period. At this the maximum LED s′ stability growth occurs within the first period as a consequence of a relative contribution of this period into the general radiant power decrease when radiated, and owing to damage coefficient decrease. Stability during the second period of radiant power decrease grows as well, but damage coefficient remains unchanged.

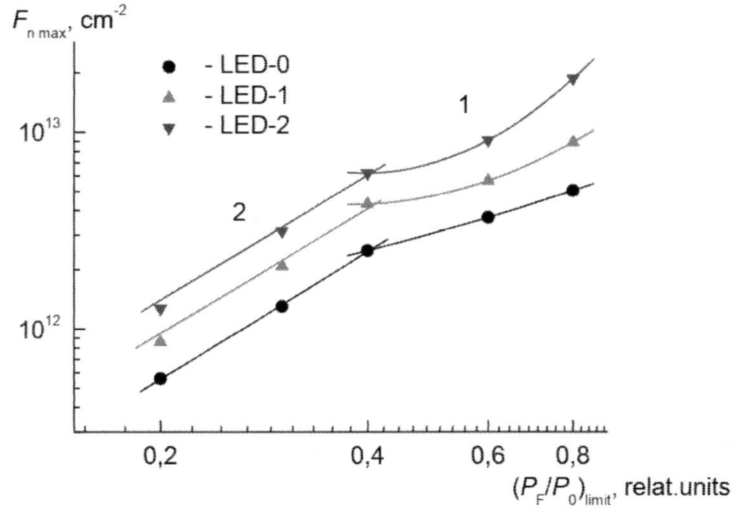

Figure 3. Relations of neutron bearable transfer and maximum permissible radiant power decrease for different LED lots: symbols – experimental data; lines - determined relations; 1 – radiant power change during the first period; 2 – radiant power change during the second period

4. Conclusion

1. For the first time a hypothesis of the "memory effect" of radiation exposure in semiconductor materials and instruments was formulated.
2. For the first time the "memory effect" existence during fast neutron radiation in doubled AlGaAs heterostructures, aimed to be used for IRED production, was experimentally proved.
3. The "memory effect" causes fast neutron influence stability growth. At this LED′s fast neutron radiation stability growth within the first period of radiant power decrease is preconditioned by relative contribution reduction of this period into the general radiant power decrease during radiation and damage coefficient depletion. Stability during the second period of radiant power decrease grows as well, but damage coefficient remains unchanged.
4. The "memory effect consideration" allows increasing a prognostication quality of LEDs′ operation in the conditions of complex (simultaneous) and combined (spread out over a period of time)

factors of long running time and fast neutrons transfer.

5. The "memory effect" application allows increasing the quality of AlGaAs heterostructured LEDs' running properties.

References

[1] Srour J R and McGarrity J M 1988 *IEEE. Proceeding,* **76** 1443-1469
[2] Claeus E and Simoen E 2002 *Radiation effects in advanced semiconductor materials and devices* Berlin, Springer-Verlag
[3] Gradoboev A V Surzhikov A P 2005 *Radiation stability of GaAs microwave devices* Tomsk, Izdatelstvo TPU 277p
[4] Skakov M K and Melikhov V D 2001 *News of Higher Educational Institutions. Phisics* **44** 38
[5] Gradoboev A V and Orlova K N 2014 *Advanced Material Research* 880 237
[6] Gradoboev A V, Orlova K N, Arefyev K P, Asanov I A 2013 *News of Higher Educational Institutions. Phisics* **56** №11/3 116

IOP Conf. Series: Materials Science and Engineering **81** (2015) 012008 doi:10.1088/1757-899X/81/1/012008

Transition in AlGaInP heterostructures with multiple quantum wells during fast neutron radiation

A V Gradoboev, K N Orlova
National research Tomsk polytechnic university, Tomsk, Russia

E-mail: gradoboev1@mail.ru

Abstract. Radiation exposure causes degradation of semiconductors' structures as well as different semiconductors based on these structures. The purpose of the research work is to study transitions in AlGaInP heterostructures with multiple quantum wells during fast neutron radiation Objects of the research are 590 nm and 630 nm LEDs based on AlGaInP heterostructures. It is proved that LEDs' radiant power decrease occurs within three periods: during the first period radiant power decrease is caused by radiation stimulated structural adjustment of a primary defect structure; during the second period the decrease is results from radiative defects introduction; with further enhancement of radiation exposure the second period develops into the third period, where LEDs evolves into the mode of electrons low injection into an active region. Empirical relations explain radiant power changes within each period are presented. Region of transitions between the first and the second periods that cause radiant power partial recovery are specified. Transitions occur both directly and indirectly for heterostructures. Potential causes of transitions occurrence are being discussed.

1. Introduction

Microelectronic articles are utilized in different conditions. More specifically, they serve in outer space, in upper atmosphere and at nuclear energy centers [1-3] being subjected to different kinds of ionizing radiation. As ionizing radiation takes place, different radiation effects are introduced into active layers of microelectronic articles, which change articles specifications and finally cause their break-down. These result in academic interest toward ionizing radiation effect on microelectronic articles.

AlGaInP heterostructures with multiple quantum wells are widely used in design and serial production of visible LEDs for different wavelengths. The fact that volume of active layers of heterostructured LEDs with quantum wells decreases makes it reasonable to suppose that they are more stable to ionizing radiation effect. This in conjunction with expected higher photoresponse determines academic interest to heterostructures.

Radiant stability of AlGaInP heterostructured LEDs with multiple quantum wells is being investigated and there are some recent papers with experimental findings [4-7]. In particular some researchers [7] present their experimental findings in the problems transitions in AlGaInP heterostructures with multiple quantum wells being subjected to gamma-quantum radiation. Those experimental findings allow anticipating alike effects in other conditions.

The purpose of the research work is to study transitions in AlGaInP heterostructures with multiple quantum wells during fast neutron radiation.

Corresponding author: Gradoboev Alexander Vasilievich, professor, Doctor of Technical Science, tel. +7 913 866 84 05, e-mail: gradoboev1@mail.ru, 634050, Tomsk polytechnic university, 30 Lenina av., Tomsk, Russia.

IOP Conf. Series: Materials Science and Engineering **81** (2015) 012008 doi:10.1088/1757-899X/81/1/012008

2. Research objects and methods

Several lots of LEDs produced on the base of different AlGaInP heterostructures with multiple quantum wells and 1,8-2,0 μm thickness of active layers which were obtained from different sources and meant to produce LEDs with 590 nm wavelength (LED 1) and 630 nm wave mode (LED 2).

LEDs were fabricated through the standard sandwich technology where deposition techniques and layer metallization to form ohmic contacts are applied; photolithography and chemical etching to form crystals (chips) as well as scribing to divide wafers into separate chips were used. LEDs had corpses and lens elements to form necessary diagram of light flux directivity from optical compound. According to the preliminary research data optical compound used to fabricate lenses does not change its optical properties even at $F_n = 2\cdot10^{14}$ cm^{-2} fast neutron radiation. From there all changes of LEDs' optical properties resulted from radiation within the mentioned above fast neutrons transfer might be considered as conditioned by changes of diodes active region optical properties.

LEDs in some lots had an identical structure of a crystal and approximately alike sectional area of an active region. Each lot contained 20 LEDs. When battery mode was continuous the intensity of forward operating current of LEDs under review made I_{op} = 100 mA, for this purpose voltage supply did not exceed U_{op} = 3 V.

Watt-ampere specification of each LED before and after radiation were measured. Measurements in a blob do not depend on LEDs' orientation. This fact allowed to provide each individual LED with radiation measurement repeatability with less than ± 2,5% deflection for (1-200) mA.

Data obtained were analyzed by mathematical statistics technique. Each lot of LEDs was classified by average measured values. Dispersion in the radiation power in primary LEDs did not exceed ±10%, but it increased after fast neutron radiation up to ±15%.

Fast neutron radiation was delivered in passive battery mode without operating current, while fast neutron impact level was characterized by F_n (cm^{-2}) particle transfer.

Each lot of LEDs was provided with a sequential number of fast neutron transfers with each definite pitch up to full break out of more than 80% of LEDs in a lot. Results from the analysis of different lots of LEDs with different fast neutron transfer and watt-ampere specification measurements make it possible to prevent any radiation defects annealing during measurement. Therefore an observable change of radiant power is determined by fast neutrons impact only.

3. Experimental findings and discussion

Observation of radiant power decrease of different lots of subjected to fast neutron radiation LEDs proves that each lot under consideration undergoes transitions identical to those that were observed earlier in the conditions of gamma-quant radiation [7] at the boundary of the first and the second periods of radiant power decrease. All AlGaInP heterostructures can be precisely divided into two specific groups. The first group is formed by heterostructures with direct transitions. Consequently, the second group is formed by heterostructures with indirect transitions. Alike results were observable in both LED-1 and LED-2 lots.

In the first instance it is necessary to study the findings obtained during the investigation of indirect transitions. Figure 1 shows a relative radiant power change of LED-1 lot measured at 100 mA operating current during fast neutron radiation. The same results occurred with LED-2 lot. .

Radiation model [8] is the most relevant to describe experimental findings. Data presented in figure 1 show that process of power decrease can be described through three periods.

During the first period radiant power decrease occurs as a result of radiation-stimulated reconfiguration of present defect structure what allows explaining the period finiteness (radiant power decrease process saturation). In this case radiant power decrease is described with the following equation:

$$\frac{P_F}{P_0} = 1 - A\cdot\left[1-\exp\left(-C_1\cdot F_n\right)\right] \qquad (1)$$

IOP Conf. Series: Materials Science and Engineering **81** (2015) 012008 doi:10.1088/1757-899X/81/1/012008

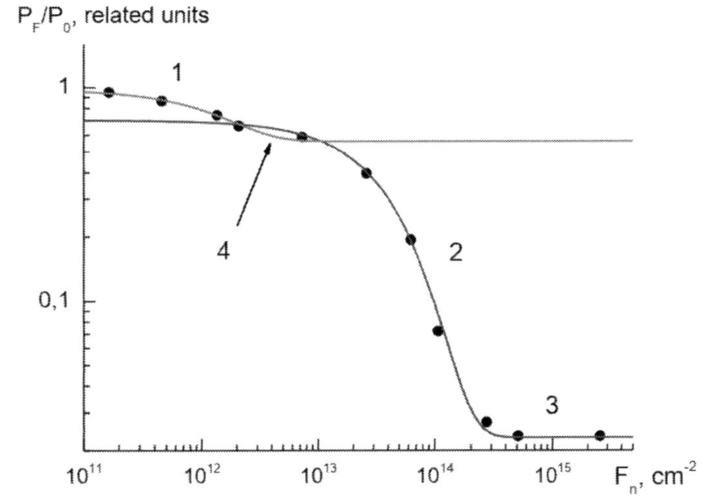

Figure 1. LED-1 (I_{op}=100 mA) radiant power change when radiated with fast neutrons: symbols – experimental data; lines – determined relations; 1, 2, 3 – determined periods of radiant power decrease; 4 – transitions region

where P_F, P_0 – radiant power with I_{op} = 100 mA, measured before and after radiation consequently; A – coefficient of proportionality that characterizes the first period input into the overall radiant power decrease when radiated with fast neutrons; C_1 – damage coefficient that specifies the speed of radiant power decrease during the first period at neutron transfer growth [cm^2].
The second period is characterized by LEDs′ radiant power change with neutrons transfer growth that occurs due to the radiation defects and is described with the equation

$$\frac{P_F}{P_0} = \frac{P_{min}}{P_0} + B \cdot \exp(-C_2 \cdot F_n) \quad (2)$$

where P_{min} – LED s′ radiant power with low injection of electrons into its active region without any dependence from current flow; B – coefficient of proportionality that characterizes the second period input into the overall radiant power decrease when radiated with fast neutrons; C_2 – damage coefficient that specifies the speed of radiant power decrease during the second period at neutron transfer growth [cm^2].
The third period corresponds to LED s′ fast neutron radiation limiting value. LEDs occur in a mode of low injection of electrons into its active region. Radiant power is described in the equation

$$P_F = P_{min} \quad (3)$$

During the third period radiant power decrease, the majority of LEDs catastrophically break down.
More precise comparison of 1-3 formulae and results, presented in Table 1, makes some discordance observable. The second period starts with some higher level of LED s′ radiant power, than it comes from formula (1).
In particular, from formulae (1,2) goes the following formula

$$A + B = 1 \tag{4}$$

Actually, results from Table 1 and equations we have

$$\delta\left(\frac{P_F}{P_0}\right) = (A + B) - 1 \geq 0 \tag{5}$$

Earlier investigations [7] in AlGaInP heterostructured LEDs′ radiant power decrease when radiated with gamma-ray quantum proved that at the edge of the first and the second periods of power decrease some transitions occur. These processes result in partial recovery of radiant power. At this LEDs′ radiant power at the edge of the first and the second periods were observable in an explicit form.

In this research radiant power growth $\delta\left(\frac{P_F}{P_0}\right)$ on the edge of the first and the second periods can also be explained by alike transitions with partial recovery of radiant power which are not observable in an explicit form. Statistical analyses of experimental findings showed that observable value is statistically significant and reproducible in series of experiments.

Same results were obtained at different volumes of operating current, with other heterostructures (LED-2). At this the higher operating current the more explicit transitions at the boundary of the first and the second periods.

Earlier alike transitions at the boundary of the first and the second periods of radiant power decrease during gamma-quant radiation were noticed among identical heterostructures [9].

It is necessary to study transitions, observable directly in heterostructures subjected to fast neutron radiation (second group of heterostructures). Figure 2 shows the way radiant power of LED-1 and LED-2 changes during fast neutron radiation in case of the above described direct transition. Experimental data of radiant power measured at different values of operating current are presented.

It is clearly seen that at the boundary of the first and the second periods of radiant power decrease transitions, which are characterized with partial radiant power recovery against general decrease explicitly occur. It is also seen that radiant power recovery degree depends on operating current volume. Such dependence was observed earlier during the study of heterostructures from the first group.

Consequently, the existence of transitions between the first and the second periods of radiant power decrease during fast neutron radiation for all AlGaInP heterostructures with multiple quantum wells meant to be used in 590 nm и 630 nm wave mode LEDs production is proved. Transitions can occur both directly and indirectly.

Experimental findings prove that transitions relate with the final period of radiation stimulated restructuring of a primary defect structure in a LED′s active layer. It is possible to suppose that in the region of fast neutron transfer action present defects transform into some complexes with radiation defects, which are not the centers of radiationless recombination. In this case, partial transformation of primary defects (centers of radiationless recombination) into such complexes might explain partial recovery of LEDs′ radiant power in the region of transitions under consideration. Dependence of the observed effect of LEDs′ radiant power recovery from operating current volume also makes it possible to suppose that the obtained complexes are electrically active.

Further research in the problems of deep centers spectrum changes in the region of observed transitions during fast neutron and gamma-quant radiation are of great importance.

Comparative study of radiant power recovery in the region of observed transitions and deep centers spectrum changes might give answers to the questions of found transitions nature.

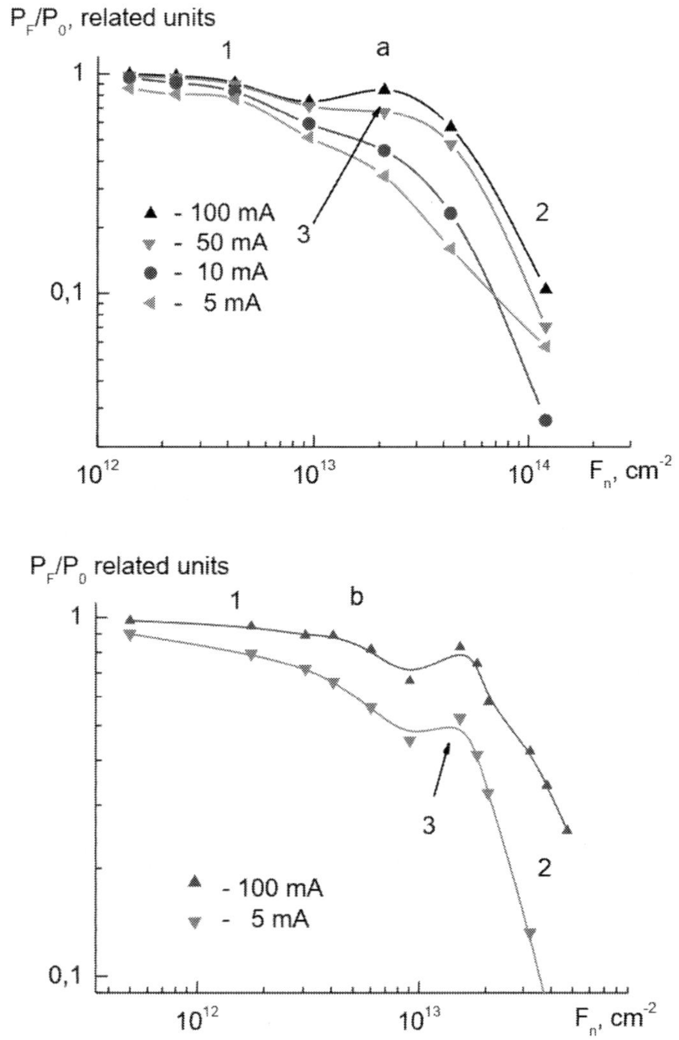

Figure 2. Radiant power changes of LED-1 (a) and LED-2 (b) with explicit transitions during fast neutron radiation: symbols – experimental data; lines – determined relations; 1, 2 – determined periods of radiant power decrease; 3 – transitions region

4. Conclusion

Radiant power decrease of LEDs based on AlGaInP heterostructures with multiple quantum wells during fast neutron radiation in a passive powering mode undergoes three periods: during the first period radiant power decrease occurs due to radiation stimulated restructuring of primary structural defects; during the second period it results from radiation defects introduction; during the third period LEDs go into the electron low injection mode and its radiant power does not depend on operating current volume. Within the third period of radiant power decrease, the majority of LEDs catastrophically break down during watt-ampere characteristic measurements.

Relations that allow describing radiant power decrease within each period are determined. These relations might be used to forecast LEDs' radiation stability.
For the first time it was experimentally proved that during fast neutron radiation in a passive mode transitions occur at the boundary of the first and the second periods. These transitions are characterized by partial recovery of radiant power against its general decrease. Radiant power recovery degree depends on operating current volume.
Radiant power recovery in the region of transitions might be both direct and indirect depending on AlGaInP heterostructure utilized to produce LEDs with different radiation wave mode.

References

[1] Srour J R and McGarrity J M 1988 *IEEE. Proceeding,* **76** 1443-1469
[2] Claeus E and Simoen E 2002 *Radiation effects in advanced semiconductor materials and devices* Berlin, Springer-Verlag
[3] Gradoboev A V Surzhikov A P 2005 *Radiation stability of GaAs microwave devices* Tomsk, Izdatelstvo TPU 277p
[4] Ryzhikov I V, Seleznyov D V 2005 *Instrument engineering techniques* **4** 11
[5] Nikiforov S G *Dissertation ... candidate of engineering* 2006 Moscow 158 p
[6] Gradoboev A V, Orlova K N, Asanov I A 2013 *J. Chem. Chem. Eng.* **7** 409
[7] Gradoboev A V, Orlova K N, Arefyev K P, Asanov I A 2013 *News of Higher Educational Institutions. Phisics* **11/3** 116
[8] Gradoboev A V and Orlova K N 2014 *Advanced Material Research* **880** 237
[9] Gradoboev A V, Orlova K N 2015 *Phys. Stat. Sol. C* **12** No.1-2 35-38

Modification of the surface layer of the system coating (TiCuN)/substrate (A7) by an intensive electron beam

Yu F Ivanov[1, 2], A I Potekaev[2], E A Petrikova[1], O V Ivanova[3], I A Ikonnikova[3], V V Shugurov[1], O V Krysina[1], and A A Klopotov[2,3]

[1] Institute of High Current Electronics SB RAS, Tomsk, 634055, Russia
[2] National Research Tomsk State University, Tomsk, 634050, Russia
[3] Tomsk State University of Architecture and Building, Tomsk, 634003, Russia

E-mail: yufi55@mail.ru

Abstract. In order to study the conditions of modification of the surface layer of the system coating (TiCuN)/substrate (A7) an analysis of processes occurring in the surface layer of the system wear-resistant coating/substrate irradiated by an intensive pulsed electron beam at a submillisecond exposure time has been carried out on the example of aluminum and titanium nitride. Irradiation has been carried out under conditions ensuring melting and crystallization of the surface layer of the material by a nonequilibrium phase diagram. It has been experimentally established that irradiation of the system coating (TiCuN)/substrate (A7) by an intensive electron beam is accompanied by changes in the phase composition of the material.It is evident that nanostructuring of the aluminum layer adjacent to the coating, and formation in it of nitride phase particles will contribute to hardening of the surface layer of the material, creating a transition sublayer between a solid coating and a relatively soft volume. The carried out analysis shows that binary nitrides based on TiN_{1-X} are most likely to form under nonequilibrium conditions, since the homogeneity range of this compound is rather large. On the other hand, formation of the ternary compound Ti_3CuN, which can be formed after an arc plasma-assisted deposition of titanium nitride of the composition TiCuN and by the subsequent intensive pulsed electron beam exposure, cannot be excluded.

1. Introduction

In most cases, surface properties of materials and products define the life of machines, mechanisms, and aggregates in general. Thus, formation of coatings for various purposes on the surface of the material allows to significantly improve physicochemical and performance properties of products, not significantly changing properties of the volume of the part, as well as its size and shape [1]. The necessary condition for performance of the system coating/substrate is a high level of adhesion forces achieved using various methods. One such method is the irradiation of the system coating/substrate by an intensive pulsed electron beam for the purpose of melting the coating into the substrate [2]. As compared to the widespread laser technology, the electron-beam technology has great potential in monitoring and controlling of the amount of input energy, differs in the locality of energy distribution in the near-surface layer of the processed material, and a high efficiency coefficient [3−5]. Ultra high rates (10^8 – 10^{10} K/s) of heating up to melting temperatures and the subsequent cooling of the thin near-surface layer of the material ($10^{-7} - 10^{-6}$ m), a very small ($10^{-6} - 10^{-3}$ s) exposure time to high temperatures, formation of limiting temperature gradients (up to $10^{7} - 10^{8}$ K/m) that provide cooling of the near-surface layer due to heat dissipation into an integrally cold volume of the material at a rate of 10^{4}–10^{9} K/s, create conditions for formation in the near-surface layer of an amorphous, nano- and submicrocrystalline structure. The purpose of this work is to analyze the processes occurring during modification of the surface layer of a wear-resistant coating/substrate irradiated by an intensive pulsed electron beam with a submillisecond exposure time.

2. Materials and methods of study

Commercially pure aluminum of the brand A7 was chosen as the modified material [6]. Titanium nitride was chosen as the wear-resistant hard coating. Electric-arc plasma-assisted deposition of titanium nitride of the composition TiCuN was carried out using the installation "KVINTA" (ISE SB RAS). The thickness of the coating was ≈0,5 μm. The high-speed heat treatment under the mode of the surface layer melting was carried out by an intensive pulsed electron beam (installation SOLO, ISE SB RAS), varying the energy density of the electron beam in the range of 10 J/cm^2 up to 30 J/cm^2; the pulse duration of the electron beam exposure was (50–200) μs; the pulse count was 3–5. The study of the phase and elemental composition, the defect substructure of the surface layer were carried out using methods of optical and scanning electron microscopy and X-ray diffraction analysis.

3. Model representation and evaluation

Numerical evaluations of the range of energy parameters of irradiation by an intensive pulsed electron beam of aluminum samples of the solid coating TiN ensuring melting and crystallization of the surface layer of the material by a non-equilibrium phase diagram with formation of a multiphase structure have been carried out: (1) temperature profiles of preheating of the surface layer of aluminum and the coating TiN, depending on the energy density of the electron beam and the duration of irradiation; (2) a comparative analysis of the calculated temperature profiles has been carried out, and an evaluation of the range of optimal values of irradiation parameters has been carried out for processing samples of aluminum and the system coating (TiN)/substrate (aluminum).

The problem of finding the temperature field in a certain range of the energy density of the electron beam is reduced to solving the heat equation. A one-dimensional case of heating and cooling of a plate with a thickness d is considered. The coordinate system is chosen so that the x axis is directed depth ward the sample. At $x = 0$ a heat flux is set, but on the back side of the plate the heat transfer is absent. In the coordinate form the heat equation takes the following form (1):

$$c\rho \frac{\partial T}{\partial t} = \lambda \frac{\partial^2 T}{\partial x^2}. \tag{1}$$

where c is the specific heat, ρ is the density, and λ is the thermal conductivity of the material.

Boundary conditions for the pulsed electron beam exposure have the form:

$$-\lambda \frac{\partial T}{\partial x} = q(t) \tag{2}$$

where the heat flux works its way depth ward the sample from the surface in the following form:

$$q(t) = \begin{cases} q_0, & 0 < t \le t_0; \\ 0, & t > t_0. \end{cases} \tag{3}$$

where q_0 is the average value of the heat flux during the exposure time by the electron beam t_0. The initial temperature $T(0, x) = T_0$ throughout the depth of the sample $0<x<d$.

For numerical solution of the problem the difference grid in increments τ in time was used: $t_j = j\tau$ and in increments h in space: $x_i = ih$, where j is the number of the layer in time, i is the number of the layer in space. The temperature value is determined in grid points. Approximation of the differential equation of heat conductivity with the first order in time and the second in space is used, an explicit scheme is chosen:

$$T_{j+1,i} = T_{j,i} + \frac{\tau(T_{j,i+1} - 2T_{j,i} + T_{j,i-1})}{c\rho h^2} \tag{4}$$

Based on the numerical solution of the heat equation the modeling of temperature fields during the electron-beam treatment of aluminum sample surfaces was carried out. Calculations were performed for an intensive electron beam with an energy density in the range from 5 J/cm^2 up to 40 J/cm^2, and a pulse duration of 50, 100, 150, and 200 μs. The thickness of the surface layer for thermal calculations $d = 10^{-3}$m, the observation time 300 μs.

Accounting of melting and vaporization of the material was carried out from the following physical considerations [3]. When the cell reached the melting temperature T_m, its temperature was fixed and assumed equal to the melting temperature, and all the input heat

$$q = \frac{\left(T_{j,i} - T_{пл}\right)c}{q_{пл}}, \tag{5}$$

where $T_{j,i}$ is the cell temperature, was supposed to go towards melting of the sample; $q = 1$ means transition of the substance into a liquid state. The crystallization process was simulated in the same manner, only the value q was decreased from 1 to 0. The process of vaporization and condensation was simulated similarly. When the cell reached the boiling point T_{boil} the temperature of the cell was fixed and assumed equal to the boiling point, and all the input heat

$$q = \frac{\left(T_{j,i} - T_{кип}\right)c}{q_{кип}} \tag{6}$$

goes towards vaporization of the substance; $q = 1$ indicates transition of the substance into vapor.

The numerical solution was carried out for thermophysical values of aluminum and titanium nitride given in the table 1.

Table 1. Thermophysical values of aluminum and titanium nitride, taken at T = 800 K [7-10] [a].

E_S, J/cm^2	Thickness of the film, μm	V, 10^{-6}, mm^3/(N*m)	V_0/V	<μ>	<μ$_0$>/<μ>
0,0 (initial A7)	0,0	7590		0.48	
15	0,5	6080	1,3	0.39	1,5
15	1,0	6570	1,4	0.40	1,2
20	0,5	5470	1,4	0.35	1,4
20	1,0	4790	1,6	0.31	1,6

[a] λ is the coefficient of thermal conductivity, c_{av} is the specific heat, ρ is the density, T_m is the melting temperature, T_{vap} is the vaporization temperature, q_m is the melting heat, q_{vap} is the vaporization heat.

The thickness of the vaporized layer of titanium nitride of the composition TiN was evaluated using the formula (7):

$$H = V_{исп.} \cdot t / \rho \tag{7}$$

where V_{vap} is the vaporization rate of the titanium nitride film equal to 1.168 10^{-5} g/(cm sec) at a temperature of 1987 K and 54.589 g/(cm sec) at a temperature of 2241 K [10], ρ is the density of titanium nitride, *t* is the time during which the temperature on the surface of the titanium nitride film, or on the surface of aluminum on which the film is deposited, is not lower than 1987 K.

The calculation results of the condition of the studied materials are presented in Figure 1 – Figure 3. It has been established that the melting temperature of the aluminum surface is achieved at E_s = 18 J/cm^2 for t_0 =150 μs (Figure 1, curve 1) and E_s = 15 J/cm^2 for t_0 = 100 μs (Figure 1, b, curve 1). The melting point of titanium nitride of the composition TiN under the specified in the work parameters of the electron beam has not been reached (Figure 1, curve 2); melting of titanium nitride with the composition $TiN_{0.84}$ is observed only at the density of the beam energy of 30 J/ cm^2 with a pulse duration of 100 μs (Figure 1b, curve 3). The maximum heating temperature of the aluminum surface (Figure 1, curve 1), attained at completion of pulse exposure, is always lower than of titanium nitride of both compositions (Figure 1).

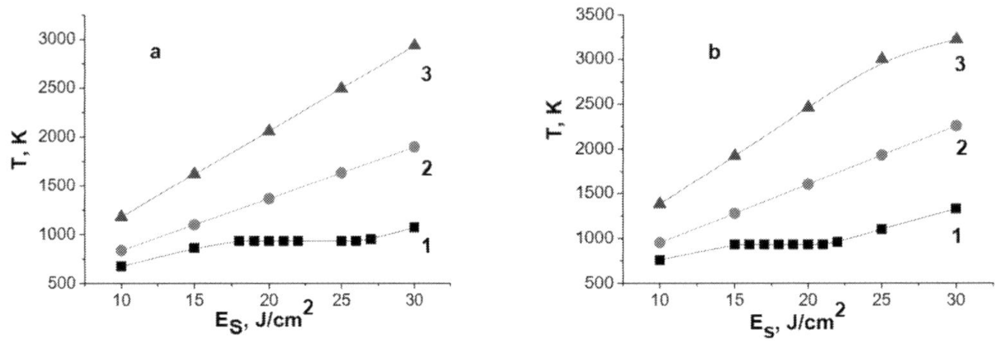

Figure 1. The dependence of the maximum heating temperature (considering melting and vaporization of the material) of the aluminum surface (curve 1) and titanium nitride of compositions TiCuN (curve 2) and $TiN_{0.84}$ (curve 3) irradiated by an electron beam on the density of the beam energy; a – t_0 = 150 μs, b – t_0 = 100 μs

Thus, irradiation of the system coating/substrate will lead to formation of thermoelastic stresses. The difference in heating temperatures ΔT for titanium nitride is 300 K under the energy density of the electron beam E_s = 10 J/cm^2 and 1000 K under E_s = 30 J/cm^2 for t_0 = 150 μs (Figure 1, a) and a small difference for t_0 = 100 μs (Figure 1, b).

It is expected that the value of thermoelastic stresses in the system coating/substrate may decrease during formation of a certain transition layer comprising a liquid phase (the two-phase condition melt – solid body). Figure 2 represents the results showing that such a two-phase layer under the pulse length of the electron beam t_0 = 100 μs begins to form at the electron beam energy density of 15 J/cm^2 (the lifetime of such a condition Δt = 6 μs (Figure 3, curve 1)); for t_0 = 150 μs – at 18 J/cm^2 (Δt = 7 μs (Figure 3, curve 2)). A single-phase transition layer (liquid phase) (Figure 2, *b*) begins to form at the electron beam energy density of 23 J/cm^2 for t_0 = 100 μs (Δt = 6 μs (Figure 3, *b*, curve 1)) and at 28 J/cm^2 for t_0 = 150 μs (Δt = 10 μs (Figure 3, *b*, curve 2)).

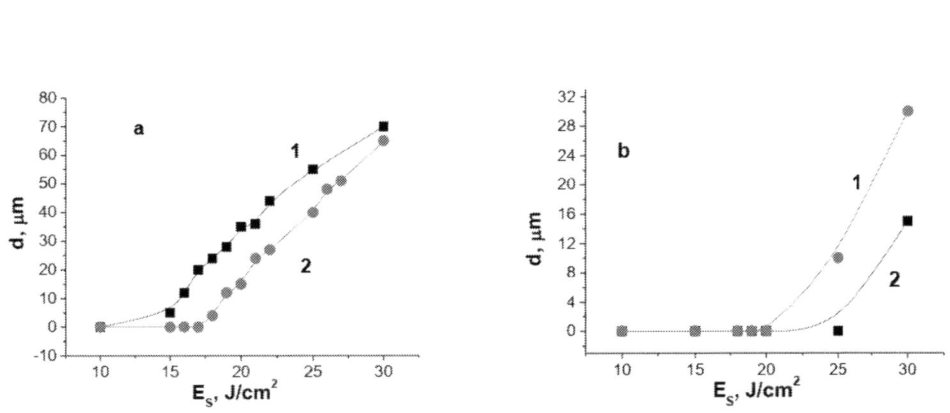

Figure 2. The dependence of the maximum thickness of the layer of the two-phase range (liquid solid + solid phase) d (a) and the melt layer d (b) on the surface of titanium on the energy density of the electron beam E_S at the exposure time of the beam 100 μs (curve 1) and 150 μs (curve 2).

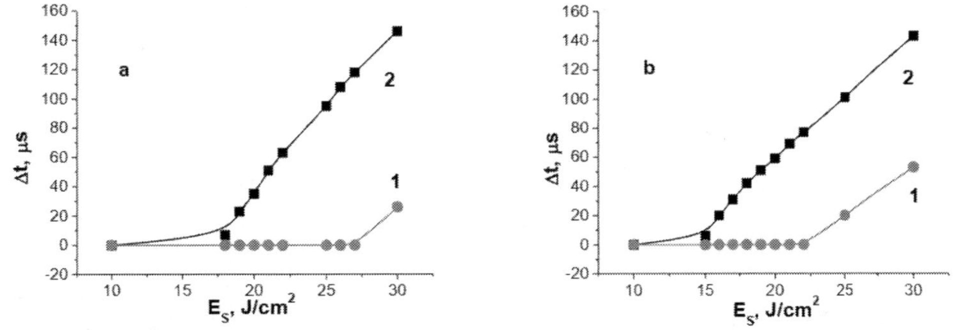

Figure 3. The dependence of the lifetime of the two-phase condition (liquid + solid phase) (curve 2) and the melt (curve 1) on the surface of aluminum on the energy density of the electron beam at the exposure time of the electron beam of 150 μs (a) and 100 μs (b).

Thus, the carried out numerical evaluations of the range of energy parameters of irradiation by an intense pulsed electron beam of the surface of aluminum and titanium nitride providing for melting and crystallization of the surface layer of the material by a non-equilibrium phase diagram allowed to construct temperature profiles of heating of the surface layer of aluminum depending on the energy density of the electron beam and the duration of irradiation pulses; to carry out a comparative analysis of the calculated temperature profiles; to assess the values of irradiation parameters of aluminum samples and the coating TiN.

4. Experimental results and their discussion

Irradiation of the system coating (TiCuN)/substrate (A7) by an intensive electron beam, as shown by thermal calculations, allows both to melt the coating into the substrate for improvement of adhesion forces and to destruct the coating (to remove the coating worn out during the operation of the part). Figure 4 shows the analysis results of the structure of the system coating (TiCuN)/substrate (A7), treated by an intensive electron beam in the mode of coating destruction. The coating is fragmented by a system of microcracks, delamination of the coating from the substrate is revealed. Melting of the coating into the substrate is not detected.

The carried out calculations show that the cooling rate of the surface layer of aluminum subjected to a pulsed electron beam irradiation at the energy density of (10–30) J/cm^2, varies in the range of (2,7–8,1)*10^8 K/s during the electron beam pulse duration of 50 μs and (0.7–2.1)*10^8 K/s during the electron beam pulse duration of 200 μs. Thus, formation of a nanoscale substructure in the surface layer of aluminum can be expected. Indeed, the studies carried out using electron microscopy methods revealed in the surface layer of aluminum, treated by an intensive electron beam (20 J/cm^2, 50 μs, 3 pulses, 0.3 s^{-1}), a structure with sizes of crystallites varying in the range of (60...100) nm (Figure 4, *d*, inset).

Irradiation of the system coating (TiCuN)/substrate (A7) by an intensive electron beam is accompanied by changes in the phase composition of the material. The studies carried out using X-ray analysis revealed formation in the system coating (TiCuN)/substrate (A7) of aluminum nitrides AlN, the volume fraction of which, depending on parameters of the electron beam, varies from 2 % up to 16 %. It is obvious that nanostructuring of the aluminum layer adjacent to the coating and formation in it of the nitride phase particles will contribute to hardening of the surface layer of the material, creating a transition sublayer between a solid coating and a relatively soft volume.

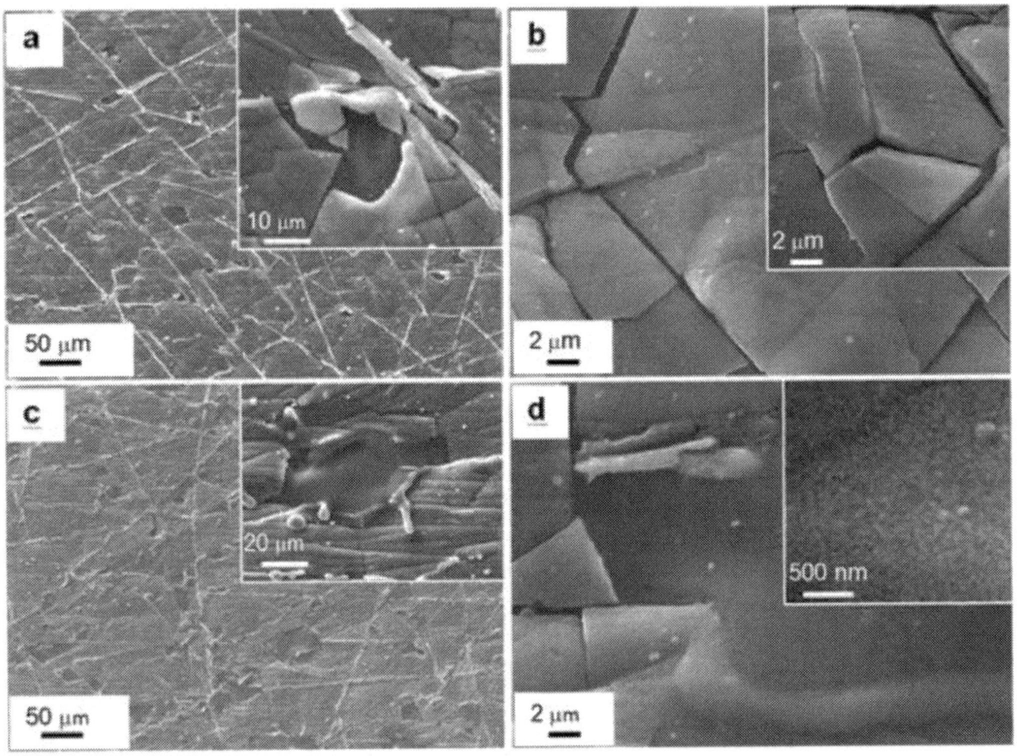

Figure 4. The electron microscopic image of the surface of the system coating (TiCuN)/substrate (A7) treated by an intense electron beam: a, b – 20 J/cm^2, 200 μs, 3 pulses; c, d – 20 J/cm^2, 50 μs, 3 pulses.

In the system Cu–Ti–N titanium belongs to transition metals [11, 12]. Titanium is in the middle portion in the row of electronegative metals, and formation of limited solid solutions and compounds with many elements is characteristic to it. For copper the value of electronegativity is located at the boundary between transition metals and electronegative metals (Zn, Cd, Al, Be). This ensures that copper equally well forms solid solutions and compounds with other elements. Formation of six intermetallic compounds takes place at melting of titanium with copper, two of which melt congruently (Figure 5) [13]: γ–$CuTi_2$ (prototype $MoSi_2$, Pearson symbol *tI6*, space group *I4/mmm*) at a temperature of 1010 °C and δ–CuTi (CuTi, *tP4, P4nmm*) – at 984 °C. Four other compounds: β–Cu_4Ti ($ZrAu_4$, *oP20, Pnna*), α–Cu_4Ti ($MoNi_4$, *tI10, I4/m*), λ–Cu_2Ti (VAu_2, *oC12, Amm2*), Θ–Cu_3Ti_2 (Cu_3Ti_2, *tP10, P4/mmn*) and ε–Cu_4Ti_3 (Cu_4Ti_3, *tI4, I4/mmm*) are formed by peritectic reactions [13]. Compounds λ and Θ exist in narrow temperature ranges and decomposition by eutectoid reactions. Compounds ζ, ε, δ and γ have narrow homogeneity ranges in composition. In addition, a number of studies revealed the existence of the compound Cu_7Ti_3, which melts congruently at a temperature of 905 °C, but at a temperature of 865 °C decompositions by eutectoid reaction on Cu_7Ti_2 and Cu_3Ti_2 (Figure 5).

The third element of the system Cu–Ti–N, nitrogen, is significantly different in chemical properties from metals forming this compound. Nitrogen with transition metals can form compounds with an ordered structure.

Nitride TiN_{1-X} has a cubic structure B1 (tprototype NaCl, Pearson symbol *cF*8, space group *Fm*3*m*). Compound TiN_{1-X} belongs to the class of disordered nonstoichiometric nitrides with a wide range of homogeneity (Figure 5) [14]. In compound TiN_{1-X} atoms of Ti form a FCC metal sublattice, the octahedral interstices of which may contain interstitial atoms N or vacancies. Retention of the type of the crystal structure of a nonstoichiometric disordered compound when changing the concentration of structural vacancies determines the presence of the homogeneity range. The homogeneity range of the binary compound TiN_{1-X} is wide.

The condition diagram of the system Cu–N has not been found in the literature. It is known that there are two nitrides based on copper: CuN_3 with a cubic system (prototype ReO_3, *cP*4 and space group $Pm\bar{3}m$) with the lattice parameter a=0,3814 nm, and the compound Cu_3N with a tetragonal structure ($I4_1/a$) with lattice parameters a=0,8653 nm and c=0,5594 nm [14].

The ternary nitride system Cu–Ti–N is characterized by only one ternary compound Ti_3CuN (Pearson symbol *tP**) with lattice parameters a=1,11968, c = 0,30217 nm [15].

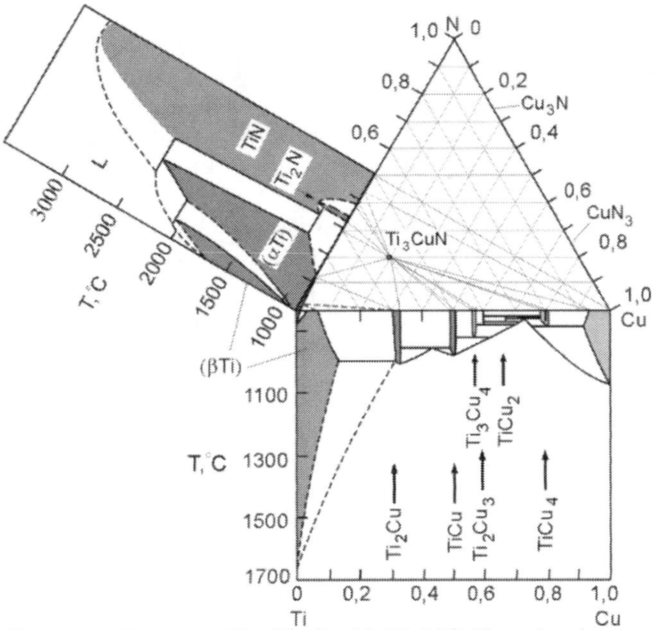

Figure 5. Binary diagrams of systems Cu–Ti, Cu–N, Ti–N [13], and an isothermal section of the ternary system Cu–Ti–N at 1000 °C [15]. Single-phase ranges on diagrams are highlighted in grey color.

Thus, the presented phase diagrams indicate that under nonequilibrium conditions formation of binary nitrides based on TiN_{1-X} is most likely to happen, since the homogeneity range of this compound is rather wide. On the other hand, it is impossible to exclude formation of the ternary compound Ti_3CuN which may be formed after an arc plasma-assisted deposition of titanium nitride of the composition TiCuN and the subsequent exposure to an intensive pulsed electron beam.

5. Conclusion

In order to study the conditions for modification of the surface layer of the system coating (TiCuN)/substrate (A7) by an intensive electron beam within the framework of a one-dimensional model the problem on finding the temperature field formed by irradiation of aluminum and titanium nitride by an intensive pulsed electron beam has been solved. Based on the numerical solution of the heat equation the modeling of temperature fields during the electron-beam treatment of sample surfaces has been carried out. The numerical solution has been carried out for thermophysiacal values of aluminum and titanium nitride. It has been shown that irradiation of the system coating/substrate will lead to formation of thermal stresses. The difference in heating temperatures ΔT for titanium nitride is 300 K at the electron beam energy density of $E_s = 10$ J/cm^2 and 1000 K at $E_s = 30$ J/cm^2 for t_0 =150 μs and rather big difference for $t_0 = 100$ μs.

It is expected that the value of thermoelastic stresses in the system coating/substrate may decrease during formation of the transition layer comprising a liquid phase (the two-phase condition melt – solid body). The presented results demonstrate that the two-phase layer at the pulse duration of the electron beam of $t_0 = 100$ μs begins to form at the energy density of the electron beam of 15 J/cm^2 (the lifetime of such a condition is $\Delta t = 6$ μs); for $t_0 = 150$ μs – at 18 J/cm^2. The single phase transition layer (liquid phase) begins to form at the energy density of the electron beam of 23 J/cm^2 for $t_0 = 100$ μs ($\Delta t = 6$ μs) and at 28 J/cm^2 for $t_0 = 150$ μs ($\Delta t = 10$ μs).

It has been experimentally established that irradiation of the system coating (TiCuN)/substrate (A7) by an intensive electron beam is accompanied by changes in the phase composition of the material. The carried out studies using X-ray analysis revealed formation in the system coating (TiCuN)/substrate (A7) of aluminum nitrides AlN, the volume fraction of which, depending on parameters of the electron beam, varies from 2 % up to 16 %. It is obvious that nanostructuring of the aluminum layer, adjacent to the coating, and formation in it of the nitride phase particles will contribute to hardening of the surface layer of material, creating a transition sublayer between the solid coating and the relatively soft volume.

The carried out analysis shows that under nonequilibrium conditions the formation of binary nitrides based on TiN_{1-X} is most likely to happen, since the homogeneity range of this compound is rather wide. On the other hand, formation of the ternary compound Ti_3CuN, which can be formed after an arc plasma-assisted deposition of titanium nitride of the composition TiCuN and by the subsequent intensive pulsed electron beam exposure, cannot be excluded.

The ranges of irradiation parameters of the system coating (TiCuN)/substrate (A7) by a pulsed electron beam, allowing to destruct the coating by means of a fragile scabbing and to melt the coating into the substrate creating a transitional reinforced sublayer, have been established.

This work was supported by RFBR grant (number 13-08-00416_a) and the Programme of improving the competitiveness of TSU among world's scientific and educational centers.

References

[1] Musil J Baroch P and Zeman P 2008 *Plasma Surface Engineering Research and its Practical Applications ed R Wei (Kerala: Research Signpost) 1–34*

[2] Laskovnev A P, Ivanov Yu F, Petrikova E A and *et al.* 2013 Modification of the structure and

properties of the eutectic silumin with an electron-ion-plasma treatment (Minsk: Belarus. Navuka)
[3] Rotshtein V, Ivanov Yu, Markov A 2006 *Surface treatment of materials with low-energy, high-current electron beams* (Materials surface processing by directed energy techniques) Ed Y Pauleau (Elsevier) charter **6** 205-240
[4] Belov A B, Bytsenko O A, Krainik A V and et al. 2012 *High current pulsed electron beam for aviation engine building* (Moscow: Dipak)
[5] Koval N N, Ivanov Yu F 2008 Nanostructuring the surface of metal-ceramic and ceramic materials under pulsed electron-beam treatment *Izvestiya Vuzov. Fizika* **5** 60-70.
[6] Stroganov G B, Rothenberg V A and Gershman G B 1977 *Alloys of aluminum with silicon* (Moscow: Metallurgy)
[7] Babichev A P, Grandma N A, Bratkovsky A M and et al. 1991 *Physical quantities: Directory* (Moscow; Energoatomizdat)
[8] Prokhorov A M 1998 *Physical encyclopedia* **5** (Moscow, Great Russian Encyclopedia)
[9] Samsonov G V and Vinnytsia I M 1976 *Refractory compounds* (Moscow: Metallurgy)
[10] Vol A E 1959 *Structure and properties of binary metal systems* **1** (Moscow: State Publishing House of physical and mathematical literature)
[11] Kornilov I I, Matveeva N M, Pryakhina L I, Polyakov R S 1966 *Metallochemical properties of the periodic system elements* (Moscow: Nauka)
[12] Kornilov I I 1964 Metallides and the interaction among them (Moscow: Nauka)
[13] Lyakishev N P 1996-2000 *Diagrams of binary metal systems* **1-3** (Moscow: Mechanical engineering)
[14] Gusev A I and Rempel A A 2001 *Non-stoichiomclrv, disorder and order in solids* (Ekaterinburg: UB RAS)
[15] Durlu N and Schuster J C 1994 *8-th Int. Conf. High Temp. Mater. Chem* (Vienna: Austria)

The study of the ferroelectric properties of lithium-titanium ferrite

A V Malyshev[1], E V Nikolaev[1], A P Surzhikov[1] and A N Zarubin[2]
[1]National Research Tomsk Polytechnic University, Tomsk, Russia
[2]National Research Tomsk State University, Tomsk, Russia

E-mail: malyshev@tpu.ru

Abstract. Loop-shaped dependences of the electric polarization on the electric field strength (the dielectric hysteresis) are registered for the first time for polycrystalline Li-Ti ferrite. Temperature evolution of the hysteresis loop parameters is investigated for ferrite samples. A thermal Barkhausen effect is detected during heating and cooling of ferrite specimens prepolarized in an electric field. The results obtained can be interpreted from the viewpoint of the Maxwell-Wagner relaxation polarization or induced ferroelectric-like state in the electric ferrite subsystem.

Keywords: Ferrite; Relaxation polarization; Residual polarization; Coercive field; Barkhausen pulse.

1. Introduction

Ferrite ceramics of the lithium group are the materials with well-established magnetic properties [1-3]. On the other hand, in [5, 6] it was shown that a ceramic material of the following composition $Li_{0.649}Fe_{1.598}Ti_{0.5}Zn_{0.2}Mn_{0.051}O_4$ under certain conditions exhibits dielectric behavior peculiar to ferroelectric materials.

It was shown that the experimental temperature dependences of the complex dielectric permittivity at the frequencies 10^2, 10^3, 10^4 and 10^6 Hz demonstrate a number of anomalies when measured at a high-level of the test signal (E_{test}=120 V/cm) at the frequency 10^2 Hz [2]. The anomalous behavior consisted in a sharp drop of the values of the real part of the permittivity ε' in the vicinity of the magnetic Curie temperature (575 K). When these dependences are measured at a low level of the test signal (E_{test}=2.3 V/cm) with a continuous voltage bias, the anomalies are manifested as sudden variations in the values of ε' and ε'' within narrow temperature intervals. Based on the analysis of the data obtained, the authors put forward a hypothesis that an additional relaxation polarization process induced by the electric field might occur in the ferrite, which is characterized by ferroelectric behavior [5, 6].

The present work reports the results of investigations of the feasibility and conditions of formation of properties similar to those of ferroelectric materials; in particular, the results on non-linearity and hysteresis of the electric-field-induced relaxation polarization dependences P = f (E_p), where E_p is the polarizing electric field strength, applied to the specimen.

2. Experimental techniques

The sintering process and specimen preparation method were the same as reported in [3, 4]. The specimens were manufactured using a standard ceramic production process from an industrially synthesized mixture (Russian designation of the grade "3ZSC-18") by thermal sintering of the compacts in air in a laboratory resistance furnace at the temperature 1283 K for two hours. The mode of sintering was selected to follow that used in the industrial production of the 3ZSC-18 ferrite. In order to remove the sub-surface layers having non-uniform distribution of electric properties over the depth, the surfaces of two sides of the specimens were ground using powdered Al_2O_3 (with different grain size) and a polishing paste.

The specimens were shaped as tablets measuring 13 mm in diameter and 0.24 mm in thickness, with silver electrodes deposited onto their surface using a PVD technique. The diagnostic electrode diameter was 5 mm.

The main phases and structural parameters of the examined ferrite were determined at room temperature by the x-ray phase analysis method (using an X'TRA diffractometer). The crystal lattice was cubic (with the lattice parameter a = 8.367 Å) and had Fd3m type spatial symmetry group. The method of differential scanning calorimetry (using an STA 449C Jupiter Netzsch thermal analyzer) did not show any noticeable exo- or endothermic reactions of the sintered ferrite ceramics at temperatures in the range 300-525 K which would be indicative of any phase structural transitions in the ferrite ceramics. In this case, the magnetocaloric effect with magnetic Curie temperature of 550 K was registered.

Experimental investigations of the temperature-field dependence of the ferrite ceramics polarization were performed using the classical Sawyer-Tower circuit [7].
Measurements of the polarization field dependences were performed at the electric field frequency changing from 20 Hz to 3 kHz and the electric field strength in the range 0-8300 V/cm with the ferrite samples whose temperatures changed in the range 300-600 K. The field dependences were registered with a Tektronix TDS-2012B electron storage oscillograph.

Repolarization current pulses (Barkhausen pulses) were monitored using a setup whose block diagram is given in Figure 1. The measurements were carried out during slow heating of (2 deg/min) of specimens in the temperature interval T=300-600 K. A DC voltage bias was applied to the specimens. Electric pulses were detected using a Velleman PCS-500 computer-automated oscilloscope.

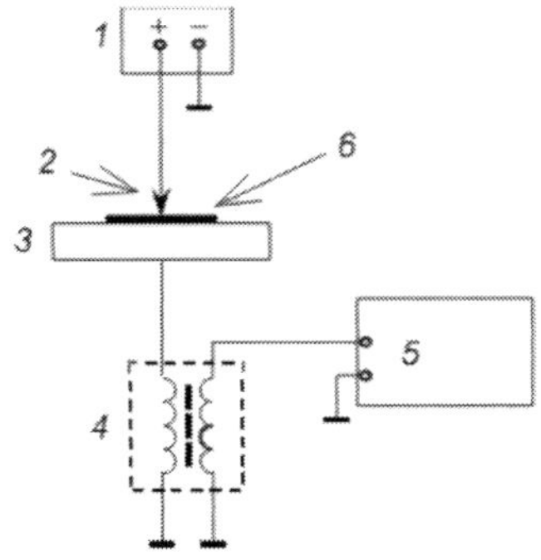

Figure 1. Layout of the Barkhausen pulses experiment: DC power supply (1), upper electrode (2), lower electrode with heater (3), transformer (4), oscilloscope (5), and specimen under study (6).

3. Result and discussion

Figure 2 presents a selection of electric field induced dependences of ferrite relaxation polarization at the temperatures 410 K.

Our investigations demonstrated the presence of dielectric hysteresis loops of classical shape for temperatures in the range 300-450 K and electric field frequencies f = 300-800 Hz (Figure 2). The hysteresis loop shapes indicate the possibility of the induced ferroelectric character of the ferrite polarization under the given conditions. It should be noted that similar hysteresis loops were observed only for the ferrite ceramic samples subjected to cooling from 600 K to room temperature in a constant electric field with strength higher than 500 V/cm. In this case, the heating degree was limited by the probability of thermal breakdown. It was believed that upon cooling of the ferrite samples in the electric field, the polarization was induced due to the orientation of electrical domains in each crystallite along the field direction, and that the residual polarization remained after the field removal. Similar processes were observed for dielectrics with the metastable electric polarization [9]. It is well known that the hysteresis loops of similar shapes are observed for a number of ferrohard materials in which internal fields are a serious obstacle to the efficient polarization [10].

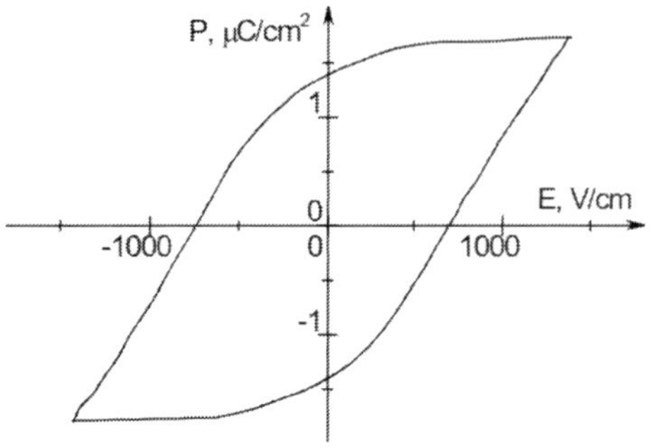

Figure 2. Dependence of the dielectric polarization P on the electric field strength E for the ferrite sample having a temperature of 410 K in the electric field having the frequency f = 300 Hz.

Beyond the indicated electric field frequency range for samples whose temperature is higher than 450 K, the loop-shaped field dependences are transformed into oval-shaped or linear dependences. It seems likely that in this case, the dielectric losses on the through conductivity are much greater than the losses on the relaxation polarization, thereby masking loops.

On the other hand, the similar hysteresis loops could be observed for nonlinear dielectrics with losses (the Maxwell-Wagner type relaxation polarization) [11]. The narrow frequency range f = 300-800 Hz in which the hysteresis loops are similar to those of ferrites testify to the Maxwell-Wagner polarization. However, a consideration of the given polarization mechanism demonstrates that processes of electron transfer and relaxation polarization are identical. Hence, the corresponding activation energies of these processes must be close in values. However, in [2] it was demonstrated that the activation energy of the process of electron transfer calculated from the temperature dependences of the conductivity for alternating current was approximately twice as great as the activation energy of the polarization process calculated from the temperature dependences of the parameter ε'.

Figure 3 shows the temperature dependences of the polarization loop parameters (the coercive field strength E_c and the residual polarization P_r) measured for an electric field frequency of 300 Hz. We note that their character is similar to the corresponding characteristics of the ceramic ferroelectric materials in which the smooth decrease in the P_r value with increasing ferrite sample temperature is caused by the nonuniformity and imperfection of the crystal structure.

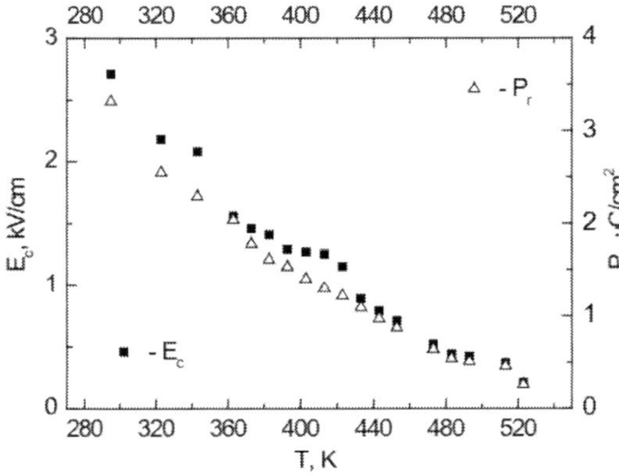

Figure 3. Temperature dependence of the coercive field strength E_c and residual polarization P_r at f = 300 Hz.

The E_c value decreases with increasing temperature due to the increased mobility of electric domains. Some special features of the E_c temperature dependence at T=410 K can be due to the influence of the heterogeneous microstructure of the ferrite ceramics on the mobility of electric domains. Thus, the character of the obtained temperature dependences is in agreement with the existing concepts about the ferroelectric materials [8].

Together with this, in the literature there are no experimental data on the temperature dependences of the hysteresis loop parameters of heterogeneous materials in the case of treatment of the results in the context of the Maxwell–Wagner model of the relaxation polarization.

Typical Barkhausen pulseforms are shown in Figure 4. Depicted in Figure 5 is the temperature dependence of the maximum Barkhausen pulse height U_m for ferrite specimens pre-polarized by an electric field. The figure implies that Barkhausen pulses are observed in the temperature interval T=465-575 K at the polarizing field strength E_p > 80 V/cm.

For E_p<80 V/cm, only electromagnetic noise is detected with a pulse height lower than 100 mV.

A sharp drop in re-polarization current pulses at 575 K might have been caused by a structural transition from ferroelectric to paraphrase state [8].

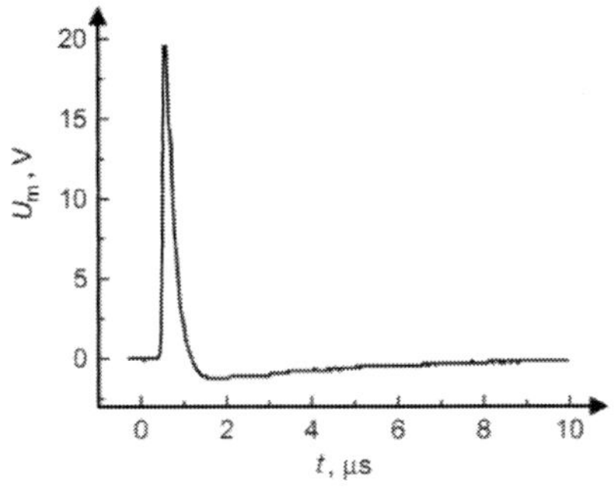

Figure 4. Typical electric pulse waveform: T=510 K, E_p=125 V/cm.

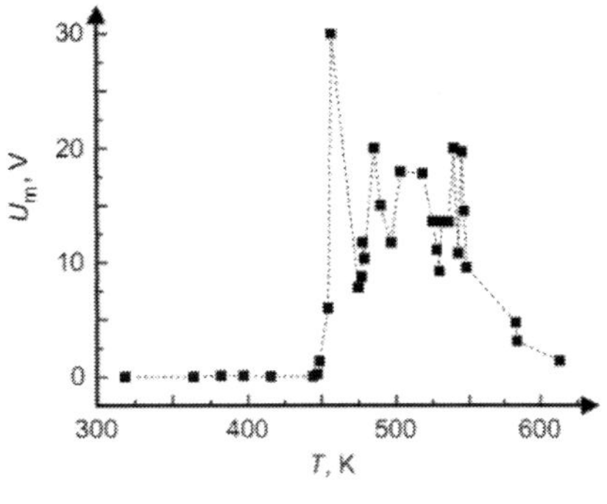

Figure 5. Temperature dependence of the maximum Barkhausen pulse height U_m for ferrite specimens pre-polarized by an electric field.

The ambiguity of treatment of the results obtained does not allow us to identify the relaxation polarization mechanism in ferrite ceramics and calls for further research in the field of relaxation phenomena in solid bodies including, for example, piezoelectric effect, temperature dependences of the lattice parameter [8]. The results obtained will allow the field of technical application of the lithium-titanium ferrite in modern electronics to be expanded together with the field of experimental and theoretical studies of multiferroics.

4. Conclusions

For the first time, the dielectric hysteresis loops have been established for the lithium-titanium ferrite ceramics, and the temperature dependences of the polarization loop parameters have been obtained.

The results obtained can be treated both as the mechanism of the Maxwell-Wagner relaxation polarization and as the mechanism that under definite conditions gives rise to the induced ferroelectric state in ferrite ceramics.

5. Acknowledgements

This work was supported by The Ministry of Education and Science of the Russian Federation in part of the "Science" program.

References

[1] Venevtsev Yu N, Gagulin V V and Lyubimov V N 1982 Ferromagnets [in Russian] (Moscow: Nauka) p 130
[2] Malyshev A V, Peshev V V and Pritulov A M 2004 *Fiz. Tverd. Tela [in Russian]* **46** No 1 185-188
[3] Malyshev A V and Peshev V V 2007 *Russ. Phys. J.* **50** No 2 161-164
[4] Surzhikov A P, Lysenko E N, Malyshev A V et al 2013 *Russ. Phys. J.* **56** No 6 681-685
[5] Belov K P, Goryaga A M and Sheremetev V V 1986 *Fiz. Tverd. Tela [in Russian]* **30** 314-320
[6] Danil'kevich M. I. and Al-Sharr D 1992 *Bull. Byelorussian Univ. [in Russian]* **3** 71-78
[7] Lines M and Glass A Principles and Applications of Ferroelectrics and Related Materials 1979 (Oxford: Clarendon Press) p 736
[8] Zheludev I S 1968 Physics of Crystalline Dielectrics [in Russian] (Moscow: Nauka) p 464
[9] Gridnev S A 1997 *Soros. Obshcheobraz. Zh. [in Russian]* **5** 105-115
[10] Panich A E and Kupryanov M F 1989 Physics and Technology of Ferroceramics [in Russian] (Rostov-on-Don: Publishing House of Rostov-on-Don State University) p 389
[11] Ioffe A F 1957 Physics of Semiconductors [in Russian]. (Moscow: Academic Press) p 758

Fractography of the fatigue fracture surface of silumin irradiated by high-intensity pulsed electron beam

Yu F Ivanov[1, 2], K V Alsaraeva[3], E A Petrikova[1], A D Teresov[1], V E Gromov[3] and S V Konovalov[3]

[1] Institute of High Current Electronics Siberian Branch of the RAS
[2] National Research Tomsk State University
[3] Siberian State Industrial University

E-mail: yufi55@mail.ru

Abstract. Irradiation mode has been revealed allowing to increase the silumin fatigue service life in three times, it was established that this fact is caused by the formation of multi-modal, multi-phase, submicro - and nano-structure able to lead to a significant increase in the critical length of crack.

1. Introduction

Fatigue fracture of the material [1] is the most frequent cause of failure of equipment, machinery, vehicles and construction. In this respect, it is important to prevent fatigue fracture (increase of service life) of critical parts, especially in those industries where occurrence of accidents leads to disastrous consequences. Fatigue cracks originate usually in the surface layer of a part [2]. A highly effective method of surface modification of parts and, increase of their fatigue life [3-5] consequently, is a material treatment by intensity pulsed electron beam allowed to modify the structure of the surface layer of tens of micrometers in thickness, turning it into a multi-modal structural-phase state and almost without changing the structural-phase state of the main volume of the alloy. [6-10]. The aim of this work is the analysis of regularities of a silumine structure modification with high-intensity pulsed electron beam, determination of mechanisms responsible for the fracture of silumin subjected to high-cycle fatigue testing.

2. Materials and methods of study

Silumin of a grade AK12 [11] was used as a material for study. Fatigue tests were carried out on a special installation according to the scheme of an asymmetrical cantilever bending [12]. The samples had the shape of a parallelepiped with dimensions 8x14x145mm. Simulation of crack was carried out by a cut in the form of a semicircle of radius 10 mm. Test temperature is of 300 K, the frequency of sample loading by bending – 15 Hz with a load of 10 MPa. Irradiation of the sample surface, prepared for fatigue testing was carried out at the device "SOLO" (IHCE SB RAS) [8] with the following parameters: electron energy of 16 keV; the pulse frequency 0.3 s^{-1}; duration of the electron beam pulse of 50 μs and 150 μs; the energy density of the electron beam (10...25) j/cm^2; the number of the exposure pulses– 1, 3, 5. The front surface of the samples was irradiated, that is, the sample surface located over the incision simulating the crack. Each irradiation mode tested at least 5 samples. Examination of electron-beam exposure and fracture surface was carried out by methods of optical (device Microvisor metallographic μVizo - MET-221) and scanning electron microscopy (device LEO EVO 50).

3. Experimental results and their discussion

The structure of silumin before electron beam irradiation (the structure of the original state) is characterized by the presence of relatively large (from tenths to tens of micrometer) inclusions of silicon predominantly of a lamellar morphology (Figure 1).

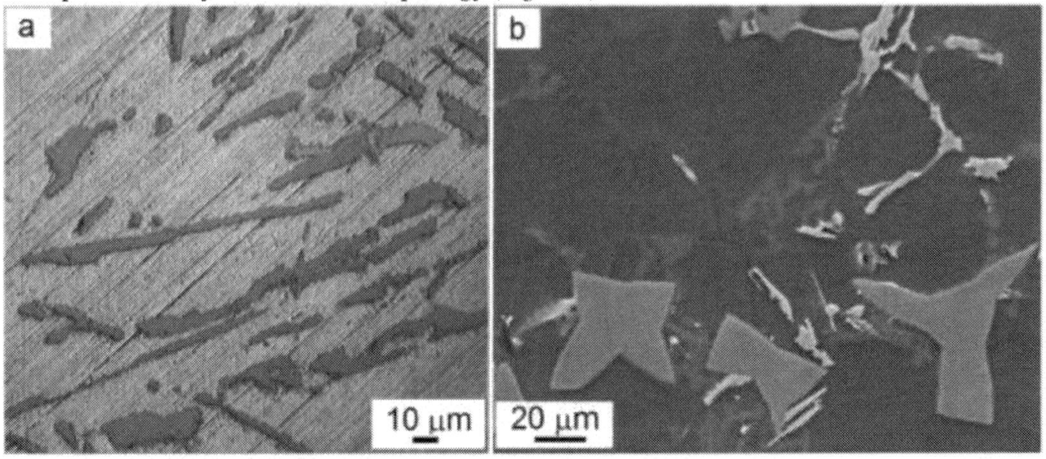

Figure 1. Structure of silumin of the original state (before intense electron beam irradiation); a - optical microscopy; b – scanning electron microscopy.

The irradiation of silumin surface by intensive pulsed electron beam with a submillisecond duration of exposure depending on the electron beam energy density is accompanied by melting of the sample surface (Figure 2, a, b), or by melting of the material layer surface of some thickness (unit - tens of micrometers) (Figure 2, c, d).

In the first case, the modification process of silicon plates is accompanied by the formation of numerous micropores along the boundary plate/matrix, and micro-cracks located in the silicon plates (Figure 2, b). It is evident, while subsequent fatigue testing silicon plates will be stress concentrators (sources of microcracks). In the second case the structure of the surface layer is fundamentally different on morphological grounds from the structure of the original sample (Figure 1) and the sample irradiated in mode of the surface melting (Figure 2, a, b). Homogeneous grain structure (grain size of eutectic varies within 30...50 μm) (Figure 2, c, d) is formed on the irradiation surface. Silicon in the form of layers or separate particles of globular form is situated at grain boundaries, the transverse dimensions of which do not exceed 10 μm. Stress concentrators which can cause the sample fracture are not detected on the irradiation surface.

Indeed, our tests revealed a significant dependence of the fatigue durability value on the irradiation mode of the silumin samples surface. While the irradiation mode leading to a partial melting of the sample surface (Figure 2, a, b), the fatigue life was below the fatigue durability of the original samples in some cases. When the irradiation of the surface layer in a stable melting mode with a forming on the surface of the grain of a structure with globular inclusions of silicon (Figure 2, c, d), fatigue life of silumin samples exceeded the fatigue life of the source material more than in 3 times.

Made assumptions about possible causes of low fatigue life of the samples irradiated by electron beam in mode 15 j/cm^2; 150 μs; 3 pulses, were confirmed while analyzing of a silumin surface fracture by scanning electron microscopy.

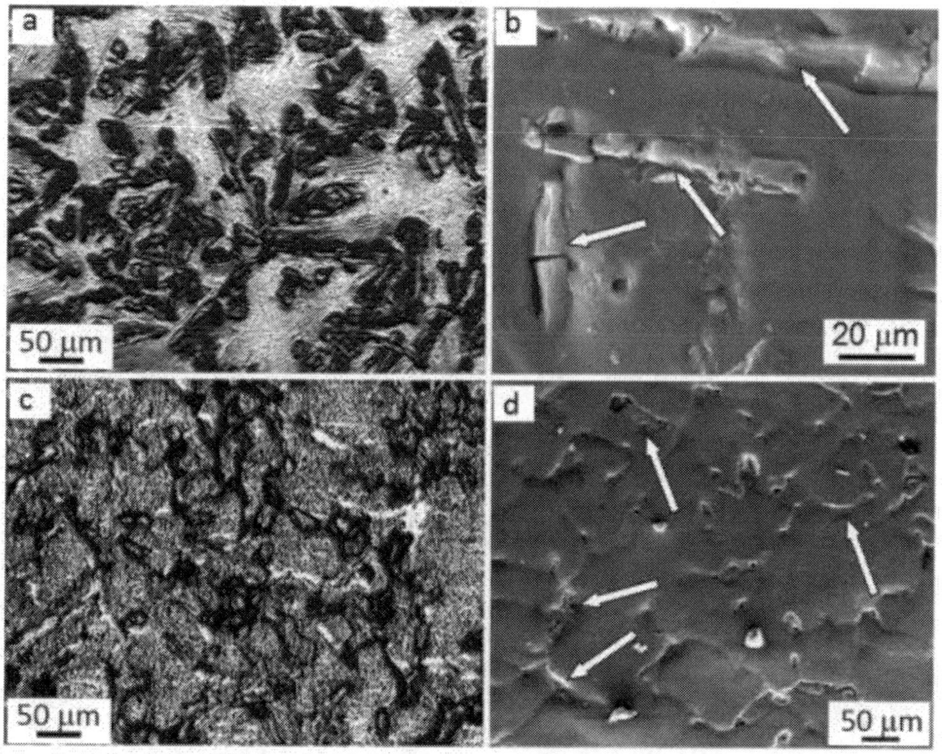

Figure 2. Surface structure of silumin irradiated by electron beam in modes: 15 j/cm^2; 150 µs; 3 pulses (a and b) and 20 j/cm^2; 150 µs; 5 pulses (c, d); a, c – optical microscopy; b, d – scanning electron microscopy; silicon particles are arrowed on (b, d).

The electron-microscopic image of the fatigue fracture surface of the silumin sample treated by an electron beam according to mode 15 j/cm^2; 150 µs; 3 pulses with fatigue life below the fatigue durability of the source material is shown on the Figure 3. It is impressive to see (Figure 3, a, b, the area of the crack formation is framed), that the fatigue crack is formed on the sample surface. Coarse inclusions of silicon located on the sample surface (Figure 3 c, the area of the material with a broken silicon plate is arrowed) are reasons for the formation of fatigue cracks. As was mentioned above, original layer structure of silicon on the treated surface does not change practically when the specified irradiation mode.

Fatigue fracture surface analysis of the sample irradiated by the electron beam in the mode of 20 j/cm^2; 150 µs; 5 pulses, showed that the thickness of the melted layer varies in the range up to 20 µm (Figure 4 a). Subsequent to the melting high-speed crystallization leads to the formation of multimodal structures represented at the macro level by the aluminium based grains, sizes of which varies within 30...50 µm with located on the boundary silicon particles, whose dimensions do not exceed 10 µm (Figure 2, c, d). Two-phase (silicon and solid solution on the basis of aluminum) crystallization cells detected on the fatigue fracture surface form meso-level of modified layer. Crystallization cell sizes vary from 100 nm to 250 nm (Figure 4, b).

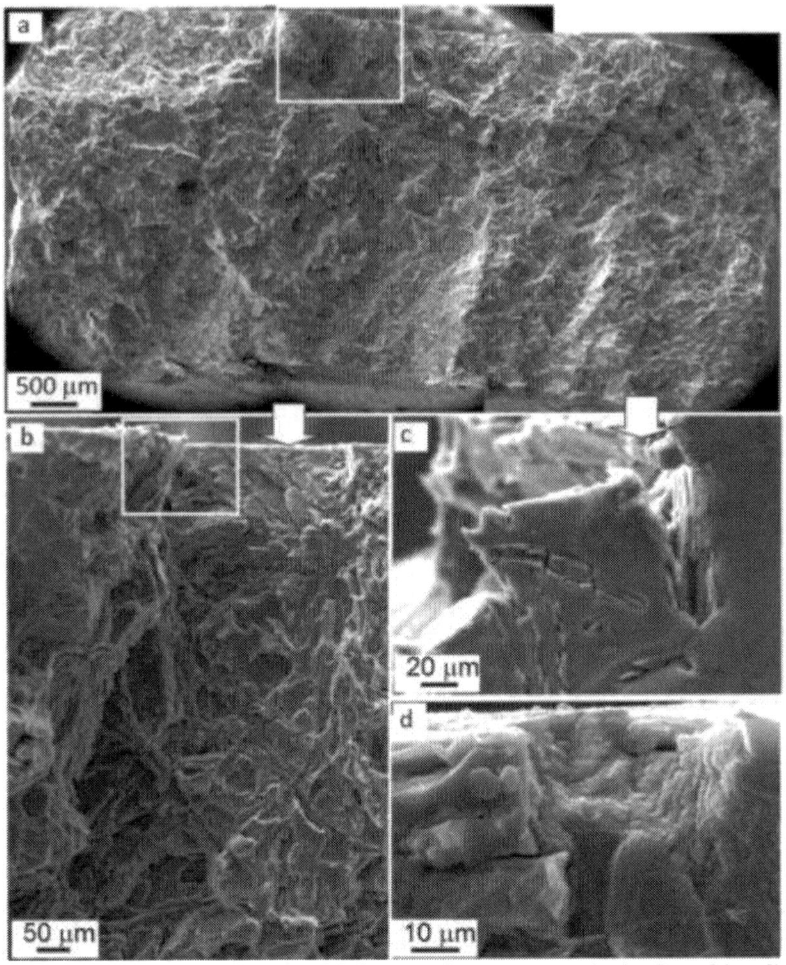

Figure 3. An electron microscope image of failure surface structure (a, b, d) and irradiation surface (b) of silumin treated by electron beam in mode 15 j/cm^2; 150 μs; 3 pulses; on (a) and (b) the area of the crack formation is framed, on (b) – irradiation surface and on (c) – site of the fatigue crack formation are arrowed.

It is evident, that the formation of such submicro - and nano-phase structure is a determinant cause providing multiple increase of fatigue durability of silumin. The physical significance of the formation of multi-level structural-phase states consists in decrease of the scaling localization level of a plastic deformation in the surface layer, leading to a more even distribution of elastic stresses in a more significant volume of material when an external mechanical or thermal effects on the surface. Consequently, the generation energy in the surface layer of stress concentrators increases greatly, and the probability of formation in the surface layer of the defective substructure decreases. In other words, the formation of multi-level structural-phase state determines the manifestation in the surface layer of the modified silumin of damping properties with regard to the base material by external mechanical and thermal influences, preventing premature nucleation of fragile microcracks and their propagation from the surface into the main volume of material leading to the formation of the main cracks and fracture of the base material [13].

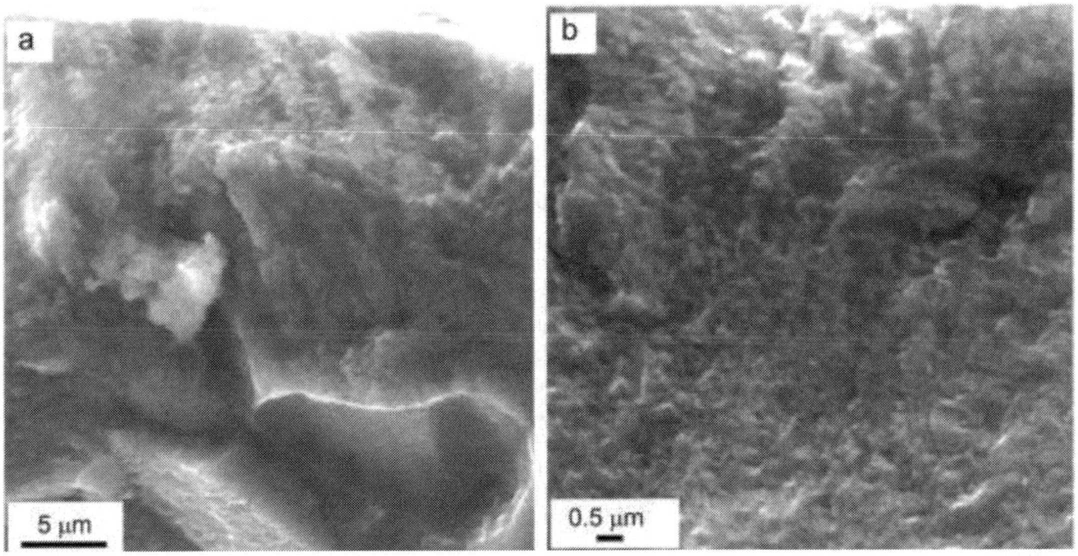

Figure 4. An electron microscope image of fatigue fracture surface of silumin treated in mode of irradiation by high –intensity pulsed electron beam 20 j/cm^2; 150 μs; 5 pulses

4. Conclusion

The surface modification of the eutectic silumin by high-intensity pulsed electron beam has been carried out. High-cycle fatigue testing was performed and exposure mode allowed increase the material fatigue life more than three times was determined. Studies of the structure of the surface irradiation and surface fatigue fracture of silumin in the initial (unirradiated) state and after modification by intense pulsed electron beam was carried out. It is shown, that in mode of partial melting of the irradiation surface modification process of silicon plates is accompanied by the formation of numerous large micropores along the boundary plate/matrix and microcracks located in the silicon plates. A multi-modal structure (grain size within 30...50 μm with located on the boundaries silicon particles up to 10 μm) is formed in stable melting mode, as well as subgrain structure in the form of crystallization cells in size from 100 μm to 250 μm). An assumption was made that formation of a multi-modal, multi-phase, submicro-and nano- sized structure assisting to a significant increase in the critical length of the crack is a main reason for the increase of the silumin fatigue life.

This work was carried out at the expense of Russia research foundation (Grant no. 14-29-00091)

References

[1] Kotsan'da S 1990 *Fatigue cracking of metals* (Moscow: Metallurgiya) p 622
[2] Fellouz Dzh. ed. 1982 *Fractography and atlas of fracture patterns* (Moscow: Metallurgiya) p 489
[3] Ivanov Yu F, Gromov V E, Grishunin V A, Teresov A D, and Konovalov S V 2013 *Fizicheskaya mezomekhanika* **16** 47–53
[4] Gromov V E, Ivanov Yu F, Grishunin V A, Raikov S V and Konovalov S V 2013 *Uspekhi fiziki metallov* **14** 67–80
[5] Konovalov S V, Atroshkina A A, Gromov V E and Ivanov Yu F 2010 *Materials Science and Engineering: A* **527** 3040–3043
[6] Vorob'ev S V, Ivakhin M P, Ivanov Yu F, Konovalov S V and Gromov V E 2008 *Structural-scale levels of high-cycle fatigue of steel when the pulse current impact* (Novokuznetsk:

SibGIU) p 270
[7] Laskovnev A P, Ivanov Yu F, Petrikova E A *et al* 2013 *Modification of structure and properties of eutectic silumin by electron-ion-plasma treatment* (Minsk: Belorusskaya nauka) p 287
[8] Koval' N N and Ivanov Yu F 2008 *Izvestiya Vuzov. Fizika* **5** 60–70
[9] Rotshtein V, Ivanov Yu and Markov A 2006 *Surface treatment of materials with low-energy, high-current electron beams*. Charter 6 in Book "Materials surface processing by directed energy techniques". pp. 205-240. Ed. by Pauleau (Y.: Elsevier) p 763.
[10] Belov A B, Bytsenko O A, Krainikov A V *et al* 2012 *High-current pulsed electron beams for aircraft engine building* (Moscow: Dipak) p 292.
[11] Stroganov G B, Rotenberg V A and Gershman G B 1977 *Alloys of aluminum with silicon* (Moscow: Metallurgiya) p 271
[12] Sosnin O V, Gromov V E, Kozlov E V and Ivanov Yu F 2004 *Fatigue of steels under pulsed current impact* (Novokuznetsk: SibGIU) p 464
[13] Psakhie1 S, Ovcharenko V, Yu B, Shilko E, Astafurov S, Ivanov Y, Byeli A and Mokhovikov A 2013 *J. Mater. Sci. Technol.* **29** 1025–1034

The role of the vacancy complexes of nanocompacted aluminum particles on the formation of structure during heat treatment

Yu Ya Gafner[1], **V S Baidyshev**[1]

[1]Abakan, pr. Lenina 90, 655000, Russia, Katanov State University of Khakassia

E-mail: ygafner@khsu.ru

Abstract.

The boundaries of thermal stability of the initial face-centered cubic (fcc) phase in perfect aluminum clusters with a diameter up to 3 nm have been investigated by the molecular dynamics method using a modified tight binding (TB-SMA) potential. Based on the performed computer analysis, it has been concluded that, in most cases, for aluminum clusters with sizes up to D = 2.5 nm, the most stable configuration is the structure with pentagonal symmetry. With a further increase in the number of atoms, the fcc structure becomes more stable. The influence of the degree of disorder of nanocompacted aluminum particles up to 4 nm in diameter on the formation of a crystal structure during heat treatment has been analyzed. It has been shown that, under the effect of the temperature factor, the clusters undergo a transition from the initial fcc phase to other structural modifications, including those with pentagonal symmetry, even in the case of clusters with fairly large sizes.

1. Introduction

Investigation of the physical properties of metal nanoparticles with characteristic nanometer-scale sizes is an important problem from both the fundamental and applied points of view due to great prospects for their practical applications. In particular, the technological base for structural and functional nanomaterials, in many cases, includes ultra-fine metal powders [1–3], which can be considered as the main raw materials for producing compact nanocrystallites. Among the basic methods for synthesizing nanoparticles that form the base of nanopowders are the technique using gas phase condensation, precipitation from colloidal solutions, low-temperature plasma, cathode sputtering, molecular beams formed by different sources of heating, electric explosion, mechanical dispersion, and some others.

Despite the great variety of methods used for synthesizing nanocrystalline particles, the experimental investigation of their structure and properties is still very complicated [4]. One of the main reasons for these difficulties is the high reactivity of nanoparticles due to their highly developed surface, which can contain up to 50% of all the atoms of the particle. In this respect, of great scientific and technical interest are currently not single nanoparticles (nanoclusters) but compacted nanocrystalline materials, which are often more convenient for the investigation and application [1].

The vast majority of the currently available technologies used for synthesizing these materials have employed various types of compaction, often with the subsequent high temperature sintering [1]. Avoiding contact with the environment in the production of nanopowders and their compaction makes it possible to exclude the contamination with different impurities. However, particles of the compacted nanomaterial prepared by this method contain numerous stacking faults, the main of which are vacancies and pores. Investigations of the density of compacted samples showed that the overall degree of structural disorder in such materials due to the presence of different types of vacancy defects can vary in the range from 5 to 30%. There is no doubt that these single defects and their complexes can have a significant influence on the formation of the internal structure of nanomaterials at the atomic level. The use of high pressures and temperatures during the compaction of nanopowders makes it possible to decrease their residual porosity and to increase the relative density of the synthesized materials. However, this method cannot completely solve the problem of the production of defect-free nonporous compact nanopowders.

Among the nanoparticles used now or in the near future, special attention should be paid to small particles, nanoobjects with different types of spatial structure, which allow us to determine the dependence of the observed physical and chemical properties of their structure. Thus, in order to more accurately determine the real structure of this material, it is of interest to investigate the dynamics of the influence of heat treatment on individual nanoparticles of the material. Indeed, for the understanding of the specific features of the formation of the structure and properties of nanomaterials, it is very important to consider the structural phase transformations occurring in these objects during heating and cooling. The thermal stability threshold of the nanostructure is highly dependent on many factors. Some of them are well known (for example, the state of grain boundaries), but some yet to be determined. Since it is very difficult to perform real experiments of this level, the main purpose of the present work is to investigate the formation of the internal structure in individual structural units of nanocompacted aluminum with the use of computer simulation techniques.

2. The computer model

For the simulation of clusters, there are two basic numerical methods that complement each other, namely, the Monte Carlo method and the molecular dynamics (MD) method. The Monte Carlo method is very effective for studying macroscopic properties of a system. This method is successfully used to investigate the order–disorder of atoms, crystal growth, annealing of radiation defects, dislocation slip through a system of obstacles, etc.

The molecular dynamics method is the most efficient for the calculation of structural characteristics of point defects, dislocations nuclei, and grain boundaries, as well as for the investigation of the diffusion of atoms. Since the molecular dynamics method allows one to determine dynamic characteristics of a simulated system, it is widely used to investigate structural phase transformations in metals and alloys. It is this technique that is the most adequate to determine the structures of metal clusters with the diameter of our interest, which are stable over a wide temperature range. Other simulation techniques, such as the Monte Carlo method or *ab initio* methods, because of the restrictions on particle sizes, calculations of temperature factors, and others, cannot completely solve the problem. Therefore, for our simulation, we chose the classical molecular dynamics method.

It is well known that in the description of metal nanoclusters is more reliable with the use of a many-body representation of the potential energy function. The main advantage of the many body representation as compared to the simpler pair interaction potential is that it better reproduces some of the main features of metal systems. The many body scheme of constructing the interaction potential overcomes many difficulties encountered in constructing pair interaction potentials by taking into account the band character of metal bonds. In this case, the relatively simple technique for the description of the atomic and electronic structure without recourse to complex *ab initio* calculations is the tight binding (TB) method [5 – 7], in which the ion–ion interaction is described taking into account the band character of the band and the short range repulsive pair potential.

Therefore, after the analysis of different types of representation of the potential energy, the heating of aluminum nanoclusters was simulated using the modified tight binding (TB-SMA) potentials [8] with a fixed cutoff radius corresponding to the fifth coordination sphere. The tight binding potentials are widely used in computer simulation of both macroscopic and nanometer scale objects. Moreover, the potentials proposed by Cleri and Rosato were thoroughly and successfully tested with many thermodynamic parameters [8, 9] and, currently, are among the main potentials used for calculating properties of metal nanoparticles.

The heating of clusters was performed in a canonical ensemble with the Nose thermostat [10]. In the simulation of the canonical ensemble, for which the number of particles N, volume V, and temperature T remained constant (NVT ensemble) and the total angular momentum of the system p was zero, the system was immersed in a thermal reservoir by introducing a random force simulating collisions with virtual particles. During the simulation, the temperature was determined by the average kinetic energy of the atoms, which was calculated using the Verlet velocity algorithm with a time step $h = 1$ fs.

In order to most accurately investigate the influence of any of the parameters on the change in the structure of clusters, it is necessary, first of all, to minimize the interaction between the particle and its environment. Based on this assumption, we considered the so-called free or isolated clusters, which were placed in thermal reservoir, filled with virtual particles and did not interact with the substrate. The initial objects were spherical fcc aluminum clusters obtained by cutting a sphere from the perfect fcc lattice.

The computer molecular dynamics simulation was carried out with the MDNTP software developed by Dr. Ralf Meyer (Universitat Duisburg, Germany).

3. Results and discussion

In the simplest case, a nanocrystalline material consisting of atoms of one type contains two structurally different components: crystalline nanograins and grain boundaries with the width ranging from 0.4 to 1.0 nm [1]. The X-ray diffraction and neutron diffraction investigations of compacted metals demonstrated that the structure of interfaces in nanomaterials is similar to that in conventional polycrystals. Pores were found only in triple junctions, rather than along the entire length of the interfaces, and the density of atoms in the grain boundaries proved to be almost the same as in the crystalline grains [11].

For a physically adequate comparison of the obtained results, at the first stage, we investigated a group of clusters with the internal structure corresponding to the perfect fcc lattice without structural defects. In order to determine the boundaries of thermal stability of different isomeric modifications in aluminum nanoclusters, first of all, we examined the influence of the size of particles and the temperature on the formation of a particular crystal modification. We studied clusters containing 13, 19, 55, 135, 141, 147, 177, 201, 225, 309, 321, 555, 561, and 923 atoms (the diameter D was varied in the range from 0.5 to 3.0 nm). In general, based on the performed computer analysis of the heat treatment processes, it can be argued that, for small defect free aluminum clusters with sizes up to 350–400 atoms, in the majority of the cases, the most stable configuration is the structure with pentagonal symmetry. With a further increase in the number of atoms, the fcc structure becomes more stable in complete agreement with the laws of thermodynamics. However, in the case of aluminum nanoclusters, the geometric magic numbers play a very essential role, which was not observed by us so obviously in the simulation of nickel and copper clusters [12, 13].

At the second stage, we investigated the influence of defects on the formation of the internal structure of clusters. In order to simulate the structure with a desired degree of disorder, some atoms were removed in a random way from the initial cluster. This approach is similar to that described in the work [6], where the role played by pores in nanoparticles during the melting of clusters was analyzed also using the molecular dynamics method (*NPH* ensemble) for perfect fcc alumina clusters with diameters from 2 to 9 nm, which contained pores of different sizes. We also investigated clusters that did not contain pores at the initial stage but with the fractions of the removed atoms (single vacancies) equal to 5, 10, 20, 25, and 30%, which corresponded to real metal nanomaterials produced by the compaction method. For each case of the fixed fraction of vacancies in the cluster, we considered an ensemble of clusters of different sizes. All the clusters were subjected to relaxation at the initial temperature $T = 60$ K for 1 ns. Then, these clusters were heated with a step of 10 K and an exposure time of 0.5 ns.

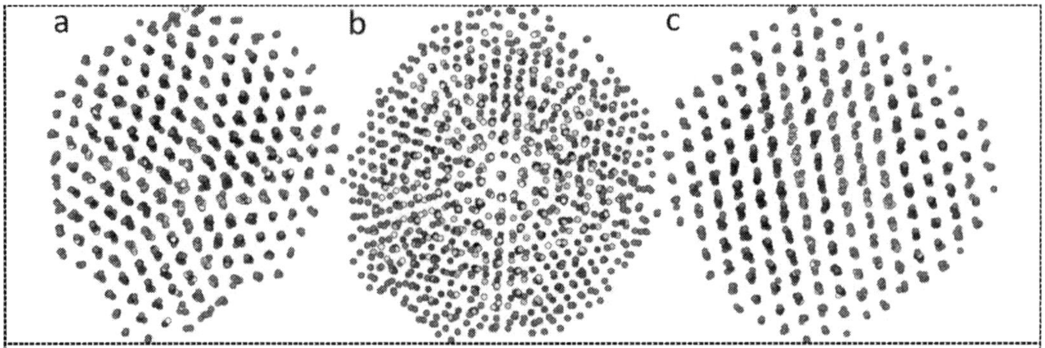

Figure 1. Aluminum clusters with the diameter D = 4 nm: (a) decahedral structure, (b) icosahedral structure, and (c) twinning in the fcc structure.

The results of the simulation demonstrated that single vacancies with the content equal to 5% of the total number of atoms in the cluster led to the formation of small defect regions even at T = 60 K. With an increase in the temperature, the size of these regions increased, although, in general, the clusters retained the initial fcc structure. When the fraction of vacancies reached 10% of the total number of atoms in the cluster with the diameter D = 3 nm, the fcc structure remained to be the most stable, but with some violation of the order of stacking of atomic planes. With an increase in the cluster size to D= 4 nm, the heating resulted in the formation of a nearly perfect fcc structure in only 20% of the computer experiments, the sufficiently defect free fcc structure with the twinning plane was observed in 60% of the numerical experiments, and model clusters had the internal structure with inclusions of different decahedral fragments in 20% of the simulated cases at temperatures T = 350–460 K.

For the fractions of the removed atoms equal to 25 and 30% of their total number in the cluster, at the stage of relaxation all the clusters had an amorphous like structure. Upon further heating of model particles, an increase in the primary disorder of clusters with single vacancies up to 25% of the total number of atoms in the cluster led to the growth of structures with pentagonal symmetry. In this case, for the clusters with D = 3 nm, the decahedral structure was stabilized in the temperature range T = 60–220 K in 30% of the model clusters, whereas the formation of the icosahedral structure was observed in the temperature range T = 220–340 K in 50% of the model clusters. Nonetheless, despite the fact that the initial state was amorphous-like, the fcc structure with twinning was observed in 20% of the computer experiments in the temperature range T = 160–240 K. With an increase in the cluster diameter to D = 4 nm, a sufficiently regular decahedral structure was stabilized in 60% of the computer experiments at temperatures T = 400–500 K (Figure 1a), the structures containing only decahedral fragments were found in 30% of the simulated clusters in the temperature range T = 100–150 K, and the formation of the defect icosahedral structure was observed in 10% of the model experiments at temperatures T = 250–300 K (Figure 1b).

An increase in the cluster diameter to D = 4 nm significantly (to 20%) decreased the probability of finding a cluster with the icosahedral structure. Moreover, particles with the decahedral structure were observed in 40% of the performed experiments, and the fcc phase was observed in 20% of the computer experiments (Figure 1c). The remaining 20% of the model clusters had a more complex internal structure with inclusions of several different fragments of the decahedral structure, which competed with each other. As in the case of clusters with the diameter D = 3 nm, the formation of the decahedral structure was observed at significantly lower temperatures in the range T = 80–200 K, the fcc structure was revealed in clusters at T = 140–180 K, the formation of the icosahedral structure proceeded the temperature range of 240–380 K, and the most complex structure with several decahedral fragments was observed at temperatures of 280–380 K. In our opinion, these differences in the temperatures of the formation of structural modification in aluminum clusters are associated with different binding energies for different ways of constructing atomic clusters. As is known, it is the icosahedral structure that has the highest binding energy, which could be judged from the higher melting temperature of the icosahedral clusters as compared to the fcc clusters of the same size. Most likely, it is for this reason that the icosahedral clusters require significantly higher kinetic energies for the transformation of the initially existing amorphous cluster into a well defined icosahedral structure.

4. Conclusions

Based on the performed computer analysis of the heat treatment processes with the use of the molecular dynamics simulation using the modified tight binding (TB-SMA) potential, it can be argued that, in small aluminum clusters with sizes up to D = 2.5 nm and without structural defects, in many cases, the most stable configuration is the structure with pentagonal symmetry. With a further increase in the number of atoms, the fcc structure becomes more stable. It was also shown that, in the case of aluminum nanoclusters with the perfect internal structure, the geometric magic numbers play a very significant role.

Thus, the performed molecular dynamics simulation of the heating of aluminum clusters with the content of vacancies from 5 to 30% of the total number of atoms in the cluster demonstrated that the formation of structures with pentagonal symmetry is possible in disordered nanoclusters with a diameter D = 3–4 nm, unlike defect free clusters in which, for this size of particles, the fcc structure is more stable. It was found that, for clusters with the diameter D = 3 nm, an increase in the number of vacancies leads to an increase in the number of clusters with the icosahedral structure, whereas for clusters with the diameter D = 4 nm, the decahedral structure remains predominant. The pentagonal clusters formed during heating, in most cases, have defect structures, in contrast to the fcc particles that are substantially free of packing defects, which was confirmed by the experimental data [14].

ACKNOWLEDGMENTS

This study was performed within the framework of the State Task of the Ministry of Education and Science of the Russian Federation.

References:

[1] Gusev A I 2005 *Nanomaterials, Nanostructures, and Nanotechnologies* (Moscow: Fizmatlit) [in Russian]
[2] Romanovskii V B and Makshina E V 2004 *Soros. Obraz. Zh.* **8** 50 [in Russian]
[3] Shpak A P, Pogosov V V and Kunitskii Yu A 2006 *Introduction to the Physics of Ultradispersed Media* (Kiev: Akademperiodika) [in Russian]
[4] Xing Y and Rosner D E 1999 *J. Nanopart. Res.* **1** 277
[5] Sutton A R, Finnis M W, Pettifor D G and Ohta Y 1988 *J. Phys. C: Solid State Phys.* **21** 35
[6] Chadi D J 1979 *Phys. Rev. B: Condens. Matter* **19** 2074
[7] Willaime F and Massobrio C 1989 *Phys. Rev. Lett.* **63** 2244
[8] Cleri F and Rosato V 1993 *Phys. Rev. B: Condens. Matter* **48** 22
[9] Li H, Wang G, Zhao J and Bian X 2002 *J. Chem. Phys.* **116** 10809
[10] Nose S 1984 *J. Phys. Chem.* **81** 511
[11] Ishida Y, Ichinose H, Kizuka T and Suenaga K 1995 *Nanostruct. Mater.* **6** 115
[12] Duque F and Mananes A 1999 *Eur. Phys. J. D* **9** 223
[13] Gafner S L, Redel L V and Gafner Yu Ya 2007 *Phys. Met. Metallogr.* **104** 180
[14] Abrosimova G E and Aronin A S 2002 *Phys. Solid State* **44** 1003

Degradation of austenitic steel 12X18H10T after electron beam impact

S V Plotnikov, N K Erdybaeva, E O Tleukenov

D. Serikbayev East Kazakhstan State Technical University, 23/2-54 Serikbayev Street, 070003, Ust-Kamenogorsk, Kazakhstan

Email: SPlotnikov@ektu.kz

Abstract. The surface structural phase changes by continuous electron beam impact on construction steel 12X18H10T were studied in this paper. As a generator of plasma-beam discharge was used device tokamak. The main objective of the work on the simulation bench was to obtain data on the structural materials samples' surface degradation degree under the influence of exposure to the electron beam and plasma discharge. Scanning electron microscope was used to study the surface morphology after continuous electron beam impact of the steel. As the matter of fact, the energy of electrons was quite low (less than 5 keV), consequently radiation defects were not formed and the entire beam energy went into heating of the target. The built-in modules of transient thermal and steady-state thermal temperature were found to be used numerically as the solution of the problem connecting with arbitrary geometry samples.

1. Introduction

The main goal of the Kazakhstan Material Research tokamak (KMT) is to conduct experimental research and testing of materials and structural solutions to protect the first wall, divertor plates and receiving nodes divertor in the mode of operation close to the ITER [1].

In support of the research program at the Kazakhstan Material Research tokamak KMT was created simulation test bench (STB). A generator is used to generate plasma, which operates on the basis of plasma-beam discharge. Experiment on irradiated samples by continuous electron beam was conducted at the current facility [2]. Tokamak has demonstrated excellent plasma confinement capability because of their symmetry but has an intrinsic drawback because of their pulsed inductive operation. Efforts have been made in the past 20 years to realize steady-state operation, the most successful utilizing a bootstrap current. In this review, progress in understanding tokamak physics related to steady-state operation is described to investigate the scientific feasibility of a steady-state tokamak fusion power system.

It is well known that radiation may significantly affect molecular materials even for relatively small doses [3]. Inter and intra-molecular reactions may promote crosslink and reticulate network in a macromolecular structure, either natural or synthetic, which has been irradiated. For instance, ionizing irradiation of a polymeric material is able to break chemical bonds and introduce modification in the material behavior [3-5]. Several works [6-11] have shown that macromolecular products such as natural fibers, biomaterials and polymers may improve their properties by ionizing irradiation as a result of crosslinking and scission processes, in which radicals are formed throughout the chains.

2. Experimental details

18-10 type steel, comprising about 18% chromium and 10% nickel (12H18N9, 10H18N10T, 108 12X18H10T at alias) is the main group. Alloying nickel - element extending γ region, leads to the fact that steel 18-10 become austenitic.

Table 1. Chemical composition in % of steel 12X18H10T.

C	Si	Mn	Ni	S	P	Cr	Cu	-
to 0.12	to 0.8	to 2	9-11	to 0.02	to 0.03 5	17-19	to 0.3	(5 C-0.8) Ti, others are Fe

In the simulation stand KMT uses plasma generator based on beam-plasma discharge (BPD) of low pressure at which the electron beam along the magnetic field is injected into the drift chamber interaction.

Compensated electron beam is transported in the electronic mode (without gas overlap, at extremely low residual pressure) on the target in a longitudinal magnetic field. It can vary within wide limits as the total power (30 kW) and the heat flux on the surface of the test materials. This provides a very wide range of thermal experiments, as well as the possibility of annealing of the samples and studying their thermal desorption.

A sample of steel 12X18H10T fastened with special plates for uncooled graphite target. To control the temperature of the sample on the back side was set XA type thermocouple. Facing the beam of the sample's surface was polished mechanically.

Table 2 shows the parameters of the unit under test of sample.

While irradiating of steel sample maximum power electron beam was equal to 75 Watts of heating power at the cathode-heated unit (CHU) 130 Watts. The maximum temperature of the steel sample registered by installed thermocouple reached 920 °C.

Appearance of the sample after irradiation is shown in Figure 1.

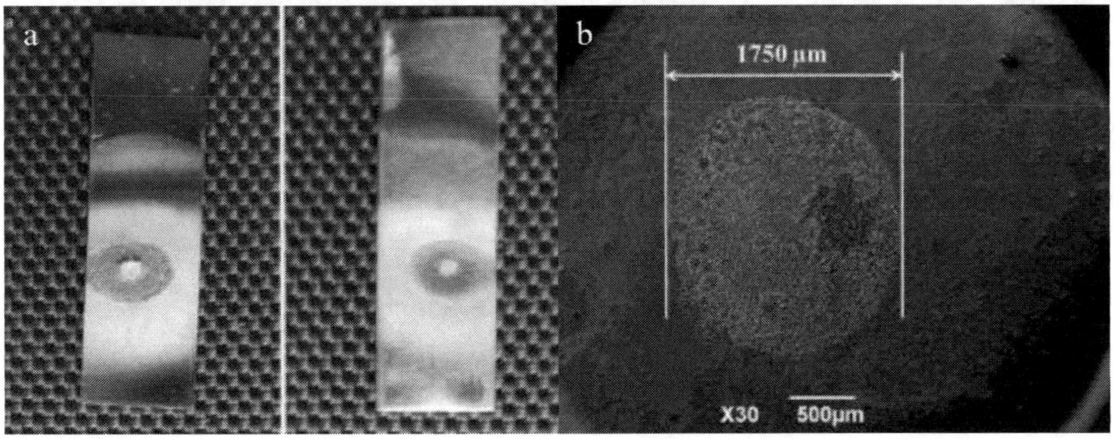

Figure 1. a) Photo of the sample after irradiation, b) micrograph of the footprint of the beam (SEM) (center).

Table 2. Parameters of the experiment at the 130W heating power CHU.

CHU			Electron beam			Duration of exposure (min)	The target temperature t(°C)
Voltag eU(V)	Curre ntI(m A)	Power N(W)	Voltag eU(kV)	Curren t I(mA)	Power W(W)		
650	200	130	3	25	75	2	920

Studies of samples' surface morphology after continuous electron beam impact were carried out on a scanning electron microscope JEOL JSM-6390LV with the prefix energy dispersive microanalysis INSA Energy.

Four heat-affected zones are clearly distinguished in the micrograph (Figure 2). In the central zone (Figure 3) precipitates titanium plates are clearly visible. Table 3 shows the analysis of the elemental composition of titanium plates and areas without isolation. The experiment shows that in the scope of the electron beam is formed thin films islet Ti. The film thickness must be less than 1.3 microns, as microanalysis shows the titanium's content is of about 98 %. If the film was thinner, microanalysis would show the content of other elements due to the "highlight" of the matrix.

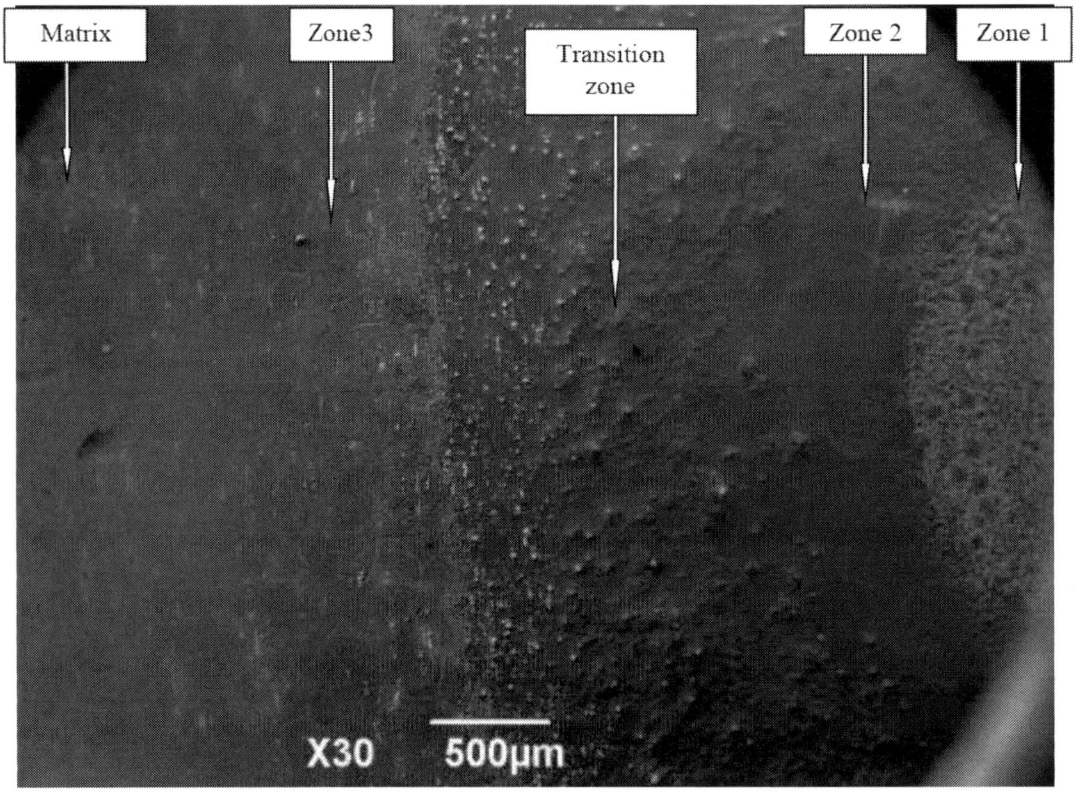

Figure 2. Microphotography of sample after irradiation, SEM x30.

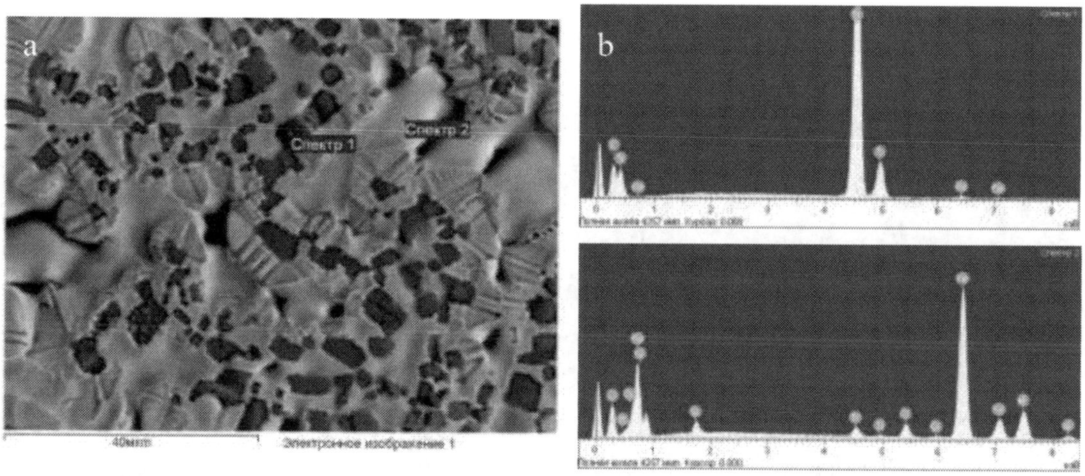

Figure 3. Microphotography of the 1st zone (center) x1500.

Table 3. Elemental analysis, in%.

Spectrum	Si	Ti	Cr	Fe	Ni	Total
Spectrum 1		98.49		1.51		100.00
Spectrum 2	2.29	2.13	4.02	70.56	21.00	100.00

Some islands have a complicated structure as nuclei crystallites. Segregation also occurs at the boundary of the 2nd and 3rd zones of Ti based alloy but in a form of cubic isolation on the surface. Table 4 shows the elemental analysis of the domains identified in Figure 6 b.

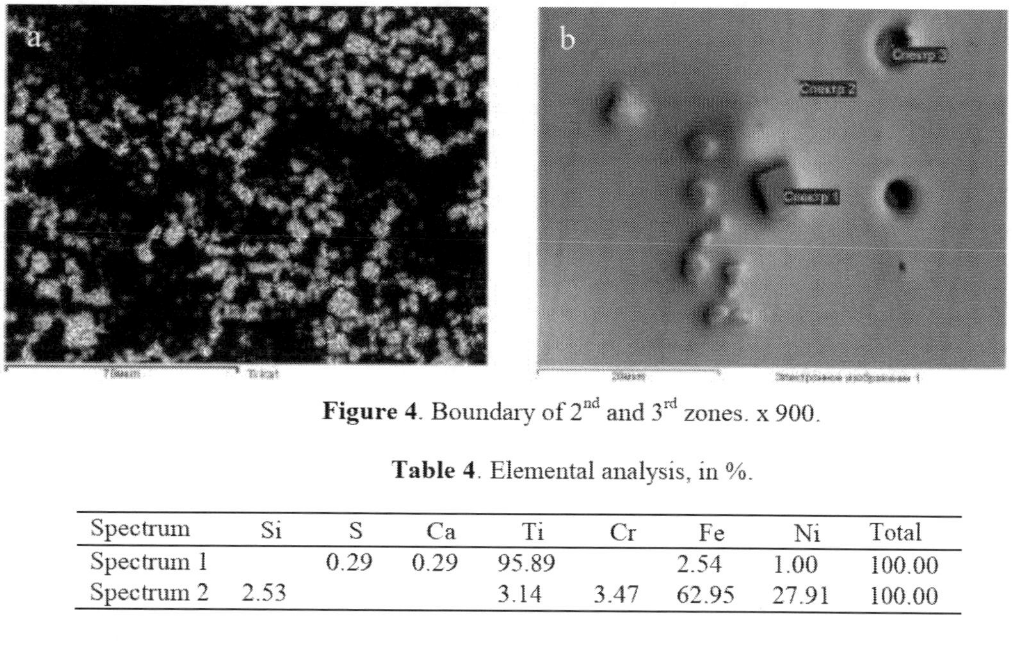

Figure 4. Boundary of 2nd and 3rd zones. x 900.

Table 4. Elemental analysis, in %.

Spectrum	Si	S	Ca	Ti	Cr	Fe	Ni	Total
Spectrum 1		0.29	0.29	95.89		2.54	1.00	100.00
Spectrum 2	2.53			3.14	3.47	62.95	27.91	100.00

Recrystallization occurs in the third zone (Figure 6), the temperature was - 1300 ^{0}C. Also, there are crystals Ti, but much less and smaller in size than on the boundary 2nd and 3rd zones and in the 1st zone.

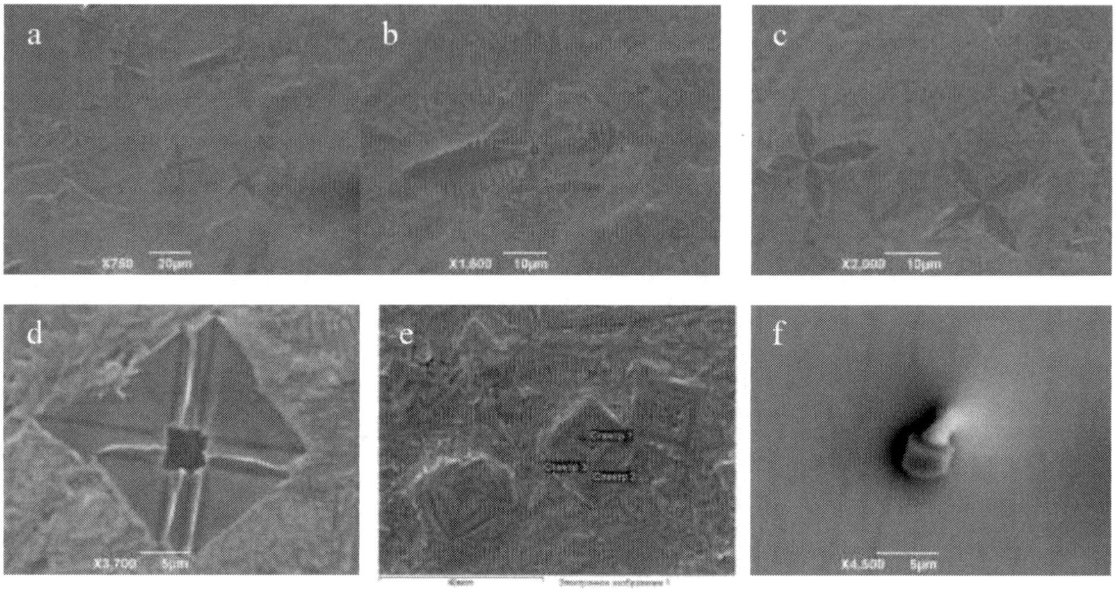

Figure 5. Microphotography of third zone from 750 to 4500-fold magnification.

The figure 5 shows the boundary of the third and fourth zones. Table 5 shows the analysis of the spectra indicated in the figure.

Table 5. Element analysis, %.

Spectrum	Si	P	Ti	Cr	Fe	Ni	W	Total
Spectrum 1	0.81		42.21	4.85	41.20	10.93		100.00
Spectrum 2			55.24	4.75	29.79	7.71	2.52	100.00
Spectrum 3	1.30	0.59	21.82	6.21	53.36	16.72		100.00
Max.	1.30	0.59	55.24	6.21	53.36	16.72	2.52	
Min.	0.81	0.59	21.82	4.75	29.79	7.71	2.52	

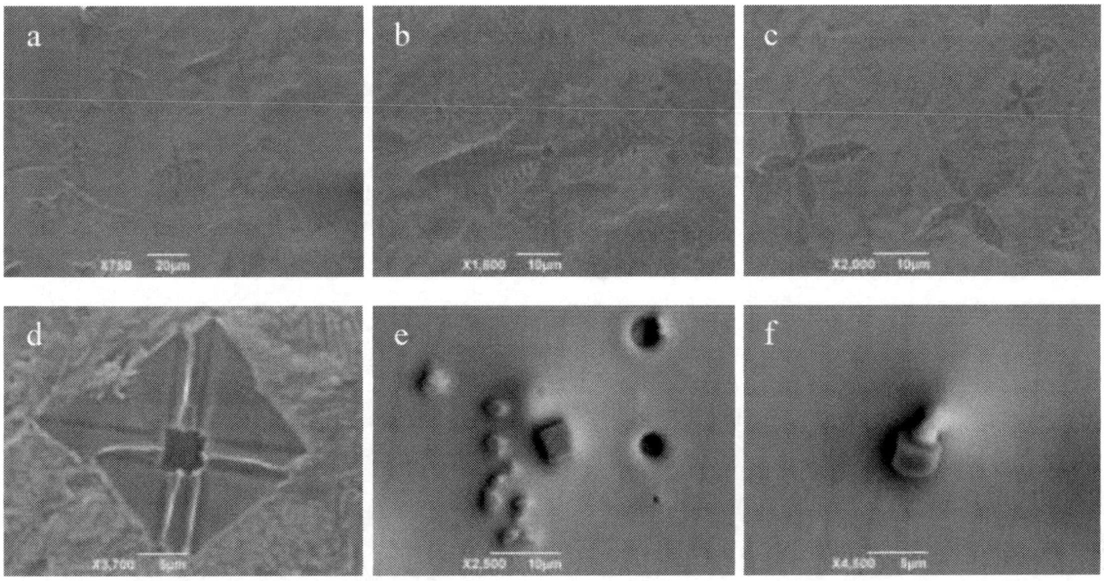

Figure 6. Photography of third zone from 750 to 4500-fold magnification.

The sample was cut at the center and the elemental analysis was carried out on depth. Figure 7 shows transverse section border of 2nd and 3rd zones photograph, graph of the intensity of element concentrations depending on the distance was plotted. From this it follows that towards the surface and at grain boundaries, the concentration of Ti significantly increases.

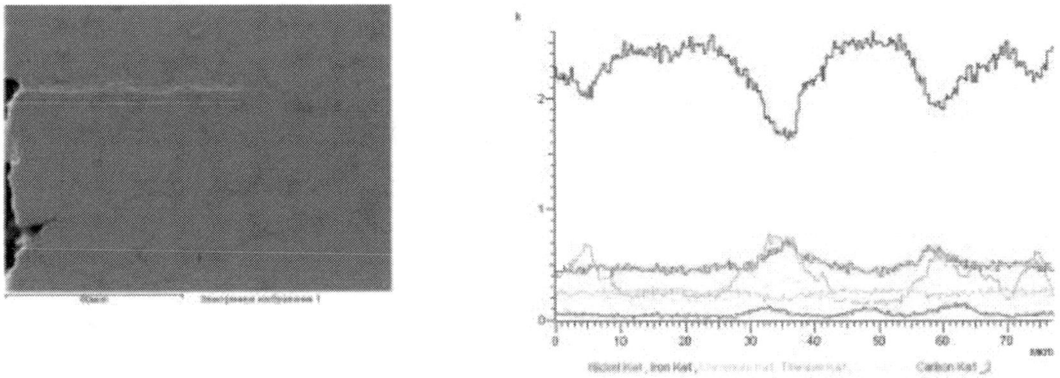

Figure 7. The boundary of the 2nd and 3rd zones. The intensity of the elemental concentrations depending on a depth.

3. Discussion of results

Due to the fact that the energy of the electrons was quite low (less than 5 keV), radiation-induced defects are not formed, and the entire beam energy goes into heating of the target. Thus structural phase changes the surface of the samples - the result of heat exposure of the electron beam.

Software system of finite-element analysis ANSYS 13.0 was used to build the model, working on the basis of the geometric core Parasolid.

Solution of the problem for an arbitrary geometry samples was found numerically using the built-in modules Transient Thermal and Steady-State Thermal. Material samples were 12X18H10T steel. The dimensions of the finite element model and the physical and technical data in both versions were introduced in Appendix Mechanical. Characteristics of the model are shown below.

Table 6. Characteristics of the model.

Dimensions of the sample:	50×10×1 mm;
Density of steel:	8055 kg/m^3;
Thermal conductivity:	13,8 W/m °C;
Specific heat:	480 J/kg °C;
Weight of steel sample:	40,2·10^{-3} kg;
Initial temperature:	22 °C;
Radius of the spot heating:	1·10^{-3} m;
Heat flux density:	~1,9·10^7 W/m^2.

Figure 8 shows a fragment of the program showing the simulation grid. The grid was generated automatically, mesh size was set manually to 1 mm. Mesh was obtained consisting of 4724 elements.

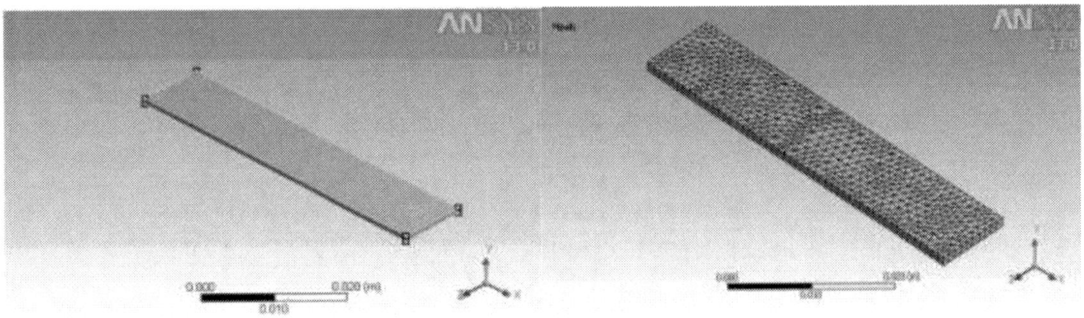

Figure 8. Generation of grid model in Ansys 13.0.

Each step of the pattern corresponds with one second of heating, 1000 iterations were taken for decision and construction of the temperature distribution during heating.

Figure 9 shows the parts of the program, showing the graphical results of calculations. Figure 9 designated temperatures at random points located around the spot heating are illustrated on the figure. The temperature scale is shown on the left.

Figure 9 shows the result of calculation with regard to the embodiment of the contacting surfaces. It is seen that the spot diameter is ~ 3 mm. Unbalanced spot, the temperature distribution with strong edge effects are shown.

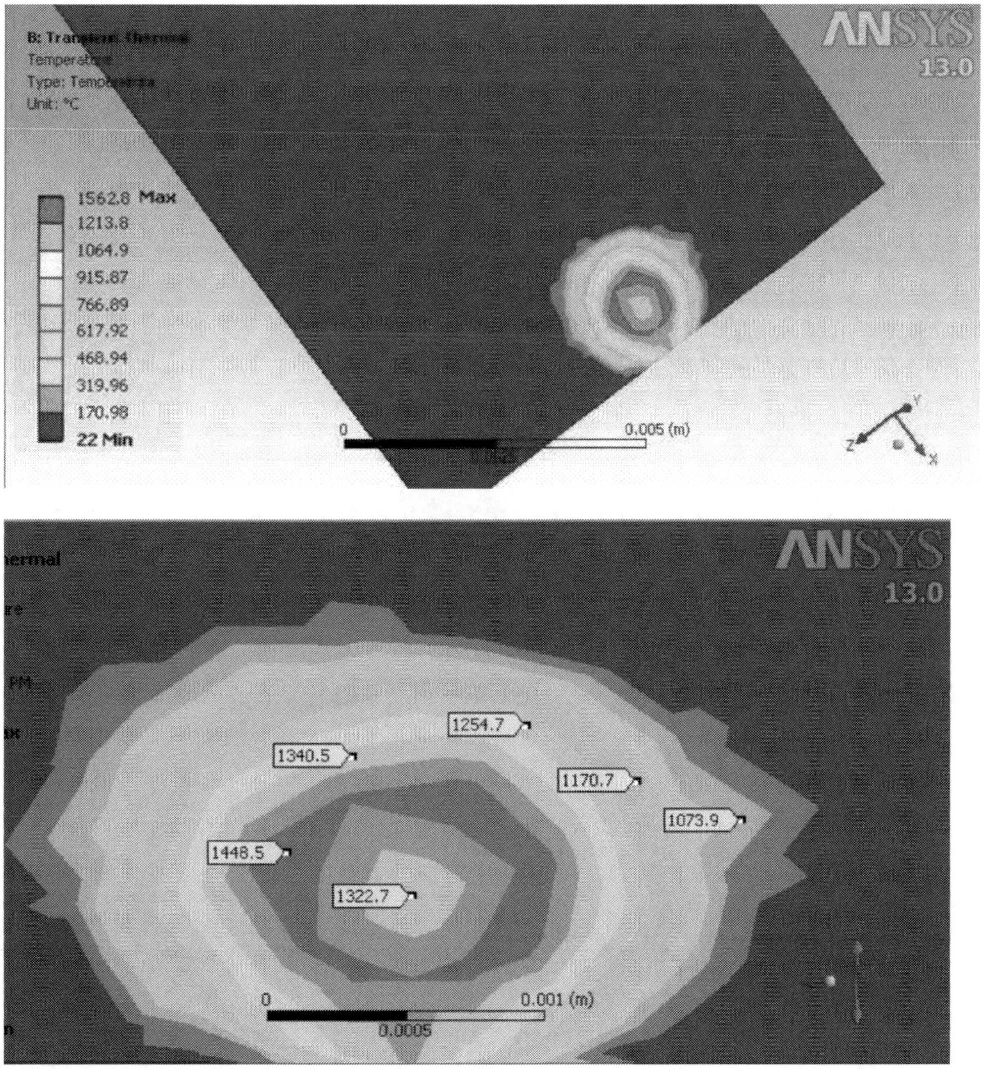

Figure 9. The results of temperature field calculation with the contacting surfaces.

Figure 10 shows a fragment of sample geometry with temperature distribution through the thickness of the sample in the embodiment with the contact surfaces. The cut is made at a distance of 0.1 mm from the center of heating spot. It is seen that the presence of the contacting surfaces leads to diameter reduction, and the overall appearance of the temperature distribution also changes.

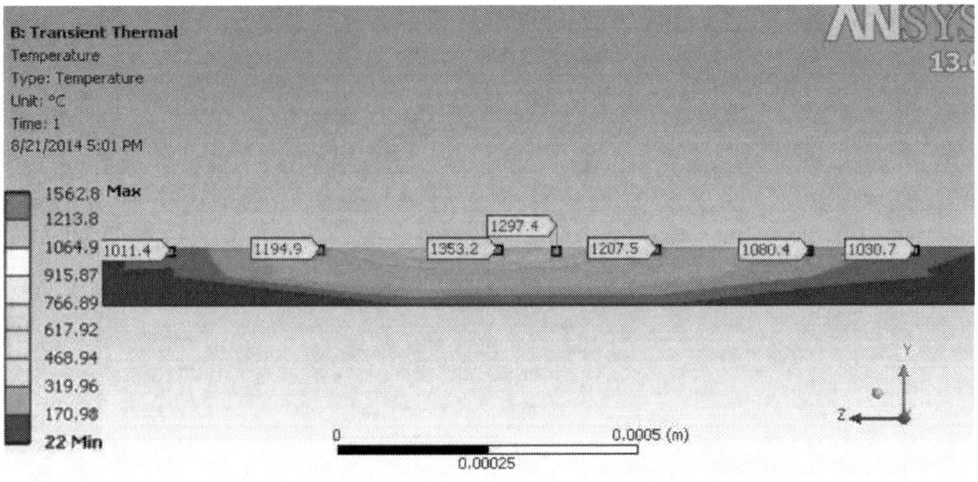

Figure 10. Temperature field of the sample cross-section in the presence of the contact surfaces.

Figure 11 is a graph, which shows the temperature change in the center of the heating spot for 20 steps, which corresponded to 20.

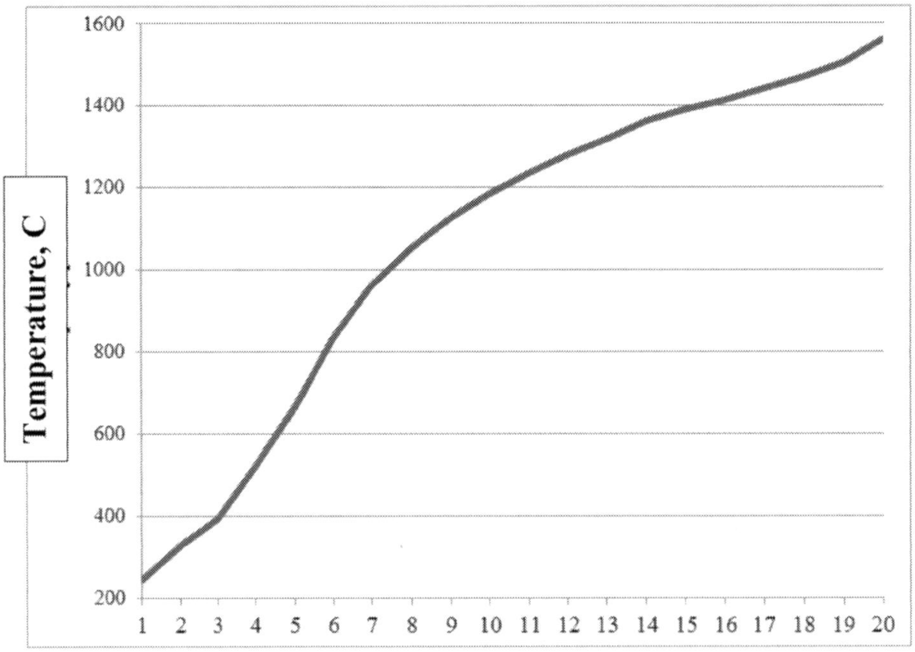

Figure 11. Dependence of temperature spots in the center on time.

Figure 12 and Figure 13 shows the temperature distribution along the height of the sample for two embodiments. The distance is measured from the center of the spot on the surface into the sample.

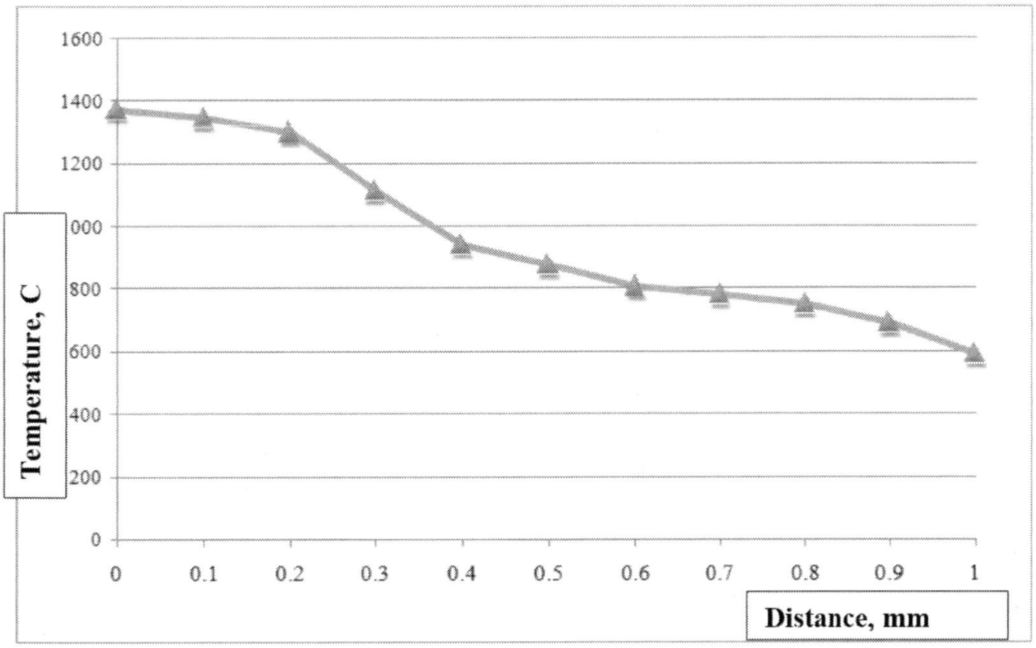

Figure 12. The temperature distribution in height of the sample in the second embodiment.

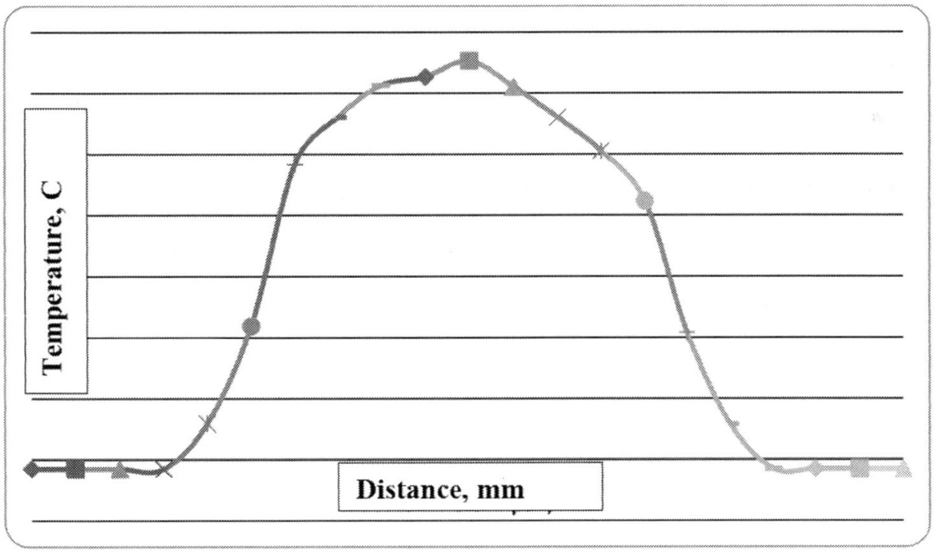

Figure 13. The temperature distribution along the length of the sample in the second embodiment.

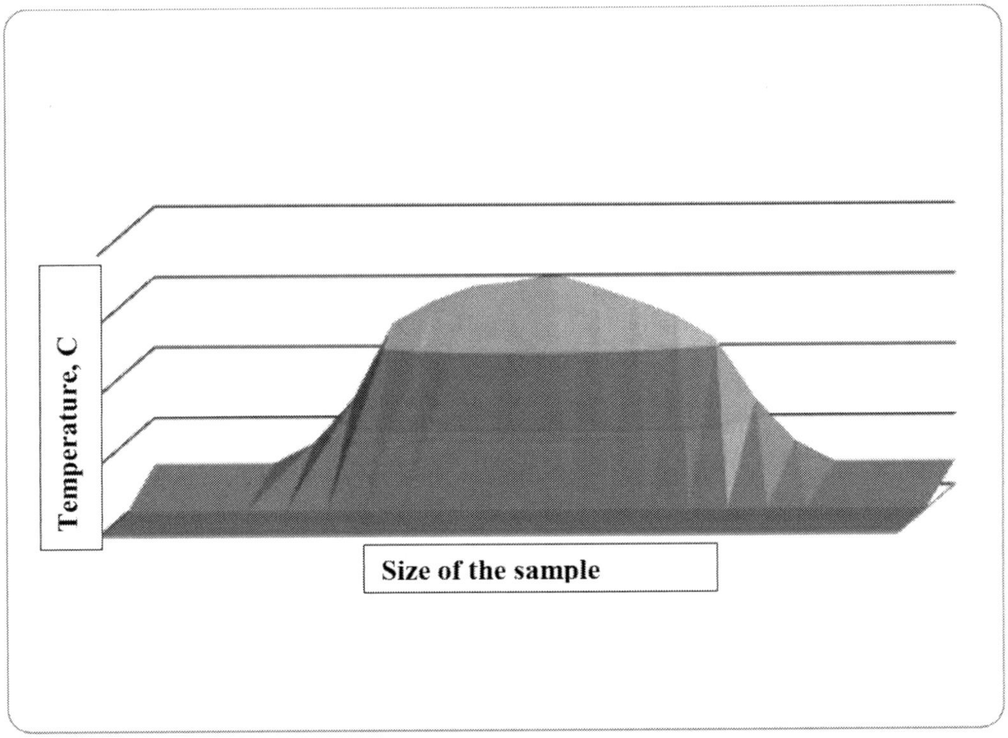

Figure 14. The temperature distribution in the second embodiment.

Isolation of impurities can be explained by the presence of gradients of point defects near the border. Due to the intense heat after turning off the electron beam occurs excess thermodynamic equilibrium vacancies. As a result, the flow of vacancies occurs, which is directed perpendicular to the surface and from the scene of the beam. The metal border is intensive sink of vacancies, so the average profile of the distribution of vacancies at the border changes its shape.

The kinetic equation is used for the calculation of the distribution of impurities in the diffusion gradients of temperature and vacancies (1) [3].

$$\frac{\partial C}{\partial t}=\frac{\partial}{\partial x}\left(-D\frac{\partial C}{\partial x}\right)+\frac{\partial}{\partial y}\left(-D\frac{\partial C}{\partial y}\right)+\frac{C(1-C)C_v}{\kappa T^2}\left(E_A D_A^* - E_B D_B^*\right)\frac{\partial T}{\partial x}$$
$$-C(1-C)\left(D_A^{'} - D_B^{''}\right)\frac{\partial C_V}{\partial x} - C(1-C)\left(D_A^{'} - D_B^{''}\right)\frac{\partial C_V}{\partial y} \qquad (1)$$

Where, C - the concentration of atoms of type B, $D_A^* = D_{0A}e^{-E_A/kT}; D_B^* = D_{0B}e^{-E_B/kT}$ - self-diffusion coefficients of atoms of type A and B; $D=(C_A D_B^* + C_B D_A^*)C_v$ - the mutual diffusion coefficient E_A, E_B - the activation energy of the atoms A and B.

The first two terms describe the concentration diffusion, thermal diffusion process is third on the nonequilibrium vacancies. Such flow is directed in a direction opposite the temperature gradient, i.e. a movable component accumulates at the cold end of the sample. The fourth and fifth terms on the right-hand side describe the process of redistribution of elements caused by non-uniform thickness distribution of nonequilibrium vacancies. In this case, a movable component must accumulate on the surface of the sample. There is the presence of competition and the effect of thermal diffusion process "vacancy wind":

Equation (1) was solved by numerical methods, as analytically solve this equation is not possible. Calculations show that the titanium concentration increases at the surface of the sample (Figure 15).

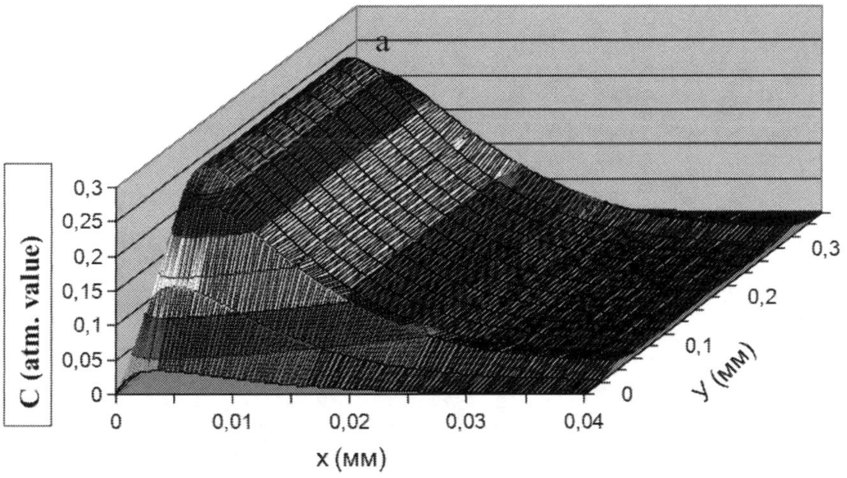

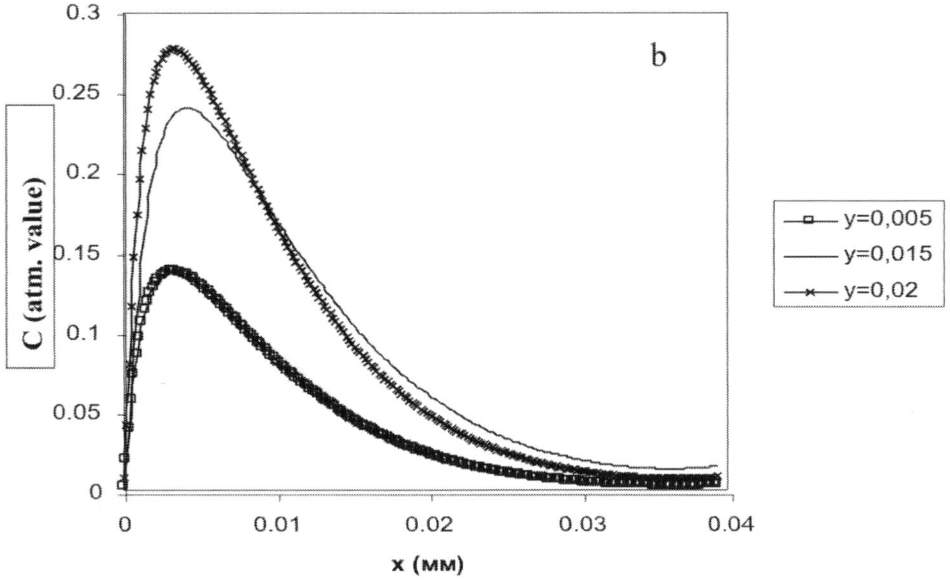

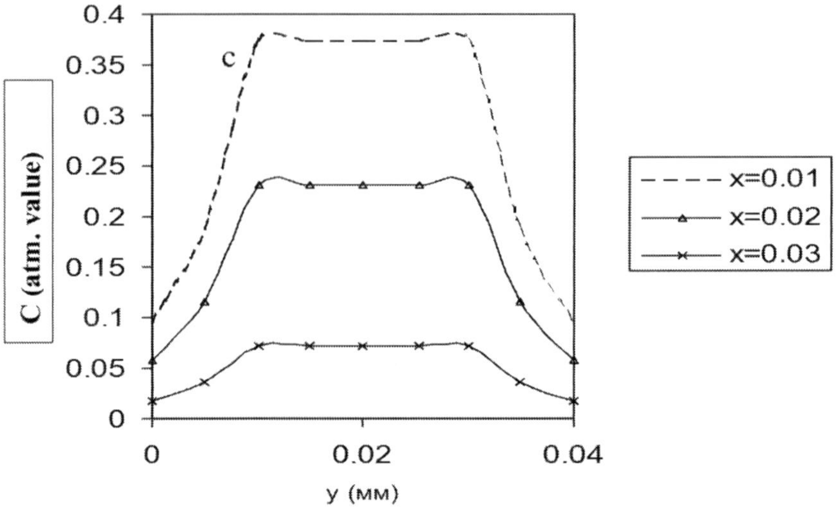

Figure 15. Distribution of titanium steel 12X18H10T after electron beam irradiation. a) space distribution b) cross section in the x-axis, c) cross-section along the y axis.

The results of calculation are in good agreement with the experimental data.

Figure 6 shows the results of measuring the distribution of the alloy components in depth. It is seen that the titanium concentration increases at the sample surface and at grain boundaries.

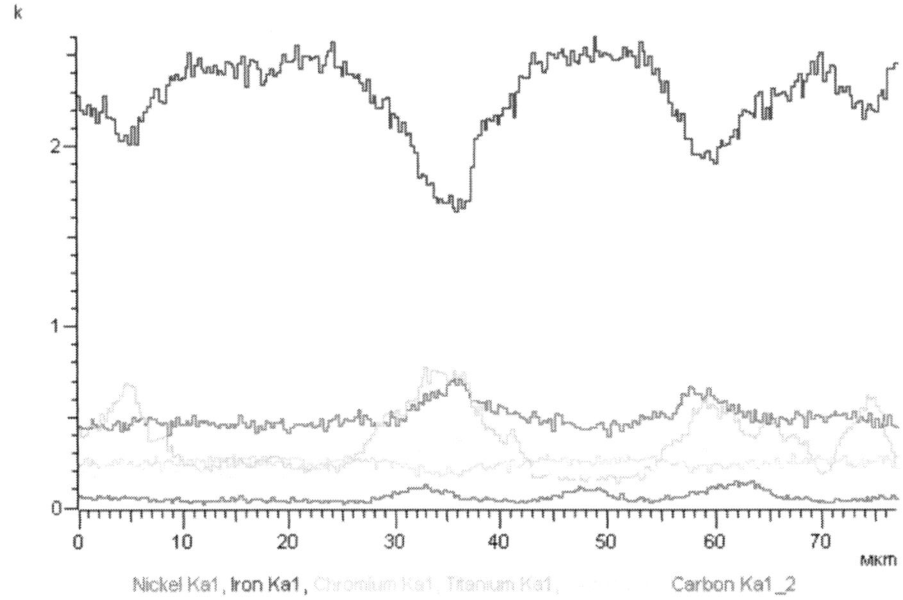

Figure 16. Distribution of the alloy components in the depth of the sample.

4. Conclusion

Electron beam irradiation of materials leads to abnormal redistribution of elements in alloys. That allows creating a predetermined impurity profile and providing the required performance of the product.

The experimental data shows that in the scope of the electron beam is formed thin films islet Ti. The film thickness must be less than 1.3 microns, as microanalysis shows the titanium's content is of about 98 %. If the film was thinner, microanalysis would show the content of other elements due to the "highlight" of the matrix.

The energy of electrons was quite low (less than 5 keV), consequently radiation defects were not formed and the entire beam energy went into heating of the target. The built-in modules of transient thermal and steady-state thermal temperature were found to be used numerically as the solution of the problem connecting with arbitrary geometry samples.

References

[1] Kurnaev V A 2005 *Stend imitacionnih ispitani v soprovojdenie issledovani na tokamake KMT* (SII-KMT), MIFI

[2] Plotnikov S V, Postnikov D V 2002 Vliyanie oblucheniya puchkom electronov na pereraspredelenie primesi binarnix metallicheskih sistemah *Izv. Vuz. Physics* **8** p 124-129

[3] Clegg D W, Collyer A A 1991 *Irradiation effects on polymers* (London, Elsevier Science Publishers)

[4] Choi H Y, Han S O, Lee J S 2008 Surface morphological, mechanical and thermal characterization of electron beam irradiated fibers *Applied Surface Science* V. **255**, p 2466-2473

[5] Takinami P Y I , Shimazaki, Colombo M A 2010 Gelatin/piassava composites treated by electron beam radiation *Revista Matéria* V. **15**, n. 2, p 339-344

[6] Buchalla R, Schüttler C, Bögl K W 1993 Effect of ionizing radiation on plastic food packaging materials: a review, Part 1. Chemical and Physical Changes *Journal of Food Protection* V. **56**, n. 11, p 991-997

[7] Machi S 1996 New trends of radiation processing applications *Radiation Physics and Chemistry* V. **47**, n. 3, p 333-336

[8] Dorschner H, Lappan U, Lunkwitz K 1998 Electron beam facility in polymer research: radiation induced functionalization of polytetrafluoroethylene *Nuclear Instruments and Methods in Physics Research B* V. **139**, p 495-501

[9] Khan F, Ahmad S R, Kronfli E 1999 Stability of jute fibers on exposure to ionizing radiation *Polymer Degradation and Stability* V. **63**, p 79-84

[10] Rouif S 2005 Radiation cross-linked polymers: Recent developments and newpplications *Nuclear Instruments and Methods in Physics Research B* V. **236**, p 68 - 72

[11] Ratnam C T, Raju G, Yunus W Z W 2007 Oil palm empty fruit bunch (OPEFB) fiber reinforced PVC/ENR blend-electron beam irradiation *Nuclear Instruments and Methods in Physics Research B*. V. **265**, p 510-514

IOP Conf. Series: Materials Science and Engineering **81** (2015) 012014 doi:10.1088/1757-899X/81/1/012014

Calculations of the heat capacity of Cu clusters synthesized by condensation from the gas phase

I V Chepkasov[1], **L V Redel**[1]

[1]Abakan, pr. Lenina 90, 655000, Russia, Katanov State University of Khakassia

E-mail: ilya_chepkasov@mail.ru

Abstract.

On the basis of our former simulations, we conclude that the heat capacity in the case of isolated free clusters can exceed that of a bulk material. It was found that at T=200K the increase in the Cu nanocluster heat capacity (D = 6nm) was only 10%, decreasing with growing nanoparticle proportionally to the reduction in the fraction of surface atoms. Thus, the considerably larger heat capacities of copper nanostructures observed in the experimental works cannot be related to the characteristics of free clusters. In our view, these properties of a nanomaterial can be associated with the degree of agglomeration of its constituent particles, i.e., the interphase boundaries and the increase in the root-mean-square displacements of atoms on the combined surface of the interconnected nanoclusters can have a strong effect. To test the above hypothesis, we took copper clusters of various sizes (4071–15149 atoms) that we produced when simulating the synthesis of Cu nanoparticles. Thus, in our molecular-dynamics experiments using a tight-binding potential at high temperatures, we failed to properly assess the role of the interphase boundaries in calculating the heat capacity of nanoparticles. The reason was the mass diffusion of Cu atoms to impart an energetically more favorable shape and structure to the cluster. At low temperatures, the heat capacity of the clusters exceeded that of the bulk same by a value from 10% to 17%. Consequently, the Cu clusters produced in direct experiments cannot be immediately applied in devices using the thermal energy of such clusters, because their external shape and internal structure are nonideal.

Introduction

Current, advances in new constructional and functional materials are associated, first of all, with the development of compactificated nanomaterials. In particular, the practical application of nanocrystalline metals and their various compounds has caused intensive studies of their mechanical, thermal, electrical, magnetic and other properties. For theoretical interpretation of experimental data obtained on compactificated nanomaterials it is, first of all, necessary to separate surface effects and those relating to small volume of particles the nanomaterial consist of. The problem is far of its solution as studying compactificated nanomaterials is yet at its prior stage of the result collection [1].

Thus, according to [2], the heat capacity of coarse-grained copper clusters of about 50 nm in size is higher than the normal heat capacity of the bulk copper in 1.2-2.0 times at temperatures from 200 K to 450 K. In [3] similar results were obtained for nickel nanoparticles of 22 nm in diameter: their heat capacity was twice higher than that of the bulk nickel at temperatures 300-800 K. A study of the heat capacity of nanostructures consisting of nickel clusters with of the diameter 10 nm by the fast neutron scattering method [4] also demonstrated the increase in the heat capacity in 1.5-2.0 times compared to that for conventional materials. However, according to [5], for nanocrystalline palladium (D = 6 nm) and copper (D = 8 nm) in the temperature range 150-300 K the heat capacity increasing effect should be much smaller (29-53% and 9-11%, respectively, compared to conventional polycrystalline values for Pd and Cu).

To sum up, obvious discrepancies take place in experimental data on the heat capacity of the compactificated metallic nanomaterials. All works show an excess of the heat capacity for a nanostructured material compared to a bulk body, but the values of this difference vary greatly. In fact, not individual nanoclusters but a nanostructured material composed of particles with a diameter of at least 6 nm was studied in these works. In this case, the sample under consideration is similar to colloidal crystals, because the developed surface and its fraction in determining the atomic mobility of the nanostructure are the main factor of an increase in heat capacity. Therefore, to fill in the existing gap, we undertook computer simulations of the heat capacity for copper nanoclusters produced through the simulation of condensation from the gas phase. This chemical element is typical representatives of the transition metals and the data obtained for him can also be largely extended to nanoparticles of other transition metals.

The computer model

The study of clusters plays a very important role in understanding the transition from a microscopic to a macroscopic structure; however, it is difficult to determine many physical properties for clusters experimentally [6]. Computer experiments may be considered as another way to investigate thermal properties of nanoparticles. In the molecular dynamics (MD) simulations presented here, Newton's equations of motion are solved numerically for each atom in the force field of the remaining ones by using a particular interatomic potential.

After analysis of different forms of the functional presentation of the potential energy, we have chosen the modified tight-binding potential TB-SMA [7] with the fixed cut-off radius corresponding to the 5-th coordination sphere, inclusively for following calculations of interatomic forces. This method is based on the fact that a significant group of transition metals can be fully described from the density of states of outer-shell d electrons. Despite a simple functional form, the tight-binding model well describes the elastic properties, the characteristics of defects, and the melting of a wide range of fcc and hcp metals.

As a test, this scheme of potential energy is tested using various thermodynamic characteristics of copper clusters, such as the melting and solidification temperatures, the hysteresis width, the convergence of a potential in the range of small clusters, and the binding energy. On the whole, when analyzing the results of various authors, we conclude that the Cleri–Rosato TB-SMA potentials [7] satisfactorily describe the behavior of a number of thermodynamic quantities of metallic fcc clusters on qualitative and quantitative levels. The differences between the simulation results and the data obtained by other collective potentials do not exceed 3–5%.

By computer simulations with molecular dynamics method, we used the MDNTP computer code developed by Dr. Ralf Meyer (Universität Duisburg, Germany). During simulation, the temperature was determined by the average kinetic energy of atoms calculated with the rapid Verlet algorithm at a time step h = 2 fs [8]. Modeling occurred in frameworks a canonical ensemble for which the number of particles N, volume V, and temperature T are constant (NVT ensemble) and total momentum p of the system is zero, we placed the system in a thermal reservoir (Nose thermostat [9]) by introducing a random force to simulate collisions with virtual particles. The advantage of this technique is a very good stabilization of a fixed temperature: the deviations from this value are at most one percent.

Results and discussion

Taking into account the mentioned above significant scatter in values of the heat capacity of nanoclusters and nanostructured materials, the heat capacity of ideal separate Ni and Cu nanoparticles was investigated in our former work [10] in the temperature range 200-800K. It was found that at T = 200K the increase in the Cu nanocluster heat capacity (D = 6nm) was only 10%, and 13% for Ni nanoclusters of the same diameter.

Having analyzed the results of our former simulations using a modified tight-binding potential, we conclude that the heat capacity in the case of isolated free clusters can exceed that of a bulk material, decreasing with growing nanoparticle proportionally to the reduction in the fraction of surface atoms. Thus, the considerably larger heat capacities of copper and nickel nanostructures observed in the experimental works [2–4] cannot be related to the characteristics of free clusters but are determined by different factors that are yet to be studied.

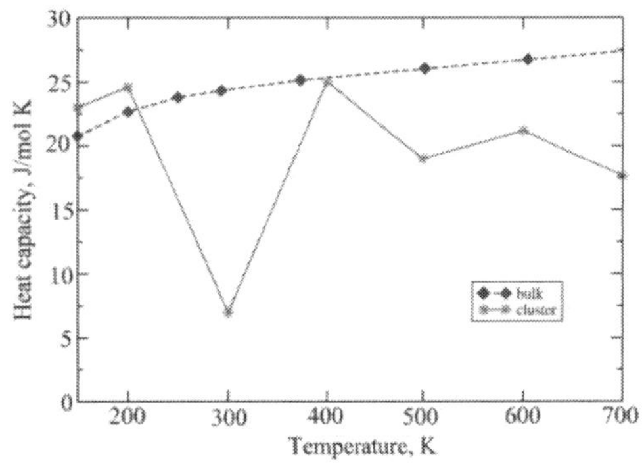

Figure. 1a. Temperature dependence of the heat capacity for the bulk copper and Cu_{4071} cluster produced by condensation from the gas phase.

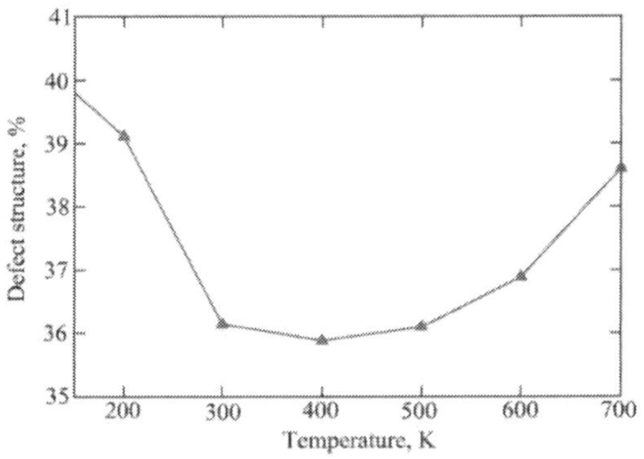

Figure. 1b. Temperature dependence of the percent of atoms with coordination number less than 12 (defective structures) in Cu_{4071} cluster.

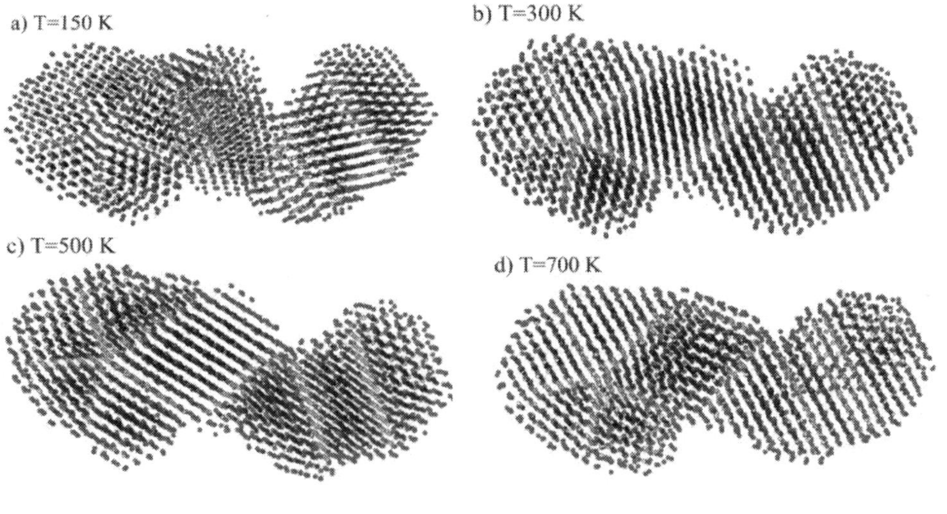

Figure. 2. Evolution of the Cu_{4071} cluster as a function of the heating temperature: a) T = 150 K, b) T = 300 K, c) T = 500 K, d) T = 700 K.

In our view, these properties of a nanomaterial can be associated with the degree of agglomeration of its constituent particles, i.e., the interphase boundaries and the increase in the root-mean-square displacements of atoms on the combined surface of the interconnected nanoclusters can have a strong effect. Thus, the increase in the heat capacity of nanostructures can be affected significantly by the appearance of low-frequency atomic vibrations with increasing amplitudes attributable to the increased surface of the agglomerated clusters and the interphase boundaries in the nanostructures.

To test the above hypothesis, we took copper clusters of various sizes (4071–15149 atoms) that we produced when simulating the synthesis of Cu nanoparticles. Therefore, at the first (preliminary) stage, we simulated the condensation of copper nanoparticles from a high-temperature gas phase for a physically and technologically proper representation of the computer sample of synthesized nanoparticles. At the second (main) stage of our studies, we calculated the heat capacity of the copper clusters with an already real shape and internal structure obtained from the gas phase to find the main patterns of the accumulation of thermal energy by them.

As example, consider a copper cluster of 4071 atoms. This cluster was ultimately formed when the Cu_{1385} particle collided with the Cu_{2686} particle. The latter turned out to be the result of the agglomeration of two particles containing 1856 and 830 atoms, respectively. In the long run, the cluster of 4071 atoms turned out to be the result of the agglomeration of more than 20 individual primary particles with typical sizes from 1.0 to 2.0 nm. However, most of these particles subsequently combined between themselves, so that only four particles separated by well-defined interphase boundaries remained distinguishable in the internal structure of the Cu_{4071} cluster at the final stage. One of the clusters with N = 2686 had a seed of the icosahedral phase. The second part of the Cu_{4071} cluster corresponded to the primary Cu_{1385} particle with a fairly regular fcc arrangement of atoms.

To properly estimate the heat capacity of the Cu_{4071} cluster, we compared it with the data obtained for a bulk copper. In Figure. 1a, we see a significant difference in the results for the bulk copper and the cluster that we produced through the simulation of real synthesis. The heat capacity of the cluster turned out to be comparable to that of the bulk sample only at low temperatures. For example, at T = 150 K it exceeded the heat capacity for the bulk copper approximately by 10.26%, but the heat capacity of the cluster was ~30% of its value when the temperature rose to 300 K. To find the cause of these discrepancies, let us examine Figure. 2 that shows the external view and internal structure of the Cu_{4071} cluster at various temperatures.

A characteristic feature of the condensation of a dispersed material is known to be a steep temperature gradient. At cooling rates of 105-108 K s^{-1}, Al, Ni, and Cu powders with particle sizes of 0.5–50 nm and specific surface areas of (70-30)·10^3 м^2 kg^{-1} can be produced [11]. Because of such a high cooling rate, the diffusion processes turn out to be essentially blocked at the final synthesis stages (T = 77 K). As a result, the synthesized nanoparticles clearly have no time to optimize their internal structure, as is clearly seen from Figure. 2a.

To minimize the energy-consuming diffusion processes at higher temperatures and to separate only the contribution from the interphase boundaries to the heat capacity, the clusters were heated at chosen temperatures for only 0.02 ns. As we see from Figure. 2, even this short time turns out to be quite sufficient for a partial rearrangement of the structure of the cluster. Thus, the heat supplied to the Cu_{4071} cluster is often expended not in increasing the store of internal energy but on the massive diffusion of atoms to impart an optimal shape and structure to the cluster. In massive reorganization of structure specifies also decrease in percent of defective structures in Cu_{4071} cluster with rising temperature to 300K (Figure. 1b).

The assumption that the cluster with the greatest difference in heat capacity with respect to the ideal nanoparticle was simultaneously also the spatially most extended one appears more plausible. The large linear length of the cluster probably allows the low-frequency part of its photonic spectrum to manifest itself by reducing the mean vibration frequency, which can lead to an increase in the cluster heat capacity.

Consequently, the Cu clusters produced in direct experiments cannot be immediately applied in devices using the thermal energy of such clusters, because their external shape and internal structure are nonideal. First of all, a preliminary stage of optimization, for example, by the thermal relaxation method, will be necessary here, which, in some cases, allows clusters with an ideal spherical shape and a single internal structure to be produces [12], or it is necessary to make changes to the cluster synthesis procedure itself to produce such ideal clusters [13].

1. Conclusions

Development of contemporary science has resulted in the necessity to use new devices comparable in size with the atomic scale. Then, up to the present time, it becomes clear that properties of nanomaterials the devices in question consist of do significantly depend on specific features of their constituents. In this connection, intensive studies have been beginning of small particles (clusters) in the size range from 10 to several thousand atoms. However, experimental study of nanoclusters faces definite difficulties relating, first of all, to their small size. For this reason, one of possible approaches to study nanosized objects may be the treatment of their computer models. Most promising for studying physical-chemical properties of metallic clusters is, in our sight, molecular dynamics method which makes it possible to understand effects of different factors on properties of particles at the atomic level.

When nanotechnologies are introduced into industry the main up-to-date problem is rather not the production of new materials itself but the investigation of thermal effects and force fields (radiational, deformational and others) in the course the usage of nanomaterials. All the above factors should affect physical-chemical properties and, as a consequence, maintenance resources of nanomaterials. The thermal properties of nanoparticles and, in particular, the capability of a cluster to accumulate the thermal energy are among the important peculiarities of their behavior. At present, active works are being performed to solve a number of problems for a more efficient use of such energy in various fields of science and engineering. Given the already revealed experimental and theoretical facts, our analysis of the heat capacity for Cu nanoparticles allows us to generalize the previously reached conclusions and to determine the direction of further research on revealing all of the mechanisms responsible for the growth of the heat capacity in nanostructured copper.

ACKNOWLEDGMENTS
This study was performed within the framework of the State Task of the Ministry of Education and Science of the Russian Federation

References:

[1] Kelsall R, Hamley I, Geoghegan M 2005 *Nanoscale science and technology* (Chichester:Wiley)
[2] Chen Y Y, Yao Y D, Lin B T 1995 *Nanostruct. Matter.* **6** 597
[3] Yao Y D, Chen Y Y, Hsu C M 1995 *Nanostruct. Matter.* **6** 933
[4] Trampenau J, Bauszus K, Petry W and Herr U 1995 Nanostruct. Matter. **6** 551.
[5] Rupp J and Birringer R 1987 *Phys. Rev. B.* **36** 7888
[6] Cleveland Ch L, Landman U, Schaaff Th G, Shafigullin M N, Stephens P W and Whetten R L 1997 *Phys. Rev. Lett.* **79** 1873
[7] Cleri F and Rosato V 1993 *J. Phys. Rev. B.* **48** 22
[8] Tao Pang 2006 *An introduction to computational physics* (Cambridge:University Press)
[9] Nosé S, 1984 *Phys J Chem.* **81** 511
[10] Gafner S L, Redel L V, Gafner Yu Ya, Samsonov V M 2011 *Journal of Nanoparticle Research* **13** 6419
[11] Ryzhonkov D I, Levina V V and Dzidziguri E L 2008 Nanomaterials (Moscow: BINOM) [in Russian]
[12] Gafner S L, Redel L V and Gafner Yu Ya 2009 *JETP* **108** 784
[13] Gafner Yu Ya, Gafner S L and Chepkasov I V 2010 *JETP* **111** 608

Influence of an initial chemical state of bivalent cation impurities on their diffusion activity in alkali-halide crystals

S A Ghyngazov[1], T S Frangulyan[1] and A V Chernyavskii[1]

[1] National Research Tomsk Polytechnic University, Tomsk, Russia

E-mail: ghyngazov@tpu.ru

Abstract. Influence of an initial chemical state of bivalent cation impurities on their diffusion activity in alkali-halide crystals is investigated. The effect of an intensive electron beam on diffusion of magnesium impurity in crystals of fluoride of lithium is studied.

1. Introduction

Diffusion phenomena in alkali-halide crystals (AHCs) have been addressed in a large number of experimental and theoretical works [e.g., 1–3]. Most of classical experiments on diffusion in AHCs have been carried out using films of halide compound impurities deposited onto crystals. The thus obtained data on the ion transport characteristics of the base material and foreign particles are well reviewed in numerous studies. Furthermore, now we are aware of most physical concepts of the mechanisms of diffusion mass transfer in AHCs. This fundamental knowledge largely predetermines the use of these crystals as model materials in the solution of an urgent and practically important problem of stimulating high-temperature diffusion in dielectric structures with ion-type bonding by ionizing radiation.

We have to admit, though, that the studies on diffusion in ionic dielectrics under radiation are very sparse [e.g., 4-7]. Impurity diffusion investigations in those works were performed from halide impurity compounds using low-power ionizing-radiation sources. No acceleration of high-temperature diffusion was observed.

The present work deals with a critical issue of the influence of the initial chemical condition of bivalent cations of the impurity on their diffusion activity and its variation in the field of a high-power beam of accelerated electrons.

2. Material and experimental methods

To study the influence of the initial chemical condition of bivalent cations of the impurity on their diffusion activity, the following experiments have been performed.

1. Diffusion was achieved following a conventional pattern from films of halide salts.
2. A thin metal film was deposited onto the surface of crystals, which was followed by thermal oxidation of this film. The resulting sandwich system was then subjected to annealing in air. A principal difference from the experiment of Type 1 consisted in that a diffusant was in a strongly bound state (in the oxide molecule).
3. The specimens with deposited metal films were subjected to a thermal diffusion annealing treatment in vacuum (the pressure being 10^{-3} Torr), which eliminated oxidation of the metal film.

Used as sources of the impurity thin films were deposited by thermal evaporation under vacuum. Their thickness was about 400 nm. The diffusant distribution over the depth was measured by the SIMS technique described in [8]. The Pb ion distribution profiles with respect to the depth of the crystal for the diffusion pair KBr-$PbCl_2$ were measured by an optical method with a microspectrophotometer.

3. Experimental results and discussion

The diffusion coefficients D_T obtained from the diffusion profiles are given in table 1. The data listed are indicative of a significant influence of the chemical state of diffusants on the values of diffusion coefficients.

Table 1. Diffusion coefficients for thermal annealing

Diffusion pair	Temperature and type of annealing	D_T, cm^2/s
KBr-$PbCl_2$ *	773K in air	$(2.5\pm0.6)\times10^{-9}$
KBr-MgO	883K in air	$(2.9\pm1)\times10^{-12}$
KBr-Mg	883K in vacuum	$(3.5\pm1)\times10^{-10}$
KBr-Al_2O_3	883K in air	$(3.5\pm0.7)\times10^{-12}$
LiF-MgF_2	1073K in air	4×10^{-10}
LiF-MgO	1073K in air	1.5×10^{-11}

In particular, for the KBr-Mg specimens annealed in vacuum the diffusion coefficients are by two orders of magnitude higher than their value after annealing in air. It is also obvious that in the case of diffusion from halide films the volume diffusion coefficients of bivalent impurities are more than 1-2 orders of magnitude higher than those obtained in diffusion experiments with oxidized metal films. Thus we may conclude that the diffusion process is sharply retarded if the impurity cation is located on the surface as an oxide compound.

We can assume this to be associated with the fact that when the impurity diffuses into the crystal it does so in a bound state. The molecules of the oxide are quite stable in the temperature interval under study. It is, therefore, very likely that the impurity cations would penetrate into the defective subsurface layers of the crystal in oxygen-bound complexes, thus limiting their mass transport.

Having penetrated into the bulk, the impurity molecules could either decompose in the course of interactions with lattice vibrations or migrate as a single whole within the crystal lattice. In the former case, the bivalent cation would start diffusing at its own characteristic rate. The diffusion coefficients measured under the given experimental conditions were found to be much lower than those for the typical values for bivalent cations in AHCs known from the literature [2]. This circumstance could be accounted for by the fact that in our experiments the impurity continues its diffusion into the bulk of the crystal in a bound state as an oxide molecule, and dissociation of molecular complexes seems very unlikely.

In order to support this supposition, in this work we investigated the composition of the surface layer after a diffusion annealing treatment of a diffusion pair "lithium fluoride crystal – magnesium oxide film", using the x-ray fluorescence method (XRF). This semi-quantitative analysis was performed with an S4 Pioneer spectrometer.

The tests were carried out as follows. A plate measuring $4\times40\times40$ mm^3 rived from crystal lithium fluoride was thoroughly polished. An XRF analysis of the initial specimens demonstrated no oxygen or magnesium in quantities exceeding the sensitivity of the method, 10^{-3} (Mg) and 10^{-2} (O) wt. %. Following this, a metal magnesium film ~400 nm thick was formed on one of the sides of the plate by the method of thermal evaporation in vacuum. This film was then subjected to thermal oxidation at T=650 K, and the diffusion pair LiF-MgO was thermally annealed in air at T=1073 K for 24 hours. The calculations demonstrated that the diffusion layer depth after this treatment was 20 – 25 μm. Then both sides of the specimen were analyzed by XRF.

The spectra from the side with a diffusant film exhibited quite strong lines characteristic of magnesium and oxygen. The concentrations of these elements were 0.186 and 0.34 wt % for magnesium and oxygen, respectively. These values have been averaged with respect to the probing depth, being 3 μm for magnesium and 4 μm for oxygen. It should be borne in mind that a certain contribution into the line strength comes from the oxide film partially maintained after diffusion annealing. In order to rule out the effects of this factor and provide an unambiguous answer as to a possibility of diffusion-induced penetration of oxygen into the specimen together with magnesium, a thin layer of 1 – 2 μm was removed from the specimen by mechanical grinding, which guaranteed a total absence of the oxide film. The removed-layer thickness was measured by an IVK-1 altimeter to a 1 μm accuracy. Then the XRF analysis was repeated on the ground-off side.

The results showed a lower strength of the lines in question as expected. The spectrum, however, still had a strong peak from oxygen. The concentration of the elements for this analysis was as follows: Mg - 0.0824 and O - 0.17 wt %.

The spectra taken from the opposite side of the specimen did not contain any lines typical of magnesium and had a weak line from oxygen, whose strength was lower than that in the impurity diffusion zone and corresponded to a content of oxygen of 0.092 wt %.

It is evident that the weight composition of the impurities Mg and O in the diffusion zone does not correspond to that of the components in magnesium oxide. Hence, the diffusion-induced penetration of one magnesium ion is accompanied by a concurrent diffusion of more than one oxygen ions. This fact is of critical importance.

Thus, the results of XRF support our hypothesis that diffusion of heterovalent cation impurities being part of oxides in the initial state occurs together with oxygen. A two-valent magnesium ion is very likely to be located in the cation site of the lattice. The compensation of the excessive, with respect to the lattice, positive ion charge occurs via two-valent oxygen located in a neighboring anion site. The titanium-oxygen complexes, which are the luminescence centers, were found in crystal lithium fluoride [9]. This shows that our supposition of an impurity being present in the lattice as oxide molecules has a certain physical meaning.

The next stage of investigation consisted in identifying the effects of a high-intensity electron beam heating the crystal on magnesium diffusion in lithium fluoride crystals, which occurred from MgF_2 and MgO films. The coefficients of radiation-thermal diffusion D_{RT} were determined for a fixed annealing temperature T=1073 K. For the radiation-thermal annealing of diffusion pairs, use was made of an ELV-6 accelerator at INP, SB RAS (E= 1.4 MeV, continuous beam current density up to j=3 $\mu A/cm^2$). The following results were obtained.

In the case of magnesium diffusion from an oxidized metal film into lithium fluoride crystals, the radiation action gave rise to a marked increase in the diffusion rate. The value of $D_{RT}=7.5\cdot10^{-11}$ cm^2/s measured for this diffusion pair was by far higher than $D_T=1.5\cdot10^{-11}$ cm^2/s obtained under the thermal annealing mode of specimens in the resistance furnace. On the other hand, the high-intensity electron beam used did not result in any change in the rate of the process if diffusion occurred from an MgF_2 film. The latter is attributed to a higher initial diffusion mobility of cations.

Acknowledgements
This work was supported by The Ministry of Education and Science of the Russian Federation within the framework of the government "Science" program.

References

[1] F Beniere and K V Reddy 1986 Diffusion of heterovalent ions in ionic crystals *J. Phys. Chem. Solids* **47(1)** pp 69–77

[2] F Beniere 1972 Diffusion in ionic crystals *Physics of Electrolytes* vol 1 ed J Hladik (London-New York:Academic Press) chapter 6

[3] A P Surzhikov, A V Chernyavskii, S A Ghyngazov and T S Frangulyan 2002 *Russian Physics Journal* **12** pp 1190-94

[4] D Mapother 1953 *Phys. Rev.* **89(6)** pp 1231-32

[5] Ya E Gegusin and Yu L Boiko 1967 *Dokl. Acad. Nauk USSR* **172(4)** pp 820-822

[6] V I Gotlib, V N Trofimov and K K Shvartz 1970 *Izv. AN Latv. SSR, Ser. Fis. and Tekh. Nauk* **6** pp 121-122

[7] S B Zahryapin, G E Gladyshev and L A Gromov 1983 *Fizika tverdogo tela* **25(4)** pp 1152-54

[8] S A Gyngazov, A P Surzhikov, T S Frangul'yan and A V Chernyavskii 2002 *Russian Physics Journal* **8** pp 753-758

[9] A I Nepomnyashchikh, E A Radzhabov and A V Egranov 1984 *Color Centers and Luminescence of LiF Crystals* (Novosibirsk: Nauka) [in Russian]

Changes in metal properties after thermal and electric impulse processing

N A Shaburova
Assistant Professor, «South Ural State University» (NRU). Chelyabinsk, Lenin Avenue, 76, Russia

E-mail: shaburovana@susu.ac.ru

Abstract. Abstract – The results of the experiments on processing metal melts by powerful electromagnetic impulses are given. The generator used in the experiments has the following characteristics: pulse height – 10KV, duration – 1ns, leading edge – 0,1ns, repetition rate – 1KHz, the output – 100KWt. The duration of the processing is 10-15min. The comparative analysis of the processed and unprocessed samples results in the changes of structure, increase of density, solidity, plasticity and resilience of cast metal. The result analysis of different external physical processing methods on alloys shows full agreement with the results of the ultrasonic processing of metals. The hypothesis of ultrasonic shock wave formation at the pulse front was adopted as the main mechanism of the electromagnetic impulse impact on alloys. The theoretical part of the research describes the transformation process of electromagnetic impulses into acoustic ones.

1. Introduction

The cast iron structure is eligible to be changed by various methods of external melt processing methods that can be subdivided into two main groups: chemical and physical methods. Chemical methods (impurity doping, modification, finishing) make changes in the chemical composition of metals. Physical methods (thermal, barometrical, gravitational, mechanical, electromagnetic, high-energy) provide interaction between the metal and the environment. Electromagnetic mixing, vibration and ultrasonic processing are the examples of the most common methods of external physical processing of metals. Each of the above methods possesses its own advantages and disadvantages but all of them are able to specifically change the structure and properties of cast metal. Our research makes use of a new external melt processing method by powerful electromagnetic impulses [1].

2. Experiment samples and methodology

Aluminum alloy Al–6 mass% Si (about 3kg of the processed melt), copper alloy BrAZh10-3 (300 kg of the melt) and 35L steel (1.9 tons) are taken as the samples for the experiment. The chemical composition of the alloys is given in Table 1.

All the alloys underwent two melt operations under the identical temperature-time conditions. During the first melt operation the temperature exceeded the melt temperature by 100°C and the alloys were left for homogenizing annealing. Then the furnace was switched off and the melts were exposed to powerful electromagnetic impulses for 10-15 min. The generator of impulses used in the experiments has the following characteristics: pulse height – 10 kV, duration – 1 ns, leading edge – 0,1 ns,

IOP Conf. Series: Materials Science and Engineering **81** (2015) 012016 doi:10.1088/1757-899X/81/1/012016

repetition rate – 1 kHz, output – 100 kW. One of the generator electrodes was put into the melt; another one was connected to the graphite crucible.

Table 1. The chemical composition of the alloys, mass %.

Alloy	Element composition
Al–6 mass % Si	6.44 % Si, 0.06 % Cu, 0.01 % Mn, 0.27 % Mg, 0.02 % Ti, 0.04 % Zn, 0.27 % Fe, the rest is Al.
BrAZh10-3	9.4 % Al; 3.4 % Fe; 0.01 % Sn; 0.09% Si; 0.26 % Ni; 0.05 % Pb; 0.83 % Zn, the rest is Cu.
35L steel	0.32–0.4 %C, 0.2–0.52% Si, 0.5% Mn, до 0.3% Ni, до 0.045% S, до 0.04% P, до 0.3% Cr, the rest is Fe.

The second melt operation was done under the same temperature-time conditions but instead of electromagnetic processing the melt was exposed in the furnace for 10-15 min before casting. During both operations the temperature was registered before casting and sand-clay molds were used.

The samples for mechanical testing and microstructure researches were cut out of the obtained moldings.

3. The impact of electromagnetic impulse processing on alloy structure

The following differences were identified in the macro and microstructure of the initial and processed Al–6 mass % Si alloy. The initial metal template has two well-defined crystallization regions: a region of columnar crystals near the mold and a region of equiaxed coarse grains in the center of the ingot. The region with crust columnar crystals in the processed ingot template is less defined. The presence of bigger equiaxed grains should also be noted in the center of the ingot. The microstructure of the cast silumin samples is shown in figure 1.

It is known that in the state of equilibrium with slow cooling of silumin ingots the erratic eutectic appears: α-solid solution, silicon crystals and lamellar structures being a compound of AlSiFe [2]. The crystals of the initial α-phase in both samples have a tree-like form. The amount of the eutectic in the structure after the electromagnetic impulse processing reduces from 30–35% to 10–15%. The particles of eutectic silicon become of a needle-like form with the length up to 25–30 μm. For unprocessed metal the eutectic silicon phase is dispersed with mean linear parameters up to 5–10 μm. Due to the decrease of eutectic release at the boundaries of dendrite cells α-phase dendrites increase insufficiently in size from 66 to 85 μm (linear parameters were define by the secant method) and become more of a round shape.

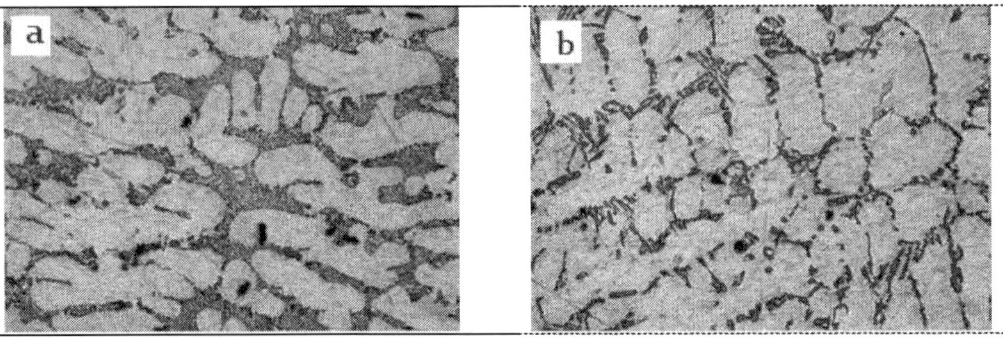

Figure 1. Microstructure of Al–6 mass % Si alloy, x100: a – initial sample, b – processed with electromagnetic impulses

The microhardness of the α-phase grains in the sample after electromagnetic impulse processing increased by 10-15% to 750 MPa. The reduction of the eutectic amount and the increase of the α-phase microhardness reflect the greater silicon content at this phase. The growth of silicon content is proved by the series of experiments carried out with the help of the JEOL JSM-6460LV scanning electron microscope with Oxford INCA X max 80 dispersion spectrometer for element analysis. According to the obtained data, the maximum silicon concentration at the α-phase of the initial metal reaches 1,36–1,66 mass %, whereas in the processed sample – 1,48–1,80 mass % (with the absolute error 0,04 %).

The microstructure examination of BrAZh10-3 bronze alloy samples on the scanning electron microscope showed the identical phase composition of the samples. Eutectic and lead phases of both samples are inhomogeneously distributed along the section plane and have irregular form. The microstructure examination of samples on the optical microscope showed the presence of dendrites of different sizes in the unprocessed sample. The dendrites in the processed sample are smaller and have more uniform sizes.

Both 35L steel samples are characterized by the presence of ferrite-pearlite structure. The samples microstructures after etching are shown in figure 2.

Figure 2. 35L steel microstructure: a – initial sample, b – processed with electromagnetic impulses

The average size of grains in the initial sample is 700μm, in the processed with electromagnetic impulses 450μm. With the evident growth of the overall length of grain boundaries in the processed sample the grain-boundary ferrite phase becomes continuous along all grain boundaries and less defined (see figure 2).

Note that there are no deformation regions in all of the examined processed metal samples, whereas they are present on the initial metal sample.

4. The impact of electromagnetic impulse processing on mechanical properties of alloys

The hardness measurement of Al-Si cast alloy samples displays that preliminary electromagnetic impulse processing of alloys increases hardness. Initial samples have 51 HB hardness, test samples – 63 HB. Using the method of tearing a ring plate off the alloy surface we measured surface tension. At 700°C it was 0.81 N/m and 0.74 N/m for initial and processed samples correspondingly. Castability defined by the State Standard 16438-70 was 195 and 295 mm for initial and processed samples correspondingly. Tensile strength was 170 MPa and 210 MPa, specific elongation 4.8 and 18.4 % for initial and processed metal correspondingly [3].

Density measurement of the bronze sample was carried out using hydrostatic weighing method that showed the density of 8.642 and 8.573 for initial and processed metal correspondingly. The slight hardness growth of 70 and 65 HRB was also noted for initial and processed metal correspondingly. The results of tensile strength tests are given in Table 2.

Table 2. Mechanical properties of BrAZh10-3 bronze samples.

Sample	Ultimate strength, MPa	Conventional yield strength, MPa	Specific elongation, %
Initial	50,6	11,8	21
Processed	50	14,4	23,4

The hardness growth for 35L steel samples after impulse processing of the alloy was also noted: average value of initial samples – 22 HRC, processed – 25 HRC.
The results of mechanical tests for tensile strength and resilience are given in Table 3.

Table 3. Mechanical properties of BrAZh10-3 bronze samples.

Sample	Yield strength, MPa	Ultimate strength, MPa	Specific elongation, %	Contraction ratio, %	Resilience at room-temperature, J/cm^2	Resilience at -60°C, J/cm^2
Initial	386	520	6	12	12	9,8
Processed	454	772	15	39	45	13

5. Discussion of results

Thus, the obtained results on electromagnetic impulse processing of metal alloys indicate the impact of this processing on the process of grain structure formation, composition and morphology of phases and mechanical properties of metals. The mechanism of this impact is not clear yet. Nonetheless, having compared the obtained results with various data of outer impact on alloys, we found out some similarities. The most similar results were obtained using ultrasonic processing of alloys (US).

The absolute majority of works on ultrasonic processing of aluminum, magnesium alloys and steels states that the form of initial α-phase products changes from tree-like to rosette and roundish. The paper [4] shows that in the crystallization phase of 0.35 kg of Al-5 mass% Si alloy 4 kW ultrasonic processing of the metal stimulates grain refinement from 1600 μm to 100 μm. The α-phase grain morphology changes to more roundish. The eutectic silicon particles become of the same rough shape as in the case of electromagnetic processing. The authors showed the roughening of eutectic for Al-11 mass %Si and Al-17 mass % Si alloys. Similarly to the paper [5] 1.5kW and 20 kHz US processing at temperatures close to those of crystallization results in the roughening of eutectic silicon.

At the same time the results of the experiments on 1.2 KW US processing of 0.4 kg AlSi9Cu alloy [6] exhibit the reduction in inter-lamellar space and refinement of eutectic silicon particles.

The controversial data obtained by the scientists on the morphology of eutectic silicon particles can be explained with the choice of processing temperature. The refinement of silicon particles takes place with the processing at an early stage of crystallization as a result of cavity phenomena. At lower temperatures the thickened melt does not allow cavity flows and thus prevents the fragmentation of phases. Input energy heats the alloy and changes the growth condition of eutectic roughening it.

The reference to the increase of basic alloying elements in the initial α-phase at US processing is also made in the paper [7].

The growth of strength and plasticity characteristics with 600 W and 19.5 kHz US processing and retention of rough needle-like form of eutectic is shown in the paper [8]. The above mentioned papers [4-7] indicate the growth of ultimate strength and specific elongation simultaneously with the grain and eutectic refinement.

As the phenomena occurring at the impulse processing are similar to those of the US processing, there is a possibility of the acoustic fields produced by the electromagnetic impulse generator to exert pressure on the alloy.

The papers [9-11] indicate that ultrasonic oscillations are generated in metal melts exposed to electromagnetic oscillation. The calculations for the induction melting of aluminum [12] in a 300 mm diameter crucible with a constant $5 \cdot 10^4$ A/m field show the oscillation pressure on the alloy to be 2 atm. This pressure is considered sufficient to achieve useful metallurgical results.

Stimulation of the mechanical oscillation is practically produced by the contact method due to the specific design of the apparatus in use. The specific placement of the oscillator makes current flow from the oscillator along the surface of an alloy and, possibly, the efficiency of the generated oscillations is greater than stated in the above mentioned papers.

To support our hypothesis theoretically, we make comparative calculations for oscillation intensity. Some papers on US impact on alloys have data making it possible to calculate generated mechanical pressure.

For plane and spherical waves acoustic pressure and displacement of particles in metals are bound through a ratio [9]:

$$p=\rho c\omega\xi=z\omega\xi, \quad (1)$$

where the product of metal density and speed of sound $\rho c=z$ is the acoustic impedance (resistance), ω - is the angular frequency ($\omega=2\pi f$); ξ is the displacement of particles from equilibrium.

The paper [13] on the oscillator in the aluminum alloy states: f=21 kHz, ξ=25 μm, the generator output is 1 kW, oscillation intensity is 100 W/cm^2, the processed metal mass is not mentioned. The speed of the longitudinal wave in solid aluminum is 6260 m/sec [14]. The oscillation speed in the alloy can reach 70% of the solid metal [14]. The density of the melted aluminum is 2390 kg/m^3 [15]. The calculations with the equation (1) give the value of pressure in the oscillator equal to 29 MPa or 290 atm.

The paper [16] on the oscillator in the aluminum alloy states: f=20 kHz, ξ=4 μm, the generator output is 150 kW, the processed metal mass is 200 g. The calculations of the acoustic pressure result in 4.6 MPa (4.6 atm.).

The paper [8] on the oscillator in the aluminum alloy states: f=19.5 kHz, ξ=30 μm, the generator output is 600 kW, oscillation intensity is 109 W/cm^2, the processed metal mass is 210 g. The calculations of the acoustic pressure result in 34 MPa (340 atm.).

The paper [12] indicates that the electrodynamic pressure in the alloy should be 1-4 atm or $4 \cdot 10^5$ Pa to produce useful metallurgical results.

The calculation equation for the surface wave pressure can be applied to calculate the oscillation pressure:

$$p=E(1+R)/c, \quad (2)$$

where p is the wave pressure, N/m^2; E is the incident wave force attributed per unit area and time, W/cm^2; R is the reflection coefficient (R=0 at total absorption, R=1 at total reflection); c is the wave velocity, m/sec. The wave velocity in smelts is $4 \cdot 10^3$ m/sec [14].

At the impulse stimulation of oscillations the incident impulse force can be approximately calculated with the equation:

$$P=\frac{U^2}{r}, \quad (3)$$

where r is the cable wave resistance of 50 Om; U is the generator voltage of 10 kW. When calculating the value of the incident impulse force with the equation (3) we obtain $P=2 \cdot 10^6$ W. The area of the metal free plane in the 80 mm crucible is $5 \cdot 10^{-3}$ m^2. Consequently, the impulse force per unit area is $4 \cdot 10^8$ W/m^2.

If we insert the obtained value of the impulse force in the equation (1), we get $P_{имп}=1{,}3 \cdot 10^5$ Pa (or 1,3 atm). This value is close to the values of the applied pressure stated in the papers [12] and [16].

6. Conclusion

The obtained results testify for the similarity between the mechanisms of the impact on metals by electromagnetic impulse and ultrasonic processing. Though the produced results are similar, the impulse type of processing is void of disadvantages peculiar to ultrasonic processing (complicated, expensive equipment, harmful effects on human health, etc.) and more efficient.

7. References

[1] Belkin V, Bukharin V and Dubrovin V 2001 *Nanosecond electromagnetic impulses and their application* (Chelyabinsk: Tatiana Lur'ye).
[2] Aristova N and Kolobnev I 1977 *Thermal treatment of casting aluminum alloy (*Moscow: Metallurgy)
[3] Znamensky L, Krymsky V and Koryagin Yu 2005 *Proceedings of VI International Conference «Electromagnetic field impact on material plasticity and strength»* 227–28
[4] Zhang L, Eskin D, Miroux A and Katgerman L 2012 *Light Metals* 999–04
[5] Jian X, Xu H, Meek T and Han Q 2005 *Materials Letters* **59** 190–93
[6] Puga H, Costa S, Barbosa J, Ribeiro S and Prokic M 2011 *Journal of Materials Processing Technology* **211** 1729–35
[7] Li Xin-tao, Li Ting-ju, Li Xi-meng and Jin Jun-ze 2006 *Ultrasonics Sonochemistry* **13** 121–25.
[8] Khalifa W, Tsunekawa Y and Okumiya M 2010 *Journal of Materials Processing Technology* **210** 2178–87
[9] Krautkremer J and Krautkremer G 1991 *Ultrasonic control of materials* (Moscow: Metallurgy).
[10] Pohl R 1962 *Einführung in die Physik* (Berlin: Springer)
[11] Grubin H 1970 *IEEE. Trans SU* ***17*** 2207–32
[12] Arkin M, Gonchrova I and Mirotvorsky V 1968 *Acoustic journal* **XIV** 344–50
[13] Yanfeng Han, Ke Li, Jun Wang, Da Shu and Baode Sun 2005 *Materials Science and Engineering* **A 405** 306–12
[14] Kikoyen I 1976 *List of physical quantities: reference book* (Moscow: Atomizdat)
[15] Amely J 1993 *Elements* (Moscow: Mir)
[16] Yu S, Feng H, Li Y and Gong L 2009 *Journal of Alloys and Compounds* **48** 360–64

The formation of stable hydrogen impermeable TiN-based coatings on zirconium alloy Zr1%Nb

E B Kashkarov[1], N N Nikitenkov[1], Yu I Tyurin[1], M S Syrtanov[1], Zhang Le[1]

[1]Tomsk Polytechnic University, 30 Lenin Avenue, Tomsk, Russian Federation

E-mail: egor_kashkarov@mail.ru

Abstract. TiN coatings were deposited by DC reactive magnetron sputtering (dcMS) method on Zr1%Nb substrates with different film thickness. The influence of crystalline structure and thickness of the coatings on hydrogen permeation was investigated. The results revealed that the increase in thickness of the film reduced hydrogen permeability. 1.54 μm TiN deposited in N_2/Ar gas mixture with a ratio of 3/1 reduces hydrogen permeation in more than two orders of magnitude at 350 °C. Adhesion strength decreased with increasing film thickness (0.55 to 2.04 μm) from 7.92 to 6.65 N, respectively. The Ti underlayer applied by arc ion plating (AIP) leads to the formation of stable Ti/TiN coatings on Zr1%Nb under thermocycling conditions up to 800 °C. Meanwhile, hydrogen permeation rate of Ti/TiN deposited by combination of AIP and dcMS remains at the same level with TiN deposited by dcMS.

1. Introduction

The problem of protection of structural and functional materials operating in hydrogen-containing ambient in nuclear reactors and fuel cells is very important. In Russian Water-Water Energetic Reactor (WWER) and High Power Channel Reactor (RBMK) the Zr1%Nb alloy is used as a cladding material. Zirconium and its alloys are susceptible to hydrogen corrosion and fretting wear. It is known that the certain concentrations of hydrogen in the bulk of the material leads to embrittlement and subsequent degradation of mechanical properties [1–6]. Titanium nitride (TiN) deposited by vacuum ion-plasma methods is promising in protection zirconium alloys from hydrogen corrosion [7]. Moreover, TiN coatings have an excellent erosion resistance, which favorably affects the fretting resistance of the coatings [8]. There is no scientific data in information about hydrogen sorption kinetics by TiN coated zirconium alloy by dcMS method. At the same time, such studies are of great practical importance for the development of TiN coating technology, allowing to work in extremal conditions.

One of the most important tasks in the development of coating technology is to ensure good adhesion of the coating under thermal cycling. The differences between temperature expansion coefficient of the coating and substrate materials leads to the coating degradation and varios accidents can occur. In this paper the problem of stability of the coatings under thermal cycling was solved by creating a Ti underlayer deposited by AIP between the zirconium alloy Zr1%Nb and magnetron sputtered TiN.

The aim of this work is to create a stable hydrogen-impermeable TiN-based protective coating under thermocycling conditions using the AIP and dcMS methods.

2. Experimental Procedure

The Zr1%Nb alloy is used as the substrate that fixed size is 0.5 mm thickness and 20 mm diameter. The samples were previously polished using sandpaper to a roughness 0.1 μm. The Ti underlayer and TiN coatings were deposited in a hybrid «Arc spectrum» technique which was developed in Tomsk Polytechnic University. Vacuum arc evaporator with coaxial plasma filter (designed by authors [9,10]) was used for AIP method. Firstly, the chamber was evacuated to a base pressure of lower, than 2.5×10^{-3} Pa. Before deposition, samples were subjected to ion bombardment in an argon glow discharge at 1500 V for 5 min.

Process parameters for AIP of Ti underlayer: arc current – 110 A; pulsed substrate bias – (-150 V); pulse repetition frequency – 100 kHz; peak pulse current – 3 A; Ar pressure – 2×10^{-1} Pa.

The TiN coatings were reactively sputtered in mixed N_2/Ar gases with different ratio (1/1, 2/3, 1/3, 1/4). Pressure in the chamber remained constant 1.54×10^{-1} Pa. Sputtering power and current was 2.6 kW and 4 A, respectively. The deposition time was varied from 10 min to 40 min.

Hydrogenation of the samples was performed by Sievert method (from hydrogen atmosphere) during 60 min at temperature 350 °C (heating rate 6 °C /min) and 2 atm hydrogen pressure. The hydrogenation temperature was chosen due to working temperature of zirconium fuel cladding in the operation process of nuclear reactors.

Hydrogen sorption kinetics was measured by Sievert method with automated complex Gas Reaction Controller. The phase constitutions of the deposited films were identified by X-ray diffraction (Shimadzu XRD-7000) using Cu Kα radiation (1.5410 A wavelength) generated at 40 kV and 30 mA with the fixed angle (θ=3°). The film thickness was measured by simple ball-cratering method performed with Calotest CAT-S-0000. The adhesion strength was measured by scratch method using Micro Scratch Tester MST-S-0000.

3. Results and Discussion

3.1. Crystalline structure and hydrogen permeation of TiN coatings deposited at the different N_2/Ar ratio

Titanium nitride thin films were deposited by reactive dcMS method in mixed N_2/Ar gases with different ratio. The process parameters are shown in Table 1.

Table 1. Process parameters for TiN coatings deposition

Sample	Pressure P, 10^{-1} Pa	DC current I, A	Power W, kW	Deposition time t, min	N_2/Ar ratio
1	1.54	4	2.6	30	1/1
2	1.54	4	2.6	30	3/2
3	1.54	4	2.6	30	3/1
4	1.54	4	2.6	30	4/1

Figure 1 shows the XRD patterns of TiN coatings applied on Zr1%Nb substrates as a function of the N_2/Ar ratio. It is seen that the difference of the TiN coatings with increasing N_2/Ar ratio is not significant. In all cases, the crystalline structure of the TiN coatings is cubical (type NaCl) and it is possible to see that the orientation in the plane (200) is the reflection of major intensity and it remains as the most intense from 1/1 to 4/1 N_2/Ar ratio. Similar results in preferred TiN orientation in plane (200) have also been observed in report [11]. The less intensive (220) and (222) reflections have been observed. The reflections in plane (111) are difficult to resolve due to the merger with substrate (Zr) reflections. Plane (111) and plane (200) have the lowest strain energy and the lowest surface energy in NaCl-type fcc nitrides, respectively. The fact that the preferential direction is (200) can be explained from investigations that Pelleg et al. made [12], which were later confirmed by Oh et al. [13,14].

The more intensive (200) reflection have been observed for the sample which was deposited at 3/1 N_2/Ar ratio compared to other samples. Furthermore, the intensity of TiN in the plane (111) and zirconium reflections decreased. It could be connected with the fact that the crystalline lattice of this sample is more preferentially oriented in the plane (200) or associated with zirconium substrate texture.

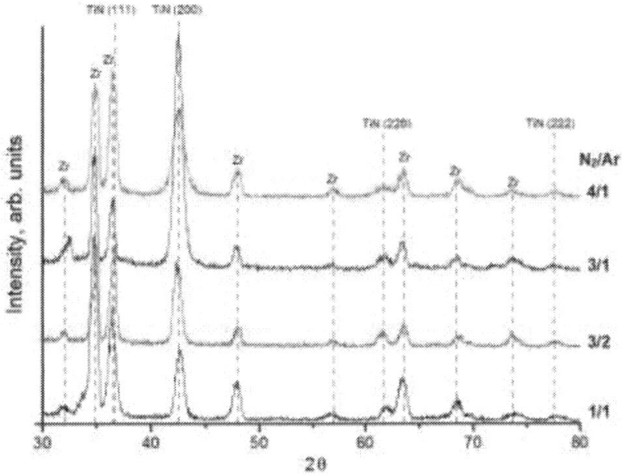

Figure 1. XRD analysis of TiN coatings deposited at different N2/Ar ratio

Hydrogenation of the samples was performed at temperature 350 °C and hydrogen pressure 2 atm during 60 min. The TiN coatings deposited with 1/1 and 3/2 N_2/Ar ratio were destroyed as a result of hydrogen embrittlement of zirconium alloy. There are no discontinuities for the TiN coatings deposited with 3/1 and 4/1 N_2/Ar ratio have been observed under optical microscope.

The decrease in the hydrogen pressure in the chamber indicates the hydrogen absorption process by the samples. The slope of the «pressure-time» curve measured in the saturation process of the samples characterizes the intensity of the hydrogen absorption. The hydrogen absorption rate is calculated by the following formula:

$$q = \frac{V}{t \cdot S} \ln(P / P_0) \qquad (1)$$

where V – chamber volume (175 cm^3), t – saturation time, S – effective surface area of the sample, P и P_0 – final and initial hydrogen pressure in the chamber, respectively.

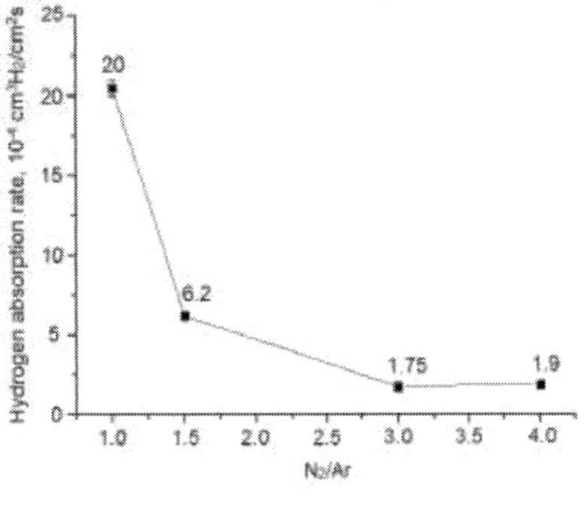

Figure 2. Hydrogen absorption rate by TiN deposited at different N2/Ar ratio (numbers near the points correspond to exact values)

The hydrogen absorption rate which characterizes hydrogen permeation was calculated by the formula (1). Figure 2 shows that the hydrogen absorption rate decreased with increasing N2/Ar ratio up to 3/1 and then insignificantly increase in the hydrogen absorption rate was observed. Therefore, TiN coatings deposited on our hybrid technique by reactive dcMS method with 3/1 N2/Ar ratio have the lowest hydrogen permeability. It appears that the highest hydrogen absorption rate for the samples with TiN deposited at gas mixture N2/Ar<3/1 has a high structural defects formed during deposition.

3.2. The influence of TiN thickness on hydrogen permeability and adhesion strength of the coatings

To study the effect of TiN coating thickness on the hydrogen permeability hydrogen absorption rate have also been measured. Deposition time was varied from 10 to 40 min to change the film thickness. The N_2/Ar=3/1 ratio was defined in terms of lowest hydrogen permeability of TiN. The thickness of the deposited coatings was measured by simple ball-cratering method using Calotest CAT-S technique. Figure 3 shows the linear dependence between the thickness of TiN coatings and deposition time. Deposition rate is 3 μm in this magnetron sputtering parameters.

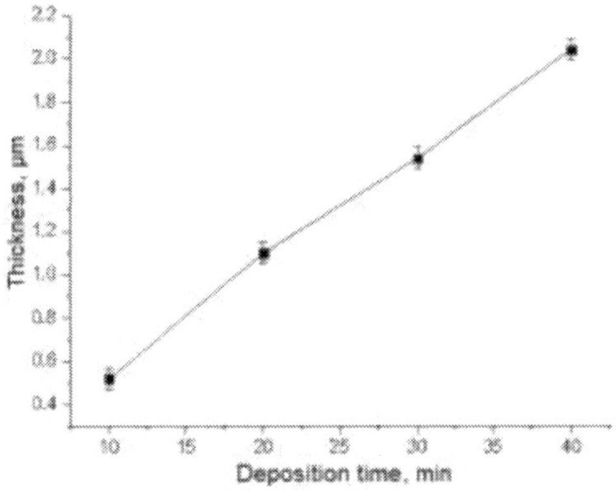

Figure 3. Thickness of the TiN

The hydrogen absorption rate was calculated for the samples with different TiN thickness and is shown in Figure 4. It can be seen that increasing the coating thickness leads to a reduction of the hydrogen absorption rate.

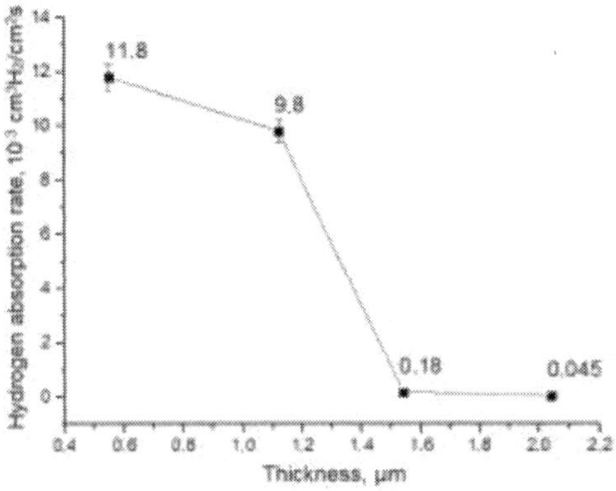

Figure 4. Hydrogen absorption rate by TiN deposited with different thickness (numbers near the points correspond to exact values)

The hydrogen absorption rate is 12×10^{-3} $cm^3H_2/(cm^2\cdot s)$ for the Zr1%Nb alloy without coatings. Thus, the hydrogen differently interacts with the TiN coatings of various thickness. It can be seen that the hydrogen absorption rate insignificantly reduced for the 0.5–1 μm TiN coatings compared to uncoated zirconium alloy. The reduction of hydrogen absorption rate in 50–60 times and more than two orders of magnitude was observed for 1.54 μm and 2.04 μm TiN, respectively. The hydrogen absorption rate for the TiN coatings has also been evaluated in reports [15], where the reduction of more than 2 orders of magnitude was observed only for 7–9 μm TiN coatings and the hydrogen absorption rate was negligible for 1–2 μm TiN.

The measurement of adhesion strength was performed with MicroScratchTester MST-S-AX-0000 technique by scratch method using a diamond indenter. Scratching parameters: initial load – 0.01 N; ultimate load – 20 N; scratching speed – 9.63 mm/min; scratch length – 10 mm. The measuring results are shown in Table 2.

Table 2. Adhesion strength of TiN

Thickness, μm	0.55	1.12	1.54	2.04
Adhesion strength, N	7.92	7.67	7.51	6.65

Table 2 shows that the adhesion strength of the TiN coatings deposited by dcMS method decreases with increasing coating thickness. The adhesion strength decreased from 7.92 to 6.65 N for the 0.55 to 1.54 μm TiN thickness, respectively.

Low adhesion strength of TiN coatings (<6 N for scratch method) along with the difference in temperature expansion coefficients of titanium nitride and zirconium alloy leads to coating delamination (detachment) during thermal cycling. Thus, a further increase in TiN thickness is impractical from the viewpoint of thermal stability and adhesion strength of the coatings.

3.3. The formation of stable Ti/TiN deposited by AIP and dcMS methods

In this paper the problem of low adhesion strength of TiN coatings and the difference between temperature expansion coefficients for TiN and Zr1%Nb was solved by applying a Ti underlayer using AIP technique.

The basis for this approach includes the following general considerations. It is known [17] that AIP of one metal to another metal may lead to the formation of the various intermetallic phases. In this report the phase composition of the surface after AIP has not been investigated. However, the formation of different Ti-Zr phases at AIP is possible, for example [18]. It seems that it will promote the balancing temperature expansion coefficient between the substrate and the coating and improve the adhesion strength.

The Ti underlayer was deposited using AIP at 110 A arc current and -150 V pulsed (100 kHz) substrate bias. Then, the 1.54 μm TiN coating was deposited by dcMS method with $N_2/Ar=3/1$ ratio at the same parameters shown in Table 1.

The adhesion strength and the hydrogen permeability of Ti/TiN were investigated before and after thermocycling.

Thermocycling of the samples was performed with a water cooled vacuum furnace. The residual pressure in the chamber was (6–7) × 10^{-3} Pa. Heating and cooling rates were 200 and ~ 2000 ° C/min, respectively. Three cycles of heating and cooling the Zr1%Nb samples coated with TiN and Ti/TiN were performed. After cooling, the samples were removed from the vacuum furnace.

The surface of the sample without underlayer contained peeling of the coating and cracks formed during thermal expansion, which was not observed in the sample with a Ti underlayer.

Hydrogen saturation of the samples with Ti/TiN coatings was performed after thermocycling at following conditions: T = 350 °C, P = 2 atm, t = 60 min. The hydrogen absorption rate of Zr1%Nb alloy coated with TiN (before thermal cycling) and Ti/TiN (after thermal cycling) was calculated. The results are shown in the Table 3.

Table 3. Hydrogen absorption rate of TiN and Ti/TiN coatings

Sample	Zr1%Nb+TiN	Zr1%Nb+Ti/TiN
Hydrogen absorption rate, $\times 10^{-4}$ cm^3 $H_2/(cm^2 s)$	1.75	1.64

After thermocycling the hydrogen absorption rate of Ti/TiN is lower than TiN without underlayer. This is evidence of an minimal reduction in the hydrogen permeability induced by additional Ti underlayer diffusion barrier or changes in the structure of the TiN coating.

Before and after thermocycling three measurements of the adhesion strength were performed for each samples using the same method as described above. The adhesion strength of the TiN deposited by dcMS after thermal cycling is reduced from 7.75 N to 4.46 N; at the same time for the samples with Ti underlayer the adhesion strength is remained at the same level before and after thermal cycling. This result demonstrates that the TiN coating with Ti underlayer is resistant when exposed to rapidly changing temperatures in the range of 20–800 °C.

4. Conclusion

The study has demonstrated that the vacuum ion-plasma TiN coatings deposited by reactive magnetron sputtering lead to the reduction of hydrogen permeation through zirconium alloys. The decrease in the hydrogen absorption rate in more than two order of magnitude was observed for 2 µm TiN coating deposited at $N_2/Ar = 3/1$ ratio. The magnetron sputtered TiN has a cubic lattice with the (200) preferred orientation.

The adhesion strength of TiN was improved by applying a Ti underlayer using arc ion plating technique. The combination of AIP and reactive dcMS methods leads to the formation of Ti/TiN coatings which are resistant during thermal cycling in the temperature range of 20–800 °C. The hydrogen permeation rate of Ti/TiN remains at the same level with TiN deposited by dcMS. The investigated coatings are perspective to protect zirconium alloys from the hydrogen embrittlement.

References

[1] Chernyaeva T P, Ostapov A V 2013 *Voprosy atomnoy nauki i tehniki* **87** 16
[2] Silva K R F, DosSantos D S, Robeiro A F, Almeida L H 2010 *Defect and Diffusion Forum* **297–301** 722
[3] Zielinski A, Sobieszczyk S 2011 *Intern. Journ. of Hydrogen Energy* **36** 8619
[4] Bai J B 1993 *Scripta Metallurgica et Materialia* **29** 1259
[5] Mani Krishna K V, Sain A, Samajdar I, Dey G K et al. 2006 *Acta Materialia* **54** 4665
[6] Singh R N, Kishore R, Singh S S, Sinha T K, Kashyap B P 2004 *Journ. of Nuclear Materials* **325** 26
[7] Ivanova S V, Glagovskiy E M, Khazov I A, Orlov V K, Shlepov I A, Nikitin K N, Dubrovskiy Yu V, Denisov E A 2008 *Proc. IV Int. Conf. on Interaction of Hydrogen Isotopes with Structural Materials (Nizhny Novgorod)* p 51
[8] Aleksandrov D A, Gorlov D S, Zhuravleva P L, Lutsenko A N, Muboyadzhyan S A 2011 *Metally* **4** 91
[9] Ryabchikov A I, Ryabchikov I A, Stepanov I B 2005 *Vacuum* **78** 331
[10] RyabchikovA I and Stepanov I B 2009 *Surface and Coating Technology* **203** 2784
[11] Chun S Y 2010 *Journal of the Korean Physical Society* **56** 1134
[12] Pelleg J, Zevin L Z, Lungo S, Croitoru N 1991 *Thin Solid Films* **197** 117
[13] Oh U C, Je J H 1993 *J. Appl. Phys.* **74** 1692
[14] Oh U C, Je J H, Lee J Y 1995 *J. Mater. Res.* **10** 634
[15] Selezneva L V, Bushmin B V, Dubrovskiy Yu V, Khazov I A, Denisov E A, Kurdyumov A A 2008 *Voprosy atomnoy nauki i tehniki* **2** 108
[16] Kurzina I A, Bozhko I A, Kalashnikov M P, Fortuna S V, Batyreva V A, Stepanov I B, Sharkeev Yu P 2004 *Bulletin of Tomsk Polytechnic University* **307** 30
[17] Bovda A M, Dmitrenko A E, Malyhin D G, Onishchenko L V, Pelyh V N 2007 *Voprosy atomnoy nauki i tehniki* **4** 173

Study of the plasma immersion implantation of titanium in stainless steel

N N Nikitenkov[1], A N Sutygina[1], I A Shulepov[1], D O Sivin[1], E B Kashkarov[1]
[1]Tomsk Polytechnic University, 30 Lenin Avenue, Tomsk, Russian Federation

E-mail: nikitenkov@tpu.ru

Abstract. The results of the study of the pulsed plasma-immersion ion implantation of titanium in steel Cr18Ni10Ti depending on the time (dose) implantation are presented. It is shown that the change of the element and the phase composition of the surface layers and their microscopic characteristics and mechanical properties (hardness, wear resistance) depending on the implantation time is not monotonic, but follows to a certain rule. The possibility of interpretation of the obtained results in the thermal spike concept of the generation on the surface by the stable (magic) clusters is discussed. This concept follows logically from the recent studies on the plasma arc composition and from a polyatomic clusters-surface interaction.

1. Introduction

The plasma-immersion implantation is a comparatively new method of the properties modifying of the near surface layers of solids, is widely used recent times. It is connected with the modification of the physical, chemical and mechanical properties of the surface layers, formation new compounds including intermetallic compounds (obtaining and study of which is constitute the separate wide sector in the chemistry and metallurgy) [1–3]. All this is very important for many areas of modern technologies (from micro- and nano-electronics to metallurgy) [4].

Plasma immersion implantation at the 1–10 keV energies of implanting ions is realized from metal plasma arc, when the implanted metals uses as the cathode of the arc discharge source. Impulse voltage is served to implantable sample (target for implanted ions) to accelerate the positive ions of the plasma without overheating the target. In this work the implantation was carried out on the installation developed in Tomsk Polytechnic University by the authors of [5]. The main advantages of the installation called by the authors "Raduga-5" is a high speed set-dose implanted ions [6]. In this work the objects of the study was Cr18Ni10Ti steel is used in equipment for oil industry, chemical plants and equipments, which are resistant to high temperatures. The strength and the wear resistance increase during the implantation titanium ions into the steel [7,8] which leads to growing operating time.

The purpose of the work was to study the microstructure, the distribution of elements and phases formed in the surface layers of the modified material depending on the dose of implanted ions and the study of the mechanical properties obtained by implantation of surface layers.

2. Materials and Methods

The Cr18Ni10Ti steel was used as the modified material (target). Implanted samples surface has a size 25x25x1 mm, which was polished. Titanium (VT1-00, technically pure, rus) has been used as the cathode of the arc plasma source. The arc current was I = 90 A. The amplitude of the voltage pulses applied to the target was U = -2 kV relative to the ground (anode of the arc evaporator); the pulse length was t_{imp}=7 μs; the pulse frequency was ν =100 Hz; the ion current density per pulse was J= 6.5 mA⊙cm^{-2}; the temperature of the implanted surface rises to 900 °C for 5 minutes, which was measured using a pyrometer. The Auger electron spectrometer "Shuna-2" and the glow discharge spectrometer GD-PROFILER-2 were used for researching distribution of elements in the depth of the modified layers. The X-ray diffractometer Shimadzu XRD 7000 was used for researching phase composition of the surface layers. Photographs of the microstructure were obtained by using the three-dimensional non-contact profilometer MicroMeasure 3D Station optical microscope with a camera 30 KB S Pruftechnik GmbH and scanning electron microscope-3000 TM. Mechanical properties of the surface were investigated using the nanohardness NANOHardness Tester NHT-S-AX-000X and the high temperature tribometer PC-Operated HighTemperature Tribometer TNT-S-AH0000.

3. Results and discussion

The concentration profiles of Fe (a) and Ti (b) are shown in fig. 1, which are depending on the implantation time of titanium in steel 18Cr10NiTi (numbers about the curves show implantation time). There are profiles of Cr, Ni, C, N and O which were obtained on a glow discharge GD-PROFILER 2, and for determining the concentration were calibrated according to the Auger electron spectrometer "Shuna-2."

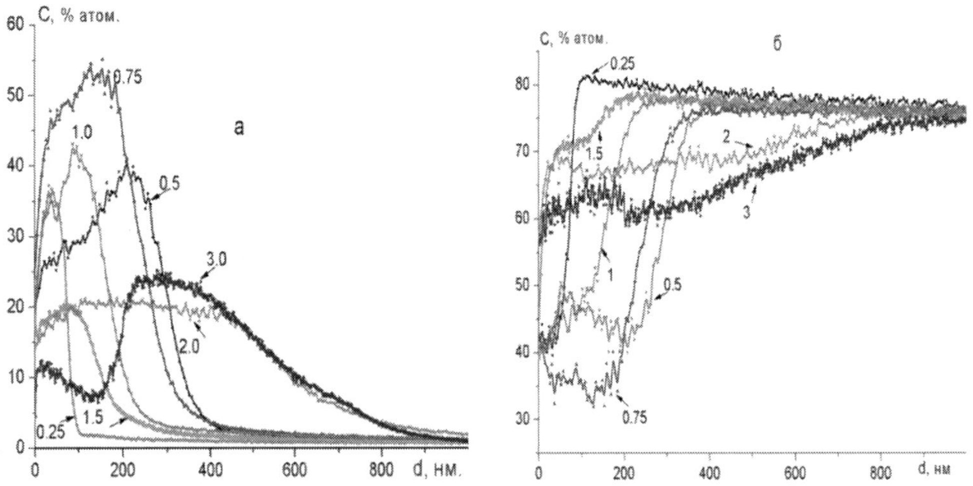

Figure 1. Concentration profiles of titanium (a) and iron (b) after the implantation of titanium in steel Cr18Ni10Ti. The numbers on the curves show the time of implantation in minutes.

From the fig. 1a follows that the modification profile of titanium can be described as the accumulation in the surface layer (up to 400 nm) with increasing irradiation time from 0.25 to 0.75 min. Broadening and shifting the profile to 800 nm with increasing the irradiation time of up to 2 or 3 minutes is observed. The curves at 2 and 3 minutes of irradiation are very similar to the thermal diffusion profiles. Profiles with irradiation time 1 and 1.5 min fall from the described trends. Note that the same fallings of the trends are observed in the study of other methods (see below): phase analysis, microhardness, and friction coefficient. Therefore, you should not talk about the experimental error in the measurement but about the features of implantation technology.

Fig. 2 shows the phase composition of the surface-modified stainless steel after 3 minute of irradiation. It is seen that the intermetallic compound are formed on the surface. However, the composition of the compounds changes from irradiation time. The TiNi and Fe_2Ti intermetallics absent at the 0.5 min and 0.75 min implantation time. The TiNi compound fixed only at the 1 min and 3 min implantation time. Line intensity of these phases changes with increasing implantation time without any regularity.

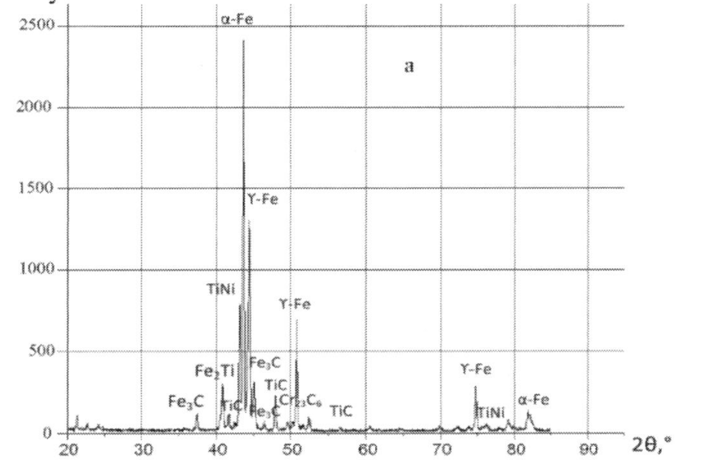

Figure 2. The surface diffractograms of the Cr18Ni10Ti implanted steel

The optical pictures of a perpendicular metallographic grinding of the steel samples are shown on the fig. 3. Metallographic were made simultaneously on all samples (including unimplanted) using epoxy resin which was poured in a metallic cylindrical shell.

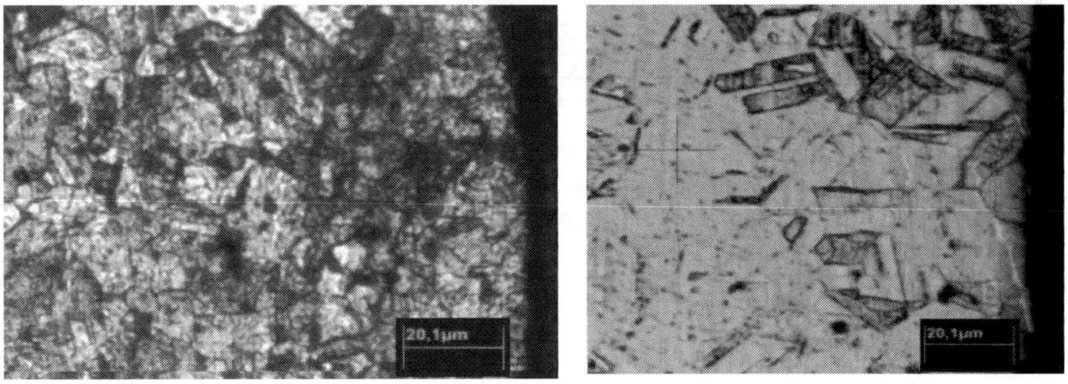

a **b**

Figure. 3. Optical images of thin surface layer of metallographic sections of steel after titanium implantation: a) initial; b) implantation time 3 min.

Therefore, the dark layers on the photos are the hardened epoxy resin has penetrated between samples. From Fig. 3 it is seen that grain structure of the near surface layer of steel varies, grains are elongated needle appearance with the increasing of the implantation time. This change is traced in all modified samples independent of dose.

The initial and modified samples were taken for the study of wear resistance of the modified surface on the tribometer «High Temperature Tribometer». High-temperature tribometer is used for measuring friction coefficient, wear resistance, wear rate in different temperature conditions, in a gas atmosphere and lubricating liquids. The velocity of the contact zone along the track was 2 $cm \cdot s^{-1}$, number of circles was 1000. The measurement results of the friction coefficient are shown in Table 1.Wear track area was measured using the three-dimensional profilometer Micro Measure 3D Station. Nine track squares and the arithmetic mean values were obtained for each sample. Results are presented in Table 2.

Table 1. The dependence of the friction coefficient from the implantation time titanium in steel.

Implantation time, min	Friction coefficient
0	0.437
0.25	0.320
0.5	0.506
1	0.490
1.5	0.494
2	0.236
3	0.168

The friction coefficient decreases at the 2 and 3 min implantation time. The wear track square for samples with the 0.25, 0.5, 0.75, 2, and 3 min implantation time is less than the initial sample. Accordingly the wear resistances of these samples are increased.

Table 2. Dependence of the arithmetic mean value of the wear track square from the implantation time of titanium.

	Implantation time, min							
	Initial	0.25	0.5	0.75	1	1.5	2	3
The arithmetic means of the square track wear μm^2.	605	84	126	266	667	747	53	49

From the above it follows that the optimal time of 2 and 3 min for obtaining the lowest friction coefficient and wear track square. The study of the microhardness and the wear resistance of the samples repeated the trends described above, i.e. the trend change of the surface microhardness occurs, but it is not monotonic.

Fig. 4 shows the example of the microhardness changing from the penetration depth of the indenter for the modified samples (the numbers near color curvers show the implantation time in minutes). These regularities may be related to effects such as the presence cluster and (or) microdroplet fraction in a flow of charged metal plasma; the effects of erosion and growth surface during implantation (thermal peaks, oversputtering layers created in the previous steps) and others. It is clear that it is impossible to evaluate the role of each of these effects at this study step for this technology.

However, it is known for example that the formation time of the thermal peak is 10^{-10}–10^{-9} s, and the its relaxation time is 10^{-10} s for monocharged and monoatomic ions [9]. The surface cannot be warmed up when the irradiation conditions such as at the pulse frequency of 100 Hz, the time between the arrival of pulses to the surface is ~10^{-2} s. The formation process of intermetallic compounds during the existence of the thermal peak requires high temperatures and pressures. The temperature in the heat peak can achieve ~1000 ^{0}C when the surface is bombarded such ions (monocharged and monoatomic) so the ion energy should be considerably higher than in the experiments presented here. Therefore it is

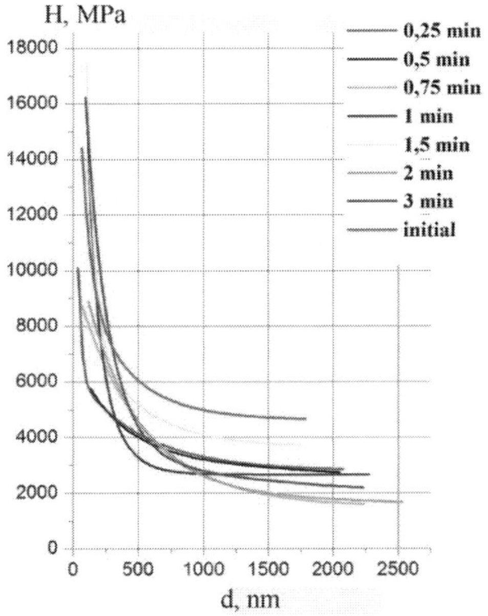

Figure 4. Dependence nanohardness from the indenter depth in Cr18Ni10Ti steel .

necessary to pay attention to lack of sufficient information about the presence of clusters with the N ≥100 number of atoms in the composition of the arc plasma. The fact that in [10] in studying the interaction of such clusters with the surface, provided that the collision energy per atom is $E_{col} \geq 10$ eV (this condition holds in our experiment) the following was established. In the collision zone for a short time (≤500 fs) from the beginning of the interaction creates a medium characterized by the temperature of ~10^4–10^5 °C, the density of atoms in the 4–5 times higher than the density of the solid and the pressure P≥1 Mbar [10]. Many physical and chemical processes that are impossible in equilibrium conditions are possible in such conditions (the formation of intermetallic compounds and even nuclear reactions). Argue that the plasma of arc discharge under certain conditions (especially the presence in it of an inert "buffer" gas; in our experiments such gas is argon) is a cluster plasma allow the results [10–13]. In these works the review of theoretical and experimental studies of the plasma clusters (including plasma of arc discharge [13]) and original theoretical results presented by the authors. In our experiment, we are dealing with beam of the cluster plasma can be argued on the basis of these studies. The hot clusters [11] play a crucial role in the modification of the surface. Cluster sizes can vary between a few atoms up to several thousand atoms (microdroplets fraction). Но до мишени долетают только стабильные ("магические" [12]) кластеры. Only stable ("magic" [12]) clusters reach the target with a high probability. In the case of titanium (densely packed hexagonal lattice) magic numbers of atoms in a cluster can be 1, 13, 57, 153, 321, 581, etc. [12]. Metal magic) cluster can carry a positive charge, at least from the 1e to 3e. Charged cluster becomes even more stable [13]. Cluster mechanism of surface modification confirmed by Fig. 5, which shows the image of the surface after plasma immersion implantation. Fig. 5a shows the optical image obtained on the microscope 30 KB S PruftechnikGmbHso 100X lens after implantation in our apparatus (in this experiment). Fig. 5b is taken from [14] in which the experiment is completely similar to ours, but the installation Platit π -80.This image obtained by the scanning electron microscope JEOL JSM6060A (Japan). Fig. 5a and 5b clearly views as microdroplets and the craters dimensions from a few microns to a disappearing small. We did not have the technical capabilities to make pictures with the increase of greater than in Figure 5 at the time of writing.

The images that are presented in Fig. 5, completely correspond to the described above concept of a cluster surface modification. They allow the following interpretation.

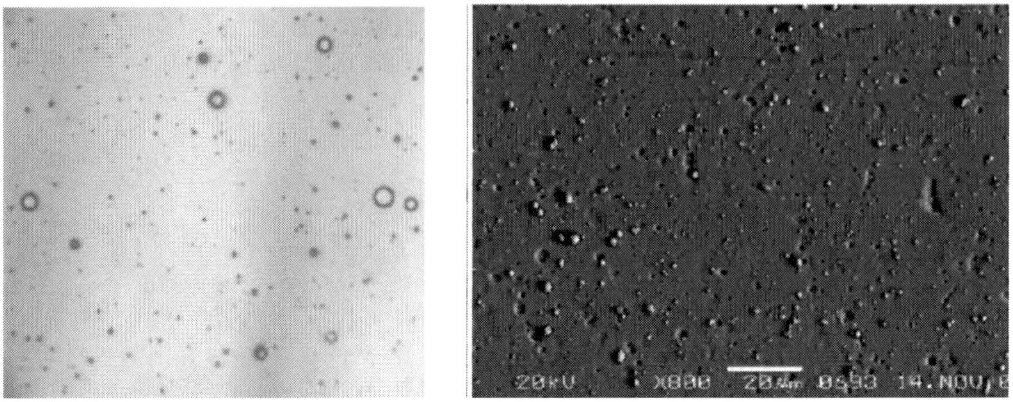

a b

Figure 5. Microscopy image of the surface of stainless steel after the plasma-immersion implantation in plants "Raduga-5" (a) и Platit π -80 (б) [14].

Microdroplets are the result of effects on the surface of the hot liquid clusters with number more than 1000 atoms; craters are formed by Magic clusters with the number of atoms less than 1000.

Conclusion

The increase in the friction coefficient and microhardness, reducing the surface wear resistance and the formation of intermetallic compounds show the great prospects for the pulse plasma immersion implantation for the surface modification. On the other hand, the data show that the charged clusters are the major contributors to the modification.

Further experiments on plasma immersion implantation should be aimed at clarifying the dependence of plasma flow on the parameters of the arc discharge.

References

[1] Conrad J R, Dodd R A, Worzala F J, Qui X 1988 *Surface and Coatings Technology* **36** 927–937.
[2] Struts V K, Zakoutaev A N, Matvienko V M 2002 *Surface and Coatings Technology* **158** 494–497.
[3] Liu J B, Iyer S S, Gronsky R, Cheung N W 1995 *Appl. Phys. Lett.* **67** 2361–2367.
[4] Gupta D 2011 *International Journal of Advancements in Technology* **2** 471–490.
[5] Ryabchikov A I, Ryabchikov I A, Stepanov I B 2005 *Vacuum* **78** 331–336.
[6] Ryabchikov A I, Stepanov I B 2009 *Surface and Coating Technology* **203** 2784–2787.
[7] Kurzina I A, BozhkoI A, Kalashnikov M P, Fortuna S V, Batyreva V A, StepanovI B, SharkeevYu P 2004 *Izvestiya of Tomsk Polytechnic Univ.* **307** 30–35.
[8] Zeng Z M, Zhang T, Tiana X B, Tanga B Y, Kwoka T K,Chu P K 2000 *Surface and Coatings Technology* **129** 236-239.
[9] Kolesnikov A S 2005 Dissertation of the candidate of Phys. and Math. Sci., St. Petersburg 176 p.
[10] Makarov G N 2006 *Uspekhi Fizicheskikh Nauk* **176(2)** 121–174.
[11] Makarov G N 2011 *Uspekhi Fizicheskikh Nauk* **181(4)** 365–387.
[12] Smirnov B M 1992 *Uspekhi Fizicheskikh Nauk* **162(1)** 119–138.
[13] Smirnov B M 2000 *Uspekhi Fizicheskikh Nauk* **170(5)** 496–543.
[14] Beregovsky V V, Duchopelnikov D V, Marachtanov M K, Shurenkova S A 2008 *Vestnik MSTU* **4** 29–32.

IOP Conf. Series: Materials Science and Engineering **81** (2015) 012019 doi:10.1088/1757-899X/81/1/012019

Effect of por-SiC buffer layer on the parameters of thin Er_2O_3 layers on silicon carbide substrates

Yu Yu Bacherikov[1], R V Konakova[1], O B Okhrimenko[1*], N I Berezovska[2], L M Kapitanchuk[3], A M Svetlichnyi[4], LA Svetlichnaya[4]
[1] *Kyiv, 03028, Ukraine*, V. Lashkaryov Institute of Semiconductor Physics, National Academy of Sciences of Ukraine
[2] *Kyiv, 01601, Ukraine*, Faculty of Physics, Taras Shevchenko National University of Kyiv
[3] *Kyiv, 03680, Ukraine*, Paton Institute of Electric Welding, National Academy of Sciences of Ukraine,
[4] *Taganrog, 347928, Russia*, Institute of Nanotechnologies, Electronics, and Electronic Equipment Engineering, Southern Federal University

E-mail: olga@isp.kiev.ua

Abstract. Using optical absorption and Auger spectrometry techniques, we studied the effect of rapid thermal annealing (RTA) on the properties of erbium oxide films deposited onto a porous silicon carbide buffer layer formed on 4H-SiC substrates. An analysis of atomic composition of the films under investigation as a function of RTA duration was performed. It is shown that phase composition of erbium oxide films on silicon carbide substrates with a porous SiC layer can be changed by varying RTA duration.

1. Introduction

High thermal stability, big values of permittivity and relative simplicity of growing process for films of a number of rare earths oxides make them promising for preparation of isolating layers in microelectronic circuits. So development of high-quality stable dielectric layers (in particular, those based on erbium and dysprosium oxides) makes an up-to-date problem in silicon carbide-based high-temperature electronics.

It is known that, in the course of further high-temperature technological process, structural defects of a semiconductor substrate penetrate a thin film growing on that substrate. This can substantially impair the device characteristics. One of the ways for reduction of the above effect is formation of a porous interlayer between the substrate and epitaxial layer [1, 2]. To illustrate, layers of porous silicon carbide (por-SiC) are used to decrease defect concentration at the porous layer-epitaxial film interface in composite structures [2-6], and porous InP and GaAs layers are applied to obtain quite unstrained autoepitaxial films on them [7, 8].

The object of the present work is investigation of the features of formation of 4H-SiC/por-SiC/Er_2O_3 structure in which por-SiC serves as a buffer layer between Er_2O_3 film and silicon carbide substrate.

* Corresponding author.

2. Samples and measurement techniques

As a first step in obtaining SiC/por-SiC/Er_2O_3 structure, we formed a por-SiC layer on the silicon carbide substrate. Porous surface of silicon carbide was obtained using electrochemical etching in the HF:C_2H_5OH = 1:1 solution for 10 min.; the current density was 10 mA/cm^2. After this, before deposition of rare-earth element, pore opening was made in KOH melt at a temperature of 550°C. Formation of Er_2O_3 film was performed in several technological steps. Firstly an erbium film was deposited onto the por-SiC surface with thermal spraying. Then the specimens were annealed in the vacuum at a temperature of 800°C for 8 min. After this the specimens were subjected to rapid thermal annealing (RTA) at 400°C for 1, 3 and 5 s.

An analysis of atomic composition of the structures under investigation was made using the Auger electron spectroscopy technique at layer-by-layer etching of the specimens by 1 keV argon ions. The specimens were studied using an instrument system LAS-2000 (Riber, France) with a cylindrical mirror spectrometer. The transmission spectra were registered with a ДФС-24 spectrometer at room temperature.

3. Results and discussion

Earlier [9-10] we considered the features of erbium oxide formation on crystalline 6H-SiC substrates by RTA at 350°C for 1, 3 and 5 s. It was found [9-10] that a thin (~100-150 nm) Er_2O_3 layer is formed on the 6H-SiC substrate in the course of RTA. An increase of RTA duration leads to increase of layer thickness and promotes formation of stoichiometric erbium oxide. The specimen surface was investigated with a scanning electron microscope (SEM) S-4800 (Hitachi, Japan).

Figure 1 presents a SEM image of porous surface. One can see that diameter of pores in the buffer por-SiC layer is ~30 nm.

Figure 1. SEM image of porous surface .

Shown in Fig. 2 are SEM images of Er_2O_3 films subjected to RTA for 1 and 5 s. One can see from Fig. 2a that, at RTA duration of 1 s, the erbium oxide film is of island type. The characteristic island sizes lie within 110-500 nm. As RTA duration is increased, the film structure becomes more uniform (Fig. 2b).

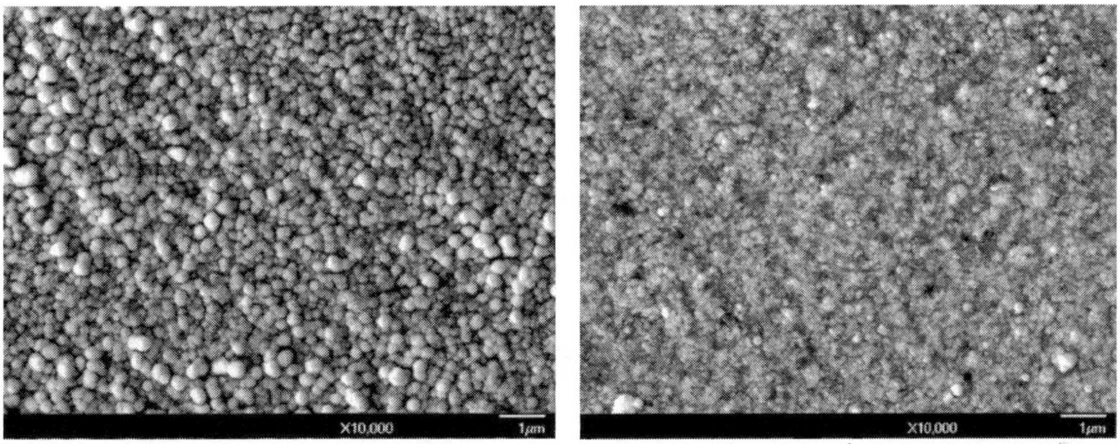

a *b*

Figure 2. SEM images of erbium oxide film surface. RTA duration: a –1 s, b - 5 s

Figure 3 presents elemental composition of the SiC/por-SiC/Er_2O_3 structures. An analysis of the elemental concentration depth profiles in those structures obtained with Auger electron spectroscopy showed that the oxide film composition is approaching to stoichiometric one ($N_O/N_{Er} \approx 1.6$–1.7) as RTA duration increases. The oxide film thickness was 80-100 nm.

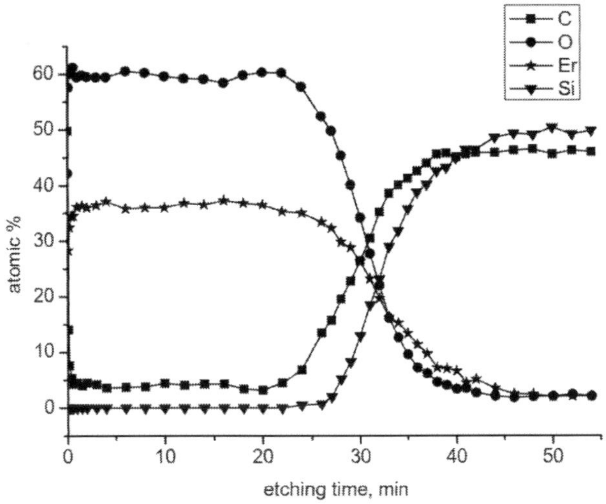

a

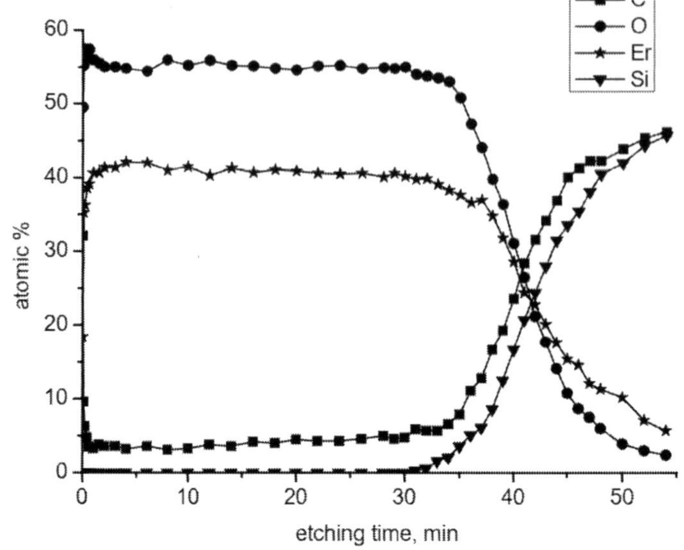

b

Figure 3. Elemental contents in the SiC/por-SiC/Er_2O_3 specimens. RTA duration: *a* - 1 s, *b* - 5 s.

Shown in Fig. 4 are optical transmission spectra for the SiC/por-SiC/Er_2O_3 structures. One can see that for those structures (as well as for the 6H-SiC/Er_2O_3 structures [9]) increase of RTA duration results in growth of transmission in the 400-800 nm spectral range. This effect is due to increase of the part of erbium oxide film of stoichiometric composition relative to sub-oxidized metal layer, which is supported by the results of Auger electron spectroscopy.

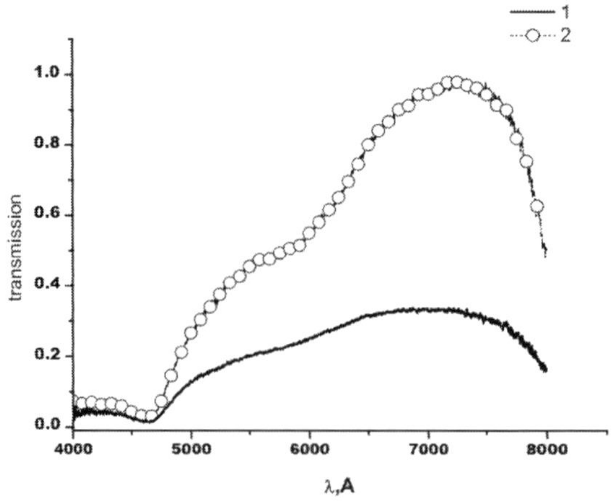

Figure 4. Optical transmission spectra of the SiC/por-SiC/Er_2O_3 structures. RTA duration: *a* - 1 s, *b* - 5 s.

4. Conclusions

So the experimental results obtained by us show that application of RTA makes it possible to form thin layers of stoichiometric erbium oxide on the surface of por-SiC. As RTA duration grows, a clearer por-SiC/Er_2O_3 interface is formed.

References:

[1] Mynbaeva M G, Lavrent'ev A A, Kuznetsov N I, Kuznetsov A N, Mynbaev K D, Lebedev A A 2003 *Semiconductors* **37** 594
[2] Savkina N S, Ratnikov V V, Rogachev A Yu, Shuman V B, Tregubova A S, Volkova A A 2002 *Semiconductors* **36** 758
[3] Sorokin L M, Savkina N S, Shuman V B, Lebedev A A, Mosina G N, Hutchison G 2002 *Technical Physics Letters* **28** 935
[4] Polupan G, Torchynska T V 2010 *Thin Solid Films* **518** S208
[5] Torchynska T V, Díaz Cano A, Dybic M, Ostapenko S, Mynbaeva M 2006 *Physica B* **376–377** 367
[6] Ashutosh Sagar, Feenstra R M, Inoki C K, Kuan T S, Koleske D D 2004 *Published in Mat. Res. Soc. Symp. Proc.* **798**, Y9.6.1.
[7] Sitnikova A A, Bobyl A V, Konnikov S G, Ulin V P 2005 *Semiconductors* **39** 523
[8] Sedova I V, L'vova T V, Ulin V P, Sorokin S V, Ankudinov A V, Berkovits V L, Ivanov S V, Kop'ev P S 2002 *Semiconductors* **36** 54
[9] Bacherikov Yu Yu, Dmitruk N L, Konakova R V, Kondratenko O S, Lytvyn O S, Milenin V V, Okhrimenko O B, Kapitanchuk L M, Svetlichnyi A M, Polyakov V V, Shelcunov A A 2007 *Technical Physics* **52** 253

Nanodefect substructures in crystal phosphors

V Lisitsyn[1,3], **L Lisitsyna**[2], **E Polisadova**[1]
[1] National Research Tomsk Polytechnic University 634050, Tomsk, Lenin Avenue, 30, Russia
[2] Tomsk State University of Architecture and Building, Tomsk, Soljanoj st., 2, Russia

E-mail: lisitsyn@tpu.ru

Abstract. We investigated spectral and kinetic characteristics of the luminescence in alkali halide crystals with polyvalent impurities, crystal phosphors based on metal tungstates in the form of nano-, micro- and bulk crystals. The possibility of the existence of complex defects (nanodefects) in materials containing high defect concentration is shown.

1. Introduction

The currently existing concepts of the stimulated processes in crystal phosphors with simple structure, with low concentration of the incorporated radiation defects (radiation-induced defects) are considered to be sufficiently consistent. These concepts are used to interpret the processes in complex crystal phosphors with high concentration of initial imperfections (widely used as phosphors, scintillators, and active and dosimetry media). It has been established that the processes in the area under radiation in complex materials with high concentration of imperfections cannot be interpreted within the concepts formulated for simple phosphor crystals. This impedes optimization of the technology for their synthesis and increase in their radiative characteristics under excitation by hard radiation.

A new approach to interpret radiation-induced processes in complex materials with high concentration of imperfections has been proposed in [1-3]. It is assumed that in these crystals, the nanodefects are formed in the process of synthesis. A nanodefect is a dopant ion, codopants entered the crystal to compensate for its charge and elastic tension, intrinsic lattice defects (vacancies, interstitial ions of the base), or a complex of defects of the structure in the crystal with low stoichiometry. A luminescence center may be the structural element of the nanodefect as well.

Nanodefects can cause significant change in the processes of radiation-induced processes in solids. Electronic excitations in the region of the nanodefect are expected to decay under laws different from those acting in the lattice. The probability of capturing electronic excitation by nanodefect, transfer of the excited energy to the luminescence center are supposed to be different from those for perfect crystals with point defects.

Determination of the structure and properties of nanodefects is a complicated problem There are no methods for experimental and theoretical studies of complex nano-sized defects. The composition and structure of the nanodefect depends on the composition and properties of the matrix and material synthesis technology.

[3] This work was supported by the Ministry of Education of the Russian Federation, government contract "Science", theme 3.1329.2014 (2152).

This paper presents the results of the study on the structure of simple scintillation crystals with a high level of imperfection indicating the possible existence of nanodefects in these crystals and their composition.

2. Samples and experimental techniques affiliations

We tested the following crystals:

• LiF-MeO (Me: Li, Fe) grown by Stockbarger method in fluorinating atmosphere in STIMS SC "S.I. Vavilov State Optical Institute" (St. Petersburg).

• LiF-MeO (Me: Ti, W) grown by Kyropoulos method in air in ISMA NAS of Ukraine (Kharkiv).

• macrocrystal ZnWO4 and ZnWO4: Eu grown by Czochralski method in air in ISMA NAS of Ukraine (Kharkiv).

• $ZnWO_4$ and $ZnWO_4$ nano-sized crystals (nanocrystals): Eu, synthesized in ISMA NAS of Ukraine (Kharkiv) by hydrothermal microwave method. Nanocrystals were two types of particles: grains with a length of ~ 25 nm and rods with a length of 100–200 nm and a diameter of about 5–10 nm. Samples to study the luminescent properties were polymer films of heat-resistant and low-molecular organic silicone rubber with incorporated $ZnWO_4$:Eu nanocrystals.

The initial imperfection in the tested crystals was estimated from the absorption spectra in the range of 13–0.1 eV. The spectra in the UV range (13–6 eV) were measured in S.I. Vavilov State Optical Institute (St. Petersburg) with a vacuum monochromator-based spectrometer; the spectra in the range of 6–1 eV were measured with SF-256 UVI and SF-256 NIR spectrometers. IR spectra were measured with FT-IR spectrometer Nicolet 5700 in the range of 0.5–0.1 eV with a spectral resolution of 2 cm^{-1}. Spectral and kinetic characteristics of pulsed cathodoluminescence (PCL) and pulsed photoluminescence (PPL) were measured in the spectral range of 4.0–1.6 eV. The characteristics of the exciting electron pulse (EP) were as follows: the length of the pulse was 10 ns; the average electron energy was 250 keV. PCL was excited by radiation of the fourth harmonic of a neodymium laser (photon energy was 4.66 eV) with a pulse duration of 5 ns and the energy per pulse of 40 mJ. The kinetic curves of PCL and PPL at a chosen wavelength were recorded with oscilloscope LeGroy-WP-6030a in the time interval of $1 \cdot 10^{-8}$–$1 \cdot 10^{-3}$ s after the end of the excitation pulse.

3. Results and discussion

3.1. The composition of nanodefects in LiF crystals with polyvalent cation-substituent impurities

LiF crystals doped with Fe, Ti, U and W are efficient converters of hard excitation energy into the luminescence. The valence of these ions is different from the valence of Li ions which they replace. Obviously, intrinsic defects, interstitial ions and vacancies enter for compensation of charge and elastic tensionin the region of the impurity ion. The crystals with multivalent ions are grown from the mix with metal oxides as added dopants, which means that oxygen ions enter the crystal as well. When oxygen enters LiF crystal, characteristic absorption occurs in the UV spectrum [4-7] in the region from 12.0 (transparency cutoff) to 4.6 eV. If the concentration of oxygen is sufficiently high, the crystal is not transparent in the region above 4.6 eV.

When LiF crystals doped with Fe, Ti and W are exposed to hard radiation pulses: electrons, laser in the region of 4.66 eV, the luminescence is excited in the region of 2.0–3.5 eV. When measured after 10 ns after the end of the excitation pulse, the luminescence spectra in all the crystals have a luminescence band at 3.1 eV (Fig.1). The luminescence at 3.1 eV in LiF crystals is caused by O^{2-} ions not bound with the dopants [8]. Over time after the end of the excitation pulse in the nanosecond range, this band decays. The bands in the long wave-length region become dominant in the spectra. As can be seen in Figure 1, the position of the luminescence bands in the spectra measured after 10 ms depends on the type of the dopant. It is assumed that in LiF-O, LiF-Fe, Ti and W the luminescence center contains O^{2-} ion as a basic structural element [1, 9-11]. It may be assumed that the luminescence center contains a fluorine vacancy. Dependence of the luminescence band on the type of the dopant indicates that the center is in close vicinity to the dopant ion. It can be concluded that the luminescence in the region of 2.5...3.0 eV in LiF-Me_nO_m crystals is caused by Me-nO^{2-} complexes [12].

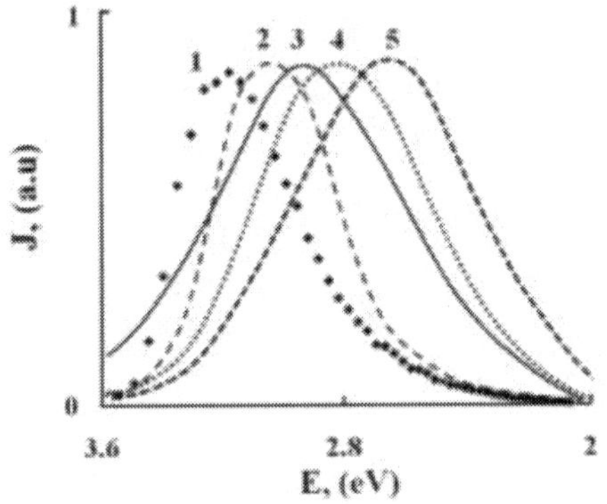

Figure 1. Photoluminescence spectra at 300 K measured with 10 ns (1) and 10 ms (2–5) delay in non-irradiated LiF crystals doped with Li_2O (1, 2), Fe_2O_3 (1, 3), WO_3 (1, 4) and TiO_2 (1, 5).

In IR absorption spectra of LiF doped with W, Ti and Fe, there is a characteristic absorption band in a free state at 3730 cm^{-1} caused by OH^- ions [13,14]. In the spectra of LiF and LiF-Li_2O, the absorption bands caused by OH^- ions are not observed. In addition to the band at 3730 cm^{-1} in IR spectra, in crystals doped with W, Ti and Fe, the bands are found at 3620–3560 cm^{1}. The position and the number of the bands depend on the type of the dopant cation. The observed dependence of the bands on the type of the dopant cation can be attributed to the effect of the cation on the absorption state of OH^- ion. The results indicate that during synthesis of the crystal, OH^- ions are located in the vicinity of the dopant ion. We can assume that during of the crystal, Me-n OH^- complexes are created in the region of the dopant.

In growing, when polyvalent cations of the impurity enter LiF crystals, anion vacancies enter the crystal and they are located in the area of the impurity ion. The experimental results to prove our conclusion are as follows. Fig. 2 shows the results of the pre-study for the kinetics of the accumulation of F_2 complex color centers. In pure LiF and LiF crystals with oxygen, the accumulation of F_2 centers occurs with an induction perio, which is characteristic of their accumulation in alkali halide crystals. In crystals with polyvalent ions, the accumulation is significantly different. The dose dependence of the accumulation of F_2 centers is a curve with saturation. The saturation level of the accumulated F_2 color centers in the dose dependence is defined by the concentration and type of the dopant cation.

The observed differences in the accumulation of F_2 centers in these crystals can be explained as follows. In pure LiF crystals, the formation of F_2 centers is the result of a complex set of processes which are generally defined by the probability of the formation of elementary F centers in spatially close lattice sites. Therefore, the kinetics of F_2 centers accumulation is described by a parabolic function.

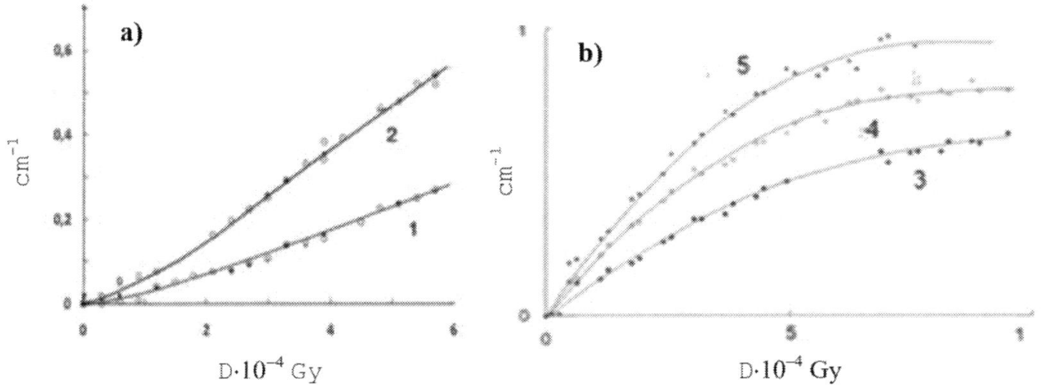

Figure 2. Dose dependence of the absorption coefficient in the maximum of F_2 band under irradiation by electrons at 300 K for different crystals: non-doped LiF (1), LiF-Li_2O (2), LiF-TiO_2 (3), LiF-WO_3 (4) and LiF- Fe_2O_3 (5).

The formation and accumulation of F_2 centers in doped crystals occurs in a different way. Incorporation of the dopant ions (Ti, W and Fe) in LiF crystal is accompanied by the entry of anion vacancies which like other lattice defects (intrinsic and extrinsic) compensate the charge and elastic tension in the region of the incorporated dopant cation. The existence of the anion vacancies in the crystal changes the kinetics of accumulation of complex centers. The accumulation of F_2 centers in doped crystals is expected to occur with a statistical probability proportional to the concentration of the dopant ions. Therefore, the dose dependence of the accumulation of F_2 color centers in crystals doped with heavy ions is a curve with a rapid initial increase in the concentration and saturation as the concentration of free all-in vacancies decreases.

Thus, in LiF crystal doped with ions of polyvalent metal, we can find oxygen ions which entered the crystal with dopants during growing, and Me-nO^{2-} and Me-n OH^- complexes. In the region of the dopant ions, there are ready vacancies (vacancy complexes) that are involved in the formation of F_2 centers.The results obtained show that during synthesis of LiF crystals doped with ions of polyvalent metals, complexes containing oxygen ions, hydroxyl groups and vacancies are formed in the region of the dopant. This type of the complex lattice defects can be regarded as a nano-sized region of the crystal with its spatial and energy structure. This region of the crystal can be called a nanodefect.

3.2. Possibility of the existence of nanodefects in scintillation crystals based on tungstates

It has been found that spectral characteristics of the radioluminescence of $PbMoO_4$ and $CaMoO_4$, $CaWO_4$, $PbWO_4$,$CdWO_4$; $CdWO_4$-Li_2CO_3; $CdWO_4$- Li_2CO_3-Bi_2O_3, LiF:WO_3, $Li_2W_{0.95}Mo_{0.05}O_4$ and suchlike crystals are similar. In the luminescence spectra of the crystals we observe the band with a maximum at 2.4–2,7 eV with a half-width at ~ 0.6–0.7 eV, and the decay time at room temperature of several tens of microseconds [12,15–20]. Close values of the spectral and kinetic parameters of the excited luminescence in oxygen-containing materials, different both by the type of the crystal lattice, and by the type of co-dopants, suggest that the structure of radiative centers have something in common.

Crystals of tungstates and molybdates are grown typically by Czochralski method from the molten mixture made by mixing the oxides of Zn, Pb and Cd and oxides of tungsten, molybdenum in proportions required to make a stoichiometric composition. It is obvious that during this growing the crystal contains a great amount of lattice defects that must have an effect on the characteristics of the material. In [12], it is suggested that in crystals with high level of imperfection, which are tungstates, molybdates, as well as in LiF crystals doped with polyvalent ions, we can find complex defects (nanodefects) as a set of intrinsic lattice defects, impurity ions entered during growth, among which we can find OH^- ions. The luminescence centers in metal tungstate crystals, as well as in the doped LiF crystals, are oxygen-containing complexes.

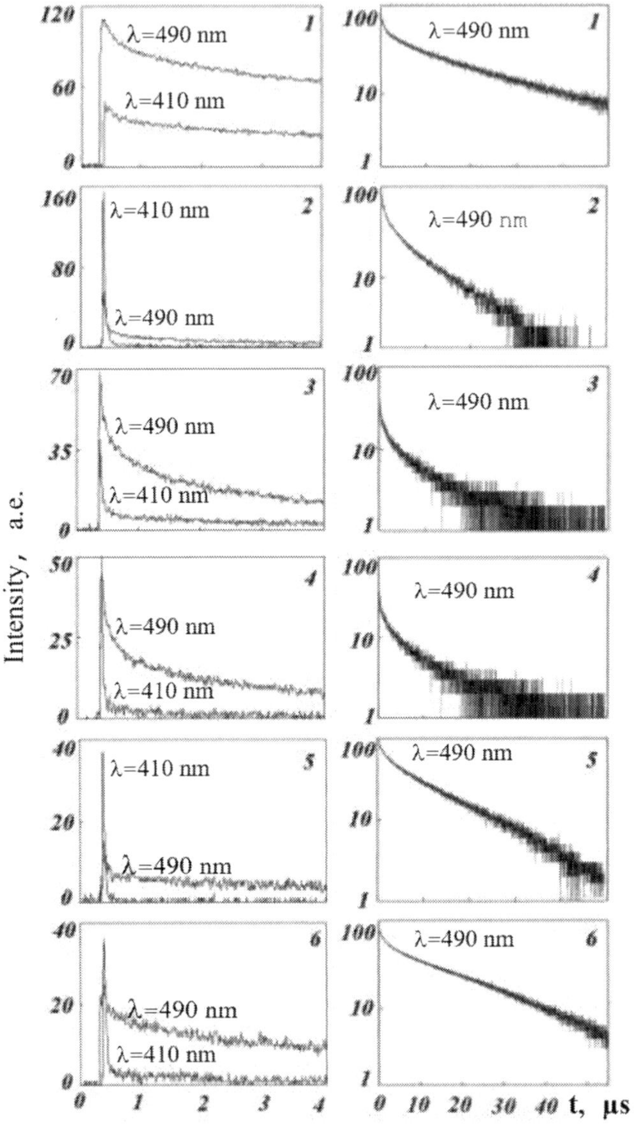

Figure 3. Kinetic curves of PCL decay for $ZnWO_4$ crystals of different dispersion incorporated into the polymer matrix. 1 is $ZnWO_4$ macrocrystal; 2 is with nanocrystals in the form of grains with a size of 25 nm, 3 is with nanocrystals in the form of rods with a length of 50–100 nm and a diameter of 5–7 nm; 4 is with nanocrystals in the form of rods with a length of 150–200 nm and a diameter of 10 nm; 5 is with crushed crystals, the size of the dominant particles being 250 nm; 6 is with crushed crystals with the size of the dominant particles being ≤ 250 μm.

We have studied the kinetic curves of attenuation of pulsed cathodoluminescence in $ZnWO_4$ crystals of different dispersion incorporated into the polymer matrix, and single crystals. The luminescence decay kinetics at 490 nm (2.53 eV) has two distinct stages of decay: in the range of 1–2 µs and 5–30 µs in different crystals. There is a significant difference in the kinetic curves of the luminescence decay in samples with synthesized nanocrystals and samples with crushed single crystals in a microsecond range. The characteristic decay time of the luminescence in samples 1, 5 and 6 (single crystal and crushed crystals) are 26, 13 and 15 µs, respectively. The characteristic times of the kinetics of the luminescence decay in crystals with synthesized nanocrystals with the size that does not exceed 200 nm are 5, 7 and 7 µs (samples 2, 3 and 4, respectively). The distinction in the characteristics of the luminescence decay kinetics may be due to the relationship between the size of the nanodefect, which contains the luminescence center, and the crystal. A sharp change in the characteristic times of the luminescence decay as the crystal size reduces to the units of nanometers can be attributed to the fact that the sizes of the crystals and nanodefects become commensurable. Note that short time luminescence can be observed in the region at 410 nm (3.03 eV) which decays after 100 ns, and it is more distinct in fine-dispersed crystals. Apparently, other centers possibly related to the surface are responsible for the luminescence in this area [12].

Fig. 4 shows the results obtained for the kinetics of the luminescence in $ZnWO_4$ macro- and nanocrystals doped with europium. The kinetic curves indicate the luminescence decay in the band at 2.6 eV (480 nm), which is caused by the oxygen complex luminescence centers in the nanodefect, and in the band at 614 nm caused by europium. The spectral and kinetic characteristics of emission from europium are weakly dependent on the size of the crystal: the spectra in macro- and nanocrystals are identical. The characteristic decay time of the luminescence at 614 nm differs slightly: in the macrocrystal it is 0.5 ms, and in the nanocrystal it is 0.3 ms.

The spectral characteristics of the luminescence in the band at ~ 480 nm, which is caused by the centers with an oxygen ion being the structural element, are identical in all the tested crystals, and this indicates the identical nature of the luminescence centers responsible for this band. However, the kinetic characteristics of the luminescence in macrocrystals and nano-sized crystals are significantly different; they are 15 µs and 3 µs, respectively. The possible cause of the difference can be as follows. The oxygen center responsible for the luminescence is a component of the nanodefect. The size of the nanodefect is several nanometers, which is comparable to the size of the nanocrystal. The size of the europium ion even coupled with the donor is less than a nanometer. Therefore, the effect of the crystal surface with the sizes of 10 ... 25 nm on the radiative properties of the oxygen luminescence center in the nanodefect is much larger than relaxation of the excitation in the europium center.

The presented results support the hypothesis that in zinc tungstate crystals having a high level of initial defects due to non-stoichiometry, complex defects with sizes of several nanometers are formed. A nanodefects in this crystal is the region of the crystal with different spatial and energy structure.

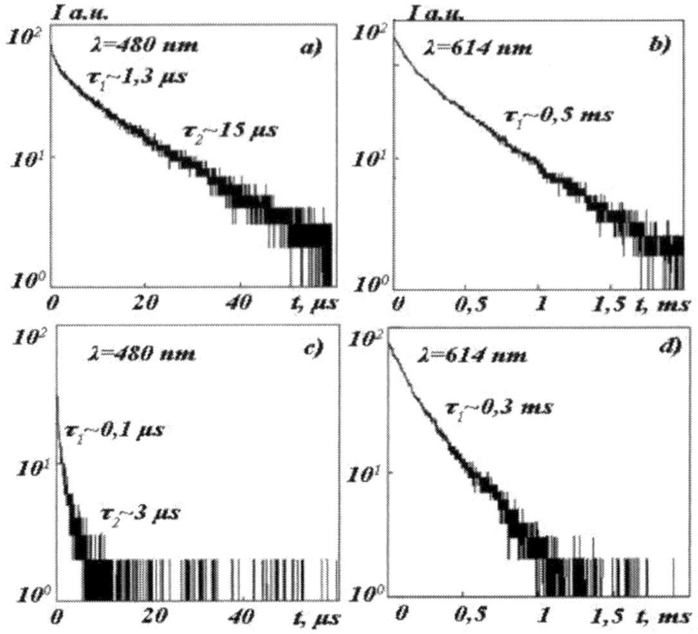

Figure 4. Decay kinetics of ICL in the bands at 480 and 614 nm for macrocrystals (a,b) and nanocrystals (c,d) doped with 3 mol% Eu under excitation by electron beam.

4. Conclusion

The nanodefect substructure in crystal phosphors can lead to sufficient increase in the efficiency of the luminescence under excitation. Nanodefects must have a very large capture cross section of the carriers of charge, energy. With a radius of nanodefects equal to 3...4 interatomic distances, the capture cross section of the hole in the region of the nanodefect 200 ... 500 times exceeds that of the random site in the crystal.

In the case of the recombination character of the energy transfer to the luminescence center, the probability of recombination in the region of the nanodefect can be much higher than that in the crystal. The luminescence through the recombination mechanism of excitation in the nanodefect can be as follows. Radiation-induced electrons and holes are found in the area influenced by nanodefects due to their large capture cross sections. A hot mobile hole, when it appears in the relaxed region of the nanodefect, is immediately self-trapped, it captures an electron. The energy of the near-impurity self-trapped exciton is transmitted to the luminescence center. If the lifetime of the inear-impurity exciton is much greater than the lifetime of the luminescence center in an excited state, the probability of the energy transfer to the luminescence center is supposed to be close to unity. This ratio is always fulfilled for scintillators: the lifetime of the near-impurity self-trapped exciton is of the order of $10^{-4}...10^{-6}$. The lifetime of the near-impurity excitons in KCl-J crystal is about 10^{-5} s [21–23].

The approach based on the existence of the nanodefect substructure opens up new possibilities for the analysis of radiation-induced processes in crystals with a high level of imperfection which are used as functional materials.

Acknowledgments
This work was supported by the Ministry of Education of the Russian Federation, government contract "Science", theme 3.1329.2014 (2152).

IOP Conf. Series: Materials Science and Engineering **81** (2015) 012020 doi:10.1088/1757-899X/81/1/012020

[1] Lisitsyna L A, Oleshko V I, Putintseva S N and Lisitsyn V M 2008 *Optics and Spectroscopy. 105,***40** 531
[2] Lisitsyna L A, Korepanov V I at al. 2012 *Optics and Spectroscopy.* 112 **6** 175
[3] Lisitsyna L A, Lisitsyn V M 2013 *Japan. J. Appl. Phys.* 55 **11** 2297
[4] Lisitsyna L A, Putintseva S N, Lisitsyn V M, Oleshko V I 2008 *Nonorganic materials.* **6** 754
[5] Freytag E 1964 *Z. Phys.* **177** 206
[6] Shvarts K K, Kristapson Ya Zh, Lusis D Yu, Podlin A V *Proc.: Radiation Physics.V,* ed. K K Shvarts, Lushina A F. Zinatne. Riga, 1967 179.
[7] Korzhik M V. 2003 *Physics of scintillators based oxygen single crystals* (Minsk: Belarusian State University)
[8] Lammers M J J, Blasse G, Robertson D S 1981 *Phys. Stat. Sol. (a)* **63** 569
[9] Egranov A V, Radjabov E A 1992 *Spectroscopy of oxygen and hydrogen impurity centers in alkali halide crystals* (Novosibirsk: Nauka)
[10] Jain V K 1986 *J. Phys. D: Appl. Phys.* 19 **9** 1791
[11] Rebane L A 1968 *Proc. IAA AS ESSR.* **37** 14
[12] Lisitsyna L A, Korepanov V I, Lisitsyn V M at al. 2011 *Optics and Spectroscopy.* 4 **110** 568
[13] Stoebe G 1970 *J. Phys. Chem. Solids.* **31** 1291
[14] Wedding B, Klein M 1969 *Phys. Rev.* **177** 1274
[15] Kolobanov V N, Kamenskikh I A, Mikhailin V V at al. 2002 *Nuclear Instruments and Methods in Physics Research A* **486** 496
[16] Nagirnyia V, Kirm M, Kotlov A at al. 2003 *Journal of Luminescence* **102-103** 597
[17] Barinova O P, Kirsanova S V, Kolobanov V N, Mikhailin V V, Spassky D A 2008 *Perspective materials* **4** 34
[18] Spassky D, Ivanov S, Kitaeva I at al. 2005 *Phys. Stat. Sol.* **2** 65
[19] Groenink J A, Blasse G 1980 *J. of solid state chem.* **32** 9
[20] Mikhaik V B, Kraus H, Itoh M, Iri D, Uchida M 2005 *J.Phys.: Condens. Matter.* **17** 7209
[21] Lisitsyn V M, Malyshev A A, Yakovlev V Yu 1983 *Russian Phys. J..* 25 **11** 3356
[22] Lisitsyn V M, Korepanov V I, Strezh V V, Bochkanov P V, Malushev A A 1985 *Russian Phys. J.* 27 **10** 3052
[23] Korepanov V I, Lisitsyn V M, Lisitsyna L A 1997 *Russian Phys. J.* 39 **11** 1082

The structure and properties of the nanocomposite films Nb-Al-N

V I Ivashenko[1], A D Pogrebnyak[2], P L Skrinski[1], V N Rogoz[2], S V Plotnikov[3], N K Erdybaeva[3], E O Tleukenov[3]

[1] *Kiev-142, 03680, Ukraine,* Institute of Materials Science problems, Kryzhanovskaya Street 3
[2] *Sumy, 40000, Ukraine,* Sumy State University, Rimsky-Korsakov Street, 2
[3] *Ust-Kamenogorsk, 070003, Kazakhstan,* D. Serikbayev East Kazakhstan State Technical University, 23/2-54 Serikbayev Street

E-mail: alexp@i.ua

Abstract.

Nanocomposite films Nb-Al-N produced by magnetron sputtering were researched in this work. Two stable crystalline structural states were found in the films: NbN_{ch} and solid solution $B1-Nb_xAl_{1-x}N_yO_{1-y}$, and also an amorphous component associated with aluminum oxynitride with reactive magnetron sputtering. Sensitivity of substructural characteristics was set up to the current supplied to Al target and their relationship with the characteristic nanohardness and Knoop hardness. Recent changes in the range of 29-33.5 GPa and 46-48 GPa, respectively. Initial principle calculations of phases NbN and Nb_2AlN and also heterostructures of NbN/AlN were carried out for the interpretation of the results. The work was performed as a part of two complex state programs: "Development of nanostructured superhard coatings formation foundations with high physical-mechanical properties" (number 0112u001382) and "Physical principles of plasma technologies for complex processing of multicomponent materials and coatings" (number 0113u000137c).

Introduction

Films based on NbN show many interesting properties such as high hardness and electrical conductivity, thermal stability and chemical inertness [1]. NbN films are used as the cathode material for field electron emission in vacuum of microelectronic devices [2]. It was shown that the introduction of Al atoms into the lattice led to the formation of solid solution $Nb_{1-x}Al_xN$. For solid solutions $Nb_{1-x}Al_xN$, B1 (type - NaCl) structure is more preferable for x below 0.45. In the range of x = 0.45-0.71, the mixture of structures B1 and B4 was observed, while when x> 0.71, B4 structure (type - wurtzite with a hexagonal structure) was formed [6,7]. Films Nb-Al-N composed of solid solution $Nb_{1-x}Al_xN$ with B1 (rare B_K) and B4 structures or mixtures thereof [3-7]. On the other hand, hitherto, nanocomposite NbN/AlN films were not investigated. Thus, in the present study, we set out to research the Nb-Al-N film, arguing that under certain conditions can be formed nanocomposite structure of the films, which may have improved mechanical properties compared to films composed of the substitutional solid solutions $Nb_{1-x}Al_xN$.

Experimental details

Films Nb-Al-N deposited on the mirror-polished wafer Si (100) using a DC magnetron sputtering Nb (99.9%, Ø72 x 4 mm) and Al (99.999%, Ø72 x 4 mm) under argon and nitrogen with the following parameters of deposition: the substrate temperature T_S = 350°C; substrate bias voltage U_B = -50 V; The flow rate (F) F_{Ar} = 40 sccm; F_{N2} = 13 sccm; operating pressure P_C = 0.17 Pa. The current supplied to the target Al (I_{Al}) was 100, 150, 200, 250 and 300 mA, which corresponds to the discharge power density P_{Al} = 5.7, 8.6, 11.4, 13.7 and 17.1 W/cm^2, respectively. Current supplied to the target Nb (I_{Nb}) was 300 mA (P_{Nb} = 17.1 W/cm^2). The base pressure in the vacuum chamber was better than 10^{-4} Pa. The distance between the target and the substrate holder was 8 cm. The dihedral angle between the targets was ~ 45^0. Substrates were ultrasonically cleaned before they were placed in a vacuum chamber. Furthermore, prior to deposition, the substrates were etched in a vacuum chamber in the hydrogen plasma for 5 min.

Results and discussion

Structural and mechanical properties were analyzed depending on I_{Al}-values. The structure of the coatings was researched by X-ray diffraction (XRD, diffractometer DRON-3M) in CuK$_\alpha$ radiation. The program of authoring profiles sharing was used with the application of complex diffraction profiles. Substructural characteristics (crystallite and microstrain size) were determined by approximation using the approximation function as - Cauchy function. The spectrum of Fourier spectroscopy (FTIR) was measured at room temperature in the range of 400–4000 cm^{-1} by spectrometer "TSM 1202" LTD «Infraspek». Knoop hardness (HK) was assessed using microhardnessmeter Microhardness Tester Micromet 2103 BUEHLER LTD under a load of 100 mN, and by nanoindentation using nanohardnessmeter G-200, equipped with a Berkovich indenter. The loads were chosen in such conditions that the indenter penetration wouldn't exceed 10-20% of the film hickness. The film thickness was determined by optical profilometer "Micron-gamma". The thickness of the Nb-Al-N coatings (d) depends weakly on I_{Al}. D values were in the range of 0.7-0.9 μm.

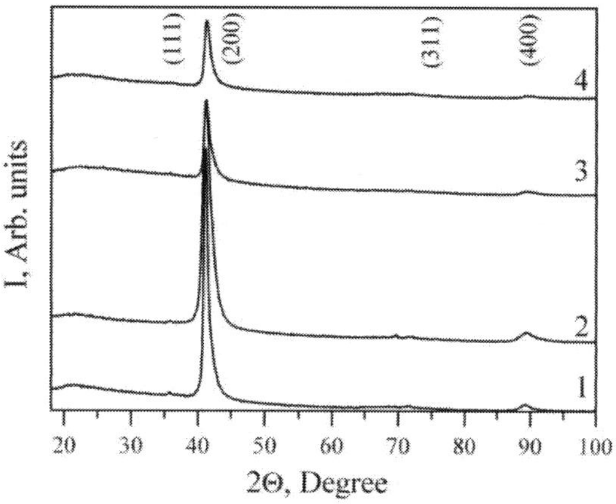

(a)

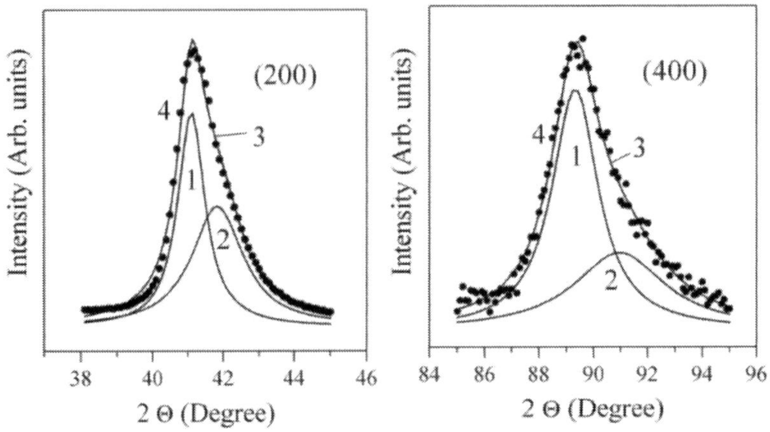

(b)

Figure 1. XRD spectra of Nb-Al-N coatings deposited at different I_{Al}: 1 – 100 mA, 2 – 150 mA, 3 – 250 mA, 4 – 300 mA (a) portion and on the division of the diffraction profile of coating components Nb-Al -N, deposited at I_{Al} = 150 mA (b): 1 – NbN, 2 – $Nb_{0,67}Al_{0,33}N$, 3 - total approximating curve, 4 - starting point of the data array.

Figure 1 (a) shows the X-ray diffraction spectra of the Nb-Al-N films at various I_{Al}. Marked peaks correspond to the structure of the planes B1–NbN_x [8]. Halo component of the amorphous phase was manifested at the same time in the range of diffraction angles 2Θ = 18–30°, based on previous studies halo component can be identified as the amorphous phase of aluminum nitride. It is seen that the reflection (200) is the main comparing with others. Crystallites were formed for all I_{Al} with a relatively small constant potential of bias on the substrate -50 V with preferentially oriented growth with [100] axis perpendicular to the plane surface. The X-ray (200) and (400) reflections are asymmetrical to larger angles. Separating these reflexes showed the presence of two components with similar lattice type, but with two typical periods. Figure 1 (b) shows the results of the deconvolution of (200) and (400) peaks in the Gaussian for the film deposited at 150 mA (curve 1 in Figure 1 (a)). Curves 1 in figure 1 (b) correspond to the cubic niobium nitride with lattice constant a = 0.439-0.438 nm. Gaussian curve 2 can be attributed to a cubic NbN with low aluminum content and the type of substitution = 0.428-0.429 nm, which is characteristic of the system Ti-Al-N with ratio of aluminum and niobium atoms in the lattice as 1/2 (approximate composition is $Nb_{0,67}Al_{0,33}N$). The latest one was determined based on the Vegard's rule for the same type of crystal lattices by substitution of atoms with different radius [9]. Along with this, lattice parameters B1-NbNx and B1-AlN were used as the basic, respectively 0.4393 nm and 0.4120 nm [5]. With increasing I_{Al} Gaussian peak position does not change, and the intensity of the peaks associated with the solid solution increases. For higher currents, the phase relationship $NbN_x/Nb_{0.67}Al_{0.33}N$ is in proportion close to 5.3.

We deposited AlN films at various I_{Al}. X-ray studies showed that all AlN films were amorphous (a-AlN, is not shown in the paper). Infrared absorption spectra of AlN films indicate that the number of Al-N bonds increases with I_{Al} enhancement: (zone of absorption at 667 cm^{-1}, associated with fluctuations of Al–N [10] becomes more noticeable).

On the basis of these results, it can be assumed that two stable crystalline structural states were found in the films: B1–NbN_x and solid solution with a composition close to the B1–$Nb_{0.67}Al_{0.33}N$. The films also contain an amorphous component associated with aluminum nitride. Thus, the films have a nanocomposite structure which is nanocrystallites- B1–NbN_x и B1–$Nb_xAl_{1-x}N$ embedded in a matrix of a-AlN (nc–B1–NbN_x/nc–B1–$Nb_xAl_{1-x}N$/a–AlN).

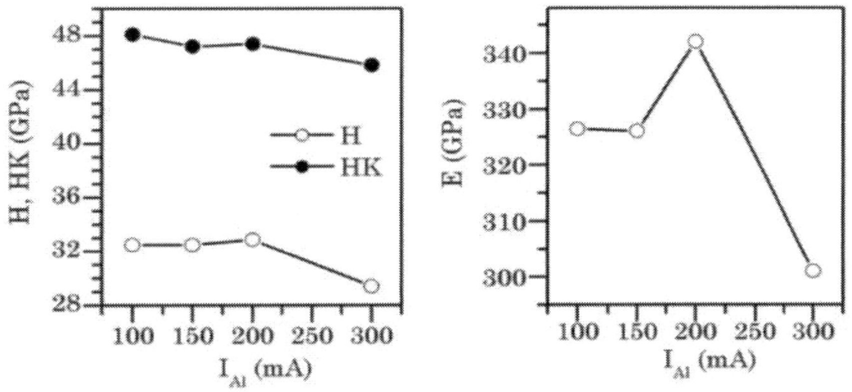

Figure 2. Nanohardness (H), the Knoop hardness (HK)(a) and modulus of elasticity (E)(b) depending on I_{Al}.

The results of nanoindentation and microindentation deposited films are shown in Figure 2. Comparison of the results in Figure 2 shows that there is a correlation between the mechanical properties and microdeformations in the Nb-Al-N films. Also nanohardness, elastic modulus and Knoop hardness (HK) are maximum for films Nb-Al-N with a grain size in the range of 30-40 nm. Increase of nanohardness from 28 GPa for films NbN [11] to 32 GPa for the Nb-Al-N film is obviously connected with the formation of a nanocomposite tipped Nb-Al-N film. We observed that the Knoop hardness higher than nanohardness ~ 50%. This may be due to the fact that the nanoindentation occurs dynamically, while the Knoop hardness is determined in a static mode. In order to verify our conclusions on the structure of Nb-Al-N films we performed initial principal calculations of B1 -NbN, solid solutions B1−$Nb_xAl_{1-x}N$, heterostructures B1−NbN(001)/B1−AlN and the ordered phase Nb_2AlN. Terms of calculation are described in detail in the work [12]. Here we note that the calculations were performed using the computational code [13]. A generalized gradient approximation was used for exchange correlational potential [14]. Simulation of molecular dynamics was performed using NVT ensemble at 1400 K with subsequent cooling to 0 K and with static relaxation [12].

96 atomic structures built by translating 8-nuclear-cell B1 as (2×2×3) were considered [12]. The compositions of solid solutions and heterostructures were selected equally. Nb_2AlN cell (space group $P6_3/mmc$, No. 194) consists of eight atoms. The above structures include all the possible configurations of the system $Nb_xAl_{1-x}N$. XRD spectra were calculated using the software PowderCell-2.4 [15].

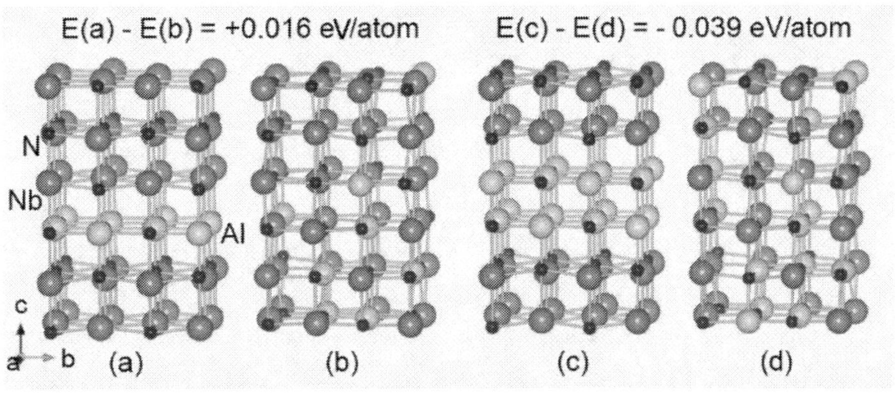

Figure 3. Atomic configuration B1−NbN(001)/1 ML B1−AlN of heterostructure (a), of solid solution $Nb_{0.83}Al_{0.17}N$ (b), B1-NbN(001)/2 ML B1−AlN of heterostructure (c), of solid solution $Nb_{0.67}Al_{0.33}N$ (d).

Composition of structures (a) and (b) is the same; a structure composition (c) is equal to the composition respectively (d). The inscription above the figure represents the difference between the total heterostructure energies and the corresponding solid solution with a random arrangement of atoms in the metal lattice. Figure 3 shows the atomic configuration of heterostructures B1−NbN(001)/ 1 ML AlN and B1−NbN(001)/ 2 ML B1−AlN, ML-monolayer, as well as solid solutions B1−$Nb_xAl_{1-x}N$. Analysis of the total energies shows that solid solutions B1−$Nb_xAl_{1-x}N$ for x <0.67 should decay and nanocomposite structure with B1-AlN interface can be formed for these concentrations.

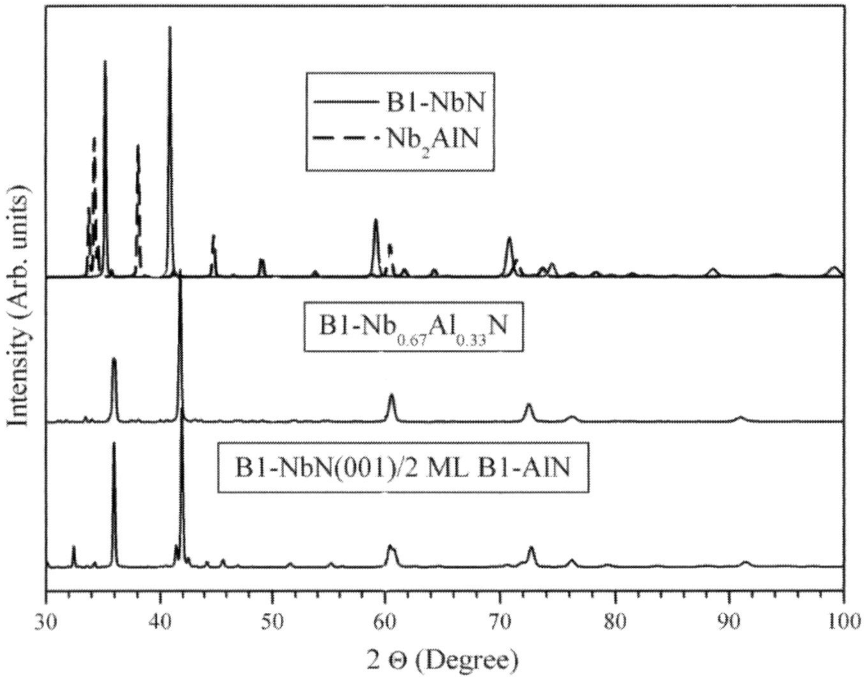

Figure 4. The calculated X-ray diffraction patterns.

Using atomic configurations resulting from first-principles calculations, we calculated diffraction patterns for the B1−$NbN_ч$, B1−$Nb_xAl_{1-x}N$, x ~ 0.67 и Nb_2AlN. The calculated XRD spectra are shown in Figure 4. Comparison of the calculated and experimental spectra (Figure 1) shows that reflexes about $2\Theta \sim 32^0$ associated with the heterostructure, and about $2\Theta \sim 38^0$, resulted by phase Nb_2AlN and does not occur at the experimental spectra. Therefore, we can assume that our films contain neither Nb_2AlN, nor any epitaxial layers B1-AlN, but rather consist of crystallites B1−NbN_{ch} and B1−$Nb_xAl_{1-x}N$, x ~ 0.67. As the matter of the fact the difference between the peak positions $\Delta 2\Theta = 2\Theta\cdot$(B1−$NbN_ч$)−$2\Theta\cdot$(B1−$Nb_xAl_{1-x}N$) for each diffraction peak (200) and (400) on experimental and theoretical diffraction patterns is almost identical. We should also note that films of niobium nitride are prone to accumulate small amounts of oxygen [5]. Oxygen can replace part of the nitrogen in solid solution and in the amorphous matrix [16,17]. Therefore, structure $Nb_xAl_{1-x}N_yO_{1-y}$, x~0.67, 1-y << 1will be more realistic for the solid solutions, as for the amorphous matrix - a-AlNO, which was shown as a result of elemental analysis, obtained by SIMS, RBS and EDS in these films.

Conclusion

Films Nb-Al-N were deposited on silicon substrates by magnetron sputtering targets of Nb and Al at different discharge powers at the target of aluminum. Experimental and theoretical studies show that the films obtained at selected deposition parameters have the nanocomposite structure that represents the nanocrystals B1−NbN_x and B1-$Nb_xAl_{1-x}N_yO_{1-y}$, embedded in a-AlNO matrix (nc−B1−NbN_x/nc−B1−$Nb_xAl_{1-x}N_eO_{1-y}$/a−AlNO). Nanocomposite coating with high microstrain action due to the difference in atomic radius of the crystal lattices metal components shows high hardness values (up to 32 GPa). Deposited nanocomposite films may be recommended with the given mechanical properties as wear resistant or protective coatings.

References:

[1] Barnett S A, Madan A, Kom I, Martin K 2003 *Mrs Bulletin*. 28, pp. 169.
[2] Gotoh Y, Nagao M, Ura T, Tsuji H, Ishikawa J 1999 *Nucl. Instr. Methods Phys. Res.* Vol. **148** pp. 925.
[3] Selinder T I, Miller D J, Gray K E 1995 *Vacuu* vol. **46** pp. 1401.
[4] Makino Y, Saito K, Murakami Y, Asami K 2007 *Solid State phenomena* vol. **127** pp. 195.
[5] Barshilla H C, Deepthi B, Rajam K S. Mater J 2008 *Res.* Vol. **23** pp. 1258.
[6] Franz R, Lechthaler M, Polzer C, Mitterer C 2010 *Surf. Coat. Technol.* Vol. **204**, 2447.
[7] Holec D, Franz R, Mayrhofer P H, Mitterer C 2010 *J. Phys.* Vol. **41**, 145403.
[8] X-ray powder diffraction file [038-1155].
[9] Umanski Y S, Skakov Y A 1978 Physica metallov. Atomnoe stroenie metallov i splavov *Atomizdat M.* pp. 352.
[10] Jadannadham K, Sharma A K, Wei Q, Kalyanraman R, Narayan 1998 *Vac. Sci. Technol* vol. **16**, 2804.
[11] Ivashchenko V I, Veprek S, Scrynskyy P L, Lytvyn O, Butenko O O, Sinelnichenko O K, Gorb L, Hill F, Leszczynski J, Kozak A O 2014 *Superhard Materials* vol. **36**, pp. 1.
[12] Ivashchenko V I, Veprek S, Turchi P E A, Shevchenko V I 2012 *Phys. Rev.* Vol. **85,** 195403-1.
[13] Giannozzi P, Baroni S, Bonini N, Calandra M, Car R, Cavazzoni C, Ceresoli D, Chiarotti G L, CococcioniI M. Dabo, Dal Corso A, de Gironcoli S, Fabris S, Fratesi G, Gebauer R, Gerstmann U, Gougoussis C, Kokalj A, Lazzeri M, Martin-Samos L, Marzari N, Mauri F, Mazzarello R, Paolini S, Pasquarello A, Paulatto L, Sbraccia C, Scandolo S, Sclauzero G, Seitsonen A P, Smogunov A, Umari P, Wentzcovitch R M 2009 *Phys.: Cond. Matter* vol. **21,** 395502-19.
[14] Perdew J P, Burke K, Ernzerhof M 1996 *Phys. Rev. Lett.* **77,** 3865.
[15] Kraus W, Nolze G 2000 PowderCell for Windows, version 2.4.
[16] Ivashchenko V, Veprek S, Pogrebnjak A,Postolnyi B 2014 *Sci. Tech. Adv. Mater.* Vol **15,** 025007-11.
[17] Pogrebnjak A D, Bagdasaryan A A, Yakushchenko I V, Beresnev V M 2014 *Rus. Chem. Rev.* Vol **83**, 1027.

IOP Conf. Series: Materials Science and Engineering **81** (2015) 012022 doi:10.1088/1757-899X/81/1/012022

Computer simulation of the single crystal surface modification and analysis at grazing low-energy ion bombardment

F F Umarov[1], A A Dzhurakhalov[2], K M Mukashev[3]

[1]Almaty, 050000, Kazakhstan Kazakh-British Technical University, 59, Tole bi str.
[2]Antwerpen, 2020, Belgium Antwerpen University, 1,Middelheimlaan
[3]Almaty, 050100, Kazakhstan, National Pedagogical University after Abai, 13, Dostyk Avenue

e-mail:farid1945 @yahoo.com

Abstract.

The specific peculiarities of 0.5-10 keV Ne^+, Ar^+, Kr^+, Be^+ and Se+ ions scattering, sputtering and implantation processes at grazing ion bombardment of GaAs (001), Si(001), SiC(001) and Cu(100) surfaces and their possible application for the surface modification and analysis have been studied by computer simulation. Sputtering yields in the primary knock-on recoil atoms regime versus the initial energy of incident ions (E_0 = 0.5-5 keV) and angle of incidence (ψ = 0-30°) counted from a target surface have been calculated. The colliding particles mass ratio influence on the energy losses of scattering particles, sputtering yields and near-surface depth distributions of implanted particles is established. It was shown that in the case of grazing ion bombardment the layer-by-layer sputtering is possible and its optimum are observed within the small angle range of the glancing angles near the threshold sputtering angle. Comparative studies of layer-by-layer sputtering for Si(001) and SiC(001) surfaces versus the initial energy of incident ions as well as an effective sputtering and sputtering threshold are discussed.

Introduction

The ion scattering, sputtering and implantation processes has been the subject of both scientific investigations for a long time and recent rapid developing micro- and nanotechnologies. Processes such as plasma etching and sputter deposition that involve ion bombardment at relatively low (~ 100 eV) ion energies are widely used in semiconductor processing. However, using glancing-angle ion bombardment for surface modification rather than conventional near-normal incidence ions allows expanding the energy range up to ~ 10 keV and has the advantages of reducing damage (such as crater formation) and preferentially removing surface asperities leading to flat surfaces. This is due to the peculiarities of sputtering processes at grazing incidence [1].

Si and SiC crystals have a great importance because of their use in semiconductor technologies. Especially, silicon carbide exhibits a large band gap, a higher break down field, a higher thermal conductivity, and a higher saturation velocity, compared to widely used silicon. Besides, SiC is a promising shielding material in nuclear fusion systems such as limiters in Tokamak devices, where the surface erosion is also important [2, 3].

In [4] atomically clean and flat Si(100) surfaces suitable for nanoscale device fabrication were prepared by wet-chemical etching followed by 0.3-1.5 keV Ar ion sputtering. It was found that wet-chemical etching alone cannot produce a clean and flat Si(100) surface which can be achieved by subsequent 300 eV Ar ion sputtering at room temperature followed by a 700 °C annealing. Application of grazing angles of incidence of ions on the solid surface opens new perspectives in the investigation of composition, structure and topography of real surfaces and their modification and polishing by ion beams. Sputtering yields of crystalline silicon carbide and silicon have been experimentally determined and results have been compared with Monte Carlo simulations for Ne^+, Ar^+ and Xe^+ ion bombardment in the energy range of 0.5-5 keV under 60° sputtering with respect to the surface normal [5]. The simulation results depend strongly on the input parameters which are not well known especially for SiC. The TRIM simulation fits the experimental results very well.

The (001) surface of III-Y semiconductors is one of the most widely used semiconductor surfaces in both homo- and hetero-epitaxial growth for the manufacturing of electronic devices. Implantation of Be and Se ions into GaAs allows to make the acceptor and donor impurities and ultra shallow junctions in this semiconductor. The evolution of surface morphology during ion beam erosion of Si(111) at 500 eV Ar^+ ion bombardment (60° from normal, 0.75 mA/cm^2 collimated beam current) was studied over a temperature range of 500-730 °C [6]. Keeping ion flux, incident angle, and energy fixed, it was found that one-dimensional sputter ripples with wave vector oriented perpendicular to the projected ion beam direction form during sputtering at the lower end of the temperature range. For temperatures above approximately 690° C, growth modes both parallel and perpendicular to the projected ion beam direction contribute to the surface morphological evolution.

Thus, though sputtering and surface modifications of single crystals are widely studied, there are not sufficient data in the case of grazing incidence. In the present paper, grazing ion scattering, sputtering and implantation processes on GaAs (001), Si(001), SiC(001) and Cu(100) surfaces at 0.5-10 keV Ne^+, Ar^+, Kr^+, Be^+ and Se+ ions bombardment have been studied by computer simulation.

Computer simulation method

The theoretical investigation of atomic collision processes in crystals caused by ion irradiation is usually done using computer simulation because real physical conditions (e.g. complicated interatomic interaction potential, surfaces, interfaces, defects) can be taken into account much easier than it is possible by using analytical methods. The simulation used in our calculations to construct the trajectories of the ions or projectile scattered by target atoms is based on the binary collision approximation [7] with two main assumptions: (1) only binary collisions of ions with target atoms or between two target atoms are considered; and (2) the path which a projectile goes between collisions is represented by straight-line segments. For the description of the particle interactions the repulsive Ziegler-Biersack-Littmark potential [8] was used. The inelastic energy losses were regarded as local depending on the impact parameter and included into the scattering kinematics.

Sputtering has been simulated in the primary knock-on regime. Only the primary knock-on recoil (PKR) atoms ejected from first, second and third layers have been considered. The presence of planar potential energy barrier on the surface was taken into account. A parallel, uniform, mono-energetic ion beam impinges on an impact area on a surface of a crystal. The angle of incidence of primary ions ψ was counted from a target surface. It is assumed that the incident beam is of small density, so the ions of the beam do not hit twice at the same place. The number of incident ions is $4x10^4$. Each new particle is incident on a reset, pure surface. The incident ions and the recoil atoms were followed throughout their slowing-down process until their energy falls below a predetermined energy: 25 eV was used for the incident ions, and the surface binding energy was used for the knock-on atoms. In order to consider simultaneous and nearly simultaneous collisions of a particle with the atoms of the adjacent chains, the special procedure proposed in [9] was used. The calculations were performed on the crystals comprising up to 120 atomic layers. The simulations were run with the crystal atoms placed stationary at equilibrium lattice sites. The initial energy of incident ions was varied from 0.5 to 10 keV, an angle of incidence ψ counted from target the surface was 0-30° and an azimuth angle of incidence ξ realized by rotating the target around its normal and counted from the <100> direction was 0-180°.

Results and discussion

In the case of Be^+ and Se^+ ions bombarded the GaAs(100) surface at grazing incidence the ions move in the surface semichannels and channels which is parallel to the surface and transfer the small portions of their energy to a great number of atoms which form of channels and semichannels walls. The μ values for Be^+ ions colliding with Ga and As atoms are equal to 7.74 and 8.31, correspondingly, (μ>1: direct mass ratio) and for Se^+ colliding with Ga and As target atoms - to 0.88 and 0.95, correspondingly, (μ<1: inverse mass ratio). Here, μ is the ratio of target atom mass m_2 to mass of ion m_1. In Figure 1a,b the number of 1keV Be^+ and Se^+ ions implanted into GaAs(100) surface versus the incidence angle Ψ for <110> (a) and <$\bar{1}$10> (b) directions are compared.

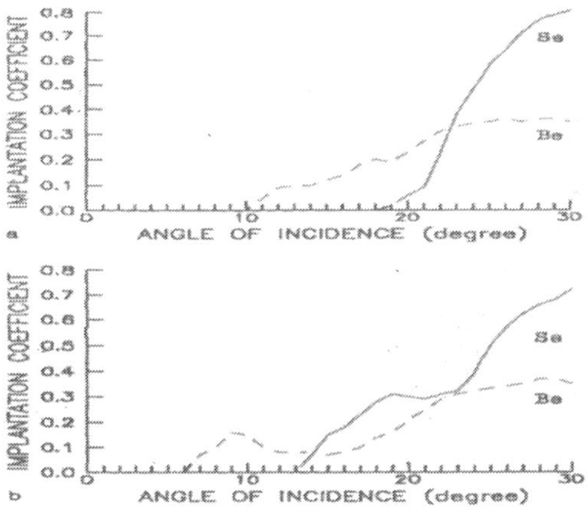

Figure 1. Number of 1keV Be^+ and Se^+ ions implanted into GaAs(100) surface versus the incidence angle Ψ for <110> (a) and <$\bar{1}$10> (b) directions

It is established that some difference of these dependencies is explained by different shapes of surface semichannels in these directions. It is clear that at grazing angles smaller than some critical angle, ion implantation does not take place. The value of the critical angle decreases with mass of incident ion decreasing. The inverse mass ratio of colliding particles in case of Se^+ ions is lead to considerable increase of incidence angle Ψ_{in} from which the ions penetration into GaAs surface layers are begin in comparison with Be^+ ions (for <110> direction: $\Psi_{inBe} = 10^0$ and $\Psi_{inSe} = 18^0$).

The number of implanted Se^+ ions is much greater than the number of Be^+ ions. At the angles of incidence where the minimum of the dependence is observed (Figure 1,b) the incident ions are intensively reflected by surface semichannels because of the ion focusing effect [7].

Figure2 a, b shows implanted depth distributions of 1keV Be^+ (a) and Se^+ (b) ions for <$\bar{1}$10> direction calculated for three incidence angles values. The oscillatory behavior of depth distributions is explained by shape of channel in the <$\bar{1}$10> direction. At grazing incidence the primary beam ions penetrate in several nearest to surface atomic layers during the process of their movement in channels of low index directions along the surface. Therefore, in this case the main peak of the implanted depth distributions is considerable shallow. Note that 1 layer is 0.136nm. The implanted depth profiles depend on the mass ratio of colliding particles, the crystal orientation and the angle of incidence. At $\psi = 30^0$ the depth profile for Se^+ ions is slightly shifted in the deep layers region (Figure 2,b). The comparison of calculated depth distributions for Be^+ and Se^+ ions shows that the range for Se^+ ions is shallower than that for Be^+ ions and the half-width of profile for Se^+ is narrow. Because the small mass and size of Be^+ ions allows them to penetrate into deep layers region.

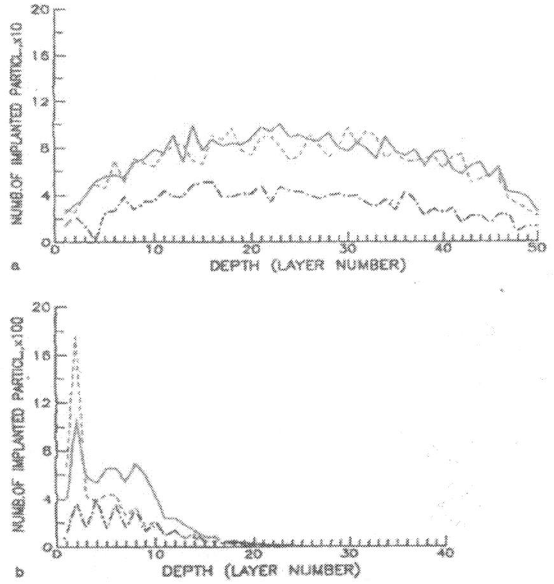

Figure 2. Implanted depth distributions of 1keV Be^+ (a) and Se^+ (b) ions for <$\bar{1}$10> direction at $\psi = 19^0$ (dash-dott), 26^0 (dash) and 30^0 (solid line)

In Figure 3 the depth distributions of 5keV Ar^+ (a) and Kr^+ (b) ions implanted into Cu(001) <110> surface have been presented for three values of angle of incidence. It is clear that the implanted depth distributions of bombarded particles contain a maximum.

With the angle of incidence ψ increasing the maximum of distribution is slightly shifted on the field of deep layers and a half-width of distribution increases. In a case of Kr ions ($\mu \approx 0.76$: inverse mass ratio of colliding particles) for the incidence angle range up to ψ=16^0, the main part of implanted ions is arranged in four-fife layers nearest to the surface. The calculated results show that at ion bombardment of Cu(001) surface along *a* <110> directions the ions are implanted much more deeply due to planar channeling and their depth distributions have not a sharp maximum in the near surface area.

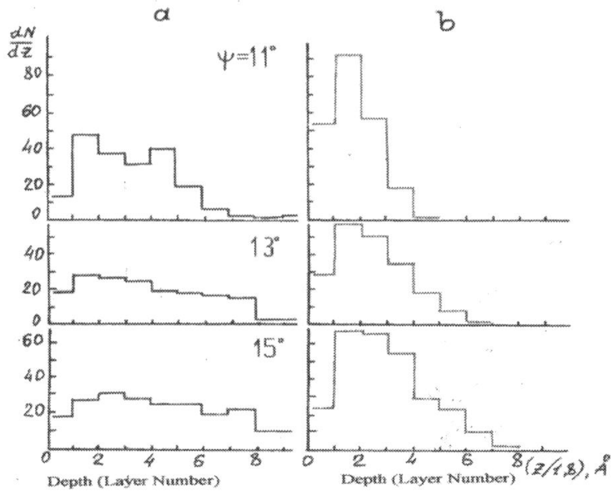

Figure 3. Depth distributions of 5keV (a) Ar^+ and (b) Kr^+ ions implanted into Cu(001) <110> surface at various grazing angles. Note that 1 layer is 1.81Å

In Figure 4a,b the angular dependences of the sputtering yield for Si(001) and SiC(001) surfaces subdivided into sputtering by the first three surface layers at 0.5 keV Ne^+ ion bombardment are compared. Note, the angle of incidence ψ is counted from the surface [10].

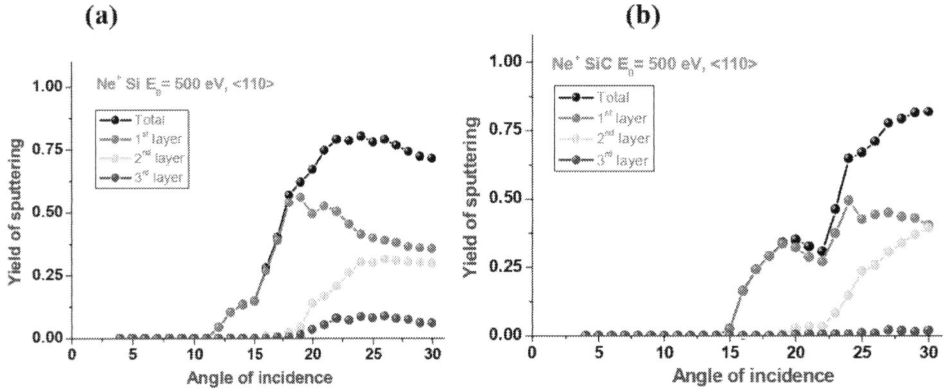

Figure 4. Sputtering yield of Si(001) (a) and SiC(001) (b) versus angle of incidence at Ne^+ ion bombardment

It is seen that there is a threshold angle of sputtering in all dependences. At angles of incidence less than the threshold angle the incident ions cannot penetrate into the crystal and cannot eject target atoms. The threshold angle shifts to the lower values of angle of incidence with increasing the energy of incident ions. At ψ large than a threshold angle, with increasing ψ the number of PKR at first rises and achieves its maximum. There is a plateau (shorter for Si and wider for SiC) near the threshold angle because of insufficient ion energy for both long moving ions within surface semichannels and their penetration to deeper layers. With increasing the initial energy this plateau disappears and the sputtering yield decreases at large ψ. This decreasing of PKR yields is explained by partial penetration of ions in deeper layers and domination of the cascade sputtering mechanism. It is clear that the relative contributions of each layer to the total PKR yield strongly depend on the angle of incidence. In the angular range of ψ = 11-20° for Si and 15-23° for SiC, the sputtering occurs only from the first layer. It is seen that the threshold angle is a bit smaller in the case of Si than for SiC. As results for high initial energy show that, in general the sputtering yield is large in the case of SiC. These dependences allow choosing an angle of incidence for an effective sputtering at given initial energy.

In Figure 5a,b the sputtering yields of Si(001) and SiC(001) surfaces subdivided into sputtering by the first three surface layers versus the energy of incident Ne^+ ions are shown at ψ =10°. It should be noted that the elastic energy losses of the scattered particles are considerably smaller than the inelastic ones in a region of glancing scattering [9].

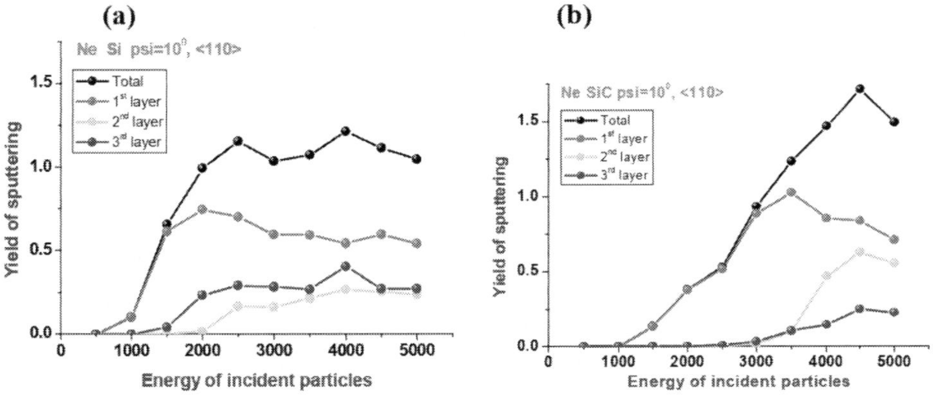

Figure 5. Sputtering yield of Si(001) (a) and SiC(001) (b) versus energy of incident Ne^+ ions at ψ =10°

The threshold energy of sputtering is about 1 keV for these cases. There is more drastic increase of sputtering yield in the beginning of dependences for Si than for SiC. It is seen that the main contribution to the total sputtering comes from the sputtering of the first layer. Moreover, in the energy range of 0.5-1.5 keV for Si and 1-3 keV for SiC, the sputtering occurs only from the first layer. Further increasing of the ion energy results in increasing the contribution from second and third layers.

The contribution to sputtering from the third layer is larger than the one from the second layer as the atomic rows in the second layer lies directly under the one of the first layer in the <110> direction. Two local maxima at 2.5 and 4 keV are observed in the total sputtering yield dependence in the case of Si. Sputtering from the first layer gives a basic contribution to the first maximum while the second maximum is formed by atoms ejected from the second and third layers. In the case of SiC the maximum of total dependence is formed by atoms ejected from the second layer.

These results show that by choosing an angle and an energy of incidence one can produce layer-by-layer sputtering of Si(001) and SiC(001) surfaces. For the realization of the layer-by-layer sputtering mechanism it is nessesary that a part of the ion energy, corresponding to the normal components of its velocity, will be lower than a threshold of sputtering of atoms of a layer, next to the surface one, i.e. $E_i \sin^2\psi_i < E_d$, where E_i is the ion energy before the ith collision, ψ_i is the angle between the ion movement direction and the semichannel axis before the ith collision, E_d – the energy of displacement of atoms of a second layer (in the case considered the bottom chain of semichannel).

In Figure 6 the azimuthal angular dependences of the sputtering yield are compared for Si(001) and SiC(001) surfaces at 5 keV Ne^+ ion bombardment, $\psi = 10°$. Main maxima and minima of dependences are observed in low crystallographic directions and near them. They are caused by the existence of original semichannels and channels in these directions. Thus, there is a good correlation between the sputtering yield dependences and crystallographic structures of studied crystals.

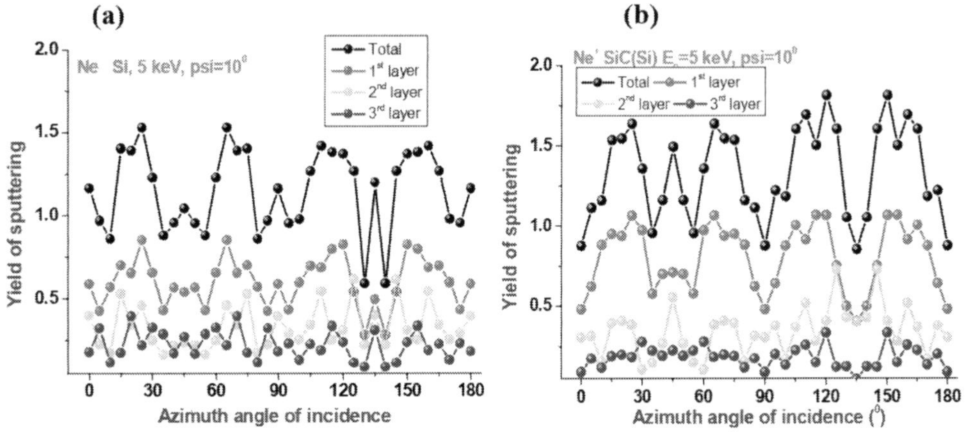

Figure 6. Sputtering yield of Si(001) and SiC(001) versus azimuth angle of incidence at $E_0 = 5$ keV, $\psi = 10°$.

From the comparison of the two curves it is seen that at some values of azimuth angle, instead of the peaks of sputtering yield of Si(001) the minima of sputtering yield of SiC(001) are observed. This difference is caused by differences of binding energies, lattice parameters and, of course, compositions of these single crystals. In most range of the azimuth angle of incidence the sputtering yield for SiC is larger than the one for Si.

Conclusion

It was shown that in case of Be^+ and Se^+ ions implantation into GaAs(001) the main peak of the implanted depth distributions is considerable shallow, the range for Se is shallower and the half-width of profile for Se^+ is narrow than that for Be^+ ions. The obtained results allow to selecting the optimum for implanted depth distributions with demanded shape in narrow near-surface area (5-10 atomic layers) of crystals production.

At sufficiently small grazing angles ($\psi < 9^0$) the sputtering processes do not take place due to impossibility of ion penetration through surface in the case of perfect surface. In the angular range $9^0 \leq \psi \leq 15^0$ the monotonous increasing of the total sputtering yield is observed.The peculiarities of formation of PKR atoms at grazing ion incidence beam on an atomically smooth surface of a single crystal promotes its layer-by-layer sputtering. In these conditions it is possible to achieve successive removal of layers without disturbance of the next layer at removal of the previous one. Ion bombardment at grazing angles reduces considerably the influence of effect of crater walls and ion mixing on results of layer-by-layer analysis and increase its accuracy and sensitivity.

IOP Conf. Series: Materials Science and Engineering **81** (2015) 012022 doi:10.1088/1757-899X/81/1/012022

References

[1] Umarov F. F., Dzhurakhalov A. A., Teshabaeva N. A. 1998 Formation of recoil atoms and layer-by-layer sputtering of the single crystal surface under ion bombardment at grazing angles *Appl. Surf. Sci.* vol. **125** pp. 221-226.

[2] Roth J., Bohdansky J., Poschenrieder W., Sinha M. K. 1976 Chemical Effects in Sputtering *J. Nucl. Mater.* vol. **63** pp. 213-222.

[3] Bischoff L., Teichert J., Heera V. 2001 Focused Ion Beam Sputtering Investi gations on SiC *Appl. Surf. Sci.* vol. **184**, pp. 372-376.

[4] Kim J. C., Ji J.-Y., Kline J.S., Tucker J.R., Shen T.C. 2003 Preparation of atomically clean and flat Si(100) surfaces by low-energy ion sputtering and low-temperature annealing *Appl. Surf. Sci.* vol. **220** pp. 293-298.

[5] Ecke G., Kosiba R., Kharlamov V., Trushin Yu., Pezoldt J. 2002 The estimation of sputtering yields for SiC and Si .*Nucl. Instr. and Meth. in Phys. Res.* vol. **B196** pp. 39-50.

[6] Brown A., George H. B., Aziz M. J., Erlebacher J.: 2004, *Mat. Res. Soc. Symp. Proc.*, vol. **792**, R7.8.1.

[7] Parilis E. S., Kishinevsky L. M., Turaev N. Yu., Baklitzky B. E., Umarov F. F., Verlrger V. Kh., Nizhnaya S. L., Bitensky I. S.: 1993, Atomic collisions on Solid Surfaces, North-Holland, Amsterdam 664p..

[8] O'Connor D. J., Biersack J. P.: 1986, Comparison of theoretical and empirical interatomic potentials *Nucl. Instr. and Meth. in Phys. Res.* vol. **B15**, pp. 14-19.

[9] Robinson, M. T. and Torrens, I. M.: 1974, Computer simulation of atomic displacement cascades in solids in the binary collision approximation *Phys. Rev.*, vol. **B9**, pp. 5008-5024.

[10] Umarov F.F. and Dzhurakhalov A.A.: 2013, Computer simulation of layer-by-layer sputtering at grazing low-energy ion-surface interactions *Surface and Interface Analysis*, **46**, pp. 83-86.

Influence of absorbed energy distribution along beam radius on ignition threshold of condensed explosives

A V Khaneft[1,2], G A Ivanov[1] and E V Duginov[3]

[1]Kemerovo State University, Krasnaya, 6, Kemerovo, Russia
[2]National Research Tomsk Polytechnic University, 30, Lenin Avenue, Tomsk, Russia
[3]Kemerovo State Agricultural Institute, Markovtseva, 5, Kemerovo, Russia

E-mail: avkhaneft@mail.ru

Abstract. The criterion of the condensed explosive ignition by the electron beam, which takes into account Gaussian distribution of electron density along the beam radius, has been obtained. It has been shown that radial heat removal of the absorbed energy leads to the increase in critical ignition energy if the effective track length of electrons in a solid is commensurate with the beam radius. The critical energy of PETN initiation by the electron beam has been calculated.

1. Introduction

The criterion of condensed explosive ignition by the short-duration electron beam was determined in the article [1]. The criterion was obtained on conditions that the rate of the chemical reaction on the border of the reaction layer is e times as small as the rate of the chemical reaction at the maximum of electron beam absorption, and the heat gain rate as a result of the chemical reaction in the solid Q_+ is to exceed the rate of heat removal from the reaction layer deep into the solid Q_- [2]. The obtained criterion didn't take into account the influence of the electron beam radius on the ignition threshold and the melting of the explosives. The track length of the explosives in the solid is known to depend on the initial kinetic energy of electrons. So, to estimate the critical energy of explosive ignition by the electron beam it is necessary to know at what electron energy the track length of electrons is commensurate with the beam radius as in this case radial heat removal from the absorption area needs to be taken into consideration [3]. The object of the current paper is to obtain the criterion of explosive ignition by the electron beam, which would take account of the melting of explosives and the beam radius, and the estimation of kinetic energy of electrons for PETN ($C_5H_8N_4O_{12}$) when the effective length of electrons is commensurate with the beam radius.

2. Problem statement

Let us analyze the following model. We consider the short-duration pulsed electron beam to fall normally on the surface of the solid parallel to axis z. As the short duration of the pulse we understand the duration when the diffusion of the heat from the area of electron absorption in the solid during the pulse duration can be neglected. We also assume the distribution of intensiveness along the beam cross-section to have Gauss shape [4]. In this case the temperature distribution along the crystal at the pulse decay is to be defined by the distribution of the absorbed energy:

$$T(\xi, r) = T_0 + \frac{W}{c\rho R_{ef}} \Lambda(\xi) \exp\left(-\frac{r^2}{r_0^2}\right), \tag{1}$$

where T_0, T – the initial and current temperatures of the sample; W – the density of absorbed energy of the electron beam; r – the distance from the beam axis; r_0 – the beam radius; c, ρ – the heat capacity and the density of the sample; R_{ef} – the effective track length of the electrons; $\Lambda(\xi)$ – the distribution of the density of the absorbed beam energy along the thickness of the crystal; $\xi = z / R_{ef}$ – the dimensionless coordinate.

3. The definition of the criterion of explosive ignition by the electron beam with Gaussian distribution of energy along the beam cross-section

Based on the condition that as the result of the chemical energy in the solid, the rate of heat gain on the border of the reaction volume is equal to the rate of heat removal from the reaction layer deep into the explosive $Q_+ = Q_-$, according to paper [3], the heat balance equation is:

$$\pi r_1^2 \Delta z_1 \rho Q Z \exp\left(-\frac{E}{RT_m}\right) = -\lambda \pi r_1^2 \left(\frac{\partial T_1}{\partial z}\right)\Bigg|_{\substack{r=0 \\ z=z_1}} - 2\lambda \pi r_1 \Delta z_1 \left(\frac{\partial T_1}{\partial r}\right)\Bigg|_{\substack{z=z_m \\ r=r_1}}, \tag{2}$$

where T_1 – the temperature on the border of the reaction volume; Δz_1 – the thickness of the reaction volume; r_1 – the radius of the reaction volume; Q – the thermal effect of the reaction on the mass unit of the explosive; Z – the frequency factor; E – the energy of the reaction rate activation; R – the gas constant.

Fig. 1*a* and Fig. 1 *b* illustrate the qualitative dependence of normal (*a*) and radial (*b*) distribution of the density of the absorbed beam energy in the sample. The thickness of the reaction volume, according to Fig. 1 *a*, is $\Delta z_1 = |z_1 - z_m|$, and it is defined by the equation [1]:

$$\Delta z_1 = R_{ef} \left(-\frac{2RT_m T_1 \Lambda_m}{E \Delta T_m \Lambda''_m}\right)^{1/2}, \tag{3}$$

where $\Lambda''_m = \partial^2 \Lambda_m / \partial \xi^2$ – the second derivative of $\Lambda(\xi)$ at $\xi = \xi_m$ ($\Lambda''_m < 0$).

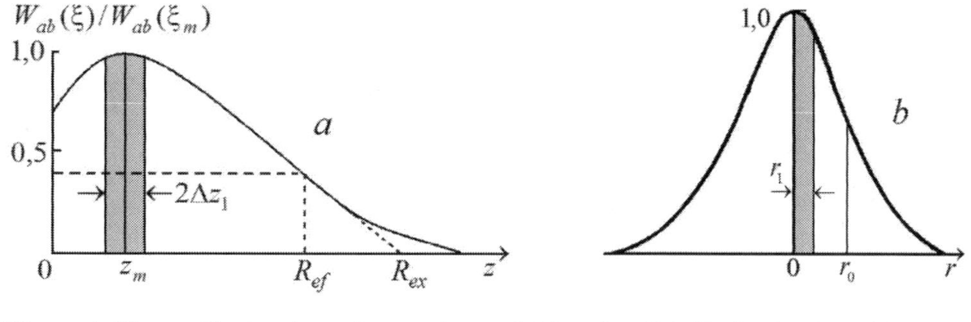

Figure 1. The qualitative dependence of normal (*a*) and radial (*b*) distribution of the density of the absorbed beam energy in the sample.

The temperature on the border of the reaction volume, defined on the condition that $Q_+(0)/Q_+(z_1) = e$, is [2]:

$$T_1 = \frac{T_m}{1 + RT_m / E}. \tag{4}$$

The distribution of the temperature along the sample at the pulse decay is defined by the expression

$$T(\xi, r) = T_0 + \frac{W}{c\rho R_{ef}} \Lambda(\xi) \exp\left(-\frac{r^2}{r_0^2}\right). \quad (5)$$

According to paper [1], the first derivative on the right side of the formula (2) is

$$\left.\left(\frac{\partial T_1}{\partial z}\right)\right|_{\substack{r=0 \\ z=z_1}} = \frac{W}{c\rho R_{ef}} \frac{\Delta z_1}{R_{ef}^2} \Lambda''_m = \frac{\Delta T_m}{\Lambda_m} \frac{\Delta z_1}{R_{ef}^2} \Lambda''_m . \quad (6)$$

It is assumed that

$$\Delta T_m = T_m - T_0 = \frac{W}{c\rho R_{ef}} \Lambda_m . \quad (7)$$

Let us define the radial derivative on the right side of the formula (2). From expression (5) and (7) we have:

$$\left.\left(\frac{\partial T}{\partial r}\right)\right|_{z=z_m} = -\frac{2r}{r_0^2} \frac{W}{c\rho R_{ef}} \Lambda_m \exp\left(-\frac{r^2}{r_0^2}\right) = -\frac{2r}{r_0^2} \Delta T_m \Lambda_m \exp\left(-\frac{r^2}{r_0^2}\right).$$

Hence,

$$\left.\left(\frac{\partial T_1}{\partial r}\right)\right|_{\substack{z=z_m \\ r=r_1}} = -\frac{2r_1}{r_0^2} \Delta T_m \Lambda_m \exp\left(-\frac{r_1^2}{r_0^2}\right). \quad (8)$$

Let us define the radius of the reaction volume. According to (5) and (7), the expression for the temperature T_1 on the perimeter of the reaction volume at $\xi = \xi_m$ and $r = r_1$ is given by

$$T_1(\xi_m, r_1) = T_0 + \Delta T_m \exp\left(-\frac{r_1^2}{r_0^2}\right). \quad (9)$$

According to (9), the expression for the radius of the reaction volume is

$$r_1 = r_0 \ln\left(\frac{\Delta T_m}{T_1 - T_0}\right)^{1/2} = r_0 \ln\left(\frac{1+\gamma}{1 - \gamma T_0 / \Delta T_m}\right)^{1/2} . \quad (10)$$

By substituting the derivatives (6) and (8) into the expression (2) and taking the account of the expression (10) for r_1, after some transformation we get the criterion of explosive ignition by the pulsed electron beam, which includes the correction for radial heat removal from the area of electron absorption:

$$\rho Q Z \exp\left(-\frac{E}{RT_m}\right) = \frac{\lambda \Delta T_m}{R_{ef}^2} \left[-\frac{1}{\Lambda_m} \frac{\partial^2 \Lambda_m}{\partial \xi^2} + \frac{4}{F(\gamma)} \frac{R_{ef}^2}{r_0^2}\right]. \quad (11)$$

where $\gamma = RT_m / E$, and the function $F(\gamma) = (1+\gamma)/(1 - \gamma T_0 / \Delta T_m)$. It should be noted, that due to the scattering the beam radius in the solid is to depend on z coordinate. That's why, the obtained criterion of explosive ignition by the electron bean, depending on the beam radius, is approximate.

The critical energy density for explosive ignition by the electron beam, taking into account the melting heat, is defined, based on the energy conservation law, by the formula:

$$W^* = \rho R_{ef} (c \Delta T_m + H_f). \quad (12)$$

4. The estimation of the effective track length of electrons

Major energy loses at the beam deceleration in the solid are ionizing. The curve of energy losses in the solid on the unit of track is normalized to the energy losses on the unit of track with energy E_0, where E_0 is the initial kinetic energy of electrons.

The experimental curve, which reflects the distribution of absorbed energy density by the solid $W_{ab}(z)$, is often approximated by the cubic [5-6]:

$$W_{ab}(x) = f(z/R_{ex})W, \; f(z/R_{ex}) = a + b(z/R_{ex}) + c(z/R_{ex})^2 + d(z/R_{ex})^3. \tag{13}$$

where R_{ex} – the extrapolated track length of the electrons; a, b, c – adjustable coefficients, which are defined on the condition of the best description of the experimental curve of the absorbed energy distribution. The function $f(z/R_{ex})$ has the maximum ($f_{\max} > 1$), and the integral is

$$\int_0^{R_{ex}} f(z/R_{ex})dz = 1.$$

If the experimental curve of the absorbed energy distribution is normalized to the value of the density of the energy, absorbed at the maximum $W_{ab}(z_m)$, the expression (13) can be transformed as

$$\frac{f(z/R_{ex})}{R_{ex}} = \frac{f_{\max}}{R_{ex}}\left[a' + b'\left(\frac{z}{R_{ex}}\right) + c'\left(\frac{z}{R_{ex}}\right)^2 + d'\left(\frac{z}{R_{ex}}\right)^3\right], \tag{14}$$

where $a = a'f_{\max}$, $b = b'f_{\max}$, $c = c'f_{\max}$, $d = d'f_{\max}$. Assume that $R_{ex}/f_{\max} = R_{ef}$. According to the agreed notations, the formula (14) goes as

$$\frac{f(z/R_{ex})}{R_{ex}} = \frac{1}{R_{ef}}\left[a' + \frac{b'}{f_{\max}}\left(\frac{z}{R_{ef}}\right) + \frac{c'}{f_{\max}^2}\left(\frac{z}{R_{ef}}\right)^2 + \frac{d'}{f_{\max}^3}\left(\frac{z}{R_{ef}}\right)^3\right].$$

We write the function in brackets in dimensionless coordinates:

$$W_{ab}(\xi)/W_{ab}(\xi_m) = \Lambda(\xi) = a'' + b''\xi + c''\xi^2 + d''\xi^3.$$

Where constants $a' = a''$, $b'' = b'/f_{\max}$, $c'' = c'/f_{\max}^2$, $d'' = d'/f_{\max}^3$.

Let us calculate the effective track length of electrons for PETN, depending on the kinetic energy of electrons. PETN is known to be a wide-band gap insulator [7]. The empiric formula for the evaluation of the extrapolated track length of electrons in explosives is given in paper [8]:

$$R_{ex} = \frac{a_1}{\rho}\left[\frac{1}{a_2}\ln(1 + a_2 e_0) - \frac{a_3 e_0}{1 + a_4 e_0^{a_5}}\right], \tag{15}$$

Here, ρ – the medium density, g/cm^3; $e_0 = E_0/mc^2$ – the medium relative kinetic energy of electron in the beam of high-current accelerator (mc^2 is the rest energy of the electron, equal to 511 keV); a_i – constants, depending on the effective atomic mass of the substance A_{ef} and the effective atomic number Z_{ef}. Constants a_i are calculated by the empiric formulae: $a_1 = 0{,}2335 A_{ef}/Z_{ef}^{1{,}209}$, $a_2 = 1{,}78 \cdot 10^{-4} Z_{ef}$, $a_3 = 0{,}989 - 3 \cdot 10^{-4} Z_{ef}$, $a_4 = 1{,}468 - 1{,}18 \cdot 10^{-2} Z_{ef}$, $a_5 = 1{,}232/Z_{ef}^{1{,}209}$.

The effective atomic mass and the effective atomic number are defined by the following formulae

$$Z_{ef} = \sum_{i=1}^{n} f_i Z_i, \; A_{ef} = \frac{Z_{ef}}{(Z/A)_{ef}}, \; (Z/A)_{ef} = \sum_{i=1}^{n}(f_i Z_i / A_i), \; f_i = A_i / \sum_{i=1}^{n} A_i.$$

Here, Z_i, A_i – the atomic number and the atomic mass of the i element; f_i – its weight fraction; n – the number of weight fractions.

The experimental curve of distribution of the absorbed energy for PETN [9] is satisfactorily described by the expression [10-12]:

$$W_{ab}(\xi)/W_{ab}(\xi_m) = \Lambda(\xi) = 0{,}7 + 1{,}57\xi - 2.31\xi^2 + 0{,}61\xi^3. \tag{16}$$

For PETN the electron track at the electron energy $E_0 = 250\,\text{keV}$ is $R_{ef} = 173{,}6 \cdot 10^{-4}$ cm, and $R_{ex} = 1{,}44 R_{ef} = 250 \cdot 10^{-4}$ cm. The evaluation of R_{ef} for PETN, using the formula (15), at $\xi_{ex} = 1{,}44$ showed, that the calculated value of R_{ef} is approximately 14% smaller than R_{ef}, calculated on the experiment. That's why, the calculated values of R_{ef} for PETN at $E_0 = 450$, 600 and 1000 keV were increased by 14%, respectively. The calculation results are given in the table.

It should be noted that in reality the spectrum of electrons in the beam of a high-current accelerator is not monochromatic. There are electrons with energy, ranging from tens of keV to the value, corresponding to the amplitude of accelerating voltage [13], in the energy spectrum. So, the extrapolated track length of electrons is to be defined by the most high-energy electrons, and not by the medium kinetic energy. Naturally, the obtained criterion of condensed explosive ignition (11) doesn't take the account of nonmonochromaticity of the electrons in the beam of the high-current accelerator.

Table 1. The dependence of the effective track length of the electrons in PETN on the kinetic energy of electrons

E_0 (keV)	250	450	600	1000
$R_{ef}, 10^{-4}$ (cm)	173,4	414,7	624,6	1247

5. Calculation results

The calculations of the critical energy of PETN ignition by the electron beam were carried out at the following thermophysical and kinetic parameters: $c = 1255{,}2$ J/(kg·K), $\rho = 1{,}77 \cdot 10^3$ kg/m^3, $\lambda = 0{,}2508$ W/(m·K), $H_f = 193$ kJ/kg, $E = 196{,}6$ kJ/mol, $Z = 6{,}3 \cdot 10^{19}$ s^{-1}, $Q = 1{,}26$ MJ/kg [14-16]. Fig. 2 *a*. shows the results of calculations for the critical density of the electron beam energy W^* for PETN, depending on the beam radius r_0 and the initial kinetic energy of electrons. According to the calculation data, presented in fig. 2 *a*, radial heat removal is to influence the critical energy density of explosive ignition by the electron beam only if the electron energy is $E_0 \sim 1000$ keV and the beam radius is $r_0 \sim 1$ mm. Fig. 2 *b* illustrates the calculation results for the critical energy density, determined using the formulae (11) and (12) depending on the initial electron energy in the pulse.

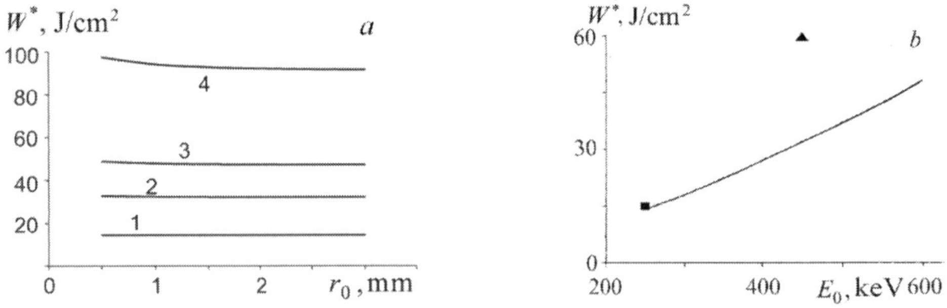

Figure 2. The dependence of critical energy density on the beam radius at different values of the beam energy for PETN: 1 – $E_0 = 250$; 2 – $E_0 = 450$; 3 – $E_0 = 600$; 4 – $E_0 = 1000$ keV (*a*) and the dependence of the critical density of the absorbed electron energy while PETN initiation on the initial electron energy in the beam: line – the calculation on the criterion (11) at $r_0 = 1$ mm; ■ – the experiment [9]; ▲ – the threshold energy of PETN detonation [17] (*b*).

As it is shown in fig. 2 *b*, the experimental value of the energy density of PETN initiation by the electron beam at $E_0 = 250$ keV [9] is in good agreement with the calculation. The density of the electron beam energy W_D ($E_0 = 450$ keV), at which the detonation of the explosive takes place [17], exceeds twice the critical value. In general, the calculation values of the density of the electron beam energy for PETN initiation by the electron pulse agree with and the experimental data quite satisfactorily, which proves the thermal mechanism of PETN initiation by a nanosecond electron beam.

The research has been carried out within the framework of the government contract № 3.1235.2014 K in the domain of science.

6. References

[1] Khaneft A V 1998 Criteria for ignition of condensed materials by an electron pulse *Chem. Phys. Reports*. **17**(8). 1573

[2] Kovalskii A A, Khlevnoi S S and Mikheev V F 1967 The ignition of ballistite powders *Combust., Expl., Shock Waves*. **3**(4). 323

[3] Khaneft A V 1998 Effect of light distribiution in the laser beam on the critical ignition energy of condensed materials *Chem. Phys. Reports*. **17**(10). 1889

[4] Morozov V A, Savencov G G, Bragin V A, Kats V M and Lukin A A 2012 Problems in initiating detonation of disruptive explosives by a high-intensity electron beam *Technical Physics* **57**(5). 706

[5] Harbeke G 1985 *Polycrystalline semiconductors. Physical Properties and Applications* (Springer-Verlag, Berlin, Heidelberg, New York, Tokyo)

[6] Shiller Z, Gayzich U and Pancer Z 1980 *Electronno-luchevaja tecnologija* (Moscow: Energija)

[7] Zhurova E A, Stash A I, Tsirelson V G 2006 Atoms-in Molecules Study of Intra- and Intermolelecular Bonding in the Pentaerythritol tetranitrate Crystal *J. Am. Chem. Soc.* **128**. 14728

[8] Tabata T and Ito R 1974 An algoritm for the energy deposition by fast electrons *Nuclear Science and Engineering*. **53**. 226

[9] Aduev B P, Belokurov G M, Grechin S S and Shvaiko V N 2007 Investigation of the early stages of explosive decomposition of PETN crystals initiated by pulsed electron beams *Izv. Vyssh. Uchebn.* Zaved., *Fiz*. **50**(2). 3

[10] Khaneft A V, Duginov E V and Ivanov G A 2012 Modeling of PETN initiation by an electron beam with nanosecond duration *Khimicheskay Fizika I Mezoskopiya*. **14**(1). 28

[11] Khaneft A V, Dolgachev V A, Duginov E V and Ivanov G A 2013 Criteria of energetic materials ignition by short laser and electron pulses *Bulletin of Kemerovo State University* **3**(55). 31

[12] Ivanov G A and Khaneft A V 2013 Radiation-thermal mechanism of initiation of PETN in the region of absorption of a nanosecond-duration electron beam *Russian Journal of Physical Chemistry B*. 2013. **7**(6). 741

[13] Kulikov V D 2004 Electrification of high-resistance materials in high-power pulsed radiation fields (Tomsk: TGU)

[14] Orlenko L P 2002 *Explosion Physics* (Moscow: Nauka)

[15] Borisov A 1981 *Detonation and Explosives* (Moscow: Mir)

[16] Baum F A, Derzhavec A S and Sanasaryan N N 1969 *Thermoresistant explosives and their effect in deep wells* (Moscow: Nedra)

[17] Oleshko V I, Korepanov V I, Lisitsyn V M, Skripin A S and Tsipilev V P 2012 Electric breakdown and explosive decomposition of PETN monocrystals initiated by electron beam *Technical physics letters*. **38**(5). 415

Spectral and kinetic characteristics of the luminescence center in LiF-WO3 and ZnWO4 crystals

L A Lisitsyna[1], I A Tupitsyna[2] and L N Trefilova[3]
[1]Tomsk State University of Architecture and Building, Tomsk, Russia
[2] Institute for Scintillation Materials, Kharkov, Ukraine
[3]National University of Civil Protection of Ukraine, Kharkov, Ukraine

E-mail: lisitsyna@mail.ru

Abstract The results of the comparative analysis of the spectral and kinetic characteristics of the luminescence centers in the non-doped $ZnWO_4$ crystals and in crystals LiF doped with tungsten oxide. It was found that the intrinsic characteristics of the scintillation transition at 2.6–2.7 eV in these crystals do not depend on the energy and symmetry of the crystal lattice, have comparable values of the light output and can be attributed to the luminescence centers of the same impurity nature. The crystal lattice determines the efficiency of the creation and accumulation of radiation defects, presence of which leads to a distortion of the spectral and kinetic characteristics of scintillation.

1. Introduction

Scintillator materials based on oxygen compounds are currently widely used as laser materials, as radiation detectors in medical diagnosis and customs control. Self-radioactivity of these materials is low, the light yield is high, and they are nonhygroscopic and chemically inert. Since the density of these materials is high, they can be used as detectors to conduct research in high energy physics.

Scintillating materials based on oxygen compounds may be divided into two types: scintillators of a doped type and natural scintillators. The radiative characteristics of scintillators of a doped type are determined by the type of the ion-dopant (occupancy of its electron shells). They can be independent of the local symmetry of the dopant position, the type and energetics of the crystal lattice (e.g. intracenter transition in Ce ion in Y_2SiO_5, LuBO, and $LiLuSiO_4$). The role of the matrix is to transfer energy to the dopant.

In natural scintillators (e.g. Y_2O_3, BeO, tungstates and molybdates of Group II metals [1]) the luminescence centers are matrix forming ions or anionic complexes, the properties of emission transitions in which depend on the symmetry and energy parameters of the crystal lattice.

In this paper, the objects under study were oxygen-containing $ZnWO_4$ crystals and LiF crystals doped with tungsten oxide (LiF-WO_3). According to [1-6], $ZnWO_4$ refers to natural scintillator and the luminescence excited in it at 2.6 eV is considered to be intrinsic, caused by self-trapped excitons in oxyanion WO_6 complex as a component of the crystal lattice. In unbroken wolframite lattice, each W^{6+} ion is in the center of an asymmetric octahedron formed from six O^{2-} oxygen ions due to three different W-O distances (1.79, 1.91 and 2.13 Å)).

LiF crystal is the scintillator of a doped type. It luminesces when being doped with tungsten oxide, in particular. Therefore, in both crystals, the radiative complex includes a tungsten ion and an oxygen ion. The comparative analysis of the luminescence parameters of the two crystals will help to understand the structure of the luminescence centers, to find out the effect of the lattice type on emission characteristics, to determine the range of doses of ionizing radiation to use the crystals as detectors of ionizing radiation.

2.Object and Methods

The investigated crystals were grown at the Institute for Scintillation Materials, NAS of Ukraine (Kharkiv). LiF crystal doped with tungsten oxide was grown by automated Kyropoulos method in nitrogen atmosphere in the form of a boule with a diameter of 240 mm and a height of 400 mm. The concentration of tungsten measured spectrophotometrically was varied from top to bottom of the boule in the range of 0.01–0.5 wt%. $ZnWO_4$ crystals were grown by Czochralski method from a mixture prepared by solid-phase synthesis of ZnO oxides (99. 995%) and WO_3 oxides (99.995%).

The initial imperfection of the crystals was evaluated by the absorption spectra in the range of 13–0.1 eV. In the range (6–13 eV), absorption spectra were measured in STIMS SC "S.I. Vavilov State Optical Institute" (St. Petersburg) with a vacuum monochromator-based spectrometer at 300 K. In the VUV region, the absorption spectra of LiF crystals doped with WO_3 contained bands at 6.2, 7.3, 8.9, 11.2 eV. The same spectrum could be observed in LiF-O crystals (figure 1), in which these bands are attributed to transitions in oxygen ions in different complexes: OH^-, O^{2-}, O_2^-, O^{2-}-V_a [7–11]. The intensity of these bands in LiF-WO_3 crystals increases as the amount of the doped oxide increases. At a concentration of more than 0.2 W wt%, the crystals were not transparent above 5 eV.

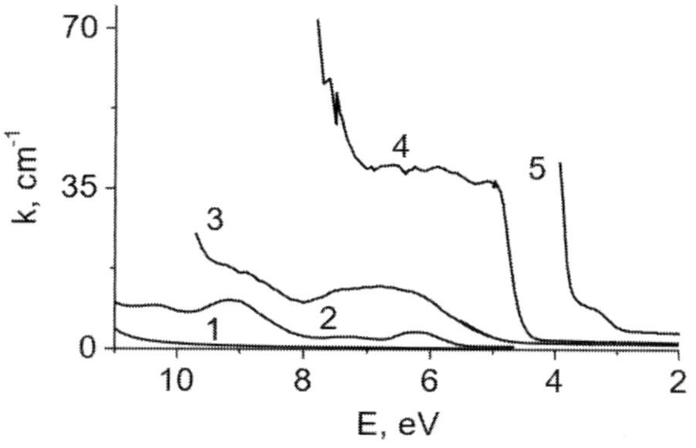

Figure 1. The absorption spectra of non-doped crystal LiF (1), LiF-Li_2O (2), LiF-WO_3, containing of W 0.1 (3) and 0.31 wt % (4) and $ZnWO_4$ crystal (5).

The non-transparency of as-grown $ZnWO_4$ crystals in the region of E> 4 eV (figure 1) is assumed [2–6] to be due to intrinsic absorption. However, this assumption does not agree with the evaluation of the band gap from 4.6 eV to 5.8 according to various authors [2,12,13]. This is one of the reasons why the authors of an alternative point of view [14–16] attribute the non-transparency of $ZnWO_4$ crystals above 4 eV to high levels of initial imperfection of the material and presence of impurity absorption states at the edge of the band gap.

In the range (0.5-0.1) eV, the absorption spectra were measured in Institute for Scintillation Materials (Kharkov) on a Fourier spectrometer "Spectrum One". IR absorption spectrum of crystals LiF-WO_3 contains two bands at 3728 and 3620 cm^{-1} caused by free ions OH^- and complexes Me-nOH^- respectively. The appearance of a proton (H^+) as an interstitial probably is due to the presence of moisture in the atmosphere of growth.

Scintillation properties of crystals (pulsed photoluminescence (PL) and cathodoluminescence (CL)) were investigated using a measurement technique with nanosecond time resolution in the range of 10^{-8} –10^{-1} s after the end of pulsed excitation. The intensity of the luminescence was detected with monochromator MDR-206 and photomultiplier PEM-97. The spectra were recalculated for PMT sensitivity.

Pulsed CL was excited by electron pulse with the parameters: pulse duration of 10 ns; the average electron energy of 250 keV. The dose absorbed by the crystal for the duration of a single electron pulse was equal to 5.10^1 Gy.

Pulsed PL was studied under excitation by the fourth harmonic of Nd-GNY (4.66 eV) laser with a pulse duration of 5 ns and an energy of 40 mJ per pulse. The excitation spectra of the luminescence at 2.6 eV were measured in a steady state in the range of the photon energy of 3–6.2 eV with spectrofluorimeter CM-2203. The luminescent properties of the crystals were studied before irradiation and after irradiation by electron beams in the range of absorbed doses of 0–10^5 Gy and temperature range of 15–300 K. To clarify the effect of the surface on the luminescence we analyzed the emission spectra of crystals of different prehistory: crystals freshly cleaved from single block and those stored for 10 hours at 900 C in oxygen atmosphere.

3.Results and discussion

In crystals of both types (LiF-WO_3 and $ZnWO_4$) the luminescence is excited by both helium-neon laser (4.66 eV), and electron beams. We have established that the parameter of the luminescence decay depends on the radiation wavelength. Figure 2 shows an example of the kinetics of the luminescence decay for $ZnWO_4$ crystal. A short decay component only can be clearly seen in the UV region of the spectrum. Its contribution decreases as the wavelength of the emitted energy grows, and the

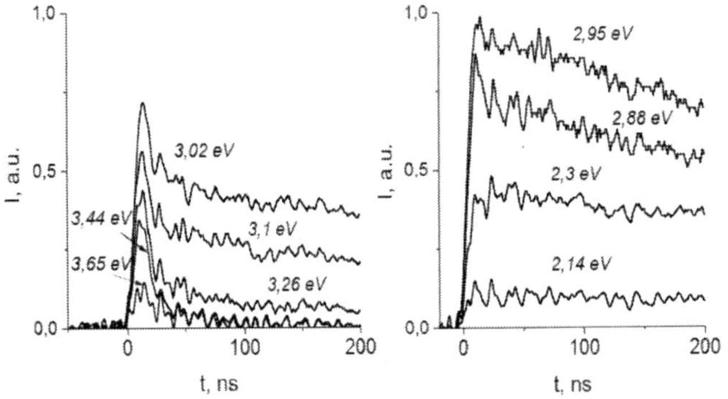

Figure 2. Decay of CL kinetics vs. different values of radiation energy for $ZnWO_4$ crystal at 300 K.

contribution of the long component decay increases. In the region of energies <2.95 eV (to the right), a short luminescence decay component is not observed in the decay kinetics.

The spectra of the decay components in crystals of both types ($ZnWO_4$ and LiF-WO_3) are a band at 3.6 eV with the decay parameter of τ<20 ns at 300 K, and the band in the region of 2.6–2.7 eV with decay parameter of 30 μs at 300 K. Figure 3 shows an example of the spectrum of pulsed PL crystal for $ZnWO_4$.

The results obtained in the study of the properties of the bands in the luminescence spectra for both crystals are as follows.

•The short-lived band at 3.6 eV is present in all $ZnWO_4$ crystals, and its intensity increases during crystal storage in air at 300 K, and when heated at 900 C in oxygen atmosphere.

As we showed in [17] the band at 3.6 eV is not found in undoped LiF crystals grown by Stockbarger in fluorinating atmosphere, and it is observed in all LiF crystals doped with different oxides: Li_2O, WO_3, TiO_2, and Fe_2O_3. Neither the spectral position of this band nor the lifetime (<20 ns) of the corresponding luminescence center depends on the type of cation-co-dopant.

•The ratio of the intensities of the bands at 3.6 and 2.6 eV (J(3.6)/J(2.6)) in both LiF-WO_3 and $ZnWO_4$ crystals depends on the conditions of luminescence excitation. In $ZnWO_4$ crystal this ratio is two orders of magnitude greater under excitation of the luminescence by photon flux with the energy of 4.66 eV than under excitation of luminescence in the same crystal by electron beams. The penetration depth of 4.66 eV photons in the crystal is less than 100 nm, while the depth of electron penetration into the crystal is equal to 200 μm. Therefore, the reduction of the ratio (J(3.6)/J(2.6)) with the increase of the excitation radiation penetration depth in the same crystal indicates that centers

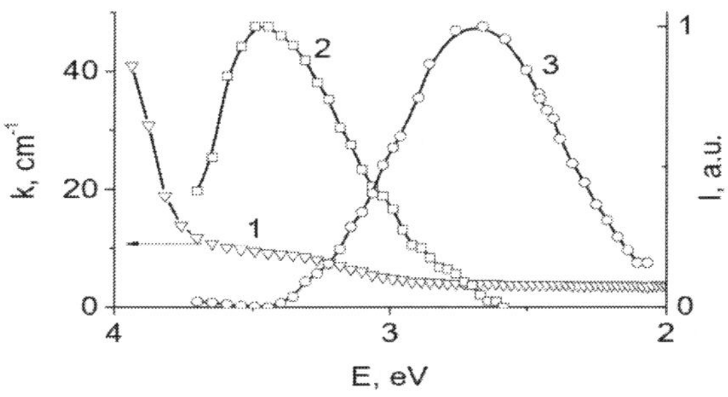

Figure 3. Absorption (1) and CL spectra (2,3) for $ZnWO_4$ crystal measured at 300 K with a delay of 10 ns (2) and 50 μs (3) after the end of excitation by electron beam. CL spectra are normalized.

responsible for the band at 3.6 eV are preferably located in the surface layer of the crystal. The fact that the band at 3.6 eV increases in heating the crystals at 900 C for 10 hours discovered by us indicates that the luminescence centers responsible for this band are not intrinsic defects of the crystal. The second band in the luminescence spectrum of LiF-WO_3 and $ZnWO_4$ crystals is the band at 2.6–2.7 eV. Its properties are as follows.

•The light yield in intensity of the long-lived band at 2.7 eV increases as the concentration of tungsten in LiF crystals grows (figure 4) and at concentration W equal to 0.32 wt % is 18% with respect to the light yield of CdWO crystal.

•For LiF crystals doped with different MeO oxides we showed [17] that the position of long-lived luminescence band at 3–2 eV and the value of the longest component of the luminescence decay at 300 K depend on the type of metal ions in the oxide. Table 1 shows the results indicating the shift in the long wavelength region of the maximum position of the dopant band from 3 to 2.7 eV (by 10%) and decrease in the decay constant at 300 K from 400 to 30 μs by (92%) in series Me: Li, Fe, Ti and W. These results irrefutably testify that the parameters of the emission transition in LiF crystals doped with oxides of different metals are affected by the cation co-dopant, which is in the vicinity of the oxygen ion. Increase in the probability of emission transition, as the atomic weight of the cation-dopant increases, indicates the so-called heavy-atom effect, first observed for organic molecules in [18]. In our case, we assume that under excitation, the wave functions of oxygen and metal

Table 1. Spectral and kinetic parameters of the luminescence at 300 K for $ZnWO_4$ and doped LiF crystals.

Crystal	Type of dopant	Band position (eV)	Component of emission decay (ns)	Band position (eV)	Components of the emission decay (μs)
$ZnWO_4$	-	3.6	<20	2.65	0.2; 3; 30
LiF					
	Li_2O	3.6	<20	3.6	400
	Fe_2O_3	3.6	<20	3.2	180
	TiO_2	3.6	<20	3	120
	WO_3	3.6	<20	2.7	0.2; 3; 25

overlapping, and the electron is transferred from excited oxygen ions $(O^{2-})^*$ to metal ion Me^n+ to form an unstable impurity complex $Me^{(n-1)+}$- O^-, which lifetime is entirely determined by Me ion. The existence of a correlation between the value of the light yield and the intensity of an absorption band at 3620 cm^{-1} in the spectrum of activated crystals with increasing of tungsten concentration in the crystal (figure 4) does not contradict with this assumption,

•The charge state of the dopant ion changes from +1 (Li) to +6 (W) and determines the number of coordinated O^{2-} ions, as well as the type and amount of cation or anion vacancies ensuring local compensation of the excess charge. (For example, in substituting the Li^+ ion by W^{6+} ion, the excess charge can be compensated in different ways: in form of $5O^{2-}$ or $(4O^{2-} + V_a)$ or $(6O^{2-}+ V_c)$ and etc. Weak dependence of the emission energy in Me-O complex on the charge state of Me (shift at 0.3 eV in series from Li^+ to W^{6+} (table 1) indicates insignificant impact of the point symmetry of the center in the lattice on the value of transition.

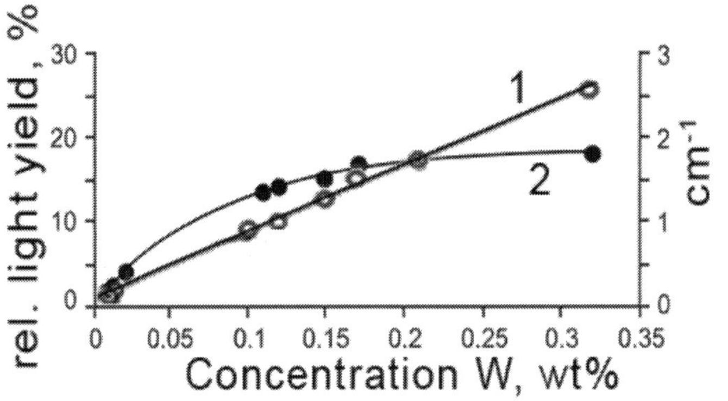

Figure 4. Dependence of absorption coefficient at 3620 cm^{-1} and the relative light yield at 300 K in the band at 2.7 eV on the concentration of W in LiF-WO_3 crystals

•To find out the nature of the luminescence band at 2.6 eV in the $ZnWO_4$ crystal we studied the spectrum of its excitation by photons in the energy range of 3–6.5 eV. The obtained data presented in figure 5 indicate: 1) the existence of the excitation energy threshold equal to 4.1 eV; 2) insignificant (10%) decrease in the band intensity as the photon energy grows in the range of 4–6.5 eV; 3) decrease in the intensity of the band at 2.6 eV as the energy of exciting photons exceeds the threshold in crystals subjected to heating at 900 C for 10 hours (by 80% for the photon energy equal to 6.5 eV).
The results show that the threshold energy (4.1 эB) to excite the band at 2.6 eV in $ZnWO_4$ does not coincide with the estimated value of the transparency cutoff 4.6–5.8 eV, as reported in [2,12,13]. This fact in [19] is attributed to the ability to create self-trapped excitons in WO_6 complexes responsible for the luminescence in the band at 2.6 eV if the photon energy falls into the Urbach "tail" on the edge of the transparency cutoff.
However, the mismatching of value of the threshold energy (4.1 эB) to excite the band at 2.6 eV with position of edge of intrinsic absorption can be considered as direct proof of the impurity nature of the luminescence at 2.6 eV excited in the impurity absorption region in the transparency cutoff. (Maintaining the stoichiometry of the melt during crystal growth is known to be difficult [20]).
Decrease in the efficiency of excitation of the band at 2.6 eV, as the photon energy increases in the heat-treated crystals (figure 5), directly indicates a change in the properties of the crystal surface during heating and indicates sufficiently high chemical activity of the material.
To determine the effect of the lattice type on the properties of emission centers, we conducted a comparative analysis of characteristics of the radiative transition in the region of 2.6 eV in the tested crystals. We obtained the following results.

In $ZnWO_4$ crystal at 300 K, FWHM of the band at 2.6 eV is equal to 0.6 eV, and the luminescence decay is described by a set of three exponential functions with values of decay constants of 30 μs, 3 μs

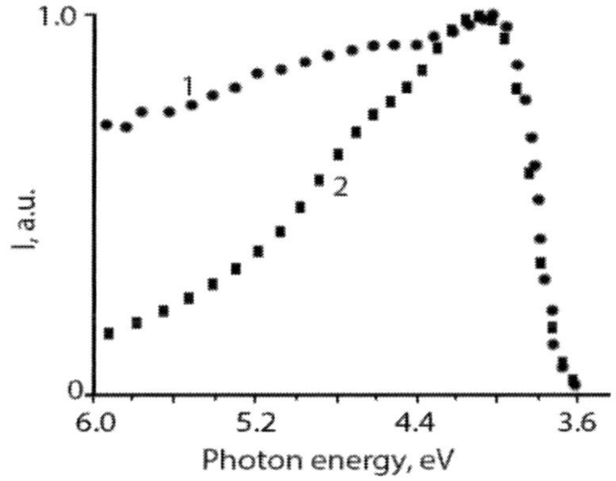

Figure 5. Spectra of luminescence excitation at 2.6 eV in different $ZnWO_4$ crystals: freshly cleaved (1) and heated at 900 C (2).

and 200 ns at 300K. In LiF-WO_3 crystals: the band is found at 2.7 eV, and the decay is described as a set of three exponential functions with values of decay constants of 200 ns, and 3 and 30 μs at 300 K similar to that in LiF-WO_3. The spectra of all the decay components are identical. They represent are a band with a maximum at 2.6 and 2.7 eV in $ZnWO_4$ and LiF-WO_3, respectively.

The results obtained show that as temperature increases in the range of 15–250 K in LiF-WO_3 crystals, the intensity of the band at 2.7 eV increases, and the duration of the decay reduces by an order of magnitude from 230 to 35 μs (figure 6). According to [21], in $ZnWO_4$ crystals, increase in the intensity and decrease in the decay time from 160 to 35 K occur in a narrow temperature range of 5–15 K. In the region of 15–250 K, the luminescence parameters remain virtually unchanged.

The light yield in the band at 2.6 eV at a temperature range of 15–250 K does not depend on the temperature in crystals of both types, but it is achieved in different ways: in $ZnWO_4$ crystal both J=const and τ=const in this temperature range, whereas in LiF-WO_3 crystal J=f(T), and it increases as the temperature of the crystal grows under excitation; τ=f(T), and it decreases as the temperature rises. However, the result of multiplying Jτ remains constant and independent of temperature.

Thus, the only distinguishing feature of the luminescence in the region of 2.6–2.7 eV in $ZnWO_4$ and LiF-WO_3 crystals is the difference in the temperature range of change in decay parameters: 4–15 K (in $ZnWO_4$) and 15–250 K (in LiF-WO_3 crystals). Within the approximation of an adiabatic potential emission center in the ground and excited states as a functions of the generalized configuration

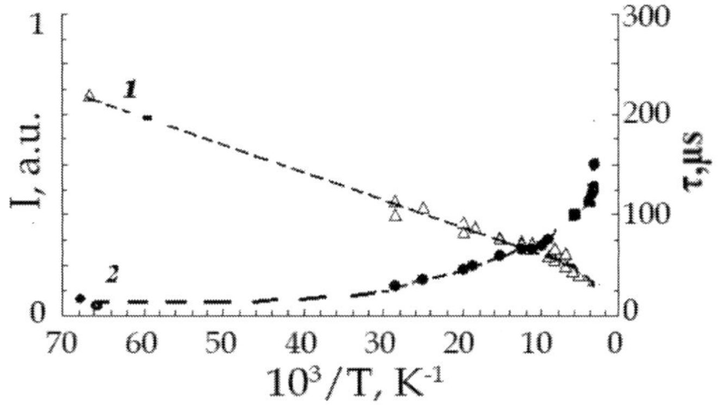

Figure 6. Temperature dependence of the decay time (1) and the intensity of the CL at 2.7 eV (2) in LiF-0.2 wt % W.

coordinate it implies different barrier height for transition of the emission center from an excited state to the ground state in these crystals. This can be due to either different degree of lattice relaxation in the luminescence center after its transition to an excited state or significant change in the elastic constants of the lattice for the center in an excited state. This is to be investigated.

The light yield in the long-lived luminescence band in the spectrum of $ZnWO_4$ crystal is 76% relative to $CdWO_4$ crystal [20]. According to our results the light yield in the LiF-0.32 wt % W is 18% relative to a reference $CdWO_4$ crystal. This means that the light yields in the bands at 2.6 ($ZnWO_4$) and 2.7 eV (LiF-WO_3) are of the same order of magnitude. The latter fact contradicts the assumption about the intrinsic character of the luminescence in the band at 2.6 eV in $ZnWO_4$ crystal since the number of a matrix-forming emission W-O complexes per unit of the volume is at least four orders of magnitude greater than activator centers in LiF-WO_3 crystal ($\sim 10^{17}$ cm^{-3}).

Thus, the parameters of the emission that is excited optically with photon energy> 4.66 eV and under ionizing radiation for luminescence centers in LiF-WO_3 and $ZnWO_4$ crystals are comparable. We studied the range of the absorbed doses within which the emission characteristics of the crystals remain constant.

We have established the following. In LiF-WO_3 crystals at 300 K under electron beams irradiation the luminescence parameters are independent of the absorbed dose in a limited range of 10^1–10^3 Gy. Further increase of the dose is followed by degradation of spectral and kinetic parameters of the luminescence, as we described in [22]. The degradation is exhibits itself as a distortion of the luminescence band (figure 7) followed by decrease in the amplitude of the band (figure 7, insert), and as an increase in the afterglow time. The degradation is caused by overlapping of the scintillation band and absorption and luminescence bands of radiation defects accumulated during electron beam irradiation. The failure in the emission band (figure 7) is caused by reabsorption of the emission by F_2 color centers. Increase in the afterglow period we attribute to post-radiation interaction of radiation-induced defects [22].

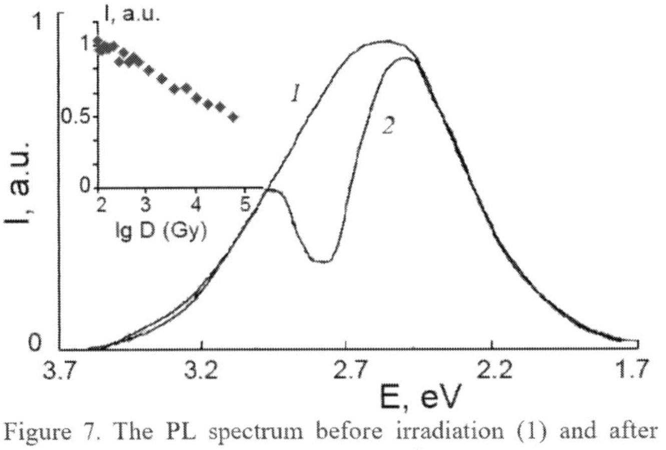

Figure 7. The PL spectrum before irradiation (1) and after irradiation by electron beams (D=2.10^4 Gy) in LiF-0.31wt % W (2). The insert shows the dose dependence of PL intensity at 2.7 eV at 300 K.

In contrast to LiF crystals, in $ZnWO_4$ crystals, the parameters of the emission at 2.6 eV remain constant in a wide range of the absorbed doses of 10^1–10^5 Gy at 300 K. In the spectral region of the luminescence we did not observe bands (emission or absorption) due to short-lived radiation defects.

The absence of radiative and absorption transitions in the region of the luminescence band at 2.6 eV in stationary measurements is reported in [14].
It is obvious that these differences in the dose dependence of the luminescence parameters in the investigated crystals are caused by the properties of the crystal lattice that determines the type and efficiency of formation and accumulation of radiation defects.

4.Conclusions

1) Emission transitions at 3.6 and 2.6 eV in both $ZnWO_4$, and LiF-WO_3 crystals belong to oxygen-containing centers of different types. When the crystal is heated in oxygen atmosphere, the band at 3.6 eV increases due to the destruction (transformation) of the centers responsible for the band at 2.6–2.7 eV.
2) In $ZnWO_4$ and LiF-WO_3 crystals, the following spectral and kinetic parameters of the emission transition in the region of 2.6–2.7 eV are found to coincide:
• the probability of emission transition at 300 K (table 1);
• the number and value of the components of the luminescence decay at 300 K (table 1);;
• independence of the light output on the temperature in range of 15–250 K;
• independence of the light output at emission bands at 2.6 and 2.7 eV on the dose values in the range of 10^1–10^3 Gy under irradiation of crystals by electron beams.
The obtained results indicate identical structure of the emission centers in the tested crystals, and show no effect of the crystal lattice and the point symmetry of W-O emission complexes on the properties of the emission transition in the region of 2.6–2.7 eV.
3) In $ZnWO_4$ crystals, W-O emission complex is not a regular matrix-forming component, and it is located near the lattice defects. The following facts can be used in favor of this hypothesis:
• the absorption in the region of 3.5–4.6 eV (the edge on the transparency cutoff) indicates the presence of defects in the crystal structure of the initial crystal (figure 3);
• the excitation energy threshold (4.1 eV) of the luminescence band at 2.6 eV (figure 5) is found in the region of the impurity absorption, it does not coincide with value of the transparency cutoff crystal and indicates the impurity nature of the excited luminescence;
• the concentration of W-O emission centers in doped LiF-WO_3 crystals does not exceed 10^{18} cm^{-3}. Since the light yield in $ZnWO_4$ and LiF-WO_3 crystals is of the same order, the number of emission centers in $ZnWO_4$ crystals is four orders of magnitude less than regular lattice sites.
4) The type of the crystal lattice, its energy parameters determine the effectiveness of formation and accumulation of radiation defects in the crystal as the absorbed dose increases. The spectral overlapping of the bands of absorption and emission of radiation defects and the luminescence center makes the emission parameters depend on the value of the absorbed dose, i.e. it leads to degradation of scintillation luminescence (figure 7). The latter, according to our estimates, does not allow use of LiF-WO_3 crystals in hard radiation fields as radiation detectors in a dose range of $\geq 5\ 10^3$ Gy.
In $ZnWO_4$ crystals, the luminescence parameters remain unchanged in a wide range of the absorbed doses of 10–10^5 Gy at 300 K. This indicates the possibility of their use as detectors of ionizing radiation in a wide range of doses and temperatures.

References

[1] Korzhik M V *Physics of scintillators based oxygen-containing single crystals*. BSU 2003 263 p.
[2] Nagirnyi V, Feldbach E, Jonsson L, Kirm M, Kotlov A, Luschik, Nefedov V and B Zadneprovski 2002 *NIM A* **486** 395.
[3] Kolobanov V N, Kamenskikh I A, Mikhailin V V, Shpinkov I N, Spassky D A, Zadneprovski B I, Potkin L I and Zimmerer G 2002 *NIM A* **486** 496.
[4] Mikhailik V.B. and Kraus H 2010 *Phys. Stat. Solidi B* **247** 1583.
[5] Krutyak N, Mikhailin V V, Vasil'ev A N, Spassky D A, Tupitsyna I A, Dubovik A M, Galashov E N, Shlegel V N and Belsky A N 2013 *J. Luminesc.* **144** 105.

[6] Ovechkin A E, Ryzhikov V D, Tamulaitis G and Zukauskas A *Phys. Stat. Sol. (a)* **103** (1987) 285.
[7]. Schwartz K K,. Kristapson Y ZH, Lucys D Y and Podin A.V *Proceed. Radiation Physics*, **V** Riga. 1967, p.179-235.
[8] Shakhnovich M I and Chubenko A I *Proceed. Radiation Physics, 1V, ionic crystals,* Riga.1966 P.125-132.
[9] Egranov AV, Radjabov E A *Spectroscopy of oxygen and hydrogen centers in alkali halide crystals.* Science, N. (1992). 161 p.
[10] Meistrich M L 1968 *J. Phys. Chem. Solids* **29** 1119.
[11] Watterich A, Foldvary I and Voszra K 1980 *J de Physique Colloque*. **41** 111.
[12] Mikhailik V and Kraus H 2006 J. Phys. D: *Appl. Phys*. **39** 1181.
[13] Itoh M, Katagiri T, Aoki T.and Fujita V 2007 *Radiation Measur*. **42** 545.
[14] Watterich A A, Kovacs L, Wurz R, Schon F, Hofstaetter A. and Scharmann A 2001 *J. Phys. Condens. Matter.* **13** 1595.
[15]. Lammers M J, Blasse G and Robertson D S 1981 *Phys. Stat. Sol. (a)* **63** 569.
[16] Lisitsyna L and Lisitsyn V 2013 *Physics of the Solid State* **55** 2297.
[17] Dauletbekova A, Lisitsyna L, Korepanov V, Lisitsyn V. and Kassymkanova R 2013 *Phys. Status Solidi C* **10** 263.
[18] McGlynn S P, Azumi T and Kinoshita M *Molecular spectroscopy of the triplet state.* New Jersey 1969 450 p.
[19] Nagirnyi V, Kirm M, Kotlov A, Lushchik A and Jonsson L 2003 *J. Lumin.* **102-103** 597.
[20] Atroschenko L, Burachas S, Galchinetsky L, Grinyov B, Ryzhikov V and Starzhinsky N *Scintillator crystals and detectors of ionizing radiation based on them.* Kiev, Naukova Dumka, 1998, p. 310.
[21] Nagirnyi V, Jonsson L, Kirm M., Kotlov A, Lushchik A, Martinson I, Watterich A and.Zadneprovski B 2004 *Radiation Measur*. **38** 519.
[22] Lisitsyna L A 2014 *Optics and Spectroscopy* **116** 257.

Impact of defectiveness on the parameters of the acoustoelectric transformations in heterogeneous non-metallic materials

T V Fursa[1], B A Lyukshin[2], G E Utsyn[1] and D D Dann[1]

[1] *Tomsk, 634050, Russia,* National Research Tomsk Polytechnic University

[2] *Tomsk, 634050 Russia,* National Research Tomsk State University

E-mail: fursa@tpu.ru

Abstract.

The article studies acoustoelectric transformations of concrete with a crack. The research presents three-dimensional modeling and 3D visualization of wave processes in a concrete sample with a surface crack. The parameters of the electrical response are found to reflect the processes of interaction between the acoustic wave front and the defect and boundaries of the sample.

1. Introduction

The problem of detecting cracks in engineering structures made from heterogeneous non-metallic materials (such as concrete) is of great importance. Surface cracks are the most common defects found in such structures. NDT methods to detect and characterize defects are of current interest for practical applications for civil engineers [1]. NDT methods such as ultrasonic and acoustic are widely used [2], as well as surface wave methods [3]. Extensive research on the development of non-contact NDT methods of concrete has been conducted recently [4-6]. The existing methods are characterized by poor accuracy and low sensitivity to internal defects and inhomogeneities.

The method to increase the accuracy of defectiveness determination taking into account the phenomenon of acoustoelectric transformations in heterogeneous dielectric materials under elastic pulse excitation is being developed [7-10].

The aim of the paper is the search of informative parameters of non-destructive testing to determine the defectiveness of heterogeneous materials by the parameters of the electrical response to mechanical pulse excitation.

2. The research results and their interpretation

The experimental research of electrical responses from a concrete sample with a natural crack was conducted. The sample size was 100×100×100 mm.

The natural crack was made in the following way. The sample was cut 5 mm deep to the center. Then a metal wedge was inserted into the cut. The system sample-wedge was loaded with a hand press. Figure 1 shows a photograph illustrating the process of crack growing in the sample of 100×100×100 mm. The sample was made from the mixture of cement, sand and gravel in the ratio 1:2:4 with the maximum fraction size of 20 mm.

Figure 1. Photograph of the loading system.

The experimental research was done using laboratory software and hardware allowing the production of pulsed mechanical excitation of the samples and recording of electrical responses.

Figure 2 shows a photograph illustrating the process of measuring the electrical response from the concrete sample with a crack located perpendicularly to the direction of the shock and in parallel to the plane of the electrical receiver.

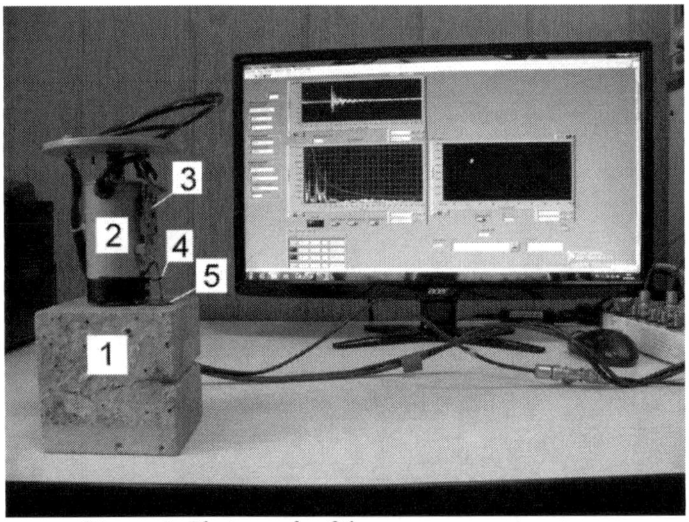

Figure 2. Photograph of the measurement system.

Pulsed mechanical excitation of the sample (1) is performed with the help of electromechanical impact devices with standardized impact force (2). To record the electrical response, a differential electrical sensor is used. Electrical measuring receiver (5) is located at the distance of 2 mm from the surface. Thus, the conditions for the noncontact reception are created. Compensation receiver (4) is located at the height of 30 mm from the measuring receiver. After that these signals are sent to the input of the differential amplifier (3), subtracted and amplified. This measurement design enables to improve significantly the signal-to-noise ratio. The signals from the electrical sensor are recorded with Multi-I/O-card "NI PCI-6251", combined with the computer, allowing the digitization of time realization of an electrical signal.

More detailed methodology of measuring the electrical response to shock excitation is reported in [11].

Figure 3 shows the electrical response from the concrete sample with a crack. The crack passes through the center of the side surface of the sample in parallel to the upper surface. The crack depth is 50 mm, length is 100 mm.

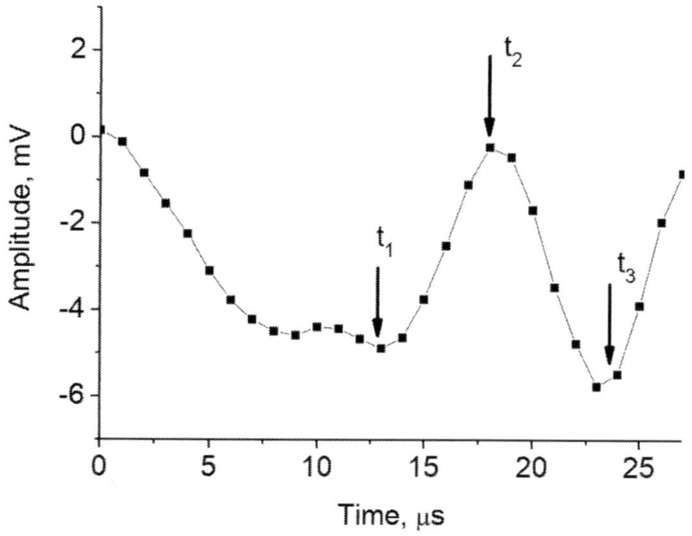

Figure 3. The initial section of the electrical response from the concrete sample with a crack.

Figure 3 shows significant deformation of the electrical response at 13 μs – (t_1), 18 μs – (t_2), and 24 μs (t_3).

3D modeling of wave processes was used to form the fields of longitudinal displacement in the sample. For calculation, the following concrete values were used: density was 2.3×10^3kg/m3; elasticity modulus was 3.3×10^{10}H/m^2; Poisson's ratio was 0.2; longitudinal wave velocity was 4×10^3 m/s. The calculation was carried out for the case when the shock was made towards the sample center. Figure 4 shows 3D visualization of the longitudinal displacements fields carried out with Voxler 3 program.

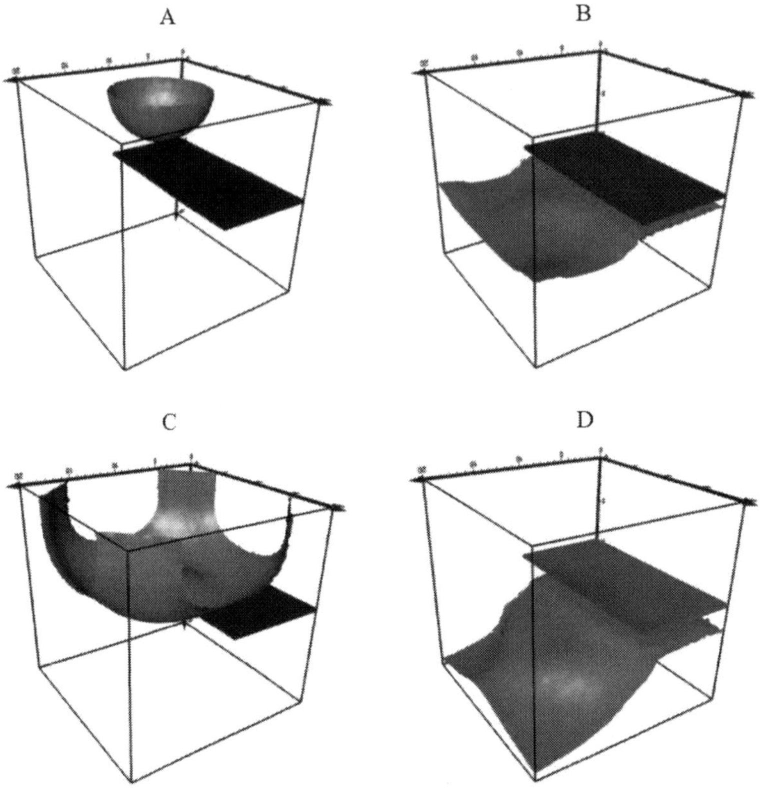

Figure 4. Longitudinal displacement fields in the concrete sample over the time intervals equal to: A is 5 μs; B is 13 μs; C is 18 μs; D is 24 μs after loading.

At the initial stage (see Figure 4A) we can see wave propagation without distortion by defects and boundaries of the sample. The crack is shown in dark in Figure 4. Within this time interval it can be seen that the shape is spherical which corresponds to the wave theory.

In Figure 4A the wave reaches the side boundaries of the sample. After 13 μs, when the wave reaches the side boundaries of the sample, the acoustic wavefront is distorted due to the reflection of the wave from the boundary. This instant time is characterized by the distortion of electrical response which corresponds to t_1 in Figure 3. The increase of the electrical response magnitude in the time interval from 13 to 18 μs occurs due to the contribution of the reflected waves. In papers [7, 8], it is shown that in heavy concrete, the main contribution to the electrical response is made by piezoelectric sources of acoustoelectric transformations. The sources are distributed throughout the sample volume. The electrical signal occurs due to the appearance of charges on the faces of the piezoelectric quartz contained in fine and coarse aggregate (river sand and gravel) under its deformation by the acoustic wave, and due to the shift of these charges relatively to the electrical receiver. Consequently, the increase in the magnitude of the electrical response in the time interval from 13 to 18 μs is connected with the contribution of the acoustoelectric transformations reflected from the side surfaces of the waves.

At 8 μs, the wave overcomes the level of the crack. In other words, it goes over it, undergoing the maximum distortion in comparison with the previous stage (Figure 4C). The change in the shape of the wavefront affects the level of the electrical response (point t_2, Figure 3).

At 24 μs (Figure 4D), the wave front is reflected from the opposite face. In Figure 3, this point is marked as t_3. The reflection from the opposite face causes signal increase.

The change in the wave process causes the change of the electrical response. The analysis of the results of mathematical modeling and obtained electrical responses further the development of non-destructive testing with dynamoelectric transformations phenomenon.

3. Conclusion

The purpose of the research is theoretical and experimental studies of the impact of the defectiveness on the parameters of acoustoelectric transformations in concrete. We received the fields of longitudinal shift in the concrete sample with a surface crack on the basis of the results of three-dimensional modeling of wave processes, formed in the sample under shock excitation. 3D visualization was used to trace the interconnection between the characteristics of the electrical response and spatial location of the defect in the sample and its orientation relatively to excitation and reception points. At the instants when the acoustic waves are reflected from the boundaries of the sample and the wave front bends the defect, the electrical response is distorted.

On the basis of the obtained results, new options of non-destructive testing of concrete defectiveness can be developed.

The work is done under the State task “Science”.

References:

[1] Carino N.J. 1994 Nondestructive testing of concrete: history and challenges in ACI SP-144 *Concrete Technology – Past, Present and Future, P.K. Mehta Ed* pp 623-678.

[2] Qasrawi H.Y. and Marie I.A., 2003 The use of USPV to anticipate failure in concrete under compression *Cement Concrete* vol. **33** pp 2017-2021.

[3] Gudra T. and Stawiski B. 2000 Non-destructive strength characterization of concrete using surface waves *Ndt&E Int.* vol. **33** pp 1-6.

[4] Algernon D., Gräfe B., Mielentz F., Köhler B. and Schubert F. 2008 Imaging of the Elastic Wave Propagation in Concrete Using Scanning Techniques: Application for Impact-Echo and Ultrasonic Echo Methods *J Nondestruct Eval* vol. **27** pp 83-97.

[5] Chekroun M., Marrec L. Le, Abraham O., Durand O. and Villain G., 2009 Analysis of coherent surface wave dispersion and attenuation for non-destructive testing of concrete *Ultrasonics* vol. **49** pp 743-751.

[6] Sung Woo Shin, Adeel Riaz Qureshi, Jae-Yong Lee and Chung Bang Yun, 2008 Piezoelectric sensor based nondestructive active monitoring of strength gain in concrete *Smart Materials and Structures* vol. **17** 055002.

[7] Fursa T.V. and Dann D.D., 2011 Mechanoelectrical transformations in heterogeneous materials with piezoelectric inclusions *Tech. Phys.* Vol. **56** pp 1112-1117.

[8] Fursa T. V., Osipov K. Yu., Dann D. D. 2001 Development of a Nondestructive Method for Testing the Strength of Concrete with a Faulted Structure Based on the Phenomenon of Mechanoelectric Transformations *Russian Journal of Nondestructive Testing* vol. **47** pp 323-328.

[9] Fursa T. V., Osipov K. Yu., Dann D. D. 2011 Developing a Metod Based on Mechanoelectric Transformation for Monitoring Concrete Defectness Variation under Influence of Freezing-Thawing Cycles *Technical Physics Letter* vol. **37** pp 291-293.

[10] Fursa T. V., Lyukshin B. A., Utsyn G. E. 2013 Relation between the Electric Response and the Characteristics of Elastic Waves under Shock Excitation of Heterogeneous Dielectric Materials with Piezoelectric Inclusions *Technical Physics* vol. **58** 263-266.

[11] Fursa T. V., Osipov K. Yu., Lyukshin B. A., Utsyn G. E. 2014 The development of a method for crack-depth estimation in concrete by the electric response parameters to pulse mechanical excitation *Meas. Sci. Technol.* vol. **25** 055605.

The electrical conductivity of the surface layers of oxide polycrystalline semiconductors irradiated by accelerated ions

T S Frangulyan[1], S A Ghyngazov[1] and S M Kaz[2]

[1] Institute of Non-Destructive Testing, National Research Tomsk Polytechnic University, 30 Lenin ave., Tomsk, Russia.

[2] Tomsk State University, 34 Lenin ave., Tomsk, Russia

E-mail: ghyngazov@tpu.ru

Abstract. The effect of ion irradiation on the electrical conductivity of low resistance and high resistance subsurface layers of polycrystalline Li-Ti ferrites is studied. Irradiation was carried out with accelerated Ar^+ ions with energy E=150 keV and the fluence F = 10^{16} ions/cm^2. It is found that exposure of high-resistance ferrites significantly decreases the activation energy E_σ and considerably increases the electrical conductivity of the surface layers. The effect of the ion beam on the stated low-resistance characteristics of the samples is much weaker. The observed decrease in the numerical values of E_σ is due to the decrease in the values of the intergranular potential barrier caused by the exposure. The decrease in the potential barrier difference is due to the decreased degree of the grain boundary oxidation which is caused by preferential desorption of oxygen under the action of the ion beam. The thermal stability of the electrical characteristics of the investigated ferrites subjected to ion radiation-induced modification is determined.

1. Introduction

In [1–3], it is shown that exposure to accelerated ions of a broad class of ionic oxide dielectrics (crystals of magnesium, quartz and lithium niobate oxides, as well as various types of electric-grade ceramics) causes a huge increase in the electrical conductivity of the surface layers. The maximum effect (more than 9 orders of magnitude) is achieved in irradiation with the fluence F = (1016–5□1016) ions/cm2. Transition of the high-resistance dielectric surface from an insulating state to a conducting one is followed by a sharp decrease in the activation energy from 1.5 eV to several hundredths of eV, which indicates the change of the conduction mechanism. The observed effect is fundamental and virtually does not depend on the nature of the accelerated ions. The analysis of the obtained results together with the data characterizing the structural state of the implanted layers suggests that distortion of the stoichiometry of the dielectric material surface caused by the preferential oxygen desorption under ion bombardment determines the formation of the resistive state [1]. It can be argued that the ion irradiation technique to modify the electrical conductivity can be extremely promising with respect to a sufficiently wide class of virtually significant oxide polycrystalline semiconductors. The considerations are based on the following assumptions. The most characteristic feature of oxides, as a special class of semiconductors, is significant deviations from the stoichiometric ratio of metal and oxygen ions. As a result, electrically active centers, donors or acceptors, are formed in the crystal lattice of oxides. The concentration of the centers depends on temperature, the nature of the gaseous medium, and other process factors. Structures with important practical properties can be created by shifting the stoichiometry of the metal-oxygen ratio. In this regard, testing of new and innovative techniques to produce an energy impact on this class of materials can help to solve this kind of problems.

The research aims to find out the nature and effectiveness of ion-irradiation impact on the conductivity of the oxide polycrystalline semiconductor surface layers, a sufficiently large group of spinel-type ceramic among them.

2. Experimental procedure

The object to be studied was ferrite lithium-titanium ceramic, which can be referred to this class of materials. The ceramics was made using ceramic technology (T_S =1280 K, the sintering time was 2 hours) from a mechanical mixture of Li_2CO_3-11.2%, MnO-2.7%, TiO_2-18.65%, ZnO-7.6%, Fe_2O_3-59.81% composition (by wt.%). Its chemical composition was $LI_{0.649}$ $Fe_{1.598}$ $Ti_{0.5}Zn_{0.2}Mn_{0.051}O_4$. The samples were made in the form of pellets of 15 mm diameter and 1.5 mm thick. Due to the heterogeneity of ceramics oxidation in sintering, the nonuniform depth distribution of the electrical conductivity and its activation energy in the thin surface layers were normally of the order of 200–300 μm. The nonuniform layers formed in sintering were removed by mechanical sanding. The samples with uniform depth distribution of the electrical characteristics were tested. The samples were irradiated with accelerated Ar ions, the energy E = 150 keV, the fluence F = 10^{16} ions/cm^2 using the ion-beam setup "Vesuvius 5M". The temperature in irradiation did not exceed 400 K. The static electrical conductivity was measured in the temperature range of (300–600) K with two-probe spreading resistance technique [4] to determine the resistance of the thin layers.

3. Experimental results

Two types of specifically prepared samples with different initial values of the electrical conductivity and electromigration activation energy were tested. The temperature dependences for the electrical conductivity are shown in Figures 1 and 2.

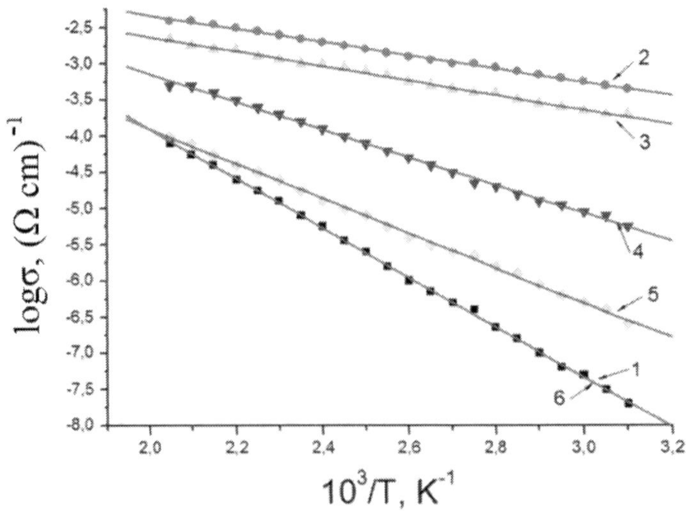

Figure 1. Temperature dependences of electrical conductivity for A-type ferrite ceramics before ion irradiation (Curve 1), after ion irradiation (Curve 2). Curves 3–6 are for temperature dependences after annealing of the irradiated sample at T = 540, 570, 590, and 630 K, respectively.

IOP Conf. Series: Materials Science and Engineering **81** (2015) 012026 doi:10.1088/1757-899X/81/1/012026

The first A type, high-resistance samples ($\rho \approx 4.5 \cdot 10^{6}$ Om·cm at T = 360 K), indicated an increased and uniform value Eσ = 0.68 eV throughout the depth. The second B type, low-resistance samples ($\rho \approx 4 \cdot 10^{3}$ Om·cm at T = 360 K), according to the data obtained by electrophysical measurements, showed low activation energy of conduction $E_\sigma = 0.22$ eV.

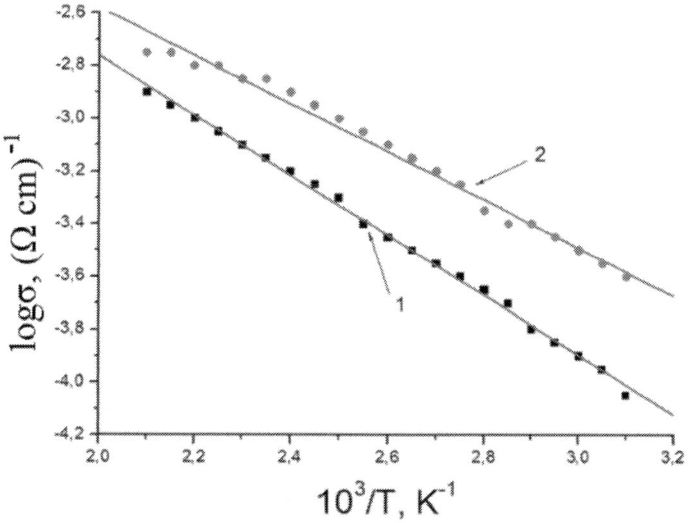

Figure 2. Temperature dependences of the electrical conductivity for B-type ferrite ceramics before ion irradiation (Curve 1), after (Curve 2) ion irradiation.

Activation energy Eσ is an essential characteristic of the electromigration in the matter. In contrast to single crystals in polycrystalline oxide semiconductors, which are an inhomogeneous conducting medium, the activation energy of electromigration is strongly affected by the grain boundaries (GB). As is well known, in heterogeneous structures, a potential barrier occurs at the grain boundaries. This barrier is caused by the excessive content of acceptors in GB, compared to the amount of the grain volume. When electrons are trapped by acceptors, the concentration of charge carriers decreases and a negative surface charge occurs, which causes increase in the electrostatic potential at the grain boundary. Oxygen atoms can be considered the most probable acceptors. In sintering of ferrite, oxidation-reduction reactions actively proceed and affect the boundary properties. Due to higher diffusion permeability, GBs are actively enriched with oxygen [5]. As a result, it becomes more oxidized; this causes decrease in Fe^{+2} ion concentration in its vicinity, and hence, charge carrier concentration.

The impact of potential barriers on electric charge transfer in heterogeneous structures is manifested through the mobility of charge carriers, which, as reported in [6], can be described by the expression:

$$\mu_{п} = \mu_{з}\ \mu_{б}\,(\mu_{з} + \mu_{б})^{-1}, \qquad (1)$$

where μ_s, μ_g, μ_b are the mobility in polycrystalline semiconductors, grains and barrier mobility, respectively. At this, $\mu_b = AT^{-1}\, d_g \exp(-\varphi_b/kT)$, where φ_b is the potential barrier difference, A is the constant, d_g is an average grain size.

Theoretically, the connection between the potential barrier at the grain boundaries and the conductivity is considered in [7], where the potential barrier height is determined by the relation:

$$\varphi_b = F_1 - F_2 = k\,T \ln n_1/n_2, \qquad (2)$$

where F_1 and F_2 are Fermi energies; n_1 and n_2 are charge carrier concentrations. Indices 1 and 2 are the volume and boundary of the crystallite, respectively.

It is obvious, that μ_s value will be defined by the smallest mobility value. If $\mu_s >> \mu_b$, then $\mu_g \approx \mu_b$. Typically, at relatively low temperatures, the barrier mobility is much lower than the intragranular one. Under these conditions, the activation growth of the ferrite ceramic electric conductivity with temperature will be determined by the potential barrier difference. The studies of the static conductivity of the Li-Ti ferrite ceramics in the temperature range of (350–500) K reported in [8] are fully consistent with these considerations.

Thus, it can be argued that the difference in the activation energy of conduction for the two tested types of ceramic samples is caused by the difference in GB oxidation. The more oxidized GB, the greater is the potential barrier difference, and hence the activation energy of conduction. In light of this understanding, we analyze the data on the effect of radiation treatment on the parameters of the electromigration in the ferrite ceramic. For the general case, the change in the value of the intergranular potential barrier under irradiation may be due to ionization processes in ceramics, which are not followed by change in its composition, and due to the oxygen exchange with the environment. In the latter case, the stoichiometric metal ion-oxygen ion ratio changes.

First, we conducted test experiments to study the effect of gamma rays irradiation with radioactive ^{60}Co isotope. The exposure dose was 10^6–10^7 r-units. This type of radiation initiates ionization processes and creation of atomic displacements by secondary electrons in the crystal lattice. No significant change in the conductivity and activation energy of conductivity was observed within these doses. Consequently, these processes cannot lead to the change in the GB state and the value of the intergranular potential barrier. Figures 1 (Curve 2) and 2 (Curve 2) show the temperature dependences of the electrical conductivity for the samples after irradiation with accelerated Ar^+ ions. Table 1 shows the values of the activation energy $E\sigma$ which characterizes the electromigration in the ion-modified layers of ferrite.

Table 1. Effect of ion irradiation and subsequent thermal annealing on the value of the activation energy of the electrical conductivity in the ferrite surface layers

Ceramics type	E_{σ}, eV					
	before irradiation	after irradiation	after annealing of the irradiated ceramics			
			T=540 K	T=570 K	T=590 K	T=630 K
A	0.68	0.18	0.19	0.32	0.49	0.68
B	0.22	0.19	-	-	-	-

The obtained results indicate that irradiation of the high-resistance ferrite with ions causes a significant decrease in the activation energy of the electrical conductivity and increases the conductivity of the modified layers. The action of the ion beam on the low-resistance samples is much weaker. At this, the activation energies of the conductivity for both ionirradiated low-resistance and high-resistance ferrite samples are found to be virtually identical. The obtained results show that ion irradiation can cause decrease in the potential barrier difference in the surface layers of lithium-titanium ferrite ceramics.

In the experiment, the sample irradiation was performed with accelerated ions in vacuum, the oxygen pressure being decreased. The temperature of the samples did not exceed T = 400 K due to radiative heating. The control experiments showed that thermal annealing of the ferrite samples in vacuum at this temperature had no significant effect on their electrical conductivity and E_{σ} value. Hence, its change after ion irradiation with these parameters is of purely radiative nature.

It is known that in interaction of ion beams with multi-component materials preferential sputtering of one of the components occurs. Bombardment of the oxide material surface with accelerated ions and the cascade processes developing therein lead to breakage of the bonds between oxygen and metal ions and preferential desorption of oxygen, its partial pressure being low, and cause stoichiometric deviation. This process should be especially efficient in the weakened areas of ceramics, grain boundaries in particular. When oxygen atoms escape, the valence balance is achieved by transition of the part of iron ions localized in GB from a threevalence charge state to a bivalent one. The decreased excess concentration of oxygen ions in the region of the grain boundary caused by ion bombardment reduces the potential barrier difference, and therefore, it decreases the activation energy of electromigration. These considerations are in agreement with the data obtained for the effect of oxidative annealing on the conductivity of the ion-modified layers.

The experiments showed that the initial electrical conductivity of the irradiated ferrites can easily be recovered by thermal annealing in air. Annealing was carried out in cyclic heating. The hold time at a fixed temperature was 10 minutes. The study showed that oxidizing annealing at $T < 540$ K has no effect on the conductivity of the modified layers. As can be seen from the data shown in Figure 1, with further increase in temperature in a sufficiently narrow range of T=(570–630) K, the electrical conductivity of the irradiated sample decreased to the values corresponding to those of the unirradiated one. The decrease in the conductivity due to thermal annealing is attended by an increase of the activation energy of electromigration (Table 1).

The described behavior of the conductivity in the ion beam-modified polycrystalline ferrites in annealing in air can be satisfactorily explained by oxidation due to chemisorption on the atmospheric oxygen surface and its diffusion to grain boundaries [9, 10]. Thus the original oxidation state of GB and potential barrier difference recover.

In conclusion, practical perspectives of the obtained results should be emphasized. A number of elements of electronic devices, which are widely used in practice, were developed based on the barrier effects in polycrystalline oxide semiconductors, [11]. They are, in particular, different types of thermistors, posistors and varistors. The results of the study show that ionbeam processing, in principle, can be successfully used as a new effective tool for effective control of the potential barrier difference.

4. Conclusions

1. Irradiation of the polycrystalline ferrite by accelerated ions of an inert argo gas causes changes in the electrical conductivity and the activation energy of the electromigration defined by the potential barrier difference. This can be attributed to activation of regenerative processes due to preferential sputtering of oxygen from GB by accelerated ion beams.
2. The electrical characteristics of the surface layers of polycrystalline ferrites subjected to ion radiation-induced modification do not depend on temperature and remain stable in heating the samples in air in the temperature range of T = (300–520) K.
3. In thermal annealing of the irradiated samples in the temperature range of T = (540–630) K, the original state of the material completely recovers. This occurs due to the oxidative process caused by chemisorption on the surface of the atmospheric oxygen and its diffusion to the region of grain boundaries.

Acknowledgements
This research is supported by The Ministry of Education and Science of the Russian Federation in part of the "Science" program.

References
[1] Pichugin V F and Frangulyan T S 2000 Effect of ion irradiation on the structure and properties of oxide dielectrics. *Perspective materials* **6** p 26

[2] Kabyshev A V, Konusov F V, Lopatin V V and Kurakov A G 2001 Properties of oxide ceramics and nitride ceramics after ion-thermal modification. *Perspective materials.* **1** p 70

[3] Pichugin V F, Frangulyan T S, Kryuchkov Yu Yu and Feodorov A N 1993 Formation of conductive layers on dielectric substrates by ion bombardment. *Nuclear Instruments and Methods in Physics Research B.* **80/81** p 1203.

[4] Mazur R G and Dickey D H 1966 A spreading resistance technique for resistivity measurement on silicon. *J. Electrochem. Soc.* **113** (3) p 255.

[5] Ghyngazov S A, Lisenko E N, Petyukevich M S and Frangulyan T S 2007 Interaction of lithiumtitanium ceramics with air during sintering. *Rus. Phys. J.* **50** (2) p 134

[6] Jerhot J and Snejdar V. 1978 Hall effect in polycrystalline semiconductors. *Thin solid films.* **52** 379.

[7] Anderson J C 1973 Barrier-limited mobility in thin semiconductor films. *Thin solid films.* **18** 239.

[8] Surzhikov A P, Peshev V V and Ghyngazov S A 2000 Electrical properties of Li-Ti ferrite. *Advanced Materials.* **6** p 66.

[9] Surzhikov A P, Frangulyan T S, Ghyngazov S A and Lysenko E N. 2010 Investigation of oxidation processes in non-stoichiometric lithium-titanium ferrites using TG analysis. *Journal of Thermal Analysis and Calorimetry.* **102** (3) p 883

[10] Surzhikov A P, Lysenko E N, Ghyngazov S A and Frangulyan T S 2002 Determination of the oxygen diffusion coefficient in polycrystalline Li-Ti ferrites. *Rus. Phys. J.* **45** (10) p 989.

[11] Valeev H S and Kvaskov V B 1983 *Nonlinear Metal Oxide Semiconductors* (M.: Energoatomizdat).

IOP Conf. Series: Materials Science and Engineering **81** (2015) 012027 doi:10.1088/1757-899X/81/1/012027

Stability of the GaAs based Hall sensors irradiated by gamma quanta

A V Gradoboev[*,**], **G F Karlova**[**]
[*]Tomsk polytechnic university, Tomsk, Russia
[**]Joint-Stock Company "Research Institute of Semiconductor Devices", Tomsk, Russia

E-mail: gradoboev1@mail.ru

Abstract. The present work is aimed at investigation of the stability of the GaAs based Hall sensors (pickups) to irradiation by gamma quanta. The examined objects are the gallium arsenide based Hall sensors manufactured on thin active layers by the methods of vapor phase epitaxy (VPE), molecular beam epitaxy, and ion implantation. Our research methodology involves measurements of the volt-ampere characteristics (VACs) of all sensors for different values of the supply voltage polarity and electron concentration and mobility by the Van-der-Pau method as well as investigations of the noise properties of the sensors before and after irradiation. The sensors are irradiated by gamma quanta of Co^{60} at room temperature in the passive mode, that is, without imposition of an electrical bias. As a result of investigations, it is established that a part of the active layer of finite thickness adjoining the substrate plays an important role in the charge carrier transmission process depending on the concentration of deep-level centers in the substrate. Irradiation by high doses leads to degradation of VACs and increase in the spectral density of the sensor noise. Low gamma radiation doses have a stabilization effect on the sensors. Periodic relaxation processes are observed for a part of the structures manufactured by the VPE method. The assumption is made that they can be caused by the deep-level centers in GaAs.

1. Introduction

Progress of methods of data processing, especially with application of microprocessors, has increased the role of sensors being a primary chain in perception of information from the object being measured. Nowadays to provide safe operation of transport systems, robotics devices, dual application navigation systems, and space technology, new sensors are required to measure magnetic fields with increased sensitivity, accuracy, and radiation stability. The Hall sensors of magnetic field possess sufficient degree of reliability and can operate under extremely adverse conditions, including irradiation. However, their radiation properties are studied insufficiently. Planar high-sensitivity gallium arsenide based sensors of magnetic field have been developed at the Tomsk Joint-Stock Company "Research Institute of Semiconductor Devices" [1]. Since the Hall voltage is inversely proportional to the sensor thickness ($1/d$), the main condition of implementation of the high-sensitivity Hall pickups is manufacture of semiconductor films with small thickness that have some specific features.

Corresponding author: Gradoboev Alexander Vasilievich, Professor, Doctor of Technical Sciences,
+7 913 866 84 05, e-mail: gradoboev1@mail.ru,
634050, Tomsk polytechnic university, 30, Lenin Ave., Tomsk, Russia.

This work presents results of experimental investigation of the influence of irradiation by gamma quanta on the electrophysical characteristics of the magnetic field sensors based on the Hall effect (the Hall pickups or sensors) arranged on thin active gallium arsenide layers and produced by different methods.

2. Objects of research and experimental procedure

Sensors manufactured from gallium arsenide layers arranged on a semi-insulating substrate were used as objects of research. The employed epitaxial structures of n^+-n-n_{bi}-n_i type were manufactured by the methods of vapor-phase (VPE) and molecular beam epitaxy (MBE). In this case, the thickness of the n^+ layer changed from 0.2 to 2 μm. A specially grown LT-layer characterized by very high resistance and small lifetime was used as a buffer n_{bi} layer. In addition, the n^+-n-n_i type structures manufactured by the ion implantation method (IIM) were used to manufacture sensors; the thickness of the n^+ layer was about 0.1 μm.

The parameters of the epitaxial structures were measured with an electrolytic profilometer. The application of the electrolyte-barrier having a mobile barrier and zero or close to zero bias moving through the film depth due to chemical etching allows us to exclude the restrictions characteristic for the metal-semiconductor barrier caused by a breakdown and to measure structures with complicated implantation profile.

The etcher based on H_2SO_4 and H_2O_2 was used as an electrolyte. The typical profile of the electron concentration distribution over the epitaxial structure used for manufacture of the Hall pickups was registered with the profilometer having a probe diameter of 3 mm at a measuring signal frequency of 10^4 Hz [2].

It was established that the profile of concentration of charge carriers providing low residual stress of the Hall pickup (HP) should be sharp on the film-substrate interface.

The concentration of charge carriers and their mobility were measured for the manufactured n^+-n-n_i type structures (here n is the concentration of charge carriers in the active layer, n^+ is the concentration of charge carriers in the contact layer, and n_i is the concentration of charge carriers in the substrate). The influence of the structure of the deep-level centers on the film-substrate interface on the conductivity was also investigated using a microwave resonator of reflection type operating at a frequency of 38 GHz [3]. The change in the conductivity of the structure irradiated by light-emitting diodes of the visible range from the side of the film (or substrate) with bias applied to the n^+-n or n-n_i transition was determined from the proportional change of the microwave power reflected from the resonator. The structures were rejected by the photoresponse shape. Its typical shape is shown in figure 1.

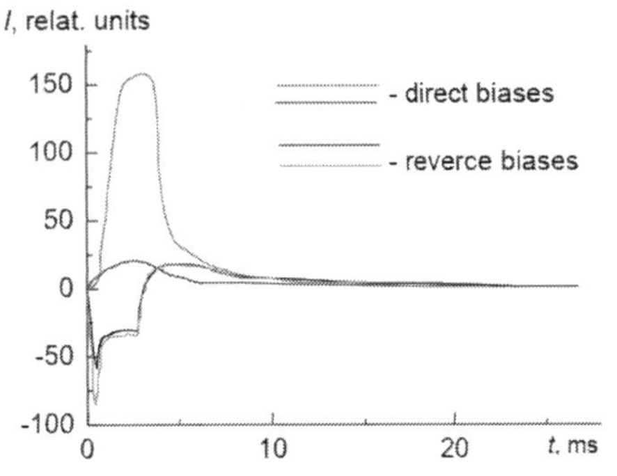

Figure 1. Photoresponce of the structure for different voltage of direct and reverse biases

From figure 1 it can be seen that the longterm relaxation of the photoconductivity depends on the recharge of the deep level (DL) localized in the n-i transition. Irradiation (injection of holes into the n-n_i transition) causes its recharge.

Crystals of sensors were prepared using traditional technology for GaAs field effect transistors. Ohmic contacts were formed by deposition of the Au–Ge–Ni alloy with subsequent alloying. The active regions for ion-implanted structures were insulated using the methods of ion implantation; for other structures, chemical etching was used. Laser scribing was used to separate the plates into individual crystals. Crystals were arranged on the ceramic metalized substrate and sealed with a compound after unwelding of the terminals.

For all sensors, the static volt-ampere characteristics (VACs) were measured for different values of the supply voltage polarity and electron concentration and mobility by the Van-der-Pau method; the noise properties of sensors before and after irradiation were also investigated. The sensors were irradiated by gamma quanta of Co^{60} at room temperature in the passive mode, that is, without imposing electrical bias. In this case, the irradiation dose was successively accumulated. The modes of VAC measurements used in our investigations did not lead to annealing of the incorporated radiation defects, which was confirmed by the identity of experimental data obtained under single irradiation of samples by preset doses and by subsequent accumulation of the dose with intermediate VAC measurements.

Let us consider the main characteristics of the manufactured Hall pickups. The VACs of sensors were in most cases symmetric (figure 2). Several characteristic regions can be identified on the forward and reverse VAC branches: region I with a linear dependence of the current on the voltage ($I \sim U$), region II with a sublinear dependence $I \sim U^k$ ($k < 1$), and region III of current saturation. Moreover, the saturation region started the earlier, the smaller is the quantity ($n{\cdot}d$). This can be explained by the effect of reverse control through the substrate as well as by the effect of traps on the active layer surface [4]. The layer of finite thickness depleted of the charge carriers was formed on the interface between the active layer and the substrate. The thickness of this layer depended on the concentration of deep-level centers in the substrate.

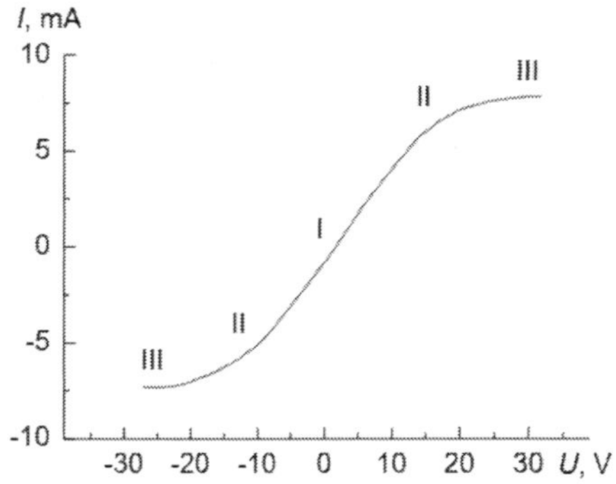

Figure 2. Volt-ampere characterisric of the Hall picup

We consider the MDS structure to be a model of this system [5]. When voltage is applied to the current contacts, some of the electrons diffuse from the active layer through the substrate into the metal, which is charged negatively. The negative charge of the metal is compensated by positive ions in the semiconductor that forms a depletion region extending deep into the semiconductor. In this case, the higher the input voltage, the larger is the depletion region and the higher is the sample resistance. The maximum width of the depletion region, disregarding the electron current through the reverse-biased Schottky barrier, can be estimated from the formula [4]

$$d_{\max} = \left[4kT\varepsilon\varepsilon_0 \ln\left(N_d / n_i\right) / qN_d\right]^{1/2} \quad (1)$$

where N_d is the impurity concentration for homogeneous implantation, ε is the dielectric constant of GaAs, ε_0 is the absolute dielectric constant, k is the Boltzmann constant, and n_i is the concentration of charge carriers. For $N_d = 6{\cdot}10^{15}$ cm^{-3} and $T = 300$ K, $d_{\max} = 0.5$ μm.

The saturation current in this case is

$$I_{sat} = N_d q A_{act} U_{sat} \quad (2)$$

where A_{act} is the thickness of the active part of the channel. From Eqs. (1) and (2) it follows that $d_{\max}$ increases and the saturation current decreases with decreasing impurity concentration in the active region, as was observed experimentally. For samples with impurity concentration of the order of $6{\cdot}10^{15}$ cm^{-3} and $(n{\cdot}d) < 6{\cdot}10^{11}$ cm^{-2}, $I_{sat} = 2$ mA, and for samples with $(n{\cdot}d) > 6{\cdot}10^{11}$ cm^{-2}, $I_{sat} > 5$ mA in agreement with our experimental results.

3. Experimental results and discussion

For irradiation doses in the range ($1.4{\cdot}10^2$–$7{\cdot}10^5$) Gy, the sensors behaved ambiguously [6]. An analysis of the results obtained allowed us to identify some characteristic groups of sensors.

In the first group of sensors manufactured by the MBE and some sensors manufactured by the VPE, the VACs remained unchanged for irradiation doses in the range ($1.4{\cdot}10^2$–$4.2{\cdot}10^5$) Gy. For higher irradiation doses, a decrease in the current was observed that was described sufficiently well by the relationship established in [7] for the change of the electron concentration upon exposure to gamma quanta, which was the quite expected typical behavior of the sensors under irradiation.

After irradiation by small doses ($1.4\cdot10^2$ Gy) in the second group comprising a part of structures manufactured by the VPE, the current increased up to 40% of its initial value and remained unchanged when the irradiation dose subsequently increased up to $4.2\cdot10^5$ Gy; in this case, the direct measurements also demonstrated the corresponding increase in the electron mobility. For higher irradiation doses, a decrease in the current was observed, which was completely described by the change of the electron concentration under irradiation by gamma quanta. That is, the main peculiarity of the second group was the restoration of the electron mobility under irradiation by small doses of gamma quanta due to the relaxation of mechanical stresses (the effect of small doses) [7, 8].

In the third group of ion-implanted structures (as well as for some VPE structures), the current after irradiation by small doses ($1.4\cdot10^2$ Gy) decreased down to 40%. Our measurements demonstrated that the electron concentration in the active layer of the sensor decreased. With further increase in the irradiation dose, the current remained unchanged up to a dose of $4.2\cdot10^5$ Gy. For higher irradiation doses, a decrease in the current was observed that was completely described by the change of the electron concentration under irradiation by gamma quanta. From the foregoing we can conclude that under irradiation of ion-implanted structures (as well as of some VPE structures) by small doses, the local mechanical stresses on the implanted layer - semi-insulating substrate interface are relaxed, which is accompanied by radiation-stimulated diffusion of defects from the substrate into the active layer, thereby decreasing the electron concentration. For low doses, the absence of the radiation effect on the structures with the LT layer in which the substrate was separated from the active layer testified in favor of this assumption.

In addition, the change of the contact resistance of the samples upon exposure to small doses can contribute to VAC changes [8]. The technology of electrode material deposition and its further annealing form an inhomogeneous electrode layer on the plate surface, thereby leading to chaotic Ge and Ni distributions over the surface area. It seems likely that the effect of small doses leads to accelerated diffusion of these elements, whereupon the resistance of the contact changes. The resistance of the contact can also be caused by the increased inhomogeneity of the *n*-layer because of the penetration of defects from outside into this layer in the process of radiation exposure. Among them are dislocations generated by thermoelastic stresses in the crystal.

And finally, the fourth group of sensors. Under irradiation by sufficiently small doses ($1.4\cdot10^2$–$7\cdot10^5$) Gy, periodic relaxation processes were observed for a number of structures manufactured by the VPE method: the first irradiation with a dose of $1.4\cdot10^2$ Gy led to the increase of the current by 50% of its initial value, and the subsequent irradiation by a dose of $1.4\cdot10^2$ Gy led to the restoration of the initial current. Under subsequent irradiation by a dose of $1.4\cdot10^2$ Gy, the current increased, and then the process repeated once again. A change of the irradiation dose during intermediate exposures in a sufficiently wide range did not change the amplitude of current oscillations.

In [9] the mechanism of their origin was discussed. As a result of investigation of deep-level centers of growth in GaAs [10] it was revealed that some deep-level centers can have several quasi-stationary states. In one of the states, this center captures the current carriers, and in the other state, it donates the captured carriers. That is, in this case we can say that each of the quasi-stationary states is characterized by its own energy level, and these states are separated by an energy barrier. Then the first radiation exposure leads to the activation of the quasi-stationary state either with the capture of a charge carrier or with its liberation depending on the initial state of the examined defect thereby leading to the observed sharp change of the current (either its increase or decrease). The subsequent exposure activates the second quasi-stationary state of the examined defect leading to a decrease (increase) of the current. This process is periodically repeated. In this case, we should note especially that the amplitude of changes of the working current is independent of the dose of intermediate exposures to gamma quanta. This has been proved experimentally by changing the intermediate irradiation doses in a wide range from $1.4\cdot10^2$ to $7\cdot10^5$ Gy.

The foregoing results allowed us to state that under irradiation with a dose of $1.4\cdot10^2$ Gy (low doses), the stabilization effect of gamma radiation on the sensors was observed.

The noise properties of the sensors were measured at frequencies in the range 5–100 kHz [11]. The band width Δf corresponded to 3 %. The bias was applied to the samples that provided the preset current. The noise was estimated from the noise temperature T_n given by the formula [12]

$$T_n = u_n / \Delta f / u_{nR} / \Delta f \qquad (3)$$

where u_n is the noise voltage measured at the output from the Hall pickup, u_{nR} is the noise voltage measured with the same ohmic resistance as for the alternating signal, and Δf is the frequency band of the noise meter. The dependences of the noise temperature on the frequency measured for the epitaxial sample and current equal to 1, 1.5, and 2.0 mA [11] demonstrated that T_n and hence, the noise voltage increased with increasing current running through the sample. The dependences measured for $I = 1$, 1.5, and 2 mA were qualitatively similar. They were similar to the curves presented in [13].

To analyze the results obtained, we note that the main components of the intrinsic noise for the Hall pickups are thermal, shot (current), and $1/f$ (flicker) noise components. In the initial segment of the curve, a linear dependence of T_n on the reciprocal frequency characteristic for the flicker noise was observed. The $1/f$ noise was manifested through fluctuations of the concentration of charge carriers when current ran through the sample. We now present quantitative estimates of the thermal and shot (current) noise components for the concrete sample with sizes $0.4 \times 0.25 \times 0.0011$ mm^3, charge carrier concentration of $1.5 \cdot 10^{16}$ cm^{-3}, and resistance of 1 kΩ.

For the thermal noise, the average squared noise voltage u^2 is determined by the Nyquist formula [12]:

$$u^2 = 4kT \cdot R \cdot \Delta f, \qquad (4)$$

where R is the active resistance of the measured source of noise. The thermal noise estimated from formula (4) yields $u^2 = 1.6 \cdot 10^{-17} \Delta f$ [V^2·s]. The shot noise is determined by the mechanisms responsible for the concentration of current carriers (largely by the lifetime of the minority current carriers) and is independent of the frequency [9].

To estimate the shot noise i, we take advantage of the formula for the shot current presented in [13]:

$$i^2 = 4I^2\tau_1^2\tau_2^2\Delta f / N_{av}(\tau_1 + \tau_2)^2(1 + \omega^2 T^2), \qquad (5)$$

where τ_1 is the average lifetime of charge carriers, τ_2 is the average lifetime of charge carriers captured in traps, N_{av} is the average number of carriers participating in charge transfer (it is determined by the product of N and the volume of the active region), and $1/T = 1/\tau_1 + 1/\tau_2$.

The average squared shot current i^2 is equal to $2 \cdot 10^{-27} \Delta f$ [A^2·s], and the average squared voltage is $2 \cdot 10^{-21} \Delta f$ [V^2·s], which is much less than the average squared voltage of the thermal noise. The flicker noise is manifested only at low frequencies. The main role is played by the thermal noise.

The spectral noise density increased approximately by one and a half orders of magnitude in comparison with S_u of the unirradiated sample. The increase in S_u can be caused by the following reasons: the decrease of the lifetime (τ) of charge carriers caused by radiation defects and liberation of charge carriers from trapping centers causing additional recombination of charge carriers. To elucidate a reason for increase in S_u, we measured the lifetime of charge carriers for unirradiated and irradiated samples by the method of injection of minority charge carriers. The measurements demonstrated that τ of an irradiated sample was 0.3 ns, and that of an irradiated sample was 0.5 ns. Since according to Eq. (5), the shot noise changed insignificantly with increasing τ, this implies that the active resistance should change.

Conclusions

1. The gallium arsenide based Hall sensors possess sufficiently high radiation stability.
2. The technology of manufacture of device structures affects significantly the behavior of sensors irradiated by gamma quanta.
3. Significant role in the processes of current transport is played by the part of the active layer of finite thickness adjoining the substrate that is depleted of the main charge carriers. The layer thickness depends on the concentration of the deep-level centers in the substrate.

4. Irradiation leads to the VAC degradation. Changes of the VACs of the sensors after irradiation with a dose of $1.5 \cdot 10^2$ Gy are caused either by decreased or increased thickness of the depletion layer near the active layer - substrate interface that changes due to reorganization of the defect structure upon exposure to radiation.
5. For irradiation dose of $1.4 \cdot 10^2$ Gy (for small doses), the stabilizing effect of gamma irradiation on the sensors was detected.
6. For some samples grown by the VPE method periodic relaxation processes were observed. It was demonstrated that they could be caused by the deep-level centers in GaAs and manifestation of 2 quasi-stationary states of the electron traps.
7. The investigation of the noise properties of the irradiated samples demonstrated that the spectral noise density (S_u) for the irradiated samples increased approximately by one and a half orders of magnitude in comparison with S_u of the unirradiated samples. The change of the spectral noise density of sensors before and after irradiation is in agreement with explanations of relaxation oscillations in the samples manufactured by the VPE method.

References

[1] Karlova G F, Porokhovnichenko L P, Umbras L P 2005 *Patent No 2262777 Russia*
[2] Gradoboev A V, Peshev V V, Kustov V G 1978 *Electronnaya Tekhnika Materialy* **9** 122-124
[3] Karlova G F, Porokhovnichenko L P 2007 **2** *Physical and Chemical Processes in Inorganic Materials* Kemerovo Kuzbassvuzizdat 84-86
[4] Shur M S *GaAs devices and circuits* 1987 Plenum New York
[5] Gradoboev A V, Karlova G F, Belopolova T Yu 2009 *Izvestiya Vysshikh Uchebnykh Zavedenii. Fizika* **52** No 8/2 515-518
[6] Gradoboev A V, Karlova G F 2011 *Proceedings of the XXI International Conference Radiation Solid State Physics* Sevastopol **2** 514-521.
[7] Gradoboev A V Surzhikov A P 2005 *Radiation stability of GaAs microwave devices* Tomsk, Izdatelstvo TPU 277
[8] Mamontov A P, Chernov I P *Effect of small doses of ionizing radiation* 2001 Moscow Energoizdat 286
[9] Karlova G F, Gradoboev A V 2013 *Izvestiya Vysshikh Uchebnykh Zavedenii. Fizika* **56** No 8/3 192-192
[10] Cho HoonYoung, Kim Fun Kyu, Min Suk-ki, et al 1991 *Appl. Phys. Lett.* **58** 1866-1868
[11] Karlova G F, Gradoboev A V 2006 *Proceedings of the 9th Conference Gallium Arsenide and Semiconductor Compounds of III–V Groups* Tomsk 19-21
[12] Van der Ziel A 1973 *Noise* Moscow Sovetskoe Radio 418
[13] Burckhardt C B and Strutt V J O 1964 *IEEE Trans. El. Dev.* **11** 47-52

IOP Conf. Series: Materials Science and Engineering **81** (2015) 012028 doi:10.1088/1757-899X/81/1/012028

Development of method of evaluation of concrete damage in the process of cyclic freeze-thawing

T V Fursa[1], A A Demikhova[1] and A P Surzhikov[1]

[1] *Tomsk, 634050, Russia, National Research Tomsk Polytechnic University*

E-mail: fursa@tpu.ru

Abstract.

The research presents a new method for determination of concrete damage under cyclic freeze-thawing by the parameters of the electrical response under impact excitation. It is established that the attenuation coefficient of the energy of the electrical responses under impact excitation can be used as a diagnostic criterion for crack process determination. A higher accuracy of evaluation of crack processes by the parameters of the electrical signal in comparison with the acoustic method is shown.

1. Introduction

Concrete constructions are often operated under significant seasonal variations of temperature. Continuous operation under these conditions leads to the development of crack processes due to the wedging action of water resulting from its freezing in the pores and microcracks.

Acoustic methods are usually used to determine damage in concrete [1-5]. These methods use direct contact with the piezoelectric transducers to generate the ultrasound and to detect the signal. The main problem with these methods is the need for an ideal acoustic contact between the transducer and the test construction.

In recent years many scientists have conducted research to develop non-contact ultrasonic methods [4–6]. Non-contact acoustic methods are based on the recording of surface waves. Therefore these methods depend on the surface roughness and are insensitive to deep cracks.

The method based on the use of characteristics of an electrical signal occurring in concrete under pulsed mechanical excitation can be used to evaluate the extent of damage in concrete [7–10]. The principle of this method consists of the excitation of the sample by a short mechanical impact which leads to the appearance of spherical acoustic waves. The shift of double electrical layers at the boundaries of the components in the heterogeneous material and the emergence of the piezoelectric effect in the piezo-containing materials occur under the influence of the acoustic waves. This gives rise to an external electromagnetic field. This field is detected by the signal receiver placed near the test object. Piezoelectric inclusions are contained in fine and coarse aggregates in concrete. The electrical signal parameters are determined by the processes of acoustic wave scattering on the structural inhomogeneities and defects. Therefore, these parameters can be used to test structural inhomogeneities and defects.

The purpose of this research is development of the criteria for the evaluation of damage in concrete in the freeze-thawing process based on the use of the parameters of the electrical response under impact excitation.

2. Materials and testing procedures

For the research two groups of heavy concrete samples with a size of 100 × 100 × 100 mm were made. For one group of the samples gravel was used as coarse aggregate, for the other group crushed stone was used. After manufacturing, the samples were kept at +25°C and 60% relative humidity for 28 days. Then the samples were exposed to cyclic freeze-thawing. Freezing was carried out in a climatic chamber at a temperature of –30 °C, thawing at a temperature of (+20±5) °C and 95% relative humidity.

The research was carried out with the laboratory hardware–software complex that enables production pulsed mechanical excitation of materials and recording of the electrical response. Figure 1 shows the image illustrating the measurement process of the electric signal from concrete samples.

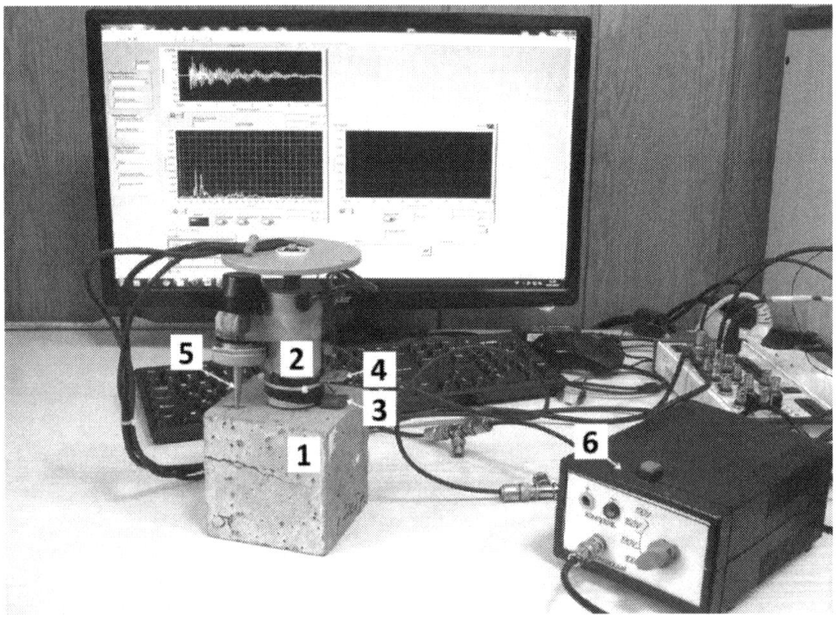

Figure 1. Laboratory hardware–software complex: (1) sample, (2) electromechanical impactor, (3) measuring receiver, (4) compensating receiver, (5) piezosensor, (6) power supply.

The pulsed mechanical excitation of samples is conducted with the electromechanical impact device with a normalized impact force. The differential electrical sensor is used to record the electrical signal. The differential electrical sensor has a measuring and a compensating receiver. The measuring receiver is placed at a distance of 2 mm from the sample surface and perceives both a useful signal and a noise signal. The compensating receiver is placed at a height of 30 mm from the measuring receiver and perceives only a noise signal. The signals from the receivers are inputted to the differential amplifier, where they are subtracted and amplified. This measuring scheme enables reliable detection of the useful electrical signal. The signals from the electrical transducer are recorded using multifunction input-output board "NI PCI 6251" combined with the PC. The multifunction I/O board allows the digitization of time realization of the electrical signal. Then, using a special program in LabVIEW, the signals and spectra are recorded and processed. The measurement technique of the electrical signal under mechanical excitation is described in more detail in paper [11].

The measurement of the acoustic signal by means of a piezosensor, with a tip in the form of a cone (the diameter of the cone is 3 mm), was made in parallel with the measurement of the electrical signal.

3. Experimental research and discussion

Cyclic freeze-thawing leads to the occurrence of cracks in the concrete samples. The scattering of the elastic waves on cracks changes the behavior of the attenuation of the electrical signal. Time-frequency analysis was used to study the behavior of the energy attenuation coefficient of the electrical signal from the concrete samples, because it allows tracking the behavior of change of the spectral energy of elastic waves as a function of time.

The experimental data were processed using a special program in LabView. The program analyzes the changes of the energy of the signal using the sliding window technique.

The main part of the energy of the electrical signals appearing in the concrete samples under elastic impact excitation is in the range of 10 to 25 kHz. Therefore, this frequency range was chosen for the analysis. A detailed technique to determine the attenuation coefficient of the energy of the electrical signal is presented in paper [12].

Figure 2 shows the dependence of the attenuation coefficient of the energy of the electrical signal on the number of freeze-thawing cycles.

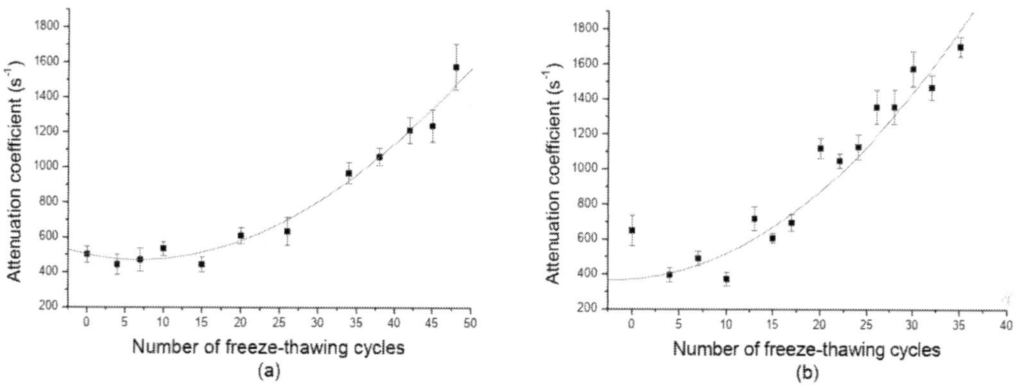

Figure 2. Dependence of the attenuation coefficient of the energy of the electrical signal on the number of cycles: a) for samples with gravel; b) for samples with crushed stone.

As can be seen from Figure 2 the attenuation coefficient of the energy of the electrical signal at the initial stages of cyclic freeze-thawing is virtually unchanged. The attenuation coefficient for samples with gravel (Figure 2.a) increases after the 25th cycle. For samples with crushed stone (Figure 2.b) the attenuation coefficient increases after the 15th cycle. The appearance of visible cracks on the surface of the samples coincides with the increase of the attenuation coefficient. Thus, the attenuation coefficient of the energy of the electrical signals can be used as a diagnostic criterion for evaluation of concrete damages in the process of freeze-thawing.

In a similar way, the acoustic signals were processed from the samples. Figure 3 shows the dependence of the attenuation coefficient of the energy of acoustic signal on the number of freeze-thawing cycles.

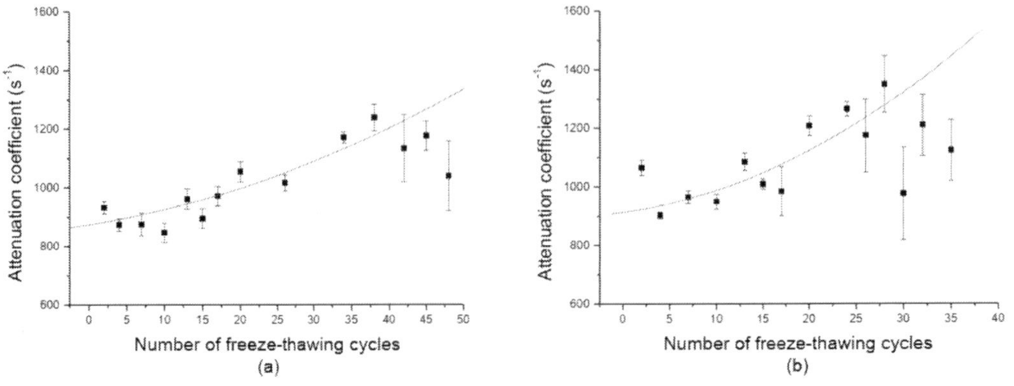

Figure 3. Dependence of the attenuation coefficient of the energy of the acoustic signal on the number of cycles: a) for samples with gravel; b) for samples with crushed stone.

As can be seen from Figure 3, at the initial stage of cracking process for samples with gravel after 25th cycle (Figure 3.a) and for samples with crushed stone after the 15th cycle (Figure 3.b) the attenuation coefficient increases. However, at the stage of the intensive cracking for samples with gravel after the 40th cycle, and for samples with crushed stone after the 27th cycle, a downward bend in the dependence is observed and the attenuation coefficient decreases again and becomes close to that of non-defective samples.

The attenuation coefficient of the energy of the acoustic signal (Figure 3) during the process of freeze-thawing increases by 1.5 times and has a large statistical dispersion. While the attenuation coefficient of the energy of the electrical signal (Figure 2) during the same number of freeze-thawing cycles increases by 3–4 times.

Therefore, the proposed method for evaluation of concrete damage in the process of cyclic freeze-thawing is preferable in comparison to the acoustic method.

4. Conclusion

The purpose of the research was the development of a non-destructive method to test the processes of cracking in concrete by the parameters of the electrical response under elastic impact excitation in the process of cyclic freeze-thawing.

The increase of the attenuation coefficient of the energy of the electrical responses can be used as a criterion for concrete damage evaluation in the process of cyclic freeze-thawing. It is shown that in concrete with different types of coarse aggregates, the difference in the dynamics of crack process is observed. The appearances of visible cracks in the sample with crushed stone occurs after the 15th cycle and in the sample with gravel they occur after the 25th cycle. The attenuation coefficient of the energy of the electrical responses reliably reflects the development of crack processes in concrete samples with different types of coarse aggregates.

A comparison of dependences of the attenuation coefficient of the energy of the electrical and acoustic signals in the process of cyclic freeze-thawing is made. It is shown that the attenuation coefficient of the energy of the acoustic signals at the expansion in the number of freeze-thawing cycles changes slightly and it cannot be used to determine the development of crack processes.

The obtained results show that the parameters of the electrical response under impact excitation more reliably reflect the crack processes compared to the acoustic signal.

The research was supported by the State task "Science".

IOP Conf. Series: Materials Science and Engineering **81** (2015) 012028 doi:10.1088/1757-899X/81/1/012028

References:

[1] Qasrawi H.Y. and Marie I.A 2003 The use of USPV to anticipate failure in concrete under compression *Cement Concrete Research* vol. **33** pp 2017 – 2021.
[2] Van Den Abeele K.E.-A., Sutin A., Carmeliet J. and Johnson P.A 2001 Micro-damage diagnostics using nonlinear elastic wave spectroscopy (NEWS) *Ndt&E Int* vol. **34** pp 239 – 248.
[3] Bui D., Kodjo S. A., Rivard P. and Fournier B 2013 Evaluation of Concrete Distributed Cracks by Ultrasonic Travel Time Shift Under an External Mechanical Perturbation: Study of Indirect and Semi-direct Transmission Configurations *J Nondestruct Eval* vol. **32** pp 25 – 36.
[4] Gudra T. and Stawiski B 2000 Non-destructive strength characterization of concrete using surface waves *Ndt&E Int* vol. **33** pp 1 – 6.
[5] Smith R.L 1984 The use of surface scanning waves to detect surface-opening cracks in concrete *Ndt&E Int* vol. **17** pp 273 – 275.
[6] Kee S.H. and Zhu J 2010 Using air-coupled sensors to determine the depth of a surface-breaking crack in concrete *J Acoust Soc Am* vol. **127** pp 1279 – 1287.
[7] Fursa T.V 2010 Mechanoelectric Transformation in Cement-Sand Composites in the Course of Crack Formation Caused by Freezing-Thawing Cycles *Technical Physics Letters* vol. **36** pp 348 – 350.
[8] Fursa T.V., Osipov K.Yu. and Dann D.D 2011 Developing a Metod Based on Mechanoelectric Transformation for Monitoring Concrete Defectness Variation under Influence of Freezing-Thawing Cycles *Technical Physics Letter* vol. **37** pp 291 – 293.
[9] Fursa T.V., Surzhikov A.P. and Osipov K.Yu 2007 Development of an Acoustoelectric Method for Determining the Porosity of Dielectric Materials *Russian Journal of Nondestructive Testing* vol. **43** pp 95 – 99.
[10] Fursa T.V., Osipov K.Yu. and Dann. D.D 2011 Development of a Nondestructive Method for Testing the Strength of Concrete with a Faulted Structure Based on the Phenomenon of Mechanoelectric Transformations *Russian Journal of Nondestructive Testing* vol. **47** pp 323 – 328.
[11] Fursa T.V., Osipov K.Yu., Lyukshin B.A. and Utsyn G.E 2014 The development of a method for crack-depth estimation in concrete by the electric response parameters to pulse mechanical excitation *Meas. Sci. Technol.* vol. **25** 055605.
[12] Osipov K.Yu. and Fursa T.V 2013 Evaluating the Depth of Open Cracks in Concrete from Parameters of Electric Response to Elastic Impact Excitation *Technical Physics Letters* vol. **39** pp 481 – 483.

IOP Conf. Series: Materials Science and Engineering **81** (2015) 012029 doi:10.1088/1757-899X/81/1/012029

The method for determination of electron-beam characteristics in the dense gaseous medium

R N Rizakhanov[1], A A Barmin[1], R I Rudshtein[1,2] and S Kh Djanibekova[1,3]

[1] SSC FSUE Keldysh Research Centre, 8 Onezhskaya ul., Moscow, 125438, Russia
[2] National Research University Higher School of Economics, 20 Myasnitskaya ul., Moscow, 101000, Russia
[3] Moscow Institute of Physics and Technology (State University), 9 Institutskiy per., Dolgoprudny, Moscow Region, 141700, Russia

E-mail: nanocentre@kerc.msk.ru

Abstract. The work has to do with optimization of vacuum-free material working processes with using the concentrated electron beam. A method for direct experimental determination of primary electrons local characteristics in the dense gaseous medium was proposed. The meter design and the measurement procedure were described. The analytical method for evaluation of the electron beam local parameters in the dense gaseous medium was developed. The method was used for evaluation of kinematic characteristics of detector displacement device.

1. Introduction

The electron beam ejection in the dense gaseous medium (including the air atmosphere) essentially enhances the technological capabilities of an application of electron beams [1]. Vacuum-free realization of such processes, as the surface hardening, the welding, the cutting of metals, raises their productivity in comparison with the vacuum implementation. New possibilities to apply the concentrated energy fluxes of this kind to solving tasks of plasma chemistry, flue gas cleaning of toxic impurities etc are opened up. It is worthy of note, that one of the currently most popular areas of technological vacuum-free application of the electron beams is the hardening of near-surface layers of railway rails aimed at lowering of their wear and an increase of their service life.

The optimization of vacuum-free material working processes with using the concentrated electron beam, which allows one to select an appropriate operational regime and parameters of a generator used, needs a knowledge of such characteristics, as distributions of electron beam current and power densities on the axis and the beam radius. Of equal importance thereat is the proficiency in methods for both analytical, and experimental determination of the magnitudes mentioned. Elaboration of the given methods is the main objective of the present work.

2. The analytical evaluation of electron beam local parameters in the dense gaseous medium

The method for analytical evaluation of the beam current density was described in [2] in the context of the phenomenological model. Using the relationships obtained, we express the power

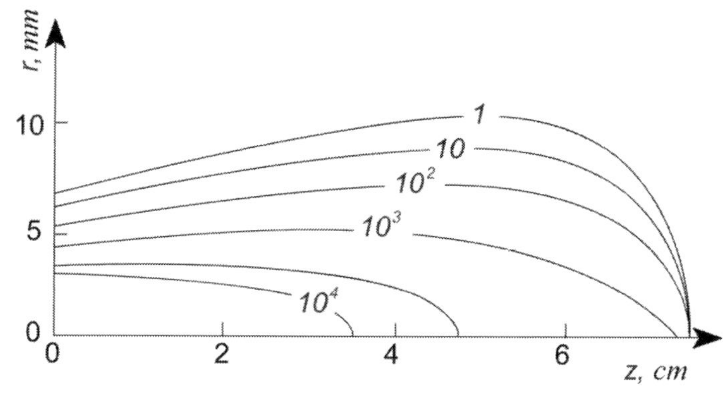

Figure 1. The isodensity lines of the primary electron beam power, W/m^2. The values were obtained for the atmospheric air in the normal conditions, $E_0 = 80$ keV, $r_0 = 2$ mm.

density of the primary electron beam at an arbitrary point at a distance of r from the axis of the beam symmetry and the axial distance z from the beam ejection window through the current density $J(r,z)$ and the mean electron energy at the cross-section $E(z)$:

$$P_{pr}(r,z) = \frac{J(r,z)}{e}E(z) = E(z)\frac{j_0}{e}\exp\left\{-\frac{z}{a} - \left[\frac{r}{b(z)}\right]^2\right\},$$

where j_0 is the axial density of the primary electron current at the initial section; a is the magnitude defined by the absorption coefficient; e is the electron charge.

In [2] is shown, that the function $b(z) = r_0\sqrt{f(z)e^{z/a}}$ refers to the envelope of the beam, which defines, for the given section z, a distance from the beam axis, where the value of the current density is e times less than the axial value, r_0 is the typical radial size of the beam ejection area.

The function $f(z)$ defines the law of degradation of the beam full current, as it recedes from the ejection window, and usually is approximated by the linear dependence $f(z) = 1 - z/L$, where L implies the extrapolated range of electron.

The mean electron energy at the section in hand z [3]:

$$E(z) = \left(E_0^{5/3} - \frac{\rho z}{C}\right)^{3/5},$$

where E_0 is the mean electron energy at the initial section, ρ is the medium density, C is the constant specific for the medium substance.

Presented in Fig. 1 is the pattern of the power density distribution for the primary electron beam on the coordinates r and z.

3. The experimental method for determination of electron-beam local parameters in the dense gaseous medium

For direct experimental measurement of primary beam parameters, the method of moving collector with a small aperture is proposed. It consists in a sequential expansion of the electron beam cross-section (with the help of a moving aperture) into laterally small matrix elements and in measurement of currents of these elements. A so found dependence of the current, passed through the aperture, on a location of this opening (the coordinates of its centre) will define the function of the current density distribution along the line of a travel of the aperture.

It should be recorded, that initially the electron gun generates a monoenergetic electron beam, but because of the passage through the dense gaseous medium a beam spreading in the spectrum takes place. In doing so, the concentration of secondary electrons produced exceeds significantly that of primary electrons ($10^4 \div 10^5$ times), and therefore, the plasma current density is considerably in excess of the primary beam current density. This circumstance results in a difficulty for measurement of the primary electron current and poses the problem to separate fast electrons with typical energies 20 to 80 keV and slow (plasma) electrons with typical energies 1 to 3 eV. As a solution of the problem, a passive separation of the indicated classes of particles may be proposed when using a thin dielectric layer on a basis of the energy dependence of the depth of electron penetration in a substance [3]. In the case the thin dielectric layer fulfils the role of a peculiar filter, allowing the high-energy electrons to easily pass and the low-energy electrons to be confined (to be "cut off").

For the typical energies of the plasma electrons at hand the effective range of electrons in the substance with charge Z and the mass number A is related to the electron energy by the relationship Kanaya-Okayama [4]

$$R_0 = 2.76 \cdot 10^{-6} \cdot \frac{AE^{1.67}}{Z^{0.889}},$$

where the energy is expressed in keV. In the above formula the law of energy loss was chosen in such a manner, that the electron range obtained would provide a most precise approximation to the size of the interaction area in respect to the depth. The magnitude found may be interpreted as the radius of semicircle with the centre at the point of incidence of the beam, which defines the envelope of electrons trajectories. Given in Tables 1 and 2 are the characteristic sizes of the depth of the electron free range in different dielectric (Ta_2O_5, Al_2O_3, TiO_2) and conducting (Al, Cu, W) materials, respectively.

Table 1. The depth of the electron penetration into dielectric material, nm.

E_0, keV	0.4	0.7	1	2	3	4	5
Ta_2O_5	2.5	6.4	11.5	36.7	72.3	116.8	169.6
Al_2O_3	4.0	10.1	18.3	58.3	114.7	185.4	269.1
TiO_2	3.9	10.0	18.1	57.4	113.1	182.8	265.4

Table 2. The depth of the electron penetration into conducting material, μm.

E_0, keV	40	50	60	70	80	90	100
Al	13.4	19.4	26.3	34.0	42.5	51.8	61.7
Cu	4.7	6.8	9.2	11.9	14.8	18.1	21.6
W	2.7	3.9	5.3	6.9	8.6	10.5	12.5

The scheme of the measuring probe and the direction of its motion by scanning procedure are illustrated in Fig. 2. The arrangement of the probe presented makes it possible to measure parameters of beam primary electrons in a wide energy range.

IOP Conf. Series: Materials Science and Engineering **81** (2015) 012029 doi:10.1088/1757-899X/81/1/012029

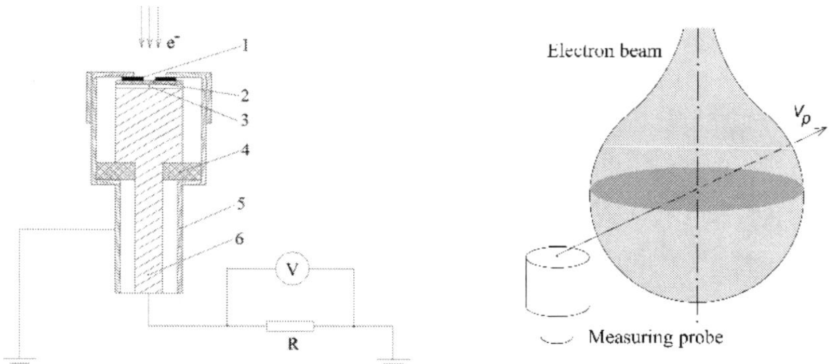

Figure 2. The measuring probe scheme (to the left) and the scheme of the regime of the beam scanning (to the right): *1* — the replaceable metal diaphragm; *2* — the dielectric screen for secondary electrons "cutoff"; *3* — the target (the current collector); *4* — the dielectric washer; *5* — the probe housing; *6* — the inner collector.

The probe is disposed in space, so that its axis is parallel to the beam ejection axis, i.e. the electron flux is chiefly directed on the normal to the probe working area. The motion of the probe is realized at a plane, perpendicular to the beam longitudinal axis. At the incidence of the electron beam to the probe working area the diaphragm *1* "cuts off" the electron beam in its cross-section, i.e. a small area total current component to be measured is selected. Then, the flux passes through the dielectric screen *2*, blocking off the slow secondary electrons, whereupon it strikes the metal target *3* (the current collector). The current collector transfers an arrived charge through the collector *6* to the meter for the following recording and processing.

The filtering out screen *2* and the current collector *3* are structurally a single part in the form of metallic disk, to which a thin coat of the dielectric (e.g., aluminum oxide) was applied. The thickness of the coating layer is about 100 nm. Starting from adaptability to fabrication, the current collector and the collector are made as individual parts, between which a good electric contact should be provided.

4. The evaluation of kinematic characteristics of the detector displacement unit and a measurement error of the method proposed

A rather long thermal action of the electron beam can lead to a fusion of the probe parts and, as a result, to a loss of operability. The most vulnerable (concerning this aspect) part of the probe is the current collector.

To reduce thermal load on the detector construction down to an acceptable level it is worthwhile to use the regime of scanning, when the probe intersects the electron beam with rather a large constant velocity v_p at some distance from the ejection window z.

We estimate the displacement velocity of the probe, necessary that the sensing element of the current collector not fail. As the velocity is perceived to be high, heat conduction and radiation from the surface may be ignored. In the process of measurements the current collector material temperature T is not to exceed the critical temperature T_{max}, at which irreversible phase changes in material start to take place.

Obviously the most stressed thermophysical conditions will come into being with detector moving along a chord, going through the axis of symmetry of the electron beam. We restrict ourselves to integration down this trajectory of motion as an upper bound for current collector heating. The energy, gained by unit area of the current collector at a distance z from the exit

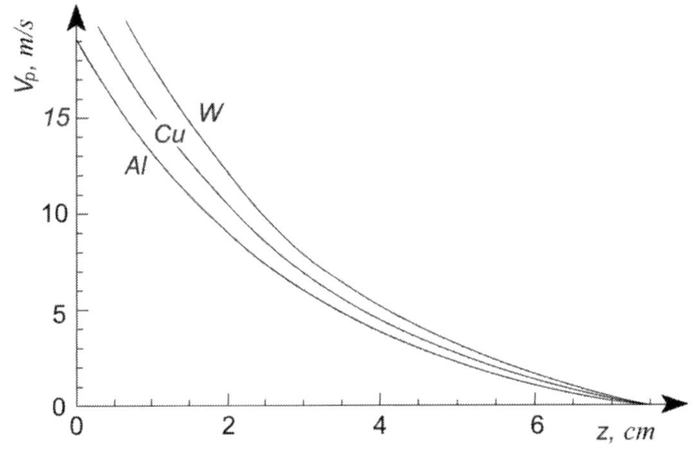

Figure 3. The minimal linear velocity of the probe displacement as a function of the distance between the ejection window of the concentrated electron beam generator and the probe for different materials of the current collector. The values were obtained for the atmospheric air in the normal conditions, the initial electron energy is 80 keV, the beam radius at the initial section 2 mm, the beam full current 60 mA.

section, equals:

$$q(z) = 2\int_0^{\infty} \frac{j(r,z)}{e} E(z)\frac{dr}{v_p} = \sqrt{\pi}\frac{E(z)}{v_p} \cdot \frac{j(z)}{eb(z)}.$$

The final estimate for determination of the detector minimum velocity appears as follows (Fig. 3):

$$(1-\eta)q(z) \le \rho C_p \Delta T_{max} \Delta z_{ep},$$

or finally

$$v_p \ge \sqrt{\pi}\frac{(1-\eta)E(z)}{\rho C_p \Delta T_{max}\Delta z_{ep}} \cdot \frac{j(z)}{eb(z)},$$

where ρ, C_p are relevant thermophysical constants of the current collector material; Δz_{ep} is the depth of penetration of primary electrons in the current collector material; $\Delta T_{max} \equiv (T_{max} - T)$ is a difference between the maximum allowable and the initial temperatures of the current collector; η is the reflection coefficient of primary electrons from the current collector surface.

For a structure of the "film-on-massive-substrate" type the reflectance η is calculated with using the formula [5]:

$$\eta = \eta_s + \eta_f(D)\left[1 - \frac{\eta_s}{\eta_{f\infty}}\right].$$

In the above expression η_s is the reflectance of electrons from a massive target-substrate, defined by the equation [6]

$$\eta_s(Z,E) = E^{0.1382-0.9211\cdot Z^{-0.5}} \cdot \left(0.1904 - 0.2236 \cdot \ln Z + 0.1292 \cdot \ln^2 Z - 0.01491 \cdot \ln^3 Z\right),$$

where Z is the atomic number of element, E [keV] is the electron energy. This formula is applicable at $5 \le Z \le 80$, $4 \le E(\text{keV}) \le 40$.

The magnitude $\eta_f(D)$ defines the reflectance of the dielectric film with the thickness D [5]

$$\eta_f(D) = \frac{a-1}{a+1}\left[1 - \left(1 - \frac{2a}{a-1}\frac{D}{R_0}\right)\cdot\left(1 - \frac{D}{R_0}\right)^a\right],$$

where R_0 is the range of electron in the film material, $a = 0.045Z$, $10 < Z < 40$, $D \ll R_0$.

The magnitude $\eta_{f\infty}$ is defined as the reflectance from a film of identical composition for a case of its infinite thickness.

When using a material representing a uniform chemical compound out of a few elements, the following rule, allowing for the reflection factors η_i and the mass fractions C_i of components, can be applied:

$$\eta_{tot} = \sum_i \eta_i C_i.$$

As is seen from Fig. 3, aluminum and copper, in regard to a totality of thermophysical characteristics, are more preferential for manufacture of the current collector, than refractory metals (tungsten, tantalum and others), as they require lesser kinematic loads on the detector displacement unit. This is caused by a circumstance, that a characteristic volume of the area of the electrons interaction with a substance for copper and aluminum exceeds substantially that for refractory metals due to a considerable difference between the depths of electron penetration in the materials indicated (Table 2).

The measurement error is a sum of the random and systematic errors. The random error depends on measurement conditions and in most cases it may be minimized at a level no more than 5%. The systematic error, stemming from imperfection of the meter design, may be decreased to aperture error, i.e., an error, correlated with terminal sizes of the analyzing aperture in the diaphragm.

The aperture error depends on the current density distribution at the cross-section of the electron beam and the radius of the diaphragm aperture:

$$\delta_a = \left[1 - \frac{j_{avr\ min}}{j(x,y)}\right]\cdot 100\%,$$

where $j_{avr\ min} = \int_{S_a} j(x,y)dxdy/\pi r_a^2$ is the minimum current density, averaged over the detector diaphragm aperture area. For lowering of this error, it is required to lessen a difference between real and measured values of the current density, which calls for a decrease in the radius of the diaphragm aperture (as compared with the radius of the electron beam).

The least value of the current, that can be measured with a specified accuracy, will be determined by a meter sensitivity [7]. The latter is restricted to the level of fluctuation noises at the meter exit. The sensitivity curves, defined as dependences of the minimum value of the electron beam current density $j_{avr\ min}$ at the plane of the diaphragm aperture, on the radius of the diaphragm aperture are shown in Fig. 4. The curves are given for different values of the magnitude β, equal to the detected signal-to-noise level ratio.

At a prescribed configuration of the detector (i.e., the diameter of the diaphragm aperture) the limiting minimum value of the current density, that the detector is capable of recording, is defined by a curve, relevant to $\beta = 1$ — the case, when the level of desired signal may be compared with the noise level. The indicated minimum value of the current density controls the maximum size of the working area in the direction of the beam ejection axis, i.e. the limiting distance from the ejection window, at which measurements are possible to be conducted. The curve, corresponding to $\beta = 10$, may be considered to be a level of stable detection of a signal. By the removal on the plot in the direction of the curve $\beta = 10$ a transition to a region of the stable detection of signal with reference to the noise level is occuring.

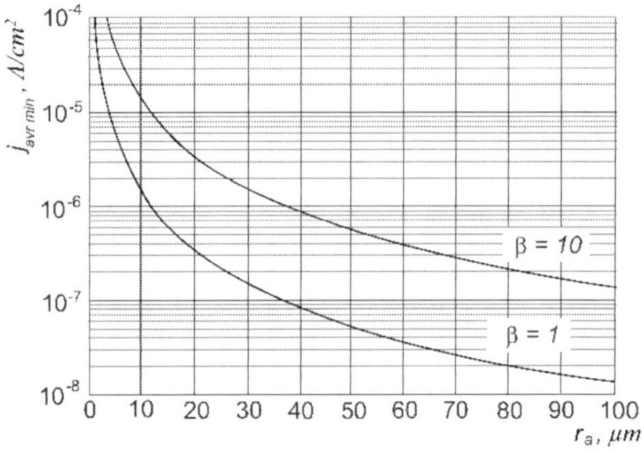

Figure 4. The curves of the meter sensitivity.

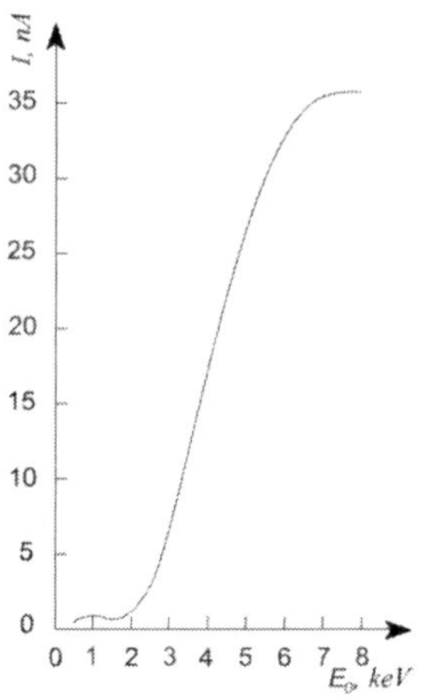

Figure 5. The experimental dependence of the current, passed through the sample with dielectric film, on the electron beam energy.

5. Results and discussion

For verification of functional performance of a current collector prototype with the target, screened with dielectric coating, the experimental investigations were conducted with using the electron source of the scanning electron microscope FEI Quanta 600 FEG.

The samples prepared had a two-layer structure. The copper plates with characteristic dimensions $10 \times 15 \times 2$ mm were used as a substrate. The samples were put through preliminary cleaning of possible impurities and oxides by way of boiling in alcohol, after which the dielectric aluminum oxide Al_2O_3 films with the thickness of about 100 nm were applied to their surface with the atomic layer deposition on the MIR-7 facility.

In the course of the experiment various points of the coating surface were subjected to the action of the electron beam at different values of the accelerating voltage. All samples were in good electric contact with the objective table of the electron microscope vacuum chamber, from which values of the current, passing through the test two-layer sample, were taken. Measurements were conducted with the help of a digital electron picoamperemeter Keithley 6485, connected to the table.

The typical dependence of the current, passed through the sample, on probing electron beam energy is shown in Fig. 5. On the plot presented, a sharp increase in the current with a rise of the accelerating voltage from 2 to 5 kV is clearly defined. This is consistent with the design data, given for aluminum oxide.

Thus, the results of the experimental investigations into dielectric coatings accord well with the calculation results and verify an applicability of this technique for secondary electrons cutoff and feasibility of similar coatings in the structure of the projected meter.

Notice, that the direct measurement of the beam current density is a favourable distinguishing feature of the technique proposed in comparison with similar techniques with a stretched moving collector, as it does not need a subsequent mathematical processing of experimental data base and introduction of supplementary assumptions, implying the axial symmetry of a plasma structure.

6. Conclusions

(i) In the work the analytical method for evaluation of electron beam local parameters in dense gaseous medium was developed and applied. The distribution of the power density for primary electrons on the axis and the radius of the beam was obtained.

(ii) The technique for the direct experimental determination of local characteristics of beam primary electrons in the dense gaseous medium was proposed. The meter design and the measurement procedure with using a passive separation of high-energy and plasma electrons were described. Kinematic characteristics of the detector travel unit were evaluated with using the mathematical model. The error of the technique proposed was estimated.

(iii) The model experimental investigations into separating properties of the dielectric film of aluminum oxide, applied to the copper substrate, at different energies of incident electrons were conducted.

References

[1] Yong-feng D, Rong L, Xian-wei H, Chang T, Rizakhanov R N and Barmin A A 2011 *Journal of Rocket Propulsion* 60–66
[2] Barmin A A and Rizakhanov R N 2007 *Applied physics* 115–118
[3] Schumacher B W 1965 *Electron and Ion Beam Science and Technology* ed Bakish A (John Wiley and Sons)
[4] Goldstein J I, Newbury D E, Echlin P, Joy D C, Fiori C E and Lifshin E 1984 *Scanning Electron Microscopy and X-Ray Microanalysis* (New York and London: Plenum Press)
[5] Saparin G V 1988 *Introduction to scanning electron microscopy* (Moscow: Moscow State University Press)
[6] Hunger H J and Küchler L 1979 *Phys. stat. sol. (a)* **56** K45–K48
[7] Balenkin V I and Ivanov A N 1978 *Methods for analysis of electron beams* (Leningrad: LETI)

IOP Conf. Series: Materials Science and Engineering **81** (2015) 012030 doi:10.1088/1757-899X/81/1/012030

The possibility of using phase transition in Ni and Cu nanoclusters for information recording processes

Yu Ya Gafner[1], S L Gafner[1]

[1]Abakan, pr. Lenina 90, 655000, Russia, Katanov State University of Khakassia

E-mail: ygafner@khsu.ru

Abstract.

The opportunity of transition metals nanoclusters' usage as a data bits in memory devices for the recording transition "order-disorder" has been analyzed. Therefore, with the help of the molecular dynamics method on the basis of TB-SMA potential the simulation of metal nanoparticles (D = 1.6 - 5.0 nm) crystallization processes have been studied. Influence of various conditions of crystallization on formation of internal structure in metal nanoclusters is investigated. The stability boundaries of various crystalline isomers are analyzed. The obtained dependences are compared with the corresponding data obtained for copper and nickel nanoparticles having similar sizes. The limiting size of nanoparticles is determined, for which a structural "order-disorder" transition necessary for the data recording is still possible.

1. Introduction

The origin of many nanotechnologies is the natural answer to the rather limited possibility of solving some difficult technical problems to methods and approaches which are typical of microelectronics. One of these typical problems is the problem of development of technical characteristics of memory devices. Besides, from the technical view point, the most interesting decisions are those that allow multiple information rerecording. From the point of view of the data storage system development, magnetic, optical and structural-phase environment characteristics used during the recording are the most interesting ones. We will examine only one of these methods, exactly the information recording methodology with using systems that can change phase state of the carrier. The central idea is to modify the phase state of the carrier material locally, to store this phase as long as it is possible, to discover and read the recorded bit and if it is necessary to delete it many times, i.e. to return the material in the initial phase state.

The most appropriate for this phase transition is the transition from crystalline state into amorphous one, though other variants are possible too. The recording principle is the principle that some carrier is in the crystalline state; by the information recording the laser beam heats a small area with a short-time pulse to the temperature that is higher than its melting temperature. After laser action cancellation of the heated area with the typical size about 100 nm the temperature falls very fast with the heat retraction speed more than 10^9 K/s. At such extreme cooling rate the area is being crystallized into amorphous stage, having other in comparison with crystallized areas optical, electrical and some other properties. Thus, the recording of one data bit occurs under such conditions. It is necessary to melt this carrier area again up to the lower recrystallization temperature and wait for full transition from amorphous stage into crystallized one in order to erase this bit.

Based on the above stated, the main purpose of this work is a detailed study of possibility to use clusters of different transition metals as a single bit of information in long-term memory devices, based on principle changing of phase memory of the carrier. As the objects of this research the clusters of copper and nickel were chosen, as exactly these clusters of the given metals are successfully synthesized of different physical-chemical methods now.

2. The computer model

Internal structure of clusters plays an important role in understanding technology of creation of newest informational resources with phase-change memory of the carrier, however, determination of a cluster structure with the help of experimental methods deals with considerable difficulties [1]. Because of the practically complete lack of true experimental data according to the definition of the single metal clusters structure of a small size (units of nanometers) many researchers deal with methods of computer simulation [2]. On this basis we have investigated computer simulation of structure formation processes for copper and nickel clusters conditioned upon ultrahigh-speed hardening in order to find out existing basic regularities and analysis of arising differences.

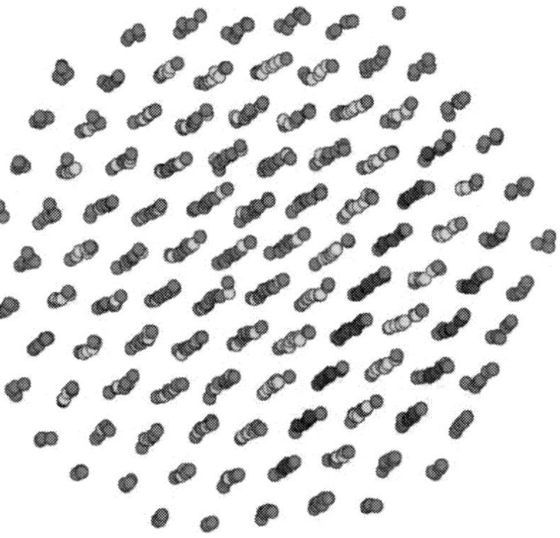

Figure 1. Ni nanocluster (D = 2.5 nm) with a hcp structure.

For molecular dynamics simulation we used the modified tight-binding potential with a fixed cutoff radius corresponding to the fifth coordination shell inclusively [3] in order to calculate the interatomic forces. On the whole, when analyzing the results of various authors, we conclude that the Cleri-Rosato potentials [3] satisfactorily describe the behavior of a number of thermodynamic quantities of metallic fcc clusters on qualitative and quantitative levels. The differences between the simulation results and the data obtained by other collective potentials do not exceed 3–5%.

During simulation, the temperature was determined by the average kinetic energy of atoms calculated with the rapid Verlet algorithm at a time step h = 2 fs [4]. The formation of small metallic particles from a liquid phase was studied along a constant energy E line (microcanonical ensemble). A microcanonical ensemble was simulated by the Andersen method [5]. This molecular dynamics (MD) method is used for an isolated system with a fixed number of particles N, system pressure P, and total energy E (NPE ensemble). The advantage of this approach consists in the fact that the Andersen thermostat can imitate the technique of cooling of nanoclusters at various rates.

3. Results and discussion

In order to study the effect of kinetics on the formation of the structure of nickel and copper clusters, the process of melting of nanoparticles with their subsequent cooling into a crystalline phase was imitated by the molecular-dynamics method. The examination of the received results we should start with nanoclusters of nickel that are already extensively used in different kinds of nanotechnological applications. The main result of simulation is that while heating of nickel nanoclusters of different diameter (D = 1.6 – 3.6 nm) the melting temperature is getting higher with their following gradual cooling the process of structure formation strongly depends on the speed of the thermal energy removal.

In [6] with the help of MD simulation method the question of structural ordering has been partially investigated but only in melted polycrystalline nickel. For this purpose, the macroscopic system at temperatures corresponding to the liquid (T = 1850 K) and supercooled state (T = 1600 and 1430 K) was simulated. The analysis was conducted according to the following several characteristics: structural factor, the angular distribution relations and CNA analysis [6]. The received in [7] results show that in the molten bulk Ni, in most cases, the areas corresponding to icosahedral and distorted icosahedral phases are being formed. However, during the system supercooling the situation is getting more difficult and other structural combinations are developing.

The analysis of the received results concerning the atomic dynamics in case of nanosizes has shown that for the Ni clusters with diameter not more than 2.8 nm the following tendency is clearly seen: if cooling time t = 0.2 ns; it means that icosahedral (Ih) phase is prevailed in these structures. With the increase of cooling time up to t = 2.0 ns the part of the clusters with icosahedral construction of atoms will uniformly decrease. As for the contention of fcc and icosahedral (Ih) phase with reducing of cooling rate, the investigated studies have shown the influence of not only the cooling rate, but also the size effects.

Thus, the frequency of the formation of the icosahedral phase decreases continuously with increasing cluster size, being accompanied with a corresponding increase in fcc phases. And if for the clusters not more than 1.6 nm in diameter at crystallization time $t = 2.0$ ns the fcc structure was not observed, the percentage of fcc phase for the nanoparticles with 2.54 nm in diameter reached about 40%. The occurrence probability of Ih modification if the sizes of nanoclusters are the same decreased more than two times (from 90 to 40%), as for the large particles fcc structure is the most suitable.

Figure 2. The structure of amorphous type of Ni cluster ($D = 2.9$ nm) with a nucleus of icosahedral phase, received during the cooling of the particle from 1500 K to 300 K during the time $t = 0.2$ ns.

At high cooling rate ($t = 0.2$ ns) for Ni clusters with $D < 2.8$ nm only in 10% of experiments the fcc structure was being formed. Here we can observe the reconstruction of the clusters structure towards the icosahedral phase. The observing situation can be explained with the following reason: the process lead time is not enough for clusters structuring with the fcc lattice. The occurrence of hexagonal close-packed configuration was registered only once at the cooling of nanocluster with diameter 2.54 nm ($N = 791$ atom) at a cooling time $t = 2.0$ ns (Figure 1). Note that for all the clusters we simulated, the probability to meet the cluster with a decahedral (Dh) symmetry, that we can practically examine as intermediate between fcc and Ih, at any cooling rate fitted in a range 20 – 40%, gradually increasing with the increase of the cluster's size.

The observed tendency was violated only for clusters with a diameter more than 2.8 nm. Clusters of such size at the high rate of crystallization corresponding to the time of cooling from 1500 to 300K for 0.2 ns frequently have no time to properly reorganize the atomic surroundings for attainment of a state with the least possible energy with the result that an amorphous phase begins to manifest itself to a great extent.

Thus, at $D = 3.6$ nm under the conditions of such continuous cooling the amorphous modification was observed in 30% of the experiments performed, and the icosahedral modification, only in 10%. When, however, the time of the process reached $t = 0.5$ ns, the amorphous phase was noted in 10% of the cases with a threefold growth of the probability of the formation of five-particle (icosahedral) symmetry. When the crystallization proceeded most slowly (cooling time $t = 2.0$ ns), the amorphous structure was no longer realized and the icosahedral structure took place in 40% of the simulations. It is necessary to mention that many observing structures of the amorphous type had the attempt of the transition to Ih modification (Figure 2).

Recent MD simulations of Ni nanoclusters [8, 9] indicate that the icosahedral structure must be prevailing one to cluster sizes on the order of 1000 atoms. However, from our studies presented in this, it follows that the situation proves to be not so clear. When the real cluster structure of nickel is formed, it is necessary to take into account both thermodynamic aspects of cluster forming and kinetics of the process. On top of that, the size of the synthesized particles is of great significance.

The pointed above regularities were checked on nanoclusters of copper. Also the methodology of the gradual cooling of Cu clusters from the molten state at a room temperature $T = 300$ K with several fixed speeds of cooling was simulated. Similarly to Ni nanoclusres the process of the structure formation depended very strongly on the cooling conditions, namely, the frequency of the formation of the icosahedral phase increased fairly uniformly with increasing rate of the process. From our data it can be seen that at a cooling time of $t = 0.2$ ns the face-centered cubic structure is formed in 5–10% of the performed experiments and the icosahedral structure is formed in 60–70% of the experiments (Figure 3). As the cooling time increases to $t = 2.0$ ns, the frequency of occurrence of the face-centered cubic modification increases to 30–40%, whereas the frequency of occurrence of the icosahedral modification decreases by a factor of 2, i.e., to 30%.

Figure 3. Cu nanocluster ($D = 3.2$ nm) with an icosahedral structure.

The above tendency is violated only for clusters with diameters larger than 3.5 nm. The clusters of this size at a high crystallization rate ($t = 0.2$ ns) have no time to form a regular structure, and, hence, the amorphous structure is dominant. In particular, the clusters at $D = 3.7$ nm upon smooth cooling have the amorphous structure in 60% of cases and the icosahedral structure in only 20% of cases, whereas the face-centered cubic structure is not formed at all. However, when the cooling time increases to $t = 0.5$ ns, the amorphous phase is formed in only 30% of experiments. At the cooling time $t = 2.0$ ns, the amorphous structure is not observed.

The model of gradual cooling (NPE ensemble) of small metal particles allows to compare the results of MD simulation for Ni clusters with data for Cu nanoparticles, researched in the same conditions. The comparative analysis of computer experiments results shows the presence of the same regularities in formation of inner clusters structure: at slow passing process of crystallization the close-packed structures are being formed, in case of quick cooling the icosahedral phase is mainly being formed. At high cooling rate the prevail structure of Cu and Ni clusters with diameter more than 3.5-3.7 nm will be amorphous one.

Our results are in agreement with the data obtained by Liu et al. [10], who also used the Cleri-Rosato many-body potential in the study of the sample containing 500 copper atoms distributed in the cubic cell with periodic boundary conditions. In [10], the authors investigated the influence of the cooling rate on the formation of different structural modifications. Although the simulated cooling rate Q in [10] differed by two orders of magnitude ($Q_0 = 8.05 \cdot 10^{13}$ K/s, $Q_5 = 3.62 \cdot 10^{11}$ K/s), the chosen cooling rates seem to be unreasonably high and cannot be reached using available technical methods. Nonetheless, the performed simulation clearly confirmed the role of the cooling conditions in the formation of a specific structure.

In particular, the data obtained in [10] make it possible to draw a similar conclusion that the face-centered cubic phase is predominantly formed at a low cooling rate and that the rapid crystallization leads to a considerable increase in the content of the icosahedral structural modification. Moreover, the sample simulated in [10] at a high cooling rate is characterized by a high percentage of amorphous and defect structures. In [10], the authors did not uniquely determine the role of the cooling rate in the development of the hexagonal close-packed modification; however, it was noted that the face-centered cubic and hexagonal close-packed modifications are frequently observed in combination and form a mixed structures.

4. Conclusions

In this represented work that deals with molecular dynamics method on basis of modifying potential of tight binding, the processes of internal structure in nanoclusters of copper and nickel on conditions of ultrahigh-speed hardening typical of memory devices with phase-change memory of the carrier have been studied. To analyze the formation processes, we used the structureless clusters formed upon melting of primary fcc nanoparticles and cooled them to room temperature. Various structural modifications were shown to form in cooling from a liquid phase, and some criteria of their stability were determined. In simulating, we revealed the role of the size factor and the heat removal rate in the structure formation in metallic clusters.

The comparative analysis of computer experiments results shows the presence of the same regularities in formation of clusters structure of copper and nickel: during the passing process of crystallization within several nanoseconds the close-packed structures are being formed, in case of quick cooling the icosahedral phase is mainly being formed. Let's notice, that such clusters in overwhelming majority of cases had quite regular forms. If we increase the cooling rate in 10 times the prevail structure of Cu and Ni clusters with diameter more than 3.5-3.7 nm will be amorphous one.

It is necessary to take into consideration all these peculiarities when using small nanoclusters Ni or Cu in devices with phase-change memory of the carrier. From the technical view point the transition from crystallized condition into amorphous one is more appropriate, but the use of transition of the FCC - icosahedrons type is also possible that is more reasonable at sizes of one bit metal domain to 3 nm. The structural transition "order-disorder" with the particle size of more than 3.5 nm for a data recording can be applied.

ACKNOWLEDGMENTS
This study was performed within the framework of the State Task of the Ministry of Education and Science of the Russian Federation.

References:

[1] Cleveland Ch L, Landman U, Schaaff Th G, Shafigullin M N, Stephens P W and Whetten R L 1997 *Phys. Rev. Lett.* **79** 1873
[2] Frenkel A I, Nemzer S, Pister I, Soussan L, Harris T, Sun Y and Rafailovich M H 2005 *J.Chem.Phys.* **123** 184701
[3] Cleri F and Rosato V 1993 *Phys. Rev. B: Condens. Matter* **48** 22
[4] Tao Pang 2006 *An introduction to computational physics* (Cambridge: University Press)
[5] Andersen H C 1980 *J. Phys. Chem.* **72** 2384
[6] Honeycutt J D Anderson H C 1987 *J. Chem. Phys.* **91** 4950
[7] Bertsch G 1997 *Science* **277** 1619
[8] Mannien K and Mannien M. 2002 *Eur. Phys. J. D.* **20** 243
[9] Qi Y Cagin, T Johnson W L, Goddard W A 2001 *J. Chem. Phys.* **115** 385
[10] Liu C S, Xia J, Zhu Z G and Sun D Y 2001 *J. Chem. Phys.* **114** 7506

Research of the thermal-tension condition and the elemental composition gradient changes of binary systems produced by combined ion-plasma method

A I Blesman[1,2], D V Postnikov[1,2] and D A Polonyankin[1,2]
[1]Department of Physics, Omsk State Technical University (OmSTU), 11 Mira avenue, Omsk, 644050, Russia

[2]Scientific-educational center of nanotechnology, Omsk State Technical University (OmSTU), 11A Mira avenue, Omsk, 644050, Russia

E-mail: blesm@mail.ru

Abstract. To increase the life cycle of the constructional steel products working at high temperatures in air environment the combined ion-plasma method of surface modification was elaborated. Using the method described in the paper, constructional steel J24056 samples with different refractory metal coatings such as molybdenum and tantalum and film-thickness up to 10 microns were prepared. The calculations of the temperature distribution and the tension on the depth of the specified sample brand coated steel were performed. The research of oxygen distribution in the surface layer after high temperature annealing in an air atmosphere has been conducted. An estimation model of the oxygen distribution within the grains in a binary steel-coating system is proposed in the paper.

1. Introduction

To increase the life cycle of constructional steel products [1], working in the conditions of the high temperatures in an oxygen atmosphere, a combined ion-plasma method of surface modification that includes ion implantation and ion-plasma magnetron sputtering was developed. Ion implantation is performed before magnetron sputtering to improve an adhesion of the constructional steel surface [2-4]. It should be noted that the ion implantation of modifier material before its sputtering improves an adhesion from 5 to 10 times, which has a significant influence on the physical, mechanical and performance properties of the products [5]. The main problems discussed in the paper are the temperature and tension distribution in a binary «steel-coating» system calculation, mass transfer of small impurities inside the grain model creation, the combined ion-plasma method elaboration, explanation of modifier material selection, as well as study of the structure and composition of the binary «steel-coating» system after annealing in air atmosphere.

2. The temperature distribution

For the research of the thermal-tension condition of constructional steel J24056 with the coating, as modifier materials were selected the following elements: titanium, nickel, chromium, vanadium, molybdenum, tantalum, tungsten and niobium [6, 7]. The selection of these elements as modifier materials was determined by their protective properties [8-11]. For the development of the constructional steel products modification method and determination the conditions of their subsequent operation, the information about the dynamic field of temperature distribution is required. Application of both the energy conservation laws and Fourier's law to analysis for heat conductivity in the motionless isotropic environment led to the differential equation of heat conductivity, which connects the temporal and spatial changes of temperature [12]. During the incidence of heat flow onto a flat plate, the problem of the temperature distribution can be considered as one-dimensional because of the uniform distribution of heat flow on the surface area of the plate and the plate area being much greater than the thickness of the layer under study (surface area is much greater than the thickness of the modifier layer).

3. Distribution of internal tensions caused by the temperature gradient

Consider the tensions occurring in the material during thermal heating. Initially, consider the two-layer plate which is heated by gas flow unevenly on one side. The wave of dynamic tension along the OX axis quickly fades; however, there is tension in the cross direction of the OX axis. This tension is caused by the temperature gradient along the OX axis and the heterogeneity of the properties in multi-layered samples. The corresponding components can be determined from the following expression (1) [13]:

$$\sigma_z = \sigma_y = \frac{\alpha(T)E(T)}{1-\mu} T(x) + C_1 + C_2 x \tag{1}$$

Here $\alpha(T)$ is the coefficient of thermal expansion, E is the elastic modulus, μ is the Poisson's coefficient, C_1 and C_2 is the constants that are defined from a condition that the resultant force and the resultant moment are equal to zero [13]. It is necessary to take into account the boundary conditions, similarly, σ_z [13]:

$$\sigma_z = \sigma_y = -K(x)T(x) + \frac{K(x)}{h}\int_0^h T dx + \frac{3K(x)(x-\frac{h}{2})}{\frac{h^3}{4}}\int_0^h T(x-\frac{h}{2})dx \tag{2}$$

Here $K = \frac{\alpha(x)E}{1-\mu}$. This type of tension is responsible for a small grid of cracks formation on a sample surface. Moreover, the change of linear expansion coefficient renders a significant influence on the cracks formation. It takes place due to the process of heating and temperature variation that is also considered in the calculation scheme. Thus, σ_y and σ_z tensions are the most destroying. As the calculations of tensions in the cross direction versus the sample depth indicated, the lowest tensions occur in the following materials: tungsten, molybdenum, chromium and tantalum. Additional tests in terms of chemical hardness narrowed the choice of materials to molybdenum and tantalum.

4. The mass transfer of small impurities inside the grain

As a result of the binary «steel-coating» system heating in an air atmosphere the oxygen diffuses from the gas phase into the surface layers of the product. In contrast with the iron atoms and the other metals the oxygen atoms are small-sized and their diffusion is carried out mainly by interstitial mechanism.

To explain and quantify the elemental composition is used the kinetic equation (3) which determines the concentration of atoms of type B in the binary system of the atoms of type A and B under the influence of the temperature gradient and internal tensions.

$$\frac{\partial c_B^i}{\partial x} = -\frac{\partial}{\partial x} D_B^i \frac{\partial c_B^i}{\partial x} \; [1 + (\frac{P}{E})^2]$$
$$+ \frac{\partial c_B^i}{\partial x} (D_A^i \frac{\partial c_A^i}{\partial x} - D_B^i \frac{\partial c_B^i}{\partial x}) \; [1 + (\frac{P}{E})^2] \; + \tag{3}$$
$$2 \frac{\partial}{\partial x} \frac{c_B^i c_A^i}{kT^2} (\frac{D_B^i E_B^i}{c_A^i} - \frac{c_B^i D_B^i E_B^i}{c_A^i} - D_A^i E_A^i)[1 + (\frac{P}{E})^2] \frac{\partial T}{\partial x}$$

Here c_A^i and c_B^i is the concentration of atoms of type A and B respectively, $D_B^i = D_{0B}^i \exp(-E_B^i / kT)$ and $D_A^i = D_{0A}^i \exp(-E_A^i / kT)$ is diffusion coefficients of atoms of type B and A (see table 1). P is the internal tensions in the material, E is the Young's modulus, T is the temperature, k is the Boltzmann constant, E_B^i and E_A^i is the activation energy of interstitial diffusion mechanism. The first term in equation (3) describes the concentration mechanism of interstitial diffusion mechanism, the second – counter flows of components A and B, the third term describes the thermal diffusion on the interstitial mechanism. The diffusion coefficients of nitrogen and oxygen atoms ($D_0 \exp(-E_A / kT)$) borrowed from [14].

To determine the distribution of elements concentrations, equation (3) must be supplemented by the influence of the mutual recombination mechanisms, when vacancy and interstitial atom disappear, as well as sinks of point defects on dislocations and grain boundaries. Basic mechanisms of elements redistribution in the surface layers associated with the point defects flows: vacancies and interstitials atoms. To calculate the concentration of vacancies and interstitial atoms consider the grain disposed on the surface as the model. Let the section of grain be rectangular. To calculate the concentration of point defects in this case, we use the following system of equations (4-5):

$$\frac{\partial C_{iB}}{\partial t} = \exp(-E_i / kT) - ZC_V C_{iB} + \frac{\partial}{\partial x}(-D_i \frac{\partial C_{iB}}{\partial x}) + \frac{\partial}{\partial y}(-D_i \frac{\partial C_{iB}}{\partial y}) - K_i C_{iB} \tag{4}$$

$$\frac{\partial C_V}{\partial t} = \exp(-E_V / kT) - ZC_V C_V + \frac{\partial}{\partial x}(-D_V \frac{\partial C_V}{\partial x}) + \frac{\partial}{\partial y}(-D_V \frac{\partial C_V}{\partial y}) - K_V C_V \tag{5}$$

Here $C_{iB} = \exp(-E_i / kT)$ and $C_V = \exp(-E_V / kT)$ is respectively the concentration of thermodynamic equilibrium vacancies and interstitial atoms at high temperature [15], $Z = 4\pi r_V v_i Z$ is the proportionality coefficient in the recombination; D_i and D_V is respectively the diffusion coefficients of interstitials and vacancies, $r_i = (\rho D_{0i} \exp(-E_i / kT)) / 2$ and $r_V = (\rho D_{0V} \exp(-E_V / kT)) / 2$ is the he rate of interstitials and vacancies annihilation on dislocation sinks [16], v_i is the hopping frequency.

The most intense annihilation mechanisms are the mutual recombination and annihilation of the dislocation at sinks. The next type of sinks is the grain boundary, which is limited. Vacancy concentration inside the grain will be higher than at the sinks. Therefore, point defects will move to the sinks, i.e. to the surface and grain boundaries. Schematically, the vacancies flows are represented in figure 1. Since the surface and grain boundaries are effective sinks of vacancies, the concentration of equilibrium vacancies decreases at the sinks. As a result of this process there are significant gradients vacancies, that stimulates the atoms diffusion. If the alloy contains several types of atoms, the equilibrium vacancy gradient leads to a flow of atoms, which in turn brings to a redistribution of the elements in the alloy.

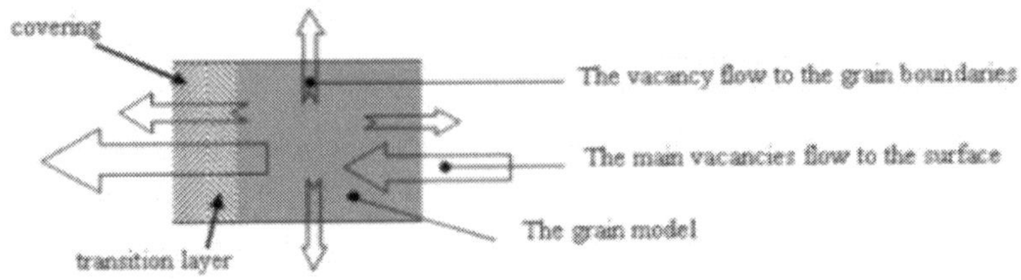

Figure 1. The scheme of the vacancies flows distribution.

Redistribution of impurities can be explained by the presence of point defects gradients at the grain boundary. Grain boundary is an intense vacancies sink, so the average profile of the vacancies distribution in the grain boundary changes its shape. As a result, two streams of impurities occur: one is directed transversely to the surface, and the second parallel to it, i.e. is directed to a grain boundary (see figure 1). The oxygen concentration in the grain area (see figure 2) was calculated according to the basis of equations (4) and (5). The theoretical studies indicate that the oxygen concentration leads to increasing in the interior area of the grain. The calculation on grain rectangular shape shows an increase in oxygen concentration of 35%.

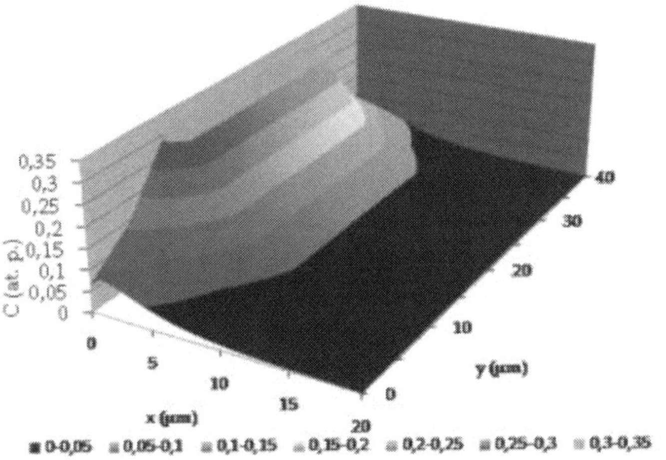

Figure 2. The oxygen distribution in steel J24056 after 60 minutes annealing, the calculation data of rectangular shape grains.

5. The oxygen influence on the structure and composition of the binary «steel-coating» system surface after heating in air atmosphere

To study the oxygen influence on the life cycle of the coating, the samples were annealed for 30-70 minutes at a temperature of 900°C in air atmosphere. As an annealing result, the oxygen diffusion in the surface layers and its redistribution within microcrystal grains [17], which is known in the literature as the segregation elements term, takes place. At high temperatures occurs the segregation of small impurities inside the grain. The calculations have shown that the oxygen concentration in the interior area of the grain is increased by reducing the concentration at the boundary. In addition, the concentration of oxygen can be increased to 40%. As it is shown by experimental SEM studies, the oxygen concentration increases within the grains and reduces to the boundaries (see figure 3). To compare the results of a numerical calculation with the experimental results, the analysis of the surface layer composition with the aid of the SEM microprobe in a straight line was carried out. The results of calculation correlate well with the experimental results.

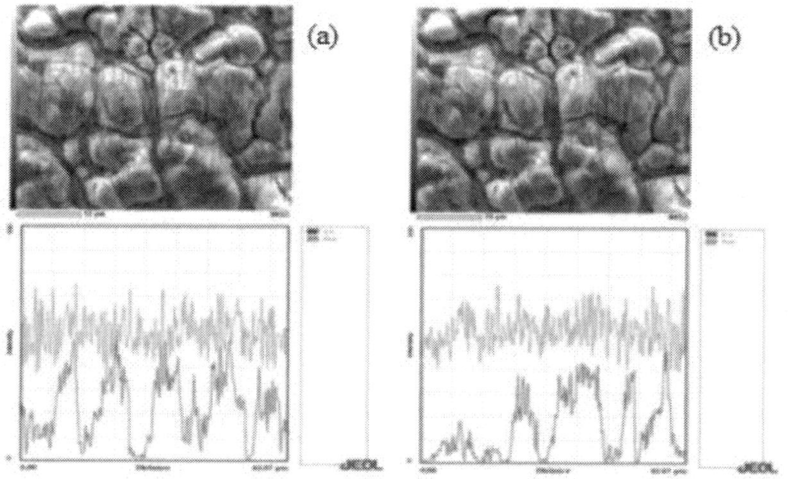

Figure 3. The elements distribution in the molybdenum-steel system after annealing at 900°C (a) within the grains (b) along the grains boundary

6. The combined ion-plasma method of surface modification

The combined ion-plasma method was realized on the NNV-6 installations together with the ADVAVAC VSM-200 installation. Technological process of ion implantation with the subsequent magnetron sputtering is performed step-by-step and includes the samples preparation, ionic cleaning, ion implantation, ion-plasma sputtering, cooling. The modes of combined ion-plasma method were defined experimentally for each type of coating material [18]. Superficial structural modification was performed according to the described method. Transition layers between the coatings and the matrix were analyzed with the scanning electron microscope JEOL JCM-5700 (see figure 4).

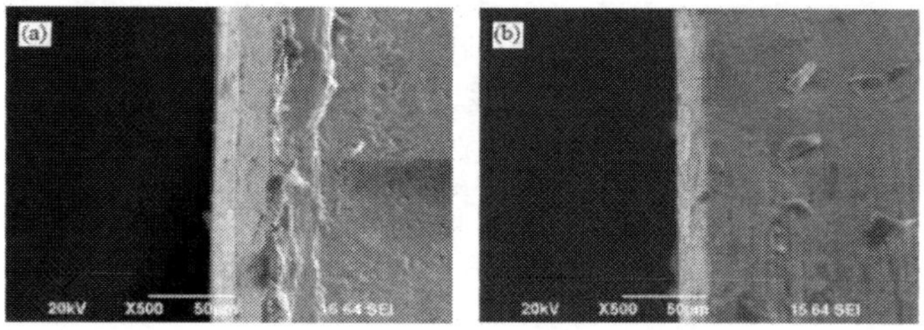

Figure 4. SEM micrograph of a cross-section of the binary «steel-coating» system with molybdenum (a) and tantalum (b) as a modifier material

During the analysis of different areas between the coatings and the matrix on the occurrence of micro-cracks, inclusions and heterogeneity regions, the best quality of the molybdenum adhesion was revealed. In accordance with the combined ion-plasma method of surface modification were obtained calibration plates for laser cutting «LaserMat-4200» installation. The plates modified by combined ion-plasma method were subjected to the field tests which have been carried out at plant of a bridge metalwork of NPO Mostovik LLC in the Omsk city. The estimated tests demonstrated at least the twofold increase in the term of plates operation, and in some cases it increased from 2 to 4 times, the enhance of periods of between-repairs interval.

7. Conclusion

To reduce the tensions caused by rapid heating and cooling, molybdenum and tantalum are the most appropriate coatings from the point view of its thermal characteristics.

The experimental and theoretical studies indicate that the segregation of small impurities (oxygen) leads to increasing in its concentration in the interior area of the grain. The experimental data of oxygen concentration in the grains increased to 40%, calculated on grain rectangular shape shows an increase in oxygen concentration of 35%, which indicates good correlation developed model. The combined ion-plasma method of surface modification including ion implantation with the subsequent magnetron sputtering was elaborated. The empirically optimized coating mode provides the improving of constructional steel products efficiency in a high-temperature exposure, thus increasing the life cycle of the calibration plates for the laser cutting «LaserMat-4200» installation twice.

Acknowledgments

This work was supported by the Russian foundation for basic research (RFBR), grant 13-08-98063.

References

[1] Frolov K V 2008 *Modern tribology: results and prospects analysis* (Moscow: LKI)
[2] Hirvonen J K 1985 *Ion implantation* (Moscow: Metallurgiya)
[3] Meyyappan M 2011 *J. Phys. D: Appl. Phys.* **44** 174002
[4] Liang J H 1999 *Japan. J. Appl. Phys.* **38** 286
[5] Palmer D W 1980 *Advances ion implantation: Ion implantation and semiconductors and other materials* (Moscow: Mir)
[6] Khatri H and Marsillac S 2008 *J. Phys.: Condens. Matter* **20** 055206
[7] Salonen E, Järvi T, Nordlund K and Keinonen J 2003 *J. Phys.: Condens. Matter* **15** 5845
[8] Yasuda K and Murota J 1983 *Japan. J. Appl. Phys.* **22** L615
[9] Navas I, Vinodkumar R, Lethy K J, Detty A P, Ganesan V, Sathe V and Mahadevan V P Pillai 2009 *J. Phys. D: Appl. Phys.* **42** 175305
[10] Uthanna S, Jagadeesh S V Chandra, Sreedhara P Reddy, and Mohan G Rao 2008 *J. Phys.: Conf. Ser.* **114** 012035
[11] Zhou Z, Pintsuk G, Linke J, Ma Y and Rödig M 2009 *Phys. Scr.* **2009** 014058
[12] Grigor'ev V A and Zorin V M. 1982 *Heat and mass transfer: thermal-engineering experiment* (Moscow: Energoizdat)
[13] Bleicher G A, Krivobokov V P and Paschenko O V 1999 *Heat and mass transfer in a solid under the influence of powerful beams of charged particles* (Novosibirsk: Nauka)
[14] Larikov L N 1985 *Thermal properties of metals and alloys* (Kiev: Naukova Dumka)
[15] Bokschtein B S 1978 *Diffusion in metals* (Moscow: Metallurgiya)
[16] Shalaev A M 1988 *Radiation-induced processes in metals* (Moscow: Atomizdat)
[17] Lee Y S, Yu E K, Shim D H, Kong H S, Bie L and Kanicki J 2014 *Japan. J. Appl. Phys.* **53** 121101
[18] Blesman A I, Postnikov D V, Polonyankin D A and Rogachev E A 2014 The combined ion-plasma method for the superficial modification of the products of constructional steel operating under the conditions of short-term extreme thermal conditions *Dynamics: Proc. Int. Conf. (Omsk, Russia, 11–13 November 2014) (Institute of Electrical and Electronics Engineers, Series: Materials and technology. Nanotechnologies and nanomaterials* vol 3) ed A V Kosykh (Omsk: OmSTU Press) pp 1-7

Some specific features of spectral bands location in the infrared spectra of crystals and liquids

G Melnikov[1], N Ignatenko[1], V Melnikov[1], E Cherkasov[1]

[1]Southwest State University (SWSU), st. 50 years October, 94, 305040, Kursk, Russia

E-mail: inmkstu@bk.ru, melnikovga@mail.ru

Abstract. A cluster model of condensed matter has been used to demonstrate that the observed peaks in the IR spectra of crystals and liquids correspond to the most stable cluster structures with the number of particles forming series of Fibonacci numbers. The relationship between the frequencies obeys the golden section rule.

1. Introduction

Recently the efforts to interpret the patterns forming in the location of spectral lines in different spectrum areas by applying the mathematical golden section rule have become more intense [1-6]. Moorman C. M. and Goff J. E. in their paper (2007) [1] studied a mechanical model of the oscillations of two physical bodies connected by weightless elastic springs in horizontal direction and obtained the resulting quadratic equation of motion of a system in the classical case.

$$\left(\omega^2\right)^2 - 3\omega_0^2\left(\omega^2\right) + \left(\omega_0^2\right)^2 = 0 \tag{1}$$

The solution of equation (1) yields two positive roots for oscillation frequencies

$$\omega_\pm = \frac{\omega_0}{2}\left(\sqrt{5} \pm 1\right) \tag{2}$$

that are related with the well-known rule of golden section: $\varphi = (\left(\sqrt{5}+1\right)/2) = 1.61803...$, $\varphi^{-1} = \varphi - 1 = 0.61803...$ and can be presented in a compact form

$$\omega_\pm = \varphi^\pm \omega_0 \tag{3}$$

i.e. natural frequencies of the considered mechanical system are related with the golden proportion in a simple manner.

In their paper Hopkins A, B., Stillinger F H and Torquato S. (2010) [2] investigated the densest local packaging (DLP) of not intersecting single diameter spheres in Euclidean space. It was shown that in any measurement region the radius of the spherical area on the surface which accommodates the centers of fifteen non-intersecting spheres and the radius of this sphere as such are in golden proportion relationship. In [3] Heyrovska R. (2013) based on experimental data on the wavelengths for hydrogen atom spectral lines demonstrated that the position of these lines in the spectrum can be represented by simple functions of the golden proportion. The review of studies where golden proportion is applied to interpret spectral properties of condensed matters can be found in [4-6].

2. Cluster model and IR spectra of liquids

Studies of condensed matter spectra in the IR region show that spectral bands appear due to rotations and librations of dimers in the structure of clusters. The more particles a cluster contains, the higher frequency of dimer's libration motions corresponds to this cluster in the IR spectrum. In the context of dynamic equilibrium of a cluster formation in classical approximation the equality will be

$$\frac{J_{\dim}\omega^2}{2} = \frac{\hat{Z}}{2}\varepsilon_{eff} \tag{4}$$

where $J_{\dim}$ - is the moment of inertia relative to the principal axes for different configurations of the dimer, ω - is cyclic frequency librations, $\hat{Z}$ - is a most probable number of particles in the cluster, ε_{eff} - the effective depth of the potential well of the interaction potential.

Following the model developed earlier in [7-9, 19], we shall write the density of clusters distribution according to the number of particles contained in their composition in the form of gamma function with the integer value of the parameter $m = 4$

$$f(Z) = \frac{\lambda^4}{6} \cdot Z^3 \cdot e^{-\lambda Z} \tag{5}$$

The most probable number of particles in cluster $\hat{Z}$ corresponds to the maximum value of the function $f(Z)$ and is defined by [7-9, 19]

$$\hat{Z} = \frac{m-1}{\lambda} = (m-1)\theta = \frac{3\pi^2}{4}\eta\exp(\eta) \tag{6}$$

where η - is the cluster structure packing factor which is determined in context of the model of hard spheres by formula $\eta = \eta_c\rho_*$, ρ_* - is the reduced density of the liquid, η_C - the coefficient of atomic packing in a critical point.

The depth of the potential well of the effective interaction potential is determined by the depth of the potential well of the pair interaction potential ε_0 and the most probable number of particles $\hat{Z}$ in a cluster [10, 19]

$$\varepsilon_{eff} = \varepsilon_0 + \Delta\varepsilon = \varepsilon_0 + \left(\frac{\hat{Z}-1}{\hat{Z}+1}\right)^2 kT \tag{7}$$

In their paper Piech M, at al. (2002) [11] successfully used a similar effective potential to account for the nearest-neighbor interaction in the system of interacting particles; Wang Dingdi, at al. (2013) in [12] simulated rotational-vibration motion of a dimer in closest neighboring environment based on the example of a dimer of iodine atoms in a rigid nano-channel and demonstrated that the set of frequencies lies in the far infrared region of the spectrum.

Based on the formulas (4) and (7) the set of frequencies in the IR spectrum is defined by [10]

$$\omega_i = \sqrt{\frac{2\Delta H_{\dim}}{J_{\dim}}}\sqrt{Z_i}\sqrt{\left[1+\left(\frac{\hat{Z}-1}{\hat{Z}+1}\right)^2\frac{kT}{\varepsilon_0}\right]} \tag{8}$$

Equation (8) implies the existence a of minimum frequency in the IR spectrum, and this frequency is determined by

$$\omega_{\min} = \sqrt{\frac{2\Delta H_{\dim}}{J_{\dim}}} \tag{9}$$

where $\Delta H_{\dim}$ - is the enthalpy of dimer configuration formation.

In the first approximation, Eq (8) considering (9) yields a simpler expression for predicting the set of frequencies in the IR spectrum of a liquid:

$$\omega_i = \omega_{\min}\sqrt{\hat{Z}_i} \tag{10}$$

where $\hat{Z}_i$ - is a set of "magic" numbers that specify the number of particles in the most stable clusters.

Reduced frequency $\omega_* = \omega / \omega_{\min}$ is a universal function of the square root of the number of particles in a cluster. In light of (10) the function of clusters distribution in accordance with the number of particles (5) contained in their composition permits us to find the distribution function of frequencies in the IR spectrum of a liquid:

$$f(\omega_*) = \frac{\lambda^4}{3}\omega_*^7 \cdot e^{-\lambda\omega_*^2} \tag{11}$$

The most probable reduced frequency $\hat{\omega}_*$ corresponds to the maximum function $f(\omega_*)$ and is equal to

$$\hat{\omega}_* = \sqrt{\frac{7}{2}\theta} = \sqrt{\frac{7}{6}\hat{Z}} \tag{12}$$

Figure 1 shows a) curves of frequency allocations (11) for liquid argon at various temperatures on the saturation line, which indicates natural displacement of frequency maximum ω_* depending on the number of particles in the cluster and b) a similar function taken for liquid CCl_4 to illustrate the golden rule in the IR spectrum.

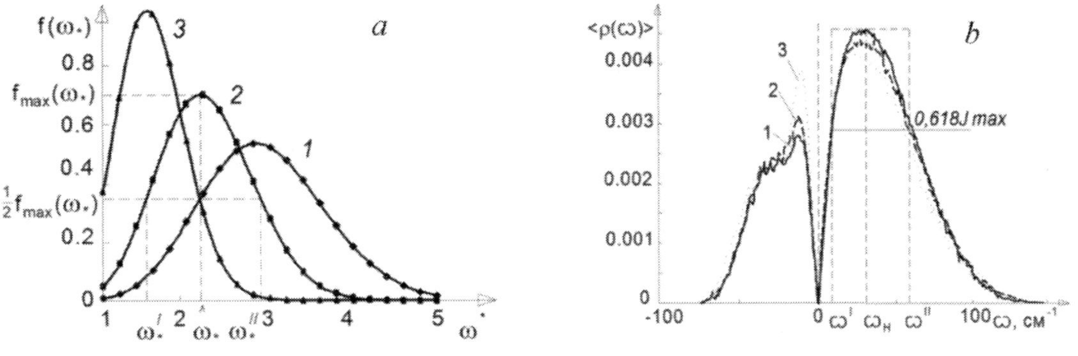

Figure 1. a) The distribution of frequencies in the IR spectrum of liquid argon (5), and b) a similar function for CCl_4 obtained by means of pair interaction potential [Moore et. al. 1995 *J. Chem. Phys.* **103**, 9. 3327].

The study of patterns of the relative allocation of the most probable frequencies in the IR spectra of liquids revealed that there is a clear functional relationship between the frequencies $\hat{\omega}$, ω'' and ω', which is defined by mathematical golden rule

$$\frac{\omega''-\omega'}{\omega''-\hat{\omega}}=\frac{\omega''-\hat{\omega}}{\hat{\omega}-\omega'}=\varphi=\frac{1+\sqrt{5}}{2}=1.6180339... \quad (13)$$

Fr calculation purposes relation (13) can be written in a more convenient form

$$\omega''=(1+\varphi)\omega_{н}-\varphi\omega' \quad (14)$$

Equation (14) implies the existence of the most stable cluster formations in liquids with the most probable number of particles represented by Fibonacci series: 0; 1; 1; 2; 3; 5; 8; 13; 21; 34; 55; 89; 144; ..., wherein the monatomic inert gases and organic liquids are characterized by small clusters with the most probable number of particles: 1; 2; 3; 5; 8; 13. According to formula (10) such set of "magic" numbers corresponds to a set of frequencies in the IR spectrum:

$$\omega_{\min},\ \omega_{\min}\sqrt{2},\ \omega_{\min}\sqrt{3},\ \omega_{\min}\sqrt{5},\ \omega_{\min}2\sqrt{2},\ \omega_{\min}\sqrt{13},\ \omega_{\min}\sqrt{3}\sqrt{7},\ \omega_{\min}\sqrt{2}\sqrt{17} \quad (15)$$

For small water clusters with $\omega_{\min}$ = 85.8 cm^{-1}, we shall obtain the following a set of frequencies: 121, [120] cm^{-1}, 149 [145] cm^{-1}, 192 [190] cm^{-1}, 242 [240] cm^{-1}, 310 cm^{-1}. Experimental frequency values [13-15] are given in square brackets. They prove the validity of applying the golden section to the analysis of the IR spectra of liquids.

The results of comparison between frequency values obtained by formula (10), Raman and IR spectroscopy experimental data available for crystals and liquids with different structures, different nature and intensity of the intermolecular interaction (such as 1-nitro-4- (trifluoromethoxy) benzene, water, ice, 4-tert butyl benzyl selenocyanate) are given in table 1. These data show good agreement between the calculations according to the rules of golden section and spectroscopy experimental data in the far region of IR spectra of a liquid.

Within the width of spectrum $\Delta\omega_*$ at the level $\frac{1}{2}f_{\max}(\omega_*)$ it is possible to estimate the lifetime of the most probable cluster with some selected conditions of a liquid state (see. figure 1).

The relationship to estimate the lifetime of a cluster in a fluid with the most probable number of particles in its structure will have the form of [19]

$$\Delta\tau=\frac{\tau_0}{\sqrt{(\hat{Z}-1)}} \quad (16)$$

If the value $\hat{Z}=2$ the formula will correspond to the dimer's lifetime in a liquid $\Delta\tau_{\dim}=\tau_0$, which can be determined by independent spectroscopic methods. For a free stable particle $\hat{Z}=1$, formula (16) corresponds to the passage to the limit and presumes an infinite lifetime.

For liquid argon near its melting point at $T=85\,K$ the lifetime of a cluster with the most probable number of particles $\hat{Z}\approx 7$ estimated by the formula (16) will yield the value $\Delta\tau=2{,}5\cdot 10^{-12}\,c$.

Time of life of cluster systems in liquids cannot be defined by experimental methods, this value can be estimated only indirectly in the context of specific assumptions or theories, that is why the relation (16) obtained through the cluster model is very important for understanding formation and disintegration processes running with cluster systems in liquids. For instance, the lifetime of hydrogen bonds in water is of the same order as the time of dipole relaxation and agrees with the estimations of a cluster's lifetime ($\Delta\tau=3{,}7\cdot 10^{-12}\,c$) in accordance with the formula (16).

Table 1. IR spectra of ice crystals modifications and some liquids with Fibonacci series.

Far –infrared spectrum of ice II at 1 bar and 100 K									
$F_i = \hat{Z}$	ω_{min} =75 cm^{-1}		ω_{min} = 106.8 cm^{-1}		ω_{min} = 139 cm^{-1}		D_2O , ω_{min} = 103.7 cm^{-1}		
	Calc.	[16]	Calc.	[16]	Calc.	[16]	Calc.	[16]	$\omega_{H_2O}/\omega_{D_2O}$
2	106	106.5	151	151.4	197	197	146.5	146.6	1.031
3	130	136.1	185	186.4	241	237	179.6	180.4	1.030
5	168	–	239	237.1	311	325	232	247.5	1.030
8	212	–	302	294.8	393	–	293	279	1.031
13	270	–	385	–	–	–	373	373	1.032
21	344	335	489	–	–	–	–	–	
Far –infrared spectrum of ice IX at 1 bar and 100 K									
$F_i = \hat{Z}$	ω_{min} = 52 cm^{-1}		ω_{min} = 97 cm^{-1}		ω_{min} = 176.8 cm^{-1}		D_2O , ω_{min} = 95.3 cm^{-1}		
2	73.5	72.7	137	137.2	250	250.6	134	134.7	1.022
3	90	–	168	163.8	306	302	165	159.9	1.018
5	116	–	216	207.4	–	–	213	201.5	1.014
8	147	–	274	282.7	–	–	270	270.2	1.015
13	187	–	350	–	–	–	–	–	–
21	238	230	–	–	–	–	–	–	–
Far –infrared spectrum of ice V at 1 bar and 100 K									
$F_i = \hat{Z}$	ω_{min} = 67.2 cm^{-1}		ω_{min} = 89.1 cm^{-1}		ω_{min} = 95.5 cm^{-1}		D_2O , ω_{min} = 87.7 cm^{-1}		
2	95	95	126	126	135	135	124	124 2	1.016
3	116	–	154	152	165	165	152	155	1.013
5	150	152	199	200	214	200	196	185	1,015
8	190	200	252	–	270	260	248	245	1.016
13	242	–	321	–	344	340	316	310	1.016
21	308	–	–	–	–	–	–	–	–
Far –infrared spectrum of 1-nitro-4-(trifluoromethoxy) benzene									
$F_i = \hat{Z}$	ω_{min} = 24 cm^{-1}		ω_{min} = 87.7 cm^{-1}		ω_{min} = 99 cm^{-1}		ω_{min} = 181.8 cm^{-1}		–
	Calc.	[17]	Calc.	[17]	Calc.	[17]	Calc.	[17]	–
2	34	–	124	124	140	140	257	257	–
3	42	–	152	–	171	–	314	–	–
5	54	58	196	198	221	–	406	418	–
8	68	69	248	240	280	296	514	499	–
13	86	82	316	–	357	347	–	–	–
21	110	113	402	398	453	–	–	–	–
Far –infrared spectrum of 4-tert butyl benzyl selenocyanate									
$F_i = \hat{Z}$	ω_{min} = 20 cm^{-1}		ω_{min} = 135 cm^{-1}		ω_{min} = 148 cm^{-1}		ω_{min} = 202 cm^{-1}		–
	Calc.	[18]	Calc.	[18]	Calc.	[18]	Calc.	[18]	–
2	28	28	191	191	209	209	286	286	–
3	35	37	234	241	256	265	350	348	–
5	45	–	302	320	331	329	–	–	–
8	56	52	382	–	419	400	–	–	–
21	92	92	–	–	–	–	–	–	–
34	117	118	–	–	–	–	–	–	–

3. Conclusion

In disordered condensed media (dense gases, liquids and polymers) clusters are distributed according to the number of particles in their composition, which can be described by two-parameter gamma function. Notable is that the parameter that specifies the order of distribution is equal to $m = 4$, the shape parameter is the function particle packing ratio in the structure of a cluster $\theta = f(\eta)$.

In their composition the most stable clusters contain such number of particles that forms Fibonacci series. These clusters determine the position of the maxima in the IR spectra of liquids. It has been proved that reduced frequencies, that correspond to the maxima of the IR spectra, are proportional to the square root of the average number of particles in the most stable clusters. Theoretical suppositions have been verified by experimental data obtained for liquids with different molecular structures.

References

[1] Moorman C and Goff J 2007 *Eur. J. Phys.* **28** p 897
[2] Hopkins Adam B, Stillinger Frank H and Torquato S 2010 *Journal of Mathematical Physics* **51** p 043302-1
[3] Heyrovska R 2013 *International Journal of Sciences* **2** p 28
[4] Hendi S and Sharifzadeh M 2012 *J. Theor. Phys.* **1** p 37
[5] Omotehinwa T and Ramon S 2013 *International Journal of Computer and Information Technology* **02** p 4
[6] Monsoriu J, Calatayud A, Remorn L, Furlan W D, Saavedra G and Andrers P 2013 *IEEE Photonics Journal* **5** p 768
[7] Melnikov G, Verveyko V, Polyansky A 2011 *International Journal of Thermophysics* **32** p 901
[8] Melnikov G, Verveyko V, Melnikov V, et. al. 2011 *Moscow University Physics Bulletin* **66** p 432
[9] Melnikov G, Verveyko V, Melikhov Yu F, et. al. 2012 *High Temperature* **50** p 214
[10] Melnikov G, Verveyko V, Melikhov Yu F, et. al. 2011 *Vestnik MGTU. Estestvennye Nauki* **3** p 108
[11] Piech M, Weronski P, Wu X and Walz J 2002 *Journal of Colloid and Interface Science* **247** p 327
[12] Wang Dingdi, Guo Wenhao, Hu Juanmei et al. 2013 *Scientific Reports* | 3: 1486 | DOI: 10.1038/s rep01486.
[13] Bosma W, Fried L and Mukamel S 1993 J. *Chem. Phys.* **98** p 4413
[14] Kang D, Dai J, Hou Y and Yuan J 2010 *J. Chem. Phys.* **133** p 014302-1
[15] Protasevich A, Starikov V 2003 *Russian Physics Journal* **46** p 1077
[16] Bertie J, Labbe H and Whalley E 1968 *Reprinted from J. Chem. Phys.* **49** p 775
[17] Arivazhagan M and Jeyavijayan S 2011 *Indian J. of Pure @ Applied Physics* **49** p 516
[18] Kumaria U, Beeguma M, et al. 2008 *RJC RASĀYAN. J. of Chemistry* **1** p 110
[19] Melnikov G, Ignatenko N, Melnikov V, et. al. 2014 *Proceedings of the Southwest State University* **3(54)** p 111

IOP Conf. Series: Materials Science and Engineering **81** (2015) 012033 doi:10.1088/1757-899X/81/1/012033

Analysis of thermal effects on copper nanoparticles synthesized from the gas phase

I V Chepkasov[1], **Z I Popov**[2]

[1]Abakan, pr. Lenina 90, 655000, Russia, Katanov State University of Khakassia

[2]Krasnoyarsk, 660036, Russia Kirensky Institute of Physics SB RAS

E-mail: ilya_chepkasov@mail.ru

Abstract.

Molecular dynamics method using the tight-binding potential to carry out simulation of ultrafast heating of the synthesized particles from the gas phase to a temperature T= 600K and T= 900K, at which the particles were kept about 10 ns. As a result of the simulation revealed that the method of ultrafast heating the particles to high temperatures virtually eliminates the possibility of a clusters of defective education, but as a result of the heat treatment, the some of investigated particles can disconnect (burst) into smaller clusters.

Introduction

The recently grown interest in the development of new and improvement of the available methods for fabrication of nanodispersed materials is due to the fact that it is this class of processing substances that finds an ever broadening application as the initial raw material in the production of the most diverse ceramic and composite compounds, superconductors, data storage media, filters, catalysts [1, 2] and nanoelectronic devices [3]. In the first place, this is conditioned by the fact that during a transition from the bulk phase to a state with a particle size of below 100 nm a number of fundamental properties of a material change, such as thermal and electrical conductivity, thermal capacity, chemical reactivity, and many others.

In the whole range of the metallic nanoparticles in use today, copper clusters hold a special place. These particles possess many unique properties and are comparatively inexpensive to manufacture. In the context of their electrically conductive properties, nanometer-sized copper particles can compete even with silver and can be applied as contacts in high-temperature electronics [4].

Currently, there are many methods to produce metal nanoparticles, such as mechanochemical grinding methods, plasma-chemical methods, some variations of chemical, photochemical and radiation reduction, the method of condensation from the gas phase [5]. Among a great variety of methods for fabrication of metallic nanoparticles, one of the promising synthesis methods is evaporation-condensation of metals in an inert gas atmosphere. A simple scalability to industrial standards and a high cleanliness of a material with the required properties (electrical conductivity, strength, ductility) favorably distinguish particles synthesized by the gas-phase method from particles produced by other methods, such as mechanical milling.

However, in most methods for fabrication of nanoparticles such conditions may be created when particles of a non-spherical shape form as a result of synthesis. Such random-shaped formations may consist of tens of united primary entities, formed at the early stages of synthesis. The non-spherical cluster formations, produced in large quantities, are not rejects in the production of perfect spherical particles, and they can also be used in various areas of nanotechnology as catalysts and battery electrodes [6-8].

However, if in the process of synthesis it is necessary to produce perfect spherical particles, actively applied, for instance, in data storage media [9], heat treatment is used at the final stage of production. The produced agglomerates and aggregates are subjected to thermal sintering, leading to compaction of nano-formations to spherical particles. But to create the most effective process of thermal action, precise information is required about the morphological changes in the investigated particles throughout the whole process of heating.

IOP Conf. Series: Materials Science and Engineering **81** (2015) 012033 doi:10.1088/1757-899X/81/1/012033

To obtain detailed information about the processes of synthesis and subsequent thermal sintering of nanopowders, computer simulation techniques are increasingly used. In the present work, to identify the main features of heat treatment of particles, condensed from the high-temperature gas phase, molecular dynamics simulation using the tight-binding potential [10] was performed, of the heating processes of copper clusters in the framework of the Nose-Hoover thermostat. In our opinion, this thermostat suits best for simulation of the processes of heating and cooling of isolated particles.

Results and discussion

Copper nanoparticles, taken for identification of the most effective post-condensation methods of heat treatment, were synthesized from the gas phase at the first stage of simulation. The condensation process was implemented with three different cooling rates U= 0.005 ps^{-1}, U= 0.025 ps^{-1} and U= 0.05 ps^{-1}, and two final temperatures T_f= 373 K and T_f= 77 K. More detailed information about the method of simulation of the gas-phase synthesis process and the obtained simulation results is given in [11, 12].

Two main techniques of heat treatment of synthesized nanopowders were studied, namely step-by-step heating and ultrafast heating. The results of the effect of step-by-step heating were presented earlier in [13]. Let us address only the main conclusions made from the results of the step-by-step heating study. As a result of a stepwise increase in the particles temperature from 100 to 1200 K (with a step of 100 K), perfect ordering of the internal structure and external shape in the process of heat treatment was observed in only 70% of the cluster array under study. The remaining 30% of the clusters in the process of thermally-activated relaxation could not form a unified shape and structure, and constituted competing parts of one particle with comparable surface energies and chemical potential energies.

In order to exclude such a behavior of conglomerates during thermal action, for this work the method of ultrafast heating of particles to relatively high temperatures of T = 600K and T= 900K was selected. In our opinion, this method, owing to high temperatures and, correspondingly, short sintering times (t = 10 ns), can lead to a rather quick formation of a particle's unified external shape and ordered internal structure, without formation of significant defects within the cluster.

All the particles produced during simulation of copper atoms condensation from the gas phase were subjected to the procedure of ultrafast heating and thermal relaxation. During the simulation of six condensation variants (with various final temperatures and cooling rates), in total 136 copper clusters of various shapes and structures with a size of 81 to 37140 atoms were synthesized. As a result of the conducted simulation, it was proved that the method of ultrafast heating of particles to the heat treatment temperatures does practically exclude the possibility of emergence of defect formations in clusters subjected to heating. For instance, the cluster Cu_{1771}, already investigated earlier and displayed in Figure.4 in the paper [13], which had not formed a united particle in the course of step-by-step heating, was subjected to ultrafast heating to high temperatures (T = 600 K and T = 900 K).

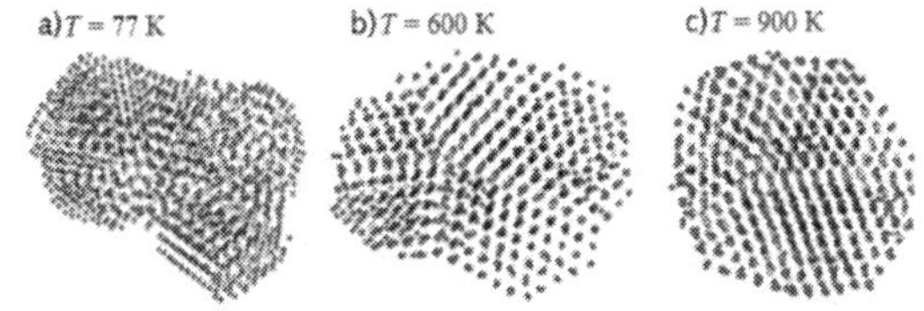

Figure. 1. Images of the cluster Cu_{1771} in the initial state (a) and after heat treatment at T= 600 K (b) and T= 900 K (c)

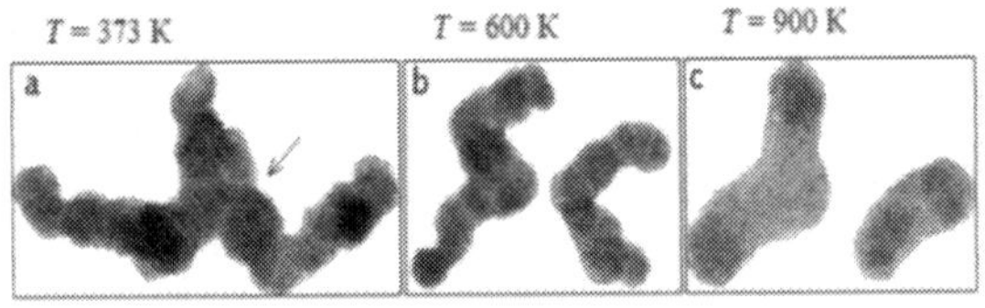

Figure. 2. Images of the cluster Cu_{17916} in the initial state (a) and after heat treatment at T= 600 K (b) and T= 900 K (c).

From the images presented in Figure.1 displaying Cu_{1771}, it is clearly visible that after the conducted heat treatment this cluster assumed a practically perfect spherical shape and completely reconstructed its structure. As a result, from a disordered conglomerate that had several local competing structures leading to emergence of defect formations during step-by-step heat treatment, the cluster Cu_{1771} turned into a perfect particle, suitable for application in high-precision instruments, in which the main criterion of efficiency is the degree of order of the internal structure and the degree of perfection of the external shape.

However, the process of ultrafast heating of particles to the selected heat treatment temperatures does not always result in formation of a united, perfectly ordered particle. As a result of heat treatment, some of the particles broke down (burst) into smaller clusters not connected with each other. The number of clusters, which were subjected to heat treatment and then «burst», varied from 5% to 83%, depending on the parameters of synthesis of these particles from the gas phase.

Figure 2 shows an example of a particle which «burst» during thermal action. Such a behavior of particles synthesized from a gas atmosphere and subjected to heat treatment at rather high temperatures is observed also in experimental works on sintering of silver clusters. For instance, in the work [14] the authors investigated the variations in the external shape of primary silver particles as a result of sintering at various temperatures. In the course of the conducted procedure, the primary particle divided into separate clusters, which subsequently formed several perfect spherical particles not connected with each other.

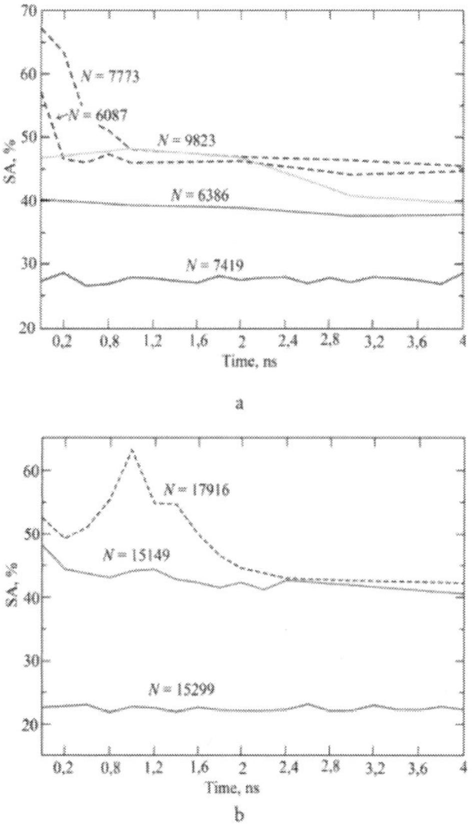

Figure. 3. Time dependencies of the percentage of surface and "defect" atoms (*SA*) for copper clusters from the group of medium particles (6000 to 10000 atoms) (a) and clusters from the group of large particles (15000 to 18000 atoms) (b). The dependencies of the burst clusters are denoted by a dashed line. Cu_{7419} and Cu_{15299} are perfect spherical copper clusters cut out from an fcc-lattice.

In order to account for such a behavior of the particles during heat treatment, a detailed analysis was performed, of the most typical representatives of the united and «burst» particles. As the most typical examples, this work mentions four particles

Table 1. Structure and shape of Cu nanoclusters synthesized from the gas phase, before and after heat treatment

Cooling rate (U, ps^{-1})	Temperature (T_f, K)	Number of clusters	Initial shape (at T_f, K)		Final shape (at T = 900 K)		
			Sphere	Agglomerate	Sphere	Agglomerate	Bursting
0,05	77	46	23,92%	76,08%	63,04%	28,26%	8,70%
0,025	77	35	25,72%	74,28%	74,28%	20,00%	5,72%
0,005	77	17	41,17%	58,83%	58,82%	35,30%	5,88%
0,05	373	19	10,53%	89,47%	42,10%	21,05%	36,85%
0,025	373	13	15,39%	84,61%	46,15%	15,39%	38,46%
0,005	373	6	16,66%	83,34%	0%	16,66%	83,84%

from a group with a particle size from 6000 to 10000 atoms and two particles from a group with a particle size of 15000 to 18000 atoms. All the particles taken for comparative analysis were produced in systems with a final condensation temperature of T_f= 373 K. In the selected examples, the absence of clusters synthesized at a final temperature of T_f= 77 K is conditioned by the fact that at such a condition of gas-phase synthesis the percentage of the «burst» clusters was only 5-9%, which could be caused by some stochastic processes.

With a view to define the various scenarios of cluster rearrangement, in the considered particles the time dependencies were studied, of the surface atoms fraction and of the fraction of atoms referring to local defect structures (the coordination number below 12). The values of these quantities were summed and represented by a single quantity *SA* whose time dependencies for the first four ns of heating at T = 600K are given in Figure. 3. As can be seen from the graphs, at the initial stage of heating for the «burst» and united particles with a medium size (6000—10000 atoms) the *SA* values differ by 17-21%. In about 0.6 ns after the start of heating, the *SA* values for the «burst» particles decline rather sharply and settle in the vicinity of the *SA* values for the united clusters. Such a behavior is related to the fact that by that time the primary particles under investigation, poorly connected with each other, had already disintegrated into secondary clusters which, in their turn, optimized the structure rather quickly.

Such a divergence in the *SA* values at the initial stage of heat treatment is significant and may lead, due to intracluster restructuring, to a complete dissociation of particles, since in the process of thermal action each cluster, guided by the law of minimum energy, tries to optimize its structure, which, in the case of particles having a large quantity of surface atoms and atoms with a local disordered structure, leads to competitive struggle between growth centers of local atomic organizations. A slightly different situation is observed in the case of the clusters Cu_{17916} and Cu_{15149}. Small differences in the *SA* values at the initial stage of heating (4.5%) for these particles, apparently, were not supposed to lead to large differences in cluster restructuring (Figure.3b). However, in the process of thermal action in the particle Cu_{17916} occurs structure destruction of a rather large layer connecting two main cluster particles, which results in a sharp rise in the *SA* values during the first ns of heating. The restructuring of this disordered area subsequently, at 2.2 ns, leads to the dissociation of the particle into secondary clusters with a size of 5760 and 12156 atoms.

Also, as a result of the conducted instant heating to the heat treatment temperatures, 136 copper nanoparticles synthesized from the gas phase were investigated in detail in the context of the shape of these particles at the final stage of heating at T= 900 K. Detailed information on the initial and final shape of the clusters under study is presented in table 1. It can be clearly seen that in all the condensation systems, heat treatment considerably improved the shape of the synthesized particles, and the percentage of spherical particles increased by 17% - 48% depending on the parameters of particle synthesis. Although in the system with a final condensation temperature of T_f= 373 K and a cooling rate of U = 0.005 ps^{-1} the number of spherical particles after heat treatment is equal to zero, this kind of a result may not be regarded as a complete disappearance of spherical clusters in the system, since a large percentage of the particles were subjected to «bursting», and spherical clusters were present in the system, but in the form of secondary particles (products of «bursting») of the primary clusters.

It also follows from analysis of table 1 that the large percentage of agglomerates (worm-like and dumbbell-like clusters), present in the systems at the initial condensation stage, diminishes considerably after heat treatment. As a result of the thermally-activated cluster restructuring, these agglomerates pass into the fraction of spherical and «burst» particles. In the systems with a final temperature of T_f =77 K a tendency is observed for a widespread integration of agglomerates into spherical particles, with formation of a small percentage of «burst» clusters. On the other hand, the opposite tendency is observed in the systems with a final temperature of T_f = 373 K, *i.e.* a widespread «bursting» of agglomerates during heat treatment, with formation of a small percentage of united spherical particles, or even their absence.

Conclusions

From the results of the performed simulation, it can be concluded that the exceeding of the 50 % threshold quantity of surface atoms and atoms having a local disordered structure may lead to the burst of a particle into secondary particles in the process of instant heating, due to rapid cluster restructuring. Thus, the use of these clusters in high-temperature chemical reactions may be severely hampered due to a rapid dissociation of the used particles, and, consequently, a sharp decline in the catalytic efficiency of the used nanomaterial.

ACKNOWLEDGMENTS
This study was performed within the framework of the State Task of the Ministry of Education and Science of the Russian Federation

References:

[1] Weber A P, Davoodi P, Seipenbusch M and Kasper G 2003 *Journal of Nanoparticle Research* **5** 293
[2] Weber A P, Seipenbusch M, Thanner C and Kasper G 1999 *Journal of Nanoparticle Research* **1** 253
[3] Fissan H, Kennedy M K, Krinke T J and Kruis F E 2003 *Journal of Nanoparticle Research* **5** 299
[4] Yiying Yao, Zheng Chen, Guo-Quan Lu and Khai Ngo D T 2012 *IEEE Transactions on* Components. Packaging and Manufacturing Technology **2** 539
[5] Binder K 1997 *Rep. Prog. Phys.* **60** 487
[6] Pratsinis S E 2010 *AIChE Journal* **56** 3028
[7] Ogawa H, Nishikawa M, Abe A 1982 *Journal of Applied Physics* **53** 4448
[8] Chew S Y, Patey T J, Waser O, Ng S H, Buechel R, Tricoli A, Krumeich F, Wang J, Liu H K, Pratsinis S E, Novak P 2009 *Journal of Power Sources* **189** 449
[9] Golovin Yu I 2007 Introduction to nano-engineering (Moscow: Mashinostroenie) [in Russian]
[10] Cleri F and Rosato V 1993 *J. Phys. Rev. B.* **48** 22
[11] Chepkasov I V, Gafner Yu Ya, Gafner S L 2011 *Letters on Materials.* **1** 107 [in Russian]
[12] Chepkasov I V, Gafner Yu Ya 2012 Basic Problems of Material Science **9** 353 [in Russian]
[13] Chepkasov I V, Gafner Yu Ya, Khalikov A R 2014 *Letters on Materials* **4** 205 [in Russian]
[14] Karlsson M N A, Deppert K, Karlsson L S, Magnusson M H, Malm J O, Srinivasan N S 2005 *Journal of Nanoparticle Research* **7** 43

Features of radiation damage of Ni-Ti alloy under exposure to heavy ions of gaseous elements

V P Poltavtseva[1], S B Kislitsin[1], D A Satpayev[1], T S Mylnikova[2] and A V Chernyavskii[2]

[1] Institute of Nuclear Physics, Almaty, Republic of Kazakhstan

[2] National Research Tomsk Polytechnic University, Tomsk, Russia

E-mail: poltavtseva@inp.kz

Abstract. The consistent patterns of changes in structural and phase state, hardening and temperature ranges of martensitic transformations in Ni-Ti alloy with the shape memory effect after implantation of heavy ions ${}_{16}O^{3+}$, ${}_{40}Ar^{8+}$ and ${}_{84}Kr^{15+}$ under comparable parameters have been experimentally studied. It is found that under the impact of ${}_{84}Kr^{15+}$ ions, a two-layer surface structure with radiation-hardened second layer is formed, radiation-stimulated phase transformation B19'→B2 occurs in the near-surface layer and out-range area, and the martensitic transformation temperature increases toward higher values after implantation of ${}_{40}Ar^{8+}$ and ${}_{84}Kr^{15+}$ ions.

1. Introduction

Modification of materials by heavy ion implantation has become a promising technology. It is important to know which heavy ions are most effective in terms of creating radiation-resistant titanium nickelide alloys to be applied in nuclear power engineering or coatings for medical supplies.

Previously it was found [1–3] that modification by heavy krypton ions of low and high energy in the two-phase Ni-Ti alloy with the shape memory effect does not cause amorphization of Ti-Ni phases. Radiation-stimulated phase transformation B19'→B2 occurs, which is the major cause of deterioration of its physical-mechanical and functional properties, and temperatures ranges of martensitic transformation extend. In the case of implantation of high-energy krypton ions, a globular structure is formed on the Ni-Ti alloy surface and the degree of its homogeneity decreases as the fluence increases, radiation hardening is found to occur in the out-range area at low (~ 10^{13} ion/m^2) fluences due to the reduced size of the structural fragments of Ni-Ti phases.

In [4,5], it is also shown that consistent high-energy krypton ion implantation and application of high-current electron beam increases the quality of the Ni-Ti alloy surface. In the out-range area, it causes formation of primary martensite phase B19' responsible for the shape memory effect, formation of nano-sized particles of NiTi R-phase, increase in the temperature interval of the martensitic transformation and further softening. In [6,7], it is found that the distinguishing feature of a purely thermal effect on the structure of Ni-Ti alloy formed by ion implantation is hardening caused by the ordering of the radiation defect structures (phases).

The paper presents the experimental results to show the effect of implantation of different heavy ions of gaseous elements on the structural and phase state, and physical and mechanical properties of Ni-Ti alloy with the shape memory effect.

2. Experimental methods and material

We studied Ni-Ti alloy with Ni of 53.46 wt.% and Ti of 46.54 wt.%, preferably consisting of NiTi with the B2 structure (austenite), NiTi with B19 structure (martensite) and a minor content of Ti, excess Ni in the form of solid solution and process particles similar in composition to $Ti_2Ni(C)$ [8]. Before implantation, a proven technology was used to prepare the surface of the samples: cutting out by spark cutting across the massive forged plate of semiindustrial Ni-Ti alloy, mechanical polishing with skins of different grits and buffing with GOI paste. The sample size was $15.000\times\sim3900\times3500$ μm^3.

The data of preliminary testing were used to choose Ni-Ti alloy with the titanium nikelid phase ratio B19'/B2 equal to ~ 0.7 which showed the highest values of martensitic transformation characteristic temperatures compared to those for higher values of the phase ratio.

Implantation of ${}_{16}O^{3+}$, ${}_{40}Ar^{8+}$ and ${}_{84}Kr^{15+}$ ions under relatively equal parameters A/Z = ~ 5.3; energies E_{ion}=1.75 MeV/a.e.m. (28.70 and 147 MeV, respectively), fluence $\Phi=10^{19}$ ions/m^2 and beam current J_{beam}=1.0–0.85 μA was performed with DC-60 cyclotron (Astana, Kazakhstan). The damaged area was ~ 1×10^{-4} m^2.

Testing of the samples before and after implantation was performed by X-ray diffraction analysis, scanning electron microscopy, measuring microhardness and the shape memory effect with D8 ADVANCE diffractometer, JSM-7500F (JEOL) microscope, microhardness tester PMT-3M and an apparatus for measuring the temperature hysteresis of the electrical resistance.

3. Experimental results and discussion

SEM studies at low magnifications revealed that in the case of ${}_{16}O^{3+}$ion implantation, traces of sputtering, i.e. tuberosity, pits, remnants of $Ti_2Ni(C)$ process particles are observed on the Ni-Ti alloy surface. As the ion mass increases 2.5 times as much (ions ${}_{40}Ar^{8+}$), the surface becomes smoother with little roughness. Part of the particles protrudes over the surface which indicates that ion etching occurs alongside with sputtering. In case of heavier ${}_{84}Kr^{15+}$ ions, etching completely dominates.

Higher magnification of SEM-images indicates formation of bright particles of a round shape, bubbles of about 1 nm, and tracks of 8 to 25 nm or conglomerates made up of several tracks with a dome, and circular bubble chains after implantation of ${}_{16}O^{3+}$, ${}_{40}Ar^{8+}$ (figure 1 a,b) and ${}_{84}Kr^{15+}$ (figure 1 c), respectively.

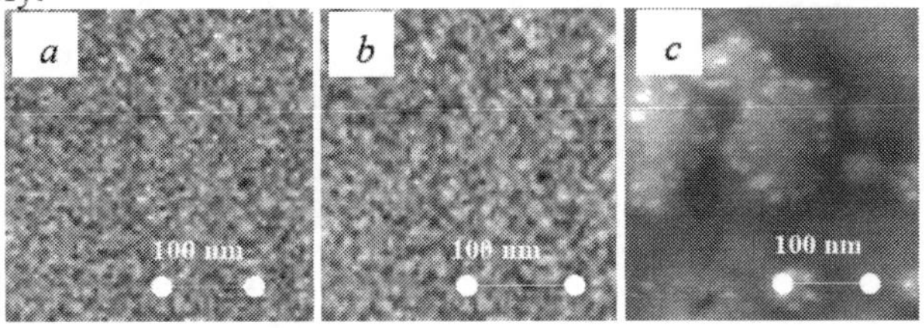

Figure 1. SEM images of the surface implanted with ${}_{16}O^{3+}$ (a), ${}_{40}Ar^{8+}$ (b) and ${}_{84}Kr^{15+}$ (c) ions.

The comparison of the results shows that formation of tracks on the Ni-Ti alloy surface requires high energy of the heavy ion (≥150 MeV). However, this is not sufficient according to previous studies with high-energy krypton ions used [5], when hemispherical convex structures (globules) with a size of up to 200 nm were observed after implantation up to the fluences of 5×10^{15} and $\geq5\times10^{19}$ion/m^2. The probability of track formation depends not only on the number of incident ions, but, apparently, on the ratio of B19'/B2 phases of titanium nickelide. Note that in this research we studied the samples of the two-phase Ni-Ti alloy with higher (~ 0.7) ratio of these phases than that of the samples studied previously.

In figure 2 the data of X-ray diffraction analysis obtained for a divergent beam mode shows that the phase composition nearby the out-range area (the depth of the analysis is comparable with the calculated ion range), as well as the structure of the implanted surface depend on the ion mass. In the case of $_{16}O^{3+}$ ion implantation (figure 2, curve 2), partial radiation-stimulated phase transformation B19'→B2 can be observed. The martensite content ~3.5 times decreases compared to unimplanted Ni-Ti alloy (figure 2, curve 1). However, $_{40}Ar^{8+}$ ion implantation (figure 2, curve 3) and $_{84}Kr^{15+}$ ion implantation (figure 2, curve 4) are characterized by complete radiation-stimulated phase transformation B19'→B2 followed by dissolving Ti phase and Ni_3Ti_3O compound. As a result of this implantation, Ni-Ti alloy becomes single phase with the B2 structure (austenite).

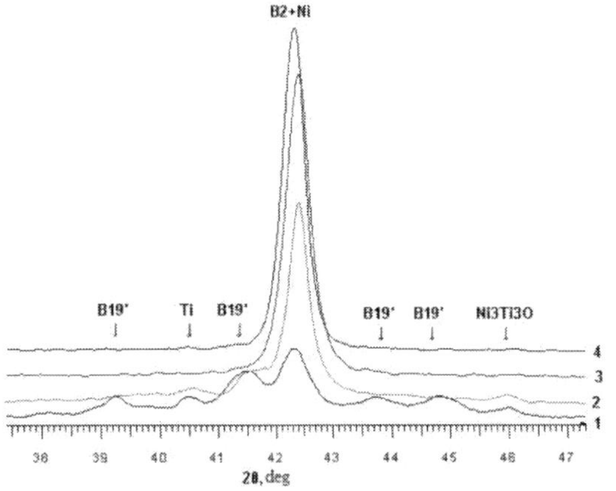

Figure 2. Parts of diffraction patterns for Ni-Ti alloy before (curve 1) and after $_{16}O^{3+}$ ion implantation (curve 2), $_{40}Ar^{8+}$ ion implantation (curve 3) and $_{84}Kr^{15+}$ ion implantation (curve 4).

Note that as the ion mass increases, the intensity of the primary X-ray line (110) of the B2 structure grows. The parameter of the body-centered cubic structure compared to unimplanted Ni-Ti alloy reduces, and this indicates approaching to the stoichiometric composition of titanium nickelide.

Similar patterns of change in the phase composition of the surface layer after $_{16}O^{3+}$, $_{40}Ar^{8+}$ and $_{84}Kr^{15+}$ ion implantation were obtained by shooting in grazing (2°) beam (analysis depth ~ 1 μm). In addition, firstly, in contrast to the out-range area, we detected slight splitting of the X-ray reflection line (110) of the B2 structure. This, as reported in [5,9,10], confirms the formation of nanoparticles of the R-phase during $_{16}O^{3+}$ ion implantation.

Secondly, the content of Ni_3Ti_3O compound does not virtually change in the out-range area and considerably reduces in the surface layer. This implies that the strain-hardened layer, formed on the Ni-Ti surface due to the technology used for its preparation, is partially sputtered during $_{16}O^{3+}$ ion implantation, its base being the Ni_3Ti_3O compound.

Earlier, the data obtained in SEM-studies of the Ni-Ti alloy surface etched prior to implantation [5] were used to attribute the surface layer hardening to titanium oxide formation during sample preparation. To make it more exact we conducted additional research in the geometry of the grazing beam at an angle of 1° to the etched surface of Ni-Ti alloy by X-ray structural analysis. The surface was found to get enriched with titanium due to release of nickel during chemical etching. This confirms the data obtained in X-ray analysis, especially as the technology of sample preparation was based on mechanical methods only.

Data on microhardness of Ni-Ti alloy measured after $_{16}O^{3+}$, $_{40}Ar^{8+}$ and $_{84}Kr^{15+}$ ion implantation indicates that, in contrast to the unimplanted alloy, distinct indentation appears even under very low (0.098 N) load. This is another evidence for the sputtering of the strain-hardened layer in interaction of heavy ions with the polished surface.

The data obtained in microhardness measurement indicates that the double-layer structure of the near-surface layer is formed with different degrees of hardening regardless of the ion mass. Radiation hardening of the second layer with a thickness of about 1.1–1.2 μm is 71, 50 and 39% for $_{16}O^{3+}$, $_{40}Ar^{8+}$ and $_{84}Kr^{15+}$ ions, respectively. Radiation hardening in the out-range area is significantly lower (7.2, 8.9 and 16%) and, vice versa, as the weight of the heavy ion grows, it increases.

The data for the temperature dependence of the resistivity of Ni-Ti alloy after $_{16}O^{3+}$, $_{40}Ar^{8+}$ и $_{84}Kr^{15+}$ ion implantation is shown in figure 3. In comparison to the corresponding R (T)-curves, measured before implantation, the martensitic transformation (MT) temperature range shifts and widens, the shape and area of the electrical resistance hysteresis loop changes. All the loops are straight, and the degree of variation depends on the ion mass, as in the case of the structure.

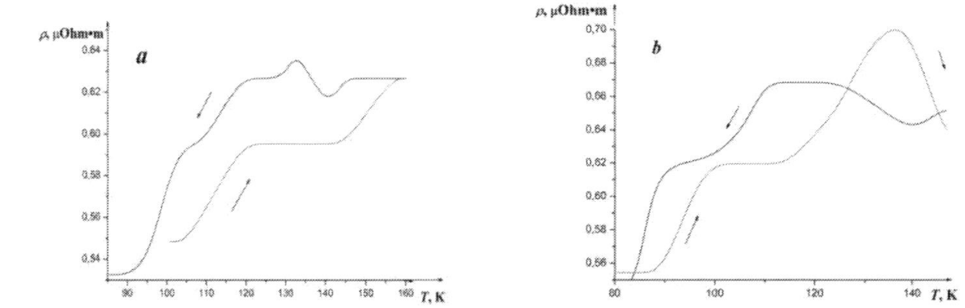

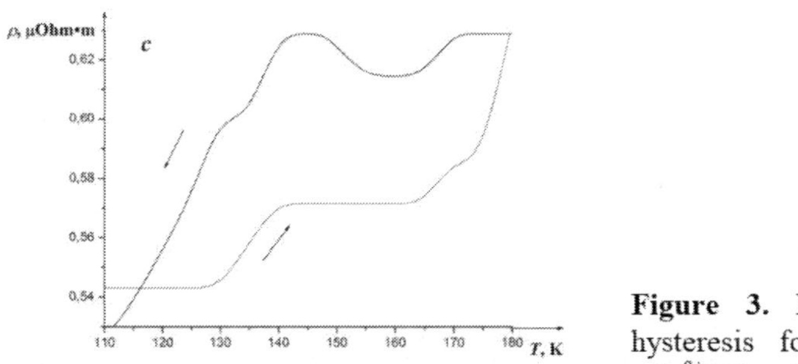

Figure 3. Loops of electrical resistivity hysteresis for Ni-Ti alloy after $_{16}O^{3+}$ (a), $_{40}Ar^{8+}$ (b) and $_{84}Kr^{15+}$ (c) ion implantation.

It was found that under the impact of $_{16}O^{3+}$ ions the temperature range shifts towards lower MT temperatures and 3.6 times increases mainly due to the greater degree of decrease in the temperature of M_{end} (figure 3 a). Martensitic transformation is preceded by pre-martensitic phase transformation B2→R, which indicates positive values of dρ/dT coefficients [11]. Changing of the shape and area of the electrical resistance hysteresis loop is related to the pre-martensitic transformation and extending temperature range reverse to MT due to a significant increase in temperature when it ends A_{end}.

As the mass of inert gases increases, the MT temperature range shifts, on the contrary, towards higher temperatures, and its widening is due to the greater degree of the increase in the temperature when MT starts M_{st}. In the case of $_{40}Ar^{8+}$ ion implantation (figure 3 b), firstly, direct MT is of distinctly stepped character compared to $_{16}O^{3+}$ ion implantation (figure 3 a) and $_{84}Kr^{15+}$ ion implantation (figure 3 c).

Secondly, in the R(T)-curve of the inverse martensite transformation we can find the electric resistance scale maximum exceeding by 0.033 μOhm·m compared to the temperature of the start M_{st} of the direct MP, and then a sharp decline as the temperature keeps increasing. A similar effect has been previously observed after heat treatment of nickel-titanium alloys doped with the third component [12].

Note that for both ions of inert gases the pre-martensitic transformation area is more extended and more pronounced. It contains the region where electrical resistance increases as the temperature falls, and the "plateau" with ρ = const in the R-curve of direct MT (figure 3 b and c). According to [12], two processes are responsible for this "plateau". The first one is related to crystal lattice distortions in transition of the R-phase, which increases electrical resistance, and the second one caused by the decrease in the amplitude of thermal vibrations (phonon component of the electric resistance) contributes to electrical resistance reduction.

The process of phase transformation B2→R in heavy ion implantation to a relatively low fluence determined by measuring the temperature dependence of the resistivity is consistent with its greater sensitivity as compared to the results of X-ray analysis.

4. Conclusion

In the experimental study, we found the features of the impact of implantation of Ni-Ti alloy with $_{16}O^{3+}$, $_{40}Ar^{8+}$ and $_{84}Kr^{15+}$ heavy ions on its structure, phase composition, and physical and mechanical properties under relatively similar parameters A/Z, E_{ion}/a.e.m., Φ and J_{beam}. The direct relationship between increase in the ion mass and radiation damage of Ni-Ti alloy with the shape memory effect is not found, but it is manifested in the degree of the processes, namely, sputtering, track formation, phase transformations and changes in the temperature ranges of martensitic transformations.

The preliminary testing of Ni-Ti alloy revealed the following features of radiation damage when exposed to $_{16}O^{3+}$, $_{40}Ar^{8+}$ and $_{84}Kr^{15+}$ ions:

- the process of sputtering prevails in the case of lighter $_{16}O^{3+}$ ions, and the process of ion etching dominates under the effect of $_{84}Kr^{15+}$ heavy ions. The strain-hardened layer based on Ni_3Ti_3O compound is dispersed regardless of the ion mass;
- for ions with M≤40, the formation of ~ 1 nm-sized bubbles can be observed, whereas, in the case of $_{84}Kr^{15+}$ ions, we observe tracks with a dome surrounded by bubble chains;
- in implantation, partial ($_{16}O^{3+}$) and complete (M≥40) radiation-induced phase transformation B19'→B2 occurs both in the near-surface layer and nearby the out-range area. The phase composition of titanium nikelid consists mainly of the B2 structure (austenite);
- regardless of the ion mass, the near-surface layer is a double layer structure. As the mass of the ion increases, the degree of the second layer hardening decreases from 71 to 39%. In the out-range area, on the contrary, it grows from 7 to 16%;
- the martensitic transformation temperature range with pre-martensitic phase transformation B2→R is found to shift towards lower and higher temperatures for $_{16}O^{3+}$ ions and M≥40, respectively.

The results obtained suggest new possibilities for modification by different heavy ions of gaseous elements to make titanium nickelid alloys for medical use which are radiation-resistant to ionizing radiation and retain the ability of the shape memory effect.

Acknowledgements

This work was partially supported by The Ministry of Education and Science of the Russian Federation within the framework of the government "Science" program.

References

[1] Kislitsin S B, Poltavtseva V P and Antoniuc V I 2009 *Proc. 8 Int. Conf. on the Interaction of Radiation with Solids* (Minsk: Publishing House Center BSU) pp 92-94
[2] Poltavtseva V P 2010 *Proc. of the 10th Int. Conf. on Modification of materials with particle beams and plasma flows* (Tomsk: Publishing House) pp 441–442
[3] Poltavtseva V P, Antoniuc V I and Kireev V V 2011 *Proc. VIII Int. Conf. in Fundamental and Applied Materials Science* (Barnaul: Publishing House AltSTU) pp 65–69
[4] Poltavtseva V P, Kislitsin S B, Koval N N and Oskomov K V 2012 *Izv. Vyssh. Uchebn. Zaved. Fiz.* **12/3** 41–43
[5] Poltavtseva V P, Kislitsin S B, Koval N N and Antoniuc V I 2013 *Proc. 10 Int. Conf. on Interaction of Radiation with Solids* (Minsk: Publishing House of BSU Center) pp 211–213
[6] Poltavtseva V P, Kislitsin S B and Antoniuc V I 2014 *Abstracts of Int. Congress on Energy Fluxes and Radiation Effects* (Tomsk: Publishing House of IAO SB RAS) p 265
[7] Poltavtseva V P, Kislitsin S B, Antoniuc V I and Satpaev D A 2013 *Proc. 9 Int. Conf. on*

Nuclear and Radiation Physics (Almaty: RSE INP) pp 103–104
[8] Poltavtseva V P, Kislitsin S B and Antoniuc V I 2009 *Izv. Vyssh. Uchebn. Zaved. Fiz.* **12/3** pp 41–43.
[9] G.E. Gunther, V.N. Khodorenko, Y.F. Yasenchuk et al. 2006 *Titanium Nikelide. Medicalapplience of New Generation* (Tomsk: MEC)
[10] Lotkov A I, Grishkov V N, Anokhin S V and Kuznetsov A I 1982 *Izv. Vyssh. Uchebn. Zaved. Fiz.* **10** pp 11–16.
[11] Potekaev A I, Klopotov A A, Kozlov E V and Kulagin V V 2004 *Weak-stable Pretransitional Structures in Ttitanium Nickelide* (Tomsk: STL)
[12] Honma Toshio and Takei Hiromi 1975 *Nippon Kinzoku Gakkaishi/Journal of the Japan Institute of Metals* **39(2)** pp 175–182

Radiation thermal processes in Cr13Mo2NbVB steel – the material of the fuel assembly shell in reactor BN-350 under mechanical tests

A S Larionov[1], A S Dikov[1], V P Poltavtseva[1], S B Kislitsin[1], M V Kuimova[2] and A V Chernyavskii[2]

[1] Institute of Nuclear Physics, Almaty, Republic of Kazakhstan

[2] National Research Tomsk Polytechnic University, Tomsk, Russia

E-mail: poltavtseva@inp.kz

Abstract. Regularities of changes of structural-phase state and mechanical properties of steel 13Mo2NbVB – the material of the fuel assembly shell in reactor BN-350 after various mechanical tests at 350°C are experimentally studied. The formation of microprecipitations FeMo, enriched or depleted with molybdenum was found in the short-time mechanical tests, which is the cause of thermal hardening of irradiated Cr13Mo2NbVB steel and its destruction by the ductile-brittle mechanism. On the basis of long-time creep tests it was shown that the material of the spent fuel assembly shell has sufficient resource for long-time storage in the temperature and force conditions simulating long-time storage of spent nuclear fuel.

1. Introduction

The majority of countries using nuclear energy, due to the limited capacity at the present for the reprocessing of spent nuclear fuel (SNF), have chosen direct disposal [1]. This period of exposure depends on the fuel burn-up (at higher burn-up the holding period can last for decades). At the second stage SNF may be directed to the ultimate disposal [1-3] or temporary dry storage for a long period (10^4-10^5 years), during which occurs the reduction of activity, heat input and the radioactivity of removed waste to a safe level [4].

The experience of recent decades shows that storage of spent fuel in dry conditions is becoming more popular, especially for new storages on off-site new nuclear power stations.

In order to justify the concept of dry storage of spent nuclear reactor designs, we proposed the studies of structural materials of the fuel assembly shells to long-time strength in the temperature and force conditions simulating long-time storage of spent nuclear fuel, and forecasting on their basis the behavior of the material during long-time storage. Conducted studies of austenitic and ferritic steels of different fuel assembly shells on fast neutrons [5, 6] have shown the prospectivity of such direction.

The article presents the experimental results of studies of changes in mechanical properties and structure of ferritic Cr13Mo2NbVB steel – material of the fuel assembly shell in reactor on fast neutrons BN-350 after various mechanical tests at a temperature of 350°C.

2. Material and experimental methods

As an object for the research we chose ferritic Cr13Mo2NbVB steel – the material of hexagonal fuel assembly shell of reactor fuel assembly (RFA) of reactor on fast neutrons BN-350 after prolonged exposure to fast neutron irradiation. At the mark of "500 mm" from the center of the core at conditions: T ~ 365°C; dose ~ 6.3 dpa; velocity of dose rate ~ 2×10^{-8} dpa/s.

According to the certification at the initial state we have Cr13Mo2NbVB steel with body-centered cubic lattice (a = 0.2866 nm) as a solid solution of Cr and Mo in iron, with carbide (Nb (V) C) hardening. Elemental composition (wt.%): Cr: – 13.2; Mo – 1.5; C – 0.11; Mn – 0.6; Nb – 0.5; V – 0.2; Si – 0.5; Fe – all the rest.

Due to the high activity of the irradiated material of hexagonal fuel assembly shell of spent RFA, the size of flat specimens for mechanical testing was $0{,}020 \times 0{,}002 \times 0{,}0003$ m^3. The samples of the irradiated and non-irradiated material were cut by the method of spark cutting from the sheet of face of the fuel assembly shell transversely with respect to its axis direction. The surface areas of the samples near the break before the investigation of the structure were subjected to mechanical grinding and polishing, chemical etching in the electrolyte: 70% C_2H_5OH; 17% HF; 13% HCl, the etching time is 20 minutes.

Long-time mechanical strength tests at various applied strains of duration 1500-1800 hours and short-time mechanical testing (at a rate of 0.5 mm/min) were performed at 20 and 350°C on a universal testing machine LR5K Plus in air environment. The studies of changes in the structure and the hardening/softening of materials after mechanical testing were performed by X-ray diffraction, optical and scanning electron microscopy and microhardness measurement with the use of diffractometer D8 ADVANCE, microscopes AxioObserver and JSM-7500F (JEOL), microhardness PMT-3M.

3. Experimental results

Figure 1 shows the results of the calculation with the steps of 1 hour creep rate on the basis of the test chart on the long-time strength at 350°C Cr13Mo2NbVB steel – the material of the fuel assembly shell in reactor BN-350. The choice of test temperature and applied strain 166 MPa was due to the temperature and force conditions typical for storages of long dry storage of spent nuclear fuel.

Clearly, the creep rate increases twofold with the increase of applied strain from 166 to 642 MPa. Regardless of the magnitude of the applied strain, the curves of the creep rate, depending on the duration of the test, three stages are observed, the duration of which is reduced with the increase of applied strain. In the studied range of test duration the destruction of samples did not happen and there were no visible traces of external damage.

The degree of reduction of the creep rate on the linear portion of the first stage is 2.5 times higher for higher applied strain (figure 1, curve 2). With further increase in the duration of the test at this stage, it decelerates in ~ 3 and ~ 5 times, respectively, for low (figure 1, curve 1) and high (figure 1, curve 2) of the applied strain.

In the second stage it is characteristic the increase in the creep rate of 1.4 and 2 times respectively, for high and low applied strain, while in the third stage creep rate increases two times, regardless of its magnitude. However, with further increase in the duration of the test, we noticed a sharp decrease in the rate of creep, with the exception of the second stage in the case of low applied strain. A distinctive feature of this stage is the presence of a time interval of about 400 hours at a constant rate of creep. It is important to emphasize that regardless of the applied strain, the creep rate is compared with the duration of almost 1,500 hours.

The obtained results of long-time creep tests at 350°C, which is only slightly lower than the temperature of irradiation of the material of the fuel assembly shell, indicate the presence of at least two types of the most effective stoppers of glide dislocations of radiation origin when exposed to high and low applied strain, respectively. It can be concluded that the Cr13Mo2NbVB steel – material of the fuel assembly shell has sufficient resources for long-time storage in the temperature and force conditions simulating long-time storage of spent nuclear fuel.

IOP Conf. Series: Materials Science and Engineering **81** (2015) 012035 doi:10.1088/1757-899X/81/1/012035

To determine the mechanical properties of Cr13Mo2NbVB steel were held short-time mechanical tests (SMT) at a temperatures of 20 and 350°C. Stress-strain diagrams at 20°C exposed and unexposed Cr13Mo2NbVB steel are presented in figure 2.

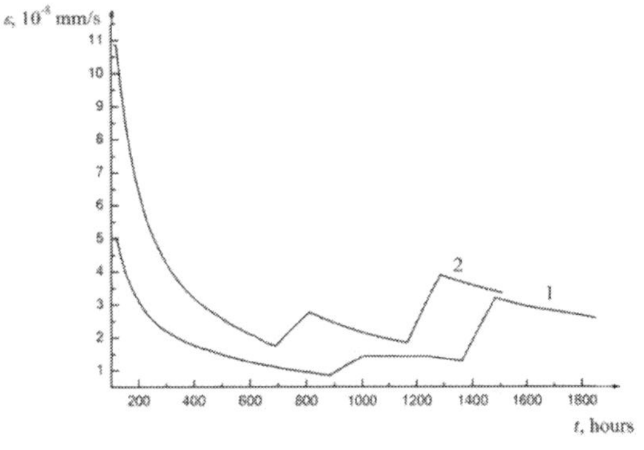

Figure 1. The dependence of creep rate Cr13Mo2NbVB steel at 350°C from the time at the strain of 166 MPa (curve 1) and 642 MPa (curve 2)

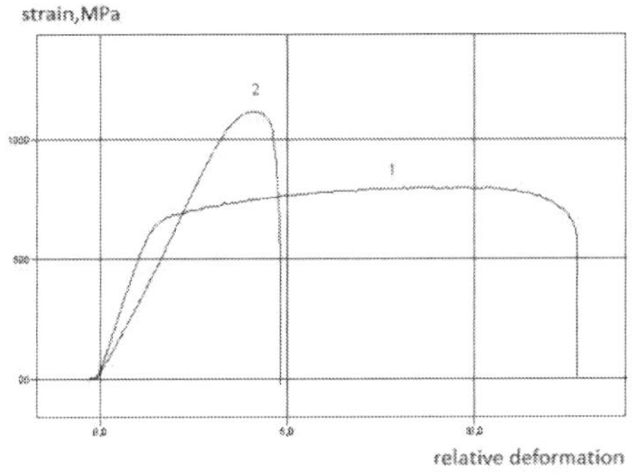

Figure 2. Stress-strain diagram at 20°C non-irradiated (curve 1) and irradiated Cr13Mo2NbVB steel (curve 2)

In the initial state the investigated steel has high ductility and a wide plasticity region necessary for the operation of the material in the fields of alternating power reactors (curve 1). Contrast with the non-irradiated steel, the material of the fuel assembly shell after long-time operation in the BN-350 radiation-hardened (curve 2). With the temperature increase of the tests to 350°C, the region of ductility of non-irradiated steel decreases 1.7 times and the elongation curve of irradiated steel does not change virtually, which is apparently due to similar values of the test temperature and radiation.

Using scanning electron microscopy it was revealed the presence in the structure of fracture non-irradiated Cr13Mo2NbVB steel after the SMT at 20°C the presence of knife-shaped cut with cup-shaped elements and practically complete absence of fractographic zone because of high ductility of the tested material. With the increase of the test temperature, the form of the cut was virtually unchanged. Irradiated Cr13Mo2NbVB steel is also characterized by knife-shape cut, but with elements of cleavage. Moreover, the cleavage area increases with the increase of test temperature. Therefore, the destruction of the irradiated Cr13Mo2NbVB steel occurs by the ductile-brittle mechanism.

Based on the analysis of mechanical characteristics of the Cr13Mo2NbVB steel indicated in table 1, it is shown that with the increase of the test temperature from 20 to 350°C, conventional yield strength $\sigma_{0.2}$ of non-irradiated and irradiated Cr13Mo2NbVB steel decreases to 4 and 22% respectively.

Table 1. Mechanical characteristics of Cr13Mo2NbVB steel

Material	Temperature of test, °C	$\sigma_{0.2}$, MPa	$\sigma_{в}$, MPa	Δl, mm	H_{μ}, MPa
Non-irradiated	20	729.0	803.6	0.18	320
	350	690.1	731.2	0.08	380
Irradiated	20	875.6	1116.8	0.05	320
	350	677.6	1005.2	0.05	350

Consequently, the main contribution to reducing $\sigma_{0.2}$ of irradiated steel is mainly due to radiation-induced processes that occur during long-time operation of the material of the fuel assembly shell. Further, we note that the contribution to the reduction of modulus of rupture of the irradiated Cr13Mo2NbVB steel is attributed to the thermal exposure during mechanical tests, whereas a zero change of elongation Δl at the increase of the test temperature, as noted above, is due to close test temperatures and radiation.

According to the microhardness measurements, irradiated Cr13Mo2NbVB steel is softer by ~ 3% (340 MPa) as compared to the initial state (350 MPa). As a result of SMT at 20°C both steels are softened; at higher temperature, on the contrary, hardened, where non-irradiated steel hardens more (table 1).

Figure 3 shows the diffraction patterns of non-irradiated Cr13Mo2NbVB steel taken with the pace of 0.07 and storage time of 7 s. before and after short-time tests at 350°C. From the comparison of diffraction patterns it can be seen that the phase composition of non-irradiated steel changes as a result of tests at 350°C (curve 2). Compared with the initial state (curve 1), three X-ray reflection lines are observed, which are, according to data files, identified as a combination close to FeMo with - body-centered cubic lattice (a = 0.3119 nm).

The amount of FeMo in the non-irradiated steel after the tests at 350°C, determined on the base of relations I_{FeMo} / I_{Fe}, is ≥ 50%. At that, the main line of FeMo is a superposition of 2 or more lines belonging to combinations of similar composition, indicating a prolonged shoulder from the side of higher angles 2θ. Further, it should be noted that in the initial state, this combination is also present, though in a small amount, according to the slightly outstanding line above the background at 2θ=40,87°.

According to the research data, by electron scan microscopy (figure 4) in the microstructure of non-irradiated Cr13Mo2NbVB steel after SMT at 350°C, one can clearly observe light (1) and darker microprecipitations (2) and microcracks of 5 to 25 nm, mainly located at the boundaries of these microprecipitations data. From the comparison of the elemental composition (table 2) it follows that the difference is determined mainly by the ratio of Fe and Mo in them. Thus, for light microprecipitations, approximately equal amount of both elements is typical, whereas for darker microprecipitations strong enrichment of Mo is typical.

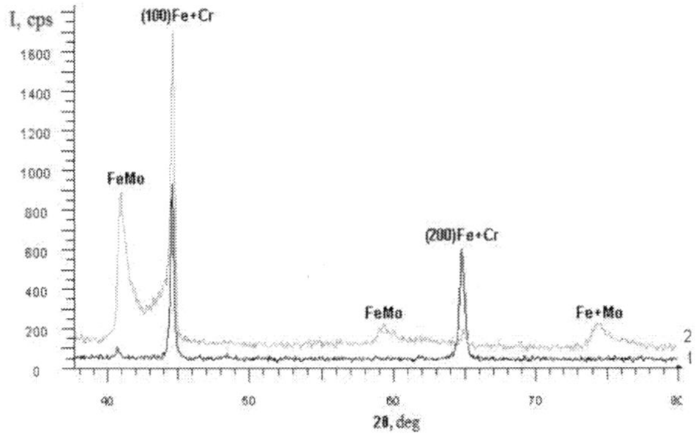

Figure 3. Sections of diffraction patterns of non-irradiated Cr13Mo2NbVB steel before (curve 1) and after SMT at 350°C (curve 2)

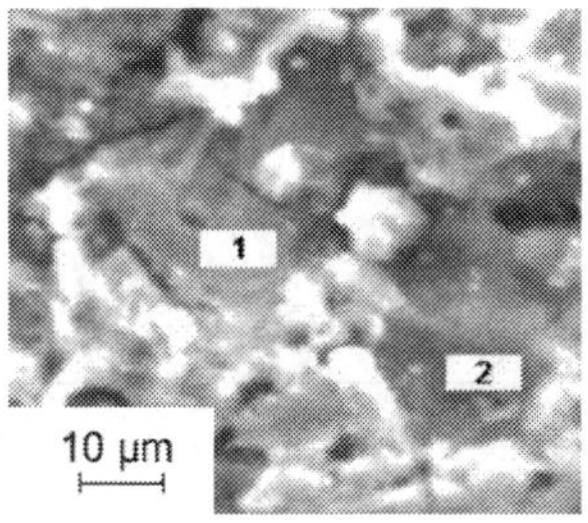

Figure 4. SEM image of the structure of Cr13Mo2NbVB steel after SMT at 350°C: 1 – $_{36}Fe_{40}Mo$; *2* – $_{17}Fe_{74}Mo$

However, such release is not detected after the test at 20°C. Only a slight reduction of carbides was found in comparison with microstructure obtained before the tests. Consequently, during the SMT at 350°C the formation occurs as a result of the cleavage of microprecipitations, enriched or depleted with Mo respectively [7].

Table 2. The elemental composition of non-irradiated Cr13Mo2NBVB steel after SMT at 350°C

№	Elemental composition (in weight %)					
microprecipitations	Fe	Mo	Cr	O	Al	Si
1	36.55	40.30	6.89	3.78	0.38	0.34
2	16.82	73.89	2.72	6.58	-	-

Conducted studies of the structure by the method of optical metallography of irradiated and non-irradiated Cr13Mo2NbVB steel also indicate the formation of microprecipitations after SMT at 350°C, but their amount is little and the size is a bit more in the irradiated steel than for the initial state. It is also shown that there are virtually no microprecipitations after tests at 20°C, whereas an inconsiderable proportion of small microprecipitations presents in the microstructure of the non-irradiated steel before the test, which is consistent with the results of X-ray diffraction analysis. Further, is should be noted that in the material after long-time operation in the BN-350 reactor, primary carbides and microprecipitations are virtually unobserved, and this is the reason for the softening of the material.

4. Conclusion

As a result of the experimental research, were found the peculiarities of changes in the structure, phase composition and mechanical characteristics of Cr13Mo2NbVB steel – the material of the fuel assembly shell in reactor at fast BN-350 neutrons after various mechanical tests at 350°C. On the basis of long-time creep tests it is shown that the material of the spent fuel assembly shell has sufficient resources for long-time storage in the temperature and force conditions simulating long-time storage of spent nuclear fuel.

It is established that the decrease in conditional limit of ductility $\sigma_{0.2}$ and modulus of rupture σ_b is mainly due to radiation-induced processes that occur during long-time operation of the material of the fuel assembly shell, and thermal effect in the short-time mechanical tests at 350°C respectively.

The formation of FeMo microprecipitations is observed, enriched or depleted with molybdenum during mechanical tests at 350°C, which is the reason for additional thermal hardening of irradiated Cr13Mo2NbVB steel and its destruction by the ductile-brittle mechanism.

Acknowledgements
This work was partially supported by The Ministry of Education and Science of the Russian Federation within the framework of the government "Science" program.

References

[1] Isaev A I 2008 *Atomnaya tehnika za rubezom* **4** pp 14-20
[2] Smidts O 2005 Rev. *E: Soce'te' royale belge des electricians* **121(2)** pp 20–29
[3] Moore R 2006 *Nucl. Future* **2(6)** pp 289-294
[4] Yamashita T 2007 *IAEA* **2** pp 197-203
[5] Dikov A S, Kireev V V, Kislitsin S B and Nesterova A Yu 2010 *Proc. XI Int. Conf.on Solid State Physics* (Ust-Kamenogorsk) pp 595-600
[6] Kislitsin S B, Gortsev A P, Dikov A S and Nesterova A Yu 2012 *Proc. XX Int. Conf. on Physics of radiation effects and radiation material* (Alushta) pp 270-271
[7] Chuistov K V 1985 *Aging of Metal Alloys* – (Kiev: Naukova dumka)

Characterization of porous glass-ceramic material as absorber of electromagnetic radiation

O Kazmina[1], **V Suslyaev** [2], **M Dushkina** [1], **B Semukhin**[3]
[1]National Research Tomsk Polytechnic University, Tomsk, Russian Federation,
[2]National Research Tomsk State University, Tomsk, Russian Federation
[3]Institute of Strength Physics and Materials Science of Siberian Branch Russian Academy of Sciences, Tomsk, Russian Federation

E-mail: kazmina@tpu.ru

Abstract. Investigations of a foam glass-ceramic material synthesized from raw siliceous earth material by the two-stage method at temperatures below 950°C have demonstrated the improvement of its physic mechanical properties in comparison with foam glass synthesized from glass cullet. This material actively interacts with microwaves and can be used for the development of protective screens reducing the adverse effect of microwaves on biological objects, anechoic chambers, and rooms with low level of electromagnetic background noise. Spectra of the transmission and absorption coefficients and of the complex dielectric permittivity for frequencies in the range 26–260 GHz are presented. The observed effects demonstrate the existence of regions with partial and total reflection arising on the glass–pore boundary and of the microwave interaction with ultradisperse carbon particles that remain after foaming with incomplete frothier transition from the soot to the gas phase.

Introduction

The increased interest to environmentally safe heat insulation materials is observed at the present time. Such materials are fire proof, durable, do not evolve harmful substances at operation and do not rot. Foam glass is one of such materials. Problems of widening of raw material base are solved both in Russia and in abroad. European researches are directed mainly on use of various kinds of broken glass, for example glasses of electron-beam tubes, computer screens and so on [1–3]. Home researches use as feed composition both broken glass and various kinds of natural and technogeneous raw materials [4 – 6]. This work considers foam glass materials obtained using two stage low temperature technology. On the first stage a frit is synthesized at temperatures not more than 900 °C. On the second stage the frit powder with additions of gas-forming agent is used to prepare foam forming mixture that is foamed at 800 – 850 °C.

The foam glass material reduces the level of reflected and transmitted electromagnetic radiation [7] and has high heat engineering characteristics, in flammability, moisture resistance, and long lifetime [8, 9]. The materials of this group can be used as heat- and noise-insulating and structural ones. The foam glass radio absorbing granules, possessing rather low weight, are promising for application in aircraft engineering [10], since their placement inside the aircraft fuselage reduces the levels of heat losses and electromagnetic and radiation exposure inevitable at high altitudes and in building industry [11]. The urgency of such multifunctional materials is caused by their fire safety and ecological compatibility, adaptability to and convenience of manufacturing of elements of various geometric shapes.

Results of comprehensive investigation of the composition of the foam glass-ceramic material and of its mechanical and electromagnetic properties are discussed in the present study to reveal functional relationships that determine the possibility of obtaining the required physical and chemical characteristics.

Experimental part

The foam glass-ceramic material have rather high strength and low density in comparison with foam glasses. This is caused not only by high-quality macrostructure of the material, but also by the structural features of the interporous barrier. In the electron photomicrographs of the FGCM interporous barrier, nanosized structural elements are observed that can be represented as quartz or cristobalite particles on the surface of which one-dimensional Si,O chains are localized (Fig. 1). In our opinion, microglobules that are absent from the foam glass synthesized from the glass cullet are formed in this glass phase. Obviously, amorphous nature of the initial siliceous earth component promotes the formation of microglobular structures, which is finally manifested through an increase in the mechanical strength. The complex structure consisting of individual spheroids and their groups with sizes from 60 to 160 nm is clearly seen in the photomicrograph.

Figure 1. Electron microscopic image of nanoglobules of the interporous barrier of the foam material: a) scanning electron microscopy and b) high-resolution transmission electron microscopy.

The IR spectral analysis demonstrated that the main differences were observed in the spectral ranges 1100–200 and 2800–3000 cm^{-1}. For the foam glass crystal material, the band lying in the first spectral range was slightly broadened, and the new band arose at 1088 cm^{-1} that corresponded to non-bridge Si–O–Si bond vibrations and was absent from the foam glass synthesized from the glass cullet. In the second spectral range, the absorption was recorded that could be attributed to valence vibrations of the OH-groups. Hydration of the silicon-oxygen skeleton due to braking of siloxine bridge bonds leads to loosening of the structures. It is obvious that this data with allowance for the photomicrographs of the structure obtained by the method of electron microscopy demonstrated the globular structure of the amorphous component of the interporous barrier.

To investigate the electromagnetic response, flat samples with sizes of 30 × 30 mm were used. The thickness of the sample from commercially available glass cullet was 3.2 mm, and the thickness of the sample from foam glass crystal material was 6.2 mm. The absorption coefficient was calculated from the formula $A = 1 - R - T$, where R is the reflection coefficient and T is the transmission coefficient. Our measurements demonstrated that the reflection coefficients of the samples from two materials were almost equal to zero. This was due to the absorbing properties of the materials and the scattering properties of the diffuse foam glass surfaces whose allowance was troublesome; therefore, values of the coefficient A were calculated with a certain error.

From Figs. 2 and 3 it can be seen that both materials interact sufficiently actively with electromagnetic radiation. The electromagnetic energy absorption by the foam glass material was most likely caused by the presence of regions of partial and total reflection that arose on the glass-pore boundary. This problem was considered in ample detail in [12] where it was demonstrated that the radio absorbing properties depended on the glass dielectric permittivity, the content of pores, the degree of system porosity, and the pore diameter.

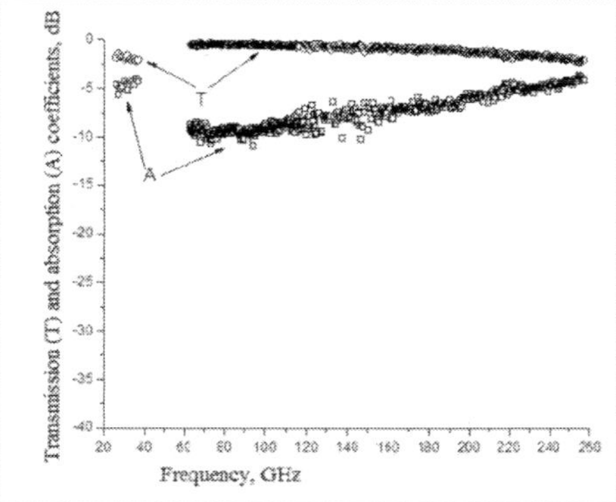

Figure 2. Parameters of the electromagnetic response of the sample synthesized from commercially available glass cullet.

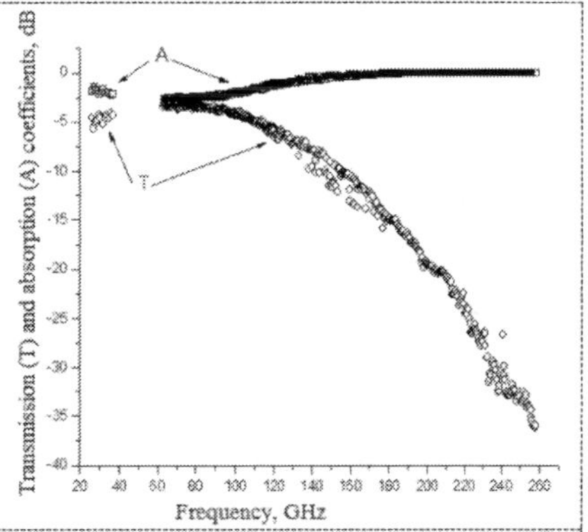

Figure 3. Parameters of the electromagnetic response of the sample synthesized from the foam glass-ceramic material.

Korolenko [12] formulated requirements to the foam glass material effectively protecting from electromagnetic radiation: ε of the foam glass skeleton must be $\varepsilon_g \geq 6$ rel. units (in this case, ε_g of the pore content must tend to 1); the degree of system porosity must be less than 0.9 rel. units; and the maximum pore diameter must be $d \leq 1.2$ mm.The dielectric permittivity of the examined samples calculated from the measured electromagnetic responses (Figs. 4 and 5) demonstrated that the foam crystal material better satisfied to these requirements.

The other physical mechanism that explains the absorbing properties of the foam glass materials is the interaction of an electromagnetic field with carbon nanoparticles remained after foaming due to incomplete transition of the carbon frother to the gas phase. The localization of carbon nanoparticles and the configuration of conglomerates they formed are mainly determined by micropore sizes and structure of the interporous barrier. The formation of coordinated carbon groups leads to the formation of Rayleigh scattering structures, and the radio absorbing properties of the foam glass increase [11].

The carbon nanoparticles can form a volume conductive skeleton. In [11] it was demonstrated that values of the EMR reflection and absorption coefficients of a composite material can be regulated by changing the relative content of the ultradisperse carbon in the composition. In this case, quantum effects arose in the foam glass material, that provide the broadband radiation absorption on a set of the formed energy levels of the transition.

Conclusions

Our investigations have demonstrated that the foam crystal material synthesized from raw siliceous earth materials by the two-stage method at temperatures not exceeding 950°C possesses the improved physic mechanical properties in comparison with the foam glass synthesized from the glass cullet, including its lower density, smaller heat conductivity, and higher compression strength.

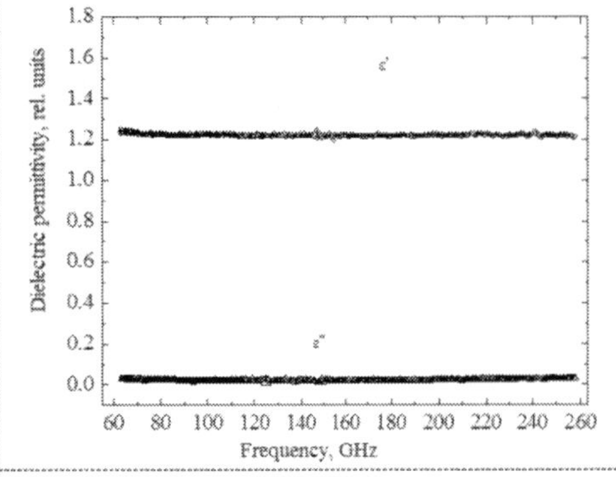

Figure 4. Spectra of the complex dielectric permittivity of the sample synthesized from the commercially available glass cullet.

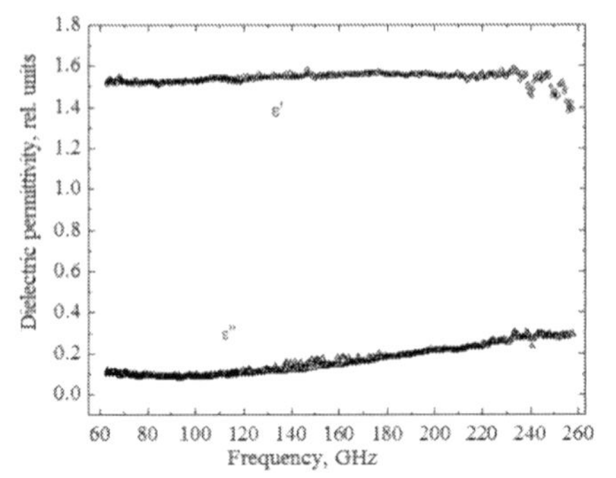

Figure 5. Spectra of the complex dielectric permittivity of the sample from the foam glass-ceramic material.

This material actively interacts with electromagnetic radiation and can be used for the development of protective screens reducing the adverse EMR effect on biological objects, unechoic chambers, and rooms with low level of electromagnetic background noise. The material interacts most actively with EMR whose frequency exceeds 60 GHz. The observed effects are explained by the existence of regions of partial and total reflection arising on the glass-pore boundary or by the EMR interaction with ultradisperse carbon particles remaining after foaming due to incomplete transition of the frother from soot to the gas phase. The electromagnetic characteristics can be controlled by changing of the composition of the foam glass crystal material and forming the shapes and sizes of micropores, the composition of interporous barriers, and the dielectric permittivity of the skeleton. The electromagnetic characteristics can change significantly after incorporation of carbon nanostructures, including single- and multiwall carbon nanotubes, bulbous structure, and fullerenes [13, 14].

Acknowledgment
The work has been performed at financial support of the state task “Science" no. 1235.

References

[1] Guo H, Gong Y, Gao SY 2010 *Preparation of high strength foam glass-ceramics from waste cathode ray tube* (Materials letters vol **64**) pp 997 – 999
[2] Chul-Tae Lee 2013 *Production of alumino-borosilicate foamed glass body from waste LCD glass* (J. of Industrial and Engineering Chemistry vol **19**) pp 1916 – 1925
[3] Fernandes H, Andreola F, Barbieri L, Lancellotti I, Pascual MJ, Ferreira JMF 2013 *The use of egg shells to produce Cathode Ray Tube glass foams* (Ceramics international 39) pp 9071 – 9078
[4] Volland S, Vereshchagin V, Kazmina O, Dushkina M 2014 *Recycling of sand sludge as a resource for lightweight aggregates (*Construction and building Materials vol **52**) pp 361-365
[5] Yatsenko E, Smoliy V, Kosarev A, Dzyuba E, Grushko I, Goltsman B 2013 *Physico-chemical properties and structure of TPP ash-based Foamglass* (Glass and Ceramics vol **1**) pp 3 – 6
[6] Damdinova D, Pavlov V, Alekseeva E 2012 *Foam Glass as the Base for Facing Materials with Controlled Porous Structure* (Construction Materials vol **1**) pp 44 – 45
[7] Suslyaev V, Kazmina O, Semukhin B 2012 (Izv. Vyssh. Uchebn. Zaved. Fiz. vol **55**) pp 312–314
[8] Bernardo E, Scarinci G, Hreglich S 2005 (Glass Sci. Technol.: Int. J. German Soc. Glass Technol DGG vol **1**) pp 7 – 11
[9] Wu J, Boccaccini A, Lee P, Rawlings R 2007 (Eur. J. Glass Sci. Technol. A 48 vol **3**) pp 133 – 141

[10] Gulbin V, Mikheev V, Kolpakov N, and Aleksandrov Yu 2012 (Proc. IX Int. Sci. Conf. on Hydroaviation "Gidroaviasalon-2012" Part II in Russian Moscow pp 108–117
[11] Gulbin V, Mikheev V, Kolpakov N, Cherdyntsev V, Inzhin Yad 2013 vol **4** No 6 pp 597 – 604
[12] Korolenko A 2004 (Elektr. Zh. Avtom. Avtomatiz. Elektrotekh. Kompl. Sist. vol **2**) pp 37 – 45
[13] Mazov I, Rudina N, Ishchenko A 2012 (Fullerenes Nanotubes and Carbon Nanostructures vol **20**) pp 510–518
[14] Suslyaev V, Kuznetsov V, Zhuravlev V 2013 (Russ. Phys. J vol **55** no 8) pp 970 – 975

Particularities of Radiation Defect Formation in Ceramic Barium Cerate

I V Khromushin[1], T I Aksenova[1], T Tuseev[2], K K Munasbaeva[1], Yu V Ermolaev[2], V N Ermolaev[2], A S Seitov[2]

1 Institute of Nuclear Physics, 050032 Almaty, Kazakhstan
2 Kazakh National Technical University, 050013 Almaty, Kazakhstan

Email: khromushin@mail.ru

Abstract. The effects of irradiation with electrons, ions of noble gases (Ne, Ar, Kr) and oxygen on the structure and properties of neodymium-doped barium cerate have been studied using the methods of X-ray diffraction analysis, scanning electron and atomic force microscopy, thermal desorption spectroscopy. It was shown that irradiation by low-energy ions of noble gases stimulates the blistering processes on the sample surface, while the high-energy ions contribute to formation of the structures on the irradiated surface that resemble the various stages of spherulite growth. The similar structures were not observed in the case of irradiation with high-energy oxygen ions. According to the data on thermal desorption of water and oxygen molecules from the irradiated barium cerate it was supposed that irradiation by the noble gas ions promotes neodymium oxidation state change. It was noticed that the electron irradiation leads to the formation of the nano-sized acicular structures on the cerate surface.

1. Introduction

Recently it was noticed that the conducting properties of the solid-oxide electrolytes with the ABO_3 perovskite structure can, in some cases, be improved by their radiation modification [1-4]. The results are explained by the oxygen vacancy formation as well as an increase in concentration and mobility of charge carriers caused by irradiation. However, the information on radiation modification of perovskite materials is still limited. The aim of this paper is to study the effect of different kinds of irradiation on the structure and properties of the neodymium-doped barium cerate.

2. Specimens and experimental procedure

The plates of the ceramic barium cerate doped with neodymium $BaCe_{1-x}Nd_xO_{3-x/2}$, (x=0.15) in the size of 10x5x1 mm were preliminary annealed in air at 650° C for 7 hours. Heavy ions irradiation was performed on the accelerator DC-60 in the Institute of Nuclear Physics in Kazakhstan (Astana city). All samples were irradiated to the same dose of $10^{16}cm^{-2}$ and ion current was less than 1 μA. The samples were mounted on a stainless steel water-cooled flange. The sample temperature did not exceed 385K at irradiation. Irradiation was carried out in a chamber evacuated at a pressure of less than 1.33×10^{-4} Pa. The irradiation parameters are presented in the Table 1. The ion ranges and the vacancy formation rate were calculated using the SRIM - 2013 software [5].

Table 1. Sample irradiation parameters

High energy				Low energy			
Ion type	Energy [MeV]	Range [μm]	Vacancy/ ion	Ion type	Energy [keV]	Range [μm]	Vacancy/ Ion
O	28	13.0	1740	O	40	0.080	243
Ne	35	12.7	2700	Ne	40	0.065	320
Ar	70	13.2	8100	Ar	100	0.080	800
Kr	147	15.3	31000	Kr	260	0.108	2500

Electron irradiation was performed on the electron accelerator ELV-4 in the Institute of Nuclear Physics (Almaty city) up to dose 10 MGy. The electron energy was 1.3 MeV providing the electron range in the material more than the sample thickness (1 mm).

The studies of heavy ion and electron effect on structure of doped barium cerate were performed using the X'Pert MPD PRO diffractometer with Cu–$K_{\alpha 1}$ radiation (λ=1,54060 Å). The morphology and chemical composition of local fragments of the specified surface areas were investigated using atomic force microscope (NT-MDT Probe Nanolaboratory Integra) and scanning electron microscope-microanalyzer (JSM-6490 LA (JEOL)) with energy dispersion extension. The special experimental setup with radio frequency spectrometer MX-7304 was used for gas thermodesorption study. This method allows to study gas desorption processes in vacuum while sample is heated with a definite rate. A personal computer with an Advantech PCI-1710HG module was used to select and control sample heating, choice of gas masses to be monitored, temperature and thermodesorption spectrum measurement. The sample heating rate was 42 K/min in all cases; pressure in working chamber was ~ 10^{-5} Pa.

3. Experimental results and consideration

The X-ray diffraction analysis showed that the structure of the non-irradiated barium cerate corresponds to the perovskite $BaCeO_3$ structure with an orthorhombic symmetry Pbnm. Fig. 1(a) shows the diffraction pattern of the non-irradiated neodymium-doped (x=0.15) barium cerate. It was found that the perovskite structure is maintained after irradiation with low-energy ions (O, Ne, Ar, Kr) and electrons. The amorphization of cerate and impurity inclusions (Fig. 1(b)) were detected in the case of barium cerate irradiation with high energy ions. It should be noted that degradation of the crystal lattice was also observed on the non-irradiating side of the ceramics, indicating that the changes in the oxide lattice, irradiated with high energy ions, occur up to the distances many times larger than the ion range in the material.

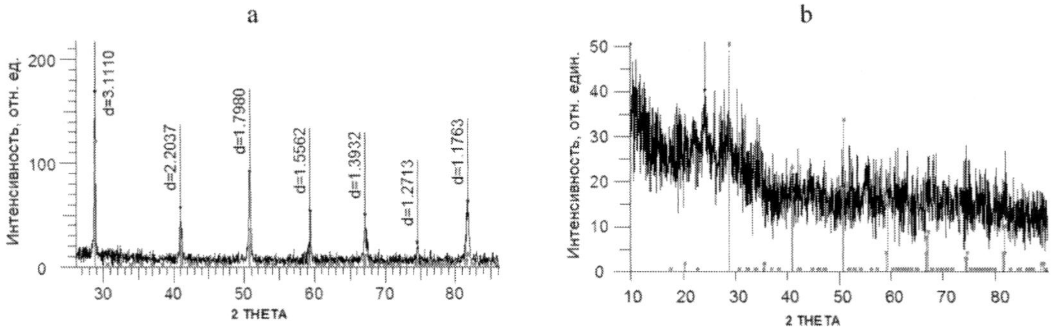

Figure 1. The diffraction patterns of non-irradiated (a) and irradiated with high energy argon ions (b) Nd-doped barium cerate.

IOP Conf. Series: Materials Science and Engineering **81** (2015) 012037 doi:10.1088/1757-899X/81/1/012037

Fig. 2 illustrates SEM images of the surface morphology of the barium cerate irradiated with the low energy ions of neon (a), argon (b) and krypton (c), made at the same magnification. According to the Fig. 2, the blistering can be observed on the surface of cerate irradiated with ions of neon and argon; it can be observed more clearly in the samples irradiated with the neon ions. There was no blistering observed under irradiation of barium cerate with the krypton ions (Fig. 2(c)).

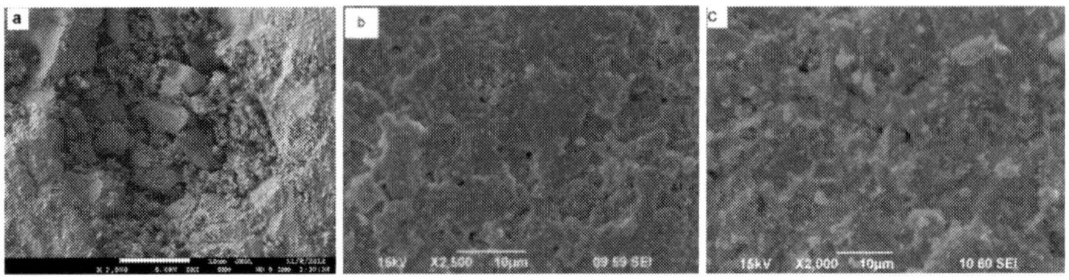

Figure 2. Electron micrographs of barium cerate surface irradiated with low energy ions of Ne (a), Ar (b) and Kr (c) to the dose of $10^{16}cm^{-2}$.

Irradiation of barium cerate with the high energy ions of noble gases caused the changes resulting, apparently, from the solid phase structural transformations on the surface (Fig. 3), and in the Ne, Ar, Kr series the barium cerate surface resembled the stages of spherulite growth: nucleation (a), growth (cauliflower view) (b), formation of spherulitic cover (c). Irradiation of this material with high energy ions of oxygen did not cause such changes of the surface (Fig. 3(d)).

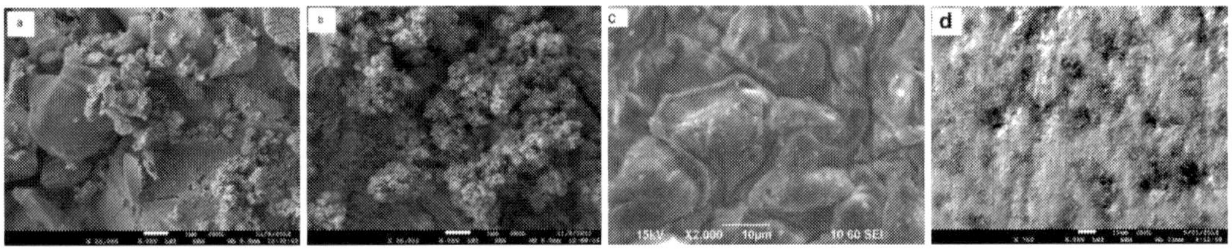

Figure 3. Electron micrographs of barium cerate surface irradiated with high energy ions of Ne (a), Ar (b), Kr (c) and oxygen (d) to the dose of $10^{16}cm^{-2}$.

Data of atomic force microscopy (AFM) are in good agreement with the results of scanning electron microscopy and also indicate the solid-phase structural transformations on the surface of cerates under irradiation with high-energy ions of noble gases. Fig. 4(b) shows the AFM images of barium cerate surface after irradiation with high-energy krypton ions. After irradiation the surface relief looks "smoother" compared with the "bumpy" surface of the non-irradiated cerate.

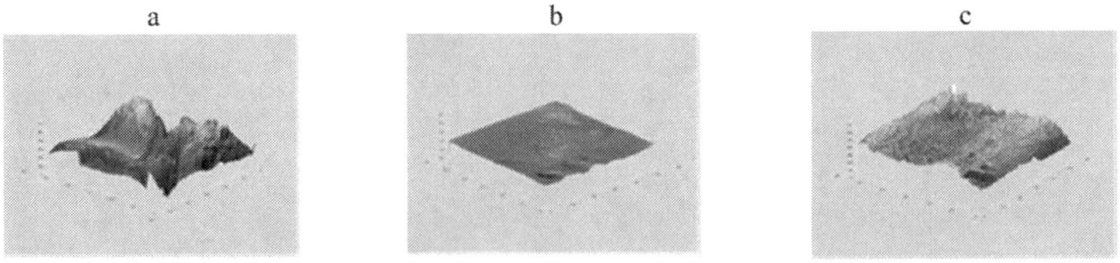

Figure 4. AFM images of Nd-doped barium cerate non-irradiated (a), irradiated with Kr (147 MeV; 10^{16} cm^{-2}) (b) and electrons (1.3 MeV; 10 MGy) (c)

The oxygen ions regardless of energy did not cause similar changes in the structure and state of the barium cerate surface, pointing to the peculiarities of defects formation in case of cerate irradiation with the ions of noble gases and oxygen. The revealed difference in the formation of the cerates surface structure under irradiation with high-energy ions of noble gases and oxygen is probably associated with their different solubility. It is assumed that in the first case the super-saturation of the material with the lattice defects takes place through their stabilization by the ions of noble gases with formation of the complexes "inert gas - vacancies". These complexes are characterized by high stability due to low solubility of noble gases in the materials. This is confirmed by the lack of noble gases release from the samples in thermal desorption experiments up to 1000°C.

In the case of high-energy oxygen ion irradiation of cerate at the elevated temperatures, the equilibrium concentration of the defects should be much lower. Indeed, the solubility of oxygen in cerate is significantly higher and under irradiation at elevated temperatures the oxygen may leave the material as a result of its diffusion to the surface and subsequent desorption. In this case, stabilization of the defect structure does not occur and the major part of defects, formed by irradiation, is annealed during irradiation.

Electron irradiation, as noted earlier, did not cause any changes in the perovskite structure of cerate; however, it resulted in smoothing of the surface relief. The peculiarity of barium cerate surface changes after electron irradiation is the formation of small pyramidal (acicular) structure (Fig. 4(c)) on it. The maximal height of the individual "pyramids" was about 50 nm; the "pyramids" effect was observed more clearly with the increase of the radiation dose. Dimensions and alternation of the newly formed structures suggest that electron irradiation promotes the nano-structuring of the cerate surface.

The studies of the processes of gases thermal desorption from the irradiated doped barium cerate indicate the dependence of these processes on the type and energy of ions. It is shown that the oxygen release was not observed from Nd-doped barium cerate irradiated with high-energy neon ions. (Fig. 5(a)).

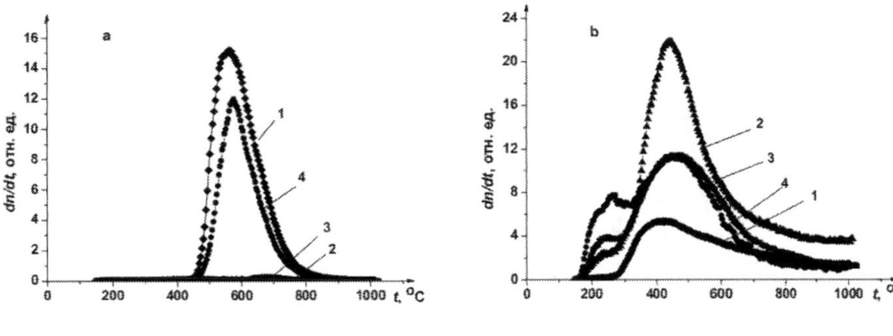

Figure 5. Oxygen (a) and water (b) release from Nd-doped barium cerate, non-irradiated (1) and irradiated with high-energy ions of Ne (2), Ar (3) and Kr (4) up to the dose 10^{16}cm^{-2}.

However, the increase in the amount of desorbed oxygen was observed with the increase of ions energy and mass, but the amount of oxygen released from the irradiated materials in all cases was less than from the initial ones (Fig. 5(a)). At the same time, in the spectra of water molecules release we observed the opposite picture compared with oxygen, i.e. the amount of the desorbed water decreased with the increase of ions energy and mass, but in all cases it was larger than that from the non-irradiated cerate (Fig. 5(b)). In addition, the low-temperature shoulder was observed in the spectra of water caused probably by the release of the physically adsorbed water.

Irradiation of barium cerate with low-energy ions of noble gases, in contrast to high-energy ions, called a significant increase in the amount of desorbed oxygen and the simultaneous decrease in the amount of desorbed water (Fig. 6(a, b)), and this effect was practically independent of ion mass and energy.

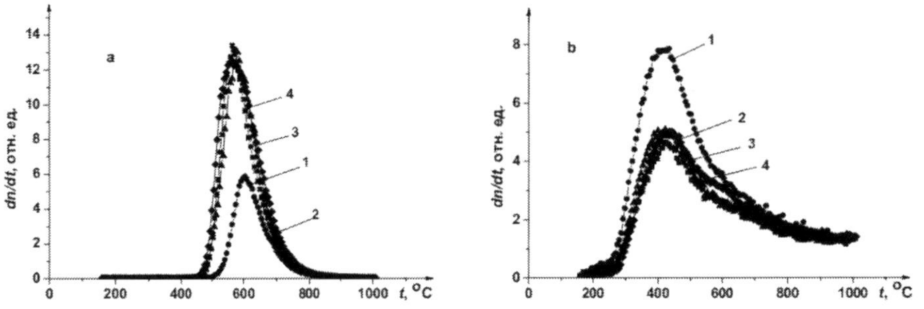

Figure 6. Oxygen (a) and water (b) release from Nd-doped barium cerate, irradiated with low-energy ions: 1 – non-irradiated sample; 2 – after irradiation with the ions of Ne; 3 – Ar; 4 – Kr (10^{16}cm^{-2}).

This behavior can be explained as follows. In the study of oxygen non-stoichiometry in barium cerate doped with various dopants it was shown that the value of the oxygen index greater than the expected value of $\delta=3-x/2$ was observed on the neodymium doped barium cerate [6, 7]. The analysis of the experimental results and their comparison with literature data allowed the authors to conclude that neodymium in the cerate lattice has the mixed valence of 3+/4+. This for the same degree of cerate doping with neodymium and other dopants, for example, Sm, Gd with the valence 3+, provides the visibility of the three-valent cation doping of less degree in the first case. According to [7] neodymium reduction takes place in cerate under the temperature increase. This, apparently, provides the release of the oxygen molecules from neodymium-doped barium cerate and reduced release of the water molecules in comparison with other dopants in the thermal desorption spectra. This pattern was observed in the thermal desorption spectra of water and oxygen from the doped barium cerate (Fig. 7).

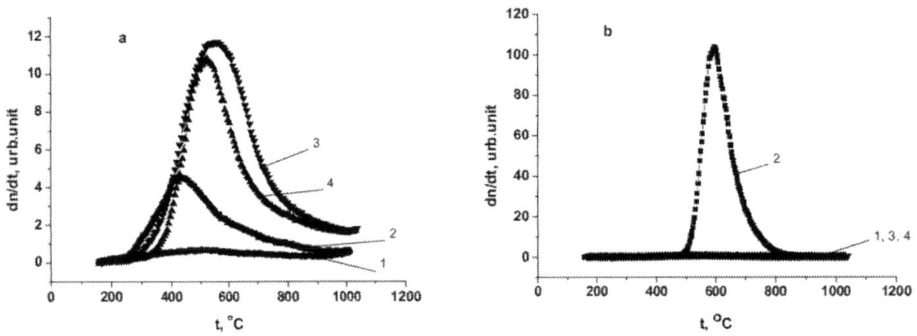

Figure 7. Water (a) and oxygen (b) release from undoped (1) and doped with Nd (2), Sm (3) and Gd (4) barium cerate.

It can be seen that the oxygen release is observed only from neodymium-doped cerate, and the amount of desorbed water from the neodymium doped cerate, is less than from Sm and Gd doped cerates with the same degree of doping. This confirms the conclusions of [7] on the mixed valence of neodymium in cerate and its reduction under the temperature increase.

Considering above, the observed behavior of the thermal desorption spectra of the neodymium doped barium cerate samples irradiated with high-energy and low-energy ions of inert gases, can be explained by the change in the valence state of neodymium during irradiation.

Thus, irradiation with low-energy ions of inert gases under these conditions apparently makes a significant part of neodymium transfer to the tetravalent state, accompanied by the increased release of oxygen and the reduced release of water compared with the non-irradiated cerate, and independence of the amount of released gases from the ions type.

Irradiation with high-energy ions of inert gases contrary causes the increase in the proportion of trivalent neodymium in the material. This, as expected, led to the reduced release of oxygen and the increased release of water from the irradiated materials in comparison to the non-irradiated ones. It should be noted that after irradiation with high-energy ions of neon practically the whole neodymium is in the trivalent state, as evidenced by the lack of oxygen release from cerate and the maximum release of water in the thermal desorption spectra.

The established patterns of the ions energies effect on the processes of oxygen and water thermal desorption are stipulated, apparently, by the peculiarities of defect formation in neodymium doped barium cerate under irradiation with low-energy and high-energy ions of inert gases. It should be noted that irradiation with oxygen ions and electrons had no appreciable effect on the desorption processes of oxygen and water molecules from the neodymium doped barium cerate.

4. Conclusion

The effects of electron and ion irradiation on the structure and properties of neodymium-doped barium cerate have been studied. Some peculiarities of irradiation type influence on the structure of doped barium cerates were revealed. On the analysis of gas desorption data it was supposed that irradiation by the noble gas ions promotes neodymium oxidation state change.

References

[1] Kim J-H, Tsuchiya B, Nagata S and Shikama T 2008 Effect of ion beam surface modification on proton conductivity of $BaCe_{0.9}Y_{0.1}O_{3-y}$ *Solid State Ionics* **179** 1182-1186

[2] Kim J-H, Choi H and Shikama T 2011 Effect of Ar+ irradiation on the electrical conductivity of $BaCe_{0.9}Y_{0.1}O_{3-y}$ *Applied Surface Science* **257** 8876-8882

[3] Khromushin I V, Aksenova T I, Tuseyev T, Munasbaeva K K, Ermolaev Yu V, Ermolaev V N and Seitov A.S 2013 Modification of barium cerate by heavy ion irradiation *Advance Materials Research* **781-784** 357-361

[4] Tsuchiya B, Morono A, Hodgson E R, Nagata S and Shikama T 2008 Change in hydrogen absorption characteristic of $SrCe_{0.95}Yb_{0.05}O_{3-\delta}$ by electron beam modification *Solid State Ionics* **179** 909-912

[5] Ziegler J F "Particle interactions with matter" http://www.srim.org

[6] Stevenson D A, Jiang N, Buchanan R M and Henn F E G 1993 Characterization of Gd, Yb and Nd doped barium cerates as proton conductors *Solid State Ionics* **62** 279–285

[7] Oishi M, Yashiro K, Mizusaki J, Kitamura N, Amezawa K, Kawada T and Uchimoto Y 2008 Oxygen nonstoichiometry of perovskite-type oxides $BaCe_{0.9}M_{0.1}O_{3-\delta}$ (M= Y, Yb, Sm, Tb and Nd) *Solid State Ionics* **179** 529-535

The nature of glow arising in PETN monocrystals' explosion initiated by a pulsed electron beam

B P Aduev[1], **G M Belokurov**[1], **S S Grechin**[1], **I Yu Liskov**[1], **A V Kalenskii**[2], **A A Zvekov**[1]
[1]Institute of Chemistry of Coal and Material Science, Sovetsky av., 18, Kemerovo, Russia, 650000
[2]Kemerovo State University, Krasnaya, 6, Kemerovo, Russia, 650043

E-mail: lesinko-iuxm@yandex.ru

Abstract. The explosive decomposition of pentaerythritol tetranitrate monocrystals under the influence of a high-current electron beam (0.25 MeV, 20 ns, 15 J/cm2) was researched with the approach of high temporal resolution optic spectroscopy. We measured kinetics and emission spectra in real time scale. The thermal nature of the explosive glow was proven with the method of spectral pyrometry. The estimated temperature of the explosion is T ≈ 3000 K..

1. Introduction

Pentaerythritol tetranitrate (PETN) is a secondary explosive that has extensive practical applications. There are numerous experimental studies where its decomposition under different initiation stimulus was researched [1-3]. One of the most fertile approaches applied in the field of explosives' decomposition research is pulse radiolysis. [4-6]. This method is believed to be highly informative as it is able to give insight into the nature of intermediate species and make it possible to estimate the temperature if it is high enough. In the present paper the kinetics of PETN monocrystals explosion initiated by pulse electron beam is studied with high temporal resolution optic spectroscopy. This way we determined the nature of glow arising during the decomposition.

2. Experimental methods and results

A pulsed electron accelerator GIN-600 was used as a source of high-current electron beam. The effective energy of an electron in the beam was 250 keV, the pulse duration was 20 ns, and the current density was up to 3 kA/cm^2.

The PETN monocrystal samples having thickness 1.5 – 2 mm and width 3 – 4 mm were installed on a copper crystal holder in a vacuum chamber of the accelerator. The distance between anode and the sample was 3 mm. The crystal surface made the angle of 45° with the beam direction and the axis of optic measurement system. The anode was made of a steel plate with a round central hole having diameter 2 mm, which was used as a discharge outlet of the vacuum diode. The distance between the steel cathode with diameter 2 mm and the anode was 3 mm. In the conditions described the incident

[1] Institute of Chemistry of Coal and Material Science, Sovetsky av., 18, Kemerovo, Russia, 650000, lesinko-iuxm@yandex.ru.

energy density of the high-current electron beam influencing the sample was 15 J/cm^2. The experiments were performed at the temperature 300 K and pressure 0.01 torr.

The kinetic and spectral dependencies of the glow arising during the PETN explosion were measured with a spectral streak-camera as it is described in [7]. The main parts of the apparatus were a polychromator used for decomposition of the glow into spectral components and a streak-camera working in the mode of linear timebase deflection. The image from the output screen of the streak-camera was read by a PCC-device and was transmitted to a computer where it was processed. The apparatus makes it possible to measure the spectral dependencies of the glow stimulated by a single pulse influence on the sample in the range 350 – 700 nm in the timebase deflection mode. The temporal resolution of the apparatus was 10 nm, the spatial resolution was 10 ns. The spectral response of apparatus was measured in the preliminary experiment where the spectral dependence of the correction coefficient k(λ) was determined with the standard incandescent lamp TRSh 2850-3000 which color temperature is known.

The kinetics of the glow at the wavelength λ = 590 nm obtained from the processed spectral-temporal image is shown on the fig. 1. The temporal dependence has two peaks. The first one observed in the nanosecond range is concerned on the sample's glow during the high-current electron beam influence, while the second one in the microsecond range is linked to the sample's decomposition.

The spectral dependence of the glow observed at the moment of the second intensity maximum (1 µs) is presented on the fig. 2. One observes a continuous spectrum with the intensity increasing in the red spectral range. The spectral lines of the excited nitrogen molecules [8] formed as a result of the decomposition are seen on the background of continuous spectrum.

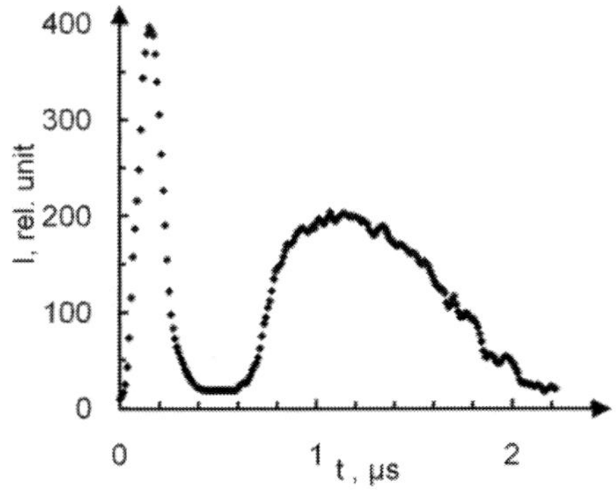

Fig. 1. The kinetics of PETN glow stimulated by the high-current electron beam pulse with energy density 15 J/cm^2 measured at the wavelength λ = 590 nm.

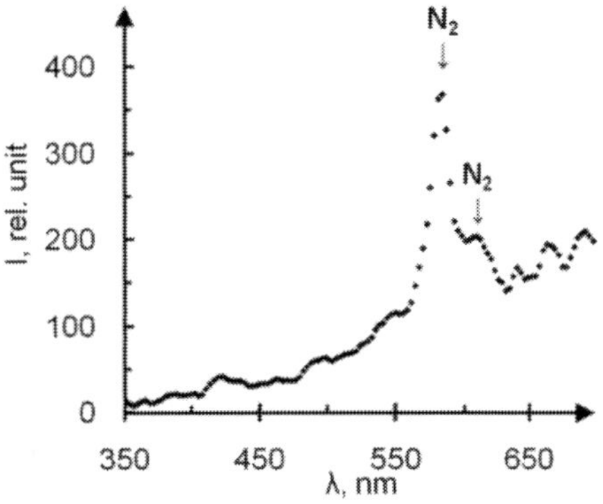

Fig. 2. The spectral dependence of the PETN monocrystal glow at 1 µs after the influence of high-current electron beam pulse.

3. Discussion

The nature of the glow arising during the high-current electron beam influence on the sample corresponding to the first peak on the temporal dependence (fig. 1) was studied thoroughly and interpreted in [4,5]. The nature of the glow observed on the stage of chemical decomposition corresponding to the second peak on the glow kinetics (fig. 1) has not got a proper understanding in [4,5]. It was suggested that the glow is linked to the NO_3-radicals luminescence, though the arguments for were unsound. Moreover, it was not determined whether the nature of the glow is the luminescent or thermal. The luminescent point of view stated in [4,5] was based on the fact that the spectral dependence in the range 600 – 1000 nm (λ_{max} = 840 nm) could not be fitted with the Plank's equation.

However, the possible spectral dependence of the emissivity factor, which distorts the thermal spectral dependencies of real objects comparing to the Plank's one, was not taken into account.

In the present paper the spectral dependencies were measured in the range 350 – 700 nm, so we used the approach of spectral pyrometry utilizing the Wien's coordinates for the results processing. This approach is as follows. The spectral dependence of the real object thermal emission is described by the Plank's equation with the emissivity factor ε:

$$I = \varepsilon C_1 \lambda^{-5}/[\exp(C_2/\lambda T) - 1] \quad (1)$$

where $C_1 = 37418$ W·μm^4/cm^2; $C_2 = 14388$ μm·K.

The problem arising when one tries to find out whether the spectrum of an object is thermal or not is concerned on the emissivity factor ε. The emissivity factor could be a smooth function of the wavelength in the case of so called gray body or may vary significantly with the wavelength. This obstacle could be overcome when one carries out the measurements in the Wien's range:

$$C_2/\lambda T >> 1 \quad (2)$$

In this case the equation (1) simplifies:

$$\ln(\lambda^5 I) - \ln(\varepsilon C_1) = -C_2/(\lambda T) \quad (3)$$

One is able to elucidate the thermal nature of the glow with plotting the function $\ln(\lambda^5 I)$ against $x = C_2/\lambda$ and checking whether the curve is linear. This way the distorting influence of the ε is eliminated. The temperature of the sample could be estimated as the slope of the linear curve.

The fig. 3 shows the spectrum of the PETN monocrystal glow at the stage of the chemical decomposition in the Wien's coordinates. The dependence is almost linear, so it proves the thermal nature of the glow. The temperature corresponding to the slope is $T = 3030$ K.

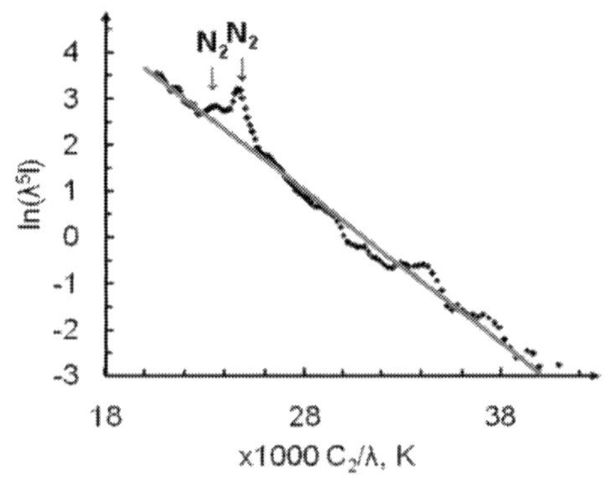

Fig. 3. The spectral dependence of the PETN monocrystal measured after 1 μs after the high-current electron pulse influence plotted in the Wien's coordinates.

4. Conclusion

The spectral-temporal dependencies of the glow arising during the explosion of PETN stimulated by the high-current electron pulse were measured. The spectral dependencies at the stage of chemical decomposition were processed in the Wien's coordinates and the thermal nature of the glow was determined. It could be considered as an argument for the thermal nature of the reaction acceleration at this stage.

Acknowledgments
The work was supported by Russian Foundation for Basic Research (grant No. 13-03-98032 r_sibir_a). The authors thank the Kemerovo Scientific Center for the apparatus used in the work.

5. References

[1] Dreger Z A, Gruzdkov Yu A et al. 2002 *J. Phys. Chem.* B **106** (2) 247
[2] Tarzhanov V I, Zinchenko A D Sdobnov V I et al 1996 *Combust. Expl. Shock waves* **32** 454
[3] Korepanov V I, Lisitsyn V M, Oleshko V I et al. 2003 *Tech. Phys. Letters* **29** 669
[4] Aduev B P, Belokurov G M, Grechin S S, Tupitsin E V 2004 *Tech. Phys. Letters* **30** 774
[5] Aduev B P, Belokurov G M, Grechin S S, Shvayko V N 2007 *Rus. Phys. J.* **50** 99
[6] Aluker E D, Aduev B P, Krechetov A G, Mitrofanov A Yu, Zakharov Yu A 2006 *Early stages of explosive decomposition of energetic materials* (Focus on Combustion Research) Ed. S Z Jiang (New York: Nova Publishers) chapter 2 pp 55 – 88.
[7] Shvayko V N, Krechetov A G, Aduev B P et al. 2005 *Tech. Phys.* **50** 732
[8] Pierce R and Gaydon A 1949 *Identification of molecular spectra* (London: Chapman and Hall, 1950) 240 p.
[9] Magunov A N 2009 *Instruments and Experimental Techniques*. **52** 451

Formation of the surface alloys by high-intensity pulsed electron beam irradiation of the coating/substrate system

Yu F Ivanov[1, 2], **E A Petrikova**[1, 2], **A D Teresov**[1, 2], **O V Krysina**[1, 2], **M E Rygina**[3]

[1] Institute of High Current Electronics SB RAS, Tomsk, 634055, Russia
[2] National Research Tomsk State University, Tomsk, 634050, Russia
[3] National Research Tomsk Polytechnic University, Tomsk, 634050, Russia

E-mail: yufi55@mail.ru

Abstract. The results of the analysis of the structure and properties of the surface layer of aluminum A7 subjected to alloying by the intense pulsed electron beam melting of the film / substrate system. Fold increase in strength and tribological properties of the modified surface layer due to the formation of submicro - nanoscale multiphase structure have been revealed.

1. Introduction

The possibility of synthesis the nanoscale metastable phases with unique physicochemical and performance characteristics in the surface layer of treated material is the main advantage of the approach based on the formation of surface alloys. This approach implemented in a number of methods: formation of coatings by electroexplosive method, welding powders using CW laser and compression plasma flows, high dose ion implantation, diffusion saturation in plasma of gas discharges and many others [1-3]. Along with the obvious advantages, these methods have and disadvantages.

For example, electroexplosive method and welding deposition method leads to the formation of inhomogeneous microstructure, the large grain size and high porosity, due to the high dimensional heterogeneity of the original powder coating and the relatively small rates of the melt quenching [3, 4].

The aim of this work is the developing of a method of alloying the surface layer of metals and alloys based on formation and irradiation of the film/substrate system by high-intensity pulsed electron beam submillisecond exposure time.

2. Materials and methods of study

Aluminum A7 was used as a material for study [5]. Formation and liquid-phase mixing of the film / substrate (A7) system was carried out in a single vacuum cycle by a pulsed electron beam on the "SOLO" setup (IHCE SB RAS) [6]. Silumin of a grade AK12 (Al-12%Si) was used as electron beam sputtering target (disk with diameter of 30 mm, a thickness of 6 mm) [7]. Aluminum substrate (plates 15*15*5 mm), located at the distance ≈ 50 mm from the target is attached to a flat plate installed on the shaft of the step motor (the last one located parallel to the plane of the manipulator) to allow rotation of the sample between the sputtering and mixing processes. At the initial moment, the modifiable surface of aluminum substrate was faced to the target and it located at an acute angle to the surface of the table. The target, fixed on a water-cooled table, irradiated by a pulsed electron beam with the following parameters: the energy density of 40 J/cm^2, pulse duration of 100 μs, the pulse frequency 1 Hz. Under these conditions, the deposition rate ≈5 nm/pulse. The coating thickness varied from 500 to 1000 nm, and determined by the number of irradiation pulses. After sputtering the plate with the sample, using a step motor was located parallel to the plane of the table and moved to the area of influence of the electron beam. Changing the mode of operation of the electron-beam setup "SOLO" (electron energy of 16 keV; the pulse frequency 0.3 s^{-1}; the pulse duration of electron beam of 50 μs and 150 μs; the energy density of the electron beam (10 ... 25) j/cm^2; the number of pulses 1, 3, 5), was conducted liquid phase mixing of the film/substrate (A7) system to form the surface alloy. All processes have been carried out in a vacuum chamber at a pressure of the working gas (Ar) ≈ 3,5 • 10^{-2} Pa. Thus, the formation and liquid phase mixing of the film / substrate system have been performed in a single vacuum cycle. Examinations of the structure of the surface alloy was carried out by optical (device Microvisor metallographic μVizo - MET-221) and scanning electron (device LEO EVO 50) microscopy. The elemental composition of the surface layer was studied using the methods of electron microprobe analysis (device LEO EVO 50). Estimation of elastoplastic properties and hardness of the surface alloy were conducted on a dynamic ultramicrohardness tester (nano indenter) Shimadzu DUH-211S and micro hardness tester PMT-3. Studies of wear resistance were carried out on a tribometer «CSEM Tribometer High Temperature S / N 07-142»; as a counterbody were used a ball of a hard alloyVK8. The measurements were carried out in a «rotation of the sample at a fixed counterface» scheme. Measurements of the profile of friction groove of the surface of the samples with the numerical determination of the depth of the friction groove and the area of its cross-section have been conducted at the end of the test using 3D-profilometer MICRO MEASURE 3D station French company STIL.

3. Experimental results and their discussion

There is rather multiplicity of the methods, which allow changing purposefully the properties of a detail at modification of structural and phase state of its surface layer [1-3]. One of the methods is the formation of surface alloys. That based on the use of the concentrated fluxes of energy (CFE) for irradiation of the film/substrate system. In the present work as it was above, the alloying of aluminum surface layer as carried out in a uniform vacuum cycle, using an intensive pulse electron beam as CFE. The results of the change of friction coefficient (μ), wear rate (V) of A7 aluminum in an initial state and after a surface alloying are presented in the table. The estimates of wear rate were carried out by equation (1):

$$V = \frac{2\pi RA}{FL}, \tag{1}$$

where R – the radius of a track of wear [mm], A – the area of cross-section of wear track [mm^2], F – the normal load [N], L – the distance [m] passed by a ball.

The analysis of the data presented in the table 1 showed that the best result is observed at deposition of a film with thickness of 1 μm and the subsequent irradiation of film/substrate system by intensive pulse electron beam at the following parameters: 20 J/cm^2, 150 μs, 5 pulses, 0,3 s^{-1}.

Table 1. Wear rate and the friction coefficient of the modified surface layer of aluminium A7[a].

E_S, J/cm^2	Thickness of the film, μm	V, 10^{-6}, mm^3/(N*m)	V_0/V	<μ>	<μ_0>/<μ>
0,0 (initial A7)	0,0	7590		0.48	
15	0,5	6080	1,3	0.39	1,5
15	1,0	6570	1,4	0.40	1,2
20	0,5	5470	1,4	0.35	1,4
20	1,0	4790	1,6	0.31	1,6

[a] The mode of treatment of the film / substrate system by electron beam: 150 μs, 5 pulses., 0.3 s^{-1}; E_S - energy density of the electron beam; V_0 and μ_0 - wear rate and friction coefficient of aluminum A7 in the initial state (before modification).

Formation of surface alloy is followed by change of strength characteristics of material. In Figure 1 the results of hardness change for the alloyed aluminum surface layer are presented. It is clear that the hardness and the Young's modulus of surface layer smoothly decreases from the values which is repeatedly exceeding strength characteristics of initial material to the values corresponding to A7 alloy. Therefore, the modified layer smoothly passes into the main volume of a sample. The boundary, which can contain concentrators of tension, is absent.

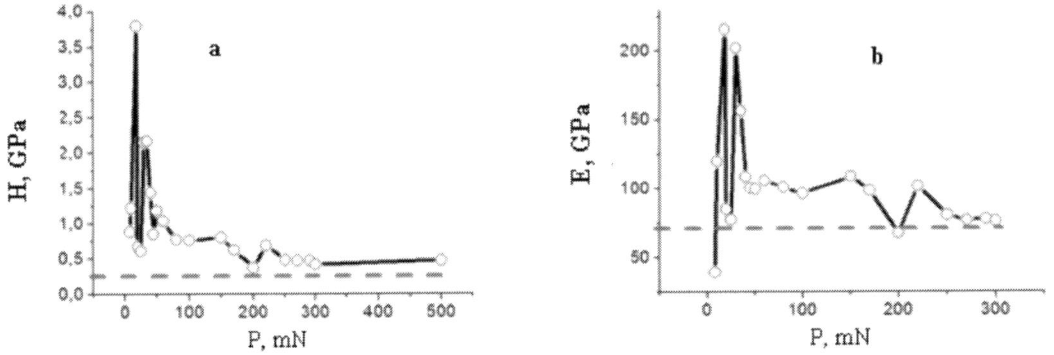

Figure 1. Dependence of hardness (*H*) and Young's modulus (*E*) on normal load (nanoindentation method) of aluminium surface layer modified by silumin deposition (thickness is 1 μm) and the subsequent treatment by an electron beam on the mode of 20 J/cm^2, 150 μs, 5 pulses, 0,3 s^{-1}. The dash-dotted lines designated the characteristics of initial aluminium A7 alloy.

Fold increase in strength and tribological properties of the surface layer of aluminum indicates a significant transformation of elemental, phase and structural state of the material. Figure 2 shows electron microscopic images of the surface structure formed by irradiation of the film/substrate system by a pulsed electron beam. Clearly shows that in the surface layer is formed submicrocrystalline structure, crystallite sizes vary from 0.3 μm to 0.6 μm.

The crystallite size of the surface layer depends on the thickness of the deposited film and on the electron beam energy density reaches the minimum sizes when irradiated by an electron beam with an energy density of 20 J/cm^2 (Figure 2, c, d).

The elemental composition of the modified layer was determined by electron microprobe analysis. Typical images of the energy spectrum obtained with this method for studying of the elemental composition is shown in Figure 3, b. The results showed that the surface layer is characterized by complex elemental composition, the main element of which, except aluminum, is oxygen and carbon. The concentration of silicon in a surface layer of the substrate repeatedly (in 25-30 times) lower than in the deposited sample. Consequently, the hardening of the modified layer and the improving of its tribological properties are due to the formation of oxycarbide phases based on aluminum.

Obviously, that sputtering of the target by a pulsed electron beam is accompanied by modification of a material surface layer. Figure 4 shows the electron microscopic images of the surface structure of AK12 silumin before irradiation (Figure 4a, b) and after treatment with a pulsed electron beam with the parameters of 40 J/cm^2, 100 μs, 100 pulses, 1 s^{-1} (Figure 4, c, d). Initially, the structure of the polished surface of the target is characterized by the presence of coarse inclusions of plate and globular forms of silicon (Figure 4, a, b). A contrast of silicon inclusions is slightly different from the contrast of the aluminum substrate when capturing an image of the surface structure of silumin in back-scattered electrons, as aluminum and silicon has similar atomic masses. Inclusions with acicular and skeletal shape in addition to silicon are detected (Figure 4, a). Comparatively bright contrast of these inclusions indicates that they are enriched with the heavier elements relative silicon and aluminum. Indeed, electron microprobe analysis revealed the presence in such inclusions of atoms of iron, copper and manganese in addition to aluminum and silicon. In [8, 9] are shown that the inclusion of such forms may be phases of Al_5SiFe (β-phase) and $Al_{15}(FeMn)_3Si_2$ (α-phase).

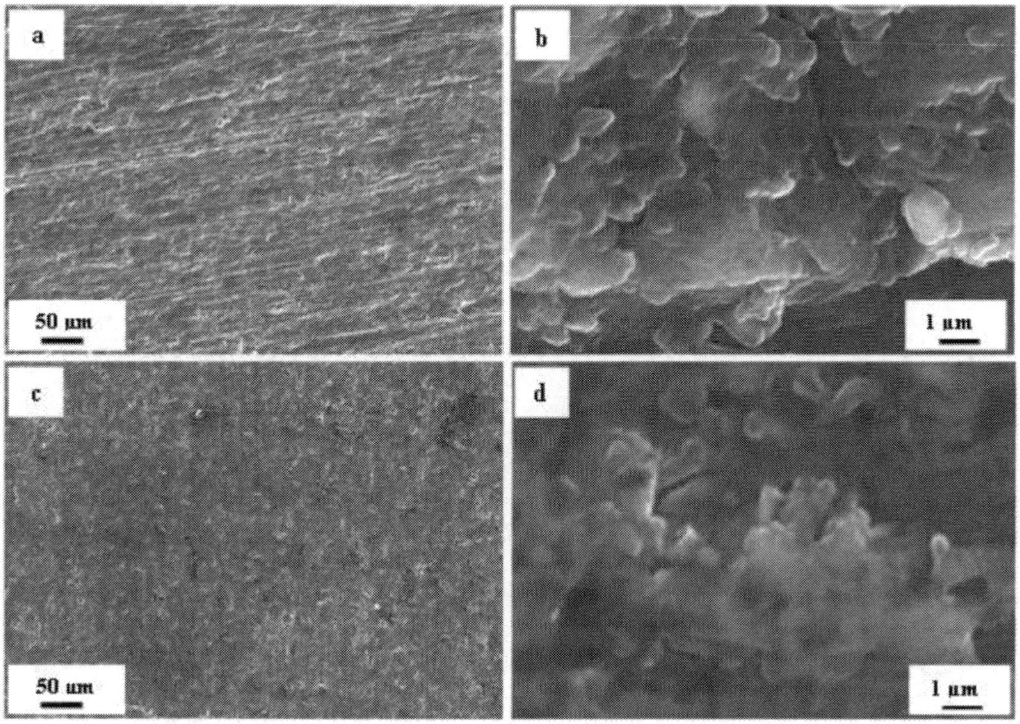

Figure 2. Electron microscopic image of the surface structure that is formed by irradiation of the film/substrate system by intense pulsed electron beam with the parameters: a, b - 15 J/cm^2, 150 µs, 5 pls., 0.3 s^{-1}; c, d - 20 J/cm^2, 150 µs, 5 pulses, 0.3 s^{-1}; a, b - film thickness of 0.5 µm; c, d - 1 µm.

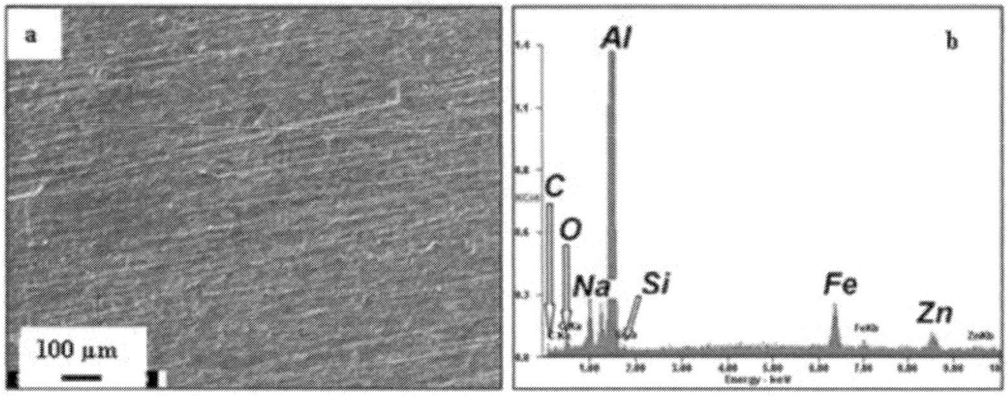

Element, at%							
Al	Si	C	O	Mg	Fe	Zn	Na
77.6	0.4	7.6	4.6	3.0	1.8	1.0	4.0

Figure 3. An electron microscopic image of the surface (a) formed upon irradiation of film/substrate system electron by pulsed electron beam; b – the energy spectrum obtained from the plot shown in (a); table elemental composition specified the area shown in (a). The thickness of the deposited film is 1 µm, electron beam irradiation mode: 20 J/cm^2, 150 µs, 5 pulses, 0.3 s^{-1}.

Figure 4. The structure of the silumin target surface in the initial state (a, b) and after irradiation by pulsed electron beam (40 J/cm^2, 100 μs, 100 pulses; 1 s^{-1}) (c, d). The arrows in (a, b) are shown the silicon particles.

After sputtering of target by pulsed electron beam the surface layer structure is changing radically. Firstly, there is the formation of a dendritic-cellular structure with high speed crystallization (Figure 4, c, d). The cell dimensions are vary from 1 to 3 μm. Inclusions of the second phase with size of 100 ... 300 nm are located at the cell boundaries. The electron microscopic images of the structure obtained in back-scattered electrons are shown that the nano-sized particles have a relatively more bright contrast, indicating the content of elements with high atomic weight in them, along with aluminum and silicon. Indeed, electron microprobe analysis of these particles revealed atoms of iron and copper in addition to aluminum and silicon. Secondly, the element and, obviously, the phase composition of the surface layer are changes: it is determined by methods of microprobe analysis that the average volume surface layer concentration of silicon is much lower (≈7 at.%) relative to the initial state (≈13 at.%). Therefore, the sputtering of target (AK12 silumin) is accompanied by a change not only the structure (formed submicro- nanoscale homogeneous structure with high speed crystallization), but also the elemental composition of the surface layer (the concentration of silicon reduces substantially in ≈2 times).

4. Conclusion

A liquid phase mixing and formation of film/substrate system by intense pulsed electron beam in a single vacuum cycle is carried out. The mode of deposition (40 J/cm^2, 100 μs, 1 s^{-1}, 1000 pulses) and irradiation (20 J/cm^2, 150 μs, 5 pulses, 0.3 s^{-1}) of the film/substrate system by intense pulsed electron beam, allowing to multiple increase in hardness and wear resistance of the surface layer of A7 aluminum is identified. It has been established that the increase of the strength and tribological properties of the modified surface layer caused by formation of submicron-nanoscale multiphase structure.

This work was supported by RFBR grant (project No. 13-08-00416_a).

References

[1] Gribkov V A, Grigoryev A I, Kalin B A *et al* 2001 *Perspective radiation-beam technology of metal treatment.* – (M. Krugluy Stol) 528
[2] Kurzina I A, Kozlov E V, Sharkeev Yu P, Fortuna S V, Koneva N A, Bojko I A 2008 *Nanocrystalline intermetallic and nitride structures formed during ion-beam exposure* (Tomsk STL) 324
[3] Bagautdinov A Ya, Budovskih E A, Ivanov Yu F, Gromov V E 2007 *Physical basis of electroexplosive alloying of metals and alloys* (Novokuznetsk: SibGIU) 301
[4] Solonenko O P, Kudinov V V, Smirnov A V *et al* 2004 *Thermal physics of plasma deposition and related technologies* (Novosibirsk: Nauka) 401
[5] Hetch Dj E, Fridlyander I N ed. 1989 *Aluminum: properties and physical metal science: Reference book* (M: Metallurgy) 423
[6] Koval' N N, Ivanov Yu F 2008 Nanostructuring the surface of metal-ceramic and ceramic materials under pulsed electron-beam treatment *Izvestiya Vuzov. Fizika* **5** 60-70.
[7] Stroganov G B, Rotenberg V A, Gershman G B 1977 *Aluminum and silicon alloys* (Moscow: Metallurgy) 271
[8] Zolotarevskiy V S, Belov N A 2005 *Metal science of cast aluminum alloys* (MISIS) 376
[9] Belov N A 2009 *Phase composition of aluminum alloys* (MISIS) 392

Mathematical modelling of the spread of contamination during fires in forests exposed to radioactive contamination

Valeriy Perminov
Department of ecology and basic safety, Tomsk Polytechnic University

E-mail: perminov@tpu.ru

Abstract.
The paper suggested in the context of the general mathematical model of forest fires [1] gives a new mathematical setting and method of numerical solution of a problem of a radioactive spread above the forest region. Numerical solution of problems of radioactive smoke spread during crown fire in exemplified heat energy release in the forest fire front was found. Heat energy release in the forest fire front was found to cause further radioactive particles spread by the action of wind. In the absence of wind, radioactive smoke particles deposit again on the underlying surface after a time. As a wind velocity increases, these particles are transferred in the ground layer over distances proportional to a wind velocity.

Introduction

The most intense and significant transfer of radionuclides to the forest area, exposed to radioactive contamination occurs as a result of the combined action of wind and forest fires [2-8]. Moreover, large forest fires are closely related to the vortex atmospheric phenomena. Large-scale forest fires, under appropriate conditions, generate a cloud of gas that generates atmospheric tornado, and the latter may intensify the spread of fire in mass due to increased flow of oxygen to the fire. That is, a firestorm, which supports the existence of the gas cloud and ensure that the products of combustion in these conditions. Thus, for mathematical modeling is necessary to formulate the problem of heat and mass transfer surface air with forest fires. Using the general mathematical model of forest fires [1], we can construct a model of the spread of radioactive contamination repeated as a result of wind and forest fires. In addition to physical and chemical factors that arise from forest fires must take into account the physical and chemical processes and the chemical components that arise during the radioactive decay of substances. In particular should be considered the dispersed radioactive particles containing uranium, plutonium, cesium and other radioactive chemicals. Experimental data suggest that as a result of radiolysis of forest fuel (FF) the rate of the drying and pyrolysis FF are reduced.

Physical and mathematical setting

Source separation of radionuclides in forest fires, is a source of heat and mass, located in a region of space [7]. Its temperature is determined by a function of time and coordinates. The basic assumptions adopted during the deduction of equations, and boundary and initial conditions: 1) the forest represents a multi-phase, multistoried, spatially heterogeneous medium; 2) the flow has a developed turbulent nature and molecular transfer is neglected; 3) gaseous phase density doesn't depend on the pressure because of the low velocities of the flow in comparison with the velocity of the sound; 4) the binary gas-dispersed mixture made of particles of condensed phase, including radioactive aerosols and gas phase - the components oxygen, inert gases and the combustible components. Let the coordinate reference point x_1, $x_2 = 0$ be situated at the centre of the surface forest fire source at the height of the roughness level, axis $0x_1$ directed parallel to the Earth's surface to the right in the direction of the unperturbed wind speed, axis $0x_2$ directed upward (Figure 1).

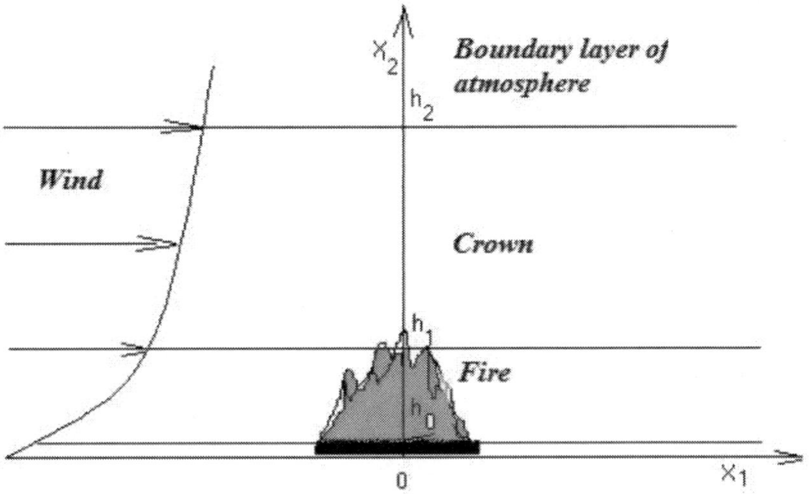

Figure 1.

For the atmospheric boundary layer stated above, the problem reduces to the solution of the following system of equations:

$$\frac{\partial\rho}{\partial t}+\frac{\partial}{\partial x_j}(\rho v_j)=0,\ j=1,2,\ i=1,2; \quad (1)$$

$$\rho\frac{dv_i}{dt}=-\frac{\partial p}{\partial x_i}+\frac{\partial}{\partial x_j}(-\rho\overline{v_i'v_j'})-\rho g_i; \quad (2)$$

$$\rho c_p\frac{dT}{dt}=\frac{\partial}{\partial x_j}(-\rho c_p\overline{v_j'T'})+q_5R_5+k_g(cU_R-4\sigma T^4); \quad (3)$$

$$\rho\frac{dc_\alpha}{dt}=\frac{\partial}{\partial x_j}(-\rho\overline{v_j'c_\alpha'})+R_{5\alpha}-\dot{m}c_\alpha,\ \alpha=\overline{1,6}; \quad (4)$$

$$\frac{\partial}{\partial x_j}\left(\frac{c}{3k_g}\frac{\partial U_R}{\partial x_j}\right)-k_gcU_R+4k_g\sigma T^4=0; \quad (5)$$

$$\sum_{\alpha=1}^{6}c_\alpha=1,\ p_e=\rho RT\sum_{\alpha=1}^{6}\frac{c_\alpha}{M_\alpha},\vec{v}=(v_1,v_2),\vec{g}=(0,g).$$

IOP Conf. Series: Materials Science and Engineering **81** (2015) 012040 doi:10.1088/1757-899X/81/1/012040

In the area located between the low level of roughness and upper boundary of forest canopy we have the system of equations averaged over the height of the canopy [1] expressing conservation laws for multiphase multicomponent reactive continuum:

$$\frac{\partial\rho}{\partial t}+\frac{\partial}{\partial x_1}(\rho v_1)=Q-(\dot m^- - \dot m^+)/h; \tag{6}$$

$$\rho\frac{dv_i}{dt}=-\frac{\partial p}{\partial x_i}+\frac{\partial}{\partial x_j}(-\rho\overline{v_i'v_j'})-\rho s c_d v_i|\vec v|-\rho g_i-Qv_i+(\tau_i^- -\tau_i^+)/h,\ i=1,2; \tag{7}$$

$$\rho c_p\frac{dT}{dt}=\frac{\partial}{\partial x_1}(-\rho c_p\overline{v_1'T'})+q_5R_5-\alpha_v(T-T_s)+(q_T^- - q_T^+)/h+k_g(cU_R-4\sigma T^4); \tag{8}$$

$$\rho\frac{dc_\alpha}{dt}=\frac{\partial}{\partial x_1}(-\rho\overline{v_1'c_\alpha'})+R_{5\alpha}-\dot m c_\alpha+(J_\alpha^- - J_\alpha^+)/h,\ \alpha=\overline{1,6}; \tag{9}$$

$$\frac{\partial}{\partial x_1}\left(\frac{c}{3k}\frac{\partial U_R}{\partial x_1}\right)-kcU_R+4k_S\sigma T_S^4+4k_g\sigma T^4+(q_R^- - q_R^+)/h=0,k=k_g+k_S; \tag{10}$$

$$\sum_{i=1}^{4}\rho_i c_{pi}\phi_i\frac{\partial T_S}{\partial t}=q_3R_3-q_{2R_{2s}}+k_s(cU_R-4\sigma T_S^4)+\alpha_V(T-T_S)\ ; \tag{11}$$

$$\rho_1\frac{\partial\phi_1}{\partial t}=-R_1,\ \rho_2\frac{\partial\phi_2}{\partial t}=-R_2,\ \rho_3\frac{\partial\phi_3}{\partial t}=\alpha_C R_1-\frac{M_C}{M_1}R_3,\ \rho_4\frac{\partial\phi_4}{\partial t}=0\ ; \tag{12}$$

$$\sum_{\alpha=1}^{6}c_\alpha=1,\ p_e=\rho RT\sum_{\alpha=1}^{6}\frac{c_\alpha}{M_\alpha}; \tag{13}$$

$$Q=(1-\alpha_c)R_1+R_2+\frac{M_c}{M_1}R_3+R_{54}+R_{55},$$

$$R_{51}=-R_3-\frac{M_1}{2M_2}R_5,R_{52}=\nu(1-\alpha_c)R_1-R_5,R_{54}=\alpha_6R_1, \tag{14}$$

$$R_{55}=\frac{\alpha_4\nu_2}{\nu_2+\nu_{2*}}R_3,\ R_3=k_3\rho\varphi_3 s_\sigma c_1\exp\left(-\frac{E_3}{RT_s}\right),h=h_3-h_1.$$

The system of equations (1)–(14) must be solved taking into account the initial and boundary conditions. At the initial time in the whole computational domain defined distribution of the unknown functions:

$$t=0: v_1=0,\ v_2=0, T=T_e, c_\alpha=c_{\alpha e}, T=T_e,\ \varphi_i=\varphi_{ie}. \tag{15}$$

Values of functions for the incoming flow in the left boundary are set as follows:

$$x_1=-x_{1e}: v_1=V_e, v_2=0, T=T_e, c_\alpha=c_{\alpha e}-\frac{c}{3k}\frac{\partial U_R}{\partial x_1}+cU_R/2=0. \tag{16}$$

On the right boundary set the following boundary conditions:

$$x_1 = x_{1e} : \frac{\partial v_1}{\partial x_1} = 0,\ \frac{\partial v_2}{\partial x_1} = 0, \frac{\partial T}{\partial x_1} = 0, \frac{\partial c_\alpha}{\partial x_1} = 0, \frac{c}{3k}\frac{\partial U_R}{\partial x_1} + cU_R / 2 = 0. \quad (17)$$

At the upper boundary of the computational domain boundary conditions will be as follows:

$$x_2 = x_{2e} : \frac{\partial v_1}{\partial x_2} = 0,\ \frac{\partial v_2}{\partial x_2} = 0, \frac{\partial T}{\partial x_2} = 0, \frac{\partial c_\alpha}{\partial x_2} = 0, \frac{c}{3k}\frac{\partial U_R}{\partial x_2} + cU_R / 2 = 0. \quad (18)$$

Values of the functions in the hearth of ignition on the ground cover are set depending on the time:

$$\rho v_2 = h_0 \dot{m}, T = T_S = \begin{cases} T_e + t / t_0 (T_0 - T_e), t \le t_0, \\ T_e + (T_0 - T_e) \exp[-k(t / t_0 - 1)], t > t_0 \end{cases} \quad (19)$$

At the interface forest canopy - the boundary layer of the atmosphere are given the following conditions:

$$\begin{aligned} &\rho v_2\big|_{x_2=h_-} = \rho v_2\big|_{x_2=h_+},\ -\rho\overline{v_1' v_2'}\big|_{x_2=h_-} = -\rho\overline{v_1' v_2'}\big|_{x_2=h_+}, \\ &v_1\big|_{x_2=h_-} = v_1\big|_{x_2=h_+},\ -\rho\overline{v_2'^2}\big|_{x_2=h_-} = -\rho\overline{v_2'^2}\big|_{x_2=h_+},\ v_2\big|_{x_2=h_-} = v_2\big|_{x_2=h_+}, \\ &-\rho\overline{v_2' c_\alpha'}\big|_{x_2=h_-} = -\rho\overline{v_2' c_\alpha'}\big|_{x_2=h_+},\ c_\alpha\big|_{x_2=h_-} = c_\alpha\big|_{x_2=h_+}, \\ &-\rho C_p\overline{v_2' T'}\big|_{x_2=h_-} = \rho C_p\overline{v_2' T'}\big|_{x_2=h_+},\ T\big|_{x_2=h_-} = T\big|_{x_2=h_+}, \\ &-\frac{c}{3k}\frac{\partial U_R}{\partial x_1}\bigg|_{x_2=h_-} = -\frac{c}{3k}\frac{\partial U_R}{\partial x_1}\bigg|_{x_2=h_+}. \end{aligned} \quad (20)$$

Here and above $\frac{\mathrm{d}}{\mathrm{d}t}$ is the symbol of the total (substantial) derivative; α_V is the coefficient of phase exchange; ρ - density of gas – dispersed phase, t is time; v_i - the velocity components; T, T_S, - temperatures of gas and solid phases, U_R - density of radiation energy, k - coefficient of radiation attenuation, P - pressure; c_p – constant pressure specific heat of the gas phase, c_{pi}, ρ_i, φ_i – specific heat, density and volume of fraction of condensed phase (1 – dry organic substance, 2 – moisture, 3 – condensed pyrolysis products, 4 – mineral part of forest fuel), R_i – the mass rates of chemical reactions, q_i – thermal effects of chemical reactions; k_g , k_S - radiation absorption coefficients for gas and condensed phases; T_e - the ambient temperature; c_α - mass concentrations of α - component of gas - dispersed medium, index α=1,2,...,6, where 1 corresponds to the density of oxygen, 2 - to carbon monoxide CO, 3 - to carbon dioxide and inert components of air, 4 - to particles of black, 5 - to particles of smoke, 6 - dust particles containing radioactive pollution; R – universal gas constant; M_α , M_C, and M molecular mass of α -components of the gas phase, carbon and air mixture; g is the gravity acceleration; c_d is an empirical coefficient of the resistance of the vegetation, s is the specific surface of the forest fuel in the given forest stratum. In equation (4) the total derivative components of the gas phase at α = 1-5 is as follows:

$$\frac{d}{dt} = \frac{\partial}{\partial t} + v_1 \frac{\partial}{\partial x_1} + v_2 \frac{\partial}{\partial x_2},$$

and for dispersed phase components at α = 6:

$\frac{d}{dt} = \frac{\partial}{\partial t} + v_1 \frac{\partial}{\partial x_1} + (v_2 - v_{2\alpha}^{(s)}) \frac{\partial}{\partial x_2}$, here $v_{2\alpha}^{(s)} = 18 g \mu / \rho_\alpha^{(s)} d_\alpha^2$ - Stokes deposition rate of the dispersed particles, μ - molecular dynamic viscosity of the gas phase, g - acceleration of gravity, $\rho_\alpha^{(s)}, d_\alpha$ - the density and diameter of the dispersed particles [6,7].

Thus, in accordance with [1], the medium is considered a two-speed, to allow for sedimentation - deposition of particles under the influence of gravity. In the system (6) - (14) the values of τ, J_α, q_T characterize the exchange of momentum, mass and energy between the boundary layer of the atmosphere and the lower layers of the forest are determined by the appropriate boundary conditions. Thus, the set of equations (1) - (14) are the ratio of the mass balance, energy and momentum, which are setting conjugate problem, whose solution is to determine the characteristics of complex, interrelated propagation process of a forest fire. Solution the problem stated above is considerable mathematical difficulties. The values of τ, J_α, q_T generally be determined in the process of solving the conjugate problem.

Reaction rates of these various contributions (pyrolysis, evaporation, combustion of coke and volatile combustible products of pyrolysis) are approximated by Arrhenius laws whose parameters (pre-exponential constant k_i and activation energy E_i) are evaluated using data for mathematical models [1].

$$R_1 = k_1\rho_1\varphi_1 \exp\left(-\frac{E_1}{RT_s}\right), R_2 = k_2\rho_2\varphi_2 T_s^{-0.5} \exp\left(-\frac{E_2}{RT_s}\right),$$

$$R_3 = k_3\rho\varphi_3 s_\sigma c_1 \exp\left(-\frac{E_3}{RT_s}\right), R_5 = k_5 M_2\left(\frac{c_1 M}{M_1}\right)^{0.25}\frac{c_2 M}{M_2}T^{-2.25}\exp\left(-\frac{E_5}{RT}\right).$$

Thermodynamic, thermal and structural characteristics correspond FF pine forests [1] and is numerically equal to: E_1/R = 9400K, k_1 = $3.36\cdot10^4$ c^{-1}, q_1 =0, E_2/R = 6000K, k_2= $6\cdot10^5$ c^{-1}, q_2=$3\cdot10^6$ J/kg, E_3/R=10^4 K, k_3=10^3 c^{-1}, q_3=$1.2\cdot10^7$ J/kg, E_5/R=11500 K, k_5 =$3\cdot10^{13}$, q_5 =10^7 J/kg, c_{p1}= 2000, c_{p2} = 4180, c_{p3} = 900, c_{p4} =1000, c_p = 1000 J/(kg ·K), s = 1000 m^{-1}, sc_d=0.1, α_C=0.06, $\rho_4\varphi_4$= 0.08 kg/m^3, ν=0.7, ρ_1 = 500, ρ_2= 1000, ρ_3= 200, ρ_e= 1.2 kg/m^3, c_{2e} = 0, φ_{3e} = 0, p_e = 10^5 N/m^2, T_e= 300 K, T_0= 1200 K, c_{1e} =0.23, c_{50} =0.09, α_4=0.66, α_0 =10^{-4}.

To close the system (1)–(14), the components of the tensor of turbulent stresses, and the turbulent heat and mass fluxes are determined using the local-equilibrium model of turbulence [1]. The system of equations (1)–(14) contains terms associated with turbulent diffusion, thermal conduction, and convection, and needs to be closed. The components of the tensor of turbulent stresses $\rho\overline{v_i'v_j'}$, as well as the turbulent fluxes of heat and mass $\overline{\rho v_j' c_p T'}$, $\overline{\rho v_j' c_\alpha'}$ are written in terms of the gradients of the average flow properties using the formulas:

$$-\rho\overline{v_i v_j} = \mu_t\left(\frac{\partial v_i}{\partial x_j}+\frac{\partial v_j}{\partial x_i}\right)-\frac{2}{3}K\delta_{ij},$$

$$-\rho\overline{v_j c_p T'} = \lambda_t\frac{\partial T}{\partial x_j}, \quad -\rho\overline{v_j c_\alpha'} = \rho D_t\frac{\partial c_\alpha}{\partial x_j},$$

$$\lambda_t = \mu_t c_p/\Pr_t, \rho D_t = \mu_t/Sc_t, \mu_t = c_\mu \rho K^2/\varepsilon,$$

where μ_t, λ_t, D_t are the coefficients of turbulent viscosity, thermal conductivity, and diffusion, respectively; Pr_t, Sc_t are the turbulent Prandtl and Schmidt numbers, which were assumed to be equal to 1. In dimensional form, the coefficient of dynamic turbulent viscosity is determined using local equilibrium model of turbulence [1]. For this case, assume that we can be neglected by non-stationary convective terms, as well as members of the diffusion of turbulent kinetic energy in the equation for the turbulent kinetic energy. Then the right-hand side of the equation, you can obtain an expression for the turbulent kinetic energy and, according to the coefficient of turbulent dynamic viscosity. To close the system of equations of the turbulent stress tensor components, turbulent flows, heat and mass are determined by the formulas. Use the local-equilibrium model of turbulence. To determine the turbulent dynamic viscosity in a two-dimensional case, we used the formula [1]:

$$\mu_t = \rho\left(\frac{c_\mu}{c_1}\right)^{3/2} l^2 \left\{2\left[\left(\frac{\partial v_1}{\partial x_1}\right)^2 + \left(\frac{\partial v_2}{\partial x_2}\right)^2\right] + \left(\frac{\partial v_2}{\partial x_1} + \frac{\partial v_1}{\partial x_2}\right)^2 - \frac{2}{3}\left(\frac{\partial v_2}{\partial x_1} + \frac{\partial v_1}{\partial x_2}\right) - \frac{g}{T\,\mathrm{Pr}_T}\frac{\partial \theta}{\partial x_2}\right\}^{1/2}, \quad (21)$$

where $\theta = T - T_e$, c_1, c_μ - constants. The length of the mixing path is determined using the formula $l = x_2 k_T /(1 + 2.5 x_2 \sqrt{c_d s / h})$ taking into account the fact that the coefficient of resistance c_d in the space between the ground cover and the forest canopy base is equal to zero, while the constants k_T = 0.4 and $h = h_2 - h_1$ (h_2, h_1 – height of the tree crowns and the height of the crown base). It should be noted that this system of equations describes processes of transfer within the entire region of the forest massif, which includes the space between the underlying surface and the base of the forest canopy, the forest canopy and the space above it, while the appropriate components of the data base are used to calculate the specific properties of the various forest strata and the near-ground layer of atmosphere. This approach substantially simplifies the technology of solving problems of predicting the state of the medium in the fire zone numerically.

Calculation method and results

The boundary-value problem Eq. (1)–(14) we solve numerically using the method of splitting according to physical processes [13]. In the first stage, the hydrodynamic pattern of flow and distribution of scalar functions was calculated. The system of ordinary differential equations of chemical kinetics obtained as a result of splitting [13] was then integrated. A discrete analog was obtained by means of the control volume method using the ***SIMPLE*** like algorithm [9]. The resulting systems of algebraic equations are solved by ***SIP*** method [6-9] for three dimensional cases. The accuracy of the program was checked by the method of inserted analytical solutions. Analytical expressions for the unknown functions were substituted in (1)–(14) and the closure of the equations was calculated. This was then treated as the source in each equation. Next, with the aid of the algorithm described above, the values of the functions used were inferred with an accuracy of not less than 1%. The effect of the dimensions of the control volumes on the solution was studied by diminishing them. The time interval was selected automatically.

Fields of temperature, velocity, component mass fractions, and volume fractions of phases were obtained numerically. Figures 2–5 *a*, b and c illustrate the distribution of relative temperatures, concentrations of radioactive products of combustion and velocity fields for different instants of time. Figures 2 a, b, c show the distribution of temperature, the concentration of the radioactive products and the combustion speed at the time of 220 seconds after the initiation of the crown fires. The wind speed was set equal to 7 m/s at a height of 14 meters above ground level. The level of contamination on the basis of the distributions of concentrations in radioactive products of combustion can be calculated, for example, per unit area according to the formula: $R = \gamma \int_0^{x_{2e}} c_6 dx_2$, where γ - quantity characterizing the level of radioactive contamination of the products of combustion per unit mass (eg, Ku/kg). Figures 2 *a*, b, c show that under the influence of the combustion source and wind velocity occur ascent of the radioactive hot combustion products and their transfer over long distances from the fire. Distribution corresponding functions after 10 minutes is shown in Figure 3. With increasing wind speed by 2 times, that is, to 14 m / c, and the concentration distribution of the isotherms pressed harder to the surface (Figure 4 *a*, b, As a result in this case combustion products more intensively move upward and in the vicinity of the fire front is a higher level of radioactive contamination in compare with previous situations.

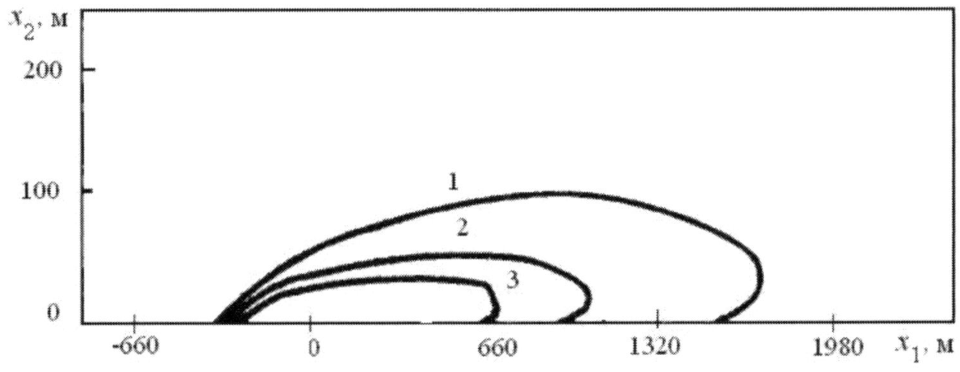

Figure. 2 *a*. $1-\bar{T}=1.01$; $2-1.05$; $3-1.1$; $\bar{T}=T/T_e$; $T_e=300\,K$.

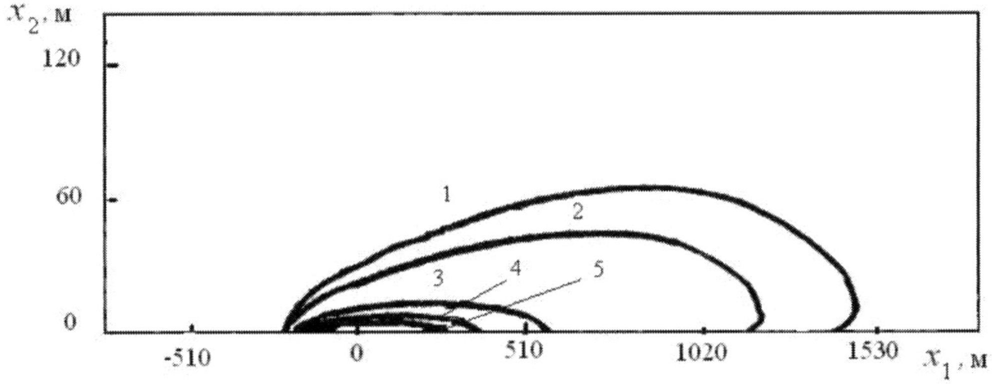

Figure 2 *b*. $1-C=0.005$ kg/m^3; $2-0.01$ kg/m^3; $3-0.05$ kg/m^3; $4-0.1$ kg/m^3; $5-0.15$ kg/m^3.

IOP Conf. Series: Materials Science and Engineering **81** (2015) 012040 doi:10.1088/1757-899X/81/1/012040

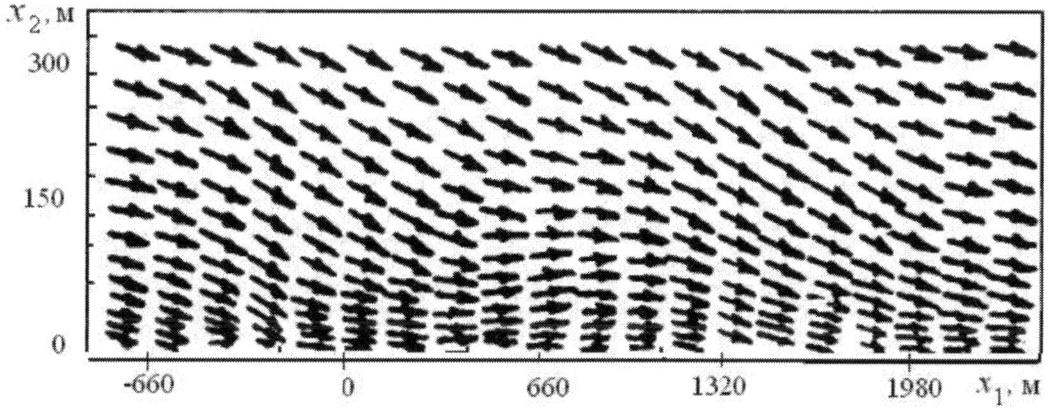

Figure 2 c. Vector flow field.

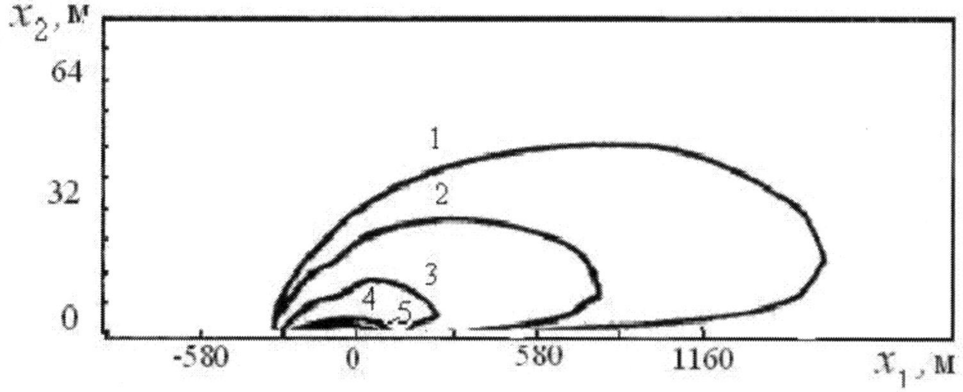

Figure 3 *a*. $1-\overline{T}=1.05$; $2-1.1$; $3-1.3$; $4-2=2.5$; $5-2.6$.

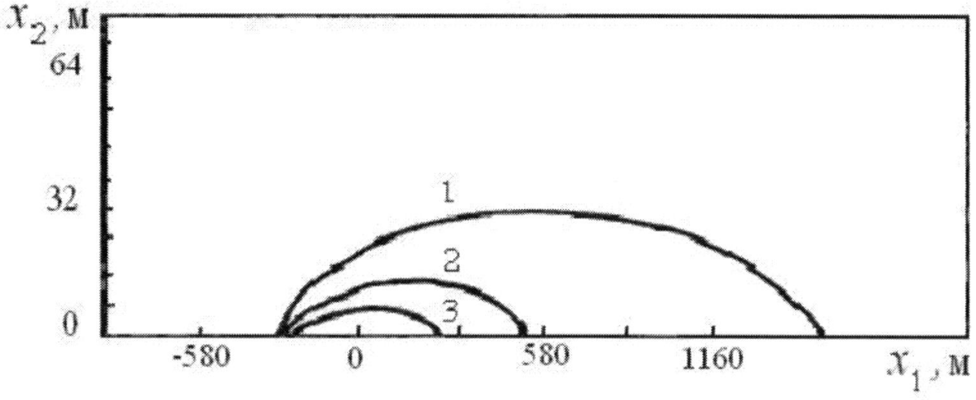

Figure 3 b. $1-C=0.02$ kg/m^3; $2-0.1$ kg/m^3; $3-0.3$ kg/m^3.

IOP Conf. Series: Materials Science and Engineering **81** (2015) 012040 doi:10.1088/1757-899X/81/1/012040

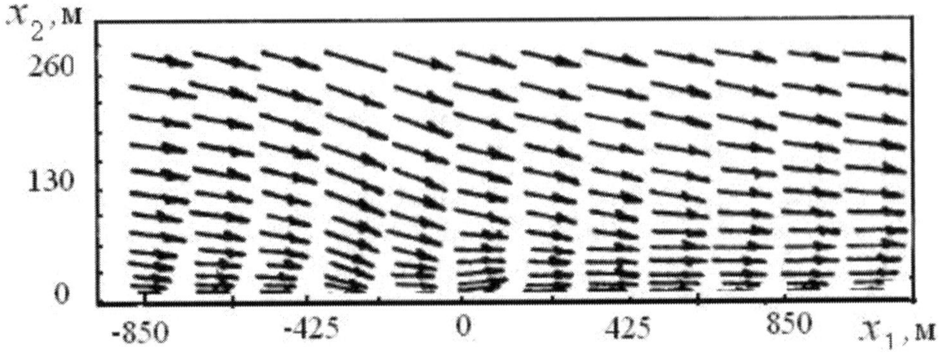

Figure 3 c. Vector flow field.

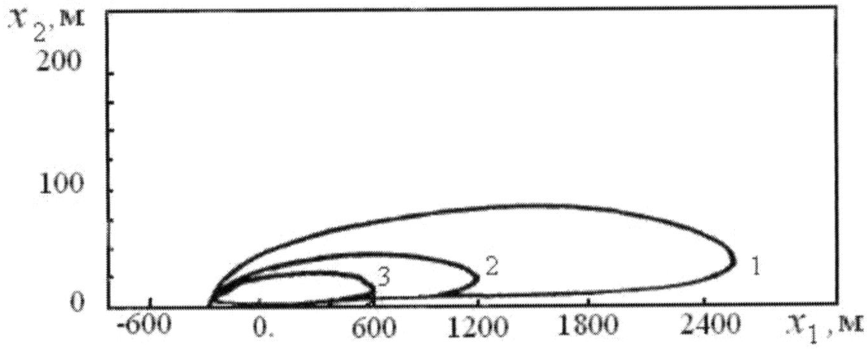

Figure 4 *a*. $1-\overline{T}=1.02$; $2-1.05$; $3-1.1$.

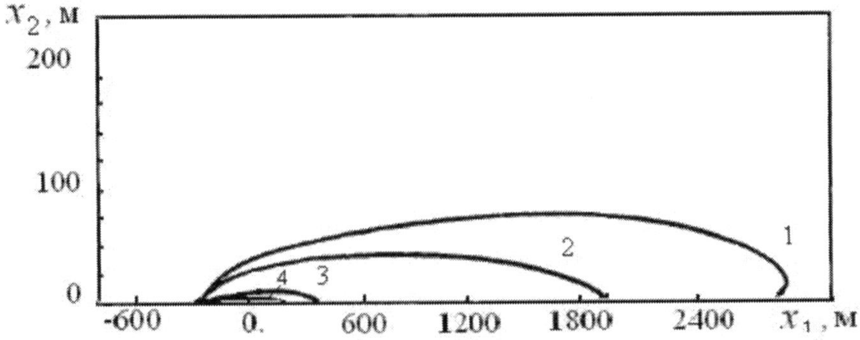

Figure 4 b. $1-C=0.005$ kg/m^3; $2-0.01$ kg/m^3; $3-0.05$ kg/m^3; $4-0.1$ kg/m^3.

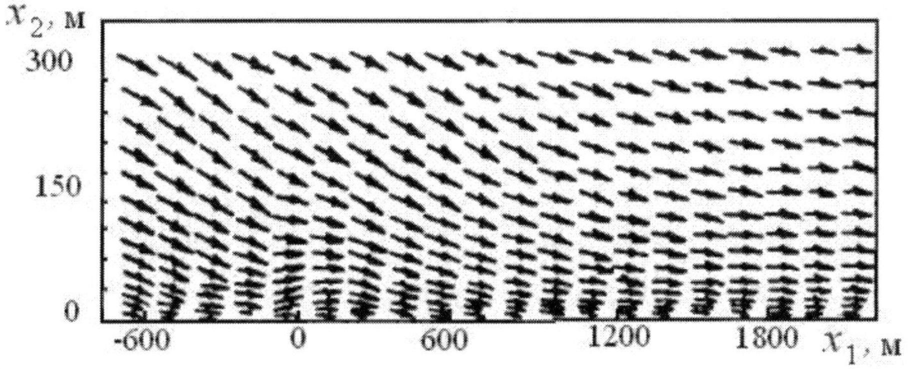

Figure 4.c. Vector flow field.

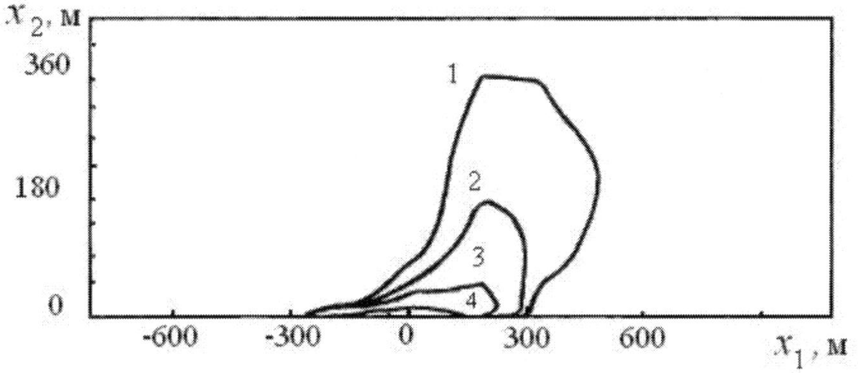

Figure 5 ***a***. $1-\overline{T}=1.03$; $2-1.05$; $3-1.1$; $4-2$.

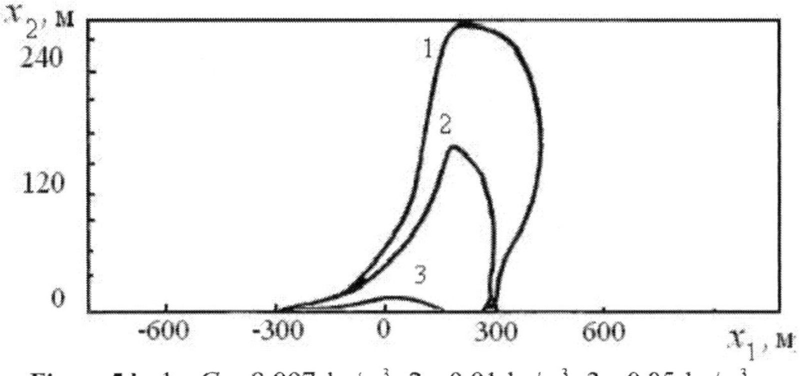

Figure 5 b. $1-C=0.007$ kg/m^3; $2-0.01$ kg/m^3; $3-0.05$ kg/m^3.

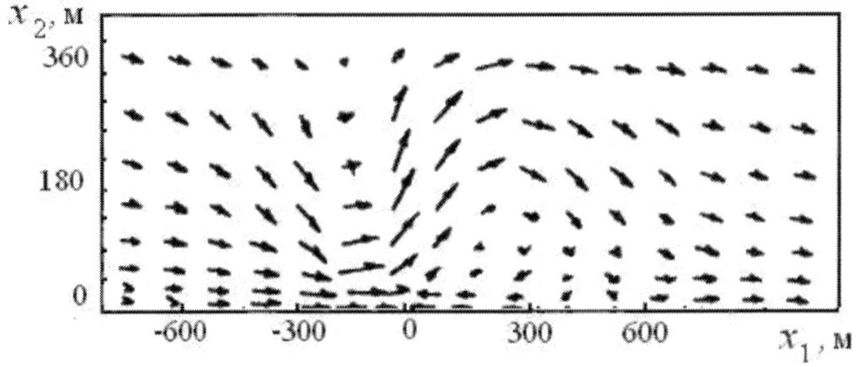

Figure 5 c. Vector flow field.

Thus, a mathematical model for prediction of the spread of radioactive contamination repeated forest fires in forests exposed to radioactive contamination as a result of accidents. In contrast to the mathematical model [6,7], where the calculated distribution of radionuclides in the vertical plane, and the front of a forest fire was set by means of boundary conditions, in this paper we consider the conjugate formulation of the problem. The system of equations to describe the crown forest fire and flow in the boundary layer of the atmosphere are solved separately, but takes into account their mutual influence with the help of (20). Obtained in this paper a mathematical model can be used in practice to determine the dynamics of the crown forest fire and to determine levels of re-contamination occurring during fires in forests exposed to radioactive contamination.

References:

[1] Grishin A M 1997 *Mathematical modelling of forest fires and new methods fighting them* (Ed.: Albini, F.) (Tomsk, Publishing house of the Tomsk University).
[2] Grishin A M 2001 Modeling and prediction of some natural and man-made disasters *Proceedings of International Conference RDAMM–2001* vol. **6** pp 134-139..
[3] Modelling the migration and accumulation of radionuclides in forest ecosystems 2002 *Report of the Forest Working Group of the Biosphere Modelling and Assessment (BIOMASS) Programme* (Vienna, Austria).
[4] Adamchikov A A, Dvornik A M 2006 The mathematical modeling of radionuclides during the forest fires in the contaminated areas *Problems of Forestry and Forestry: Proceedings of the* (Gomel: Forest Research Institute National Academy of Sciences).
[5] Paatero J, Vesterbacka K, Makkonen U, Kyllönen K, Hellen H, Hatakka J, Anttila P 2009 *Journal of Radioanalytical and Nuclear Chemistry* vol. **282** pp 473-476.
[6] Grishin A M, Perminov V A 1993 Mathematical modeling of forest communities in the conditions of natural and man-made disasters *Mathematical modeling* (Moskow: Moskow State University) pp 167-185.
[7] Grishin A M, Perminov V A 1994 Mathematical model and mathematical modeling of the propagation of aerosols during forest fires *Comp. Techn.* vol. **8** pp 72-86.
[8] Grishin A M, Merzlyakov A L, Perminov V A 1995 Mathematical modeling of radionuclide migration by the action of wind and forest fires *Proceedings of the International Conference "Forest fires: Initiation, spread and ecological consequences*, (Tomsk: Tomsk University).
[9] Patankar S V 1981 *Numerical Heat Transfer and Fluid Flow* (New York: Hemisphere Publishing Corporation).

The optical control system of dispersed phase properties in thermal spray process

A V Dolmatov[1], I P Gulyaev[2] and V I Jordan[3,4]

[1]Ph. D., Ass. Prof., Ugra State University, Khanty-Mansiysk, Russia
[2]Ph. D., Ass. Prof., Institute of Theoretical and Applied Mechanics SB RAS, Novosibirsk, Russia
[3]Ph. D., Ass. Prof., Altai State University, Barnaul, Russia

E-mail: jordan@phys.asu.ru

Abstract. The models of measuring the velocity and temperature of particles using the processing of their track images are introduced. The method of brightness pyrometry of moving objects uses the calibration procedure based on the static temperature standard. Performance of the statistical analysis of thermal data by means of optical control system of the particles properties in gas-thermal spraying flow equals to 2200-2700 particles per second. Investigation of stationary plasma spraying process allowed to obtain the distributions of velocity and temperature of particles over the volume of spraying jet. The error in determining the velocity of the particles was 1%, and the error in determining the temperature is 3%.

1. Introduction

The effective way of deposition of protective and functional coatings onto the machine parts is realized by means of the various thermal spraying technologies [1, 2]. They include plasma-arc, plasma, detonation and other types of spraying technology. Velocity and temperature of the particles are the "key" physical parameters that define the basic characteristics and quality of the coating as a whole. The control of these parameters allows to maintained of the required quality of spraying coatings, optimize the operation modes of the processing station to improve the efficiency of use of raw materials, and also to carry out the adaptation process of technological mode at the transition to processing station of another class. Modern optical measuring systems based on methods of non-destructive testing and measurements, are able to provide the information required for the technology [3, 4]. However, it should be noted that modern devices of optical control of speed and temperature of particles have certain disadvantages: low accuracy; inability to use of devices for the flows with a high concentration of the dispersed phase, too slow response. The most difficult task is to measure the temperature. The solution of this task requires the development of new methods with high resolution [5-10].

This paper presents the system for optical control and measurement of key physical parameters of the dispersed phase in a heterogeneous flow in thermal spray process. This system allows to control the operation regime stability of technological facility.

[4] To whom any correspondence should be addressed.

2. The optical control system of properties of dispersed phase in thermal spray flow

The control system is based on the idea of registration of particles radiation in the form of a series of images of their tracks and determining the parameters of the individual particles through these tracks analysis. In the stationary state of operation of the spraying facility the statistical characteristics of the dispersed phase of spraying jet in any place it should remain constant in time. Then evaluation of these characteristics can be performed through selection of particles which have been identified in video frames.

The control system comprises two data channels. The first channel is used to collect the thermal data about the brightness temperature of the dispersed phase of spraying flow, and the second channel controls its spectrum and provides the spectral temperature [3]. The sensor of first channel is the video camera with information bandwidth 1 Gb/s. Processing of this information is complicated so it is unrealizable to implement at hardware level in the physical device. Therefore, the structure of the control system was built using the concept of virtual instruments, which provides interaction between the data acquisition module and the data processing module. The data acquisition module integrates the optical subsystem, streaming camera HD1-1312-1082-G2 (PhotonFocus, Switzerland) of visible and near IR range with Gigabit Ethernet interface and high-speed spectrometer LR1-T (ASEQ Instruments, Canada). To implement the data processing unit is used MATLAB software which provides powerful tools of conversion of multidimensional signals based on the technology of parallel computations. Block diagram of the control system is shown in Figure 1.

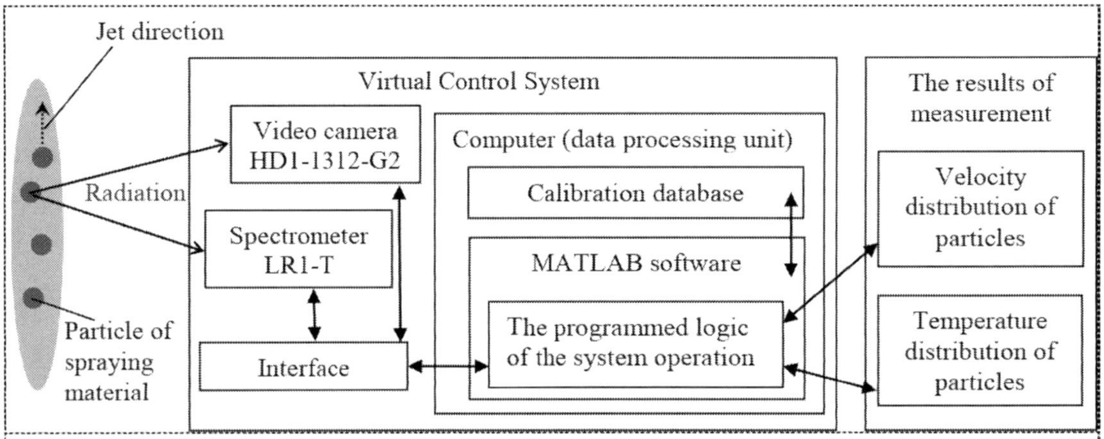

Figure 1. Block diagram of the control system.

Monochrome camera HD1-1312-1082-G2 is based on a CMOS matrix of 1312x1082 pixels with spectral sensitivity range of 400-1000 nm. The signal of CMOS matrix is represented by 12-bit digital code, and interface real-time allows transmit data to a computer with a frequency of 55 frames per second. Camera driver supports the standard MATLAB environment interface and allows to use it as a source in a pipeline processing. To provide temperature measurements by means of HD1-1312-1082-G2 video camera bandpass filter SL-575-50 (LLC "Photooptik", Russia) with a central wavelength of 575 nm and a half-width of 50 nm was added into its optical channel. In addition, on the automated test bench was developed for calibration and correction of nonuniformity of camera sensitivity. Composition and method of operation of the automated test bench presented in [4].

Digital spectrometer LR1-T is based on the linear CCD-sensor of 3648 photocells and includes thermoelectric cooler to -35 °C. Spectrometer provides measurements in the spectral range 200-1100 nm. Digital spectrometer data represented a 16-bit digital code that is transferred to the computer by the interface USB 2.0. To display spectrometer in MATLAB-software was developed the M-class of facilities "aSpectr", which represents it as a virtual instrument with a set of properties and methods. To measure the spectral temperature was performed the correction of the nonuniformity of detector sensitivity by means of the automated test bench and methods [3].

3. The method of measuring the velocity and temperature of particles of the dispersed phase in the thermal spray flow

During thermal spray process particles of size 10-100 microns are heated to temperatures of 1000-3500 K, while remaining in the condensed phase (in the form of liquid droplets of spherical shape), and spectrum of their heat emission can be registered in the range from 300 to 1000 nm.

The following method was used for velocity measuring of particles moving in the thermal spray flow. The image of the particle stream is projected onto the matrix photodetector of camera that is placed in the focal plane of the optical system. Photodetector works on the principle of charge accumulation during the exposure time τ. The value τ is chosen so that the particle can undergo a straight path, the length of which is 20-30 times greater than particle diameter. The camera registers the trajectory of particle-the track but not a particle itself. The image of a spherical particle will be the area of a circle of diameter D, and the track of a moving particle should look as rounded rectangle. The length of the movement of the circular image of the spherical particle is equal to

$$L = Z - D, \tag{1}$$

where Z - the maximum (longest) size of the rounded rectangle and velocity of the particle is equal to

$$v = \mu \cdot L / \tau \tag{2}$$

where μ - scaling factor that converts the length of the track in pixels into the real path of particle.

By means of MATLAB-software for each video-frame was created a mask and performed morphological analysis which tests the connectivity of the binary image areas and is performed the image division into objects corresponding to the individual tracks. Selected tracks that touch the frame boundaries are removed from the image, since it is impossible to find their length. Each selected object is analyzed for the definition of the apparent diameter of the particle track length, the coordinates of its center, orientation, brightness. In order to exclude overlaid tracks objects are filtered in according to the length, diameter and the ratio of maximum to average brightness. Objects which do not satisfy the filter parameters are deleted from the image and the residual objects are identified as tracks (Fig. 2).

At first, the estimates of mathematical expectations of parameters of the particle tracks are calculated with a significance level of 0.05. Tracks whose parameters fall within the appropriate confidence intervals, believed to be significant, but the residual tracks are discarded. Velocities of the particles, and hence the velocity distribution of particles, are determined by the parameters of the identified tracks according to formula (2).

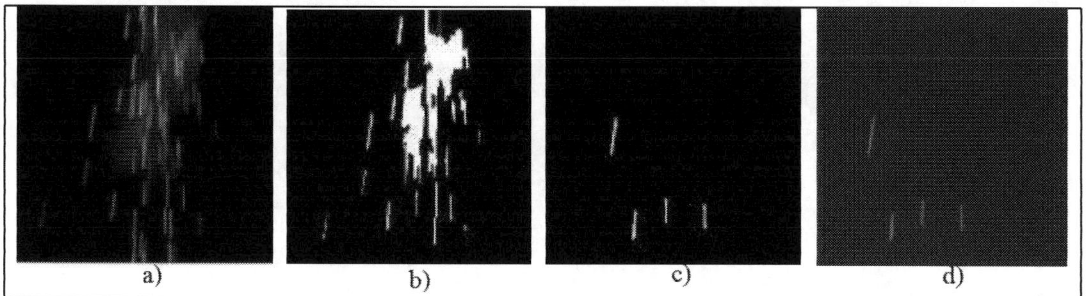

Figure 2. The video frames processing steps: a - the original image; b - the result of applying a binary mask; c - the image with probable tracks; d - the processed image with the identified tracks.

Given the range of the measured temperature 1000 - 5000 K and the "effective" wavelength of filter $\lambda_f = 575$ nm (filter SL-575-50) the Wien approximation can be used to describe the spectral brightness of the particle radiation. The parameters of the optoelectronic channel in the process of measuring are fixed, and the photodetector signal is proportional to the energy of the particle radiation during the exposure and depends only on the temperature of the observable object:

$$F(T) = K \cdot \varepsilon(T) \cdot \exp\left(-\frac{C'}{T}\right), \tag{3}$$

where K - constant that includes the coefficient of optoelectronic conversion, $C' = (h \cdot c)/(k \cdot \lambda_f)$, c – the speed of light, h and k – Boltzmann and Planck constants, $\varepsilon(\lambda_f, T)$ - spectral emissivity of the particles at the effective wavelength λ_f. In the calibration process by means of the temperature lamp TRU-1100-2350 for each signal value F of the photoelectric cell on the basis of (3) the brightness temperature T_b is calculated and these data are saved in a table:

$$(F_1, F_2, \ldots, F_{255}) \Leftrightarrow (T_{b1}, T_{b2}, \ldots, T_{b255}). \tag{4}$$

Denote F_p - signal of the sensor photocell recorded during the exposure time for motionless (stationary) particle, and S_p - the area of its image. If during the observation particle temperature remains constant, then the energy of the thermal radiation emitted by it in equal intervals of time, will be the same. Under the assumption of uniform distribution of the signal level of the photocell on the area S_{tr} of the track, must satisfy the relation:

$$F_p \cdot S_p = F_{tr} \cdot S_{tr}. \tag{5}$$

The area of the model image of motionless particle and her track respectively are equal

$$S_p = \frac{\pi \cdot D^2}{4}; \quad S_{tr} = \frac{\pi \cdot D^2}{4} + L \cdot D. \tag{6}$$

Denote T_{tr} - the brightness temperature, which corresponds to the signal level F_{tr} according to formulas (3) and (4). Then, based on (3), (5), (6) the temperature of the particle can be determined as follows:

$$\frac{1}{T} = \frac{1}{T_{tr}} + \frac{1}{C'}\left(\ln \varepsilon(T) - \ln\left(1 + \frac{4}{\pi} \cdot \frac{L}{D}\right)\right). \tag{7}$$

Formula (7) allows to measure the thermodynamic temperature of moving particles with the parameters of the tracks L, D, and with the known dependence of the spectral emissivity of the material, and the thermal calibration of photodetector can be performed on a motionless temperature standard. Otherwise, their brightness temperature is determined. Further, similar to velocity calculation, the confidence intervals of the track parameters are estimated with a significance level of 0.05 and then statistical filtering is performed.

4. Results and discussion

Research carried out by means of optical control system at the Paton Electric Welding Institute NASU was focused on the study of characteristics of the dispersed phase of the spraying jets to optimize the operating regimes of plasma-arc facility "Plazer". In the coating process the anode-wire was fed with constant rate after the plasma torch nozzle [11, 12]. Under the action of the arc and the plasma gas the wire of different composition was melted, fragmented into droplets the size of 5-50 mm and they were accelerated to the treated surface. The sight window of the control system was 2,9x2,4 cm. The distribution of particles parameters over the spraying flow in the area 16x8 cm was investigated using scan of the jet with the optical module. The spectrometer was used to ensure the absence of the plasma band-emission lines in the spectral range of thermal imaging channel and to measure the spectral temperature - maximum thermodynamic temperature of the dispersed phase in the thermal spray flow. After processing of video files recorded in a stationary operation regime of the spraying facility and merging the analysis results the frequency distributions of parameters of the dispersed phase for the entire jet were obtained (Fig. 3).

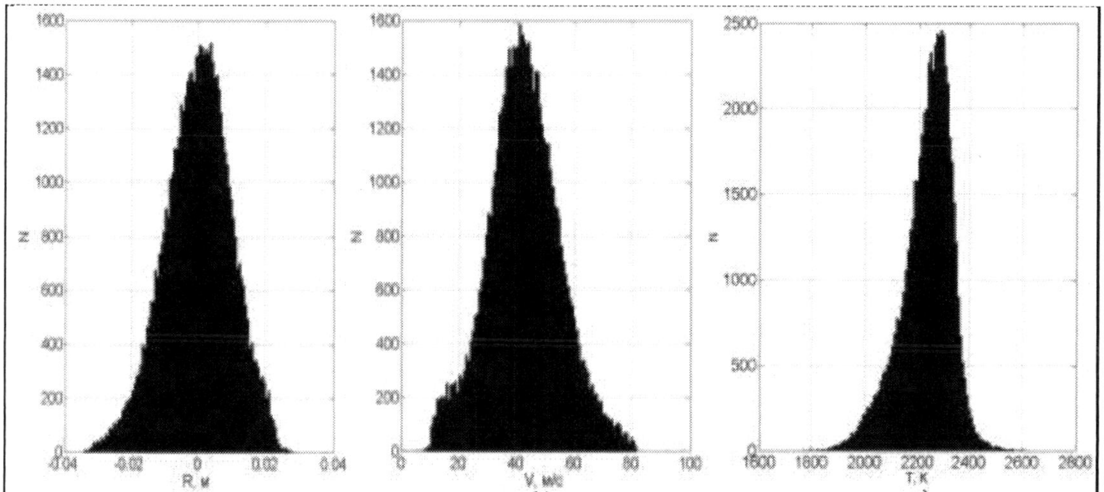

Figure 3. Frequency distributions of the dispersed phase parameters in the spraying jet (steel wire diameter 1.6 mm, arc current 220 A): a – radial coordinate (m), b – velocity (m/s), c – temperature (K).

A sample consisting of 60000 identified tracks was used to construct statistical distributions. The distribution of particles radial coordinate (Fig. 3, a) demonstrates the asymmetry of the jet and determines the spraying spot and coating deposition rate. Distributions of particles velocities (Fig. 3, b) and temperature (Fig. 3, c) allows to estimate the dynamic range and variation ratio of these characteristics providing a base for thermal and kinetic parameters of the spraying jet optimization.

The spatial dependences of the statistical estimates (average values) of temperature, velocity and direction of motion of the particles are presented in Fig. 4, a-c. The number of particles identified in various jet positions is shown in Figure 4, d.

Particles velocity and temperature spatial dependences reflect their motion and heating dynamics. The obtained results allow us to calibrate the mathematical models describing the processes of plasma-arc spraying, optimize spraying distance, to produce the transfer of technological spraying regime between facilities of different classes, as well as to optimize the operation regime of a given spraying facility and optimize its design.

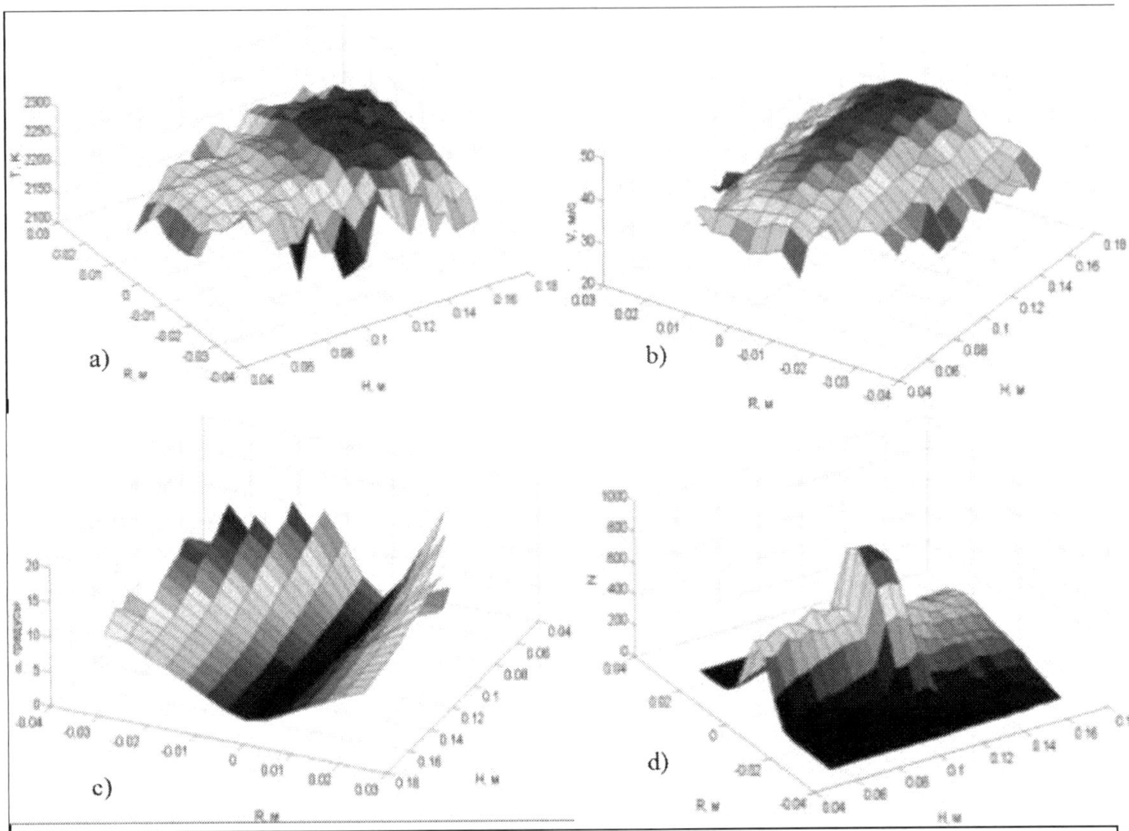

Figure 4. The spatial dependence of dispersed phase parameters in "Plazer" facility spraying jet: a – temperature (K), b – velocity (m/s), c - motion angle (degrees), d – number of identified particles of dispersed phase (pieces). Axis notations are as follows: R – radial distance from the axis of the jet (m); H – axial distance from the nozzle (m).

The above method of constructing histograms of the particles velocity (and temperature) uses the traditional statistical approach counting velocities of individual particles corresponding to different spatial and temporal positions in the jet (different particles of the accumulated statistics are being registered in large time intervals).

The following describes the different approach [13], in which the histogram of the velocities is constructed by the individual image frame (for short exposure time) and each column is obtained by averaging parameters in a certain cross-section of the jet.

Experiments were conducted in Khristianovich Institute of Theoretical and Applied Mechanics SB RAS at the spraying facility using 50 kW DC plasma torch for production of hollow ceramic spheres [14] from yttria-stabilized zirconia (YSZ) feedstock powders of 50-56 mkm diameter (204B, Sultzer Metco). The regime was as follows: plasma-forming nitrogen flow rate 30 slpm, arc current 150–200 A (voltage 180–190 V), thermal efficiency 60 %, anode diameter 10 mm, feed rate of powder 6 kg/h, double side radial powder injection.

In this technique the particles image processing unit is implementing a consistent pipelined processing consisting of five procedures [13]: the smoothing by means of the spatial Gaussian filter, the computation of the brightness profile derivative function of the image by means of the discrete differential Sobel operator, thinning of tracks boundaries, dual threshold filtering, parameterization of tracks by Hough algorithm. The particle velocity is calculated according to the formula (2) and estimated track lengths *L*. Then, in each cross section of the jet are determined the average values of the particle velocities and the histogram of particle velocity distribution (along the axis of the jet). Figures 5 and 6 demonstrate examples of image processing, calculating the particle velocity distribution and the particles density distribution in the flow.

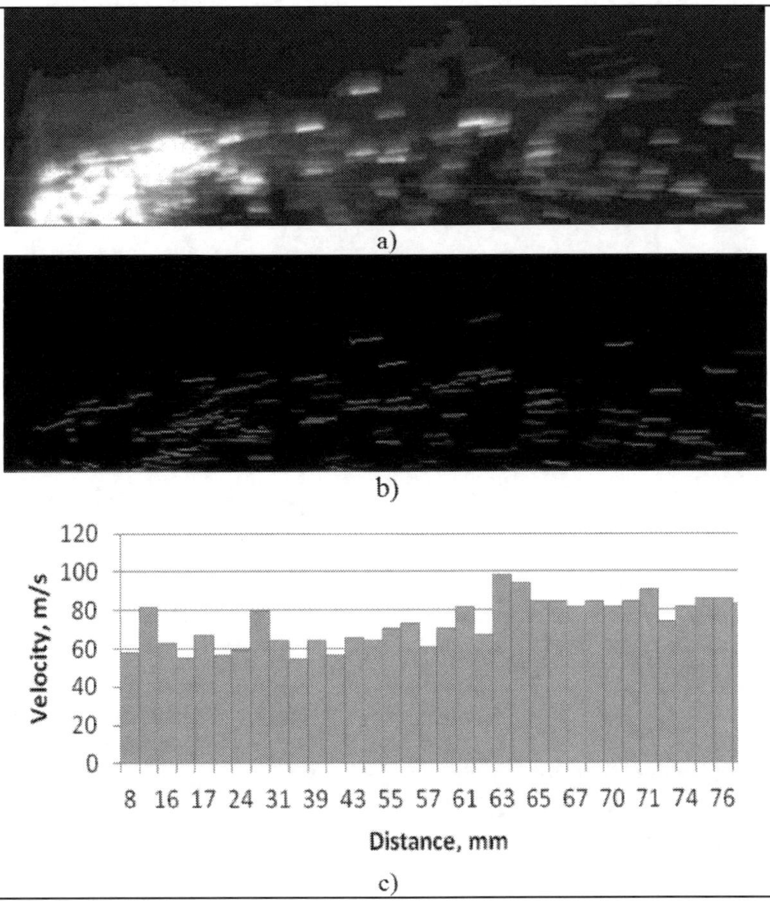

Figure 5. Processing of the single frame: a - the image frame with an exposure time $\tau = 40\ \mu s$; b - the result of pipelining frame processing; c - the velocity distribution of particles along the direction of the plasma flow.

As can be seen from fig. 5, c the particle velocity has stabilized at the distance 65 mm from the nozzle of plasma torch. For a small time exposure of frame (fig. 6, a) the tracks are presented as points (particles) and after appropriate image processing in different cross sections of the flow this assumption allows to estimate the relative portion of particles (particle flow density, fig. 6, b).

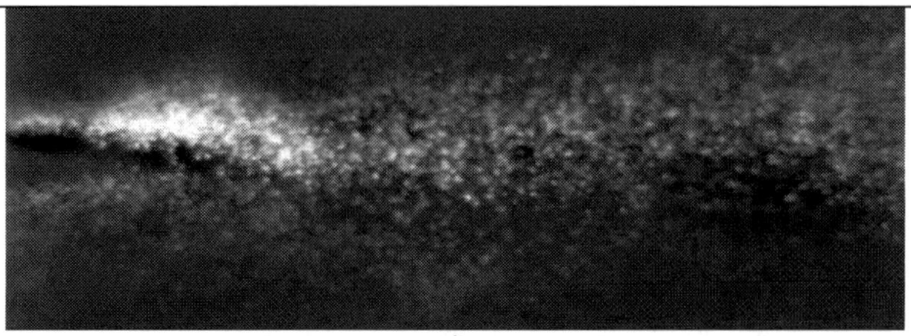

a)

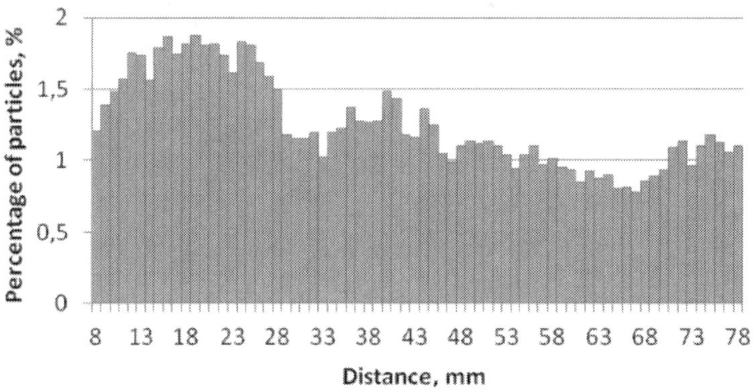

b)

Figure 6. The image frame with an exposure time $\tau = 2\ \mu s$ (a) and the particle flow density along the direction of the plasma flow (b).

Analysis of fig. 6, b uncovers the "wavelike" nature of the particle flow density along the axis of the plasma jet (sequence of "compactions" and "decompactions"), which provides a tool for of stationarity control during facility operation.

5. Conclusion

The system of optical control of velocity and temperature of the dispersed phase in the thermal spray flow was developed on the base of the camera HD1-1312-1080-G2, the spectrometer LR1-T, and signal processing module in the environment of MATLAB. The models for measuring velocity and temperature of the spraying particles from the images of their tracks are presented. The method of the brightness pyrometry of the moving objects based on a calibration by the motionless temperature standard source introduced. The error in measuring the velocity of the particles is 1%, and the error in measuring the temperature is estimated by authors to be 3%. Performance of thermal video-data analysis by means of optical control system is determined to be 2200-2700 particles per second. The developed methods for particles velocity and temperature distribution evaluation allow to investigate the stationarity of the spraying facility operating regime. The optical control system can be used to control the spraying process in the real-time, for the optimization of the technological regime of improving of the spraying facility. The work was performed as part of a joint project of Russian Fund for Basic Researches (№14-08-90428) and National Academy of Sciences of Ukraine (№ 06-08-14).

References

[1] Borisov Yu S, Kharlamov Yu A, Sidorenko S L and Ardatoskaya E N 1987 *Thermal spray with powder materials: handbook* (Kiev: Naukova Dumka) (in Russian)
[2] Korzhyk V N, Krivtsun I V, Petrov S V and Kharlamov M Yu 2009 New technology "Plazer"

for renovation of rail road components *Repair, restoration, modernization* **1** pp 20–22 (in Russian)
[3] Gulyaev P Yu and Dolmatov A V 2007 Diagnostics of temperature and velocity distribution of sprayed powder in impulse plasma flow *Russian Physics Journal* **9** p 114 (in Russian)
[4] Gulyaev P Yu, Dolmatov A V, Popov V A and Leonov G N 2012 Optical diagnostics methods of particles in high-temperature flows *Polzunovskiy vestnik* **2/1** pp 4–7 (in Russian)
[5] Boronenko M P, Gulyaev I P, Gulyaev P Yu, Demyanov A I, Dolmatov A V, Jordan V I, Korzhyk V N, Krivtsun I V and Kharlamov M Yu 2013 Methods of Control of Temperature and Velocity of Particles of the Condensed Phase in Process of Plasma-Arc Spraying *Fundamental Research* **10** pp 1194–1199 (in Russian)
[6] Gulyaev I P, Solonenko O P, Smirnov A V, Chesnokov A E, Gulyaev P Yu, Jordan V I, Milyukova I V and Dolmatov A V 2008 Method for Dispersed Phase Particles Temperature Distribution Determination in Two-Phase Plasma *Patent of Russian Federation* 2383873
[7] Gulyaev I P, Ermakov K A and Gulyaev P Yu 2014 New high-speed combination of spectroscopic and brightness pyrometry for studying particles temperature distribution in plasma jets *European researcher* **3/2** pp 564-570
[8] Jordan V I 2013 Revers Integral Transformation for Temperature Distribution of Particles in Heterogeneous flow Reduction from their Integral Heat Spectrum *Russian Physics Journal* **8/3** pp 293-299 (in Russian)
[9] Dolmatov A V, Gulyaev I P and Imamov R R 2014 Spectral Pyrometer for Temperature Control in Thermal Synthesis Processes *Ugra State University Bulletin* **2** pp 32–42 (in Russian)
[10] Dolmatov A V, Ermakov K A, Lavrikov V V and Makoveev A O 2012 Automated Thermal-Vision System Calibration Complex based on MATLAB *Ugra State University Bulletin* **2** pp 56–63 (in Russian)
[11] Kharlamov M Yu, Krivtsun I V, Korzhik V N and Petrov S V 2011 Heating and Melting of Anode Wire in Plasma Arc Spraying *Paton Welding Journal* **5** pp 2–7
[12] Kharlamov M Yu, Krivtsun I V and Korzhyk V N 2014 Dynamic Model of the Wire Dispersion Process in Plasma-Arc Spraying *J. Therm. Spray Technol.* **23** pp 420–430
[13] Jordan V I and Ryabchenko I K 2012 Determination of Velocity Field and Particles Density in Plasma Jet by Processing of Image Stream during High-Speed Video Capturing *Russian Physics Journal* **9/2** pp 169-175 (in Russian)
[14] Gulyaev I P 2014 Experience in plasma production of hollow ceramic microspheres with required wall thickness *Ceramics International* **41** pp 101-107

IOP Conf. Series: Materials Science and Engineering **81** (2015) 012042 doi:10.1088/1757-899X/81/1/012042

Gas – discharge electron sources with gas- dynamic beam output windows

L N Orlikov and N L Orlikov

Electronic devices Department, Tomsk State University of Control Systems and Radioelectronics, Russia

E-mail: oln4@yandex.ru

Abstract. The extraction of focused electron beams via the system of funnels and outlets is discussed. The increase of efficiency is achieved by the use of gas-discharge electron sources, complication of a funnel shape and consideration of a beam effect on pressure differential.

1. The present state of the problem

Electron beams, injected into gas, can find wide application in material machining, metallurgy, quantum electronics and medicine.

The devices with beams removal usually consist of an electron source and an output window that separates its vacuum volume from a chamber, filled with a gas. The window for extraction of high-energy beams (above 100 KeV) can be made of a metal or polymer film, transparent for electrons, but opaque to gases. For extraction of low-energy continuous focused electron beams the gas-dynamic windows, consisting of supersonic gas jets, metal vapor jets and liquid jets can be used. Besides, jet gates for extraction of focused beams some systems of aligned diaphragms can be used with differential gas pumping between them [1-13]. The pressure differential, obtained with such means, increases with a number of stages and pumping capacity, but drops with the increase of output hole dimensions. Gas-dynamic windows parameters can be improved by a better design of gas-dynamic windows itself and also of pumping means. The decrease of an incoming gas can be achieved by overlapping of holes in the case of beams absence, e.g. using a rotating cylinder with a channel for beams extraction, or with a disk with holes at its edge for the same purpose. With advances in vacuum techniques instead of mechanical pumping means cryogenic and ejector ones are used currently.

Now existing means for extraction of stationary beams through the system of diaphragms allow to extract beams into atmosphere from electron sources with a thermal cathode, operating at pressures of the order 10^{-3} Pa, using 5-6 pumping stages, through holes with the diameter 1-10 mm with power consumption about 5-7 KW per 1 mm of the hole area and current losses up to 75%.

Despite growing interest to application of continuous beams their application is retarded by a number of problems, evolving in the process of electron sources operation with gas-dynamic windows. The main of them is contradiction between hole dimensions for beams extraction and capacity of pumping means to maintain vacuum for electron generation.

In most cases a beam is extracted with probabilistic forecasting as to a pressure differential and removed power. However, in some cases a form and mutual arrangement of gas-dynamic windows elements together with the passing beam modes can define a pressure differential and beam losses, but not capacity of pumping means. One of the reasons of probabilistic efficiency of beams extraction is the fact that well-known calculation methods, based on relationships of vacuum techniques, can not reveal the conditions of efficiency increase, but only establish tentative correlation between the hole diameter and necessary gas pumping rate.

2. Problem solution

The increase of beams extraction efficiency is related to solution of a number of problems that can be subdivided according to the following directions: an electron source, pressure differential system, electro physical events, current passage, correlation.

3. Electron source

The employment of gas-discharge electron sources, operating in mud vacuum, is rather promising. The advances in electron - beam and plasma technologies of electron guns, based on Penning gas discharge and high-voltage glow discharge, operating in forvacuum under conditions of lesser requirements to gas-dynamic windows parameters which allows to use only mechanical pumps and to utilize gas-dynamic effects to increase a pressure differential. The development of multi-channel electron guns as electron sources that can generate electron beams at pressures of 80 Pa and more [14] was great achievement in this field. On figure 1 the scheme of electron source on the base of high-voltage smoulder discharge with gas-dynamic window is described.

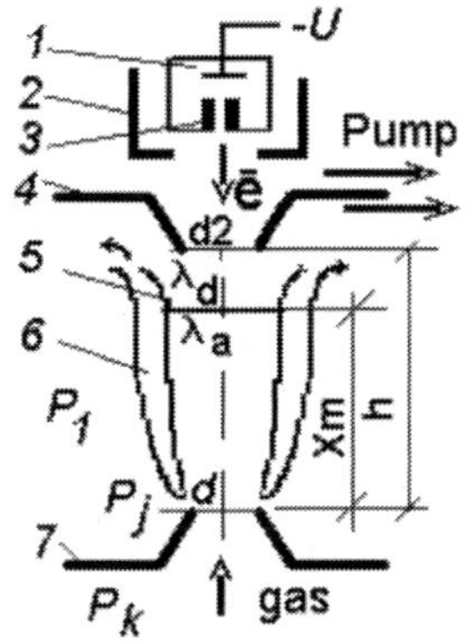

Figure 1. The scheme of electron source on the base of high-voltage smoulder discharge with gas-dynamic window. *1* – Cathode, *2* – anode, *3* – magnetic lens, *4*, *7* – gas-dynamic window, *5* – Mach disk, *6* – jumps of pressure.

The source consists of a cold aluminum cathode *1* and anodes *2*, between which the high-resistance resistor R is included. At pressure in the source of 1-10 Pa the negative potential (up to 100 kW) is given on a cathode. Between cathode and anode the high-voltage smoulder discharge is lighted. An ion from plasma of the discharge directs on cathode. From cathode are dislodged electrons under action of ion-electronic emission. The electron beam is focused by the magnetic lens *3* in the holes of elements *4,7* of gas-dynamic window and are removed in the camera. Gas flow from the working camera is formed in a kind of supersonic jet with jumps of pressure *5* and *6*. Transverse jump of pressure *5* means a Mach disk. Distance X_m up to Mach disk is described by following [15]:

$$X_m = 0{,}7d\sqrt{kP_j / P_1} \qquad (1)$$

Where k - is a factor of gas adiabatic, P_j - is pressure on the element cut, P_1 - is pressure between elements.

4. Pressure differential system

The most promising means for provision of a pressure differential is the system of aligned diaphragms of different shapes. There is an optimum distance which corresponds to the maximum pressure differential. To optimize the element shapes and the analysis of gas- dynamic effects a jet supersonic model of a gas flow has proved its efficiency. At the pressure differential above two orders at a hole an element shape does not effect pressure in an electron source, which is related to compliance of the principle of independence of flow parameters on a body shape. For accounts on figure 2 the experimental dependencies of optimum distance between elements, at which difference of pressure is maximum for geometric sharp (curve 2) and flat (curve 1) elements are submitted.

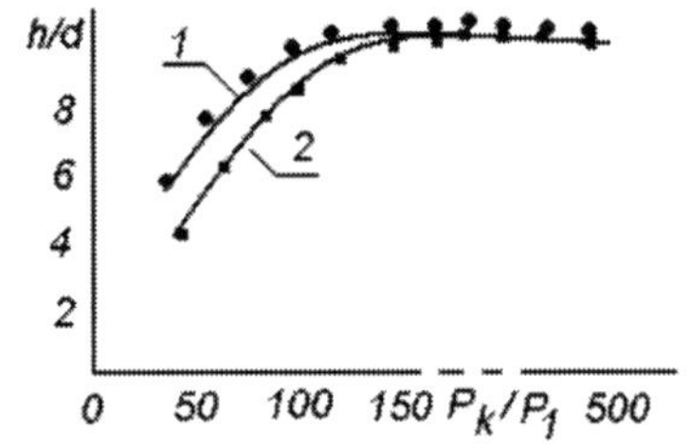

Figure 2. The experimental dependencies of optimum distance between elements, at which difference of pressure. 1 – flat elements, 2 – sharp elements

Initial area of curves coincided with known ratio for Mach disk. Stabilization and some reduction of optimum distance with increase of difference of pressure is connected with fact that at the difference of more than two orders (M>3) the principle of flow independence from the body form is executed. In this case under h/d>10 the calculation of differential pump out system can be carried out on ratios of vacuum technique. However, the employment of gas-dynamic effects under conditions of low pressure becomes problematic one due to gas rarefaction; pressure decrease causes the decrease of gas pumping means efficiency. Energy input for gas pumping out is limited by dimensions of pumping means and by capacity of vacuum communications; the complexity of element shapes is also limited by the pressure recovery limit in a gas flow by $(k+1)/(k-1)$ times.

5. Electro physical effects

To provide a pressure differential it is useful to apply the pumping effect, produced by an extracted beam, or by a discharge, ignited between elements [16-21]. Gas, flowing into the vacuum, is accelerated into the region of supersonic velocities, and expands in the form of a hollow jet with formation of the pressure compacting zone (Mach disk) in front of the output elements. Following the Mach disk the flow is decelerated and its pressure increases.

However, an extracted beam and a discharge in some cases can impair a produced pressure differential. The influence of an extracted beam on a pressure differential can be analyzed, using the scheme of a gas flow and changes of its parameters at a pressure differential (figure 3).

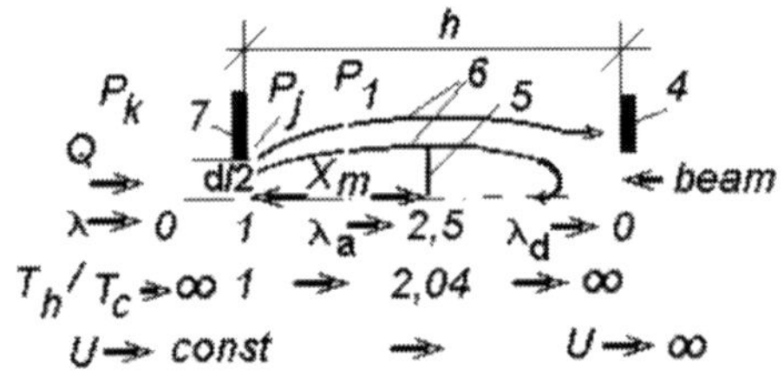

Figure 3. The scheme of a gas flow and changes of its parameters at a pressure differential. *4, 7* – gas-dynamic window, *5* – Mach disk, *6* – jumps of pressure.

The scheme of gas parameters changing is based on the temperature dependence of the holes capacity U and also on a thermal model of beam interactions with a gas, according to which the greater is a coefficient of a flow velocity the lesser are its ability to be heated and pressing pressure in a gas flow. Gas temperature change is related to the change of the flow velocity coefficient λ with relation:

$$T_h / T_c = (1+\lambda_c{}^2)/4\lambda^2{}_c \quad (2)$$

The indices "h" and "c" correspond to hot and cold gases. It is evident that the lesser is a gas velocity the greater is its heating. The gas flow rate change Q in the course of heating can be defined by the expression:

$$Q_c / Q_h = [2T_h / T_c - 1]^{0,5} \tag{3}$$

In the inter element space the velocity coefficient λ can take limiting values from 1 to 2.5. Correspondingly limiting temperature changes can take values from 1 to 2,04. Thus, a limiting gas flow drop due to gas heating by a beam cannot exceed 1,75 for the air.

Table 1 gives the values of the total pressure change function $f(\lambda)$ for gas heating in subsonic and supersonic jet regions in an output facility, taken from paper [15]:

$$f(\lambda) = (\lambda^2 + 1)[1 - (k-1)/(k+1)\lambda^2]^{1/(k-1)} \tag{4}$$

Table 1. Some parameters of a gas flow

λ	0	0,528	1	2	2,5
$f(\lambda)$	1	1,1	1,2	0,32	0

The table shows that the value of the total pressure change function f in a supersonic region decreases by more than 2 orders. In a subsonic jet region the total pressure does not almost change during gas heating. In an operating chamber a flow is motionless $\lambda \to 0$; $T_h/T_c \to \infty$. In the region of a beam input into a chamber $\lambda \to 1$; $T_h/T_c \to 1$; consequently the hole capacity remains constant.

$$U \sim (T/M)^{0,5} \tag{5}$$

Where T is temperature, M is gas molecular weight. The gas flow rate into the system of pressure differential near a beam decreases at the expense of the gas density decrease, applied to the chamber out of the beam output region.

The flow in an inter element space up to Mach disk is accelerated from $\lambda = 1$ to λ_a. In a supersonic region and decelerated from $\lambda_d = 1/\lambda_a$; to $\lambda = 1$ in a subsonic region after a shock wave. Without a beam effect the gas temperature decreases with a velocity increase, but being effected by a beam it increases. At maximum flow (air) velocity we have:

$$M \to \infty, \lambda \to 2{,}5;\ T_h / T_c \to 2{,}04\ ;\ \delta = (1) / f(\lambda_a)$$

Approaching the element of the second pumping stage in a subsonic region the jet flow is decelerated. $\lambda_d \to 0$; $T_h/T_c \to \infty$ which increases the capacity ($U \to \infty$).

The coefficient of the total pressure change is expressed by:

$$\begin{aligned} \delta &= [f(0)/f(1)][f(1)/f(\lambda_a)][f(\lambda_d)/f(1)] \approx \\ & f(\lambda_d)/f(\lambda_a) \approx f(0{,}46)/f(\lambda_a) \end{aligned} \tag{6}$$

So the total pressure change during beam passage (6) is defined by relationship of the total pressure change functions behind a shock wave and in front of it. Deterioration of the pressure differential, affected by a beam, can be explained by the fact, that at a small interval between elements the flow principally consists of a subsonic region with great gas heating and correspondingly great hole capacity.

6. Current passage

By contrast to traditional notions about current passage value, determined by relationships between diameter of a beam and an output hole, current passage can increase during formation of a concave plasma boundary in the neighborhood of elements under specific compromise between a pressure differential and an element shape. In the region of unsteady interactions of a gas flow with elements of extracting means the current passage is of unsteady character. When approaching the channels of extracting means the current passage is of unsteady character.

During discharge generation in gas-dynamic windows a part of a transported beam can be screened. The current passage from high current electron sources is defined by a geometrical form of a plasma flow as far as an extraction system.

7. Correlation

An electron source with a beam extraction means is a unified complex, in which both a volt-ampere characteristic and also the increase and decrease of a pressure and current losses depend on beam parameters, mutual arrangement and a shape of extracting elements.

The increase of a source operating pressure can be achieved at the expense of harder discharge ignition and by leak- in of light gases (He) into a source. The employment of a cold cathode with a central insertion made of a material with a high coefficient on ion-electron emission (LaB_6) in combination with an optimal focusing optical system enable to increase beams brightness and to decrease its losses during extraction. The current passage at a pressure differential can be higher, compared with the case without a differential, which is related to the effect of ion focusing in forming of a plasma boundary in the region of extracting means elements. In contrast to traditional notions a gas-dynamic windows is a multi-functional means and can be adjusted either to a maximum pressure differential (by provision of ejector events in the neighborhood of elements) or to a maximum current passage (with hole dimensions, not exceeding internal dimensions of a jet) or to generation of pressure oscillations or the fulfillment of the pressure stabilization function (in the range of Reynolds numbers 150-200). In fabrication of elements with dimensions not greater than internal dimensions of a jet cavity a flat shape is flowed around in a mode of a sharp body. Due to this the notion of the element shape has not geometrical, but a parametric sense. With the pressure drop the priority in provision of a pressure differential passes to the systems with a discharge between gas-dynamic windows elements.

Results

A new gas-dynamic method for calculation of systems for beams transfer was developed that allows to optimize calculations of required parameters together with new multi-channel and multi-functional gas-discharge electron sources at the operating pressure up to 80 Pa and higher, the power consumption of pumping means was reduced in the source of beams output into atmosphere from 10 to 1.5 KW/mm per 1 mm of a hole area at the distance between diaphragms from 2 to 5 diameters of output holes.

References

[1] Schumacher B 1962 *Trans. 8 th. Nat. Vacuum Symposium and 2-nd Int.Congress a Vacuum science and technology* 1961 (Washington) pp 1192-1200.
[2] Shapiro A 1963 *Vacuum* **13, 3**, pp 83-87
[3] Kalbfell D 1971 Patent USA 3585349 cl 219-121
[4] Hershcovitch A 1995 *Appl. Phys* **78(9)** November p 5283
[5] Bakisch D 1971 *Electron beam welding, noes Building* (Park Ring: New Jersey USA) 151 p
[6] Niedrielski R 1965 Patent USA 3 171 943 cl 219-121
[7] Niedrielski R 1965 Patent USA 3 175 073 cl 219-121
[8] Niedrielski R 1969 Patent USA 3 444 350 cl 219-121
[9] Osaka E 1974 Patent Japan 49-36 544 cl 12B11
[10] Osaka E 1973 Patent Japan 48-27 060 cl 12B11
[11] Steigerwald K 1961 Patent USA 2 987 610 cl 219-121
[12] Schiller S, Heising V, Jasch G 1971 *Schweistechnik* **21** pp 11-16
[13] Smith R and Schumacher B 1974 *Instruments and methods* **118** pp 73-91
[14] Bichkov J, Korolev J, Mesjaz G 1979 "*Izvestiya Akademii Nauk USSR, Seria fizicheskaya*, vol **43** no 2 pp 226-229
[15] Abramovich G 1976 *The Applied gas dynamics* M Nauka 808 p
[16] Orlikov L 1999 Electrophysical methods of pressure differential generation in the course of

injection of focused electron beams into gas *The Third international Simposium proceedings* (Tomsk: Russia May 18-20 Vol **2** pp 476-478
[17] Orlikov L 2002 *The theory and practical of conclusion in gas of electronic beam* (Tomsk: TGU) 150 p
[18] Orlikov L and Orlikov N 2004 *Proc. 2 int. conf. On Material and technologies XX1 age* (Rusia: Pensa) pp 169-172
[19] Orlikov L and Orlikov N 2005 *Proc. 3 int. conf. on Electronic facilities and managerial system* (Tomsk: Rusia) pp 8-12
[20] Orlikov L 2006 *Izvestiya Vyssh. Ucheb.Zaved* Series Physics **11** pp 74 - 77
[21] Orlikov L and Chugunov A 2008 Electro Physical Ways of Creating Pressure Differential in the Course of Charged Particles Output into the Atmosphere: *Proceeding 15 Int Symposium on High Current Electronics 21-26 Sept 2008* (Tomsk: IAO SB RAS) pp 85 -88

IOP Conf. Series: Materials Science and Engineering **81** (2015) 012043 doi:10.1088/1757-899X/81/1/012043

The combination methodic of diffusion and implantation technologies for creating optic wave-guided layers in lithium niobate

L N Orlikov, N L Orlikov, S I Arestov, K M Mambetova and S M Shandarov

Electronic devices Department, Tomsk State University of Control Systems and Radioelectronics, Russia

E-mail: oln4@yandex.ru

Abstract. The implantation of copper into Lithium Niobate in the prohibited crystal zone forms a definite energetic level for optic transits. This paper examines conditions of optic wave-guided layers formation on Niobate Lithium due to the method of implantation copper ions with the next diffusion. Reflect Spectrum in consequences implantation is extended. The transfer of the optical power from the primary beam into the another beam was discovered and in reverse. Photo galvanic characteristics of implantation specimen identity of crystal by traditional technology and doping CuO manufacture.

1. Present state of the problem

Copper-doped Niobate Lithium is challenging to form energetic levels in the crystal prohibited zone for the optical transitions. Local intensities of the electric field appearing due to doping are accompanied by the alteration of the refraction index, occurrence of photorefractive holograms and other optic effects. It lets create optically controlled miniature crystal optoelectronic devices. As it shows [1], the largest changes of the refraction index are observed while the copper concentration is 10^{24}-10^{26} m^{-3} using the diffusion from the scum which is made under the copper thermal evaporation. For instance, the possibility to change the refraction and absorbance index with the concentration growth of two-times ionized centers of copper Cu^{2+} playing a role of traps in the prohibited zone is shown [2]. However, the realization of some electro-optic effects, while forming films on typical vacuum unit with lube means to create vacuum, is restrained by the probabilistic representation of wave-guided layers. There is an interest to simplify the technology of forming active wave-guided layers due to the primary copper ions implantation with the next diffusion on the implanted layer. The most perspective for the similar purposes are the sources of metal ions with the generation of particles from a vacuum arc cathode stain [3].

2. Problem solution technique

A solution technique of the problem of optical waveguide synthesis is based on variation of the refraction coefficient of a crystal as a result of the implantation of ions in the near-surface field and their reallocation on necessary depth as a result of the diffusion. As a result, the effect of the ionic stream can lead to amorphization of the crystal structure, in other words to create "swelling" of atomic packing, and to change both the refractive index and the optical absorption.

3. Process modelling

3.1. Modelling of the implantation process

The implanted dose of ions of copper D_{Cu} was calculated on a ratio:

$$D_{Cu} = n_{Cu} \cdot N_A / a \cdot b \tag{1}$$

Where n_{Cu} is the quantity of copper mole; N_A=6.02·10^{23} is the Avogadro number; a=2.5 cm is the length of a sample; b=0.8 cm is width of a sample; d is the thickness of a film.

$$n_{Cu} = \rho_{Cu} \cdot a \cdot b \cdot d / M_{Cu}, \tag{2}$$

ρ_{Cu}=8,9 g/cm^3 is the copper density; M_{Cu}=81,72 g/mole is the molar mass of copper.
At the calculated thickness of the film d=50·10^{-9} m the calculated implanted dose makes: D_{Cu}=3,6885·10^{17} ion/sm^2

3.2. Modelling of the diffusion process

During the diffusion time considerably exceeding the time needed for the complete intromission of an impurity from the film into the crystal, the regularity of its concentration is featured by the Gaussian function. The thickness h of the wave-guided layer used here is related to the diffusion depth, as [4]:

$$h = 0{,}892\sqrt{4Dt}, \tag{3}$$

where D is the diffusion constant and t is its duration.

3.3. Modelling of the optics process

The absorption coefficient $k(\lambda)$ was calculated on the ratio:

$$k_\lambda = -\tfrac{1}{d}\ln\left(\frac{-(1-\frac{1}{R})^2}{2T} + \sqrt{\frac{1}{R^2} + (1-\frac{1}{R})^4 \cdot \frac{1}{4T^2}}\right), \tag{4}$$

where $T(\lambda)$ is the transmittances constant, and $R(\lambda)$is the reflexions constant. Calculations were carried out with the software package MATCAD.

4. Procedure and technique of the experiment

On fig. 1 the circuit of the ionic source with the microsecond duration with the initiation of a surface breakdown of the dielectric is presented [5]. The source consists of the cathode *1*, the dielectric interposing *2*, the ignitor electrode *3* and the anode *4* closed by the grid *5*. The device of neutralising of the ion charge is fulfilled on the basis of a plasma source of electrons.

At pressure 0.01 Pa switching of size C capacity (5 kV, 100 μF) in the chain of the transformer T. The disruption in the dielectric interposing between the cathode and the ignitor electrode takes place. In the source cavity a vacuum arc is formed. When feeding of the positive polarity speeding up impulses (U~20 kV, 250 μs, 1- 50 Hz), the ions extraction with the current under 1000 mA occurs. Ions when transiting the anode *4*, the grid *5*, and the neutralizer of charge *6* hit the specimen *7* installed on the work table *8*. The reduction of the drop wise fraction and the stability of operation was attained by feeding of the plasma gap at the moment of the ionic current takeoff. On the fig. 2(a) the circuit of Hall's source for the crystal ion clearing of Niobate Lithium and the circuit of the magnetron (fig. 2b) to form the copper film are presented.

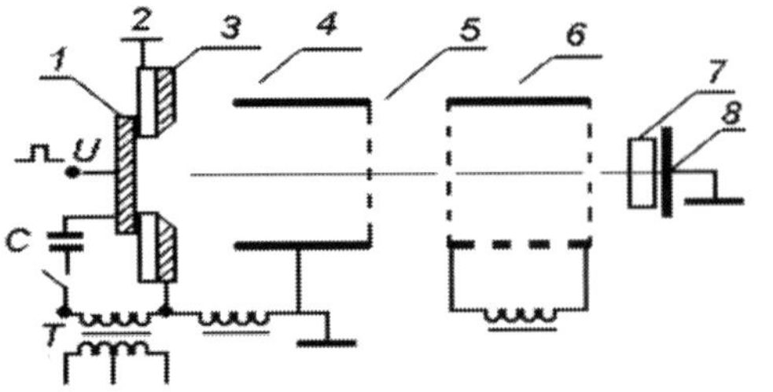

Figure 1. Pilot model. 1 – cathode, 2 – dielectric interposing, 3 – ignitor, 4 – anode, 5 – grid, 6 – neutralizer, 7 – specimen, 8 – work table.

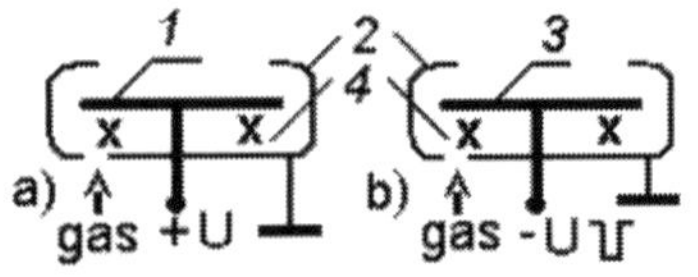

Figure 2. a) – Circuit of Hall's source; b) – magnetron's circuit, 1 – anode, 2 – box, 3 – target, 4 – magnet.

In the capacity of a dose analyser, the device with the dose increase control during exposure to radiation is used [5].

5. Preparation of samples

To increase a film adhesion the gas emission mechanism was estimated [6]. For this purpose the crystal raying is made in the low-energy plasma, and the monochromator DMR 2 with the photoreceiver FSK-1 recorded the intensity of the oxygen lines under different discharge rates. The crystals with the minimal gas extraction were chosen when the intensity of the oxygen lines was minimal and almost did not change while alternating the current voltage characteristics of the discharge.

6. Forming the wave-guided layer

The usage of the constant vacuum system wash with argon at the rate of flux of 30 sm^3 atm/hour according to mass-spectrometry by a measurer of partial pressures IPDO-1 allows decreasing the quantity of such optical modes quenchers as hydrocarbons. The direction of the synthesis reaction was estimated by the pressure change during a period of time. During the experiment, the following condition was provided: the inflow stream of inert gas Q did not exceed the pressure change of the gas component ΔP for a period of the implantation time Δt (~ 20 min).

$$Q = V(\Delta P / \Delta t)\,, \tag{5}$$

where V is the size of the evacuated chamber.

Implantation was carried out under the speeding up voltage 20 kW in the argon medium under a pressure 0.01 Pa at arc current less than 100 A. The ionic current impulse was 25-1000 mA at the duration 250 μs. The impulse repetition frequency was 1- 50 Hz. The measured implanted dose was $3 \cdot 10^{17}$ ion/sm^2 which is close to the calculated value for the film with the thickness of 50 nm.

The crystal annealing with the size of 25x8x2 mm was made in the furnace SUOL 025 with the speed of the temperature increase 2 K/s and with the temperature endurance 1350 K within 4 hours. Then, the obtained waveguide was cooled within 8 hours with the velocity of ~2 K/s.

Then the forming process of copper scum having thickness 0.5 micron on Niobate Lithium was carried out by means of the small-scale pulsed magnetron which works on a frequency of 50 kHz under a pressure of 0,1 Pa. The analysis carried out on the focus beam microscope showed that the quantity of copper in the scum amounts 4% under the thermal forming of covering in the vacuum, and it calculates 36% in argon. While forming by means of the magnetron the part of copper compounds 40% total.

The diffusion of copper is realized in the air atmosphere at a temperature of 1050 K during 2-20 hours.

7. Undertaken researches

The figure 3 presents a circuit for optic tests. The reference beam of laser *1* (LCS-DTL-317) with the power 40 mW on the wave length 532 nm through the collimator *2* falls on the beam splitting cube *3* where divides into two beams with equal intensities on 390 mW /sm^2. Then one part of the beam is aimed to the crystal *6* at Bragg angle (12,5° for the emission with the length wave 532 nm). The other part of the beam from the beam splitting cube through the prism *4* is also directed to the crystal symmetrically to the normal by its entry edge. Reading of the photorefractive diagram in the process of its forming is carried out according to the Bragg angle diffraction method of the light beam with the length wave 655 nm radiated by the laser diode *5* with the power output 6 mW. The angle of descent of the reading beam on its entry edge is equal to the. Bragg angle which amounts 7,7°. Power timing dependencies of the beam passed the diffraction and the transmitted beam *7, 8* are recorded by means of photodiodes *9* (FD-24 K). The data processing system *10* included two digital microampermeter as V7-40/1, an interface and personal computer. It is worth to point out, that the readout radiation of the wave length 655 nm weakly influences the formation of photorefractive holograms by writing beams on the wave length 532 nm due to the strong decrease of photovoltaic constants of crystals with the wave length. The system is set up for the minimal discrepancy of beams in the plane XZ. The test sample of crystal was exposed to annealing under the temperature 190°C or is kept in the darkness in order to delete recorded photorefractive holograms. On the fig. 4 specific timing the dependencies of relative powers of the beam after the diffraction (the curve 2) and the transmitted (reading) beam (the curve 1) are presented.

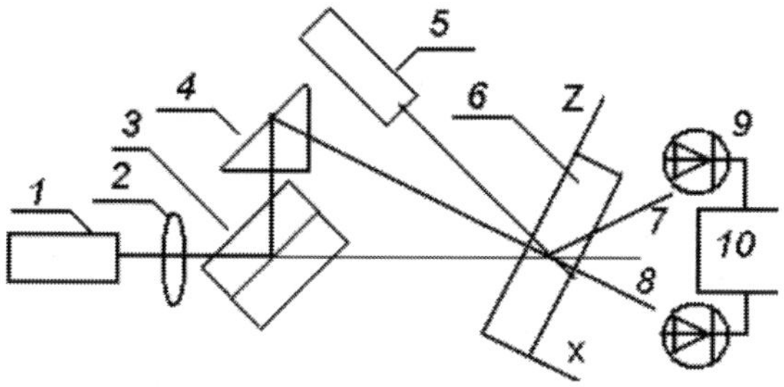

Figure 3. Circuit of optic tests. *1* – laser, *2* – collimator, *3* – splitting cube, *4* – prism, *5* – laser diode, *6* – crystal, *7, 8* – transmitted beam, *9* – photodiodes, *10* – processing system

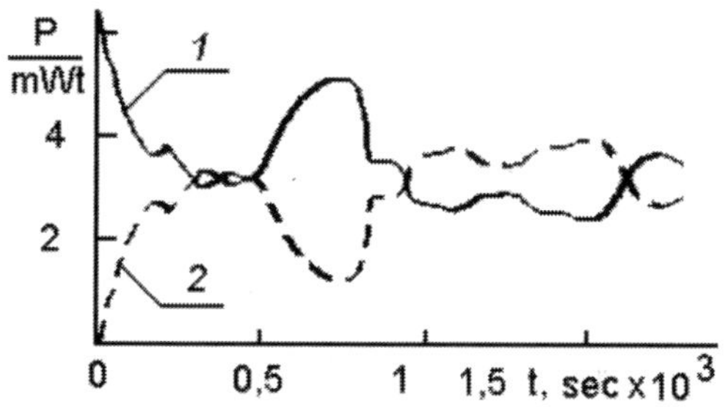

Figure 4. Power dependence of the beam after the diffraction and reading beam on time

The photo induced absorption of the non-object beam and the transfer of the optical power from the beam *7* (figure 3) into the beam *8* was discovered and in reverse. Oscillographic measurements made by the stand-by oscillograph S8-12 within the microsecond interval show discontinuous variations of intensity patterns are observed when the disruption occurs on the crystal surface. In this case, the field amplitude of the spatial charge on the crystal frontier reaches the value 30 kV/sm what exceeds the electrical surface stability of the piezoelectric.

8. Results and outputs

The biggest alteration of the diffraction pattern is observed under the maximal copper doping of the crystal. Due to the optical absorption coefficient on the waves length 477 and 1040 nm, according to the methodic [1] the concentration C of ions Cu^{2+} and Cu^{+} responsible for electro-optical characteristics of copper doping crystals was determined:

$$C_{Cu+} = 2\times10^{21}\alpha_{477}\ ;\quad C_{Cu^{2+}} = 6{,}8\times10^{22}\alpha_{1040}\ [\mathrm{m}^{-3}]; \qquad (6)$$

where α is the absorption coefficient (sm^{-1}). The index points out the wave length 477 and 1040 nm.

The table 1 presents parameters of formed wave-guided layers in a time of the diffusion t (hour) for the films of copper having thickness δ (nm).

Table 1. Parameters of formed wave-guided layers

№	δ(nm)	t(h)	$Cu^{+}(m^{-3})$	$Cu^{2+}(m^{-3})$	α_{532} (sm^{-1})
1	100	2	$7{,}9\times10^{22}$	$8{,}7\times10^{22}$	0,063
2	200	3	2×10^{23}	$1{,}4\times10^{24}$	0,615
3	260	20	$7{,}8\times10^{23}$	6×10^{24}	0,88
4	900	9	6×10^{24}	$2{,}8\times10^{25}$	7,9

Calculations show that perhaps changes of diffraction efficiency are connected with the magnitude variation of the crystal spatial field.

9. Conclusion

The process to form wave-guided layers on the typical vacuum sets requires computer maintenance to define the order and parameters of producible process. The doze of copper ions ~ $5\times10^{17}sm^{-2}$ is minimal to create an active wave-guided layer. The next increasing of the concentration is attained by the diffusion of copper from the scum formed in the inert atmosphere.

Powers changes of the transmitted beam and the beam after the diffraction are connected with the alteration of the crystal electric field caused by cracking on its surface.

The availability to conduct the intensity of photo induced absorption using the change of copper concentration in the crystal offers the challenge to create solid sources of electrons and X radiation [7]. The combination methodic of diffusion and implantation technologies is perspective in order to create light-controlled active optic wave-guided layers.

On figures 5, 6 dependences of transmittances $T(\lambda)$, absorption $k(\lambda)$ and reflexions $R(\lambda)$ on the wave length through the waveguide nanolayer before and after implantation are presented.

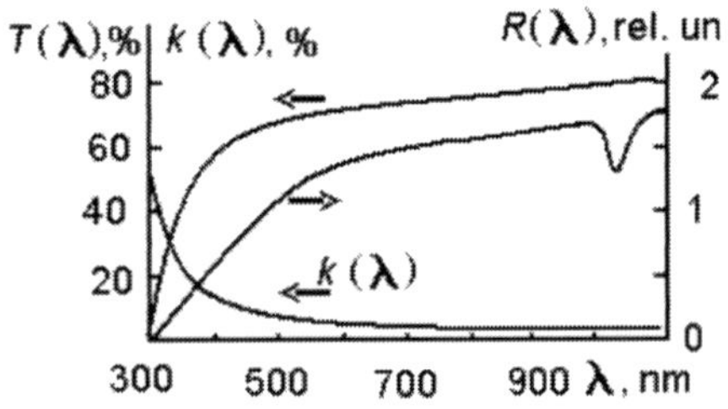

Figure 5. Dependences of transmittances $T(\lambda)$, absorption $k(\lambda)$ and reflexions $R(\lambda)$ on the wave length before implantation.

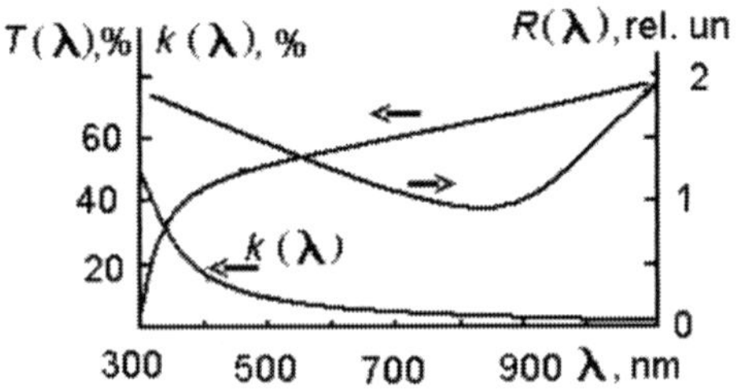

Figure 6. Dependences of transmittances $T(\lambda)$, absorption $k(\lambda)$ and reflexions $R(\lambda)$ on the wave length after implantation.

We can see that before the implantation the following effects are present: transmittance of the waveguide layer on the wave length almost does not vary, and the reflectivity increases with the wave length growth. After implantation and annealing the transmittance decreases; (the layer "clarifies"), and the reflectivity has a minimum on the wave length of 800 nm

References

[1] Peitmann K, Hukriede J, Buse K and Kratzig E 2001 *Phus Rev* vol **61** no **7** p 4615
[2] Esseling M, Holtmann F, Woerdemann M and Denz C 2010 *Optics Express* vol **18** no **16** p 17404

[3] Arestov S and Orlikov L 2013 *Izvestiya vuzov, ser. Fizika* vol **56** no **1/2** pp 178-181
[4] Danilina T, Smirnova K, Ilyushin V and Velichko A 2005 *Micro- and nanotechnology processes* (Tomsk: Tomsk University of Control Systems and Radioelectronics) 316 p
[5] Arestov S, Arestov G, Orlikov L and Shandarov S 2011 *Transactions International science-practical conferences on Problem of avtomatic and control by tekhnichal sistems* (Penza: Russia 19-22 April 2011 PGU 2011 pp 9-11
[6] Orlikov N and Orlikov L 2004 The Russian patent 2223481 MPK7 G01N25/00 Russia
[7] Orlikov L, Romanov V, Arestov S and Shandarov S 2011 *Lectures of Tomsk State University of Control Systems and Radioelectronics* no **2 (24)** P **2** pp 135-138

IOP Conf. Series: Materials Science and Engineering **81** (2015) 012044 doi:10.1088/1757-899X/81/1/012044

Production of nanostructured TiO_2 films by pulsed laser deposition

Ye Khuanbay[1], N Kh Ibrayev[2], D A Afanasyev[2] and A K Aimukhanov[2]
[1] *Karaganda, 100027, Kazakhstan*, Karaganda State Technical University,
[2] *Karaganda, 100028, Kazakhstan* Institute of Molecular Nanophotonics, E.A. Buketov Karaganda State University

E-mail: esen_d@mail.ru

Abstract. The possibility of titanium dioxide films formation by pulsed laser deposition was studied. It was found that the titanium dioxide nanorods were formed on the surface of quartz glass under certain conditions. Craters with diameter about 40-50 μm and depth about 5-10 μm were formed on the target surface exposed with the nanosecond laser pulse. Electronic structure of the samples was studied by energy dispersive X-ray spectrum. Energy transitions of electrons determined from the energy dissipating spectra.

Introduction

The titanium dioxide (TiO_2) films have unique physico-chemical properties. In accordance with these properties the TiO_2 films are used as an element in photocatalysis devices, solar photovoltaic devices and they are widely used in optical devices [1, 2]. Titanium dioxide is different from other semiconductor materials (ZnO, SnO_2, Fe_2O_3). It has a high photosensitivity and acceptable band gap about 3.2 eV [3]. Features of the physico-chemical properties of TiO_2 films associated with its nanoscale structures. There are many methods for the synthesis of TiO_2 films. A method of pulsed laser ablation is one of the most promising methods for producing TiO_2thin films.

In this paper the results of the research the structure and energy dispersive X-ray spectra of TiO_2 films produced by pulsed laser deposition were presented.

There are two methods of laser-induced transfer of matter: direct transfer of matter and reverse transfer of matter. In the direct transfer of matter the laser beam passes through the transparent substrate evaporating material deposited on it. The evaporated material condenses on the location along the laser beam substrate. In the reverse transfer of matter laser beam passes through the transparent substrate and the acceptor focuses on the target surface. The evaporated material is deposited on substrate towards the laser radiation and condenses on the substrate. The technology of reverse transport of material was used in the present work.

Methods and Results

Universal laser technological installation was used for the deposition of TiO_2 films. Technological installation included a laser emitter ILTI-407B and vacuum universal installation VUP-4. The design of the vacuum chamber allows changing the distance from the sample to the substrate. The film thickness is controlled by changing the distance between the substrate and the sample and changing

[1] Author to whom any correspondence should be addressed.

the duration of the laser ablation exposure. The deposition of films occurred by evaporating matter of target under the action of YAG:Nd^{3+} laser radiation with 10 ns duration, at $\lambda = 1064$ nm and a pulse repetition rate about 14 Hz. The angle between the laser beam and the sample surface was 45 degrees. The pressed cylindrical sample of titanium dioxide was used as the target. Target dimensions were 10 mm in diameter, 150 mm in length. The laser beam was focused on different parts of the target due to the translational motion of target during the ablation.

Quartz glass was used as substrate (dimensions about 10x7x1.5 mm). The substrate was placed on the resistive heater at a distance about 10 – 30 mm from the target surface. The pulse energy was 11 mJ. Power pulsed laser radiation was $10^5 - 10^7$ W·cm^{-2}.

The study of the obtained films microstructure was carried out using the scanning electron microscope (SEM) JEOL JSM-5910. Its work is based on the registration of secondary electrons (SE). Space resolution in the scanning mode at voltage 15kV was 3 nm. Working distance was 8 nm. Probe current was 5 nA (analysis mode). Thermal-type Schottky electron gun was mounted into the microscope.

SEM -image of TiO_2 target surface after laser ablation is shown on Figure 1 (a). Craters formed on the surface target by nanosecond laser pulse. Craters have a diameter of 40-50 μm and a depth of 5-10 μm. The absorbed laser energy, that exceeds the critical density of the radiation flux, heated the target surface till the temperature of melting and then till the temperature of evaporation.

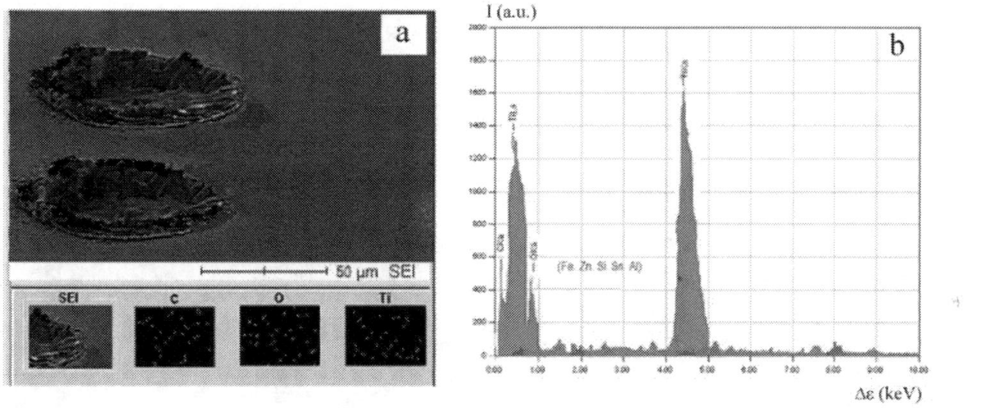

Figure 1. SEM images of the target surface (a), energy dispersive characteristic of X-ray spectrum (b).

The flux density of the laser radiation q can be represented as a function equivalent to multiplication function of time and the functions of the surface coordinate [4, 5]. The flux density of the laser radiation given by following formula (1):

$$q = A\varphi(t)q^*(x, y), \tag{1}$$

where A – the absorptivity that depends on the surface state and its temperature. The function $\varphi(t)$ describes the temporal structure of the pulse. The function $q^*(x,y)$ is the spatial distribution of the flux density of the laser radiation.

Form of melted area of sample changes during laser irradiation continuously and depends on the pulse duration. X-ray diffraction spectroscopy was used to study the chemical composition and elemental microanalysis of TiO_2 films and targets.

Energy dispersive X-ray spectroscopy technique is based on electron beam excitation of the sample followed by detection of the X-ray characteristic [6]. Energy dispersive characteristic X-ray spectra were obtained by X-ray spectrometer JXA-8200 with the energy dispersion and wave dispersion. This spectrometer was installed in the SEM microscope. Type of detector was EX54133MUK. The electron energy probe was 25 keV. The probe current was 1 nA. The measurement time was 30 seconds. Working distance was 10 mm.

Energy dispersive characteristic of X-ray spectrum of elemental constituents of the film are shown on Figure 1, (b). The spectrum has the peaks corresponding to all the chemical elements of the film. Quantitative determination of elements was performed using analytical signals OK_{α}, TiK_{α}. OK_{α} spectrum line was formed at the transition between atomic states $2p_{3/2} \rightarrow 1s_{1/2}$, for $TiK_{\alpha} - 2p_{3/2} \rightarrow 1s_{1/2}$, $TiL_{\alpha} - 3d_{5/2} \rightarrow 2p_{3/2}$, taking into account the dipole selection rules [7].

OK_{α} line shows the distribution of 2p states of oxygen atoms. TiL_{α} lines show information about d-electrons and the distribution of 3d states of titanium atoms primarily. This is due to the smallness of the contribution to the intensity of the s-electrons. Energy electron transitions corresponding analytical lines determine from the energy dissipating spectra: $\Delta\varepsilon(OK_{\alpha}) = 0.525$ keV; для $\Delta\varepsilon(TiL_{\alpha}) = 0.46$ keV; $\Delta\varepsilon(TiK_{\alpha}) = 4.54$ keV. The experimental data of the transition energy $\Delta\varepsilon$ correspond to quantum-mechanical calculations using Moseley's law on the theory of atomic spectra [7].

Concentration of TiO_2 films elements were determined by mathematical processing of energy dispersive spectra. Mathematical processing was made by the program PHI-RHO-Z using the method of Bence and Albee binary oxide systems. Comparison of atomic (*At.%*) and the weight (*G,%*) elemental concentrations at various elementary areas of the objects shows that the samples are chemically homogeneous.

The films were prepared at different repetition frequencies of the laser radiation. These films don't differ in the percent abundance of atoms and mass. The value x for TiO_x is close to 2 in all cases. It corresponds to the stoichiometric composition of TiO_2.

Images of the microstructure of titanium dioxide films are presented on Figure 2.

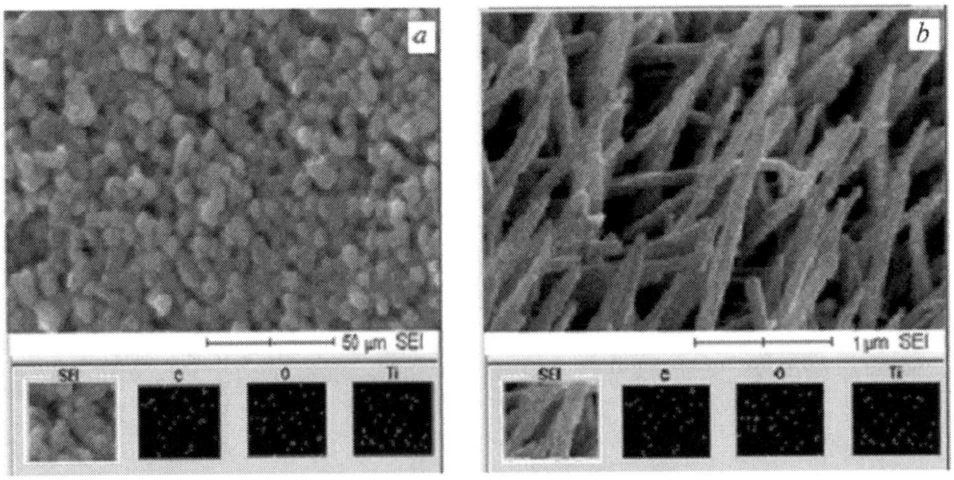

Figure 2. Microstructure of titanium dioxide film (a), a cross section of film (b).

The film has a mesoporous structure, as can be seen from the figure 2. Regularities aren't observed in the arrangement of the grains and their spatial orientation.

The average grain size is 5-8 microns. Direct forms of walls are observed for the tubes. Most of the surface of the film is smooth with a small roughness of 10-15 nm. The diameter of the tubes is about 190 nm in average. A series of experimental measurements carried out using SEM show that the film thickness and stoichiometry of TiO_2 don't depend on the parameters of laser radiation. The thickness of the film increases with deposition time.

References:

[1] Xu M, Huang N, Xiao Z and Lu Z 1998 Photoexcited TiO_2 nanoparticles through OH-radicals induced malignant cells to necrosis *Supramol. Sci.* vol.**5** pp. 449-551.

[2] Kim B., Kim D, Cho D and Cho S. 2003 Bactericidal effect of TiO_2 photocatalyst on selected food-borne pathogenic bacteria *Chemosphere* vol.**52** pp. 277-281.

[3] Kominami H, Kato J I 2003 Solvothermal syntheses of semiconductor photocatalysts of ultra-high activities *Catalysis Today* vol.**84** pp. 181-189.

[4] Rykalin N N, Uglov A A, Smurov I Y 1979 Spatial nonlinear problem of metals heating by laser radiation *Physics and Chemistry of Materials Processing* vol.**2** pp.3-13.

[5] Uglov A A, Smurov I Y et all 1981 Laser modification of gas-thermal coatings *Technical Physics* vol. **59** pp.1787-1791.

[6] Frish S E 2010 *Optical spectra of atoms* (Saint Petersburg: Lan).

[7] Armstrong J T 1984 Quantitative analysis of silicate and oxide minerals: a reevaluation of ZAF corrections and proposal for new Bence-Albee coefficients *Microbeam Analysis* pp.208-218.

IOP Conf. Series: Materials Science and Engineering **81** (2015) 012045 doi:10.1088/1757-899X/81/1/012045

Crystallographic features of magnetostatic waves spectra in ferrite films

V V Shagaev and Tun Tun Lin
Bauman Moscow State Technical University, Kaluga Branch, Russia

E-mail: shagaev_vv@rambler.ru

Abstract. Influence of crystallographic magnetic anisotropy on dispersion laws of magnetostatic spin waves is discussed. Classification of waves spectra in single crystal films of ferrites is given. The scale of magnetic anisotropy influence on spectra of volume and surface types of waves is investigated. The film model with cubic magnetic anisotropy is constructed and expressions for components of a tensor of the effective demagnetization anisotropy factors are deduced. Results of spectra calculation of magnetostatic waves in a film of yttrium iron garnet are given.

1. Introduction

One of the perspective directions of functional electronics is development of devices, in which transformation of signals is carried out by means of magnetostatic spin waves (MSWs). The greatest interest is represented by MSWs extending in films of ferrite [1, 2]. Their phase speed is some orders less than velocity of light and variations of fields in space than in time are more essential to them. The theory of such waves can be constructed on the basis of the equation of the movement of magnetization and Maxwell's equations taken in magnetostatic approach.

Important feature of films is their perfect crystal structure and as a result films possess anisotropy of magnetic properties. Crystallographic magnetic anisotropy belongs to the factors defining dispersive characteristics of MSWs. Radiation effects and changes of temperature lead to changes of anisotropy parameters and it influences on dispersion waves laws. To the present time many aspects of such an influence are studied. Private results relating to the films of the cubic ferrites are together with common approaches to a description of the MSWs spectra in the films with arbitrary magnetic anisotropy type were obtained [3 – 8]. The orientation of the investigations just to such ferrites is caused by wide use in the spin-wave electronics of the yttrium iron garnet films (YIG, $Y_3Fe_5O_{12}$), which are grown up on the single crystal gadolinium gallium garnet ($Gd_3Ga_5O_{12}$) substrates. In the present work the model, in which crystallographic orientation of a cubic ferrite film is the varied parameter, is built and investigated. Thus the considered earlier separate orientations become only special cases of the developed model.

2. General ratios

The studied model is presented in figure 1. The ferrite film thickness of d and the unlimited sizes in the plane is magnetized before saturation by an external constant magnetic field. The vector of magnetization $\boldsymbol{M}_0$ forms an angle θ with the film plane. Two systems of coordinates are entered into consideration. In xyz system the y axis is parallel to the film plane and the z axis is parallel to the $\boldsymbol{M}_0$

vector, $z \parallel \boldsymbol{M}_0$. In $\xi\eta\zeta$ system the ξ axis is directed along a normal to a film $\boldsymbol{n}$ and the η axis along a wave vector of MSW $\boldsymbol{k}$. The φ angle defines a direction of the η axis regarding the xz plane. The dispersive equations (DEs) of the magnetostatic waves are deduced by joint integration of the linearized equation of the magnetization movement without an exchange and losses and Maxwell's equations taken in magnetostatic approach and with the corresponding electrodynamic boundary conditions. As a result the DEs can be presented in the form of the equations expressed through the components of the magnetic permeability tensor of the film material taken in $\xi\eta\zeta$ system

$$\tan\left[\frac{kd\sqrt{(\mu_{\xi\eta}+\mu_{\eta\xi})^2-4\mu_{\xi\xi}\mu_{\eta\eta}}}{2\mu_{\xi\xi}}\right]=\frac{\sqrt{(\mu_{\xi\eta}+\mu_{\eta\xi})^2-4\mu_{\xi\xi}\mu_{\eta\eta}}}{\mu_{\xi\eta}\mu_{\eta\xi}-\mu_{\xi\xi}\mu_{\eta\eta}-1} \tag{1}$$

if $(\mu_{\xi\eta}+\mu_{\eta\xi})^2 \geq 4\mu_{\xi\xi}\mu_{\eta\eta}$ and

$$\exp\left[\frac{kd\sqrt{4\mu_{\xi\xi}\mu_{\eta\eta}-(\mu_{\xi\eta}+\mu_{\eta\xi})^2}}{\mu_{\xi\xi}}\right]= =\frac{\left[\sqrt{4\mu_{\xi\xi}\mu_{\eta\eta}-(\mu_{\xi\eta}+\mu_{\eta\xi})^2}-2\right]^2+(\mu_{\xi\eta}-\mu_{\eta\xi})^2}{\left[\sqrt{4\mu_{\xi\xi}\mu_{\eta\eta}-(\mu_{\xi\eta}+\mu_{\eta\xi})^2}+2\right]^2+(\mu_{\xi\eta}-\mu_{\eta\xi})^2} \tag{2}$$

at a reverse inequality.

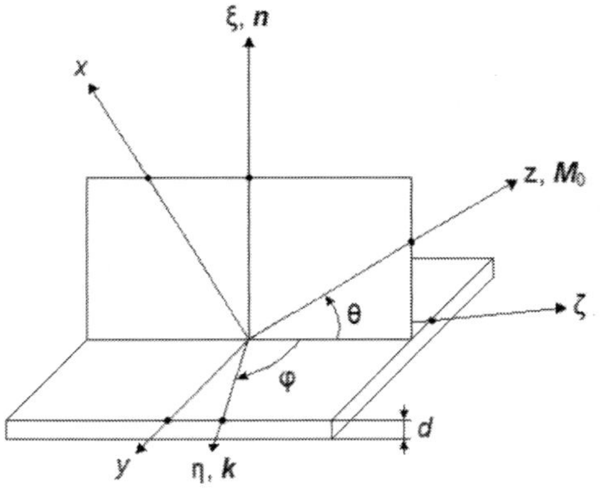

Figure 1. The coordinates systems used at a conclusion of the dispersion equations of the magnetostatic waves.

In the xyz coordinates system the tensor of magnetic permeability has a view

$$\hat{\mu}=\begin{pmatrix}\mu_x & \mu_s+i\mu_a & 0\\ \mu_s-i\mu_a & \mu_y & 0\\ 0 & 0 & 1\end{pmatrix},$$

$$\mu_x=1+\frac{4\pi M_0\omega_x}{\Omega},\quad \mu_y=1+\frac{4\pi M_0\omega_y}{\Omega},\quad \mu_s=-\frac{4\pi M_0\omega_s}{\Omega},\quad \mu_a=\frac{4\pi M_0}{\Omega}\left(\frac{f}{g}\right),$$

$$\omega_x=H_{in}+M_0\left(N_{yy}^a-N_{zz}^a\right),\quad \omega_y=H_{in}+M_0\left(N_{xx}^a-N_{zz}^a\right),\quad \omega_s=M_0N_{xy}^a,$$

$$H_{in} = \frac{(\boldsymbol{H}_e \boldsymbol{M}_0) - 4\pi(\boldsymbol{M}_0 \boldsymbol{n})^2}{M_0}, \quad \Omega = \omega_x \omega_y - \omega_s^2 - \left(\frac{f}{g}\right)^2.$$

Here $g = 2{,}8$ MHz/Oe is the gyromagnetic ratio; $\boldsymbol{H}_e$ is the external magnetizing field strength; N_{ij}^a are the tensor components of effective demagnetizing anisotropy factors [9]; f is the wave frequency. In passing to the $\xi\eta\zeta$ coordinate system, the components of tensor $\hat{\mu}$ are transformed as follows

$$\left.\begin{aligned} \mu_{\xi\xi} &= \mu_x \cos^2\theta + \sin^2\theta, \\ \mu_{\eta\xi} &= [(1-\mu_x)\sin\theta\cos\varphi + (\mu_s + i\mu_a)\sin\varphi]\cos\theta, \\ \mu_{\xi\eta} &= [(1-\mu_x)\sin\theta\cos\varphi + (\mu_s - i\mu_a)\sin\varphi]\cos\theta, \\ \mu_{\eta\eta} &= (\mu_x \sin^2\theta + \cos^2\theta)\cos^2\varphi - \mu_s \sin 2\varphi \sin\theta + \mu_y \sin^2\varphi. \end{aligned}\right\} \tag{3}$$

The analysis of the spectra of the main types MSWs is based on the equations (1), (2).

2.1. The perpendicularly magnetized layer when $\theta = \pi/2$, and it is also possible to combine y and η axes and therefore equality $\varphi = \pi/2$ is carried out.

Then $\mu_{\xi\xi} = 1$, $\mu_{\eta\xi} = \mu_{\xi\eta} = 0$ and DE has a view

$$\tan\left(kd\sqrt{-\mu_{\eta\eta}}\right) = -\frac{2\sqrt{-\mu_{\eta\eta}}}{1+\mu_{\eta\eta}}, \tag{4}$$

$$\mu_{\eta\eta} = 1 + \frac{4\pi M_0\left[H_{ez} - 4\pi M_0 + M_0\left(N_{xx}^a - N_{zz}^a\right)\right]}{\left\{\left[H_{ez} - 4\pi M_0 + M_0\left(N_{xx}^a - N_{zz}^a\right)\right]\left[H_{ez} - 4\pi M_0 + M_0\left(N_{yy}^a - N_{zz}^a\right)\right] - \left(M_0 N_{xy}^a\right)^2 - (f/g)^2\right\}}.$$

The equation (4) describes modes of direct volume MSW. H_{ez} is a projection of the vector of the external magnetizing field strength to a magnetization vector.

2.2. Tangentially magnetized layer when $\theta = 0$ and $\boldsymbol{k} \parallel \boldsymbol{M}_0$, thus equality $\varphi = 0$ is carried out.

Then $\mu_{\eta\eta} = 1$, $\mu_{\eta\xi} = \mu_{\xi\eta} = 0$ and DE has a view

$$\tan\frac{kd}{\sqrt{-\mu_{\xi\xi}}} = \frac{2\sqrt{-\mu_{\xi\xi}}}{1+\mu_{\xi\xi}}, \tag{5}$$

$$\mu_{\xi\xi} = 1 + \frac{4\pi M_0\left[H_{ez} + M_0\left(N_{yy}^a - N_{zz}^a\right)\right]}{\left\{\left[H_{ez} + M_0\left(N_{xx}^a - N_{zz}^a\right)\right]\left[H_{ez} + M_0\left(N_{yy}^a - N_{zz}^a\right)\right] - \left(M_0 N_{xy}^a\right)^2 - (f/g)^2\right\}}.$$

The equation (5) describes modes of the backward volume waves.

2.3. Tangentially magnetized layer when $\theta = 0$ and $\boldsymbol{k} \perp \boldsymbol{M}_0$, $\varphi = \pi/2$.

The equations (1), (2) after all substitutions and algebraic transformations assume the following view

$$\tan\frac{kd\sqrt{\left(f_1^2 - f^2\right)\left(f^2 - f_2^2\right)}}{f^2 - f_0^2} = \frac{2\sqrt{\left(f_1^2 - f^2\right)\left(f^2 - f_2^2\right)}}{(4\pi M_0 g)^2 + f_1^2 + f_2^2 - 2f^2} \tag{6}$$

$$\exp\frac{2kd\sqrt{(f^2-f_1^2)(f^2-f_2^2)}}{f^2-f_0^2}=1+\frac{4\sqrt{(f^2-f_1^2)(f^2-f_2^2)}}{(4\pi M_0 g)^2-\left(\sqrt{f^2-f_1^2}+\sqrt{f^2-f_2^2}\right)^2} \quad (7)$$

The equation (6) corresponds to bulk wave and (7) to surface wave (or mixed at $N_{xy}^a \neq 0$). Here

$$(f_0/g)^2 = H_{ez}^2 + H_{ez}M_0\left(N_{xx}^a + N_{yy}^a - 2N_{zz}^a + 4\pi\right) + M_0^2\left[\left(N_{xx}^a - N_{zz}^a + 4\pi\right)\left(N_{yy}^a - N_{zz}^a\right) - \left(N_{xy}^a\right)^2\right], \quad (8)$$

$$f_{1,2}^2 = f_0^2 + 2\pi M_0^2 g^2\left[\left(N_{xx}^a - N_{yy}^a\right) \pm \left|\sqrt{\left(N_{xx}^a - N_{yy}^a\right)^2 + 4\left(N_{xy}^a\right)^2}\right|\right]. \quad (9)$$

Just surface MSW represents the greatest practical interest.

In experiment orientation of the external magnetizing field but not orientation of magnetization is usually controlled. In the anisotropic and uniformly magnetized film at $\boldsymbol{H}_e \perp \boldsymbol{n}$ and $\boldsymbol{k} \perp \boldsymbol{H}_e$ the angles θ and φ defining the $\boldsymbol{M}_0$ orientation can differ from values $\theta = 0$ and $\varphi = \pi/2$. It is physically clear that this difference of subjects is less than intensity of the magnetizing field in comparison with the magnetic anisotropy effective fields is more. The detailed consideration of a given question is a separate task and goes out beyond the frames of carried out investigation of dispersion laws. It is possible to note however that on the formulas given above can be calculated angular derivative of frequency and at $\theta = 0$, $\varphi = \pi/2$ they become zero: $df/d\theta = 0$ and $df/d\varphi = 0$. So the corrections of dispersion laws connected with consideration of small deflections of the angles from the considered values will have the second order of smallness.

The example of the spectrum calculated on formulas (6) and (7) for values $(f_0/4\pi M_0 g)^2 = 1$; $(f_1/f_0)^2 = 1{,}1$; $(f_2/f_0)^2 = 0{,}95$ is given in figure 2. SMSW, FVMSWs and BVMSWs are respectively surface, forward volume and backward volume MSWs.

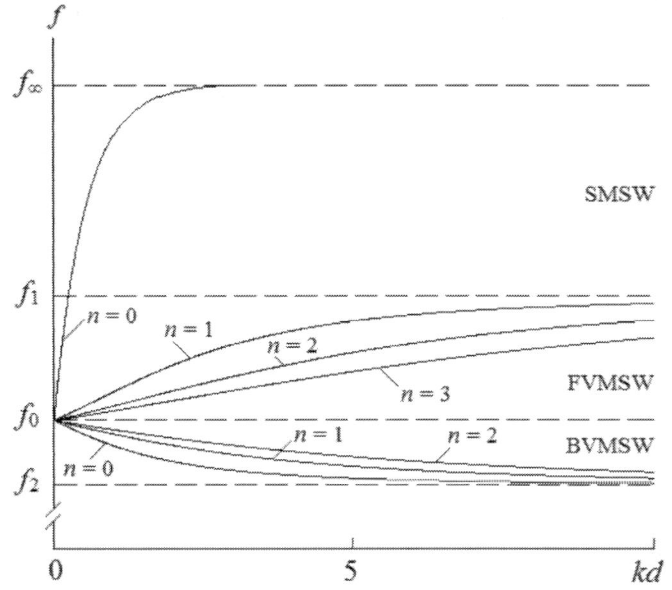

Figure 2. Spectrum of magnetostatic waves in the tangentially magnetized film and with a wave vector perpendicular to a magnetization vector

Unlike a surface wave both types of volume waves are multimode. In figure 2 are represented some modes with initial numbers "n". Spectrum of the forward volume waves is situated in interval $f_0 < f < f_1$, and in the interval $f_2 < f < f_0$ there is a spectrum of the backward volume waves. The surface MSW spectrum occupies frequency interval

$$f_1 < f < f_\infty = gH_{ez} + 2\pi M_0 g + 0.5gM_0\left(N^a_{xx} + N^a_{yy} - 2N^a_{zz}\right)$$

and the wave vectors are restricted to the values

$$kd > k_{sv}d = \frac{2\left(f_1^2 - f_0^2\right)}{\left(4\pi M_0 g\right)^2 - \left(f_1^2 - f_2^2\right)}.$$

Dispersive dependence of the main mode of a forward volume wave smoothly passes into dispersive dependence of a surface wave at $f = f_1$ and $k = k_{sv}$.

The comparative estimation of intervals of the frequencies occupied by volume waves and a surface wave looks as follows

$$\frac{f_1^2 - f_2^2}{f_\infty^2 - f_0^2} = \frac{16\pi\sqrt{\left(N^a_{xx} - N^a_{yy}\right)^2 + \left(2N^a_{xy}\right)^2}}{\left(4\pi + N^a_{xx} - N^a_{yy}\right)^2 + \left(2N^a_{xy}\right)^2}.$$

It follows from this expression that in the tangentially magnetized films of the weakly anisotropic ferrites (when $\left|N^a_{ij}\right| << 1$) the large part of MSW spectrum with wave vectors perpendicular to the magnetization will occupy by a surface type wave. Besides spectra of the volume waves arise in the considered conditions only due to crystallographic magnetic anisotropy and in isotropic model these spectra aren't present.

3. Anisotropy of the surface wave spectrum in a cubic ferrite film

In connection with the considered task we will note that YIG films with crystallographic orientation of {111} type are most widely applied now in spin-wave devices. At the same time some characteristics of devices can be improved when using films of other orientations. In particular the thermostability of SMSW spectra in the YIG films with orientations of {110} and {100} type is higher than in the {111} films.

We shall build a model, allowing executing of calculations for the films oriented along by any of the crystallographic planes passing across an axis of <110> type. Then, for example, all three mentioned orientations {111}, {110}, {100} become special cases of such model. Geometrical aspects of placed task are represented in figure 3.

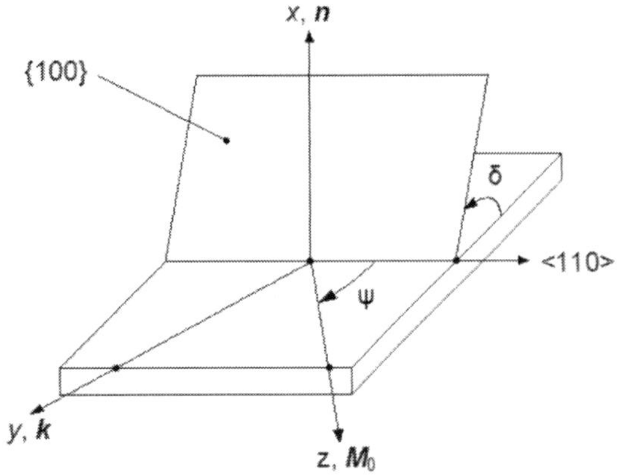

Figure 3. The investigated model of a cubic ferrite film

The crystallographic orientation of the magnetization vector is defined by two angles. The angle ψ defines the direction in the film plane concerning <110> axis and the δ angle sets an inclination of the plane {100} passing through the considered <110> axis to the film plane.

Magnetic anisotropy of ferrite is taken into account with the help of $M_0 N_{ij}^a$ components. Expressions for them can be deduced from expression for energy of cubic magnetic anisotropy [9]. As a rule it is enough to take only one constant into account in this energy (two constants are necessary for {111} orientation and this aspect is discussed below).

$$M_0 N_{xx}^{a(1)} = -2\frac{K_1^c}{M_0}\left\{1 - \frac{1+3\cos 2\delta}{16}\left[(1+3\cos 2\delta) + 3(1-\cos 2\delta)\cos 2\psi\right]\right\},$$

$$M_0 N_{yy}^{a(1)} = -\frac{K_1^c}{M_0}\left\{1 + \frac{1+3\cos 2\delta}{32}\left[(1+3\cos 2\delta) + 3(5-\cos 2\delta)\cos 4\psi\right]\right\},$$

$$M_0 N_{zz}^{a(1)} = -\frac{K_1^c}{M_0}\left\{1 + \frac{1+3\cos 2\delta}{32}\left[(1+3\cos 2\delta) + 4(1-\cos 2\delta)\cos 2\psi - (5-\cos 2\delta)\cos 4\psi\right]\right\},$$

$$M_0 N_{xy}^{a(1)} = \frac{3}{2}\frac{K_1^c}{M_0}\sin 2\delta\left[\cos 3\psi + \frac{1+3\cos 2\delta}{8}(\cos\psi - \cos 3\psi)\right].$$

K_1^c is the first constant of cubic magnetic anisotropy. In figure 4a examples of angular dependences of SMSW frequencies calculated on DE (7) with substitutions of (8), (9) and $M_0 N_{ij}^{a(1)}$ expressions are given. The crystallographic orientation of a film was defined by the values of the δ angle. Calculations correspond to the following values of magnetic parameters of YIG crystals (at 293 K): $4\pi M_0 = 1750\ \text{G}$, $K_1^c / M_0 = -43\ \text{Oe}$. It was supposed that $H_{ez}/4\pi M_0 = 1/3$ and $kd = 1$.

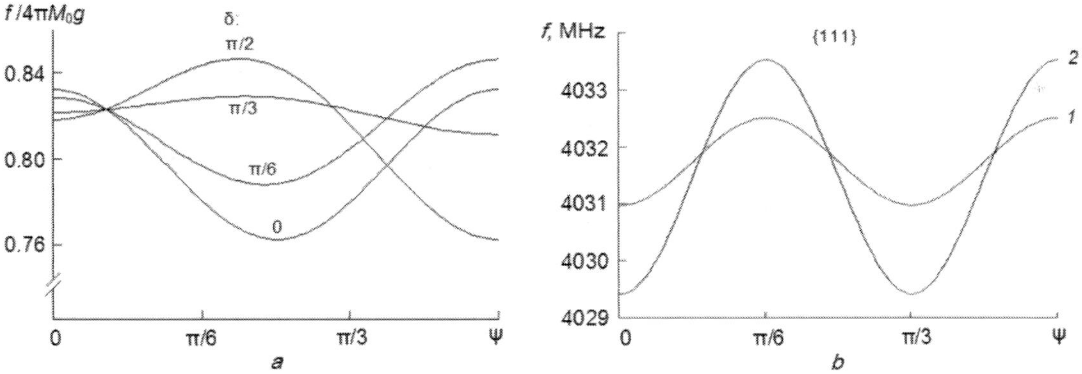

Figure 4. Angular dependences of SMSW frequencies in films with various crystallographic orientations.

Films with {111} orientation possess the following feature. Equality of $(1+3\cos 2\delta) = 0$ is carried for them and as appears from the $M_0 N_{ij}^{a(1)}$ expressions dependence on a ψ angle remains only at $M_0 N_{xy}^{a(1)}$. And unlike diagonal components, the parameter $M_0 N_{xy}^{a(1)}$ is included into DE in the quadratic kind and therefore angular dependence of the frequencies in the {111} plane will be the weakest. In this case the account not only the first, but the second constant of magnetic cubic anisotropy is necessary [5]. Calculation of the effective demagnetizing anisotropy factors corresponding to the accounting of the second constant (K_2^c) leads to the following formulas

$$M_0 N_{xx}^{a(2)} = \frac{K_2^c}{6M_0},\ M_0 N_{yy}^{a(2)} = \frac{K_2^c}{18M_0}(1+5\cos 6\psi),\ M_0 N_{zz}^{a(2)} = \frac{K_2^c}{18M_0}(1-\cos 6\psi),$$

$$M_0 N_{xy}^{a(2)} = \frac{\sqrt{2}}{6}\frac{K_2^c}{M_0}\cos 3\psi .$$

The results shown in figure 4b visually demonstrate effect of the second constant on the angular dependence of the SMSW frequency in the {111} film. .Curve *1* take into account only the first anisotropy constant while curve *2* account for the first and the second anisotropy constant. The latter was taken as $K_2^c/M_0 = -2$ Oe . Values of other parameters didn't change.

4. Conclusions

The MSW dispersion laws in ferrite films are deduced and analysed. It is shown that the accounting of crystallographic magnetic anisotropy of ferrite leads to emergence in spectra of MSW of the additional dispersive dependences which are absent in isotropic models. The surface MSW spectrum in the tangentially magnetized films was in details investigated. It is established that the greatest anisotropy of spectrum is inherent in the films with {100} and {110} orientations and smallest with {111} type.

The results can be used for developing spin-wave devices. In particular change of characteristics of these devices at radiation and thermal effects will be defined by changes of the magnetic parameters of ferrite of such as saturation magnetization and a magnetic crystallographic anisotropy field. The obtained analytical dependences allow both forecasting of these changes and minimize them promoting development of devices with thermostable and resistant to radiation characteristics.

5. References

[1] Glass H 1988 *Proc. IEEE.* **76** 151

[2] Vashkovsky A V, Stalmakhov V S and Sharayevsky Yu G 1993 *Magnetostatic waves in microwave electronics* (Saratov: Saratov State University Press)

[3] Beregov A S and Kudinov E V 1986 (No 6) *Electron. Tekh. Ser. 1: Electron. SVCh* 41 [in USSR]

[4] Beregov A S and Kudinov E V 1987 (No 6) *Electron. Tekh. Ser. 1: Electron. SVCh* 8 [in USSR]

[5] Chivileva O A, Gurevich A G and Emiryan L M 1987 *Fiz. Tverd. Tela (Leningrad)* **29** 110 [in USSR]

[6] Zavislyak I V, Talalayevsky V M and Chevnyuk L V 1989 *Fiz. Tverd. Tela (Leningrad)* **31** 319 [in USSR]

[7] Shagaev V V 1998 *Fiz. Tverd. Tela (St. Peterburg)* **40** 2089 [in Russia]

[8] Shagaev V V 2004 *Zh. Tekh. Fiz. (St. Peterburg)* **74** 108 [in Russia]

[9] Gurevich A G and Melkov G A 1994 *Magnetic oscillations and waves* (Moscow: Nauka)

Effect of morphology of ZnO nanowire arrays on photovoltaic and electron transport properties of DSSC

B R Ilyassov*, **N Kh Ibraev, D B Abzhanova**
Institute of Molecular Nanophotonics, E.A. Buketov Karaganda State University, 100028 Karagandy, Kazakhstan.

*E-mail: baurdinho@mail.ru

Abstract. ZnO nanowire arrays were synthesized by hydrothermal method for 8, 10, and 12 hours. The array synthesized for 10h had the highest surface area and aspect ratio, and DSSC based on the 10h nanowire array revealed the best photovoltaic performance. The electron transport properties of DSSC based on the ZnO nanowire arrays were investigated by Electrochemical Impedance Spectroscopy (EIS).

Keyword: ZnO nanowires, arrays, DSSC, hydrothermal growth, EIS, electron transport, electron diffusion coefficient

1. Introduction

Dye sensitized solar cells (DSSCs) based on nanostructured metal oxides attract much attention of researchers in the recent years [1-3]. Preparation technologies of DSSCs are much easier and cheaper than one of solar cells based on p-n junction [1-3]. The efficiency of DSSCs has reached a value of more than 10% [4-5], and the efficiency of organic-inorganic perovskite solar cells which are also based on metal oxides has exceeded 15% [6].

One of the key components of the DSSCs is a metal oxide semiconductor photoelectrode which performs a double function: (1) It is a scaffold with a large specific surface needed for the dye loading, and (2) at the same time it is an electron transport material [1]. Morphology and defect density of the metal oxide determine its specific surface and its electron transport properties, respectively, and also influence on electron recombination in DSSCs. Nanostructured TiO_2 currently is being studied as photoelectrode of DSSCs most of all. High efficient DSSCs have been reached on TiO_2.

ZnO nanostructures are alternative material as a photoelectrode of DSSC. The bandgap and the conduction band edge position of ZnO is approximately the same as that of TiO_2. The first advantage of ZnO over TiO_2 is a large variety of nanocrystals with different morphology that have sufficiently large specific surface area. These ZnO nanocrystals can be obtained by solution processable methods [7-10]. The second is the electron mobility in ZnO, at least in single crystals, by an order of magnitude is greater than that of TiO_2 (anatase) [11-14]. However so far, the efficiency of DSSCs based on ZnO nanostructures is lower than one based on TiO_2. The efficiency of DSSCs based on multilayer assembly of ZnO nanowire arrays has reached 7% [15].

Some reports suggest that dye adsorption is the major problem in ZnO DSSCs [12-14]. Cells with high dye loading tend to be inefficient, whereas cells with lower dye loading show good efficiencies. These problems are mainly related to the high acidity of the carboxylic acid binding groups of the dyes that can lead to dissolution of ZnO and precipitation of dye-Zn^{2+} complexes, leading to a poor overall electron injection efficiency of the dye [14].

According to the report [16], the majority of studies regarding ZnO DSSCs have been focused on two main aspects: (1) using the ease of preparing one-dimensional ZnO nanostructures with attractive transport properties and (2) searching for suitable dyes for ZnO. However, in comparison with TiO_2, less detailed studies of electron transport and recombination have been carried out on DSSCs based on ZnO nanostructures.

Vertically aligned ZnO nanowire arrays have recently attracted considerable interest because of their one-dimensional structure and unique properties. In the field of energy conversion, nanowire DSSCs have been fabricated based on the fast electron transport in single crystalline ZnO nanowires. The majority of studies regarding ZnO nanowires for DSSC have been concentrated on preparing nanowires with the high aspect ratio and density on a substrate that gives the high surface area of nanowires. However, there are no researches related to an effect of morphology of ZnO nanowires on electron transport properties of DSSCs.

In this paper, we studied the effect of morphology of ZnO nanowire arrays on the photovoltaic and electron transport properties of DSSC. ZnO nanowire arrays were synthesized by hydrothermal methods. Electron transport properties of DSSC based on the ZnO nanowire arrays were investigated by Electrochemical Impedance Spectroscopy (EIS).

2. Experimental procedure

ZnO seed layers were prepared by the sol-gel spin coating fabrication process [17]. Zinc acetate dehydrate [($Zn(CH_3COO)_2\cdot 2H_2O$, Sigma Aldrich)] was dissolved in the mixed solution of monoethanolamine (C_2H_7NO, Sigma-Aldrich) and isopropyl alcohol. The concentrations of both zinc acetate dehydrate and monoethanolamine in the resulting solution are 0.5 M. The coating solution was spin-coated onto ITO substrates at 3000 rpm for several times. The ITO substrates were subsequently annealed at 400°C in air for 30 min in order to convert zinc acetate to ZnO.

ZnO nanowires were prepared by hydrothermal deposition [7]. ZnO nanowires were grown by placing vertically the ZnO-seeded ITO substrates in solutions with 25 mM $Zn(NO_3)_2$ (Sigma-Aldrich) and 25 mM hexamethylenetetramine ($C_6H_{12}N_4$, Sigma-Aldrich) at 90°C for 10 hours. In order to obtain a constant nanowire array growth rate, the solutions were refreshed during the reaction period (solution turnover time 2 h). Subsequently, the substrates were washed with water/ethanol and annealed at 400°C for 30 min to remove any residual organics.

The ZnO-based photoanodes, each with an active surface area of 0.2 cm^2, were dyed by immersing them in a dry ethanol solution containing 0.5 mM of N719 (Solaronix) at 50 °C for 1 h. An ITO substrate coated with Pt was used as the counter electrode. Pt coat was prepared by electrodeposition methods on ITO that can be found elsewhere [18, 19] The DSSC was fabricated by placing the Pt-coated counter electrode over the dyed ZnO-based photoanode separated by a 25 μm thick thermoplastic sealing spacer (Meltonix 1170-25, Solaronix). The internal space of cell was filled with a high stable liquid electrolyte, Iodolyte Z-150 (Solaronix; Redox couple: iodide/tri-iodide; redox concentration: 150mM; additives: ionic liquid, alkylbenzimidazole, thiocyanate; solvent: 3-methoxypropionitrile).

The surface morphology of ZnO nanowire arrays was characterized on a field-emission scanning electron microscope (FESEM MIRA-3, Tescan). The photocurrent-voltage characteristics were measured using a Source Meter instruments (Keithley 2400) by irradiating with simulated solar light, that is, AM 1.5 100 mW/cm^2 (PET PHOTO Emission TECH., INC). The cell active area was 0.2 cm^2. The impedance of the cells was measured by using an impedance meter (Z-500PRO, Elins). All impedance measurements were performed under a bias light illumination of 100 mW/cm^2 from a simulated solar light (PET PHOTO Emission TECH., INC) at open circuit condition. Impedance measurement of cells was recorded in a frequency range from 0.05 Hz to 200 kHz with an *ac* amplitude of 10 mV.

3. Result and discussion

Figure 1 shows the SEM top view images of ZnO nanowire (NW) arrays synthesized for 8, 10, and 12 hours. Table 1 shows morphology parameters of NW arrays (NW density, diameter, length, and aspect ratio) calculated from the data of SEM. As shown in Figure 1 and table 1, the NW average diameter and length increase with the increasing of reaction time while NW density (number per cm^2 of projected surface area) decreased. On the basis of these calculated parameters, the surface area of the NW arrays located on a substrate area of 1 cm^2 was approximately evaluated (see table 1). At the beginning, the NW surface area increases from 31 to 47 cm^2 when the reaction time is increased from 8 to 10 h, respectively, and further at 12 h it decreased to 41 cm^2. The aspect ratio (diameter/length) also has a maximum value for the 10 h NW array.

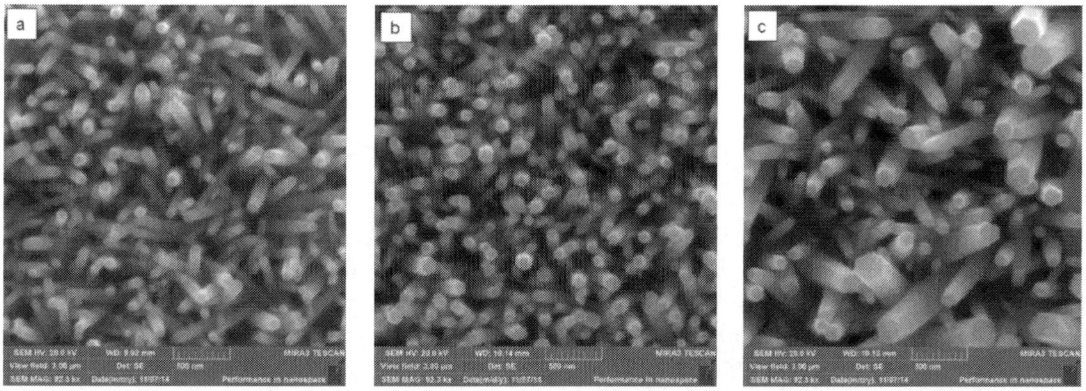

Figure 1. The SEM top view images of ZnO nanowire arrays synthesized for 8 (a), 10 (b), and 12h (c).

Table 1. Morphology parameters of ZnO nanowire array synthesized for 8, 10, and 12 h.

Reaction time, h	Nanowire density (cm^{-2})	Average nanowire diameter (d), nm	Average nanowire length (L), μm	d/L	Surface area of nanowire array, cm^2
8	$503 \cdot 10^7$	92	2.1	$4.38 \cdot 10^{-2}$	31
10	$295 \cdot 10^7$	151	3.3	$4.76 \cdot 10^{-2}$	47
12	$215 \cdot 10^7$	160	3.4	$4.59 \cdot 10^{-2}$	42

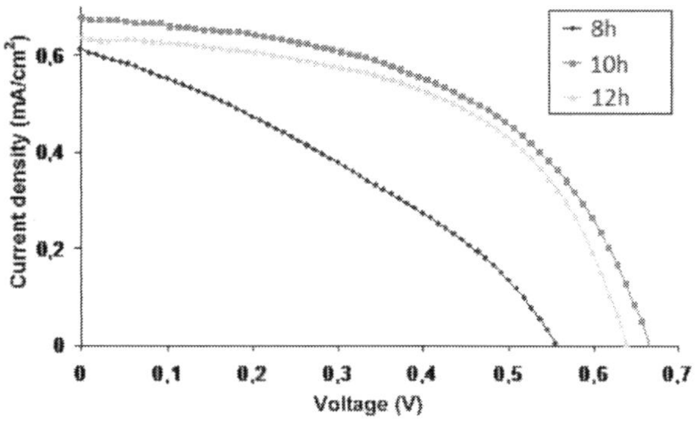

Figure 2. Current-voltage characteristics of DSSCs based on ZnO NW arrays synthesized for 8, 10, and 12 h.

Table 2. Photovoltaic performances and electron transport properties of DSSCs based on ZnO NW arrays synthesized for 8, 10, and 12 h.

	J_{sc}, µA/cm^2	U_{oc}	FF	η, %	R_w, Ohm	R_k, Ohm	R_k/ R_w	k_{eff}, s^{-1}	τ_{eff}, ms	D_{eff}, cm^2/s
8h	610	0.56	0.34	0.12	9.3	132.4	14.2	120	8.31	$7.41\cdot10^{-05}$
10h	680	0.66	0.52	0.25	7.3	85.2	11.7	275	3.64	$35.2\cdot10^{-05}$
12h	629	0.63	0.55	0.22	4.8	64.1	13.4	479	2.09	$74.2\cdot10^{-05}$

The current-voltage characteristic of DSSCs based on ZnO NW arrays synthesized for 8, 10, and 12 h. are shown in figure 2 and their photovoltaic performances are listed in table 2. The cell based on the array grown for 10h has revealed the highest conversion efficiency, short circuit current density, and open circuit photovoltage. This result confirms that the 10h grown array has the highest surface area.

Effect of nanowire morphology parameters on electron transport properties of the DSSCs were investigated using Electrochemical Impedance Spectroscopy (EIS). Figure 3 represents the Nyquist plots of the impedance data of the cells operated in the open circuit under the standard illumination. According to the diffusion-recombination model [20], a suggested equivalent circuit representing the DSSCs based on NW array shown in the figure 4 is employed to fit the impedance spectra for extracting the impedance of the nanowires and the nanowires/electrolyte interface portions. The fitted results are the solid lines shown in the Figure 3. Estimation of the electron transport parameters in the nanowires and electron transfer parameters on the nanowire/electrolyte interface of the cells was conducted from the Nyquist plots of the nanowire and the nanowire/electrolyte interface portions according to the procedure demonstrated by Adachi et al. [21]. The electron transport resistance in NWs ($R_w = r_wL$, as represented in the equivalent circuit and L is the thickness of the anode), charge-transfer resistance related to recombination of an electron ($R_k = r_k/L$), first-order reaction rate constant for electrons being lost (k_{eff}), effective lifetime of an electron in NWs (τ) estimated from the impedance analyses are also listed in Table 1.

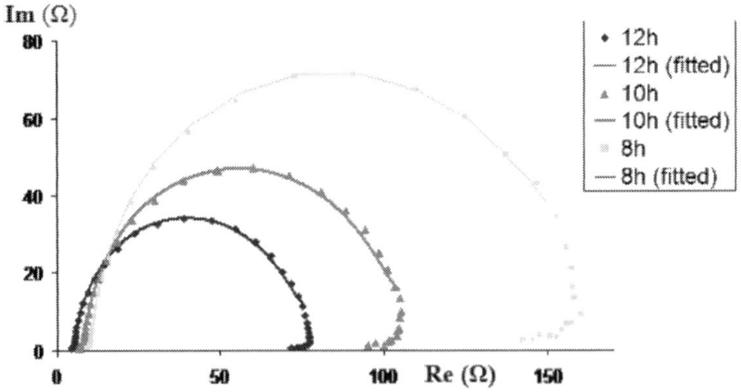

Figure 3. Nyquist plots of the impedance data of DSSCs based on ZnO NW arrays synthesized for 8, 10, and 12 h. The solid lines are the fitting results based on the equivalent circuit model shown in the figure 4.

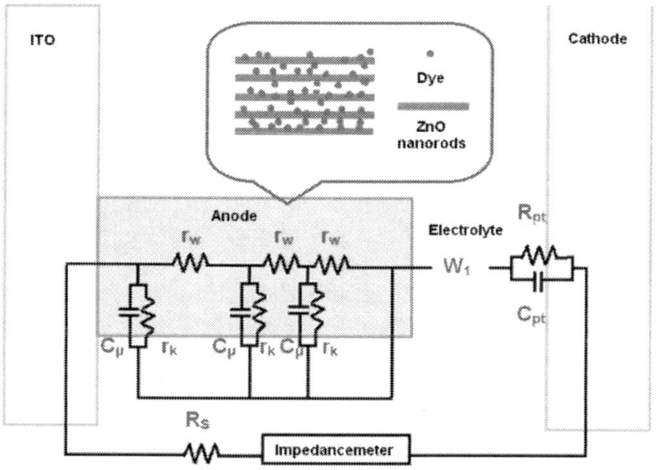

Figure 4. The equivalent circuit model of DSSCs based on ZnO NW arrays, where $C_\mu(=c_\mu L)$ is the chemical capacitance, R_s is a total series resistance for the transport resistance of ITO and all resistances out of the cell, W_1 is the impedance of diffusion of redox species in the electrolyte, and R_{Pt} and C_{Pt} are the charge transfer resistance and the interfacial capacitance at the counter electrode/ electrolyte interface, respectively [22].

As shown in Table 1, when the reaction time of NW synthesis is increased from 8 to 12 h, the electron transport resistance in ZnO NWs decreases; consequently, FF of cells improves. The charge-transfer resistance which related to recombination of electrons at the ZnO/electrolyte interface and the effective electron lifetime also decrease. The latter indicates that with the increasing of the reaction time of NW synthesis, recombination losses of electron intensify. It is can be due to the increasing a defect density on NW surface [23] and in case of 10h NW due to the increasing of the surface area. However, effective diffusion coefficient of electrons (D_{eff}) increases with the increasing of the reaction time of NW synthesis. This phenomenon indicates that bulk defects in NWs are reduced.

5. Conclusion

In summary, ZnO nanowire arrays were synthesized by hydrothermal method for 8, 10, 12 hours. The arrays synthesized for 10h had the highest surface area and aspect ratio, and DSSC based on their revealed the best photovoltaic performance. According to the Impedance Spectroscopy date, with the increasing of reaction time of nanowire arrays from 8 to 12 h, the electron transport resistance, the charge-transport resistance and the effective electron lifetime in ZnO nanowires decreased, whereas the effective electron diffusion coefficient increased.

REFERENCES

[1] O'Regan B, Gräzel M 1991 A low-cost, high-efficiency solar cell based on dye-sensitized colloidal TiO_2 films. *Nature* **353** 737–740.
[2] Kong F-T, Dai S-Y, Wang K-J 2013 Review of Recent Progress in Dye-Sensitized Solar Cells. *Adv. Opto Elect.* **75384** 13.
[3] Hamann T W, Jensen R A, Martinson A B F, Ryswyk H V, Hupp J T 2008 Advancing Beyond Current Generation Dye-Sensitized Solar Cells *Energy Environ. Sci.* **1** 66–78.
[4] Hagfeldt A, Grätzel M 2000 Molecular Photovoltaics *Acc. Chem. Res.* **33** 269–277.
[5] Grätzel M 2001 Photoelectrochemical cells *Nature* **414** 338–44.

[6] Burschka J, Pellet N, Moon S J, Humphry-baker R, Gao P, Nazeeruddin M K, Grätzel M 2013 Sequential deposition as a route to high-performance perovskite-sensitized solar cells *Nature* **499** 316–9.

[7] Baruah S, Dutta J 2009 Hydrothermal growth of ZnO nanostructures *Sci. Technol. Adv. Mater.* **10** 18.

[8] Lamia Z 2010 Sol–gel-deposited ZnO thin films *Materials Science and Engineering B* **174** 18–30.

[9] Skompska M, Zarebska K 2014 Electrodeposition of ZnO Nanowire Arrays on Transparent Conducting Substrates *Electrochimica Acta* **127** 467–488.

[10] Chen H, Zhu L, Liu H, Li W 2013 Effects of preparing conditions on the nanostructures electrodeposited from the $Zn(NO_3)_2$ electrolyte containing KCl *Thin Solid Films* **534** 205–213.

[11] Law M, Greene L E, Yang P 2005 Nanowire dye-sensitized solar cells *Nature Materials* **4** 455–459.

[12] Gao Y, Nagai M, Chang T-C, Shyue J 2007 Solution-Derived ZnO Nanowire Array Film as Photoelectrode in Dye-Sensitized Solar Cells *Crystal Growth and Design* **7** 2467–2471.

[13] Kaidashev E M, Lorenz M, von Wenckstern H, Rahm A, Semmelhack H-C, Han K-H, Benndorf G, Bundesmann C, Hochmuth H, Grundmann M 2003 High electron mobility of epitaxial ZnO thin films on c-plane sapphire grown by multistep pulsed-laser deposition *Applied Physics Letters* **82** 3901–3903.

[14] Hendry E, Koeberg M, O'Regan B, Bonn M 2006 Local field effects on electron transport in nanostructured TiO_2 revealed by terahertz spectroscopy *Nano Letters* **6** 755–759.

[15] Xu C, Wu J, Desai U V, Gao D 2011 Multilayer Assembly of Nanowire Arrays for Dye-Sensitized Solar Cells *J. Am. Chem. Soc.* **133** 8122–8125.

[16] Guill E, Peter L M, Anta J A Electron Transport and Recombination in ZnO-Based Dye-Sensitized Solar Cells *E. J. Phys. Chem. C* **115** 22622–32.

[17] Kamaruddin S A , Chan K, Yow H, Sahdan M Z, Saim H, Knipp D 2011 Zinc oxide films prepared by sol-gel spin coating technique *Applied Physics A* **104** 263–268.

[18] Lee K S, Lee H K, Wang D H, Park N, Lee J Y, Park O, Park J H 2010 Dye-sensitized solar cells with Pt- and TCO-free counter electrodes *Chem. Commun.* **46** 4505–4507.

[19] Gao Y, Chu L, Wu M, Wang L, Guo W, Ma T 2012 Improvement of adhesion of Pt-free counter electrodes for low-cost dye-sensitized solar cells *Journal of Photochemistry and Photobiology A: Chemistry* **245** 66–71.

[20] Bisquert J 2002 Theory of the Impedance of Electron Diffusion and Recombination in a Thin Layer *J. Phys. Chem. B* **106** 325–333.

[21] Adachi M, Sakamoto M, Jiu J, Ogata Y, Isoda S 2006 Determination of Parameters of Electron Transport in Dye-Sensitized Solar Cells Using Electrochemical Impedance Spectroscopy *J. Phys. Chem. B* **110** 13872–80.

[22] Wu J, Chen G, Yang H, Ku C, Lai J-Y 2007 Effects of dye adsorption on the electron transport properties in ZnO-nanowire dye sensitized solar cells *Applied physics letters* **90** 2131091–3.

[23] Kern R, Sastrawan R, Ferber J, Stangl R, Luther J 2002 Modeling and interpretation of electrical impedance spectra of dye solar cells operated under open-circuit conditions *Electrochimica Acta* **47** 4213–4225.

IOP Conf. Series: Materials Science and Engineering **81** (2015) 012047 doi:10.1088/1757-899X/81/1/012047

The peculiarities of implanted deuterium capture in Kh16N15M3T1 austenitic steel with different structure

G A Raspopova[1], **V A Arbuzov**[1]

[1]Institute of Metal Physics, Ural Branch of Russian Academy of Sciences, 18 Kovalevskaya str., Ekaterinburg, 620990, Russia

E-mail: raspopova@imp.uran.ru

Abstract. The results of studying of implanted hydrogen (deuterium) capture in austenitic steel Kh16N15M3T1 are reported. The peculiarities and possibilities of using the nuclear reactions method for measuring deuterium capture are discussed. The traps of implanted atoms and their capture by the elements of steel structure under radiation damage are considered. It is shown that the segregation intensity depends on the initial structure of the samples. A method is proposed for separating the contribution from different types of traps – structure elements – to implanted deuterium capture in irradiated volume. The mechanism of formation and evolution of radiation-induced segregation is discussed.

1. Introduction

As is known, the solubility of hydrogen in steels is very small. This causes embrittlement of articles made thereof that are used in hydrogen-containing atmosphere. If it is possible to block hydrogen in numerous isolated areas of the bulk (traps), the critical conditions of nucleation and development of pores and cracks can be changed. In unirradiated material, such traps are created in the form of precipitates of second phases, and under irradiation – as large density of radiation-induced hydrogen traps. The larger amount of hydrogen can be captured by these traps, the more effective they are. This problem is topical also for materials exposed to neutron irradiation, since radiation-induced nuclear reactions and corrosion processes result in a large concentration of hydrogen (up to 0.4 at. %) in parts made of austenitic steels after their long-term operation in reactors [1].

The behavior of hydrogen in fusion materials is usually examined with the use of hydrogen isotope – deuterium. Materials are saturated with deuterium either from gaseous phase or by implantation, and capture is measured by thermal desorption or by nuclear reactions. In the latter case, in situ analysis can be performed.

In this paper, the results of works [2 - 4] devoted to implanted deuterium capture in austenitic steel Kh16N15M3(T1) in different structural states, as well as the original findings are presented.

2. Materials and methods

2.1. Steel Kh16N15M3T1.

In this work we used specially prepared samples of Kh16N15M3(T1) steels with pre-assigned structure. Titanium-free and titanium-containing steels are designated as ss01 and ss11. Both steels contain (wt. %) 15.9 Cr; 15.0 Ni; 2.53 Mo and 0.03 C. Steel Kh16N15M3T1 contains also 1.02 wt. % Ti.

The samples in different structural states were prepared under the following conditions: hardening (without symbolic notation) – heating to 1373 K, 1 h exposure; deformation (**d**) – rolling of hardened samples at room temperature until 30 % deformation; annealing **(do)** – heating of cold-worked samples to 650 K and exposure at this temperature for 1 h. The sample ss11x was prepared from titanium-containing steel, which was aged upon hardening at 923 K for 20 h (**x**).

The choice of cold deformation and annealing conditions is explained by the following reasons:

1. 30% deformation degree provides for 10^{10}-10^{11} cm^{-2} dislocation density that is realized during standard deformation of austenitic steels.
2. The post-deformation annealing temperature of 650 K is sufficient for complete annealing of deformation vacancies (DV) [5], but it does not lead to considerable variation of the dislocation density.

The examined samples have the following structural peculiarities and possible initial hydrogen traps.

All the samples contain inclusions of metallurgical origin, namely, complex calcium oxides and other deoxidizing agents. Only some fragments of large inclusions in titanium-containing steels turned out to be carbides TiC that did not dissolve during heating for hardening (<<0.007 vol. %). No other precipitates were found in the hardened samples.

Particles of Ni_3Ti intermetallic or γ'-phase. Besides nonmetallic inclusions typical of hardened sample, the aged sample was fount to contain ~2 vol. % small (~6 nm) particles of Ni_3Ti intermetallic. In the strained steel samples, besides nonmetallic inclusions of metallurgical origin, no other particles were found either in dislocations or outside them. However this result obtained by TEM analysis with the use of dark-field images and microdiffraction does not allow one to completely exclude the existence of very small deformation-induced [18] coherent precipitates or pre-precipitates of deformation-induced Ni_3Ti.

Dislocation structure. The TEM analysis showed that the unstrained samples contained only single dislocations that are distributed rather uniformly. Their density is $\sim 1 \cdot 10^{7}$ cm^{-2}. According to data [2], such dislocations are not effective traps of implanted deuterium.

The TEM analysis for strained samples of steel revealed that their dislocation structure has a network-cellular character. The cell size is 0.2...0.4 μm and the average dislocation density in the samples is $\sim 3.4 \times 10^{10}$ cm^{-2}. Deformation twins and stacking faults are observed on the background of the network-cellular substructure in all the samples.

In the process of post-deformation annealing, the dislocation structure changes only slightly. So, the measured average cell size in annealed sample remains within the same limits that upon deformation. The density of dislocations is 3.3×10^{10} cm^{-2}, which is at the level of dislocation density values in unstrained sample. No changes were observed upon annealing in other crystal structure defects, such as twins and stacking faults.

Deformation vacancy clusters (DVC). According to work [5], upon deformation the examined steels contain more than 50 appm DV. Based on the data on the mobility of dislocations in these steels and high stability of vacancy clusters and considering the effect of multiplicity on the stability and size of DVC, the authors of work [5] make a conclusion about the size and concentration of DVC, which are respectively 0.6 nm and more than 10 appm.

2.2. Radiation-induced segregation of hydrogen

Hydrogen atoms are very labile and usually decorate defects. This occurs also in case of implantation of ions into hydrogen-containing target. Since the distribution of radiation defects in ion-irradiated target is not uniform, the distribution of hydrogen related to defects is not uniform either, i.e. radiation-induced segregation of hydrogen (RIS) takes place [6]. RIS arises also during implantation of hydrogen [7 - 10]. In the conditions when defects formed during implantation or their complexes with hydrogen atoms have small mobility, the region of enhanced hydrogen concentration (the region of segregation or RIS) is localized in ion-irradiated volume. This volume is limited to a section of the ion beam, the surface and the length of the projected range of the implanted ion.

It is obvious that the data on the amount of trapped hydrogen and on its distribution in the depth provide information about the amount and nature of radiation-induced defects – hydrogen atom traps. Such information can be received by employing the following schemes.

1. RIS is generated by implanting beam of accelerated deuterium ions (deuterons) and is analyzed by accelerated helium ions by means of nuclear reaction $D(^3He,p)^4He$ [11]. Both beams "work" simultaneously combining the ion irradiation and measuring processes in time.
2. RIS is generated and analyzed by one beam of accelerated deuterons (nuclear reaction D(d,p)T).

Each of these schemes has its own merits and demerits. The obvious advantage of both schemes is that all deuterium registered in the experiment is of implantation origin.

2.3. Nuclear reactions method

The unique prospects of using the reaction D(d,p)T for measuring the concentration of deuterium in the real time of ion irradiation process have been repeatedly considered in the literature [13]. The small number of experimental works where this nuclear reaction is used is connected, in our opinion, with the ambiguous character of the obtained results and with the difficulty of their interpretation. Since with a decrease in the energy of deuterons entering into the nuclear reaction the energy of the formed protons increases, then, with allowance for deceleration of deuterons and protons in the target, the protons coming from different depths get in one spectrometer channel, i.e. there is an ambiguous dependence between the energy of the nuclear reaction products and the depth of analysis. By setting the detection angle of 120 degrees, i.e. considerably increasing the energy losses of protons, we managed to overcome partially this ambiguity. Using the sample with constant deuterium concentration, $TiD_{1.78}$, as a standard and selecting the optimal conditions for its irradiation and measuring, we obtained the information about the concentration of deuterium in irradiated material in the *in situ* mode.

The choice of the deuteron energy value in the reaction D(d,p)T (700 keV), as opposed to numerous works where this energy does not exceed tens of keV, allowed us to expand the analysis region up to ~4 μm (steels, iron, nickel) and thereby to take it way from the sample surface and concomitant impurities [11]. Besides, this approach provided the accuracy of the experiment with respect to introduction of additional component into the target (helium) and excluded the creation of radiation-damaged areas outside the deuterium implantation region.

The choice of irradiation, detection and calculation conditions [6] allowed us to determine from the energy spectrum of protons the average concentration of deuterium in irradiated volume $\boldsymbol{C_D}$ as a function of the implantation dose $\boldsymbol{F}$. In the experiment, $\boldsymbol{C_D}$ was measured during continuous ion irradiation and discrete registration of the energy spectrum of reaction products – protons. All the experiments with the use of reaction D(d,p)T were carried out at room temperature.

As distinct from heavier ions, during implantation of deuterium, its real distribution at room temperature differs essentially from the estimated values. Diffusion redistribution of deuterium takes place incompletely. Some implanted atoms (captured in traps) remain in irradiated volume. This phenomenon is characterized by trapping coefficient α. It is calculated as a fraction of all implanted deuterium remaining in the irradiated region in the moment of measuring, i.e. $\alpha = C_D / C_D^{theor} \times 100\%$, where C_D^{theor} is the average concentration of deuterium calculated on condition that all implanted atoms remain in the irradiated volume. It is evident that calculations with the SRIM software yield α = 1, while for real distribution of implanted deuterium in irradiated material at 300 K $\alpha \ll 1$.

3. Results and discussion

3.1. Capture of deuterium in irradiated volume

The reason why implanted atoms partially remain in irradiated volume at temperatures providing diffusion redistribution of deuterium in the target is not only radiation-induced hydrogen traps in the radiation region (RIT), but also initial traps (IT). RIT are unevenly distributed in the depth of the radiation region (leading to RIS), whereas the distribution of IT in the target is uniform.

The elements of the microstructure of the examined steel identified as IT affect hydrogen capture in RIS both directly and indirectly (they can be point defect sinks). Hydrogen capture in the radiation damage region depends immediately on their total capacitance. If IT do not undergo evolution under irradiation, their capacitance during irradiation is a constant value.

Both single radiation-induced point defects and their clusters can be RIT. In the displacement cascades, vacancy clusters of cascade origin, VCc, can be formed. When radiation-induced structural phase transformations take place in the target under the action of radiation ([13]), the products of these transformations (precipitates of second phases) can also become RIT.

The dose dependence of total deuterium capture consists of dose dependences of capture on RIT and IT, i.e. $\alpha_{tot}(F) = \alpha_{IT}F) + \alpha_{RIT}(F)$. At a relatively small density of RIT, when the fields of interaction of hydrogen traps do not overlap, $\alpha_{RIT}(F) \sim F$. If the capacitance of IT is denoted as $\boldsymbol{Q_{IT}}$, after occupation of IT $\alpha_{IT}(F) = Q_{IT} / C_D^{theor}(F)$. Since $C_D^{theor} \sim F$, then $\alpha_{IT}(F) \sim 1/F$.

Analysis of total capture of implanted deuterium showed that there is a real possibility to separate the contributions to hydrogen capture by IT and RIT. So, at descending dependence $\alpha_o(F)$, the dependence $\alpha_{IT}(F)$ can be determined from its first points. Otherwise in the beginning of measuring the IT can be considered to be already filled, and for estimation of $\alpha_{IT}(F)$ the first point of the dependence $\alpha_o(F)$ should be used.

This approach makes it possible to determine the capacitance of each type of IT and their influence on the formation of RIT. In Figure 1, the results of examination of RIS in titanium-containing austenitic steel Kh16N15M3T1 are presented as an example [12].

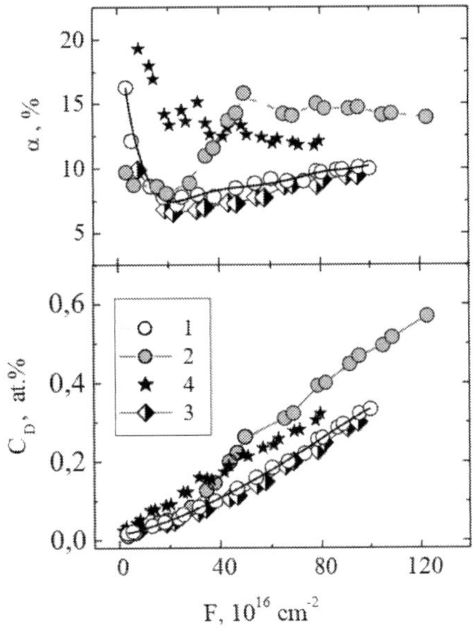

Figure 1. The effect of radiation dose F on CD and α for Kh15N16M3T1 steel samples.

The numerals 1 - 4 denote samples ss11, ss11d, ss11do and ss11x, respectively.

Special preparation of the samples that provided a wide range of microstructures makes it possible to separate the contribution of each type of IT to total capture of implanted deuterium. So, for determination of the contribution of deformation vacancies to total deuterium capture, samples with the same dislocation density – after cold deformation without annealing and annealed sample – have been obtained. Samples with different dislocation density (hardened sample and sample annealed after deformation) were prepared to establish the trapping role of dislocations. In order to determine the contribution of initial Ni_3Ti particles to total capture of implanted deuterium, along with hardened sample, we prepared an aged sample with Ni_3Ti particles. The capacitances of IT of the examined samples derived from $\alpha_{IT}(F)$ measurement results can be written as

$Q^{ss01} = 0;$

$Q^{ss01d} = Q^{ss01}{}_{DV} + Q^{ss01}{}_{disl};$

$Q^{ss01do} = Q^{ss01}{}_{disl};$

$Q^{ss11} = Q_{TiC};$

$Q^{ss11d} = Q_{TiC} + Q^{ss11}{}_{DV} + Q^{ss11}{}_{disl};$

$Q^{ss11do} = Q_{TiC} + Q^{ss11}{}_{disl};$

$Q^{ss11x} = Q_{TiC} + Q_{Ni3Ti},$

where Q^{ss01} is the total capacitance of IT of a sample of hardened titanium-free steel, which consists of capacitances of oxide inclusions and small-density dislocations. It is obvious that the mentioned structural elements are not effective deuterium traps. The DV, disl., TiC and Ni_3Ti indices are related respectively to deformation vacancy clusters, dislocations and titanium carbide TiC and Ni_3Ti particles.

Note that the $C_D(F)$ dependences of all the examined samples (Fig. 1a) are parallel lines when the implantation doses are greater than 50×10^{16} cm^{-2}. Equal inclination of these dependences to the F axis shows that the RIT in all the samples are of one type, whereas different level of C_D values is indicative of the difference in the Q values. The total values of Q can be determined with a high degree of accuracy either from the initial area of dose dependences of total deuterium capture in RIS or from the ratio of Q values from $C_D(F)$ dependences.

The types of the initial traps and their capacitances Q obtained from analysis of $C_D(F)$ and $\alpha(F)$ and the results of TEM studies of the structure of the samples [2, 3] are listed in the Table. The asterisk * in the Table corresponds to Ni_3Ti particles formed during cold plastic deformation.

Table. Capture of implanted deuterium on initial traps in Kh16N15M3T1 steel

Sample	Capacitance of different types IT, appm				
	Q_Σ	Q_{TiC}	Q_{Ni3Ti}	Q_{disl}	Q_{DVC}
hardened	170	170	0	0	0
aged	770	170	600	0	0
strained	450	170	180*	250	30
annealed	420	170	165*	250	0

The fact that carbides are effective hydrogen traps agrees well with the results of work [14].

The total deuterium capture for all the examined samples was decomposed according to the scheme presented in Figure 2 for sample ss11 (hardened steel with titanium). The indices tot, IT, RIT and Ni3Ti correspond to total deuterium capture, to capture on IT, RIT and radiation-induced particles of Ni_3Ti ($\alpha_{tot} = \alpha_{IT} + \alpha_{RIT}$ and $\alpha_{RIT} = \alpha_{VCc} + \alpha_{Ni3Ti}$). The notation VCc corresponds to deuterium capture on vacancy clusters of cascade origin. Since deuterium capture on IT in ss01 sample is absent, for this sample $\alpha_{tot} = \alpha_{RIT} = \alpha_{VCc}$

Comparison of $\alpha_{RIT}(F)$ of Kh16N15M3(T1) steel samples revealed that high-density dislocations with coherent deformation-induced precipitates of Ni_3Ti are effective vacancy sinks. Radiation-induced vacancies, in their turn, are building materials for effective RIT – vacancy clusters. The initial Ni_3Ti precipitates produced by aging of unstrained steel are also effective vacancy sinks and deuterium traps.

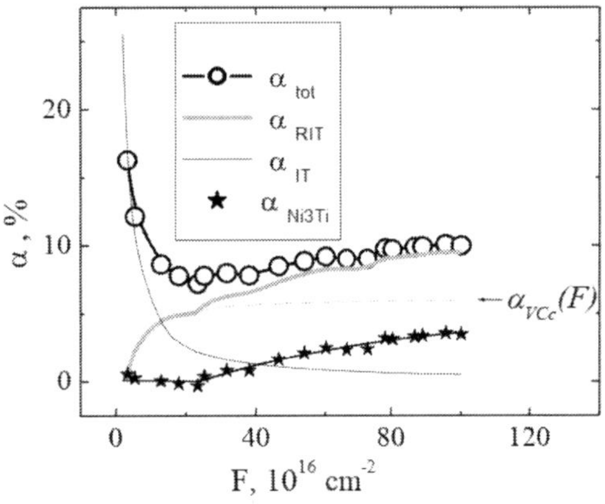

Figure 2. The scheme of decomposition of total deuterium capture αo into trapping capture in hardened sample of steel Kh15N16M3T1

The efficiency of vacancy sinks manifests itself first of all in displacement of deuterium capture onset on VCc towards larger radiation doses. The implantation doses, at which deuterium capture on RIT begins, have been determined. For a hardened sample this dose is ~3×10^{16} cm^{-2}, and for ss11d, ss11do and ss11x samples it is ~18×10^{16} cm^{-2}.

4. Conclusions

1. The scheme employed in this work for studying radiation-induced segregation allowed us to separate the contributions from initial and radiation-induced traps to the total capture of implanted deuterium in irradiated volume. With special preparation of the initial structure of the samples, it has been possible to determine the type and capacitance of initial deuterium traps in them, as well as their effect on the number of traps appearing under irradiation.

2. It was found that cold deformation of steels at room temperature is accompanied by the formation of deuterium traps related to dislocations. Presumably, these traps, effectively capturing implanted deuterium, are precipitates (or rather pre-precipitates) of Ni_3Ti on dislocations.

3. It was shown that the initial traps (carbide particles, Ni_3Ti and clusters of strained vacancies) are effective sinks of vacancies that are mobile under experiment conditions. The quantity of clusters of radiation-induced vacancies decreases, and consequently implanted deuterium capture on radiation-induced traps lowers.

The work was done within RAS Program (project «Spin» № 012001463330, and «Flux» № 01201463334), with partial support of Russian Foundation for Basic Research (grant № 14-03-00359-a) and Ural Branch of RAS (grant № 15-17-2-3)

REFERENCES

[1] Garner F A, Oliver B M, Greenwood L R, et al. 9th Meeting on environmental degradation of materials, Tahoe, Utah, august 2001 54–72
[2] Arbuzov V L, Raspopova G A, Sagaradze V V, et al. 2004 *FMM* **97** No. 6 66–74 [in Russian]
[3] Arbuzov V L, Raspopova G A, Sagaradze V B, et al. 2009 *FMM* **108** No. 1 81–92 [in Russian]
[4] Raspopova G A, Arbuzov V L 2009 *Problems of atomic science and technology* Iss. 2 14-19
[5] Druzhkov A P, Arbuzov V L, Perminov D A 2005 *J. Nucl. Mater.* **341** 153-163
[6] Arbuzov V L, Vykhodetz V B, Raspopova G A 1995 *Fusion Technology* **28** 1127–31

[7] Arbuzov V L, Raspopova G A, Danilov S E, et al. 2000 *J. Nucl. Mater.* **283-287** 849–853
[8] Arbuzov V L, Raspopova G A 1999 *Advances in Science and Technology.* **24** 507–514
[9] Raspopova G A, Arbuzov V L 2009 *Physics of Metals and Metallography* No. 107 58-67
[10] Arbuzov V L, Raspopova G A 1999 *J. Nucl. Mater.* **271-272** 340–344
[11] Neklyudov I M, Tolstolutskaya G D, et al. 2004 *Fizika i khimiya obrabotki materialov (Physics and chemistry of material working)* No. 2 50-55 [in Russian]
[12] Kawachi N, Katabuchi T, Yamaguchi M and Tagishi Y 2002 *Nucl. Instr. and Meth.* **190** Iss. 1-4 195-198.
[13] Borodin O V, Bryk V V, Voyevodin V N, Neklyudov I M 1999 *Metallofiz. noveishiye tekhnol. (Metal physics and high technologies)* **21** No. 6 51–65 [in Russian]
[14] Zouev Yu N, Podgornova I V, Sagaradze V V 2000 *Fusion Engineering and Design* **49-50** 971–976

IOP Conf. Series: Materials Science and Engineering **81** (2015) 012048 doi:10.1088/1757-899X/81/1/012048

Model of cratering by single ions in heavy metals

A I Kalinichenko, S S Perepelkin[1], V E Strel'nitskij
National Science Center "Kharkov Institute of Physics and Technology",
1 Akademicheskaya Str., Kharkov, 61108, Ukraine

E-mail: sperepelkin@kipt.kharkov.ua

Abstract. In the model of the nonlocal thermoelastic peak (NTP) of ion the theoretical investigation of nanocratering in targets of heavy metals at bombardment by ions Xe^+ with energy from 25 eV up to 30 keV is carried out. Simulation using program package SRIM2008 has shown possibility of formation of subsurface singly connected cascade with energy capacity sufficient for arising of the droplet sputtering for Ag targets. Calculations have shown opportunity of cratering on surface of Ag flat targets with yield strength <0.1 GPa at bombardment by Xe^+ ions with energy E > 10 keV. Craters are not formed on Pt flat targets. Possible influence of nanoscale roughness on cratering by single ions is discussed.

1. Introduction

Interest in formation of defects on the surface of structural materials at ion bombardment is related to the development of nanotechnology, demanding for its implementation tools that are able to create and/or modify nanoscale objects. To use fast heavy ions bombarding solid surface as such a tool, it is necessary to investigate related phenomena and, above all, the processes of droplet/cluster sputtering and nanocratering by single ions [1].

The sputtering of metal targets by heavy ions may exhibit as a macroscopic effect, to obey the laws of thermodynamics and continuum mechanics. High temperatures and pressures arising in cascades of excited atoms after thermalization can lead to melting, evaporation and subsequent ejection of target material in the form of vapor, polyatomic clusters or droplets, which provides sputtering factors $S >> 10$ [2]. The surface is subjected to increased erosion with the formation of nanometer-sized craters. [1-3].

The evaporation model [4,5], the shock wave model [6,7], the brittle fracture model [8,9] were offered for description of nonlinear sputtering processes. In last years, except for analytical research methods, methods of molecular dynamics are widely used allowing to trace the spatial - temporal evolution of the atom ensemble near the ion path during the transition from nonequilibrium state to state, obeying thermodynamic laws.

The model of cratering associated with arising of subsurface overheated nanoscale regions containing the molten material flowing away under action of thermoelastic stress was proposed in [10]. However, the above model can not explain the experimentally observed cases of nanoparticle ejection at the ion bombardment of heavy metals (gold) [3]. In addition, the analytical methods used in [1-10], do not have sufficient universality and do not give complete understanding of cratering and cluster sputtering. Also, the use of the molecular dynamics for cratering analysis is associated with

[1] E-mail: sperepelkin@kipt.kharkov.ua

very time-consuming calculation.It is therefore of interest to develop a simple physical model for analysis the process of cratering and for determination main parameters that influence on cratering.

In [11, 12] the model of the droplet sputtering based on the concept of the nonlocal thermoelastic peak (NTP) of the ion has been proposed. In this model, the minimal permissible energy $E_{\min} \sim 10$ eV of the ion is determined by opportunity of thermodynamic description of physical processes in NTP. The maximal permissible energy E_{NTP} of the ion is determined by a qualitative structural change of the cascade, resulting in formation of several weakly connected overheated regions determined by parameters of subcascades forming them. The value E_{NTP} depends on ion species and target material. To determine E_{NTP} the detailed statistical analysis of overheated region was required which has permitted estimating probability of acceptable parameters of the region [13].

In this paper, the cratering on surface of Ag and Pt targets with nanoparticle ejection at bombardment by single Xe^{+} ions with energy from 25 eV to 30 keV is investigated in the NTP model. The range of ion energies causing droplet sputtering and cratering is determined. The influence of target surface roughness on cratering is discussed.

2. Mathematical model

Cratering is related to the transition of the target material in a liquid state in the NTP volume and ejection of molten mass. At $E \le E_{NTP}$ the simulation results obtained by the well-known program package SRIM2008 have shown that cratering of a heavy metal can be related to arising of the ion NTP [11,12]. The resulting peak can be approximated by a spherical segment adjacent to the target surface and containing the phonon loss of ion energy E_{ph}. Initial temperature and phase state of the material in the NTP are determined by the average density of thermal energy:

$$\varepsilon(E,d) = \frac{E_{ph}}{V(E,d)}. \tag{1}$$

Here $V(E,d)$ is the NTP volume, depending on ion energy E and effective size d of target crystallite.

Melt ejection from the NTP is due to elastic unloading of stressed state arising into its volume. Stress in NTP is mainly determined by the sum of the thermal stress caused by thermal expansion of target material in the NTP volume, and some intrinsic stresses having other nature. Provided occurrence of phase transition the thermoelastic stresses arising in the NTP can be written as:

$$\sigma_T(E,d) = \int_0^{\varepsilon(E,d)} \Gamma(\varepsilon)\,d\varepsilon, \tag{2}$$

where $\Gamma(\mathrm{E})$ is the effective Gruneisen parameter connecting $\sigma_T(E,d)$ and $\varepsilon(E,d)$.

The Gruneisen parameter of the target material varies significantly with density of thermal energy in the NTP. At σ_T calculation the following components of heat expansion were taken into account: 1) material expansion at its heating to melting point; 2) expansion at melting; 3) melt expansion at heating up to temperature corresponding to energy density in the NTP. Further calculations were performed assuming that Gruneisen parameters for these processes are constants. This approximation is acceptable for materials under consideration. In σ_T calculation the values of effective Gruneisen parameter of the target material is equal to Γ_s and Γ_l in the solid and liquid state, respectively, and Γ_m in transient state "solid → liquid". Here $\Gamma_m = \zeta K/(q_m \rho)$, where ζ is the specific volume change at target material melting, q_m is the specific heat of melting, K and ρ are the effective bulk modulus and density of target material, respectively. Values ε_{m1}, ε_{m2} and ε_{b1} are the energy densities of the melting beginning, the melting end and the boil beginning, respectively. In this case, the general expression (2) transforms to:

$$\sigma_{\mathrm{T}}(\mathrm{E},\mathrm{d})=\Gamma_{\mathrm{s}}\varepsilon\,\Pi(\varepsilon;0,\varepsilon_{\mathrm{m1}})+\left[\Gamma_{\mathrm{s}}\varepsilon_{\mathrm{m1}}+\Gamma_{\mathrm{m}}(\varepsilon-\varepsilon_{\mathrm{m1}})\right]\Pi(\varepsilon;\varepsilon_{\mathrm{m1}},\varepsilon_{\mathrm{m2}})$$
$$+\left[\Gamma_{\mathrm{s}}\varepsilon_{\mathrm{m1}}+\Gamma_{\mathrm{m}}(\varepsilon_{\mathrm{m2}}-\varepsilon_{\mathrm{m1}})+\Gamma_{\mathrm{l}}(\varepsilon-\varepsilon_{\mathrm{m2}})\right]\Pi(\varepsilon;\varepsilon_{\mathrm{m2}},\varepsilon_{\mathrm{b1}}) \quad (3)$$
$$+\left[\Gamma_{\mathrm{s}}\varepsilon_{\mathrm{m1}}+\Gamma_{\mathrm{m}}(\varepsilon_{\mathrm{m2}}-\varepsilon_{\mathrm{m1}})+\Gamma_{\mathrm{l}}(\varepsilon_{\mathrm{b1}}-\varepsilon_{\mathrm{m2}})\right]\Pi(\varepsilon;\varepsilon_{\mathrm{b1}},\infty),$$

where

$$\Pi(x;a,b)=\begin{cases}0, & x\in(-\infty,a)\cup[b,\infty);\\ 1, & x\in[a,b),\ (b>a).\end{cases}$$

Cratering with ejection of molten material is possible if the elastic energy W_{el} of the melt spent on drop ejection exceeds energy W_s spent on formation of drop and crater surfaces. Therefore, the criterion of droplet sputtering and cratering has the form:

$$W_{el}\geq W_s. \quad (4)$$

Critical parameters of the NTP corresponding to droplet sputtering initiation (the critical volume V_c of the NTP and the critical number $Y_c=V_c\rho/M$ of atoms in the NTP, where M is the atom mass of target material) are determined from equation $W_{el}=W_s$.

The elastic energy W_{el} of the melt spent on droplet ejection was estimated by formula:

$$W_{el}(E,d)=\lambda\frac{\left[\sigma_T(E,d)+\sigma_i\right]^2}{2K}V(E,d). \quad (5)$$

The parameter λ determines the relative fraction of the elastic energy spent on drop formation and its detachment from target surface. Precise λ calculation is possible by solving equations of continuum mechanics, but it lies beyond the scope of this paper. Therefore, λ parameter is the free model parameter, which depends on surface geometry and the NTP position. Let's estimate λ value basing on the following considerations.

Assuming one-dimensional approximation we conclude that only half-wave propagating to surface involves in drop formation and detachment. Therefore, the value $\lambda=0{,}5$ may be taken for fraction of the elastic energy of the NTP spent on drop formation in the case of flat target surface. However, the estimate λ should change in case of a non-planar surface if its radius of curvature is comparable to the NTP radius R_{NTP}. In case of a convex surface, fraction of elastic energy in the wave interacting with the surface increases ($\lambda>0.5$). Analogously, in case of a concave surface, fraction of elastic energy going on drop formation and detachment decreases $\lambda<0.5$. In case of rough surface the droplet sputtering process becomes the probabilistic one since the fraction of the elastic energy spent on drop formation varies depending on surface geometry near ion incidence point.

Analysis showed that molten material ejection in the form of single spherical drop is the least energy-consuming. It allows estimating the minimum energy needed for surface formation of drop and crater, using the formula:

$$W_s(E,d)=\sqrt[3]{36\pi}\,\beta\delta\left(V(E,d)\right)^{2/3}, \quad (6)$$

where δ is the surface tension of the target material in liquid state, β is the growth factor which is equal to the ratio of additionally formed square to the square of the spherical drop. In the case of the flat surface the value of β lays in range from 1 (drop ejection from liquid or plastic material without cratering, ($\mu << K$ where μ is the average shear modulus of target material) up to ~ 2 (drop ejection from a rigid material with formation of the cavity replicating the NTP shape). For arbitrary combination of "ion-target" the value of β is determined by model stated in [14]. According to this model the crater is represents as an axially-symmetric hollow which surface is formed by rotation along vertical axis of two conjugate arcs with radii $R_1 = \delta / \sigma_Y$ and $R_2 = 2R_1 + (R_{NTP} - 2R_1) \cdot \Theta(R_{NTP} - 2R_1)$. Here σ_Y is the yield strength of the target material, $\Theta(x)$ is the unit Heaviside function. Factor β, the shape, diameter and the depth of the crater are determined on the assumption that the crater forms in the complex process of drop ejection by the action of two opposing forces - surface tension and elasticity/stiffness of the target material. Details of determining the value of β for flat target surface are given in [14]. In case of a rough surface the value β varies depending on the surface curvature at the point of ion incidence.

3. Results and discussion

The simulation performed using the program package SRIM2008, has allowed to determine the shape and geometric parameters of NTPs generated in the heavy metal targets (Ag, Pt) by ions Xe^+ with energy $E \le E_{NTP}$ bombarding target surface, as well as the energy phonon loss E_{ph} of ions [12]. Value E_{NTP} increases with increase of the crystallite size d. The maximum admissible energies E_{NTP} for polycrystalline, nanocrystalline (d =2 nm) and amorphous target of Ag (Pt) have the following values: 20 (25) keV, 16 (21) keV and 13 (17) keV, respectively. The obtained values of NTP parameters have allowed to estimate thermal energy density in the resulting NTP, to check fulfillment of the criterion of droplet sputtering and cratering (4), to determine minimal E_1 and maximal E_2 values of ion energy ensuring droplet sputtering, to calculate the yield of droplet sputtering and crater dimensions.

Figure 1 shows the functions of number of atoms in the NTP $Y(E,d)$ and the critical number of atoms $Y_c(E)$ satisfying starting condition $W_{el} = W_s$ of droplet ejection from NTP of Xe^+ ion for targets of crystalline Ag (figure 1(a)) and amorphous Pt (figure 1(b)). Calculation of functions $Y_c(E)$ for bulk Pt and Ag targets was performed at $\lambda = 0.5$ and β values calculated by expressions from [14].

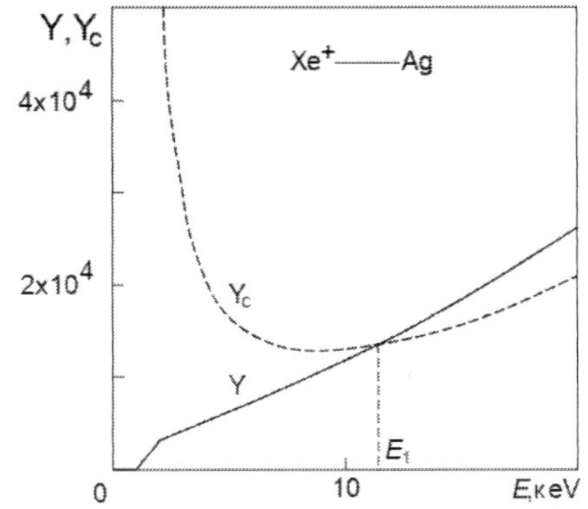

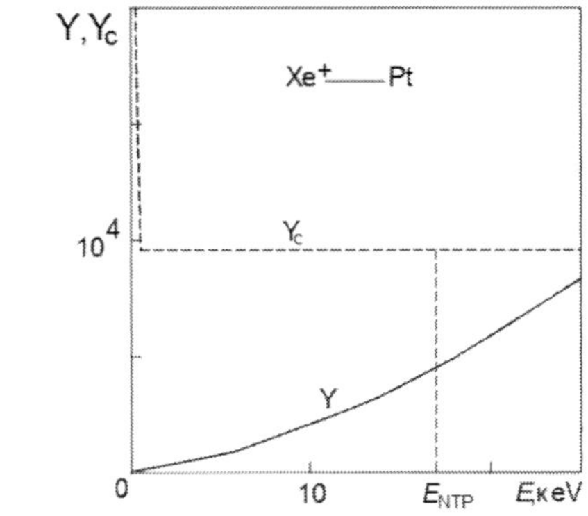

Figure 1. The number Y of atoms in the NTP of Xe^+ ions and critical number Y_c of atoms, corresponding to condition (4) in silver polycrystalline target (a), and platinum amorphous target (b) with flat surfaces ($\lambda = 0.5$). The threshold energies for Ag target are equal to: $E_1 \simeq 11.6$ keV; $E_2 = E_{NTP} = 20$ keV. For platinum amorphous target the criterion (4) is not satisfied at $E < E_{NTP}$.

As one can see from figure 1, for polycrystalline ($\beta = 1.016$) target of Ag the droplet sputtering is possible at $E \geq E_1 \approx 11.6\,\text{keV}$. The largest value of sputtering yield is determined by the value $Y(E_2, d)$ where $E_2 \sim E_{NTP} \approx 20$ keV. For Pt targets with flat surface the droplet sputtering criterion (4) is not carried out even in the most favorable case of amorphous ($\beta = 1.012$) target. However, the last conclusion may changes at ion bombardment of a rough surface. Thus, at ion incidence on convex section of a rough surface, the curvature radius of which is comparable to R_{NTP}, the fraction of elastic energy spent on drop detachment increases ($\lambda > 0.5$) in comparison with a case of a flat surface, and droplet sputtering may be possible [12]. At the same time, the fraction of elastic energy spent on drop detachment decreases at ion bombardment of a concave section of surface, and droplet sputtering is impossible. One should note, that such selective action of ion bombardment on convex sections of a rough surface should lead to surface smoothing, i.e. it appears as original «ionic polishing».

Figure 2 shows dependences of energy density ε in Xe^+ ion NTP on ion energy for amorphous (the curve 1), nanocrystalline (the curve 2, $d = 2$ nm) and polycrystalline (the curve 3) silver targets with flat surface, $\lambda = 0.5$ (see figure 2(a)), and platinum targets with a rough surface, $\lambda = 0.7$ (see figure 2(b)).

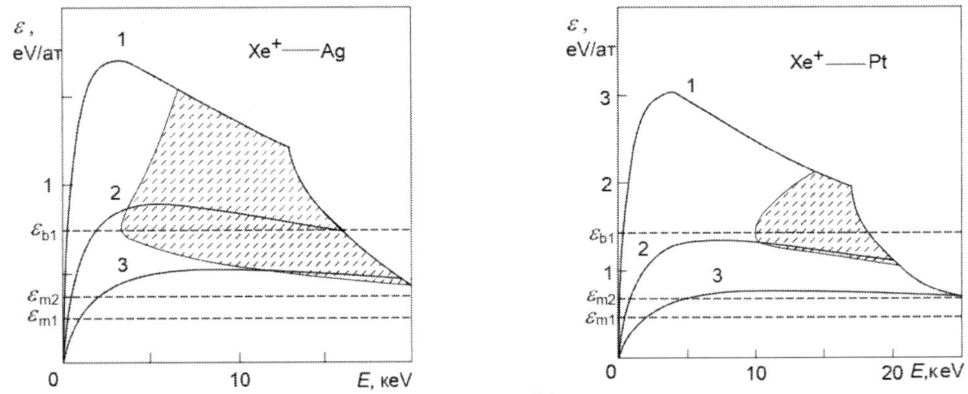

(a) (b)

Figure 2. Thermal energy density ε in Xe^+ ion NTP depending on ion energy in amorphous (curve 1), nanocrystalline d = 2 nm (curve 2), polycrystalline (curve 3) targets: (a) silver target with flat surface (λ = 0.5); (b) platinum target with rough surface (λ = 0.7). Dashed lines correspond to the energy density of melting start point (ε_{m1}),melting end point (ε_{m2}), and boiling start point (ε_{b1}). Droplet sputtering and cratering are possible within hatched areas.

As one can see from figure 2, thermal energy density in the NTP $\varepsilon(E)$ decreases with d increasing, its maximum shifts to higher energies. For all Ag targets there is ion energy range, in which $\varepsilon(E)$ exceeds the value ε_{m2} to be necessary for the material full melting in NTP. Criterion of droplet sputtering (4) is met in hatched areas. The use of the equations derived in [14] enables to determine the size and the shape of forming craters which considerably depend on surface tension coefficient δ and yield strength σ_Y of target material. As a result of calculation, we obtain the following values of diameter D and depth H of crater. For Ag polycrystalline target (λ = 0.5, δ = 0.93 J/m^2, σ_Y = 0.02 GPa) we have D = 36 nm, H = 1.2 nm. For Pt nanocrystalline (d = 2 nm) target (λ = 0.7, δ = 1.74 J/m^2, σ_Y = 0.07 GPa) we obtain D = 24 nm, H = 1 nm.

Analogous calculations [11, 12] have shown opportunity of droplet sputtering and cratering at bombardment of Au and U targets by single Xe^+, Au^+, U^+ ions of low energies.

The results can be used in nanotechnology for smoothing of heavy metal surfaces by heavy low-energy ions.

Conclusions

1. A simple model of cratering as a result of heavy metal droplet sputtering at low-energy ion bombardment is suggested. The criterion of droplet sputtering and cratering is obtained with a glance of surface irregularities.

2. It is shown that the criterion of droplet sputtering and cratering by Xe^+ ions is met for amorphous, nanocrystalline and polycrystalline Ag targets with yield strength σ_Y <0.1 GPa , with flat surfaces or in the presence of nanoscale irregularities.

3. It is shown that the effect of the droplet sputtering is absent for platinum targets with flat surfaces. The cratering is possible in amorphous and nanocrystalline platinum targets with nanoscale irregularities.

References

[1] Donnelly S E and Birtcher R C 1997 *Phys. Rev.*B **56** 13599
[2] Bouneau S, Brunelle A, Della-Negra S, Depauw J, Jacquet D, Le Beyec Y, Pautrat M, Fallavier M, Poizat J C, and Andersen H H 2002 *Phys. Rev.*B **65** 144106
[3] Birtcher R C, Donnelly S E and Schlutig S 2000 *Phys. Rev. Let.* **85** 4968
[4] Kelly R 1977 *Rad. Effects* **32** 91
[5] Johnson R E, Evatt R 1980 *Rad. Effects* **52** 187
[6] Vorobyova I V, Geguzin J E, Monastyrenko V E 1984 *Physics of the Solid State* **26** 1964 (in Russian)
[7] Bitensky I S, Parilis E S 1987 Nucl. Instr. Meth. Phys. Res. **B21** 26.
[8] Davydov A A, Kalinichenko A I 1985 *Problems of atomic science and technology, series: "Physics of radiation damages and radiation material science"* **3/36** 27 (in Russian)
[9] Kalinichenko A I, Strel'nitskij V E 2010 Nucl. Instr. Meth. Phys. Res. **B268** 2246
[10] Averback R C and Mai Ghaly 1994 J. Appl. Phys. **76** 3908
[11] Kalinichenko A I, Perepelkin S S, Strel'nitskij V E 2011 *Russian Physies Journal* **54** 324 (in Russian)
[12] Kalinichenko A I, Perepelkin S S, Strel'nitskij V E 2013 *Journal of Kharkiv National University. Physical series "Nuclei, Particles,Fields"* **3/59/** 57 (in Russian)
[13] Kalinichenko A I, Perepelkin S S, Strel'nitskij V E 2012 *Proc.VIII Int.Scien.Conf."Radiation-thermal effects and Processes in Inorganic Materials" (Tomsk)* p 516 (in Russian)
[14] Kalinichenko A I, Perepelkin S S, Strel'nitskij V E 2013 *Problems of atomic science and technology, series: "Physics of radiation damages and radiation material science"* **2/84** 128 (in Russian)

Calculation of intrinsic stresses in the diamond-like coatings produced by plasma ion deposition in modes of DC and pulse bias potentials

A A Kalinichenko, S S Perepelkin[1], V E Strel'nitskij
National Science Center "Kharkov Institute of Physics and Technology",
1 Akademicheskaya Str., Kharkov, 61108, Ukraine

E-mail: sperepelkin@kipt.kharkov.ua

Abstract. The formula derivation for calculation of intrinsic stress in diamond-like coatings deposited from the ion flux in modes of continuous and pulsed potentials in view of process of defects formation is given. The criterion of applicability of obtained formula allowing to determine critical parameters of the pulsed potential mode is suggested. Results of calculation of stresses in diamond-like coatings at deposition of low-energy ions C^+ from filtered vacuum arc plasma are adduced. The influence of the bias potential, repetition frequency and pulse duration, on the value of intrinsic stress is discussed. Qualitative agreement of calculated stress and experimental data is stated. The important role of deposition temperature in control of intrinsic stress in deposited coating is noted.

1. Introduction

Intrinsic stresses σ arising in diamond-like (ta-C) coatings at deposition of C^+ ions substantially determine the physical properties and operational characteristics of the obtained coatings. In [1] formula for σ calculation using the point thermal spike (PTS) model of the ion has been suggested. The PTS model was used for determination of the number of thermally activated transitions specifying the relaxation rate of intrinsic stress.

However, the use of the PTS model for description of stress relaxation contradicts to nonlocality of energy transition from ion to target matter. Consequently, the qualitative agreement with experimental data is achieved at values of migration energy $u = 3 - 11$ eV, exceeding in many times known values for processes of defect migration. Therefore, u is the fitting parameter in the Davis formula [1], and its physical interpretation is difficult. Also, one should note the internal contradiction of PTS model which was developed only for the case of zero target temperature and for constant thermal capacity at the same time. The last condition contradicts to the Debye theory as well as to experimental data. Hence, Davis model can not explain experimentally observed dependence of intrinsic stress on deposition temperature T_0.

In [2] modification of Davis formula for calculation of intrinsic stresses σ was proposed, taking into account the deposition temperature T_0:

$$\sigma(E,u,T_0) \sim \frac{E_Y}{1-\Pi} \cdot \frac{\sqrt{E}}{\xi + w(E,u,T_0)}, \tag{1}$$

[1] E-mail: sperepelkin@kipt.kharkov.ua

where E is the ion energy, E_Y and Π are the Young's modulus and the Poisson's ratio of a target material, ξ is the ratio of fluxes of deposited atoms R and ion j. The model of nonlocal thermoelastic peak (NTP) was used for calculation of the number of thermoactivated transitions $w(E,u,T_0)$:

$$w\left(E,u,T_0\right)=n_0\nu\int_0^{\tau_c}V\left(t,E\right)e^{-\frac{u}{k_BT(t,E,T_0)}}dt, \quad (2)$$

where n_0 is the concentration of atoms, ν is the frequency of atomic oscillations, $V(t,E)$ is the NTP volume, $\tau_c \sim R_{NTP}{}^2/(4\kappa)$ is the life time of NTP with radius $R_{NTP}(t, E)$, κ is the thermal diffusivity of target material, T is the temperature in NTP, and k_B is the Boltzmann constant [3].

The modified formula (2) permitted to explain some regularities observing at deposition of carbon and BN coatings in mode of direct current (DC) bias potential [2]. However expressions (1), (2) cannot explain experimental results of paper [4] concerning deposition of diamond-like coatings in the DC bias potential mode. Furthermore the implementation of new deposition technologies, in particular, the pulse bias potential mode requires generalization of theory of intrinsic stress formation in deposited coatings.

The purpose of the work is derivation of the formula for calculation of intrinsic stress in coatings deposited in DC and pulse bias potential mode and its application for calculation σ in ta-C coatings at deposition from C^+ ion fluxes.

2. Mathematical model

Deriving formula for intrinsic stress we followed the reasonings used by C.A. Davis [1]. Intrinsic stresses in films and coatings deposited from ion fluxes arise as the result of two opposite processes: 1) subsurface ion implantation, leading to formation of defects and origin of compressive stress, and 2) decrease of defect number due to their thermal migration in ion NTPs that result in stress relaxation.

In the pulse potential mode the ions of two different energies $E_2 = E_0 + ie(U + U_0)$ and $E_1 = E_0 + ieU_0$ are alternately deposited. Here E_0 is the initial ion energy, U is the potential supplied to the substrate, U_0 is the floating potential, e is the proton charge, and i is the ion charge in e units. If the duration and the repetition frequency of high-energy pulses are equal to t_p and f, respectively, then the duration of bombardment of target by low-energy ions is equal to $f^{-1} - t_p$. In this case the intrinsic compressive stress which is established in the coating depends on the type of the formed coating. If the pulse width is sufficiently large, then material deposited during t_p can be regarded as a solid layer of macroscopic thickness. The intrinsic stress in each layer should be evaluated using the expression (1), in which E is the energy of the ions forming the deposited layer. In this case, the coating can be regarded as multi-layer sandwich, and the equilibrium stress can be calculated taking into account the stress and the thickness of each layer [5].

However, if the effective thickness of each layer does not exceed the interatomic distance a, then it is senseless to talk about the stress formation in each of these 'layers', because the size of the ion NTP considerably exceeds the layer thickness [2]. In this case, we can assume that the coating is formed by a mixture of ions of two different energies (the approximation of mixed beam), and relative fractions of ions with energies E_2 and E_1 are ft_p and $1 - ft_p$, accordingly. The condition of realization of the last case has the form:

$$f = Ra^2. \quad (3)$$

Taking for estimations R= 5 mA/cm^2 ~ $3\cdot10^{20}$ ions·m^{-2}s^{-1}, a =$2.5\cdot10^{-10}$ m, we get from (3): $f > 20$ Hz. In known experiments on coating deposition in the pulse potential mode this requirement is surely satisfied.

The considered model, as well as the Davis model [1] is based on the hypothesis of a linear relation between volumetric deformation of solid films bombarded by ions, and density of defects formed as a result of ion scattering on the target atoms. In [6] based on the linear theory of cascade collisions it is shown that a generation rate of defects on unit area $\dot{n}_i$ is related to the flux density of bombarding ions j and ion energy E by relationship $\dot{n}_i \sim jE^{1/2}$. One should note that the above dependence arising from model assumptions is only analytical approximation presenting satisfactorily process of a defect formation only at enough high energy of the bombarding ion $E >> E_d$, where E_d is the energy of defect formation. However direct calculation of the part of ion energy spent on formation of vacancies $\upsilon(E)$ can result in the change of the dependence $\dot{n}_i(E)$. We assume that each vacancy corresponds to an interstitial, composing the Frenkel pair with it. According to definition of function $\upsilon(E)$, number of point defects (Frenkel pairs) forming by the primary ion is equal to $E\upsilon(E)/E_d$. It is necessary to consider also the contribution Υ to volume strain of a primary ion which is constant for all ions with energy E, exceeding surface energy $E_s \sim 10$ eV. As the result, expression for generation rate of defects can be presented as:

$$\dot{n}_i \sim j\left[\frac{E\,\upsilon(E)}{E_d} + \Upsilon\right], \quad (E > E_s). \tag{4}$$

If ions of two various energies E_1 and E_2 are presented in flux, then (4) can be written in the form:

$$\dot{n}_i \sim j\left[ft_p\zeta(E_2) + (1 - ft_p)\zeta(E_1)\right], \tag{5}$$

where we introduce the notation $\zeta(E_i) = \left(E_i\upsilon(E_i)/E_d + \Upsilon\right)$.

According to expression (2) stress relaxation is determined by the number of thermally activated transitions of atoms in the ion NTP. The relaxation rate $\dot{n}_R$ per unit area is proportional to the number of thermally activated transitions of atoms in the ion NTP $w(E,u,T_0)$, to relative fraction of defects n/n_0, and to the flux of implanted ions j:

$$\dot{n}_R = \frac{n}{n_0} jw(E,u,T_0), \tag{6}$$

where n and n_0 are the defect and atom concentrations, accordingly.

If the ion flux consists of ions of two different energies they produce NTPs of two different types, making various contributions in forming processes of thermally activated transitions. Then the expression for $\dot{n}_R$ takes the form:

$$\dot{n}_R = \frac{n}{n_0} j\left[ft_p w(E_2,u,T_0) + (1 - ft_p)w(E_1,u,T_0)\right]. \tag{7}$$

At plasma-ion deposition intrinsic stress in the coating can be calculated from the assumption that there is equilibrium between processes of defects generation by ion implantation and their loss due to migration, therefore the density of the implanted atoms does not depend on time.

The rate per unit area with which the defects are implanted into the film, is equal to $R(n/n_0)$ where R is the total rate per unit area of attachment of atoms to growing film. On the other hand, the resultant rate of defects formation is given by the difference between the rate of defect generation due to ion implantation and the rate of their loss due to thermally activated migration. Consequently, the condition of stationarity (i.e., constancy of defect density) leads to the relation [1]:

$$\dot{n}_i - \dot{n}_R = R(n/n_0). \tag{8}$$

Substituting $\dot{n}_i$ and $\dot{n}_R$ in (8), we obtain the part of the implanted ions in a film n/n_0 :

$$\frac{n}{n_0} \sim \frac{ft_p\zeta(E_2)+(1-ft_p)\zeta(E_1)}{\xi + ft_p w(E_2)+(1-ft_p)w(E_1)}. \quad (9)$$

For thin coating the compressive stress acting in coating plane is related to volumetric strain υ by expression $\sigma \sim \sigma_Y \upsilon$, where $\sigma_Y = E_Y/(1-\Pi)$. Considering that volumetric strain is proportional to the relative fraction of the implanted atoms in film n/n_0, we obtain:

$$\sigma(U) = B\frac{E_Y}{1-\Pi}\frac{ft_p\zeta(E_2)+(1-ft_p)\zeta(E_1)}{\xi + ft_p w(E_2)+(1-ft_p)w(E_1)}, \quad (10)$$

where B is the normalizing constant depending, generally, on a material and technology of coating. Assuming $ft_p = 1$ in (10) we obtain expression (1) for intrinsic stress in the case of DC bias potential mode.

3. Results and discussion

Expression (10) allows to determine the value of intrinsic compressive stress in ta-C coating deposited from C^+ ion flux in modes of DC and pulse bias potentials at energies of ions E <1 keV.

In the figure 1 dependences of internal compressive stresses in deposited ta-C coating on potential U in modes of DC (the curve 1) and pulse (curves 2, 3) bias potential are presented. Calculation of stresses was carried out by formula (10) at following values of parameters: U_0 = 16 V, E_0 = 20 eV, u = 0.3 eV, ξ =1; curve 2: t_p= 5 μs , f= 24 kHz; curve 3: t_p= 20 μs , f= 1.2 kHz. Calculation of function $\zeta(E_i)$ and $w(E_i)$ for combination C^+ - ta-C was carried out using program SRIM2000 [7]. In calculations E_d = 35 eV and E_s = 10 eV were accepted.

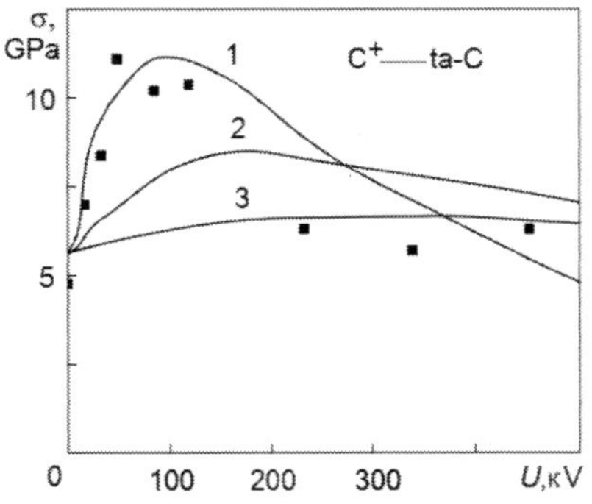

Figure 1. Dependences of intrinsic compressive stresses in deposited films ta-C on DC bias potential (the curve 1) and pulse bias potential (curves 2, 3), calculated at deposition temperature T_0= 370 K. ■ - the experimental data obtained in DC bias potential mode from work [4].

At calculation of intrinsic stress in deposited coating it is necessary to consider deposition temperature T_0 which can essentially change with energy of deposited ions because the ion flux heats up coating surface. One can show that in approximation of linear heat equation with a constant heat conductivity and in steady conditions deposition temperature T_0 is associated with potential U by a linear relation: $T_0(U) = \lambda[q(U+U_0)+E_0] + T_{00}$, where T_{00} is the temperature of unirradiated substrate. The charge q of the ion C^+, according to experimental data [8], with adequate accuracy is equal to proton charge.

The value λ depends on technical parameters of the installation for coating deposition. It was chosen from requirement of equality of deposition temperature to its experimental value at known energy of deposited ions. In accordance with experimental data given in [4] we accepted $T_0(0\ V) \approx 300$ K and $T_0(450\ V) \approx 380$ K. That gives an estimation of $\lambda \approx 0.1$ K/eV.

Dependence of intrinsic stress in deposited ta-C film, obtained at deposition temperature $T_0 = 293$ K in mode of DC bias potential (curve 1) qualitatively agrees with experimental data from [4]. Curve 1 is normalized on maximum of experimentally observed stress ~ 11.1 GPa that determines the scale of all other stress curves. In DC bias potential mode the maximum of calculated compressive stress σ_m ~ 11.1 GPa is situated at U ~ 100 V (the curve 1). The maximum experimental value σ in ta-C coatings also was observed at U = 100 V [4]. Also, calculation has shown that intrinsic stress sharply decreases at increase of deposition temperature.

In mode of pulse bias potential for ft_p =0.12 (0.024) the maximun of intrinsic stress σ_m ~ 8.5(6.6) GPa is placed at U ~ 170 (350) V (the curve 2(3) in fig. 1) at the same deposition temperature, i.e. the maximum shifts to higher energies. One should be noted that at U <270 (370) V (the curve 2 (3) in figure 1) the deposition in mode of pulse potential results in smaller stresses than deposition in mode of DC bias potential. The obtained results qualitatively agree with the experimental data from [9,10].

Noted features remain at other deposition temperatures.

CONCLUSIONS

1. In model of the nonlocal thermoelastic peak with a glance of defect formation the expression for the internal stress in the coating deposited in the plasma environment with simultaneous bombardment of ion flux is derived. The expression is valid in regimes both DC and pulsed bias potential.

2. Comparison of the calculation results of internal stresses in ta-C coating deposited from C^+ ion beam in DC bias potential with the experimental data demonstrates their qualitative agreement.

3. It is shown that in the mode of pulse bias potential the deposition of ta-C coatings for low-energy ions (E < 500 eV) leads to a lower value of compressive stress than the deposition in DC bias potential at the same temperature. The maximum internal stress in the coating increases with ft_p parameter. The bias voltage corresponding to stress maximum decreases with ft_p parameter. Pulse potential mode allows to obtain ta-C coatings with low internal stress at appropriate choice of the deposition temperature.

References

[1] Davis C A 1993 *Thin Solid Films* **226** 30
[2] Kalinichenko A I, Perepelkin S S, Strel'nitskij V E 2010 *Diam. & Rel. Materials* **19** 996
[3] Kalinichenko A I, Perepelkin S S, Strel'nitskij V E 2006 *Diam. & Rel. Materials* **15** 365
[4] Fallon P J, Veerasamy V S, Davis C A, Robertson J, Amaratunga O A J and Milne W I, Koskinen J 1993 *Phys. Rev.*B **48** 4777
[5] Kalinichenko A I, Perepelkin S S, Strel'nitskij V E 2012 *Journal of Kharkiv National University Physical series "Nuclei, Particles,Fields"* **4/56/** 102 (in Russian)
[6] Windischmann H 1987 J. Appl. Phys. **62** 1800
[7] Ziegler F, Biersack J P, Littmark U 1996 *The Stopping and Range of Ions in Solids* (New York: Pergamon Press)
[8] Aksenov I I, Andreev A A, Belous V A, Strel'nitskij V E, Khoroshikh V M 2012 *Vacuum Arc: Plasma Sources, Coating Deposition, Surface Modification* (Kyiv: Naukova Dumka) (in Russian)
[9] Zhang Y B, Lau S.P, Sheeja D,Tay B.K 2005 *Surface & Coatings Technology* **195** 338
[10] Bilek M M M, McKenzie D R, Moeller W 2004 *Surface & Coatings Technology* **186** 21

IOP Conf. Series: Materials Science and Engineering **81** (2015) 012050 doi:10.1088/1757-899X/81/1/012050

The study of the structural and superconducting properties of YBaCuO films produced by laser ablation method

Ye Khuanbay[1], K M Makhanov[2], D B Abzhanova[2], G B Turebaeva[1], N S Smakova[1]
[1] *Karaganda, 100027, Kazakhstan*, Karaganda State Technical University,
[2] *Karaganda, 100028, Kazakhstan* E.A. Buketov Karaganda State University

E-mail: esen_d@mail.ru

Abstract. Microstructure and oxygen stoichiometry of YBaCuO films manufactured by laser ablation are investigated. Dependence of the specific electrical resistivity and superconducting parameters of films on microstructure and oxygen stoichiometry is demonstrated. Conditions of deposition are determined, under which films with minimum microstructure defects and good conducting properties are formed. It was found that the concentration of oxygen atoms in the deposited film on the $SrTiO_3$surface changes non-linear with the substrate temperature increasing. Also the stoichiometry of the films was calculated and their chemical composition was determined.

Introduction

A perspective material of the present-day cryoenergetic is high-temperature superconductor (HTSC) $YBa_2Cu_3O_{7-x}$ (YBaCuO) in form of films on a variety of dielectric substrates [1]. So the investigation of the relationship between microstructure and superconducting characteristics of YBaCuO films has the practical interest. It is quite clear that the main superconducting properties of the films (critical temperature T_c, width of superconducting transition ΔT_c, and density of critical current J_c) depend on how ideal the microstructure of the resulting films is. This fact is mainly determined by the pinning force on the microstructure imperfections. Planar defects of YBaCuO films microstructure favor the formation of grain interlayers with distortion of the element stoichiometry, including primarily that of oxygen. On the other hand, critical temperature T_c and width ΔT_c of the superconducting transition in YBa2Cu3O7–x are functions of the oxygen stoichiometry (7–x) [2,3]. Therefore, an important issue in the formation of superconducting films is the solution of a problem of possible variation of composition, microstructure and superconducting properties of a multiphase composite material.

Methods and Results

In this paper the structural properties and superconducting characteristics of YBaCuO films produced by laser ablation on a $SrTiO_3$ (100) single crystal substrate are studied.

For the films deposition a universal laser facility was used. It consists of two lasers, a universal vacuum station, a system of laser emission control, and a system of substrate temperature control T_s control. The films were deposited by evaporation of the targets by two crossed YAG-Nd^{3+} lasers at the wavelength of λ= 1060 nm, pulse duration 10 ns, and pulse repetition frequency 10 Hz. The radiation pulse energy was 0.09 J. Two hot-pressed specimens of HTSC $YBa_2Cu_3O_7$ ceramics were used as the targets. The spacing between the substrate and the target was 50 mm. The deposition time was 4 min.

The working chamber pressure was varied from 10^{-5} to 10^{-2} mm Hg. The substrate temperature was varied within 1053–1123 K.

Crystallographic characteristics of the films and substrates were measured at a DRON-6 diffractometer using CuK_{α}-radiation. Structural quality of the films was characterized by the angles of misorientation of the *a*, *b*, and *c* crystallographic axes. The X-ray diffraction analysis revealed that the diffraction patterns from the YBaCuO films on the surface of $SrTiO_3$ contained series of reflections (001) up to high orders, evidencing of the crystal lattice orientation along the *c*-axis and of the film epitaxial character. The *a*- and *b*- axes of crystal lattices of the YBaCuO films were found to be parallel to the substrate plane. The specific electrical resistivity $\rho_{ab}(T)$ in the *a* and *b* planes, and the critical superconducting parameters of the films were measured using a four-wire method.

The superconducting parameters were determined from the analysis of the dependence of the normalized specific electrical resistivity $\rho_{ab}(T)/\rho_{ab}(300)$ in the plane *ab* of the films formed at different substrate temperatures and the working chamber pressure 10^{-3} mm Hg. The experimental dependences $\rho_{ab}(T)/\rho_{ab}(300)$ for the films formed at different substrate temperatures are shown at Figure 1. The curve shapes are practically the same. It is evident that apart from a sharp change in the specific electrical resistivity of these films there is a weak temperature dependence $\rho_{ab}(T)$.

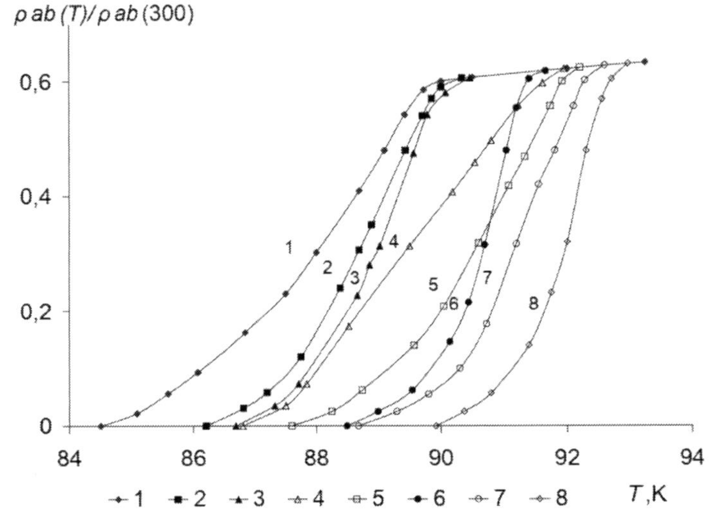

1-T_s = 1053; 2 – 1063; 3 – 1073; 4 – 1123; 5 – 1113; 6 – 1083;7 – 1103; 8 – 1093 K

Figure 1. Temperature dependences of the normalized specific electrical resistivity of the films formed at different substrate temperatures T_s.

In the general case, these curves could be described by a well-known formula that was used in the analysis of the temperature dependence of the specific electrical resistivity of superconducting cuprates in the *ab* plane [4]:

$$\rho_{ab} = \frac{\rho_c}{AT\cosh^2\frac{T_0}{T}}, \tag{1}$$

where ρ_c – specific electrical resistivity along the *c*-axis.

The values of superconducting parameters and $\rho_{ab}(300)$ of the films obtained from the curves shown in Figure 1 and $\rho_c(300)$ calculated by the formula (1) are listed in Table 1. When calculating A parameter, the experimental results on oxygen stoichiometry (7–x) were used. The critical temperature of the superconducting transition T_c is determined by the level of $0.5\rho_{ab}(T)$, and the width of the superconducting transition ΔT_c – by the difference of the levels $0.9\rho_{ab}(T)$ and $0.1\rho_{ab}(T)$.

Table 1. Values of superconducting parameters of YbaCuO films.

T_S, K	$\rho_{ab}(300)$, mΩ·cm	$\rho_c(300)$, mΩ·cm	T_C, K	ΔT_C, K
1053	0,259	7,771	88,0	4,0
1063	0,300	9,000	88,7	2,6
1073	0,157	4,553	89,0	2,0
1083	0,190	5,130	90,7	1,7
1093	0,230	4,830	92,0	1,8
1103	0,260	6,240	91,2	2,3
1113	0,280	7,560	90,6	3,0
1123	0,330	9,240	89,5	3,5

It follows from Table 1 that for all films there is high-temperature superconductivity at $T_c \geq 88$ K. The films with T_s=1093 and 1083 K demonstrate a significantly sharp transition into superconducting state, while the samples with Ts=1053, 1113 and 1123 K show a comparatively gradual transition into superconducting state. The lower substrate temperature bound, where the YBaCuO film transits into superconducting state at the temperature 88 K, corresponds to the substrate temperature about 1053 K. The upper bound, at which formation of the YBaCuO film is still possible, corresponds to the substrate temperature about 1123 K. For YBaCuO films in the substrate temperature interval 1083–1103 K, the superconducting parameters slightly vary. It should be noted that there is a narrow substrate temperature interval at which optimal superconducting parameters are achieved for YBaCuO films. The different character of variation of YBaCuO films properties is worth mentioning, when the films are formed at substrate temperatures decreasing and increasing with respect to their optimal values.

The calculation of oxygen stoichiometry (7–x) of the investigated films was based on microanalysis of their elemental composition using a special PHI-RHO-Z code via a mathematical processing of the characteristic of X-ray energy dispersion spectra. The microstructural analysis and the measurement of energy dispersive spectra from the YbaCuO films were carried out on a JEOL JSM-5910 scanning electron microscope (SEM). Figure 2 shows the characteristic of X-ray energy dispersive spectra from elemental components of the films with T_c= 89 K and ΔT_c= 2.0 K. The peaks corresponding to all elemental components of the films are evident in these spectra. The probe electrons energy is 20 keV, the probe current is 0.1 nA, the duration of measurements is 50 s. The obtained peaks are: OK_α, YL_α, YL1, BaL_α, BaL1, BaL_β, CuL_α, CuL1, CuK_α, and CuK_β. According to the dipole selection rules [5], the peaks OK_α, CuK_α result from an electron transition $2p_{3/2}\rightarrow 1s_{1/2}$ and correspond mainly to the valent p-electrons of oxygen and copper atoms; YL_α, BaL_α, CuL_α – $3d_{5/2}\rightarrow 2p_{3/2}$ result from small contributions of s-electrons into the electron intensity and bear information on d-electrons of the atoms of yttrium, barium, and copper; BaL_β– $3d_{3/2}\rightarrow 2p_{1/2}$; CuK_β– $3p_{3/2}\rightarrow 1s^{1/2}$; YL1, BaL1, CuL1 – $3s_{1/2}\rightarrow 2p_{3/2}$ represent the 3s-electrons of the atoms of yttrium, barium, and copper, which do not participate in the formation of chemical bonding.

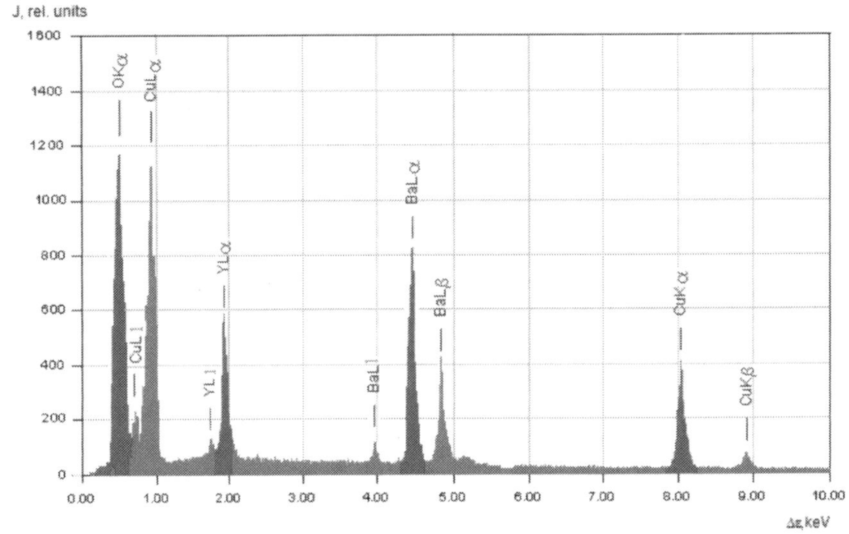

Figure 2. Characteristic X-ray energy dispersive spectra from an YBaCuO film at T_c= 89 K.

A microscopic analysis of the elemental composition was performed using OK_α, YL_α, BaL_α, and CuL_α analytical signals, since the intensity of these signals (J, rel. units) is higher than that of the remaining peaks. The data presented in Figure 2 were used to determine the X-ray quantum energy: for YL_α $\Delta\varepsilon$= 1.92 keV, for BaL_α $\Delta\varepsilon$= 4.46 keV, for CuL_α– $\Delta\varepsilon$= 0.930 keV, for OK_α– $\Delta\varepsilon$= 0.525keV.

Based on the performed calculations using the data of the energy dispersive spectra, it was found out that within the selected substrate temperature interval the oxygen stoichiometry in YbaCuO films lies within the range 6.8–6.94. At substrate temperature about T_s= 1093 K there is a maximum concentration of oxygen in the film [6].

The established dependences of the critical temperature and the width of the superconducting transition on the oxygen stoichiometry 7–x based on the Figure 1 and the Table 1 are shown in Figure 3.

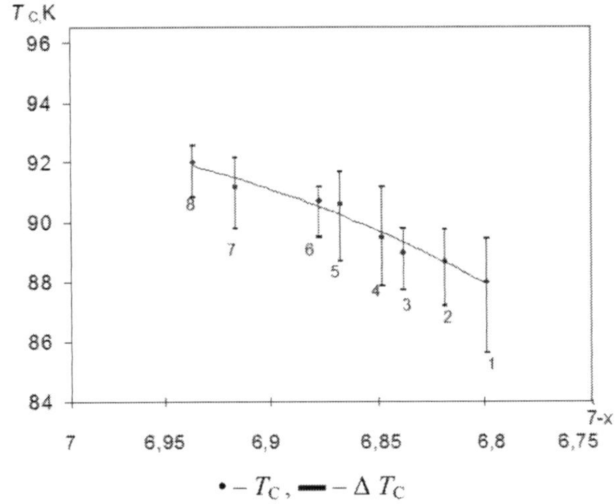

Figure 3. Dependences of superconducting parameters of the films on oxygen stoichiometry

At the oxygen stoichiometry 6.94, the film has a maximum T_c, while at 6.8 – T_c is minimal. The microstructure of the YBaCuO film with T_c= 88 K and ΔT_c= 4 K formed at T_s= 1053 K possesses a pronounced granular structure with a predominance of spherical grains (Figure 4, a). Neither regular patterns in grain arrangement nor predominant direction in their orientation are observed. For the films with T_c= 92 K and ΔT_c= 1.8 K formed at T_s= 1093 K a higher microstructure density, smoothness and uniformity are observed compared to those films with T_c= 88 K. There are spherical regularly shaped microdrops on the surface (Figure 4, b). The works [7, 8] showed that high values of the superconducting transition critical temperature (T_c= 92 K) and optimal values of ΔT_c result from the uniformity and dense low-defect microstructure of the YBaCuO films.

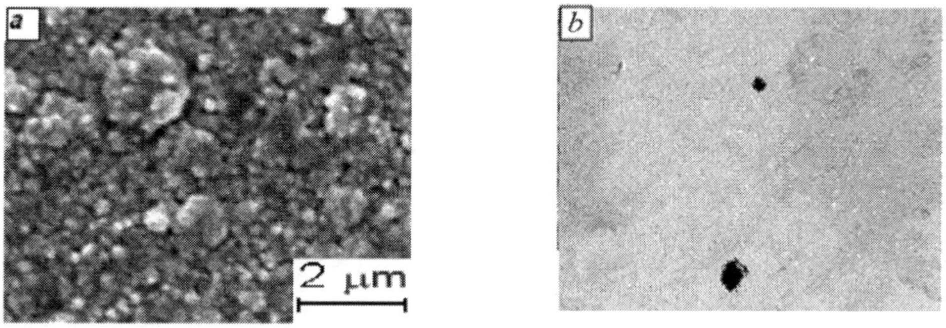

a – T_C=88 K, ΔT_C= 4 K b – T_C=92 K , ΔT_C = 1,8 K

Figure 4. Microstructure of the films with different superconducting parameters

The observed dependence of the film structure on the deposition temperature could be interpreted in the following way. Structural formation of YBaCuO films occurs directly in the process of deposition. As the material is supplied onto the substrate, there is the formation and growth of crystal nuclei with maximum growth rate anisotropy. Due to migration of the adsorbed particles, there is a physical contact between the granules. If the interaction between the granules is stronger than that between the granules and the substrate, the crystallites are aggregated into big clusters. This process would be enhanced with the temperature increase.

References:

[1] Cukauskas E J, Alen L H 1999 Critical current characteristics of composite thin films of Au and $YBa_2Cu_3O_7$ *Physica C* vol. **313** pp. 11–20.

[2] Khoryushin A V, Mozhaev P B, Mozhaeva J E, Grivel J C, Hansen J O B, and Jacobsen C.S 2013 Yttrium-enriched $YBa_2Cu_3O_x$ thin films for coated conductors fabricated by pulsed laser deposition *Physica C: Superconductivity* vol. **485** pp. 39–46.

[3] Golman E K, Ilyin R N, Plotkin D A, Razumov SV, Sakharov V I, Serenkov I T 2000 Growth of ultrathin $YBa_2Cu_3O_{7-x}$ films on $SrTiO_3$ substrate *Physics of the Solid State* vol. **42** pp. 397–400.

[4] Abrikosov A A 1998 Resonant tunneling in high-T_c superconductors *Uspekhi Fizicheskikh nauk* vol.**168** pp. 683–695.

[5] Frish S E 2010 *Optical spectra of atoms* 2 Edition (St-Petersburg: Lan).

[6] Khuanbai E, Ibraev N Kh, Kargin D B 2004 Microstructure and superconducting characteristics of YBaCuO thin films obtained by laser ablation *The KazNU Journal* vol. **2** pp. 99–106.

[7] Khuanbai E, Ibraev N Kh 2011 Influence of structure on the electrical properties of superconducting YBaCuO films *Russian Physics Journal* vol. **54** pp. 294–295.

[8] Khuanbai E, Ibraev N Kh, Kargin D B 2005 Microstructure of $YBa_2Cu_3O_{7-x}$ *Russian Physics Journal* vol.**48** pp. 663–664.

IOP Conf. Series: Materials Science and Engineering **81** (2015) 012051 doi:10.1088/1757-899X/81/1/012051

Effect of electric pulse processing on physical and chemical properties of inorganic materials

S E Sakipova[1], B R Nussupbekov[1], D Ospanova[1], A Khassenov[1], Sh E Sakipova[2]

[1] E.A. Buketov Karaganda State University,

28, Universitetskaya Str., Karaganda, 100028, Kazakhstan

[2] Kazakh National Agrarian University,

8, Abay avenue, Almaty, 050010, Kazakhstan

E-mail: sesaule@mail.ru

Abstract.

This article analyzes various aspects of the practical application of electric pulse technology of industrial raw materials processing as a result of a spark electric discharge in a liquid solution of the raw material under processing. The object of the study are samples of technogenic materials from a deposit in Central Kazakhstan, which are crushed and ground to particles with a preset degree of fragmentation. The electric pulse processing is performed by using different numbers of discharges. The effect of electric pulse processing with different electrical parameters is carried out on the basis of comparison of the properties and structure of metal-containing and industrial raw materials after machining and electric pulse processing. The X-ray spectral microanalysis was performed using a scanning microscope. The researchers obtained data on changes in the microstructure and elemental composition of inorganic material samples as a result of electric pulse processing. It was established that the technology of electric pulse crushing and grinding of inorganic materials makes it possible to obtain not only a final product with desired size of dispersed particles, but also to change their physical and chemical properties.

1. Introduction

Kazakhstan has large deposits of ferrous and non-ferrous metals and is a major producer of iron and steel products. Increasing demand for metal-containing raw materials, on the one hand, and the requirements strengthening for environmental protection, on the other hand, required both improvement of ores and concentrates parameters and higher degree of utilization of secondary energy resources. Achieving high comprehensive utilization of raw materials is not possible without the development and implementation of methods and equipment to ensure the optimization of technological processes with high reliability [1-4].

The role of the enrichment processes increases in connection with processing of low-grade metal ores and accumulation of final tailings of mining and metallurgical production. The degree of metal content in the processed raw materials determines the index numbers of a metallurgical enterprise: productivity; labour costs; consumption of fuel, power, and auxiliary materials; loss of recoverable metals and cost of final products, etc. The higher the metal content in the processed raw material, the more economical is its metallurgical processing net result. As a rule, ore enrichment is carried out by machining when only the ratio of the quantitative values between valuable minerals and waste rock changes in the feedstock and the final product.

In recent years, more and more studies are carried out on the development of technologies for inorganic materials enrichment aimed at increasing the degree of extraction of valuable components. For example, the development of technology for the extraction of gold and non-ferrous metals from technogenic structures composed of mature final tailings using gravitational, flotation, hydrometallurgical methods of enrichment [5, 6].

The works of the Institute of Geological Sciences named after K.I. Satpaev are well known as well. One of them is deals with the development of the device with mechanical and hydrodynamic high-frequency pulsating rotary aerator generating fluctuations within a narrow range of 2,4-3,4 kHz that makes possible almost 1.5-fold increase in the extraction of lead in lead concentrate.

In hydrometallurgy leaching procedure, electrolysis, chemical precipitation, adsorption, ion exchange, evaporation and flotation are used to extract metals from aqueous solutions and for the selective extraction of non-ferrous, rare and precious metals from ores, sludges, settlings or solutions. These methods are widely used in the process of waste recovery and wastewater treatment as well [4]. These data confirm the importance of the development and implementation of technologies for processing of inorganic materials not only to increase the output of precious and non-ferrous metals without the cost of developing new fields, but also to restore favorable environment in the areas of accumulation of process plants waste. This article discusses various aspects of use of a high-voltage spark discharge in a liquid medium.

2. Statement of the problem

Modern technologies of generating materials with desired properties a high-voltage electric discharge in a liquid medium is used as a powerful source of mechanical energy. The essence and distinctive feature of the electric pulse technology is that processing of ore and technogenic raw materials using strong energy pressure makes it possible to grind quickly and clear raw material from impurities and get enriched product with desired degree of fineness. That product can then be used directly in the production processes of metalworking. The electric pulse processing technology based on the unique effect of instantaneous energy release at the moment of electric discharge has huge unknown potentialities and unexpected effects [7]. Release of enormous energy in nearly pinpoint space of incompressible liquid medium is accompanied by a sharp rise of temperature and pressure and causes various physical and chemical processes, phase transformations and thermal kinetic phenomena. As a result, the structure and properties of the processed inorganic materials change.

The electrical discharge parameters and properties of liquids are the factors that determine the speed and amount of kinetic and thermal energy released in the area of electrical discharge. In this case, due to the intensive evaporation of the liquid in the area of electrical discharge and the steam expansion in the electric arc gap, a liquid compressional wave arises. The liquid compressional wave can be caused both by a single powerful electric discharge impulse and a series of successive impulses. The electric pulse fragmentation is an efficient method for grinding of various materials in order to obtain a product with a desired degree of fineness of a specific granulometric composition and has a high selectivity for fragmentation [8]. Possibility to control the energy level in the discharge channel and the duration of the energy release, i.e. changing the electrical pulse repetition rate in combination with electrode systems of various design make it possible to change the processing mode depending on the properties of the raw material. It is possible to change the granulometric composition of the product by varying the parameters of a single pulse or series of them during electric pulse disintegration of the material. [9] shows the possibility to affect the size of the resulting particulate product for mineral raw materials. The authors chose for the latter the copper-sandstone ore from Central Kazakhstan.

3. Experimental results and discussion

To ensure identical conditions, samples of the same raw material were ground by machining and electric pulse processing to a diameter of fragmented particles (0.1-0.3)* 10-3m. Within the framework of this project a vibrating cone crushing mill VCCM-6 (St. Petersburg) was used for mechanical crushing and grinding of samples of raw materials.

The operational blocks, functional principle and electro technical parameters of the electric pulse plant are described in detail in [8-10]. The electric pulse processing was carried out by changing the inter electrode distance within the range $l_d = 3 \div 14 \cdot 10^{-3}\ m$ while adjusting the values of supplied pulse voltage $U = 8 \cdot 10^3 V \div 40 \cdot 10^3 V$. The capacitance value of the capacitor bank varied within $C = (0{,}25; 0{,}5; 0{,}75) \cdot 10^{-6}\ F$. In order to ensure, the processing of each sample was carried out under the same conditions at least 5-7 times. In the course of the experiments the optimal capacitance value equal to $C = 0{,}5 \cdot 10^{-6}\ F$ was determined that provided maximum yield of the final product, crushed to desired sizes. For samples of raw material from the Nurkazgan mine the yield of the final particulate product was 37%.

Analysis of changes in the microstructure and properties of the start material composition after the electric pulse processing compared to machining was carried out using a scanning electron microscope Philips SEM 515 on the basis of the Scientific Center of Tomsk State University. The studies were performed in accordance with the standard procedure STR TSU N041-2009 "Procedure of research on the structure of a solid surface by scanning electron microscopy." Data on the elemental composition were obtained on the basis of the measurements of energy spectra, the mass concentration W,% and the atomic concentration A,%.

Figures 1 and 2 show photographs of the microstructure at a magnification by101 times and energy spectra of the samples of industrial raw materials from Nurkazgan mine after treatment in a different ways.

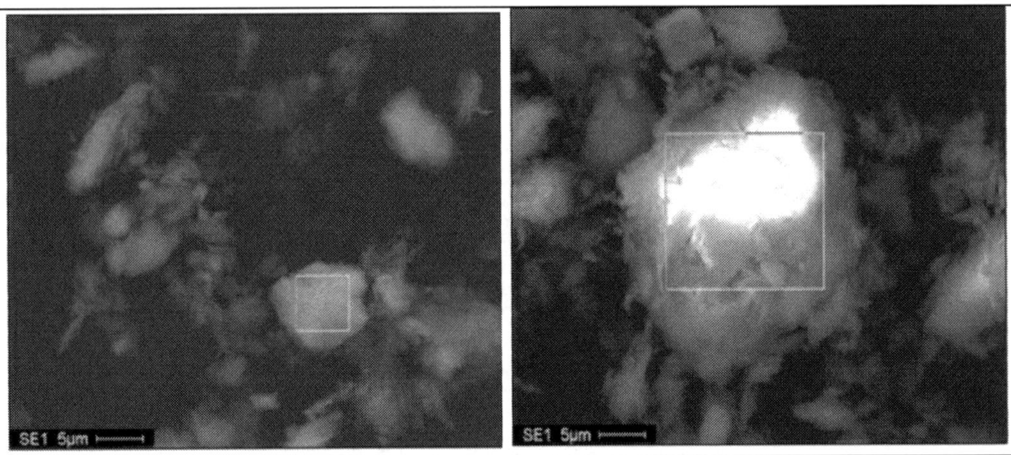

a) b)

Figure 1. Photographs of the microstructure of technogenic raw materials from Nurkazgan mine after treatment: a) machining; b) electric pulse processing

The results of X-ray spectral microanalysis were obtained by using a rather known sufficiently accurate ZAF method, which includes three types of corrections in the data retrieval [11]. These corrections are represented in the title and indicate: Zi - correction for difference in an average atomic number between the sample and reference standard, it is due to reflection and deceleration of electrons; Ai - correction for the absorption of X-rays in the sample, it is determined by the energy of the probe, emergent X-ray angle and mass absorption coefficient for the element of interest in the sample; Fi - correction for fluorescence due to excitation by the secondary X-ray emission of the i element by the radiation of other elements and by deceleration radiation.

In the experiments the experimenters used X-ray spectral microanalysis based on the same principles as the "conventional" emission X-ray spectral analysis. The X-ray spectral microanalysis differs in that the excitation of the primary radiation is in a relatively small volume of the sample by very narrow electron probe. Therefore, the basis for the X-ray microanalysis is an electron-optical system of a scanning electron microscope. Table 1 presents data on changes in elemental composition.

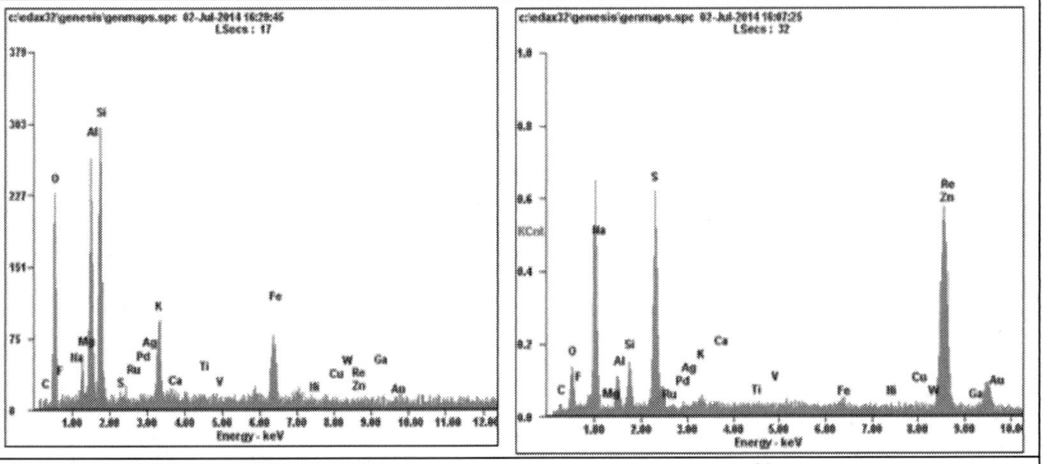

a) b)

Figure 2. The energy spectra of technogenic materials from Nurkazgan mine after treatment: a) machining; b) electric pulse processing

Table 1. The elemental composition and its change for samples from Nurkazgan mine

Element	W_m , %	W_{ep} , %	K= W_{ep}/W_m
F	0.08	0.80	10.00
Mg	0.08	2.12	26.50
Al	1.11	14.34	12.92
Si	1.14	16.29	14.29
S	9.65	0.45	0.05
K	0.40	5.24	13.10
Ti	0.04	0.59	14.75
Fe	0.36	10.58	29.39
Ni	0.15	0.52	3.47
Cu	0.20	0.84	4.20
W	0.43	2.30	5.35
Re	2.50	57.61	23.04
Ga	0.35	0.89	2.54
Au	2.43	3.41	1.40

The coefficient K is defined as the ratio of the mass concentration value of the element after the electric pulse processing W_{ep}% to the mass concentration values of the element after machining W_m%. For these samples of the industrial raw materials from Nurkazgan mine it was found that as a result of electric pulse processing the percentage of magnesium increased by 26.50 times, the same of aluminum grew by 12.92 times, of iron – by 29.39 times, of copper – by 4.2 times.

More clearly the results are shown in the diagram of changes in the elemental composition of samples of raw material from Nurkazgan mine, Figure 3, which confirms the enrichment of the original product and possibility to extract some valuable components more efficiently.

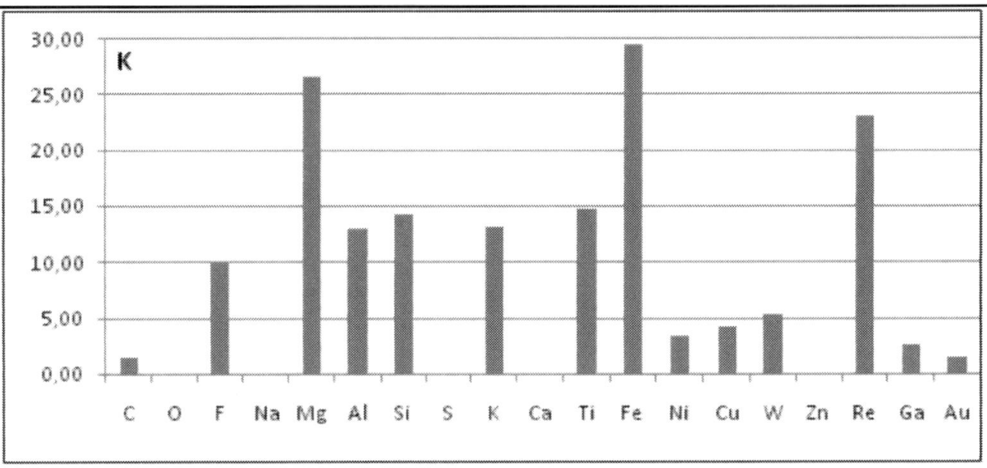

Figure 3. Change in the elemental composition of raw materials Nurkazgan mine.

The analysis confirms various changes both in the structure and elemental composition of the processed samples after the mechanical grinding and after electric pulse processing. Similar results were obtained for samples of industrial raw materials and "poor" ore from Annensk and Kushmurun deposits in Kazakhstan [9, 10].

These data confirm that the developed electric pulse technology for disintegration and grinding of metal-containing raw materials makes it possible to obtain a final product with desired properties such as parameters of granulometric composition and the degree of fineness, and to increase the percentage of particular elements.

4. Conclusion

This paper discusses the results of the use of electric pulse technology for enrichment of inorganic materials using the example of technogenic raw materials from Nurkazgan mine in Central Kazakhstan. The authors studied the effect of the electric pulse processing on the properties of inorganic materials on the ground of research on their structure and elemental composition after treatment.

The experimenters obtained data on the concentration rise of particular elements by several times. The experimental results provide an opportunity to draw a conclusion that the electric pulse processing makes it possible not only to crush and grind metal-containing and technogenic raw materials effectively, but also to enrich it, thus increasing the possibility to extract valuable components, including rare metals.

The results of these experiments provide an opportunity to elaborate the foundations for a new, knowledge-based electric pulse technology for processing of inorganic materials and mineral structures to implement them further at the processing industry enterprises.

The work was performed as part of the project №508-Ph 14 within the program No 055 of the Kazakhstan Republic Ministry of Education and Sciences

References:

[1] Batrakova L H, Egeubaev B S, Edenbaev SS 2003 *Complex processing of mineral raw materials of Kazakhstan (state, problems, solutions)* Almaty 390
[2] Usov AF 2001 Prospects technologies of destruction of rocks and ores. *Proceedings of the Academy of Sciences, Energy* vol. **1** pp. 54 – 62
[3] Mamaev YA 1996 Scientific and methodological and technological basis for the rational development of gold-mining: Author. dis. Doctor. techn. Science MGGA Moscow 40
[4] Kramarenko SA 2005 Processing of mature tailings of CMF "Ilyich MMP " *IMS NAS of Ukraine*, Kiev pp. 130-132
[5] Kazakova Y, Novikov O N 2006 Patent 2006104513/02 (004898) RU. The method of extraction of metals.
[6] Ivasishin OM, Oshkaderov SP, Spivakovsky VB, Makovskaya GV et al. 2003 Patent of Ukraine 59062. Pulsation reactor.
[7] Yutkin LA, Goltsova LI 1983 Method of crushing hard materials **Bi**. 8
[8] Nussupbekov BR, Kussaiynov KA, Sakipova SE, Khassenov AK 2014 On the issue of improving the technology of complex extraction of rare and trace metals by electro-method. *Metal Physics and Advanced Technologies*. vol. **2 (36)** pp. 275-286
[9] Kussaynov K, Sakipova SE, Nussupbekov BR, Aitpaeva ZK 2013 *Eurasian Physical Technical Journal* vol. **1(19)** pp.52-55
[10] Kussaiynov K., Sakipova SE, Nussupbekov BR, Khassenov AK, Akhmerova KE 2014 Electro-pulse processing metal containing raw materials for extraction of valuable components. Int. Scientific Colloquium "*Modeling for Electromagnetic Processing*"- MEP 2014, Hannover pp.289-295.
[11] Physical bases of X-ray spectral microanalysis. http://lab.bmstu.ru/rem/index

IOP Conf. Series: Materials Science and Engineering **81** (2015) 012052 doi:10.1088/1757-899X/81/1/012052

The calculation method of interaction between metal atoms under influence of the radiation

S N Yanin[1]
[1]*Tomsk, 634050, Russia,* Tomsk Polytechnic University
E-mail: yanin@tpu.ru

Abstract. A method of calculation of interatomic interaction potentials in the presence of ionized states has been developed. They have been obtained for the atoms with different ionization degree on example of aluminum. The Heine-Abarenkov-Animalu model potential form factors was employed. The form factor parameters of ionized atoms was determined on the base of the quantum defect method using the atomic-spectroscopy data. The potential of interatomic interaction for different charged states with different degree of ionization were determined.

1. Introduction

It is known that intensive radiation generates ionized atoms within the track of a fast charged particle as well as near the solid surface. The potentials of interactions between these atoms and surrounding particles change considerably. The knowledge of corresponding potentials of interatomic interaction is necessary for studying the behavior of partially or entirely ionized matter.
Therefore, aluminum was taken as an example studying the interatomic potentials in metal containing ionized states. Calculations were based on the method of pseudopotentials using Heine-Abarenkov-Animalu model potentials with parameters which were determined from spectroscopic terms of free ions following the method of quantum defect.

2. Quantum defect method

The energetic levels E_{nl} of a single electron in the area of positive ions with the electron shells being similar to the spherically symmetric electron shells of inert gases, in particular Na^{+}, Mg^{2+}, Al^{3+}, Si^{4+}, P^{5+},S^{6+} , Cl^{7+}, are known to satisfy the relation [1]:

$$E_{nl} = -Z^2 / (n - \delta_{nl})^2 + \Delta_{nl}, \qquad (1)$$

where Z is the ion-core charge, δ_{nl} and Δ_{nl} denotes quantum defects characterizing the $|Ze|$–charge induced deviation of an atomic potential in such a configuration of inert gas shells from a Coulomb ones. Figure1 shows a linear dependence of the value of spectroscopic terms of the atomic series considered on a square of the ion-core charge.

It turns out that there also are similar dependences for additionally ionized cores (figure2). Here the following series Na^{2+} Mg^{3+}, Al^{4+}, Si^{5+}, P^{6+},S^{7+} , Cl^{8+} will be realized.

Using this empirical law we have determined the mode potential parameters of the additionally ionized cores with respect to the energies $\varepsilon_{nl} = E_{nl} - \Delta_{nl}$, as it was done for transition metals in [1].
Radial wave equation with a model potential describing the electron motion is written as

$$\left[-\frac{1}{2}\frac{d}{dr^2}+\frac{l(l+1)}{2r^2}+v_l\right]\chi(r)=\frac{1}{2}\varepsilon\chi(r), \tag{2}$$

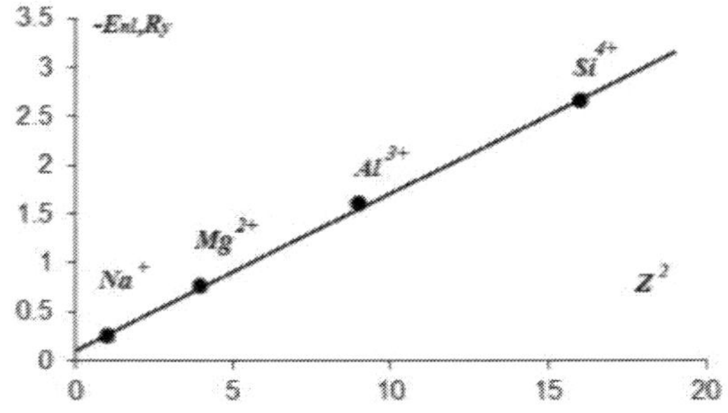

Figure 1. Spectroscopic term values E3p as function of the chemical valence.

where χ(r) is the radial wave function and l is the angular momentum. With such a representation ε is given in rydbergs and vl in atomic units.

The model potential of the electron ion-core interaction has a form

- $V_l = -A_l(\varepsilon)$ for $r \le R_m$
- $V_l = -Z/r$ for $r > R_m$, (3)

here R_m is a model radius closed to the ion-core radius which provides for joining the inside and outside solutions of the wave equation.

After making a substitution of

$\rho = Zr$ and $\bar{\rho} = \mathrm{r}\sqrt{2\mathrm{A}_1 - |\varepsilon|}$ into expression (2) it will be written as

- $\left[-\frac{1}{2}\frac{\partial^2}{\partial\bar{\rho}^2}+\frac{l(l+1)}{\bar{\rho}^2}+1\right]\chi = 0$ for $r \le R_m$, (4)
- $\left[-\frac{\partial^2}{\partial\rho^2}-\frac{l(l+1)}{\rho^2}+\frac{2}{\rho}\right]\chi + \varepsilon'\chi = 0$ for $r > R_m$, (5)

where $\varepsilon' = \varepsilon / z^2$.

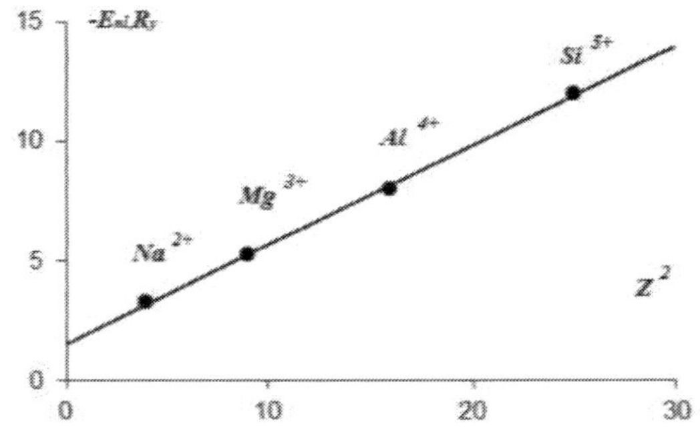

Figure 2. Spectroscopic term values E_{2p} as function of the ion-core charge.

Solution of equation (4) are the spherical Bessel functions and equation (5) is the Coulomb wave one which was solved according to the procedure described in [2].

By setting the logarithmic derivatives from solutions of equation (4) equal to those in equation (5) at $r=R_m$ one obtains an expression for determining the parameters A_l :

$$\frac{xj_{l-1}(x)}{j_l}-l=\frac{{}^0D^l(ZR_m)+\gamma(N,l)^1D^l(ZR_m)}{{}^0U^l(ZR_m)+\gamma(N,l)^1U^l(ZR_m)}, \tag{6}$$

where $x=R_m\sqrt{2A_l-|\varepsilon_{nl}|}$, $j_l(x)$ denotes the spherical Bessel functions, ${}^0U^l$, ${}^1U^l$ are regular and irregular parts of the Coulomb wave function respectively, $N=n-\delta_{nl}$ expresses an effective quantum number, $D^l = \rho(dU^l/d\rho)$,

$$\gamma(N,l)=\frac{\Gamma(N-l)N^{2l+1}tg\,\pi(N\text{-}l\text{-}1)}{\Gamma(N+l+1)}, \tag{7}$$

where Γ is a gamma-function [3].

The values of spectroscopic terms $\varepsilon=\varepsilon_{nl}$ were calculated from tables [4]. Al values as function of energy $\varepsilon=\varepsilon_{nl}$ at fixed l.

Al may be calculated precisely only for the energies which are consistent with the lines observed in the electron emission spectra of a core potential. There in the function $A_l(\varepsilon_{nl})$ is chosen so that the model potential will give eaten values consistent with ε_{nl}. Therefore, first, we estimate A_l for as many energy values as experimental lines exist, after wards linear interpolation add extrapolation to the Fermi energies is made.

Here the question arises, what value is a real energy of the conduction electrons. In contrast to isolated ion (equation2) Schrodinger equation for the conduction electron in metal has a form

$$\left[-\frac{\hbar^2}{2m}\nabla^2+(V+\Sigma)_{ion}+(V+\Sigma)_{rest}\right]\Psi=\varepsilon\Psi, \tag{8}$$

where $(V+\Sigma)_{ion}$ is the potential of this ion and $(V+\Psi)_{rest}$ is the potential caused by other ions and conduction electrons.

Expression (8) may be written as

$$\left[-\frac{\hbar^2}{2m}\nabla^2+(V+\Sigma)_{ion}+(V+\Sigma)_{rest}\right]\Psi=\varepsilon\Psi, \tag{9}$$

Equation (9) is similar to equation (2) for an isolated ion with electron energy

$$E_F=eF-(V+S)_{rest}, \tag{10}$$

where

$$\varepsilon_F=-|I_i|-|I_c|+\mu_x+\mu_c+\frac{2}{5}\left(\frac{\hbar^2k_F^2}{2m^*}\right)-E_x-E_c+\frac{0{,}6Ze^2}{R_a}, \tag{11}$$

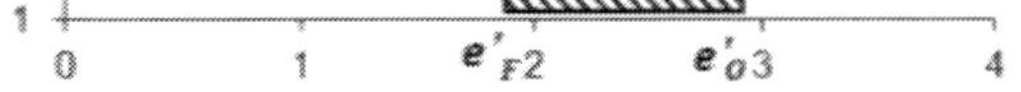

Figure 3. Energetic dependence of the parameters Al for $l=0$ and $z=3$. Dashes region corresponds to the energies of the occupied conduction electron states.

I_i is the ionization energy of outside electron in a free atom, I_c is the electron cohesive energy, k_F is the Fermi radius, m^* is the effective electron mass, E_x is the exchange energy, E_c is the correlation energy of a free electronic gas per an electron, μ_x and μ_c are respectively, exchange and correlation potentials, and R_a is an atomic radius.

The potential $(V+\Sigma)_{rest}$ induced by the rest residual ions and conduction electrons may be written as [5]:

$$\Sigma_{rest} = \mu_x + \mu_c, \tag{12}$$

$$V_{rest} = \frac{Ze^2}{R_a}\left[3 - \frac{3}{4}\left(\frac{R_m}{R_a}\right)^2\right]. \tag{13}$$

Figure3 shows an energetic dependence of parameter A_0 for an electron in the field of a three-charged ion. The energy levels εnl in Al^{3+}-ion, the Fermi energy ε_F' and that of the bottom of conduction band $\varepsilon_0' = \varepsilon_F' - \hbar^2 k_F^2 / 2m$ are shown in terms of the ion energy [6].

The ion-core radius R_c is an important value for calculating interatomic potential. Its procedure estimation is following.

According to [7,8] the wave electron function outside the core has a form

$$\Psi = -\frac{2}{n}\sqrt{\frac{(n-l-1)!}{[(n+l)!]^3}}\exp(-\rho^*/2)\rho^* L_{n+1}^{2l+1}\rho^*, , \tag{14}$$

where L is the added Lagerr polynomial, $\rho^*=2Zr/na_0$, a_0 is a Bohr radius.

Then the wave function will be

$$\Psi_{nl} = -\frac{2}{n}\sqrt{\frac{(n-2)!}{[(n+l)!]^3}}\exp(-\rho^*/2)\rho^* \sum_{k=0}^{n-2}\frac{(-1)^{k+l}[(n+1)!]^2\rho^{*k}}{(n-2-k)!(k+3)!k!}. . \tag{15}$$

Inside the core it is

$$\Psi_{nl} = Arj_l(\tilde{\rho}). \tag{16}$$

Here A is a normalization factor, $\tilde{\rho} = yr$, $y = \sqrt{2A_l - |\varepsilon_{nl}|}$.

Expression (15) included the numbers up to $(n\text{-}2)$ order, ρ^{*n-2} and ρ^{*n-3} are the major ones. If keep only these two members, one may think that n adopts any fractional values. At $r = R_c$ the logarithmic derivatives from expressions (15) and (16) may be set equal :

$$\left[-\frac{1}{2} + \frac{1}{\rho^*} + \frac{n-2}{\rho^* - (n+1)(n-2)}\right]\frac{2Z}{na_0} = \left(\frac{\tilde{\rho}}{1-\tilde{\rho}ctg\tilde{\rho}} - \frac{2}{\tilde{\rho}}\right)y. \tag{17}$$

By solving graphically equation (17) we hand found the ion-core radius of additionally ionized atom.

As it should be guess R_c value seemed to be not very sensitive to the degree of ionization.

As for an effective mass and charge of electron conduction, we counted $m^*/m = 1$, as it was done in [1], while the parameter inserted in expression to define the effective charge $e^* = e(1+\alpha)$, which results from orthogonality correction, was calculated from a formulae

$$\alpha = (R_c/R_a)^3/2. \tag{18}$$

Table 1. Parameters of model potential

Z	A_0	A_1	A_2	R_m	R_c	α
3	1,38	1,64	1,92	2	1,08	0,024
4	1,35	1,756	1,46	2	1,05	0,022

Table 1 illustrates the parameters of a model potential for usually charged aluminum, taken from reference [1], and those obtained by us for additionally ionized core for comparison. All the measured values are in atomic units.

3. Results and discussion

Figure 4 represents the potential of interatomic interaction for different charged states in aluminum which were estimated using the above data. Calculation was done for a situation when the ionized atom concentration is small and consequently the conduction electron concentration does not differ from an usual one.

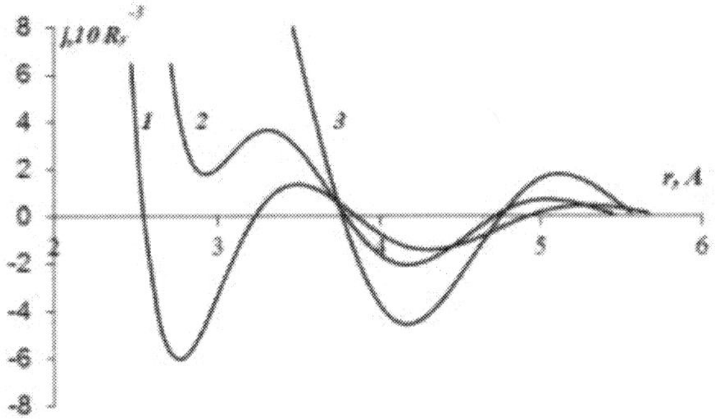

Figure 4. Interatomic potential: 1 – Al^{3+} and Al^{3+}; 2 – Al^{4+} and Al^{3+} ;3 – Al^{4+} and Al^{4+}.

This figure shows that ionization leads to the strong decrease in the depth of the first minimum of the potential function corresponding to the distance area between the nearest neighbors for three- or four-charged ions. Moreover, for the pair of particles with four charges the first minimum disappears absolutely. Atoms fall to the repulsion branch of interaction potential. As a result, the crystal lattice changes to the state of nonequilibrium.

On the basis of the similar potentials of interatomic interactions, applying the method of molecular dynamics, we modeled the dispersion of aluminum surface within the track of charged particles. It was shown that ionization greatly affects the dispersion process of surfaces of metals. However, the molecular-dynamic dispersion modeling in many cases is qualitative. For the more thorough investigation of destruction processes of surfaces we should solve the equation of thermal elasticity, where in its turn we should know the equation of substance condition.

References

[1] Animaly A O E 1973 *Phys. Rev.* B **8** 3542
[2] Kertis A R 1969 *The wave function of Coulomb* (Moscow: Computer Center of USSA Academy of Sciences) p 14
[3] Ham F S 1955 *Solid State Physics* vol 1 (New York) p 127
[4] Moore C E 1949 *Atomic Energy Levels* vol 1 (Washington National Bureau of Standarts) p 130
[5] AnimalyA O E and Heine V *Phys. Mag.* 1965 **12** 1249
[6] Ziesche P and Lehman G 1984 *Progress of the Electronic Theory of Metals* vol 1 (Moscow: Mir) p 134
[7] Landau L D and Lifshits E M 1989 *Theoretical Physics* vol 3 (Moscow: Nauka) p 153
[8] Harrison W 1989 *Electronic Structure and Properties of Solids* vol 2 (Moscow: Mir) p 114

IOP Conf. Series: Materials Science and Engineering **81** (2015) 012053 doi:10.1088/1757-899X/81/1/012053

Structural features of the hydride phase formation in nanostructured α-titanium

Yu R Kolobov[1,2], V I Torganchuk[1,2], V N Fokin[1] and B P Tarasov[1]

[1] Institute of Problems of Chemical Physics of RAS, Chernogolovka, Russia
[2] Belgorod State National Research University, Center of Nanostructured Materials and Nanotechnology, Belgorod, Russia

E-mail: torganv@mail.ru

Abstract. Features of structurally-phase transformations in nanostructured and coarse grained (recrystallized) titanium in process of diffusion saturation by hydrogen in a wide interval of temperatures are studied.

1. Introduction

Nowadays titanium and its alloys [1] are the most popular metal materials used worldwide in production of medical implants. And intentional alloying is the traditional way to achieve the complex of necessary mechanical properties in such kinds of materials. As a rule, the most common alloying elements in the titanium alloys, such as Al, V, Mo, Ni, Nb, etc. are harmful or even toxic for living organisms. This can also be referred to a famous titanium alloy Ti-6Al-4V (BT6 in Russian terminology, Grade 5 alloy – its international analogue) that is widely used as a material for medical implants. However, a high level of toxicity of V has been recently found in experiments for cells survival in biological fluids [2]. There is a number of confirmations that Al can cause Alzheimer's disease and lead to bone softening (a systematic disease that is characterized by the lack of bone tissue mineralization) [3-5]. These drawbacks are not found in the nanostructured non-alloying titanium that has been got with the method of mechanical-thermal treatment under special technological regime of plastic deformation (severe plastic deformation); so it has got structural characteristics after quasistatic and periodic loading at the level of BT6 alloy. The idea of severe plastic deformation (SPD) use for such a titanium preparation was proved in the articles earlier [1, 6, 7].

It is well knows that the change of metals and alloys structure with the help of SPD can lead to the major modification of mechanic and other properties, as well as of SPD mechanisms [6, 8]. When we use this method to achieve the formation of submicrocrystalline structure (SMC) or nanostructured (NS) condition we can achieve tensile strength duplication of commercial pure titanium VT1-0 that due to its bio inertness is used in medicine as implant material.

It is known that the intensity of mechanical and other properties deterioration in medical metal materials, when in operation, depends heavily on interreacting with biological tissue and fluids. The increase of hydrogen concentration in implants can be the result of such an interreacting. When the concentration of hydrogen in metals with hcp lattice is high, there starts the formation of fragile hydride phases in the volume and on the boundaries of grains succeeded by the beginning and rise of cracks in the phase interface [9-14]. It is stated that hydrogen diffusion saturation of titanium alloys starts in the biological environments containing hydrogen ions, shown on the example of dental implants.

This leads to grain boundary destruction on the surface of the volumes as the result of hydrides formation over there. It has been published earlier in the articles that such a situation leads to preterm implant destruction when in operation [13,15]. Probably the origin of hydrogen embrittlement of titanium alloys used in medicine is grain boundary segregation of iron that is a foreign matter in titanium alloys of commercial pure. It is known that this chemical element stabilizes titanium β-phase, mainly on grain boundaries. We can also watch the release of intermetal combinations of Ti_xFe types on the grain boundaries. This leads to the decrease of metal corrosion resistance because the particles of intermetallides serve as the ways for activating hydrogen diffusion inside the metal. Intermetallides of iron group interreact with hydrogen at room temperature. The diffusuvity of hydrogen in titanium alloys within these characteristics increases greatly [15]. However, notwithstanding the characteristics mentioned above, titanium alloys are widely used in medicine [16-19].

2. Material and experimental procedures

The coarse grained (99, 99%) and commercial pure titanium VT1-0 (total concentration of foreign matters 0.283 wt. % (N – 0.003, C – 0.004, H – 0.0008, Fe – 0.12, O – 0.143, Al – 0.01, Si – 0.002)) in nanostructured (NS) and coarse grained (recrystallized) conditions were used to determine the effects of how the structure influences hydrogen absorbtion in a wide temperature range. NS condition was formed by the impact of plastic deformation in screw and lengthwise rolling combination [20]. The average size of grain and subgrain structured elements in NS condition is d ~0.2 micrometer. Recrystallized alloy VT1-0 had the grain average size of d ~12 micrometers, iodide titanium – d ~240 micrometers. Microstructure and the bar chart of size distribution of grain and subgrain structured elements are shown on figure1.

Hydrogen diffusion saturation of titanium was performed on a special laboratory machine of high pressure (Siverts method) 60 ml. The sampling material was placed into the reactor-digester, then it was vacuumized to the residual pressure of ~1 Pa within 1 hour at different temperatures from 300 to 500 °C. After vacuum condition was achieved, we added hydrogen under the pressure of 2-2.5 Pa and shut off the heat. If there was no hydrogenation, the heat was resumed and it continued before the pressure in the system stopped falling down (2-10 hours). Temperature and time interval of hydrogen saturation was put together taking into account the necessity to maintain NS condition.

Titanium cut was used as an accelerant for hydrogenation temperature decrease and hydrogen absorbation process increase. As it is known, the dissociation of the molecular hydrogen on the atomic one takes place on the free surface of such a titanium cut. In this case the molecular and atomic hydrogen mixture is formed near the contact place of titanium cut with the surface in the reactor-digester. We have to mention that the atomic hydrogen has got a high level of diffusion mobility in the metal crystal lattice. This lets us reduce the temperature of hydride formation beginning. The speed of dissociation also depends on gas pickup conditions (temperature, pressure). The decrease of hydrogen charging temperature is a major factor for experimentation in the research of regularity of hydrogen influence on the NS materials structurally-phase condition, as it helps to maintain this condition.

All the works, preceding the hydrogen diffusion saturation of titanium, were performed in argon atmosphere in a dry box with the controlled inert atmosphere of MBraun company (Germany). Highpurity hydrogen (99.9999%) from a metal hydride accumulator on the basis of $LaNi_5$ was used for hydration.

The structure of hydride phases was identified with the help of gas-volumetric, chemical (CHNS/O elementary analyzer «Vario Micro Cube») and X-ray phase analyses. Thermal analysis was performed with the use of synchronic TG-DTA/DSC (Netzsch STA 409 Luxx) in argon atmosphere (the heating rate of 10 °C/min).

Structural research was made on a scanning electron microscope Quanta 600 FEG with the field emission and in a transmission electron microscope Tecnai. The samples were prepared with the help of mechanic grinding-polishing installations LaboPol-5 (Struers), electropolishing of the sample surface was made on LectroPol-5 (Struers) installation in 60 ml $HClO_4$ + 600 ml CH_3OH + 360 ml butylglycol solution at a temperature of +5 °C and voltage of 23 V.

3. Results and discussion

Commercial pure titanium VT1-0 in its nanostructured and coarse grained (recrystallized) conditions (the average size of a grain and subgrain mixture is 0.2 and 12 micrometers respectively) and coarse grained iodide titanium (the grain size is 240 micrometers) (figure1) were used to determine the effects of a structurally-phase condition on the process of hydrogen absorbtion by titanium. Hydrogen saturation of iodide and NS commercial pure titanium was made in one reactor-digester with a metal separating baffle, because the speed of titanium oxide layer (TiO_2) in oxygen containing environments is very high and it is a strong barrier for hydrogen penetration into titanium matrix. It was discovered that initial structure of the sample influence the temperature of intensive hydrogen diffusion saturation of titanium.

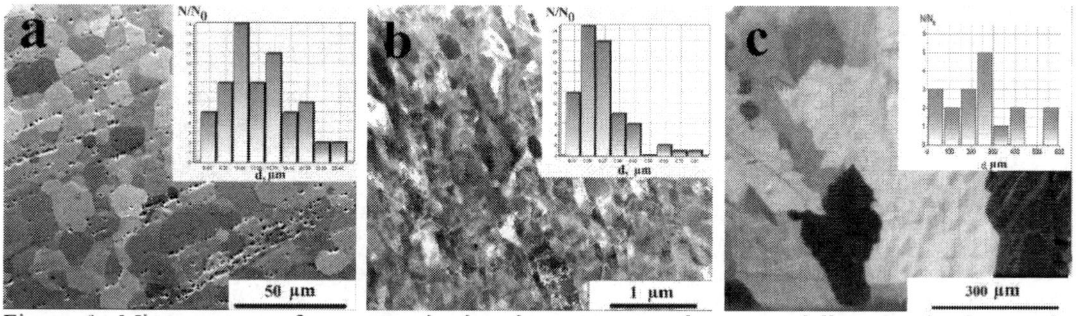

Figure 1. Microstucture of coarse-grained and nanostructured commercially pure titanium and histograms of grain size distribution: a – coarse grain (annealed) titanium alloy VT1-0, b - nanostructured titanium alloy VT1-0, c - iodide titanium.

As we can see on the graphs (figure2) that have been obtained with the help of differential scanning calorimetry, there are differences in quantity and location of peaks for iodide and commercial pure titanium in its NS condition after diffusion saturation in hydrogen atmosphere. After hydrogen diffusion saturation of titanium at a temperature of 400 °C we can observe the peaks displacement to high temperature for NS titanium. Besides, there are additional peaks in the interval between 400 and 500 °C. This type of graph is similar to the one for titanium hydride in coarse grained (iodide) titanium [21]. The DSC curves, obtained for compact specimens, differ from the respective graphs for chipped samples. This is due to several processes. Firstly, heterogeneous processes located on the boundary line of solid/vapor influence seriously on specific surface value (surface area per unit volume), and when the latter rises, the processes proceed intensively. Secondly, the difference is in accelerating of gas pickup in the chipped samples. This happens due to the fact that hydride phase starts its formation in uncrystallized initial grain, but the grain has enough time to grow in the compact specimen under long term hydrogen saturation before the hydride phase begins. The article [21] shows a complex character of phase transformation of titanium hydride under annealing; this proves a possible structural change of the latter into one of its crystal modifications.

There is an explanation for the presence of two peaks on DSC curves of titanium dihydride, that correspond to nonequivalent position of hydrogen in hollows: one peak is in tetrahedral hollow, the other one is in octahedral [22]. For certain reasons [23] the additional peak formation, the second one exactly, is due to the presence of bct titanium hydride that transforms in to a cubic structure while heated. When the temperature of hydrogen diffusion saturation of NS titanium decreases the grain rate of growth goes down. As the result, hydride phase appears in a smaller grain. And when the temperature of hydrogenation is 400 °C the graph shows three distinct peaks. This indicates a double displacement of the structure or the presence of hydride mixture of different crystal lattice modification.

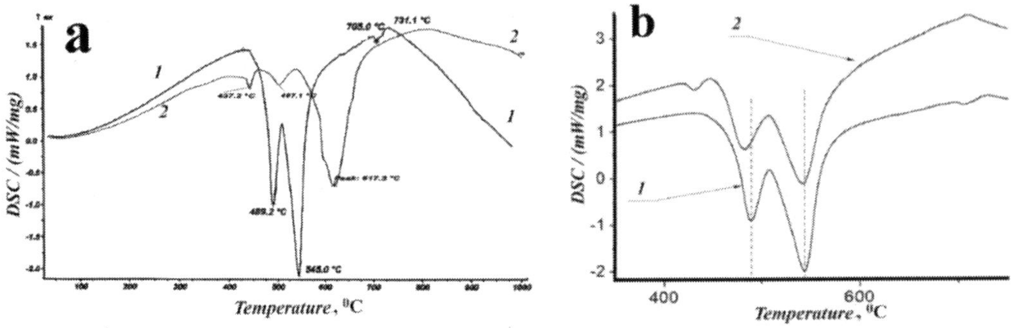

Fugure 2. DSC curves of the titanium shavings after hydrogen absorption at 400 °C: (a) the iodide titanium (1) - 3.81 weight. % H; commercially pure NS titanium VT1-0 (2) - 3.71 weight. % H; b) DSC curves of the compact samples after hydrogen absorption at 400 °C: (1- 3.84 weight. % H; 2 - 3.83 weight. % H).

A high concentration of iron in VT1-0 alloy can influence the hydride formation kinetics significantly, as intermetallide TiFe serves an accelerator in the processes of titanium hydrogenation. Together with other mixtures that are often out of purposeful control and that are a part of VT1-0 alloy, iron can influence the morphological features of hydride phase formation in α-titanium.

To exclude the influence of foreign elements on hydride formation kinetics there was done a comparative research of commercial pure VT1-0 titanium diffusion saturation in nanostructured and coarse grained titanium that was formed as the result of recrystallization annealing from NS condition of this alloy.

Figure 3 shows the differences in DSC graphs for VT1-0 alloy after diffusion saturation in hydrogen atmosphere and for iodide titanium. Hydrogen charged samples in the form of titanium cut are characterized by a more complex nature of a graph with a number of endothermal peaks (figure 3a), a distinctive feature of NS condition is a less integral area between the peaks. It is important to note a complex way of DSC graphs, this can show structurally-phase transformations of hydride phase as well.

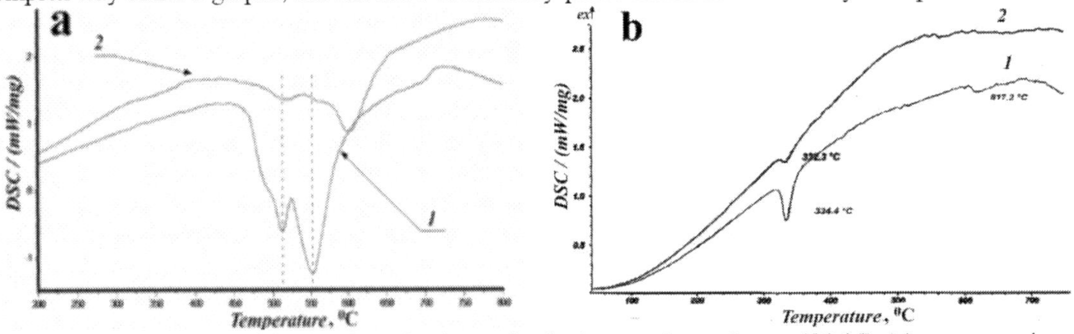

Fugure 3. DSC curves of the titanium shavings after hydrogen absorption at 420 °C: (a) coarse grain commercially pure titanium VT1-0 (1) - 3.65 weight. % H; commercially pure NS titanium VT1-0 (2) - 3.44 weight. % H; b) after hydrogen absorption at 400 °C (b): 1- 0.84 weight. % H; 2 – 0.54 weight. % H).

At hydrogen diffusion saturation of NS and coarse grained titanium (400 °C, 4 hours, 2MPa) the hydrogen concentration in the samples in the form of cuts is 0,54 and 0.84 % respectively. It is known that the plasticity of particle hardened metal alloys, for titanium alloys with hydride particles here (they are formed as the result of diffusive annealing in hydrogen atmosphere), directly depends on a form and character of hydrides distribution. Plasticity and impact hardness can increase at spherical and globular form of such particles under certain conditions. Long lamellar precipitate (here, hydrides) are a stress raiser that lead to decrease of material plasticity.

Morphological features of hydride formation in NS titanium are a significant result of our research.

Figure4 shows that zigzag configuration hydrides are formed in such a material. Probably, this fact states that a significant role in hydride extraction in metals and alloys in SMK and NS condition belongs to residual inner stress that is common for such materials obtained under SPD [1]. A partial de-stress takes place in the process of formation and growth of zigzag configuration hydrides. The above described type of hydride extraction, called «hydride platelets» was observed earlier for some zirconia alloys (e.g: Zr702, (Hf 4.5%; Fe + Cr 0.2%; H 0.005%; N 0.025%; C 0.05%; O 0.16%) [25]. Titanium hydride extraction typical for pure titanium and some its alloys in coarse grained condition of test titanium alloy is observed on the grain boundaries [10, 14, 24].

Nanosized platelets extractions of 10 nm thick and 600 nm long that present inside a grain were discovered while investigating the structurally-phase condition of NS titanium through transmission electron microscopy (figure 4). The length of such extraction is limited with the size of grains. The given morphology of extractions is typical for hydride phase. It is known that depending on hydride phase type the lamellar precipitate are located on the boundaries and inside a grain [10]. Besides, particles of globular form were discovered (figure 4c), but due to their small size it is impossible to obtain their electron diffraction pattern.

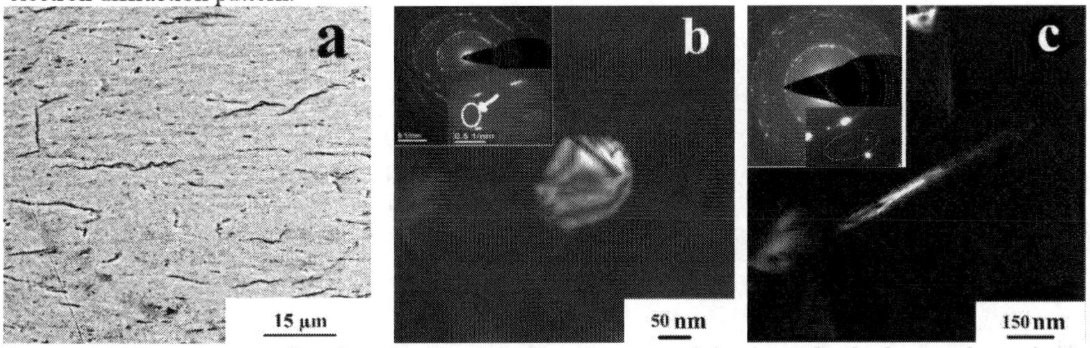

Fugure 4. Microstructure of surface the samples of nanostructured titanium after hydrogen absorption at 400 °C (a - scanning electron microscopy; b, c - transmission electronic microscopy).

The electron-diffraction pattern and dark-field figures, obtained from different sides of structure (b-longitudinal section, c- cross-section) of VT1-0 NS titanium after hydrogen saturation at the temperature of 400 °C is shown on figure 4b. This electron diffraction pattern shows an additional reflex circle that corresponds neither to hcp lattice (α-titanium) nor to body centered cubic lattice (β-titanium). While identifying the pattern there were some difficulties because of a large number of reflexes that correspond to titanium α-matrix. However these reflexes prove the hydride extraction of $TiH_{1.971}$ and/ or $TiH_{1.924}$ type ($TiH_{1.971}$ (PDF – 2 №7-370, cubic system, space group I4/mmm), $TiH_{1.924}$ (PDF – 2 №25-982, cubic system, space group Fm3m)). Relative data of interplanar distance for hydride phases that have a cubic lattice as well as the combination of other factors didn't let determine the hydride phase structure precisely. Probably, $TiH_{1.924}$ hydride phase is presented by globular extractions, while $TiH_{1.971}$ has got sheetlike extractions. The presence of at least two crystal modifications of titanium hydride is confirmed indirectly by the presence of additional plateau and peaks on DSC graphs for NS condition.

4. Conclusion

Comparative examinations of structurally-phase transformations in nanostructured and coarse grained (recrystallized) titanium under its hydrogen diffusion saturation at a wide temperature interval are conducted. The differences in location and quantity of peaks on the graph are discovered, this proves more complex processes of hydride phase dissolution. The phase was formed after nanostructured titanium hydrogenation in comparison with the one fit with coarse grained (recrystallized) titanium.

It is stated that hydride formation in nanostructured titanium takes place in the form of zigzag productions (by type of cracks formation and expansion) that look like the colonies of finely dispersed particles of submicrometer dimension compared to long particles typical for recrystallized condition.

Acknowledgments
The work was financially supported by the Ministry of Education and Science of the Russian Federation under contract No.02.G25.31.0103 and support from the Ministry of Education and Science of the Russian Federation through the contract No. 3.1492.2014/K

References
[1] Kolobov Yu R 2009 *Rossiyskie nanotekhnologii* (Nanotechnologies In Russia) **4** (11-12) 69-81 (in Russian).
[2] Okazaki Y, Rao S, Ito Y and Tateishi T 1998 *Biomaterials* **19** 1197-1215.
[3] Ham K N, Brown D, Dawborn J, Johnson C, Nelson S and Xipell J 1985 *Pathology (Phila.)* **17** 458-463.
[4] Rodriguez M, Felsenfeld A, Samara S, Pederson J and Llach F 1985 *ASAIO J.* **31** 655-659.
[5] Balakrishnan A, Lee B C, Kim T N and Panigrahi B B 2008 *Trends Biomater. Artif. Organs.***22** 54-60.
[6] Kolobov Yu R, Valiev R Z and Grabovetskaya G P 2001 Zernogranichnaya diffuziya i svoystva nanostrukturnykh materialov [Grain boundary diffusion and properties of nanostructured materials]. (Novosibirsk: NAUKA) p. 232 (in Russian).
[7] Kolobov Yu R, Grabovetskaya G P, Dudarev Eu F and Ivanov KV 2004 *Voprosy Materialovedeniya – Problems of Materials Science* **1 (37)** 56-63 (in Russian).
[8] Dudarev Eu F, Golosov Eu V, Kolobov Yu R, Pochivalova G P, Bakach G P and Torganchuk V I 2010 *Bull. Russ. Acad. Sci., Physics* **53** 31-37 (in Russian).
[9] Murzinova M A and Salishchev G A 2010 *Adv. Eng. Mater.***12**, 765-768.
[10] Zwicker U 1979 Titan i ego splavy [Titanium and Titanium Alloys] (Moscow: Metallurgiya) p.511 (in Russian).
[11] Adva N, Alam M, Ravindranath T, Mubeen A and Saluja B 2005 *J. Indian Prosthodont. Soc.* **5** 126-131.
[12] Yokoyama K, Kaneko K, Miyamoto Y, Asaoka K, Sakai J and Nagumo M 2004 *J. Biomed. Mater. Res. A* **68** 150-158.
[13] Yokoyama K, Ichikawa T, Murakami H, Miyamoto Y and Asaoka K 2002 *Biomaterials* **23** 2459-2465.
[14] Kolachev B A 1985 Vodorodnaya khrupkost' metallov [Hydrogen embrittlement of metals] (Moscow: Metallurgiya) p.216.
[15] Diffusion of hydrogen in titanium by Abdul-Hamid. O.S. TH: Ph.D. Thesis; PBD: 1993.
[16] Rodrigues D C, Urban R M, Jacobs J J and Gilbert J L 2009 *J. Biomed. Mater. Res. B, Appl. Biomater.***88B** 206-219.
[17] Yamamoto Y and Shinohara K 2002 *Anat. Rec.* **269** 217-223.
[18] Lütjering G and Williams J C 2007 Titanium. 2nd edition. (Berlin–New York: Springer) p.454
[19] Corrosion 1987 **13** (Handbook 9th Ed. M.: ASM Int. ISBN.) 871700190.
[20] Ivanov M B, Kolobov Yu R, Golosov Yu V, Kuz'menko I N, Veynov V P, Nechaenko D A and Kungurtsev Yu S 2011 *Rossiyskie Nanotekhnologii* [Nanotechnologies In Russia] **6 № 5-6** 108-114 (in Russian).
[21] Spivak L V and Malinina L N 2011 *Vestnik permskogo universiteta* [Bulletin of the University of Perm] **69** (in Russian).
[22] Vonzeppelin F, Hirscher M, Stanzick H and Banhart J 2003 *Compos. Sci. Technol.* **63** 2293-2300.
[23] Kennedy A R 2002 *Scr. Mater.* **47** 763-767.
[24] Metallography and microstructures 2004 Ed. GF Vander Voort, Handbook A. ASM Intenational. **9.**
[25] Danielson P E, Sutherlin R C and Chang W 2004 Metallography and Microstructures of Zirconium, Hafnium, and Their Alloys, Materials Park, OH: ASM International. 942–958.

Computer simulation of processes of radiation-thermal heating[1]

V A Zhuravlev*, V A Meshcheriakov*, V I Suslyaev*

*Tomsk State University, Lenina av., 36, Tomsk, 634050 Russia

E-mail: mva@mail.tsu.ru

Abstract. In this work we present results of computer simulation of radiation-thermal heating of magnetic materials to change their structure. The energy source is a pulsed electron accelerator. The simulation results are presented in the form of the temperature distribution in the irradiated material, depending on the pulse duration, pulse repetition rate and the parameters of the material.

1. Introduction

Synthesis of ferrites beams of accelerated electrons is discussed in [1-3]. The radiation source produces a stream of electrons perpendicular to the flat object specified thickness. The distribution of energy losses of electrons on the thickness of the object is described by a Gaussian function. Thermal radiation and heat transfer is discussed in [4]. A mathematical model of a computer program is the heat equation [5, 6]. For the numerical solution of the equations used explicit template finite-difference schemes for one-dimensional case:

$$T_i^{n+1} = \frac{\lambda\,\Delta t}{\rho\,c\,h^2}\left(T_{i+1}^n - 2\cdot T_i^n + T_{i-1}^n\right) + T_i^n + \frac{P_i^n\,\Delta t}{\rho\,c\,h^3},$$

where P_i^n - is the power density of the radiation source in n - including the temporary layer, h is the grid spacing, $T=T(x,t)$ - the temperature pressure, λ - the thermal conductivity, ρ - the density of the heated substance, c - is the heat capacity of the substance.

2. The main parameters system computer simulation

The mathematical model is implemented in a software system with the interface presented in figure 1. The main window of the program includes tools for setting and changing parameters of the model. The main parameters of the system are: E_e - the energy source; Z - the number of the main chemical element; A_e - its atomic mass; R_e - extrapolated mileage of electron flow; J_{imp} - current source; d_{ti} - pulse duration; F_i - pulse frequency; N- number of pulses; G - the pulse energy; D - the power absorbed dose; T_{max} - maximum temperature; T_{cp} - average temperature; D_x - the thickness of the heated material. The simulation results can be observed in the main window. The received data can be saved in files.

3. The results of computer simulation

[1] Work carried out in the framework of the increase of competitiveness of Tomsk State University

The purpose of computer simulation is to obtain the temperature profile in the radiation-thermal heating in the process of synthesis and sintering of magnetic materials, and comparison with experimental results. The objective of the simulation - optimization procedures, radiation-thermal heating, energy and time costs and the devices used. This task can be solved by variation of the variables in the parameter space of the system.

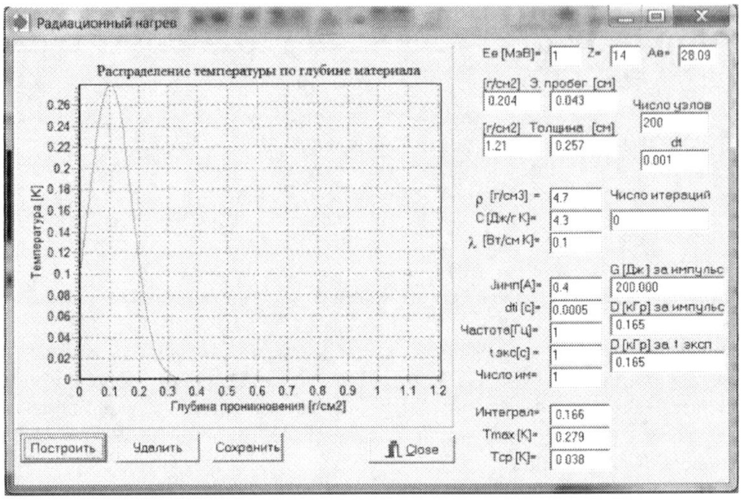

Figure 1. A fragment of the main window

An important characteristic is the distribution of energy losses of the incident electrons, the depth of the material from one pulse (with specified energy E_e and pulse duration $J_{имп}$). Figure 2 presents the distribution of energy losses of electrons.

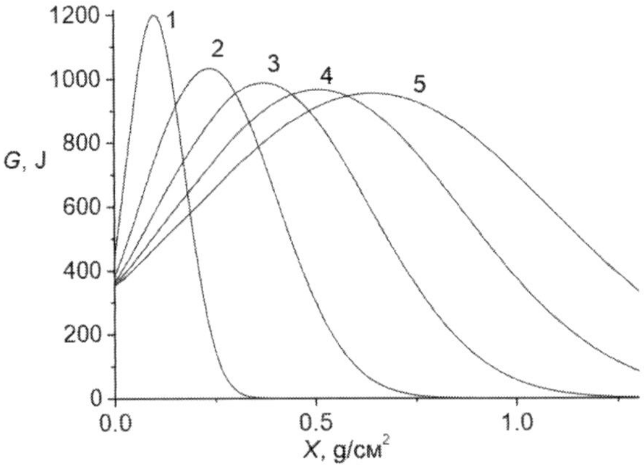

Figure 2. The distribution of energy losses of electrons in the bulk material. Curve 1 corresponds to E_e = 1 MeV, 2 - 2 MeV, 3 - 3 MeV, 4 - 4 MeV, 5 - 5 MeV.

Figure 3 presents the temperature distribution in the material for different time intervals of exposure, after the action of one pulse. The results are presented for the energy flow E_e = 1 MeV, for a material with physical characteristics: R_e = 0,204 g/cm^2, D_x = 0,204 g/cm^2 (is equal to the extrapolated mileage), λ = 0.1 W/(cm * K), ρ = 4.7 g/cm^3, c = 4.3 J/(g * K), $J_{имп}$ = 0.4 A, d_{ti} = 500 μsec, the time t_{exp} was changed from d_{ti} to 0.4 seconds.

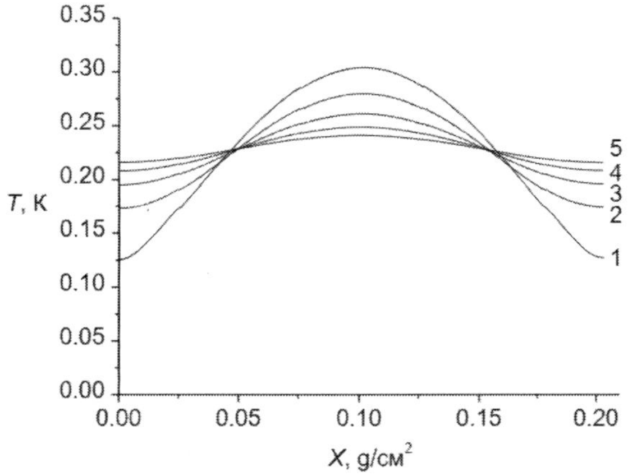

Figure 3. The temperature distribution along the depth of the sample from the time of the observation. Curve 1 corresponds to t_{exp} = 500 μsec; 2 - 0.1 sec; 3 - 0.2 sec; 4 - 0.3 sec; 5 – 0.4 sec. The thickness of the material equal to the length of the extrapolated path.

There is interest in studying the effect of electron energy on the temperature distribution along the thickness of the material. As we know with increasing beam energy increases the extrapolated path of the electron. At a small thickness of material, you can get a situation where a small fraction of the energy loss of the beam will occur in the material. Figure 4 shows the temperature distribution along the depth of the material for different beam energy and a fixed time interval of exposure. The results are presented for the following parameters: material thickness varied from D_x = 0.3 g/cm^2 to D_x = 1.1 g/cm^2 . Exposure time t_{exp} = 0.1 sec.

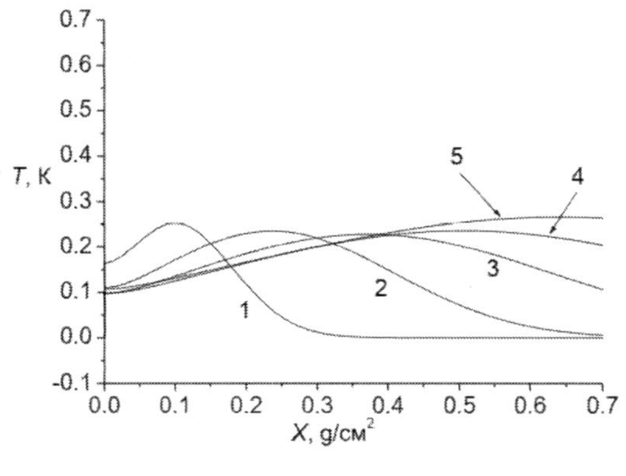

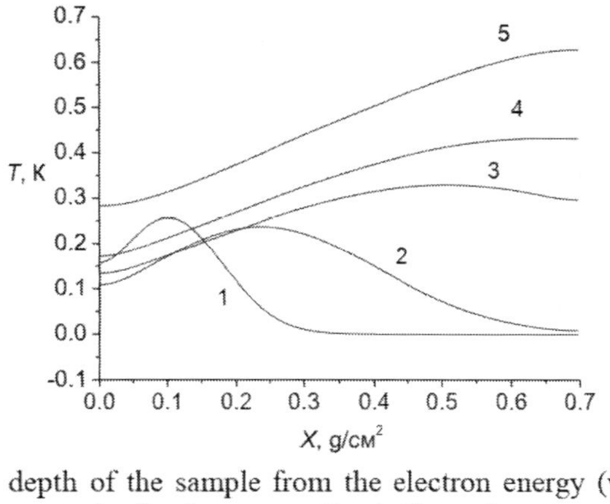

Figure 4. The temperature distribution along the depth of the sample from the electron energy (when the beam current pulse is 0.4 A, a pulse duration of 500 μsec and a fixed exposure time). Curve 1 corresponds to E_e = 1 MeV; 2 - 2 MeV; 3 - 3 MeV; 4 - 4 MeV; 5 - 5 MeV. Sample thickness: D_x = 0.7 g/cm^2 - left graph; D_x =1.1 g/cm^2 - right graph.

Values extrapolated path of the flow of electrons R_e and the average temperature T_{cp} presented in the table 1:

Table 6. Values extrapolated path.

E_e [MeV]	$D_x = 0.3$ g/cm^2		$D_x = 0.7$ g/cm^2		$D_x = 1.1$ g/cm^2	
	R_e	T_{cp}	R_e	T_{cp}	R_e	T_{cp}
1	0.204	0.155	0.204	0.066	0.204	0.042
2	0.474	0.310	0.474	0.133	0.474	0.085
3	0.744	0.465	0.744	0.199	0.744	0.127
4	1.014	0.620	1.014	0.266	1.014	0.169
5	1.284	0.775	1.284	0.332	1.284	0.211

The main task of computer simulation is to obtain the temperature profile in the radiation-thermal heating in the process of synthesis and sintering of a number of magnetic materials and the determination of the temporal parameters of the heating process.

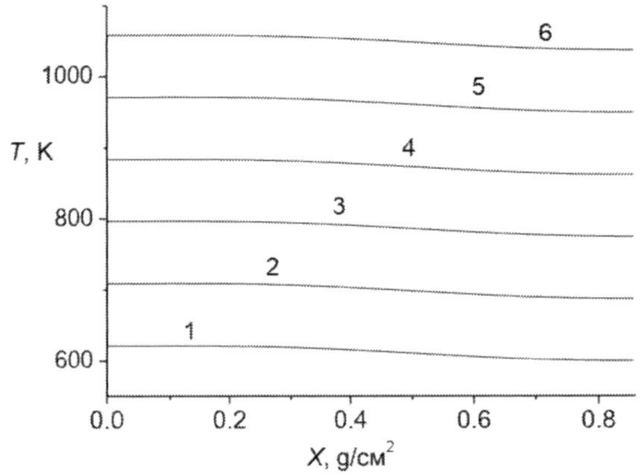

Figure 5. The temperature distribution along the depth of the sample from the time of exposure (when the beam current pulse is 0.25 A, the pulse duration of 500 μsec and pulse frequency of 10 Hz). Curve 1 corresponds to t_{exp} = 140 sec, 2 - 160 sec, 3 - 180 sec, 4 - 200 sec, 5 - 220 sec, 6 - 240 sec.

For the experimental studies were calculated source parameters for heating materials with the following characteristics: $\lambda = 0{,}063$ W/ (cm * K), $\rho = 3 \div 4.7$ g / cm^3, $c = 0.8$ J / (g * K). These settings correspond to the ferromagnetic material which was heated by the electron accelerator. The beam energy $E_e = 2.4$ MeV, current $J_{имп} = 0.25$ A, $d_{п} = 500$ μsec. Material thickness 2 mm. Figure 5 shows the temperature distribution along the depth of the material for different time intervals of exposure, frequency source $F_i = 10$ Hz. Temperature values when changing the frequency of the source is presented in the table 2:

Table 2. The temperature dependence of the frequency.

F_i [Hz]	t_{exp} [sec]	T_{cp} [K]
11	180	870
13	180	1040
15	180	1170
17	180	1360

4. Conclusion

Created a computer program to simulate the heating of the object stream of accelerated electrons. We have compared the simulation results with experimental results. In real experiments the temperature in 1200^{0}C was obtained when F_i = 10 Hz, with an error of less than 3% corresponds to the results of computer simulation. The program can be effectively used to solve optimization problems of the processes of obtaining materials with specified characteristics.

Acknowledgments

The authors thank the staff of Tomsk Polytechnic University Gyngazova S.A., Lysenko E.N. for their valuable advice.

5. References

[1] Surzhikov A P, Sokolovskiy A N, Vlasov V A, Vasendina E A. Synthesis of lithium-zinc ferrite in bunch of accelerated electrons. // Rare Metals, 2009.-T.28-C. 418-420.
[2] Itin V I, Kirdyashkin A I, Minin R V, et al. 2006 Izv. Vyssh. Uchebn. Zaved. Tsvetn. Metallurg. 5 83.
[3] Minin R V. 2008 Technology of synthesis of hexagonal oxide ferrimagnetic materials with W-, M-, and Z-compositions by the method of self-propagating high-temperature synthesis, Cand. Tech. Sci. Dissert., Tomsk Polytechnic University, Tomsk).
[4] Siegel, John R. Howell, Robert; Howell. John R. Thermal radiation heat transfer. New York: Taylor & Francis, Inc. ISBN 978-1-56032-839-1. Retrieved 2009-07-23.
[5] Samarskii A A. The Theory of Difference Schemes. DJVU. New York: Marcel Dekker, Inc., 2001, 762 p.
[6] Tihonov A N, Samarskii A A. Equations of mathematical physics. M.: Nauka, 1977. – 735 p.

Interconnection between parameters of rock samples electromagnetic signals and content of magnetite in the samples

L V Yavorovich[1], A A Bespal'ko[1], P I Fedotov[1], A I Popelyaev[1]

[1] Problem Research Laboratory of Electronics, Dielectrics and Semiconductors, Nondestructive Testing Institute, Tomsk Polytechnic University, Lenin Avenue, 30, Tomsk, 634050, Russia

E-mail: pif@tpu.ru

Abstract. The paper describes the results of studies of electromagnetic response parameters of samples of bearing strata and ore rocks from the Tashtagol mine on acoustic effects. Patterns of changes in amplitude-frequency parameters of electromagnetic signals for rock samples with different content of magnetite are experimentally found. The conducted research shows that the maximum amplitudes of electromagnetic signal spectral components increases in samples of the same mineral composition with the reduction of their ultimate strength. This is caused by heterogeneities, and defective areas which contribute to active transformation of mechanical energy into electromagnetic energy. For rocks, which contain magnetite, the emissivity depends not only on heterogeneities and defective areas, but on the quantity of high-conductivity minerals in their composition.

1. Introduction

When developing the method to monitor changes in the stress-strain state of rocks and control bump hazard by characteristics of electromagnetic emission, knowledge of basic regularities in mechanoelectric transformations (MET) occurring in the array is required [1]. Physical modeling in laboratory and natural conditions is considered to be available [2-4]. Research of mechanisms of Ems and its sources in rocks at their straining and application of investigated parameters Ems for the control of destruction process also was spent in works of authors [5-16].

A fundamental property of the rocks is heterogeneity in structure, texture, composition and physico-mechanical properties. The author of the paper [17] divides the degree of heterogeneity into four groups:

1) large scale heterogeneity, including facies variation, tectonic breaks, the zone of weathering and unloading, technological heterogeneity;
2) heterogeneity of structure and composition of rocks within a single layer, including the fractured zone, presence of small tectonic dislocations);
3) heterogeneity of rocks within the elementary volume (sample), difference in chemical and mineral composition, shape and sizes of grains, cracks;
4) heterogeneity of real crystals, defects of the crystal lattice, and dislocations.

Thus, when studying rock samples, 3 and 4 groups are important factors of heterogeneity to be considered while analyzing the obtained results.

The only fundamental force which integrates all the substances, including rocks, is the electromagnetic force. If mechanical stresses in the rock change, transformation of mechanical energy into electromagnetic one depends on petrographic and physical-and-mechanical properties of the interacting systems which carry energy. Real rock samples differ both in mineral and phase composition, the size and shape of grains, and in textures (pores, cracks and other stress raisers). Charged particles are more likely to accumulate on heterogeneities, electric field changes and, as a consequence, mechanical energy transforms into electromagnetic one [18]. MET causes generation of electromagnetic signals (EMS) by the rock which can be recordered with special instruments [19–20]. The method for monitoring changes in stress-strain state and forecasting rock burst being developed in Tomsk Polytechnic University (TPU) is based on the study of the properties and characteristics of MET.

In our earlier papers we presented the study of relation between parameters of electromagnetic signals and electrical properties of rocks under acoustic impact [21]. It is found that the amplitude of generated EMS depends on the conductivity of the rocks samples under study. In this regard, the task was to identify the dependence of EMS parameters on the amount of minerals with low electrical resistance in the rock.

2. Experimental technique and objects of research

Skarns of different petrographic composition containing no magnetite and ore with different content of magnetite from the Tashtagol iron-ore deposit were studied. The samples were cut from the core in the form of a cylinder (42 ± 1) mm in diameter and (80 ± 2) mm in height. The end faces of the samples were polished to make a parallel-sided sample with the discrepancy of (0.5 ± 0.1) and the axis of the sample and its ends were arranged under the angle of $(90 \pm 1)^{o}$.

The acoustic impact was made with a piezoelectric transformer (PET), by method developed in the Problem Research Laboratory of Electronics, Dielectrics and Semiconductors (PRL EDaS), Tomsk Polytechnic University (TPU) [22]. The duration of the acoustic signals to excite the sample changed discretely in the interval of 5 to 100 μs. The piezoelectric irradiator was excited by the electric pulse of a rectangular shape at a voltage of 800 V. The electromagnetic signals from rock samples were recorded with a digital storage oscilloscope Tektronix TDS2024B and transferred to a personal computer. Then, the amplitude-frequency spectra of EMS were calculated using fast Fourier and the data obtained were analyzed.

3. Experimental results

At the first stage of the experiments the amplitude-frequency parameters of the EMS for the samples with different content of magnetite were studied. Figure 1a shows a typical electromagnetic signal from a host rock sample under the duration of the PET electric excitation pulse of 10 μs. The amplitude-frequency spectrum of EMS is shown in figure 1b.

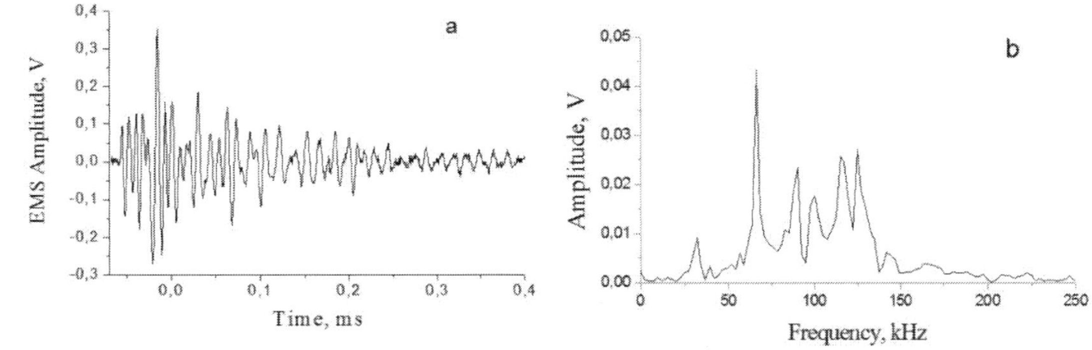

Figure 1. Electromagnetic signal (a) and its amplitude-frequency spectrum (b) in excitation by the acoustic pulse of 10 μs duration for bearing strata rock samples.

To simplify the analysis of the calculated spectra for each sample rock the frequency bands with the maximum amplitude values were determined. The obtained values were used to plot graphs for amplitudes of main spectral components of EMS (figures 2a,b). The figures show the graphs for samples from one group bearing strata rocks with different duration of the excitation pulse and different ultimate strength. Under uniaxial compression with the press (Testing Machine 500) the ultimate strength of the samples of bearing strata of rocks was defined:

- №0 is 188 kN;
- №4 is 320 kN;
- №3 is 234 kN.

Figure 2a shows that for the sample with the least value of the ultimate strength (sample №0) the maximum amplitude of EMS spectral components is observed under the excitation pulse of 5 μs. If the duration of the acoustic excitation pulse increases to 10 μs (figure 2b), the values of the maximum amplitudes of EMS decrease by 1 to 2 orders of magnitude if compared with acoustic excitation of 5 μs.

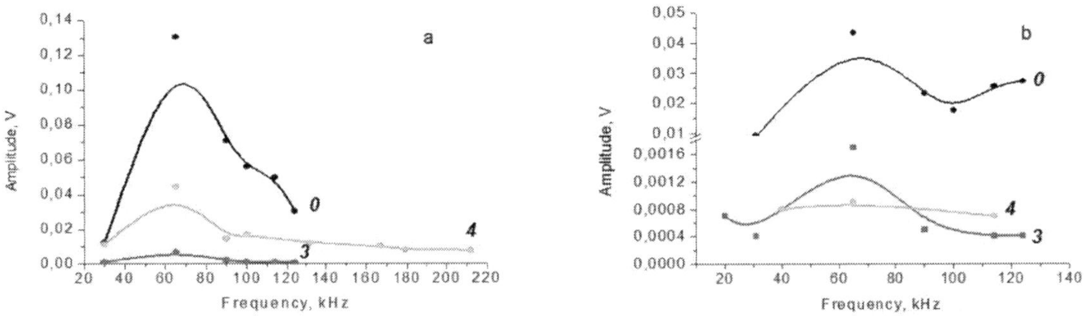

Figure 2. Graphs for the maximum amplitudes of EMS spectral components in excitation by the acoustic pulse of 5 μs (a) and 10 μs (b) for bearing strata rock samples.

Thus, it was experimentally found that as pulse duration of acoustic excitation increases from 5 to 10 μs, the maximum amplitude of EMS spectral components decreases. In samples of the same mineral composition decrease in their ultimate strength causes increase in the maximum amplitudes of EMS spectral components. This testifies to heterogeneities and defective areas that encourage active transformation of mechanical energy into electromagnetic energy due to increase in the number and extent of the charged electric double layers.

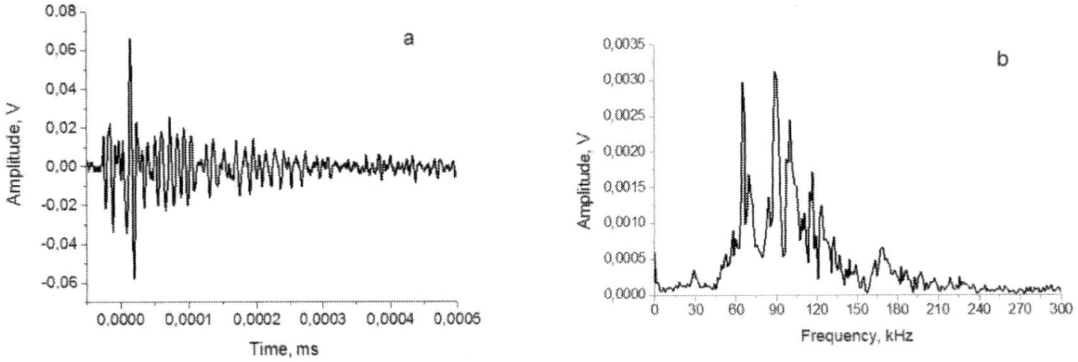

Figure 3. Electromagnetic signal (a) and its amplitude-frequency spectrum (b) in excitation by the acoustic pulse of 10 μs duration for ore samples.

The second group under study was ore samples with different content of magnetite. At the first stage, the ore samples as well as samples of bearing strata were excited by acoustic signals of 5 and 10 μs duration using the same acoustic transducer. Figure 3 shows a typical electromagnetic signal and the amplitude-frequency spectrum of one of the samples from this group.

The graphs with the maximum amplitudes of spectral components for this group of samples are presented in figure 4. The values of ultimate strength recorded under uniaxial compression for ore samples are 258 kN for №3, 317 kN for №4, 193 kN for №2, and 234 kN for №1. Figure 4 shows that the maximum amplitude of EMS spectral components in samples №3 and №4 are 2–3 orders of magnitude less than the amplitude of EMS in samples №1 and №2. These samples differ in their petrographic composition from the samples of the first group by magnetite content. Thus, the emission ability is affected not only by heterogeneities and defects of the composition, but the content of minerals with different electric and magnetic properties. It is known that the electric properties of rock samples depend on the electrical properties of minerals forming the rock. Therefore, they can change from sample to sample. For example, electrical resistivity (p) of bearing strata from the Tashtagol iron-ore deposit, presented by syenites, skarns of different composition, and diorites equals 10^3-10^6 Om•m. Electrical resistivity of magnetite mineral as a part of ores equals 10^{-2}-10^{-5} Om•m. Resistivity of the magnetite ore increases up to 10 Om•m due to inclusion of minerals with small value of ρ [23]. Thus, the more magnetite in the sample, the less its electrical resistivity, and therefore, emissivity of the rock decreases. The amount of magnetite affects emissive ability of ore samples. In our experiments the amount of magnetite in the samples was determined by weighing. It is known [24] that the specific gravity of magnetite is about 5 g/cm^3, for ore the proportion varies from 3.2 to 4.5 g/cm^3, hence, samples with a large amount of magnetite will have a higher weight. Considering that in our experiments, the volume of samples is equal, the specific gravity equals 3.8 g/cm^3 for №3, 3.7 g/cm^3 for №4, and 3.4 g/cm^3 for №2 and №1. Comparing the regularity presented in figure 4 and the calculated specific gravity of samples we can argue that the amount of high-conductive minerals affects the emission ability of the tested samples.

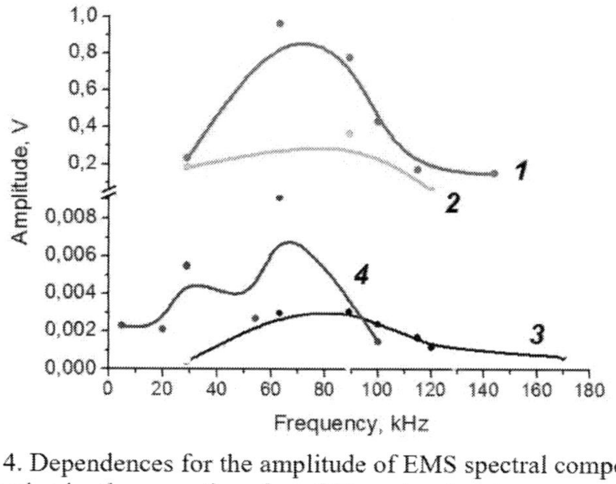

Figure 4. Dependences for the amplitude of EMS spectral components in excitation by the acoustic pulse of 10 μs duration for ore samples.

4. Conclusion

Thus, the conducted research of parameters of electromagnetic signals from the samples of bearing strata and ore rocks from the Tashtagol iron-ore deposit under acoustic excitation shows that in samples of equal mineral composition the maximum amplitudes of EMS spectral components increase if their ultimate strength decreases. This is caused by presence of heterogeneities and defective areas which contribute to active transformation of mechanical energy into electromagnetic energy. For rocks with high-conductivity minerals (such as magnetite) the emissivity depends not only on presence of heterogeneities and defective areas, but on the number of high-conductivity minerals as well.

Acknowledgements
Work is performed with financial support of a grant of the Russian Federal Property Fund No. 14-08-00395 and within the state task of the Ministry of Education and Science of the Russian Federation.

References
[1] Bespal'ko A A, Yavorovich L V, Viitman E E, Fedotov P I and Shtirts V A 2010 Dynamoelectric energy transfers in a rock mass under explosion load in terms of the Tashtagol mine *J. Min. Sci.* **46** No 2 136-42
[2] Bespal'ko A A, Gol`d R M, Yavorovich L V, and Datsko D I 2002 Influence exerted by siltstone lamination on the electromagnetic signal parameters during acoustic excitation of samples *J. Min. Sci.* **38** No 2 124-8
[3] Bespal'ko A A, Gol`d R M, Yavorovich L V, and Datsko D I 2003 Excitation of electromagnetic radiation in laminated rocks under acoustic influence *J. Min. Sci.* **39** No 2 112-7
[4] Bespal'ko A A, Surzhikov A P, Yavorovich L V 2007 Study of Mechanoelectrical Transformation in Rocks Under Dynamic Impact *Rus. J. Non Ferr. Met.+* **1** 9-11
[5] Lokajıcek T, Sikula J and Vasina P 1997 Detection of crack position by AE and EME effects in solids *Proc. 8th Int. Symp. on Nondestructive Characterization of Materials* (Boulder, Co, June 15–20, 1997) 517
[6] Sikula J, Koktavy B, Vasina P and Lokajıcek T 1998 Cracks and quality testing by acoustic and electromagnetic emission *Proc. 23rd European Conf. on Acoustic Emission Testing* (TUV Austria, Vienna, May 6–8, 1998) 130
[7] Frid V, Bahat D, Goldbaum J and Rabinovitch A 2000 Experimental and Theoretical Investigations of Electromagnetic Radiation Induced by Rock Fracture Israel *J. Earth Sci.* **49** 9-19
[8] Koktavy P, Sikula J and Koktavy B 2001 A model for an electromagnetic emission signal generated by crack creation *Proc. NDT in Progress* (Trest, June 20–22, 2001) *Czech Society for Non-Destructive Testing* 283
[9] Sikula J, Lokajicek T and Mori Y 2002 Cracks characterisation by electromagnetic and acoustic emission *Proc. 8th ECNDT, AEND* (Barcelona, June 17–21, 2002) 290
[10] Bahat D, Frid V, Rabinovich A and Palchik V 2002 Exploration electromagnetic radiation and fractographic methods of fracture properties induced by compression in glass-ceramic *International Journal of Fracture* **116** 179–94
[11] Goldbaum J, Frid V, Bahat D and Rabinovitch A 2003 An analysis of complex electromagnetic radiation signals induced by fracture *Meas. Sci. Technol.* **14** 1839–44
[12] Frid V, Rabinovitch A and Bahat D 2003 Fracture induced electromagnetic radiation *J. Phys. D: Appl. Phys.* **36** 1620–8
[13] Koktavy P, Pavelka J and Sikula J 2004 Characterization of acoustic and electromagnetic emission sources *Meas. Sci. Technol.* **15** 973–7
[14] Xueqiu He, Wenxue Chen, Baisheng Nie, Hani 2011 Electromagnetic emission theory and its application to dynamic phenomena in coal-rock *Int. J. Rock Mech. Min. Sci.* **48** 1352–8
[15] Wang Chao, Xu Jiankun, Zhao Xiaoxia, Wei Mingyao 2012 Fractal characteristics and its application in electromagnetic radiation signals during fracturing of coal or rock *Int. J. Min. Sci. Tech.* **22** 255–8
[16] En-Yuan Wang, En-Lai Zhao 2013 Numerical simulation of electromagnetic radiation caused by coal/rock deformation and failure *Int. J. Rock Mech. Min. Sci.* **57** 57–63
[17] Glushko V T, Yamshikov V S, Yalanskii A A 1987 Geophysical monitoring in mines and tunnels (Moskva: Nedra) 278

[18] Vorobyev A A 1975 Physical conditions of occurrence and properties of deep substances (Tomsk: Publishing house of TPU) 296

[19] Bespal'ko A A, Fedotov P I, Yavorovich L V 2008 Registrar of electromagnetic and acoustic signals to control the strength and fracture of materials and rocks *TPU Bulletin* 2 **312** 255-8

[20] Bespal'ko A A, Bombizov A A, Loshchilov A G, Yavorovich L V 2011 Registrar of electromagnetic signals for tracking changes in the stress state of the rock *Control. Diagnostics* **11** 14-7

[21] Bespal'ko A A, Yavorovich L V, Fedotov P I 2005 Dependence of parameters of electromagnetic signals on electric properties of rocks under acoustic and quasistatic impact *TPU Bulletin* 7 **308** 18-23

[22] Bespal'ko A A, Yavorovich L V, Surzhikov A P 2011 Dependence of electromagnetic signal parameters on electric properties of rocks under the acoustic and quasi-static impact (Tomsk: Publishing house of TPU) 120

[23] Vachromeev G S, Erofeev L Y, Kanaikin V S 1997 *Petrophysics: Textbook for universities.* (Tomsk: P. University)

[24] Physical properties of rocks and minerals *Petrophysics* 1976 Reference Geophysics (Moskva: Nedra) 527

IOP Conf. Series: Materials Science and Engineering **81** (2015) 012056 doi:10.1088/1757-899X/81/1/012056

Research on modes of plasma generation in low-pressure discharge for themal radiation processes

T V Koval[1], I V Lopatin[2], Nguyen Bao Hung[3]

[1] Professor, National Research Tomsk Polytechnic University, 30, Lenin Ave., Tomsk, 634050, Russia.
[2] Junior researcher, Institute of High Current Electronics of Siberian Branch of Russian Academy of Sciences, 2/3, Akademichesky Ave., Tomsk, 634055, Russia.
[3] Post-graduate student, National Research Tomsk Polytechnic University, 30, Lenin Ave., Tomsk, 634050, Russia.

E-mail: lopatin@opee.hcei.tsc.ru

Abstract In this work a theoretical research on modes of low-pressure glow discharge generation and its parameters in a large area hollow cathode are carried out. The relations describing the dependence of the burning voltage on the gas pressure and the geometry of the system were obtained. It is shown that it is possible to adjust the ion current density and burning voltage by external current injection independently of the surface area and material of the processed parts. This work also carries out a numerical study of the influence of plasma parameters on the heating of processed parts. Theoretical results are compared with experimental data.

1. Introduction

Thermal radiation processing enhances the surface properties and lifetime of machine parts, while decreasing their cost. A target for plasma treatment is immersed in gas plasma in which ions are accelerated toward the target in the electric double layer formed between the target and discharge column. The saturating element (gas atoms) in the treated target volume propagates due to heat-activated diffusion whose rate depends mostly on the treatment temperature and saturating element gradient. The main parameters of plasma treatment are the gas composition or gas mixture ratio, temperature and time of the process, operating pressure, discharge parameters, dissociation and ionization degrees of the working gas, ion energy, and ion current density at the surface of a treated target [1]. Most of these parameters are interrelated. So, the gas composition or gas mixture ratio affects the discharge initiating voltage and hence the ion energy, and the voltage and current density are limited by the temperature allowable for treatment and by the gas pressure. Nitriding at 260–510 °C provides the formation of hard nitride layers of thickness ranging up to 200–300 μm [2, 3].

The vacuum plasma technology used for surface modification of materials and large-size objects is based on the generation of low-temperature glow discharge plasmas [4–9] for which it is required to provide an ion current density of ~1 mA/cm^2 to a treated surface and a discharge operating voltage of hundred volts [6–9]. One of the important problems is to control the discharge current and discharge operating voltage irrespective of the gas kind, gas pressure, and treated surface area.

The paper presents numerical simulation results on plasma generation modes (self-sustained and non-self-sustained) in a low-pressure discharge with a large-area hollow cathode. The effect of the plasma parameters, material and geometry of treated parts on their heating rate and temperature is investigated. The theoretical research results are compared with experimental data.

2. Hollow-cathode gas discharge

The cylindrical hollow cathode used in experiments on plasma generation is shown in Figure 1 [6, 7]. Anode *1* in the form of two tubes of total area S_a is located at the side wall of the cathode. The anode area S_a is varied by displacing the anode deep into the cathode cavity the volume of which is $V_c = 2\times10^5$ cm^3, $S_a = 200...500$ cm^2. The plasma is generated both in the main (self-sustained) mode and in the mode in which the discharge is sustained by an electron beam extracted from the plasma of auxiliary arc discharge *3*. The plasma is formed inside the volume of the hollow cathode. The cathode potential fall U_c is almost equal to the discharge voltage U: $U_c = U - U_a \approx U$, where U_a is the negative anode potential fall. The plasma ions accelerated in the cathode layer provides ion-electron emission from the cathode surface. The electrons accelerated and set in oscillatory motion in the hollow cathode (with about the same probability of ionization at any point of the cathode cavity) lose their energy in collisions with neutrals and ensure a self-sustained discharge. This provides high homogeneity of the ion current density, which is defined by the properties of the discharge plasma.

The hollow-cathode discharge is described by equations of energy balance of charged particles and current continuity which allow us to relate the internal parameters (temperature, plasma potential) to the external parameters (working gas pressure, discharge current and voltage, system geometry). From the balance equation for fast electrons we can obtain the discharge operating condition. In the general case of auxiliary discharge current injection I_{ext}, the balance equation for fast electrons is written in the form

$$\frac{\gamma I_i}{e} + \frac{\delta I_i}{e} = \frac{n_f V \nu_i}{u} + \frac{S_a}{4} n_f \langle \mathrm{v} \rangle \tag{1}$$

Here γ is the ion-electron emission coefficient which depends on the discharge voltage and gas kind; $I_i = e\,\nu_i n_f V$ is the ion current which is defined in terms of energy loss by fast electrons in gas ionization in the cathode cavity $V = (V_c - V_a)(1 - h/2D)$, h is the anode height, D is the length and diameter of the cathode cavity; $\delta=(1-\alpha)(I_{ext}/I_i)$, α is the fraction of fast electrons not involved in ionization; n_f and $\langle v \rangle$ are the density and average velocity of fast electrons; $\nu_i = n_g \sigma_i v$ is the ionization rate; and e is the electron charge. From equation (1) we can obtain the condition of the main discharge in the non-self-sustained mode:

$$u = \frac{P}{P(\gamma+\delta)-1}, \tag{2}$$

which gives a relation between the dimensionless discharge operating voltage $u = e\,U_c/W$, gas pressure $P = (p/kT_e)\sigma_i L$, and effective length of the cathode cavity $L = 4V/S_a$; W is the total electron energy expended in gas ionization. At high pressures $p >> (kT_e/\sigma_i L)(\gamma+\delta)^{-1}$, the discharge operating voltage does not depend on the effective length of the cathode cavity; the glow discharge condition is defined by the relation $(\gamma+\delta)u = 1$, and at $S_a << S_c$ it coincides with the relation derived from energy balance [8] and at $\delta=0$ with the relation obtained elsewhere [10].

To determine the spatial distribution of the electron temperature and plasma density and to study the effect of the geometry, size, and material of targets placed in the cathode cavity, we used a hydrodynamic model which describes the density of charged particles and their average energy as a function of time and space [7]. The transport coefficients (discharge coefficients) dependent on the electron energy distribution function are output parameters for the hydrodynamic model and are calculated using the BOLSIG+ program [11]. The plasma generation in the hollow cathode is simulated without considering the cathode region in which emission electrons are accelerated. When the current continuity condition is valid at the computational domain boundaries, the output parameters for the hydrodynamic model are the discharge characteristics obtained from (1)–(2).

Self-sustained mode. In the self-sustained mode, the high voltage of the main discharge is a shortcoming for the hollow cathode, because it leads to intense ion etching of a treated surface [3]. Figure 2 shows calculated (formula (2)) and experimental dependences [6, 7] of the discharge operating voltage on the gas pressure (argon, nitrogen) in the self-sustained mode at different effective lengths of the hollow cathode $L = 4V/S_a$; the discharge current is 30 A; the experimental values correspond to S_a = 500 cm^2. Decreasing the operating pressure of the main discharge causes a substantial increase in discharge operating voltage at a pressure below 0.65 Pa. At these pressures, a large fraction of ionizing electrons whose energy is only partly expended in ionization reaches the anode. The decrease in ionization efficiency is compensated by a corresponding increase in the coefficient of secondary electron emission $\gamma = \gamma(u)$ due to the increase in discharge voltage; in the range of voltages (200…1000) V, γ_{Ar} = 0.03…0.11 and γ_{N_2} = 0.04…0.2, ($\sigma_{i\,Ar}/\sigma_{i\,N_2} \approx$ 1.5).

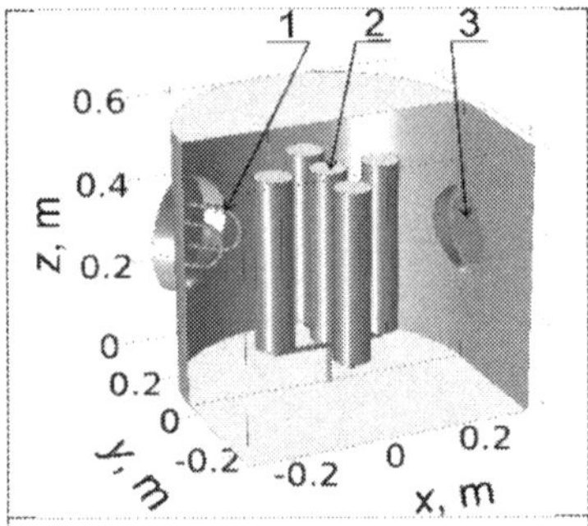

Figure 1. Hollow cathode model: *1* – anode; *2* – targets; *3* – auxiliary discharge window.

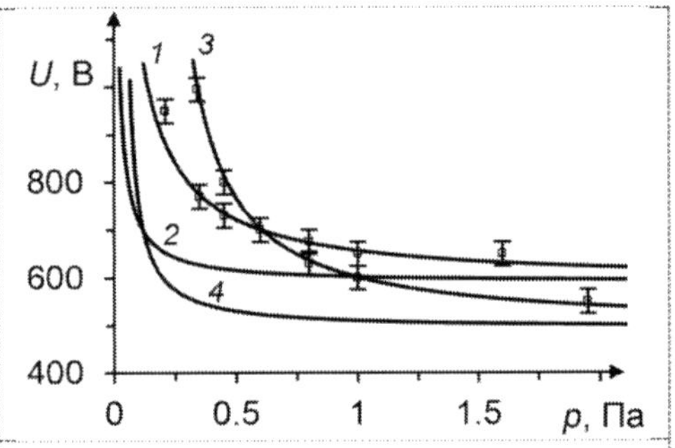

Figure 2. Discharge operating voltage vs the gas pressure in the self-sustained mode for nitrogen (*1*, *2*) and argon (*3*, *4*) at L=1.6×10^3 cm (*1*, *3*) and L=6.4×10^3 cm (*2*, *4*); I – experiment, I = 30 A.

From relation (2) and Figure 2 it is seen that decreasing the effective length of the hollow cathode $L = 4V/S_a$ (increasing the anode area S_a or decreasing the cathode volume due to targets, V_c-V_d) increases the discharge voltage. This owes to capture of fast electrons by the anode, and the self-sustained discharge current is kept at the specified level due to the emission current (the coefficient γ), i.e., due to the voltage in the cathode layer.

The current to the anode is defined by plasma electrons and anode fall. Figure 3 shows calculated gas pressure dependences of the argon plasma density in the center of the cathode cavity, plasma temperature, and plasma potential in the self-sustained mode at a constant discharge current of 30 A. The discharge current and current density at the cathode are kept constant due to the increase in anode fall with decreasing plasma density (S_a = 220 cm^2).

Non-self-sustained mode. Additional electron injection into the cathode cavity is through the window of the auxiliary discharge (Figure 1). As can be seen from relation (2), the auxiliary discharge shifts the characteristics of the main discharge toward lower voltages and gas pressures. This decreases the energy gained by an electron in the cathode layer of the discharge and the number of its ionization events.

Studies of the main discharge operating voltage U depending on the relative volume of targets (Figure 4) show that in both modes, the voltage U varies slightly at a nitrogen pressure of p_{N2}>0.65 Pa. In the non-self-sustained mode at an operating pressure of p_{N2}<0.65 Pa and discharge voltage of U>350 V, the main discharge operating voltage can be stabilized by varying the auxiliary discharge current.

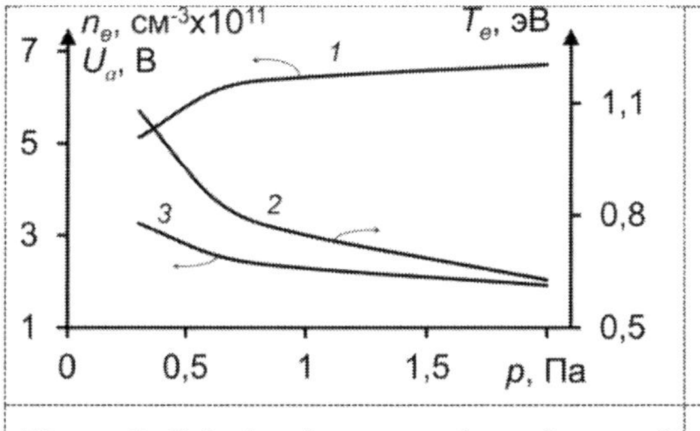

Figure 3. Calculated pressure dependences of the argon plasma in the self-sustained mode: *1* – plasma density; *2* – plasma temperature; *3* – plasma potential; $S_a = 220$ cm^2.

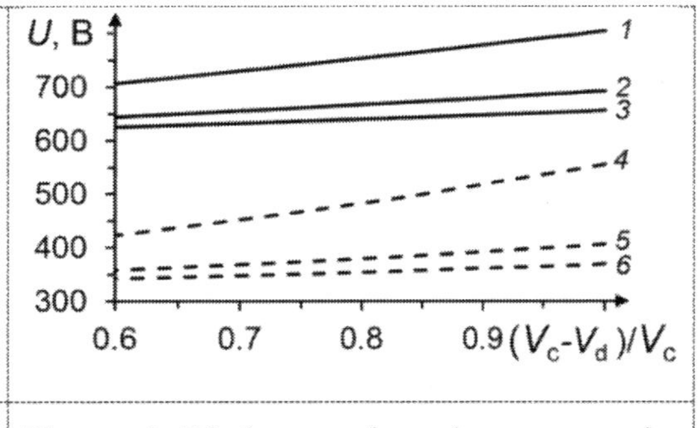

Figure 4. Discharge voltage in argon vs the relative volume in the self-sustained (solid lines) and non-self-sustained (dashed lines) modes at different pressures *p*: 0.35 (*1*, *4*), 0.65 (*2*, *5*), and 1 Pa (*3*, *6*); $\delta = 0.11$.

The external injection current and plasma potential vs the number of targets N (shaped as cylinders of dimensions ∅10 cm × 40 cm) for $p_{N2} = 0.65$ are shown in Figure 5. Stabilization of the discharge voltage at 370 V and ion current density with increasing the number of targets (N=1÷8) takes place as the current of additional emission is increased from 9 to 14 A. The plasma potential increases slightly, because the relative volume $(V_c\text{-}V_d)/V_c$ varies in the range of unity up to 0.9 (Figure 4), the total volume of targets is (0.25÷2)×10^4 cm^3, and their relative surface area is S_d/S_c=0.071÷0.57.

Using the hydrodynamic model, we obtained plasma density distributions in the hollow cathode for nitrogen. Figure 6 shows calculated lines of the plasma density in the non-self-sustained mode at I_{ext} = 14 A and U = 370 V in two sections of the hollow cathode with targets. The pressure gradient due to gas supply through the electron source increases the plasma density gradient in the region of the emission window.

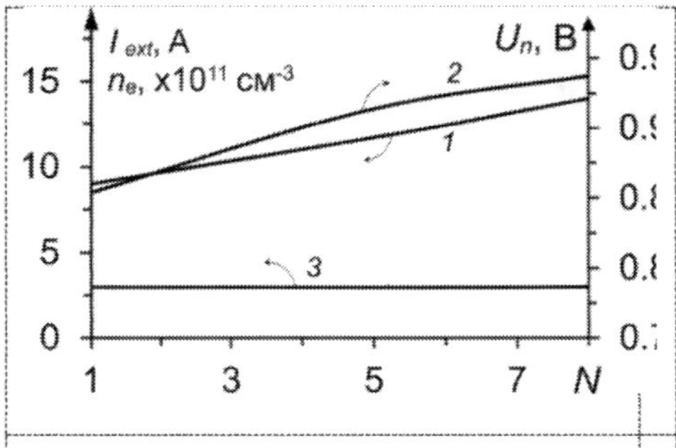

Figure 5. External injection current (*1*), plasma density (*3*), and plasma potential (*2*) vs the number of targets; nitrogen.

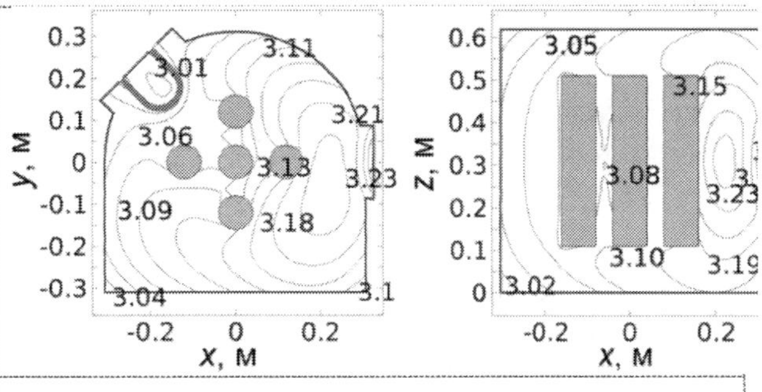

Figure 6. Plasma density distribution for nitrogen (cm^{-3}×10^{11}) in the lateral and longitudinal sections of the hollow cathode with targets; non-self-sustained mode.

The properties of the cathode material, configuration of a treated target, and its size can affect the discharge characteristics and the degree of treatment [12]. When treated, a target serves as a cathode; between the plasma and treated target, a space charge layer arises and its electric field provides acceleration of plasma ions. Due to ion bombardment, the target is heated. The heating time depends on the material and dimensions of a treated target. The heat conduction problem with regard to water cooling of the hollow cathode walls (Figure 1) was numerically solved for titanium and copper parts of dimensions ∅5÷13 cm and 20÷40 cm. The calculations show that with no shield, the specimen temperature is no greater than 330 °C, which is lower than 530 °C optimal for nitriding [3].

With an active shield surrounding the treated targets, the shield creates active particles and ensures additional heating of the targets. The times of heating to 400°C for targets of different geometries with an active shield placed at a distance of 2 cm from the cathode are presented in Table 1. The heating time of targets depends on their material and dimensions.

Table 4. The times of heating to 400°C for targets of different geometries with an active shield placed at a distance of 2 cm from the cathode.

Dimension of a part	Heating time, min Fe	Heating time, min Ti
∅10cm×40 cm	123	86
∅10cm×20 mm	113	75
∅ 5cm×40 mm	66	43
∅13cm×40 mm	155	108
∅13cm×20 mm	140	93

It is seen from the table that the difference in heating time for the parts is more than one hour. This defines the difference in nitriding time and hence in nitrogen penetration depth. The energy expended in heating several targets occupying the entire cathode volume can be the same and even lower than the energy expended in heating a single target. This owes to a decrease in discharge voltage and mutual thermal radiation of targets, which impairs heat removal from them.

The ion current density and ion energy are not decisive factors for nitriding, and the process efficiency is defined to a greater extent by the dose of implanted nitrogen ions at an ion current density to a treated target higher than 1 mA/cm^2 [3]. In this case, thermal diffusion dominates and ion bombardment is secondary. Experiments show that there are two competitive processes: penetration of nitrogen through diffusion and reduction of its depth due to sputtering [14]. An increase in nitrogen concentration correlates well with an increase in surface microhardness, and an increase in nitrided layer thickness provides an increase in wear resistance [3].

The Arrhenius curve characterizes the rate of increase in nitride layer thickness $\ln(d^2/t)$ depending on the temperature T [3]. The linear form of this dependence evidences that the growth of nitride layers is controlled by diffusion of nitrogen atoms. The diffusion is described by the formula

$$\frac{d^2}{t} = \exp(-\frac{E_a}{kT}), \quad (3)$$

where d is the nitride layer thickness, t is the treatment time, E_a is the diffusion activation energy determined from experimental data, T is the treatment temperature.

3. Conclusion

Thus, we studied the plasma generation in a low-pressure gas discharge with a large-area hollow cathode in the self-sustained mode and in the mode with additional electron injection.

Using the theoretical model in which fast electrons are responsible for the birth of charged particles in the hollow cathode plasma, dependences of the discharge operating voltage on the gas pressure and on the geometry of the hollow cathode and treated targets were obtained. It is shown that the discharge operating voltage decreases more than two times due to the current of additional electron injection. It is demonstrated that whatever the surface area and material of treated targets, it is possible to control the ion current density and discharge operating voltage by varying the current of additional electron injection.

Using the hydrodynamic model, it is shown that at a discharge operating voltage of 300–400 V, gas pressure of 0.65 Pa, and plasma temperature of about 1 eV, the plasma density for nitrogen and argon reaches 3×10^{11} and 6×10^{11} cm^{-3}, respectively.

The active shield provides uniform heating of targets of different shapes and dimensions to a temperature required for nitriding. The heating time of targets depends on their material and dimensions and decreases with increasing the number of targets due to their mutual thermal radiation. The temperature conditions of treated targets are controlled by varying the auxiliary discharge current and gas pressure.

The theoretical and numerical models agree with experimental data [6, 7].

This work was carried out at the expense of Russia research foundation (Grant no. 14-29-00091)

References

[1] Hosseini, Ashrafizadeh F 2009 Accurate measurement and evaluation of the nitrogen depth profile in plasma nitrided iron *Vacuum* **83** 1174–1178

[2] Zagonel L F, Figueroa C A, Droppa Jr. R, Alvarez F 2006 Influence of the process temperature on the steel microstructure and hardening in pulsed plasma nitriding *Surf. Coat. Tech.* **201** 452–457

[3] Berlin E V, Koval N N, Seidman L A 2012 *Plasma thermochemical surface treatment of steel parts* (Moscow: Tekhnosfera) p 462 (in Russian)

[4] Lopatin I V, Akhmadeev Yu Kh, Koval N N, Schanin P M, Yakovlev V V 2010 Generation of plasma in non-self-sustained glow discharge with hollow cathode *Modification of Materials with Particle Beams and Plasma Flows: Proc. of 10th Int. Conf. (Tomsk: Russia, 19-24 september 2010) (Publishing house of the IOA SB RAS)* pp 35–38

[5] Lopatin I V, Akhmadeev Yu Kh, Koval N N, Schanin P M, Shnaider A V 2008 Plasma Production in a Low-pressure Hollow-cathode Non-selfsustained Discharge *Proc. of 9 Int. Conf. on Modification of Materials with Particle Beams and Plasma flows* (*Tomsk: Russia 21-26 september 2008) (Publishing house of the IOA SB RAS)* pp 312–315

[6] Lopatin I V, Schanin P M, Akhmadeev Y H, Kovalsky S S, Koval N N 2012 Self-sustained low pressure glow discharge with a hollow cathode at currents of tens of amperes *Plasma Phys. Rep.* **38** 7 583–589

[7] Koval T V, Lopatin I V, Ogorodnikov A A, Nguyen Bao Hung 2014 Research on low-temperature plasma generation with a large area hollow cathode glow discharge *Izv. Vussh. Uchebn. Zaved. Fiz.* **57** 3/2 122–125

[8] Korolev Y D, Frants O B, Landl N V, Shemyakin I A, Geyman V G 2013 High-Current Stages in a Low-Pressure Glow Discharge With Hollow Cathode *IEEE Trans. Plasma Sci.* **41** 8 2087–2096

[9] Korolev Yu D, Frank K Discharge formation processes and glow-to-arc transition in pseudospark switch1999 *IEEE Trans. Plasma Sci.* **27** 5 1525–1537

[10] Metel A S, Grigoriev S N 2005 *Glow discharge with electrostatic confinement of electrons: physics, engineering, applications.* (Moscow: MSU IC STANKIN, Janus-K)

[11] Hagelaar G J M, Pitchford L C 2005 Solving the Bolrzmanneguation to obtain electron transport coefficients and rate coefficient for fluid models *Plasma Sources Sci. Technol.* **14** 722

[12] BOLSIG+ Electron Boltzmann equation solver: http://www.bolsig.laplace.univ-tlse.fr.

[13] de Ataide A R P, Alves Jr. C, Hajek V, Leite J P 2003 . Effects during plasma nitriding of shaped materials of different sizes. - Surface and Coating Technology *Surf. Coat. Tech.* **167** 52–58

[14] Galdikas A, Moskalioviene T 2011 Modeling of stress induced nitrogen diffusion in nitrided stainless steel *Surf. Coat. Tech.* **205** 3742–3746

Study of Sorption Properties of Aluminium Oxides and Hydroxides Powders Obtained by Electro-Impulse Methods

S P Zhuravkov[1], G L Lobanova[1], D V Martemiyanov[1] and L V Nadeina[2]

[1] Institute of High Technology Physics, Research Tomsk Polytechnic University, 2A, Lenin Ave, Bld. 11, 634028, Tomsk, Russia
[2] Institute of Natural Resources, Research Tomsk Polytechnic University, 2A, Lenin Ave, Bld. 5, 634028, Tomsk, Russia

E-mail: zhursp@yandex.ru

Abstract. Experimental results of physicochemical and sorption properties of material samples obtained by electro-spark dispersion in water and by conductor electric explosion in argon are shown in the paper. Due to comparison of investigated samples sorption activity under static conditions, it was able to determine the most effective samples in the process of extraction of Cu^{2+} ions from aqueous solutions.

1. Introduction

The development of new kinds of nanomaterials and their properties investigation are of the main interest in recent years [1-8]. Oxide and hydroxide forms of aluminium obtained by water and aluminium electro-explosive powder interaction are of primary concern among well-known nanosized materials [9]. The advantage of this technology is that there are no alkali metal or ammonium cations when aluminium hydroxide precipitates. And this, in turn, provides high purity of product derived from electro-explosive powders. However, different conditions of the process of conductor electric explosion and post passivation of obtained nanosized metal powders result in, even in the same laboratory, the original metal powders vary in particle size and shape, and they also differ in surface condition, they vary in a fraction of metal phase in powder composition etc. Hydroxides are formed in the metal oxidation zone as a result of nanosized metal powder of aluminium and water interaction. Due to the fact that hydroxides have slight water solubility they are capable of producing high oversaturated solutions when the metal is oxidized intensively. That is why obtained products contain not only crystalline mass but cryptocrystalline and colloid ones. In spite of way of hydroxides formation in time especially under the air-dry condition, they lose capillary and adsorbed water forming chemically-combined hydroxide group compounds. Therefore, anhydrous oxides are formed which have property stabilized surfaces. Thus, comparative testing of their use as an active component of filter and sorption materials in order to clean water from microbiological pollution [10] and to treat sewage from heavy metals [11] is of the main interest.

Electric discharge metal dispersion is another method to obtain metal nanopowders using electric discharges. This method has been presented in sufficient detail in the papers [12-13]. Electric discharge metal dispersion is carried out in metal electrode when between metal electrodes there is a metal load (in the form of granules or arbitrary shape particles size 5-10 mm) submerged in liquid (for example, in water). Therefore, dispersion of metal and producing fine metal particles and water interaction are not separated from each other as in the use of electro-explosive powders. In such kind of system electric breakdowns of some sections (as power fluid layers and thin oxide films on the surface of granules) between metal granules are due to impulses of voltage energizing electrodes. The breakdowns can be shown by way of multiple sparking (discharge channels). When applying practically voltage impulse parameters from 0,5 to 25 kV, range of sizes of finely dispersed metal particles is sufficiently wide within the interval from nanometer units to tens of micrometers. Being to the left of hydrogen in the line of metal activities the finely dispersed metal particles, forming and accumulating in investigated system, are oxidized and displace hydrogen from water [14]. Hydraulic shocks, accompanying electric discharges, maintain the finely dispersed metal particles in suspension. They help the diffusion processes to be easy at "surface of metal particle – water" boundary. They also provide accelerated development of oxidation-reduction reactions in the bulk suspension. At the same time the large-sized granules of metal loading remain at the bottom of reaction chamber. In electroerosion chamber the metal oxidation undergoes partially in the periods of electrodes voltage supply, partially it undergoes in the periods between impulses, and it continues after disconnection of voltage source when there is no electrode voltage. In this period the metal particle oxidation is carried out by chemical method at uncontrolled potential [14].

Therefore, according to the process for making products of metal load erosion – polydisperse metal particles – they, on the one hand, are acted upon by factors originated with a discharge channel both in solid and in gas and in liquid. But, on the other hand, the particles interact with water or solution at uncontrolled potential [12].

Hence, metal nanopowder method can effect on composition and properties of products which have been formed.

The aim of work is to compare and to estimate structural characteristic and sorption properties of both aluminium hydroxides, obtained by oxidation of aluminium electro-explosive and electroerosion powders and water interaction products, and active aluminium oxide for chromatography by containing Cu^{2+} cation aqueous solution contact.

2. Experimental

Active aluminium oxide (sample 1 Table 1) was used as a comparison standard [15].

Sample 2 (Table 1) was obtained by aluminium electroerosion powder oxidation in water. Experimental apparatus reported in [12] was used to carry out electroerosion dispersion of aluminium load (granulated aluminium chemically pure grade) in water. After the treatment of aluminium load by electric discharges had been finished, suspension of eroded materials was separated from load by decantation and it was holding in touch with water until hydrogen stopped giving off at room temperature. Then the product was dehydrated and dried at 105^0C.

Electro-explosive aluminium nanopowder, the initial particles of which had a size from 20 nm to 100 nm, was used to obtain aluminium oxyhydroxide (sample 3 Table 1). The initial particles are combined into aggregates with the size to 500nm and they are combined into loosely coupled aggregate agglomerates with the size to 5 μm . Average surface diameter of aluminium nanopowder particles is 110 nm. Aluminium nanopowder was previously passivated in air, and hereby the metal aluminium content 20nm was min 91 % mass. Then aluminium nanopowder interacted with water according to the method of AlOOH nanofiber process used in the paper [9].

The thermal nitrogen desorption method was used to estimate structural characteristics of investigated materials. Samples of pores were tested with a help of analyser «SORBTOMETR-M», and specific surface (S) and specific volume of pores (P) were estimated according to measurement data.

Cu^{2+} ion sorption was carried out under static conditions. Samples of 0,2 g materials were added to 20 sm^3 solutions (V) with different initial concentration (C_0) of adsorbed substances from 2 to 40 mg/dm^3. After carrying out this process during 150 minutes the solution was separated from sorbent with a help of centrifugation rotating at speed of 20000 r/min, and adsorbed substances equilibrium concentrations (C_p) were determined. Cu^{2+} ion concentration was determined by method of stripping voltammetry with a help of a device with analyser TA-07. Cu^{2+} ion sorption was carried out in activated aluminium oxide – chromatographic standard sorbent – under similar conditions to quantify sorption properties of derived products [15].

Sorption capacity (mg/m^2) was calculated according to formula:

$$A_p = \frac{(C_0 - C_p) \cdot V}{m \cdot S}$$

Where: C_0– initial concentration, mg/dm^3; C_p - equilibrium concentration, mg/dm^3; V– volume of solution, dm^3; m – mass of sorbent, g; S – specific surface of investigated products, m^2/g.

3. Discussion of Results

Data of phase composition, specific surface, specific volume and average surface diameter of investigated sample pores are presented in Table 1.

Table 1. Specific surface, specific volume of pores and average size of investigated sample pores

Number of sample	Sample	$S_{уд}$, m^2/g	P, cm^3/g	Average diameter of pores, nm
1	Standard (Al_2O_3) 100%	194,33	0,083	1,722
2	AlO(OH) 24,1% +$Al(OH)_3$ 66,1% $Al_{2,66}O_4$ 9,7%	85,07	0,037	1,720
3	AlO(OH) 89,5% +10,5% $Al(OH)_3$	220,57	0,095	1,717

As shown in Table 1, the sample №3 possesses maximum specific surface and specific volume of pores. It is aluminium oxyhydroxide synthesized by water and aluminium powder interaction. This powder was obtained under argon with a help of conductor electric explosion technology consisting of two phases. Larger portion of product includes AlO(OH), whose formation energy is less than $Al(OH)_3$. Sample №2 contains similar substances in its composition but different ratio. This sample makes up 66% of $Al(OH)_3$ phase content. According to XRD data, nonstoichiometric oxide $Al_{2,66}O_4$ was found out together with hydroxides in sample №2. Similar structure oxide was previously described in the paper [16]. Discharges resulting in erosion of metal are most likely to take part in hydrate formation process. Probably, this oxide formation is possible in vapour-gas bubble, when the discharge channel degenerates into the vapour-gas bubble. Oxide phase formation was observed due to electrical erosion of zinc load in water [17]. In comparison with sample №3, enumerated special characteristics of aluminium hydroxides obtained by electrical erosion in water have determined not only the change of phase composition of product but the change of structural characteristics – specific surface and volume of pores have decreased. Therefore, the conditions of nanodispersed powder obtaining influence on its interaction with environment, in particular, with water. So, the authors of paper [18] noted significant influence of aluminium oxyhydroxide synthesis conditions (the temperature of process, pH medium, time and temperature of precipitation ageing, temperature and time of samples treatment) on composition and properties of precipitations being formed when oxidizing conductor electric explosion aluminium nanopowder. Thus, investigated samples, obtained from aluminium with a help of electric discharges of different stages technology process, differ from each other both composition and structural characteristics, and this, without doubt, must have an effect on properties of their surfaces.

Adsorption isotherms of Cu^{2+} ions of investigated products and active aluminium oxide taken by us as a standard according to adsorption amount are presented in Figure .

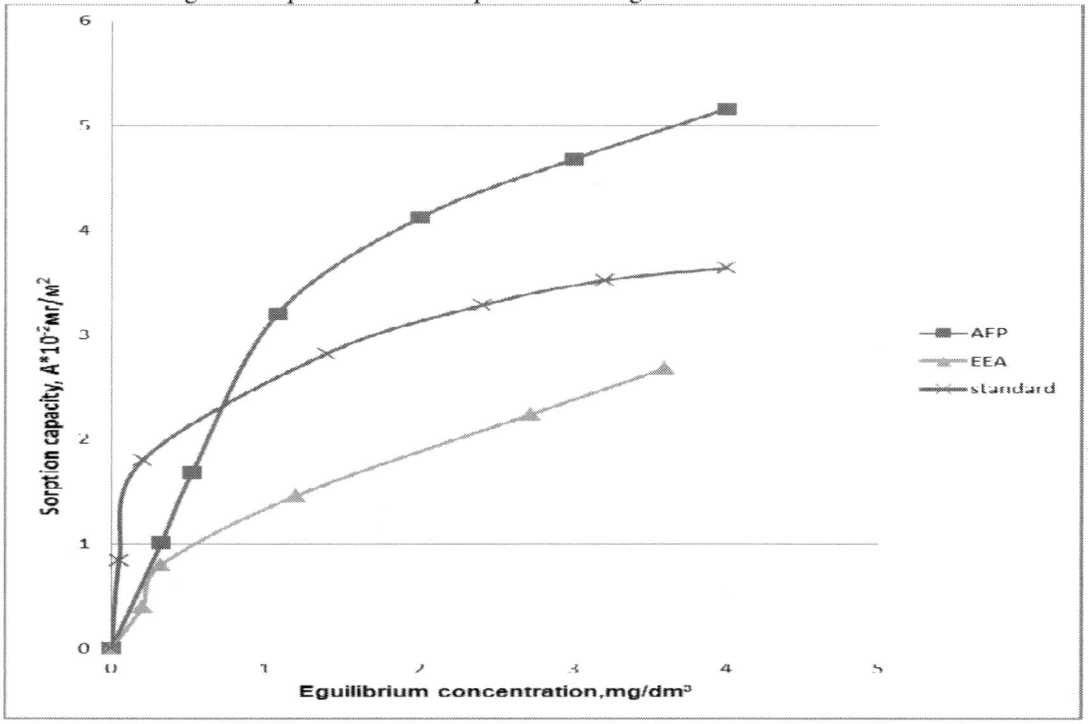

Figure . Adsorption isothere of Cu^{2+} ions.

Figure indicates that ions of Cu are the most actively sorbed on the surface of sample №2, which is a mixture of aluminium oxide and hydroxides. Sorption properties of sample №3 are close to standard.

4. Conclusions

1. It is found that according to structure and phase composition, the products of aluminium powders oxidized by water obtained by various electro-impulse methods are different.
2. Comparison tests of material samples obtained under static conditions in the process of extraction of Cu^{2+} ions from aqueous solutions were carried out. It was shown, that the sample consisting of mixture of aluminium oxide and hydroxides, which was obtained under conditions of electrospark aluminium-in-water dispersion, is the most effective.

References:

[1] Lobanova G L, Zhuravkov S P and Pustovalov A V 2011 Izv. Vys. Ucheb. zav. Fizika. B.54 N.11/3 pp 168-171

[2] Romanov A N, Romanova N A, Yavorovskiy N A, Zhuravkov S P and Pustovalov A V 2012 Izv. Vys. Ucheb. zav. Fizika. B. 55 N 6/2 pp 225-230

[3] Yavorovskiy N A, Pustovalov A V, Lobanova G L and Zhuravkov S P 2012 Izv. Vys. Ucheb. zav. Fizika. B. 55 N 6/2 pp 236-243

[4] Fedushchak T, Vosmerikov A, Uimin M and Zhuravkov S Proceedings – 2012 7th International Forum on Strategic Technology, IFOST 2012 2012. C. 6357498.

[5] Fedushchak T A, Petrenko T V, Vosmerikov A V, Kanashevich D A, Velichkina L M and Zhuravkov S P 2012 Russian J. of Physical Chemistry A. B. 86. N3 pp 375-379.

[6] Zhuravkov S, Plotnikov E, Martemiyanov D, Yavorovsky N, Hasse U and Zander S 2014 Adv. Mater. Res. V 872 pp 74-78

[7] Lysenko E N, Surzhikov A P, Zhuravkov S P, Vlasov V A, Pustovalov A V and Yavorovsky N A 2014 J. of Ther. Anal. and Cal. V 115 Issue 2 pp 5-11

[8] Zaikovskii V I, Anufrienko V F, Kodenev E G, Echevskii G V, Vasenin N T, Ismagilov Z R, Parmon V N, Vosmerikov A V, Korobitsyna L L and Zhuravkov S P 2005 Dokl. Phys. Chem. V 404 N4-6 pp 201-204

[9] Pustovalov A V, Shiyan L N and Zhuravkov S P 2010 Sbornik trudov 3-y Vserossiyskoy nauchnoy konferentsii s mezhdunarodnym uchastiem. Nanomaterialy i tekhnologii.Nanostruktuirovannye sistemy v fizike kondensirovannogo sostoyaniya. Tekhnika i tekhnologiya nanomaterialov - Ulan-Ude, 26-28 avgusta 2010.- Ulan-Ude: Izd-vo Buryatskogo gosuniversiteta pp 126-131

[10] Martemiyanov D V, Korotkova E I and Galanov A I 2012 Vestnik Karagandinskogo universiteta : Seriya: Khimiya V 67 N 3 pp 61-65

[11] Martemiyanov D V, Galanov A I and Yurmazova T A 2013 Fundamental'nye issledovaniya N 8/3 pp 666-670

[12] Sergey Zhuravkov, Alexey Pustovalov, Galina Lobanova, Olga Kvashnina and Nikolay Yavorovsky 2014 Proceedings International Conference on Mechanical Engineering, Autornation and Control Systerns (MEACS) 978-1-4799-6221-1/14/$31.00 ©2014 IEEE.

[13] Evgenii Plotnikov, Sergey Zhuravkov, Andrew Gapeyev, Vladimir Plotnikov and Dmitry Martemiyanov 2014 Adv. Mater. Res. V 1040 pp 65-69 Online available since 2014/Sep/24 at www.scientific.net

[14] Sharlo G. Metody analiticheskoy khimii. Izd-vo «Khimiya», M. - L., 1965 pp 152 - 205

[15] Oksid alyuminiya aktivnyy AOA-1 (GOST 8136-85)

[16] Shirasuka K, Yanagida H and Yamaguchi G. 1976 J. of the Ceramic Assoc. of Japan V 84 pp 610-613

[17] Galanov A I, Lobanova G L, Zhuravkov S P, Saprykin F E, Kornev Ya. I and Yavorovskiy N A 2013 Russian J. Perspektivnye materialy N 4 pp 64-71

[18] Volkova G I, Sedoy V S 2008 Russian J. Zhurnal prikladnoy khimii V81 Vyp5 pp 721-725

IOP Conf. Series: Materials Science and Engineering **81** (2015) 012058 doi:10.1088/1757-899X/81/1/012058

The energy dependence of refractory metals and alloys radiation damageability

K M Mukashev[1], F F Umarov[2]

[1]Almaty,050100, Kazakhstan, National Pedagogical University after Abai, 13, DostykAvenue,

[2]Almaty,050000, KazakhstanKazakh-British Technical University, 59, Tole bi str.

e-mail:mukash_kanat@mail.ru

Abstract.

In this work the systematical investigation of the radiation defects distribution profile energy dependence in three different materials - tantalum, molybdenum and 10X18H10T–VD stainless steel irradiated by high energy protons has been performed. It has been shown that in stainless steel and tantalum, independently of proton energy, the vacancy complexes related by configuration appear which are described by the slightly expressed elastic channel. The defects are recovered in one annealing stage with different migration activation energy. At the same time the molybdenum radiation damageability is composed of two components in each of which its specified defects formation mechanism takes action. For high energy protons the inelastic channel of interaction is the basic and subcascades appearance is created by primary knocked-on atoms of considerable energies. For low energy protons the processes of elastic interaction with lattice atoms and atomic hydrogen in the end of run creation are major.

Introduction

As is well known, the effects of nuclear radiation on matter is accompanied by a number of new phenomena. The most important among them are nuclear reactions and change associated with these reactions in the elemental composition, formation and the emergence of clusters of point defects, damage to the matrix caused by cascade atom - atom collisions, etc., and as a result, violation of the integrity of the crystal.For the study of phenomena related to changes in the crystal structure of the material in the active zone of reactor, it is often sufficient to conduct simulation experiments at accelerators of charged particles. It is very important task of studying the profile of the defect distribution along the depth of the damaged layer. At one time, for this purpose on the basis of theoretical research program has been developed for calculations on computer of the displaced atoms profile in depth passage of heavy ions in the material [1]. But any program, as it was neither universal, yet cannot take into account all aspects of the complex process of interaction of charged particles with the real crystal lattice, the more it cannot be acceptable when the object of study is the multicomponent alloys.

Charged particles during moving in matter lose energy. The energy loss of the incident particle can occur in various ways, including ionization and excitation of the electron shells, the atoms' polarization of the medium, the radiation losses and nuclear stopping, whose role in the formation of structural defects may be different. Consequently, the profile of the defect distribution in the depth may also be different, the location of which depends on the type and parameters of the bombarding particles of the target material, and the irradiation temperature. In this regard, experimentally obtained parameters of the defect structure may differ considerably from the theoretically calculated.

The average path traversed by a charged particle in matter to a full stop, called its range - R. The latter depends on the particle energy andmaterial properties of the target.The range of the particle is usually expressed through the length of the path - d and density - ρ material:

$$R = d \cdot \rho \; [g \cdot cm^{-2}] \qquad (1)$$

To assess the interaction of particles with the matter in the directories are: energy particles - E in the lab coordinates, expressed in MeV; particle ranges R (E), expressed in g • cm-2; stopping power S (E) in MeV • cm-2 • g-1; derivatives of stopping power by energy D (E), used as a correction factor [2,3]. To calculate the range of a particle whose energy lies between the tabular data, use the formula:

$$R(E)\Delta E) = R(E) + \frac{\Delta E}{S(E)} - \frac{1}{2}\left|\frac{\Delta E}{S(E)}\right|^2 \cdot D(E), (2)$$

where E - the nearest table value of energy. On the contrary, for the calculation of energy corresponding to range, the value of which lies between the values in the table, you can use the formula:

$$E(R + \Delta R) = E(R) + S(E) \cdot \Delta R - \frac{1}{2} S^3(E) \cdot D(E) \cdot (\Delta R)^2 \qquad (3)$$

To find the intermediate values of S (E), are absent in the table, is using a linear interpolation relationship:

$$S(E + \Delta E) = S(E) + D(E) \cdot \Delta E, (4)$$

where$S(E)$ - the closest value of the stopping power. Similarly, absent in the table intermediate value $D(E+\Delta E)$ can be found by the relation:

$$D(E + \Delta E) = D(E) + \frac{\delta D}{\delta E} \cdot \Delta E, (5)$$

Under these conditions, the accuracy of calculation of the tabulated data is ~ 1%.

Experimental

Because as the passage of charged particles in matter, there is a consistent discharge its energy, the study of the profile of the defect distribution in depth, in principle, is the task of the study of the energy dependence of metal radiation damageability. To solve this problem, you can go the way of studying of metal defect structure by sequentially etching the surface, or on the basis of variable thickness of the absorber method. Naturally, the most acceptable is the second, non-destructive method of research, the essence of which is to irradiate by the charged particles of high energy and studying the stack of foils, the total thickness of which is greater than the mean free path of the particles in the material. By using such a technique, each foil is irradiated by particles of different energy and contains the respective structural damages characteristic forgiven depth of the material. The test objects were used polycrystalline Mo and Ta, and stainless steel-VD 10H18N10T as a foil thickness of 100 μm and a diameter of 17 mm each. The initial state of metals achieved by annealing at T = 1200°C and steel - at 1050°C for 1.5 hours in a vacuum of 10^{-5} Pa. The thickness of each foil Δd defines a path element $\Delta x_i = \Delta d_i \rho$, on which there is a loss of energy of protons $\Delta E_i = s_i(E)\Delta x_i$; the mean energy of the protons on the other side of each foil will be $E_{xi}=E_i-\Delta E_i$.. Consequently, each successive foil is irradiated with protons of different energies, studying the degree of damage that can be installed its energy dependence. Irradiation was carried out by protonsflow $1{,}2 \cdot 10^{13}cm^{-2}c^{-1}$up to $2 \cdot 10^{17}cm^{-2}$.fluence.

The main research tool in this case is the electron-positron annihilation (EPA)method. Being one of the major nuclear-physical methods of studying the state of the condensed medium, the method of electron-positron annihilation is a very sensitive tool to various kinds of damages of crystal structure. Spectrum shape of the angular distribution of annihilation photons (ADAP), resulting from the annihilation of positrons with electrons of the material changes significantly at localization of positrons near the defects in the crystal lattice, as well as from the atomic environment of defective regions. Slow positrons react also to change the order of the structure. Therefore positron probe is an ideal tool for studying the electronic states of the metallic materials local microregions. The study of structural defects of materials produced by the spectrometer with a linear-slit geometry with an angular resolution of 0.5 mrad.As a positrons source the isotope ^{22}Na of with 10 mCiactivity was used. The measurement ofADAPspectrum allows to determine the relative contribution to the annihilation of positrons process with the conduction electrons and ion core electrons. For this experimentally measures the intensity of the annihilation gamma rays as the dependence of counting rate of pulses coincident in time 2 photons detected by opposing detectors on thedisplacement angle of the movable detectorθ.ADAP spectra measured for different states of the material, are normalized to a single space. Not difficult to establish that the spectrum for the defective material has a higher intensity at the maximum and the narrow width at half maximum (Figure 1).

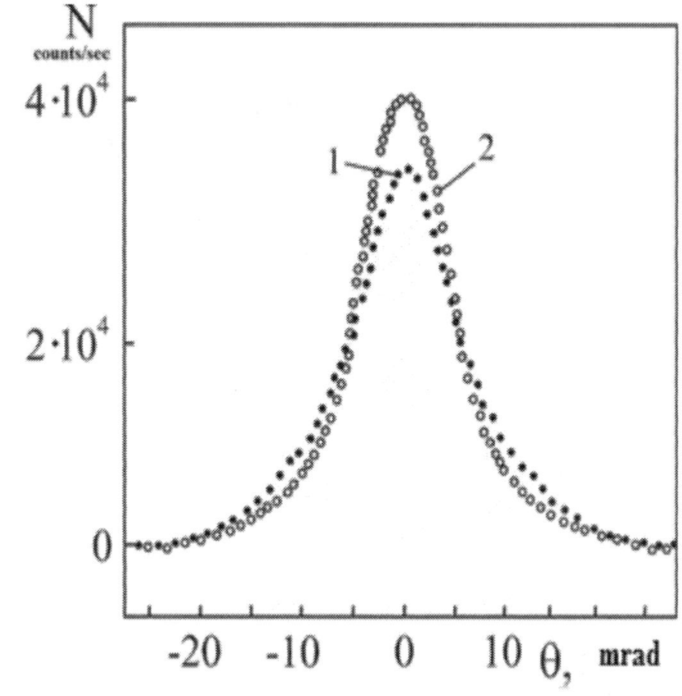

Figure 1. Experimental ADAP spectra: 1 - for the source;
2- irradiated materials

Results and discussion

To interpret the results of studies were used the following structure-sensitive annihilation parameters: F - positron annihilation probability redistribution between the conduction electrons and bound electrons, as well as its corresponding increment ΔF of relative values for the initial state is recovered by processing the spectrum of the angular correlation of annihilation radiation [4]. Results of studies of stainless steel according to these conditions by the annihilation parameters change dataare summarized in Table 1.

Table 1. Parameters annihilation became the depth of passage protons$E_{initial}$ = 30 MeV (F = 2 • 10^{17}cm^{-2})

Sample number	E_P, MeV	X, µm	$F=S_p/S_g$	$f=N(0)/N(8)$	FWHM, mrad.
annealed	-	-	0.15	3.1	6.1
19	0	-	0.27	3.9	5.6
18	2.33	1750	0.24	3.9	5.6
17	6.40	1650	0.24	3.8	5.7
16	9.09	1550	0.28	3.4	6.0
15	11.29	1450	0.25	3.7	5.7
14	13.19	1350	0.31	3.6	5.8
13	14.91	1250	0.26	3.4	5.9
12	16.49	1150	0.23	3.2	5.9
11	17.97	1050	0.27	3.5	5.8
10	19.35	950	0.28	3.8	5.7
9	20.67	850	0.27	3.5	5.7
8	21.92	750	0.29	3.6	5.7
7	23.12	650	0.27	3.5	5.9
6	24.27	550	0.21	3.4	5.9
5	25.39	450	0.25	3.6	5.8
4	26.47	350	0.23	3.3	5.9
3	27.51	250	0.25	3.4	5.9
2	28.52	150	0.24	3.7	5.7
1	29.50	50	0.26	3.7	5.7
error±	0.05	1.00	0.02	0.1	0.1

High energy protons irradiation is changeconsiderably the spectrum shape of annihilation photons angular distribution. The spectrum is narrow in half-width and maximal intensity raiseconsequently the redistribution of positrons annihilation probability with conduction and ion core electrons. These factors are visible in annihilation parameters change. If for annealed sample F = 0.15, whereas after irradiation it increases almost twice as much. The spectrum half-width (FWHM) for initial state is equal 6.1mrad. The protons irradiation is reduce to it decreases at the average up to 5.6mrad. In spite ofexisting parameters considerable change the evident regularity between them and protons energy does not visible in this case. Though, the certain tendency in the annihilation parameters behavior yet it can be established. Thus, for example the positrons annihilation relative probability F with particles energy increase is steadily decrease at the average. It can testify that the main contribution in the steel radiation damageability carried in the low energy protons are suffered the elastic collisions with steel components atoms.

Therefore it can suppose that the structural damages arising meanwhile in the steel samples are not differ practically between itself both by configuration and by positrons capture efficiency and correspond to the traps of one type. The last is confirmed by shape of isochronal annealing curves from the stack of the samples irradiated by different energy protons (Figure 2). At E_P= 29.5 ÷ 13.2 MeV protons energy the materials basic properties recovery is ended in the 350-600°C temperature interval and at $E_P \leq$ 6 MeV - in the field of 250 - 550°C, that is the specified relationship between protons energy and defects annealing temperature existence is present.

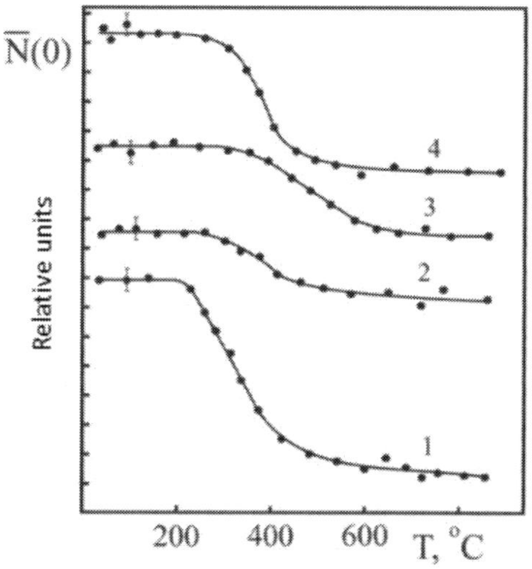

Figure 2. The energy dependence by protons irradiated 10X18H10T – ВД steel annealing kinetics
1. E_P = 4.6 MeV; 2. E_P = 31.2 MeV;
3. E_P = 23.1 MeV; 4. E_P = 29.5 MeV

Apparently of the defects migration activation energy value, consequently irradiation by high energy protons the dislocation loops with E_a= 2,1 – 2,2 eV are appeared. In the low energy protons case the vacancy complexes as a small subcascades or connected vacancy-impurity complexes with E_a= 1,7 – 1,8 eV are created evidently. The connected vacancy-Cr atom state which difficultly than other decomposed at annealing is most likely[5].

Unlike of stainless steel the polycrystallinetantalum irradiation at the same conditions is lead to same characteristic changes of the annihilation parameters. It is very difficult to separate out for this metal the spectrum parabolic component in consequence of insignificant part of the freecharge carriers. Therefore the basicequivalent (instead of F) parameter is the counting rate in the maximum of spectrum relation to its value at the angle θ = 8 mrad, that is f= $N(0)/N(8)$. The dependence of this parameter on protons energy is given in Figure 3a. The maximal spectrum narrowing and parameter f= $N(0)/N(8)$ considerable increase correspondingly is observed at low energies ~5 - 8 MeV. With particles energy increase these parameters are possessed a steadilyincrease or decrease character.

Such changes of parameters characterized of the spectrum shape are possible only at corresponding changes of capture efficiency of positrons by defects, created by protons. As far as the effectivenessmaximum is fitted on field of low energy protons, that whichsuppose that the largest damageability tantalum is suffered consequently of elastic interactions. The nuclear reactions influence activated by inelastic interactions in the high energy region, here is negligible. The actual picture of the structure damages in tantalum is can be determined consequently isochronal annealing of individual samples from the stack irradiated by different energy protons (Fig.3b).

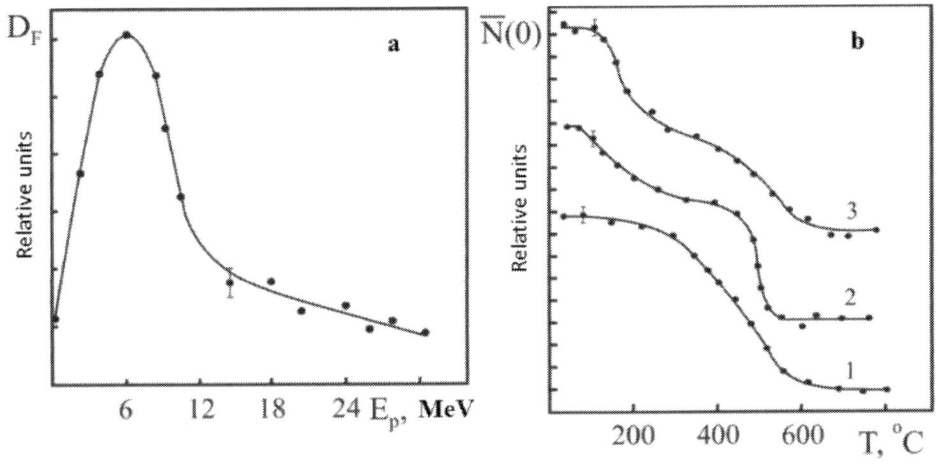

Figure 3. The energy dependence of damageability (a) and annealing kinetics (б) of tantalum, irradiated by protons with 30 MeV initial energy: 1. E_P = 6 MeV; 2. E_P = 25 MeV; 3. E_P = 30 MeV

One can see that the sample irradiated by low energy protons is recover in one stage in 250 - 600°C temperature interval with curve amplitude~15 % (curve 1), whereas at high energy irradiation the structure damages in Ta are appearing with two recovery stages (curves 2,3). And with particles energy decreasing from 30MeV to 25MeV the point defects migration beginning is shifted sideways to lower temperatures with simultaneous increasing of vacancy defects part. This confirm the decisive contribution of elastic interactions in the process of defect formation. So, at protons irradiation with E = 30 MeV the vacancy stage part is compose 41% from the damages common level and with energy decreasing up to 25MeV it increasing up to 47%. Meanwhile the recovery second stage is receive the more relief appearance than in the first case. The migration activation energy for vacancy components is meanE_{a1} = 1,41–1,45 эВ, and for dislocation and more complicated components are annealed pending second stage - E_{a2} = 2,33–2,35 eV.

Still large interest in these experiments are can present the damageability distribution profiles of molybdenum irradiated at the same conditions study results. For molybdenum in annealed state the annihilate photons angular distribution is approximated by one Gaussian, as far as the probing positrons are annihilated basically with ion core electrons. The parabolic component in spectrum is does not succeeded. But consequently protons irradiation the nascent structure damages arepromoted to the parabolic component responsible for positrons annihilation with free charge carriers appearing in the spectrum. Meanwhile the energy dependence of the relative annihilation probability $F = S_P/S_g$have a more complicated and nonmonotoniccharacter

(Figure 4). After certain energetic region 5 - 10 MeV in such the relative probability *F* is stay constant, in the sequel it is sharply decreased and achieve the minimum at about 18 - 20 MeV is increasing again up to E_P= 30 MeV. Such annihilation parameter dependence is can be activated by the appropriate defects distribution profile which is determine the defects both concentration and configuration along the trajectory of charged particles in matter.

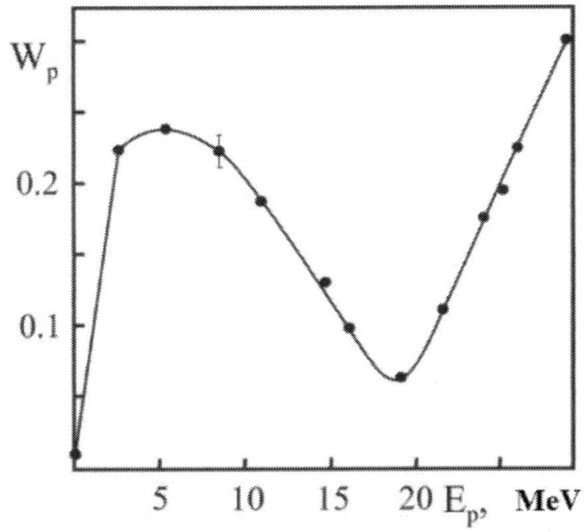

Figure 4. The energy dependence of the Mo damageability irradiated by protons, with initial energy 30 MeV

The probablereason of such dependence can be the different contribution in the radiation defects generation process of the elastic interactions at E_P<18 MeV energies and subcascade regions formation atE_P>20 MeV on the one hand and also the nuclear reactions specifiedroleon the other hand. But in both cases the defects structure must be vacancy that is confirmed by isochronal annealing results. Regardless of energy the proton irradiation consequently the vacancy complexes are appeared in Mo and in annealing temperature region up to 800°C the only one incomplete return stage have been observed. Meanwhile as far as proton energy increasing the defects migration temperature threshold is gradually shift to more high temperatures sideways - from 300 - 350°C up to 450 - 500°C that testify about appearance in the metal the more stable temperature concerning defects. But in work [6] is stated that the full annealing of irradiated by protons Mo completed at temperature about 300 - 320°C. But this temperature is considerable lower than Mo recrystallization temperature, that give rise to doubt in above mentioned conclusion [6]. But annealing temperature reached in this investigation (850°C) is also does not sufficient for completioneven the first stage of recovery (III-stage), whereas the recrystallization temperature of Mo is higher than 1000°C[7].

Conclusion

Thus the defects distribution profiles investigation by charged particles passing depth in three different materials is shown that in stainless steel and Mo regardless of protons energy similar to configuration vacancy complexes are appeared which recoverable in one annealing stage with migration activation energy E_a= 1,7 – 1,8 eV. The steel total damageability in the whole of protons energy interval is formed basically by one low-grade elastic channel, whereas for molybdenum it is added from two components in each of which its defined mechanism of defects formation is work. If for high energy protons the inelastic channel of interactions and subcascades appearing are the basic, that which the major processes for low energy protons are the elastic interactions with lattice atoms and atomic hydrogen formation in the end of run. The basic mechanism of the radiation defects formation in Ta is the evidently expressed elastic interactions. The nuclear reactions role in the structure damages creation are low-grade in this case. The defects created by low energy protons are annealing in one stage, whereas at high energy particles influence the radiation defects appearing which are recovery in two stages possessed the vacancy and dislocation nature.

References:

[1]Antonov A.G. 1993 The investigation of tritium entry in environment at liquid leady- lithium irradiation eutectic through channel wall from stainless steel // The liquid metal application in the national economy. - Obninsk, pp 181 - 183.

[2]Myllyia R., Xostomaova J. 1985 A time-to amplitude converter with constant-fraction timing discriminators for short time interval measurements.*Nucl.Jnst. and Methods* A **239**pp. 568 - 578.

[3]IbragimovSh.Sh., Reutov V.F., Abdurashitov I. Yu.1977Radiation damageability of Mo at high temperature irradiation by protons with initial energy 30 MeV.*VANT, ser. FRP and RM*B.**1(21)**pp. 76 – 79.

[4]Mukashev K.M. 2007Low Positrons Physics and Positron Spectroscopy Almaty508 p.

[5]Berestetsky V.B., Livshits E.M., Litaevsky A.P. 1968The relativistic quantum theory. MNaukaPart.1 480p.

[6]Dextyar I.Ya.,Mukashev K.M., RustamovSh.A., at all.1984 The positron annihilation in irradiated by chromium- ruthenium alloys *Ukr. Fis.Journal* V.**29** N**11**pp.1679 - 1681.

[7]Belyaev V.N., Zemlerub P.A., KovalevV.Yu. at all.1980 The positrons lifetime spectrometer Preprint ITEF 95 M 26 p.

Use of Simulation Methods of Wave Processes for Non-Destructive Testing of Concrete Products of Different Geometrical Size

T V Fursa[1], B A Lyukshin[2], G E Utsyn[1], D D Dann[1] and M V Kuimova[1]

[1] *Tomsk, 634050, Russia,* National Research Tomsk Polytechnic University

[2] *Tomsk, 634050, Russia,* National Research Tomsk State University,

E-mail: fursa@tpu.ru

Abstract.

Theoretical and experimental research of electrical responses under pulsed mechanical excitation of different size concrete products was carried out. Comparison of calculated and experimental data shows very close agreement. The obtained results will help to develop new opportunities to improve the method of non-destructive testing of concrete by creating a bank of calculated electrical responses from products of different geometrical arrangement and elastic response and using them as standard samples for defectiveness evaluation of products according to previously established criteria.

1. Introduction

While in operation, constructions of concrete are subjected to mechanical and thermal impacts that often lead to the development of cracks in the material. Hence, there is a need for current defects monitoring in constructions of concrete. To solve this problem various methods of non-destructive testing have been developed. Mostly, acoustic and ultrasonic testing methods are used for these purposes.

Research is being done to determine the starting time of crack formation and its location with the help of acoustic emission [1-3]. These methods enable only observation of the process of crack formation.

Common methods of ultrasonic testing [4, 5] require a perfect acoustic coupling between the transducer and the surface of the concrete.

Currently, research is being done on the development of non-contact ultrasonic methods. The registration of ultrasonic signals through the air and the use of laser vibro-meters is based on the measurement of surface waves [6-9]. However, surface waves are not sensitive to deep cracks and depend on surface undulation of concrete.

To solve the problem of non-destructive testing, the method based on the use of the phenomenon of emergence of alternating electromagnetic fields under pulsed mechanical excitation of non-metallic materials is developed [10-12].

As a result of the previous research, it was established that the coefficient of cross-correlation of spectra of electrical responses from the tests of defect-free samples can be used as a diagnostic criterion for defectiveness determination [13].

In addition it is shown that the use of a reference spectrum enables one to obtain a very high coefficient of correlation with a specially made, defect-free basic reference standard. In our work we used samples of 50×50×100 mm, and the shock impact was imposed in the direction of the major flat surface. Therefore, for the samples of this geometrical arrangement, freely damped harmonic motions were offered to be used as a reference standard. In this paper, theoretical and experimental research is conducted on the development of a similar approach to solving the problem of non-destructive testing of different size products.

2. Methods of experimental and theoretical research

The principle of the method of the research is that under pulsed mechanical load, a transversal wave propagates across the concrete volume. The transversal wave affects electric double layers at the boundary of phase interface and piezo inclusion. Piezoelectric inclusions are contained in sand and gravel which are used for concrete manufacture. The electromagnetic field occurs due to the appearance of charges on the faces of the piezo-quartz under its deformation and owing to the displacement of charge of the electric double layers relative to the electrical receiver. The electrical measuring receiver is located in close proximity to the sample, and is in the coverage area of this field. As the result of the appearance on the surface of the receiving electrode of free charge carriers induced by an electric field, current flows through the input resistance of the measuring circuit.

The experimental research of the electrical response to pulsed mechanical excitation was conducted with the use of a hardware-software complex. The complex includes an electro-mechanical impact device and a differential electrical sensor. The impact device enables it to produce a single impact of 60 μs duration. The electrical sensor consists of a receiver, differential amplifier, input-output board and a computer. The receiver has two receiving metal plates: measuring and compensating. The measuring plate is located at a distance of 2 mm from the sample surface, and the compensating plate is located at a distance of 30 mm from the surface. This construction design of the receiver provides significant reduction of the level of external electromagnetic interferences, and enables it to make measurements of an electrical signal without the use of additional shielding systems. In more details the experimental method is described in [14].

On the basis of the previously performed research, the physico-mathematical model of mechanoelectrical transformations in heterogeneous non-metallic materials is developed [15]. On the basis of the developed model with the use of numerical methods for the simulation of wave processes, the calculation of electrical responses to shock impact is performed.

The numerical implementation was carried out with the use of non-central differencing scheme of second-order accuracy relative to the steps in space and time. The correctness of the numerical results was evaluated both by intrinsic convergence of the results under the change of the parameters of a finite-difference grid and step of integration in time and by calculations of model problems.

The numerical algorithm is based on the Runge-Kutta method for solution of ordinary differential equations.

For the calculation, the following values for the concrete properties were used: density is 2.3×10^3 kg/m^3, modulus of elasticity is 2.1×10^{10} N/m^2; Poisson's ratio is 0.2; longitudinal wave velocity is 4×10^3 m/s. The calculation was performed for the case where the impact is made on the center of the sample and the receiving electrical sensor is located at a distance of 20 mm from the point of impact at a height of 2 mm above it.

To improve the accuracy and convergence the size of the elements in the finite element model was 2.5×2.5 mm. The calculation contains 512 points at sampling frequency with the interval of 1 μs. The finite element model consisted of 1024 elements.

The calculated time signal was transformed into the frequency domain using fast Fourier transform.

3. The results of theoretical and experimental research

The concrete products having a size of 100×100×100 mm, 100×100×200 mm and 100×100×300 mm were manufactured for the research.

Figure 1a shows the calculated electrical responses, and Figure 1b shows the experimental responses of concrete of different geometrical size.

As the images show the spectra of theoretical and experimental results coincide.

The conducted research shows that theoretically calculated electrical responses can be used as standard samples in the procedure of non-destructive testing.

This approach enables to avoid manufacturing of special defect-free samples and to exclude the impact of the original structural inhomogeneity in complex heterogeneous materials.

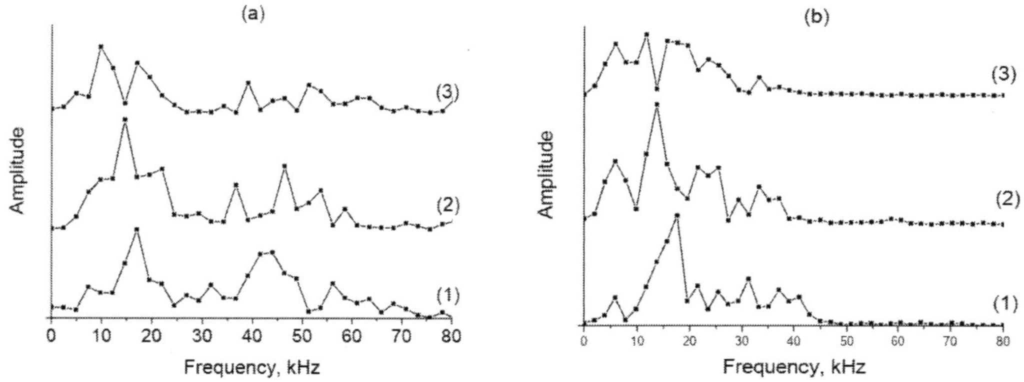

Figure 1. Spectra of the experimental and calculated electrical responses for concrete Products of different geometrical size: (1) 100×100×100 mm, (2) 100×100×200 mm and (3) 100×100×300 mm.

4. Conclusion

The purpose of this work is to carry out theoretical and experimental research of the influence of geometrical dimensions of concrete products on the electrical response and development of applications of numerical methods for non-destructive testing.

The electrical responses to the shock impact of the concrete samples of various sizes were calculated based on the modeling of wave processes with the use of continuum mechanics. The amplitude spectra of the signals were calculated using fast Fourier transform; the comparison with experimental results was made. It is shown that the spectrum of calculated electrical signals is in good agreement with the experimental results. The creation of a bank of theoretically calculated electrical responses from products of different geometrical arrangement and elastic properties will enable using them as standard samples for defectiveness evaluation of products of any geometrical dimensions.

On the basis of the obtained results, new additional possibilities for non-destructive testing of defectiveness can be developed.

The work is done under the State task "Nauka".

References:

[1] Carpinteri A., Xu J., Lacidogna G., Manuello A. 2012 Reliable onset time determination and source location of acoustic emissions in concrete structures *Cement and Concrete Composites* vol. **34** pp 529-537.

[2] Tittelbooma K. V., Beliea N. D., Lehmannb F., Grossec C. U. 2012 Acoustic emission analysis for the quantification of autonomous crack healing in concrete *Construction and Building Materials*

vol. **28** pp 333-341.
[3] Ohno K., Ohtsu M. 2010 Crack classification in concrete based on acoustic emission *Construction and Building Materials* vol. **24** pp 2339-2346.
[4] Bray D E, Stanley R K 1997 *Nondestructive evaluation* CRC Press, New York;
[5] Hsiao C M, Cheng C C, Liou T H and Juang Y T 2008 Detecting flaws in concrete blocks using the impact-echo method *Ndt&E Int* vol. **41** pp 98-107.
[6] Kee S-H, Zhu J 2011 Surface wave transmission measurements across distributed surface-breaking cracks using air-coupled sensors *J. Sound Vib.* vol. **330** pp 5333-5344.
[7] Dhital D, Lee J R 2012 A Fully Non-Contact Ultrasonic Propagation Imaging System for Closed Surface Crack Evaluation *Exp. Mech.* vol. **52** pp 1111-1122.
[8] Kee S-H, Zhu J 2010 Using air-coupled sensors to evaluate the depth of a surface-breaking crack in concrete *J. Acoust. Soc. Am.* vol. **127** pp 1279-1287.
[9] Aggelis D G, Leonidou E Matikas T E 2012 Subsurface crack determination by one-sided ultrasonic measurements *Cem. Concr. Compos.* **34** 140–146
[10] Fursa T.V., Surzhikov A.P., Osipov K.Yu. 2007 Development of an Acoustoelectric Method for Determining the Porosity of Dielectric Materials *Russian Journal of Nondestructive Testing* vol. **43** pp 95-99.
[11] Fursa T.V. 2010 Mechanoelectric Transformation in Cement-Sand Composites in the Course of Crack Formation Caused by Freezing-Thawing Cycles *Technical Physics Letters* vol. **36** pp 348-350.
[12] Fursa T.V., Osipov K. Yu., Dann D. D. 2011 Developing a Metod Based on Mechanoelectric Transformation for Monitoring Concrete Defectness Variation under Influence of Freezing-Thawing Cycles *Technical Physics Letter* vol. **37** pp. 291-293.
[13] Fursa T.V., Osipov K.Yu., Dann D.D. 2011 Development of a Nondestructive Method for Testing the Strength of Concrete with a Faulted Structure Based on the Phenomenon of Mechanoelectric Transformations *Russian Journal of Nondestructive Testing* vol. **47** pp 323-328.
[14] Fursa T. V., Osipov K. Yu., Lyukshin B. A., Utsyn G. E. 2014 The development of a method for crack-depth estimation in concrete by the electric response parameters to pulse mechanical excitation *Meas. Sci. Technol.* Vol. **25** 055605
[15] Fursa T.V., Lyukshin B. A., Utsyn G. E. 2013 Relation between the Electric Response and the Characteristics of Elastic Waves under Shock Excitation of Heterogeneous Dielectric Materials with Piezoelectric Inclusions *Technical Physics* vol. **58** pp 263-266.

Features of the Amplitude-Frequency Characteristics of Electromagnetic Emission during Uniaxial Compression of Dielectric Composites

V P Surzhikov[1]and N N Khorsov[1]

[1] *Tomsk, 634050, Russia,* Tomsk polytechnic university
E-mail: iva1954@ mail.ru

Abstract. We have studied the electromagnetic emission from samples of epoxy resin filled with sand subjected to uniaxial compression . Capacitive sensor measures the electrical component of the response when excited electromagnetic emission of ultrasonic pulse using a differential amplifier. It was shown the influence of the load on the spectral signal characteristics: with increasing pressure, the formation of bands at frequencies corresponding to possibly quasi-Rayleigh waves generation. It is concluded that the use of the experimental geometry studies the main contribution to the response of electromagnetic emissions create born normal vibrations, which are damped standing waves.

1. Introduction

Investigation of the state of the object under load is one of the tasks of non-destructive testing. For this it is developed a variety of methods based on different physical phenomena, such as an acoustic emission [1], which allows to evaluate the concentration of defects on the parameters of acoustic signal, accompanied by their formation and development. Moreover, the character of energy distribution of acoustic emission may serve as a measure of estimate [2].

To study the dynamics of crack it is also used electromagnetic emission (EME), due to the formation of an alternating electromagnetic field in the separation of charges in the mouth developing cracks, at the interfaces of heterogeneous media, the interaction of acoustic waves with inclusions having piezoelectric properties.

For the first time this phenomenon has been used in the Tomsk Polytechnic University in developing methods for forecasting geodynamic phenomena (earthquakes, rock bursts, landslides). In the future, efforts were focused on the phenomenon EME in developing methods for non-destructive testing of defects and strength [3-5].

It was investigated the effect of the stress-strain state of the sample on the parameters of electromagnetic emission during pulsed acoustic excitation, the effect of the volume defects on space-time characteristics of the electromagnetic response due to the mechanoelectrical conversions in dielectric samples, determined the nature of the EME energy distribution of the level of operating loads and possible criteria for monitoring the stress strain state [6-8]. However, it was not carried out analyzing the EME frequency response, depending on the load until the destruction of the samples.

2.Experiment results and discussion

For the experiment the sample was prepared of an epoxy resin with sand filling comprising quartz. Specimen dimensions were $60\times80\times100$ mm^3.

The edge $60\times80mm^2$ served as basis. Hammering device on the basis of piezoelectric transducer was pressed by the end edge 100x60mm2. Symmetrically the plates with capacitive sensors were located at the edges $100\times80mm^2$. The sample was placed in the press and subjected to a stepwise uniaxial compression .

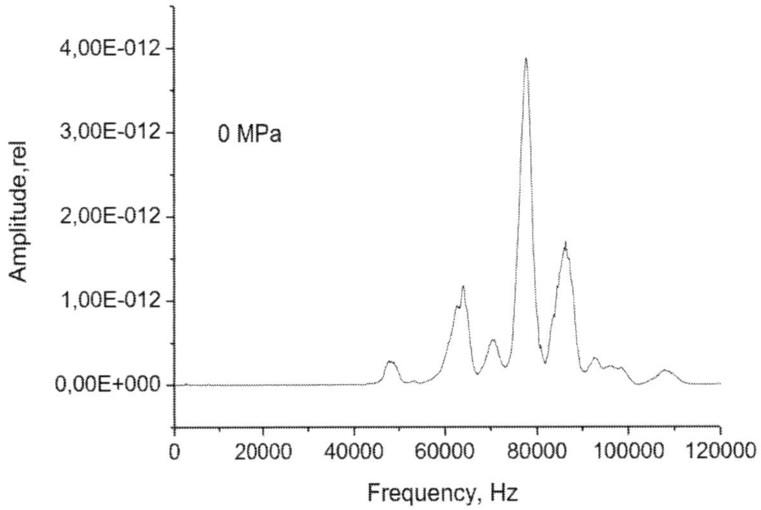

Figure 1. EME frequency response power of the sample without load.

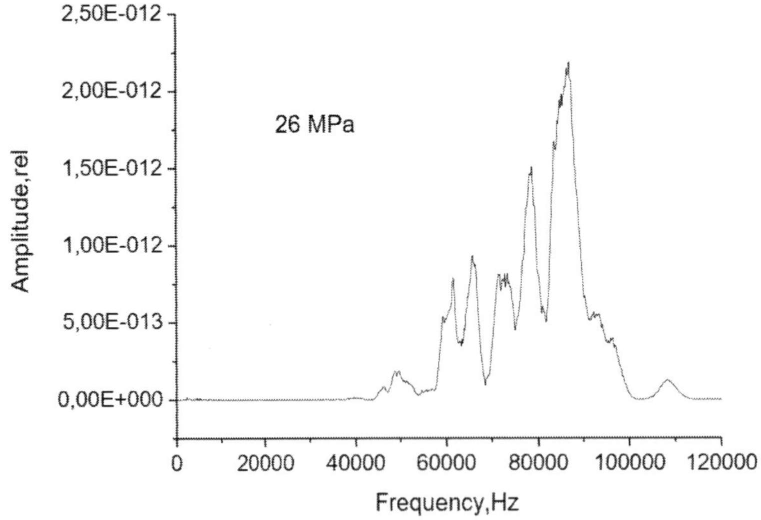

Figure 2. EME frequency response power of the sample under 26 MPa load

The step size was about 4 MPa. At each the step of the pressure the sample was excited by a series of 80 mechanical pulses with the registration responses using the capacitive receiver connected to an input of the differential amplifier.

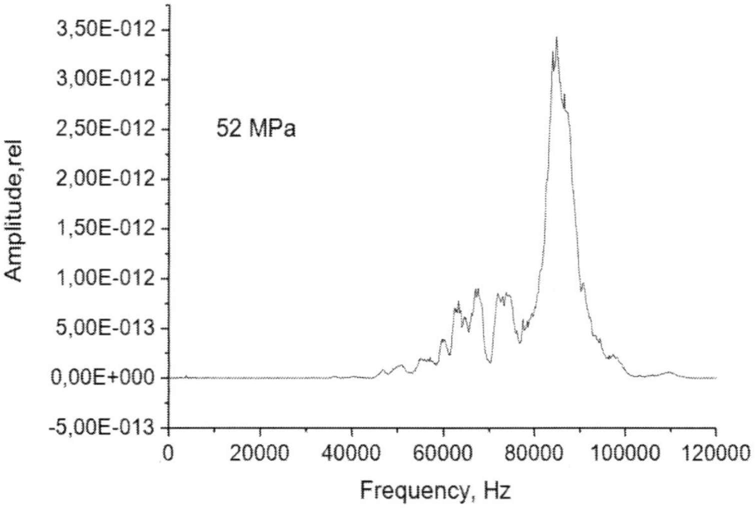

Figure 3. EME frequency response power of the sample under 52 MPa load.

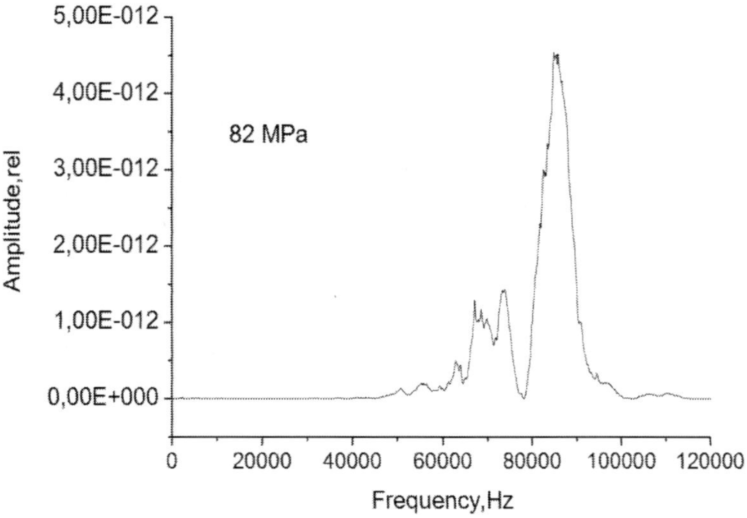

Figure 4. EME frequency response power of the sample under 82 MPa load.

Amplitude-frequency response characteristics were calculated using the fast Fourier transformation. The EME spectra calculated for different values of the applied pressure are shown in figures 1,2,3,4. The figures show that the increase in load leads to the transformation of the spectrum: decrease in the intensity of the low-frequency peaks and the appearance of a broad intense higher frequency peak with a maximum at a frequency of 88.5 kHz.

Analysis of the amplitude-frequency characteristics was conducted for the cut EME responses in which portions from the beginning of the response to a certain time tn excluded. The presence of a broad spectrum band with maxima at frequencies of 87.85 kHz and 89.74 kHz has been revealed. Peak amplitudes at these frequencies at tn of 48 μs and 64 μs reached the maximum value that is significantly greater peak amplitudes for other frequencies. This ratio of peaks in the spectrum with minor changes remains up to tn is 105 μs. With a further increase tn amplitude of the peaks at the frequencies 87.85 kHz and 89.74 kHz begin to decrease and at tn equal 160 μs almost become indistinguishable. The amplitude of the excitation pulse at the same time was decreased by 46 times. When tn is 1500 μs and the pressure is 82 MPa in the EME spectrum more low-frequency peaks at 1.9 kHz, 3.8 kHz, 7.9 kHz, 11 kHz appear.

Values of tn 48 μs and 64 μs coincide with periods of natural oscillations of the sample excited longitudinal and transverse waves. The sample can be excited vibrations of ultrasonic waves with a frequency of 88,5 kHz with wavelength, propagating at a speed 1325 m/s, equal to 1/4 of the sample thickness d, that corresponds to the speed of the surface wave in the sample.

The analysis used the speed of the longitudinal wave, measured in the experiment, and the velocities of transverse and surface waves obtained from the estimates.

Analysis of the EME spectra indicates the major role in the formation of the eigen oscillations of the sample frequencies given geometrical dimensions of the sample and the rate of propagation in it longitudinal, transverse and surface waves.

The appearance of size d/4 may also indicate the existence of a quasi-Rayleigh waves in the sample [9] of length equal to d, at which the maximum displacements and maximum stresses are created on the surface of the component parallel to the surface and that at a depth approximately equal to ¼ wavelength changes according to the direction and sign [10].

2. Conclusion

In the experimental geometry studies the main contribution to the response of EME create born normal vibrations, which are damped standing waves.

This work was financially supported by The Ministry of Education and Science of the Russian Federation in part of the science activity program.

References

[1]. Dunegan H.L., Harris D.O. 1969 *Ultrasonics*,. **7**.
[2]. Damasinskaja E.E., Kadomzev A.G. 2013 *Technical Physics Letters in Russian*. **39**. p. 29
[3]. Surzhikov V. P. , Khorsov N. N. 2012, *Control. Diagnostics in Russian* **11**(173), p. 69
[4]. Surzhikov V. P. , Khorsov N. N. , Khorsov P. N. 2012 *Russian Journal of Nondestructive Testing* , **48**, p. 85
[5]. Surzhikov V. P. , Khorsov N. N. 2013, *Proceedings of the universities. Physics in Russian* **56**. p. 261
[6]. Surzhikov V. P. , Khorsov N. N. 2015, *Technical Physics*, **60**., p.148.
[7]. Surzhikov V. P. , Khorsov N. N. 2014, *Modern problems of science and education in Russian*. **5**; URL: http://www.science-education.ru/119-15035
[8]. Surzhikov V.P., Khorsov N.N. 2011 *Russian Journal of Nondestructive Testing*, **47**, p 687.
[9]. Viktorov V.I., Grigorjan R.A. 1959 , *Acoustic journal in Russian* **5**. p. 366
[10].Viktorov V.I. 1966 The physical basis for the use of ultrasonic Rayleigh and Lamb waves in engineering (Moscow :Nauka) 164 p.

Changing the properties of metals under conditions of exposure to ionizing radiation

S N Yanin[1]
[1]*Tomsk, 634050, Russia,* Tomsk Polytechnic University
E-mail: yanin@tpu.ru

Abstract. The equation state for metals were provided on the basis of the pseudopotential method when part of the ionic cores further ionized. The isotherms were obtained for different degrees of the atoms ionization on the example of aluminium. The change in the physical properties of metals were analyzed under conditions of excitation by ionizing radiation of the electron subsystem. The interatomic interaction potentials were calculated for the case of a local increase concentration of conduction electrons.

1. Introduction

While studying the radiation-stimulated processes it is necessary to learn out not only the interatomic coupling potential, but the consistent relations considering the ionized particles, as well. Such relations can be provided on the basis of binding energy.

According to the pseudopotential, the binding energy in the model of the second order looks as follows:

$$E(\Omega) = E_e^{(0)} + E_e^{(1)} + E_e^{(2)} + E_i \,. \tag{1}$$

Where Ω -atomic volume, $E_e^{(0)}$-the energy of homoheneous electron gas, including the kinetic, exchange and correlative energy of conduction electrons, $E_e^{(1)}$-the energy, caused by the fact that the ions are not the points and have some dimensions, $E_e^{(2)}$-energy of zonal structure, E_i–Evald energy.

The registration of differently ionized atoms was being made on the base of technique, been developed for the alloys of replacement [1].

The lattice is placed in homogenious negative charge with density $\bar{Z}/\Omega$ in concordance with the technique.

Here

$$\bar{Z} = Z_3 C + Z_4 (1 - C)\,, \tag{2}$$

Z_3 is the charge of three-, Z_4 is the four – charged ions, C is the concentration of Z_3 ions.

The calculations was being defined assuming that δ-electrons were leaving the observed area, that is $\bar{Z}=3$.

The energy of homogeneous electron gas per rydberg on electron has a form [2]:

$$E_e^{(0)} = Z\left(\frac{2{,}21}{r_s^2} - \frac{0{,}916}{r_s} + E_c\right), \text{Ry} \tag{3}$$

Here $r_s = (3/4\pi n_0)^{1/3}$ is the usual radius giving the volume per electron, $n_0 = \bar{Z}/\Omega$, E_c is the correlation energy. Expression for E_c may be written as Nozieres – Pines approximation [2]:

$$E_c = -0{,}115+0{,}031\ \ln(r_s)\,. \tag{4}$$

The energy connected to the not Coulomb part of the potential has a form [3]:

$$E_e^{(1)}=2\left(\frac{3\ Z\ \bar{Z}}{2\ R_a}\left(\frac{R_m}{R_a}\right)^2-\bar{A}\ \bar{Z}\left(\frac{R_m}{R_a}\right)^3\right),\ \text{Ry} \tag{5}$$

$\bar{A}$ is the average level between A_l of the potential well.

The energy of zonal structure looks as follows:

$$E_e^{(2)}=-\Omega\sum_{\vec{\tau}}\frac{\pi(\tau)}{\varepsilon(\tau)}\left|V(\vec{\tau})\right|^2, \tag{6}$$

where $\vec{\tau}$ is the reciprocal lattice vector. Here the vector $\vec{\tau}=0$ is absent.

The static dielectric permeability of the conduction electrons is written as:

$$\varepsilon(q)=1+\frac{4\pi}{q^2}\pi(q)\,, \tag{7}$$

where q is the wave vector.

The polarization operator has a form:

$$\pi(q)=\frac{\pi_0(q)}{1-4\pi(G(q)/q^2)\pi_0(q)}\,. \tag{8}$$

Hartre permittivity it is:

$$\pi_0(q)=\frac{3\ n_0}{2\ \varepsilon_F}\left[\frac{1}{2}+\frac{1\text{-}x^2}{4x}\ln\left|\frac{1+x}{1\text{-}x}\right|\right], \tag{9}$$

where $x=q/2k_F$, n_0 is the concentration of the conduction electrons.

Ewald energy has a form:

$$E_i=-\frac{\bar{Z}^2e^2}{r_0}a_M\,, \tag{11}$$

where $r_0=(3\Omega/4\pi)^{1/3}$, a_M is Madelung constant, $a_M=1.7917$ for face-centered cubic lattice.

Static component of pressure- $P_{st}(\Omega)$ can be defined by equation

$$P_{st}=-\frac{\partial E}{\partial\Omega}\,. \tag{12}$$

Differentiation E on volume Ω was made under the formula:

$$\frac{\partial}{\partial\Omega}=\left(\frac{\partial}{\partial\Omega}\right)_{n_0,\boldsymbol{\tau},\mathbf{R}}+\frac{\partial\mathbf{R}}{\partial\Omega}\left(\frac{\partial}{\partial\mathbf{R}}\right)_{n_0,\Omega,\tau}+\frac{\partial\boldsymbol{\tau}}{\partial\Omega}\left(\frac{\partial}{\boldsymbol{\tau}}\right)_{n_0,\Omega,\mathbf{R}}+\frac{\partial n_0}{\partial\Omega}\left(\frac{\partial}{\partial n_0}\right)_{\boldsymbol{\tau},\Omega,\mathbf{R}} \tag{13}$$

The energy is function of volume. The volume depends from n_0, $k_F=(3\pi^2n_0)^{1/3}$, r_s, $\mathbf{R}=a\mathbf{m}$, $\boldsymbol{\tau}=\mathbf{n}2\pi/a$, where a is the lattice constant, **R** is lattice vector.

The contribution caused by harmonic phonon pressure and zero oscillation energy has been negligible.

The $E_e^{(0)}$ was defined assuming that δ-electrons were leaving the observed area and the concentration of conductivity electrons was constant [4].

Calculating the $E_e^{(1)}$ and $E_e^{(2)}$, the model potential parameters (after quantum defect method [4]) had been taken. Figure 1 shows the aluminium isoterms at the 0 °K temperature for the different grades of ionization. So the 10% ionization results in the 15 kbar internal pressure appearance.

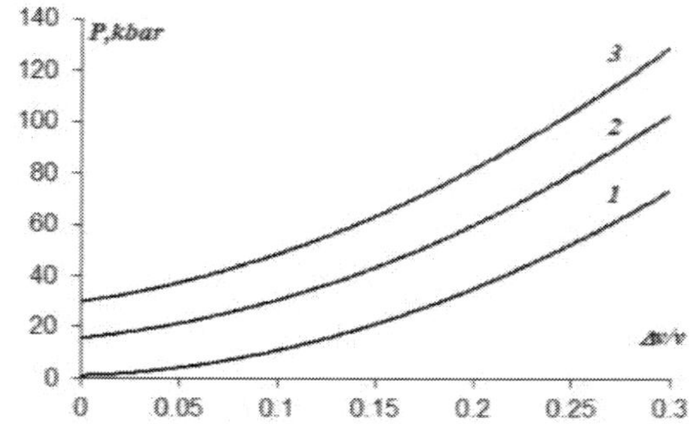

Figure 1. Aluminum isotherms for 0 K temperature: 1 – par, 2 – 10% of ionized atoms; 3 – 20% of ionized atoms.

Thus, the equation of the matter states for the case of ionization were investigated by means of the given method.

The consistent relations, resultant, are proposed to be engaged for studying the destruction of surfaces under the charged particles sturdy beam bombardment.

It is known that intensive radiation generates ionized atoms within the track of a fast charged particle as well as near the solid surface. The potentials of interactions between these atoms and surrounding particles change considerably. The knowledge of corresponding potentials of interatomic interaction is necessary for studying the behavior of partially or entirely ionized matter.

2. Change of Interatomic Forces upon Excitation of Electronic Subsystem

Essential part of energy of fast charged particle in metal is spent on ionization of atoms and excitation of conduction electrons. Electrons accelerating in the field of charged particle slow down transferring their energy to conduction electrons in the space limited by the radius of about 10 interatomic distances from track. For example, density of electronic energy reaches 10^{12} erg/cm^3 [5] in the area of fission fragment particle trajectory, which corresponds to approximately double increase in the energy of free electrons in aluminum over that of Fermi.

Though the fact of change of the interatomic interaction nature upon excitation of electronic subsystem does not raise any doubt, the experimental observation of it is quite a task. In paper [6] there is an analysis of this phenomenon on the basis of experimental data on the threshold energy of the atom displacement from the equilibrium state E_d, which is determined by change of the value of some physical quantity, which is dependent on concentration of point defects (electric conductance for example) under changing energy of falling particles.

The displacement threshold was determined in many papers. In different experiments different values were obtained for one and the same elements even for polycrystalline materials. Comparison of experimental results indicates the existence of some dependence of E_d value on the interval of energies of bombarding particles.

The analysis of work enumerated in [6], enable to conclude that the lower the applied down limit of bombarding particle energy (0.1 – 3 MeV in the considered energy range), the smaller threshold energy of displacement. Peculiarities of E_d behavior depending on the energy of incident particle can be explained by change of the nature of interatomic interaction in metal under conditions of excitation of electronic subsystem. During movement the fast charged particle loses energy transferring it to free electrons and to electrons of the ion core, and in case of collision with the core ion – directly to it. It can be assumed that excitation wave moves in front of the accelerated particle. Thus, already before the collision with the fast charged particle, the initially knocked-on atom and surrounding ions are in the state other than equilibrium.

In order to find out whether the existence of such area is possible let us consider the times of energy re-distribution in metal. According to [7], the characteristic time of excitation relaxation of a separate electron is about 10^{-15}s, and characteristic time of excitation transfer from the electron gas to the ion subsystem is 10^{-12}s (characteristic time of the lattice temperature change from the moment of fast charged particle track occurrence associated with the energy exchange between Fermi electrons and ion cores). Time of equilibration between the electron gas and the lattice is about 10^{-12}s. Thus, if the period of atom transition from one equilibrium state to another is inversely proportional to Debye frequency which is about 10^{-13}s, then quite a large number of transitions of ions amid the electrons with enhanced energy is possible during the time of excitation state existence.

In figure 2 the displacement threshold energies obtained by extrapolation taken from paper [6] are presented in comparison with the graphs of primary electrons energy losses for ionization and excitation of free electron gas

.

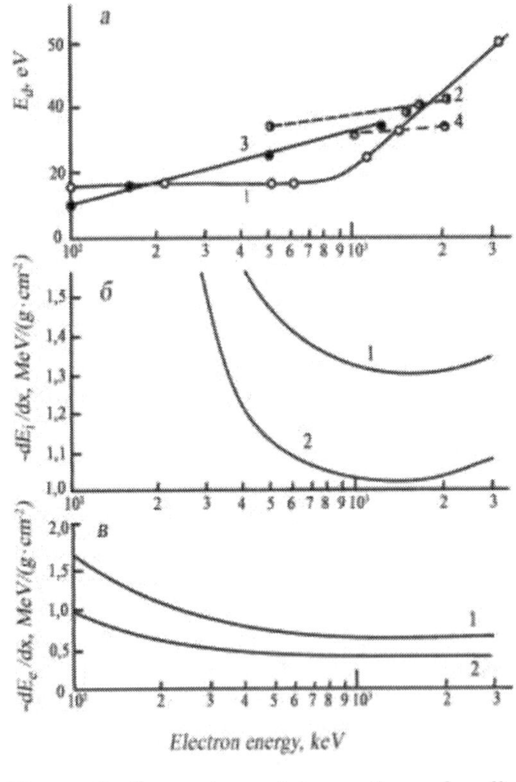

Figure 2. Parameters of interaction of radiation with metals depending on energy of bombarding electrons: a – threshold displacement energy; б – energy losses for initial ionization; в – energy losses for electronic excitation: 1- Al; 2 – W; 3 – Cu; 4 – Mo..

In experiments for determination of displacement threshold the target thickness is much smaller than the length of electron free path, therefore the energy absorbed by electrons is proportional to $dE\,/\,dx$. As is seen from the figure, the correlation between losses for electronic slow down and the value of displacement threshold is observed. The more energy is transferred to the electronic sybsystem, the lower is experimentally observed value of the displacement threshold. The effect of the displacement threshold dependence on the energy introduced to the electronic subsystem is also observed upon investigation of monocrystalline samples depending on the crystallographic direction.

Phenomena associated with non-elastic sputtering of metals by ion bombardment are examined in paper [8]. The local heating of electronic subsystem which results in interatomic interaction forces change was considered as one of the reasons for abnormally high fission fragments sputtering factor.

Besides, it was discovered experimentally that gamma and electron radiation of comparably weak intensity have an essential influence on phase conversions in metals and alloys [9]. In particular, under the influence of gamma radiation the ordering temperature of β - Cu – Zn alloy is shifted. A hypothesis was stated that gamma radiation causes "softening" of the crystalline lattice, i.e. the decrease in elastic moduli which is manifested in the ordering temperature shift. Since the density of point defects created by radiation is extremely small and observed effect depends not so much on the dose as on the radiation intensity, it can be assumed that this phenomenon is conditioned by deviation of function of distribution of electrons from thermodynamically equilibrium distribution. The deviation was caused by radiation effect.

Thus, the growth of conduction electrons temperature in metals causes change of interatomic interaction potentials and consequently to reduction of potential barriers for ion movement. In our view the nature of this phenomenon consists in strengthening of screening of ion cores by Fermi electrons.

In paper [9] an attempt was made to calculate the function of electron distribution in nontransition metals under conditions of gamma radiation effect on the basis of kinetic equation and using this function to calculate the metal shear modulus change in pseudopotential approximation. Deviation of the function of electron distribution from that of Fermi was directly taken into account in changing of polarization operator through some component leading to decompensation of direct Coulomb interaction between the ions by means of conduction electrons. Then elastic moduli were calculated based on changes of interatomic interactions nature. The conclusion has been made that change of elastic moduli is not big due to small number of over-Fermi electrons.

We made an attempt to find out the influence of effect of thermal smearing of Fermi surface on potentials of interatomic interaction by way of direct calculations. For this purpose we updated the Lindhard function obtained in the approximation of Fermi sphere, having included the function of conduction electron distribution against pulses into the expression under integral sign [10].

Calculations of interatomic interaction potentials with this function showed that temperature has weak influence on the screening properties of conduction electrons.

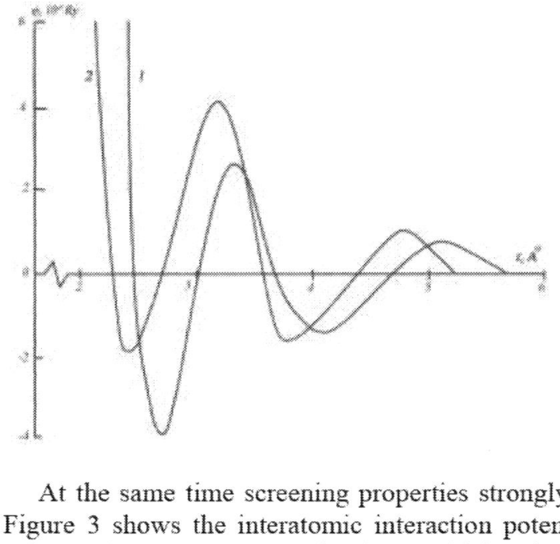

Figure 3. Potentials of interatomic interaction for aluminum: 1- in normal state; 2- the conduction electron pulse exceeds that of Fermi by 10%.

At the same time screening properties strongly depend on concentration of conduction electrons. Figure 3 shows the interatomic interaction potentials for normal state, and for the case, when the valence electrons concentration is 30% higher than the normal one (in this case the electron pulse on the Fermi surface grows by 10%). Such local growth of concentration can occur in the area of fast charged particle track and lead to "softening" of the crystalline lattice.

3. Conclusion

Thus, the method of calculation of pseudopotential formfactor parameters for additionally ionized ion cores has been developed on the basis of the method of quantum defects. The parameters of Heine-Abarenkov-Animalu formfactor have been calculated and pair potentials of interatomic interaction have been obtained on the basis of these parameters. It was discovered that depth of the first minimum in the potential function of the atom in normal charge state and additionally ionized core decreases. Moreover, for the pair of ionized particles the first minimum in the potential function disappears at all, as a result the particles get on the repulsion branch.

On the basis of pseudopotential approach, the behavior of interatomic interaction potentials under conditions of electronic subsystem excitation has been analyzed. It was discovered that thermal smearing of Fermi surface does not cause any essential change of interatomic interaction forces. At the same time local growth of conduction electron concentration causes "softening" of crystalline lattice when configuration with smaller interatomic distance becomes the equilibrium one. The potentials of interatomic interaction have been calculated for this case.

Thus, the change of interatomic interaction potentials change conditioned by ionized radiation causes different physical phenomena, such as intensification of diffusion processes, sputtering of the surface from the area of fast charged particle track etc. These phenomena can be studied by means of mathematical simulation, particularly, by means of the molecular dynamics method.

4. References

[1] Portnoi K I, Bogdanov V I and Fuks D L 1981 *Calculation of Interaction and Stability of Phase* (Moscow: Metallurgy) p 78
[2] Baryahtar V G, Zarochentsev EV and Troitskaay E P 1990 *Methods of Computing Physicsin Theory of Solids*. vol Atomic Properties of metals (Kiev: Naukova Dumka) p 120
[3] Heine V, Cohen M and Weaire D 1973 *Pseudopotential Theory* (Moscow: Mir) p 306
[4] Krivobokov V P and Yanin S N 1992 *Russian Izvestiya RAN* 56 **6** 12
[5] Borin I P 1978 *Reactor Materials Science* vol 1 (Moscow: TsNII atominform) p 123
[6] Shalayev A M and Adamenko A A 1977 *Radiation stimulated change of the electron structure* (Moscow: Atomizdat) p 110
[7] Anisimov S I, Imas Y A and Romanov G S 1979 *High power radiation effect on metals* (Moscow: Nauka) p 272
[8] Baranov I A Martynenko Y V Tsepelevich S O and Yavlinsky Y N 1988 *Russian Uspekhi fizicheskikh nauk* 156 477
[9] Bakay A S, Sergeyeva G G and Fateyev M P 1983 *Russian VANT ser. Fizika radiatsionnykh povrezhdenii I radiatsionnoye materialovedeniye*, 3 **26** 32.
[10] Krivobokov V P and Yanin S N 2005 *Eurasian Physical Technical Journal* 2 **2** 39.

IOP Conf. Series: Materials Science and Engineering **81** (2015) 012062 doi:10.1088/1757-899X/81/1/012062

The separation process of methane-butane fraction from natural gas before transport

O V Galtseva[1], S V Bordunov[2]
[1] Tomsk, 634050, Russia, National Research Tomsk Polytechnic University
[2] Tomsk, 634055, Russia, Public Corporation Scientific - Innovation Enterprise "Echteh"

E-mail: bvv@academ.tsc.ru

Abstract. A series of experiments of adsorption of vapor-air mixtures of some individual hydrocarbons on fibrous sorbents out of polyethylene and polypropylene is made. The calculated changes of concentrations of hydrocarbons in the vapor-air mixture base on experimental data. It is shown that concentration of hydrocarbons decreases after column by 1.4 times and that sorption capacity of fibers reaches 1 g of hydrocarbons on 1 g of a fibrous sorbent. This result was assumed as a basis of calculation of model of the separator for separation of hydrocarbons C_{1-4} from water and fractions C_{5-n}, the degree of separation of the first stage gas of the Verkhne – Tarskoe's field was 98 %.

1. Introduction

The damage by harmful aerosols in countries with the developed industry forces them to take the measures for reduction of volumes of emissions and for providing a normal sanitary state of the atmosphere.

The choice of the method of air purification depends on the specific conditions and it is determined by technical and economic calculations. One of the main methods of purification of air from pollutants is a filtration through the filter nozzles of different types and multicomponent composition of air pollutants causes the application of materials for filtering nozzles with polyfunctional action.

An attempt was made to use polymer fibers the resulting by a technology from waste of domestic and industrial thermoplastic [1, 2] for vapor recovery of hydrocarbons from the air.

2. Experimental

Experiments were made on laboratory installation for research of absorption of vapors of a heptane, a hexane, benzene and toluene on fibers of polyethylene and polypropylene at 20, 40 and 60 ^{0}C. Average diameter of the fibers used in this work made 40 microns, density of packing of the filter is 150 kg/m^3, fiber layer height in a column is 100 mm.

Experiments were made as follows. Into the flask placed in the thermostat we poured 100 cm^3 of hydrocarbons, turned on the thermostat and kept a flask and a column with fiber connected to a neck flask up to achievement of the set temperature. After this we turned on compressor with a constant consumption of air 1.8 of dm^3/min. Through certain periods of time a column with fiber was weighed and an additional weight was defined.

Installation is supplied with the refrigerator on exit from a column for measurement of losses of the hydrocarbons with air (V_4), which passed through fibrous material.

3. Results and considerations

The experiment results of sorption capacity for polypropylene and polyethylene fibers are shown figures 1 and 2.

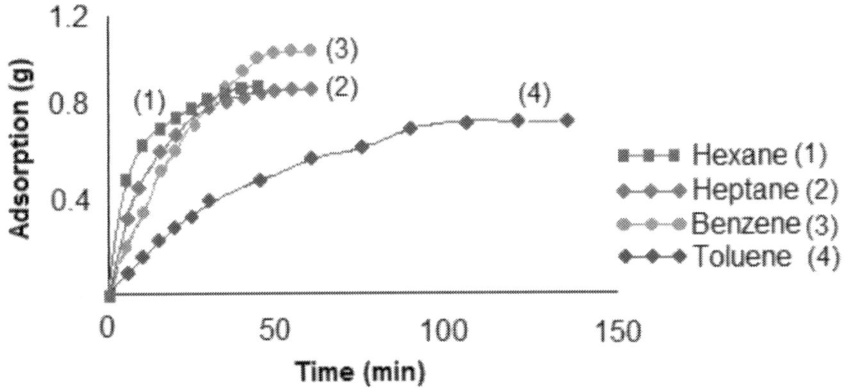

Figure 1. The adsorption of hydrocarbons on the polypropylene fiber (g/g of fiber).

Process of adsorption of these hydrocarbons can be described by:

$$A = A_{\infty} \cdot kt/(1 + kt) \qquad (1)$$

where A is an amount of adsorbed substance, A_{∞} is a maximum quantity of adsorbed substance at fiber saturation, k – constant of adsorption process, t – time of adsorption.

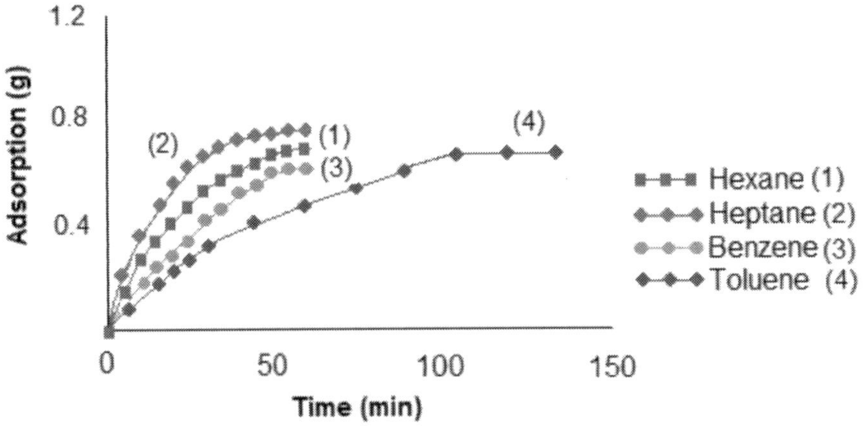

Figure 2. The adsorption of hydrocarbons on the polyethylene fiber (g/g of fiber).

In table 1 results of mathematical processing of the schedules represented in figures 1 and 2 on the equation (1) are given. In table 2 results material balance of experiments of adsorption of hydrocarbons on the column with a mass of 1 gram of fiber are given. Balance is calculated based on the volume of hydrocarbon filled into flask (V_1), residue in flask volume (V_2), on fiber weight hydrocarbon calculated for volume (V_3), and losses of hydrocarbons with air (V_4).

Table 1. The calculation constants of hydrocarbons by equation (1).

Hydrocarbon	Temperature (^{0}C)	Polyethylene A_∞	Polyethylene K	Polypropylene A_∞	Polypropylene K
Hexane	40	1.122	0.029	1.032	0.172
	60	1.175	0.210	1.281	0.223
Heptane	40	1.063	0.049	1.088	0.085
	60	0.799	0.332	0.907	0.839
Benzene	40	1.103	0.019	2.010	0.022
	60	3.472	0.047	1.917	0.097

According to table 2 changes of concentration of hydrocarbons after a column are calculated with a constancy of concentration before and after a column. For example, concentration for hexane and polypropylene fibers varies from 0.22 cm^3/dm^3 to 0.155 cm^3/dm^3 and for octane and polypropylene fibers varies from 0,018 cm^3/dm^3 to 0,013 cm^3/dm^3, i.e. concentration of hydrocarbons is reduced by 1.4 times. These results were assumed as a basis at design and production of the model of the separator which is shown in figure 3.

Table 2. The material balance for adsorption of hydrocarbons on polypropylene (A) and polyethylene (B) fibers at a temperature of 60 ^{0}C and at an air flow rate of 1.8 dm^3 / min.

Saturation time (min) / Hydrocarbon	(V_1) A	(V_1) B	(V_2) A	(V_2) B	(V_3) A	(V_3) B	(V_4) A	(V_4) B
10 / Hexane	10	10	5.9	5.6	1.3	1.31	2.8	3.09
10 / Heptane	10	10	7.6	7.8	1.097	0.88	1.303	1.32
120 / Octane	10	10	6.1	6.2	1.12	1.38	2.78	2.42
12 / Benzene	10	10	6.35	7.4	1.46	1.13	2.19	1.47
50 / Toluene	10	10	7.8	7.8	1.09	0.70	1.11	1.5

At preparation of gas for transport it is necessary to extract water and hydrocarbons $C_{5\text{-}n}$ from it, because water vapors and hydrocarbons $C_{5\text{-}n}$ at temperature decrease are condensed, forming liquid, ice and hydrate corks in pipelines.

For implementation of objectives the following operations are carried out: cleaning, dehydration, gas compressing and extraction from sour gas of petrol fractions [3-4]. These are multistage power-intensive processes. Various devices for separation and purification of gases are known: system from several adsorbents [4], removal of impurity $C_{5\text{-}n}$ by their absorption [5-7], gas-liquid separators [8-11], filter nozzles on columns [12, 13]. Disadvantage of all these devices, in our opinion, is lack of possibilities for simultaneous separation of water and hydrocarbons $C_{5\text{-}n}$ from faction $C_{1\text{-}4}$.

The most promising, in our opinion, is a device for gas purification from impurities in the form of a vertical cylindrical housing with a tangential inlet pipe, a gas outlet nozzle, a nozzle of liquid discharge, a horizontal wall and a rotor in the form of a perforated cup [12]. In perforated glass two shells are installed coaxially and with a gap, one of them is adjacent to the perforated glass and this shell is executed by the continuous, but other is perforated. Thus a both shells don't reach to bottom of perforated glass and adjoin respectively to external and internal diameters of ring tray established under them. End of drain tube is removed into cavity of vertical cylindrical case. Disadvantage of device is a low degree of gas purification $C_{1\text{-}4}$ from other hydrocarbons because shells are not working as an adsorbent but for dehydration the diethylene glycol is recommended.

We made the separator model which is schematically given on figure 3.

The separator consists of the case cooled by tap water in which the rotating cartridge. It consists from internal and external perforated shells which connected by the upper distributor of cooling water and the lower tank of cooling water. The space between shells is filled by adsorbent from polyolefin fibers which is cooled by water flowing in pipes from the distributor to tank. The lower part of the shell is executed in the form of continuous pipe and the bearing with consolidation is pressed on it. The same bearings are pressed on a pipe of tank and branch pipe of input into cartridge of the cooling water connected with an engine shaft by coupling. Thus, rotation of the cartridge with adsorbent is provided and cameras of input of initial mixture of hydrocarbons and water from the camera of separation of liquid and gaseous hydrocarbons are hermetically separated. On an internal surface of case are established the regiments at an angle 50 – 70 ^{0} in relation to an internal axis of installation, which reduce a spraying of the liquid leaving the cartridge under the influence of centrifugal forces and the direction of this liquid into branch pipe of outlet of condensate.

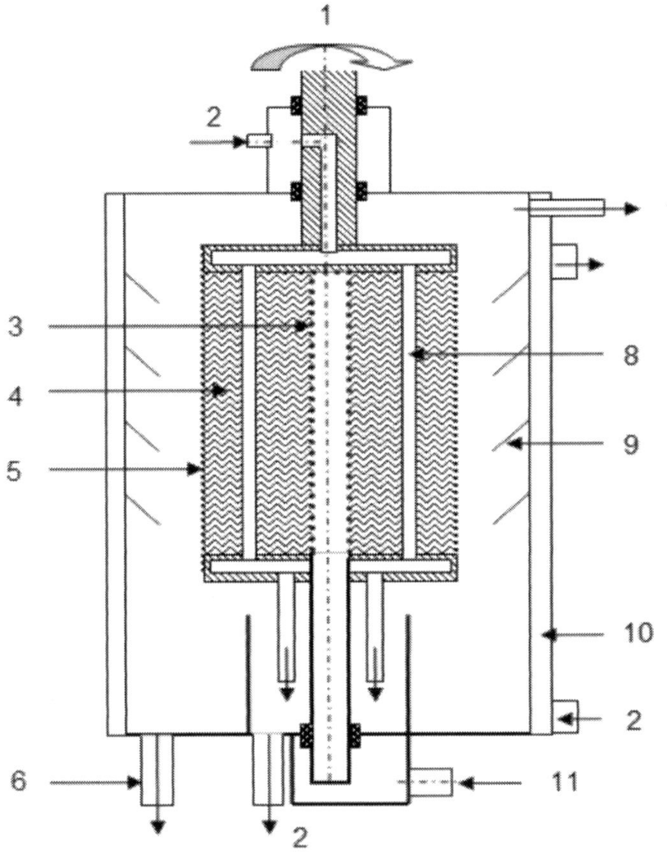

Figure 3. Scheme of model of separator for separation of hydrocarbons C_{1-4} from water and fractions C_{5-n} : 1 – rotation drive; 2 –coolant; 3 – punched shaft; 4 – fibrous adsorbent; 5 – punched rotor shirt; 6 – pipe output of hydrocarbons C_{5-n}; 7 – gases C_{1-4}; 8 – tubes for coolant; 9 – reflectors drops; 10 – cooling shirt; 11 – initial mixture of gases on separation.

Tests of separator showed that extent of division at the first stage of sour gas of the Verkhne – Tarskoe's field makes 98%. Results of experiment are presented in table 3.

Table 3. Material balance of preparation of initial raw gas of the Verkhne – Tarskoe's field to transport according to technological scheme (fig. 3).

Feed gas composition		Contents of components after separator (mass %)		Contents of components after column (mass %)	
Component	Mass (%)	Gas	Condensate	Gas	Condensate
Carbon dioxide	0.859	0.859	-	0.859	-
Nitrogen	1.402	1.402	-	1.402	-
Methane	32.684	32.684	-	32.684	-
Ethane	6.506	6.506	-	6.506	-
Propane	22.451	22.451	-	22.451	-
Isobutane	11.044	11.044	-	11.044	-
N-butane	13.679	13.679	-	13.679	-
Pentanes	8.548	1.2604	7.2876	-	1.2604
Hexanes	2.338	0.2361	2.1019	-	0.2361
Heptanes	0.439	0.0135	0.4285	-	0.0135
Octanes	0.045	0.0004	0.0446	-	0.0004

4. Summary

- It is shown that fibers from junk and waste of thermolayers can be used for extraction of hydrocarbons from steam-air mixes in the adsorptive columns with a nozzle from fibers of thermolayers, for example, in processes of catching of hydrocarbons from gas fractions deleted from oil at its preparation for transport.
- The separator model for division of natural gases and gas condensate into methane-butane fraction and fraction of other heavier hydrocarbons into one stage on the rotating fibrous nozzle with regulation of its temperature is developed; degree of separation at the first stage of sour gas of the Verkhne – Tarskoe's field makes 98%.

References

[1] Bordunov V V *et al* 2002 RF Patent 2179600 *Byull. Izobret.* **5**
[2] Galtseva O V and Bordunov S V 2014 *Testing. Diagnostics* **13** 68
[3] Uzhov V N and Myagkov B I 1970 *Purification of industrial gases by filters* (Moscow)
[4] Churakaev A M 1983 *Processing of oil gases* (Moscow)
[5] 1987 *Pat. Appl.* **0248720** (Russia)
[6] 1987 *Pat. Appl.* **0247585** (Russia)
[7] Mutin F I and Makarov E P 1981 RF Patent 835461 *Byull. Izobret.* **21**
[8] Guzhov A I and Bogachev S P 1984 RF Patent 1068141 *Byull. Izobret.* **3**
[9] L'vov V M 1977 RF Patent 585856 *Byull. Izobret.* **48**
[10] Kazaryan V A *et al* 1981 RF Patent 801855 *Byull. Izobret.* **5**
[11] Baranov V N and Barinov B V 1981 RF Patent 860829 *Byull. Izobret.* **33**
[12] Milshtein L M *et al* 1984 RF Patent 1068142 *Byull. Izobret.* **2**
[13] Manfred Sh *et al* 1985 RF Patent 232835 *Byull. Izobret.* **2**
[14] Ribakin V F 1988 RF Patent 263650 *Byull. Izobret.* **4**
[15] Diarov R K *et al* 1997 RF Patent 2095122 *Byull. Izobret.* **31**
[16] Bordunov V V *et al* 2000 RF Patent 2160332 *Byull. Izobret.* **34**

IOP Conf. Series: Materials Science and Engineering **81** (2015) 012063 doi:10.1088/1757-899X/81/1/012063

Magnetic iron oxides in the cementation technology of the boron-containing radioactive waste

M A Fedotov[1], O A Gorbunova[2], O V Fedorova[3], G E Folmanis[4] and L V Kovalenko[5]

[1] Graduate Student, A.A.Baykov's Institute of Metallurgy and Material Science RAS, Russia, 119991 Moscow Leninsky Ave 49

[2] Head of Department, FSUE "RADON" of State Corporation "ROSATOM", Russia, 119121 Moscow 7th Rostovsky lane 2/14

[3] Head of Laboratory, FSUE "RADON" of State Corporation "ROSATOM", Russia, 119121 Moscow 7th Rostovsky lane 2/14

[4] Leading Researcher, A.A.Baykov's Institute of Metallurgy and Material Science RAS, Russia, 119991 Moscow Leninsky Ave 49

[5] Head of Laboratory, A.A.Baykov's Institute of Metallurgy and Material Science RAS, Russia, 119991 Moscow Leninsky Ave 49

E-mail: mikle_fed@mail.ru; olg-gorbunova@yandex.ru

Abstract. Two ways of synthesis of non-detachable dispersed particles of magnetic materials useful for the boron-containing waste cementation process regulation were developed. Powder XRD showed that the method of carbothermic recovery of nanoscale iron hydroxide allows obtaining a mixture of iron oxides with content of the magnetic phase up to 70%. Method of low-temperature hydrogen reduction of the raw materials allows obtaining various compositions of α-iron and iron oxides with the possibility to change the size of the final particles in a wide range. The possibility of using composites of magnetic iron oxides and metal oxide compositions instead of ferromagnetic rods with VEP of boron-containing liquid radioactive waste in the fluidized field was studied. It was shown that the use of fine and nano particles of the iron oxides in the pre-treatment of the boron-containing LRW increases the strength of the final compounds and accelerates the cement setting compounds from 13 to 5-9 days.

1. Introduction

At the nuclear industry enterprises various of radioactive waste (RW) has been formed. A significant amount of liquid and solid radioactive waste has been already accumulated in Russian Federation [1]. The long-term controlled storage and final disposal of the radioactive waste (RW) require their preservation in the form that prevent their interaction with the environment and be sustainable throughout the storage time. And it is important to minimize the volume of the conditioned RW. In Russia, the liquid radioactive waste (LRW) primarily goes to cementation [2]. This is the most cost-effective solution for disposal of low and intermediate level radioactive waste, although the technology of cementation increases the total volume of the stored waste. To obtain a cement compound that meets the requirements of the normative documents, it is important to determine the recipe of the cementation. The optimal compound recipe provides the maximum inclusion of waste. The minimum amount of stored compounds must consider the chemical properties of different types of waste and as well as the established national standards.

The boron-containing radioactive waste with low pH<6 values cannot be cemented without alkaline additives that neutralizes the acidic form of the borate compounds. The introduction of the alkaline additives increases the volume of the final cement compound several times, which significantly increases the financial and operational costs.

The process of hardening of the borate cement matrix can be accelerated by reducing the effects of the borates in electromagnetic processing of the LRW in the vortex field with ferromagnetic rods. But this leads to the formation of secondary LRW after the decontamination of the reusable rods. Their replacement by non-separable dispersed particles of the magnetic iron oxides allow to accelerate the cement compounds setting, to improve their mechanical properties and to prevent the formation of secondary LRW [3].

The aim of this work is to develop the technology of preparation of dispersed magnetic iron oxide particles suitable for using instead of ferromagnetic rods in the process of the vortex of electromagnetic processing (VEP) boron-containing liquid radioactive waste, and testing obtained dispersed magnetic iron oxide particles for changing the state of aqueous solutions of borates after processing in the vortex electromagnetic field.

2. Magnetic iron oxides and methods for their preparation

Two ways of synthesis of non-detachable dispersed particles of magnetic materials useful for the boron-containing waste cementation process regulation were developed. The resulting materials are composites of dispersed magnetic iron oxides or composites with the presence of dispersed α-iron. The process of obtaining such materials is based on reduction of nanoscale raw material in the form of iron hydroxide by solid or gaseous reducing agents. Magnetic iron oxide phases are γ-Fe_2O_3 (maghemite), Fe_3O_4 (magnetite) and dispersed iron. The geometrical dimensions of the magnetic particles are controlled by reduction temperature.

Dispersed iron oxide magnetic particles synthesis process is separated into two phases. (Process for preparing the particulate magnetic iron oxide particles is separated into two phases). The first step is preparing a raw in the form of iron hydroxide by precipitation of iron chloride $FeCl_3$ or iron nitrate $Fe(NO_3)_3$ with ammonia aqueous solution. The result is a colloidal solution of ferric hydroxide $Fe(OH)_3$. In the second step the ferrous hydroxide is reduced with activated carbon. For this purpose, ferric hydroxide is mixed with crushed carbon in various proportions and heated to a temperature of 500 - 600 °C. The result of reduction is obtained nanoscale iron oxide powder with an average particle size of about 50 nm, which are composed of iron oxide phase α-Fe_2O_3, γ-Fe_2O_3 and Fe_3O_4, there is also soot carbon, sedimenting on the iron oxide particles. Reflections of magnetite and hematite are presented on the powder XRD (figure 1). The content of hematite (α-Fe_2O_3) in the resulting composite of iron oxides is not more than 2%. Maghemite (γ-Fe_2O_3) was detected by using Mössbauer spectroscopy.

Another process that allows obtaining composites of magnetic iron oxide is low-temperature hydrogen reduction of nanosized iron hydroxide. This process also involves the preparation of nanosized raw material as a first step. The method allows obtaining composites of magnetic iron oxides γ-Fe_2O_3, Fe_3O_4, as well as various iron oxide phase composites with metallic iron. It is possible to vary the geometrical dimensions of the obtained particles in a wide range. As an example, the Mössbauer spectrum of nanosized particles of different iron phases is shown on figure 2 [4].

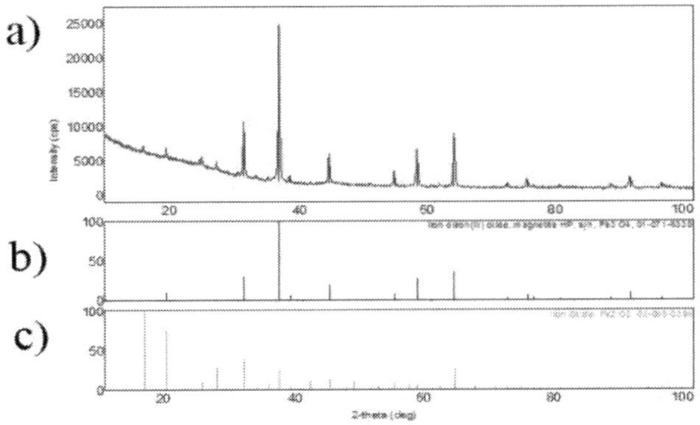

Figure 1. Powder XRD of nanoscale iron oxides: a – composite of α-Fe_2O_3, γ-Fe_2O_3 and Fe_3O_4; b – reflections of the magnetite Fe_3O_4; c – reflections of the hematite α-Fe_2O_3

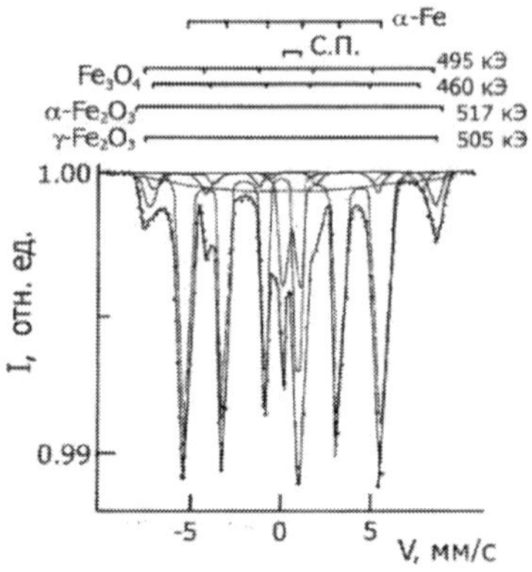

Figure 2. Mössbauer spectrum of nanosized metal oxide particles composition

The possibility of using composites of magnetic iron oxides and metal oxide compositions instead of ferromagnetic rods with VEP of boron-containing liquid radioactive waste in the fluidized field was studied.

3. The effect of iron oxide mixtures on the properties of the boron-containing liquid radioactive waste which goes to cementation

The properties of the boron-containing LRW are regulated using mixtures of dispersed magnetic iron oxides during the electromagnetic processing in the vortex field. Figure 3 shows the IR spectra of boron-containing LRW before and after VEP.

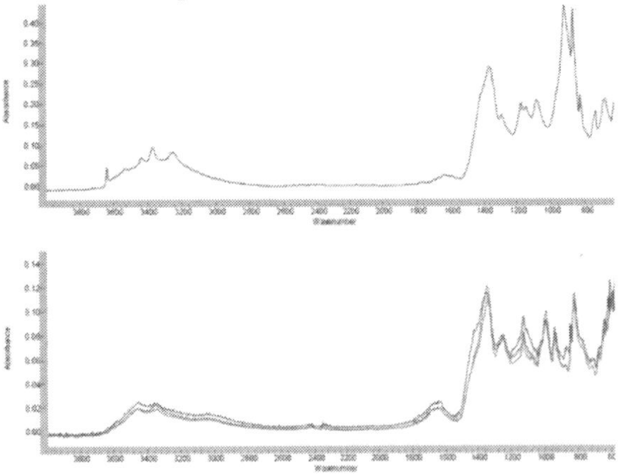

Figure 3. Infrared spectra of the boron LRW samples: a - before processing, b - IR spectra overlaid after 30, 60, and 180 - second VEP with Fe_3O_4 and γ- Fe_2O_3 magnetic powders

The spectra confirm the changes of the forms of the hydrate oxygen boron compounds. Due to the electromagnetic interference in the hydrogen bonds dissociation of the crystal hydrates and characteristic boron bonds (B-O-B), the magnetic and electromagnetic water activation may change the crystal phases and the pH of the solutions [5-8]. Thus, in the IR spectra peaks of 3640 cm^{-1} and 920 cm^{-1} disappeared. Vibration of 3640 cm^{-1} corresponds to stretching vibration of the free single end bond -O-H, and the vibration of 920 cm^{-1} correspond to the deformation vibration of -A-O-H bond (where A is the atom of boron). Moreover, peaks in the range of 1450 cm^{-1} to 750 cm^{-1} have changed, reflecting the state of the bonds of boron atoms. During the electromagnetic processing of the LRW the free end -O-H bond has disappeared, and there was a partial destruction of the crystal hydrates, which increases pH.

To determine the effect of the magnetic iron oxides obtained from the initial mixture of different composition, samples were selected of the o-boron acid solution to determine the change of the pH values, that were prepared using iron chloride and iron nitrate, and as well as with various proportions of carbon. The pH changes and conductivity of the solutions are given in table 1. The processing was performed within 180 seconds. Using nano particles of iron oxides as activators in the electromagnetic processing of the solutions of o-boric acid, pH value increased to 1.6 - 1.7 times, and the conductivity increase of 10-20 times depending on the composition of the initial mixture. The use of raw material of the various salts of iron (NC-1 and NC-2) has practically no effect on pH, while increasing the amount of carbon in the initial mixture NC-2 compared to NC - 3 has a positive effect on the increase of pH. So, using powders of iron oxides as activators in electromagnetic processing it improves the reactivity of the cement minerals in the presence of a solution of o-boric acid.

Table 1. Changes in pH and conductivity of the aqueous solutions of o-boric acid after the VEP depending on the type of activator

Activator type	pH	Electrical conductivity γ (μkS/cm^3)
No VEO processing	3.35	119
NC-1 (the ratio of the iron chloride $FeCl_3$ to the activated charcoal 1:1)	5.73	2410
NC-2(the ratio of iron nitrate $Fe(NO_3)_3$ to the activated carbon 1:1)	5.71	1700
NC-3 (the ratio of iron nitrate $Fe(NO_3)_3$ to the activated carbon 1:0.5)	5.43	1260

In the structure of the hardened cement compound with a boron-containing LRW the magnetic particles can act as reinforcing materials. Figure 4 shows the results of the cement compounds strength measurements. It is obvious, that the use of fine and nano particles of the iron oxides in the pre-treatment of the boron-containing LRW increases the strength of the final compounds. This is confirmed by the scanning electron microscopy data of the cement compounds on the basis of the VEP boron containing LRW (figure 5), indicating of a denser and orderly structure of the cement matrix already in the early stages of the hardening (10-14 days).

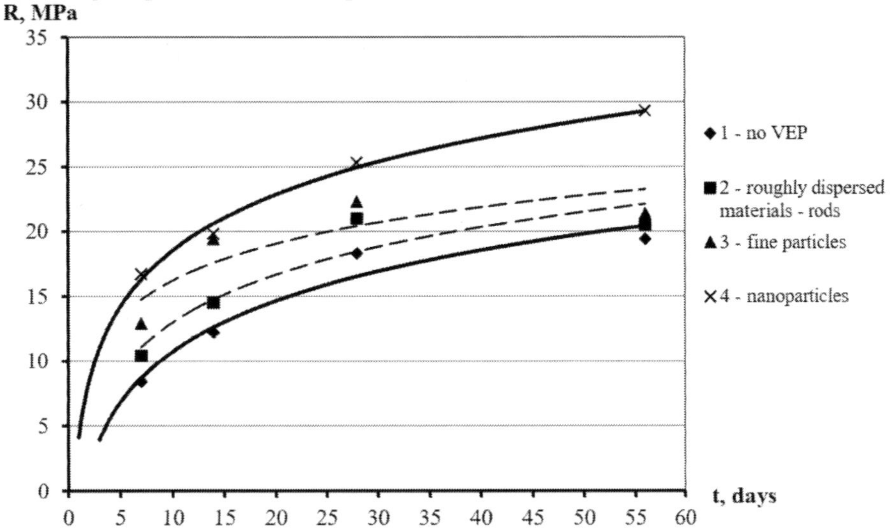

Figure 4. Change of compressive strength of the cement compounds

In cementation process using VEP the hardening dynamics of the compounds varies depending on the used activator. The use of a mixture of iron oxides instead of the separable ferromagnetic rods accelerates the cement setting compounds from 13 to 5-9 days. The iron oxide nanoparticles show better results after the boron containing LRW cementation using VEP, observed early setting and uniform strength. The mixture of nanoscale iron oxides act as well as dispersion strengthened elements that increase the mechanical strength of the cement compounds.

Figure 5. Ordered structures of the new cement formations in the presence of magnetic iron oxides and VEP boron containing LRW

4. Conclusions

There are two ways to obtain non separable dispersed particles of magnetic materials suitable for regulation of the cementation processes of the boron-containing LRW.

Method of carbothermic reduction of nanoscale iron hydroxide allows obtaining a mixture of iron oxides with content of the magnetic phase up to 70% and particles with an average size of 50 nm.

Method of low-temperature hydrogen reduction of the raw materials allows obtaining various compositions of α-iron and iron oxides with the possibility to change the size of the final particles in a wide range.

It is shown that a result of the electromagnetic processing (VEP) in the vertical field of magnetic iron oxides through changes in the boron atoms bonds in hydrated compounds, the pH values increase in 1.6-1.7 times in the borate aqueous solutions, which leads to improvement of the properties of the cement compounds with boron containing LRW.

Replacement the ferromagnetic rods with dispersed mixture of magnetic iron oxides in the vortex electromagnetic processing of the boron LRW does not lead to the loss of positive effect and helps to simplify the technological process of processing the boron containing LRW using the method of cementation and to avoid the formation of secondary liquid radioactive waste.

References

[1] Aparkin F M 2010 The concept of radioactive waste management in JSC Concern Rosenergoatom *Materials of the Int. scientific technical Conf. "Safety, efficiency and economics of nuclear energy" (Moscow 26-27 May 2010)* - http://pptonline.ru/slide/id/74022

[2] Kozlov P V and Gorbunova O A 2011 *Cementation as a method of radioactive waste immobilization* (Ozersk: FSUE "PA Mayak")

[3] Gorbunova O A, Kamaeva T S, Tananaev I G, Folmanis G E, Myasoedov B F and Kulyako U M 2014 Method of conditioning the liquid radioactive waste: Patent RU N2516235 C2, G21 F 9/16, priority 12.04.2012, issued on 20.05.2014

[4] Tananaev I G, Baldohin U V, Korneev V P, Kovalenko L V, Suzdalyov I P, Prusakov V E, Arsentyeva I P and Folmanis G E 2011. Study of nanostructures obtained in the process of low-temperature hydrogen recovery of the iron hydroxide *Academy of Sciences reports* vol. **436** N **4** 474–7

[5] Gorbunova O A, Fedotov M A, Kovalenko L V and Folmanis G E 2013 Magnetic phases of iron oxides for LRW cementation using the electromagnetic processing of the vortex field *RW* management *Materials of the Int. scientific technical seminar for young scientists and PhD students FSUE "RADON" and Russian Academia of Science (Sergiev Posad, 8 April 2013)* ed A V Safonov and O A Gorbunova (Moscow: Russian Academia of Science) pp 61–64

[6] Gorbunova O A, Fedorova O V, Krupskaya V V and Fedotov M A 2014 The formation of a cement compound based on boron containing LRW and nanoscale ferromagnetic activators *6th Russian youth school on radiochemistry and radiochemical technologies: abstracts (Ozersk, 8-12 September 2014)* (Ozersk: FSUE "PA Mayak") pp 101–105

[7] Gorbunova O A Cementation of liquid radioactive waste with high content of borate salts *2nd International Symposium on cement-based materials for nuclear waste (June 3-5 Palais des Papes Avignon June 6 CEA-Marcoule, France)* pp 742–755

[8] Gorbunova O A 2015 Cementation of liquid radioactive waste with high content of borate salts *Journal of Radioanalytical and Nuclear Chemistry* vol. **303** N **2** February 2015, DOI:10.1007/s10967-014-3886-3. – Springer: Akademiai Kiado, Budapest Hungary, 2015. – http://www.link.springer.com/article/10.1007/s10967- 014-3886-3

Data on granulometric composition of calcium phosphate obtained by dispersion method

O A Golovanova[1], E S Chikanova[1], A V Malyshev[2] and T S Mylnikova[2]
[1] Omsk State University, Omsk, Russia
[2]National Research Tomsk Polytechnic University, Tomsk, Russia

E-mail: golovanoa2000@mail.ru

Abstract. The kinetics of calcium phosphate crystallization from model solutions of saliva and liquid phase of dental plaque has been studied by the dispersion method. It was found that the composition of the saliva model system is favorable for the growth of larger crystals. The size of the particles in crystallization varies nonlinearly. As supersaturation grows, the amount of formed particles increases, however, the average rate of crystallite growth decreases.

Keywords: Biological fluid; Saliva; Microcrystallization; Mathematical model; Self-assembly; Nano- and micro-level analysis of variance.

1. Introduction

Investigation of phase formation in biological environment is one of the current issues in the field of crystallization from solution [1–3].

Crystallization of low-solubility compounds (LSC) from model solutions of biological fluids is difficult to study since they are multicomponent, and the process is multi-factorial. In addition, the formation of LSC most often occurs under non-equilibrium conditions and is controlled by kinetic factors [4]. By now, there is not much information on the nature of LSC crystallization in physiological solutions of complex composition [5, 6].

One of the new and promising optical methods to study crystallization is the dispersion analysis which allows obtaining information on particle size distribution [7]. The particle size distribution is directly related to the crystallization parameters that determine the mechanism of nucleation and crystal growth. This technique is not commonly used in the experimental study of crystallization, though it helps to accumulate different information about the origin and growth of crystals and is easy to use.

Thus, the aim is to study the kinetics of LSC (calcium phosphate) crystallization from model solutions of saliva and liquid phase of dental plaque (LPDP) by the dispersion method.

2. Experimental techniques

To study the process of crystallization we used the model solutions with the concentration range of basic inorganic components and saliva pH and LPDP of an average healthy adult [4].

For each series of the experiments, the solutions containing cations and anions were prepared since in these conditions, soluble compounds are not formed. In each of the solutions, pH was adjusted to the physiological value (6.93-7.00 ± 0.05) by adding 20% of NaOH or HCl solution (conc.). The equivalent volumes of the solutions were mixed to make a solution with a designed supersaturation (S, to determine the effect of supersaturation on the parameters of nucleation the following values were modeled S = 5, 10, 15, ..., 50 for both systems) and calculated concentration of the components. The prepared solution was analyzed at different time intervals with a laser diffraction analyzer. The diagrams of particle size distribution were taken, and the average size of the particles formed was calculated.

The dispersion analysis was performed with laser diffraction particle size analyzer Shimadzu SALD-2101 (Laser Diffraction Particle Size Analyzer). For more reliable results, the 4-5 fold replicated analysis of the samples was carried out. The relative standard deviation for the measurement data was S_r = 0.02-0.04.

3. Result and discussion

In the study of crystal formation and growth in model solutions, the differential curves for particle size distribution at different intervals of crystallization (Figure 1, 2) were obtained for all supersaturations under investigation with the diffraction analyzer.

The analysis of the dependencies indicated differences in particle size distribution for the two tested systems. The LPDP system is characterized by monomodal particle distribution throughout the crystallization time, and for the system modeling the saliva composition, the distribution is polymodal.

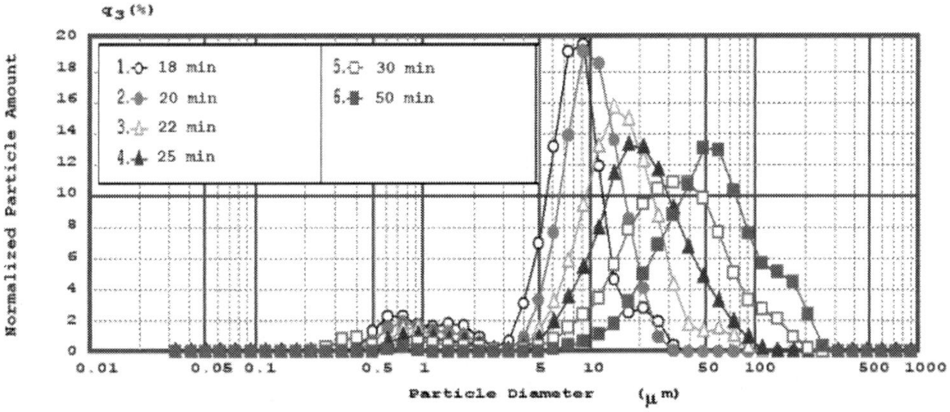

Figure 1. Diagrams for particle size distribution at different time intervals in the LPDP system with S = 15.

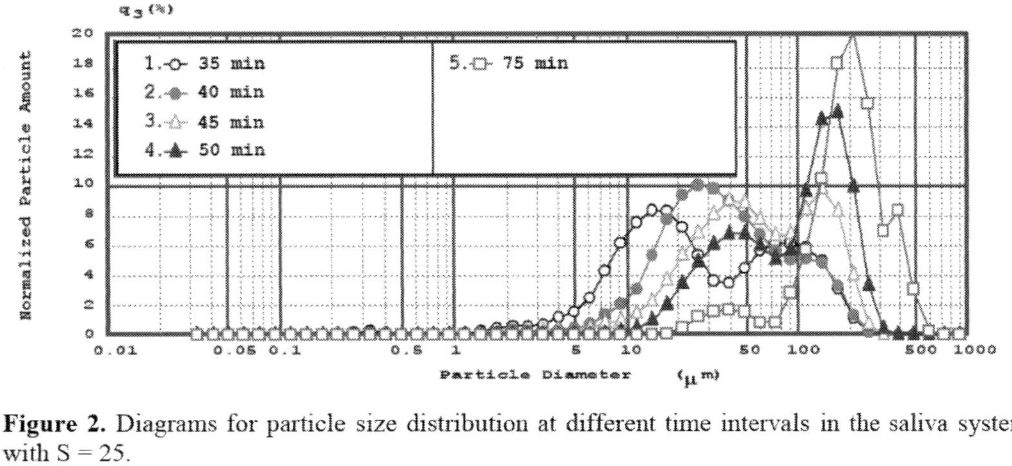

Figure 2. Diagrams for particle size distribution at different time intervals in the saliva system with S = 25.

This type of particle size distribution can be explained if assume that in crystallization, calcium hydrogenphosphate (brushite) is a largely dominating compound in the LPDP, and in the model saliva solution, the dominating compounds are calcium hydrogen phosphate and octacalcium phosphate [4]. At that, in the process of solid phase crystallization, the following processes can be observed: nucleation of solids (nucleation), primary crystal growth to a critical size, aggregation of the crystals and growth of the formed nano-sized particles, aggregation of the particles and formation of crystals with a diameter of several micrometers, and their further growth. As a result, nucleation and crystal growth in the model system of saliva occurs simultaneously in several centers of different critical size. This makes the obtained distribution curves polymodal.

After appropriate mathematical treatment of the differential curves, an average size of the particles formed in crystallization was determined (table 1 and 2).

Table 1. Average particle size for LPDP solution in time under varied supersaturations.

Supersaturation values					
S=15		S=20		S=25	
t, min	d, μm	t, min	d, μm	t, min	d, μm
18	7.724 ± 0.288	5	8.358 ± 0.334	2	9.042 ± 0.274
20	9.699 ± 0.285	7	15.180 ± 0.338	3	19.324 ± 0.323
22	14.52 ± 0.309	10	20.492 ± 0.303	5	25.808 ± 0.384
25	19.32 ± 0.339	15	25.439 ± 0.341	8	27.189 ± 0.418
30	30.621 ± 0.361	20	30.324 ± 0.409	10	26.438 ± 0.426
35	40.056 ± 0.402	25	37.463 ± 0.443	15	25.224 ± 0.442
40	45.619 ± 0.468	27	38.315 ± 0.380	20	25.189 ± 0.427
45	49.287 ± 0.216	30	**39.206 ± 0.,374**	22	26.384 ± 0.449
50	**52.648 ± 0.373**	-	-	25	26.779 ± 0.446

Table 2. Average particle size for saliva solution in time under varied supersaturations.

Supersaturation values					
S=25		S=30		S=35	
t, min	d, μm	t, min	d, μm	t, min	d, μm
35	23.388 ± 0.489	5	20.216 ± 0.423	5	21.696 ± 0.454
40	34.775 ± 0.374	10	36.124 ± 0.420	10	30.682 ± 0.353
45	56.738 ± 0.370	15	49.138 ± 0.333	15	35.737 ± 0.392
50	105.759 ± 0.334	20	66.761 ± 0.316	20	37.050 ± 0.398
55	140.910 ± 0.360	25	76.418 ± 0.281	25	40.577 ± 0.433
60	168.842 ± 0.549	30	84.962 ± 0.417	30	**41.824 ± 0.438**
65	176.731 ± 0.316	35	90.112 ± 0.312	-	-
70	192.877 ± 0.293	40	92.540 ± 0.288	-	-
75	**198.350 ± 0.256**	45	**94.831 ± 0.349**	-	-

It is evident that during crystallization the particles increase in size, and the confidence intervals for different periods of time do not overlap. Despite the similarity of the pattern, the size of the solid phase particles in the saliva model solution and that of the particles in the LPDP solution are different. In the saliva model solution, it varies from 42 to 198 μm in transition from S = 25 to S = 35, and in the LPDP solution, it ranges from 52 to 26 μm for supersaturations S = 15 and S = 25, respectively. These results are consistent with the data obtained previously, that is, the composition of the "saliva" model system is favorable for growth of larger crystals [7]. The analysis of the table data shows that the size of the particles in crystallization changes nonlinearly with the maximum extremum and indicates similar character of the obtained patterns for the LPDP and saliva model solutions (Figure 3). However, for the "saliva" system, the crystal growth is observed before supersaturation equal to 30, while in crystallization of the LPDP, the crystals grow at supersaturation equal to 20, and these dependencies do not change over time. This behavior can be attributed to higher concentrations of the basic inorganic components of the LPDP solution if compared to the saliva model system [4]. In addition, decrease in the crystal size under increased supersaturation (saliva > 30, LPDP > 20) may be caused by transition from a heterogeneous nucleation of the crystallite to a homogeneous one or by secondary processes (aggregation, dissolution, recrystallization, etc.). The higher the value of the initial supersaturation, the larger is the aggregation. The calculated values of the particle size enabled to find the dependence of this parameter on the supersaturation in the modeled solutions (Figure 3).

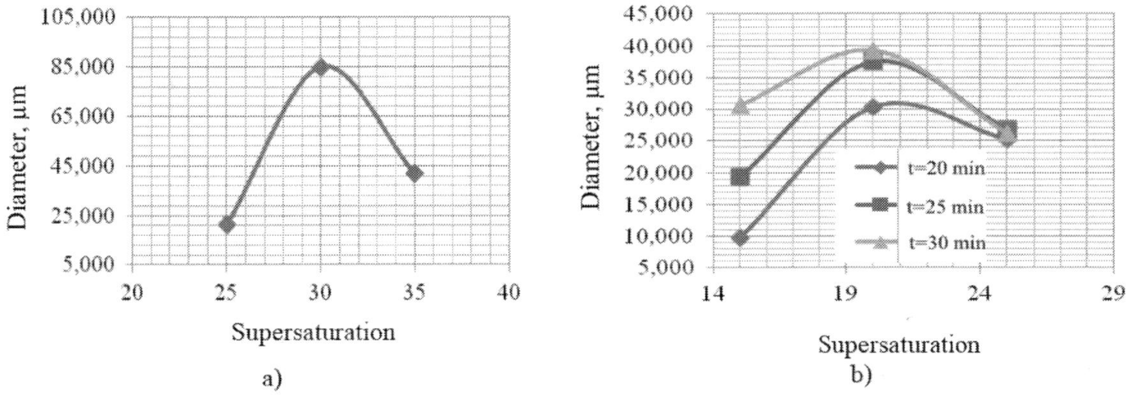

Figure 3. Dependence of the particle size on supersaturation in the model systems: a) is Saliva and b) is LPDP.

As the supersaturation increases, the number of the formed particles grows, however, the average rate of crystallite growth reduces (table 3).

Table 3. Average rate of the crystallite growth in the model systems.

Supersaturation (saliva)	Average growth rate μm/min	Supersaturation (LPDP)	Average growth rate μm/min
25	4.38	15	1.38
30	1.85	20	1.24
35	0.81	25	0.74

The obtained values of the particle diameter depending on the time of crystallization (except for the LPDP model system with the supersaturation of 25) are linearly correlated in the coordinates $d^3 = f(t)$

IOP Conf. Series: Materials Science and Engineering **81** (2015) 012064 doi:10.1088/1757-899X/81/1/012064

(Figure 4). Therefore, in terms of the Lifshitz-Slyozov-Wagner theory [9], the particle growth in these systems occurs due to isothermal distillation (Ostwald ripening), that is the substance is transferred from small to large particles since the chemical potential of the latter is smaller (Kelvin effect). As a result, fine particles gradually dissolve and larger particles grow. Thus, the growth of particles is caused by the diffusion substance transport.

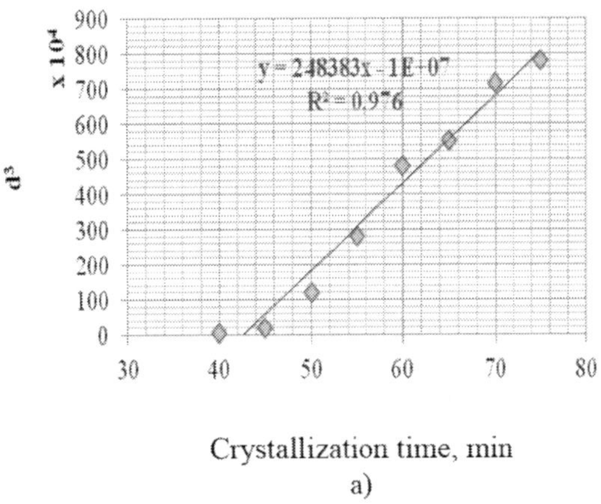

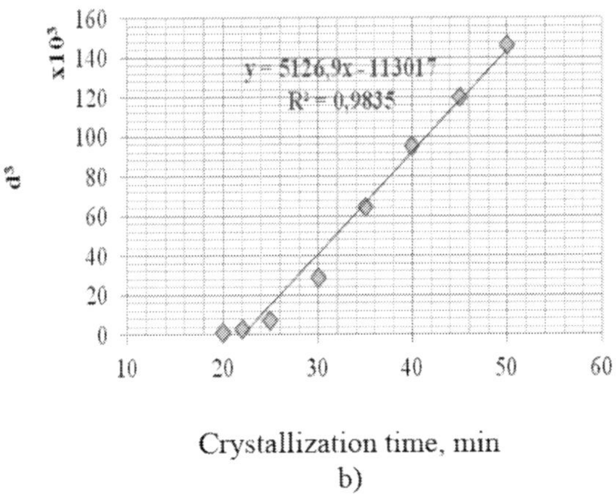

Figure 4. Dependence of the particle size on crystallization time, a) is saliva system S = 25, b) is LPDP system S = 15.

It was found that it (the correlation?) is not linear, it has the maximum extremum, and it is of the similar character for both LPDP and saliva solutions. The obtain interrelation indicates that two-dimensional nucleation is characteristic of the modeled liquids since it can be described by the equation $\frac{dR}{dt} = k_n S^n$ (where n is the order of the reaction surface [10]), and the calculated value is equal to n = 2.00 ± 0.06. These patterns are in good agreement with the data in [9] and with previous results [4].

The shape of the curve "size-time" for the LPDP system with the supersaturation equal to 25 is different. This may be due to a different mechanism of nucleation (transition from heterogeneous to homogeneous nucleation mechanism as the supersaturation increases) and crystal growth. It can be assumed that the particle growth in the system occurs due to coagulation, i.e. aggregation (coacervation) of the particles in a dispersed phase. Since a large number of nuclei of crystallization are formed in this system, the density of the solid-phase particles is higher, they are close to each other, and this enhances their approach and mutual aggregation.

4. Conclusions

Thus, in the research by means of the dispersion analysis it was found:

1. For the liquid phase of dental plaque, the monomodal particle distribution can be observed throughout the crystallization time, however, for the system modeling saliva it is polymodal.
2. In crystallization, the particle size increases and the confidence intervals for different time periods do not overlap.
3. The composition of the saliva model system is favorable for growth of crystals larger in size.
4. The particle size in crystallization varies nonlinearly, and it has the maximum extremum. As supersaturation increases, the number of formed particles grows, however, the average rate of the crystallite growth decreases.

5. Acknowledgements
The study was partially supported by the Ministry of Education and Science of the Russian Federation, within the public task of universities in terms of research works on the 2014-2016 years, (project No 2953).

References
[1] Nagata Y, Higashi M, Ishii Y 2006 *J. Life Science.* **78** No 15 1677
[2] Van Wuyckhuyse B C, Perinpanayagam H E, Bowen W H 1995 *J. Of Dental Research.* **74** No 2 686
[3] Zappacosta B, Manni A, Persichilli S et al 2003 *J. Clinica Chimica Acta.* **338** 57
[4] Golovanova O A 2007 Pathogenic minerals in the human body (Omsk) p 395
[5] Tanaka M, Matsunaga K, Kadoma Y 2000 *J. Of Medical And Dental Sciences* **47** No 1 55
[6] Matsuo S, Lagerlof F 1991 *J. Archives of Oral Biology* **36** No 7 525
[7] Golovanova O A, Achkasova E Y, Puning J O, Zhelya E V 2006 *J. Crystallography* **51** No 2 376
[8] Kutsarev I P 2003 Handbook for physicians and clinical laboratory technicians. Indicators of fluid systems of the man in health [in Russian] (Rostov on/D: Phoenix) 59
[9] Larichev T A, Sotnikova L V, Sechkarev B A et al 2006 Bulk crystallization in inorganic systems. Textbook. [in Russian] (Kemerovo) p 177
[10] Timofeev V A 1978 Crystal growth from solutions and melts [in Russian] (Moscow: Nauka) p 268

Effect of cations ($Mg^{2+,}$ Zn^{2+}, Cd^{2+}) on formation of the mineral phase in $Ca(NO_3)_2$-$Mg(NO_3)_2$-Na_2HPO_4-H_2O system

O A Golovanova[1], R M Shlyapov[2], Sh K Amerkhanova[2], A S Uali[2] and V A Vlasov[3]
[1] Omsk State University, Omsk, Russia
[2] Karaganda State University, Karaganda, Kazakhstan
[3]National Research Tomsk Polytechnic University, Tomsk, Russia

E-mail: golovanoa2000@mail.ru

Abstract. Synthesis of hydroxyapatite in the system $Ca(NO_3)_2$-$Me(NO_3)_2$-$NaHPO_4$-H_2O with pH in the range of 6-12.2 is performed, and hydroxyapatite of B-type is formed. The effect of magnesium, zinc and cadmium ions on the properties of hydroxyapatite is studied. It is shown that as the concentration of these ions increases, the crystallinity of hydroxyapatite and the Ca/P ratio decreases. It is found that in all the experiments Zn^{2+} cations affect the hydroxyapatite structure to a lesser extent which results in less structural defects, wherein the Ca/P ratio attains the highest values. It is shown that the solubility of the samples containing zinc ions is greater than that of the samples with other additives.

Keywords: Hydroxyapatite structure; Dissolution; Crystallization; Acidity; Cationic substitution.

1. Introduction

Millions of people are susceptible to bone tissue lesions as a result of pathological diseases, such as osteomyelitis, osteosarcoma, osteoporosis, or injury. Since ancient times health professionals has been searching for techniques to reconstruct the bone tissue by implanting the defect to allow normal functioning of the damaged organ, for example, to withstand physiological loading. The problem is to find the material for the implant. Ideally, the material should be biocompatible with the tissue, that is, cause an adequate response from the body, to be nontoxic. It should not cause negative immune and other types of body reactions, and not be rejected by the body as a foreign body. It should be biologically active, that is, form interaction with the biological system to form bone tissue on it or replace it by the bone tissue and, preferably, and to induce bone formation processes. The implant must maintain its functional qualities over a certain period of time without change in its structure and mechanical properties. A great number of materials have been tested for this purpose [1–10]. However, bioinert materials were limited in application in reconstructive surgery because of the inevitable rejection reactions [1–9].

In recent years, interest in biogenic apatite has significantly increased. One of the reasons is the fact that hydroxyapatite is close in chemical composition to the inorganic component of the bone, and therefore, it has similar physicochemical and mechanical properties. Therefore, it can be used as an implant in the form of ceramics, cements and composites [7, 8].

Another reason of great interest in studying these minerals is due to the fact that apatite is the part of the majority of pathogenic biominerals, and the cause and mechanism of their formation has not yet been found. Unfortunately, data on the composition and structural features of biogenic apatite are very few in number. At the same time, the literature data [7–10] suggests a complex isomorphism which often causes considerable non-stoichiometry of the substance and significant variations in the parameters of the crystal lattice.

The aim of the research is to study the effect of metal ions on patterns of formation and nature of the solid phase crystallized from aqueous solutions of specified pH and ionic composition.

2. Experimental techniques

The crystals of the substituted calcium phosphate were synthesized from aqueous solution at room temperature through spontaneous crystallization according to the chemical reaction:

$$10\ CaX_2 + 6\ M_2HPO_4 + MOH \rightarrow Ca_{10}(PO_4)6(OH)_2\downarrow + 20\ MX + 6\ H_2O \quad (1)$$

Crystallization was conducted at room temperature (22–25 °C) and under varying concentration of the metal cation dopant in the range of 0.0017–0.0255 mol/l (for Mg^{2+}) or 0.0009–0.0102 mol/l (for Zn^{2+} and Cd^{2+}). The crystallization time was 48 hours. In the liquid phase, the content of calcium and magnesium ions was determined by complexometric titration with two indicators (Eriochrome Black T and murexide, according to RD 52.24.403-2007); the final content of the metal ion dopant (Zn^{2+} according to GOST 18293-72 and Cd^{2+} by the photometric dithizone method) was found; the concentration of phosphate ions was evaluated by molybdene blue spectrophotometric method. The resulting solid phase was studied by physical and chemical analyses [8].

StatSoft Statistica 6.0 was used to perform statistical data processing.

3. Result and discussion

Hydroxyapatite was synthesized from the solution at varying pH and ionic strength. The results of X-ray diffraction analysis of the solid phase (Figure 1) indicated the formation of hydroxyapatite. Other calcium phosphates in the sediment were not found.

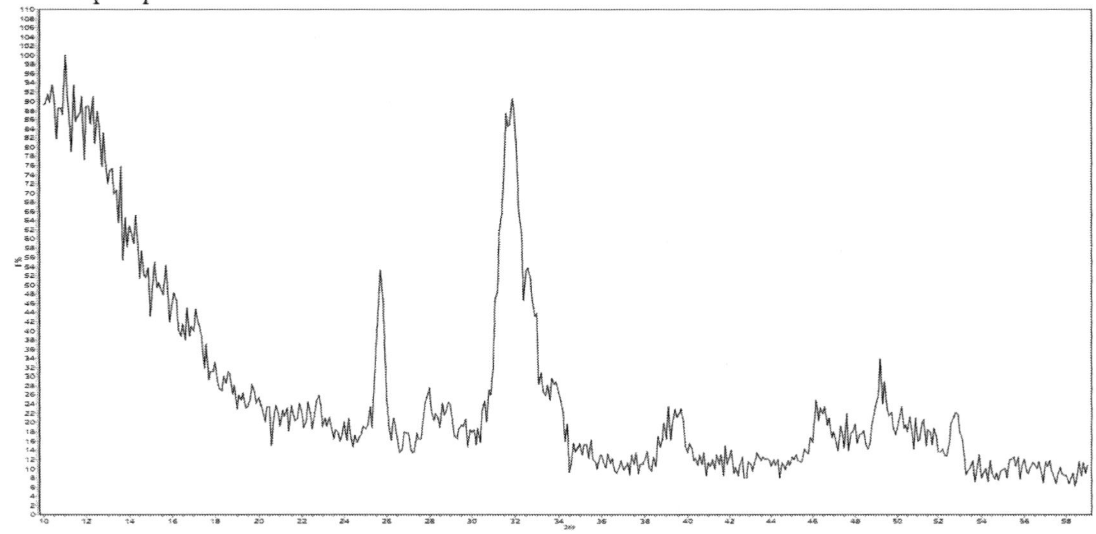

Figure 1. X-ray phase analysis of the samples produced by deposition: at pH = 9.0.

The characteristic main absorption bands caused by vibrations of O-P-O bonds in PO_4^{3-} tetrahedras are observed in the IR spectra (Figure 2). The maximum absorption in these bands is at 1087, 1040,

962, 601, 575 and 474 cm^{-1}. The peaks at ν = 1040 and 1087 cm^{-1} are caused by triply-degenerated antisymmetric valence vibrations of the O-P-O bonds, ν3. The absorption band at 962^{-1} refers to a non-degenerated symmetric valence mode, ν_1. The intensities with the maximum absorption at 601 and 575 cm^{-1} are caused by doubly-degenerated valence vibrations of the O-P-O bond, ν_4. Faint peak at 472 cm^{-1} is the component of the degenerated valence vibrations of the ν_2 mode. The peak with ν = 640 cm^{-1} and the intensity at 1650 cm^{-1} in the near spectrum region correspond to OH-group variations; the broad band at 3150–3400 cm^{-1} in the far spectrum region can be caused by the absorption of the molecular water adsorbed by apatite.

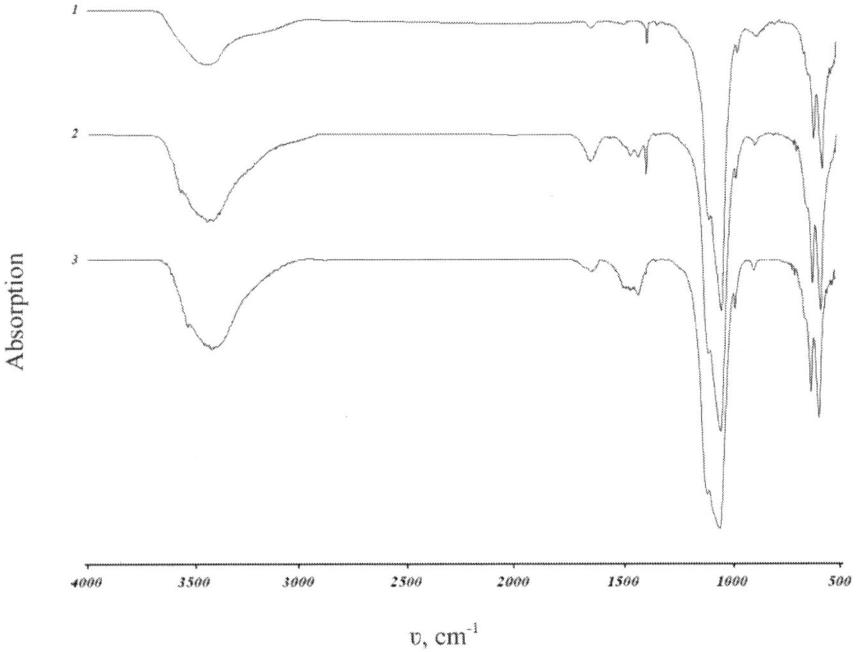

Figure 2. IR spectra of samples prepared by precipitation: 1 is at pH = 6.00, 2 is at pH= 9.00, 3 is pH = 12.00 (hydroxyapatite phase).

In the spectra of the synthesized hydroxyapatite, we can also observe the bands of the ν3 mode of CO_3^{2-} vibrations at 1420 and 1450 cm^{-1} and those of the ν2 mode at 873 cm^{-1}. These absorption bands in the spectra indicate partial substitution of RO_4^{3-} by carbonate ions in the hydroxyapatite structure of B-type. Thus, all the samples prepared at pH ≥ 7.00 are found to be carbonate -substituted hydroxyapatite of B-type [8].

The analysis data (table 1) showed that pH growth increases the precipitated mass of the resulted phase. At pH=12, precipitation is the largest. This indicates more favorable conditions for crystallization of the mineral phase.

Table 1. Data of the analysis of ions in the solution.

pH initial	m ave., g	Ca/P ave.
6.00±0.05	1.195±0.3563	1.35
9.00±0.05	1.335±0.0497	1.58
12.00±0.05	1.567±0.3476	1.68

An important characteristic of hydroxyapatite is stoichiometry of its composition which is usually expressed by the Ca/P ratio. To explain the deviations from the ideal stoichiometry Ca/P = 1.67 the hydroxylapatite composition is written by the formula $Ca_{10-x}(HPO_4)_x(PO_4)_{6-x}(OH)_{2-x}$ (1.5 <Ca/P <1.67, i.e. 0<x<1).

The chemical analysis showed that the Ca/P ratio in the prepared samples is from 1.35 to 1.68 depending on the initial concentrations. Table 1 shows the change in Ca/P at varying pH. These changes indicate that the data obtained at pH = 12 (Ca/P = 1.68) is the closest to the ideal stoichiometry. This data is in agreement with the thermodynamic results obtained earlier [8].

Comparison of possibility for crystallization of the mineral phase in $Ca(NO_3)_2$-$Mg(NO_3)_2$ - Na_2HPO_4-H_2O system containing (Mg^{2+}, Zn^{2+}, Cd^{2+}).

Cationic substitutions are known to have a significant impact on biological behavior of hydroxyapatite and are one of the known techniques to improve reabsorption (solvability) of hydroxyapatite materials. In addition, the elements of the apatite crystal lattice can exchange with ions of the solution surrounding the crystal and change due to the ions in this solution. In living systems, this property makes apatite highly sensitive to the ion composition of biological fluids. The ion exchange in the hydroxyapatite crystal lattice changes its properties, affects the strength and size of crystals.

For example, it is known that strontium apatite is formed in substitution of Ca by Sr^{2+} in the apatite crystal lattice. This disturbs the crystal structure. Kashin-Beck disease is reported to be accompanied by osseous lesion, meromicrosomia of limbs in humans and animals. In the radionuclide-contaminated areas, the negative effect of strontium apatite on a human being is caused by the possibility of radioactive strontium deposition.

In the case of filling the cation site, the determining factor is the nature of the chemical bond formed by the divalent cation in a specific position of the crystal structure. The factor that the substituting cation and calcium cation are similar in size is of less relevance.

We studied the effect of magnesium, zinc and cadmium ions as dopants in hydroxyapatite synthesis since the ionic radii of these cations are close in values to those of calcium ions (table 2). It was found that the highest values for Ca/P are obtained for zinc ion dopants (1.4), and the minimum values are typical of cadmium ion dopants (0.8). In the presence of magnesium ions in the hydroxyapatite structure, the Ca/P ratio takes intermediate values between zinc ion and cadmium ion dopants. It can be assumed that Zn^{2+} ions have a lesser effect on the hydroxyapatite structure and cause less imperfection of its structure.

Table 2. Ion and metal radius.

Cation	Ion radius, Å
Ca	0.99
Mg	0.66
Zn	0.83
Cd	0.97

The analysis of the resultant solid phase weight indicated that Zn^{2+}, Mg^{2+} and Cd^{2+} ions present in the starting solution increase the precipitation mass. The minimal effect was observed for magnesium ions as dopants.

It is important to emphasize that the biological activity of hydroxyapatite samples doped with magnesium, zinc and cadmium ions is similar, but the greatest bioactivity is observed for precipitation containing zinc ions. All the solid phases with dopants exhibit higher solvability than hydroxyapatite samples without additives.

IR spectroscopy indicated that doping with ions and slight increase in their concentration in the starting solution does not virtually affect the IR spectra form (Figure 3 a–d); the B-type hydroxyapatite is observed in all the cases. In turn, sharp increase in dopant concentration (up to 75%) causes formation of the amorphous phase (Figure 4 a–b). Magnesium ions, whose ionic radius is significantly smaller as compared to calcium, may destabilize the apatite structure to a greater extent.

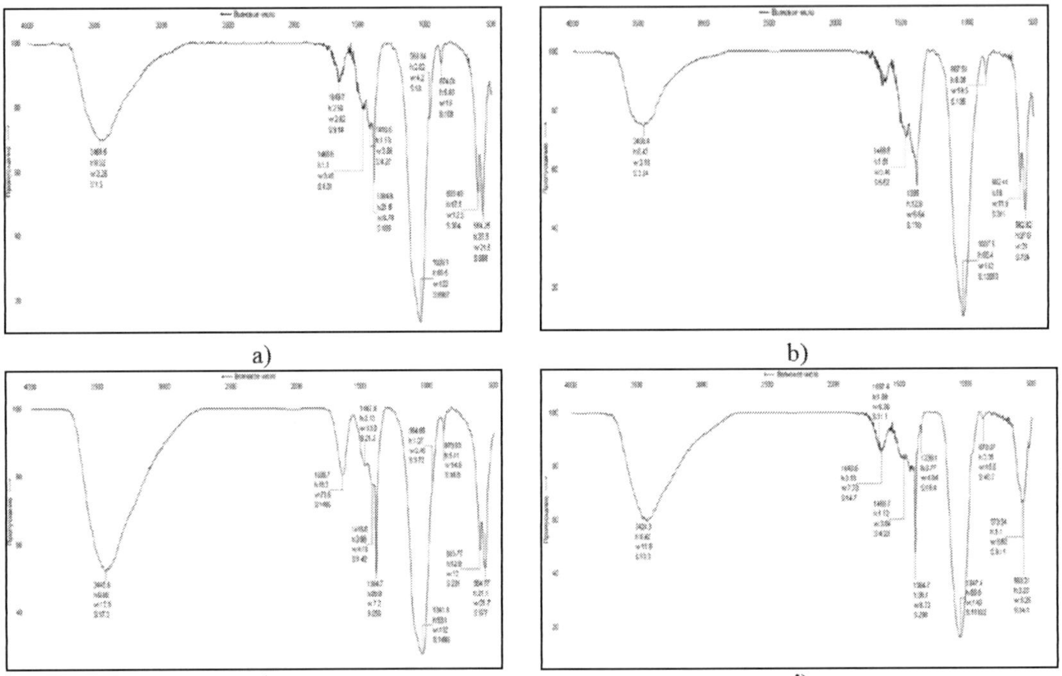

a) b) c) d)

Figure 3. IR spectrum of the sample at pH=12: a) C(Cd)=2.5%; b) (Cd)=30%; c) C(Zn)=2.5%; d) C(Zn)=30%

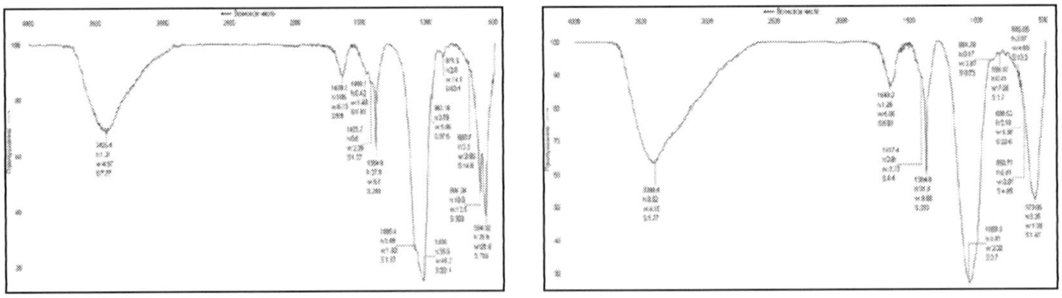

a) b)

Figure 4. IR spectrum of the sample at pH=9 c a) C(Mg)=5%; b) C(Mg)=75%

4. Conclusions

1. Synthesis of hydroxyapatite in the $Ca(NO_3)_2$-$Me(NO_3)_2$-$NaHPO_4$- H_2O system with the initial ratio Ca/P = 1.70, with pH in the range of 6-12.2 has been performed.

2. The Ca/P ratio in synthetic samples is found to be from 1.35 to 1.68 depending on the initial concentration of the doped ions. The value for Ca/P, closest to the hydroxyapatite stoichiometry (1.67), can be obtained at pH = 12 with hydroxyapatite of B-type being formed.

3. The effect of magnesium zinc and cadmium ions on the properties of hydroxyapatite is studied. As the concentration of these ions increases, the hydroxyapatite crystallinity and Ca/P ratio are shown to decrease.

4. It is established that in all the experiments Zn^{2+} cations lead to less imperfection of the hydroxyapatite structure, and the Ca/P ratio attains its highest values.

5. It is shown that the bioactivity of the samples containing zinc ion dopants is greater than that of other dopants.

It is found that ion doping does not affect the IR spectra form.

5. Acknowledgements
The study was partially supported by the Ministry of Education and Science of the Russian Federation, within the public task of universities in terms of research works on the 2014-2016 years, (project No 2953).

References
[1] Surmeneva M A, Surmenev R A, Chaikina M V et al 2012 *J. Physics and chemistry of material treatment* [in Russian] No 3 51
[2] Sventskaya N V 2011 Slicophosphate biocomposite materials with controlled pore structure for osteoplastic surgery, thesis abstract [in Russian] (Moscow) p 23
[3] Gomes Sandrine, Nedelec Jean-Marie Jallot, Edouard et al 2011 *J. Crystal Growth Design.* **11** 4017
[4] Dorozhkin S V, Agatopoulus Simeon 2002 *Chemistry and Life* [in Russian] No 2 8
[5] Carlisle E M 1970 *J. Science* **167** 279
[6] Wang Yongsheng, Zhang Sam, Zeng Xianting et al 2007 *J. Materials Science and Engineering.* **27** 244
[7] Barinov S.M 2010 *Russian Chemical Reviews* [in Russian] **79** No 1 15
[8] Solonenko A P, Golovanova O A 2013 *J. Inorganic Chemistry.* **58** No 12 1420
[9] Veresov A G, Putlyaev V I, Tretyakov Y D 2004 *Russian Journal* [in Russian] **48** No 4 52
[10] Golovanova O A 2007 Pathogenic minerals in the human body [in Russian] (Omsk) p 395
[11] Larichev T A, Sotnikova L V, Sechkarev B A et al 2006 Bulk crystallization in inorganic systems. Textbook. [in Russian] (Kemerovo) p 177
[12] Timofeev V A 1978 Crystal growth from solutions and melts [in Russian] (Moscow: Nauka) p 268

Physics and chemistry of producing silicon-hydroxylapatite-titanium composite materials

O A Golovanova[1], A V Zayts[1], T V Panova[1] and T S Frangulyan[2]
[1] Omsk State University, Omsk, Russia
[2]National Research Tomsk Polytechnic University, Tomsk, Russia

E-mail: golovanoa2000@mail.ru

Abstract. In the research, Si-HA poweders have been synthesized using SBF solution with different content of silicon. It was found that all the samples synthesized from the model solution of extracellular fluid under varying concentration of silicate ions are single-phase and repesent hydroxyapatite. The nature of the reagent containing SiO_4^{4-} ions does not affect the hydroxyapatite structure. In the study of the surface and morphological characteristics of the phosphate coatings modified by silicon ions, it was found that as the degree of phosphate calcium substitution by SiO_4^{4-} ions increases, the surface wettability on the titanium substrates deteriorates and cohesive energy decreases. It is shown that the titanium coating is formed in three stages. After irradiation of titanium substrates coated with Si-HA, the crystals can keep growing and the surface can keep regenerating.

Keywords: Calcium phosphates; Apatite nanocomposite synthesis; Crystallization; Modifying; Bioavailability; Biological fluids titanium coating.

1. Introduction
Development of calcium phosphate biomaterials is one of the important areas of modern material science which is at the interface of chemistry, medicine and physics. Synthetic hydroxyapatite $Ca_{10}(PO_4)_6(OH)_2$ (HA) is a widespread material used in medicine for treatment of damaged bone tissue due to its chemical and phase similarity to the inorganic component of the bone tissue [1–9].

The bioactivity of materials is considered to be determined by their chemical composition, morphology of the crystals, and surface properties [9–14]. The main techniques to modify these characteristics are chemical modification of calcium orthophosphates and variation of the synthesis conditions. Since silicon is vital in the physiological processes of growth and remodeling of the bone and cartilaginous tissue, research in the properties of silicon-substituted hydroxyapatite (Si-HA) synthesized under conditions close to physiological ones is a promising and relevant physical and chemical problem.

The literature review shows that synthesis of thermally stable silicon-containing hydroxyapatite-based ceramic materials is currently of great interest since silanol (Si-OH) groups on the material surface contribute to splicing of the implant with the bone [10–14], which indicates their high practical significance to make progress in orthopedics and implantology.

It can be argued that silicon has a beneficial effect and improves the characteristics of non-substituted hydroxyapatite, and after annealing, the properties of Si-HA samples are similar to those of a normal human bone. This fact indicates that silicon-substituted hydroxyapatite samples may be used in medicine as a ceramic coating on titanium implants.

Most bioactive materials, when applying them to bioinert (titanium) substrate, lose their characteristics, therefore, it is important to produce materials which exhibit high stability, and retain their bioactive properties.

In biomimetic synthesis, it is essential to find optimal conditions to produce powders of a highly dispersed phase, with high resorption and biocompatibility with the human body. When depositing these calcium phosphates on a titanium substrate, it is necessary to find a stepwise mechanism of Si-HA deposition to use these samples in practice, to observe in vitro the processes occuring, possible destruction or degradation of the prepared ceramics. Therefore, research in this area is considered to be of current importance in modern implantology [15–17].

Thus, the aim of the research is to study the possibility of producing coatings on the titanium substrate by applying the modified silicate ions of calcium phosphates prepared via deposition from a model solution of extracellular fluid (SBF).

2. Experimental techniques

Synthesis of the substituted calcium phosphate crystals is carried out by deposition from aqueous solution at room temperature through spontaneous crystallization by the chemical reaction:

$$10\ CaX_2 + 6\ M_2HPO_4 + MOH \rightarrow Ca_{10}(PO_4)6(OH)_2\downarrow + 20\ MX + 6\ H_2O \ (1)$$

The main compounds for silicate groups were Na_2SiO_3 and tetraethoxysilane (TEX) as the most frequently used compounds. The weight content of silicon in the starting solution was 0.5, 2.5 and 5.0 %. The chosen ratio Ca/P = 1.67 corresponded to that for preparing stoichiometric hydroxyapatite [10]. The PH value was 7.40, pH characteristic of the extracellular fluid, and it remained constant.

The duration of the synthesis was 48 hours. Upon completion of the solid-phase crystallization, pH and concentration of Ca^{2+} ions were measured in the supernatant fluid by potentiometric method. The colorimeter CPC-2 was used to measure the residual concentration of phosphate-ions [RD 52.24.382-2006], silicate-ions (RD 52.24.433-2005) in the solution. The solid phase was filtered, dried and analyzed with FT-IR spectroscopy and X-ray diffraction analysis (XRD). IR spectra were recorded with FT-801 spectrometer (samples made in the form of pellets with KBr). XRD was performed with the Bruker D8 Advance X-ray diffractometer (Germany) under monochromatic Cu-k_α radiation.

Statistical data processing was carried with StatSoft Statistica 6.0.

To deposit hydroxyapatite on the surface of VT1-0 titanium the titanium plates with a size of 15 mm ∗15 mm ∗1.2 mm were pretreated: polished; etched (we chose the pickling mixture HNO_3: HF 1: 1 to enlarge the surface of the titanium substrates); exposed to powerful ionization beam (the setup"Temp" was used for irradiation); subjected to laser ablation; perforated (etched, not subjected to mechanical impact and chemical aggression, subjected to laser ablation), and plates which were not subjected to mechanical impact and chemical aggression.

After that, the plates were immerged into the prototype of the model solution of human interstitial fluid [18]. The system was kept for 3, 6 and 12 days (SBF base solution being changed and being not changed). After that, the plates were removed from the test solution, dried at room temperature and analyzed using optical microscopy.

3. Result and discussion

The X-ray analysis showed that all the samples synthesized in the medium of the model solution of extracellular fluid under varying concentration of silicate ions are single-phase and represent hydroxyapatite. In the diffraction pattern, the HA reflexes are found at 25.8°(002), 31.7°(211), 32.2°(112), and 32.9°(300) (Figure 1). It should be noted that the diffraction patterns corresponding to the solid phases with different sources materials for Na_2SiO_3/TEL silicate groups are found to be similar. Thus, the nature of the reagent containing SiO_4^{4-} ions does not affect the hydroxyapatite structure.

The size of the prepared unmodified HA crystallite was 6.3 nm, and the size of Si-HA was 6.0÷7.1 nm, which indicates the formation of nanocrystalline compounds. The parameters of the crystal lattice of the formed solid phase grew larger as compared to the unmodified hydroxyapatite. This confirms the possible substitution of the orthophosphate-ion by the silicate-ion of different size (distance Si-O = 1.66 Å, P-O = 1.55 Å [18]).

In the IR spectra of the samples formed in the presence of the silicate ions, the absorption bands characteristic of hydroxyapatite can be observed (Figure 2). The HA structure is defined by the following bands: the bands at 1040–1080 (ν_3), 960, 840 (ν_2), 602, 574 (ν_4) and 473 (ν_2) cm^{-1} corresponding to the vibrations of the phosphate groups, the broad band at 3440–3570 cm^{-1} which corresponds to deformation vibrations of OH^--groups and valence vibrations of the adsorbed water. High intensity of this peak in the modified HA is related to the fact that SiO_4^{4-} ions have high affinity for water molecules which results in hydration. Since HA is synthesized in air, during its formation carbon dioxide is adsorbed from atmospheric air, and carbonate ions are localized in the lattice in the position of the phosphate ion, and in the IR spectra, the bands absorbing deformation vibrations of C-O bonds of the CO_3^{2-} ion at 875 cm^{-1} and valence vibrations at 1422–1457 cm^{-1} correspond to these ions.

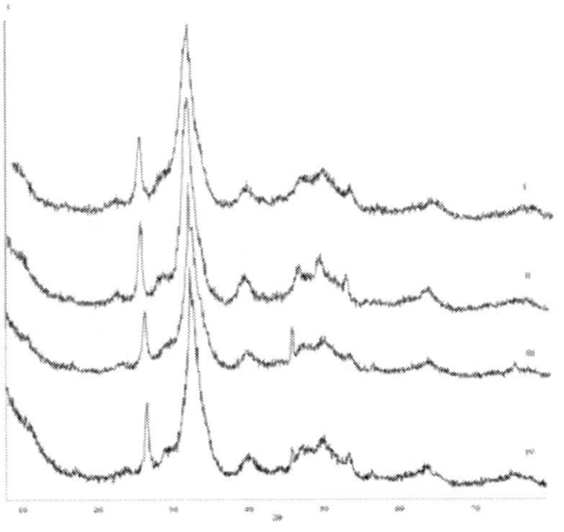

2θ, deg

Figure 1. The diffraction patterns for hydroxyapatite samples modified by Na_2SiO_3: *I* is C Si init. 0.5%; *II* is C Si init. 2.5%.; TEX: *III* is C init. 0.5%; *IV* is C Si Si init. 2.5%.

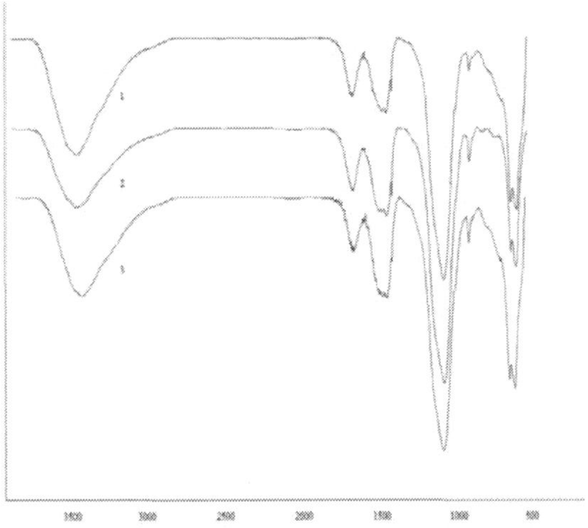

υ, cm^{-1}

Figure 2. IR spectra for hydroxyapatite samples modified by: *1* is Na_2SiO_3C Si init. 0.5%; *2* and *3* are TEX C Si init. 0.5 and 2.5%.

It has been observed that IR spectra contain a low intensity band of absorption of the silicate group bond vibrations at 511 cm^{-1} due to the deformation vibrations of SiO bonds. The absorption band of the Si–O bond vibration at 945 cm^{-1} is overlapped with the absorption band which corresponds to P–O bond vibrations at 960 cm^{-1}. Occurrence of these signals in the spectrum indicates the substitution of PO_4^{3-} by SiO_4^{4-} in the HA structure. This leads to distortion of the PO_4^{3-} tetrahedra, and as a result, the parameters of the cell change. Thus, the data obtained by FT-IR spectroscopy are in agreement with that obtained by XRD.

An essential problem in the production of calcium phosphate coatings is their long-time resistance to resorption which depends on the adhesive strength of the phase composition and the degree of the structure crystallinity. Hydroxyapatite synthesized from the SBF (Simulated Body Fluid) solution was used as adhesive material.

The values obtained in measuring the limiting wetting angle and the surface tension of the prepared suspensions on titanium substrates were used in the Young-Dupre equation [19] to calculate the adhesion energy of the calcium phosphate suspension applied to the solid surface of the titanium samples (Table 1).

Table 1. Adhesive characteristics of the titanium-based calcium phosphate suspension

Additive Si	Composition	σ01,MJ/m^2	cosΘ	W_a, MJ/m^2	W_k, MJ/m^2	W_a/W_k
	1.00 %	85.12	0.73	147.26	170.24	0.87
Na_2SiO_3	2.50 %	76.16	0.26	115.76	152.32	0.76
	5.00%	76.82	0.44	110.62	153.34	0.72
	1.00 %	66.04	–1	0	132.08	0
TEX	2.50 %	64.57	0.26	81.36	129.14	0.63
	5.00%	66.67	–0.76	16.00	133.34	0.12

It was found that as the concentration of silicate ions increases in the samples synthesized with inorganic ion carrier SiO_4^{4-}, the surface tension and limiting wetting angle decrease. This may be caused by partial dissolution of the resulting calcium phosphate and formation of inorganic surface-inactive substances in the solution. For samples with TEX as a SiO_4^{4-} ion, change in the concentration of silicon in the suspension leads to ambiguous results which require further research. The adhesion-cohesion energy ratio is an important characteristic of the produced coatings (Table 1). The calculated data allow the conclusion that the suspension containing 1.00 wt% of the silicate ion in the form of SiO_4^{4-} exhibits good adhesive interaction between silicon-containing hydroxyapatite and titanium samples.

In the study of the surface and morphological characteristics of the phosphate coatings modified by silicon, it was found that more complete deposition of Si-HA on the titanium substrate surface occurs in the etched samples (Figure 3). This technique of surface treatment provides uniform, dense, and finely-dispersed coating, and HA crystals are found to grow in the form of dendrites. The crystal growth becomes more structured. Increase in the time of keeping the titanium sample in the model solution results in discontinuous columnar crystal growth which indicates the start of Si-HA surface structuring. At this stage, the height of the formed aggregates is found to be different (Figure 3b). Further increase in the time of keeping the sample in the model solution for 6 days causes formation of non-uniform coating. The crystals are found to grow in the form of dendrites. The start of the crystal growth in the form of cylindrical columns can be observed, that characterizes the beginning of Si-HA structuring on the substrate surface (Figure 3 c). For longer period of keeping the sample in the solution without changing the model solution, the crystals are found to accumulate in the form of aggregates, and the crystal growth goes beyond the substrate (Figure 3 d).

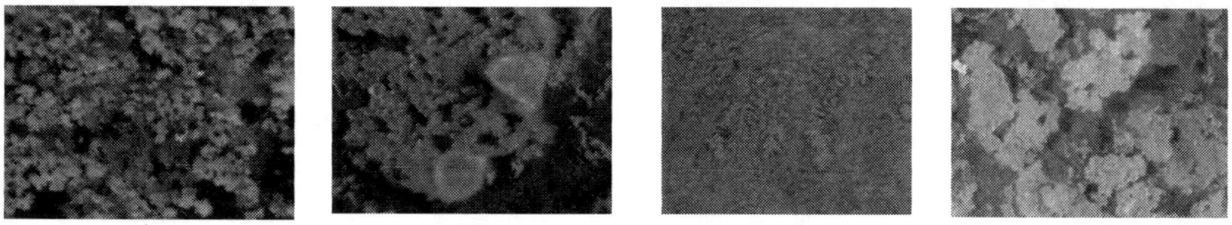

a) b) c) d)

Figure 3. Deposition of Si-HA on titanium substrates (etched surface).Duration a) 3 days; b) 3+3 days, the model solution being changed; c) 6 days, the model solution being not changed; d) 3+3+6 days, the model solution being changed.

Different techniques are used to produce biocoatings on titanium implants, for example, plasma, and magnetron or laser deposition of the bioactive material. High temperature and high-energy action, which are typical of these techniques, cause partial degradation of the bioactive material and decrease its efficiency. Development of new methods for microcrystalline hydroxyapatite synthesis, low-cost and technological techniques to form bioactive titanium coatings from composites is considered to be of current relevancy and in great demand.

Coatings made on titanium substrates subjected to laser ablation (Figure 4a) are a non-uniformly distribution of the deposited phase. The deposition occurs in the craters formed under the exposure and in the near-crater areas. The crystals grow upwards, and columnar structures characteristic of the Si-HA phases are formed. Increase in the number of pulses results in a more uniform coating over the entire surface (Figure 4b). The deposition is dense and the thickness of the coating is significantly different. In some areas, thick coating is found to be with typical cracking. The deposition occurs in craters (more dense deposition), and in the near-crater areas. Further increase in the number of pulses causes deposition in the near-crater areas. Virtually no crystal growth can be observed in craters which may be due to hindered diffusion of the electrolyte components.

Another technique to treat the titanium substrate surface is to expose to a powerful ion beam (PIB). After this treatment, the surface coating is uniform, the deposition is dense and finely-dispersed (Figure 4a).

At the next stage, after synthesis of the Si-HA layer on the titanium substrates, the samples were exposed to PIB with j= 0 A/cm^2 (Figure 4c-d). As a result, coating was observed to be uniform, dense, with melted edges. Dendritic crystals were found to grow on the substrates, and this enhanced further growth and renewal of the surface under physiological conditions.

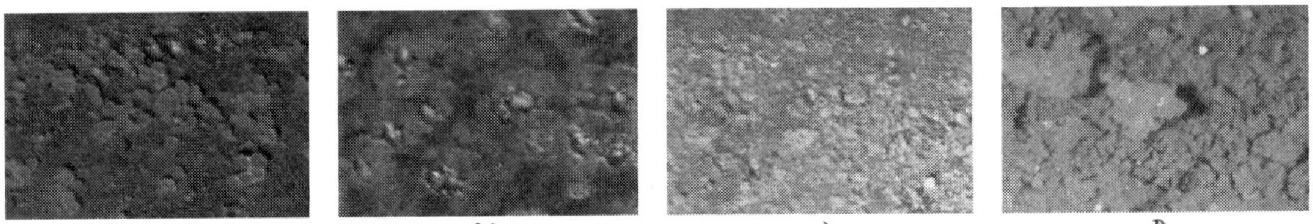

a) b) c) d)

Figure 4. Deposition of Si-HA on titanium substrates after subjecting to laser ablation, deposition time is 3 days: a) the number of pulses is 1: b) the number of pulses is 3; c) the number of pulses is 5; d) the number of pulses is 3, the deposition time is 3 + 6 days, the model solution being changed.

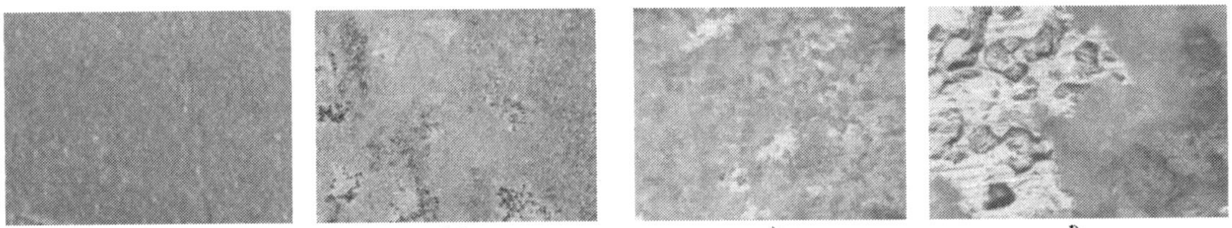

a) b) c) d)

Figure 5. The morphology of the hydroxyapatite crystal surface grown on the titanium surface 1-0: a) surface treatment by exposure to PIB before deposition; b–d) after irradiation by PIB with j =50 A/cm^2.

When measuring the coated area (Table 2), it was found that the highest percentage of deposition onto the substrate occurs within the first three days. This is attributed to the fact that three days are the basic time for Si-HA dissolution. After that, Si-HA dissolution stops. To attain better characteristics for the coated area the suspension is to be changed. As a result, the coated area of the substrate is not only larger, but the coating is more uniform both in size and in homogeneity of the crystals.

Table 2. The area of the coated surface depending on the technique of titanium substrate treating

Treatment	$S_{coating}$, %				
	3 days	6 days	3+3 days	3+6 days	3+3+6 days
Etching	72	75	79	-	85
Not treated	60	67	70	-	78
Perforation etching	30	-	-	73	-
Perforation without treatment	25	-	-	79	-
PIB	-	60	-	-	-
Laser n=1	20	-	-	30	-
Laser n=2	23	-	-	35	-
Laser n=3	25	-	-	40	-
Laser n=5	27	-	-	44	-
Laser perforation laser n=1	19	-	-	26	-

The analysis of the data from Table 2 shows satisfactory coating of the Si-HA surface for non-treated titanium samples. This confirms that titanium substrates are highly biocompatible, non-toxic, resistant to corrosion, and their characteristics are similar to the mechanical properties of the bone tissue [8]. The coating appears to be uniform, dense, with characteristic cracking, which indicates sufficient thickness of the coating; large and small Si-HA aggregates occur (Figure 6 a–d). As the time of keeping the sample in the model solution increases, discontinuous columnar growth of crystals can be observed and this indicates the start of Si-HA surface structuring.

No significant difference is found in the height of the newly grown crystals (Figure 6 b). The crystal growth does not go beyond the substrate surface, and the Si-HA lower layer takes the form of the titanium crystal lattice. In long-time keeping the titanium substrates in the SBF model, the coating appears to be uniform over the entire surface, dense, no Si-HA agglomerates can be detected (Figure 6 c, d).

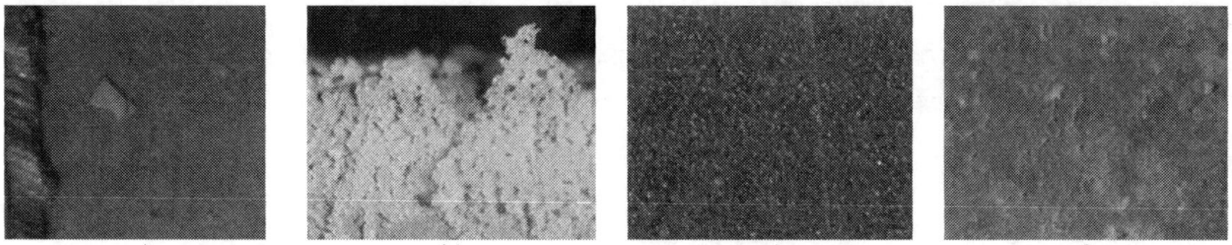

a) b) c) d)

Figure 6. Deposition of Si-HA on titanium substrates without surface treatment: a) 3; b) 3+3 days, the model solution being changed; c) 6; d) 3+3+6 days, the model solution being changed.

In addition to the above techniques of the substrate surface treatment, we used perforation of the titanium samples. The Si-HA coating on these substrates is not uniform; virtually no deposition occurs (Figure 7a). Repeated treatment of the surface (etching, laser ablation) results in partial depositing of Si- HA crystals in laser craters and on creater cuts (Figure 7b). As the time of keeping the titanium samples in the model solution increases, with the solution being changed, glassy aggregates are formed in the process of perforation hole closing with the following growing of crystals upwards in the shape of cylinders. In the near-perforation areas, the crystals grow in the form of dendrites (Figure 7 a–d).

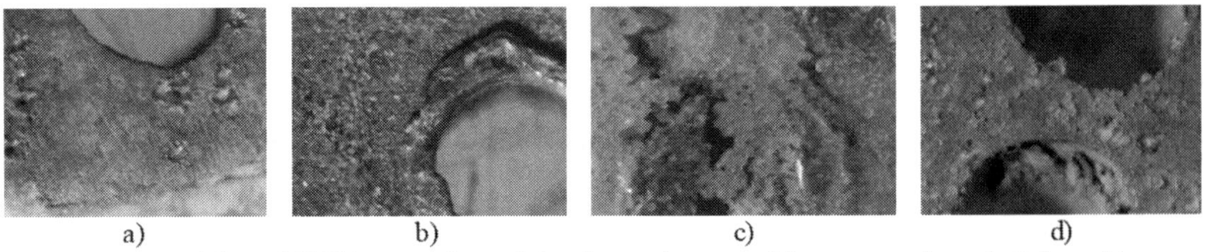

Figure 7. Deposition of Si-HA on perforated titanium substrates a) laser-treated n = 1, 3 days; b) etched, 3 days; c) unetched 3+6 days; d) laser-treated n = 1 3+6 days.

4. Conclusions

Si-HA powders have been synthesized from SBF solution with the silicon content of 0.50; 1.00; 1.50; 2.50; 5.00 wt.%. It was found that all the samples synthesized in the model solution of extracellular fluid under varying concentration of silicate ions are single-phase and represents hydroxyapatite. The nature of the reagent containing $\mathbf{SiO_4^{4-}}$ ions does not affect the hydroxyapatite structure.

In the study of the surface and morphological characteristics of the produced phosphate coatings modified with silicon ions, it was found:

As the degree of the substitution of calcium phosphate by SiO_4^{4-} ions increases (inorganic silicate ions promoter was used in the synthesis) the surface wettability, limiting wetting angle and cohesion energy are found to deteriorate. The samples with organic silicon ion dopants are mainly characterized by nonwettability of the surface and low cohesive energy.

5. Acknowledgements

The study was partially supported by the Ministry of Education and Science of the Russian Federation, within the public task of universities in terms of research works on the 2014-2016 years, (project No 2953).

References

[1] Surmeneva M A, Surmenev R A, Chaikina M V et al 2012 *J. Physics and chemistry of material treatment* [in Russian] No 3 51.
[2] Sventskaya N V 2011 Slicophosphate biocomposite materials with controlled pore structure for osteoplastic surgery, thesis abstract [in Russian] (Moscow) p 23
[3] Gomes Sandrine, Nedelec Jean-Marie Jallot, Edouard et al 2011 *J. Crystal Growth Design.* **11** 4017
[4] Wang Yongsheng, Zhang Sam, Zeng Xianting et al 2007 *J. Materials Science and Engineering.* **27** 244
[5] Barinov S M 2010 *Russian Chemical Reviews [in Russian]* **79** No 1 15
[6] Solonenko A P, Golovanova O A 2013 *J. Inorganic Chemistry.* **58** No 12 1420
[7] Golovanova O A 2007 Pathogenic minerals in the human body [in Russian] (Omsk) p 395
[8] Stillman M I 2010 *High-molecular compounds, Series A* **52** No 9 1551
[9] Astakhova I N 2008 Surgery [in Russian] No 12 34
[10] Putlyaev V I 2004 *Soros Educational Journal* **8** No 1 44
[11] Fomin A S, Komlev B C, Barinov S M et al 2006 *Perspective materials [in Russian]* No 2 51
[12] Kosjakov A N 2010 *Orthopedics, traumatology and prosthetics [in Russian]* No 4 105
[13] Gnedenkov S V et al 2011 *Perspective materials [in Russian]* No 2 49
[14] Sidel'nikov A I 2000 *Info-Dent [in Russian]* No 5 10
[15] Ferro D, Barinov S M, Rau J V et al 2005 *Biomaterials* **26** 805
[16] Ulasevich S A, Poznyak S K, Kulak A I et al 2012 *Sviridov readings* Iss. 8 168
[17] Bharati S, Sinha M, Basu D 2005 *Bull. Mater. Sci.* **28** No 6 617
[18] Berdinskaya M V, Golovanova O A, Zaits A V 2014 *J. Structural Chemistry [in Russian]* No 5 **55** 992

[19] Izmailov R R, Golovanova O A, Zaits A V 2014 *Inorganic Materials [in Russian]* **50** No 6 641

IOP Conf. Series: Materials Science and Engineering **81** (2015) 012067 doi:10.1088/1757-899X/81/1/012067

Pulsed non-self-sustained glow discharge with a large-area hollow cathode for nitriding of iron-based alloys

V V Denisov[1], Yu H Akhmadeev[1], I V Lopatin[1], P M Schanin[1], S S Kovalsky[1], N N Koval[1], V V Yakovlev[1]

[1] Institute of High Current Electronics SB RAS, Tomsk, 634055, Russia

E-mail: volodyadenisov@yandex.ru

Abstract. The paper reports on a study of a pulsed non-self-sustained glow discharge with a hollow cathode area of 2.3 m^2 and pulse current of up to 370 A. The ignition and operation of the discharge at low pressures (0.4–1 Pa) was sustained through electron injection from the plasma of an arc discharge based on a cold hollow cathode. The current-voltage characteristics of the pulsed glow discharge were measured. Specimens of 12 Cr18Ni10Ti, Cr12, and 40Cr steels were nitrided in the glow discharge plasma with double thermal shields at a temperature of 520 °C and pressure of 0.9 Pa for 4 h. The nitrided layer thickness for 40Cr, Cr12, and 12Cr18Ni10Ti steels was 240, 90, and 50 µm, respectively.

1. Introduction

Research in low-pressure discharges for surface modification of materials is assuming more and more significance. New avenues for surface treatment are opened up by pulsed discharges [1] which allow operation at increased powers (one-two orders of magnitude higher than those for steady discharges) and control of discharge pulse durations [2]. Among the methods of ion nitriding as a tool of chemical heat treatment is the use of a pulsed glow discharge which allows control of its operating voltage and treatment of different metal materials [3]. In this type of discharge, the average ion current density and temperature of treated parts are controlled by varying the discharge pulse duration at a constant pulse repetition frequency [4]. Despite much progress made in ion nitriding technologies, there still remains a series of problems such as surface roughness of treated parts due to etching, their nonuniform heating, and initiation of microarcs on their surface. These problems are urgent for both pulsed and steady glow discharges. It has been shown that a steady low-pressure non-self-sustained glow discharge based on a hollow cathode have a high potential for technological applications, including ion nitriding of Fe- and Ti-based alloys [5–10]. A special feature of this discharge is the possibility of its operation at decreased pressures 0.4…1 Pa which enables efficient surface cleaning during nitriding. Moreover, the discharge is initiated at low voltages 100…350 V, making it possible to greatly decrease surface etching and to attain minimum surface roughness. In pulsed mode, this type of discharge provides a possibility to independently control the main parameters of the process: the energy of ions bombarding the surface of treated parts, their temperature, and pressure in the chamber.

By now, no theoretical and experimental research has been performed on the formation of a low-pressure (0.4–1 Pa) non-self-sustained glow discharge with a large-area hollow cathode (>2 m^2) in pulsed mode at a pulse current more than 100 A and pulse power higher than 30 kW. Also no experimental studies have been reported on nitriding of Fe-based alloys in this type of discharge under the conditions specified above.

The objective of the present study was to investigate the characteristics of a pulsed non-self-sustained hollow-cathode glow discharge and the possibilities of its use for nitriding of 12Cr18Ni10Ti, Cr12, and 40Cr steels.

2. Experimental setup

The experimental setup used in the study was built on the basis of NNV-6.6-I1 commercial ion plasma unit (Fig. 1). The vacuum chamber was pumped with a TMN-500 turbomolecular pump to a pressure of 1×10^{-2} Pa. The operating pressure was controlled in the range 0.4…1 Pa by varying the gas (nitrogen) flow rate.

The glow discharge was ignited between a hollow cathode (inner surface of the chamber) and a plane anode. The cathode area was 2.3×10^{4} cm^2, the anode area was 1350 cm^2, and the ratio of these areas was thus 17/1. The main discharge was powered by a stabilized voltage source with an output voltage of 30…320 V, output pulse current of 0…500 A, and maximum average output power of 25 kW; the output voltage instability was ±5%. The power supply was equipped with a protection system which precluded initiation of microarcs at the cathode surface. The discharge current was measured using a Honeywell CSNJ481 Hall sensor with a current output. The voltage was measured using a probe with a divider of 1:100. The signals were transmitted to a TDS2014C oscilloscope and averaged over 16 points. To provide a flat peak during the entire voltage pulse, a capacitor bank with a capacitance of 7.2 mF was used.

For easier ignition and stable operation of the glow discharge at low pressures, an electron source based on an arc discharge with a cold hollow cathode was used [6]. The working gas to the discharge chamber was supplied through this electron source arranged at the chamber gate. Electrons were emitted to the hollow cathode (chamber) through an electrode shaped as a truncated cone and covered on all sides with a fine grid of dimensions 0.4×0.4 mm and geometric transparency 45%. This electrode served as the anode of the auxiliary discharge and was simultaneously at the potential of the main discharge cathode. The plasma generator was powered by a welding rectifier which ensured a constant arc current of up 200 A at an arc voltage of 30…40 V; the current of the magnetic coil in all experiments was 0.18 A. The current and voltage of the auxiliary discharge were measured using pointer instruments with recording of their effective values.

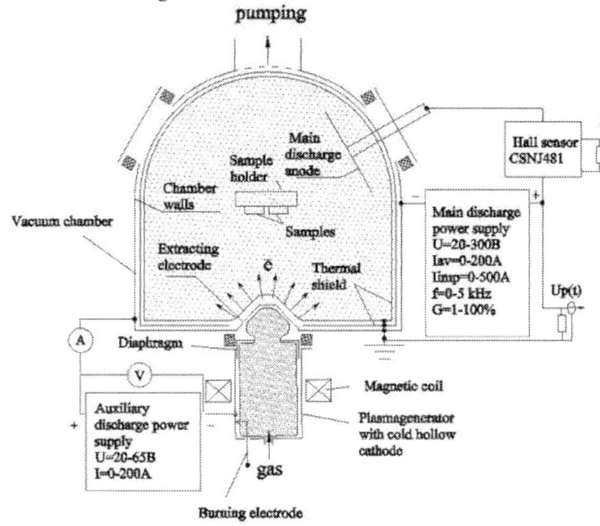

Figure 1. Schematic of the experimental setup (top view).

The current-voltage characteristics of the pulsed non-self-sustained glow discharge were taken from oscillograms of the current and voltage at 320 μs of the discharge pulse during the operation with a frequency of 1 kHz and duty factor of 34 %. When these characteristics were recorded, the discharge voltage varied in the range 40…320 V. The pressure dependence of the main discharge current was determined at an auxiliary discharge current of 40…120 A. Before the measurements, the chamber walls were subjected to ion cleaning for which a pulsed non-self-sustained glow discharge was initiated in the working chamber through nitrogen supply and its walls were bombarded by ions for 30 min at a voltage of 300 V and average current of 25 A.

To decrease thermal loss at the water-cooled walls of the vacuum chamber, double shields made of polished stainless steel were used in nitriding. In this experiment, the thermal shield walls facing the discharge gap served as the hollow cathode of the glow discharge.

For nitriding, the 12Cr18Ni10Ti, Cr12, and 40Cr steel specimens made to dimensions 15×15×5 mm were fixed in a holder placed at the potential of the hollow cathode. The working gas mixture contained 90 % of nitrogen and 10 % of helium. The addition of helium to nitrogen was made to increase the nitriding rate [9]. The specimens were nitrided for 4 h at a gas pressure of 0.9 Pa, discharge operating voltage of 280 V, and average ion current density of 1.7 mA/cm^2. The specimen temperature was 520°C and was controlled at a constant ion energy by varying the pulse duty factor which was 40 % at a pulse repetition frequency of 1 kHz (pulse duration 400 μs, pause 600 μs). The ion current density to the specimens reached 4.4 mA/cm^2 in a pulse. Heating and cleaning of the specimens was through bombardment by ions accelerated in the cathode layer of the non-self-sustained glow discharge. Before arrangement into the vacuum chamber for nitriding, the specimens were grinded and polished, washed with benzene and acetone in an ultrasonic bath to remove organic impurities, and wiped with ethanol-wetted calico.

The nitriding efficiency in the pulsed non-self-sustained glow discharge plasma was estimated by measuring the Vickers microhardness on lateral sections in depth from the specimen surface with a step of 10 μm at an indenter load of 0.2 N. The measurements were performed on a PMT-3 device. The in-depth microhardness distribution was measured for each specimen until its initial microhardness was reached, whereupon the result was averaged over three measurements at each point.

3. Results and discussion

Waveforms of the main discharge current at two values of the auxiliary discharge current are presented in Fig. 2. The current rise and fall times, regardless of the pulse duration, are rather long due to a gradual increase in plasma density when voltage is applied to the discharge gap and to deionization of the discharge gap when it is removed. The discharge current in the experiments reached 370 A in a pulse. The discharge pulse power was about 90 kW.

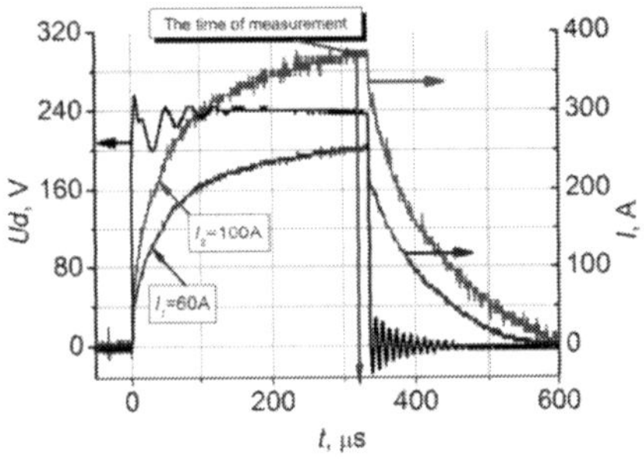

Figure 2. Waveforms of the current and voltage of the non-self-sustained glow discharge at two auxiliary discharge currents: I_1=60 A and I_2=100 A; discharge operating voltage U_d=240 V, pulse repetition frequency f=1 kHz, duty factor γ=34 %, pulse duration t_p=340 μs, operating pressure p=1 Pa.

Figure 3a shows a current-voltage characteristic of the discharge at an operating pressure of 0.65 Pa.

The discharge current is the cathode current and equals to the product of the amount of injected electron current from the auxiliary discharge and γ-electron current formed as a result of the secondary ion-electron emission from hollow cathode walls to the average number of ionizations produced by these electrons after acceleration in the discharge cathode layer. In turn, γ-electron current is equal to the product of the ion current to the secondary-ion-electron emission factor γ. Increasing the discharge voltage increases the average number of ionization events realized by each of the electrons accelerated in the cathode layer. Because the voltage in the near-cathode layer at the initial portion of the current-voltage characteristic is low, the secondary-ion-electron emission factor γ is small and γ –electrons have low influence on the characteristic. Beginning with a discharge voltage of about 200 V, the rate of rise of the current increases which can be explained by a considerable increase in the number of γ –electrons [11].

The dependence of the main discharge current on the auxiliary discharge current (Fig. 3b) suggests that the efficiency of the hollow cathode increases with increasing discharge operating voltage. The characteristics are linear because the cathode current is proportional to the amount of auxiliary discharge electron current and current of γ –electrons. When U_d = const the secondary ion-electron emission factor remains constant and the growth of the cathode current is carried out by increasing of the electron current from the auxiliary discharge and the ion current to the cathode. Increasing the discharge operating voltage increases the rate of rise of the discharge current due to an increase in the fraction of electrons formed in secondary ion-electron emission. Moreover as noted above increasing the discharge voltage increases the average number of ionization events realized by each of the electrons accelerated in the cathode layer.

The dependence of the main discharge current on the operating pressure at a constant discharge voltage (Fig. 3c) demonstrates that the discharge current has a slight decrease in the investigated range of operating pressures.

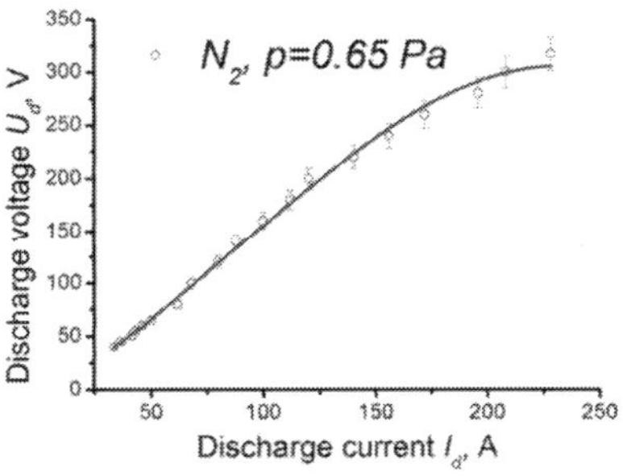

Figure 3a. Current-voltage characteristic of the non-self-sustained glow discharge at an auxiliary discharge current of I_a=40 A.

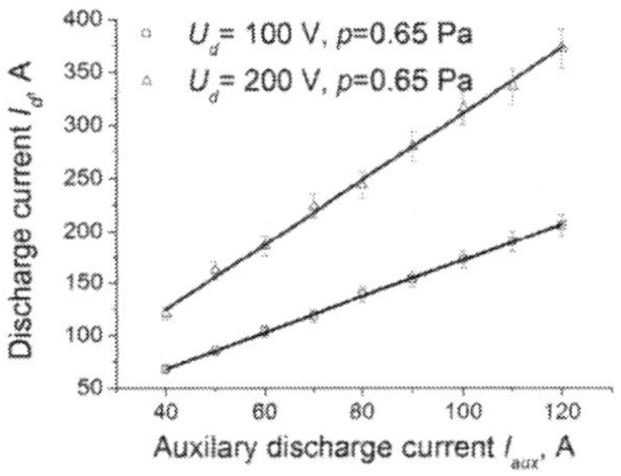

Figure 3b. The main discharge current vs the auxiliary discharge current at a constant discharge voltage.

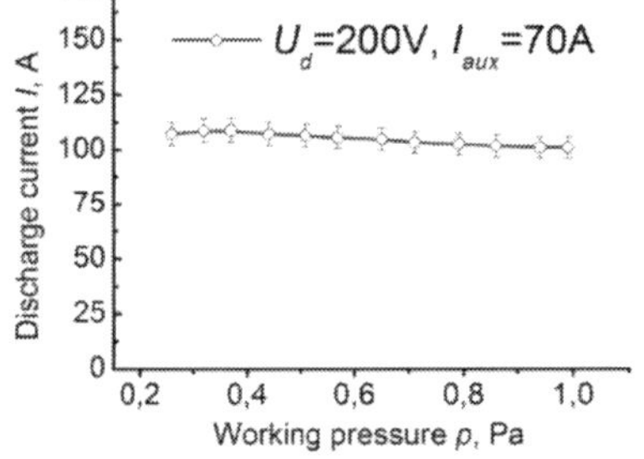

Figure 3c. The main discharge current vs operating pressure at a constant discharge voltage.

This can be explained by the fact that for a given area ratio of anode and cathode the upper limit of the pressure range within which there will be a constant value of the discharge voltage is 0.33 Pa. Above this value, according to [12] there will be an increase of the discharge voltage at a constant current discharge. The dependences in Figure 3c shows that at the constant discharge voltage in the pressure range (0.3 - 1) Pa discharge current decreases slightly.

Figure 4 shows in-depth microhardness distributions for 12Cr18Ni10Ti, Cr12, and 40Cr steels nitrided in the glow discharge plasma. The result of nitriding in the nitrogen-helium mixture for 4 h is a substantial increase in surface layer microhardness.

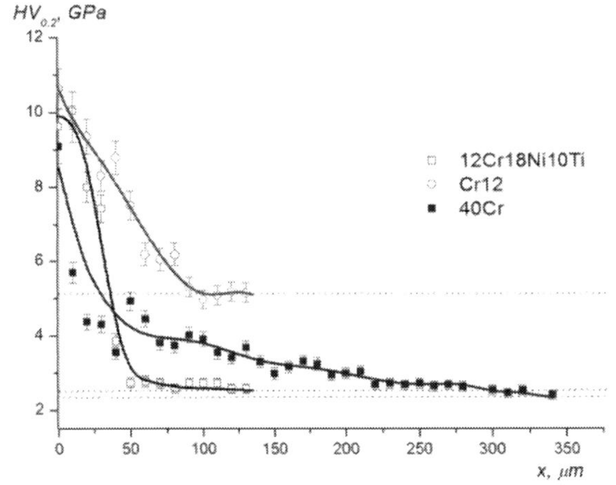

Figure 4. In-depth microhardness distributions for 12Cr18Ni10To, Cr12, and 40Cr steels nitrided in the glow discharge plasma for 4 h at a temperature of 520 °C and nitrogen-helium pressure of 0.9 Pa. Dotted lines – initial microhardness.

In-depth microhardness distributions show that the nitrided layer depth for 40Cr steel is 240 μm; for Cr12 steel, it is 90 μm; and for 12Cr18Ni10Ti, it is 50 μm. This result evidences that the surface of the steels is nitrided with a sufficiently high efficiency. In comparing the nitriding processes in pulsed and steady modes, it should be noted that the nitriding rate for 40Cr steel in a steady non-self-sustained hollow-cathode glow discharge was about the same [9]. Despite this, an important advantage of the pulsed non-self-sustained glow discharge is the possibility to control the temperature of treated parts by varying the pulse duty factor at a specified discharge voltage amplitude over a rather wide range of operating pressures.

4. Conclusions

Thus, we realized a pulsed non-self-sustained glow discharge with a large hollow cathode area (>2 m^2), pulse current of up to 370 A, and pulse power of up to 90 kW. The measured current-voltage characteristics of the discharge demonstrate stability of the discharge current in the operating pressure range 0.4–1 Pa, showing promise for chemical heat treatment of materials and parts with a large surface area.

Nitriding of 12Cr18Ni10Ti, Cr12, and 40Cr steels in the non-self-sustained glow discharge plasma at a temperature of 520 °C for 4 h was realized. The nitrided layer depth for 40Cr, Cr12, and 12Cr18Ni10Ti steels was 240, 90, and 50 μm, respectively.

The research results suggest that pulsed non-self-sustained glow discharges with a large-area hollow cathode have a high potential for increasing the efficiency of nitriding of steels and alloys.

This work was carried out at the expense of Russia research foundation (Grant no. 14-29-00091).

References

[1] Anders A 2004 Fundamentals of pulsed plasmas for materials processing *Surf. Coat. Tech.* **183** 301–311

[2] Conrads H, Schmidt M 2000 Plasma generation and plasma sources *Plasma Sources Sci. Technol.* **9** 441–454

[3] Berlin E V, Koval N N, Seydman L A 2012 *Plasma thermochemical surface treatment of steel parts* (Moscow: Tekhnosfera) p 464

[4] Biro A S 2013 Trends of nitriding processes *Production Processes and Systems* **6** 1 57–66

[5] Lopatin I V, Schanin P M, Akhmadeev Y H, Kovalsky S S, Koval N N 2012 Self-sustained low pressure glow discharge with a hollow cathode at currents of tens of amperes *Plasma Phys. Rep.* **38** 7 585–589

[6] Lopatin I V, Akhmadeev Y H, Koval N N , Schanin P M 2011 Plasma generator on the base of non-self-sustained low pressure glow discharge with a hollow cathode of big volume *Prib. Tekh. Eksp.* **1** 151–156

[7] Schanin P M, Koval N N, Akhmadeev Y H, Grigoryev S V Arc disharge with cold hollow cathode in crossed electric and magnetic fields 2004 *J. Tech. Phys.* **74** 5 24

[8] Lopatin I V, Akhmadeev Yu Kh, Koval N N, Schanin P M, Yakovlev V V 2010 Generation of plasma in non-self-sustained glow discharge with hollow cathode *Modification of Materials with Particle Beams and Plasma Flows: Proc. of 10th Int. Conf. (Tomsk: Russia, 19-24 september 2010) (Publishing house of the IOA SB RAS)* pp 35–38

[9] Akhmadeev Yu H, Ivanov Yu Ph, Koval N N et al. 2012 Influence of plasma gas composition on the nitriding in non-self-sustained glow discharge with hollow cathode of large size *Poverkhnost. Rentgen. Synchrot. Neitron. Issled.* **2** 62–67

[10] Gavrilov N V, Kamenetskikh A S, Khatmullin I G 2014 Generation of polyenergetic beam in the electrone source with plasma cathode *Izv. Vyssh. Uchebn. Zaved. Fiz.* **3/2** 74–77

[11] Ulyanov K N 1999 Superdense glow discharge: theoretical model of the cathode region *Teplofizika Vysokikh Temperatur* **37** 3 363 – 373

[12] Metel A S, Grigoryev S N, Melnik Yu A, Panin V V 2009 Filling the vacuum chamber of a technological system with homogeneous plasma using a stationary glow discharge *Plasma Phys. Rep.* **35** 12 1140-1149

Investigation of Massive Catalyst based on Molybdenum Disulphide by Simultaneous Thermal Analysis and Mass Spectrometry Methods

A S Akimov[1], **M A Morosov**[1], **T A Fedushchak**[1], **M A Uyimin**[2], **A V Vosmerikov**[1], **S P Zhuravkov**[3], **E N Lysenko**[3]

[1] Institute of Petroleum Chemistry, 4 Akademichesky Ave, 634021, Tomsk, Russia
[2] Institute of Metal Physics, 18 S. Kovalevskaya Str., 620990, Yekaterinburg, Russia
[3] Tomsk Polytechnic University, 30 Lenin Ave, 634050, Tomsk, Russia

E-mail: zerobox70@mail.ru

Abstract. The paper presents the results of experimental studies of massive sulfide catalysts by simultaneous thermal analysis and mass spectrometry. It is found that the STA/MS methods are quite informative for testing the catalyst systems based on MoS2 and are useful in identification of the reference features that could be used to predict their activity. It is also shown that the defect structure of molybdenum disulfide formed during mechanical activation is reflected on the DSC curves.

1. Introduction

The development of methods for the synthesis of domestic massive sulphide catalysts for hydrotreating of petroleum fractions is a strategic priority in the field of oil refining to produce high quality distillate fuels. Conventional western technologies allow producing bulk catalysts with a high activity and selectivity for all types of hydroprocessing procedures (hydrogenation, hydrodesulfurization, hydrodenitrogenation, hydrodemetallization, hydrodearomatization, hydroisomerization, hydrodewaxing, hydrocracking) [1]. These multi-stage and energy-intensive methods require the use of large amounts of aqueous solutions of salts-precursors as the active components, which is inevitably accompanied by the formation of substantial amounts of wastewater with toxic anions [2-4]. The use of mechanical activation method to prepare the hydrotreating catalyst systems has a number of advantages, as it allows reducing the number of stages and preparation time, and eliminating the use of aqueous solutions of reagents and organic solvents.

We propose a new promising approach to manufacturing 4-, 3-, 2-, and 1- component massive sulfide catalysts, which is implemented in a single-stage mixing of initial solid-phase reagents under mechanical activation (MA). In this case, MoS_2 is a dimensional inorganic precursor of the active component only and it has to be subjected to dispersion [5]. We have previously shown [6, 7] that the catalysts of this series are capable of providing an ultra-low level of sulfur (about 10 ppm) in a model reaction of hydrogenolysis of dibenzothiophene (DBT) and in the course of hydrorefining the straight-run diesel fraction (S_{initial} = 2.08%).

For 1-component catalysts based on MoS_2, a unique experimental fact was discovered – their activity in the model reaction of dibenzothiophene hydrodesulfurization can be significantly increased in case of using the trace amounts of polar liquids (S_{res} = 1-3 ppm) during MoS_2 MA-dispersion. There is no answer to the question of the reason for such a high catalytic activity and diagnostics of such systems so far.

It is well known that thermal analysis methods are based on detection of thermal effects which reflect the physical and chemical transformations occurring in the analyzed sample. For example, a simultaneous thermal analysis (STA) allows us to monitor the change in mass, oxidation, decomposition of the sample substance, temperature and enthalpy of the phase transition, the thermal effects of chemical reactions, and to study the kinetics of the reactions [8-10]. A combination of the simultaneous thermal analysis and mass spectrometry (MS-STA) allows us to obtain a lot of additional data on the composition of volatile products.

The aim of this study was to investigate massive sulfide catalysts based on MoS_2 via a simultaneous thermal analysis and mass spectrometry.

2. Experimental

A number of sulfide catalysts were prepared by the solid phase method. The commercial coarse-grained powder of molybdenum disulfide (the content of molybdenum disulfide is 99.72%) was used as a precursor of the active catalyst component. The powder was subjected to mechanical activation in a vertical vibratory mill in the presence of trace amounts of polar liquids (in this case, methanol) and in their absence.

The activity of the catalyst system was evaluated by the residual content of sulfur in the course of hydrogenolysis of a model dibenzothiophene compound which represents the most stable sulfur structures presented in the diesel fraction. The model reaction experiments were performed in an 'AUTOCLAVE ENGINEERS' autoclave with the reactor volume of 100 ml, at a pressure of 3.4 MPa, and a temperature of 340 °C ($S_{initial}$ = 500 ppm).

The samples were analyzed by the STA-MS method using a STA-449C ("Netzsch") apparatus combined with a quadrupole QMS 403C mass spectrometer. Heating was carried out in air (flow rate of 20 ml/min) to 650 °C at a heating rate of 10 °C/min. The results of thermogravimetry (TG) and differential scanning calorimetry (DSC) were processed using the 'Proteus Analysis' program.

3. Discussion of Results

It was found out experimentally that in the presence of mono-component catalysts prepared by the solid phase method (Table 1), a complete hydrodesulfurization of dibenzothiophene was observed which had not been reported in the literature. For example, the coarse-grained molybdenum disulfide (catalyst system AK-0) used as a commercial reagent (DMI-7) exhibits certain catalytic activity resulting in a 20% conversion of DBT and a 100 ppm decrease in the residual sulfur content in the reaction mixture (from the initial 500 ppm). The sulfur content in hydrodesulphurizate was reduced by a factor of 5 (down to 100 µl) in the presence of mechanically activated MoS_2. In the presence of trace amounts (100 µl) of a polar liquid (methanol) the MoS_2 monocatalyst exhibits its maximum activity in the model reaction of DBT hydrogenolysis (residual sulfur content 1-2 ppm; AK-60).

Table 1. Composition of catalysts, conditions of their MA and their activity in a model reaction.

Number of sample	Catalyst composition	Time of MA (h)	Residual sulfur (ppm)
AK-0	Commercial MoS_2	-	400
AK-27	Commercial MoS_2	8	105
AK-60	MoS_2+100 µl CH_3OH	8	1-2
AK-71	MoS_2+ 200 µl CH_3OH	8	50
AK-74	MoS_2 + 200 µl CH_3OH	12	6

A slightly higher volume of CH_3OH added to MoS_2 (up to 200 μl) requires longer (by about 4 hours) MA time (Table 1), which is necessary to provide ultra-low sulfur content in the reaction products.

On the other hand, we have to answer the questions on the reason for the threshold effects in the activity of MoS_2 in the presence of trace amounts of CH_3OH and the origin of active centers in the above catalysts. The method of simultaneous thermal analysis and mass spectrometry allowed us to obtain the following additional information on the AC-27, AC-60 and AC-74 catalysts *before* and *after* their participation in a model reaction.

Figures 1 and 3 show typical TG, DTG, and DSC curves using the AC-74 catalyst system as an example. First of all, it should be noted that the mass loss occurs in the temperature region up to 170-280 °C. The calculated content of methanol in the sample is 0.026% which could be hardly measured without special techniques. According to MS spectra, the ions with the following masses: m/e = 18(H_2O^+), 12(C^+), 2(H_2^+), 34(H_2S^+), etc. have been identified in gas atmosphere in this temperature region, and an interpretation of their presence in the spectra is difficult. The mass loss of the AC-74 catalyst sample *before* the reaction at the temperature up to 300 °C is 11.19% while *after* the reaction it is 8.66%. The mass changes in MoS_2 within the temperature range above 300 °C are due to its oxidation.

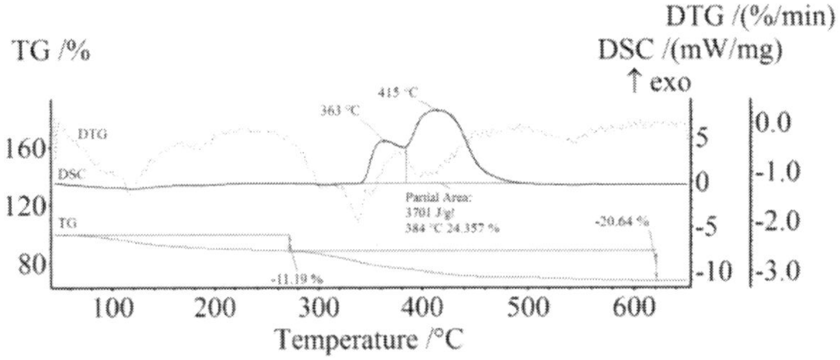

Figure 1. Results of thermal analysis for AC-74 catalyst *before* to its use in the model reaction; $m_{initial}$ = 5.61mg.

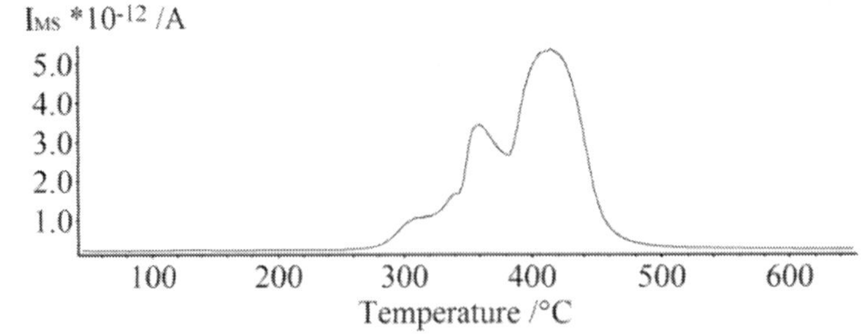

Figure 2. MS spectra with *m/e* = 64 (SO_2^+) for the gas atmosphere during the heating of AC-74 catalyst *before* to the reaction.

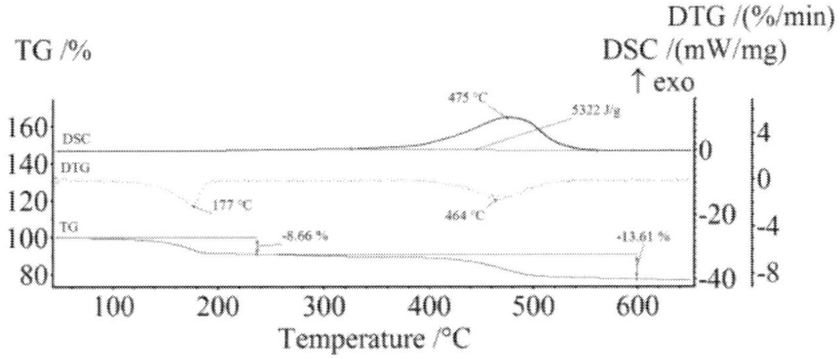

Figure 3. Results of thermal analysis for AC-74 catalyst *after* its use in the model reaction; $m_{initial} = 4.54$ mg.

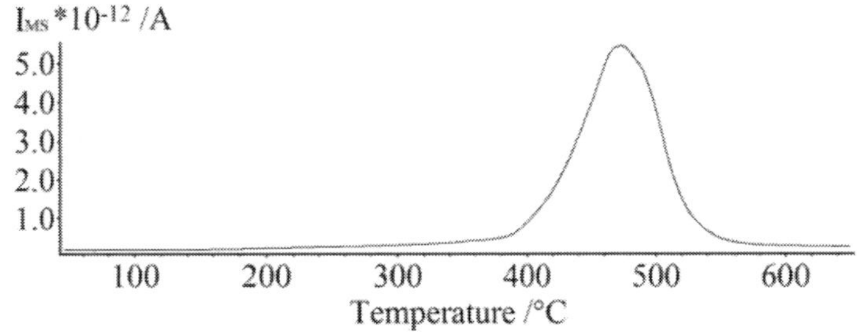

Figure 4. MS spectra with *m/e* = 64 (SO_2^+) for the gas atmosphere during the heating of AC-74 catalyst *after* to the reaction.

According to thermal analysis for all investigated samples, the identical changes of the shape in DSC curves are observed for the catalysts *before* and *after* their use in the model reaction. For example (see Fig. 1, 3), the DSC curve for AC-74 catalyst *before* the reaction is a double-humped curve of heat emission with the temperature maxima of its major peaks at 415 °C and 363 °C (Table 2), while for the AC-60 and AC-27 catalysts, the DSC curves are polymodal. The first low temperature peak in the thermograms for these catalysts disappears *after* the reaction.

Table 2. Temperature peaks on the DSC curves for monocomponent sulfide catalysts.

Catalyst	The temperature of the first peak on the DSC curve *before* the reaction (°C)
AC–27	354
AC–60	377
AC–74	363

One more specific feature of the formal external analysis of the curves should be noted: the peaks of heat emission in the temperature range 300-550 °C (Fig. 1, 3) in the DSC curves repeat the shape of the mass spectral peaks in the MS curves (Fig. 2, 4). Moreover, these changes are accompanied by mass loss (DTG and TG curves in Fig. 1 and 3). The ratio m/e = 64 is identified as the molecular SO_2^+ ion formed by oxidation of MoS_2. According to the literature data, oxidation of a coarse-grained MoS_2 begins at about 400 °C. For nanodisperse states, the initial oxidation temperature of the substrate can be much lower.

It is well known from the literature [11, 12] that the mechanical activation of the reaction mixture is accompanied by an increase in the number of point contacts, generation of defects, their constant regeneration, multiplication, and migration throughout the entire volume of the reaction mixture. Hence, the broadening of the MS peaks corresponds to a multiple set of different discrete states produced during mechanical activation of MoS_2, which are due to varying degree of bonding of sulfur atoms with Mo and imperfection of its nanocrystallites. The shift of the MS peak temperature from 415 to 475 °C in Fig. 4 might testify in favour of the initial activity of low-temperature catalytically active centers of MoS_2 and their functioning during the model reaction. In addition to that, there is a reasonable question about the effective lifetime of mono-component catalysts based on molybdenum disulphide in the reactions of DBT hydrodesulfurization in the absence of such centers.

4. Conclusion

Thus, for the first time we have shown a possibility of identifying structure defects formed during mechanical activation using the DSC curves. The experimental results obtained show that the STA/MS method is quite informative for testing the catalyst systems based on MoS_2 and is useful in identification of the reference features that could be used to predict their activity.

References:

[1] Eijsbouts S, Mayo S W, Fujita K 2007 Unsupported transition metal sulfide catalysts: From fundamentals to industrial application *Applied Catalysis A: General* **322** 58.

[2] Hinnemann B, Norskov J K, Topsoe H A 2005 A density functional study of the chemical differences between type I and type II MoS_2-based structures in hydrotreating catalysts *J. Phys. Chemistry* B **109** 2245.

[3] Tomina N N, Nikul'shin P A, Tsvetkov V S, Pimerzin A A 2009 Thiophene hydrodesulfurization and diesel fuel hydrorefining activities of $XMo_6(S)/\gamma\text{-}Al_2O_3$ and Ni-$XMo_6(S)/\gamma\text{-}Al_2O_3$ (X = Al, Ga, In, Fe, Co, and Ni) catalysts *Kinetics and Catalysis* **50** 220.

[4] Nikul'shin P A, Mozhaev A V, Ishutenko D I, Minaev P P, Lyashenko A I, Pimerzin A A 2012 Influence of the composition and morphology of nanosized transition metal sulfides prepared using the anderson-type heteropoly compounds $[X(OH)_6Mo_6O_{18}]$n- (X = Co, Ni, Mn, Zn) and $[Co_2Mo_{10}O_{38}H_4]^{6-}$ on their catalytic properties *Kinetics and Catalysis* **53** 620.

[5] Kouzu M, Uchida K, Kuriki Y, Ikazaki F 2004 Micro-crystalline molybdenum sulfide prepared by mechanical milling as an unsupported model catalyst for the hydrodesulfirization of diesel fuel *Applied catalysis A: General* **276** 241.

[6] Fedushchak T A, Uyimin M A, Akimov A S, Morozov A S, Vosmerikov A V, Zhuravkov S P 2013 Direct Synthesis of Sulfide Multifunctional Catalysts for Hydrogenolysis Molecular Aspects of Catalysis by Sulfides *Proc. VI International Symposium* (France, Lion) 84.

[7] Feduschak T A, Uimin M A, Ermakov A E, Akimov A S, Schegoleva N N, Petrenko T V, Zhuravkov S P, Vosmerikov A V 2013 Mechanical activation for obtaining sulfide catalysts of hydrogenolysis *Chemistry for sustainable development* **6** 683.

[8] Surzhikov A P, Frangulyan T S, Ghyngazov S A, Lysenko E N 2010 Investigation of oxidation processes in non-stoichiometric lithium-titanium ferrites using TG analysis *J. Therm. Anal. Calorim.* **102** 883.

[9] Surzhikov A P, Frangulyan T S, Ghyngazov S A, Lysenko E N 2012 Investigation of structural states and oxidation processes in $Li_{0.5}Fe_{2.5}O_{4-\delta}$ using TG analysis *J. Therm. Anal. Calorim.* **108** 1207.

[10] Lysenko E N, Surzhikov A P, Zhuravkov S P, Vlasov V A, Pustovalov A V, Yavorovsky N A 2014 The oxidation kinetics study of ultrafine iron powders by thermogravimetric analysis *J. Therm. Anal. Calorim.* **115** 1447.
[11] Knyazheva O A, Baklanova O N, Lavrenov A V, Drozdov V A, Leont'eva N N, Trenikhin M V, Arbuzov A B, Likholobov V A 2011 Mechanochemical synthesis of nanocrystalline nickel-molybdenum compounds, their morphology and application in catalysis. I. Effect of the Ni:Mo atomic ratio on the structure and properties of nickel-molybdenum compounds prepared under mechanochemical synthesis conditions *Kinetics and Catalysis* **52** 886.
[12] Knyazheva O A, Baklanova O N, Lavrenov A V, Buluchevskii E A, Gulyaeva T I, Leont'eva N N, Drozdov V A, Likholobov V A, Vasilevich A V 2012 Mechanochemical synthesis of β-$NiMoO_4$ as a precursor of bulk highly dispersed catalyst for the hydrocracking of oil fractions *Catalysis in Industry* **4** 179.

Thermal shock removal of defective glass-enamel coating from cast-iron products

A D Aleutdinov[1], S A Ghyngazov[1], T S Mylnikova[1], P A Luchnikov[2]

Tomsk Polytechnic University, 634050, 30 Lenin ave, Tomsk, Russia
Moscow State Institute of Radio Engineering, Electronics and Automation, 119454, Vernadsky ave, Moscow, Russia

E-mail: a_aleut@mail.ru

Abstract. A setup for light beam exposure has been developed. The setup was used to consider the technology of thermal shock destruction of the coating by pulsed-periodic exposure to powerful focused light from the xenon arc lamp DKsShRB-10000. It is shown that this type of exposure can effectively remove the glass-enamel coating from iron products. The optimal mode of setup operation to efficiently remove the defective glass-enamel coating is found: the diameter of the focused light beams is 2.5–3.5 cm; the lamp arc pulse current is 350–450 A; pulse duration is (0.5–1) s and pulse repetition frequency is (0.15–0.5) s^{-1}.

1. Introduction

Glass-enamel coatings (GEC) for metal products are the most effective protection against corrosion and, in addition, they provide the metal surface with a number of useful properties inherent in oxide glasses such as hardness, smoothness, acid- and alkali-resistance and aesthetic appearance, and so on [1]. While being manufactured and operated, GEC on metal enameled products may contain different defects [1]. One of the stages to recover GEC is to remove its defective part from the surface of the enameled product. Steels can be treated locally with subsequent recovering or re-enameling the enamel coating on the top of the product defective part [2]. For cast iron products, the problem is fundamentally different. Due to specific properties of iron (high gas evolution rate upon heating) [2, 3], this technology can not be implemented. For cast iron products, recovery of the enamel coating should be performed after complete removal of the old. Conventional cleaning methods based on mechanical and chemical cleaning are labor-intensive and time consuming, and in addition, they are not environmentally friendly [3]. Thermal shock destruction of GAC with a powerful thermal effect can be suggested as an alternative method free of these drawbacks [4, 5].

The paper aims to study the features of pulse-periodic action of a powerful focused light of the xenon arc lamp on GEC on iron products.

2. The experimental procedure

Xenon arc lamp DKsShRB-10000 was used as a light source. The configuration of the setup to implement a pulse-periodic action mode is shown in Figure 1. The apparatus comprises a switched power supply (1), radiant energy generator (3), ignition unit (2), column of "T-31021" type gun-welder (4), bogie (6) on which the workpiece to be treated is attached (5). The pulsed mode of the xenon lamp operation is achieved via switched power supply (1), which includes a three-phase transformer, rectifier, the device to control the lamp current and reference generator to regulate the duration and repetition rate of current pulses.

The features of arcing in the xenon lamp do not allow implementation of the pulsed mode with completely interrupted glow between pulses, since DC is required to provide stable arcing. In this setup, the pulsed mode of lamp operation can be achieved via pulse-periodic energy supply of the lamp with direct background dark current therein. The dark current was regulated in the range of 15–200 A.

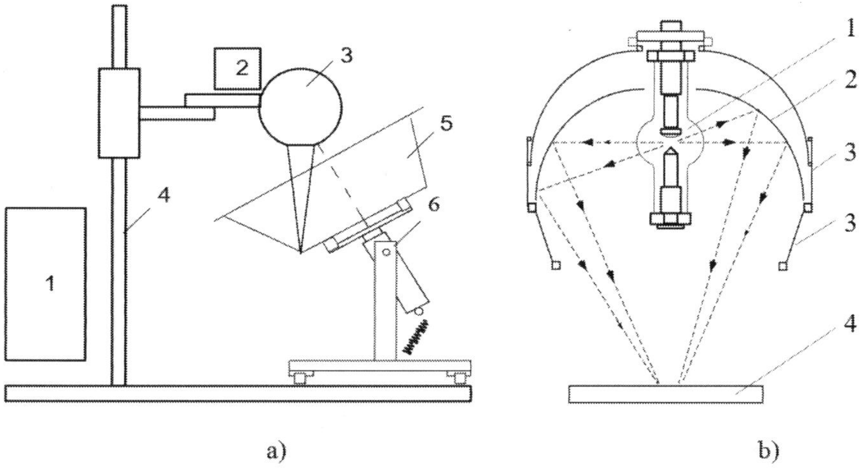

Figure 1. The configuration of the setup (a) and (b) are the radiant energy generator (3): 1 is a xenon lamp; 2 is a reflector; 3 is a protective casing; 4 is the target.

The power density applied to the target in the light spot is directly proportional to the current through the lamp and is inversely proportional to the squared diameter of the light spot (the area of the latter). The valid estimated values of the power density in the treatment region for the setup are shown in Fig. 2. It should be emphasized that the current exceeding 250 A is permissible only under pulsed mode. The integral power dissipation for the lamp should not exceed the maximum rated capacity of the arc source.

The heat flow from the radiant energy generator with the xenon lamp DKsShRB-10000 has lower and upper bounds. The lower bound is due to the minimum current for steady lamp arcing (dark current I_0 is 15–20 A), and the upper bound is determined by the capacity of the power supply and current capacity for the lamp.

For the lamp DKsShRB-10000, the maximum permissible mode to maintain its performance is as follows: pulse current I of 750 A, pulse duration τ of 0.5 s, pulse repetition rate υ of 0.5^{-1}. The luminous flux is directly proportional to the lamp current. In terms of the rated conversion efficiency, the light flux per pulse is 2.25 kW, and if the exposed area is 5 cm^2, the power density of light exposure (per pulse) will be 450 W/cm^2.

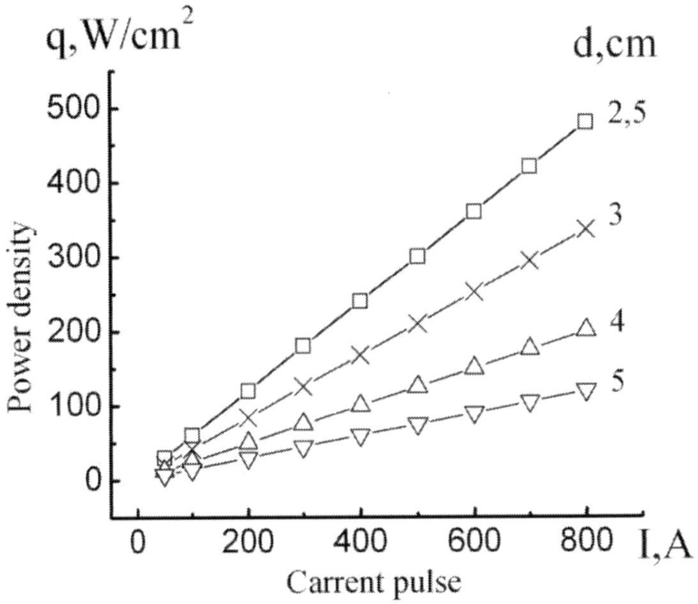

Figure 2. The estimated pulse power density of the light output in the exposed area versus the pulse current amplitude and the spot diameter. The diameter varies in the range of (2.5–5) cm.

A measure of exposure effectiveness was chosen the rate of GEC removal (destruction), i.e. the area of the surface region of the cast iron product with GEC removed within one minute (the area of the light spot being fixed). In the experiment, the dependence of the rate of GEC removal on the thickness of the coating, pulse current, duration and pulse repetition rate was investigated. The dark current was maintained at a minimum level. The diameter of the light spot was 3 cm.

3. Results and discussion

The experimental results obtained for the light beam exposure by means of the designed setup are shown in Table 1.

As can be seen from the Table, for the area of the focused light emission with a diameter of 3 cm, the maximum removal speed is achieved within the following modes: the pulse current of the arc lamp I of (350–450) A; pulse duration τ of (0.5–1) s, and pulse repetition rate υ of (0.15–0.5) s^{-1}.

Under the focused incoherent light of the arc source, the coating on the cast iron product absorbs radiation and heats up. Due to the temperature difference between the GEC and cast iron, significant thermal stresses arise and cause separation of the coating. Under exposure with the designed parameters, the destruction occurs at the coating-iron interface which is the optimal solution for the problem of GEC removal from the cast iron product surface.

The destruction of GEC on cast iron under intense heating by light flux can occur under continuous exposure mode, however, it is typically followed by microcracks, exfoliation and fusion of small enamel fragments. In the concave parts of the surface, fusion may not be followed by further destruction of the coating.

Table 1 Parameters for glass-enamel coating removal

Coating thickness mm	Pulse current I, A	Pulse duration τ, c	Pulse repetition rate υ, c^{-1}	Removal speed cm^2/min	Effect
1–1.5	400	1.5	0.15	160	GEC partial fusion
	450	1.5	0.15	200	
	450	1.0	0.25	300	Shearing
	500	1.0	0.25	320	
	500	0.5	0.5	320	
	550	0.5	0.5	140	GEC partial fusion
	600	0.5	0.5	80	
2.5–3.5	400	1.5	0.15	60	Shearing
	400	1.5	0.25	80	Fusion
	450	1.5	0.15	120	Shearing
	450	1.0	0.25	140	
	500	1.0	0.15	145	
	500	1.0	0.25	140	
	500	0.5	0.5	140	
	550	0.5	0.5	80	Partial fusion
	600	0.5	0.5	30	GEC fusion
1.5–2.5	400	1.5	0.15	80	Partial fusion
	400	1.5	0.25	120	GEC fusion
	450	1.5	0.15	200	Partial fusion
	450	1.0	0.25	240	Shearing
	500	1.0	0.25	260	
	500	1.0	0.5	280	GEC fusion
	500	0.5	0.5	280	Shearing
	550	0.5	0.5	150	Partial fusion
	600	0.5	0.5	80	GEC fusion

The pulsed mode allows significant increase in the exposure effectiveness without fusion of the surface, as it enables to implement the greatest thermal-gradient tensions on the GEC-cast iron boundary. In addition, this boundary, apparently being the tension concentrator, will inevitably accumulate microdamages induced by repeated propagation of thermal deformation waves which finally result in separation of the coating. Thus, removal of an old or defective GEC from cast iron products with pulse-periodic heating by the focused radiation of the arc source is, apparently, fatigue destruction.

It should be noted that increase in the pulse repetition rate (if the duration is maintained or insufficiently reduced) not specified in the above table, is unpractical, since this results in overheating of the surface which causes or may cause not shearing but fusion.

Based on the experimental data, the technological procedure to recover the defective GEC for cast iron products can be as follows: complete removal of the glass-enamel coating through exposure to the light-beam and subsequent enameling of the cast iron products according to the proven conventional technology: treatment of the surface, application of the ground enamel slip; base coat drying, ground layer sintering, powdering and sintering of the enamel coating.

It should be noted that the setup can be successfully applied for local restoration of the defective glass-enamel coating on cold steel products.

4. Conclusions

The developed setup for light beam exposure allows effective removal of enameled cast iron products.

The process modes for thermal shock destruction of the glass-enamel coating on cast iron by pulse-periodic action of the powerful focused light flux from a xenon arc lamp.

Acknowledgements

This work was financially supported by The Ministry of Education and Science of the Russian Federation in part of the science activity program.

References

[1] Vargin V V 1972 *Enameling of Metal Products* (Moscow: Mechanical Engineering)
[2] Kholodilin N N 1962 *Enameling of Steel And Iron Products* (Moscow: Gosstroiizdat)
[3] Petzold A and Peshma G 1990 *Enamel and Enameling: Reference Textbook* (Moscow: Metallurgy)
[4] Romanov B P, Artish A S, Zuev A M and Aleutdinov A D 1989 Method for removing glass-enamel coating on metal products: *AS 1791466. RF. Application No 4710609* Publ. 30.01.1993, B.E. No 4
[5] Romanov B P 1999 Processes of self-organisation on strukture-formation of glassenamel coating under the effect of electromagnetic emission from electrophyssical heating sources *V. Russion Chinese International Simposium «Advanced materials and processes»* p 246

STUDY OF TYPES OF DEFECTS IN WOOD CHIPBOARD PRODUCTION

Elena Vasendina[1], Inna Plotnikova[1], Ludmila Red'ko[1] and Natalia Zyablova[2]

[1] Institute of Non-Destructive Testing, National Research Tomsk Polytechnic University, Lenin Avenue 30, 634050 Tomsk, Russia

[2] Institute of Physics and Technology, National Research Tomsk Polytechnic University, Lenin Avenue 30, 634050 Tomsk, Russia

E-mail: inna@tpu.ru

Annotation: Types of defects in wood chipboard production and losses connected with the defects were studied. Faulty products were studied by statistical methods of control; analysis of causes of faulty products at wood chipboards production was carried out. Economical appraisal of losses according to peculiarities of production was given.

Key words: defect, faulty products, statistical methods of control, wood chipboards.

Introduction

Choice of technical measures allows increasing effectiveness of functioning of wood chipboards production, which is connected with launch of quality and competitive products for broad utility in building, special purpose furniture production, and constructions working in atmospheric conditions [1, 2]. The main purpose of commercial organization is to get a profit [3], provided that one of effective ways of increasing profit is decreasing expenses. Furthermore, expenses need to be identified, evaluated and controlled. Nowadays not all expenses can be decreased, for instance, it is not recommended to decrease expenses on quality materials and salary. However, some part of expenses, which is called losses [4], can and must be decreased for increasing effectiveness of activity, competitiveness, and profitability of an enterprise. Losses are expenses for mismatches; classification of expenses for quality was fully presented in [5]. It is possible to decrease expenses for mismatches by identification of faulty products, analysis of causes of faulty products and exclusion or minimizing their influence. Most enterprises keep a record of characteristics of quality of production in the form of a report of different types of faulty products. This allows evaluating losses from faulty products in monetary equivalent without additional studies. Statistical methods of quality control of production and processes can be effective methods of managing losses [6, 7].

Purpose of the work

The present work aims to show how application of statistical methods can help identify sources of losses in the form of faulty products, analyze their causes at wood chipboards production.

Methods and objects of the study

Technological process of wood chipboard production is a coordination of a lot of stages (fig. 1) which have a possibility of mismatches.

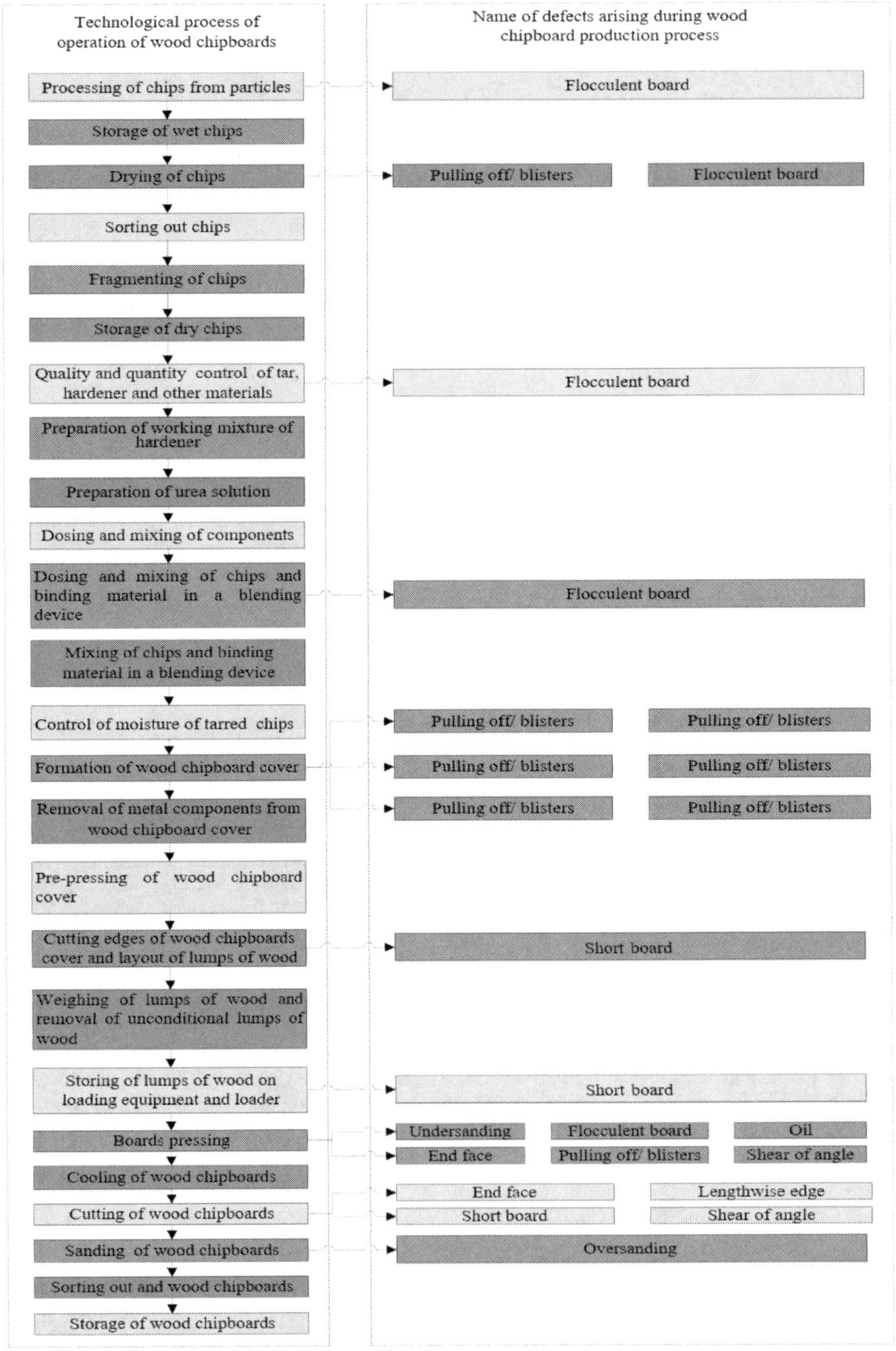

Figure 1. Scheme of a technological wood chipboard production process with stages and types of defects arising on them more often.

To get an overview of the quality of products a share of unconditional products was defined from the whole quantity of wood chipboards launched by a factory. A chart of the p-card type for the share of defective products was made to visualize the results, fig. 2.

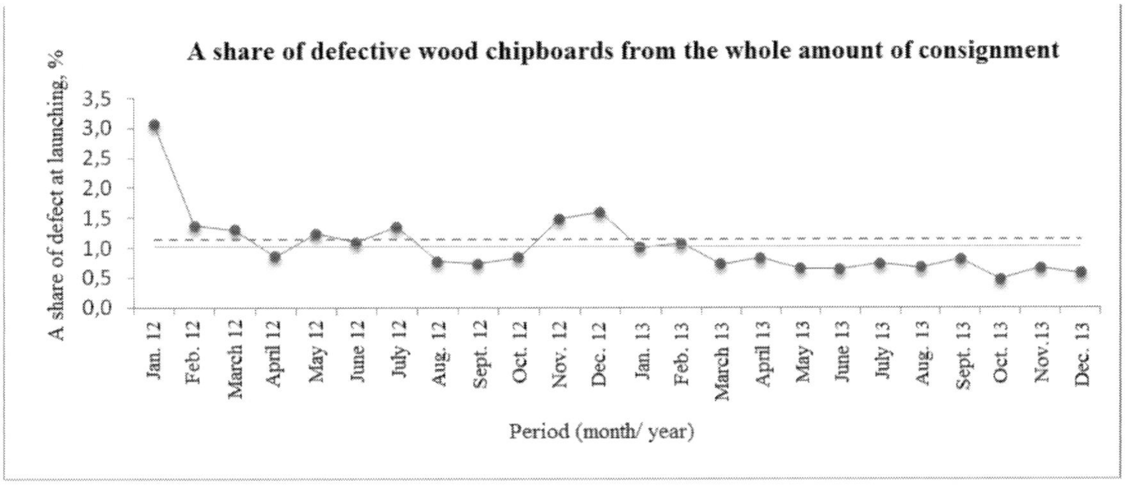

Figure 2. Chart with defective products share.

Charts help in setting periods with high and low level of defectiveness and elicit productive situations which enable best quality of data [4].
On the basis of the chart it was stated that:

- a share of defective products in 2013 decreased in comparison with 2012. The cause of decrease of level of defectiveness could not be identified which was connected with a large period of time passed since the changes in the process;
- a process of wood chipboards production in 2013 became stable;
- a share of defectiveness at an enterprise is very low. It was on the average less than 1 % in 2013.

Financial losses of an enterprise for 2012-2013 were defined for to detect losses from defectiveness in monetary equivalent (in RUB.). Types of defects revealed during wood chipboards production and fixed monthly in a quality control department in reports of defectiveness registration were detected for this reason. They are: sanding defects; dust and tar pitch; end face; lengthwise edge; pulling off; shear of angle; short and flocculent board; oil. Ways of use of wood chipboards in dependence on the type of a defect were identified. Prices of wood chipboards having a defect were set. Losses connected with the launch of a defect made 13,8835 mln. RUB.

Results

Analysis of frequency of defects by means of their names with the help of the diagram [5], fig. 3, was completed.

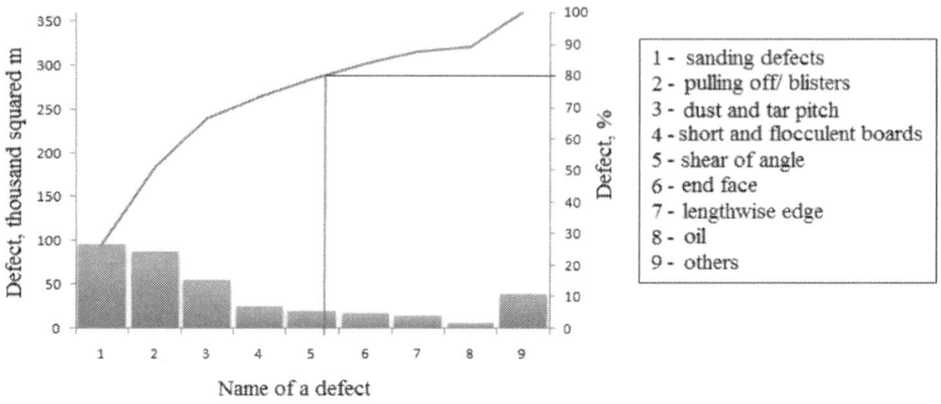

Figure 3. Frequency of defects of wood chipboards by means of their names.

The diagram shows the most frequent defects: sanding defects, pulling off/ blisters, dust and tar pitch, short and flocculent boards.

Having drawn an analogy between the diagram and financial losses because of defects of launch a conclusion can be made about small losses for an enterprise in case of a 'short and flocculent boards' defect.

A 'short and flocculent boards' defect at an enterprise is registered as one type of the defect. In reality these are two types of defects – 'flocculent boards' and 'short boards'.

The main reasons of 'flocculent boards' defect are defined as: high moisture of dry chips; incorrect mode of tarring; low quality of tar, insufficient pressure on the press; small weight of cover; short cycle of pressing; low temperature of pressing; incorrect reduce of pressure.

Most reasons of defects are connected with operator or technologist faults because of insufficient experience.

The diagram in fig. 4 was made for detecting operations of technological production process of wood chipboards where a lot of defects take place (80 %).

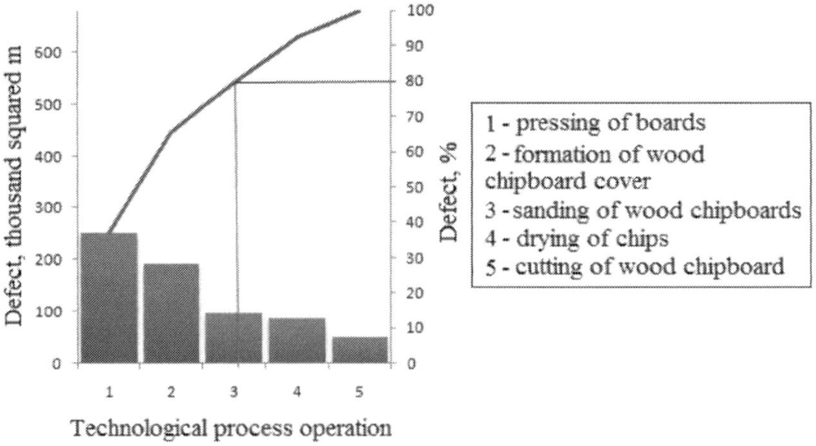

Figure 4. Diagram of detection of operations of technological production process of wood chipboards where a lot of defects take place.

As it can be seen from fig. 4, most defective boards take place at the stage of pressing of boards and formation of wood chipboard cover. Therefore, these operations were described in detail by means of diagrams in fig. 5 and fig. 6, where quantity proportion of types of defects in these operations is shown.

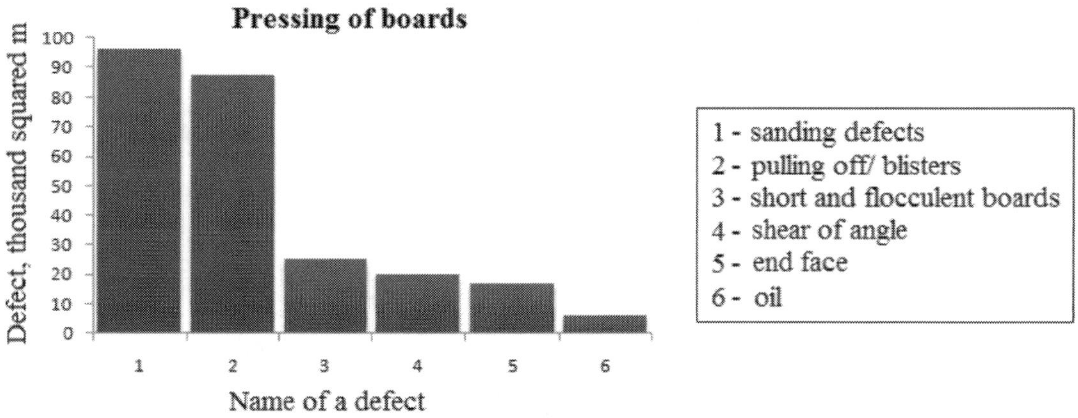

Figure 5. Types of defects at the stage of pressing of boards.

Sanding defects and 'pulling off/ blisters' defects take place more often at the stage of pressing boards, the least quantity refers to 'oil' defects.

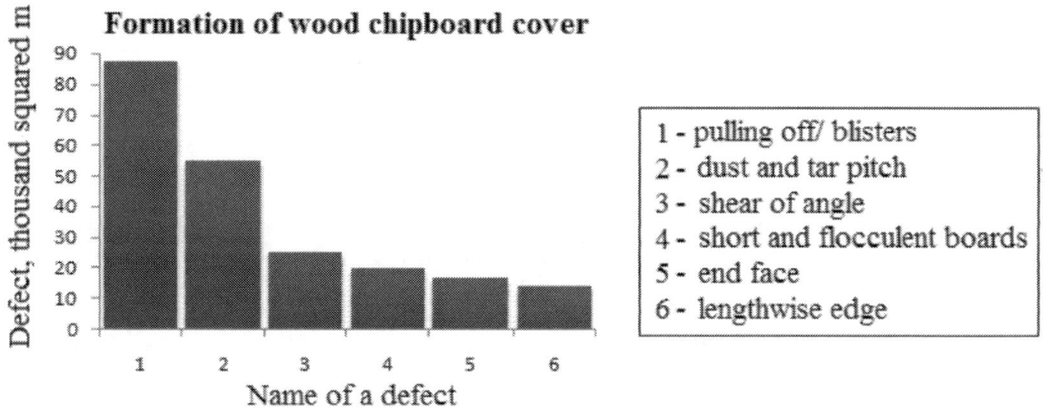

Figure 6. Types of defects at the stage of formation of wood chipboard cover.

'Pulling off/ blisters' defects take place at the stage of formation of wood chipboard cover more often, 'lengthwise edge' defects take place less.

Charts of individual data and moving range towards every type of the detected defect were made for detection of variations of production process of wood chipboard launch. The chart of sanding defects was presented in fig. 7 as an example.

IOP Conf. Series: Materials Science and Engineering **81** (2015) 012070 doi:10.1088/1757-899X/81/1/012070

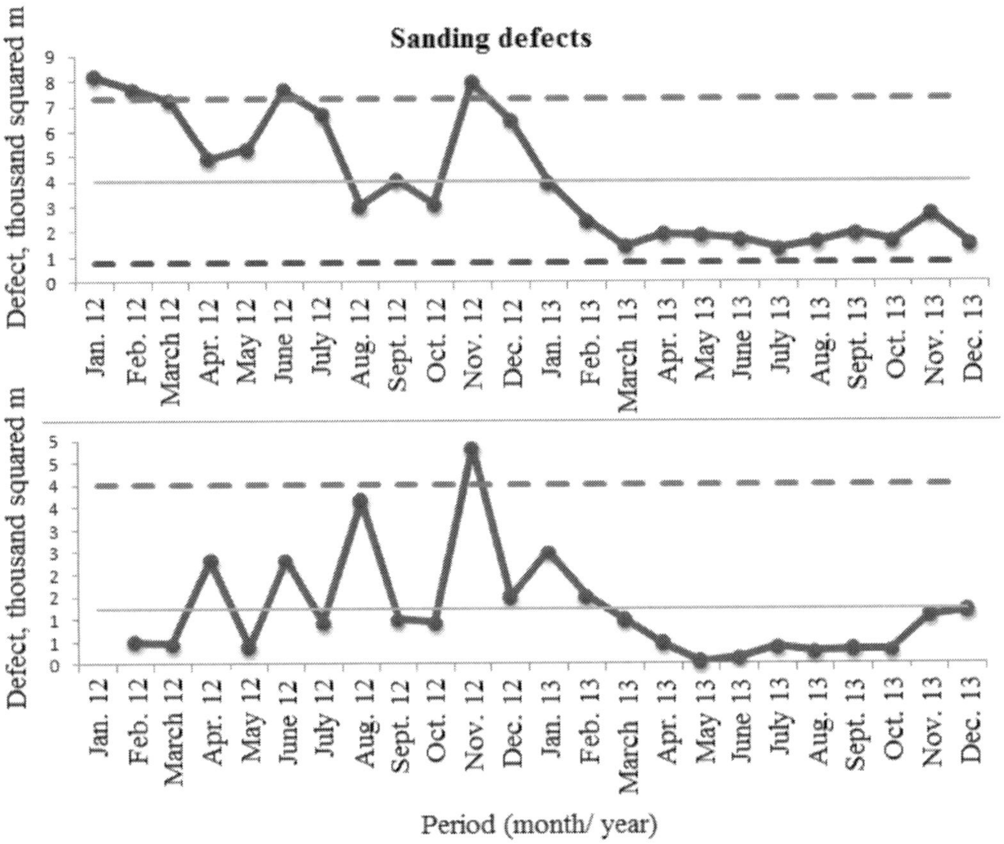

Figure 7. Chart of sanding defects.

The upper chart depicts a number of wood chipboards with sanding defects. The lower chart shows how data of quantity of defective boards change in relation to the previous or next period.

The chart demonstrates instability of production process of launch of boards with sanding defects in 2012. The process has a statistically controlled (stable) character in 2013.

Conclusion

Thus, the use of statistical methods allowed detecting neck stages and defining priorities for quality improvement. Frequent types of defects which cause a lot of monetary losses at enterprises were defined and their causes analyzed. Technological operations with a lot of defects and their types were defined. It should be noted that such kind of analysis was not carried out earlier. The sum of losses caused by defective details was calculated; it comprises about 1 % or a share of 1 % from the whole volume of the ready-made production. However, as a rule the sum of losses in monetary equivalent caused by faulty products attracts attention of managers to the quality of products.

A conclusion about the necessity of further study of factors of technological process which influence the parameters of quality of wood chipboards was made on the basis of the study for reasonable fixation of their optimal data in normative and technical documentation.

References:

[1] Plotnikova G P, Denisov C B , Chelysheva I N 2010 Povyshenie jeffektivnosti proizvodstva drevesnostruzhechnyh plit *Krasnojarsk: Vestnik KrasGAU* **7** pp 152-158

[2] Beljaev A 2012 Drevesnye plity dorozhajut *Lesnaja industrija* **10** pp 42-43

[3] Oblivin A N, Lopatnikov M V, Brancev V A et al 2011 Nanotehnologii i nanomaterialy v lesnom komplekse *M: FGBOU VPO MGUL* p 221

[4] Fejgenbaum A V 1986 Kontrol' kachestva produkcii *M.: Jekonomika* p 470

[5] Isikava K 1988 Japonskie metody upravlenija kachestvom *M.: Jekonomika* p 215

[6] Kumje H, Iizuka J, Takahasi T et al 1991 Statisticheskie metody povyshenija kachestva *M.: Finansy i statistika* p 301

[7] Grodzenskij S J, Grodzenskij J S 2012 Ob jeffektivnosti primenenija semi prostyh instrumentov kachestva *Metody menedzhmenta kachestva* **2** pp 46-49

[8] GOST 10632-2014 Plity drevesno-struzhechnye. Tehnicheskie uslovija. EN 312:2010, NEQ

[9] Volynskij V N 2010 Tehnologija drevesnyh plit i kompozicionnyh materialov *SPb.: Izdatel'stvo "Lan'"* p 366

[10] Ishhenko T L 2009 Izmenenie prochnosti drevesnostruzhechnyh plit v period ih poslepressovoj vyderzhki *Derevoobrabatyvajushhaja promyshlennost'* **2** pp 7-9

[11] Panov N G 2010 Povyshenie prochnostnyh svojstv drevesnyh materialov na osnove primenenija nanotehnologii *Tehnologija i oborudovanie himicheskoj pererabotki biomassy dereva. Himija drevesiny. SPb: SPbGLA.* pp 173-176

[12] Panov N G, Pituhin A B, Rozhkov S S, Cvetkov V E, Sanaev V G, Firjulina O V 2012 Drevesno-struzhechnye plity na osnove karbamidoformal'degidnoj smoly, modificirovannoj nanorazmernym shungitom *Lesnoj vestnik MGUL* **2(85)** pp 135-139

Ignition of wood subjected to the decreasing radiant energy flux

V T Kuznetsov[1], A I Filkov[1,3,4], Yu N Isaev[2] and V O Guk[1]

[1] National Research Tomsk State University, 36, Lenina avenue, Tomsk, 634050, Russia
[2] National Research Tomsk Polytechnic University, 30 Lenina avenue, Tomsk, 634050, Russia
E-mail: aifilkov@gmail.com

Abstract. In this paper we analyze the ignition of wood samples subjected to the decreasing heat flow. The experimental setup was created on the base of the optical wave "Uran-1". The intensity of the heat flow was changed during the experiment by moving the test sample along the optical axis of the elliptic reflector in the setup. Pine wood was used as the test samples. We received the delay times for ignition of pine wood during heating by the decreasing heat flow. The received data were compared with the data for a static heat flow.

1. Introduction

In [1] it is reported that the ignition process is essentially dependent on the type of the heat flow, whether it is static or variable (when the ignition of fuels is initiated by the time-dependent heat flow). In recent years the dynamic regimes of ignition are mostly used in solving applied problems in the field of fire safety. In the real conditions the dynamic regimes are observed during anthropogenic and forest fires. At the same time, the ignition of fuels under variable conditions has been insufficiently studied. There are only some works [2–5], where the ignition of fuels was carried out under variable heat flux. Bilbao et al. [4] examined the ignition of wood exposed to different conditions of constant and variable heat fluxes. A time-decreasing heat flux was employed in their work, corresponding to the thermal radiation of a flame of natural gas or any other hydrocarbon produced in an accidental release of such material. However, the external heat flux always varies with the growth of a fire. Lizhong et al. [5] carried out the theoretical and experimental investigation on pyrolysis and spontaneous ignition of wood exposed to heat fluxes linearly increasing with time. Experiments were conducted for different wood species subjected to several increasing rates of heat fluxes. These results have shown that the characteristics of pyrolysis and auto-ignition of woods under variable heat flux differs from those under constant heat flux. However in examined works decreasing and increasing heat flux investigated under low heat fluxes. Taking into consideration that the variable type of the heat flow dominates in real conditions, the urgency of this problem is obvious.

2. Methods

[3] To whom any correspondence should be addressed.
[4] This work was supported by the Government Assignment in the Field of Scientific Activity (number 1624).

To study the ignition of wood under variable heat flux, we created an experimental setup on the basis of the optical furnace "Uran-1" (figure 1*a, b*) with a radiation source (xenon lamp with a power of 10 kW) [6].

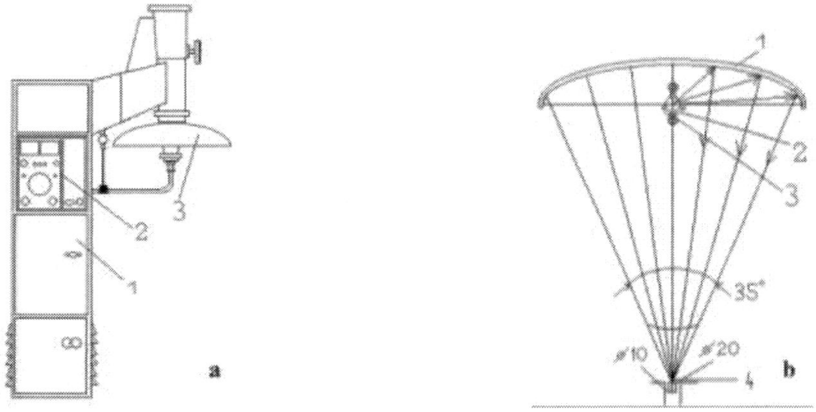

Figure 1. The scheme of experimental installation: a) optical furnace URAN-1: 1–power supply, 2–control panel, 3–radiator; b) optical radiator: 1–reflector; 2–counter reflector; 3–lamp; 4–action spot.

Optical radiation accounted for only 5.5 kW out of 10 kW generated by the xenon lamp. The radiation spectrum distribution was as follows: 0.5 kW (9%) in the ultraviolet region; 2 kW (36%) in the visible region, and 3 kW (55%) in the infrared region. The diameter of the focused flux was about $2{\cdot}10^{-2}$ m.

The recording unit of the experimental setup provided the measurements of the radiant flux density, exposure time, initiation of flame and controlled experimental conditions. It included a radiant flux density sensor, photodiodes, time relays and an N-117 light-beam oscillograph.

The density of a radiant energy flux in a focal spot was controlled by using radiation attenuators and changing electric power applied to the xenon lamp. The attenuators of the radiant energy flux were a metal grid that was placed perpendicularly to the optical axis at a distance of 0.3-0.4 m from the focal plane. The system of gates protects the test sample against early heating by the radiant flux and dosed supply of energy radiation.

The variable flow was modeled by the movement of the sample along the optical axis. For the movement of the sample it was used the following installation (by the example of the decreasing radiant energy flux, figure 2).

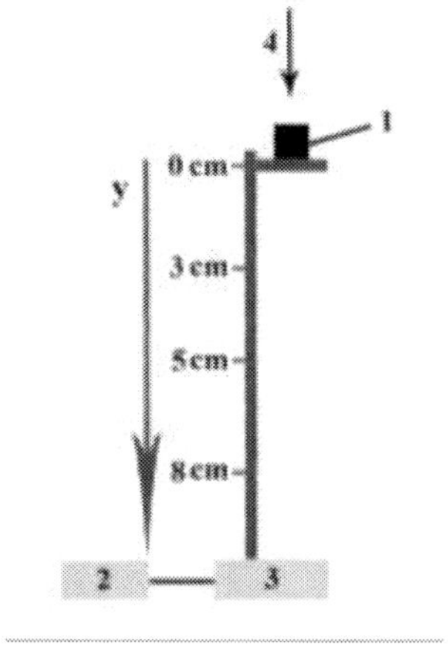

Figure 2. Installation for the movement of the sample: 1 – sample, 2 – control panel for motion of the site with a sample; 3 – mechanism that moves the sample; y = 0 cm is the focal plane; 4 – radiant heat flux; 3, 5, 8 cm are the distance from the focal plane.

Simultaneously with the opening of the gate, the transporter was turn on and the sample started moving with a constant speed of 0.117 m/s. The opening (closing) time of the gate was $4 \cdot 10^{-2}$ sec.

The test procedure was as follows. The intensity of the radiant flux was measured by a calorimeter fixed to the holder at the upper and lower points of the optical axis. Then the test sample was located on the place of the calorimeter and tested. The registration unit recorded the opening time of the gate and the initiation of flame. Ignition of the condensed substance was determined by the initiation of flame on the sample surface.

The calorimeter was a copper disk $1 \cdot 10^{-2}$ m in diameter and $3 \cdot 10^{-3}$ m in thickness with a thermocouple calked in the center of the disk to a depth of $1.7 \cdot 10^{-3}$ m. The measurement error of the radiation intensity did not exceed 10%. To determine the density of the radiant flux, we experimentally measured the heating rate of the copper disc subjected to radiation supplied to a blackened surface. The heat flux density was measured in the range q = 20–110 W/m^2.

The delay time of ignition was determined by a photoelectric method. The ignition time was considered to be a time interval from the exposure to the radiant energy flux up to the initiation of flame above the surface of the sample. Radiation of flame was registered using photodiodes. The H-117/1 oscilloscope was used for recording electrical signals. The measurement error for the ignition time is not more than 4% and is determined mainly by the physicochemical properties of the material studied.

Pine wood was used as the test samples. The samples were cylinders $1.9 \cdot 10^{-2}$ m in diameter and $1.5 \cdot 10^{-2}$ m in height. The surface of the samples, absorbing radiation, was covered by lampblack. The light radiation reached the sample perpendicularly to the wood fibers. The initial temperature of the samples corresponded to room temperature (297 K). The moisture content of the samples was 1.8 % and determined by using an A&D MX-50 moisture content analyzer with an accuracy of 0.01%. Unchangeable moisture content of the samples was provided by a desiccator filled with silica gel.

In this paper we analyze the ignition of wood samples subjected to the decreasing heat flow. To compare the delay time of ignition under static and variable conditions, the heat flux was averaged under variable conditions. For this purpose, the following method was used. The intensity of the radiant energy flux was measured with a microcalorimeter at fixed distances from the focal plane of the reflector.

Knowing the speed of the sample along the optical axis of the elliptical reflector, we can turn from fixed values of the distance to the time. In this case, the dependence of the heat flow on time is well enough (correlation R^2 = 0.997) approximated by the function (figure 3):

$$q = a\,e^{-0.7t} - 2 \tag{1}$$

where q is the heat flow, W/m^2; t is the time, s; a is the nondimensional coefficient.

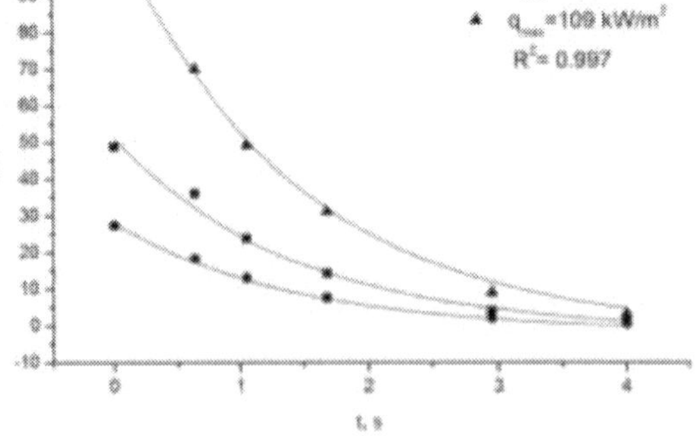

Figure 3. Heat flow versus time when the sample is moved from the focal plane.

Thus, depending on the value of heat flow in the focal plane (t=0) we can determine a coefficient *a* and changes of the heat flow at the surface of the moving sample by using the formula (1). The delay time of ignition was measured using a photodiode determining the initiation of flame. Knowing the ignition delay time of the sample we can find a heat flow in the ignition point using a distribution curve q(t). Since the sample of pine ignites within one second in the tested range of the heat flow, then from figure 3 it can be seen that the curve is quasilinear in this range. In this connection, the average heat flow was determined as the arithmetic mean value for variable conditions.

3. Results and discussion

Figure 4 shows the ignition delay times of the pine samples for variable (q=f(t)) and static (q=const) [6] heat supply.

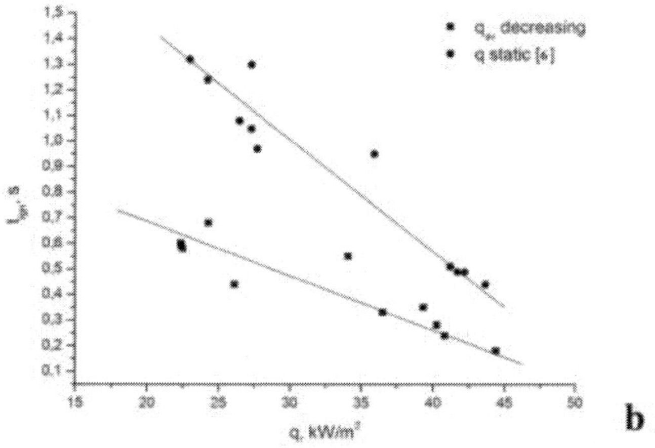

Figure 4. Ignition delay times of the pine samples versus heat flow for variable and static conditions [6].

When the radiation flux decreases, the delay time for ignition of the pine samples is 2-2.5 times less than that for the constant radiation flux. At the same time, figure 4 shows that the decrease in value of heat flux leads to the fact that the delay time of ignition increases faster under static conditions than under variable ones. It should be noted the fact that when we compared the time of ignition for the variable and static heat flow in this work, the variable mode was determined by the average heat flow during the experiment.

The physical considerations imply the following facts. If the ignition time is substantially less than the thermal relaxation time for the material under study in the experiment, the variable supply of radiant energy results in the formation of the heated layer on the sample surface. Heating of this layer requires less energy for ignition of wood.

Estimation of the thermal relaxation time of pine wood gives the following values

$$t_r \approx \frac{a}{u^2} = 1.84 \cdot 10^3 \ s \qquad (2)$$

where a is the thermal conductivity of pine wood, $1.84 \cdot 10^{-7}$ m²/s [7]; u is the normal combustion rate, 10^{-5} m/s [8]. It is seen that the relaxation processes of heat distribution in wood is much slower than the ignition process for the considered heat flows. At the same time, the temperature gradient on the sample surface is generally higher in a variable mode than that in the static mode, i.e. the surface temperature reaches a critical value faster than in the static mode. Therefore, for the same heat flow, the ignition time in the variable mode should be smaller than in the static mode. The increase in flow of radiant energy leads to the fact that the difference in the ignition delay decreases for variable and static conditions (figure 4), which confirms this assumption.

These results do not contradict the paper [9], where the mathematical modeling shows that in the case when the heat flow decreases, the delay time for ignition of wood is less than that for the constant heat flow.

The next step of this work is to study the properties of ignition for the increasing heat flow and the data analysis. The received results can be used to improve the prediction of dangerous zones for ignition of wood during wildland and anthropogenic fires.

References

[1] Filkov A I, Kuznetsov V T, Guk V O, Telyakov E Sh and Enalejev R Sh 2013 *Proc. of the 4th Fire Behavior and Fuels Conf.* (Saint-Petersburg, Russia, July 1-4, 2013), (Missoula: IAWF) p 57

[2] Rosenband R Sh, Barzykin V V and Merzhanov A G 1968 *Combust. Explo. Shock.* **4(2)** 171-5

[3] Enaleev R Sh, Mateosov V A, Sinaev K N, Dinavetsky V D and Gainutdinov R Sh 1973 *Physics of Combustion and Methods of Research* 80-6

[4] Bilbao R, Mastral J F, Lana J A, Ceamanos J, Aldea M E, Betran M 2002 *J. Anal. Appl. Pyrol.* **62** 63-82

[5] Yang L, Guo Z, Zhou Y, Fan W 2007 *J. Anal. Appl. Pyrolysis* **78** 40–45

[6] Kuznetsov V T and Filkov A I 2011 *Combust. Explo. Shock.* **47(1)** 74-9

[7] Aseeva R M, Serkov B B and Sivenkov A B 2010 *Combustion of Wood and Fire Behavior* (Moscow: Academy of the State Fire Service of the Russian Ministry of Emergency Situations) p 262

[8] Abduragimov I M, Androsov A S and Bartak M 1986 *Combust. Explo. Shock.* **22(1)** 10-3

[9] Enaleev R Sh, Krasina I V, Gasilov V S, Chistov Yu S and Tuchkova O A 2013 *Bulletin of Kazan Technological University* **16(10)** 99-106

Characterization of copper nanopowders after natural aging

Yu A Amelkovich, O B Nazarenko, A I Sechin and P M Visakh

Tomsk Polytechnic University, Lenin av. 30, 634050 Tomsk, Russia

E-mail: obnaz@mail.ru

Abstract. Copper nanopowders after storage under natural conditions can oxidize and change their properties. In this work, the phase composition, morphology and thermal properties of copper nanopowders after natural aging of 10 years were studied. The copper nanopowders were produced by the method of electrical explosion of wires in different gaseous ambient: argon and carbon dioxide. The nanopowders were characterized by X-ray diffraction, thermal analysis, scanning electron microscopy, Fourier transform infrared spectroscopy. The content of copper oxides phases was found to be increased in comparison with freshly synthesized powders. This transformation results in the change of the thermal characteristics of the nanopowders. The effect of the synthesis conditions on the composition and thermal characteristics was shown.

1. Introduction

In recent years, copper nanopowders (Cu NPs) have attracted much attention due to their catalytical, optical, tribological and electrical properties [1–4]. The problem in Cu NPs application is their high activity with oxygen atoms and rapid oxidation during handling and storage under ambient conditions that results in changes phase composition, morphology and functional properties [5–7].

At present, Cu NPs are prepared using a number of methods, including vacuum vapor deposition [8], metal vapor synthesis [9], thermal reduction [10], chemical reduction [11], and laser ablation [5]. Among these methods, electrical explosion of wires is known to be a promising method for the synthesis of nanoscale powders [12–14]. The metal nanopowders produced by EEW are characterized by a high chemical reactivity at high temperatures and a high stability during their storage. To achieve the high stability of NPs produced by EEW towards oxidation, a passivation of freshly synthesized powders by slow oxidation in air should be carried out [12–15]. In our previous work, the characteristics of aluminum NPs produced by EEW and stored under ambient conditions for a long time were investigated [16]. It was shown that aluminum NPs remain to be stable toward oxidation with time that was explained by the formation of the oxide-hydroxide layer on the particles surface at passivation. Cu NPs have the properties different from those of aluminum NPs.

The properties of Cu NPs freshly synthesized by EEW were intensively investigated [13, 17]. It is known that they have a specific surface area of 5-10 m^2/g, a wide particle size distribution in the range from 100 to 4000 nm. The metallic copper content in the particles can vary from 80 to 95 %. The surface of the particles is covered with oxide layer having the thickness of 5-15 nm. Cu NPs stored under ambient conditions are oxidized; during this process a growth of the oxide layer thickness on the particles surface occurs to 30-35 nm for 3 years [17] and the degree of crystallinity of oxides increases. Full oxidation of fine particles is possible.

IOP Conf. Series: Materials Science and Engineering **81** (2015) 012072 doi:10.1088/1757-899X/81/1/012072

The changes in the physicochemical properties and phase transformation occurring in Cu NPs at their long-term storage have been studied insufficiently. At the same time, the studies of the phase composition, morphology and the thermal stability of Cu NPs to oxidation during the storage in air are of interest for practical use, also in term of the safety of nanotechnology. The objective of this work is the investigation of the phase composition, morphology and thermal stability of Cu NPs produced by electrical explosion of wires in argon and carbon dioxide, after long-term storage under ambient conditions.

2. Experimental

Copper nanopowders were produced by the method of EEW in the following gaseous ambient: argon (the sample Cu(Ar)) and carbon dioxide (the sample Cu(CO_2)). After production a stabilization of the copper particles surface was carried out by a controlled slow stream of air. The Cu NPs had been stored in the closed containers under natural conditions: temperature 20-25 °C and humidity 60-80 % for 10 years.

The phase composition of the studied Cu NPs was determined by X-ray diffraction (XRD) analysis with CuKα radiation (Shimadzu XRD-7000). The size and morphology of the studied samples were investigated using scanning electron microscope (SEM) JEOL JSM-7500FA. The specific surface area of the samples was determined using a BET-analyzer Sorbi. A thermogravimetric analyzer SDT-Q600 was used to study the thermal gravimetric (TG) and the differential scanning calorimetric (DSC) behavior of the samples from 20 to 1000 °C at a heating rate of 10 °C/min in a flow of air. The infrared spectra of the samples in the spectral range 4000-400 cm^{-1} were recorded using Nicolet 5700 Fourier transform infrared (FTIR) spectrometer.

3. Results and discussion

The XRD results in figure 1 indicate the presence of crystalline phases of metallic copper and copper oxides Cu_2O and CuO in the composition of the studied powders. Cu(CO_2) NP contains more copper oxide CuO in comparison with the sample Cu(Ar) NP, that is caused by the use of oxygen containing gas CO_2 as a medium for fabrication in the process of EEW and partial oxidation of the particles directly in this process.

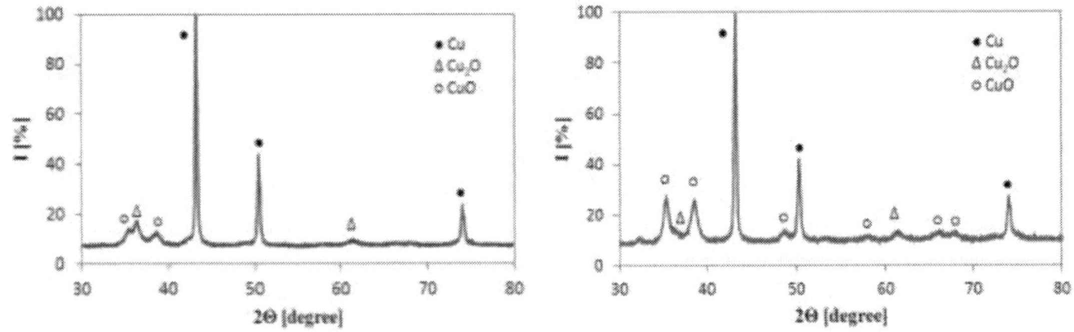

Figure 1. XRD patterns of Cu NP: (a) Cu(Ar), (b) Cu(CO_2).

The average crystallite size of the Cu nanoparticles was estimated using the Debye-Scherrer formula [18] and for the sample Cu(Ar) it was 32 nm, for the sample Cu(CO_2) - 30 nm.

The specific surface area of the samples Cu(Ar) and Cu(CO_2) was 6.72 and 5.93 m^2/g, respectively, which is not much different from that of the freshly prepared Cu NPs.

The SEM micrographs of Cu NPs are shown in figure 2. The particles of Cu NPs have a size of about 100-250 nm with a shape close to spherical with friable surface. Among the particles of nanodispersed fraction the particles having the size up to 10 mcm with a smooth surface can be found.

It is known that polymodal particles size distribution is typical for all NPs produced by the EEW [12-14]. It can be assumed that nanodispersed fraction of Cu NPs is fully oxidized during long storage in air, which was previously shown in [13, 17].

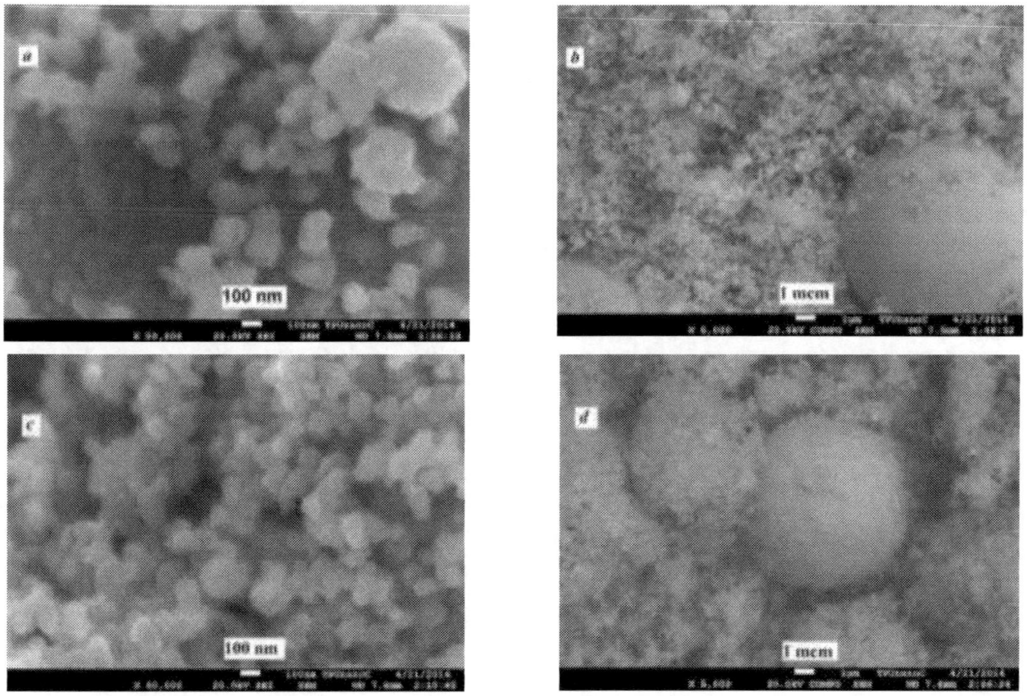

Figure 2. SEM micrographs of Cu NPs: (a), (b) - Cu(Ar); (c), (d) - $Cu(CO_2)$.

Figure 3 shows the results of thermal analysis obtained at a non-isothermal heating of Cu NPs in air. The desorption of adsorbed gases and water vapor from the surface of the particles occurs in the temperature range 20-200 °C that lead to a weight decrease by about 0.4-0.7 wt. %. This corresponds to the endothermic process on the DSC curve with a minimum at 52.6 °C for the sample Cu(Ar) and 65.7 °C for the sample $Cu(CO_2)$. Then, after further heating a sharp increase in the weight and heat release occurs as a result of oxidation of the powders. The onset temperature of oxidation is 183 °C for Cu(Ar) NP and 234 °C for $Cu(CO_2)$ NP.

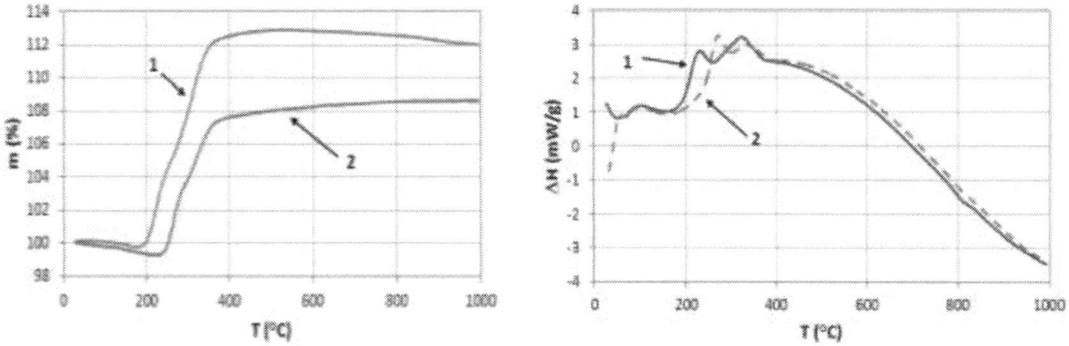

Figure 3. TG (a) and DSC (b) curves for Cu(Ar) NP - 1 and $Cu(CO_2)$ NP - 2.

The oxidation of Cu NPs occurs in two stages, with maxima in the DSC curve for the sample Cu(Ar) at 225 and 321 °C and for the sample Cu(CO_2) at 267 and 328 °C. As it was described in [17], a sequential transformation of Cu into oxide Cu_2O and then into oxide CuO occurs during the oxidation of Cu NPs. The presence of several stages in the oxidation process is typical for nanoscale metals produced by the method of EEW and it is associated with a polymodal particle size distribution and the oxidation of fine particles in the first instance [16].

At the first oxidation stage of the sample Cu(Ar) from 183 to 270 °C, metal copper is oxidized to Cu_2O: $4Cu + O_2 = 2Cu_2O$. The sample weight at this stage is increased by 5.4 wt. %. Further heating of Cu(Ar) NP to 525 °C is accompanied by oxidation of Cu_2O to CuO by the reaction $2Cu_2O + O_2 = 4CuO$ and an increase in the sample weight by 7.8 wt. %. Upon heating in the range of 525-1000 °C, a slight decrease in weight of 0.8 wt. % is observed, due to the partial decomposition of CuO.

The weight of the sample Cu(CO_2) at the heating to 1000 ° C increased by 9.4 wt. %. In the first steps of this sample oxidation in the temperature range from 235 to 300 °C the sample weight increased by 4.3 wt. %. Further heating of the sample up to 1000 °C is accompanied by an increase in the sample weight by 4.1 wt. %.

The specific heat effects during oxidation of the samples Cu(Ar) and Cu(CO_2) are 904.7 and 750.4 J/g, respectively. The larger thermal effect of the sample Cu(Ar) is caused by a higher content of metallic copper in the composition of this powder, as shown by the XRD results.

According to the thermal analysis data a metallic copper content in the powders is estimated to be 55 % in the sample Cu(Ar) and 35 % in the sample Cu(CO_2).

The FTIR spectroscopy technique allows obtaining infra spectra of the copper particles surface oxide layer and to examine their state and the composition. Figure 4 shows the FTIR spectra of the studied powders.

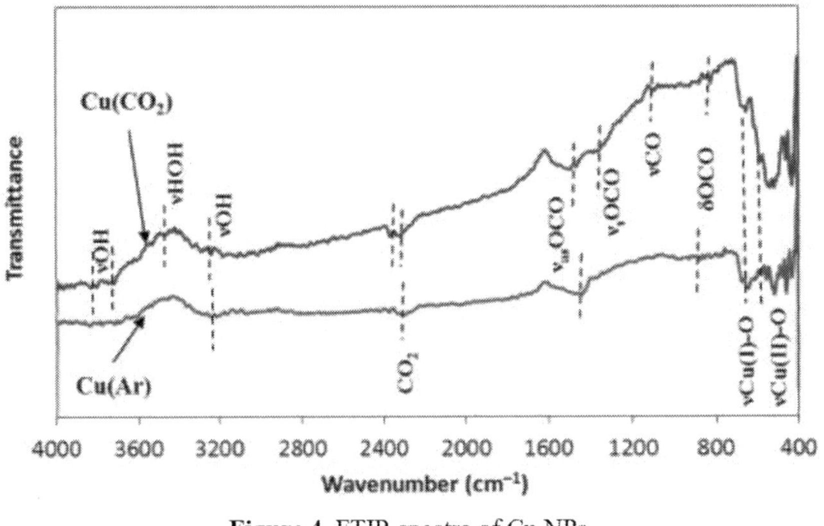

Figure 4. FTIR spectra of Cu NPs.

The spectrum of the sample Cu(Ar) is characterized by the presence of several absorption bands in the range of 592-430 cm^{-1} caused by stretching vibrations νCu(II)-O of copper oxide CuO. The absorption bands with maxima at 643 and 677 cm^{-1} can be attributed to the vibrations νCu(I)-O, indicating the presence of oxide Cu_2O in the composition of the sample [18, 19]. In the spectrum of the sample Cu(CO_2) the absorption bands in the range of 579-430 cm^{-1} typical for copper oxide CuO are observed and the band with a maximum 537 cm^{-1} appears. The absorption band at 657 cm^{-1} is associated with the presence of oxide Cu_2O in the sample Cu(CO_2).

The absorption bands of the adsorbed CO_2 molecules are observed at 2301 cm^{-1} in the spectrum of the Cu(Ar) sample and at 2369 and 2322 cm^{-1} in the spectrum of the Cu(CO_2) sample. As a result of carbon dioxide adsorption on the surface of the sample Cu(Ar) carbonate structures are formed, which are characterized by absorption bands due to asymmetric stretching vibrations ν_{as}O-CO at 1441 cm^{-1} and deformation vibrations δO-CO at 878 and 829 cm^{-1} [20]. In the spectrum of the sample Cu(CO_2) the bands of asymmetric stretching vibrations of the carbonate ion ν_{as}O-CO at 1495 and 1466 cm^{-1}, symmetric stretching vibrations ν_sO-CO at 1336 cm^{-1}, symmetric stretching vibrations νC-O at 1096 cm-$^{-1}$ and deformation vibrations δO-CO at 833 cm^{-1} appear.

The numerous low-intensity absorption bands in the range of 3800-3000 cm^{-1} in the spectra of the studied Cu NPs are caused by the stretching vibrations of the water molecules and structural hydroxyl groups.

4. Conclusion

The study of the phase composition, morphology and thermal stability of copper nanopowders were carried out after long-term storage under ambient conditions. Copper nanopowders were prepared by the method of EEW in argon and carbon dioxide and then they were stored in the closed containers 10 years. During long storage in the air nanodispersed fraction of the particles was oxidized almost completely due to their high reactivity. The sample Cu(CO_2) is more oxidized than the sample Cu(Ar). The use of oxygen-containing gas as a medium for a synthesis of powders by EEW results in the obtaining particles finer than in the case of using argon because of the passivating effect. The presence of the smaller particles leads to their more rapid oxidation and smaller metallic copper content in the sample Cu(CO_2). As a result, the thermal stability of the powder Cu(CO_2) to the oxidation at the heating is higher than that for the powder Cu(Ar).

Acknowledgments

The authors are thankful to the Scientific and Analyzing Centre of Tomsk Polytechnic University for the providing the TG, DSC and FTIR measurements, and the Innovation Centre for nanomaterials and Nanotechnologies for the providing the SEM.

References

[1] Karakhanov E A, Maximov A L, Kardasheva Y S, Skorkin V A, Kardashev S V, Predeina V V, Talanova M Yu, Lurie-Luke E, Seeley J A, Cron S L 2010 Copper nanoparticles as active catalysts in hydroxylation of phenol by hydrogen peroxide *Appl. Catal. A- Gen.* **385** 62-72

[2] Yeshchenko O A, Dmitruk I M, Dmitruk A M, Alexeenko A A 2007 Influence of annealing conditions on size and optical properties of copper nanoparticles embedded in silica matrix *Mater. Sci. Eng.: B* **137** 247-254

[3] Tarasov S, Kolubaev A, Belyaev S, Lerner M, Tepper F 2002 Study of friction reduction by nanocopper additives to motor oil *Wear* **252** 63-69

[4] Yabuki A, Arriffin N 2010 Electrical conductivity of copper nanoparticle thin films annealed at low temperature *Thin Solid Films* **518** 7033-7037

[5] Swarnkar R K, Singh S C, Gopal R 2011 Effect of aging on copper nanoparticles synthesized by pulsed laser ablation in water: structural and optical characterizations *Bull. Mater. Sci.* **34** 1363-1369

[6] Xia X P, Xie C S, Cai S Z, Yang Z H, Yang X L 2006 Corrosion characteristics of copper microparticles and copper nanoparticles in distilled water *Corros. Sci.* **48** 3924-3932

[7] Yanase A, Komiyama H 1991 In situ observation of oxidation and reduction of small supported copper particles using optical-absorption and X-ray-diffraction *Surf. Sci.* **248** 11-19.

[8] Liu Z, Bando Y 2003 A novel method for preparing copper nanorods and nanowires *Adv. Mater.* **15** 303-305

[9] Vitulli G, Bernini M, Bertozzi S, Pitzalis E, Salvadori P, Coluccia S, Martra G 2002 Nanoscale copper particles derived from solvated Cu atoms in the activation of molecular oxygen

Chem. Mater. **14** 1183-1186

[10] Dhas N A, Raj C P, Gedanken A 1998 Synthesis, characterization, and properties of metallic copper nanoparticles *Chem. Mater.* **10** 1446-1452

[11] Khanna P K, Gaikwad S, Adhyapak P V, Singh N, Marimuthu R 2007 Synthesis and characterization of copper nanoparticles *Mater. Lett.* **61** 4711-4714

[12] Kwon Y S, Jung Y H, Yavorovsky N A, Ilyin A P, Kim J S 2001 Ultra-fine powder by wire explosion method *Scripta Mater.* **44** 2247-2251

[13] Kotov Yu A 2003 Electric explosion of wires as a method for preparation of nanopowders *J. Nanopart. Res.* **5** 539-550

[14] Kwon Y S, Kim J C, Ilyin A P, Nazarenko O B, Tikhonov D V 2012 Electroexplosive technology of nanopowders production: current status and future prospects *J. Korean Powder Metall. Inst.* **19**(1) 40-48

[15] Kwon Y S, Gromov A A, Ilyin A P, Rim G H 2003 Passivation process for superfine aluminum powders obtained by electrical explosion of wires *Appl. Surf. Sci.* **211** 57-67

[16] Nazarenko O B, Amelkovich Y A, Sechin A I 2014 Characterization of aluminum nanopowders after long-term storage *Appl. Surf. Sci.* **321** 475-480

[17] Korshunov A V, Ill'in A P 2009 Oxidation of copper nanopowders on heating in air *Russ. J. Appl. Chem.* **82** 1164-1171

[18] Zhang Y C, Tang J Y, Wang G L, Zhang M, Hu X Y 2006 Facile synthesis of submicron Cu_2O and CuO crystallites from a solid metalorganic molecular precursor *J. Cryst. Growth.* **294** 278–282

[19] Prakash I, Muralidharan P., Nallamuthu N, Venkateswarlu M, Satyanarayana N 2007 Preparation and characterization of nanocrystallite size cuprous oxide *Mater. Res. Bull.* **42** 1619–1624

[20] Yermakov A Ye, Uimin M A, Galakhov V R, Mysik A A, Koryakova O V, Kharchuk V G, Gaviko V S, Robin S, Neumann M, Vykhodetz V A 2005 Structure and surface states of Cu-O based nanocrystalline powders *J. Metastable and Nanocrystalline Materials* **24-25** 43-48

Ultrasonic inspection of spent nuclear fuel casks

I Bolotina[1], A Bulavinov[2], A Lider[3], D Sednev[4], A Shtaynbreher[5]

[1] Associate Professor of Department of Industrial and Medical Electronics
National Research Tomsk Polytechnic University, Russia
E-mail: bolotina@tpu.ru

[2] Director of GmbH «I-Deal Technologies», Germany
E-mail: andrey.bulavinov@i-deal-technologies.com

[3] Head of Department of General Physics
National Research Tomsk Polytechnic University, Russia
E-mail: lider@tpu.ru

[4] junior scientist of Tomsk Open Laboratory for Material Inspection
National Research Tomsk Polytechnic University, Russia
E-mail: sednev@tpu.ru

[5] Student of Department of Industrial and Medical Electronics
National Research Tomsk Polytechnic University, Russia
E-mail: aas7@tpu.ru

Abstract. The paper describes the main aspects of ultrasonic non-destructive inspection of casks with spent fuel. Digital Focus Array technique is proposed for acoustic imaging of spent nuclear fuel casks' closure weld. Scientific and engineering basics and imaging algorithms of this technique are discussed. Advantages of Digital Focus Array technique in comparison with common techniques of acoustic images processing are revealed.

Introduction

Spent nuclear fuel (SNF) is potentially the most hazardous product of nuclear energy application, it contains up to 98% of total radioactivity which is concentrated in all materials of nuclear fuel cycle. Presence of fissionable materials, hypertoxicity of radionuclies and residual heat release in SNF requires advanced safety measures as well as special technologies of its treatment. 8,500 tons of SNF out of approximately 10,500 tons of annually accumulated SNF in the world is placed in SNF storages, the rest amount is placed for long term storage and reprocessing. International Atomic Energy Agency (IAEA) does not establish any regulations related to methods or terms of storage, however IAEA states that the terms of storage of SNF in dry storage casks can reach a hundred years period. For instance, Nuclear Regulatory Commission (NRC) reached the decision about the possibility of SNF storage in dry casks for the period of 60 years after the reactor fuel uploading. However the problem of long-term storage remains the most important problem for nuclear industry of all countries. It is evident that to perform such long-term storage an effective technique of quality control of constructions is required which would not infringe the integrity of technological processes related to storage and transportation of controlled objects.

1. Ultrasonic Inspection of SNF Casks

The necessity of weld joints inspection and residual life assessment of SNF casks leads to development of new techniques in non-destructive testing [1, 2]. One of the promising inspection techniques is ultrasonic inspection.

However the transducers positioning for ultrasonic inspection represents serious problem in relation to SNF casks. At nuclear power plants manual ultrasonic inspection can be performed only in case of the shutdown situation in order to avoid unacceptable exposure of NDT inspectors. Accessibility and ecological factors do not allow to perform such inspection for SNF casks. Scanning is proposed to be performed by the means of steel rod with the ultrasonic transducer positioned at one of its ends or by the means of robot caterpillar chassis.

In one of the published papers [3], ultrasonic inspection of cask is performed by the means of phased array transducers. Preliminary the cask lid area is divided into two control zones as it is shown in Figure 1. Taking into consideration that the thickness of each zone varies, different focal depth is applied during scanning.

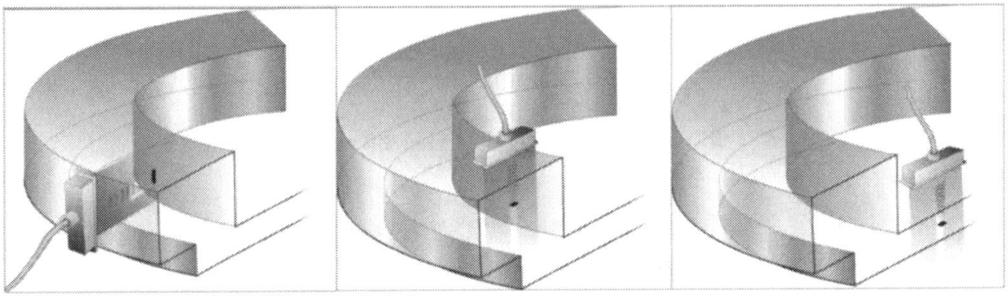

Figure 1. Ultrasonic inspection of cask lid.

In the paper [4] inspection of friction welded joint was performed by the means of ultrasonic testing and radiography (Figure 2). Linear phased array transducer is used for ultrasonic inspection; the transducer is positioned on the cask lid surface. Digital radiography is performed with 9 MeV linear accelerator in combination with linear X-ray detector.

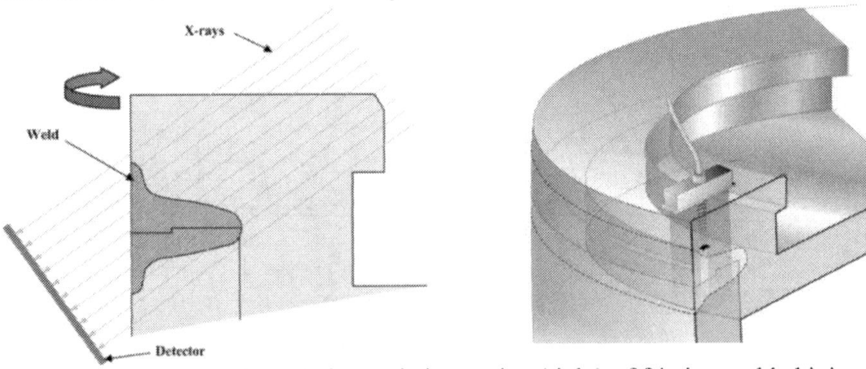

Figure 2. Radiography (left) and ultrasonic inspection (right) of friction welded joint.

2. Digital Focus Array Imaging Visualization Technique

Unfortunately the commercially available image reconstruction algorithms and inspection equipment do not allow to reconstruct 3D images of the controlled object. Reconstruction algorithms applied in X-ray inspection use Radon transform defining varying degrees of X-ray attenuation passed through the controlled object with flaw and flawless controlled object. During ultrasonic inspection almost total ultrasonic wave reflection takes place at boundaries between flaw and flawless area. Due to this, direct application of such algorithm becomes not feasible. Special algorithms for acoustic data treatment need to be developed to obtain 3D image of the controlled object. The algorithms will allow to determine the coordinates of defects locations as well as defects size enabling objective evaluation of product condition and its further service life.

A new technique of measurement and images reconstruction with application of typical multielement transducers was developed at Tomsk polytechnic university in collaboration with Fraunhofer Institute for Nondestructive Testing (IZFP) [5, 6]. Digital Focus Array Technique (DFA) is applied with only one phased array element transmitting a signal, while the rest elements are receiving it. Received time signals are saved. For example, one of the tacts of measurement is sending out *i* element, the elements from 1 to *N* receive the signal, the matrix Aij is formed (Figure 3). After N tacts the matrix is full – it contains time signals of each combination transmitter-receiver.

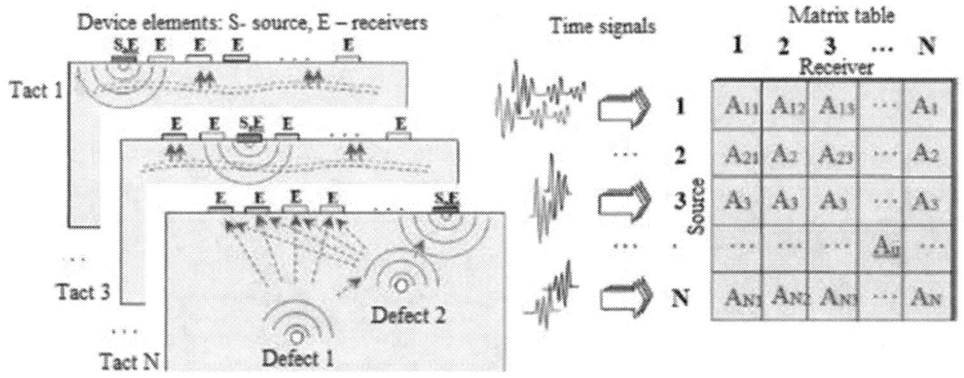

Figure 1. DFA technique principles.

In the basis of the Digital Focus Array concept is the idea that physical superposition of elementary ultrasonic waves formed in material by phased arrays can be generated artificially. Elementary waves can be presented by matrix elements using impulse multielement transducer. Having summed up the signals in relation to desired focusing it becomes evident that sound wave propagation periods from the transmitter to the point in material volume obtained by artificial net A-scan are absolutely equivalent to periods obtained by common transducers.

DFA-reconstruction principle is explained using the linear phased array example. In case of linear array processor the reconstruction area is the surface perpendicular to the controlled object surface. The corresponding image is represented by sector-scan on analogy with the common technique.

The back projection of measured time signals is realised in the following way: sector diagram is divided into separate pixels. For each pixel of the reconstruction area the wave propagation time T_{ij} to and back from the pixel (X_m, Z_m) is calculated considering actual transmitter position (X_i, Z_i) and receivers position (X_j, Z_j) with determined wave propagation speed. Further the amplitude value of wave propagation time is summed up to the already saved value for the pixel (Figure 4). This procedure is repeated for each combination «receiver-transmitter».

Whereas the multielement transducers create divergent acoustic field, separate elements do not provide data about azimuth location of reflector – geometrical location of projected signals A_{ij} answers to hemisphere with $i=j$ or to semiellipse with $i \neq j$ (Figure 5). With increase of data received (number of multielement array transducers) the defect in the material is more visible in reconstructed sector-scan (Figure 5).

Wave propagation time of each wedge element for each volume point can be calculated and recorded in so-called subsidiary table. Thus the process of image reconstruction can be accelerated. Physically realized focusing area and subsequently improved resolution power of multielement transducers techniques in wedges is limited to the area of immediate fields. This means that for components with greater thickness the advantage of the technique is only in the speed of data processing and image reconstruction. Based on the size of the area of study the conclusion can be made that frequency of images receiving and processing can reach one KHz that corresponds to the speed of about one meter per second in measurement automation.

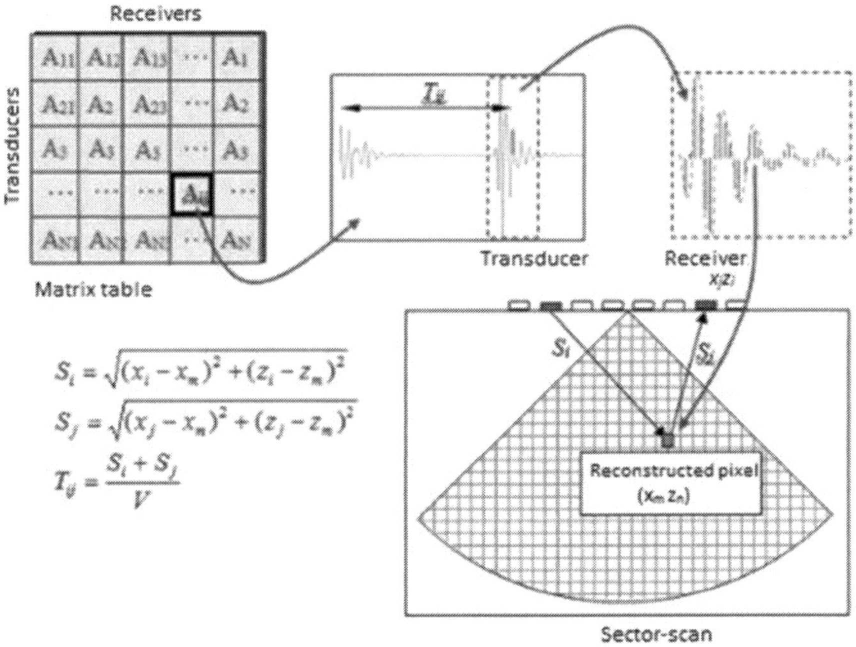

Figure 4. Sector-scans reconstruction principles.

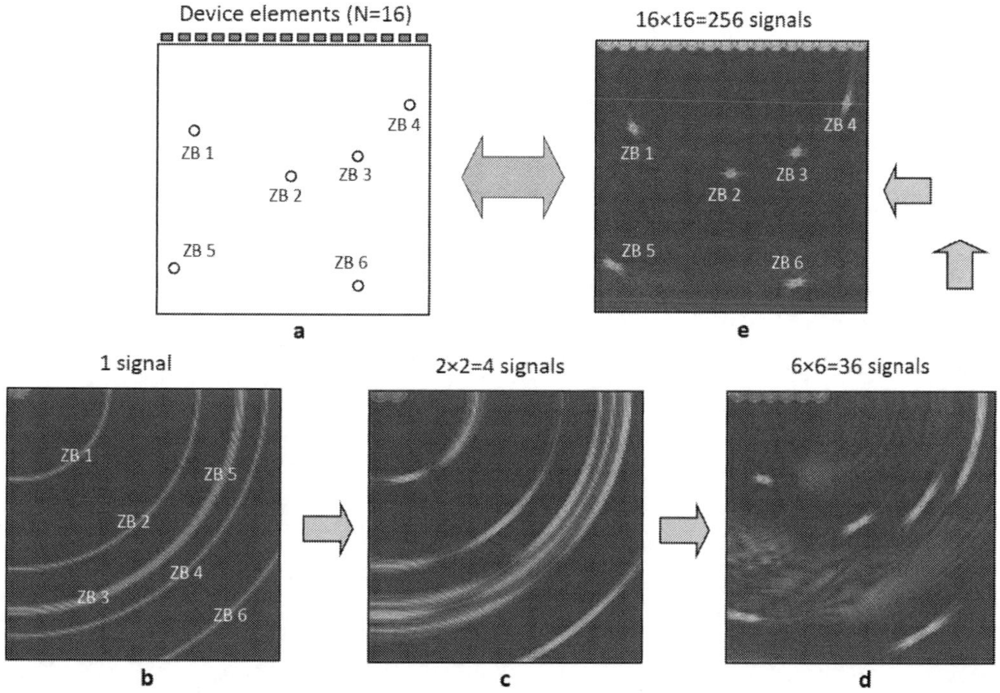

Figure 5. Example of image reconstruction for Sector-scan for 16 element transducer. Times of propagation of waves reflected by six side-drilled holes form the image according to DFA technique.

If 2-D sector diagrams (sector-scans) of neighboring transducer positions are imposed on one another the situation can be improved for materials with greater thickness thus enabling the opportunity to increase the focal area by constructing 1-D synthetic aperture (Figure 6).

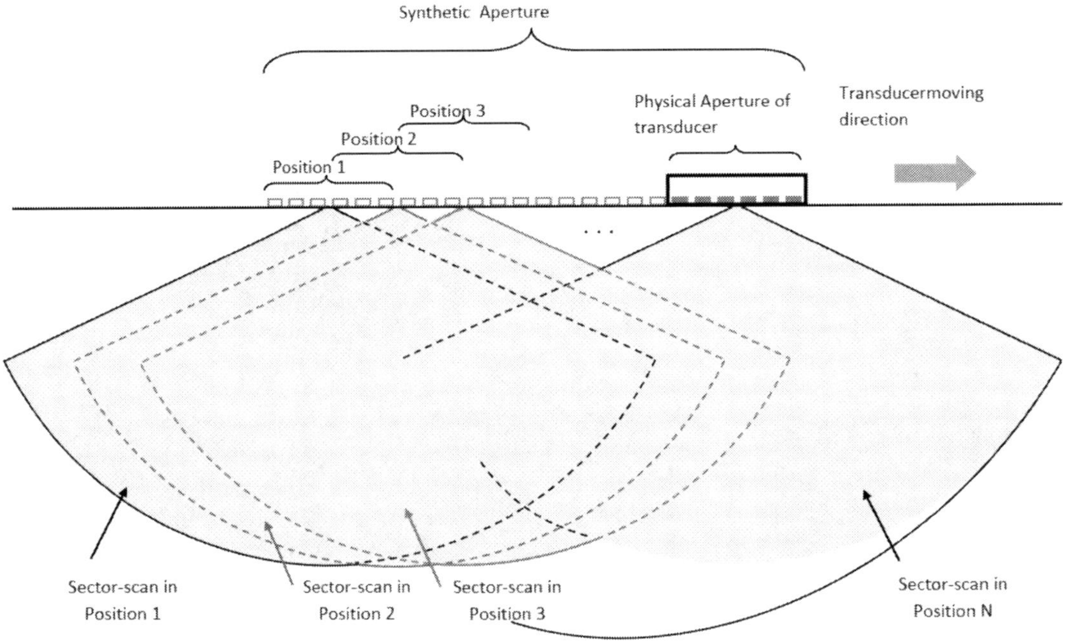

Figure 6. Construction of synthetic aperture.

In contrast with common reconstruction using SAFT algorythm, here not separate time signals are summed up but rather 2-D amplitude fields. Thereby as well as during SAFT-reconstruction physically realized focal area is increased, in which the 2-D data field (compound scan) is built in sectional plane. Among other ultrasonic reconstruction techniques DFA technique demonstrated higher effectiveness, inspection results are notable for high resolution capacity and contrast [7, 8].

3. Conclusion

Application of radiographic, thermographic, eddy current and ultrasonic inspection techniques is known in the world practice. Radiographic inspection involves additional radiation exposure for personnel, as well it does not allow to detect a number of critical defects, for example «kissing bond» defect. Thermographic inspection requires thermal dynamics of an object hence additional heating of an object is required. Eddy current inspection has fundamental limitation in the depth of defect occurrence (about 1 mm). Complexity of received data interpretation and high requirements to the level of operating personnel qualifications are referred to as the main limitations of ultrasonic inspection. Suggested advanced ultrasonic visualization and image reconstruction techniques are free from the limitations mentioned above.

Currently existing common techniques of presenting accosting data are not sufficiently informative. With development of signals digital processing the quality of inspection results became comparable with the level of data detail acquired during radiographic inspection. The most efficient direction of technologies development is considered to be SAFT and Digital Focus Array technique (developed at TPU in collaboration with Fraunhofer Institute for Nondestructive Testing (IZFP)) which represent further development of phased array techniques.

The paper reveals that the suggested new technique of measurement and ultrasonic signals reconstruction - Digital Focus Array (DFA) possesses high details of data in comparison with common methods of acoustic data representation. The technique allows to perform tomographic reconstruction of defects` images in controlled object in real-time mode, ensures high data validity, reliability, reproducibility and high efficiency of ultrasonic nondestructive inspection techniques.

Acknowledgement
Research has been conducted with financial support of Federal Target Programme «Research and Development in Priority Fields of Russian Research and Technological Complex for 2014 - 2020», project unique identifier is RFMEFI57514X0048.

Reference

[1] Lee J-R, Jeong H, Ciang C, Yoon D-J, Lee S-S 2010 *Nuclear Engineering and Design* vol 240 pp 3513–3520
[2] Ardashkin I B, Yakovlev A N, Martyushev N V 2014 *Advanced Materials Research* vol 1040 pp 912–916.
[3] Larsen E, Watkins A, McJunkin T, Pace D, Bitsoi R 2006 *5th International Conference on NDE in Relation to Structural Integrity for Nuclear & Pressurized Components*, San Diego USA pp 929–942
[4] Muller C, Pavlovic M, Takahashi K, Ewert U, Rosenthal M, Brekow G, Bohm R 2007 *6th International Conference on NDE in Relation to Structural Integrity for Nuclear and Pressurized Components* Budapest Hungary pp 1–12
[5] Bulavinov A, Kröning M, Pinchuk R, Reddy K M 2009 Echtzeit-3D-Bildgebung mittels der getakteten Gruppenstrahlertechnik für Ultraschallprüfanwendungen *Ber. der Jahrestagung der Deutschen Gesellschaft für Zerstörungsfreie Prüfung e. V. (DGZfP),* Münster Germany pp 1–12
[6] Bulavinov A, Pinchuk R, Pudovikov S, Reddy K M, Walte F 2010 Industrial application of real-time 3d imaging by sampling phased array *European Conference for Non-destructive Testing* Moscow Russia pp 1–9
[7] Bolotina I O, Djyakina M E, Krening M K, Mokhr F, Reddi K M, Soldatov A I, Zhantlesov E 2013 *Russian Journal of Nondestructive Testing* vol 49 **3** pp 145–158
[8] Bolotina I O, Kryoning K V, Kvasnikov K G, Sednev D A, Sumtsova O V 2014 *Advanced Materials Research* vol 1040 pp 959–964

IOP Conf. Series: Materials Science and Engineering **81** (2015) 012074 doi:10.1088/1757-899X/81/1/012074

Optical laser diffraction transducer for measuring single-wire electric cable eccentricity

M E Fedorov
National Research Tomsk Polytechnic University, 634028, Tomsk, Russia

E-mail: evgeny_fyodorov@list.ru

Abstract. The paper outlines some issues related to optical sensors based on the diffraction method which are used as a part of the devices to control the outer diameter and eccentricity of a single-core electric cable. The conversion function which allows calculation of the diameter of cylindrical objects regardless of their position in the tested area is considered. The calculations are based on the diffraction method to detect shadow boundaries of the object in two-dimensional optical sensors that use a divergent laser beam. The obtained results can be used to develop precise optical instruments to control the outer diameter and eccentricity of the object in a cable industry.

1. Introduction

In-process control of the diameter and eccentricity of the electric cable aims at solving the problem of practical importance to minimize rejects, to improve the quality of the manufactured cable products, to reduce raw material costs in production, and, consequently, to reduce production costs.

A wide range of instruments and devices of different types has been developed to control geometrical sizes of wires and cables. The existing foreign counterparts of similar devices with good metrological characteristics do not fully meet the production requirements. They have the following disadvantages:

- the complexity of the technical solutions, and a high cost of maintenance and purchase;
- the presence of mechanical moving parts reducing the potential reliability;
- a large size in the longitudinal direction, and some one-piece constructions imposing a restriction on their use.

The need to measure the extrusion die in the output (conductor insulation with the temperature of ~ 130 C) and the continuity of the process does not allow using the contact and destructive methods of eccentricity control. Figure 1 shows the insulated conductor cross-section, where the distance from the center of the cable conductor to the center of the cable sheath (line segment e) is the cable eccentricity, and line segments e_x and e_y are the eccentricity projections along the corresponding axes. The values of these characteristics are to be used for the automatic process control.

The eccentricity measuring technique includes an optical two-dimensional system to determine the position of the center, the diameter of the cable outer sheath, and a transformer mutual inductance magnetic sensor to measure the displacement of the cable conductor center. The data processing of optical and electromagnetic channels makes it possible to determine the eccentricity with sufficient accuracy. Figure 2*b* shows a diagram of the eccentricity measuring instrument, where *W1.1–W1.4* are the coils of the magnetic sensor which measures the displacement of the cable conductor for the first channel, *W2.1–W2.4* are the coils for the second channel, *LAZ1* is a laser and *CCD1* is a radiation detector for the first channel of the optical measuring system, and *LAZ2* and *CCD2* are those for the second channel. The performance of the mutual inductance magnetic sensor measuring the position of the cable conductor is investigated in [1– 3].

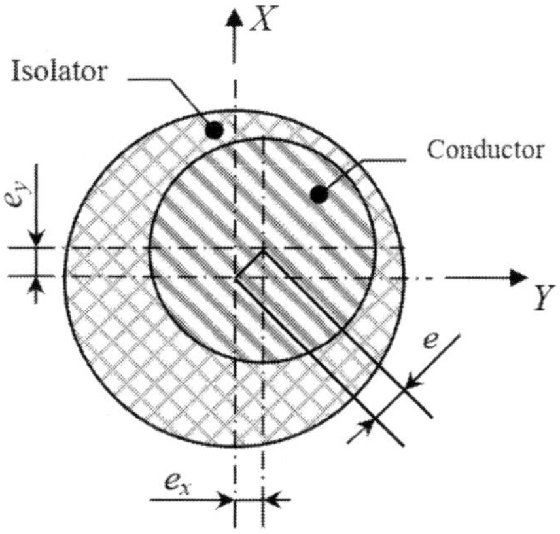

Figure 1. Wire section. Line segment e is the eccentricity under evaluation.

A range of advantages in relation to other optical measuring methods is achieved when the eccentricity and diameter of round wire materials is measured with the help of the laser beam divergence technique [4, 5]. Particularly, the lack of catadioptric optical system and movable optical components essentially simplifies the optical system and design of a primary measuring transducer. Design and production of two-dimensional diameter measuring instruments based on this method, is a promising trend in cable instrument engineering due to their reliability, relative ease of fabrication, and objective adjustment.

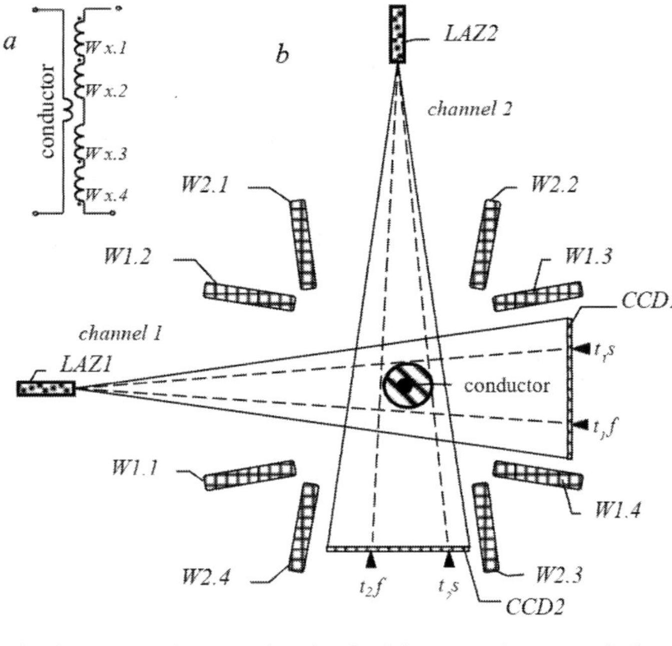

Figure 2. Optical inductive sensor: *a*) connection circuit of the magnetic sensor winding for one of the channels; *b*) the transducer design, where LAZ_1 and $LAZ_2$are point radiation sources; $CCD_1$and CCD_2 are multielement photodetectors for the1[st] and the 2nd measuring channels, respectively; the quantities t_1f, t_1s *and* t_2f, t_2s are the shadow boundaries of the workpiece under evaluation; *W1.1–W1.4* are coils of the magnetic sensor which measures the displacement of the cable conductor for the first channel, *W2.1–W2.4* are those for the second channel.

Laser beam divergence technique for diameter measurement used for long wire materials is based on detection of shadow boundaries of the object by means of multielement linear photodetectors placed in two orthogonal measuring channels. Figure 2 shows a schematic layout of the optical two-dimensional primary measuring transducer which implements this measurement technique. Traces of laser beams emitted by point radiation sources LAZ_1 and LAZ_2 are shown by dashed lines. These laser beams are directed tangentially to the work piece edges and form light–shadow boundaries t_1f, $t1s$ и t_2f, t_2s on the respective multielement photodetectors CCD_1 and CCD_2. This technique and functions of primary measuring data transformation are described in detail in works [6] and [7].

In practical application, an accurate detection of geometrical boundaries of rising and falling edges of a work piece shadows using a multielement photodetector is rather complicated. This is because the slew rate and the shape of boundaries depend on a local lighting of photodetector and a position of the work piece in a plane orthogonal to the photodetector surface. Scratches, dust, dirt and other during-operation defects of optical glass of measuring instruments affect the accuracy of shadow boundary determination. Even though these defects will be taken into account or effectively eliminated, the accuracy of optical instruments is restricted by diffraction effects occurring at the work piece boundaries that results in a blurring effect of a shadow.

In the patent [8], the principle of the shadow boundary determination is described on the basis of the extreme value distribution from the edge of the opaque object. It is a well-known technique that was investigated in the works [9] and [10]. The principle of the shadow boundary determination is widely used in science and technology [11–22]. In particular, it is applied to enhance the accuracy of geometry measurements of various wire materials. In order to improve a resolution of optical transducers based on a laser beam divergence measurement technique, the analysis of the Fresnel diffraction pattern of large-scale objects was carried out by instruments produced by Sikora and Zumbach Companies. However, in the above mentioned literature, the transformation function allowing the accurate mathematical calculation of the boundary position in measuring wire materials with diameters exceeding the wavelength is not described. This fact restricts the application of Fresnel diffraction by optical transducers based on this technique. In addition to the transformation function, the authors present research into the object movements within the gaging zone affecting the diffraction pattern that is very important for the industrial development of measuring devices.

2. Determination technique of the geometrical boundary of shadow based on diffraction.

As shown in Figure 3, the principle of Fresnel diffraction occurs on the boundary of opaque cylindrical objects. Partially the light penetrates into the shadow region while in the illuminated region it forms the system of diffraction minima and maxima, the difference between them monotonically decreases, and the intensity of light goes to the initial illumination I_0. The distance L between the point source and the multielement photodetector depends on the structural properties of the optical transducer and is constant. The distance y may vary depending on a position of the work piece under control.

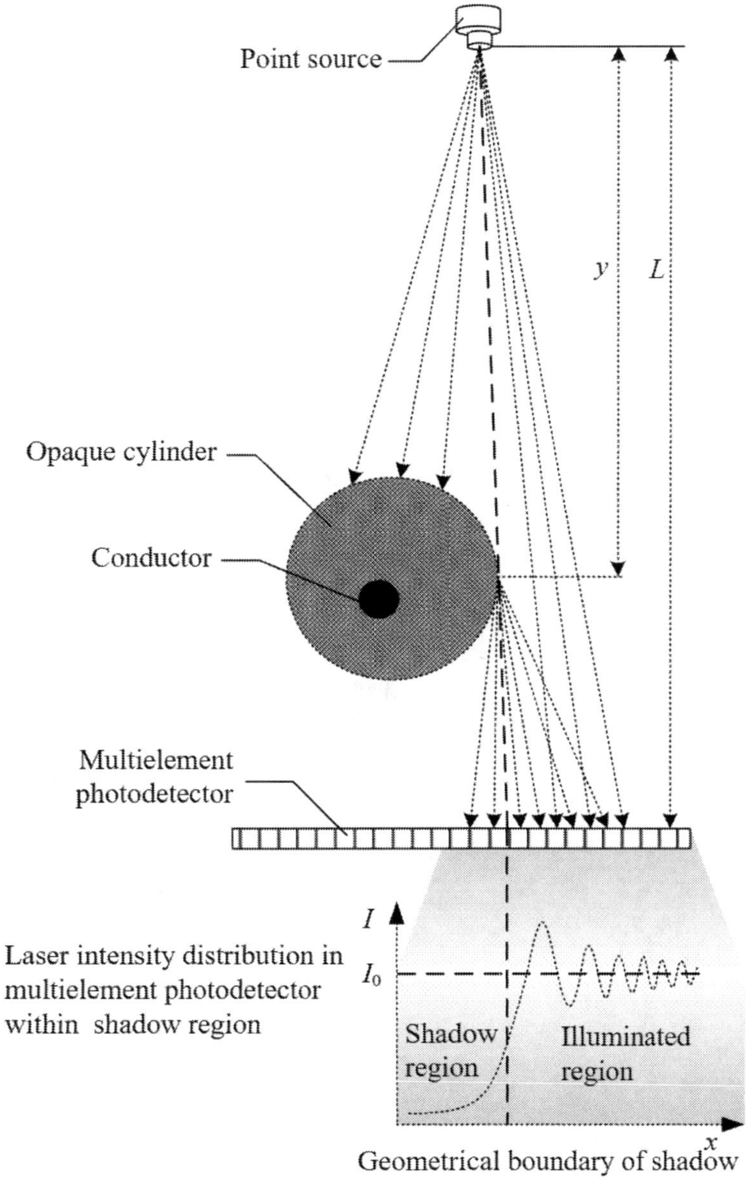

Figure 3. Fresnel diffraction at the boundary of opaque cylinder: I_0–initial illumination; L–distance between the point source and the multielement photodetector; y–distance between the point source and opaque cylinder.

Figure 4 allows the study of diffraction extremum distribution in the vicinity of geometrical boundary. In case the shadow boundary is projected orthogonally to the photodetector plane, the distance X_i from the point X_t to its respective maximum M_i and the distance x_i from the same point X_t to its respective minimum m_i are defined by formulas

$$X_i = \sqrt{\frac{\lambda L(L-y)}{2y}\left(4i+\frac{3}{2}\right)},\quad x_i = \sqrt{\frac{\lambda L(L-y)}{2y}\left(4i+\frac{7}{2}\right)}, \tag{1}$$

where i is the number of the respective maximum or minimum starting from zero; λ is the wavelength of the point source (Fig. 3).

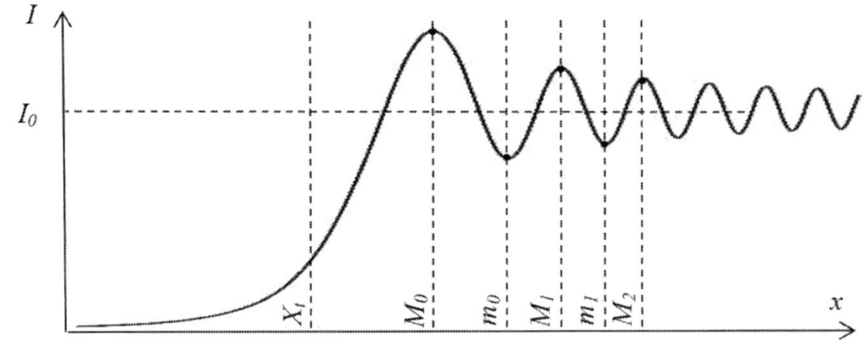

Figure 4.Diffraction extremum distribution in the vicinity of geometrical boundary: X_t– geometrical boundary of shadow; M_0,M_1, M_2–minima of the first, second and third orders respectively; m_0, m_1–minima of the first and second orders respectively.

A position of the boundary X_t on the multielement photodetector is the original value for a calculation of diameter using method presented in [6]. Having determined the distance between the first two maxima (interval M_0M_1) or minima (interval m_0m_1) shown in Figure 4, the boundary X_t can be found. Since factor $\sqrt{\lambda L(L-y)/2y}$ in (1) is fixed for all extreme values, distribution of these values will then be defined by factors $\sqrt{4i+3/2}$ and $\sqrt{4i+7/2}$ for maxima and minima respectively. Thus, the distance between the extreme values can change proportionally depending on parameters of L and y, however, correlation between them is being constant. In particular, the interval X_tM_0 correlates with the interval M_0M_1 with fixed coefficient 1,093, while a correlation between intervals X_tm_0 and m_0m_1 equals to 2,154. Thus, the formulas below can be derived to find coordinates of geometrical boundaries of rising and falling edges:

$$X_{ft} = 1{,}093(M_0 - M_1) + M_0 = 2{,}154(m_0 - m_1) + m_0$$
$$X_{st} = M_0 - 1{,}093(M_1 - M_0) = m_0 - 2{,}154(m_1 - m_0), \quad (2)$$

where X_{ft} and X_{st} are positions of geometrical boundaries of rising and falling edges; M_0, M_1, m_0, m_1 are the extreme values of diffraction distribution.

3. Determination of the shadow boundary position when the object under evaluation is placed in an arbitrary position

The test installation was designed to conduct the experiment. The block diagram of the test installation is shown in Figure 5.

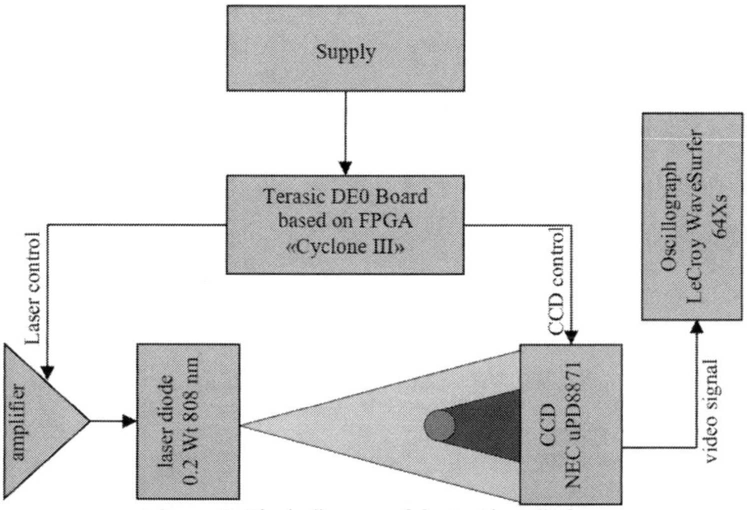

Figure 5. Block diagram of the test installation

The angle measurement was provided by the mechanical dial with 1′angle-error detection. In the centre of the mechanical dial a board with the multielement photodetector was fixed. The cylindrical object~4 mm diameter was also mounted in the centre next to the board. The linear CCD (charge-coupled device) NEC μPD8871 was used as a multielement photodetector. It has 3 rows of 10680 pixels and 4×4 μm photocell size. CCD scanning rate and exposure time were 1 kHz and about 50 ms respectively. Diode laser HLDH-808-B20001 with parameters of 808 nm wave length, 0,2 Wt optical power, and 42°beam divergence angle, was fixed on a hanger mounted to the dial. A driving pulse generation for the board with the multielement photodetector and laser emitter was performed by the Terasic DE0 Board based on FPGA Cyclone III. FPGA Cyclone III is used to accurately CCD clock and control with 20 MHz frequency observing all intervals in compliance with its datasheet. Extremum positions for the diffraction pattern were registered by LeCroy WaveSurfer 64Xs Digital Oscilloscope. The test installation was supplied from the power source.

Figure 6 presents the oscillogram of the work piece scanned by a laser beam. All the notations used in this Figure are taken from Figure 4.

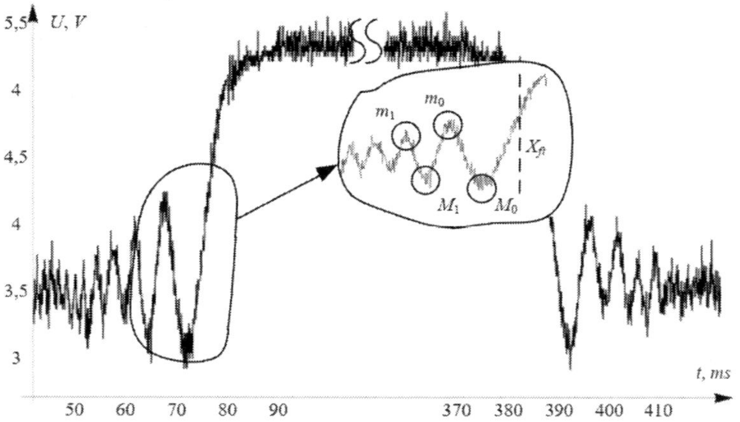

Figure 6. Oscillogram of the work piece with diffraction effects occurred at the boundaries.

Figure 7 describes the experimental dependence between coefficients k_M and k_m (experimentally equal to 2 and 1,1 respectively) and the workpiece movement within the gaging zone normal to the multielement photodetector. As shown in Figure 7, their values are found to be in good agreement with theoretical results. These values are constant within the wide range of the work piece movements that is true for formulas (2) in case when the diffraction pattern is formed by an incident rim ray normal to the surface of multielement photodetector.

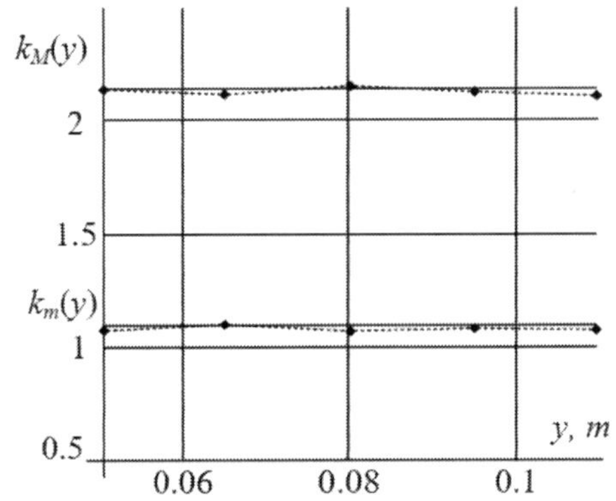

Figure 7. Dependence between scale coefficients of diffraction and the work piece movements within the gaging zone: $k_M(y)$ and $k_m(y)$ – coefficients for the first two maxima respectively.

However, in real instruments a work piece can move not only along the axis normal to the photodetector plane but also in any other direction. This results in the fact that rim rays incident at an angle α different from 90°, and the geometry of the optical system including parameters L and y, is transformed to parameters L' and y'. Diffraction extremum distributionis also transformed from X_t, M_i, m_i to X_t', M_i', m_i' states depending on the incident angle α, where X is the geometrical boundary of shadow; M and m are maxima and minima of the i-th order as shown in Figure 8.

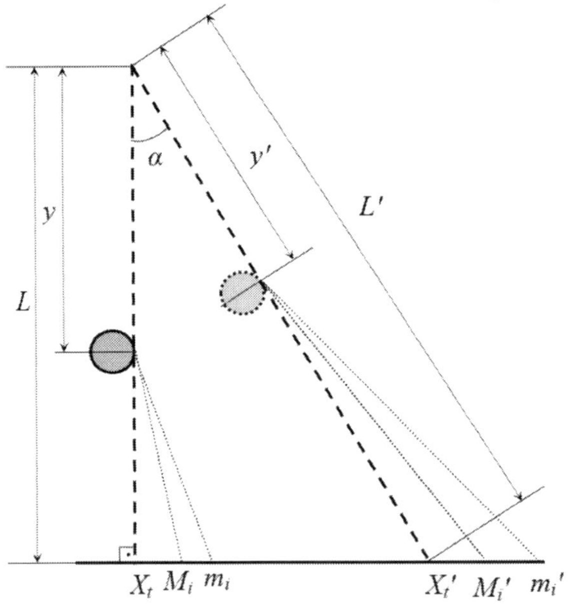

Figure 8. Formation of diffraction pattern on the multielement photodetector at the angle of incidence α.

To validate formulas (2) in case of oblique incidence of rim rays, it is necessary to clarify the manner in which distances between the principle extreme values of the diffraction pattern correlate depending on the angle of incidence. Distances between the first and the second order and between the second and the third order maxima of diffraction pattern were taken as test distances that correspond to intervals M_0M_1 and M_1M_2 shown in Figure 4.

Diagrams shown in Figure 9*a*, demonstrate the empirical relation between intervals M_0M_1 and M_1M_2 and the incident angle α. Zero corresponds to a normal incidence of a rim ray. As it was assumed, distances between the extreme values increase with the increase of beam deflection from the normal to the photodetector plane. As shown in Figure 9*b*, the dependence diagram determines a proportionality of a distance change between these extreme values. This diagram demonstrates how the coefficient $k(\alpha)$affects the correlation M_0M_1/M_1M_2depending on the angle of incidence. Fig. 9, *b* shows that coefficient $k(\alpha)$ (experimentally equal to 1,35) keeps constant under a wide range of incident angle that proves a proportional change of distances between the extreme values of diffraction pattern. This allows formulas (2) to be used for an accurate detection of the geometrical boundary of the work piece shadow in a wide range of its movements.

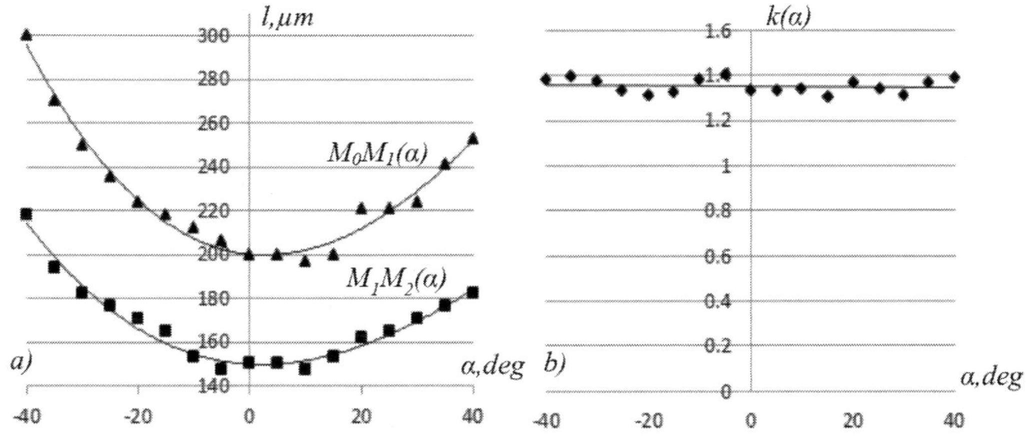

Figure 9.Diffraction extreme values depending on the rim ray angle of incidence: *a)* M_0M_1 and M_1M_2 relation depending on the angle of incidence α; *b)* coefficient *k* and incident angle α dependence.

4. Discussion

Techniques suggested in this work were tested at various intensities of laser radiation similar to a real operation of a measuring instrument. Results of investigation are given in Table 1 as compared to those obtained for a classical amplitude detector which detects shadow position by the slew rate and the shape of boundaries on the CCD picture.

Table 1 Comparison of detectors

I	Amplitude detector		Diffraction detector	
	Rising and falling edges (pixel)	Detection error (μm)	Rising and falling edges (pixel)	Detection error (μm)
$0.95 \cdot I_0$	1875.2	-10.4	1876.7	-0.4
I_0	1877.8	0	1876.8	0
$1.05 \cdot I_0$	1880.7	+11.6	1877.0	+0.8

The detection error for the shadow boundary as shown in Figure 6, comes to 10 μm at laser brightness variation ±5% from a certain initial value I_0. During operation, this error can increase multiply due to contamination of optical elements, detection errors of rising and falling edges being summed up for the diameter calculation. In suggested technique of the boundary detection (by diffraction pattern extreme values) the error is around 1 μm at the similar flare brightness. This provides high metrological characteristics of measuring instruments regardless of the optic emitter drift characteristics and purity of optics instruments.

Suggested techniques were approved on many opaque cylindrical objects with diameters ranging from 0,5 to 40 mm and made of different materials such as polypropylene, polyethylene, polyvinylchloride, rubber, metals, etc. Semiconductor diodes 808 nm length and 0,2-0,5 Wt energy were used in this study. As a rule, they possess different beam divergence along different symmetry planes, in particular $\Theta// \approx 8 \div 11°$, $\Theta\perp \approx 39 \div 48°$. In laser beam divergence technique for diameter measurement, only semiplane $\Theta\perp$ is used to provide a flare of the entire gaging zone. Therefore, other laser positions and, consequently, differences in the light beam polarization are not presented in this paper.

Depending on a configuration of the optical system, suggested techniques provide resolution for a single diameter measurement within 2÷3 μm range allowing for optical magnification of a laser-beam-divergence optical transducer (Fig. 2). Further mathematical processing of obtained data in conformance with methodology described in works [6, 7]. In the work [23] there are also considered statistical methods for increasing of resolution up to 1 μm for optical measuring devices applied to control cylindrical extended objects such as cables, wires, cords etc.

5. Conclusions

The paper shows the technique of measuring the diameter on the basis of light diffraction at the object boundary. The expression (2) to determine the exact position of the shadow geometric boundary of the object without the amplitude analysis of the image front has been obtained. The conversion functions allow determination of the shadow boundary regardless of the spatial position of the object under evaluation. The obtained results enable a precise control of the inner diameter of extended objects and eccentricity of a single cable if the described optical transducer is used together with the inductive transducer investigated in [1–3].

REFERENCES

[1] Goldshtein A.E. and Fedorov E.M. A Mutually Inductive Measuring Transducer of Transverse Displacements of a Rectilinear Conductor *Russian journal of nondestructive testing* Vol. 46 **6** 2010.

[2] Patent RF № 2010107490/28; pending 01.03.10; published 27.06.11, bulletin № 18, «Induktivno-opticheskij preobrazovatel' izmeritelja jekscentrichnosti jelektricheskogo kabelja» / A. E. Gol'dshtejn, V. V. Red'ko, E. M. Fedorov; Tomskij politehnicheskij universitet (TPU); Nauchno-proizvodstvennoe ob'edinenie "Redvill".

[3] Fedorov E.M., Gol'dshtejn A.E., Svendrovskij A.R., Red'ko V.V. «Izmeritel' diametra i jekscentrichnosti jelektricheskogo kabelja na osnove induktivnoopticheskogo metoda» *Izvestija* TPU 2010 **2**

[4] C.J Tay, S.L Toh, H.M Shang Time delay and integration imaging for internal profile inspection, *Optics & Laser Technology*, Volume 30 *8* 1998, Pp. 459–465

[5] Jinhui Lan, Jian Li, Guangda Hu, Yiliang Zeng Distance estimation using a panoramic sensor based on a novel spherical optical flow algorithm, *Optics & Laser Technology*, Volume 45, 2013, Pp.168–176

[6] Svendrovskii A.R., Raschet diametra v beskontaktnykh dvukhkoordinatnykh izmeritelyakh [Measurement of diameter by non-contact two-dimensional detectors]. Proc. 1stAll-Rus. Sci. Conf. 'Scientific and engineering problems of instrument engineering'. Tomsk, 2005. Pp. 31–33. (rus)

[7] Fedorov E.M., Edlichko A.A. Vychislenie geometricheskikh parametrov dvukhkoordinatnykh izmeritelei diametra protyazhennykh izdelii [Calculation of geometrical parameters of two-dimensional instruments used for measuring diameter of long wire materials] *Bulletin of the Tomsk Polytechnic University*, 2008. No. 2. (rus)
[8] Patent N EP 0924493 B1 «Measurement of diameter using diffraction borders and electronic soiling correction». Adrian Beining, Werner Dr.-Ing. Blohm, HaraldSikora, 2002.
[9] Sommerfeld A., Optics (German edition 1950, English translation: Academic Press, 1964), sec 37, 38.
[10] Born and E. Wolf, Principles of Optics, 3rd Edition Pergamon Press, 1965.
[11] Toenshoff, H. K., Tuennermann, A., Korthals, J., SPIE, Use of Fresnel diffraction for the measurement of rotational symmetrical workpieces. Proceedings SPIE The International Society For Optical Engineering, 3784, 1999, Pp. 334-343.
[12] Chi-Tang Li, James V. Tietz, Improved accuracy of the laser diffraction technique for diameter measurement of small fibres. *Journal of Materials Science*, 1990, Volume 25 **11** Pp. 4694-4698.
[13] Khodier S.A., Measurement of wire diameter by optical diffraction. Original Research Article Optics & Laser Technology, Volume 36, **1**, 2004, Pp. 63–67.
[14] Durgin G.D. , The Practical Behavior of Various Edge-Diffraction Formulas. *IEEE Antennas and Propagation Magazine*, Volume 51, **3**, 2009, Pp. 24–35.
[15] George S., Straight edge diffraction using a laser. *Physics Education*, Volume 7, **6**, 1972, Pp. 349–352
[16] Jinhuan Li, Osami Sasaki, and Takamasa Suzuki, Measurement of diameter of metal cylinders using a sinusoidally vibrating interference pattern. Optics Communications, Volume 260, Issue 2, 2006, Pp. 398-402.
[17] R. Jabłoński, J. Mąkowski Measurement of cylinder diameter by laser scanning, Recent Advances in Mechatronics 2007, pp 596-600
[18] Y. Chugui,V. Bazin,L. Finogenov,S. Makarov,A. Verkhogliad Optical electronic measuring systems and laser technologies for scientific and industrial applications, International Symposium on Instrumentation and Control Technology No6, Beijing , CHINE (2006) 2006 , vol. 6357 (2)[Note(s) : 2 vol.,] (13 ref.) **ISBN 0-8194-6452-X** ; 978-0-8194-6452-1
[19] Ryszard Jablonski, Pawel Fotowicz New generation of laser mike, *Proc. SPIE 4420, Laser Metrology for Precision Measurement and Inspection in Industry*, 91 (September 11, 2001); doi:10.1117/12.439198
[20] Smith S.W. "Chapter 8: The Discrete Fourier Transform". The Scientist and Engineer's Guide to Digital Signal Processing (Second ed.). San Diego, Calif.: California Technical Publishing. **ISBN 0-9660176-3-3.1999.**
[21] Luis Miguel Sanchez-Brea, Juan Carlos Martinez-Anton, Eusebio Bernabeu Effect of the refraction index in the diameter estimation of thin metallic wires, *Proc. SPIE 5858, Nano- and Micro-Metrology*, 585819 (August 26, 2005); doi:10.1117/12.612652
[22] Lars Benckert, Lars Forsberg, and Nils-Erik Molin, Fresnel diffraction of a Gaussian laser beam by polished metal cylinders, *Applied Optics*, Vol. 29, **3**, pp. 416-421 (1990)
[23] Yury A. Chursin, Evgeny M. Fedorov "Methods of resolution enhancement of laser diameter measuring instruments" *Optics & Laser Technology*, Volume 67, 2015, Pages 86-92

Study of adsorption process of iron colloid substances on activated carbon by ultrasound

K I Machekhina*, L N Shiyan, T A Yurmazova and DA Voyno

Department of General and Inorganic Chemistry, Institute of High Technology Physics, Tomsk Polytechnic University, 634050, Tomsk, Russia.

*E-mail: mauthksu@yandex.ru

Abstract. The paper reports on the adsorption of iron colloid substances on activated carbon (PAC) Norit SA UF with using ultrasound. It is found that time of adsorption is equal to three hours. High-frequency electrical oscillation is 35 kHz. The adsorption capacity of activated carbon was determined and it is equal to about 0.25 mg iron colloid substances /mg PAC. The iron colloid substances size ranging from 30 to 360 nm was determined. The zeta potential of iron colloid substances which consists of iron (III) hydroxide, silicon compounds and natural organic substances is about (-38mV). The process of destruction iron colloid substances occurs with subsequent formation of a precipitate in the form of $Fe(OH)_3$ as a result of the removal of organic substances from the model solution.

Keyword: iron colloid substances, adsorption, activated carbon, ultrasound

1. Introduction

The present work relates to the problem of high-quality drinking water supply using processes of adsorption on activated carbon with using ultrasound with the oscillation frequency 35 kHz. In papers [1, 2] it is shown that iron containing groundwater of Western Siberian can be divided into two types. The first type is the water with a high concentration of hydrocarbonate where iron presents in the form of hydrocarbonate iron. The second type is water where iron presents partly in the form of hydrocarbonate iron and partly as a complex with dissolved organic substances. The classical water treatment which includes such step as the aeration, sedimentation and filtration is successfully applied for producing potable water from the water of the first type [2]. Using traditional methods of water treatment for the second type is not always allowed to provide the necessary quality of water. This is due to the ability to form iron colloid solutions during the oxidation [3]. These solutions are stable for a long time. This complicates the water treatment process and significantly reduces the performance of the facilities used in [1].

Currently, many research groups engaged in the study of the processes of absorption for the removal colloid substances [4, 5]. In addition there is work studying the adsorption processes of organic substances on activated carbon [6, 7]. Therefore, this work is relevant.

The objective of the work is to investigate the processes of adsorption for the removal iron colloid substances on activated carbon with using ultrasound.

2. Experimental approach

For synthesis and analytical study, reagents $FeSO_4{\bullet}7H_2O$, $Na_2SiO_3{\bullet}9H_2O$ and distilled water were used. Organic substances for experiments were extracted from peat bogs of the Tomsk region. The concentration of organic substances was determined using an analyzer of dissolved organic carbon (DOC) with subsequent IR-analysis of CO_2 evolved. The molar mass of organic substances was determined by the gel-chromatography method [8]. The samples contained fractions of 200–20000 Da.

All model solutions were synthesized at room temperature. A shaker GFL 3005 (Germany) with a shaking frequency 100 rounds / min was used for mixing.

The iron and silicon concentrations in the solution were determined using an ICP-OES plasma optical emission spectrometer (Varian, USA). The pH value was measured using a WTW Multiline P4 multifunctional device (WTW GmbH, Germany). The colloid particle size distribution and the ζ-potential in the model solutions were determined using a Zetasizer Nano ZS analyzer (Malvern Instruments, UK), which allows measuring the particle size in the range 0.6…6000 nm. The particle size was determined by dynamic light dispersion (at a dispersion angle of 173°). The measurement result was in the form of numerical particle distribution s $\varphi(r)=dN/dr$. The mode δ_m was found from the maximum of this distribution and was taken as an average particle size because the distribution functions approximated normal log functions.

For the adsorption experiments activated carbon Norit SA UF was used, which was produced by Norit, table 1. The specified sorbent has high kinetic characteristics due to the superthin structure of its particles.

Table 1. Specification of the sorbate Norit SA UF

Characteristic	Value
Produce	Norit
Name	Norit SA UF
Total surface area m^2/g	1200
Particles size,μm	7
Apparent density, kg/m^3	160
Moisture, mass.,%	5

The ultrasonic bath (PSB-4035-05, Russia) was used to power source. The operating frequency was 35 ± 10 kHz. The power supply was the ultrasonic generator that connects to electricity 220 V and 50-60 Hz. Amplitude modulation is up to 90%.

3. Results and discussion

Adsorption is the accumulation of atoms or molecules on the surface of a material. The substance on whose surface there is an adsorption, is referred to as adsorbent, and the substance which is adsorbed as adsorbate. Forces of interaction between adsorbent and adsorbate, determining adsorption, are various, and usually consider two extreme cases when adsorption is characterized by physical or chemical interactions: so-called physical and chemical adsorption [9].

The equation of adsorption shows functional dependence of the following kind:

$$f(q,C,T)=0 \tag{1}$$

where q is the amount of adsorbed substance (in kmol), designed on $1m^2$ adsorbent surfaces (or on 1 kg adsorbent); C is the equilibrium concentration in the solution; T is the temperature.

If the temperature is constant, the adsorbed amount of substance is a function of the amount of adsorbent and the concentration of adsorbate. This equation is referred to as adsorption isotherm.

There is no general theory which correctly enough would describe all kinds of adsorption on various adsorbents and different interfaces. Common theories are the monomolecular adsorption theory from Langmuir and the Freundlich isotherm [9].

The Langmuir isotherm is described by the following equation:

$$q= q_m (K_l C/(1+K_l C)) \tag{2}$$

where q is the amount of the substance adsorbed on 1 m^2 of a surface (or 1g) adsorbent (specific adsorption of a substance), g/m^2 or g/g; q_m is the maximum amount of the substance adsorbed 1 m^2 of a surface (or 1g) adsorbent in g/m^2 or g/g; C is the equilibrium concentration of the adsorbent in the gas phase or the solution; K_l is the Langmuir adsorption coefficient which depends on the structure of the adsorbed molecules.

Concerning adsorption from water solutions, the Langmuir isotherm suitable for the adsorption of big molecules, like humic substances which cause the color of natural waters [10].

During adsorption two kinds of intermolecular interactions compete: hydration of molecules of the dissolved substance, i.e. their interaction with molecules of water in a solution, and interaction of molecules of adsorbed substance with atoms of a surface of a firm body. The energy difference of these two processes also represents energy from which the substance taken from a solution is kept on a surface shipped in a solution of eluate. The strongest adsorptive interaction is observed if in the structure of the molecules there are the double bonds formed with participation of π - electrons. Therefore, adsorption of aromatic compounds is much better than of aliphatic substances.

Adsorption decreases at increase in molecules of number hydroxide groups possessing the big energy of hydration due to hydrogen bonds. Adsorption also decreases with the presence of electric charge focusing around of dipoles of water.

The model solution has been prepared with a technique [11]. Different properties of the solution have been investigated: pH-value and electroconductivity were analyzed with a multipurpose device WTW Miltiline P4, particle-sizing interferometry and measurement of ζ -potential in solution have been performed. The basic parameters are shown in the table 2.

Table 2. Comparison of different properties of the model solution and groundwater

Parameter	Model solution	Groundwater
pH	6.5	6.0…7.0
Electroconductivity, μS/cm	205	-
ζ- potential, mV	-21...-32	-20…-120
Concentration of organic substances (DOC), mg/L	2.43	2.0…14.0
Concentration of iron, mg/L	6.92	1.0…25.0
Concentration of silicium, mg/L	24.13	10.0…28.0

Table 2 shows that the model solution corresponds to real groundwaters of northern regions of Tomsk area [12]. Hence, it could be used as model system for the investigation of adsorption processes as well as ultra and nano filtration in the experiments.

Figure 1 shows a distribution of particle size in the model solution.

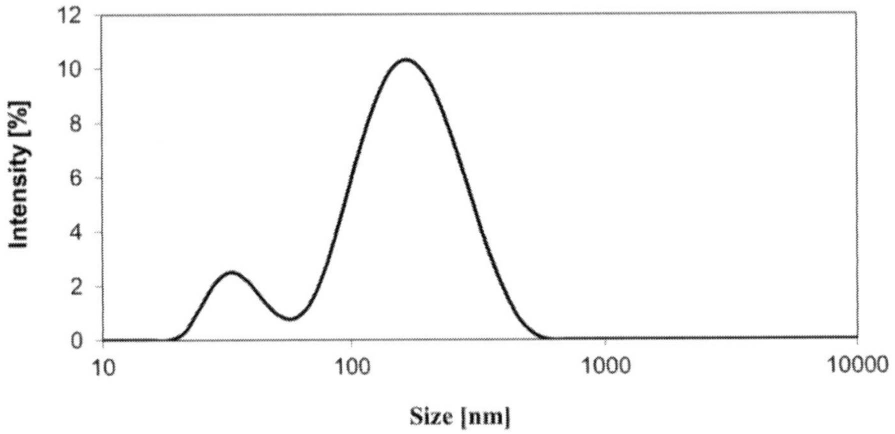

Figure 1. Size distribution of the colloid particles in the model solution.

Figure 1 shows that the solution contains basically two types of particles (maxima at 33 and 165 nanometers).

Adsorption experiment consisted of two basic parts. The first part is analysis of the adsorption kinetics and second one is analysis of the adsorption isotherm. Activated carbon was used as adsorbent for the analysis of the adsorption kinetics. The mass of activated carbon was 12.5 mg in 250 ml solution (50mg/L). The activated carbon (m=12.5mg) was put into a flask (volume 250 ml) filled with model solution. After that the model solution with activated carbon was shaked with using ultrasound. The flasks were removed after 15, 30, 60, 120, 180 and 1440 minutes after the beginning of shaking. After that solutions were filtrated through a membrane with a pore size of 0.4μm (Millpore, USA).

Concentrations of DOC, Fe and Si were measured in the solution after filtration. The results are presented in figure 2.

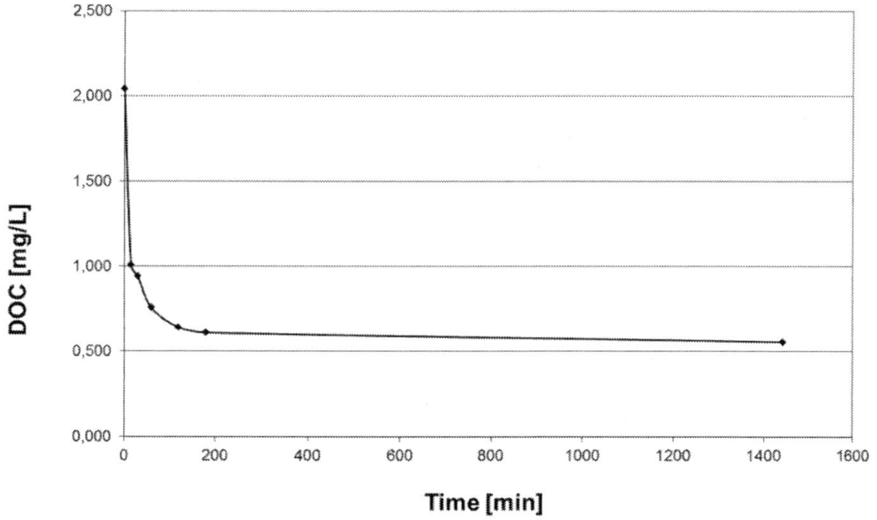

Figure 2. Adoption kinetics of the model solution on activated carbon with ultrasound.

Figure 2 shows that silicium and iron practically did not adsorb, but the concentration of organic substances decreased from 2.24 mg/L to 0.6mg/L.

The adsorption process without sonication proceeds less efficiently. Changes in the concentration of DOC are presented in Figure 3.

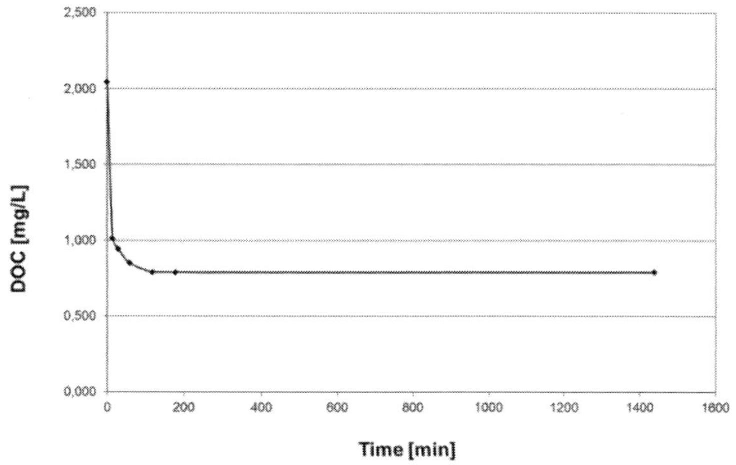

Figure 3. Adoption kinetics of the model solution on activated carbon without ultrasound.

Figure 3 shows that the concentration of organic substances decreased from 2.24 mg/L to 0.79 mg/L. This is due to the fact that the rearrangement of the organic molecules and their division into smaller occurs under the action of ultrasound. This process dominates the desorption of organic molecules from activated carbon. Thus three hours were found as minimum to achieve an adsorption-equilibrium and were hence chosen as adsorption time for the determination of the isotherm. Concentration of activated carbon remained 50 mg/L, concentration of iron and silicium in initial solutions remained at 6.73 mg/L for iron and 25.61 mg/L for silicium. The concentration of DOC increased from 1.2 mg/L to 30.1 mg/L.

The Langmuir model was applied for the calculation of the adsorption isotherm. For this purpose eq. 3. was used in the form formula:

$$\frac{C}{q} = \frac{C}{q_m} + \frac{1}{q_m \cdot K_l} \qquad (3)$$

It was not possible to fit the entire data set to the Langmuir model. For the four points of highest initial concentrations. K_l and q_m were determined, and K_l is equal to 0.26 and q_m is equal 0.25mg/mg. The bad results for the fitting of the whole data set might be due to experimental mistakes in the lower concentration range. Figure 4 shows the adsorption isotherm.

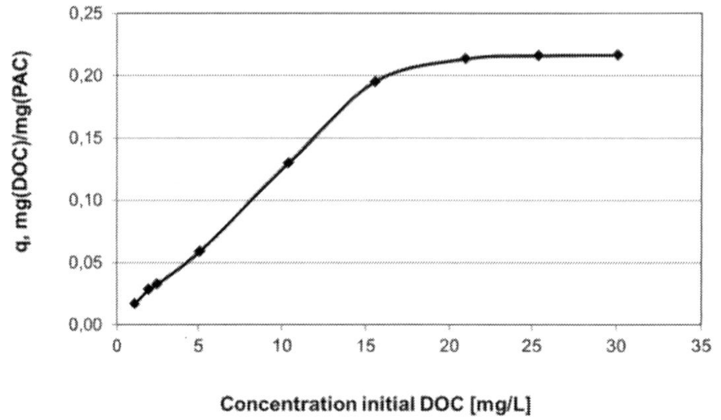

Figure 4. The adsorption isotherm of model solution on activated carbon Norit SA UF.

The maximum adsorption capacity of activated carbon to organic substances is 0.25 mg / mg. Thus, the destruction of iron colloid substances occurs due to the removal of organic substances from the model colloid solution through the adsorption process on activated carbon with formation of a precipitate in the form of $Fe(OH)_3$.

4. Conclusion

1. Experimental modelling was performed to study the chemical properties of colloid solutions containing iron ions, dissolved organic substances, and silicon ions.
2. The iron hydroxide nanoparticles size is in the range from 30 to 360 nm and characterizes the steady state of the colloid system.
3. The adsorption capacity of activated carbon was determined and it is equal to about 0.25 mg iron colloid substances /mg PAC.

Acknowledgements
The study was supported by National Research Tomsk Polytechnic University, the project No. 7.1326.2014.

References

[1] Serikov L V, Tropina E A, Shiyan L N, Frimmel F H, Metreveli G, Delay M 2009 *Journal for Soils and Sediments* **9** 103–110
[2] Machekhina K I, Shiyan L N, Tropina E N, Voyno D A 2012 *The 7th International forum on strategic technology (Tomsk)* vol 1 (Tomsk: Russia/Tomsk Polytechnic University) p 10-14
[3] Shiyan L N, Machekhina K I , Tropina E A, Gryaznova E N, An V V 2014 *Advanced Material Research* **872** 237–240
[4] Vincent B 1974 *Adv. Colloid Interface Sci.* **4** 193
[5] Pincus P A, Sandrof C J, Witten T A 1984 *J. Physique* **45** 725–729
[6] Saravia F, Frimmel F H 2005 *Proc. Of 10th Aachen Membrane Colloquim (Aachen)* vol 1 (Aachen: Germany/ American Elsevier) p 315–320

IOP Conf. Series: Materials Science and Engineering **81** (2015) 012075 doi:10.1088/1757-899X/81/1/012075

[7] Campos C, Schimmoller L, Marinas B J, Snoeyink V L, Baudin I, Laine J.-M. 2000 *Journal of American Water Works Association* **92** 69–83

[8] Perminova I V, Frimmel F H, Kudryavtsev A V, Kulikova N A, Abbt-Braun G, Hesse S, Petrosyan V S 2003 *Environ. Sci. Technol.* **37** 2477–2485

[9] Paul C H, Raj R 1997 *Principles of Colloid and Surface Chemistry* (New York: Marcel Dekker)

[10] Voyno D A, Machekhina K I, Shiyan L N 2014 *Advanced Materials Research* **971-973** 266–269

[11] Shiyan L N, Tropina E A, Machekhina K I , Gryaznova E N, An V V 2014 *SpringerPlus*. **3** 260

[12] Machekhina K I, Shiyan L N 2014 *Advanced Materials Research* **1040** 342–346

Investigation of the mechanism of microplasma impact on iron and aluminum load using solutions of organic substances

G L Lobanova[1], T A Yurmazova[2] , L N Shiyan*[2], D A Voyno[2]

[1]Laboratory 12, Institute of High Technology Physics, Tomsk Polytechnic University, Tomsk, Russia
[2]Department of General and Inorganic Chemistry, Institute of High Technology Physics, Tomsk Polytechnic University, 634050, Tomsk, Russia.

*E-mail: lshiyan@rambler.ru

Abstract. The paper reports on the study of mechanism of electroeffects on iron and aluminum and pellets with using solutions of organic substances. Methylene blue solution, furacilin and eosin were used. It is observed the reactions of the pulse at the time and after switching off the voltage source. It is shown that there are two developing process in the conditions studied. The first process depends on material of electrodes and pulse parameters. The second process occurs spontaneously and it is determined by the redox reaction and sorption processes. The products of electrode erosion and active particles react in the redox reactions. Active particles are formed in solution by the action of pulsed electric discharge in water. The highest efficiency of the process was demonstrated on an iron pellets.
Key words: *pulsed electric discharge erosion, aluminum, iron, an aqueous solution of methylene blue, eosin furatsilin.*

1. Introduction

The using of pulse electric discharge in heterogeneous mediums is the great scientific interest. For instance, for production of metal nanoparticles and their compounds and for waste and natural water purification due to a wider variety of chemical processes depends on electrode material.

The scientific interest in this field is due to the particular conditions of chemical reactions in a pulsed electric discharge (PED), their kinetics and mechanism.

On the one hand different kinds of dissolved substances (impurities) are involved in different types of reactions: redox, acid-base, ion-exchange, chemisorption, coprecipitation, etc. On the other hand, the type and behavior of reactions can depend on the type of PED and the chemical nature of electrode.

The use of various types of electric discharges for cleaning water due to their ability to initiate the formation of a wide spectrum of short-lived reactive species - radicals and ions, which form with the influence of degradation and oxidation of impurities in water [1 – 3]. Recently it is known papers which are devoted to the electric discharge wastewater treatment in reactors with aluminum and iron loadings [4 – 6]. Some developments brought to the level of patents [7]. However, there is no clear opinion of the processes.

The pulsed electric discharge in a layer of iron or aluminum granules placed in different aqueous solutions of organic substances differs from other types of discharges. It is due to the input energy influence on the electrodes. Metal pellets disintegrate to form highly dispersed materials. When using PED there are different physical and chemical processes with the active particles. Active particles have oxidizing and reducing properties. The study process of formation of active partials in the aquatic environment and chemical reactions with them are difficult due to time of life.

The mechanism of their transformation identified by chemical methods. For example, in papers [2, 3], it is investigated the formation of active particles when exposed to water an electric corona discharge with using solutions: $KMnO_4$, KJ and oxalic acid.

Most of the energy of pulsed electrical discharges expended on the dispersion of the metal by melting and evaporation.

Metal dispersed particles react with the medium to form the oxidized form metal erosion products. Products erosion of metallic granules determine the future direction of chemical reactions. There is no consensus explanation of these processes. Currently, many research groups engaged in the study of the process on electropulse with the solution of organic substances [1 – 4]

The objective of the work is to study the mechanism of physical and chemical processes occurring in reactors under the action of pulsed electrical discharges with iron and aluminum load using aqueous solutions of organic substances.

2. Experimental approach

The experimental setup is presented in [5]. The apparatus consists of a reactor volume of 1.5 L and power source. The reactor was made of a dielectric material. At the bottom of the reactor filled with aluminum or iron granules. Selection of different granules determined electrochemically active metal in an aqueous medium. The two external electrodes immersed in the metal loading was applied pulse voltage amplitude of 500 V with a pulse repetition rate of 300 s-1 and a duration of 15 ms. The maximum discharge current was 250A. Pulse energy 0.5 J / pulse.

A breakdown of intergrain space occurs with electroconductive canal formation and much energy liberation on the surface of contacting grains –"hot" spark formation. At that the main energy consumption is for metal dispersion, due to melting, evaporation, spraying.

Figure 1 shows the sketch of spark erosion process.

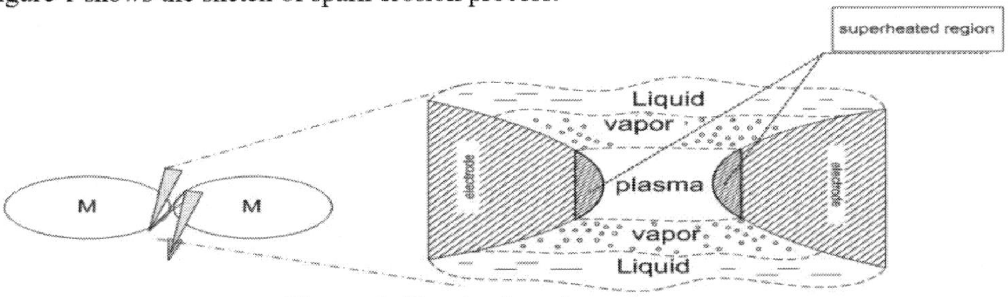

Figure 1. Sketch of spark erosion process.

On the left are two electrodes immersed in a dielectric liquid and connected to a pulsed power source. When the field strength in the gap is sufficiently high, a spark is produced. The spark results from the breakdown of the dielectric fluid as depicted in the enlarged view of the reaction zone on the right-hand side of Fig. 1.

A plasma channel of small diameter (perhaps 50 μm) is established in approximately 10 ns after the electric field is applied. The temperature in the plasma channel has been determined to be in excess of 10000 K.

The colored aqueous solutions of organic substances with well-studied properties: methylene blue, furacilin and eosin were used to determine the mechanism of physical and chemical processes in reactors. Selected organic substances can demonstrate the possibility of chemical reactions such as oxidation-reduction, and adsorption. Chemical formulas and properties of organic substances listed in the table 1.

Table 1. Chemical formulas and properties of organic substances

Name of substances	Formula	Properties
thylene blue	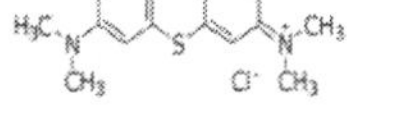 $C_{16}H_{18}ClN_3S{\cdot}HCl$	Reversible redox and cation adsorption indicator. Dissociation of the cation forms a shape in the form of $[C_{16}H_{18}ClN_3S]H^+$

osin	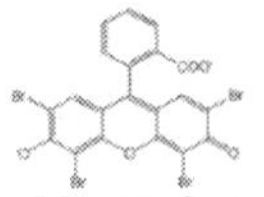 $NA_2[C_{20}H_6O_5Br_4]$	Irreversible redox and ion adsorption indicator. Dissociation of forming anionic form as $[C_{20}H_6O_5Br_4]^{2-}$
Furacilin	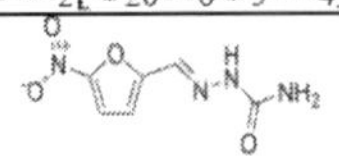 $C_6H_6N_4O_4$	Participates in redox processes and oxidized with decomposition

Treatment of aluminum or iron pellets electrical discharges in the solutions leads to the formation of a homogeneous suspension. The concentration of solids in the suspensions depends on the time of the metal processing load.The suspensions is centrifuged in a centrifuge slurry «Allegra 64R» firm «Beckman-Coulter» US at 20.000 rpm for the separation of the solid phase and the solution. Changing the intensity of the color solutions of organic substances upon the action of the pulsed electric discharge was determined after 15 min.

Absorbance of the solution was measured at the maximum of the absorption bands of the instrument ApelPD-303UV, the company «ApelCo., LTD», Japan. For methylene blue is at a wavelength of 590 nm to 440 nm at furatsilina and eosin is at 490 nm.

3. Results and discussion

The concentration of different forms of iron are measured to study the dynamics of accumulation of erosion products of an iron charge in various forms in water. The data on iron accumulation in various forms in solid erosion products are shown on figure 2.

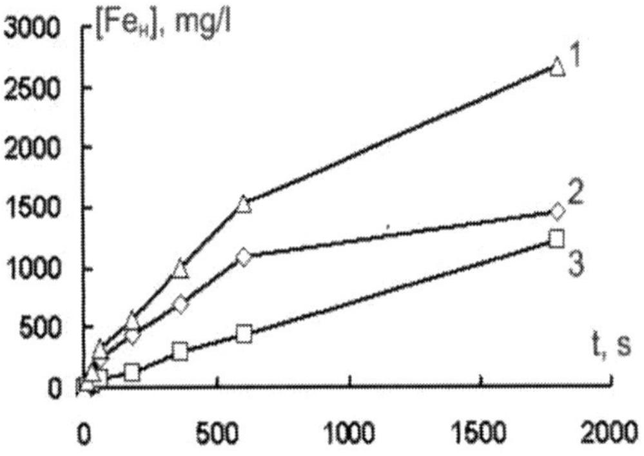

Figure 2. Data on accumulation of erosion products of the iron grains in various forms (1) – ΣFe, (2) – Fe^{2+} и Fe^0, (3) – Fe^{3+} in water.

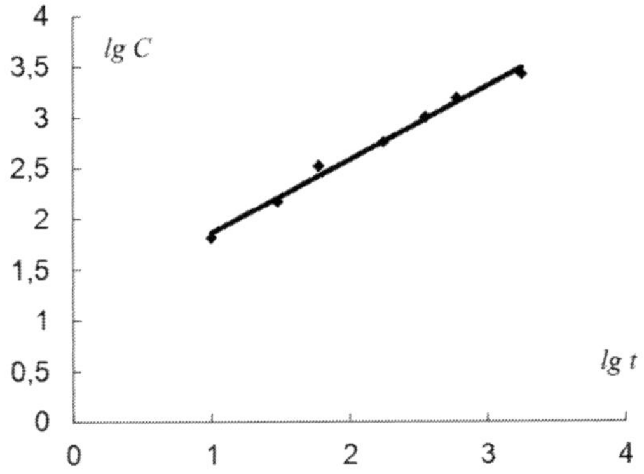

Figure 3. Concentration $\sum Fe$ of iron ions in erosion products (in suspension) vs. (versus) the treatment time *t*.

The experimental data obtained were processed and dependence plotted in the log-log coordinates (Figure 3). The total amount of iron (mg/L) removed from the electrodes (charge) is described as a function of the time of action of the discharge by the following empirical relations:

$$lg\ C_{Fe} = 0,72\ lg\ t + 1,1331 \quad (1)$$

$$or\ C_{Fe} = 13,6\ t^{0,72} \quad (2)$$

Thus, the kinetic equation describing the spark-erosion process of iron grains in water can be represented as a general formula:

$$[M_н] = k_e \cdot t^v, \quad (3)$$

where $v \leq 1$.

The index *v* was determined experimentally and it is connected with the change of contact geometry in a layer of iron grains and electrophysical characteristics of medium.

The value k_e is an empirical constant which depends on intensity and other characteristics of discharge, properties of electrodes and medium mg/(l·s^v).

The mechanism of this process can be represented by the following reactions, in which Me_n - dispersed metal particles

$$Me \xrightarrow{PED} Me_н \quad (4)$$

$$Me_н \rightarrow Me^{+n} + ne^- \quad (5)$$

$$Me_н + n\,H_2O \rightarrow Me(OH)_n + H_2 \quad (6)$$

$$Me^{+n} + n\,OH^- \rightarrow Me(OH)_n \quad (7)$$

$$4H_2O + 4\,e^- \rightarrow 2H_2 + 4OH^- \quad (8)$$

$$4OH^- + 4\,e^- \rightarrow 2H_2O + O_2 \quad (9)$$

In the above reactions 4, 8 and 9 react on the electrodes and depend on parameters of the electrical discharge pulse. Reactions 5, 6 and 7 are spontaneously, even after turning off the voltage source. As follows from equations 5-7, the reactions start at the interface "metal solution". The solutions of methylene blue, eosin and furacilin were used to identify reactions in the reactor.

Exposure to pulsed electric discharge on the methylene blue solution leads to a reduction of the intensity of its color. Metal particles in its oxidized form as oxides and hydroxides are deposited on the bottom of the reactor after switching off the voltage source.

Aluminum (III) react with water and form a sediment in the form of $Al(OH)_3$ when dealing with aluminum loading ions. The result of this process is the return methylene blue in initial color partly. Intensity of colors methylene blue in this case depends only weakly on the number of submitted pulses.

Figure 4 (curve 1) shows the change in the color intensity of methylene blue solution when pulsed electric discharge exposed for 15 s, 30 s and 60 s. The color intensity of methylene blue on the aluminum load varies slightly. This change does not reflect the actual mechanism of the process. The oxidation process proceeds to form electrons in reaction (5) due to the high electrochemical activity of aluminum.

Active particles that are obtained during discharge were involved in the oxidation of methylene blue is losing color.

The hydrogen which are formed by the reaction (6), involved the methylene blue by oxidation reactions (11). Methylene blue has a reversible indicator. It allows evaluating the development of the processes of oxidation and subsequent recovery. Reduction reactions involving hydrogen occur primarily.

$$[\text{Active particles}] + Red_{ind.} \rightarrow H_2O + Ox_{ind} \quad (10)$$

$$Ox_{ind} + H_2 \rightarrow Red_{in} \quad (11)$$

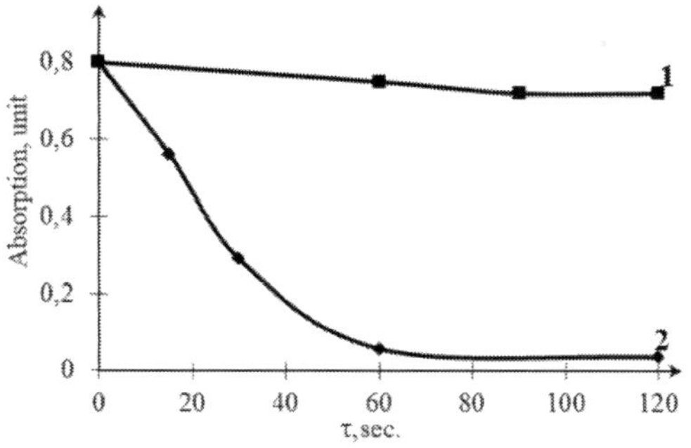

Figure 4. The dependence of the optical density of methylene blue on the time of exposure to pulsed electric discharge: curve 1–on aluminum loading curve; 2 – on iron loading.

Processes which is occurring on the iron granules, more intense than aluminum. This can be seen in Figure 4 curve 2. Intensive reduction of methylene blue staining is associated with a longer time of the process in comparison with the processes occurring on the aluminum granules. This is due to variable valence of iron and participation of free electrons in the oxidation of methylene blue and the lower electrochemical activity of iron as compared with aluminum.

IOP Conf. Series: Materials Science and Engineering **81** (2015) 012076 doi:10.1088/1757-899X/81/1/012076

A less concentration of hydrogen ions in the iron granules reduced the velocity reduction reactions Confirmation of a slow recovery of the color intensity of methylene blue can serve kinetics are shown on the Figure 5.

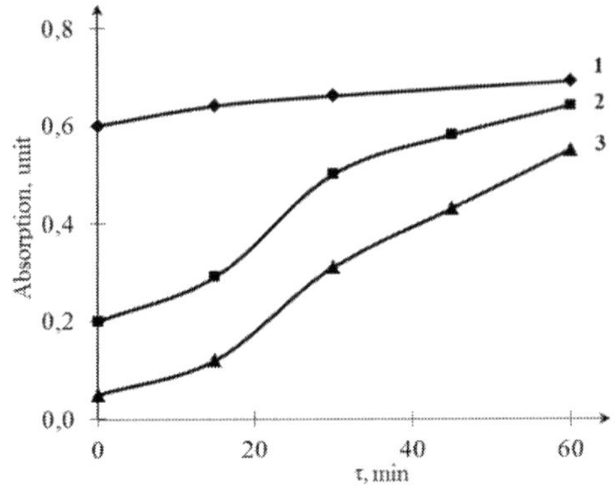

Figure 5. Restoring the color intensity of methylene blue solution after cessation of exposure PED iron loading at different time processing solution discharge (1 – 15 s., 2 – 30 s ., 3 – 60 s.).

The same experiments were performed with a solution furacilina. Furacilin is irreversible in oxidation-reduction reactions and may show only the oxidative processes. Change the color intensity of the solution furacilina under the action of pulsed electric discharge is shown in Figure 6.

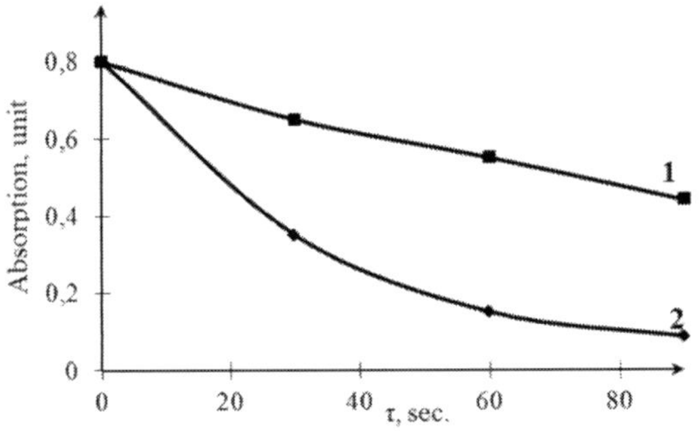

Figure 6. Dependence of the optical density furacilina from the time of exposure PED: 1-on aluminum load, 2 - on iron loading.

View kinetic dependences indicates that the oxidation furatsilina on aluminum and iron loads flows through different mechanisms. This is due to different kinetics of accumulation of products electrolytic erosion solution. In [5, 6], the dependence of the amount of metal eroded by time is described by the general formula (12):

$$[Me_{н}] = A\,t^{n}, \qquad (12)$$

where $[Me_{н}]$ –is concentration of electrolytic erosion products, mg/L;
A is empirical constant that depends on the intensity and other characteristics of the discharge, properties of the electrodes and the medium, mg/L c^{n}.

The exponent n is associated with a change in the geometry of the contacts in the layer of granules and electrical characteristics of the medium. For the aluminum loading of n equals 1, while for iron n equals 0.72.

The presence of hydroxide ions formed in reaction (8) and high local temperatures due to electrical discharges suggest to the reaction (13). This reaction is accompanied by cleavage of the furan ring to form the hydrazine and ammonia regardless of the material load.

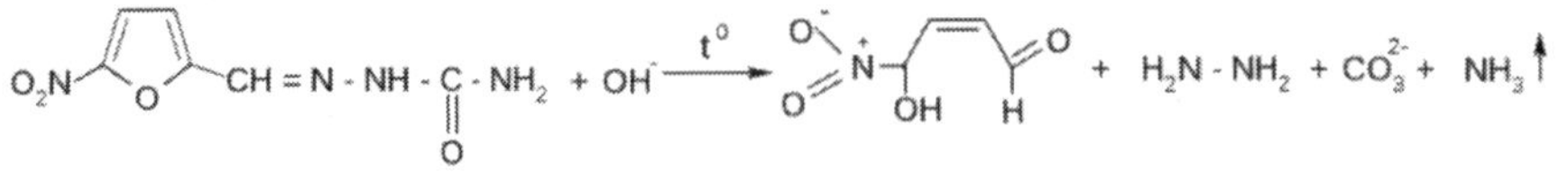

(13)

Hydrogen formed in the interaction of the metal particles with the water in this case is not involved in the processes of recovery furacilina.

However, the most intensive decrease color of the solution of furacilina is observed at iron pellet. This can be attributed to its interaction with iron ions to form insoluble compounds by reaction (14)

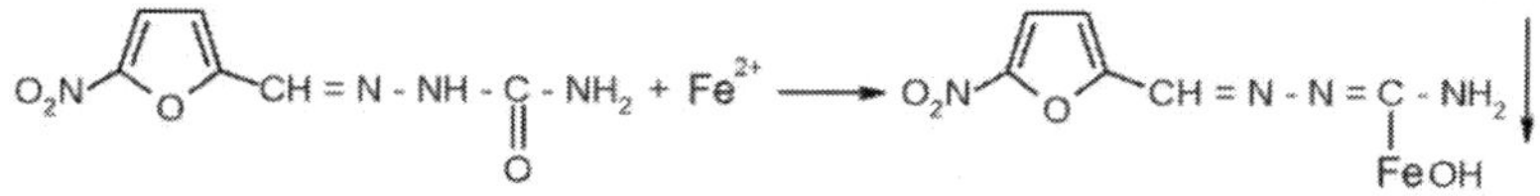

(14)

The possibility of this reaction was confirmed by reaction of furacilina solution with a model solution of $FeSO_4$. Oxidation of hydroxide Fe (II) leads to the formation of hydroxide Fe (III) with a developed adsorption of surface. This leads to a decrease in the intensity of color furatsilina due to sorption processes.

Established facts confirm the important role of material of loading in oxidation-reduction reactions during the processing of materials that are prone to oxidation in the PED reactors.

Adsorption processes have been examined for adsorption of the eosin solution. The results are shown in Figure 7.

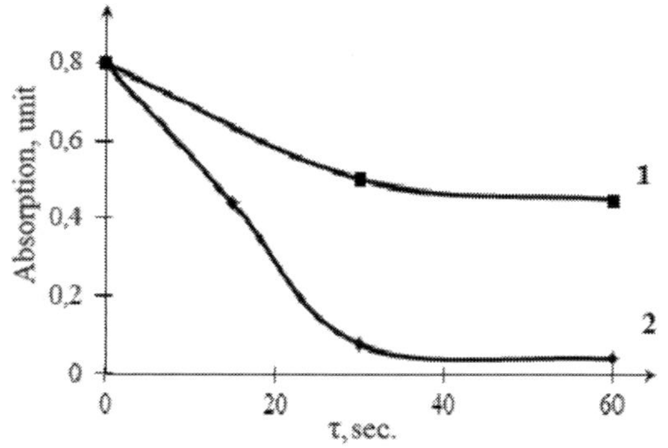

Figure 7. Dependence of the optical density of the solution of eosin from time of PED: 1–on aluminum loading; 2 – on iron loading.

Figure 7 shows the change in color of eosin at the initial time exposure pulses. It cannot be explained only by the adsorption process due to a solid phase with a minor amount.

In this case it may be assumed other reactions that lead to changes in the intensity of eosin staining. For example, a redox reaction with the active particles or destruction with the influence of the discharge.

According to figure 7 the sorption properties of erosion products of iron loading is much more efficient than aluminum. This is associated with the positive charge of the surface of the iron hydroxide having basic properties [7]. Aluminum hydroxide is amphoteric and it may contain two types of surface ions - H^+ and OH^- . Efficient sorption of eosin having anionic form ($[C_{20}H_6O_5Br_4]^{2-}$) is reduced.

4. Conclusion

1. It is shown that the main process in the reactors are redox reactions that developing during operation a pulsed electric discharge. Efficacy of redox reactions is dependent on the material load.
2. It is found that methylene blue is involved in redox reactions.
3. It is shown that furacilina simultaneous occurrence of two processes such as the destruction due to high local temperatures and the formation of insoluble compounds with heavy metal ions.
4. Flow adsorption processes are shown in eosin solution. The effectiveness of which is determined by the surface charge of the oxidized products of erosion.

Acknowledgements
The study was supported by National Research Tomsk Polytechnic University, the project No. 7.1326.2014.

Reference

[1] Piskarev I M 2002 *Journal High Energi Chemistry* **36** 360–361
[2] Aristova N A, Piskarev I M 2003 *Journal* of Physical of Chemistry **77** 723–726
[3] Parkansky N 2013 *Journal Plasma Chemistry Plasma Process* **33** 907–919.
[4] Kim J R et all 2013 *American Journal of Analytical Chemistry* **4** 115–122
[5] Jaworowsky N A, Savelev G G, Galanov A I, Shiyan L N, Yyrmazova T A, Lobanova G L 2008 *Advanced Material Research* **4** 74–78
[6] Danilenko N B, Savelev G G, Yavorovsky N A Yyrmazova T A 2008 *Russian Journal of Applied Chemistry* **81** 803–810.
[7] Shiyan L N, Yavorovskii N .A, Pustovalov A V, Gryaznova E N 2014 *Advanced Materials Research* **1040** 59–64.

IOP Conf. Series: Materials Science and Engineering **81** (2015) 012077 doi:10.1088/1757-899X/81/1/012077

Influence of the type of electric discharge on the properties of the produced aluminium nanoparticles

L N Shiyan, N A Yavorovskii, A V Pustovalov and E N Gryaznova*
Tomsk Polytechnic University, Institute of High Technology Physics, Tomsk, 634050, Russia

*E-mail: t_elena@mail2000.ru

Abstract. The effect of the method of aluminum nanopowder production on the aluminum products with water reaction is described. It has been established that the interaction of aluminum nanopowder prepared by the electric wire explosion, the phase composition of the reaction products mainly consists of boehmite (AlOOH) and has a fibrous structure. Therefore, that boehmite (AlOOH) can be used for modification of polymer membranes. The modified membranes can be used as water treatment from the impurity of formed true solutions according to adsorptive mechanism, and from colloidal nanometer and micron particles according to the mechanism of mechanical separation of particles depending on sizes.

Keywords: aluminum nanopowder, electrical explosion, modification of membranes

1. Introduction

Modern methods of producing nanofibrous aluminum oxyhydroxide (AlOOH) sufficiently diverse due to their use in the synthesis of various precursors based on aluminum, such as salts or in the form of metal powders [1–5]. The basis of these methods is the chemical synthesis of aluminum oxyhydroxide. The most common method of cooking is aluminum oxyhydroxide precipitation from solutions of salts of aluminum or aluminate solutions. More commonly, the salts used as starting chlorides, nitrates and sulfates of aluminum. With this method of producing aluminum oxyhydroxide, its composition and properties depend on the pH and temperature of the reaction medium. The reproducibility of this method is relatively low, which affects the properties of the resulting product. One way to obtain aluminum oxyhydroxide with desired properties is aluminogels hydrothermal treatment. The advantages of this method are the possibility of choosing the parameters (t, P, τ), providing the properties of good reproducibility and high quality product. A disadvantage of this method is the necessity of using high temperature and pressure, which affects the final product cost.

There are direct methods of producing aluminum oxyhydroxide, which are based on the oxidation reaction of aluminum metal. The main problem of these methods is to activate the surface of aluminum metal. It is therefore of interest to develop technologies for the production of aluminum with a thin oxide (protective film). There are methods of producing, allowing to obtain an active alumina capable of easily reacting with water to form aluminum oxyhydroxide by reacting:

These methods are the method of electric wire explosion (EWE) and the method of pulsed electric discharge (PED).

The purpose of this study is to compare the products of oxidation of aluminum, obtained by EEW and PED with water.

2. Materials and methods

In this work was used nanofibers of aluminum oxyhydroxide, which were obtained by the two methods:

1. The reaction of water and nanodispersed aluminum particles preliminary generated in the argon atmosphere by electric wire explosion.
2. The reaction of water and aluminum particles generated directly in water by electric discharge.

The main difference of these methods are difference of a precursors and epy type of discharge.

The first method is basing on the interaction of water and nanodispersed aluminum particles generated by electric wire explosion in an argon environment. Figure 1 a shows the scheme of this method.

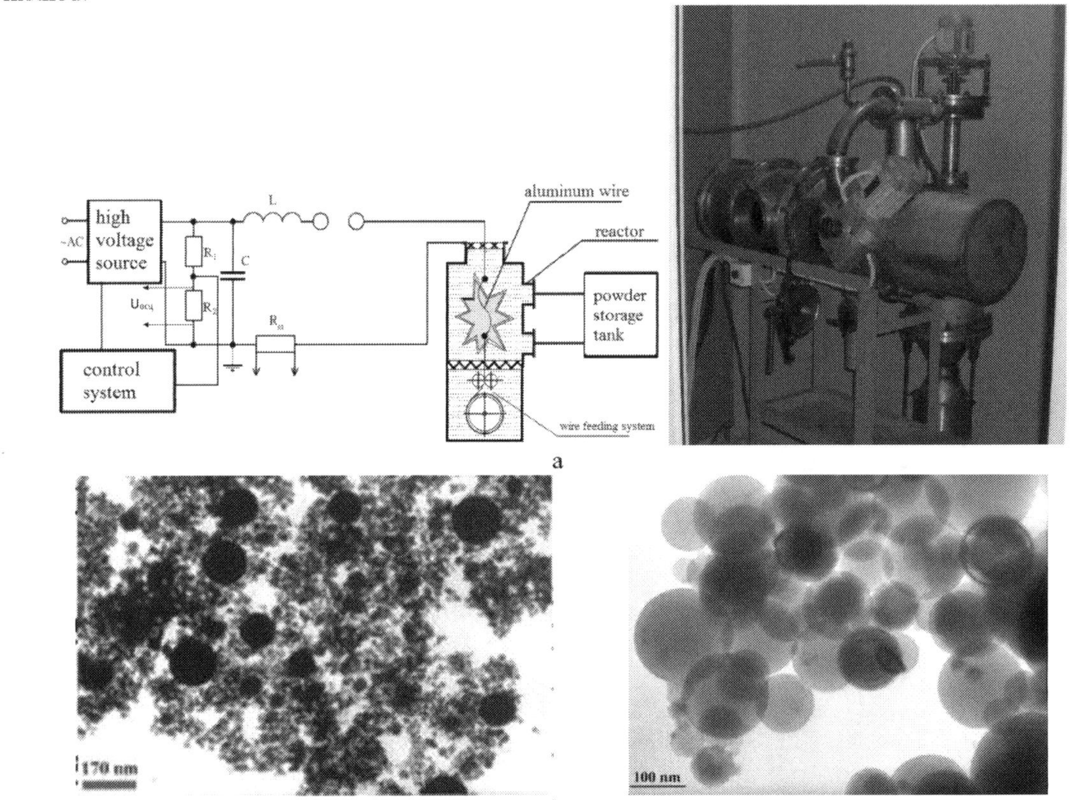

Figure 1. Setup scheme of the electric wire explosion – a;
TEM image of aluminum nanopowders – b.

This technology allows getting the particles of spherical shape with normal logarithmic distribution of particles by sizes (Figure 1*b*). The disparity of powders can be easily regulated by the change of energy applied to the wire.

The facts that there is no solid protective layer on the aluminum powders generated by this method, and the content of active metal changes in time to 100 % and less are very promising in comparison with metallic aluminum. In spite of a low content of active aluminum, the powders continue to react with water actively.

This method provides a nanosized powder of aluminum, but aluminum oxyhydroxide. So the next step was to carry out the oxidation of aluminum nanopowder with water at a constant temperature of 60 °C for 8 hours [6]. Only when the reaction of water with aluminum nanopowder aluminum oxyhydroxide can be obtained, whose the electron micrographs are shown on Figure 3*b*.

The second approach represents a generation of nanopowders directly in water. For this purpose, we used electric discharge technology [7]. The principle of this method is the following: two electrodes are placing in the chamber with the particular geometry. The space between the electrodes were filling with water and some pieces of aluminum shavings. When high voltage pulse were supplying to the electrodes, multiple micro discharges were generating between the shavings. Under the action of these micro discharges, the metal was destructing and thrown out in water as the small particles. Figure 2 shows the scheme of this method.

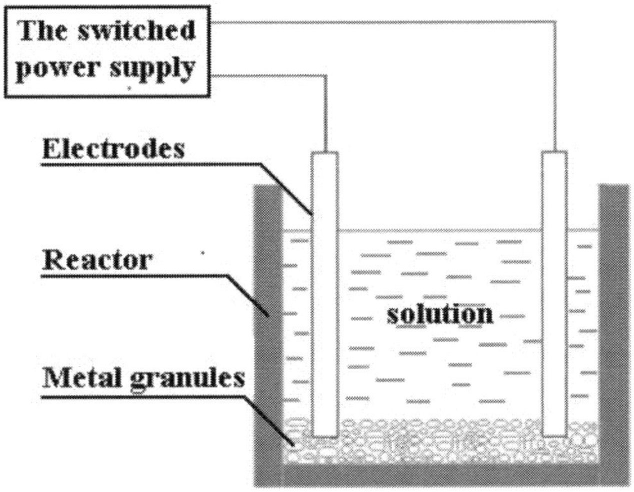

Figure 2. Setup scheme of the pulsed electric discharge.

This method allows obtained aluminum oxyhydroxide by one step. The electron micrograph of the resulting product are shown on the Figure 3*b*.

The all sample of the nanofibers of aluminum oxyhydroxide content of active (metallic) aluminum, phase content, area of specific surface and particle shape were determined. The phase composition of the samples was examined by X-ray diffraction using a Shimadzu XRD-7000 diffractometric at the angular range 2θ 10 – 100 ° and scanning rate of 1 deg min^{-1}. The specific surface area of aluminum nanoparticles were measured by the BET technique on Sorbtometer-M. The size and shape of particles were determined using the electron microscope. Morphological structure of all nanofibers were studied on transmission electron microscope - JEM-2100F, Japan.

The heat effects and changes in the phase composition during gradual heating of the samples in an inert argon atmosphere were determined on an SDT Q600 – Simultaneous TGA/DSC/DTA analyzer at 0 – 1000 °C.

3. Results and Discussion

Figure 3 shows the electron micrographs of products of oxidation of metallic aluminum with water, obtained by the two methods described above.

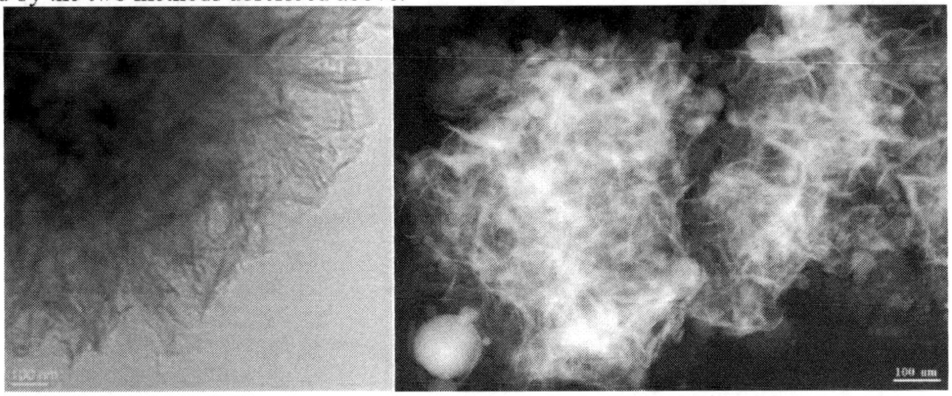

a b

Figure 4. The electron micrographs of products of the reaction of aluminum with water: a – by the electric explosion; b – by the electric discharge.

It is seen that the product obtained from aluminum nanopowder has a fibrous structure and product obtained by the electrical discharge method is part of a fibrous structure with the particles of round shape - aluminum metal. The results of the products analysis of this reaction are shown in the table 1.

Table 1. The characteristics of aluminum-water reaction products.

Type of products	Phase composition [%]	Specific surface [m^2/g]	Particle shape
Products generated by electric wire explosion	AlOOH – 85,0 $Al(OH)_3$ – 10,0 Al_2O_3 – 4,0 Al – 1,0	175,5	fibers
Products generated by electric discharge method	AlOOH – 6,0 $Al(OH)_3$ – 90,0 Al_2O_3 – 2,0 Al – 2,0	69,3	crystals of indefinite shape and plate-shape particles

Products obtained from electric explosion powders contain mainly nanofibers about 5 to 8 nm. in diameter. $Al(OH)_3$, Al_2O_3 and Al were in a small amount. The specific surface of nanofibers changes over a wide range, more than single-order and it is determined, basically, by depressiveness of an initial powder, the contents of active metal and initial water temperature.

The products obtained by electric discharge in water contain the crystals of an indefinite shape and very thin plates. The phase structure of these products includes basically Al $(OH)_3$. In some tests AlOOH, Al_2O_3 и Al are registered. The specific surface of the products depends more often on pH and water temperature.

We studied the behavior of the products reaction obtained by the different methods under heat by the method of the combined thermal analysis, using SDT Q600 Analyzer. Figure 5 demonstrates the results of analysis.

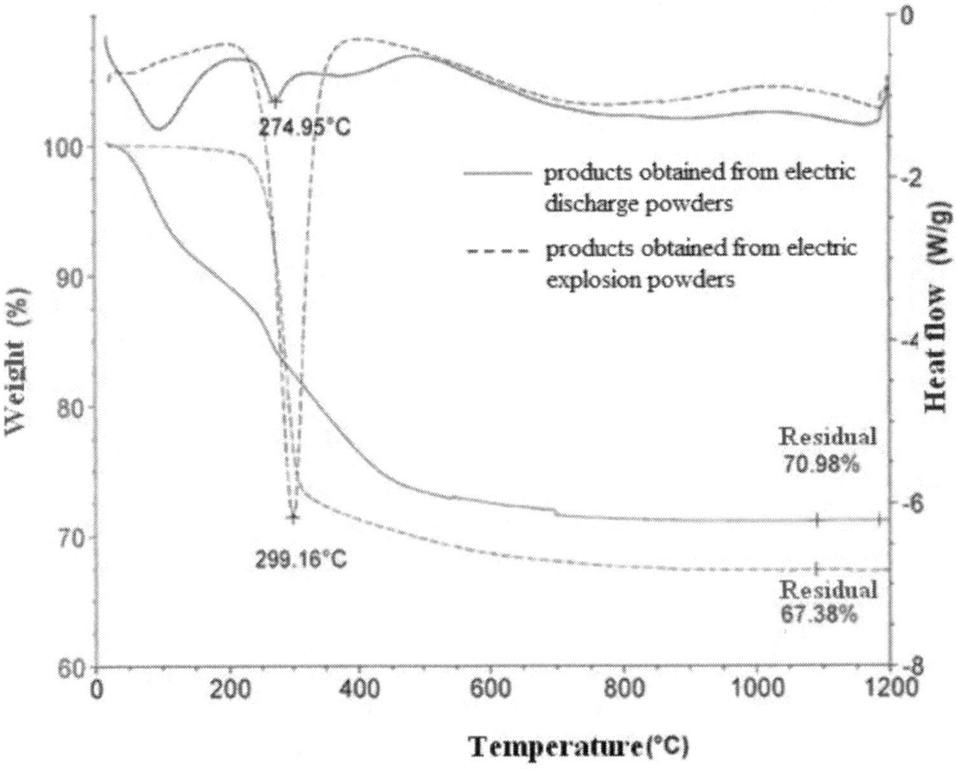

Figure 5. Thermograms of the products.

It is evident from the thermograms that the curves of weight loss and the character of phase transitions at the temperature range below 500 °C differ essentially.

These differences are the result of phase composition of analyzed products. The AlOOH phase passes into Al_2O_3 phase at 300 – 400 °C with conservation of fibrous structure. Thus, Al_2O_3 fibers get more precise shape. The specific surface increases on 20 – 25 wt. %. The $Al(OH)_3$ phase passes in oxides in wider temperature interval. So the crystals of long shape are formed.

4. Conclusion

It was demonstrated that the phase AlOOH are composed of the product by oxidation reaction aluminum nanopowder obtained by EVD, with water, and the products obtained during electroerosion composed of Al $(OH)_3$. The specific surface area of AlOOH are greatest, which makes this material a promising treatment methods for sorption of various media [8], as well as the possibility of its use as a modifier for polymeric membranes and as the carrier for catalysts [9].

Acknowledgement

The work was supported by State task of «Science» of 7.1326.2014.

References

[1] Zhang J, Shi F, Lin J,Weia S, Chen D, Gao J, Huang Z, Ding X, Tang Ch 2008 *J. Mater. Res. Bull.* **43** 1709–1715
[2] Jarayaman V, Guanasekaran T, Periaswami G *Mat. Lett.* 1997 **30** 157–162
[3] Chen M, Xiang L *Nano Biomed Eng.* 2010 **2** 121–125
[4] Bulent E. *J. Appl. Chem. Biotechnol.* 1973 **23** 803–809
[5] Wooa S, Park J-H, Rhee Ch, Lee J, Kima H *Microelectr. Engin.* 2012 **89** 89–91
[6] Gryaznova E, Shiyan L, Yavorovskii N, Korobochkin V *Rus. J. of Appl. Chem.* 2013 **86** 360–365
[7] Shiyan L, Yavorovskii N, Pustovalov A, Gryaznova E *Advanced Materials Research* 2014 **1040** 59–64
[8] Shiyan L, Tropina E, Machekhina I, Gryaznova E, An V *SpringerPlus* 2014 **3** 1–7
[9] Gryaznova E, Shiyan L 2012 The 7th International forum on strategic technology (Tomsk) vol 1 (Tomsk: Russia/Tomsk Polytechnic University) p 263–266

Thermally activated motion of dislocations in irradiated KBr crystals

M V Korovkin, Yu I Galanov
National Research Tomsk Polytechnic University, Tomsk, Russia

E-mail: mvk@tpu.ru

Abstract. Dynamics of thermally activated motion of dislocations in real time were studied by recording the pulses of electrical potential on the surface of the KBr crystals (in situ). It is shown that the radiation influence leads to crystals hardening, activation energy increasing and movement of dislocations initial stage shift to higher temperatures.

Keywords: alkali halide crystals, dynamics of dislocations, pulses of electrical potential

1.Introduction

Mechanical stresses in the crystal, powder, ceramic micro - and nanomaterials after pressing, are eliminated by prolonged thermal annealing. Residual stresses occur in these materials after radiation sintering also. Since the relaxation of mechanical stresses in crystalline dielectrics is determined by the motion of charged dislocations, it is necessary to investigate the dynamics of their thermally activated motion, which is extremely time-consuming process.

It is known that the motion of charged dislocations is accompanied by a pulse change of the electric potential on the surface of the crystal [1-3]. Starting temperature and the activation energy of their thermally activated motion is defined by recording the pulses of electrical potential (PEP) on the surface of LiF crystals [4]. In this work, the dynamics of thermally activated dislocations in the radiation-irradiated crystals in real time (in situ) was studied by recording PEP by the example of alkali halide crystals KBr.

2. Experimental techniques

Samples of KBr crystals (size 15×15×0,89…1,0 mm) were prepared from artificially grown from the melt of the crystals. The dislocations density is not more than 10^5 sm^{-2} by the etching method. Platinum electrodes, which served as a capacitive sensor antenna, is applied on the pattern surface by the method of cathode sputtering in a vacuum. The using of these type electrodes results to reduce ten times the level of background radiation, due to the near-electrode processes [5]. Electrical signals were amplified using high-frequency broadband amplifier, and then they were observed on the oscilloscope screen and were recorded by counting device counting mode pulse per unit time. Measurements were carried out in vacuum while heating at a constant rate 0,1 K·s^{-1} with a sensitivity of the 350 μV in the frequency range 100…2×10^7 Hz. The samples of crystals were subjected to x-ray irradiation (U=50 kV, I =20 mA, t=1 h).

3. Result and discussion

Figure 1 presents the temperature dependence of the counting rate of the pulses of electrical potential and synchronously measured thermally stimulated luminescence (TSL) of crystals KBr. We can observe two areas differing of the counting rate of the pulses of electrical potential. The first portion (a) is the number of the maximum PEP counting rate that are correlated with TSL. It is due to thermal annealing hole centers with subsequent recombination of holes on the F-centers [6].

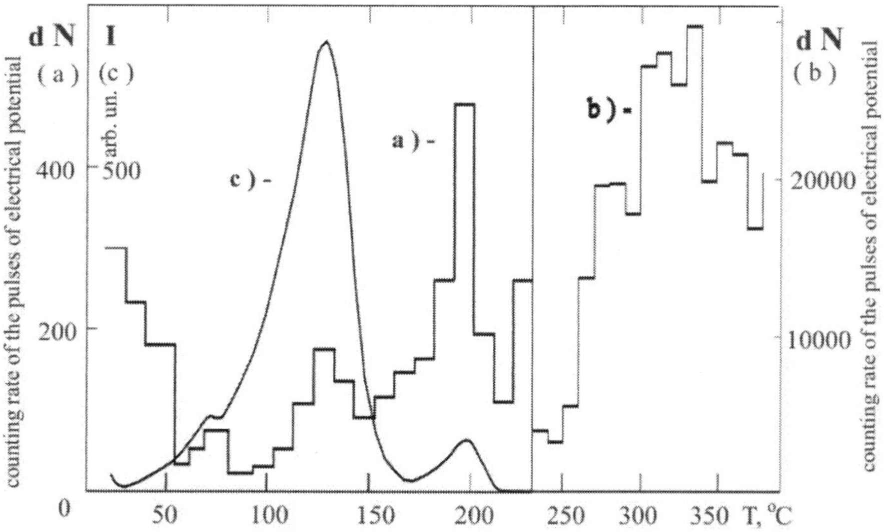

Figure 1. The counting rate changing the pulses of electrical potential (a, b) and luminescence (b) by thermal stimulation KBr crystals irradiated by x-rays.

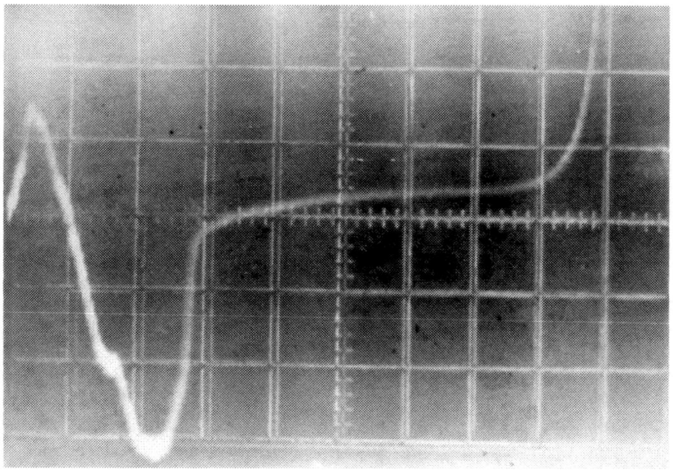

Figure 2. The parameters of the pulse shape of the electric potential of the irradiated x-ray crystals KBr in the field thermally activated motion of dislocations. The speed of the scanner 10 is seconds, the amplitude of the 500 μV /span.

The second part (b), above 220 °C, there has been exponential growth PEP exceed the background level by more than three orders of magnitude. Then the counting rate pulses of electrical potential reaches the maximum value and has some tendency to decrease but it remained high. Non-irradiated crystals have only the second high-temperature portion (b).

The pulses of electrical potential is observed on the oscilloscope screen in the form of signals, both positive and negative polarity, bell-shaped form with the amplitude of 10 mV with a duration of the leading edge 10...20 μs (Fig. 2). Because «jump» of dislocations occur spontaneously, it can be superimposed pulses of electrical potential, resulting in the observed pulses of complex shape.

The sharp increase of the counting rate of pulses on part (b) begins with some "starting" temperature, which is characteristic for various alkali halide crystals. It coincides with the known research results of dislocations dynamics.

According to modern concepts [7, 8], the motion of dislocations in the crystal becomes possible when the energy of the dislocation is sufficient to thermofluctuation overcoming the retarding its movement stoppers. Since some "starting" temperature the spontaneous dislocations movement occurs and its number increases with temperature. Surrounded by a «cloud» of charged defects the dislocations moves jumply at low speed ($<10^{-3}$sm c^{-1}) without looking up from his «cloud».

When jumping motion of a charged dislocations relatively to compensating its charge cloud of the Debye-Hückel due to the separation of the centers of gravity of the charges, an electric dipole moment. The appearance of the pulses of electrical potential on the surface of the crystal caused by the oscillations of the electric dipole moment

Caused by pulse of electric potential dipole moment P_i is determined by the expression

$$P_i = \eta \times q \times d \times \xi,$$

where η is the number of dislocations segments, making simultaneous jump,

d is the length of the segment,

q is the linear charge density per unit length of dislocations,

ξ is the offset dislocations relative to the compensating its charge cloud.

The magnitude of the electric potential φ on the crystal surface, at a distance **r** is determined by formula

$$\varphi = (\boldsymbol{P} \times \boldsymbol{r}) / 4\pi\, \varepsilon_0\, \varepsilon\, r^3$$

Given for alkali halide crystals the average evaluation values included in this ratio ($q \approx 3{,}1\times10^{-11}$ Kl/m, $d \approx 6{,}6\times10^{-5}$ m, $\xi \approx 1{,}0\times10^{-6}$ m [7, 8]) we obtain the minimum detectable value of the dipole moment $P_i \approx 1{,}0\times10^{-21}$ Kl•m and the potential on the surface ($r \approx 5\times10^{-4}$ m), corresponding to the sensitivity threshold detection scheme $\varphi \approx 350$ μV at $\eta \approx 3$.

It is quite possible the simultaneous collective movement of a large number of the dislocation segments [3].

Due to the thermally activated dislocations motion, counting rate changing the pulses of electrical potential described by the formula [4]:

$N \approx A\, N_O \exp(-E_a / kT)$,

where A is the coefficient, generally depends on the density of mobile dislocations, the dislocations in the number of jumps per unit time, concentration of stoppers, etc.

According to this formula, the dependence lg N = f (1/T) should be a straight line. Indeed, the experimental point of high-temperature portion for the temperature dependence counting rate changing the pulses of electrical potential is a straight line, its declination corresponds to the thermally activated dislocations motion energy (Fig.3). The change of activation energy reflects the dislocations dynamics and their interactions with local stoppers.

The dislocations motion in non-irradiated crystal KBr begins since ≈180 ^{O}C, but in crystals irradiated by x-rays since ≈220 ^{O}C, wherein in "start-up" stage, the activation energy in the irradiated crystals reaches 1.26 eV. Starting values of energy thermally activated motion of dislocations in all probability can not be less than the bond energy center fastening with dislocation.

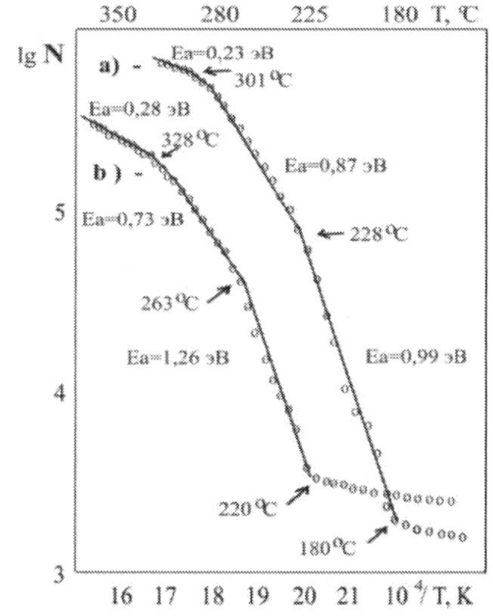

Figure 3. Changing the values of energy of thermally activated dislocations motion in non-irradiated (a) and irradiated by x-rays (b) KBr crystals

Experimentally we observe the values of the starting temperature at which breakaway of dislocations from the stopper occurs only due to the thermal motion in radiation irradiated and non-irradiated alkali halide crystals correspond to the definitions produced by the method of internal friction by extrapolation based the activation energy at zero amplitude deformation [7].

Further changes in the activation energy occurs in several stages and then reaches its minimum value of 0.23…0.28 eV. The values of activation energy, for example, in deformed crystals KBr ≈ 0,3 eV, obtained by different methods, probably correspond to the «above-barrier» dislocations motion [9] when their interaction energy with stoppers is not affected to the dislocations moving, because it is overcome by the power of external mechanical, electrical, or heat exposure in our case.

Jumping motion of charged dislocations cannot be explained only by the presence of random barriers along the path of motion. Jumping motion of charged dislocations is their own property and, therefore, it is determined by the state of the dislocations and its "atmosphere" - the clouds surrounding the charged defects [10]. Irradiation of the crystals by x-rays leads to the appearance of charged radiation-induced defects in the crystal, which also interact with dislocations, represent itself as stopper.

4. Conclusions

The temperature dependence counting rate the pulses of electrical potential of in alkali halide crystals at high temperatures reflects the dynamics of thermally activated dislocations motion.

This is confirmed by the literature and experimental data on the dynamics of dislocations in alkali halide crystals obtained by other methods (etching of the surface, internal friction, and others). Irradiation of crystals ionizing radiation leads to the increase of activation energy of their thermally activated motion. It indicates the dislocations pinning and strengthening crystal. This method can be very effective to study the influence of radiation and thermal processes at strengthening crystals (including ceramic) materials in real time (in situ).

References

[1] Urusovskaya A A 1968 *J. Uspehi physicheskih nauk.* **V 96** No 2 39

[2] Martishev Yu N 1965 *J. Crystallography.* **V 10** No 2 224

[3] Golovin Yu I et al 1985 *J. Physica tverdogo tela.* **V 27** No 2 124

[4] Korovkin M V et al 1998 *Russian Physics J.* No 11 124

[5] Korovkin M V and Galanov Yu I 1989 *Russian Physics J.* No 3 102

[6] Korovkin M V and Galanov Yu I 2011 *Russian Physics J.* **V 54** No 1/2 363

[7] Botaki A A et al 1980 Radiation physics of ionic crystals.[in Russian] (Moscow: Science) p 250

[8] Zuev L B 1989 Physics of electroelasticity alkali-halide crystals. [in Russian] Novosibirsk: Science p 120

[9] Darinskaya E V et al 1982 *Physica tverdogo tela.* **V 24** No 23 940

[10] Gekstrin S G et al 1997 *J. Crystallography.* **V 42** No 6 965

IOP Conf. Series: Materials Science and Engineering **81** (2015) 012079 doi:10.1088/1757-899X/81/1/012079

Synthesis of polymer ion-exchange hydrogels under γ - irradiation ^{60}Co

V. M. Le[1], V. D. Zhevnyak[1], V. Kh. Pak[1], V. A. Anan'ev[1] , U.V.Borodin[2]

[1]Kemerovo State University, Kemerovo, Russia,
[2]National Research Tomsk Politechnic University

E-mail: ya808@yandex.ru

Abstract. We have reported earlier about the modification of ion-exchange hydrogel under the influence of gamma radiation. The optimal absorbed dose of irradiation had been choosen for radiation modification of polymer hydrogels by ionits to produce products with a high content of the gel - fractions and sufficient mechanical properties. The dependence of the static exchange capacity of hydrogels on the type of ionit and its fractional composition had been studied. The dependence of the static exchange capacity of the quantitative composition of the ionit in the volume of the hydrogel had been investigated. The ion-exchange medical eye lenses had been made under selected conditions of synthesis. Their sorption properties had been studied.

1. Introduction

Nowadays polymeric medical supplies are widely used in many areas of medicine. Therefore polymeric materials with desired properties are especial interest when it contacts with a human, for example, hydrogels. Their creation is possible by synthesis of new or modification of known materials. The basic requirement for hydrogels is their biological inertness, which is primarily depends on the conversion degree of the monomers. One of the ways to hydrogels preparation is radiation polymerization. Polymerization of various monomers, high conversion of monomers, high purity polymers synthesized, ease of process control, are advantage of the radiation technology [1], [2], [3], [4]. Various modifications of hydrogels for soft contact lenses (SCLs) with specified properties can be synthesized by this technology.

SCLs are effective treatment of fresh chemical and thermo-chemical eye burns by neutralize the damaging agent, the acceleration of the cornea epithelialization, preventing of the simblefarona, and at a later date – after of burn corneal healing.

SCLs show the best sorption capacity than saline solutions, it is not large enough. The increasing of SCLs sorption capacity to searing chemicals, toxins, viruses and bacteria, results in the increasing of their therapeutic activity in the treatment of burns and viral eye lesions.

The goal of this paper is to study the synthesis of polymeric ion-exchange materials in the form of the soft contact lenses for the treatment of chemical and thermal burns, trauma and infectious diseases of the eye, by the radiation modification of hydrogel by ion-exchange resins. The hydrogel for SCLs had been used to prepare an ion-exchange polymer hydrogel [5].

2. Material and methods

The samples were irradiated with ^{60}Co γ-rays at 310 K. The dose rate was measured with a Fricke dosimeter, assuming the radiation yield of Fe to be equal to 15.6 (100 eV). The dose absorbed by the sample was calculated using the mass energy absorption coefficients.

The hydrogel has been prepared by radiation copolymerization of methyl methacrylate and N-vinylpyrrolidone in the presence of divinyl ether diethyleneglycol under γ-irradiation [6]. It has a high (70%) water content, plasticity and stability. Lenses from this hydrogel are used for vision correction (from -30D to +30 D), and as therapeutic lenses in diseases such as bullous keratopathy, keratitis, corneal transplants, and others.

Weakly acid cation resins (KB-2E - methacrylic acid-divinyl ether diethyleneglycol (DVEDEG), KB-4 - methyl ester copolymer of methacrylic acid, divinylbenzene (DVB), and D-113 (Granion CWP-1) - copolymer "acrylic" and DVB) were used as modifying agents for polymeric ion-exchange hydrogel synthesis [7]. Modifiers with fraction 0.25-0.35 mm and less than 0.25 mm have been investigated.

The content of the gel-fraction can be calculated as

$$X=(m_2/m_1)\times 100, \%, \tag{1}$$

where m1 is weight of the sample before the extraction; m2 is weight of the sample after the extraction. The average value of six measurement was used.

The deformation and strength characteristics of the samples were studied by tensile machine RMB 30. Strength at failure (puncture) σ is

$$\sigma=k\sigma\times(P/(h\times d)), \text{kgs/cm 2 (MPa)}, \tag{2}$$

where P is load at which the sample was destroyed, kgs; h is thickness, cm; d is diameter of the sample working area, cm (0.4 cm); kσ is normalization factor (1.1).

Deformability ε is calculated as

$$\varepsilon=k\varepsilon\times(l/d)\times 100,\%, \tag{3}$$

where l is elongation, the value of the indenter sliding from the beginning of its contact with the sample to sample destruction, mm; kε is normalization factor (2.2).

The determination of cross-linked polymer (gel-fraction) in modified ion-exchange resins and in the original hydrogel samples was performed by extraction with distilled water during 36 hours in Soxhlet apparatus. After extraction the samples were dried in a vacuum oven at 60 ° C to constant weight.

The acid-base titration has been used for estimated of sorption activity.

The absorption of alkali from 0.02 M of solution NaOH, for samples of hydrogels modified by ion exchangers in the H-form, has been determined. The absorption of the acid from 0.02 M of HCl solution for hydrogels containing ionits in Na-form has been determined. The hydrogels were placed in the successive volume of 0.02 mmol/g acid or alkali solution and it was titrated to determine static exchange capacity (SEC) of acid or alkali.

3. Results

Figure 1 displays the results of the cross-linked polymer determination in the ion-exchange hydrogel samples prepared at different absorbed doses. As seen, the increase of absorbed dose results in increase the degree of monomers. Lower content of gel-fraction in the modified by ion-exchange material vs the hydrogel may be due to the breakage of the polymer chain on the modifier, which does not react. The gel-fraction content is closed to 100% at absorbed doses high than 25 kGy, e.g. 35-40 kGy.

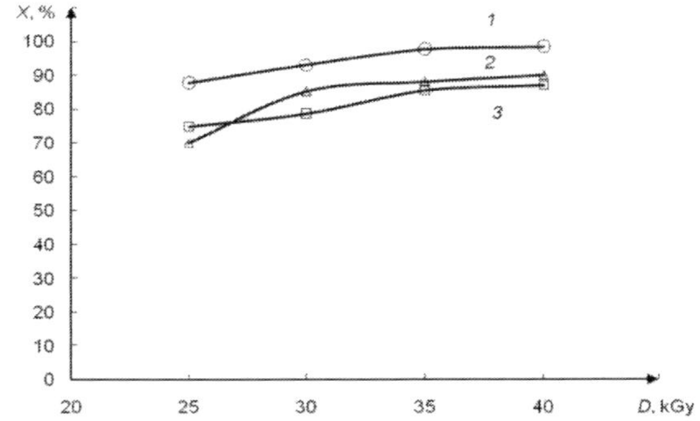

Figure 1. The content of the ion-exchange polymer hydrogels gel-fraction vs absorbed doses of irradiation. 1 - hydrogel; 2 - hydrogel +5% D-113; 3 – hydrogel +5% KB-4.

Figure 2 displays the results of the strength and elongation at puncture tests for ion-exchange hydrogel samples containing 5% resin D-113, irradiated with different doses. As seen, the tensile strength at puncture values are below at the higher deformability values for samples irradiated with dose 35 kGy. The strength values of the material increase with increase of the absorbed dose but it loses elasticity. So, the value of absorbed dose is equal to 33-35 kGy has been choosing for synthesis of ion-exchange hydrogel samples.

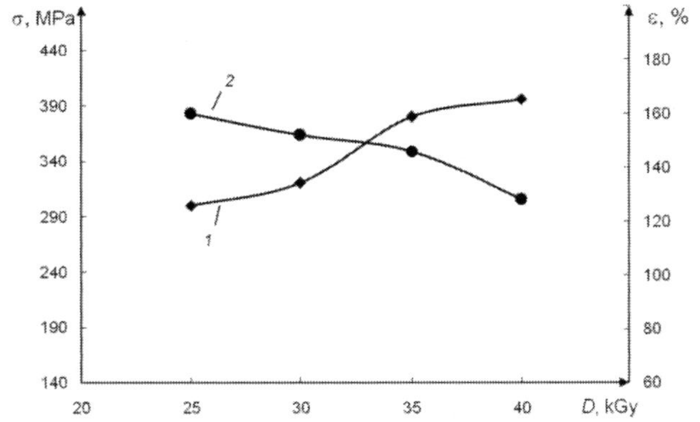

Figure 2. The strength characteristics of polymer ion-exchange material vs absorbed doses of irradiation. 1-tensile strength puncture (σ); 2-deformability (ε).

In previous studies the content of modifiers in ion-exchange samples was 5%. The SEC is not high enough for such content. Therefor it was studied SEC for higher modifiers content. Figures 3-4 displays the dependence of SEC of acid and alkali by ion-exchange hydrogels with different fraction from saturation time. As seen, the SEC to acids and alkali of the fractional composition of ion-exchange resins with fraction 0.25 $<d<$0.35 mm are higher than for the fine fraction. SEC of hydrogel modified by ionit D-113, is up to ~1.2 and 3-fold to the acid and to the alkali 2-2.5-fold, than SEC materials modified by ionits KB-4 and KB-2E, respectively. So, the cation exchanger D-113 has been choosing for subsequent studies.

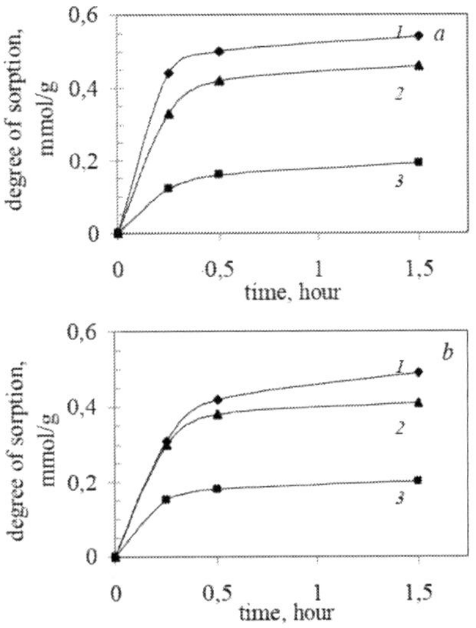

Figure 3. Absorption curves of HCl by ion-exchange hydrogels. Modifiers (5%): 1 - D-113, 2 - KB-4, 3 - KB-2E. Fraction of modifier: a) 0.25 <d <0.35 mm, b) d <0.25 mm.

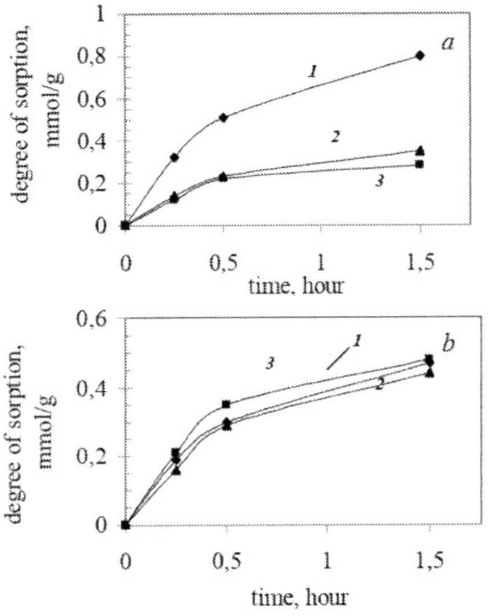

Figure 4. Absorption curves NaOH by ion-exchange hydrogels. Modifiers (5%): 1 - D-113, 2 - KB-4, 3 - KB-2E. Fraction modifier: a) 0.25 <d <0,35 mm, b) d <0.25 mm

Figure 5 displays the dependence of the SEC hydrogels containing 5, 10, 15% of the ion-exchange resin to alkali from saturation time. The dependence of the SEC from saturation time to acid is the same. As seen, the increase of the ion-exchange resin content in the hydrogel has been resulted in a significant increase of the sorption capacity.

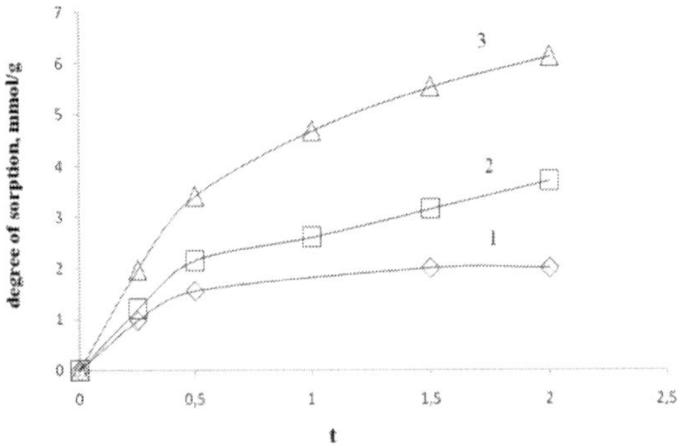

Figure 5. Dependence SEC of the IEMEL containing 5, 10, 15% of the ion-exchange resin to alkali from saturation time.

Ion-exchange medical eye lenses (IEMELs) were prepared by absorbed dose 35 kGy, the contain of cation exchanger D-113 (fraction 0.25 <d <0.35 mm) is 15 %.

Figure 6 displays the dependence of the SEC of ion-exchange medical eye lenses and hydrogel lens to alkali and acid from saturation time. As seen, the first lenses SEC value to alkali is equal to 6.2 mmol/g for two hours. IEMEL SEC to acid is increased rapidly (30 min) and then it becomes constant and equal to ~3mmol/g.

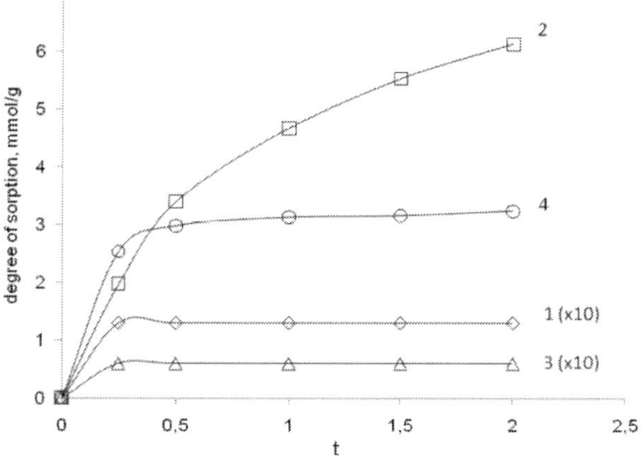

Figure 6. Dependence SEC of the IEMELs (15% of modifiers) from saturation time. SEC of alkali: 1-lens of hydrogel, 2-IEMELs; SEC of acid: 3-lens of hydrogel, 4- IEMELs.

4. Discussion

Biocompatibility is basic requirement for medical devices in contact with the human body. Biological inertness of polymer material provides a high degree of conversion of monomers, under which is not possible to disengage into the human body incomplete polymerization products. The study of depending on the content of gel fraction on the absorbed irradiation dose was performed 25kGy dose at which the monomer mixture is polymerized to the solid state. Increase of the dose to 35kGy allowed to extend the content of the gel fraction to 90%. Further increase of the absorbed dose had no significant effect on the change of the degree of conversion of the monomers.

The absorbed dose of irradiation also affects the processes of cross-linking and destruction of the polymer, which leads to change of the strength characteristics of the hydrogel. Study of the strength characteristics showed in the range of 25 to 40 kGy of the absorbed dose, that at higher absorbed doses of irradiation the breaking strength of the hydrogels had been increased, but at the same time their elasticity had been decreased. As seen from Figure 2, the strength and the deformability of the hydrogel have optimal values at the absorbed dose 33 kGy. The absorbed dose γ-irradiation 35kGy had been selected during the comparison of the results obtained for the synthesis of polymeric ion-exchange hydrogel.

SEC investigation of the dependence of ion-exchange polymer hydrogels on the type of cation resins and its fractional composition (Fig. 3 - 4) showed that as a modifier the ion-exchange resin D-113 fraction from 0.25 to 0.35 mm is preferable to the use. A smaller fraction of cation resins beds and stick together during polymerization, which leads to reduction of its active sorbent surface.

However, absorption of acids and alkalis in the ion-exchange hydrogel is determined not only by the properties of the sorbent, and the processes of diffusion of sorbate in the hydrogel matrix. The nature of the polymer and its moisture content affects on sorption and desorption of substances. In hydrogel materials diffusion rate of solutes generally increases with increasing water content. Simply increasing of the percentage of the sorbent in the hydrogel will reduce the moisture content and, consequently, to reduce the rate of diffusion of adsorbed substances. Changing the composition of the polymerization mixture in the direction of increasing the hydrophilic component and reduce the amount of crosslinking agent, helped to keep the water content, increase the amount of ionit to 15% and get ion exchange hydrogels with high adsorption activity (Fig. 5). A further increase of ion exchangers is possible provided that the basic properties of the base of the hydrogel.

Studies of the sorption properties of ion-exchange medical of eye lenses (Fig. 6) showed that IEMEL actively absorbs alkali for 2 hours or more, and the sorption of acid is almost complete within 30 minutes. The lens swells and gradually increases in size at the sorption of alkali. This is due to the fact that, in parallel with the sorption of alkali ion exchanger s in the hydrogel matrix flow processes of alkaline hydrolysis of polymethyl methacrylate to form polymethacrylic acid. Spatial grid is loosening the hydrogel water content and increase the absorption of alkali. Not appropriate to use for the sorption of alkali IEMEL more than 2 hours since the size of the lens is increased significantly distorted her boarding radius and this lens will not be comfortable to the eye. During the sorption of acid IEMEL lens gradually shrinks and diffusion of sorbate in its scope is difficult. Thus, efficient use of time to retrieve IEMEL acid is 30 minutes.

5. Conclusion

The samples of ion-exchange materials had been obtained in the result of conducted studies by the radiation modification of polymer hydrogel soft contact lenses.

The absorbed dose at which had been achieved the highest possible value of the content gel - fraction and strength characteristics selected.

The cation resins D-113 with fraction 0.25 $<d<$0.35 mm showed the best results of sorption activity in the studied range of ion exchangers.

The quantitative content of the ion exchange resin was increased to 15% for increase the sorption capacity of ion-exchange polymer hydrogels, subject to changes of the ratio of monomers and cross-linking agent in the base hydrogel.

The eye treatment ion exchange lenses made in the chosen conditions of synthesis. Sorption activity EMIEL exceed the value of static exchange capacity of contact lenses on the acid 50-fold and 47-fold on the alkali.

References

[1] Zhevnyak V, Pak V, Le V 2013 *17 th International Conference on Radiation Effects in Insulators* (Book of Abstracts)

[2] Clough RL, Gillen KT, Dole M 1991 *Radiation resistance of polymers and composites* (In. Irradiation Effects on Polimers, D.V. Clegg, A.A. Collyer (eds.) Elsevier Apple Sci. p.79- 156)

[3] Ivanov V 1988 *Radiation chemistry of polymers* p.320

[4] Shur A 1969 *Polymer science* p.656

[5] Pak V, Zhevnyak V, Dikunova T, Shraibman G 2008 *Ion-exchange polymer hydrogels for the treatment of chemical eye burns* The fundamental problems of modern materials pp.25- 28

[6] Zhevnyak V, Stalkovsky V, Fomina M 1997 *A process for preparing a polymer material for soft contact lenses* (Russia federation Patent No2119927)

[7] Le V , Zhevnyak V , Pak V , Hatminsky Y u 2012 *The new sorption materials for the treatment of chemical eye burns* (Journal for ophthalmologists and optometrists «Eye» No 2)

IOP Conf. Series: Materials Science and Engineering **81** (2015) 012080 doi:10.1088/1757-899X/81/1/012080

Method of case hardening depth testing by using multifunctional ultrasonic testing instrument

Y A Salchak[1], D A Sednev[1], M Kroening[2] and I B Ardashkin[1]

[1]National Research Tomsk Polytechnic University, 30 Lenin Ave, Tomsk, Russia
[2]Saarland University, Saarbrücken, D-66123, Germany

E-mail: yasalchak@gmail.com

Abstract. The paper describes usability of ultrasonic case hardening depth control applying standard instrument of ultrasonic inspections. The ultrasonic method of measuring the depth of the hardened layer is proposed. Experimental series within the specified and multifunctional ultrasonic equipment are performed. The obtained results are compared with the results of a referent method of analysis.

1. Introduction

Nowadays metalworking industry is growing all over the world. Its primary manufacture operations consist of fabricating or assembling metal parts. This industry produces a very wide range of products, such as: electrical machinery (motors, generators, and transformers), other machinery (engines, turbines, pumps, mining and metalworking equipment), transportation equipment (automobiles, aircraft, railroad equipment, ships and boats) and other fabricated metal products. Each of these areas contains the type of details that are used under certain conditions, such as dynamic loads. Due to the fact that the amount of various parts that work under the influence of dynamic loads is continually growing it is necessary to monitor their quality effectively, both at manufacturing and operation stages. The ability to sustain loads without damaging is provided by specific structure of heavy loaded components. Such components have hardened surface layer and mild core material. The case hardening depth is the criteria of a component quality.

The state-of-the-art principles of diagnostics are based on destructive testing methods which are costly and time consuming, and not always effective. Destructive testing is a statistical method; usually the results of inspection are formed according to the condition of certain selected details. There is no guarantee of each detail quality in a wide line of components. Such method does not provide comprehensive output control. Of course, there are some well-known nondestructive (NDT) methods too. Most of them are based on the physical properties of the material, in metalworking industry it is usually steel or another ferromagnetic material. Unfortunately, these methods of inspection have considerable limitations in use. Electromagnetic method is the most widely used method of nondestructive testing in Russia. However, this method can be applied only for monitoring the subsurface of the hardened layer. The depth of the layer that can be inspected is to be 1 mm or less. Further it is extremely difficult to automate the process of scanning while using the electromagnetic method of control.

The analysis of worldwide experience in this area has revealed that application of ultrasonic methods of control have been applied to solve this problem At Fraunhofer Institute for Nondestructive Testing (IZFP) the method of measuring the depth of the hardened surface layer of metal details was developed [1]. Furthermore this method formed the basis for development of the instrument which lately found wide application in the metallurgical industry market all over Europe. This instrument has very specific application; basic function of ultrasonic inspection is not provided. This implies that effective quality monitoring requires an operator to use several instruments. One instrument is for measuring case depth hardening and determining quality according to the technical requirements, while the other is for determining whether a component has any defects that may decrease its sustainability. It is useful to combine these two functions in one instrument to provide the effective procedure of quality control of heavy-loaded components. In this respect the main aim of the research was to determine the suitability of the ultrasonic nondestructive testing method for measuring case hardening depth applying multi-operated ultrasonic instrument.

2. Ultrasonic Attenuation in Hardened Materials

Usually an ultrasonic inspection system consists of several components, such as electronic unit, transducer and display devices. The receiver is also the transmitter; it can produce high voltage electrical pulses. Transducer with piezoelectric element can convert these pulses into high frequency ultrasonic energy. If there is a flaw such as a crack or a pore some of the energy will be reflected back from the flaw surface. The time between the moment when the signal was generated and when the echo was received, can be measured. According to this time it is quite simple to count the distance that the signal has traveled. Moreover the information about the reflector location, size, orientation and other features can also be obtained.

Ultrasound attenuation is the reduction of the amplitude and the intensity, and consequently of the acoustic wave itself as it propagates. The main reason for this effect is acoustic scattering. Ultrasound scattering occurs due to the sharp change in fluid properties (density, modulus of elasticity) at the boundaries of inhomogeneities which dimensions are comparable with the wavelength.

In general the sound waves are scattered in all directions. Some of them are also scattered back to the ultrasonic testing (UT) probe. On the one hand the effect that means the magnitude of the scattered and backscattered sound energy depends on the change of density and/or the elastic material properties within the volume range. On the other hand it depends on the relation of the geometric size of the volume range and the wavelength of the ultrasonic waves.

When inspection of the hardened surface layer is performed, it is important to pay attention to the structure of materials under inspection. The surface hardened layer is the material with fine grain size permeable to ultrasonic waves (Figure 1). At the same time unhardened material grain sizes are much bigger and this difference causes strong scattering of ultrasonic waves in different directions.

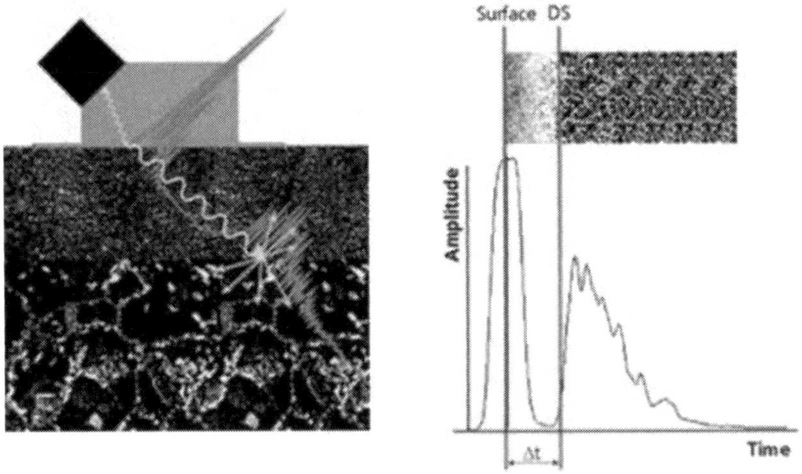

Figure 1. Attenuating in steel.

Due to the fact that magnitude of ultrasonic backscattering can be influenced by adjusting the wavelength of ultrasonic waves it is very important to find the right frequency range. Thus in case when size of a grain is larger than the wavelength the sound can be reflected by its boundary.

Ultimately the scattered signal is scattered back to the testing probe. The main item to be measured is the time of wave propagation. Time measurement process can be automated by developing specialized software. As a result of measurement the depth of the hardened layer of material in mm is displayed on the instrument screen.

Following these principles the instrument can be produced to measure the depth of hardened layer which is the main quality characteristic of hardened items.

3. Experimental Procedure

The samples were selected for the experiments to verify the use of ultrasonic NDT methods for quality control of dynamically loaded components applying multifunctional instrument. Each of the samples is a part of the slewing bearings. The choice of the component was made due to the fact that bearing is the most commonly used detail in metalworking industry. The experimental samples are of different shape and therefore have different measuring positions (Figure 2). There are two rows of rolling elements: internal and external rings that clamp together axially. Depending on the load point, each part has different points of hardening. This conditioned the choice of measuring positions for the experiment. Series of measurements were performed in each position to provide reliable results.

(a) (b) (c)

Figure 2. Experimental samples: a – sample №1, b – sample №2, b – sample №3.

Two types of equipment – specific and multifunctional were chosen to provide reliable results of the experiment. Case Depth Tester (P312), QNet was chosen as a specialized instrument. Fraunhofer Institute for Non-Destructive Testing (IZFP) has developed the ultrasonic method for fast and nondestructive case depth testing. This instrument provides fast production control and ensures consistently high quality levels of quality management.

IOP Conf. Series: Materials Science and Engineering **81** (2015) 012080 doi:10.1088/1757-899X/81/1/012080

The case depth tester P3121 consists of the following components: computer unit and sensor system (Figure 3).

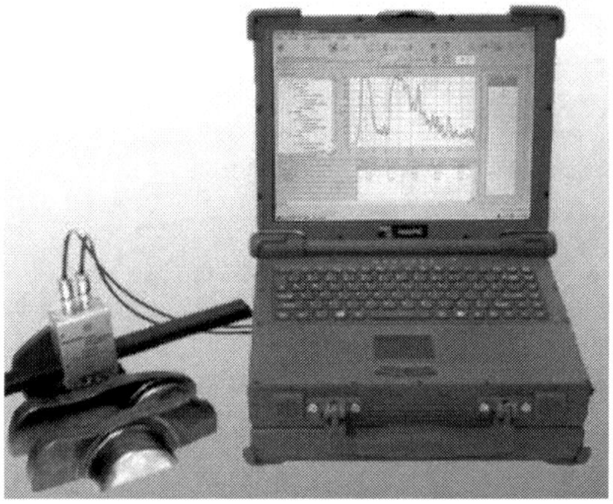

Figure 3. Ultrasonic testing instrument for testing the case depth, P3121, Case Depth Tester.

Case hardening depth values of experimental sample №2 measured by the specific ultrasonic instrument in each position are relevant:

- Rht=4.7 mm, position 1 – right flank;
- Rht=3.6 mm, position 2 – left flank (Figure 4);
- Rht=5.1 mm., position 3 – center.

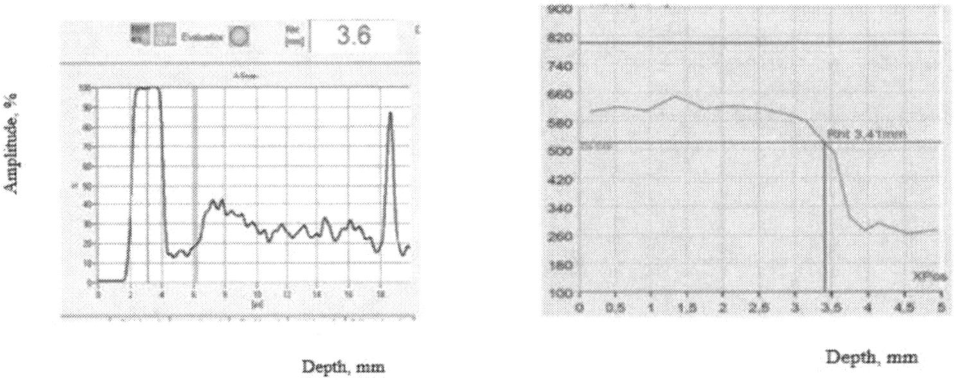

(a) (b)

Figure 4. Measurement results for the sample №2 (position 2 – left flank): a – ultrasonic testing; b – metallographic analysis.

Metallographic analysis was chosen as a reference method and the case depth according to this method constitutes 3,41 mm (test sample №2, left flank). The results are shown in Figure 4 (b).

It is worth mentioning that the case depth determined by ultrasonic backscattering does not necessarily equal to the case depth measured with metallographic-, micro-hardness- or any other established method. However it is also evident that close correlation is often found between the ultrasonic hardness depth and the reference hardness depth. The reason is that the depth of the core material and the case depth determined by the established methods are influenced by the process parameters of hardening.

The following conditions produce ideal results:

- The test parts are induction-hardened.
- The test parts are forged, not cast.
- The minimum case depth to be measured is 1,5 mm.
- There is no intermediate microstructure between the martensitic and the core material.
- The grain size of the core material is large enough to cause significant backscattering of shear waves with the frequency of 20 MHz.

SAPHIR_Quantum (AREVA) was chosen as a multifunctional instrument. It is a versatile high end ultrasonic system for phased array as well as for single or parallel multi-channel applications in the industrial UT nondestructive testing. The electronic unit contains the ultrasound transmitter and receiver channels as well as the signal processing. In standard configuration 32 channels are implemented in one electronic unit. Each channel has excellent technical characteristics and is individually adjustable.

The test stand consisting of the ultrasonic electronics (SAPHIR_Quantum), transducer (20 MHz) and a laptop constitute the experimental setup (Figure 5).

Figure 5. The experimental setup.

To achieve reliable results the experiment was performed on the same samples as in the first case. Typical results obtained by the second series of experiments by using multifunctional ultrasonic system are presented in Table 1 below.

Table 1. Experimental results obtained by the test stand (test sample № 2, left flank)

(a) Signal perfomed by the PC	(b) Digitized signal
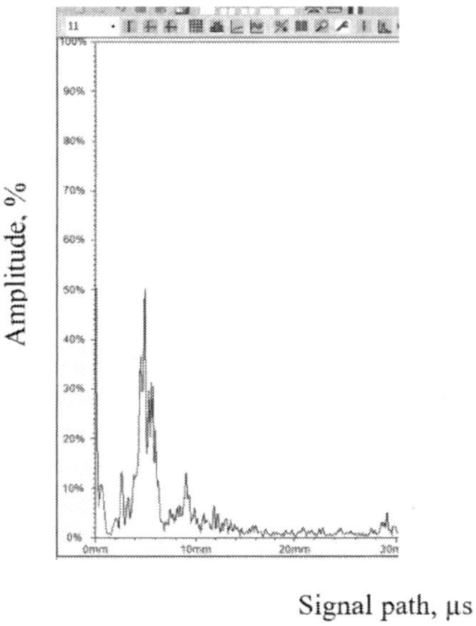	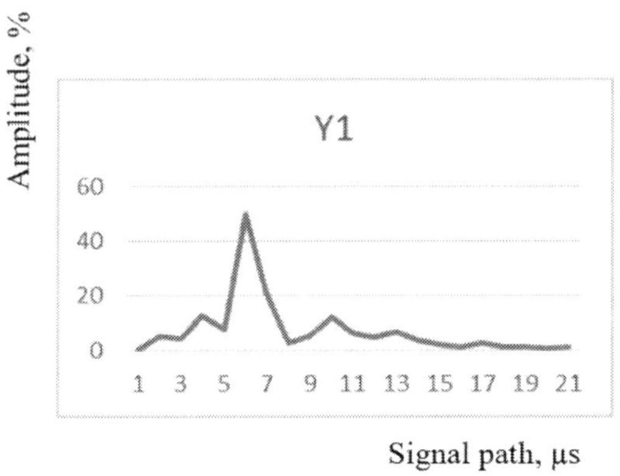
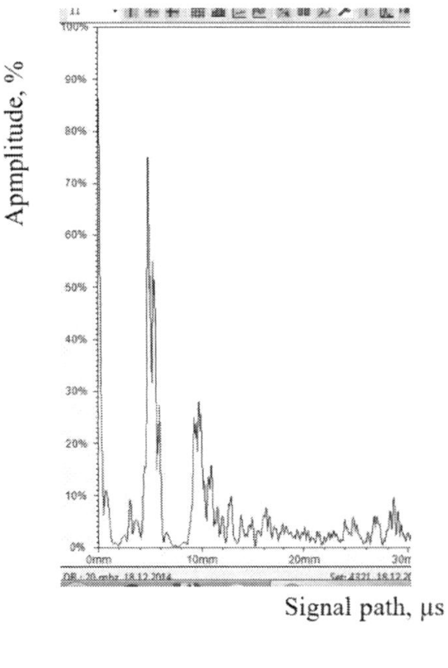	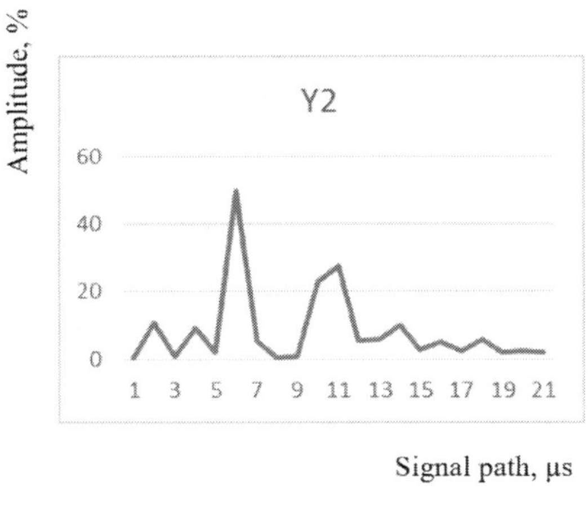

It is notable from Table 1 that the signal received during inspections (Table 1(a)) by standard equipment of ultrasonic testing does not allow to determine the depth of the hardened layer correctly, in comparison with the specialized instrument used in the first series of experiments (Figure 4). The output signal contains many peaks with close values of the amplitude and high noise levels. However the signal has the same character. It can be assumed that by further signal processing, especially, by signal averaging relevant results could be achieved. Further filtration is to be improved in order to reduce the level of electrical noise.

Averaging it is a type of filtering that is aimed at reduction of step-like changes of the signal. In this case there is a range of peaks with the same amplitudes which corresponds to point reflectors within the hardened layer (structure with different grain sizes). Relevant signal does not contain such steps. Due to a range of sudden changes but with small amplitude level difference, step signal contains high-frequency components that are not to be found in the averaged and smoothed signals. However currently there are many ways of averaging and signal processing as well as programs with appropriate algorithms to achieve this goal. Within this research the results obtained by testing samples with SAPHIR_Quantum system were averaged with the most common algorithm. The results are represented in the form of digitized and averaged signal (Figure 6). The results were digitized by Engauge program (Table 1(b)), the obtained values were summarized and the average value for each of respective measurement points was calculated. According to the obtained results case hardening depth of the left flank of the experiment sample №2 is 3.5 mm (the actual depth of the layer is 3.41 mm).

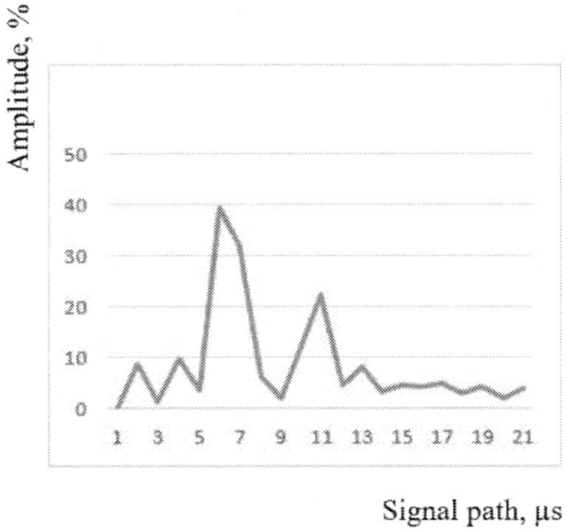

Figure 6. Digitized and averaged signal.

4. Summary

The developed case hardening depth control method discussed in this paper can be used for quality assurance testing in different areas of metalworking industry. The efficiency of the described method is proved by the experimental series with the specified equipment: QNet – P3121, Case Depth Tester. The results of these inspections correspond to the relevant metallographic analysis results. Deviation of the results merely reached 5 %.

SAPHIR_Quantum system, AREVA was used to verify the suitability of the method of ultrasonic nondestructive case hardening depth testing applying multi-operated ultrasonic instrument. According to the results of these experimental series deviation was only 2.5%. Metallographic analysis showed that case hardening depth was 3.41 mm and ultrasonic testing – 3.5 mm. Series of experiments demonstrated that the method of ultrasonic measurement of case hardening depth can be applied with the multifunctional equipment of ultrasonic testing in the case of implementing additional algorithm of signal processing.

Acknowledgments

Research has been conducted with financial support of Federal Target Programme «Research and Development in Priority Fields of Russian Research and Technological Complex for 2014 - 2020», project unique identifier is RFMEFI57514X0048.

References

[1] Lejeune Y 2001 Case Depth Tester P3121 *Quality Network (Q NET) Engineering GmbH* (Electronic Materials) p 5
[2] Yastrebova O 2011 Optical Flow Transducer Position Tracking for Quantitative Manual Ultrasonic Testing *Dissertation* p 107
[3] Drury J 2004 *Ultrasonic flaw detection for technitians* (Llandarcy: Silwerwing Limited)
[4] Paul A 2000 *Proc. 15th World Conference on Nondestructive Testing* (Rome) pp 157–159
[5] Salchak Y, Sednev D, Kroening M 2014 *Advanced Materials Research* **1040** pp 937–942
[6] Bolotina I O, Kryoning K V, Kvasnikov K G, Sednev D A, Sumtsova O V 2014 *Advanced Materials Research* **1040** pp 959–964

Effect of ion-plasma treatment on oxidation–reduction processes in lithium–titanium-zinc ferrites

A P Surzhikov[1], E N Lysenko[1], S A Gyngazov[1], T S Frangulyan[1], S A Lamonova[1]

[1]National Research Tomsk Polytechnic University, 30 Lenin Ave, 634050, Tomsk, Russia

E-mail: lysenkoen@tpu.ru

Abstract. We examined the effect of nitrogen, oxygen and argon plasma on the diffusion-controlled oxidation-reduction processes in lithium-titanium-zinc ferrite ceramics by measuring the activation energy of electrical conductivity in the depth of the sample. The experimental results show that the high-temperature treatment in polycrystalline ferrites by nitrogen or argon ion plasma greatly accelerates the oxidation-reduction processes in ferrites and changes the process direction depending on the partial pressure of oxygen.

1. Introduction

Recently, the radiation effect methods have been used to exert the effective effect on the structural condition and different physical-mechanical properties of a wide range of materials [1-5]. A particular interest has been shown in research on the modification of the surface and near-surface layers of materials based on oxide compounds in order to develop new a new optical, mechanical, electro-physical and chemical properties. [6-7]. This is due to the need to resolve a number of practical problems, such as the creation of conductive surface layers on non-metallic substrates, creation of surface layers with high adhesion and strength properties, etc.

As regards the efficiency of the effect on the surface condition, the methods of ion plasma treatment occupy one of the leading positions. By varying the parameters of the electric discharge forming plasma, as well as the type of gas it is possible to control the nature of that impact. In view of this, the applications of ionic technology methods as a means aimed at changing the properties of solids were very diverse.

The basic factor combining the ion plasma processes is the application of low-temperature gas plasma with low pressure as a source of particles for material treatment. This plasma is a weakly ionized gas (degree of ionization of a percent) consisting of a mixture of stable and excited atoms and molecules, products of the dissociation of molecules: radicals, electrons, positively and negatively charged ions. Modifying of surface properties in these cases is defined by both the radiation damage of the surface structure and physico-chemical interaction with components of the plasma particles of the irradiated material.

As regards the oxide materials, a special interest is the investigations of the effect of ion-plasma treatment in different atmosphere on the efficiency of the proceeding of diffusion-controlled oxidation-reduction processes which largely determine the stoichiometry of the surface layers and coatings, and are the basis of planar manufacturing technology of electronic devices.

Thus, in this work using the lithium-titanium-zinc ferrite ceramics as an example, we examined the effect of nitrogen, oxygen and argon plasma on the proceeding of diffusion-controlled oxidation-reduction processes having a significant role in the formation of the physical properties of the oxide materials. This investigations subject is preferred because the ferrites are characterized by the high electrical conductivity at high-temperature and, consequently, this greatly facilitates plasma treatment of the ferrites. When using high resistance oxide dielectric materials for these purposes, this procedure is complicated because of the effect of surface charging.

2. Experimental

The experiments were carried out on lithium-titanium-zinc ferrites produced by ceramic technology from a mechanical mixture with the following composition (in wt%): Li_2CO_3 – 11.2; MnO – 2.7; TiO_2 – 18.65; ZnO – 7.6; Fe_2O_3 – 59.81.

The ferrite samples were produced by one-sided pressing in the form of tablets with a diameter of 18 mm, thickness 3 mm. The sintering temperature was 1283 K. Depending on the sintering time, samples were obtained with different stoichiometric composition, which is estimated by the γ value (degree of deviation from stoichiometry) [8-9]. The chemical composition of the sintered ferrites was $Li_{0.649}Fe_{1.598}Ti_{0.5}Zn_{0.2}Mn_{0.051}O_{4-\gamma}$.

In the present work, ion-plasma treatment was carried out using a plasma generator based on a non-self-arc low-pressure discharge in the equipment of the Institute of High Current Electronics SB RAS, which was detailed in [10]. The presence of semiconductor properties in the test ferrite materials allows carrying out its processing on DC.

A vacuum chamber with a volume of 0.5 m^3 was evacuated to a pressure of 1 Pa, and then the ionized gas was supplied into the chamber. A discharge current of 80 A produced in the chamber a uniform gas-discharge plasma with a concentration in the center of the chamber of 10^{10} cm^{-3} and resulted in the density of the saturation ion current up to 10 mA/cm^2. The nitrogen, oxygen and argon gases were used. The gas was supplied through the cathode section of plasma generator, and consequently, it was possible to control the concentration of reactive gas ions in the chamber. To intensify a treatment process, a negative bias voltage of up to 400 V was supplied to the samples. The thermal heating of the samples was carried out by the stream of ions extracted from the discharge plasma and accelerated in the discharge space layer formed at the surface of the samples when submitting them negative bias. The energy of the accelerated single-charged ions corresponded to the bias voltage. The calculations show that with such energy ions penetrate into the surface layer of the ferrite samples to a depth not exceeding 20–40 Å. In order to ensure the required temperature, ion treatment was combined with electron heating. In this case, electron heating is 2/3 times of plasma treatment. Electron heating was carried out by supplying the anode potential to the samples. The discharge current was 50 A, the discharge voltage was 50 V. The procedure allows to heat the samples to temperatures of 970–1070 K.

The effect of plasma was determined by comparison with the investigation results derived from the samples subjected to heat treatment in non-ionized gas atmosphere at low pressure. The pressure of the gas (P), temperature (T) and annealing down (t) in the gas atmosphere corresponded to the pressure, temperature and time in plasma treatment, and the values for nitrogen were $P_{N2} = 10^{-3}$ mm Hg, T = 1000 K, t = 2h; for oxygen $P_{O2} = 4\times10^{-3}$ mm Hg, T = 1020 K, t = 1 h; for argon $P_{Ar} = 6\times10^{-3}$ mm Hg, T = 1020 K, t = 1 h. Measuring the thickness of the samples before and after ion-plasma treatment showed the absence of appreciable surface sputtering of samples.

The variation of the oxygen concentration and the nature of the process (oxidation or reduction) in the ferrite samples were evaluated based on the variation of the activation energy of electrical conductivity (E_a) in the depth of the sample. As shown in [11], the value of E_a is correlated with oxygen content in grain boundaries of polycrystalline ferrite, and increases with oxidation of ferrite and, on the other hand, decreases during the reduction process. The activation energy was calculated from electrical conductivity measured in the temperature range 300–550 K by two-probe method after consecutive removing of thin layers from the ferrite sample.

3. Results and discussion

The samples of two types differing in the initial degree of oxidation were investigated. The first type – the A samples were oxygen deficient and had the uniform distribution of oxygen in the sample depth before ion-plasma treatment. The initial oxygen concentration was determined using nuclear microanalysis techniques [12]. The oxygen concentration for these samples was 54.3 at%, and is lower than the value corresponding to stoichiometric composition in the ferrites equal to 57.14 at%. It is estimated that $\gamma = 0.199$ for A samples.

According to the electrical conductivity results, the A samples were a low resistance ($\rho = 10^3$ ohm·cm at $T = 300$ K) with a low value of $E_a = 0.19$–0.20 eV in the sample depth (Fig.1a, curves 1). This value of E_a for the polycrystalline ferrite samples corresponds to the minimum difference of oxygen content between grain boundaries and grains.

The second type – the B samples are characterized by 200–300 μm nonuniform distributions of $E_a(x)$ in thin subsurface layers (Fig.1b, curves 1), which were formed during ferrite sintering. The E_a=0.75 eV on the surface, where the grain boundaries are characterized by the maximum oxygen saturation, and changes smoothly up to 0.19–0.2 eV at 250–300 μm depth.

The variation of electrical properties in the course of ferrite heat treatment affects mainly the subsurface layers of the samples. Then the comparative analysis of E_a distribution formed after plasma treatment and thermal annealing was performed.

Figure 1 shows the E_a depth profiles measured before and after thermal annealing in gas and plasma atmospheres for A samples (Fig.1a) and B samples (Fig.1b). It can be seen that the plasma treatment of the ferrite greatly differs from both the annealing in gas atmosphere in the plasma absence at low pressure and the annealing in high vacuum ($P = 10^{-5}$ mm Hg).

3.1. *Ion plasma treatment in nitrogen or argon*

According to distributions of $E_a(x)$ for both A and B samples, a nitrogen plasma treatment greatly accelerates the process of ferrite reduction, i.e. activates the oxygen diffusion from the samples into the surrounding atmosphere. As a result, the E_a value reduced up to 0.1 eV on sample surface, and then gradually increased to values typical for non-annealed samples at 200–300 μm depth (Fig.1a and Fig.1b, curves 5).

In addition, the sample has been processed in inert gas argon plasma in order to ensure that the observed changes in the activation energy of conductivity are determined by oxygen content variation and not by possible nitrogen diffusion in ferrite.

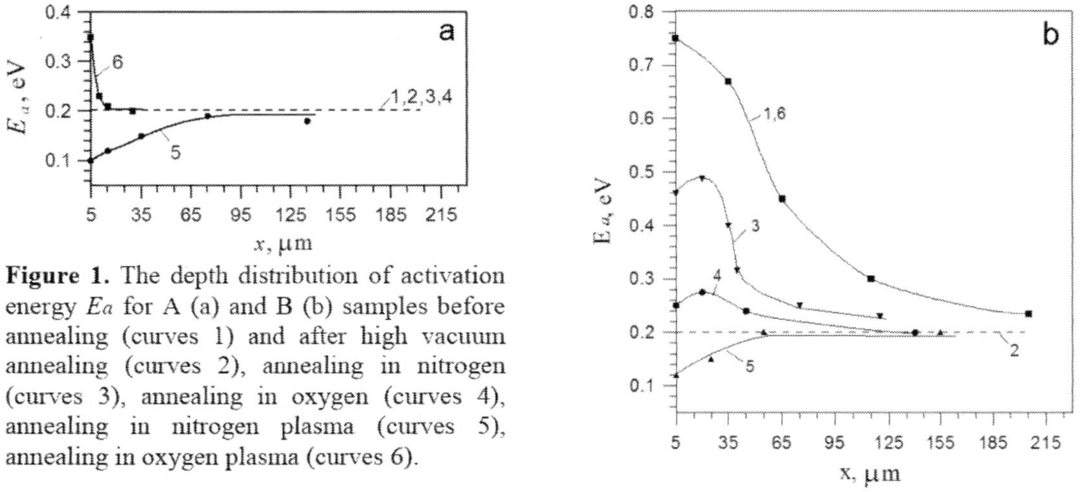

Figure 1. The depth distribution of activation energy *Ea* for A (a) and B (b) samples before annealing (curves 1) and after high vacuum annealing (curves 2), annealing in nitrogen (curves 3), annealing in oxygen (curves 4), annealing in nitrogen plasma (curves 5), annealing in oxygen plasma (curves 6).

3.2. Ion plasma treatment in oxygen.

According to results, the oxygen plasma treatment of samples affects the direction of the diffusion processes, which in the general case depends on the partial pressure in the surrounding atmosphere. Generally, for samples B, annealing in oxygen (in absence of plasma) results to ferrite reduction due to the loss of the oxygen from the sample (Fig.1b, curves 1, 4). However, oxygen plasma treatment under the same pressure does not change the initial depth profiles of E_a (Fig.1b, curves 1, 6).

This can be explained only by the fact that in the implemented experimental conditions, the oxygen diffusion from the sample surface is balanced by plasma enhanced its direct diffusion into the sample. For non-stoichiometry A samples, annealing in oxygen plasma results to oxidation of the surface layer (Fig.1a, curves 1, 6), regardless of the low oxygen partial pressure, at which the plasma is ignited.

We can explain these results as follows. Effects caused by the ion-plasma treatment may be due to two factors, namely, the state of an ionized gas environment and/or direct effect of their component on the surface layers of ferrite material. The dissociation of the gas molecules occurs in the plasma, causing the formation of reactive components which can strongly interact with the surface of the treated sample. The efficiency of this process increases as a result of the increased energy of the ions, accelerated in the electrical field. In this case, the adsorbed active oxygen at high temperature diffuses into the ferrite predominantly along grain boundaries due to the ceramic structure of ferrite samples. Then oxygen diffuses from the oxidized grain boundaries into the volume of the grains. The potential barrier formed on the boundary depends on the ratio of these diffusion flows, and determines the activation energy of electric conductivity in ferrite.

4. Conclusions

Thus, the ion plasma effect on the oxidation–reduction processes in polycrystalline ferrite has been determined. In particular, the effect is that the high-temperature treatment in polycrystalline ferrites by nitrogen or argon ion plasma greatly accelerates the oxidation-reduction processes in ferrites and changes the process direction depending on the partial pressure of oxygen. Namely, the nitrogen or argon plasma treatment of ferrites speeds up the process of ferrite reduction, i.e. activates an oxygen diffusion from the samples into the surrounding atmosphere. The oxygen plasma effect on the ferrites has a stimulating impact on the diffusion penetration of oxygen into the samples.

In conclusion, the results of this study confirm the active role of accelerated ions of nitrogen in the destruction of the oxide film on the surface of metals and alloys, and as the authors suggested [13], it was one of the reasons for increasing efficiency of nitriding. Previously, nitriding efficiency of these materials is achieved by introducing into the chamber a certain amount of hydrogen.

5. Acknowledgements

This work was financially supported by The Ministry of Education and Science of the Russian Federation in part of the science program.

References:

[1] Chernov I P, Berezneeva E V, Beloglazova P A, Ivanova S V, Kireeva I V, Lider A M, Remnev G E, Pushilina N S, Cherdantsev Y P 2014 Physicomechanical properties of the surface of a zirconium alloy modified by a pulsed ion beam *Technical Physics* **59** 535.

[2] Surzhikov A P, Vasendina E A, Lysenko E N, Nikolaev E V 2014 Kinetics of phase formation in a Li_2CO_3–TiO_2–Fe_2O_3 system during radiation-thermal synthesis *Inorganic Materials: Applied Research* **5** 102.

[3] Liu X, Mei X, Qiang J, Remnev G E, Wang Y 2014 Effects on structure and properties of $Zr_{55}Al_{10}Cu_{30}Ni_5$ metallic glass irradiated by high intensity pulsed ion beam *Applied Surface Science* **313** 911.

[4] Surzhikov A P, Pritulov A M, Lysenko E N, Sokolovskii A N, Vlasov V A, Vasendina E A 2012 Dependence of lithium–zinc ferrospinel phase composition on the duration of synthesis in an accelerated electron beam *J. Therm. Anal. Calorim.* **110** 733.

[5] Kholodnaya G, Ponomarev D, Sazonov R, Remnev G 2014 Characteristics of pulsed plasma-chemical synthesis of silicon dioxide nanoparticles *Radiation Physics and Chemistry* **103** 114.
[6] Surzhikov A P, Frangulyan T S, Ghyngazov S A, Vasil'ev I P 2014 The effect of a low-energy high-current pulsed electron beam on surface layers of porous zirconium Ceramics *Technical Physics Letters* **40** 762.
[7] Surzhikov A P, Frangulyan T S, Ghyngazov S A, Vasiliev I P 2014 Electron microscopy studies of near-surface layers of $ZrO_2(Y)–Al_2O_3$ composite ceramic modified by high current beam of low-energy electron *Inorganic Materials: Applied Research* **5** 536.
[8] Surzhikov A P, Frangulyan T S, Ghyngazov S A, Lysenko E N 2010 Investigation of oxidation processes in non-stoichiometric lithium-titanium ferrites using TG analysis *J. Therm. Anal. Calorim.* **102** 883.
[9] Surzhikov A P, Frangulyan T S, Ghyngazov S A, Lysenko E N 2012 Investigation of structural states and oxidation processes in $Li_{0.5}Fe_{2.5}O_{4-\delta}$ using TG analysis *J. Therm. Anal. Calorim.* **108** 1207.
[10] Schanin P M, Koval N N, Tolkachev V S, Gusheiiets V I 2000 Plasma-emitter electron accelerators *Russian Physics Journal* **43** 427.
[11] Surzhikov A P, Pritulov A M, Peshev V V, Malyshev A V 2001 Nonuniform distribution of electric properties in ferrite ceramics *Russian Physics Journal* **44** 1244.
[12] Surzhikov A P, Lysenko E N, Ghyngazov S A, Frangulyan T S 2002 Determination of the oxygen diffusion coefficient in polycrystalline Li-Ti ferrites *Russian Physics Journal* **45** 989.
[13] Shchanin P M., Koval' N N, Goncharenko I M, Grigor'ev S V 2001 Structural steels nitriding in low-pressure gas discharge *Fizika i Khimiya Obrabotki Materialov* **3** 16.

About accuracy of the discrimination parameter estimation for the dual high-energy method

S P Osipov, S V Chakhlov, O S Osipov, A M Shtein, D V Strugovtsev

National Research Tomsk Polytechnic University, Lenin Avenue, 30, Tomsk, 634050, Russia

E-mail: osip1809@rambler.ru

Abstract. A set of the mathematical formulas to estimate the accuracy of discrimination parameters for two implementations of the dual high energy method – by the effective atomic number and by the level lines is given. The hardware parameters which influenced on the accuracy of the discrimination parameters are stated. The recommendations to form the structure of the high energy X-ray radiation impulses are formulated. To prove the applicability of the proposed procedure there were calculated the statistical errors of the discrimination parameters for the cargo inspection system of the Tomsk polytechnic university on base of the portable betatron MIB-9. The comparison of the experimental estimations and the theoretical ones of the discrimination parameter errors was carried out. It proved the practical applicability of the algorithm to estimate the discrimination parameter errors for the dual high energy method.

1. Introduction

The different implementations of digital radiography are widely used for the custom examination and for the flaw detection [1–4]. One of the major implementation of the digital radiography is a dual energy method (DEM) [5–8], which can discriminate the testing object (TO) materials. The discrimination for the custom examination means the correlation of test object material or their constituent with one of the extensive material classes. The mentioned division of the materials to classes is carried out by two main methods [8–11] – by effective atomic number and by level line method. In any modification of DEM on the initial stage there are generated two digital radiographs for two specially selected maximal X-ray energies. One of the important quality index of the radiographic system with the option of discrimination of TO material and their constituents is a precision of the discrimination parameters. This index is essentially dependent from many factors, for example, from the range of effective atomic numbers and from TO constituent sizes, from the given error of the discrimination parameter, from the maximal high energy of X-ray radiation. The state of the art analysis shows that the estimation task of the discrimination quality of the TO material and their constituents for major implementations of the discrimination method by dual high energy method is not solved in corpore.

2. The estimation of the discrimination parameter errors for the dual high energy methoda

To solve the problem of the discrimination TO materials and their constituents by DEM there are used different versions of two major approaches [8]. The first approach estimates the effective atomic number of the TO material. The second approach [8] generates the discrimination parameter image.

The value of every image point is compared with some level lines, and the comparison result is used to correlate T material with one or another class of material. The mentioned approaches are differed essentially by the volume of used preliminary information and by the algorithms of the primary radiograph processing therefore it is necessary to consider a specific character of the mathematical models to discriminate TO materials by use the DEM for the both abovementioned approaches.

Let consider some common part of the abovementioned discrimination approaches.

2.1. Common part

On the output of the digital radiographic system with the TO material discrimination option the maximal X-ray energies E_i, i=1,2 are corresponded to the primary radiographs $\mathbf{J}_i = \{J_i(x, y) : (x, y) \in S\}$, i = 1,2, where S is a discrete set of R^2. The physical meaning of the primary radiographs is the distribution of the absorbed energy of the primary attenuated X-ray radiation along the set S.

In the introduction we emphasized that the major characteristic of the discrimination quality is the discrimination parameters error, which evidently related with noise level of the primary radiographs. The research purpose of is to detect the mutual relation of the major customer characteristic with the cargo inspection system parameters, therefore we shall make more detailed analysis of the noise transformation during the step-by-step processing of the primary images.

On first stage we shall estimate the noise levels on the original images. In the first approximation we shall consider the radiation source as a point and isotropic one. Let suppose that the sensitivity volume of the singled radiometric detector has a rectangular prism form, which the center axis is oriented to the radiation source. The radiometric detector has size hd in the propagation direction of the primary X-ray beam, but in the perpendicular plane – $a_d \times b_d$. The distance between the radiation source and the front surface of the detector is equal to F. At the first stage of the design there is known some characteristic of the X-ray radiation source – P_{0max}, for example, the absorbed dose power or exposure dose of X-ray radiation on air at 1 meter distance and ν is a frequency of X-ray radiation impulses. The maximal energy E of X-ray radiation is adjustable value and can take values in the range from E_{min} to E_{max}. The control system of X-ray source can generate the sequence of alternated impulses with maximal energies E_1 and E_2, $E_{min}<E_1<E_2<E_{max}$ according to a defined principle. We would name as an impulse package a collection of the serial X-ray radiation impulses, those need to produce a single line of the primary shadow radiographs J_1, J_2, corresponding one another.

For the pulse X-ray radiation sources the time is measured in the pulse period – 1/ν. The number of impulses n in a package, which is need to generate a single line of the identification image, is equal to the sum of impulses with energy E_1 – n_1 and impulses with energy E_2 – n_2. Here the identification image is a distribution of the calculated effective atomic number or discrimination parameter of the level lines method along the set S. The value n defines the inspection performance. The maximal performance of the radiographic system with option of the TO material discrimination and their constituents is reached for n_1=1, n_2=1. Evidently that in practice it is reasonable to advance a requirement $n_2< n_1$. Let introduce a parameter $p=n_1/n_2$. If the parameter p and the number of impulses in the package n, are known that $n_2=n/(1+p)$ and $n_1=np/(1+p)$.

For the customer it is most interesting the minimal size of constituent, for which the material is surely discriminated by DEM. Let the projection of the specified constituent on the set S is consisted from M pixels.

Note that for every X-ray radiation source with the regulated maximal energy it is known a function W, which describes the dependence of the X-ray radiation source characteristic P_0 from the current value of the maximal energy E. Let the mentioned dependence has the following form

$$P_0(E) = P_{0\max} W(E, E_{\max}). \quad (1)$$

The function $W(E, E_{max})$ in the range of the maximal X-ray energies above 2 MeV is a smooth increasing function, for example, to describe the absorbed dose power [12] there is used the power dependence $W(E, E_{max})=(E/E_{max})^3$.

The number of photons N_{0i} of the X-ray radiation with the maximal energy E_i, whose are intersected the front surface of the single detector in the time unit when the test object is absent, is executed by formula

$$N_{0i} \approx \frac{CP_{0max}W(E_i, E_{max})a_d b_d}{vF^2 \int_0^{E_i} Ef(E, E_i)\mu_{ab\ air}(E)dE}, \quad (2)$$

where C is the conversion coefficient of the X-ray radiation source characteristic to dimension MeV/sec; $f(E, E_i)$ is the energy spectrum of the X-ray radiation with maximal energy E_i; $\mu_{ab\ air}(E)$ is the linear absorption coefficient of the proton radiation with energy E in the air.

Let the X-ray radiation flow is weaken by the test object with the thickness ρH and the material's effective atomic number Z. In this case the number of photons $N_i(\rho H,Z)$ of the X-ray radiation with the maximal energy E_i, whose are not interacted with the TO and are registered by the radiometric detector in the time unit, is defined by the approximate expression

$$N_i(\rho H, Z) \approx N_{0i} \int_0^{E_i} f(E, E_i) \mathrm{e}^{-m(E,Z)\rho H}\left[1 - \mathrm{e}^{-\mu_d(E)h_d}\right]dE, \quad (3)$$

where $m(E,Z)$ is the mass attenuation coefficient of the photon radiation with energy E by the test object material; $\mu_d(E)$ is the linear attenuation coefficient of the proton radiation with energy E by the detector material.

The formula to estimate the radiometric signal beyond the test object thickness ρH and the effective atomic number Z is I_i $(\rho H, Z)$, i=1,2 without contribution of the background noises of the radiometric detectors has the form

$$I_i(\rho H, Z) = N_i(\rho H, Z) n_i \overline{E_{iab}(\rho H, Z)}, \quad (4)$$

where $\overline{E_{iab}(\rho H, Z)}$ is the mean value of the absorbed energy of the registered X-ray photon with maximal energy E_i beyond the testing object. The values $\overline{E_{iab}(\rho H, Z)}$ are calculated by the following way

$$\overline{E_{iab}(\rho H, Z)} = \frac{\int_0^{E_i} \overline{E_{ab}(E)} f(E, E_i) \mathrm{e}^{-m(E,Z)\rho H}\left[1 - \mathrm{e}^{-\mu(E)h_d}\right]dE}{\int_0^{E_i} f(E, E_i) \mathrm{e}^{-m(E,Z)\rho H}\left[1 - \mathrm{e}^{-\mu(E)h_d}\right]dE}, \quad (5)$$

where $\overline{E_{ab}(E)}$ is the mean energy value of the registered photon with energy E [13].

The dispersion of the radiometric signals $I_i(\rho H,Z)$ – $DI_i(\rho H,Z)$, i= 1,2 are calculated by formula [13]

$$DI_i(\rho H, Z) = n_i N_i(\rho H, Z) \overline{E^2_{iab}(\rho H, Z)} = n_i N_i(\rho H, Z) \overline{E_{iab}(\rho H, Z)}^2 \eta_i^2(\rho H, Z), \quad (6)$$

where $\overline{E^2_{iab}(\rho H, Z)}$ and $\eta_i^2(\rho H, Z)$ are the mean values of the absorbed energy square for the registered photons and the accumulation coefficient of fluctuations for X-ray radiation with the maximal energy E_i beyond the testing object under investigation. The formula to calculate $\overline{E^2_{iab}(\rho H, Z)}$ has the form similar to (5)

$$\overline{E_{iab}^2(\rho H,Z)}=\frac{\int_0^{E_i}\overline{E_{ab}^2(E)}f(E,E_i)\mathrm{e}^{-m(E,Z)\rho H}\left[1-\mathrm{e}^{-\mu(E)h_d}\right]dE}{\int_0^{E_i}f(E,E_i)\mathrm{e}^{-m(E,Z)\rho H}\left[1-\mathrm{e}^{-\mu(E)h_d}\right]dE},\tag{7}$$

where $\overline{E_{ab}^2(E)}$ is the mean value of the energy square of the registered photon with energy E, which can be calculated, for example, by use formulas from [13].

After substitution (3) and (7) to formula (6) we get the expression coupling the dispersion $DI_i(\rho H,Z)$ with the number of impulses n_i.

$$DI_i(\rho H,Z)=N_i(\rho H,Z)n_i\frac{\int_0^{E_i}\overline{E_{ab}^2(E)}f(E,E_i)\mathrm{e}^{-m(E,Z)\rho H}\left[1-\mathrm{e}^{-\mu(E)h_d}\right]dE}{\int_0^{E_i}f(E,E_i)\mathrm{e}^{-m(E,Z)\rho H}\left[1-\mathrm{e}^{-\mu(E)h_d}\right]dE}.\tag{8}$$

At the next stage we consider the process of the noise transformation during the generation from primary images Ji the resulting radiographs $\mathbf{R}_i=\{R_i(x,y):(x,y)\in S\}$.

The transformation of the primary images J_i to the resulting images R_i is consisted from several transformations, whose are the «black» calibration, the «white» calibration, the normalization on the reference detector signal, taking the logarithm. The «black» calibration is reduced to the subtraction of the mean values of the background noises levels of the radiometric detectors J_b. For the detectors J_b can be the determined values. During the «white» calibration the signals from the detectors are normalized on the signals without the testing object $I_{0i}=J_{0i}-J_b$ is the calibration. The transformation of the original radiometric signals $J_i(x,y)$ to I_i(x,y), and then to R_i(x,y) is described by formula

$$R_i(x,y)=-\ln\frac{J_i(x,y)-J_b}{J_{0i}-J_b}=-\ln\frac{I_i(x,y)}{I_{0i}}.\tag{9}$$

The values of R_i(x,y) are numerically equal to the ray-lengths of the testing object for the X-ray radiation with the maximal energy E_i along the line, connecting the source point and the detector point with coordinates (x,y), the measurement unit is the length of mean free path.

The dispersions R_1 and R_2 are estimated by use decomposition of (9) on small increments. The final expressions to calculate the dispersions $DR_1(\rho H,Z)$ and $DR_2(\rho H,Z)$ for the testing object thickness ρH and the effective atomic number Z have the form

$$DR_i(\rho H,Z)=\frac{DI_i(\rho H,Z)}{I_i^2(\rho H,Z)}+\frac{DI_i(0,0)}{I_i^2(0,0)},\tag{10}$$

where $I_i(0,0)$, $DI_i(0,0)$ are the signal value and the signal dispersion when the testing object is absent. With including (4), (8) the expression (10) would have the following form

$$DR_i(\rho H,Z)=\frac{\tau_i(\rho H,Z)}{n_i},\ \tau_i(\rho H,Z)=\frac{\eta_i^2(\rho H,Z)}{N_i(\rho H,Z)}+\frac{\eta_i^2(0,0)}{N_i(0,0)},\tag{11}$$

where $N_i(0,0)$, $\eta_i(0,0)$ are the number of protons registered by detector and the accumulation coefficient of fluctuations without the testing object.

Note 1. In practice the «white» calibration is carried out by the large number of the image lines k_0, therefore the parameter estimations τ_i are defined by the expression

$$\tau_i(\rho H,Z)=\frac{\eta_i^2(\rho H,Z)}{N_i(\rho H,Z)}+\frac{\eta_i^2(0,0)}{k_0N_i(0,0)},\tag{12}$$

where $\overline{N_i(0,0)}$ is the selective mean of the photon number registered by detector without the testing object. For large values of k_0 the second summand in sums (11), (12) can be not included.

At the next stage the further transformations of the noise levels are depended from the image processing algorithms.

Note 2. The expressions to estimate the dispersions $R_1(\rho H,Z)$ and $R_2(\rho H,Z)$ for the testing object thickness ρH and the effective atomic number Z (10), (11) are derived without consideration of the pulsed X-ray radiation instability. In this case the instability means that the parameters characterized the single X-ray radiation pulse are the random values. The instability influence minimization is achieved by the normalization of the detector signals $I_i(\rho H,Z)$ on the signals from the reference detector I_{pi}. The reference detector signals I_{pi} are not dependent from the testing object. The expression to estimate the dispersion of the normalized signal $I_i(\rho H,Z)/I_{pi}$ $-D(I_i(\rho H,Z)/I_{pi})$ has the following form

$$D\left(\frac{I_i(\rho H,Z)}{I_{pi}}\right)=\frac{I_{pi}^2 DI_i(\rho H,Z)+I_i^2(\rho H,Z)DI_{pi}}{I_{pi}^4}. \quad (13)$$

The expression (13) was derived for the condition of the full compensation of instability, which can be reached in that case when only the number of electrons dropped on the accelerator's target is instable [14].

From analysis of expression (13) it follows that during implementation of some requirements to the reference channel the estimation of the signal dispersions $R_1(\rho H,Z)$ and $R_2(\rho H,Z)$ can be done by the expressions (10), (12) whereas the first note. The abovementioned requirements in the formalized form look like

$$I_{pi} >> I_i(\rho H,Z). \quad (14)$$

The physical implementation of the restriction (14) is reduced to use as the reference channel the special selected radiometric detector working synchronously with the measurement channels and distinguish from them by the large amount of registered photons and by the low value of the fluctuation accumulation coefficient.

2.2. Discrimination by the effective atomic number

As result of the combined processing of the resulting radiographs R_1 and R_2 there are producing the DEM parameters images – **A** and **B**. To estimate DEM parameters $A(x,y)$ and $B(x,y)$ for every point (x,y) of set S there is solved the non-linear equation system

$$\begin{aligned}
-\ln\frac{\int_0^{E_1}\overline{E_{ab}(E)}f(E,E_1)\mathrm{e}^{-A(x,y)w_1(E)-B(x,y)w_2(E)}\left[1-\mathrm{e}^{-\mu(E)h_d}\right]dE}{\int_0^{E_1}\overline{E_{ab}(E)}f(E,E_1)\left[1-\mathrm{e}^{-\mu(E)h_d}\right]dE}&=R_1(x,y)\\
-\ln\frac{\int_0^{E_2}\overline{E_{ab}(E)}f(E,E_2)\mathrm{e}^{-A(x,y)w_1(E)-B(x,y)w_2(E)}\left[1-\mathrm{e}^{-\mu(E)h_d}\right]dE}{\int_0^{E_2}\overline{E_{ab}(E)}f(E,E_2)\left[1-\mathrm{e}^{-\mu(E)h_d}\right]dE}&=R_2(x,y),
\end{aligned} \quad (15)$$

where w_1(E), $w_2(E)$ are the energy dependencies of two main types of the photon radiation interaction with the testing object material. The first process in the system (15) is the Compton effect and the second one is the pair production.

To solve the system (15) relative to the parameters $A(x,y)$ and $B(x,y)$ it is necessary know the maximal energy values – E_1 and E_2, the energy spectra $f(E, E_1)$ and $f(E, E_2)$, the energy dependencies $\overline{E_{ab}(E)}$, $w_1(E)$ and $w_2(E)$. There are known the analytical descriptions of the energy dependencies of the mean value of the registered photon energy [13] and the interaction cross-section of the photon radiation with the material, for example [15]. Besides the analytical dependencies we can use the database of the gamma radiation interaction with the material [16, 17].

The certain complication during the organization of the preliminary information for the system (15) is related with the estimation of the energy spectra of the high-energy X-ray radiation. At the present time there is used the analytical approach to describe $f(E,E_i)$, which is based on the Shiff's formula [18–20] adjusted for the attenuation in the radiation source filters and adjusted for the registration efficiency of the photon radiation.

During solution of the system (15) the random value dispersions A, B – DA and DB, and also the covariation cov(A,B) are founded by the least increments method application. The final expressions for the dispersions $DA(\rho H,Z)$, $DB(\rho H,Z)$ and the covariation $\mathrm{cov}(A,B)(\rho H,Z)$ for the testing object thickness ρH and the effective atomic number Z have the form

$$DA(\rho H,Z)=\frac{g_{22}^2 DR_1(\rho H,Z)+g_{12}^2 DR_2(\rho H,Z)}{G^2(\rho H,Z)},$$
$$DB(\rho H,Z)=\frac{g_{11}^2 DR_2(\rho H,Z)+g_{21}^2 DR_1(\rho H,Z)}{G^2(\rho H,Z)}, \quad (16)$$
$$\mathrm{cov}(A,B)(\rho H,Z)=\frac{-g_{22}g_{21}DR_1(\rho H,Z)-g_{12}g_{11}DR_2(\rho H,Z)}{G^2(\rho H,Z)}.$$

where

$$g_{ij}=g_{ij}(\rho H,Z)=\frac{\int_0^{E_i} w_j(E)\overline{E_{ab}(E)}f(E,E_1)\mathrm{e}^{-m(E,Z)\rho H}\left[1-\mathrm{e}^{-\mu(E)h_d}\right]dE}{\int_0^{E_i}\overline{E_{ab}(E)}f(E,E_1)\mathrm{e}^{-m(E,Z)\rho H}\left[1-\mathrm{e}^{-\mu(E)h_d}\right]dE},$$
$$G(\rho H,Z)=g_{11}(\rho H,Z)g_{22}(\rho H,Z)-g_{12}(\rho H,Z)g_{21}(\rho H,Z).$$

The substitution (11) in (16) gives

$$DA(\rho H,Z)=\frac{g_{22}^2\,\tau_1(\rho H,Z)/n_1+g_{12}^2\,\tau_2(\rho H,Z)/n_2}{G^2(\rho H,Z)},$$
$$DB(\rho H,Z)=\frac{g_{11}^2\,\tau_2(\rho H,Z)/n_2+g_{21}^2\,\tau_1(\rho H,Z)/n_1}{G^2(\rho H,Z)} \quad (17)$$
$$\mathrm{cov}(A,B)(\rho H,Z)=\frac{-g_{22}g_{21}\,\tau_1(\rho H,Z)/n_1-g_{12}g_{11}\,\tau_2(\rho H,Z)/n_2}{G^2(\rho H,Z)}.$$

The final research stage of the noise transformation is connected with the estimation of the effective atomic number dispersion. In the high-energy domain [8] the effective atomic number estimation $Z_V(\rho H,Z)$ is related to the DEM parameters by the following expressions $A(\rho H,Z)$ and $B(\rho H,Z)$ [8]

$$Z_V(\rho H,Z)\approx\frac{B(\rho H,Z)}{A(\rho H,Z)}. \quad (18)$$

Appling the least increments method to (18) we get for the testing object thickness ρH and the material's effective atomic number – Z the formula to calculate the dispersion of the effective atomic number estimation $Z_V(\rho H,Z)$ – $DZ_V(\rho H,Z)$

$$DZ_V(\rho H,Z)\approx\frac{A^2DB(\rho H,Z)+B^2DA(\rho H,Z)-2AB\,\mathrm{cov}(A,B)(\rho H,Z)}{A^4}. \quad (19)$$

Let substitute the expression (17) in the formula (19). After simple conversions the expression to estimate the effective atomic number dispersion reduce to the form

$$DZ_V(\rho H,Z) \approx \frac{(Ag_{21}+Bg_{22})^2\,\tau_1(\rho H,Z)/n_1 + (Ag_{11}+Bg_{12})^2\,\tau_2(\rho H,Z)/n_2}{G^2(\rho H,Z)A^4}. \quad (20)$$

We are interesting in the relation of the dispersion of the estimated testing object parameter $DZ_V(\rho H,Z)$ with the number of pulses in the package $n=n_1+n_2$, with the parameter $p= n_1/n_2$ and with the physical TO characteristics – the mean value of the material's effective atomic number Z and with the thickness ρH. The parameters of the high-energy implementation of DEM $A(\rho H,Z)$ and $B(\rho H,Z)$ are related to Z and ρH by the following way [8]

$$A(\rho H,Z)=\rho H,\quad B(\rho H,Z)=Z\rho H. \quad (21)$$

After substitution of (21) in (20) under conditions $n_2=n/(1+p)$ and n1=$np/(1+p)$ we get

$$DZ_V(\rho H,Z) \approx \frac{(1+p)\left[(g_{21}+Zg_{22})^2\,\tau_1(\rho H,Z)/p + (g_{11}+Zg_{12})^2\,\tau_2(\rho H,Z)\right]}{nG^2(\rho H,Z)(\rho H)^2}. \quad (22)$$

The expression (22) together with formulas (11), (12) allow to estimate the measurement accuracy of the TO material's effective atomic number, in term of the time to generate lines of the primary radiographs, the TO parameters, the maximal energies of the X-ray radiation. The measurement accuracy of the effective atomic number Z can be estimated by the least-square deviation ΔZ – $\Delta Z_V(\rho H,Z)=\sqrt{DZ_V(\rho H,Z)}$.

Evidently, that there is exist the optimal value of the parameter p, for which the dispersion value $DZ_V(\rho H,Z)$ under fixed values Z and ρH is minimal. The formula to calculate p_{opt} has form

$$p_{opt}(\rho H,Z)=\frac{g_{21}+Zg_{22}}{g_{11}+Zg_{12}}\sqrt{\frac{\tau_1(\rho H,Z)}{\tau_2(\rho H,Z)}}. \quad (23)$$

The dispersion value $DZ_V(\rho H,Z)$ corresponding to p_{opt}, is defined by substitution (23) in (22).

The formula (22) can calculate the number of pulses in the package which need to generate the line's pair of the images I_1 and I_2, on the basis of the user defined maximal deviation of the effective atomic number ΔZ_{lim}. The mentioned maximal deviation of the effective atomic number ΔZ_{lim} is named the resolution by Z. The unknown expression has the form

$$n(\rho H,Z) \approx \frac{(1+p)\left[(g_{21}+Zg_{22})^2\,\tau_1(\rho H,Z)/p + (g_{11}+Zg_{12})^2\,\tau_2(\rho H,Z)\right]}{G^2(\rho H,Z)(\rho H)^2\,\Delta Z^2_{lim}}. \quad (24)$$

Note 3. The expressions (20)–(22), (24) were derived for the case when the estimation of the effective atomic number of the testing object material or their constituent $Z_V(\rho H,Z)$ are produced for one identification image point (one pixel). In practice the image of the minimal constituent defined by user consist of M pixels M>>1. Therefore it is logically to use the mean estimation value of the effective atomic number $\overline{Z_V(\rho H,Z)}$ as the identification parameter. The dispersion of mean value $\overline{DZ_V(\rho H,Z)}$ is defined by the formula

$$\overline{DZ_V(\rho H,Z)}=\frac{DZ_V(\rho H,Z)}{M}. \quad (25)$$

Including (25) the all mentioned in Note 3 the formulas are corrected by including the additional factor equal to $1/M$ into the right part.

Note 4. The inspection efficiency is essentially depended of the testing object thickness ρH. The internal structure randomness of the inspecting objects creates some dilemma before the customer.

This dilemma must give the answer to the question – what is more preferable – the high inspection efficiency with the qualified discrimination of the testing object materials and their constituents with small thickness or the low inspection efficiency with the qualified discrimination of the large thickness material.

2.3. Discrimination by the level lines method

The second implementation of the testing object discrimination and their constituents by DEM we named the level lines method [8]. In this implementation on the base of resulting images R_1 and R_2 the identification image is created

$$\mathbf{Q} = \left\{ Q(x,y) = \frac{R_2(x,y)}{R_1(x,y)} : (x,y) \in S \right\}. \tag{26}$$

The testing object material in the point with coordinates (x,y) is correlated with some material, if the following restriction is carried out

$$U_-(R_1(x,y)) < Q(R_1(x,y)) \leq U_+(R_1(x,y)), \tag{27}$$

where $U_-(x,y)$, $U_+(x,y)$ are the level lines for the correlated material.

For the testing object thickness ρH and the material's effective atomic number Z the identification parameter dispersion $Q(\rho H,Z)$ – $DQ(\rho H,Z)$ is defined by the small increment decomposition

$$DQ(\rho H,Z) = \frac{R_1^2(\rho H,Z)DR_2(\rho H,Z) + R_2^2(\rho H,Z)DR_1(\rho H,Z)}{R_1^4(\rho H,Z)}. \tag{28}$$

After substitution (10), (11) in (28) including $n_2=n/(1+p)$ and $n_1=np/(1+p)$ we get

$$DQ(\rho H,Z) = \frac{(1+p)\left[R_1^2(\rho H,Z)\tau_2(\rho H,Z) + R_2^2(\rho H,Z)\tau_1(\rho H,Z)/p\right]}{nR_1^4(\rho H,Z)}. \tag{29}$$

The collection of the expressions (3), (9), (11) and (29) allows to estimate the identification parameter accuracy by the level lines method – $\Delta Q(\rho H,Z) = \sqrt{DQ(\rho H,Z)}$.

The value of the parameter p – $p_{opt}(\rho H,Z)$, for which the value $DQ(\rho H,Z)$ is minimal, is calculated by the formula

$$p_{opt}(\rho H,Z) = \frac{R_2(\rho H,Z)}{R_1(\rho H,Z)}\sqrt{\frac{\tau_1(\rho H,Z)}{\tau_2(\rho H,Z)}}. \tag{30}$$

The minimal dispersion value $DQ(\rho H,Z)$ is defined by substitution (30) in the expression (29).

The total number of pulses n, which determine the cargo inspection performance with the material discrimination by the level lines method, is calculated from (29), starting from the limiting accuracy of the identification parameter ΔQ_{lim},

$$n(\rho H,Z) = \frac{(1+p)\left[R_1^2(\rho H,Z)\tau_2(\rho H,Z) + R_2^2(\rho H,Z)\tau_1(\rho H,Z)/p\right]}{R_1^4(\rho H,Z)\Delta Q_{lim}^2}. \tag{31}$$

The expressions (3), (9), (11), (31) allow to estimate the cargo inspection performance with the option of the material discrimination by level lines method. The notes 3 and 4 from the previous section must take into account.

3. Calculation ΔZ_V and ΔQ

To show the applicability of the above mentioned algorithms for estimation of the identification parameter accuracy by high-energy DEM there was carried out the series of the mean-square deviation calculation for the identification parameter for two pairs of the maximal values of the X-ray radiation

– E_1=4.5 MeV, E_2=7.5 MeV and E_1=4.5 MeV, E_2=9 MeV. The testing object thickness ρH was varied from 10 to 120 g/cm^2, but the effective atomic number Z take the values 6, 13, 26 and 82. The calculations were done for betatron MIB-4/9 with characteristics P_{0max}=20 R/min for and E_{max}=9 MeV, ν= 200 pulses per second. The inspection geometry was characterized by the parameters: the focal length F=4.2 m; the radiometric detectors on base of the CdWO4 scintillator with thickness h_d=50 mm and the lateral sizes $a_d \times b_d$ =5×6 mm^2. The values n_1 =3, n_2=1. The minimal number of pixels in the projection on the mapping surface for the surely identified constituent is M=16.

The tables 1 and 2 show the calculation results ΔZ_V and ΔQ for the different thicknesses ρH and the different testing object materials.

Table 1. The dependence of the mean-square deviation for the effective atomic number ΔZ_V from ρH.

Material	$E_1 - E_2$, MeV	ρH, g/cm^2											
		10	20	30	40	50	60	70	80	90	100	110	120
C	4.5 – 7.5	11.2	6.6	5.1	4.5	4.2	4.2	4.2	4.3	4.5	4.7	5.0	5.4
	4.5 – 9	5.6	3.3	2.6	2.3	2.2	2.1	2.2	2.2	2.3	2.5	2.6	2.8
Al	4.5 – 7.5	11.4	6.8	5.3	4.7	4.5	4.4	4.5	4.6	4.8	5.1	5.5	5.9
	4.5 – 9	5.8	3.4	2.7	2.5	2.3	2.3	2.4	2.4	2.6	2.7	2.9	3.2
Fe	4.5 – 7.5	12.0	7.2	5.8	5.3	5.1	5.1	5.2	5.4	5.8	6.2	6.8	7.4
	4.5 – 9	6.1	3.8	3.1	2.8	2.7	2.7	2.8	3.0	3.2	3.5	3.8	4.2
Pb	4.5 – 7.5	14.6	9.5	8.3	8.0	8.0	8.6	9.3	10.3	11.6	13.2	15.1	17.5
	4.5 – 9	8.0	5.3	4.7	4.6	4.7	5.1	5.5	6.2	7.0	8.1	9.3	10.8

The analysis of the data mentioned in the table 1, shows that the accuracy of Z is high for the small and large thicknesses. The accuracy of Z is reduced with increasing of the greater energy.

Table 2. The dependence of the mean-square deviation of the discrimination parameter ΔQ ×0.01 from ρH.

Material	$E_1 - E_2$, MeV	ρH, g/cm^2											
		10	20	30	40	50	60	70	80	90	100	110	120
C	4.5 – 7.5	1.10	0.81	0.75	0.76	0.79	0.85	0.92	1.01	1.12	1.24	1.39	1.56
	4.5 – 9	0.94	0.69	0.64	0.64	0.66	0.71	0.77	0.85	0.94	1.05	1.17	1.32
Al	4.5 – 7.5	1.10	0.82	0.76	0.77	0.80	0.86	0.93	1.02	1.13	1.26	1.41	1.58
	4.5 – 9	0.94	0.67	0.65	0.65	0.68	0.73	0.79	0.87	0.96	1.07	1.21	1.36
Fe	4.5 – 7.5	1.08	0.83	0.77	0.78	0.82	0.88	0.96	1.06	1.18	1.32	1.49	1.68
	4.5 – 9	0.93	0.71	0.66	0.67	0.70	0.76	0.83	0.92	1.02	1.15	1.30	1.47
Pb	4.5 – 7.5	0.25	0.29	0.33	0.39	0.46	0.54	0.63	0.75	0.89	1.06	1.27	1.51
	4.5 – 9	0.23	0.26	0.30	0.36	0.42	0.49	0.58	0.69	0.82	0.98	1.16	1.39

The analysis of the data mentioned in the table 2 shows that the accuracy of the identification parameter Q has less evident dependence from ρH and the larger maximal energy, than the accuracy of Z.

4. Experiments

To verification the efficiency of algorithms designed to estimate the material identification parameter accuracy by use different implementation of the DEM, there was done the series of the experiments on the cargo inspection system of the Tomsk polytechnic university. The testing object composed from the constituents with different thicknesses from the organic materials, aluminum, iron and lead, was scanned by two narrow X-ray beams E_1=4.5 MeV, E_2=9 MeV. The experimental conditions were differed from the calculation conditions only by the power of the exposition dose – P_{0max}=5 R/min. The indicated power was selected to provide the stable work of the betatron for a long time. The figures 1 and 2 show the results of the comparison the experimental and the numerical dependencies $\Delta Z_V(\rho H)$ and $\Delta Q(\rho H)$.

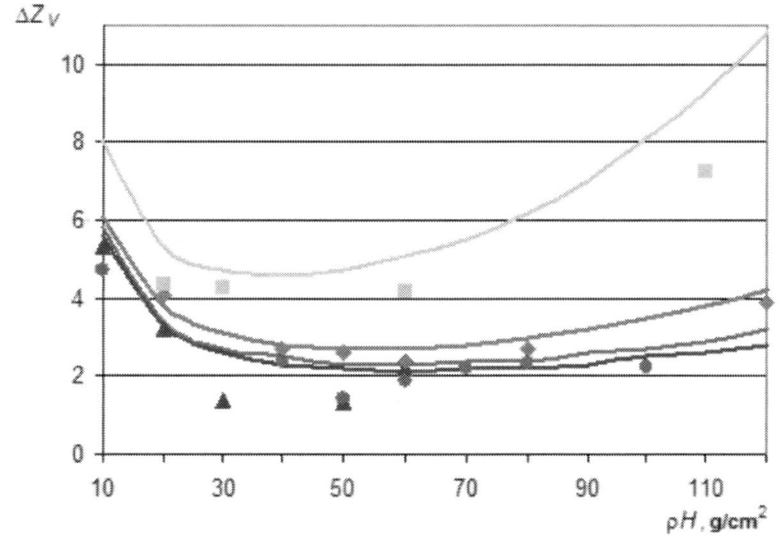

Figure 1. The calculated and the experimental dependencies $\Delta Z_V(\rho H)$:
– calculation, ■ – experiment, material – Pb;
– calculation, ♦ – experiment, material – Fe;
– calculation, ● – experiment, material – Al;
– calculation, ▲ – experiment, organics.

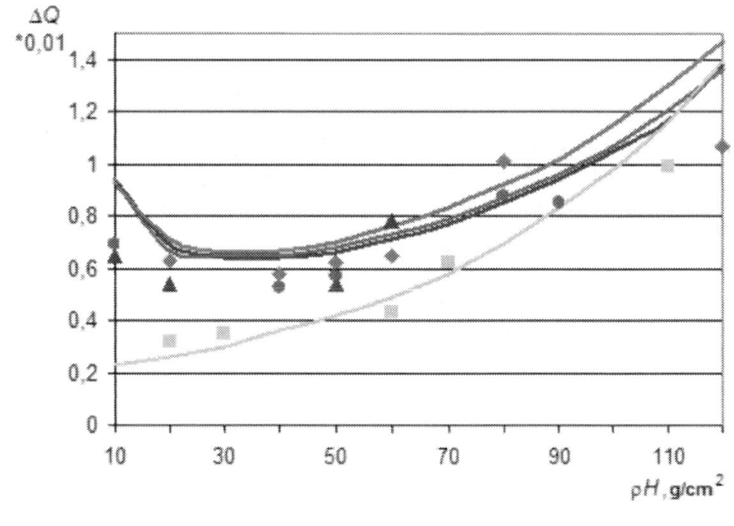

Figure 2. The calculated and the experimental dependencies $\Delta Q(\rho H)$:

– calculation, ■ – experiment, material – Pb;

– calculation, ♦ – experiment, material – Fe;

– calculation, ● – experiment, material – Al;

– calculation, ▲ – experiment, organics.

From analysis of the data presented on figures 1 and 2 it follows that in the considered range of the test object thicknesses the experimental estimations of the identification DEM parameter accuracy are closes to the numerical ones, or less than ones. It proves the applicability of the proposed algorithms to estimate accuracy of the effective atomic number and the identification parameter of the level lines method in practice.

5. Relation between impulses in the package

In section 2 it was shown that the DEM identification parameters accuracy is essentially depending from value p. The parameter p is equal to the ratio of the number of the pulses with the less maximal X-ray energy in the package to the one with the larger maximal energy. Before it was emphasized that there are exist the optimal values of the parameter p – p_{opt}, for which the dispersion values of the identification parameters are minimal.

The parameter p_{opt} for the identification parameter Z is calculated by use the formula (23), and for the identification parameter Q – with use the expression (30). The series of the calculation of the dependencies $p_{opt}(\rho H,Z)$ for the identification parameters Z and Q for the conditions of the numerical example from section 3 was carried out. The dependencies $p_{opt}(\rho H)$ for the objects from carbon, aluminum, iron and lead are given in the table 3.

From analysis of the data in the table 3 a number of conclusions can be done. The first conclusion – with increasing the object thickness ρH the values p_{opt} are increased during identification by Q and by Z. The second conclusion – the values popt for the Q identification are practically closed to the similar values popt, for the Z identification the divergence is not more than 10 %. The third conclusion – the values popt for the usage of the pair of maximal energies 4.5 – 9 MeV are always large than values popt for the usage of the maximal energies pair 4.5 – 7.5 MeV. We can make the following recommendations how to use the table 3. The number of pulses in the package n is selected by the formulas (24) or (31) including the notes 3, 4. The number of pulses of the larger energy n_2 is the maximum from two numbers – the nearest integer to $n/(1+p_{opt})$ and 1, but it is logically to get as n_1 the maximum of the nearest integer to $np_{opt}/(1+p_{opt})$ and 1.

Table 3. The optimal values p_{opt}.

Material		$E_1 - E_2$, MeV	ρH, g/cm²											
			10	20	30	40	50	60	70	80	90	100	110	120
C	Q	4.5 – 7.5	1.31	1.42	1.52	1.61	1.71	1.8	1.89	1.98	2.07	2.16	2.26	2.36
		4.5 – 9	1.45	1.61	1.76	1.9	2.05	2.19	2.34	2.49	2.64	2.79	2.95	3.12
	Z	4.5 – 7.5	1.31	1.41	1.52	1.61	1.71	1.8	1.89	1.99	2.08	2.17	2.27	2.36
		4.5 – 9	1.45	1.61	1.76	1.91	2.06	2.21	2.36	2.51	2.67	2.82	2.99	3.15
Al	Q	4.5 – 7.5	1.33	1.43	1.53	1.62	1.71	1.79	1.87	1.95	2.03	2.11	2.19	2.27
		4.5 – 9	1.47	1.63	1.78	1.92	2.05	2.19	2.32	2.45	2.58	2.71	2.84	2.98
	Z	4.5 – 7.5	1.34	1.44	1.54	1.63	1.71	1.8	1.88	1.97	2.05	2.13	2.21	2.29
		4.5 – 9	1.5	1.65	1.8	1.95	2.09	2.23	2.36	2.5	2.64	2.77	2.91	3.05
Fe	Q	4.5 – 7.5	1.35	1.47	1.56	1.64	1.72	1.79	1.86	1.93	2	2.06	2.12	2.18
		4.5 – 9	1.51	1.69	1.83	1.96	2.08	2.2	2.31	2.42	2.52	2.62	2.73	2.83
	Z	4.5 – 7.5	1.4	1.5	1.59	1.67	1.74	1.82	1.89	1.96	2.03	2.09	2.15	2.22
		4.5 – 9	1.6	1.76	1.9	2.03	2.15	2.28	2.39	2.5	2.62	2.72	2.83	2.94
Pb	Q	4.5 – 7.5	1.84	1.9	1.95	1.99	2.02	2.06	2.08	2.11	2.14	2.16	2.18	2.21
		4.5 – 9	2.3	2.38	2.45	2.52	2.57	2.62	2.66	2.7	2.74	2.78	2.81	2.85
	Z	4.5 – 7.5	1.68	1.8	1.88	1.94	2.00	2.05	2.09	2.13	2.16	2.2	2.23	2.25
		4.5 – 9	2.09	2.28	2.42	2.52	2.61	2.68	2.75	2.81	2.87	2.92	2.96	3.00

6. Results

There are given the collections of the mathematical relations whose allow to estimate the identification parameter accuracy for the two DEM implementations – the discrimination by the effective atomic number and the discrimination by the level lines method. The example of the identification parameter accuracy calculation was done for the cargo inspection system of the Tomsk polytechnic university on the base of the portable betatron MIB-4/9. The results of the comparison the theoretical and the experimental estimations of the identification parameters accuracy proved the applicability of the proposed algorithms to validate the quality of the radiographic systems with the material discrimination option for the testing object and their constituents. There are formulated the recommendations to select a structure of the package of X-ray radiation impulses, which produce a single line of the dual energy image.

7. Acknowledgement

The authors are very much obliged to the support of the Russian Foundation for Basic Research (grant 13–08–98027), National research Tomsk polytechnic university (grant VIU INK 66 2014) and the company PowerScan Technologies Ltd (Beijing, China).

References

[1] Kolkoori S, Wrobel N, Osterloh K, Redmer B, Deresch A and Ewert U 2013 High-energy radiography for detecting details in highly complex packings *MPMater Test.* **55** 683–8
[2] Mery D 2014 Computer vision technology for X-ray testing *Insight-non-destructive testing and condition monitoring.* **56** 147–55
[3] Martins M N and Silva T F 2014 Electron accelerators: History, applications, and perspectives *Radiat Phys Chem.* **95** 78–85
[4] Ryzhikov V D, Opolonin O D, Lysetska O K, Galkin S M, Voronkin Y F and Perevertaylo V L 2013 Research on improvement of receiving-detecting circuit for digital radiographic systems with advanced spatial resolution *Nondestructive testing of materials and structures. Springer Netherlands* **6** 105–9
[5] Wells K and Bradley D A 2012 A review of X-ray explosives detection techniques for checked baggage *Appl Radiat Isotopes* **70** 1729–46
[6] Rebuffel V and Dinten J M 2007 Dual-energy X-ray imaging: benefits and limits *Insight-non-destructive testing and condition monitoring* **49** 589–94
[7] Beldjoudi G, Rebuffel V, Verger L, Kaftandjian V and Rinkel J 2012 An optimised method for material identification using a photon counting detector *Nucl Instrum Meth A* **663** 26–36
[8] Chakhlov S V and Osipov S P 2013 High-energy digital X-ray imaging method for substance identification *Kontrol'. Diagnostika* **9** 9–17
[9] Ji Sung, Park and Jong Kyung Kim 2011 Calculation of effective atomic number and normal density using a source weighting method in a dual energy X-ray inspection system *J Korean Phys Soc.* **59** 2709–13
[10] Klimenov V A, Osipov S P and Temnik A K 2013 Identification of the substance of a test object using the dual-energy method *Russian J Nondestr Test.* **49** 642–9
[11] Osipov S P, Temnik A K and Chakhlov S V 2014 The effects of physical factors on the quality of the dual high energy identification of the material of an inspected object *Russian J Nondestr Test.* **50** 491–8
[12] Ananiev L M and Stein M M 1971On the question of the mathematical description of the betatron *Bulletin of the Tomsk Polytechnic Institute* **180** 3–7
[13] Zav'yalkin, F.M. and Osipov, S.P. 1985 Dependence of the mean value and fluctuations of the absorbed energy on the scintillator dimensions *At Energ.* **59** 281–3
[14] Zav'yalkin F M and Osipov S P 1989 Effect of the instability of parameters of bremsstahlung beams on the accuracy of radiometric measurements *Sov J Nondestr Test.* **25** 108–12
[15] Siegbahn K 2012 *Alpha-, beta-and gamma-ray spectroscopy* (Burlington, MA : Elsevier)
[16] Gamma data for elements. http://www.ippe.ru/podr/abbn/libr/groupkon.php
[17] Chadwick M B, Obložinský P, Herman M, Greene N M, McKnight R D, Smith D L, Young P G, MacFarlane R E, Hale G M, Franklr S C, Kahler A C, Kawano T, Little R C, Madland D G, Moller P, Mosteller R D, Page P R, Talou P, Trellue H, White M C, Wilson W B, Arcilla R,

Dunford C L, Mughabghab S F, Pritychenko B, Rochman D, Sonzogni A A, Lubitz C R, Trumbull T H, Weinman J P, Brown D A, Cullen D E, Heinrichs D P, McNabb D P, Derrien H, Dunn M E, Larson N M, Leal L C, Carlson A D, Block R C, Briggs J B, Cheng E T, Huria H C, Zerkle M L, Kozier K S, Courcelle A, Pronyaev V and Van der Marck S C 2006 ENDF/B–VII. 0: Next generation evaluated nuclear data library for nuclear science and technology *Nucl Data sheets* **107** 2931–3060

[18] Schiff L I 1951 Energy-angle distribution of thin target bremsstrahlung *Phys Rev.* **83** 252–3

[19] Aliev F K, Alimov G R, Muminov A T, Osmanov B S and Skvortsov V V 2005 Simulation of experiment on total external reflection of electron bremsstrahlung *Tech Phys.* **50** 1053–7.

[20] Mordasov N G, Ivashchenko D M, Chlenov A M and Astakhov A A 2004 Simulation of methods for a rapid determination of the energy spectrum of bremsstrahlung from electron accelerators *Tech Phys.* **49** 1213–20

Determination of sensitivity for in-process control of cable product insulation

V V Redko[1], N S Starikova[1], L A Redko[1], G V Vavilova[1]

[1] National Research Tomsk Polytechnic University, Tomsk, Russia

E-mail: nadushasns@tpu.ru

Abstract. This article presents current methods of cable insulation control. The new method which allows to improve reliability of cable insulation control was offered. The cable model with several types of defects was developed by using Comsol Myltiphysics software. Minimal sizes of defects which can be detected by using given in-process control method.

1. Introduction

Quality of insulation is one of significant parameters for cable product (continuity, external geometry, isotropy, chemical composition) [1]. This parameter influences operational characteristics and cable quality. The insulation defects lead to cable spoilage because it is cause of noise distorting a useful signal.

Nowadays, cable insulation control is implemented by several ways:

1.1. Voltage test

According to the current regulatory documentation [2], there are two categories of high voltage insulation testing: ET-1 and ET-2.

ET-1 category testing is stationary and implemented by immersing hank into a water. Immersed cable is kept under the water during defined time by applying high testing voltage. This method of control makes manufacturing cycle longer and it is also necessary to provide additional technological area in the plant [3]. One more significant disadvantages of water testing is the fact that this cable testing category is carried out in the final stage of technological process. Thus, detected defects are cause of high costs of the company.

The main idea if ET-2 testing is applying of high testing voltage to the surface of cable insulation by the special electrode during technological process while core, armor or shield is grounded. When defect part is in control area breakdown is occurred and it is registrated by automatics of electrospark detector (Figure 1) [1].

In comparison with ET-1 testing ET-2 testing has some advantages. Firstly, electrospark detector for ET-2 testing have significantly smaller size. Then, water medium is not necessary for this kind of testing. Also, ET-2 testing is provided during technological process. It allows to refine technological parameters and prevent mass spoilage [3]. Disadvantage of ET-2 testing is low defect detection sensitivity in comparison with stationary water testing (ET-1).

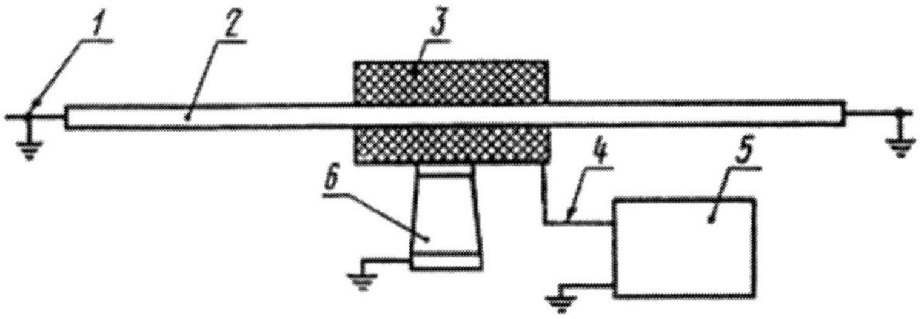

Figure 1. Basic scheme of ET-2 testing: 1 is core; 2 is insulation; 3 is electrode, 4 is high voltage output; 5 is signal processing device; 6 is isolator.

1.2. Linear capacitance control

The cylindrical electrode is used to realize this testing method. The cylindrical electrode is immersed under the water of cooling bath for measuring of cable linear capacity. When defect part is in control area linear capacitance of control area is changing in comparison with linear capacitance of defect-free part. Significant advantage of this method is carrying out of in-process control. But there are some disadvantages. Water medium is necessary for this method of control. Additional equipment and maintenance is required for correct work of electrocapacity device.

Take into account disadvantages of present method of control it is offered to explore the electrocapacity method during high voltage technological testing (ET-2) of cable insulation. The aim of this work is determination of minimal defects size which can be detected by this method.

2. Mathematical model

To provide this research the mathematical model was developed in Comsol Myltiphysics software.

The cable model is presented as two concentric cylinders which have length higher then l value (l is the length of cable area with applied high testing voltage). High testing voltage is applied to the cable area surface with l length. The voltage value (U_{test}) was set depending on the insulation thickness [4]. The air space is considered around the cable model (Figure 2).

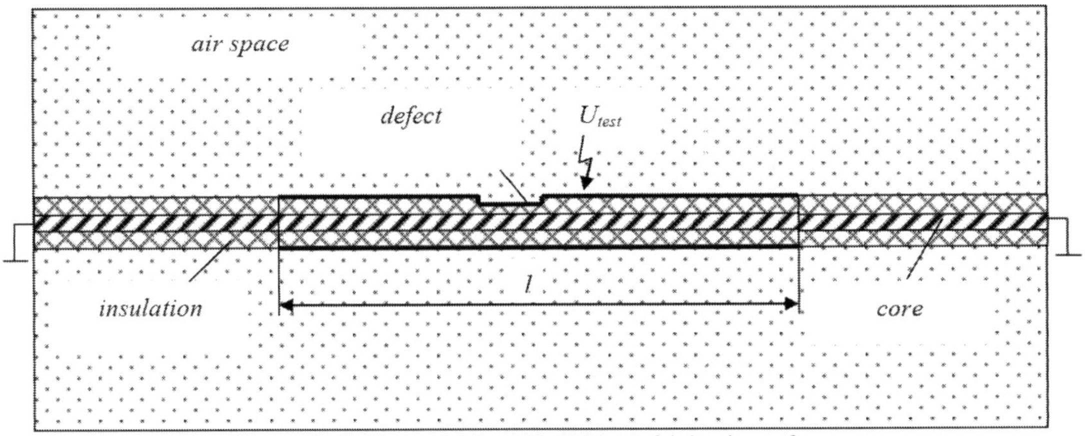

Figure 2. Cable model in COMSOL Multiphysics software

3. Defects detection conditions

During the capacity changing in-process control measurements it is impossible to detect the defect of any size. Insulation coarseness, horizontal or vertical shifting during moving of the sample through the measuring device, changing of geometrical parameters, instability of testing voltage lead to deviation of control area cable electrical capacity from defect-free zone value. Consequently, to consider any deviation of electrical capacity as a defect misoperation can take a place. To increase control reliability the deviation is considered as a defect when it is higher than 20% [5, 6].

4. Defects simulation

4.1. "Longitudinal cut" type defect

By using given special software "longitudinal cut" type defect was simulated for «NV-3» cable product (Figure 3).

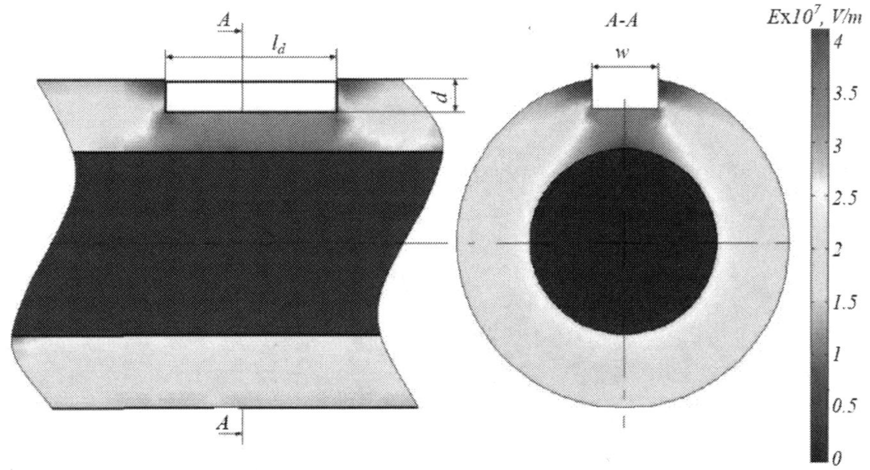

Figure 3. "Longitudinal cut" type defect: l_d is defect length, d is defect depth, w is defect width, A-A is cross-section, E is value of electric field

The dependencies of relative electrical capacity value from relative defect length with several values of l parameter (length of measuring electrode is 10 cm, 20 cm, 40 cm) are shown below (Figure 4).

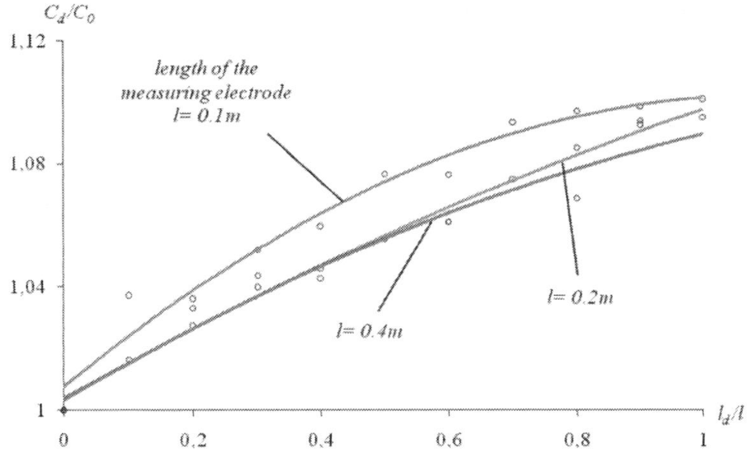

Figure 4. Dependencies of relative electrical capacity value from relative defect length: C_d is electrical capacity of control area with defect, C_0 is electrical capacity of defect-free control area, l_∂ is defect length, l is electrode length.

By analyzing obtained dependencies it can be notices that it has a similar view. As it was written above deviation is considered as a defect when deviation of cable electrical capacity from defect-free zone value is higher than 20%. This defect with constant value of *d* and *w* parameters lead to deviation less than 20%. Consequently, this kind of defects cannot be registered.

4.2. "Crack" type defect

By using the mathematical calculation minimal size of "crack" type defect which can be detected by in-process control was defined.

For calculation of "crack" type defect electrical capacitance for cable area following formulae was derived:

$$C = 2\pi\varepsilon_0\varepsilon_r \cdot \left(\frac{l-w}{\ln(R_i/R_c)} + \frac{w}{\ln((2R_i - D_d)/(2\cdot R_c))} \right),$$

where l – length of controlled area of the cable (measuring electrode);
w – defect width
R_i – external radius of insulation (defect-free part);
R_c – core radius;
D_d – depth of the defect (R_i- R_d) (Figure 5).

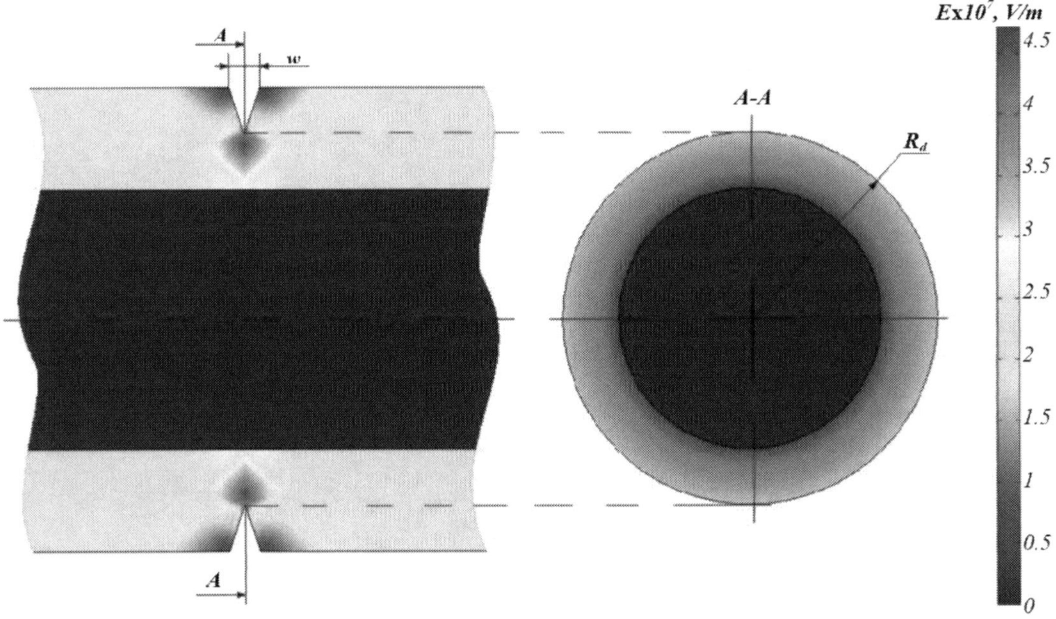

Figure 5. "Crack"-type defect: Rd – external insulation radius in defective zone.

Calculations was provided for NV-3 cable. The characteristics of given cable are:

- core cross-section is 1.5 mm^2(diameter of core is 1.4 mm);
- external insulation diameter is 2.5 mm;
- thickness of insulation is 0.55 mm;
- insulation material is PVC.

Length of cable controlled area is 0.1 m.

Dependences of relative electrical capacity of cable area (C_d/C_0, where C_d – electrical capacity of defective part, C_0 – electrical capacity of defect-free part) from "crack" type defect dimensions (d_d is ration of defect depth to thickness of defect-free insulation) was plotted on the basis of calculations results(Figure 6). It was noticed that in-process control allows to detect cracks which has depth equal to depth of defect-free insulation (relative depth d_d=1) with opening width not less than 21% of controlled area length l. It is also possible to detect defects which have lower depth but higher opening width.

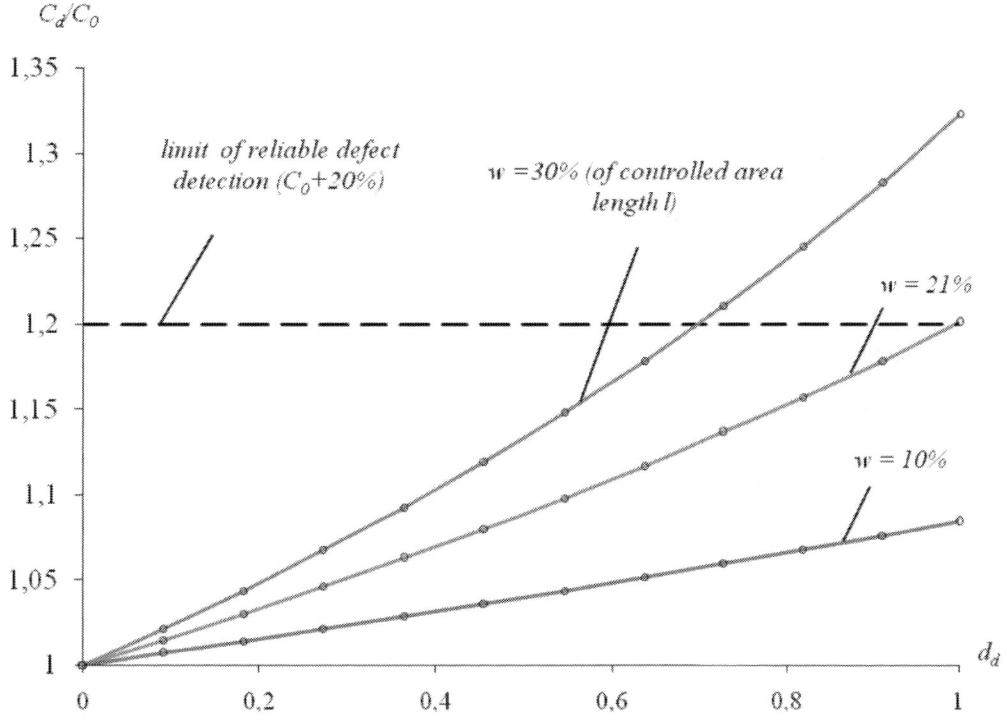

Figure 6. Dependence of relative capacity of cable part from size of "crack" type defect.

4.3. *Change in external diameter of insulation*

By using mathematical calculations minimal deviations of "local changing of external diameter of insulation" type defect (Figure 7) which can be found due to in-process control (with deviations of capacity of controlled area from nominal value higher than 20%) was found.

To find theoretical value of electrical capacity of cable part following formula can be used:

$$C = 2\pi\varepsilon_0\varepsilon_r\left(\frac{l - l_d}{\ln\frac{R_i}{R_c}} + \frac{l_d}{\ln\frac{R_i+\delta}{R_c}}\right),$$

where

l – length of controlled area of the cable

l_d— length of defective part;

Ri – external radius of insulation (defect-free part);

R_c – core radius;

δ— insulation thickness difference on defective and defect free part.

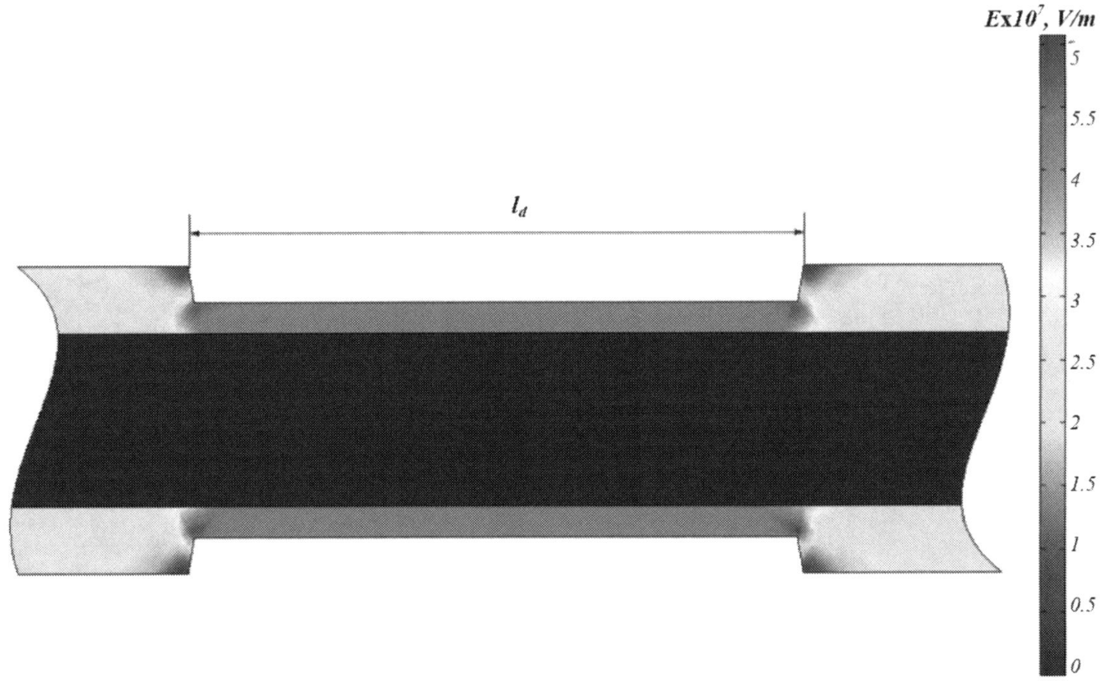

Figure 7. "Local thinning" type defect

Dependences of relative capacity of cable part from deviation of insulation thickness with l=10cm, 20cm, 40cm (Figure 8) were found and plotted. Length of defect was 20%, 50% and 100% for each l

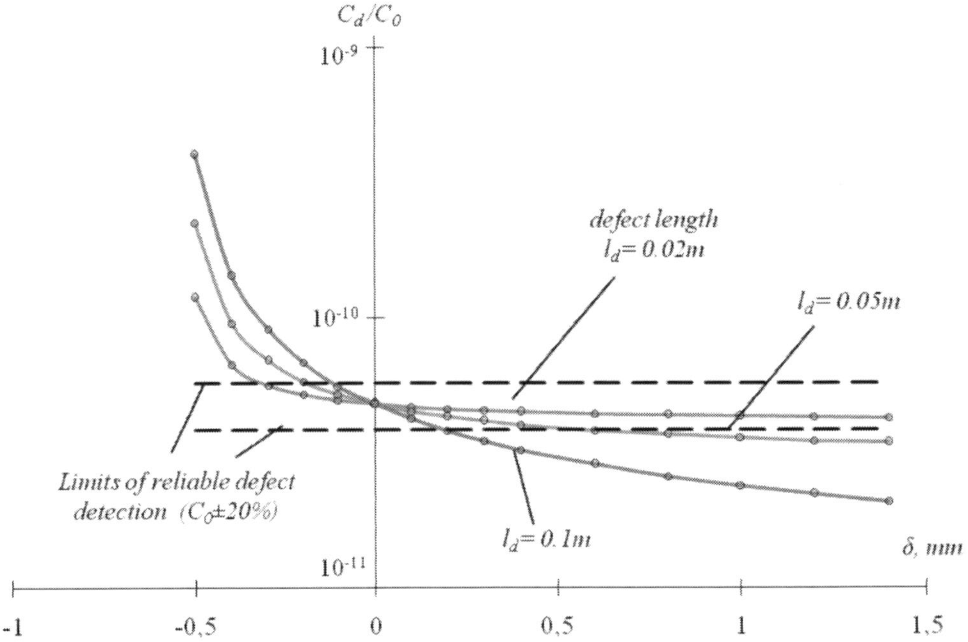

Figure 8. Dependences of relative capacity of cable part from deviation of insulation thickness (in the defective part) relatively to defect-free area with l=10 cm.

According to plotted figures it was noticed that behavior of dependences for several relative lengths of the defects (10%, 50%, 100%) is similar and sensitivity of the measurements is higher in case of higher relative lengths of the defect. Sensitivity is also improving if insulation thickness is increasing. (Figure 9).

To define capacity variation coefficient in presence of the defect (C_d) and without any defect (C_0) additional value k (sensitivity) was added:

$k = \frac{C_0}{C_d}$ for local increasing of external diameter;

$k = \frac{C_d}{C_0}$ for local decreasing of external diameter.

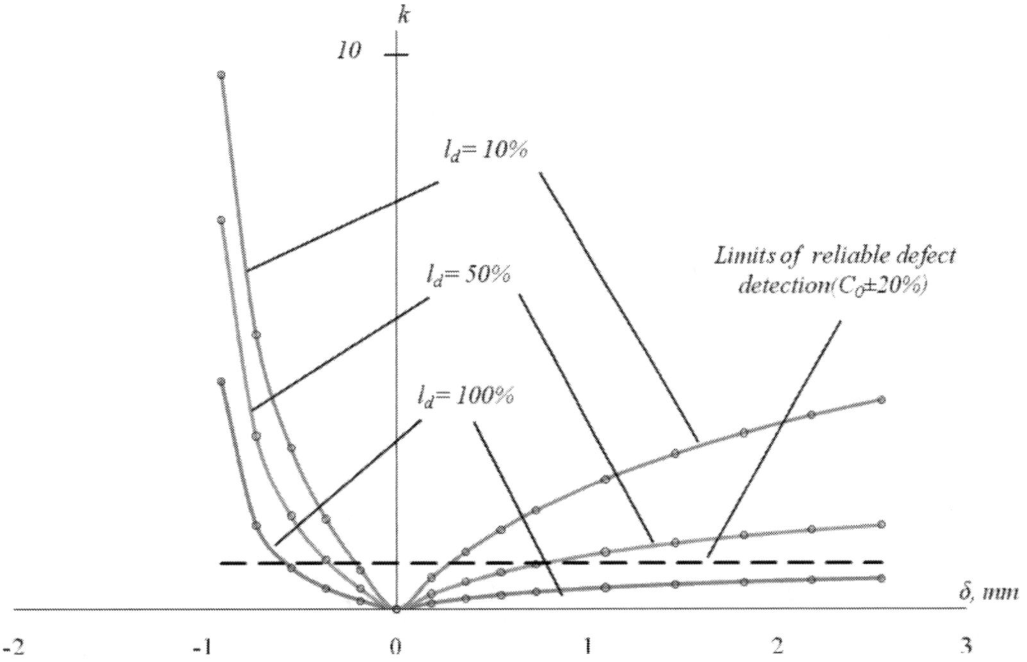

Figure 9. Dependences of measurements sensitivity from variation of insulation thickness in the defective area relatively defect free area with l_d=10%, 50%, 100%

Thus, it was defined that by changing of cable area electrical capacity it is possible to identify following changes of insulation thickness:

– thickness decrease by 57% and increase by 397% (if defect amounts 20% of controlled area length);

– thickness decrease by 35% and increasey by 109 % (if defect amounts 50% of controlled area length);

– thickness decrease by 21% and increasey by 36% (if defect amounts 100% of controlled area length).

Based on these results it is possible to conclude that size of defects which can be detected by this method considerably depends from the length of controlled area. Since detection of local variation of external diameter is possible only when defect amounts approximately 100% of controlled area length (reliability of detection is much lower in case of lower defect length) it is possible to change measurements sensitivity by variation of controlled area length.

5. Conclusion

In this research special mathematical model was developed. By using this model effect of different defect dimensions on the linear capacity value with different lengths of measuring electrodes was defined as well as ability to detect structural and geometrical defects with offered control method. A simulation of several types of cable insulation defects was provided by using COMSOL Multiphysics software.

By using simulation results it was noted that "longitudinal cut" type defect cannot be detected with given method. In case of "crack" type defect and local variation (increasing or decreasing) of insulation external diameter it is possible to evaluate the defect considering variation of capacity value of controlled area.

References

[1] Redko V V Burtceva L B and Redko L A 2010 *Bulletin of the Tomsk Polytechnic University* **4(317)** 111
[2] Starikova N S, Redko V V and Vavilova G V 2015 *Applied Mechanics and Materials* **756** 486
[3] Grigorjev M G Vavilova G V and Starikova N S 2015 *Applied Mechanics and Materials* **756** 546
[4] Leonov A P Redko V V and Soldatenko E Y 2014 *IOP Conference Series: Materials Science and Engineering* **66** 1
[5] Redko V V 2013 *Spark quality control of cable products insulation* (Tomsk: TPU)
[6] Redko V V Redko L A and Starikova N.S 2013 *Polzunovsky Gazzette* **2** 164

Evaluation of equivalent defect heat generation in carbon epoxy composite under powerful ultrasonic stimulation by using infrared thermography

D A Derusova, V P Vavilov, S S Pawar
National Research Tomsk Polytechnic University, Russia 634028, Tomsk
vavilov@tpu.ru

Abstract
Low velocity impact is a frequently observed event during the operation of an aircraft composite structure. This type of damage is aptly called as "blind-side impact damage" as it is barely visible as a dent on the impacted surface, but may produce extended delaminations closer to the rear surface. One-sided thermal nondestructive testing is considered as a promising technique for detecting impact damage but because of diffusive nature of optical thermal signals there is drop in detectability of deeper subsurface defects. Ultrasonic Infrared thermography is a potentially attractive nondestructive evaluation technique used to detect the defects through observation of vibration-induced heat generation. Evaluation of the energy released by such defects is a challenging task. In this study, the thin delaminations caused by impact damage in composites and which are subjected to ultrasonic excitation are considered as local heat sources. The actual impact damage in a carbon epoxy composite which was detected by applying a magnetostrictive ultrasonic device is then modeled as a pyramid-like defect with a set of delaminations acting as an air-filled heat sources. The temperature rise expected on the surface of the specimen was achieved by varying energy contribution from each delamination through trial and error. Finally, by comparing the experimental temperature elevations in defective area with the results of temperature simulations, we estimated the energy generated by each defect and defect power of impact damage as a whole. The results show good correlation between simulations and measurements, thus validating the simulation approach.

Keywords: composites, impact damage, ultrasonic infrared thermography, heat generation

1. Introduction

Current research on ultrasonic infrared (IR) thermography (also called "sonic IR thermography" and "thermosonics") is being conducted in two main directions: 1) powerful stimulation by using piezoelectric [1–4] and magnetostrictive [5] units, 2) low-power stimulation based on the resonant interaction between ultrasonic waves and defects due to local defect resonance [6]. Although there is lack of complete understanding of the physics governing the heat generation process, it is generally accepted that external ultrasonic stimulation of structural defects causes friction at the defect edges. The generated heat diffuses away from these defects and the corresponding temperature distributions are monitored by using an infrared (IR) camera. Analysis of equivalent heat power produced by a particular defect is of primary interest because, in combination with temperature resolution of a used IR imager, it defines threshold parameters of detected defects. For example, the so-called "kissing" defects may provide frictional heating along the whole defect edge while large defects become thermographically visible only at defect tips where friction predominantly occurs. The defect energy can be accurately evaluated by using both analytical and numerical thermomechanical models that is a challenging task. Thin delaminations which are caused by impact damage in composites subjected to ultrasonic excitation can be considered as local heat sources. Such heat sources operate during ultrasonic stimulation time and their apparent power depends on the material properties and a mechanism of heat release. This study proposes a further step in this direction to quantitatively evaluate the "defect power" under ultrasonic stimulation.

2. Experimental results

The experimental data was obtained on a 100×150×4 mm carbon epoxy sample which contained a 16 J impact damage defect. The damage was barely seen on the impacted (front) surface as a dent of the size of about 4×4 mm (Figure 1(a)), while on the rear surface the damage was consisted of two extended areas producing the known "butterfly-like" pattern, and its size was about 48×16 mm (Figure 1(b)). Thus we can interpret that the damage has grown in conical fashion through the thickness like a typical low velocity impact damage in composite materials. Note that the piezoelectric sensor shown in Figure 1(a) was not used for the measurements.

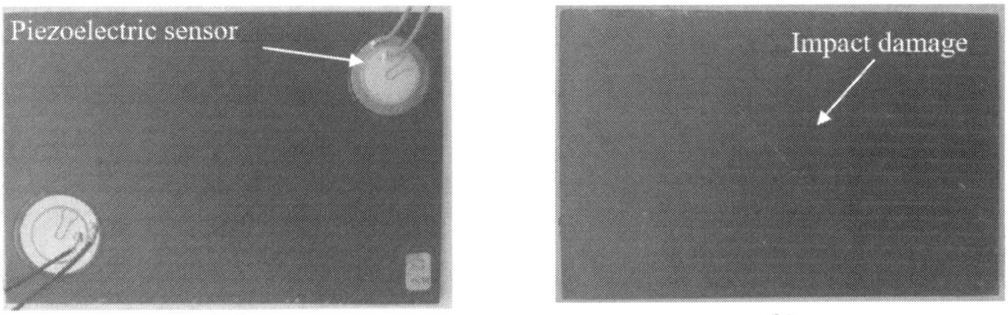

(a) **(b)**

Figure 1. Carbon epoxy composite sample with 16 J impact damage: (a) front and (b) rear surface.

The sample was ultrasonically stimulated by means of a magnetostrictive device described in [5] (electrical power up to 2 kW, frequency 22 kHz, stimulation time 5-10 s). Test results were recorded on both the sample surfaces as a sequence of IR thermograms by using a NEC TH-9100 IR camera. As can be seen from Figure 2, the maximum temperature rise, due to the application of ultrasonic stimulation for 10 seconds, was 1.04 °C at the defect location on the front surface and 1.65 °C on the rear surface. It is worth noting that on the rear surface the defect pattern revealed two sections (Figure 2(b) & (c)) thus pressing the need to model it with two independent 'heat source' delaminations located near the rear surface.

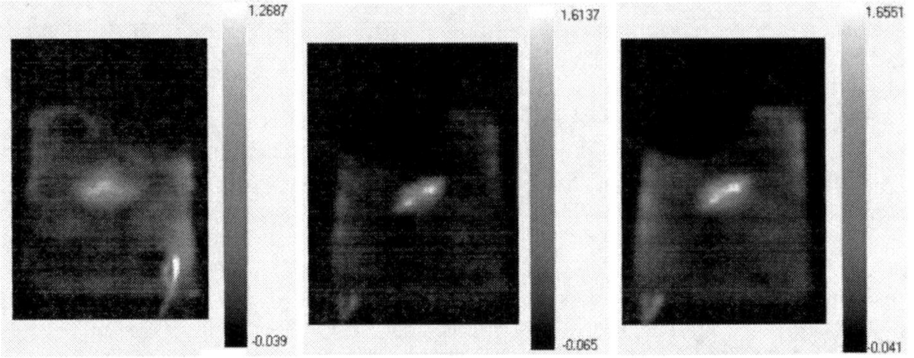

a) **b)** **c)**

Figure 2. IR thermograms of carbon epoxy composite during ultrasonic stimulation : (a) front surface after 10 seconds of stimulation (b) back surface after 5 seconds of stimulation (c) back surface after 10 seconds of stimulation.

3. 3-D modeling and simulation

As discussed in section 2, we have assumed that impact damage can be modeled as a cone, or a pyramid in the Cartesian coordinates, containing multiple delaminations. A particular cross-section of this pyramid (Fig.3) represents the delaminations which generate thermal energy under ultrasonic stimulation. This particular impact damage was modeled as a set of 9 parallelepiped-like air-filled heat sources each having thickness of 0.4 mm. The distance between the adjacent layers was 0.2 mm. Two rear-surface defects were modeled as two different sections in accordance with the visible pattern (Figure 3). Model parameters were set closer to the experimental values (sample thickness 4 mm, thermal conductivity 0.8 W/(m·°C), heat capacity 760 J/(kg·°C) and density 1560 kg/m^3).

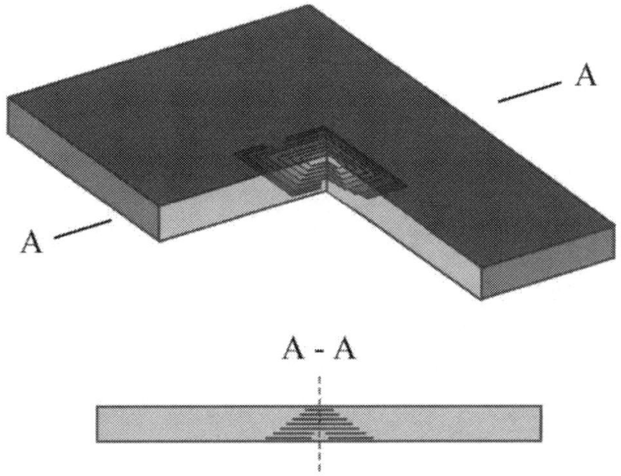

Figure 3. Modeling impact damage as a set of multiple delaminations (Cartesian coordinates, pyramid-like defect shape).

Since pixel-based temperature evolutions were obtained by analyzing experimental image sequences, the next step was the simulation of the surface temperature signal by varying the temperature responses from each delamination so as to match the simulated temperature *vs.* time graph to the experimental graph and to calculate the "defect power". The corresponding 3D thermal NDT problem of transient heat conduction was solved by using a home-made ThermoSource program. The mathematical formulation of the problem was as follows:

$$\frac{\partial T_i(x,y,z,\tau)}{\partial \tau} = \alpha_i^x \cdot \frac{\partial^2 T_i(x,y,z,\tau)}{\partial x^2} + \alpha_i^y \cdot \frac{\partial^2 T_i(x,y,z,\tau)}{\partial y^2} + \alpha_i^z \cdot \frac{\partial^2 T_i(x,y,z,\tau)}{\partial z^2} + \frac{w(x,y,z,\tau)}{C\rho}; \quad (1)$$

$i = 1 \div 16$ (*seven layers + nine defects*)

$$T_i(\tau = 0) = T_{in}; \quad (2)$$

$$K_1^z \cdot \frac{\partial T_1(x,y,z=0,\tau)}{\partial z} = h_F \cdot [T_1(x,y,z,\tau) - T_{amb}]; \quad (3)$$

$$K_3^z \cdot \frac{\partial T_3(x,y,z=L_z,\tau)}{\partial z} = -h_R \cdot [T_3(x,y,z,\tau) - T_{amb}]; \quad (4)$$

$$\frac{\partial T_i(x,y,z,\tau)}{\partial x} = 0 \text{ for } x=0, y=0 \div L_y; x=L_x, y=0 \div L_y; \quad (5)$$

$$\frac{\partial T_i(x,y,z,\tau)}{\partial y} = 0 \text{ for } y=0, x=0 \div L_x; y=L_y, x=0 \div L_x; \quad (6)$$

$$T_i(x,y,z,\tau) = T_{i\pm1}(x,y,z,\tau) \quad (7)$$

$$K_i^{q_j} \cdot \frac{\partial T_i(x,y,z,\tau)}{\partial q_j} = K_{i\pm1}^{g_j} \cdot \frac{\partial T_{i\pm1}(x,y,z,\tau)}{\partial q_j} \quad (8)$$

Here: T_i is the temperature in the *i*-th region counted from the initial object temperature (i = 1 - 7 corresponds to specimen layers (if a composite consists of 7 plies), i = 8 - 16 corresponds to 9 internal heat sources); T_{in} is the specimen initial temperature; $\alpha_i^{q_j}$, $K_i^{q_j}$ are the thermal diffusivity and the thermal conductivity in the ith region by the coordinate q_j; x, y, z are the Cartesian coordinates; q_j is one of the Cartesian coordinates x, y or z (j = 1 - 3); τ is the time; $w(x,y,\tau)$ is the heat power density designated to internal heat sources; h_F, h_R are the heat exchange coefficients on a front and rear surface respectively; T_{amb} is the ambient temperature; L_x, L_y, L_z are the sample dimensions.

Eq. (1) is the 3D parabolic equation of heat conduction; Eq. (2) is the initial condition; Eq. (3) is the boundary condition on a front surface (both heating and cooling); Eq. (4) is the boundary condition on a rear surface (cooling only); Eqs. (5) and (6) are the adiabatic conditions on side surfaces by the coordinates x and y respectively; Eqs. (7) and (8) are the temperature and heat flux continuity conditions on the boundaries between the layers and between the layers and the heat sources respectively.

The power of heat generation in each of the 9 defects was varied by performing repeated calculations to match the experimental temperature evolutions on both the sample surfaces through trial and error. Since the maximum temperature rise is fully determined by the total power of all heat sources, which can be easily determined by comparing maximum experimental and theoretical values, the main difficulty was the fitting of the temperature response curve shape which is affected by the depth location of each heat source. The fitting results are presented in Figure 4.

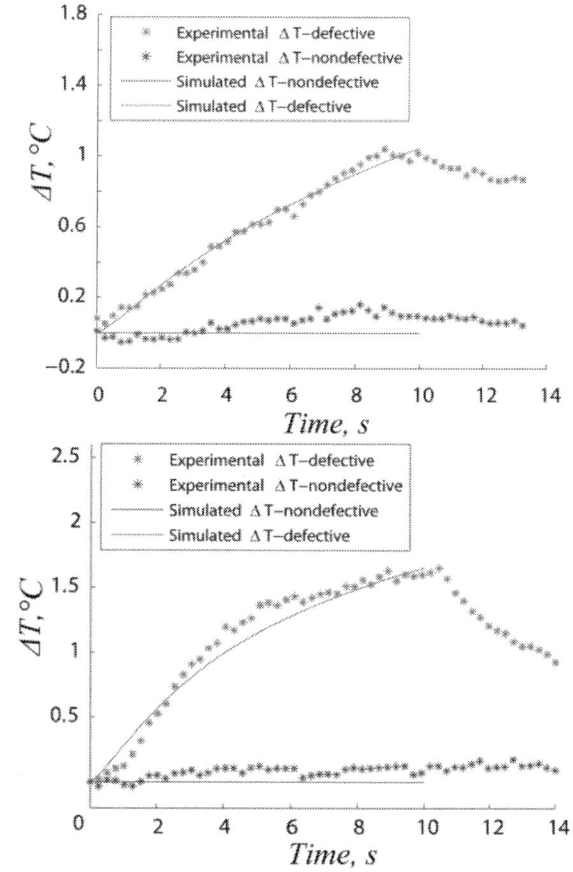

a) b)

Figure 4. Fitting experimental (a) front and (b) rear surface temperature evolutions by using the model in Figure 3 (heat source operation time 10 s).

Table 1 contains values of particular defect section volumes (V) and their depths (l), as well as estimated heat power (Q), combination of which provides the best fitting of the experimental data (Figure 4). The total "defect power" is 152.9 mW that is in accordance with estimates obtained earlier [7].

Table 1. Fitting parameters in modeling heat generation in impact damage defect (ThermoSource program, ultrasonic stimulation time 10 s)

Defect number	1	2	3	4	5	6	7	8	9
V, $\times 10^{-9}$ m^3	6.4	24	42	64	120	104.4	111.4	133.5	119.8
l, mm	0.1	0.6	1.1	1.6	2.1	2.7	2.7	3.3	3.3
Q, mW	2.75	8.88	12.6	17.28	26.4	20.88	22.28	22.03	19.8

Figure 5 shows temperature responses which would appear if subsurface heat sources are turned on consecutively starting from the inspected sample surface. It is seen that major contribution to the total temperature rise is provided by the first three defect sections which are closer to the inspected surface. The corresponding sample thickness considering these first three defect sections is 1.5 mm.

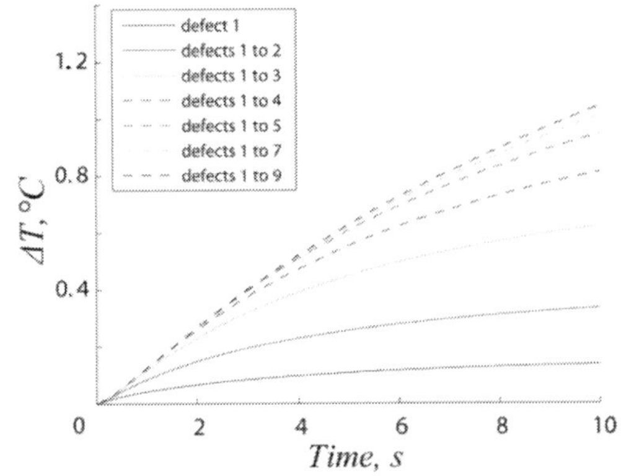

a)

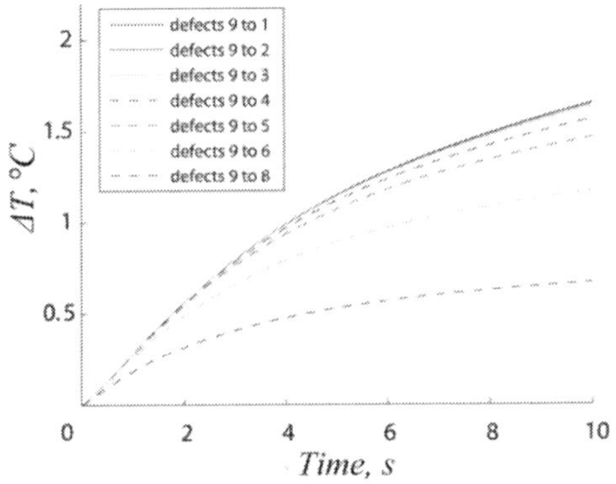

b)

Figure 5. Total temperature response obtained by consecutively turning on subsurface defects for (a) the front and (b) the rear surface (heat source operation time 10 s).

4. Conclusions

In this work, we demonstrated an approach for the evaluation of equivalent thermal power of subsurface defects which operate as heat sources under ultrasonic stimulation. We have restricted our calculations to the case of impact damage of 16 J in a carbon epoxy composite plate. This particular impact damage was modelled as a set of multiple delaminations each acting as a heat source. After the application of ultrasonic stimulation for 10 seconds, experimental measurements showed a maximum temperature rise of 1.04 °C and 1.65 °C at the defect location on the front and rear surface respectively. During the temperature simulations, the power of heat generation in each of the 9 defects was varied by performing repeated calculations, to match the experimental temperature evolutions on both the sample surfaces through trial and error. Thus, the total "defect power" calculated for the impact damage of 16 J was 153 mW that is in accordance with estimates obtained in earlier works (by taking into account the power of ultrasonic stimulation which determines differential temperature signals). It was also found that the major contributions towards the temperature rise were made by defects located within the 1.5 mm-thick material layer adjacent to the monitored surface. The main difficulty was the matching of the temperature response curve shape which was affected by the depth location of each heat source. In all future cases, the simulation approach validated in this paper will be a valuable tool to estimate the defect power and to predict the possible through the thickness positions of the major delaminations in composite materials.

This research was supported by the NIR # 445 (ONG), State order of the Russian Ministry of Higher Education for 2014-2016.

References

[1] Burke M.W., Miller W.O. Status of VibroIR at Lawrence Livermore National Laboratory. *Proc. SPIE "Thermosense-XXVI"*, vol. 5405, **313-321**, 2004.
[2] Zweschper Th., Dillenz A., Busse G. Ultrasound lock-in thermography - a defect selective method for the inspection of aerospace components. *Insight,* vol. 43, No. 3, **173-179**
[3] Favro L.D., Han X., Ouyang Z. IR imaging of cracks excited by an ultrasonic pulse. *Proc. SPIE "Thermosense-XXII"*, vol. 4020, **182-185,** 2000.
[4] Homma S., Rothenfusser M., Baumann J., Shannon R. Study of heat generation mechanism in acoustic thermography. *Rev. Quant. NDE,* ed. D.O. Thompson and D.E. Chimenti, vol. 25, **566-573**, 2011.

[5] Guo X., Vavilov V. P. Crack detection in aluminum parts by using ultrasound-excited infrared thermography. *Infrared Physics & Technology*, vol. 61, **149-155**, 2013.

[6] Solodov I., Bai J., Busse G., Resonant ultrasound spectroscopy of defects: Case study of flat bottomed holes. *J. Appl. Phys.,* vol. 113 Issue 22, **p 223512,** 2013.

[7] Vavilov V.P., Khorev V.S., Chulkov A.O. Investigating the technique of ultrasonic infrared testing of cracks in composite materials. *Control. Diagnostics,* No. 13, **197-201**, 2012 (in Russian).

Waste management through life cycle assessment of products

Yu V Borodin[1], T E Aliferova[1], A Ncube[2]
[1]Tomsk Polytechnic University, Tomsk, Russia
[2]Environmental Sciences Institute, Harare, Zimbabwe
*email: uryborodin@tpu.ru

Abstract. The rapid growth of a population in a country can contribute to high production of waste. Municipal waste and industrial waste can bring unhealthy and unpleasant environment or even diseases to human beings if the wastes are not managed properly.With increasing concerns over waste and the need for 'greener' products, it is necessary to carry out Life Cycle Assessments of products and this will help manufacturers take the first steps towards greener designs by assessing their product's carbon output. Life Cycle Assessment (LCA) is a process to evaluate the environmental burdens associated with a product, process or activity by identifying and quantifying energy and materials used and wastes released to the environment, and to assess the impact of those energy and material used and released to the environment. The aim of the study was to use a life cycle assessment approach to determine which waste disposal options that will substantially reduce the environmental burdens posed by the Polyethylene Terephthalate (PET) bottle. Several important observations can be made.
1) Recycling of the PET bottle waste can significantly reduce the energy required across the life cycle because the high energy inputs needed to process the requisite virgin materials greatly exceeds the energy needs of the recycling process steps.
2) Greenhouse gases can be reduced by opting for recycling instead of landfilling and incineration.
3) Quantity of waste emissions released from different disposal options was identified.
4) Recycling is the environmentally preferable disposal method for the PET bottle.
Industry can use the tools and data in this study to evaluate the health, environmental, and energy implications of the PET bottle. LCA intends to aid decision-makers in this respect, provided that the scientific underpinning is available. Strategic incentives for product development and life cycle management can then be developed.

1. Life Cycle of the Polyethylene Terephthalate

An analysis of the flow of energy involved in the production of a product is one aspect of a life cycle assessment, an objective process of analysis that attempts to evaluate the environmental burdens associated with a product, process, or activity by [1]:

- Quantifying the quantities of energies and materials used and the quantities of waste emissions released into the environment.
- Assessing the impact of energy and emissions releases on the environment.
- Evaluating opportunities to effect improvements in the final disposal of the product.

The assessment, if at all possible, should consider all the activities related to the manufacture of a product or operation of a process; this includes activities such as processing of raw materials, manufacturing, transportation and distribution, use/reuse, recycling, and final disposal [2]. Figure 1 shows the simplified processing and manufacturing of PET.

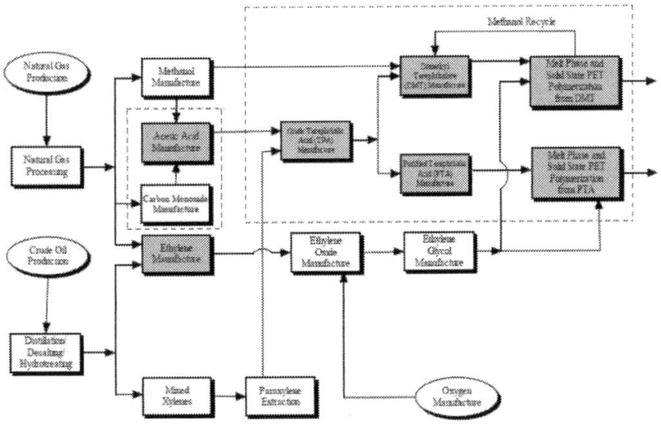

Figure 1. The simplified processing and manufacturing of PET.

2. Life Cycle Inventory

Plastic Production. This section gives an overview of the complex and varied processes involved in making the PET bottle plastic. We also look at 'additives', which enhance the performance of plastic bottle or aid its processing. These chemicals often have a greater potential impact on the environment than the polymer itself. Finally, we consider information on the impact of making plastics on the environment and human health [3].

Production Stage – Emissions and Energy Requirements for producing a 1L PET bottle. The production of different kinds of polymers has totally different energy requirements and the amount of emissions released. In this section, PET plastic bottle will be discussed in order to provide a clear result during the production stage (table 1).

Table 1. Environmental Profile for PET – PET Production Stage

Raw material	Oil & natural gas	**Main Product**	1kg PET is produced
Energy	83.8 MJ/kg	**Solid waste**	0.045130 kg/kg
Transport	0.2 MJ/kg	**Emissions**	CO_2 2330 g/kg SO_x 25 g/kg NO_x 20.2 g/kg CO 18 g/kg HCs 40 g/kg

Use of PET bottle. This section looks at the main sector that uses PET plastic bottle: packaging. During this stage, there are no energy and emissions released involved. Therefore, no data can be used to indicate the effect of PET plastic bottle during the use stage to the entire life cycle of plastic bottle.

Transport of PET bottle. In this section, the produced PET bottle will be transported to consumers and eventually to recycling centers and disposal sites once it reaches the end of life of its services. Meaning to say, consumers dump the bottles into their dustbins after the plastic product becomes useless. We assumed the maximum of transport distance to consumers, recycling centers and disposal sites to be 200 km. The energy required for transportation was 0.01 MJ/kg/km according to. We assumed no emissions released during transportation stage because the emissions released are most like to be insignificant.

Transportation Stage – Emissions and Energy Requirements for a 1Kg PET bottle

Environmental Profile – Transportation Stage of PET

Transport = 0.01*200=2MJ/Kg

Disposal of PET plastic bottle Waste. Plastic waste can be disposed-off using different kinds of methods. Recycling was also one of the improvement methods to be considered in this study. The idea being that recycling activities will reduce the waste generated to the earth and eventually lessens the risks of environmental burdens caused by the PET bottle.

PET Waste Recycle – Emissions and Energy Requirements. Where reuse is not the most environmentally sound way of extracting value from plastic bottle wastes, an alternative is to recycle them into feedstock or into energy recovery so that their intrinsic value is not lost. These two technological methods of plastic waste recovery have been developed in the industrialized countries on a large scale, mechanical recycling and incineration with energy recovery. Once the recyclates have been cleaned and shredded, the process is much the same as for the production of the PET plastic bottles from feedstock. Most plastics are recycled mechanically, but chemical recycling is at a developmental stage. Plastic bottles are the main type of plastics collected and recycled from household waste. During the process of recycling, energy is required and emissions are released to environment.

Waste Recycled – Emissions and Energy Requirements for a 1 Kg PET bottle (table 2)

Table 2. Environmental Profile for PET – Waste Recycled

Raw material	-	**Main Product**	1 kg new PET bottle is produced
Energy	27.07 MJ/kg	**Solid waste**	-
Transport	0.2 MJ/kg	**Emissions**	CO_2 163 g/kg SO_x 0 g/kg NO_x 0.081 g/kg CO 0.205 g/kg HCs 0.016 g/kg VOCs 6.95 g/kg

Waste to Landfill – Emissions and Energy Requirements for a 1Kg PET bottle (table 3)

Table 3. Environmental Profile for PET – Waste to Landfill

Energy	60.007 MJ/kg	**Solid waste**	-
Transport	0.2 MJ/kg	**Emissions**	CO_2 94.597 g/kg SO_x 0.848 g/kg NO_x 1.728 g/kg HCs 2080.609 g/kg

Waste Incinerated – Emissions and Energy Requirements for a 1Kg PET bottle (table 4)

Table 4. Environmental Profile for PET – Waste Incinerated

Energy	32.5MJ/kg	**Solid waste**	-
Transport	0.2 MJ/kg	**Emissions**	CO_2 2016 g/kg SO_x 0.609 g/kg NO_x 2.436 g/kg CO 0.609 g/kg

Summary – Emissions and Energy Requirements for a 1Kg PET bottle. In this section, we summed up and summarized the emissions released and energy requirements during the entire life cycle of a PET bottle. Table 4 summarizes the total energy required in MJ and total emissions released for the entire life cycle of 1 kg of the PET bottle for the different methods of waste disposal.

3. Life Cycle Impact Assessment

In this section we used the results from the life cycle inventory to assess the potential impacts of PET bottle treated with different disposal methods. The best way of making a comparison of the PET bottle waste disposal method from an environmental point of view was to compare and analyze the energy involved and waste emissions released in the entire life cycle of the PET bottle. Next step was to assign the air emissions to the impact of global warming [4].

Energy Use. Results for the total energy used are shown in Table 5 for PET with different treatment methods of disposal. The energy required to recycle 1Kg PET bottle is relatively lower than the energy consumed by the other two options. This is because it takes less energy to make recycled products due to its greatest environmental benefit provided that the recycled product is used to substitute virgin polymers. Besides, recycling is advantageous because every time that the material is reused, it is not produced again and, therefore, only half the waste remains in the soil than if virgin resin is used. The more times a material is recycled, the less the amount that remains in the soil and this will subsequently reduce the environmental load in the waste stream. According to Table 5, landfilling option required much more energy for the whole cycle of a kg PET bottle maybe due to vehicle fuel consumption on transportation, maintenance and the continuous monitoring on the landfill area. On the other hand, incineration is also one of the options to manage the PET bottle waste. Incinerator will burn all waste in ovens and the energy recovered is equivalent in amount to the heat of combustion of the different components.

Table 5. Energy consumed and emissions released for entire life cycle of 1kg PET bottle

	Waste Recycled	**Waste to Landfill**	**Waste Incinerated**
Total Energy (MJ)	113.27	144.2	118.7
Total Emissions (g)	2603.45	4610.98	4452.85

Global Warming. Table 6 presents the environmental impact for a 1kg PET bottle treated with different waste disposal methods. Global warming is the rising of the global temperature due to emissions of greenhouse gases. The only greenhouse gas emissions of any significance in the manufacture and disposal stages of the PET bottle are carbon dioxide and methane. Based on the results in table 6, the order of preference is that recycling is better than incineration and landfilling in terms of energy and emissions. The emissions contributing to global warming for incineration of PET bottle is rather similar to the emission for landfilling. This is because all the fossil carbon is released during incineration, as well as during landfilling.

PET can also be co-incinerated with other combustible products from the waste stream which will give even greater contributions to the reduction of greenhouse gases by the prevention of the emission of methane gas from landfills. Methane has a global warming potential of 30 times that of CO_2 [5]. This is why the prevention of waste going to landfill is a key measure to reduce the greenhouse gas emissions.

Table 6. Environmental impact for different treatment methods of the PET bottle

Environmental Impact	**Waste Recycled**	**Waste to Landfill**	**Waste Incinerated**
Global Warming (kg CO_2-eq)	3.33	47	4.3

Example of Calculation for Global Warming Value

Process X releases 2.495 kg CO_2 and 0.040016 kg CH_4

The equivalency factors are the 100-year Global Warming Potentials (GWPs):

Q global warming-CO_2 = GWPCO_2 = 1 g CO_2

Q global warming-CH_4 = GWPCH_4 = 21 g CO_2/ g CH_4. The potential contribution to global warming of methane is:

Q global warming-CH_4 x gCH_4 = 21 g CO_2/ g CH_4 x 40 g = 840 g CO_2-Eq.

The total contribution of Process X to global warming is:

Total global warming = (2.495 + 840) kg CO_2 = **3.33 kg CO_2-Eq.**

4. Summary

PET bottle plastics make a valuable contribution to the way we live, but as a society we need to find ways of using these plastics more wisely. The way we make, use and dispose of PET plastic bottles should have a minimal impact on the environment. Some of the methods to reduce the impacts on the environment are:

1) A greener plastics industry

The manufacture of plastic materials is one of the major industries with potential for serious pollution to the surrounding environment. Different types of plastics manufacturing processes and disposal methods will contribute to different effects on the environment. Therefore, government agency must ensure that industry operates in a way that minimises adverse effects on people and the environment, and contributes to the achievement of sustainable development.

2) The potential impact of chemicals leaching from PET plastic bottles on human health and the environment

The use of PET plastic bottles in beverages and packaging increases the risk of exposure. Risk assessments have been carried out for some, but not all chemicals used in plastics. Where information exists, it suggests that the quantities released during the manufacture and disposal stages of the life cycle of plastics are much greater than those lost during the use phase, when humans are likely to be most exposed. This information gap needs to be addressed.

3) Reducing the waste generation

As a society, we are generating an increasing amount of waste. As with other materials, more must be done to reduce the amount of PET plastic bottle waste we produce. There has been a shift towards a 'disposable' culture, and the lower cost of plastics may have contributed to this and the associated growth in waste.

4) Practicing recycling habit

The attitude of the public in recycling practices should be improved. The public should be well informed of the importance to recycle. Environmental influences appear particularly effective among members of the public who have a "strong belief in personal responsibility and influence, as well as the power of self-determination". The public would recycle more if they had a greater understanding of the environmental benefits of recycling.

5) Charging for waste

Although this practice is still not common in most countries, but several recycling centers in Russia are already implementing this concept, as a way to encourage more people to recycle. This concept is mainly targeting specifically at those who don't recycle, and in turn supports those who do. By charging for the waste sent to the recycling centers, this can help to encourage more recycling practices among the public.

6) Introducing more recyclable products

By replacing the PET plastic bottles for packaging beverages, with recyclable and biodegradable ones; this can initially encourage more people to recycle. By displaying the international standard *'Recycling Logo'* on recyclable products, the consumer will be fully informed whether the items can be recycled or not. When the consumer is aware that the items can be recycled, they will automatically categorize the items as recyclable products and will not dump them together with non-recyclable wastes. With such measures, it is hoped that the public will be more aware of the recyclable items available in the market.

References

[1] ISO 14040 (2006). Environmental Management. Life Cycle Assessment. Principles and Framework. ISO/FDIS.

[2] Ncube A., Borodin Y. V. Life Cycle Assessment Of Polyethylene Terephthalate Bottle // 7th International Forum on Strategic Technology (IFOST - 2012): Proceedings: in 2 vol., Tomsk, September 18-21, 2012. - Tomsk: TPU Press, 2012 - Vol.1 - **P. 64-69.**

[3] ISO 14041 (1998). Environmental Management. Life Cycle Assessment. Goal and Scope Definition and Life Cycle Inventory Analysis. ISO/FDIS.

[4] ISO 14042 (1999). Environmental Management. Life Cycle Assessment. Life Cycle Impact Assessment. ISO/FDIS.

[5] McDougall, F. R., White, P. R., Franke, M. and Hindle, P. (2001). Integrated Solid Waste Management: A Life Cycle Inventory (2nd Edition). Blackwell Science, Oxford, UK.

Conception of low-rise earthquake-resistant energy-efficient buildings

Y V Shefer[1], B S Ordobaev[2], S V Romanenko[3], O A Antonevich[4]

[1] Graduate, Tomsk Polytechnic University, Tomsk, Russia
[2] Head of Department, Kyrgyz-Russian Slavic University, Bishkek, Kyrgyzstan
[3] Head of Department, Tomsk Polytechnic University, Tomsk, Russia
[4] Associate Professor, Tomsk Polytechnic University, Tomsk, Russia
E-mail: yuri@shefer.ru

Abstract. The article proposes a new earthquake-resistant technology of low-rise building with increased energy efficiency and long-life operation. The proposed solution allows to build low-rise buildings with increased resistance to natural and man-made disasters. The building is frame (made of tube-concrete) and also completely monolithic, where foundation, all walls, floors and roof are filled of polystirolconcrete (composed of concrete and polystyrene), which forms a monolithic construction.

1. Introduction

There is a need to build buildings and constructions in seismic areas in many regions of the world. The sad experience of the past, devastating earthquakes and analysis of their effects revealed that the problem of reliable seismic protection of citizens and their homes still has not been solved.

It is seen from the analysis of the reports on the effects of strong earthquakes [1] that earthquake-resistant buildings, designed in strict accordance with the local regulations, in many cases were destroyed even under the "non-hazardous" seismic load level, which is lower than the calculated level.

In this case, it is definitely found out that in columns and walls of low-rise buildings all official seismic norms always decreases the real level of seismic stresses it many times, as well as significantly understates the real level of seismic stress and overestimates their actual seismic resistance. For example, according to official standards, these buildings have a wide margin to withstand a 10-point earthquake (by MSK-64 scale). But there are cases of their destruction in reality. This drawback of seismic norms can not be corrected, as it is a direct consequence of the basic postulates of the official theory of the direct dependence of seismic stresses in columns and walls on ranking building mass.

Seismic building codes in some cases increase the level of seismic risk for the population. For example, it is well known that buildings with walls made of bearing materials can not take expending stress (i.e., brick, stone, etc.). As a rule, they are destroyed at 9 magnitude earthquake. However, none of the normative document prohibits construction of such buildings. The main and the only argument is that the "vibrational" doctrine of seismic damage is globally accepted.

2. Experimental part and Discussion

Our research has shown that a devastating earthquake regularly raises a lot of facts and phenomena that can not possibly be caused by low-frequency vibrations and which can not be explained on the basis of the official "vibrational" model of seismic damage to buildings. For example, the low-frequency seismic ground oscillation must cause fractures of reinforced concrete columns near their clamped ends with formation of flexural plastic hinges in frame buildings. However, in reality, instead of it reinforced concrete columns only cut by a skew cracks in the span of the column at a distance from their ends always arises. It should be emphasized that for flexible reinforced concrete columns as for any flexible rod elements on quasi-static loads occurs only bending destruction with their break.

This is confirmed by numerous experiments on simulating the vibrational seismic impact on the model of frame buildings [2,3].

Earthquake-resistant buildings should not only withstand a major earthquake once, save lives, but to be able to function in the future and continue to withstand new tremors without damaging bearing and life supporting systems. We propose to combine the best native and international construction technology. Such construction technology is based on the construction of framework with tube-concrete and technology of monolithic polystyrene concrete pouring.

First articles on the use of tube-concrete were published in the USSR in 1932 by Professor A.A. Gvozdev on the method of calculation on tube-concrete as construction. However, this technology got a large spread of abroad in the United States, Japan, China, the United Arab Emirates. Tube-concrete columns from steel pipes of rectangular cross-section, filled with concrete are able to withstand considerable horizontal shear stress, which makes them indispensable in earthquake-prone areas.

The essence of the proposed concept is the following: the construction of a 'floating' building frame from tube-concrete with the possibility of nearly any layout and floor height. The building at the same time must consist of at least two frames.

Technical challenge was to increase the energy efficiency of buildings, exceeding the requirements of the regulations; increase structural strength, allowing the construction of multi-storey buildings; reduce the cost of construction.

The solution of this problem is provided by the fact that the building, (Figure1) containing the foundation, walls, building frame, floors overlap, formwork sheets the space between them is filled with light concrete. Walls contain two rigid frames: an inner frame consisting of columns, linked with girth rail with fixed roll-formed profile, which rests on the foundation and is the supporting framework of the building in its lower part. And at a distance of 0.1 to 1 meter from the inner frame light frame of the building is made, fastened, if necessary, with an internal temporal frame locks as filling with light concrete. Interfloor overlaps of building are made of interfloor girth rail fixed to the inner frame with girth rail. All components are equipped with roll-formed profile in such a way that after sheathing with formwork sheets they form a single formwork space of walls and floor overlaps.

Columns, floor girth rail and an outer frame on the upper floor of the inner frame are fixed on a formwork sheets and may be formed as mansard roof. Bearing structures can be used as a foundation of earlier constructed or rehabilitated buildings

Columns, girth rails of inner frame and beams can be made of steel tube of square and / or rectangular cross-section. If it is necessary to increase the load-carrying capacity, fire resistance and seismic resistance steel tubes can also be filled with heavy and / or light concrete.

Industrial applicability of the claimed technical solution can be explained in examples:

- For low-rise buildings (1–3 floor.) Load-bearing structures of building, columns, girth rails and beams are made, for example, of steel pipe cross-section 50 × 100 mm with a 4 mm wall, fastened together by welding.
- Middle-rise buildings (4–5 floor.) Load-bearing structures of building, columns and girth rails are formed, for example, from a steel pipe of 100 mm × 100 mm cross-section with a 4 mm wall, and the beams are made, for example, of steel pipe cross-section 50 × 100 mm with a 4 mm wall, filled with light monolithic concrete of necessary brand strength, fastened together by welding and in addition with screw connections and / or rivets.
- Outer frame is set at such a distance from the inner to provide the required parameters of thermal protection of building.

Inner bearing frame of the building: columns, girth rails and formed floor overlaps are made of metal pipes of rectangular cross-section, fastened to each other by welding and also with rivets or screw connections filled with concrete, where the pipe parameters and concrete brand are determined according to the height of the building and the length of the span.

Outer frame combines the functions of the frame for fastening permanent formwork sheets and at the same time can be used for fastening cleats with the subsequent installation of porcelain stoneware. The distance between frames may be 10 cm above the thickness filling of monolithic polystyrene concrete. In between formwork space necessary utilities are laid. In outer walls ventilation device with heat recovery is set. It serves fresh air into the living room and kitchen. In winter the air is heated and cooled in summer to a comfortable temperature. This can be called air heating in winter and air conditioning in summer. On the top floor of the building walls with thickening are transformed into a roof terrace. To this design (Figure 1) an application for the invention is filled [4] and is currently carrying out the examination on the merits.

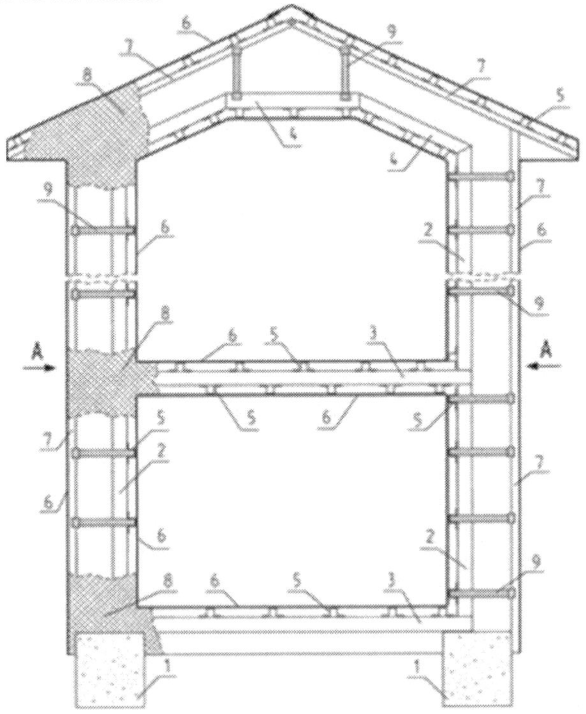

Figure 1. Cross-cross-section of the building
1 – foundation; 2 – inner frame column; 3 – girth rail inner frame; 4 – joisted floor; 5 – roll-formed profile; 6 – formwork sheet; 7 – outer frame; 8 – lightweight concrete; 9 – temporary lock.

We propose to fill constructions with lightweight concrete, for example, polystyrene concrete. Polystyrene technology was patented in Germany in 1952. In Soviet times polystyrene concrete was considered as strategic building material and was used mainly for the construction of large-scale production and storage facilities, as well as military installations. Concrete and Reinforced Concrete Research Institute (CRCRI) gave second life to PSC in our country, spent large scientific work and released All Union State Standards (P 51263-99) on polystyrene in 1999.

The main features of polystyrene. To be more precise, then it would be better to call this material: polystyrene-air-concrete. For the preparation of polystyrene cement, water, foam polystyrene balls and air-entraining resin are used. In the process of cement milk mixing fine bubbles of air are involved. Polystyrene balls are needed for homogeneous distribution of bubbles in the cement matrix. Because of this volumetric weight of polystyrene is homogeneous in whole monolithic poured volume and only slightly different in pouring height. At present, the production of polystyrene, can withstand up to 500 freeze-thaw cycles, then there is no destruction of the material under the design load.

According to our calculations for a comfortable, safe and energy efficient residence one must apply monolithic polystyrene brand D-200 200-600 millimeters thick in boundaring constructions. Thereby, the whole house (foundation, walls, floors, roof) poured monolithically, there are no "cold bridges", and the whole array of polystyrene, in fact, is an effective thermal insulation of its internal space from the environment. In such a house it is warm in winter and cool in summer as in wooden house(Table 1).

Table 1. The dependence of heat flow on the thickness of boundaring constructions

Wall thickness, мм	heat transfer resistance	$\frac{м^2 \times °C}{Вт}$	heat flow $Вт/м^2$
300	4.282		17.50
350	5.000		15,00
400	5.,714		13.12
450	6.428		11.66
500	7.142		10.50
550	7.857		9.54
600	8.571		8.75
650	9.258		8.10
700	10.000		7.50

It is also necessary to consider such characteristics of monolithic polystyrene concrete poured into the permanent formwork as flammability and fire resistance. First of all for the manufacture requires only the use of food self-extinguishing polystyrene brands. So when in contact with open flame polystyrene balls in cement matrix seem to disappear.As polystyrene balls consist by volume of 98% of air, then it is completely combusted in an excess atmosphere oxygen, in contact with air not only from the outside, but also the inside. During the oxidation reaction carbon dioxide and water appear. Carbon dioxide prevents burning and water transforming into vapor at the time of formation, prevents further destruction of the balls within the array.

It should be noted that in the monolithically poured polysterine bearing structures of the building are protected from aggressive environmental factors (moisture, air, heat, vibration, etc.), repeatedly increasing the life of the structure in whole (Figure 2).

As permanent formwork sheets (Figure3) it is useful to apply glass magnesite sheets that have 10 mm thickness, which are non-flammable materials and have fire resistance EI 90. It allows their use in buildings up to 150 meters according to the standards. It protects polystyrene from open flames and additionally protects the bearing structures of the building: columns, girth rails, joisted floors from exposure of high temperatures and destruction.

IOP Conf. Series: Materials Science and Engineering **81** (2015) 012086 doi:10.1088/1757-899X/81/1/012086

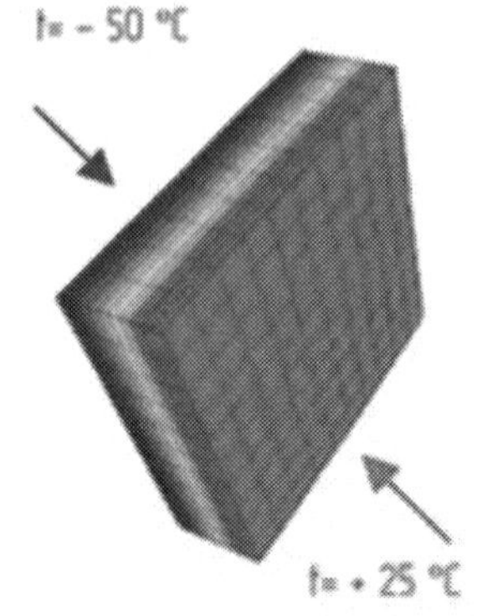

Figure 2. Temperature field.

This design concept allows to combine outside steel frame of the permanent formwork with a suspension system for fixing porcelain. The outer sheets of permanent formwork cleats are attached to the guide profiles with retractable rivets. At the same time it is possible to install windows and balcony doors. Frame of the building enclosed as if in a "shell". For engineering of exterior walls lined with porcelain a patent for utility model is received [5].

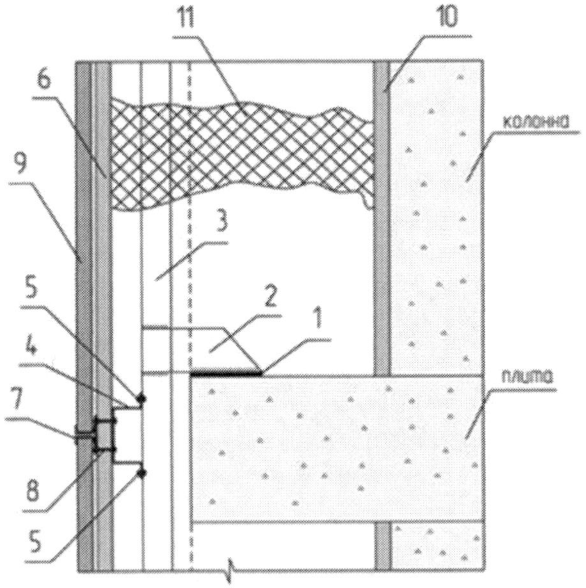

Figure 3. Multi-layer wall

1 – embedded parts; 2 – scarf ; 3 – vertical stand; 4 – horizontal profile; 5 – short rivet ; 6 – outer sheet permanent formwork; 7 – cleat; 8 – long rivet; 9 – covering element; 10 – the inner sheet of non-removable formwork; 11 – polystyrene.

The gap between the porcelain and the formwork is found to be the thickness of cleat. In such a way we get a three-layer instead of five-layer walls construction. Where porcelain and air gap appear apart from two sheets of polystyrene formwork and layer. It plays a very important complementary role in heat and fire protection of walls. In winter heat flow is coming out of the room, heats the air layer, as a result of a small gap and roughness of contacting surfaces. Warm air moves up in the

IOP Conf. Series: Materials Science and Engineering **81** (2015) 012086 doi:10.1088/1757-899X/81/1/012086

laminar regime that leads to the additional thermal insulation effect. In summer, on the contrary, the sun's rays warm porcelain and form airflow in the gap that helps to cool the surface of the wall.

3. Conclusions

So in accordance with the proposed concept of the combined use of technology tube-concrete and monolithic polystyrene allows to design and rapidly build low-rise buildings with increased resistance to earthquakes, fires, which have, in addition, exceptionally high energy efficiency. The concept allows to reduce construction costs and operating costs of maintaining a comfortable temperature indoor air in summer and winter.

References

[1] A sure report for building damages due to the 1995 Hyogo-Ken Nanbu earthquake *Building Research institute. Ministry of Construction (Japan).* 1996 March, 222 p.

[2] Smirnov S.B. Seismic shear buildings - the result of the impact of soil strata slidable deep seismic waves *Housing construction.* 2009. **№9**. P. 32-35.

[3] Smirnov S.B. The reasons for the destruction of earthquake-resistant buildings and principles of effective seismic protection *Concrete and reinforced concrete.* 1994. **№3**. P. 22-25.

[4] Y. Schaefer Prefabricated frame building energy efficient. The application for the invention RU **№ 2012131704;** appl. 24.07.12.

[5] Y. Schaefer The multilayer outer wall lined. Utility model patent RU № 97147 U1; appl. 30.04.10

Effect of particle size on the diffuse reflection coefficient of titanium dioxide powder

V A Vlasov[1], A L Astafyev[1], A N Zarubin[2]

National Research Tomsk Polytechnic University, Tomsk, Russia[1]
National Research Tomsk State University, Tomsk, Russia[2]

E-mail: vlvitan75@mail.ru

Abstract. In the present work a model of light scattering is shown which explains the result about effect of particle size on the diffuse reflection coefficient of initial titanium dioxide powders. The diffuse reflection coefficient depending on particle size for TiO_2 pigment varies on the curve with maximum. The experimental results and the model can be used for technology development of manufacturing pigment for light-reflecting temperature-control coatings of spacecraft

1. Introduction

During use device uncovered parts are exposed natural background and electromagnetic radiation, which penetrate to material surface layer. The accumulation of defects induced by radiation leads to microscopic changes related to the rupture of the individual valence bonds. These ruptures contribute to the appearance of luminescence centers. Moreover, this contributes to phase transitions and chemical reactions that lead to a change in the atomic structure of the crystal lattice and it properties. These changes are expressed in the ageing or degradation of the materials and coatings.

The main problem of maintaining the stability of the technical devices in radiation conditions is to develop composite materials and coatings that provide the desired level of their initial properties, a sufficiently high degree of protection in case of accidental drift of environmental parameters and the low degradation rate of the desired properties.

In crystals always there is some concentration of equilibrium pre-irradiation defects. As the rule, surface defects concentration increased during milling. This leads to the effect of particle size on irradiation stability. The effect of grain size on the irradiation stability of the materials is investigated in several studies [1-4] and the effect of particle size on the initial properties is shown in papers [5-8].

Titanium dioxide powders-pigments are applied for manufacturing of enamels, white paints and a reflecting temperature-control coatings (RTC) of spacecraft.

In this paper are given explanations about the results of the effect of particle size (r) on the diffuse reflection coefficient (ρ) initial powders using the developed model of light scattering.

2. Experimental techniques

Conducting experiments in real conditions using the spacecraft are connected with the serious technical and scientific challenges. That is why, in the laboratory for the study of irradiation effects in aerospace materials at Tomsk Polytechnic University to test RTC developed and manufactured special settings that simulate of space environment: vacuum, temperature, flow of charged particles (electrons and protons) and electromagnetic radiation, plasma of ionosphere, plasma "Sunny the wind".

The reflectivity of powders were measured by an absolute method of globular photometer using the "Spectrum" [9] which comprised of a source of electromagnetic radiation imitating solar spectrum in the wavelength range of 0.2 to 2.2 microns; the sources of electrons and protons imitating radiation belts of the Earth also the gas puffing system and control system their partial pressure. The investigated samples are clamped by fixed block to surface copper object stage. There are five clamps which allow to fasten 20 samples having diameter 15 mm. Object stage has a possibility to rotate around axis from magnetic drive this allows to investigate the samples by turn without breaking a vacuum. For control and maintain temperature of samples in range 80-470 K applied a clamping table-thermostat constitute copper ring with a channel for a liquid heat-transfer and electric heater. The temperature is measured by copper-constantan thermocouples caulk in substrate samples.

Absolute method of registration diffuse reflection spectra consist in measurement the illumination produced on a certain portion of the inner surface of the integrating sphere at two different positions of the light beam.

Powder factions were obtained by the sedimentation method based on Stokes' law [10]. The size of the accumulated particles is calculated by the formula from Stokes' law:

$$r = \sqrt{\frac{18 \cdot \eta \cdot U}{(d_2 - d_1) \cdot g}}$$

where d_2 – the density of material particles, d_1 – the density of dispersion liquid, g – free fall acceleration, η – the liquid's viscosity, U – the deposition rate of particles in a viscous medium.

The powders TiO_2 (collected sample + water) were applied on an aluminum substrate whereupon a suspension was dried during 24 hours at the room temperature. Samples were fixed on the object stage of setting "Spectrum", atmosphere in vacuum chamber was to 10^{-5} Pa, the initial diffuse reflection spectra were recorded in the range 360-2100 nm at 300 K and samples were irradiated by electron flux from $7 \cdot 10^{15}$ up to $4 \cdot 10^{16}$ cm^{-2} with 30 keV energy and intensity $1 \cdot 10^{13}$ cm^{-2}s^{-1}. After each irradiation step we recorded the diffuse reflection spectra of irradiated samples in vacuum. Diffuse reflection spectra ($\Delta\rho_\lambda$) were calculated by subtracting the spectra ρ_λ before and after irradiation and equated to the induced absorption spectra.

3. Result and discussion

We have previously found [11] that the dependence of diffuse reflection in different regions of the spectrum before irradiation (ρ_1), after irradiation by electrons (ρ_2) in the range of the average particle size 1-15 microns TiO_2 powder varies along a curve with a maximum and the dependence $\Delta\rho = \rho_1 - \rho_2$ varies along the curve with the minimum. In the area with the greatest reflection of initial powders observed maximum of irradiation stability. The optimal particle size lies in a range 2-4 microns. Increasing irradiation stability by varying the particle size for TiO_2 powders is over 5%.

Reducing diffuse reflection with increasing particle size explained the technological conditions of prepare samples. The increasing diffuse reflection with increasing particle size is explained by geometric factors considered powder system (scattering model). The rise $\Delta\rho$ with decreasing particle size is explained by defects emerging during formation a new surface. These defects in the initial condition are centers of gas sorption, but in the case when it located under action before-threshold irradiation it becomes potential color center. On the other hand an increase in the specific surface area decreases the concentration of defects, due to their drain on the dislocations to the surface of the crystallite, where occurs dissipation of energy electronic excitations.

Next, we consider the theoretical model of light scattering crystallites spherical form in more detail, which explains the growing part of the diffuse reflection from particle size. This model assumes existence of two light sources: 1) surface source - in a crystallite shape with size r_A; 2) volumetric source - in shape of scattering center with size r_C. For simplicity, let's consider the particles of spherical shape (Figure 1), where there are m scattering centers in volume.

Assume that scattering coefficient has constant value σ_C on the one center, on surface – σ_S. In this case, scattering coefficient of single particle can be written as $\sigma_P = W_S\,\sigma_S + W_C + \sigma_C$ or in expanded form:

$$\sigma_P = \sigma_S + \frac{r_C^2}{r_A^2} m \sigma_C \tag{1}$$

where W_S=1, $W_C = \frac{r_C^2}{r_A^2} m$ - the probability of ingress of luminous flux to the surface and the

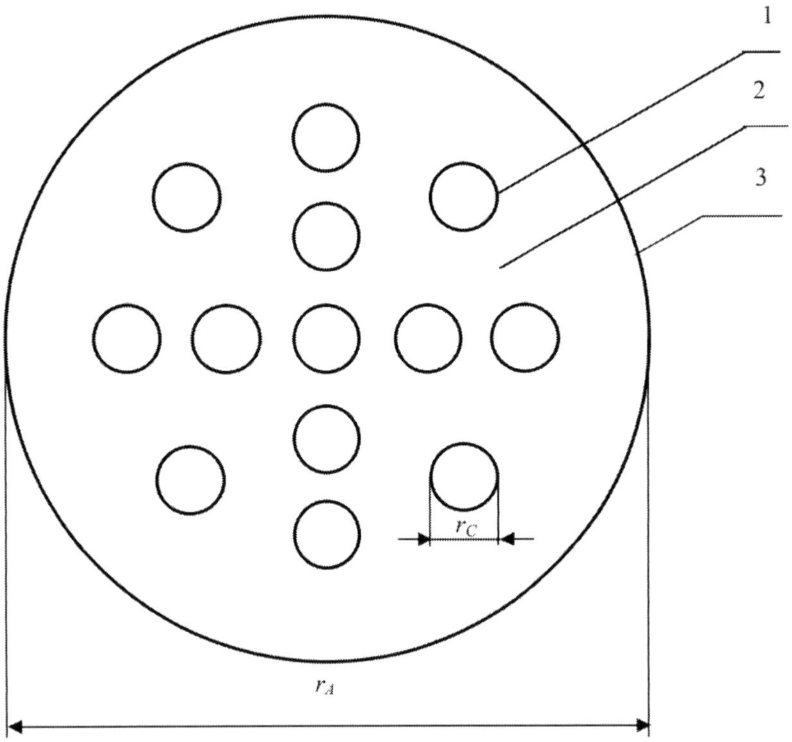

Figure 1. The scattering model: 1 – scattering center; 2 – perfect crystal lattice; 3 – scattering surface of crystallite.

scattering centers respectively; $\frac{r_C^2}{r_A^2}$ – the relative area of the one center.

Assuming that $\sigma_S \approx \frac{\pi \cdot r_A^2}{2}$ and $\sigma_C \approx \frac{\pi \cdot r_C^2}{2}$, the particle scattering coefficient is obtained from (1):

$$\sigma_P = \frac{\pi}{2} r_A^4 + \frac{\pi}{2} \frac{r_C^4}{r_A^2} m = \frac{\pi}{2}(r_A^4 + r_C^4 m)$$

Thus, the more r_A, the more σ_p.

This means, that reflecting surface of crystallite are increased with increasing particle size, and consequently diffuse reflection coefficient are increasing.

4. Conclusions

The diffuse reflection coefficient depending on particle size varies along a curve with a maximum. The highest irradiation stability is observed in an area of the greatest reflection unexposed initial powders. Particles which provide high values of diffuse reflection coefficient in sunny range of spectrum and good the TiO_2 powder stability to the action of electrons have a size in range from 2 to 4 microns. The increasing diffuse reflection with increasing particle size is explained by geometric factors considered powder system. A model of light scattering crystallites spherical shape was offered, which explains this increase.

5. Acknowledgements

This work was supported by The Ministry of Education and Science of the Russian Federation in part of the "Science" program.

References

[1] Aluker V E, Greenberg M G, Nesterov S N et al 1986 *Izv. AN Lat. SSR. Series nat. and tehn. nauk* [in Russian] No 5 23-25
[2] Mikhailov M M, Kuznetsov N I and Ryabchikova L E 1988 *Izv. SSSR Akad. nauk. Neorgan. Mater.* [in Russian] V **24** No 7 1136-1140
[3] Joshi N G, Dhoble S J, Moharil S V et al 1994 *Radiation Physics and Chemistry* V **44** No 3 317-322
[4] Dhoble S J, Sahare P D and Moharil S V 1991 *J. Phys.: Condens. Matter.* V **3** No 9 1189-1195
[5] Gurevich M M, Itsko E F, Seredenko M M 1984 *Optical properties of coatings* [in Russian] (Leningrad: Himia) p 120
[6] Lucas G 1974 *Heat transfer and thermal regime of the spacecraft* [in Russian] (Moscow: Mir) p 544
[7] Levshin V L and Ryzhikov B D 1962 *Optica i Spectroscopia* [in Russian] V **12** No 3 400-406
[8] Mikhailov M M and Utebekov T A 2013 *Modern materials and technologies* No 1 178-180
[9] Kositsyn L G, Mikhailov M M, Kuznetsov N I et al 1985 *PTE* [in Russian] No 4 176-180
[10] Figurovsky N A 1948 *Sedimentation analysis* [in Russian] (Moscow-Leningrad: Academia nauk SSSR) p 332
[11] Mikhailov M M and Vlasov V A 1998 *Russian Physics Journal* V **41** No 12 1222-1228

IOP Conf. Series: Materials Science and Engineering **81** (2015) 012088 doi:10.1088/1757-899X/81/1/012088

Application of secondary ion mass spectrometer for measuring the diffusion profiles in alkali-halide crystals

A V Chernyavskii[1] and M S Kaz[2]

[1] National Research Tomsk Polytechnic University, Tomsk, Russia

[2] National Research Tomsk State University, Tomsk, Russia

E-mail: chernyavski@tpu.ru

Abstract. Depth profiles of magnesium, fluorine and oxygen impurities was examined in the surface layers of alkali-halide KBr crystals using method of secondary ion mass spectrometry. Samples of potassium bromide, coated with a surface film of magnesium fluoride were subjected to isothermal diffusion annealing in air at various times. It is shown that the diffusion of O ions occurs from the ambient atmosphere besides the diffusion of Mg and F ions during annealing of KBr crystals. Accurate estimation of the diffusion coefficients of cationic impurity Mg requires taking into account the possible interaction of this impurity and oxygen.

1. Introduction

A significant number of solid phase reactions, which determines the real defect structure and the most important physical, chemical properties of inorganic materials, are among the diffusion-controlled. Radiation impact is one of the most promising methods and greatly accelerates the process of sintering of powders of alkali salts [1], a number of oxides [2], as well as intensifies the process of synthesis of complex oxide compounds [3, 4]. Opportunity to stimulate the diffusion phenomena in ionic dielectrics by flows of ionizing radiation, especially in the most practical interest region of high temperatures, is an actual problem. Therefore, we carried out experiments using electron beams, providing heating of the samples and the high level of excitation of the electron and nuclear subsystems of crystals, as well as radiation defect, for the study of their effect on the diffusion mass transfer model in alkali halide crystals, the results of which are given in [5-7].

Secondary-ion mass spectrometer MS7201M (PC "Electron", USSR) was used in these studies to measure the diffusion profiles. The design features of the instrument required to implement additional steps for sample preparation and measurement. More detailed methodology of research on this device is described in our previous works [5, 8]. In addition, analysis of the results of studies high temperature diffusion of Mg impurities from a variety of chemical compounds (MgF_2, Mg) in ionic crystals LiF [9] and KBr [8] showed that the calculated value of the diffusion coefficient has always been less for investigated impurities than the known values of bulk diffusion coefficients [10] and this difference was (2-3) an order of magnitude. The original chemical state of impurity cations had no significant effect on their diffusion characteristics in this works.

It has been suggested that an impurity film reacts with oxygen to form metal oxide under annealing in contact with the air. A similar problem faced by the authors of [11] in the study of the diffusion of cobalt impurity in potassium chloride, for which extremely low diffusion coefficients were also received. To substantiate this hypothesis, we studied high-temperature diffusion of Mg, F, O impurities in the surface layers of ionic crystals KBr during thermal annealing, the results of which are given in this article.

2. Material and experimental methods

Secondary-ion mass spectrometer PHI 6300 (Perkin-Elmer PHI 6300 Ion Microprobe, USA) was used to measure the depth profiles in this work. The instrument allows analysis of both positive and negative secondary ions. Thus, it is possible to conduct a study of depth profiles of cationic and anionic impurities in ionic crystals. Study of the profiles of anionic impurities could not be done by using a mass spectrometer MS7201M as the device allows analysis only the positive secondary ions. The yield of secondary positive ions is very small for the investigated anionic impurities in KBr crystals. Primary source of Cs^+ ions with energy of 7 keV was used for etching samples in PHI-6300. The sample surface is scanned by beam of focused primary ion with sides of the scan area of 500 μm. To neutralize the positive charge building up on the surface of the sample during analysis, electron gun was used, the electron beam of which is scanned sample surface. After the measurements it was determined the depth of the etching crater using the profilometer, and then there were built dependencies of the intensity of the secondary ion impurities signal from the depth. Secondary positive ions were analysed for magnesium impurities and negative ions - for oxygen and fluorine impurities. In our experimental conditions, we can assume that the signal intensity of the secondary ions is directly proportional to the concentration of the elements of interest in the sample.

Single crystals of potassium bromide were used in our experiments. Magnesium fluoride film with about 400 nm thickness was deposited by thermal evaporation in vacuum on the surface of the samples. Unlike the previously used technique [8], the film was deposited on the entire surface of the sample. Then, the film sample surface was superimposed another crystal, and they are pressed together. Thereafter, the samples were subjected to isothermal annealing in air at 650°C in an electric resistance furnace. Annealing times were 30, 90 and 600 min.

3. Experimental results and discussion

Figure 1 shows the depth profiles of the magnesium impurity ions in the crystals of potassium bromide, subjected to annealing with different duration, and depth profile of magnesium ions in the original sample before the diffusion annealing. Dependencies the signal intensity of the secondary ions of magnesium on the depth presented in relative units. Signal intensity was normalized to the intensity value at the boundary surface, the impurity film - crystal.

As can be seen from figure 1 the depth of penetration and the number of embedded impurities increases with increasing time of annealing, indicating a diffusive mass transfer of impurities in the depth of the crystal. However, the calculated value of the diffusion coefficient for magnesium impurities for these measurements is equal 3×10^{-12} cm^2/s and comparable to values previously obtained for the diffusion of impurity from the metal films [8]. Diffusion of magnesium ions previously nobody investigated in KBr crystals. In the literature [10], there are experimental data relating to the divalent cationic impurities other varieties. Typical values for their diffusion coefficients corresponding to temperature of our experiments are in the range of 10^{-8}–10^{-10} cm^2/s. The calculated value of the diffusion coefficient, determined by us for magnesium ions in potassium bromide, was considerably reduced by orders of magnitude. This result confirmed our earlier results [9].

To numerically evaluate the diffusion coefficients of magnesium ions approximation of the experimental profiles was carried out by solution of Fick's equation for the diffusion from the region with a constant concentration in a semi-infinite crystal. However, such evaluation gives approximate results, as seen in this model only diffusion of one impurity is considered and does not account diffusion of other ions, as well as their possible interactions between them. In our experimental conditions diffusion of oxygen ions and fluoride was also observed that can be seen from these results further.

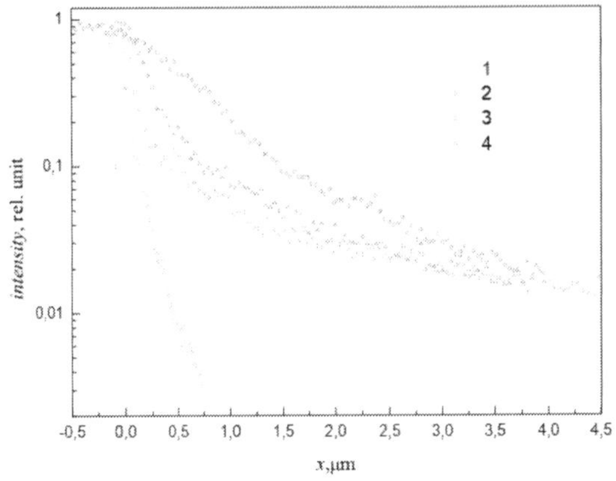

Figure 1. Magnesium depth profiles in the original sample without annealing (1) and samples annealed for various duration (2 - 30 min., 3 - 90 min., 4 - 600 min.) at 650°C

Figure 2 shows the depth profiles of F and O ions for the initial sample and the sample after annealing at 650°C for 600 min. Signal intensity of the secondary ions of oxygen (in the graph shown in absolute units) increases by an order of magnitude or more for a sample subjected to an annealing, as compared to the initial sample. Moreover, the oxygen penetrates into the MgF_2 film, which is on the sample surface and the potassium bromide crystal itself.

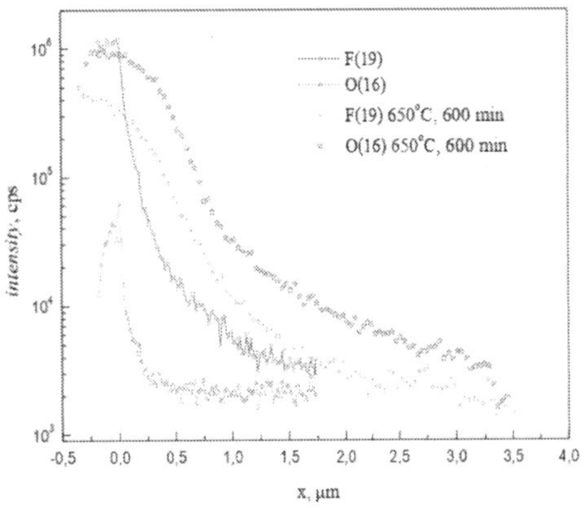

Figure 2. Depth profiles of oxygen and fluorine ions in the original and annealed samples

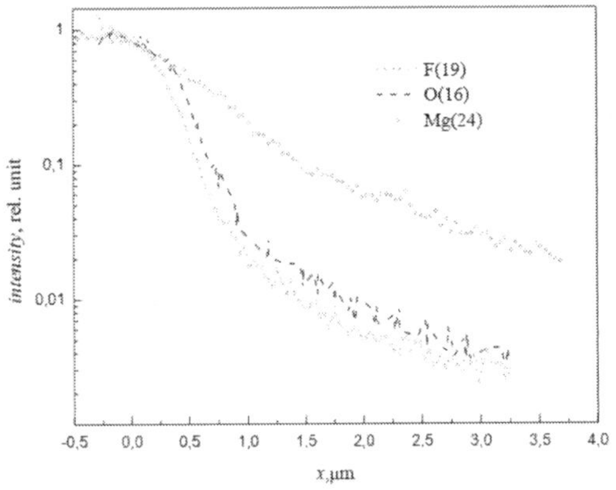

Figure 3. Depth profiles of O, F, Mg impurities in KBr crystal followed by 600 min. annealing at 650°C

The depth of penetration of impurity ions of fluorine and oxygen was essentially less compared to the depth of penetration of magnesium ions, as best seen in figure 3 (secondary ion intensity signal is given in relative units), which shows the depth profiles of indicated ions. This indicates a lower activity of the diffusion fluorine ions and oxygen ions in comparison with magnesium.

The results obtained confirm that the diffusion of atmospheric oxygen simultaneously takes place in the sample and source diffusing film in the diffusion of magnesium in KBr crystal from MgF_2 film.

In the crystal lattice of the oxygen can react with the cation impurity to form an oxide molecule. Oxide molecules in the investigated temperature range are quite stable. We can assume that because of this impurity cation penetrates into the surface layers in a bound state as inactive oxide molecular complexes. Thus, for the numerical determination of the diffusion coefficients of impurities in these experiments, it is necessary to take into account the fact of simultaneous penetration of magnesium, fluorine, oxygen impurities in the crystals as well as the information on the form of penetration of impurities in the crystals and therefore, the development of a specific diffusion model for the solution of the Fick equation are required.

4. Conclusions

According to the presented results following conclusions can be:
- oxygen diffusion occurs from the ambient atmosphere during annealing on air for samples of the alkali halide crystals with deposited on their surface of the metal halide film as a source of impurity for diffusion experiments;
- correct estimation of diffusion coefficients of cation impurities requires to consider the possible interaction of these impurities with oxygen in ionic crystals.

Acknowledgements

This work was supported by the Ministry of Education and Science of the Russian Federation within the framework of the government "Science" program.

References

[1] Yu M Annenkov, T S Frangul'yan and A V Voznyak 1991 *Soviet powder metallurgy and metal ceramics* **30(8)** pp 634–636
[2] V V Grishaev, A P Erastova, B M Lebed', Yu G Saksonovand G Ya Fedorova 1988 *Izv. Akad. Nauk SSSR. Neorg. Mater.* **24(11)** pp 1857–60
[3] V L Auslender, I G Bochkarev, V V Boldyrev, N Z Lyakhov and A P Voronin 1997 *Solid State Ionics* **101-103**(PART 1) pp 489-493
[4] V V Boldyrev, A P Voronin, O S Gribkov, E V Tkachenko, G R Karagedov, B I Yakobson and V L Auslender 1989 *Solid State Ionics* **36(1-2)** pp 1-6
[5] A P Surzhikov, A M Pritulov, S A Ghyngazov and A.V Chernyavskii 2000 *Perspekt. Mater.* **1** pp 30-34
[6] A P Surzikov, T S Frangulyan and A V Chernyavskii 2001 *Perspekt. Mater.* **3** pp 24-29
[7] A P Surzhikov, T S Frangulyan, S A Ghyngazov and A V Chernyavskii 2004 *Perspekt. Mater.* **2** pp 77–82
[8] S A Gyngazov, A P Surzhikov, T S Frangul'yan and A V Chernyavskii 2002 *Russian Physics Journal* **8** pp 753-758
[9] A P Surzhikov, A V Chernyavskii, S A Ghyngazov and T S Frangulyan 2002 *Russian Physics Journal* **12** pp 1190-94
[10] F Beniere 1972 Diffusion in ionic crystals *Physics of Electrolytes* vol 1 ed J Hladik (London-New York:Academic Press) chapter 6
[11] D Damien and K V Reddy 1983 *Indian J. Pure Appl. Phys.* **21(3)** pp 133–136

Effect of ion irradiation on tribological properties of composite coatings

E S Platonova[1], S A Guchenko[1], A Sh Syzdykova[1], V Ch Laurinas[1], V M Yurov[1], E N Lysenko[2], T S Mylnikova[2]

[1]Buketov Karaganda State University, Karaganda, Kazakhstan
[2]National Research Tomsk Polytechnic University, Tomsk, Russia

E-mail: lysenkoen@tpu.ru

Abstract. The paper discusses the results of the research in the effect of ion irradiation on tribological properties of multiphase ion-plasma coatings. It is shown that all the investigated coatings behave differently under ion irradiation: the Zn-Al, Fe-Al, Zn-Cu-Al coating is radiation-resistant and the friction coefficient does not virtually change; the Cr-Mn-Si-Cu-Fe-Al coating exhibits twofold increase in the friction coefficient, and that for Mn-Fe-Cu-Al coating decreases threefold. These changes are related to changes which occur in the coating surface under ion bombardment.

1. Introduction

Ion irradiation, considered as an unwanted phenomenon in airspace and reactor engineering, now has become one of the techniques to modify the material surfaces, including nanomaterials, to provide them with desired properties [1-8].

Ion bombardment of a solid surface implies several aspects. First, exposure of surfaces of different materials to metal ions makes the basis of the ion-implantation technique [9, 10]. The ion energy of tens or hundreds of keV allows the penetration depth of up to 1μm. Ion-implantation results in the formation of the near-surface alloy layer with different chemistry which, in its turn, changes its physical-and-mechanical properties. There is no distinct interface common to deposited layers. Ion-implantation is an active and non-equilibrium process and it allows creation of materials with structure and composition which cannot be obtained with conventional metallurgical techniques. Ion-implantation technique enables to dope any element of a desired quantity.

The second aspect of ion bombardment is related to nanostructuring of deposition directly in the production process [11, 12].

The third aspect of ion bombardment considers the effect of ions of different gases on the material surface when it has already been formed [13]. This is of current relevancy for spacecrafts in the Earth ionosphere [14].

The main effect of ion bombardment on the coating properties can be observed at the stage of their formation due to the stress relaxation in the region of ion action and restructuring of the crystal structure. Point defects, which are active adsorption centers, are formed on the surface.

The number of point defects can be increased either by increasing the ion flow energy, or by an increase in the ion beam density. Simultaneously with the formation of defects, the reverse process of their recombination, "annealing", occurs. It decreases the defect concentration. As a result of these two processes, the amount of nucleation sites becomes equilibrium which can be affected through changing the parameters of ion irradiation.

2. Experimental

In the experiments, we used the composite cathodes prepared by induction melting. These cathodes were used to perform deposition on the steel substrate with ion-plasma instollation HHB-6.6I1 under different process conditions. Atomic force microscope NT-206 was used to conduct a nanoscale examination of the coating surface.

Irradiation of the coating with argon ions was carried out by means of the multi-ampere ion source with a hollow cathode. The arc current was 1 A, and the potential of the substrate was maintained equal to 300 V.

The information-measuring system for tribological studies is described by us in [15]. It consists of two main components: the experimental setup and upper level software.

To apply the nanocoating the following basic process approaches are used: 1) ion-assisted deposition; 2) multilayer deposition with layers of nanometer thickness; 3) multiphase deposition; 4) a combination of these methods. We used ion-assisted multi-phase deposition.

3. Results and discussion

Figure 1a shows an electron microscopical image of the coating performed by simultaneous sputtering of the Cr-Mn-Si-Cu-Fe-Al cathode and a titanium cathode in nitrogen atmosphere. Figure 1b presents the same coating after argon ion irradiation.

It can be seen from Figure 1a, the average grain size is (100–150) nm. This type of coatings is referred to as submicrocrystalline coating [16]. The results of the quantitative analysis of the Cr-Mn-Si-Cu-Fe-Al + Ti coating in a nitrogen atmosphere in different parts of the surface showed that the contents of chromium, titanium and nitrogen are close to each other. This suggests that in addition to titanium nitride formation, chromium nitride is formed as well. Figure 1a also shows that the microcrystallites of titanium nitride and chromium have a preferred orientation (presumably in direction (200)), which is different from the spherical symmetry of the pure titanium microcrystallites.

a b

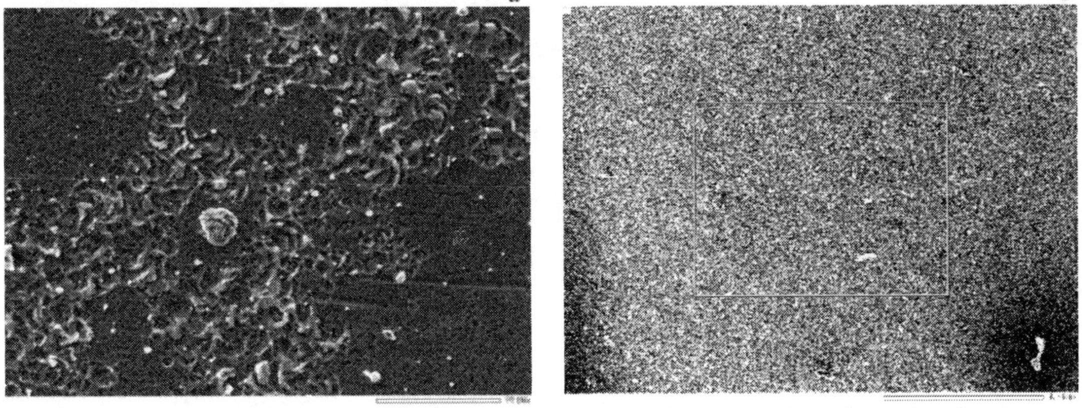

Figure 1. Electron microscopical image of the Cr-Mn-Si-Cu-Fe-Al + Ti coating in nitrogen atmosphere (a); and after ion bombardment (b).

Irradiation of the Cr-Mn-Si-Cu-Fe-Al + Ti coating with argon ions causes grinding of the grain structure (Fig. 1b), so that the grain size becomes less than 100 nm. This type of structures is called nanocrystalline structures [16].

Ion bombardment of the Fe-Al + Ti and Zn-Al + Ti surfaces does not result in grain grinding as that observed for the Cr-Mn-Si-Cu-Fe-Al + Ti coating. This is due to the weak effect of ion irradiation on the Fe-Al and Zn-Al coatings [17].

We attribute the radiation resistance of the Zn-Al coating to its pronounced globular structure. The presence of the "ball" system leads to elastic scattering of argon ions, so that the local deformation becomes negligible. This is confirmed by the behavior of the Young's modulus, which remains unchanged during irradiation. The radiation resistance of the Fe-Al coating disorder is associated with its amorphism. Strong "amorphization" of the coating makes it radiation insensitive.

The above properties of the structure of multiphase coatings are also reflected in their tribological properties. Table 1 shows the friction coefficients before and after ion bombardment.

Table 1. Friction coefficients for coatings after ion irradiation.

Coating	Friction coefficient	
	Before irradiation	After irradiation
Zn–Al+Ti	0.349	0.344
Fe–Al+Ti	0.374	0.367
Zn–Cu–Al+Ti	0.243	0.241
Cr–Mn–Si–Cu–Fe–Al+Ti	0.711	1.422
Mn–Fe–Cu–Al+Ti	0.367	0.122

As can be seen from Table 1, all the investigated coatings behave differently under ion irradiation: Zn-Al, Fe-Al and Zn-Cu-Al coatings are radiation-resistant and the friction coefficient virtually does not change; the friction coefficient for Cr-Mn-Si-Cu-Fe-Al coatings increases twofold, and that for Mn-Fe-Cu-Al coatings decreases threefold. These changes are caused by the changes which occur in the coating surface under ion bombardment.

4. Conclusion

The results obtained make it possible to draw the conclusions.

- Ion bombardment is a promising technique to regulate the structure and properties of coatings and can be used to create a variety of combined methods for deposition.
- The tribological properties of coatings, closely related to their microstructure, can be considerably changed under ion irradiation.

5. Acknowledgements

This work was financially supported by The Ministry of Education and Science of the Russian Federation in part of the science activity program.

References:

[1] Rybin V V, Zhivushkin A A, Berezhko A I et al. 2001 *Voprosy materialovedenia* **2** 166.

[2] Ivanov L I, Platov Yu M 2004 *Radiation physics of metals and its applications* (Cambridge International Science Publishing) p 356.

[3] Andrievskii R A Effect of irradiation on the properties of nanomaterials 2010 *Physics of Metals and Metallography* **110** 229.

[4] Ghyngazov S A, Vasil'ev I P, Surzhikov A P, Frangulyan T S, Chernyavskii A V 2015 Ion processing of zirconium ceramics by high-power pulsed beams *Technical Physics* **60** 128.

[5] Stepanova O M, Krivobokov V P 2010 Erosion of solid surface under irradiation by high-power submicrosecond ion beams *Bulletin of the Russian Academy of Sciences: Physics* **74** 122.

[6] Berlin E V, Koval N N, Sheydman L A 2012 *Plasma chemical-thermal surface treatment of steel products* (Moscow: Technosphere) p 640.
[7] Rogozhkin S V, Aleev A A, Zaluzhnyi A G, Kuibida R P, Kulevoi T V, Nikitin A A, Orlov N N, Chalykh B B, Shishmarev V B 2012 Effect of irradiation by heavy ions on the nanostructure of perspective materials for nuclear power plants *Physics of Metals and Metallography* **113** 200.
[8] Uglov V V, Remnev G E, Kvasov N T, Safronov I V 2014 Dynamic processes in metal nanoparticles under irradiation *Journal of Surface Investigation* **8** 703.
[9] Russell Ch, Ruge I 1983 *Ion implantation* (Moscow: Science) p 360.
[10] Pogrebnyak A D, Remnev G E, Chistyakov A S, Ligachev A E 1987 Modification of the properties of metals by high-power ion beams *Soviet Physics Journal* **30** 39.
[11] Panin V E, Sergeev V P, Panin A V 2008 *Nanostructuring of the surface layers of structural materials and nanostructured coatings* (Tomsk: Tomsk Polytechnic University Publishing) p 285.
[12] Lyakhov N Z, Psakhie S G 2008 *Surface nanoengineering. Formation of non-equilibrium states in the surface layers of materials by electron-ion-plasma technologies* (Novosibirsk: Publishing House of SB RAS) p 275.
[13] Yurov V M, Vertyagina E N, Guchenko S A 2011 *Sovremennye naukoemkie technologii* **5** 63.
[14] Nadiradze A B, Korsun A G, Kurshakov M U et al. 2011 *Electronic J. Trudy MAI* (Moscow: Moscow Aviation Institute) **43** 11.
[15] Kolesnikov V A, Baysagov Ya Zh, Yurov V M 2011 *Fundamentalnye issledovaniya* **12** 121.
[16] Reshetnyak E N, Strel'nitskij V E 2008 Synthesis of hardening nanostructural coatings *Problems of Atomic Science and Technology* **2** 119.
[17] Yurov V M, Laurynas V C, Guchenko S A, Zavatskaya O N 2013 Structure and properties of multiphase ion-plasma coatings (Karaganda: Publishing House of the Kazakh-Russian University Press) p 150.

Multiphase composite coatings: structure and properties

V M Yurov[1], S A Guchenko[1], E S Platonova[1], A Sh Syzdykova[1], E N Lysenko[2]

[1]Buketov Karaganda State University, Karaganda, Kazakhstan
[2]National Research Tomsk Polytechnic University, Tomsk, Russia

E-mail: lysenkoen@tpu.ru

Abstract. The paper discusses the results of the research into the formation of ion-plasma multiphase coatings. The types of the formed structures are found to be not so diverse, as those formed, for example, in alloy crystallization. The structures observed are basically of globular type and, more rarely, of unclosed dissipative and cellular structures. It is shown that the properties of the coating formed in deposition are largely determined by its surface energy or surface tension. Since the magnitude of the surface tension (surface energy) in most cases is an additive quantity, each of the elements of the coating composition contributes to the total surface energy. In case of simultaneous sputtering of multiphase cathodes, high entropy coatings with an ordered cellular structure and improved mechanical properties are formed.

1. Introduction

The microstructure of single-phase films can be described qualitatively well using models proposed by Movchan and Demchishin, and Thornton [1, 2]. However, if the film is doped with impurities, these models greatly change. The columnar microstructure typical mainly of single-phase films completely changes in films with an average and high impurity content. This fact is described by the model developed by Barna and Adamik [3].

To obtain nanocomposites with plasma deposition methods, the multicomponent flows are to be deposited on the substrate. In most cases, the coating of a complex elemental composition is synthesized through layering or simultaneous deposition of the flows on a substrate in reactive gas atmosphere from sources: ion [4], magnetron [5], vacuum arc [6], and their combinations thereof. [7]

The paper presents the experimental data on the structure of multi-phase coatings and their properties. The results obtained are discussed in terms of the role of the surface tension in formation of the coatings.

2. Experimental

In the research, the Zn-Al, Al-Fe, Zn-Cu-Al, Mn-Fe-Cu-Al and Cr-Mn-Si-Cu-Fe-Al composite cathodes prepared by induction melting were used to deposit coatings. The technique of simultaneous sputtering of different cathodes in argon and nitrogen atmosphere was used as well.

The thickness of the coatings and their elemental composition were measured using an electron microscope 200 Quanta 3D. The phase composition and structural parameters of the samples were studied by means of the diffractometer XRD-6000 with CuKα-radiation. The phase composition, size of the coherent scattering regions and internal elastic stresses ($\Delta d/d$) were analyzed using PCPDFWIN and PDF4+ databases and the POWDER CELL 2.4 software was used for a full-profile analysis. The nanohardness of the coatings was determined with the nano-identification system by the Oliver-Pharr method with the Berkovich indenter under a load of 1 g and the time of exposure of 15 s.

3. Results and discussion

Figures 1 and 2 show the images of Zn-Al and Fe-Al two-phase coatings obtained with an atomic force microscope. In the first case, the structure is globular, and in the second one, the structure is unclosed dissipative. We suppose that this difference in the structure of the coatings is caused by the difference in the surface tension of the basis metals of these coatings.

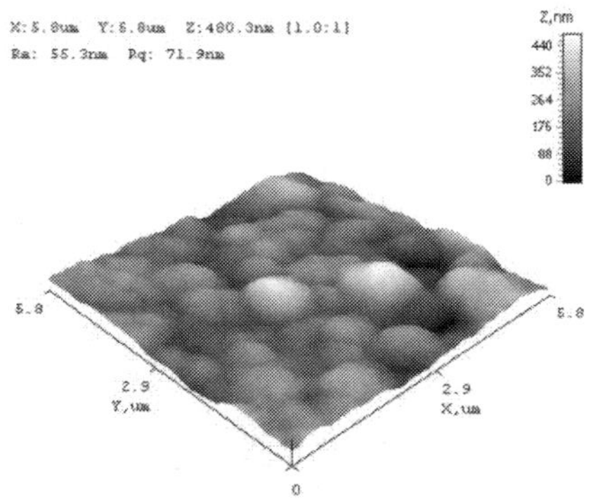

Figure 1. AFM image of the Zn-Al coating surface.

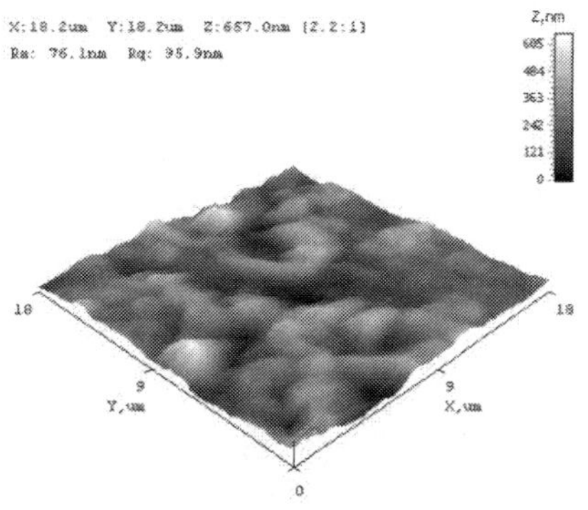

Figure 2. AFM image of the Fe-Al coating surface.

The surface tension for Zn-Al and Fe-Al coatings was determined by the technique described in [8]. The results obtained by the quantitative elemental analysis of the composite coatings were $Fe_{0.57}Al_{0.43}$ and $Zn_{0.34}Al_{0.66}$. The mean values of the surface tension were found to be $\sigma_{Fe\text{-}Al} = 0.972$ J/m^2 and $\sigma_{Zn\text{-}Al}$=0. 617 J/m^2.

Using the values of the surface tension for pure metals [9], we obtain:

$$\sigma_{Fe\text{-}Al}= 0.57\cdot\sigma_{Fe}+0.43\cdot\sigma_{Al}=1.01 \text{ J/m}^2;$$

$$\sigma_{Zn\text{-}Al}= 0.35\cdot\sigma_{Zn}+0.65\cdot\sigma_{Al}=0.551 \text{ J/m}^2.$$

It can be seen that the experimental and theoretical values of the surface tension for Fe-Al and Zn-Al coatings coincide within the experimental error. The surface tensions for Fe-Al and Zn-Al coatings are two times different.

According to the classical theory, the nucleation of a new phase in from the metastable phase is regarded as a fluctuation process [10]. In this case, the work of the spherical nucleus formation is

$$W = \Delta F_{кр} = \frac{16\pi}{3}\left(\frac{M}{\rho}\right)^2 \frac{\sigma^3 T_0^2}{q^2(\Delta T)^2},$$

where M is molecular weight, ρ is nucleus density, q is heat of fusion, T_0 is the equilibrium temperature of the two phases of infinite radius.

From the last expression it follows that the work of the globule formation is proportional to the cube of the surface tension. This indicates the importance of considering the surface tension of the components in plasma coating formation.

Figure 3 shows the structure of the coating prepared by simultaneous sputtering of Cr-Mn-Si-Cu-Fe- Al and titanium cathodes in nitrogen atmosphere.

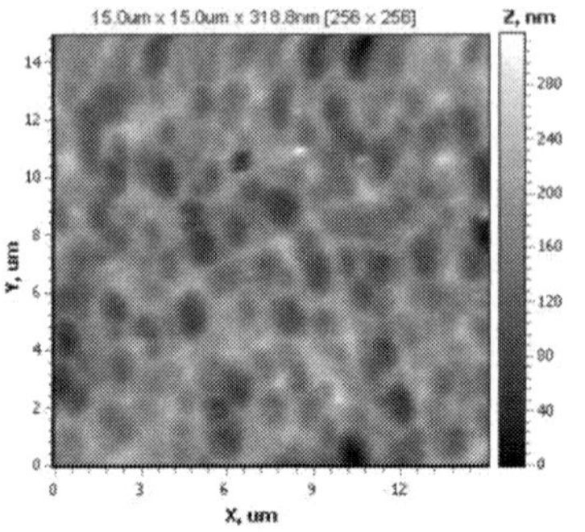

Figure 3. AFM images for the cellular structure of Cr-Mn-Si-Cu-Fe-Al + Ti coating in nitrogen atmosphere.

The results of the study of the phase composition and structural parameters of the sample are shown in Table 1. The nanohardness of Cr-Mn-Si-Cu-Fe-Al+Ti coating in nitrogen gas environment is 7.413 GPa. For 12X18H10T+Ti coating in nitrogen gas environment, the nanohardness is very high and is equal to 35.808 GPa. For comparison, Table 2 shows the nanohardness values for the most solid materials.

Table 1. Phase composition of Cr-Mn-Si-Cu-Fe-Al+Ti coating in nitrogen gas environment.

Sample	Determined phases	Phase content, vol.%	Lattice parameters, Å	Coherent scattering region nm	$\Delta d/d^*$ 10^{-3}
Cr-Mn-Si-Cu-	$FeN_{0.0324}$	60.6	a=3.598	103.37	3.460
Fe-Al+Ti	$TiN_{0.31}O_{0.31}$	39.4	a=4.211	25.6	5.143

Table 2. Properties of the materials calculated on the basis of the nanoindentation data [11].

Material	*H*	*E*	*R*
	GPa	GPa	%
Multilayer film Ti/α-C:H	8.0	128	34
Amorphous tape Zr-Cu-Ti-Ni	11.5	117	42
Silicon (100)	11.8	174	62
Thin film Ti-Si-N	28.4	295	62

The comparison shows that the nanohardness of 12X18H10T+Ti coating in nitrogen atmosphere exceeds that of all the materials in Table 2, among which the last three materials are used as abrasive and hardening coatings. The nanohardness of Cr-Mn-Si-Cu-Fe-Al+Ti coating in nitrogen gas environment is almost 2 times higher than that of the pure titanium (H = 4.1 GPa) and close to the nanohardness of the multilayer film Ti/α-C:H.

Figure 3 shows the cellular structure of the coating. The same structure can also be observed for 12X18H10T+Ti coating in nitrogen atmosphere. A cellular substructure is often formed during solidification as a result of the concentration supercooling [12]. If in formation of a cellular structure the zone of liquid melt enriched with impurities occurs at the crystallization front, the emergence of the impurity segregation at the cell boundaries can be attributed to the lateral impurity diffusion flux from the top of the growing projection. This model can be used to explain the observed cellular nanostructures with titanium nitride as a dopant. However, the cause of self-organization of the crystallizing melt on the substrate surface is to be found.

To address the issue of self-organization of structural units of the coating, consider the Bénard cell model. The Bénard cell phenomenon is the emergence of self-organization of convection cells in the form cylindrical shafts or regular hexagonal patterns in the layer of viscous fluid with a vertical temperature gradient [13]. Benard cells are one of the three standard examples of self-organization, along with the laser and the Belousov-Zhabotinsky reaction [14]. The temperature gradient is the governing parameter of self-organization.

More strictly (see, as an example, [13]) in the analysis of processes in the Bénard system, the governing parameter is the Rayleigh number:

$$\mathrm{Re} = gL^3bdT / \nu a,$$

where g is gravitational acceleration, L is the characteristic size, b is the coefficient of volumetric expansion, dT is the temperature gradient, ν is kinematic viscosity, and a is the coefficient of thermal conductivity.

Since the kinematic viscosity $\nu \sim 1/\sigma$, σ is surface tension, then the above expression for the Rayleigh number indicates that in this case (and presumably in the case of the Bénard process), surface tension is the governing parameter. Thereby, we return to the issues discussed above.

The above consideration on the formation of the plasma coating structure points out a more complicated situation than it might appear at first sight. Consider this issue at another angle, namely, using the model of dislocation cell structure (DCS). Inelastic deformation of crystals (and coatings) accompanied by the formation of the deformation relief on their surface, which reflects the process of deformation localization at meso-, micro- and nanoscale levels. Dislocation cell structure starts to form in a deformed crystal at the end of the second stage and at the beginning of the third stage of strain hardening of metals, and finishes at the end of the third stage (see [15] and references therein). In further deformation, a fragmented dislocation structure is formed (FDS) (at the fourth and fifth stages of strain hardening).

DCS is considered to be the process of self-organization of dislocations under multiple slip. For its occurrence, the specific criteria (as in the case of Bénard cells) binding together the coefficients of multiplication, immobilization, and dislocationannihilation are to be met.

In ion-plasma coating and coating cooling, stress states are formed in the latter [16]. These stress states can be the sources of dislocation propagation throughout the volume of the deposited coating. A sharp increase in microhardness of the formed film is the result of dislocation hardening the coating material. [17]. In [18], a sharp increase in microhardness was observed under ion-plasma deposition of high entropy alloys in nitrogen atmosphere.

4. Conclusion

From the results of the research into the formation of multiphase coatings we may draw the following conclusions:

- types of structures formed in ion-plasma deposition of coatings are not as diverse as, for example, in alloy crystallization. Basically, it is a globular structure and, rarely, unclosed dissipative and cellular structures;
- physical properties of the coatings can significantly vary depending on their composition of elements;
- since the value of the surface tension (surface energy) in most cases is an additive quantity, each element, being the part of the coating, contributes to the total surface energy;

- in case of simultaneous multiphase cathode sputtering, high entropy coatings are formed. They possess an ordered structure and improved mechanical properties.

Thus, the general pattern obtained is as follows: multi-phase (multi-element) cathodes should be used for superhard nanostructured coatings, preferably in simultaneous sputtering.

5. Acknowledgements

This work was financially supported by The Ministry of Education and Science of the Russian Federation in part of the science activity program.

References:

[1] Movchan B A, Demchishin A V 1969 Investigation of the structure and properties of nickel, titanium, tungsten, alumina, and zirconia thick vacuum condensates *Fizika metallov i metallovedenie* **4** 23.
[2] Thornton J A High rate thick film growth 1977 *Annual Review of Material Science* **7** 239.
[3] Barna P B, Adamik M 1977 *In Protective Coatings end Thin Films* Edited Pfleau Y, Barna P B. (Kluwer Academic, Dortrecht, The Netherlans) pp 279–297.
[4] Watanabe H., Sato Y., Nie C, Ando A, Ohtani S, Iwamoto N 2003 The mechenical properties and microstructure of Ti-Si-N nanocomposite films by ion plating *Surf. Coat. Technol.* **169–170** 452.
[5] Carvalho S, Ribeiro E, Rebouta L, Vaz F, Alves E, Schneider D, Cavaleiro A 2003 Effects of the morphology and structure on the elastic behavior of (Ti,Si,Al)N nanocomposites *Surf. Coat. Technol.* **174–175** 984.
[6] Flink A, Larsson T, Sjolen J, Karlsson L, Hultman L 2005 Influence of Si on the microstructure of arc evaporated (Ti,Si)N thin films; evidence for cubic solid solutions and their thermal stability *Surf. Coat. Technol.* **200** 1535.
[7] Li Z G, Mori M, Miyake S, Kumagai M, Saito H, Muramatsu Y 2005 Structure and properties of Ti-Si-N films prepared by ICP assisted magnetron sputtering *Surf. Coat. Technol.* **193** 345.
[8] Yurov V M, Laurinas V Ch, Guchenko S A, Zavatskaya O N 2014 Surface tension of hardening coatings *Uprochnyayushie tehnologii i pokrytiya* **1** 33.
[9] Jurov V M 2011 Superficial tension of pure metals *Eurasian Physical Technical Journal* **8** 10.
[10] Kidyarov B I 1979 *Kinetic of crystal formation from liquid phase* (Novosibirsk: Nauka) p 134.
[11] Golovin Yu I 2008 Nanoindentation and mechanical properties of solids in submicrovolumes, thin near-surface layers, and films: A Review *Physics of the Solid State* **50** 2205.
[12] Vayngard W 1967 *Introduction to the physics of metal crystallization* (Moscow: Mir) p 170.
[13] Gershuni G Z, Zhukhovnitsky E M 1972 *Convective stability of incompressible fluid* (Moscow: Nauka) p 232.
[14] Haken H 1991 *Information and self-organization: Macroscopic approach to complex systems* (Moscow: Mir) p 240.
[15] Malygin G A 2007 Simulation of the surface deformation relief of a plastically deformed crystal *Physics of the Solid State* **49** 1460.
[16] Barvinok V A 1990 *Control of the stress state and properties of plasma coatings* (Moscow: Mashinostroenie) p 384.
[17] Goryachev S E 1984 *Microscopic mechanisms of strain hardening* (Moscow: Moscow Engineering Physics Institute) p 61.
[18] Sobol' O V, Andreev A A, Gorban V F, Krapivka N A , Stolbovoi V A, Serdyuk I V, Fil'chikov V E 2012 Reproducibility of the single-phase structural state of the multielement high-entropy Ti-V-Zr-Nb-Hf system and related superhand nitrides formed by the vacuum-arc method *Technical Physics Letters* **38** 616.

Investigation of antioxidant properties of metal ascorbates and their mixtures by voltammetry

A N Vtorushina and E D Nikonova

Tomsk Polytechnic University, 30 Lenina av., Tomsk, 634050, Russian Federation

anl@tpu.ru

Abstract
The paper describes modern ways for selection of anti-radical substances. Molding of such components with a carbon-based material decreases the rate of its oxidative destruction. Addition of such a component to a carbon-based material decreases the rate of its oxidative destruction. The purpose of this study is to determine the antioxidant activity of ascorbates metals (Ca, Mg, Li, Co, Fe), used in the practice of medicine, as well as mixtures based on them together with well-known antioxidants. In this article we examine the effect of metals on the process of ascorbate oxygen electroreduction.
From these ascorbates lithium and magnesium ascorbate showed the greatest activity toward cathode oxygen reduction process.
Also mixtures with well-known examined antioxidants ascorbate (glucose, dihydroquercetin) were investigated at different concentrations of components. It is shown that the multicomponent mixtures exhibit lower activity than the individual drugs. Recommended the creation of drugs on the basis of ascorbate Mg and Li with not more than 3 number of components.

1. Introduction

Nowadays some calcium, lithium, iron, magnesium preparations, such as calcium chloride, lithium carbonate, lithium chloride and others are wildly used in clinical practice of psychoactive diseases [1, 2]. Calcium and magnesium play an important role in various physiological processes of an organism. In the fact, calcium ions are necessary for the process of transmission of nerve impulses and the activity of certain enzymes [3–6]. Magnesium plays a key role in the energy, plastic and electrolyte exchange, acts as a regulator of cellar growth, and is important component at all stages of the synthesis of protein molecules. In particular, the proper functioning of ribosomes and binding them with messenger RNA (a key mechanism of protein biosynthesis) depend on a sufficient amount of magnesium in the body. Besides, magnesium takes part in sharing phosphorus, ATP synthesis, regulation of glycolysis, the building of bone, etc. [7, 8]. Iron is an important component of a plenty processes of life. In the body's cells iron is consumed on the synthesis of gemo-containing enzymes and ferritin — the main protein containing iron stores. Iron is essential in the synthesis of hemoglobin, as well as in increasing of erythrocyte production. Lithium actively influences neurochemical processes occurring in the brain, that underlies its therapeutic activity in mental illness. It is found that lithium has the ability to stop acute manic excitement and to prevent affective attacks. However it is known that these drugs have toxic side effects on the organism [8].

Therefore creating new forms of drugs based on salts of these metals, as well as studying their bioactivity, including antioxidant properties, are current interests.

The aim of the paper is to investigate the influence of complexes of magnesium, lithium, cobalt, iron, calcium, containing ascorbic acid as bioactive ligand on the electrochemical behavior of oxygen in the aqueous medium in a semi-infinite linear diffusion.

2. Methods and materials

A simple designed in hardware, express and highly sensitive method of voltammetry (VA) was used in the work.

Investigations were carried out on a computerized voltammetric analyzer TA-2 producted by LLC Research and Production Enterprise RPE «Tomanalyt» (Tomsk).

Voltammetric curves were recorded at room temperature in a three-electrode electrochemical cell connecting to the analyzer. A working mercury film electrode (MFE), a silver-silver chloride electrodes with KCl saturated ($Ag|AgCl|KCl_{sat}$), as reference and counter electrodes, were used. An open type cell was used in this investigation.

As a supporting electrolyte, phosphate buffer 0.025 $mol \cdot l^{-1}$ (equimolar mixtures of Na_2HPO_4 and KH_2PO_4, pH 6.86) was used.

Constantly cathode-current mode VA was used; speed of potential sweep (W) was 40 mV/s, working potential range 0 to -1 V. The substances were mixed by vibration of the electrodes.

Oxygen reduction reaction was used as a model reaction. Records were in following order: record of first wave cathodic reduction of oxygen dissolved in a supporting electrolyte in the specified potential range. After substance addition, the solution was stirred about 20s. After the stirring is stopped, the potential was scanned negatively, causing oxygen reduction, giving a current first wave electroreduction of oxygen (ER O_2). Its value was proportional to the amount of oxygen in the bulk of the solution. Oxygen concentration was monitored by oxygen analyzer. Based on the ammetric measurements the concentration of oxygen in phosphate buffer at 25.0 ± 0.5°C was 2.56 ± $0.05 \cdot 10^{-4}$ $mol \cdot l^{-1}$.

Oxygen concentration in the electrolyte solutions was monitored by potentiometric oxygen sensor № 5972 produced by the center of computing automation and measuring "MERA-ELVRO", with automatic temperature compensation in the range of 0 to 40 C. Weigh sample material was carried out on laboratory analytical balance VL 210 firm "Gosmetr" with a weighing error of ± 0.0002 g. Nanopure water was used for making solutions.

3. Experimental

The list of methods of determining the antioxidant activity (AA) is wide enough: the chemiluminescent method, gas phase chromatography, and electrochemical methods, fluorimetric, etc. [9–14]. The method of cathode voltammetry (VA) in determining AA of metal ascorbates was used in the work [15].

The method is based on a process of ER O_2 that has a mechanism similar to the restoration of oxygen in the cells and body tissues (1-3). At the same time active forms of oxygen are generated on the electrode: superoxide anion of oxygen

$O_2^{\cdot -}$ and hydroperoxide $HO_2^{\cdot}$.

$$O_2 + e^- \overset{k}{\rightleftarrows} O_2^{\cdot -} \quad (1)$$

$$O_2^{\cdot -} + H^+ \rightleftarrows HO_2 \quad (2)$$

$$HO_2 + H^+ + e^- \rightleftarrows H_2O_2 \quad (3)$$

The antioxidant activity was determined by a relative decrease in the current ER of O_2 in the presence of the test components in solutions.

This paper considers the influence of complexes of magnesium, lithium, calcium, cobalt, iron, that contain ascorbic as bioactive ligand (Figure 1), on the electrochemical behavior of oxygen in the aqueous medium in a semi-infinite linear diffusion. Also authors made a comparative evaluation of the activity of the new complex of magnesium with previously synthesized metal complexes and ascorbic.

The influence of the test substances on the electrochemical behavior of oxygen is considered in a supporting electrolyte – phosphate buffer with pH 6.86.

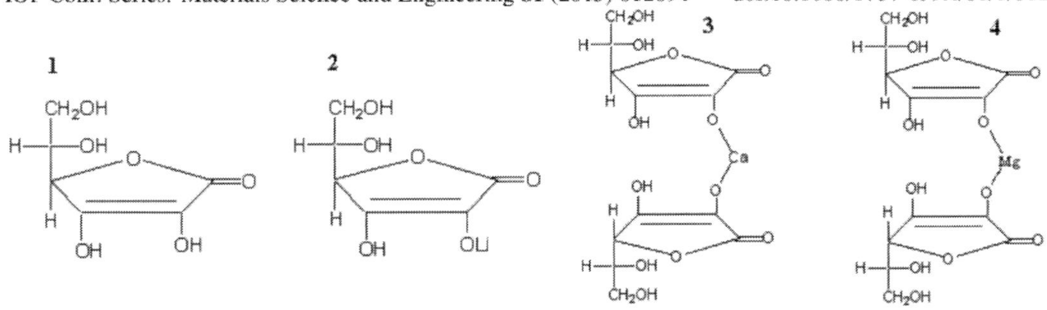

Figure 1. Structural formulas of the test substances: ascorbic (1); lithium ascorbate (2); calcium ascorbate (3); magnesium ascorbate (4)

For the study of AA of drugs voltammograms of first current wave ER O_2 in the absence and presence of investigated ascorbates were recorded. Studies showed a decrease in the limiting current ER O_2 in the presence of examined ascorbates, with the exception of iron ascorbate.

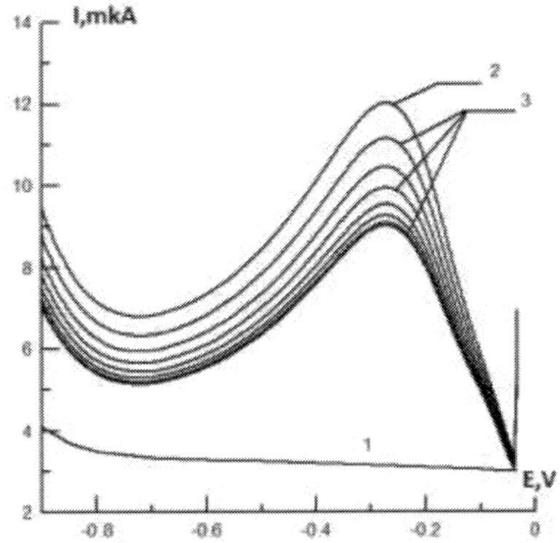

Figure 2. Voltammograms of the ER O_2 current in phosphate buffer (0.025M, pH 6.86) on the MFE without (2) and with 10^{-4} g/ml Mg ascorbate in solution (3), the residual current of supporting electrolyte in the absence of O_2 and substances in the solution (1)

Lines (3) in Figure 2 show the reduction of oxygen current in depending on the time of reaction between the active forms of oxygen and magnesium ascorbate solution. Similar dependences were obtained for all the studied ascorbates.

According to the results of obtained voltammograms graphs of dependency of relative decrease of current ER O_2 on the time of the process in the presence of the test drug were built.

The antioxidant activity of the investigated drugs was estimated by kinetic criteria of antioxidant activity – K (μmol $l^{-1}min^{-1}$), which reflects the efficiency of interaction of the sample with active forms of oxygen [14]:

$$K = (1 - \frac{I}{I_o})\frac{C^0}{t}$$

where C^0 [$\mu mol \cdot l^{-1}$] is the oxygen concentration at the electrode in absence antioxidant, I [A] is the ER O_2 current with the investigated substance addition in the solution, I_o [A] is the limiting ER O_2 current without substance in the solution, t [min] is time of the interaction between reactive oxygen species (ROS) and antioxidant at the MFE.

4. Discussion of results

Reducing the current of first wave of ER O_2 is connected with the reaction between investigated metals ascorbates and products of oxygen reduction. The most active in relation to the process of the ER O_2 is Mg ascorbate, while Fe ascorbate · is completely neutral and do not show antioxidant properties.

Table 1. AA of the samples (c = 10^{-4} g/ml) in relation to the process of ER O_2 (p=0.95, n=5).

Substance name	K, μmol $l^{-1}min^{-1}$
Asc Mg	2.125 ± 0.053
Asc Li	1.714 ± 0.024
Asc Ca	1.550 ± 0.042
Asc Co	0.879 ± 0.027
Asc Fe	–
Ascorbic acid	1.165 ± 0.053
Glucose	0.169 ± 0.037
Dihydroquercetin	0.650 ± 0.040

Most of the metal complexes and ascorbic acted more actively, then ascorbic itself, which is andoubtly antioxidant and synergist in the process of interrupting the chain radical reactions.

As AA is depends on both time of interaction of antioxidant with active forms of oxygen and concentration of antioxidant, the influence of Asc Mg on a process of ER O_2 (Figure 3) was considered. It is shown that with the increasing of concentration of Asc Mg in solution the relative change in the limiting current ER O_2 is growing, that means that the test substance is more responsive to the oxygen radicals.

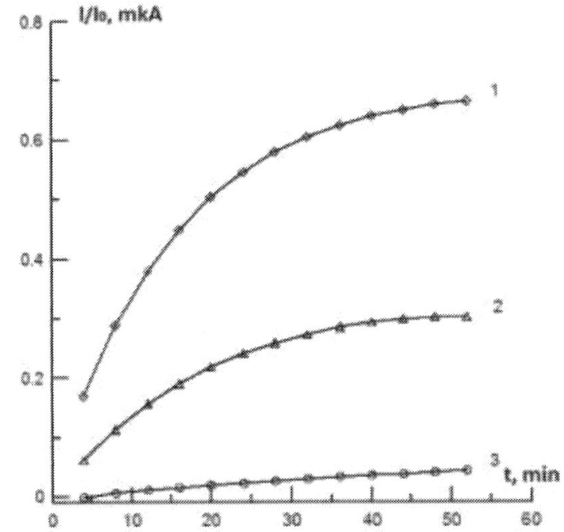

Figure 3. The dependence of the relative change of the current ER O_2 from the time of the process in the presence of $2 \cdot 10^{-4}$ (1), $1 \cdot 10^{-4}$ (2), $1 \cdot 10^{-5}$ g/ml (3) Asc Mg in the solution

Table 2. AA of the samples in relation to the process of ER O_2 in dependence on concentration (p=0.95, n=5).

Concentrarion Asc Mg, g/ml	K, μmol $l^{-1}min^{-1}$	Storage time Asc Mg
$2 \cdot 10^{-4}$	5.374 ± 0.082	fresh
$2 \cdot 10^{-4}$	2.711 ± 0.023	after 2 months
$1 \cdot 10^{-4}$	2.125 ± 0.053	fresh
$1 \cdot 10^{-5}$	0.251 ± 0.004	fresh

Therefore with the growth of storage time Asc Mg becomes less active in relation to the process of ER O_2.

In order to establish the existence of synergies the activity of mixtures of known antioxidants and ascorbates was determined. The results are shown in Table 3.

Table 3. Antioxidant activity of samples in relation to the process of ER O_2 in dependence on concentration (p=0.95, n=5).

The mixture of antioxidants	K, μmol $l^{-1}min^{-1}$	The mixture of antioxidants	K, μmol $l^{-1}min^{-1}$
Asc Mg, dihydroquercetin (1:1)	2.727 ± 0.131	Asc Li, glucose (1:1)	0.338 ± 0.094
Asc Mg, dihydroquercetin (1:3)	1.111 ± 0.128	Asc Li, glucose, dihydroquercetin (1:1:1)	0.453 ± 0.023
Asc Mg, glucose (1:1)	0.383 ± 0.019	Asc Li, glucose, dihydroquercetin (1:2:1)	0.642 ± 0.068

Due to the data it is proper to say that multicomponent mixtures are less active in relation to the process of cathode oxygen reduction than two-components mixtures. It should be noted that in the case of binary mixtures the effect of synergy can be observed, i.e. strengthening of active action of each component in the presence of another one. The same behavior of components of Mg and Li ascorbates' mixtures wasn't noted. In the mixtures with the number of components more than 3 total activity in relation to the process of ER O_2 was decreasing arithmetically. Perhaps this is because of spherical difficulties of organic residues that prevent the cathodic reduction of O_2 in the MFE. Due to the obtained data it is fair to say that for more effective influence creating of new drugs based on Mg and Li ascorbates with a number of components no more that 3 is really necessary.

Toxicity studies were also carried out on the basis of the compositions of Mg and Li ascorbates 25 mice of BALB/c with intraperitoneal administration of an aqueous solution of the drug in a wide range of doses

Also researches of toxicity of mixtures based on Mg and Li ascorbates were held on 25 mice of BALB/c line with intra-abdominal impact of aqueous solution of preparation in wide dose range (from 100 to 1000 mg/kg weight). The observations were made within 30 days. During this time no toxic side effects or death of animals were observed. Perhaps reducing toxic effects of metals is connected with their binding with ascorbic and the creation of mixtures with widely known antioxidants such as dihydroquercetin, glucose and others.

5. Conclusion

Thus, studies of antioxidant properties of ascorbate metals showed significant activity of Asc Mg and Asc Li. It is shown that with the increasing of concentration of Asc Mg its activity in relation to the process of ER O_2 increases. It was also noted that there is a synergistic effect in mixtures of dihydroquercetin with Asc Mg with the number of components no more than 2. Based on these data it is possible to recommend the establishment of pharmacological drugs on the basis of magnesium and lithium ascorbates in mixtures with small amounts of components for greater impact on the body.

References

[1] Rasmussen L, Husted S and Johnsen S 2003 *Acta Anaesthesiologica Scandinavica* **47** 1038-40
[2] Haft W and Vallejo M 2013 *Anesthesiology Clinics* **31** 517
[3] Moodley J 1998 *International Journal of Obstetric Anesthesia* **7** 73
[4] Perényi A, Rihmer Z and Bánki C 1983 *Journal of Affective Disorders* **5** 171
[5] Angelucci F, Aloe L, Vasquez P and Mather A 2003 *The International Journal of Neuropsychopharmacology* **6** 225
[6] Serefko A, Szopa A, Wlaź P, Nowak G, Radziwoń-Zaleska M, Skalski M and Poleszak E 2013 *Pharmacological Reports* **65** 547

[7] Ebel H and Gunther T 1998 *J. Clin. Chem. & Clin. Biochem* **18** 257
[8] Gosmaro F, Bagnati M, Berto S, Bellomo G and Prenesti E 2013 *Talanta* **115** 526
[9] Moyer R, Hummer K, Finn C, Frei B and Wrolstad R 2002 *Journal Agricultural and Food Chemistry* **50** 519
[10] Musa K H, Abdullah A, Kuswandi B and Hidayat M A 2013 *Food Chemistry* 141 4102
[11] Dhavamani S, Rao Y and Lokesh B 2014 *Food Chemistry* 164 551
[12] Campanella L, Bonanni A, Bellantoni D, Favero G and Tomassetti M 2004 *Journal of Pharmaceutical and Biomedical Analysis* **36** 91
[13] Blasco A, Rogerio M, Gonz´alez M and Escarpa A 2005 *Anal. Chim. Acta* **539** 237
[14] Schepetkin I, Potapov A, Khlebnikov A, Korotkova E, Lukina A, Malovichko G, Kirpotina L and Quinn M 2006 *J. Biol. Inorg. Chem.* **11** 499
[15] Avramchik O, Korotkova E, Plotnikov E, Lukina A and Karbainov Y 2005 *J. of Pharmaceutical and Biomedical Analysis* **37** 1149

Preclinical animal acute toxicity studies of new developed MRI contrast agent based on gadolinium

I F Nam [1,2] and V V Zhuk [3]

[1] Technical director, "MedContrastSynthesis" LLC, Tomsk, Russia
[2] Asssistant Professor, National research Tomsk politechnic university, Tomsk, Russia
[3] Chemist, "MedContrastSynthesis" LLC, Tomsk, Russia

E-mail: irenanam@gmail.com

Abstract. Acute toxicity test of new developed MRI contrast agent based on disodium salt of gadopentetic acid complex were carried out on Mus musculus and Sprague Dawley rats according to guidelines of preclinical studies [1]. Groups of six animals each were selected for experiment. Death and clinical symptoms of animals were recorded during 14 days. As a result the maximum tolerated dose (MTD) for female mice is 2.8 mM/kg of body weight, male mice – 1.4 mM/kg, female rats – 2.8 mM/kg, male rats – 5.6 mM/kg of body weight. No Observed Adverse Effect Dose (NOAEL) for female mice is 1.4 mM/kg, male mice – 0.7 mM/kg, male and female rats – 0.7 mM/kg. According to experimental data new developed MRI contrast agent based on Gd-DTPA complex is low-toxic.

1. Introduction

According to the rules governing medical products in Russian Federation the novel active pharmaceutical ingredient and all pharmaceuticals based on it are liable to toxicological studies. Toxicological studies consist of general toxical action studies and specific toxicities. Toxicity studies involve wide range of tests in different species with regular monitoring for physiological or biometrical abnormalities observed in long-term administration of the drugs. General toxical action of the drug determined in acute and subacute toxicity studies. The aim of acute toxicity studies is to estimate no observed adverse effect dose, maximum tolerated and lethal doses and get experimental animal mortality analysis. Acute toxicity describes the adverse effects of a substance that result either from a single exposure or from multiple exposures in a short space of time (usually less than 6 hours) during a 24 hours [2].

The aim of this research is acute toxicity studies of new developed MRI contrast agent based on Gd-DTPA complex using intravenous bolus injection to experimental mice and rats.

2. Materials and methods

2.1. Profile of Experimental Product

New developed MRI contrast agent is based on disodium salt of gadopentetic acid and used as solution for injection (0,5 mM/ml). It was produced in laboratory of "MedContrastSynthesis" LLC, Tomsk, Russia.

Predetermined therapeutic dose according to preclinical functional studies is 0,1 mM/kg of body weight.

2.2. Animals

To carry out the acute toxicity studies Mus musculus mice and Sprague Dawley rats [3] of both sexes were used according to guidelines of preclinical studies [1] as shown in table 1.

Table 1. Experimental animals

Animals	Mus musculus (mice)	Rattus sp (rets)
Stock	CD-1	SD (Sprague Dawley)
Sex	Male and female	Male and female
Age in the beginning of the research	8-10 weeks	3 – 3,5 month
Body weight in the beginning of the research (mean)	female 22,7 g male 32,0 g	female 253,0 g male 487,0 g

Groups of six animals each were selected for experiment (table 2). Administration volume and number of exposures were calculated using requirements [4].

Table 2 Grouping of the animals, the dose, administration volume and number of exposures

Group number	Dose	Animals	sex	administration volume, ml	number of exposures
Group № 1	5.6 mM/kg	mice	♀	2.2	5 (4x0.5 ml + 0.2 ml)
Group № 2	2.8 mM/kg	mice	♀	1.1	3 (2x0.5 ml + 0.1 ml)
Group № 3	1.4 mM/kg	mice	♀	0.5	1
Group № 4	0.7 mM/kg	mice	♀	0.25	1
Group № 5	0.3 mM/kg	mice	♀	0.12	1
Group № 6	5.6 mM/kg	mice	♂	3.2	4 (4x0.8 ml)
Group № 7	2,8 mM/kg	mice	♂	1.6	2 (2x0.8 ml)
Group № 8	1.4 mM/kg	mice	♂	0.8	1
Group № 9	0.7 mM/kg	mice	♂	0.4	1
Group № 10	0.3 mM/kg	mice	♂	0.2	1
Group № 11	11.2 mM/kg	rats	♀	6	3 (3x2 ml)
Group № 12	5.6 mM/kg	rats	♀	3	2 (2x1.5 ml)
Group № 13	2.8 mM/kg	rats	♀	1.5	1
Group № 14	1.4 mM/kg	rats	♀	0.75	1
Group № 15	0.7 mM/kg	rats	♀	0.4	1
Group № 16	11.2 mM/kg	rats	♂	11.5	4 (4x2.9 ml)
Group № 17	5.6 mM/kg	rats	♂	5.75	3 (3x 1.9 ml)
Group № 18	2.8 mM/kg	rats	♂	2.8	2 (2x1.4 ml)
Group № 19	1.4 mM/kg	rats	♂	1.4	1
Group № 20	0.7 mM/kg	rats	♂	0.7	1

The animals were feed laboratory chow and water ad libitum and maintained in an air-conditioned environment (20-22 °C, 45-65 % humidity) in a 12 h light/dark cycle. Animal studies conducted humanely [5].

2.3. Statistics

Data obtained in the experiments were expressed in terms of mean± SE [6]. Statistical significance of data was assessed by analysis of variance (one-way ANOVA) followed by a comparison between different groups using Dunnet's test. Differences were considered statistically significant if the p value is not greater than 0.05 [6, 7].

3. Results

3.1. Mortality and behavioral analysis

In 14-days period the behavior pattern of experimental animals was observed 1h, 5h after the administration first and then twice a day every day, registered mortality and any symptom of toxicity.

After intravenous bolus injection mortality appeared in Group №1 (mice, female) at a dose 5.6 mM/kg: 5 of 6 animals in this group died during exposure. There was no mortality in Groups № 2, 3, 4, 5 (mice, female) at 2.8 mM/kg, 1.4 mM/kg, 0.7 mM/kg and 0.3 mM/kg corespondently.

In the Group № 6 (mice, male) at 5.6 mM/kg 5of 6 species and in Group №7 (mice, male) at 2.8 mM/kg 3 of 6 animals died. Lethality observed in the day of exposure. There is no mortality in Groups № 8, 9 and 10 (mice, male) at 1.4 mM/kg, 0.7 mM/kg and 0.3 mM/kg correspondently.

Summary of mortality data after pharmaceutical administration are in table 3. Based on these information the value of LD_{50} were calculated and is equal to 4.19 mM/kg for female mice and 2.8 mM/kg for male mice.

Table 3. Summary of mice mortality data after administration at different doses

Dose, mM/kg	Number of animals in group	Number of died animals	LD_{10}	LD_{16}	$LD_{50} \pm SE$ (confidence interval)	LD_{84}
Mice, female						
5.6	6	5	2.86 mM/kg	3.12 mM/kg	4.19±0.73 (3.00 – 5.86) mM/kg	5.64 mM/kg
2.8	6	0				
1.4	6	0				
0.7	6	0				
0.3	6	0				
Mice, male						
5.6	6	5	1.33 mM/kg	1.37 mM/kg	2.80±0.88 (1.58–4.94) mM/кг	5.70 mM/kg
2.8	6	3				
1.4	6	0				
0.7	6	0				
0.3	6	0				

After intravenous bolus injection all animals in Groups №11, 12 (rats, female) at 11.2 mM/kg and 5.6 mM/kg: died during administration. In Groups № 13, 14 and 15 (rats, female) at 2.8 mM/kg, 1.4 mM/kg, 0.7 mM/kg and 0.3 mM/kg no mortality were observed.

In Group № 16 after administration at a dose 11.2 mM/kg 4of 6 (rats, male) died in the first day. In group №17 (rats, male) at 5.6 mM/kg 1 of 6 animals died due to technical reasons and was not taken into account. In Groups № 18, 19 and 20 (rats, male) at 5.6 mM/kg, 2.8 mM/kg, 1.4 mM/kg, 0.7 mM/kg and 0.3 mM/kg correspondently no mortality were observed as shown in table 4.

Calculated LD_{50} is equal to 2.81 mM/kg for female rats and 8.85 mM/kg for male rats.

Table 4. Summary of rats mortality data after administration at different doses

Dose, mM/kg	Number of animals in group	Number of died animals	LD_{10}	LD_{16}	$LD_{50} \pm SE$ (confidence interval)	LD_{84}
Rats, female						
11.2	6	6	2.79 mM/kg	2.79 mM/kg	2.81±0.01 (2.80 – 2.83) mM/kg	2.84 mM/kg
5.6	6	6				
2.8	6	0				
1.4	6	0				
0.7	6	0				
Rats, male						

11.2	6	4	6.19 mM/kg	6.71 mM/kg	8.85±1.57 (6.28 – 12.47) mM/kg	11.67 mM/kg
5.6	5	0				
2.8	6	0				
1.4	6	0				
0.7	6	0				

All animals were observed daily for clinical signs. The time of onset, intensity and duration of these symptoms, if any, were recorded. There are no significant changes in behavior of female mice after admission of the drugs in doses of 1.4 mM/kg, 0.7 mM/kg и 0.3 mM/kg and male mice in doses of 0.7 mM/kg and 0.3 mM/kg. The main signs of toxicity in mice are a violation of the frequency and rhythm of respiratory movements, depression, convulsions, weakening (in high doses until the disappearance) reactions to various stimuli. Mentioned symptoms occurred within the first 15 minutes after drug administration at doses: female mice - 5.6 mM/kg and 2.8 mM/kg in male mice - 5.6 mM/kg, 2.8 mM/kg and 1.4 mM/kg.

The main signs of toxicity in rats are a violation of the frequency and rhythm of respiratory movements, depression, convulsions, weakening (in high doses until the disappearance) reactions to various stimuli, increased skeletal muscle tone, pale mucous membranes, and constriction of the pupil. These symptoms were observed in males and females rats for the first 15 min after drug administration at doses of 11.2 mM/kg, 5.6 mM/kg, 2.8 mM/kg, 1.4 mM/kg and 0.7 mM/kg. Degree of manifestation of health problems, the number of animals with signs of ill health and long-term health was dose-dependent. By clinical signs of ill healthit can be concluded that there gender differences: mouse males and females rats are more sensitive to the toxic effects of the drug. Degree of manifestation of health condition, the number of animals with health deterioration symptoms and duration of expression was dose-dependent. It can be concluded that there are gender differences: mouse males and females rats are more sensitive to the toxic effects of the drug.

3.2. Body weight

The surviving mice body weight (males and females) after administration of the drug increased during research period within its physiological range.

Female rat body weight during 14 days period did not change or slightly increased within its physiological range. Male rat body weight significantly decreased in the first five days after administration, and then gradually increased, reaching the initial value at the end of the observation period.

3.3. Necropsy

If the animal died during the study time of death were determined and recorded as accurately as possible. Animals were weighed and autopsied as quick as possible. Dying animals were euthanized, weighed and autopsied. Euthanasia of dying animals were carried out after the decision of the head of the study. If it were possible, blood samples were taken for analysis of indicators of clinical pathology. Surviving experimental animals were sacrificed on the 15 day after administration except for animals which died in observed period using CO_2 inhalation technique followed by exsanguination. Necropsy of all animals was carried out and examined the body surface condition, the inner surfaces and holes, the cranial cavity, thoracic, abdominal and pelvic region with organs and tissues and neck with organs and tissues, skeleton and musculoskeletal system. All abnormalities were documented.

Examination of the dead mice (males and females administrated a dose of 5.6 mM/kg and 2.8 mM/kg of body weight) internal organs showed the hemorrhages in the diaphragm, adrenal glands, macro- and microbleeds in the lung parenchyma, subepicardial hemorrhages, hepatic congestion. The cause of death of these animals - acute respiratory and adrenal insufficiency on the background of stagnant plethora with hemorrhages in the vital organs, caused by toxic effects of the drug.

Pathological examination of mice (males and females), carried out on the 15 day after administration showed that all or most of all animals of all groups under the planned euthanasia have macro- and micro hemorrhages found in the lung parenchyma, increased blood filling in the mesenteric and omentum. Some females also noted an increased blood supply of the fallopian tubes. Some males found hemorrhages in the diaphragm, increasing the size of the kidneys, increased blood circulation in the cortical substance.

Pathomorphological examination of the dead rats (males and females administrated a dose of 11.2 mM/kg and 5.6 mM/kg) showed hepatic congestion, hemorrhage into the diaphragm, flatulence, subepicardial hemorrhages, large hemorrhages in the lung parenchyma, petechial hemorrhages and bloody contents in the gastrointestinal tract, adrenal glands hemorrhage, bleeding in the ovaries, brain swelling, bleeding in the pia mater, microbleeds in the lungs, focal necrosis and microbleeds in the liver, congestion of mesenteric vessels, convolution smoothness and cerebellum dislocation, subdural hematoma, in one case - thymomegalia and hemorrhages in the thymus. The cause of animal death - acute respiratory and adrenal insufficiency on the background of stagnant plethora with bleeding in vital organs, caused by toxic effects of the drug.

Pathomorphological examination of rats (males and females), conducted on the 15 day after administration of the drug showed that all animals found microbleeds in the lung parenchyma, hepatic congestion, subepicardial hemorrhages; there are cases of splenomegaly, bleeding in the ovaries (female); a small number of animals with increased blood supply vessels of the mesentery and omentum, hepatomegaly, petechial hemorrhages in the intestine and ecchymosis, bleeding in the adrenal glands, extra sex glands (male), cerebral edema, and smoothness convolutions dislocation of the cerebellum, thymomegalia, congestion of the brain vessels, increased blood circulation in the kidneys, renal ischemia, renal cysts, foci of necrosis in the lungs.

Identified during necropsy examination pathologies in mice and rats are the result of minimal toxic effect of drug administered in complex with the selected method of euthanasia, the vast number of pathological findings corresponds to euthanasia using CO_2.

4. Conclusion

As a result the maximum tolerated dose (MTD) for female mice is 2.8 mM/kg of body weight, male mice – 1.4 mM/kg, female rats – 2.8 mM/kg, male rats – 5.6 mM/kg of body weight. No Observed Adverse Effect Dose (NOAEL) for female mice is 1.4 mM/kg, male mice – 0.7 mM/kg, male and female rats – 0.7 mM/kg. According to experimental data new developed MRI contrast agent based on Gd-DTPA complex is low-toxic.

References

[1] Mironov A N et al 2012 *Guidelines for preclinical studies of drugs* **1** (Moscow: Grif and K) p 944
[2] http://www.ilpi.com/msds/ref/acutetoxicity.html
[3] http://www.spf-animals.ru/animals/rats/outbred/
[4] Makarenko I, Avdeeva O, Vanati G, Rybakov A, Khodko S, Makarova M and Makarov V 2013 Possible ways and amounts of drug administration to laboratory animals *International Veterinary Gazette* Vol **3** pp 78-84
[5] Altman D Bland J 1999 How to randomize *BMJ* **11** 319:703-704
[6] Bland M 2000 *An Introduction to Medical Statistics* (3 rd edition) (Oxford Medical Publications) p 422
[7] Directive 2010/63/EU of the European Parliament and of the Council of 22 September 2010 on the protection of animals used for scientific purposes *OJ* L 276 p 33

IOP Conf. Series: Materials Science and Engineering **81** (2015) 012093 doi:10.1088/1757-899X/81/1/012093

Simultaneous spectrophotometric determination of chromium (VI) and iron (III) in alloys

E V Larionova and K A Bulygina

634050, Russia, Tomsk, Lenin av., 34, Tomsk Polytechnic University

evl@tpu.ru

Abstract. In this work the possibility of simultaneous spectrophotometric determination of chromium (VI) and iron (III) in alloys with help of the mixed organic reagents (diphenylcarbazide and 1,10-phenanthroline, diphenylcarbazide and sulphosalicylic acid) is studied. We have applied curve fitting method to separate individual signals of chromium (VI) and iron (III) from overlapped absorption spectrum. Peak shape analysis of pure signals of complexes of chromium (VI) with diphenylcarbazide and iron (III) with the 1,10-phenanthroline and sulphosalicylic acid is previously carried out. For adequate application of curve fitting method we chose the optimal empirical models of signals taking into account preliminary information on their shape. Correctness of determination of concentration by means of the offered technique is proved by "added-found" method for a series of mixtures with different ratios of concentration of chromium (VI) and iron (III). It is founded that the error of determination of concentration doesn't exceed 11%.

1. Introduction

Many common alloys consist of chromium and iron [1]. A quantity of iron in alloys is determined by ICP-AES [2], flame atomic absorption spectrometry [3], spectrophotometry [4], X-ray fluorescence spectrometry [5], voltammetry [6]. Chromium can be analyzed by ICP-AES [7], sequential injection analysis [8], spectrophotometry [9], X-ray fluorescence spectrometry [10]. Simultaneous determination of chromium and iron can be realized by ICP-AES [11]. Analysis methods of chromium and iron often require additional stages of sample preparation, the chemical pretreatment stages, unconventional and high-cost equipments. Moreover for properties characterization of alloys it often requires to know the speciation of elements. It gives reasons for the development of simple and accessible techniques of simultaneous determination of Cr (VI) and Fe (III) in alloys.

In the industrial analysis the spectrophotometric determination of chromium (VI) with diphenylcarbazide (λ_{max} = 540 nm) and iron (III) with 1,10-phenanthroline (λ_{max} = 510 nm) and sulphosalicylic acid (λ_{max} = 500 nm) are widely applied [12]. Spectrophotometric techniques of chromium (VI) and iron (III) determination are rather sensitive and selective. The problem of simultaneous determination of Cr (VI) and Fe (III) can be solved with use of the chromogenic and mixed organic reagents [13–17]. However in this case absorption spectra of chromium (VI) and iron (III) complexes overlaps therefore direct determination of concentration of the studied substances is impossible. To separate individual signals from overlapped absorption spectrum the different mathematical approaches, such as H-point method, PLS regression, principal component analysis and artificial neural networks were applied [13–17].

In this work we applied the curve fitting approach to resolve overlapped absorption spectrum of chromium (VI) and iron (III). Spectrophotometric signals have a peak shape. The curve fitting method is based on the simulation of complex signal as sum of single peak models, using iterative least squares minimization. Application of the curve fitting method has some specific features. To signal modeling the adequate empirical models should be chose. Special attention has been called to a priori evaluation of partial peaks parameters.

Methodology of curve fitting method was thoroughly described in our previous works [18, 19]. Program Resolution with effective algorithm of resolution overlapped signals based on flexible empirical models was developed. In the Resolution program to approximate the peaks the system of modification and construction of base elementary peaks proposed by the papers [20–22] was applied.

Thus our goal is to study the possibility of simultaneous spectrophotometric determination of Cr (VI) and Fe (III) with mixed organic reagents (diphenylcarbazide and 1,10-phenanthroline, diphenylcarbazide and sulphosalicylic acid) via curve fitting.

2. Results and discussion

We obtained the individual spectra of Cr (VI) with diphenylcarbazide and Fe (III) with 1,10-phenanthroline and sulphosalicylic acid and their sum spectra with mixed organic reagents (diphenylcarbazide and 1,10-phenanthroline, diphenylcarbazide and sulphosalicylic acid). Figure 1 shows the received signals. For detection of absorption spectra we used UV Cary 60 Agilent Technology spectrophotometer in the range of wave length from 200 to 800 nm with cell length equal to 10 mm.

As shown in figure 1, spectra of Cr (VI) and Fe (III) complexes are closely overlapped in both cases. Direct measurement of signals sizes and determination of concentration of Cr (VI) and Fe (III) from overlapped spectrum is impossible. As the signals are additive the application of mathematical methods of signal resolution is valid. Signal pretreatment and resolution of overlapped signals were carried out with the help of the home-made Assayer and Resolution software (ref. figures 2, 3).

For correct application of curve fitting method the preliminary study of individual peak shapes and choice of optimal models are necessary.

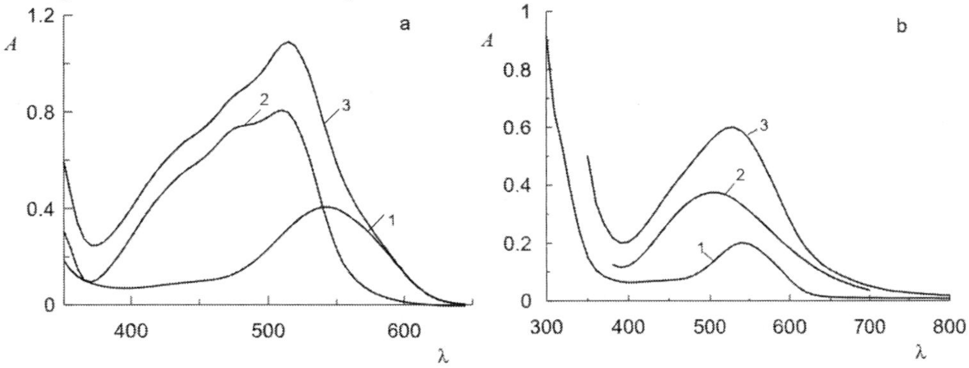

Figure 1. Absorption spectra of: a – (1) Cr(VI) with diphenylcarbazide; (2) Fe(III) with 1,10-phenanthroline and (3) Cr (VI) and Fe (III) with the mixed reagent (diphenylcarbazide and 1,10-phenanthroline); b – (1) Cr(VI) with diphenylcarbazide; (2) Fe(III) with sulphosalicylic acid and (3) Cr (VI) and Fe (III) with the mixed reagent (diphenylcarbazide and sulphosalicylic acid).

It is necessary to subtract residual current before the signals treatment. It highly significantly to give more attention to accurate base line subtraction since the method is very sensitive to even small distortions of the peak shape. Base line subtraction is provided by liner interpolation with help of Assayer program. This program was developed on the basis of statements of work [18]. Shape parameters and peak position [20, 23, 24] are also estimated by the Assayer program.

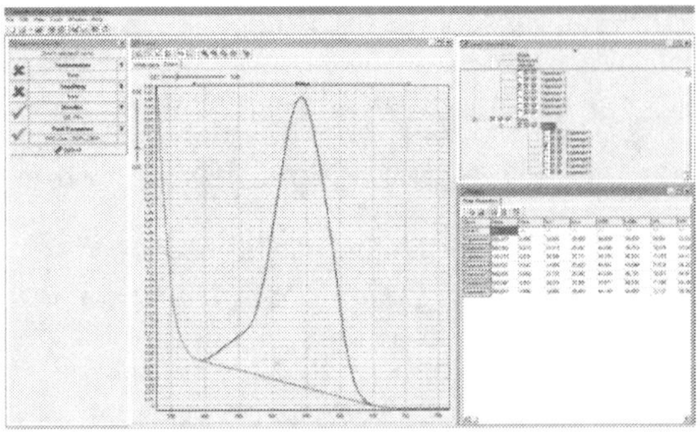

Figure 2. Interface of the Assayer program.

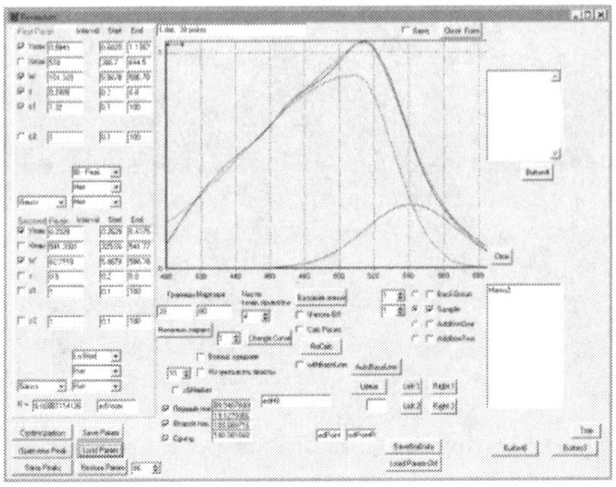

Figure 3. Interface of the Resolution program.

In table 1 shape parameters and peaks position of individual absorption spectra of Cr (VI) and Fe (III) are given. Signals parameters are used as initial values at the resolution of overlapped absorption spectra of Cr (VI) and Fe (III) with mixed reagent. Peak shape analysis is also useful for the choice of optimal empirical models of individual signals.

We chose the empirical functions for modeling of individual absorption spectra of Cr (VI) and Fe (III) which are shown in table 2. As the peak of chromium is symmetrical (s = 0,02) we offer to apply the Gaussian peak. In the case of iron the peaks have an asymmetrical shape (s = –0,50; 0,11). So we investigated several functions obtained by the modification base elementary peaks (Gaussian peak, peak of the logistic and Cauchy peak) proposed in the papers [21, 22]. Programs Grapher and Statistica are used for signals approximation.

Table 1. Parameters of individual absorption spectra Cr (VI) and Fe (III).

Parameters of signals	Cr (VI) with diphenylcarbazide	Fe (III) with 1,10-phenanthroline	Fe (III) with sulphosalicylic acid
Asymmetry s	0,02	–0,50	0,11
Asymmetry s'	0.52	0,25	0,55
Peak half-width at peak half-height W	90,38	116,61	147,38
Peakedness v	0,44	0,44	0,43
Taildness t	0,33	0,29	0,31
Maximum position λ_{max}, nm	543,15	510,38	508,99

As shown in table 2 the best approximation result is obtained by Gaussian peak (R = 0,993) for chromium spectrum and $PM_{x1s1}G$ and $PM_{s1}G$ peaks (R = 0,996; 0,997) for iron spectra. Figure 4 gives the illustration of the approximation results with help of the optimal empirical functions.

The preliminary information about peak shape and peak position (ref. Table 1) is used as initial values in the optimization process or for the decreasing the quantity of the optimized parameters. Adequate empirical functions are used for modeling signals in the curve fitting process. Figure 5 shows the results of resolution of overlapped spectra of Cr (VI) and Fe (III) with mixed reagents in Resolution program.

Table 2. Results of approximation of individual Cr (VI) and Fe (III) absorption spectra by various empirical models (PG – Gaussian peak, $PM_{s1}G$ $PM_{x1s1}G$, $PM_{s1}L$, $PM_{s1}C$ – asymmetrical modification of Gaussian peak, peak of the logistic and Cauchy peak.

Absorption spectra	Empirical models	Correlation coefficient R
Cr (VI) with diphenylcarbazide	$^{1}\, q_{PG} = 2^{-4p^2}$	0.993
Fe (III) with 1,10-phenanthroline	$^{2}\, q_{PM_{s1}G} = 2^{-\left(\frac{p-s\lvert p\rvert}{1-s^2}\right)^2}$	0.994
	$q_{PM_{s1}L} = \cosh^{-2}\left(\left(\log\sqrt{2}+1\right)\frac{p-s\lvert p\rvert}{1-s^2}\right)$	0.987
	$q_{PM_{s1}C} = \dfrac{1}{1+\left(\dfrac{p-s\lvert p\rvert}{1-s^2}\right)^2}$	0.970
	$^{3}\, q_{PM_{x1s1}G} = 2^{-\left\lvert 2\frac{p-s\lvert p\rvert}{1-s^2}\right\rvert^{2c_1}}$	0.996
Fe(III) with sulphosalicylic acid	$q_{PM_{s1}G} = 2^{-\left(\frac{p-s\lvert p\rvert}{1-s^2}\right)^2}$	0.997

[1] $q = \dfrac{y}{y_m}$, $p = \dfrac{x - x_m}{W}$, where y_m is a peak maximum, x_{max} –is a position of peak maximum, W is a peak width;

[2] s – is an asymmetry parameter;

[3] c_1 – parameter function, which changes the shape of top and tail of peak.

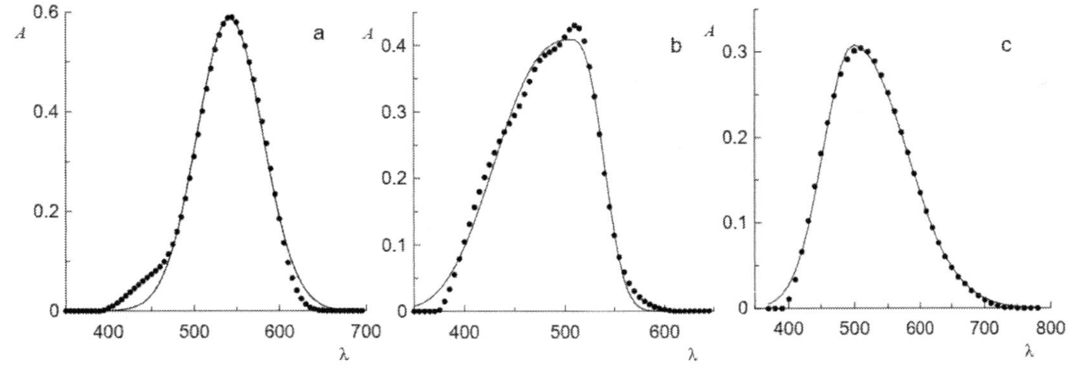

Figure 4. Absorption spectra of: (a) Cr(VI) with diphenylcarbazide and approximation by the Gaussian peak; (b) Fe(III) with 1,10-phenanthroline and approximation by the $PM_{x1s1}G$ function; (c) Fe(III) with sulphosalicylic acid and approximation by the $PM_{s1}G$ function (dotes is experimental absorption spectra, lines is approximation by the empirical function).

As it can be seen the satisfactory coincidence of experimental and theoretically calculated signals is observed.

For the evaluation of correctness of the proposed technique of simultaneous determination of Cr (VI) and Fe (III) we used several synthetic mixtures that had ratios of elements prepared according to some alloys [1]. Concentrations of Cr (VI) and Fe (III) are calculated by the individual calibration plots of Cr (VI) and Fe (III) and peaks heights obtained by the resolution of overlapped spectra. Table 3 contains the results of concentrations of Cr (VI) and Fe (III) founded by "added-found" method with help of the proposed technique. As it can be seen the error of concentrations determination does not exceed 11 percent.

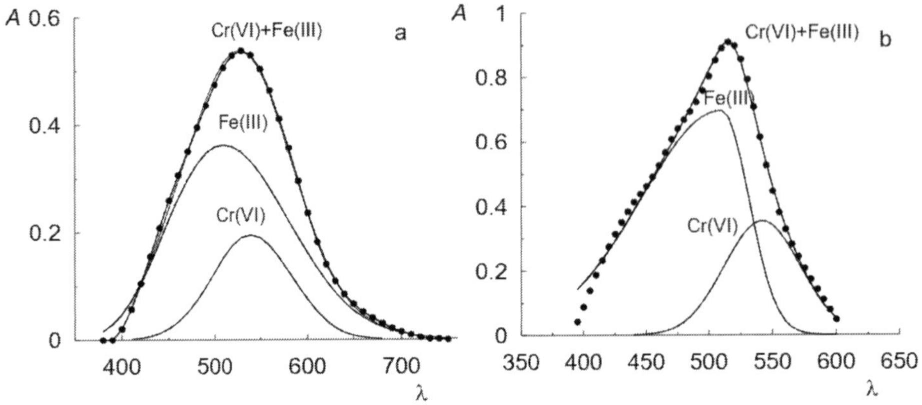

Figure 5. Overlapped absorption spectra of Cr(VI) and Fe(III) with mixed reagents and individual absorption spectra of Cr(VI) and Fe(III) obtained by curve fitting. Mixed reagents are (a) diphenylcarbazide and sulphosalicylic acid; (b) diphenylcarbazide and 1,10-phenanthroline. Dotes are the experimental absorption spectra, lines are the spectra obtained by curve fitting method).

Table 3. Results of concentration determination of Cr (VI) and Fe (III) by "added-found" method.

Added		Found		Δ, %	
C_{Cr}, mg/ml	C_{Fe}, mg/ml	C_{Cr}, mg/ml	C_{Fe}, mg/ml	C_{Cr}, mg/ml	C_{Fe}, mg/ml
mixed reagent (diphenylcarbazide and 1,10-phenanthroline)					
0,0006	0,0060	0,0006	0,0058	0	3,3
0,0009	0,0053	0,0009	0,0047	0	11
0,0010	0,0046	0,0009	0,0045	10	2,2
0,0013	0,0043	0,0012	0,0046	7,7	6,9
mixed reagent (diphenylcarbazide and sulphosalicylic acid)					
0,0005	0,0090	0,0005	0.0100	0	11

3. Conclusion

The problem of simultaneous spectrophotometric determination of Cr (VI) and Fe (III) can be solved by application of mixed reagents and curve fitting method. For effective application of curve fitting method the optimal empirical models of signals and preliminary information on their shape and position should be used. The offered technique can be used for the chromium and iron speciation in alloys. Application of mathematical treatment of signals expands the possibilities of spectrophotometric methods, allows avoiding the chemical pretreatment stages and using high-cost instrumentation and techniques.

References

[1] http://www.csudh.edu/oliver/chemdata/alloys.htm
[2] Xu J, Li M, Li S, Peng H, Zhao B and Yu D 2014 Determination of main components in nd-fe-B alloy with high precision by inductively coupled plasma atomic emission spectrometry *Zhongguo Xitu Xuebao/Journal of the Chinese Rare Earth Society* **32** 234
[3] Wang Y and Feng X 2007 Determination of micro chromium, copper, magnesium and iron in silicon aluminum alloy by flame atomic absorption spectrometry *Yejin Fenxi/Metallurgical Analysis* **27** 50

[4] Devanna N, Satheesh K P and Chandra Sekhar K B 2005 Derivative spectrophotometric determination of iron(II) using diacetylmonoxime benzoyl hydrazone *Asian Journal of Chemistry* **17** 1767

[5] Yao Q, Wang Y, Zhu Y, Yang D and Wu Q 2011 Determination of six elements in magnesium alloys by X-ray fluorescence spectrometry *Yejin Fenxi/Metallurgical Analysis* **31** 54

[6] Locatelli C and Torsi G 2003 Voltammetric determination of metals in copper alloys at trace and ultratrace concentrations *Collection of Czechoslovak Chemical Communications* **68** 1437

[7] Ma Q, Li Z and Zhai B 2013 Determination of chromium and zirconium in copper-chromium-zirconium alloy by inductively coupled plasma atomic emission spectrometry *Yejin Fenxi/Metallurgical Analysis* **33** 52

[8] Fan J, Sun Y, Wang J and Fan M 2009 An organic-reagent-free method for determination of chromium(VI) in steel alloys, sewage sludge and wastewater *Analytica Chimica Acta* **640** 58

[9] Xue C and Wang R 2009 Determination of chromium in aluminium alloy by potassium iodide-starch spectrophotometry *Yejin Fenxi/Metallurgical Analysis* **29** 75

[10] Zhang H, Ji S and Li C 2014 Determination of twelve elements coexisting with chromium, vanadium and titanium in titanium alloys by X-ray fluorescence spectrometry *Yejin Fenxi/Metallurgical Analysis* **34** 30

[11] Cheng Y 2012 Determination of the silicon, vanadium, iron, aluminum, nickel, molybdenum and chromium in titanium alloy by ICP-AES *Xiyou Jinshu Cailiao Yu Gongcheng/Rare Metal Materials and Engineering* **41** 1871

[12] Marczenko Z and Balcerzak M 1998 *Spektrofotometryczne metody w analizie nieorganicznej* (Warszawa: Wydawnictwo Naukowe PWN)

[13] Martelli P B, Reis B F, Kronka E A M, F°H B, Korn M, Zagatto E A G and Araujo A N 1995 Multicommutation in flow analysis. part 2. binary sampling for spectrophotometric determination of nickel, iron and chromium in steel alloys *Analytica Chimica Acta* **308** 397

[14] Fernandes R N, Reis B F and Campos L F P 2003 Automatic flow system for simultaneous determination of iron and chromium in steel alloys employing photometers based on LEDs as radiation source *Journal of Automated Methods and Management in Chemistry* **25** 1

[15] Abdollahi H 2001 Simultaneous spectrophotometric determination of chromium(VI) and iron(III) with chromogenic mixed reagents by H-point standard addition method and partial least squares regression *Analytica Chimica Acta* **442** 327

[16] Safavi A, Abdollahi H and Mirzajani R 2006 Simultaneous spectrophotometric determination of fe(III), al(III) and cu(II) by partial least-squares calibration method *Spectrochimica Acta - Part A: Molecular and Biomolecular Spectroscopy* **63** 196

[17] Kompany-Zareh M, Massoumi A and Pezeshk-Zadeh S 1999 Simultaneous spectrophotometric determination of fe and ni with xylenol orange using principal component analysis and artificial neural networks in some industrial samples *Talanta* **48** 283

[18] Romanenko S V 2006 *Phenomenological modeling of peak-shaped analytical signals. Ph.D. thesis.* (Tomsk: Tomsk Polytechnic University publishers)

[19] Romanenko S V, Stromberg A G, Selivanova E V and Romanenko E S 2004 Resolution of the overlapping peaks in the case of linear sweep anodic stripping voltammetry via curve fitting *Chemometrics and Intelligent Laboratory Systems* **73** 7

[20] Romanenko S V, Stromberg A G and Pushkareva T N 2006 Modeling of analytical peaks: Peaks properties and basic peak functions *Analytica Chimica Acta* **580** 99

[21] Romanenko S V and Stromberg A G 2007 Modelling of analytical peaks. peaks modifications *Analytica Chimica Acta* **581** 343

[22] Romanenko S V and Stromberg A G 2000 Classification of mathematical models of peak-shaped analytical signals *Journal of Analytical Chemistry* **55** 1024

[23] Stromberg A G, Romanenko S V and Romanenko E S 2000 Systematic study of elementary models of analytical signals in the form of peaks and waves *Journal of Analytical Chemistry* 55 615

[24] Stromberg A G, Selivanova E V and Romanenko S V 2004 Simulation of asymmetric peak-

shaped analytical signals by the frame representation of their shape using stripping voltammetry as an example *Zhurnal Analiticheskoj Khimii* **59** 832

Determination of enamel insulation corona resistance by high-frequency modulated pulses

A P Leonov[1], V V Red'ko[2], L A Red'ko[3,2], E Y Soldatenko[4]
[1] Associate Professor at the Tomsk Polytechnic University, RF
[2] Associate Professor at the Tomsk Polytechnic University, RF
[3] Associate Professor at the Tomsk Polytechnic University, RF
[4] Post-graduate student of the Tomsk Polytechnic University, RF

E-mail: leonov_ap@tpu.ru, ultratone@rambler.ru

Abstract. In the article test equipment is described for corona resistance testing of enameled winding wire samples. The primary element of equipment is generator producing test voltage with necessary waveform and magnitude according to the required PWM. Test conditions are accurately simulated by operational loads on a winding insulation (simultaneous impact of temperature and corona discharges). Obtained results of average time to breakdown show that the enamel insulation modified by silicon nanoparticles has a maximum corona resistance.

Keywords: enameled wire, corona resistance, winding, VFD, PWM, tests.

I. Introduction

Recent advances in the development field of power semiconductors and the method of pulse-width modulation allowed increasing significantly the frequency of switching operations (up to 20 kHz), minimizing power losses, and improving the performance of variable frequency drive [1]. However, increasing the switching speed reduced the rise time of the voltage pulses, which had a negative impact on the transition processes in the inverter – power cable – motor circuit. Impedance mismatches between the inverter, cable and motor excited wave processes in the cable and signal reflection phenomenon that resulted in overvoltages at the motor terminals [2–4]. Such electrical loads tightened operation conditions of VFD insulation system and, first of all, turn-to-turn insulation as the weakest element of the winding insulation. Corona discharges leading to accelerated aging of the insulation and further breakdown began appearing in voids and air gaps [5–7]. Currently, there are some standardized methods to determine the electrical, mechanical and thermal properties of the winding wire insulation. Unfortunately, there is no detailed information about any method to determine the enamel insulation resistance to the effect of corona discharges. In this regard, a reasonable choice of enameled wires for use in low-voltage machine windings with frequency regulation system is complicated. The aim of this work is to develop equipment for determining the enameled wire insulation resistance to the effect of operating loads in the VFD.

II. Theory

Total operational impacts causing aging of the insulation is varied and complex, but in many cases the main reason of aging is the high electric field. The main negative factors in the VFD resulting in overvoltage appearance in LV windings are:

- high slew rate (du/dt) – high rate of inverter output voltage rise has a negative impact on the cable insulation and its winding by small distances between inverter and motor.

- overvoltages at the motor terminals – it is the result of occurring wave phenomena being sometimes called "The long transmission line effect ", "The reflected wave effect" or "The standing wave effect".

- additional losses in a motor and feeder – the higher harmonics of voltage and current lead to additional losses in feeder, motor core and winding, especially in the rotor "squirrel cage".

The destructive effect of surface discharges is a result of heating impact, charged particle bombardment (generally, electrons), the impact of chemically active products formed during the surface discharge effect, as well as impulse waves and radiation.

The effect of these factors causes erosion (destruction of the surface with material entrainment), as well as structural changes (degradation, cross-linking of the polymer chains). Erosion can be caused by the following reasons:

a) oxidation of polymers by highly reactive radicals existing only in the discharge area;

b) reactions of polymer microradicals formed under the influence of a discharge, with oxygen of gaseous atmosphere or another active material arisen as a result of the surface discharges;

c) the dielectric surface bombardment by electrons and ions.

Ozone formation is typical of discharges in the air. Ozone is very corrosive, easily oxidize. In an enclosed space ozone formation is only a transitional phenomenon. Ozone oxidizes air nitrogen, after a while ozone disappears and nitrogen oxide stays. Various stable and unstable nitrogen bonds are formed by discharges in the void filled with air.

The mechanical and electrical insulation properties are sharply deteriorated because melting, emission, carbonization and chemical decomposition occurs in the field of surface discharges effect. Conductivity and lose factor of electrical insulating material increase because of rising the temperature that enhance the local heating.

Above factors have an effect together with increased temperature. Based on these conditions, tests of enameled wire must adequately reproduce the thermal and electrical loads on winding.

III. Test method

Sample is the standard twisted pair placed in heat chamber. Tests are carried out at a temperature of heat resistance class [7]. High-frequency electric loads are created on the twisted pair insulation using specially designed equipment (Figure 1). Criterion is the average time to insulation breakdown.

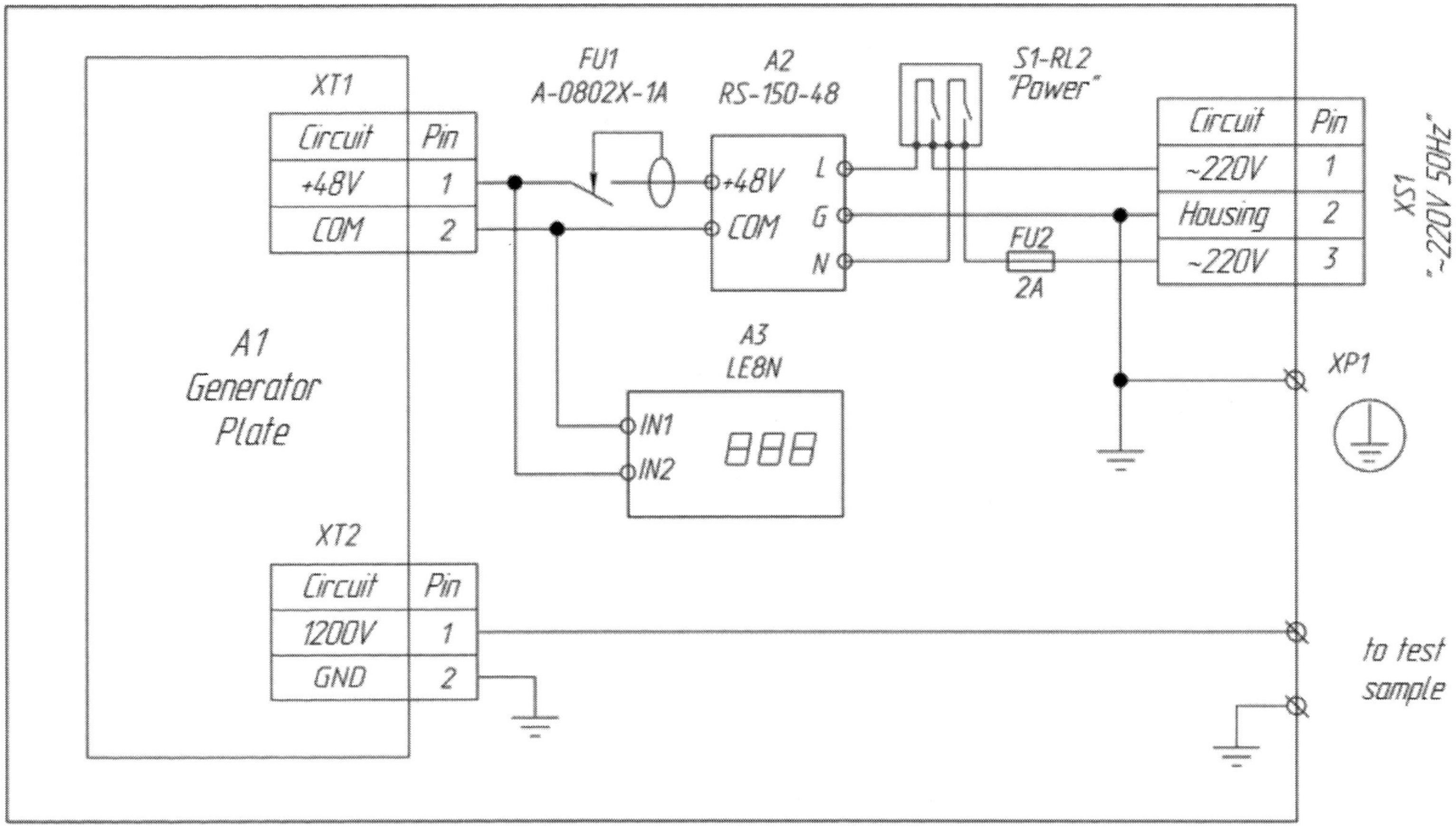

Figure 1. Schematic structure of equipment for test voltage generating, where *A1* – generator plate, *A2* – module power supply RS-150-48, *A3* – accumulated hour meter LE8N, *FU1* – circuit breaker assembly A0802X-1A, FU2 – amp fuse 2A, *S1* – network commutator RL2.

Figure 1 shows equipment schematic structure forming the test voltage. Further, the operating principle of equipment is stated. Generator plate *A1* generates a pulse signal with amplitude $Ua = \pm 1200\ V$ feeding the test sample in the heat chamber by the high-voltage wires. Generator plate is powered by the voltage $Us = 48\ V$ from module power supply *A2* through circuit breaker assembly *FU1*.

Accumulated counter timer *A3* is activated by synchronous power supply to the generator plate and test initiation. Time is calculated until control voltage of *48 V* supplies the counter. The threshold level of circuit breaker is selected so that the input current is increased in the feed network by sample breakdown and circuit breaker *FU1* breaks the circuit. Tests are broken off and the accumulated counter timer *A3* stops counting. Equipment turn on / off is made by the *S1*. The equipment construction according to the current structure provides the simplest implementation.

Generator plate schematic is shown in Fig. 2. The scheme operational principle is stated below. A rectangular voltage formation with required PWM law is realized by the processor module *A1*. This solution is the most ideal for solving single tasks in the case of laboratory research. Module is powered by a galvanically isolated power supply *DA2*. Generated signals are transmitted to the following stages via photo coupler *DA1*. The galvanic isolation of the processor module allows providing the equipment resistance to interference. Driver *DA3* separates signals on two channels and enters a pause between control pulses of 0,4 us by the power transistors. Drivers of control transistors *VT1* and *VT2* are implemented on chips *DA4* and *DA5* respectively. Switching circuits are identical and correspond to the standard ones recommended by manufacturer. Consider the example of driver work on the circuit *DA4*. Switching on is carried out by LED through a resistor R1. Chain *C10*, *R5*, *VD3* protects the transistor *VT1* against overcurrent, including short circuit. Resistor *R1* limits the switching speed of transistor and protects the output stage circuit *DA4* against overcurrent. Diode *VD1* and resistor *R7* provide protection of transistor gate against overvoltage and increase resistance to interference. Diode *VD5* provides a path flow of reverse current when the transistor *VT1* is off. The power stage is built by the scheme of circuit transformer with a common point. This scheme enables efficient operation of equipment at a low supply voltage. Transformer *T1* increases the test voltage up to $\pm 1200\ V$, and provides galvanic isolation of equipment from high-voltage circuit. The output voltage is supplied to pins 3 and 4 of the connector *X1*. Supply voltage *48V* is supplied to the generator plate through the terminals 1 and 2 of connector *X1*.

IOP Conf. Series: Materials Science and Engineering **81** (2015) 012094 doi:10.1088/1757-899X/81/1/012094

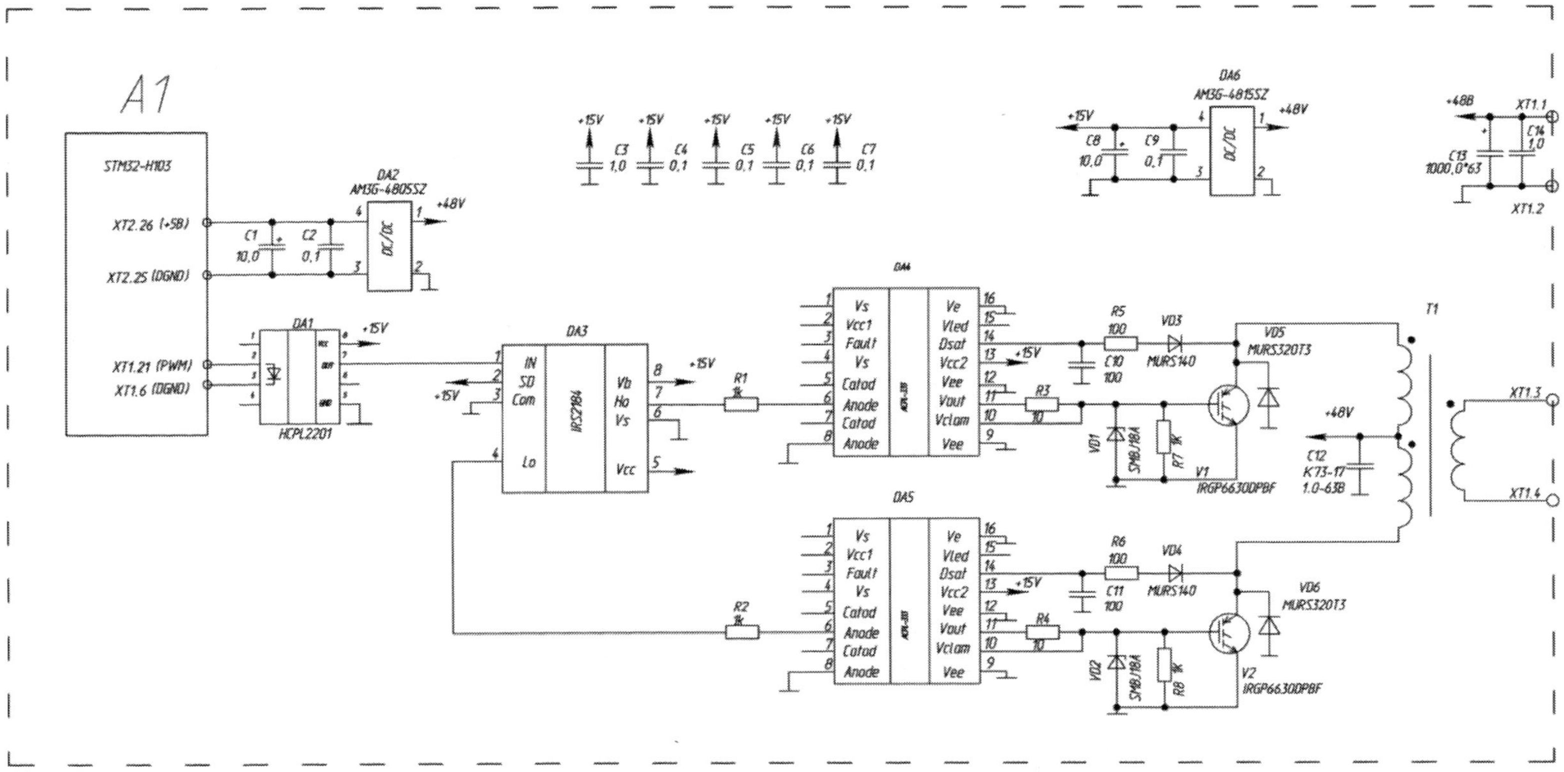

Figure 2. Generator plate schematic: *A1* – processing module STM32-H103, *DA1* – photo coupler HCPL2201, *DA2* – modulate power supply AM3G-4805SZ, *DA3* – driver IRS2184, *DA4, DA5* – drivers ACPL-333, *DA6* – modulate power supply AM3G-4815SZ, *VD1, VD2* – protective diodes SMBJ18A, *VT1*, *VT2* – transistors IRGP6630DPBF, *VD3, VD4* – diodes MURS140, *C1, C8* – tantalic capacitors TESAR CHIP 10uF x 10V type C 10%, *C2-C7, C9-C11* – ceramic capacitors 1206-X7R- 50, *C12, C14* – polymer film capacitors K73-17-1.0-63B, *C13* – electrolytic capacitor ECAP 1000/63V 1625 105C SH, *R1, R8* – CHIP resistors 1206 0.25 W 1%, *T1* – voltage transformer МП2000НМ- 4*К45× 28×8, $w_1 = w_2$=105 wire windings PEV2-0.15, *w3= 2800* wire windings PEV2-0.8.

IV. EXPERIMENTAL RESULTS

The general scheme of tests is shown in Figure 3.

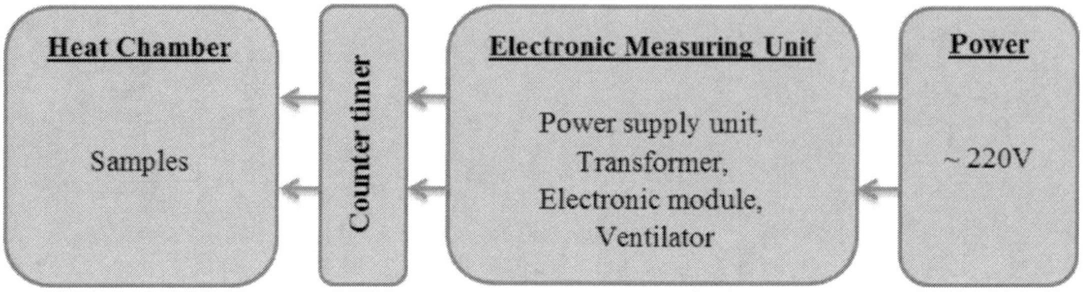

Figure 3. Block-scheme of high-voltage test equipment.

During the tests were obtained values of the mean time to breakdown samples wires with enamel isolation of various types. Samples were tested as in the initial state and after impregnation with impregnating materials. For each experiment was tested at least 20 samples. Summarized results are presented in Table 1.

Table 1. Determination of average time to wire twists breakdown by high voltage tests

Wire insulation type	Impregnating compound type	Average time to breakdown t_{br}, s
1 layer (inner) – 3-hydroxiethylcyanurate varnish, comprising at least 0.1 volume percent of silicon nanoparticles, 2 layer (outer) – polyamide imide varnish	Without IC	9509
	KO-916K	10664
	KP-50	14728
	KP -55-5	11412
1 layer (inner) – polyester varnish , 2 layer (outer) – polyamide imide varnish	Without IC	2009
	KO-916K	1286
	KP -50	1278
	KP -55-5	1192
Polyester varnish	Without IC	2337
	KO-916K	1085
	KP -50	1736
	KP -55-5	1812

V. Conclusion

1. Tests on the high voltage laboratory bench allow producing the whole range of operating loads, closest to the real operation of the VFD winding with PWM. In this case, temperature, electrical voltage, corona discharges influence on sample.
2. Developed equipment generates electrical impulses exciting appearance of stable corona discharge across the twisted part of the sample during the entire experiment. The value and shape of electrical impulses adequately reproduce overvoltages caused by the PWM operation: an alternating voltage with the amplitude of 1200 V, frequency of 400 Hz with the quantization voltage frequency (modulation) of 5 kHz (the duration of the wave front), with the rate of pulse rise 4 ms.
3. Wire with two-layer insulation modified by silicon nanoparticles has the greatest time to a breakdown. It is apparently associated with non-traditional mechanism of wire insulation destruction modified by nanoagents. In case of conventional wire aging, corona discharges cause material

IOP Conf. Series: Materials Science and Engineering **81** (2015) 012094 doi:10.1088/1757-899X/81/1/012094

destruction processes leading to the polymer chains break and the formation of free radicals and the volatile decomposition products. It results in relatively rapid insulation breakdown. But in the case of corona resistance wires, nanoparticles put into the enamel insulation have a strong interaction between the particles (bond strength) by the corona resistance development. Thus, it creates a kind of protective layer slowing the insulation breakdown during the material degradation.

References

[1] Melfi M, Sung A M J, Bell S, Skibinski G L 1998 *Industry Applications, IEEE Transac-tions on* p 766 - 775

[2] Stone G C, Culbert I M, Lloyd B A 2007 *Cement Indus-try Technical Conference Record, IEEE*, p 187 – 192

[3] Bogh D, Coffee J, Stone G, Custodio J 2006 *Industry Applications, IEEE Transactions on* p 148 – 154

[4] Fenger M, Campbell, S R, Pedersen J 2003 *Industry Applications Magazine,* IEEE p 22- 31

[5] USSR copyright certificate №536443, МПК5 G01R31/00, published 25.11.1976

[6] Pulse width modulation simulator for testing insulating materials. US Patent No. 6051980, Is-sued on April 18, 2000.

[7] Leonov A P, Red'ko V V, Soldatenko E Yu 2014 *20th International Conference for Students and Young Scientists: Modern Techniques and Technologies (MTT'2014),* Tomsk, Russia Vol. 66

Thermal properties of epoxy composites filled with boric acid

P M Visakh, O B Nazarenko, Yu A Amelkovich and T V Melnikova

Tomsk Polytechnic University, Lenin av. 30, 634050 Tomsk, Russia

E-mail: visagam143@gmail.com, obnaz@mail.ru

Abstract. The thermal properties of epoxy composites filled with boric acid fine powder at different percentage were studied. Epoxy composites were prepared using epoxy resin ED-20, boric acid as flame-retardant filler, hexamethylenediamine as a curing agent. The prepared samples and starting materials were examined using methods of thermal analysis, scanning electron microscopy and infrared spectroscopy. It was found that the incorporation of boric acid fine powder enhances the thermal stability of epoxy composites.

1. Introduction

Epoxy resins are one of well-known thermosetting polymers and used as construction material, for coating and adhesive applications, which can also be reinforced with additives for obtaining high strengths and good chemical resistances. However, they are intrinsically combustible. Since environmental concerns using the halogenated and brominated flame retardant arose from the daily consumer products, new approaches for the developments of non-halogenated flame retardants has been initialized and flourished at both academics and industries. Mineral fillers are considered to be the most suitable flame retardants.

In a study performed by Ollier et al. [1], the author incorporated 5 wt. % of bentonite in unsaturated polyester (UP) matrix. They noted that the addition of bentonite increases the thermal stability of the UP resin. Carrasco and Pagès [2] showed that, at low clay contents (up to 5 wt. %) the addition of clay had no effect on the thermal stability of the epoxy matrix, whereas for higher concentration (10 wt. %) a clear increase on this parameter was observed. In addition, Lakshimi et al. [3] reported an improvement in the thermal stability of epoxy resins with the incorporation of montmorillonite (MMT). Saitoh et al. [4] found that the phosphonium cations used to obtain the organoclays influenced the thermal resistance of the resulting epoxy/clay nanocomposites. The explanation for this behavior is that the dispersed MMT-Clay nanolayers can act as barrier protecting the epoxy polymer matrix from volatilizing gaseous products of degradation. Régnier et al. performed a kinetic study on the thermal degradation of carbon fibre/epoxy composites, both in air and in inert atmosphere. The thermal degradation of the composites occurred in three stages [5]. Brnardic et al. studied the thermal stability of nanocomposites based on organically modified MMT and an epoxy resin [6]. Compared to the neat resin, small changes in the thermal stability were observed in the case of nanocomposites.

The thermal stability of epoxy resin/TiO_2 nanocomposites was found to be dependent on the nanoparticles loading, as well as on their dispersion state [7]. At a very low TiO_2 loading into the matrix, the nanoparticles were dispersed uniformly and formed a barrier to heat and oxygen, due to their ceramic nature. The incorporation of the hybrid TiO_2-SiO_2 nanofillers into an epoxy resin increased the thermal stability of the neat resin [8]. Also the char yield increased from 0% for the neat resin to 25% for the nanocomposites. These phenomena are a consequence of the hybrid nanoparticles which acted as thermal stabilizers for the epoxy resin. The addition of multi-walled carbon nanotubes (MWCNTs) leads to a decrease of the thermal stability of epoxy matrices [9, 10]. This effect is caused by the increase of the polymer thermal conductivity as a consequence of MWCNTs addition. Kuan et al. reported that the incorporation of the MWCNTs functionalized with vinyltriethoxysylane into an epoxy resin increased its thermal stability [11]. The same effects were obtained in the case of MWCNTs grafted with triethylenetetramine [12] and MWCNTs functionalized with silane. Mehmet et al. reported [13] on their work on solid particle erosion behavior of glass fibre reinforced boric acid filled epoxy resin composites.

Boron compounds, including boric acid, are known as effective flame retardant additives [14, 15]. As a rule, boric acid is used in cellulosic products and coatings: wood, plywood, particle board, wood fibre, paper, cotton. Boron compounds reduce the flame spread of wood but may have diverse effects on hygroscopicity. Wood treated with inorganic flame-retardant salts is usually more hygroscopic than untreated wood, particularly at high relative humidity. Increases in the equilibrium moisture content of such treated wood will depend upon the type of chemical, level of chemical retention, and size and species of the wood involved.

To the best our knowledge no systematic work has been reported on the use of boric acid as reinforcement in epoxy resin, although these composites can be used for many applications. The objective of this work was the investigation of effect of the boric acid fine powder with different percentage as flame-retardant filler of epoxy resin on the thermal properties of the composites obtained.

2. Experimental

2.1. Physical properties of starting materials

The materials were used in this study: the epoxy-diane resin ED-20 (GOST 10587-84), the molecular weight of 390-430, the content of epoxy groups 20.0-22.5%, the viscosity 12-18 Pa·s; boric acid H_3BO_3, the molar mass 61.83 g/mol, the density 1.48 g/cm^3 (25 °C), the melting point 171 °C.

2.2. Preparation of composites

For the preparation of epoxy composites the epoxy-diane resin ED-20 was used and hexamethylenediamine (HMDA) as an epoxy resin curing agent. Boric acid fine powder having 45% of particles less than 40 microns was used as flame-retardant filler. The preparation of epoxy composite samples was performed as follows. The surface of boric acid was modified by a hardener to improve the adhesion of the particles with the epoxy resin. Then, the required amount of filler was added into epoxy resin. The filler content in the compositions was 1; 2.5; 5 and 10 wt. %. The epoxy resin was mixed by hand with the filler for 5 min at room temperature. After that HMDA was added into mixture and mixed again for 3 min. The ratio of epoxy resin and hardener was 10:1 by weight. The obtained mixtures were cured in the silicone molds at room temperature for 24 h. The sample coding is given in table 1.

Table 1. The compositions (wt. %) of the specimens.

Samples	ED-20	Boric acid
H_3BO_3 Boric acid powder	0	100
Epoxy resin + 1 wt. % H_3BO_3	99	1
Epoxy resin + 2.5wt% H_3BO_3	97.5	2.5
Epoxy resin + 5wt. % H_3BO_3	95	5
Epoxy resin + 10wt. % H_3BO_3	90	10
Epoxy resin	100	0

2.3. *Characterizations*
The obtained samples and starting materials were examined using methods of thermal analysis (TA instrument SDT Q600), scanning electron microscopy (Hitachi TM-3000). The parameters of thermooxidative degradation in air atmosphere and thermal decomposition in argon were investigated at a linear heating rate of 10 °C/min in the temperature range 20-1000 °C.

3. Results and discussion

3.1. *Scanning electron microscopyr*
According to the scanning electron microscopy data (figure 1), boric acid powder is polydisperse systems. The analysis of the dispersed composition revealed that 45 % of the particles have a size less than 20 mcm. The powder particles are scaly crystals composed of the planar layers of thickness about100 nm.

Figure 1. SEM micrograph of the boric acid powder.

3.2. *TGA results in air*
Figure 2 shows the TGA analysis results of boric acid powder, neat epoxy polymer and epoxy composites at different percentage of boric acid done in the presence of air.

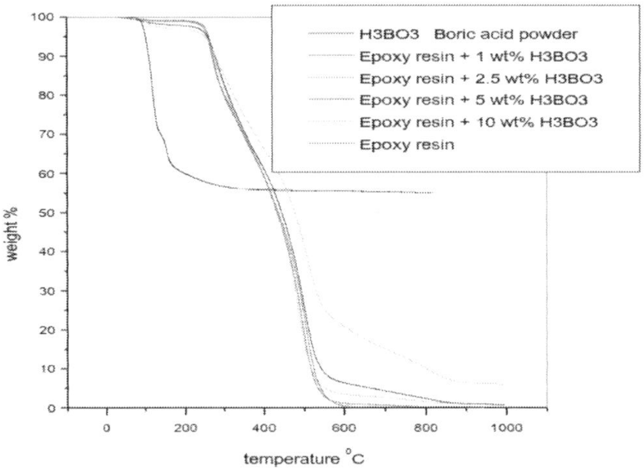

Figure 2. TGA plots of boric acid, neat epoxy and their composites at the heating in air.

Thermo-oxidative degradation of the samples occurs in several stages. The onset temperature of degradation is between 250-260 °C for all samples, steadily shifted between 550-560 °C. The process of thermooxidation for neat epoxy polymer ends at 600 °C. Using the TGA results we determined the values of temperature for a fixed weight loss - 10, 30, 50, 70 and 90 % (table 2). Introduction of the filler has a positive effect on the thermal stability the filled samples. The temperature corresponding to 50 % weight loss for the sample with a filler content of 10 wt. % is higher by 40 °C than the neat epoxy polymer.

Table 2. Degradation temperature at different weight loss levels of composites.

Samples	Degradation temperature (°C)				
	$T_{10\%}$	$T_{30\%}$	$T_{50\%}$	$T_{70\%}$	$T_{90\%}$
H_3BO_3 Boric acid powder	109	139	-	-	-
Epoxy resin + 1 wt. % H_3BO_3	269	349	433	482	516
Epoxy resin + 2.5wt% H_3BO_3	276	347	432	478	518
Epoxy resin + 5wt. % H_3BO_3	277	354	444	492	542
Epoxy resin + 10wt. % H_3BO_3	279	377	476	530	799
Epoxy resin	272	348	436	488	524

It is found that the thermal stability increases with increasing amount of boric acid content, and the decomposition temperature of the composites at different stages increased upon the addition of boric acid. Increased thermal stability of the composites with increased boric acid content may be explained by the more uniform distribution of boric acid in epoxy matrix and decreased mobility of epoxy phase in the vicinity of the boric acid particles.

3.3. TGA results in argon
Figure 3 shows the TGA analysis results of boric acid powder, neat epoxy polymer and epoxy composites at different percentage of boric acid done in the presence of argon.

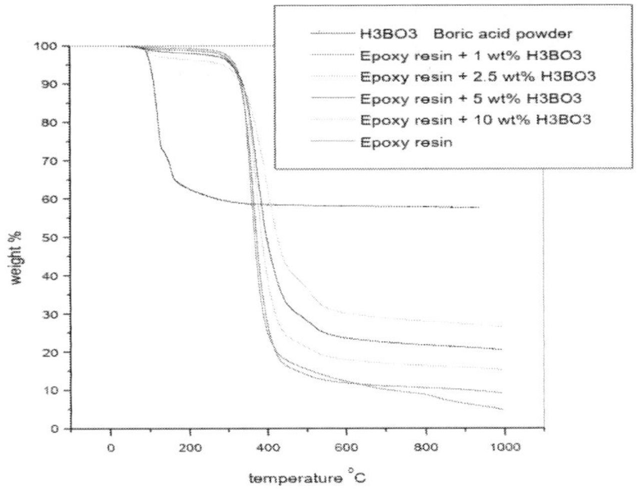

Figure 3. TGA plots of boric acid, neat epoxy and their composites at the heating in argon.

The process of thermal degradation of the studied samples occurs in several stages. We can see from figure 3, the onset temperature of degradation is between 200-400 °C for all samples, steadily shifted between 400-600 °C. From this figure also we can say that thermal stability increases with increasing boric acid content. The decomposition temperature of the epoxy composites at different stages increased upon the addition of boric acid. From the figure, neat epoxy and 1 % boric acid composites are not very difference in thermal degradation, but 2.5, 5 and 10 % boric acid based composites shows very good improvements in the thermal degradation, these results show, the boric acid content effect in epoxy composites.

The temperature values for a fixed weight loss (10, 30, 50, 70 and 90 %) at the thermal degradation are presented in table 3. It can be seen that the temperature corresponding to a fixed weight loss increases with the increasing the filler content. The best result was obtained for the sample with a filler content of 10 wt. %. The temperature of 50 % weight loss for this sample is higher by 58 °C than that for the neat epoxy polymer.

Table 3. Degradation temperature at different weight loss levels of composites.

Samples	Degradation temperature (°C)				
	$T_{10\%}$	$T_{30\%}$	$T_{50\%}$	$T_{70\%}$	$T_{90\%}$
H_3BO_3 Boric acid powder	111	148	-	-	-
Epoxy resin + 1 wt. % H_3BO_3	331	357	372	395	899
Epoxy resin + 2.5wt% H_3BO_3	338	366	385	419	-
Epoxy resin + 5wt. % H_3BO_3	338	371	397	475	-
Epoxy resin + 10wt. % H_3BO_3	334	389	425	604	-
Epoxy resin	335	354	367	388	707

The decomposition of boric acid takes place in two steps and is accompanied by the release of water. The first stage begins at temperature of 70 °C, the second stage - at 130 °C and finishes at 400 °C. Boric acid decomposition reaction is endothermic, resulting in cooling of the polymer matrix.

4. Conclusion

The effect of the addition of boric acid fine powder in the epoxy polymer on the thermal stability of epoxy composites at the heating in the atmosphere of air and argon was studied. It is shown that the influence of boric acid as a filler depends on its content. The temperature of 50 % weight loss in the process of thermo-oxidative degradation at the heating in air and thermal degradation at the heating in argon increases; the yield of the residue also increases. The results obtained demonstrate the effectiveness of using the fine powder of boric acid as the additive in the epoxy resin for reducing the flammability.

Acknowledgments

The authors are thankful to the Scientific and Analyzing Centre of Tomsk Polytechnic University for the providing the TG measurements, and prof. Sivkov A.A. for the providing the SEM.

References

[1] Ollier R, Rodriguez E, Alvarez V 2013 Unsaturated polyester/bentonite nanocomposites: influence of clay modification on final performance *Composites Part A* **48** 137-143

[2] Carrasco F, Pagès P 2008 Thermal degradation and stability of epoxy nanocomposites: influence of montmorillonite content and cure temperature *Polym. Degrad. Stab.* **93** 1000-1007

[3] Lakshmi M S, Narmadha B, Reddy B S R 2008 Enhanced thermal stability and structural characteristics of different MMT-Clay/epoxy-nanocomposite materials *Polym. Degrad. Stab.* **93** 201-213

[4] Saitoh K, Ohashi K, Oyama T, Takahashi A, Kadota J, Hirano H, Hasegawa K 2011 Development

of high-performance epoxy/clay nanocomposites by incorporating novel phosphonium modified montmorillonite *J. Appl. Polym. Sci.* **122** 666-675

[5] Régnier N, Fontaine S 2001 Determination of the thermal degradation kinetic parameters of carbon fibre reinforced epoxy using TG *J. Therm. Anal. Calorim.* **64** 789-799

[6] Brnardic I, Macan J, Ivankovic H, Ivankovic M 2008 Thermal degradation kinetics of epoxy/organically modified montmorillonite nanocomposites *J. Appl. Polym. Sci.* **107** 1932–1938

[7] Chatterjee A, Islam M S 2008 Fabrication and characterization of TiO_2–epoxy nanocomposite *Mater. Sci. Eng. A* **487** 574-585

[8] Omrani A, Afsar S, Safarpour M A 2010 Thermoset nanocomposites using hybrid nano TiO_2–SiO_2 Mater. Chem. Phys. **122** 343-349

[9] Zhou Y, Pervin F, Lewis L, Jeelani S 2007 Experimental study on the thermal and mechanical properties of multi-walled carbon nanotube-reinforced epoxy *Mater. Sci. Eng. A* **452-453** 657-664

[10] Ma P C, Kim J K, Tang B Z 2007 Effects of silane functionalization on the properties of carbon nanotube/epoxy nanocomposites *Compos. Sci. Technol.* **67** 2965-2972

[11] Kuan C F, Chen W J, Li Y L, Chen C H, Kuan H C, Chiang C L 2010 Flame retardance and thermal stability of carbon nanotube epoxy composite prepared from sol–gel method *J. Phys. Chem. Sol.* **71** 539-543

[12] Yang K, Gu M 2009 The Effects of triethylenetetramine grafting of multi-walled carbon nanotubes on its dispersion, filler-matrix interfacial interaction and the thermal properties of epoxy nanocomposites *Polym. Eng. Sci.* **49** 2158–2167

[13] Bagci M, Imrek H 2011 Solid particle erosion behavior of glass fibre reinforced boric acid filled epoxy resin composites *Tribology International* **44** 1704–1710

[14] Awoyemi L, Westermark U 2005 Effects of borate impregnation on the response of wood strength to heat treatment *Wood Science and Technology* **39** 484-491

[15] Shen K K, Kochesfahani S H, Jouffret F 2009 Boron based flame retardants and flame retardancy *Fire Retardancy of Polymeric Materials* 2nd ed C A Wilkie and A B Morgan (Baco Raton: CRC Press) pp 207-237

Effect of the anisotropy of monocrystalline silicon mechanical properties on the dynamic characteristics of a micromechanical gyroscope

T G Nesterenko, S E Vtorushin, E S Barbin and A N Koleda

Department of Precision Istrument Making, National Research Tomsk Polytechnic University, 30a Lenin ave., Tomsk, Russia

E-mail: ntg@tpu.ru

Abstract. The aim of the research was to determine the effect of temperature on mechanical properties of a micromechanical gyroscope with the sensing element mounted on a silicon wafer, with the crystallographic orientation of (100) (110) (111). The research is of current relevancy since the metrological characteristics that depend on the eigenfrequencies over the full temperature range are to be controlled. The temperature-modal analysis of the micromechanical gyroscope model was performed with ANSYS program. The temperature dependence for eigenfrequencies was obtained. The dependence of the scale factor on temperature for the most temperature-independent variant of sensor positioning on the wafer was determined. The developed mathematical model was used to find the forms of the output oscillations of the gyroscope.

1. Introduction

The perspectives of modern instrument making are related to the development of the devices of low weight, small dimensions, low cost, power consumption and high reliability. Microelectromechanical sensors (MEMS) meet these requirements. The main advantages of the devices are as follows.

- ***Small parameter spread within the finished product.*** Manufacturing of components in a single process cycle allows production of components with almost similar parameters.
- ***High manufacturability and repeatability.*** Well-proven and controlled microelectronics process technologies are used in MEMS manufacturing, which allow the product to obtain the desired characteristics. Lack of assembly operations improves manufacturability of complex systems.
- ***Subminiature.*** Chip technology makes micromechanical components much smaller than those made with traditional technologies.
- ***High functionality.*** Miniature products and the ability to manufacture sensors processing circuits and actuators in one device enable to create complete systems of sufficiently high complexity in a single chip, comparable in dimensions to the integrated circuit (IC).
- ***Improved performance of operation.*** Electronics and the electrical channels for connection between sensors and mechanics based on integrated technology and small dimensions of the chip can improve the characteristics of the gyro such as the operating frequency, signal / noise ratio, etc.
- ***Low power consumption.***

- ***Low cost.*** The cost of MEMS-based devices is much lower than that of the devices made without using integrated technology. Low cost of the MEMS-based final product depends on its serial production and MEMS components in the product.

The areas of MEMS application are expanding. The market in the field of manufacturing of MEMS inertial sensors - microaccelerometers and microgyroscopes, the most complex MEMS devices, is expected to exponentially grow in the next few years [1]. The sensing element of the micromechanical gyro (MMG) is a movable structure made of single-crystal silicon.

2. Description and architecture of MMG

The effect of the anisotropy of mechanical properties of single-crystal silicon on the dynamic characteristics of an LL-type micromechanical gyro [2] was investigated under stationary thermal fields.

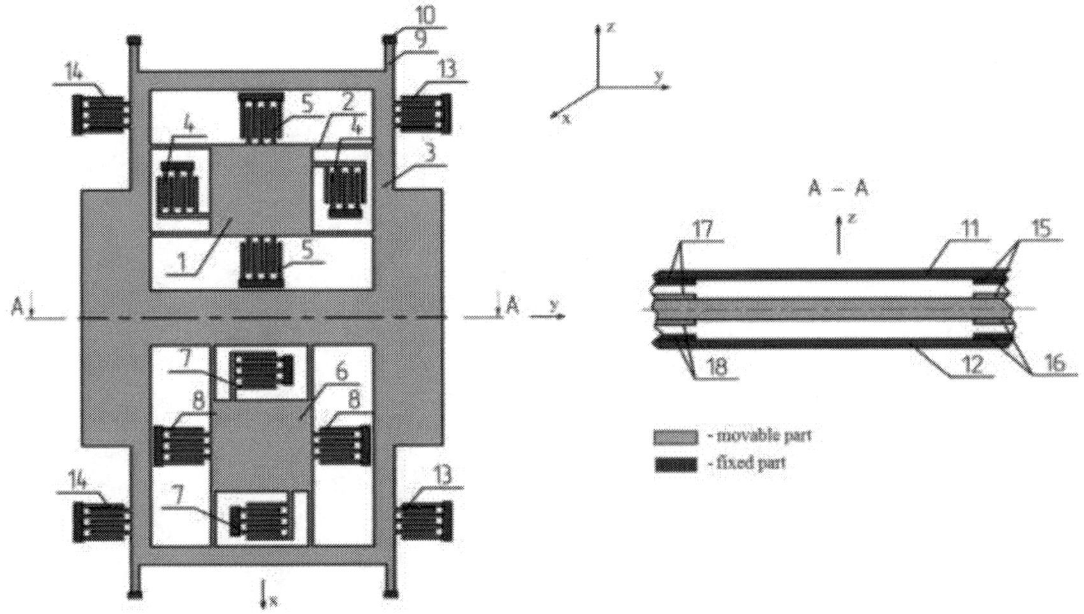

Figure 1. Schematic diagram of the silicon micromechanical gyroscope sensor.

The gyro contains an electromechanical silicon sensor (figure 1) and an electronic unit. The sensor includes frame 3 mounted on elastic suspensions 10 in base 9. The frame contains two sensitive masses 1 and 6 with mutually perpendicular suspension axes. The frame with sensitive masses produces primary oscillations along Z axis, which are excited by comb vibrodrive 13 or by a planar vibrodrivein in the form of two planar electrodes located under and above the inertial mass.

One of the electrodes in the form of metal deposition is located on the basis, and the other one is represented by frame 3. Presence of angular velocities of the object rotation causes the Coriolis forces. The forces are proportional to these angular velocities and cause secondary informative oscillations of the sensitive mass along X and Y axes. The oscillation amplitude of sensitive mass 6 along Y axis is proportional to the angular velocity of the object around X axis. The oscillation amplitude of mass 1 along sensitive X axis is proportional to the angular velocity of the object around Y axis.

Comb structures 4 and 8 are used to obtain data on the angular velocity of the object. Comb structures 5 and 7 are used to correct the errors of the gyroscope. Comb structure 14 is used to measure the amplitude of primary oscillations in the system controlling the gyroscope primary oscillations. Thus, MMG has two measuring axes, and represents a system of three oscillators.

3. Properties of single-crystal silicon

The silicon structure is a diamond-type lattice with tetrahedral bonds [3]. Each atom has four nearest neighboring atoms. These atoms interact due to forces of the covalent bonds. Packing of atoms in plane (111) (or direction (111)) in the complex silicon structure is more dense compared to that in plane (100) (direction (100)). Physical properties of silicon in different crystallographic directions are different [4-5]. Quantitative relationships for the elastic modulus, shear modulus and Poisson's ratio in different crystallographic directions are determined by the generalized Hooke's law for anisotropic media.

Metrological characteristics of MMG considerably depend on the material of the oscillator. Oscillators of high-Q and high amplitude oscillating provide high precision measurements of the angular velocity of the base. Therefore, the maximum match operating frequency of the model with a desired frequency is to be achieved, when calculating eigenfrequencies and modes of the sensor.

Silicon, like all materials changes its mechanical properties depending on temperature. This can significantly affect the dynamic characteristics of MMG which in many ways depend on the stability of the scale factor. Stable relationship between the construction eigenfrequencies is required to make the scaling factor constant. Slight deviation of the oscillator eigenfrequencies results in a significant change in the scale factor. The eigenfrequencies of the device depend on the hardness of the elastic suspension; therefore, the stability of this parameter should be maintained with high accuracy. Analytically, the natural frequency of the oscillator is related to the stiffness of the elastic suspension by the expression

$$\omega_i = \sqrt{\left(\frac{G}{m}\right)},$$

where G is the stiffness of elastic suspension, m is the inertial mass.

The temperature effect causes the drift of the silicon elastic characteristics and changes the linear dimensions of the structure. This causes the internal stresses, the geometry of the construction changes, temperature imbalance arises, and the dynamic characteristics of MMG change [6].

To estimate the changes in the eigenfrequencies of the oscillator in the operating temperature range from minus 50 to plus 80, the temperature-modal analysis of MMG was performed. The material of the construction was single-crystal silicon with anisotropy of the mechanical and thermal properties [7]. The research was carried out for five the most commonly used variants of sensor positioning on the wafer depending on the plane and the orientation of the wafer (table 1).

Table 1. Variants of the sensor orientation on the wafer.

Wafer orientation	Variant 1 (100)	Variant 2 (100)	Variant 3 (110)	Variant 4 (110)	Variant 5 (111)
Orientation of crystallographic axis	X <100> Y <010> Z <001>	X <110> Y <-110> Z <001>	X <100> Y <011> Z <0-11>	X <111> Y <-211> Z <0-11>	X <-1-12> Y <1-10> Z <111>

The mechanical and thermal characteristics of silicon, such as Young's modulus, shear modulus and temperature coefficient of Young's modulus were set for each variant of its positioning in an explicit form [8]. Due to the cubic symmetry of single-crystal silicon, the anisotropic mechanical properties can be determined by the tensor of second rank with three independent stiffness coefficients or crystal compliance coefficients. Appropriative mechanical and thermal characteristics of the crystal for each variant can be found by converting the tensor components of the initial coordinate system to the tensor components of the target coordinate system using the expression [9]:

$$T_{ij}^{'} = a_{ik} a_{jl} T_k ,$$

where T'_{ij} is tensor components of the target coordinate system, a_{ik}, a_{jl} are the matrices of direction cosines, T_k is tensor components of the initial coordinate system.

4. Research and results

The research was carried out using ANSYS. The finite element model (FE) of the test MMG is shown in figure 2.

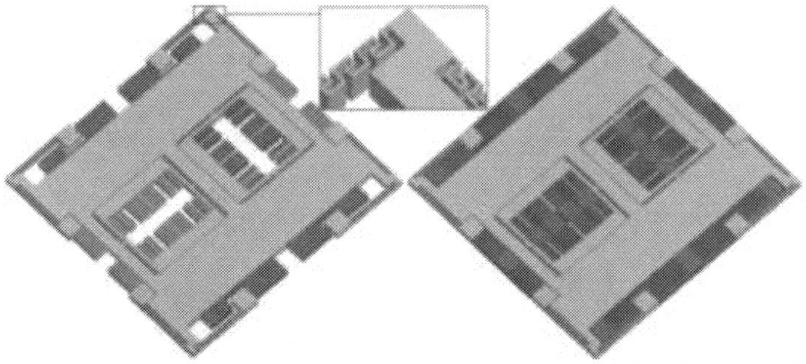

a) is without fixed electrodes b) is with fixed electrodes

Figure 2. MMG Silicon sensor.

When the sensor geometry had been designed, the modal analysis was performed. figures 3, 4 and 5 show the first vibrational modes of the sensor. As can be seen, that the movable sensor parts make the required movements along X, Y and Z axes with the eigenfrequencies which differ in no more than 5%.

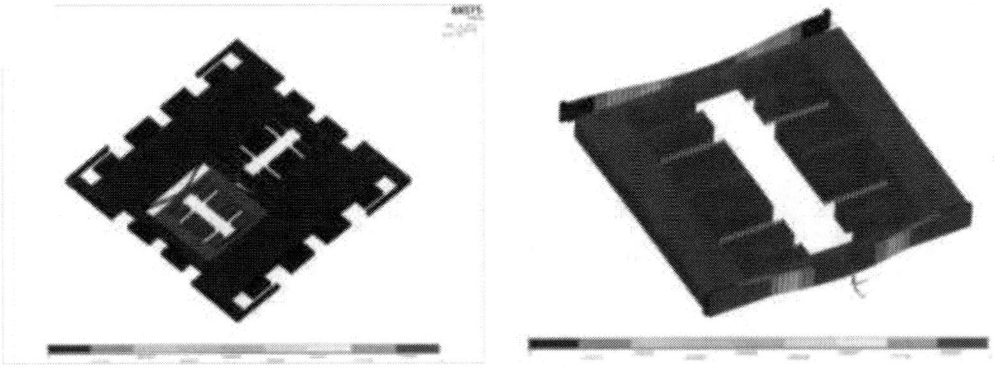

a) is sensor b) is inertial body

Figure 3. The first mode of sensor eigenfrequencies (f_y= 5032.57 Hz).

The first form of the gyro eigenfrequiency corresponds to its informative vibrations due to the presence of angular velocity Ω_x. While vibrating, the inertial body 6 makes a linear movement along Y axis (figure 3).

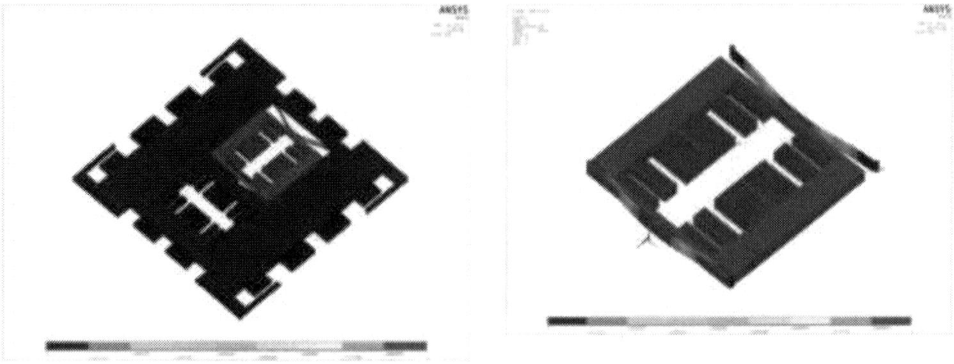

a) is sensor b) is inertial body

Figure 4. The second mode of sensor eigenfrequencies (f_x=5034.36 Hz).

The first form of the gyro eigenfrequiency corresponds to its informative vibrations due to the presence of angular velocity Ω_y. While vibrating, the inertial body 1 makes a linear movement along X axis (figure 4).

The third form of the gyro eigenfrequencies (figure 5) corresponds to its forced vibrations under electrostatic forces generated by the vibrodrive. Along Z axis, the frame and inertial bodies move as one body. The difference between the frequencies of the first three modes of vibration is no more than 1.33%.

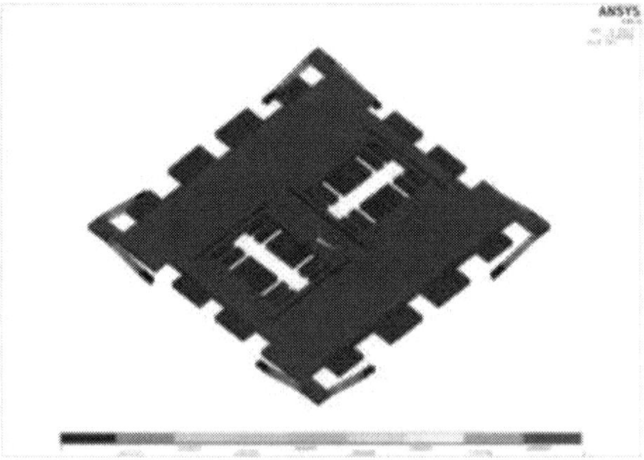

Figure 5. The third mode of sensor eigenfrequencies (f_z=5099.48 Hz).

The initial operating temperature of the sensor corresponding to the nominal dimensions of the gyro sensor and its initial mechanical properties was accepted as 20 °C. The operating temperature range was from –50 ° C to + 80 °C.

The dependences for the change in the MMG eigenfrequencies due to changes in the thermoelastic state of the structure are shown in figures 6, 7 and 8.

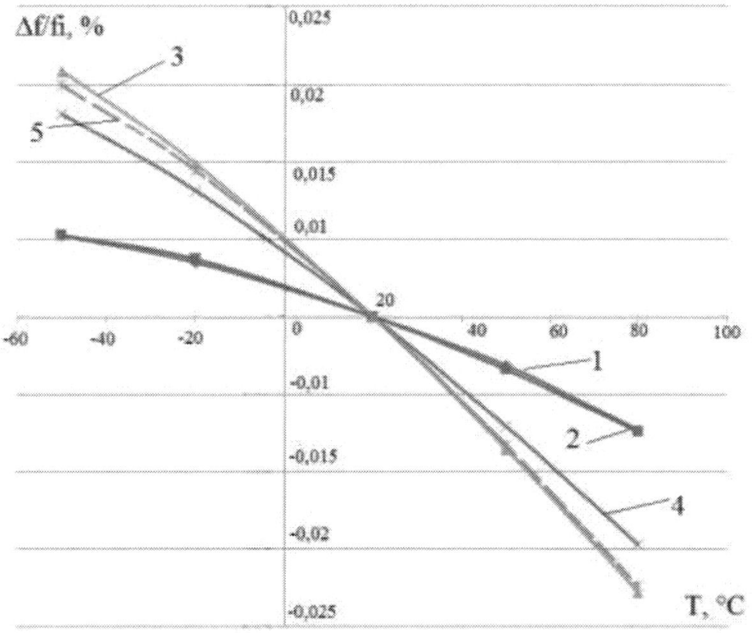

Figure 6. The first mode of eigenfrequencies (Y axis).

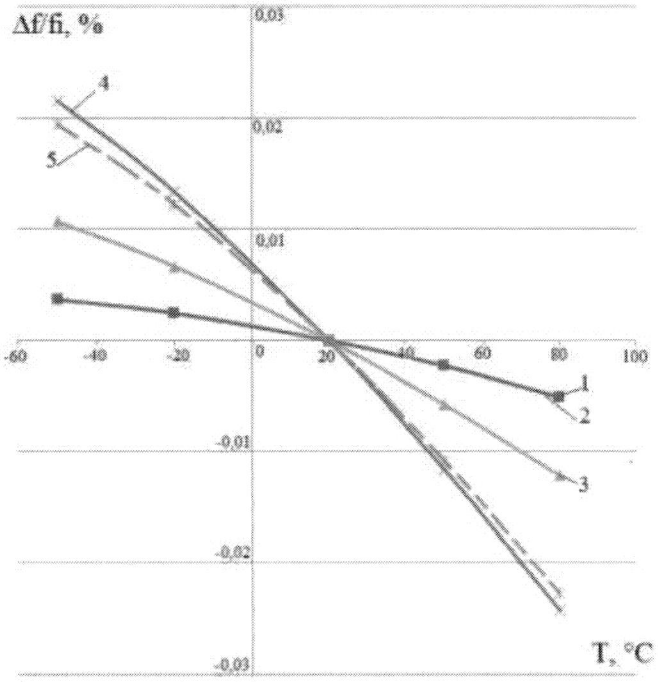

Figure 7. The second mode of eigenfrequencies (X axis).

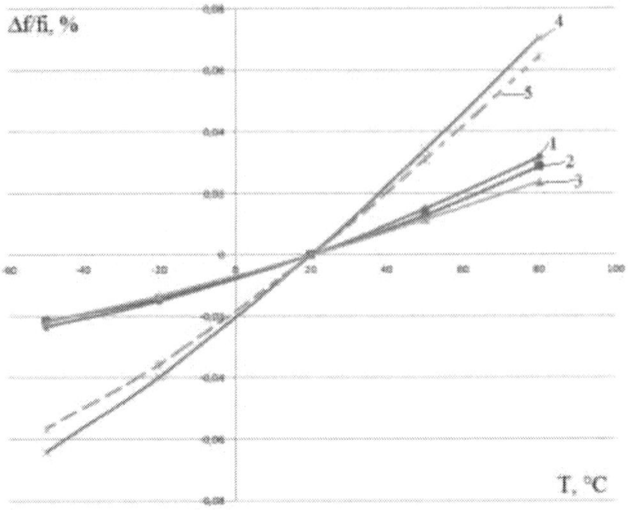

Figure 8. The third mode of eigenfrequencies (Z axis).

Absolute and relative changes in the eigenfrequencies for five variants of sensor positioning on the plate are shown in table 2 and figure 9, respectively.

Table 2. Absolute changes in gyro eigenfrequencies.

Mode	Variant 1 (100), Hz	Variant 2 (100), Hz	Variant 3 (110), Hz	Variant 4 (110), Hz	Variant 5 (111), Hz
1	0.60384	0.60384	1.18252	1.0064	1.15736
2	0.2517	0.256734	0.55374	1.10748	1.08231
3	1.63168	1.5297	1.17277	3.5693	3.31435

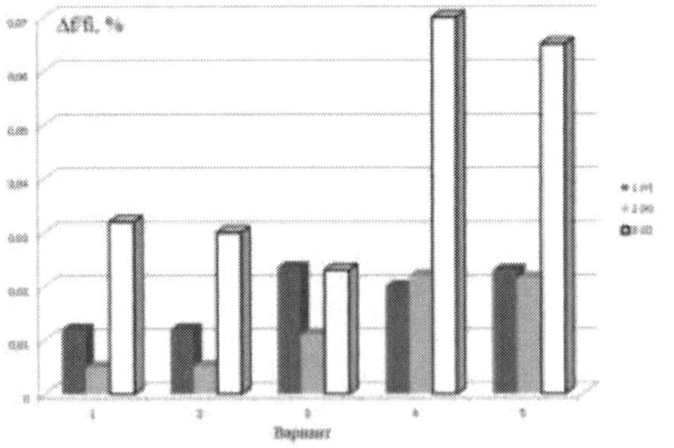

Figure 9. Relative changes in gyro eigenfrequencies.

The wafer with crystallographic orientation (100) is found to be the least sensitive over the full range of temperature change (figure 9).

To obtain the temperature dependence for scale factor the mathematical model was made. It consists of two nonlinear differential equations of second order, which reflects the dynamic behaviors of MMG. The mathematical model contains temperature dependence for eigenfrequencies, stiffness coefficients and modal mass approximated by second-order polynomials.

The resulting dependence of the scale factor vs temperature is shown in figure 10.

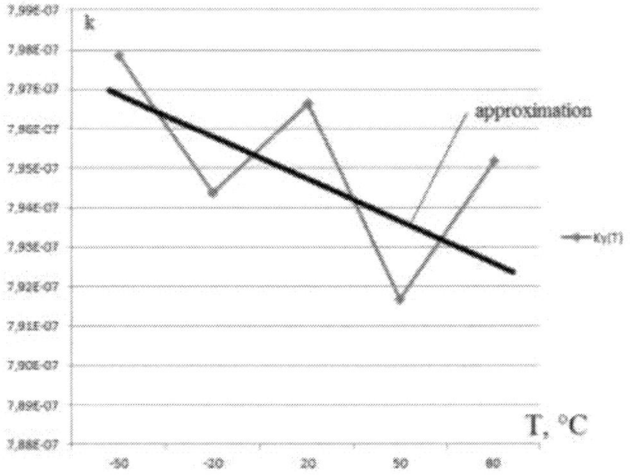

Figure 10. Dependence of scale factor vs temperature.

To reduce the effect of temperature on the accuracy of the angular velocity measurement the temperature stabilization system is to be used. The system is a combination of capacitive sensors to create positive or negative stiffness. The dependences reflecting the drift of the eigenfrequencies are virtually linear. This simplifies the thermal stabilization algorithm.

As a result of the research, direction (100) is found to be the most suitable sensor positioning on the silicon wafer. It causes the least deviation of the eigenfrequencies and the scale factor in the operating temperature range.

Acknowledgments

The research has been conducted in Tomsk Polytechnic University and it is financially supported by grant No. 14.575.21.0068 FTP "Research and development in priority directions to develop the scientific-technological complex of Russia for 2014–2020".

References

[1] http://www.yole.fr

[2] Nesterenko T G, Koleda A N, Barbin E S, Peresvetov M V Integrated micromechanical gyroscope Patent №2012122431 RUS

[3] V. Y. Raspopov 2007 *Micromechanical sensors* (Moscow: Mechanical engineering)

[4] Kim J, Cho D, Muller R 2001 Why is (111) silicon a batter mechanical material for IVIEMS? - The 11 International Conference on Solid-State Sensors and Actuators (Transducers '01), Munich, Germany 662-665

[5] Brantley W A 1973 Calculated elastic constants for stress problems associated with semiconductor devices *J. Applied Physics* **44** 534-535

[6] Nesterenko T G, Koleda A N, Barbin E S, and Uchaikin S V Temperature Error Compensation in Two-Component Microelectromechanical Gyroscope *IEEE Transactions on Components, Packaging and Manufacturing Technology* **10** 1598-1605

[7] Ikehara T, Tsuchiya T 2009 Effects of anisotropic elasticity on stress concentration in micromechanical structures fabricated on (001) single-crystal silicon films *J. Applied Physics* **105** 975

[8] Chen G S, Ju M S, Fang Y K 2000 Effects of monolithic silicon postulated as an isotropic material on design of microstructures *Sensor and Actuators* **86** 108-114

[9] Nye J F 1964 *Physical properties of crystals their representation by tensors and matrices. Lecturer in Physics in the University of Bristol* (Oxford at the Clarendon press)

PV effectiveness under natural conditions

A Yurchenko[1], V Syriamkin[1,2], A Okhorzina[1] and N Kurkan[1]

[1] National Research Tomsk Polytechnic University, Lenin Avenue 30, Tomsk 634050, Russia

[2] Department of Innovative Technologies, National Research Tomsk State University, Lenina Avenue 36, Tomsk, RU

ohra.avit@gmail.com

Abstract. Based on previous studies, the main factors which influence photovoltaic module under natural conditions were determined, In terms of hardware factors created the monitoring system prototype of photovoltaic module electrical parameters. Next, the system was upgraded and expanded by taking into account the meteorological data. The operating system results were available on the internet and presented in the article. The data analysed the efficiency of photovoltaic modules in Tomsk based on operating results of the monitoring system prototype. Finally, a conclusion was made about application efficiency of photovoltaic modules.

1. Introduction

The main factor limiting the widespread use of solar energy in practice is the cost of energy systems based on of alternative energy systems. Modern development of solar techniques can be used photovoltaic modules (PV) at latitudes previously considered unsuitable for this type of renewable energy.

The factors which may impact the work of PV were analyzed in previous papers [1, 2]. All factors affecting PV systems in natural conditions can be divided into two groups:

- Hardware factors which depend on the construction of the PV module and solar elements, and the technology of their implementation, specific arrangement of PV modules on buildings and structures (architectural factors), and PV module distribution and degradation characteristics (short-circuit current, performance coefficient and temperature coefficient of idling).
- Climatic factors which depend on different climatic parameters affecting the output energy characteristics of solar batteries (solar radiation, air temperature, humidity, wind speed, and aerosol concentration).

To determine of the PV effectiveness in Tomsk was a system of control of the solar module with the influence of climatic parameters. Main characteristic, which describe the PV work under natural conditions and during the operation, is the current-voltage characteristic.

To study the influence of factors on operating photovoltaic module and determine its effectiveness developed the electronic load, which allows measuring the current-voltage characteristics of PV.

The characteristics of PV are determined based on the measured current-voltage characteristic: performance coefficient, short circuit current and open circuit voltage.

2. The measuring of current-voltage characteristic with electron load

Figure1 shows a diagram of current-voltage characteristic measuring via electron load. The main element of a current-voltage characteristic measurement is electronic load. Electronic load changes its resistance under the influence of the signal from the DAC. The voltage of electron load is determined by the ADC. The operating principle of the electronic load is based on the change in resistance of the field effect transistor gate [2].

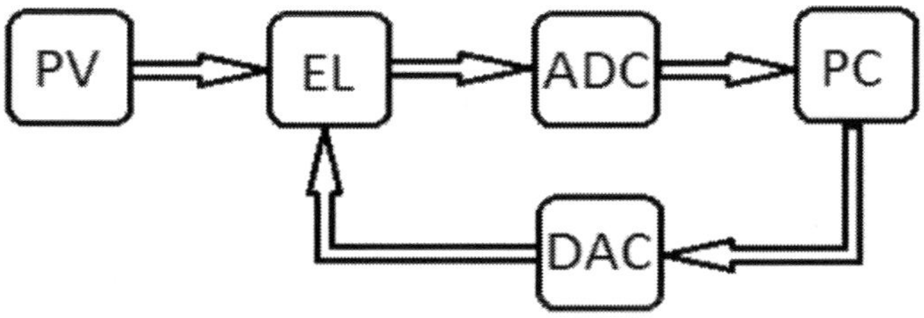

Figure 1. Scheme of measuring with electron load, PV – photovoltaic module, El – electron load, ADC – analog-digital converter, DAC – digital-analog converter, PC – personal computer.

The similar scheme implementation is not possible to measure current-voltage characteristic in low light. The number of points in the current-voltage characteristics of solar radiation at 200 W/m^2 is 2-3 points.

3. The modification measuring of current-voltage

In order to be able to measure the current-voltage characteristic in low light, has been modified the measuring scheme of current-voltage characteristic on the basis of a multi-valued measure of resistance. Multi-valued measure of resistance is controlled the computer and microcontroller (Figure 2). The number of dimensions is reaches 30, thus increasing the accuracy of measurement. [2]

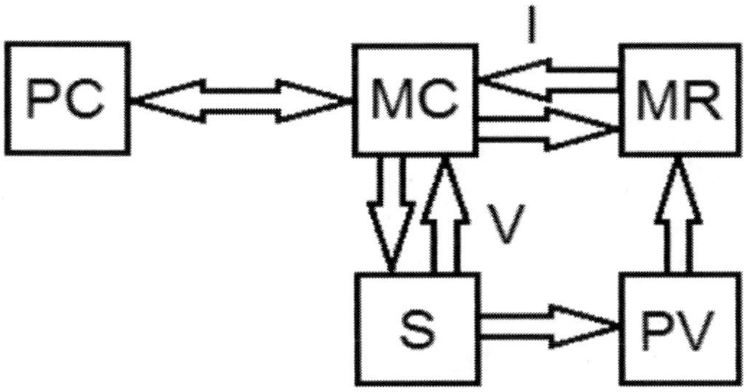

Figure 2. Modification scheme of measuring with electron load, PC – personal computer, MC – microcontroller, MR – multi-valued measure of resistance, S – switch, PV – photovoltaic module.

At startup program signal is sent to a microcontroller that provides a control signal to the switch, which connect the load. The load is a multi-valued measure of resistance from 1 Ohm to 8 Ohm that allows to measure current-voltage characteristic with high accuracy and to determine the operating point.

The time period between measurements of current-voltage characteristic of the PV modules is one of the parameters of the program. During its automatic operation the electron load resistance gradual changes from 0.1 Ohm with the simultaneous current and voltage control. Measuring stops at the moment when the PV module current approaches zero. After this the electron load is switched off and open circuit voltage is measured. Afterwards the system is in the waiting mode.

The computer contains specific software which controls the electron loading, receives and stores measurement results. The program has created in LabVIEW. The window of the program is shown in Figure 3.

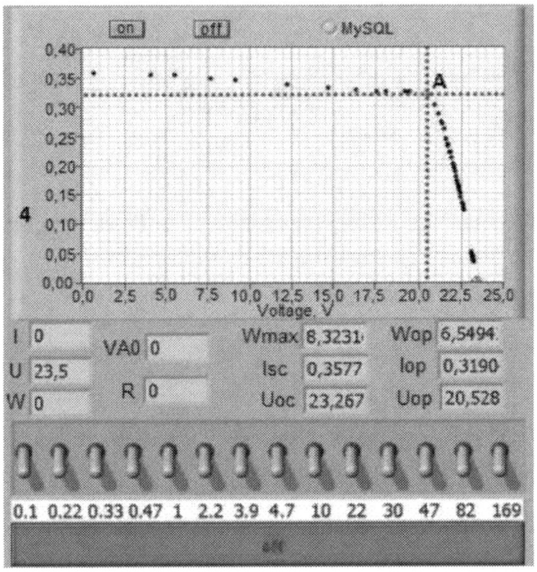

Figure 3. The window of the software.

In the lower left corner of the program window represented 14 switches that allow measurement by hand, that gives the chance most accurately evaluate the performance of PV [2].

To implement a full research combined current-voltage characteristic measuring station with climatic parameters sensors.

4. Station of on-line monitoring

The control system of PV allows to measure the current-voltage characteristic of PV, determine its parameters and to measure the main parameters of climate (temperature, pressure and humidity, the amount of incoming solar radiation), the surface temperature of the PV. This enables to determine the efficiency of the PV under different weather conditions.

Described station is set on the roof of the Institute of Atmospheric Optics in Tomsk and is connected to the PV (25 W), that is installed at an angle of 54° to the horizon to the south. [2, 3]

Measurements of current-voltage characteristics and climatic parameters are carried out once with an interval of time; recorded voltage characteristic photovoltaic module, the temperature, the air temperature, the operating point is calculated. As a result, one cycle for each parameter is collected one-dimensional array, which consists of 30 elements. Next the array is averaged. The result is stored in the data file. Moreover, the measurement results, where the short circuit current less than 0.1 A are not retained [3].

4.1. Database

As a result of the tests was collected database consisting of 20 000 entries. The following table shows structure of the database.

Table 1. Database structure

Field	Measurement units	Description
Date time	DD.MM.YYYY hh:mm:ss.	Date and time
Open-circuit voltage	V	Open-circuit voltage
Operating voltage	V	operating Voltage
Short-circuit current	A	Short-circuit current
Operating current	A	Operating current
Temperature	˚C	Air temperature
Humidity	%	Humidity
Total solar radiation	W/m2	The total solar radiation
Wind speed	m/s	Wind speed
A	Grad	Azimuth of the sun
H	grad	Declination of the sun
Pressure	mm Hg	Air pressure

4.2. On-line resource

The measurement data are automatically recorded and are shown on the site *lop.iao.ru* in January 2011 (Figure 4 and 5).

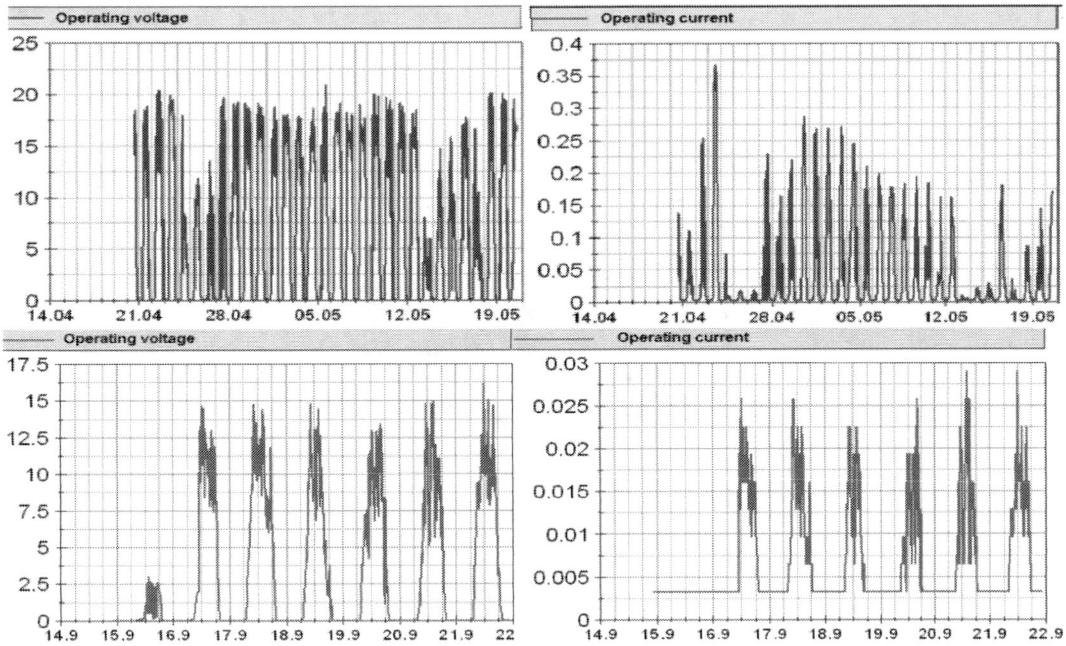

Figure 4. On-line measures.

"Solar Eye" shows incoming solar radiation throughout the year by month and hours. According to this schedule is possible to tell about the duration of sunshine during the day.

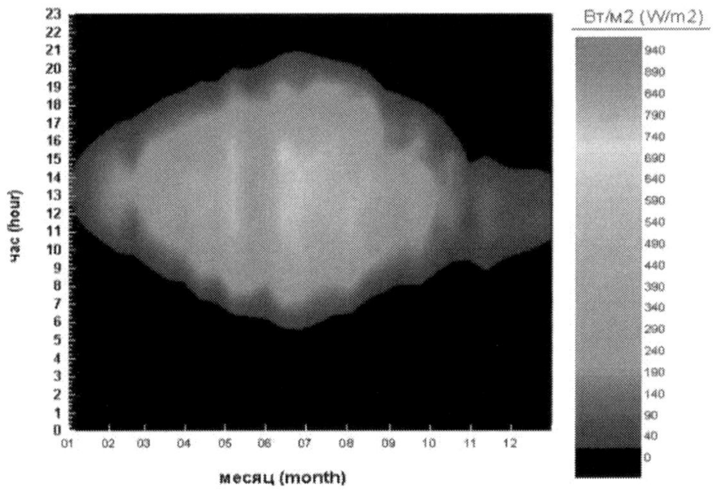

Figure 5. "Solar Eye" in 2014.

5. Effectiveness of PV module

Based on previous data, we calculated the efficiency of the solar module during the year [2]. Efficiency was calculated as the ratio of the PV module average power generated during the day to coming in for a day of solar radiation on the surface of the PV module [5]. The results are shown in Figure 6.

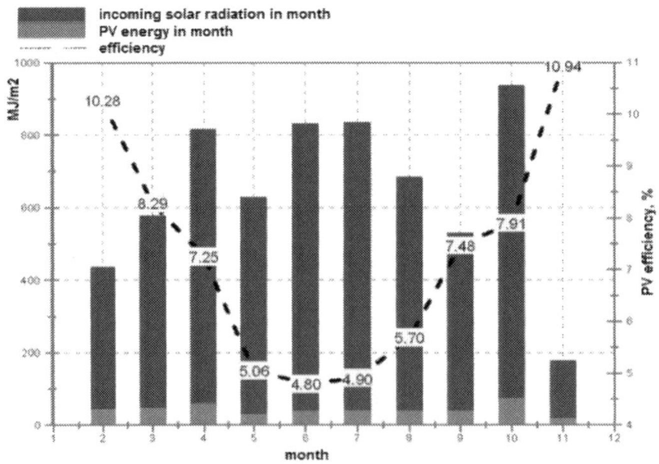

Figure 6. Effectiveness of PV module.

Figure 6 shows that the effectiveness of PV module varies greatly in different months. The maximum value is achieved in the winter months; this is due to a high proportion of reflected light from the snow in the winter, as well as due to the low angle of declination of the sun, and light falls on the solar module throughout the day. In the summer months the efficiency a PV module is significantly reduced due to the fact that the sun in the morning and evening is the surface of a PV module due to high seasonal angle.

On average the efficiency is about 6%, which is about half the maximum possible efficiency for a used the PV module (η_{PV} = 13.8%, the efficiency is a parameter specified by the manufacturer in the passport on the basis of laboratory tests PV module). Efficiency can be increased by using more modern modules, tracking or combined systems.

The calculated values of PV power during the year compared with the results obtained on the basis of a fifteen-year analysis of the PV in Siberia (Figure 7) [2, 3].

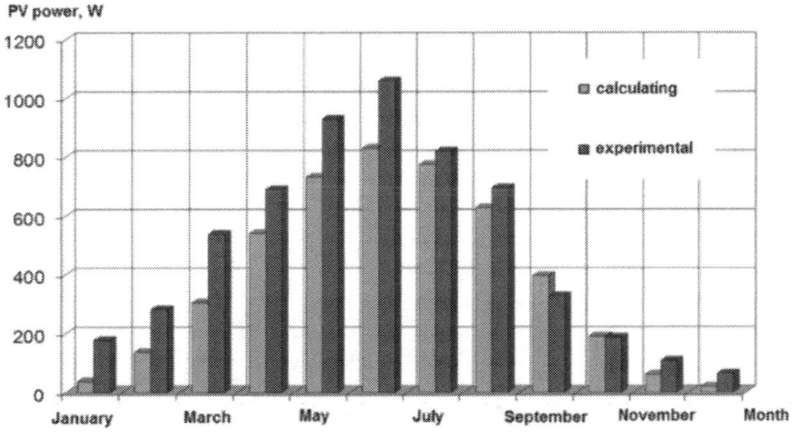

Figure 7. The PV power for a year.

As seen in Figure 7 changes the power dynamics of the PV during the year is the same. The experimental values are higher than the theoretical, which is due to the fact that the average annual solar radiation in the year under review was above average. Also influenced by the background and scattered radiation. In the winter months is a big difference due to the presence of the reflected light from the snow, which is also not taken into account in the calculations.

6. Conclusion

- The online-monitoring system of PV module operating has been designed (http://lop.iao.ru/activity/?id=sb). The system consists of a PV module, personal or portable computer and electron load. The computer contains specific software which controls the electron loading, receives and stores measurement results.
- The efficiency of the PV in Tomsk has been determined, that allows draw a conclusion on the effectiveness of solar systems on the territory of Tomsk and Siberia.

Acknowledgment
Work has been funded by President Grant MD-5352.2013.08

References

[1] Nann S 1991 *Solar Energy* **45** pp 385–393
[2] Kitaeva M, Yurchenko A, Ohorzina A 2012 *the 7th International Forum on Strategic Technology* (Tomsk: TPU Press vol 2) pp 107–110
[3] Yurchenko A, Kozlov A, Volkov A 2008 *23st European Photovoltaic Solar Energy Conference* (Valencia Spain, 4-8 September 2008: WIP Wirtschaft und Infrastruktur GmbH and Co Planungs-KG) pp 2989–2991
[4] Yurchenko A, Kitaeva M, Ohorzina A, Skorohodov A 2011 *26th European Photovoltaic Solar Energy Conference and Exhibition* (Frankfurt: WIP Wirtschaft und Infrastruktur GmbH and Co Planungs-KG) pp 4360–4362
[5] Markvart T, Castafier L 2003 *Practical Handbook of Photovoltaics: Fundamentals and Applications* (USA: Elsevier Science Ltd)

Power supply of remote and almost inaccessible settlements

O A Surzhikova
National Research Tomsk Polytechnic University
Czech Technical University in Prague

E-mail: surzhikovaoa@tpu.ru

Abstract. Problems of power supply of remote and almost inaccessible settlements are analyzed. Variants of their connection to transmission lines and the use of diesel electric power stations are considered. It is suggested to revise the policy of power supply for consumers of this category based on the experience in implementation of nontraditional energy production technologies, namely, based on natural renewable resources.

1. Introduction

The special feature of Russia characteristic primarily for regions of Siberia and Far East is a significant number of small remote and sparsely populated settlements. The problem of reliable and high-quality power supply of such settlements remains urgent in social, technical, and economic aspects [1].

Generally, these consumers can be supplied either by centralized or decentralized power supply.

Technical feasibility of power supply of an object through transmission lines (TL) depends on the installed electrical power of the object of power supply and the distance to the nearest point of connection to mains.

2. Results and considerations

The All-Union Standard on the electric energy parameters establishes restrictions on voltage deviations from the nominal value within ±5%. Hence, the maximum TL extension l_{max} of power distribution networks with standard voltages of 6 and 10 kV can be calculated from the formula

$$_{max} = \frac{\gamma F U_{nom} \Delta U}{P_{nom}} [km],$$

where γ is the wire material conductivity, m/Ω·mm^2; F is the wire cross sectional area, mm^2; U_{nom} is the nominal voltage equal to 6–10 kV; DU is a 5% voltage loss, kV; and P_{nom} is the nominal power of an electrical receiver, kW.

Rural distribution networks are typically made from aluminum wire with specific resistance of 32.2 Ωm/mm^2, and the minimum standard wire cross section for powers up to 160 kW is equal to 16 mm^2.

The maximum TL lengths are presented in Table 1 depending on the power of the supplied object for distribution networks with voltages equal to 6 and 10 kV, respectively (given that the voltage losses are 5%).

A higher voltage of the distribution network allows the distance of the centralized power supply to be increased under the same conditions.

Thus, electrification of objects with power smaller than 250 kW from the centralized electric networks is limited by distances ≤10 km.

Table 1. Maximal TL length (l_{max}) versus the object power (P) for network voltages of 6 and 10 kV.

P, kW	5	0	3	00	60	50	2
l_{max}, km (6 kV)	7	3	4				–
l_{max}, km (10 kV)	03	4	1	6	6	0	1

Construction of higher-voltage more expensive TL (for example, 35 kV) is inexpedient for low transmitted power because of the low load factor.

Hence, it would be inexpedient to estimate the economic characteristics of high-voltage TL operating in the regime close to idling one.

Thus, for the overwhelming majority of electrification objects located in regions with low population density and poorly developed infrastructure, the technical criterion for electrification by means of TL construction cannot be fulfilled.

In addition to this purely technical restriction, a question arises on the high cost of transmission lines themselves and of their maintenance and service.

All this taken together makes abundantly clear the low probability of possible electrical supply of small remote consumers through their connection to centralized electric networks.

In the second variant of power supply of sparsely populated and remote settlements, the most widespread electric power source are stationary and mobile diesel electric power stations (DEPS); more than 5 thousand stations can be counted in Russia. They produce about 1.8 billion kW·h of electric power and consume about 0.8 million ton of standard fuel annually.

The main problems of power supply of the remote and sparsely populated settlements from DEPS are:

- bad technical conditions of electric power sources,
- long-distance transport of fuel and dependence on its deliveries,
- limited terms of seasonal delivery (the fuel can be transported to some remote regions during a year and longer, with intermediate storage in storage terminals) to the most remote regions,
- poor development of transport infrastructure,
- dependence on budgetary financing.

The poor development of transport infrastructure complicates significantly fuel supply. Long transportation distances and multistage and seasonal delivery of fuel lead to large fuel losses and multiple price increase. The transport component of cost of the fuel delivered to the most remote consumers can reach 70–80%.

The low-power sources used for autonomous power supply, as a rule, have low technical and economic characteristics. In addition, the increased fuel prices and transport tariffs (especially large for remote regions) lead to high price of electric power production that is several times higher than its average value for systems of centralized power supply. Incomplete fuel delivery entails long interruptions in power supply.

The DEPS power required to supply an individual consumer is determined by the peak power consumption depending on the electrical load chart of the object. As a rule, individual consumers are characterized by extremely nonuniform electric energy consumption, thereby causing significant underutilization of the nominal DEPS power.

The majority of sources of autonomous power supply are unprofitable, since the cost of electric power production is much higher than the tariff established for population, and the absence of industrial enterprises in the overwhelming majority of small isolated settlements does not allow the payment to be compensated by the industrial tariffs. Therefore, the power supply of small individual consumers from DEPS is subsidized by budgetary funds of local and regional administrations.

Losses from sale of electric power at the price below the cost are covered by the budgetary grants to provide admissible tariffs on electric energy in remote and sparsely populated regions. The lack of budgetary financing leads to incomplete fuel delivery, thereby putting this category of consumers in extremely severe conditions of power supply.

The problems of power supply of consumers in remote and sparsely populated settlements are aggravated because of the financial situation in the country and require additional attention of management. However, design organizations develop schemes only of regional energy systems and energy centers, while problems of power supply of numerous small individual consumers are considered on the level of either consumers themselves (industrial enterprises participating in economic activities in the given territory and providing maintenance of all accompanying industrial and social infrastructure) or regional and municipal authorities [2].

The increased price of organic fuel and transport tariffs for its delivery and the inefficient regimes of operation of the existing DEPS necessitate revision of the power supply policy for consumers of this category based on the accumulated experience on the application of modern power production technologies, being guided primarily by the directions providing decreased consumption of imported fuel. One of such directions is the application of renewable energy sources (RES).

The total RES resources in the world and Russia are huge [3, 4]. The total technical potential of solar, wind, biomass, water, and geothermal energy in the world is $2.4\cdot10^5$ mln ton of standard fuel units (s.f.u.); it is $4.5\cdot10^3$ mln ton of s.f.u for Russia.

Converted into the traditional fuel and energy resources, for Russia this is equivalent to the replacement of about $3.15\cdot10^9$ ton of oil, $3.7\cdot10^{12}$ m^3 of natural gas, $5.7\cdot10^9$ ton of coal, and $1.3\cdot10^{10}$ m^3 of firewood per year.

For a comparison: Russia consumes about $1.6\cdot10^8$ ton of coal, $1.3\cdot10^8$ ton of oil, and $4.6\cdot10^{11}$ m^3 of gas per year for the activity of all types.

In the long term, the contribution of nonconventional renewable energy sources to the world energy balance is estimated from (1–2) to 10%, although already today there are countries where the contribution of these sources exceeds half the national energy balance.

3. Summary

One of the indispensable conditions for the successful development of objects of renewable power engineering is the state legislative support of institutions and persons engaged in the introduction and utilization of RES.

Thus, the active application of RES observed in the last few years in Europe is largely caused by the adoption of regional and all-European legislations and instructions on the development of renewable power engineering. Various mechanisms of RES stimulation have been developed and tested. Among the most efficient of them are recognized introduction of special tariffs on energy produced with the use of RES and commitments to the manufacture and (or) sale of energy from RES (bonds, green certificates, etc.).

Recently a number of attempts have been undertaken in Russia on the federal and regional levels to develop programs of utilization of energy sources based on renewable energy resources [5].

However, despite many adopted resolutions and developed programs, practical implementation of projects on power supply of individual consumers, including utilization of energy sources based on RES, is carried out on insignificant scales. This did not allow the problem of their power and fuel supply to be solved adequately.

Among the most *recent* guidance materials on the aforesaid issues are RF Government Order "On Utilization of Renewable Energy Sources" No. 1-r of January 8, 2009 and Federal Law "On Energy Saving, Increase in Energy Efficiency, and Amendments to Some RF Acts" No. 261-FZ of November 23, 2009.

However, no noticeable positive changes are observed in the development and expansion of works.

At the same time, the lack of the state legislation must not block the adoption by regional and municipal authorities of the corresponding standard-legislative documents and interfere with practical RES implementation of local decisions.

References

[1] Surzhikova O A and Nikulina I E 2006 *European Journal of Natural History* **2** 151
[2] Surzhikova O A 2008 *European Journal of Natural History* **3** 98
[3] Belyaev L S *et al* 2000 *World Power Engineering and Transition to Sustainable Development* (Novosibirsk)
[4] Shafranik Yu K 1995 *Novel Energy Policy of Russia* (Moscow)
[5] Nikulina I E and Surzhikova O A 2006 European Journal on Natural History **5** 69

IOP Conf. Series: Materials Science and Engineering **81** (2015) 012099 doi:10.1088/1757-899X/81/1/012099

NEM and MFEM Simulation of Interaction between Time-dependent Waves and Obstacles

V A Perminov[1], T S Rein[2], S N Karabtcev[2]

[1]Tomsk national research polytechnic university, chair of ecology and health and safety, Tomsk, Russia
[2]Kemerovo state university, UNESCO NIT chair , Kemerovo, Russia

E-mail: p_valer@mail.ru, tsrein@mail.ru, skarab@kemsu.ru

Abstract. Fundamental research challenge of structural integrity and construction resistance while interacting with fluid or gas is of high importance when estimating their efficiency and lifetime. The paper presents the simulation results of interaction between incompressible ideal fluid and an escarpment at the bottom, and interaction between viscous incompressible fluid and an obstacle above fluid surface. Flow patterns at different times and chronograms of hydrodynamic loads on solid walls of computational domain, horizontal and vertical obstacles are displayed.

1. Introduction

The paper shows solution results of the problems dedicated to fluid interacting with vertical and horizontal obstacles in the context of classical nonlinear problem statement where velocity vector component and pressure domain are unknown. This statement enables to carry out comprehensive study of wave flows during all stages of simulation experiment including wave collapsing , vortex formation, turbulent flows with possible intermixing fluid layers formation.

Nowadays there is a whole range of mesh methods. Finite elements method (FEM) [1] and control volume method (CVM) [2] are the most popular of them. However, mesh-free methods [3] are more often used. They approximate equations into partial derivatives being based on nodes set without knowing any additional information about mesh structure. Smoothed Particle method [4], Lagrangian and Eulerian particles method [5] appear to be very efficient in the context of hydrodynamics and gasdynamics dynamics problems. Relatively low accuracy and difficulties based on carrying in boundary conditions are disadvantageous features of mesh-free methods.

The facts described above conduced to new conventionally mesh-free methods which include both ideas and opportunities of mesh-free methods and grid method advantages. Mesh-free finite elements method (MFEM) and natural element method [6, 7] appeared to be the first of new generation mesh-free methods.

Their main feature is that they occur to be usual (classical) Galerkin methods in the context of non time-dependent problems. To solve time-dependent problems Lagrange approach of medium description is applied: based on every time step a mesh is created, it identifies new structure of neighbors for each node of the domain. Equation system is solved again with the help of Galerkin method on newly-created grid. Thus, NEM and MFEM have some advantages of classical Galerkin method, that are simplicity of shape function in definition domain, continuity between elements, easy boundary conditions input. On the other hand they have all advantages of mesh-free methods, which means that natural element method form functions depend on node location.

2. Problem statement

Complete nonlinear problem of ideal incompressible fluid with free boundary is stated below.

Ideal incompressible fluid flow described by Euler equation system and equation of continuity is specified in flow D computational domain represented by finite node set and limited by free surface Γ_0 and solid boundaries Γ_1, Γ_2 and Γ_3:

$$\frac{Du_i}{Dt} = -\frac{1}{\rho}\frac{\partial p}{\partial x_i} + f_i, \ \mathbf{x} \in D, \ i = \overline{1,2}. \tag{1}$$

$$\partial u_i / \partial x_i = 0, \ \mathbf{x} \in D \tag{2}$$

Here $\mathbf{x} = (x_1, x_2)$ – spatial coordinates, $\mathbf{u} = (u_1, u_2)$ – velocity vector, p – pressure, ρ – density, $\mathbf{f} = (f_1, f_2) = (0, -g)$ – external force vector. Node motion in the domain is describe by the following equation:

$$dx_i / dt = u_i, \ \mathbf{x} \in D, i = \overline{1,2}. \tag{3}$$

Dynamic condition $p = p_{atm}$ is fulfilled on free surface Γ_0, impermeability condition $\mathbf{u} \cdot \mathbf{n} = 0$ is fulfilled on solid boundaries Γ_1, Γ_2, Γ_3, where $\mathbf{n} = (n_1, n_2)$–the external normal to the fluid boundary.

General formulation of viscous incompressible fluid flow problem is presented below. Newtonian viscous incompressible fluid flow described by Navier-Stokes equations and equation of continuity (2) is considered to occur in flow domain D. Navier-Stokes equations are presented in Euler's formulation as follows:

$$\frac{Du_i}{Dt} = -\frac{1}{\rho}\frac{\partial p}{\partial x_i} + \frac{\mu}{\rho}\frac{\partial}{\partial x_j}\left(\frac{\partial u_i}{\partial x_j}\right) + f_i, \ i = \overline{1,2}, \ j = \overline{1,2}. \tag{4}$$

Pressure p and velocity vector $\mathbf{u}$ are the required functions in equation system (2),(4). Density ρ, external forces vector $\bar{f}$ and dynamic viscosity coefficient μ are the required parameters in equation system (2),(4).

Boundary conditions for Navier-Stokes equations (2),(4) are as follows: as far as fluid is viscous , no-slip boundary condition $u_i = 0, \ i = \overline{1,2}$ is fulfilled on solid boundaries Γ_1, Γ_2 and Γ_3. Dynamic condition $p = p_{atm}$ is fulfilled on free surface Γ_0 like in ideal fluid case.

It is needed to set nodes location $\mathbf{x}(0) = \mathbf{x}^0$ and distribution of unknown functions in whole flow domain $\mathbf{u}(\mathbf{x},0) = \mathbf{u}^0(\mathbf{x})$, $p(\mathbf{x},0) = p^0(\mathbf{x})$ for time-dependent problem of both viscous and ideal fluids flows.

3. Numerical computation

3.1. Mesh-free finite elements method

In 1994 L.Traversoni suggested natural elements [8] method should be used to solve plasticity theory problems. This method is a variation of Galerkin mesh-free method. Like Galerkin's method unknown functions are approximated as follows: $q = N^T Q$, where $q = (u_i, p, \rho), i = \overline{1,2}$, Q – nodal values of unknown function, N - interpolating form function. Later the method was adapted to solve fluid flow problems and became more popular as mesh-free finite elements method [9].

Integrated form of weighted residuals method is used to create discrete equations system. Integrals taken are based on elements of extended Delone triangulation. Multitude of natural elements for each node and free boundary nodes at new time step are defined with the help of «sweep line» method и «alpha-shape» method respectively [10].

The essential difference of classical finite elements method from mesh-free method includes the necessity of computational mesh recreating at every time step. Each time new Sibson and Laplace interpolating functions [11,12] are constructed and rigidity matrixes are calculated, resulting equations are set up on a newly created mesh. Solution of multidimensional problem at fractional steps results in solution of individual equations with the help of conjugate gradient method.

To define the structure of neighbors and its nodes and to calculate Sibson and Laplace form functions the domain is divided by Voronoi cells [12]. Computational mesh is needed for numerical integration while calculating the elements of rigidity matrix of resulting equations (weak form of medium motion equation is under consideration).

At first time step the domain is divided with the help of Delone triangulation. Though, there is a necessity of computational mesh quality improvement in time-dependent problems (as soon as the mesh does not meet Delone criteria). If a mesh node location changes it is necessary to interpolate the variables values from the old "bad" mesh to the new upgraded one. Sibson and Laplace form functions are considered to be interpolation coefficients while recalculating the variables values.

3.2. Sibson and Laplace Interpolating functions

Focused on natural elements concept Sibson's interpolating functions are based on Voronoi diagrams of the first and second order and identified by polygons area ratio in two-dimensional case [11]:

$$\alpha_I(\mathbf{x}) = A_I(\mathbf{x}) / A(\mathbf{x}),\ A(\mathbf{x}) = \sum_{J=1}^{N} A_J(\mathbf{x}),\ I = \overline{1, N}. \tag{5}$$

Here N – number of natural elements for node $\mathbf{x} = (x_1, x_2)$; $A(\mathbf{x})$ – Voronoi cell area of the first order, that includes node $\mathbf{x}$, $A_I(\mathbf{x})$ – crossing area of Voronoi cell area of the second order $\mathbf{x}$ with area $A(\mathbf{x})$ (refer with: Figure 1,*a*).

Laplace interpolation is based on neighbors defining by having the domain divided with the help of Voronoi cells [12]. Node $\mathbf{x}$ is related to Voronoi polygon with side number equals N. Let us set polygon sides lengths by $s_I, I = \overline{1, N}$, and the heights put down from $\mathbf{x}$ to s_I, – by h_I. Then Laplace interpolation coefficients are as follows :

$$\alpha_I = (s_I / h_I) \left(\sum_{J=1}^{N} s_J / h_J \right)^{-1},\ I = \overline{1, N}. \tag{6}$$

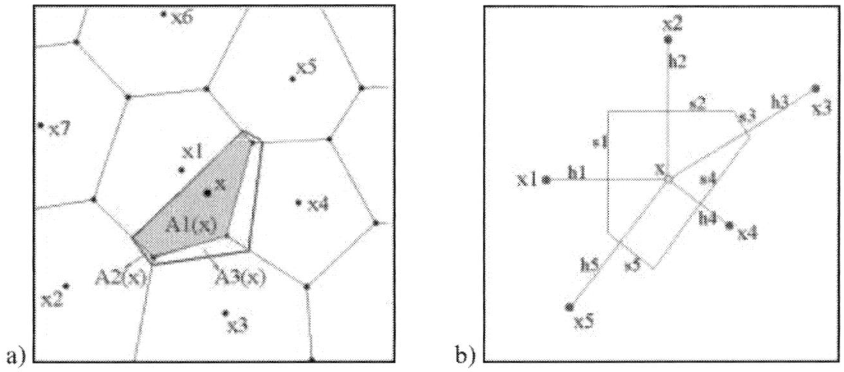

Figure 1. Interpolation by: a)Sibson; b) Laplace

This method of coefficients α_I defining is easier and cost-effective compared to Sibson's approach, as it does not require calculating polygons crossing area. Fig.1b shows Laplace interpolation architecture for two-dimensional case.

Nonsibson's interpolation has a specific feature that can be defined taking into consideration its basic properties. Many dissection polygons appear to be simplexes if the domain is divided by Voronoi cells for the specified node set. In the context of simplex polygons Laplace form functions act like linear functions. Form function behavior on some polygons is known beforehand. This fact can simplify integration problem and derivative calculation problem.

Derivative of Sibson's and Laplace's form functions can be calculated by deriving equations (5) and (6) respectively.

4. Numerical results

4.1. Solution of interaction with obstacle above the fluid surface problem

Computational domain contains reservoir with flat bottom and solid impenetrable walls filled with uniform viscous incompressible fluid and at initial time separated by thin impermeable wall that creates fluid level difference (refer with: Figure 2*a*). At initial time the wall starts moving upwards steadily with specified velocity while developing fluid column with zero velocity vector starts collapsing due to gravity.

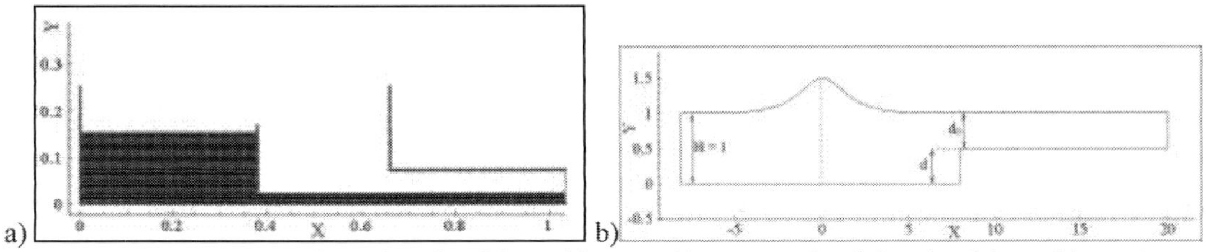

Figure 2. a) Computational domain of the problem 4.1; b) Computational domain of the problem 4.2

Peak load and maximum wave force experienced by horizontal obstacle are estimated in the context of different surface heights above free surface. Figure 3 shows flow patterns at different times for the case when height of horizontal surface quadruples layer depth. This parameter value was chosen because while dam collapsing maximum wave amplitude quadruples layer depth initial value at the bottom. It is worth noting that horizontal obstacle length was chosen in the way that moment of maximum wave amplitude coincides with the moment of wave collision with escarpment.

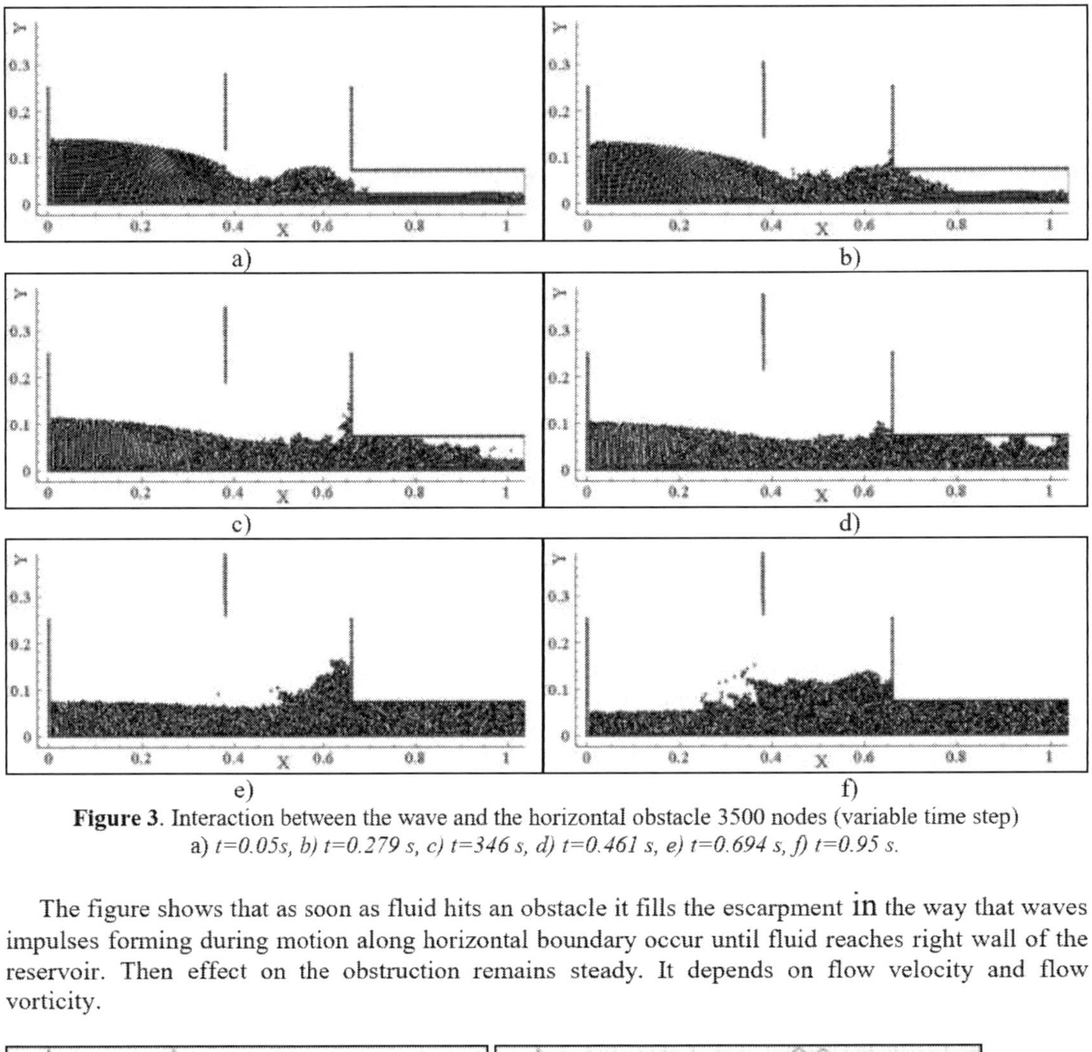

a) b) c) d) e) f)

Figure 3. Interaction between the wave and the horizontal obstacle 3500 nodes (variable time step)
a) *t=0.05s, b) t=0.279 s, c) t=346 s, d) t=0.461 s, e) t=0.694 s, f) t=0.95 s.*

The figure shows that as soon as fluid hits an obstacle it fills the escarpment in the way that waves impulses forming during motion along horizontal boundary occur until fluid reaches right wall of the reservoir. Then effect on the obstruction remains steady. It depends on flow velocity and flow vorticity.

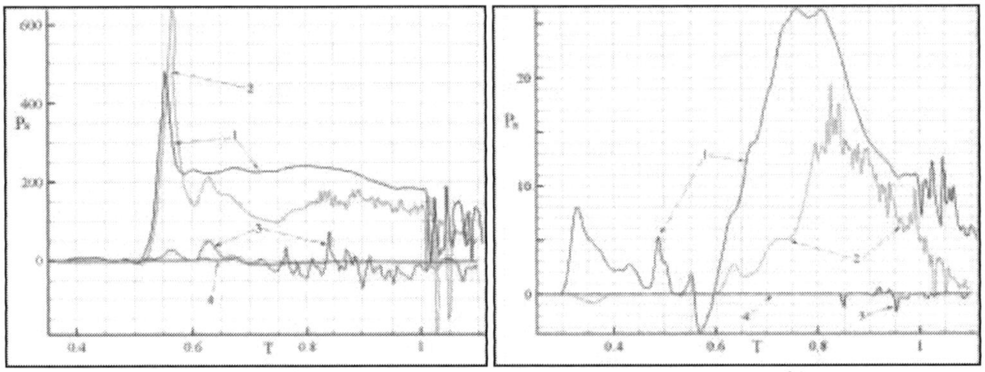

a) b)

Figure 4. Chronograms of hydrodynamic load

$1-h_g=4h$; $2-h_g=6h$; $3-h_g=10h$; $4-h_g=12h$

a) horizontal boundary of the escarpment; b) vertical boundary of the escarpment

In case horizontal obstacle is quite distant from free fluid surface fluid lapping at the obstacle is identified. As far as the fluid is viscous, no-slip condition is set at the solid boundary. While fluid flowing down the horizontal surface fluid particles are "released" from solid boundary and so called negative pressure zones occur at the "releasing" moment.

4.2. Solution of the problem of interaction with sledge located at the reservoir bottom

The problem was solved by using the natural elements method in the context of nondimensional variables. Figure 2b shows computational domain scheme, where d – underwater sledge height, d_0 – fluid depth above the sledge, A – arriving wave amplitude.

Figure 5 shows free surface evolution as time goes by. Wave blob occurs at the front boundary of the underwater sledge when the sledge experiences arriving wave uprush. Wave amplitude starts increasing, double bulge occurs on its surface, which later divides into reflected wave and transmitted wave. While moving along the channel transmitted wave shape transforms, the wave grows in the context of its amplitude and results in the clearly defined second wave that follows the first wave and drops behind the first wave because of the smaller amplitude and velocity. Principal waves are followed by disperse waves as well. It was mentioned in the experimental work [14].

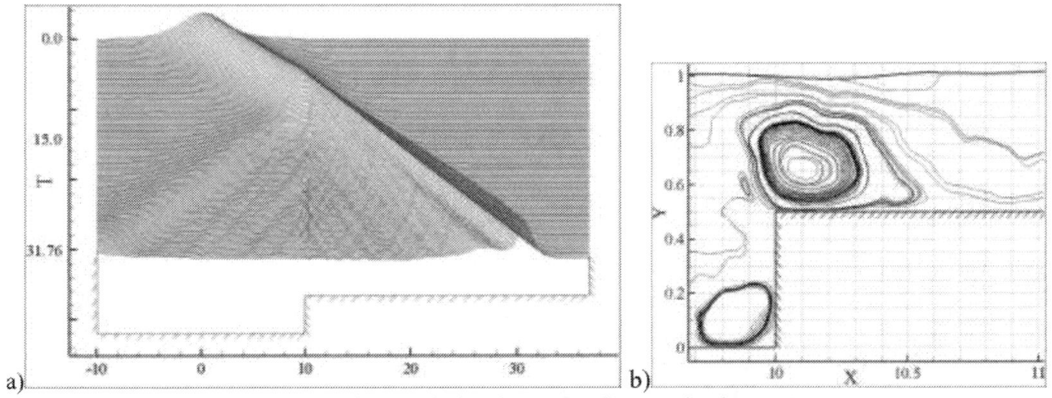

Figure 5. a) Free surface evolution; b) Vortex flows at the time *t=15.945 s*.

Table 1 shows amplitudes of transmitted waves (second and third columns) and reflected wave (forth column). The reflected wave amplitude was measured at the node with abscissa $x = 7$. The distance between the node with abscissa $x = 7$ and front boundary of the ledge equals the distance between amplitude sensor and the sledge in experiment [13]. Transmitted wave amplitude was measured at the node with abscissa $x=25$ that corresponds to last sensor location in the experiment carried out [13]. The table shows great numerical correspondence to experimental results.

Table 1. Wave amplitude

Amplitude	A_{t1}	A_{t2}	A_r
Experimental results [16]	0,2655	0,09	0,03
NEM method calculation	0,2703	0,0833	0,03036

Fig.5a shows that there is wave ripple on the surface of fluid in abscissa variation range $10 \le x \le 11$ (above the sledge). Vortex flow above the sledge (refer with: Figure 5b) results in wave ripple. Vortex existence is defined in experimental work [13] and in paper [14] though its impact on transmitted waves and reflected waves is not researched yet. Small vortex of low intensity in front of front sledge boundary occurs. Figures 5,6 show flow lines while amplitude soliton wave $A = 0.1825$ moving above the sledge $d = 0.5$. Vortex velocity rotor value above the sledge is negative, that means vortex rotating direction coincides with clockwise rotation.

5. Summary

The paper presents calculation results of the problem of interaction between viscous and ideal fluid and different obstacles. The calculation results show that mesh-free method of finite elements can be considered to be applicable to solve the problems described above. It is worth mentioning the result presenting hydrodynamic loads. Load distribution diagrams are essential values to identify contractions lifetime affected by fluid and possible destroying consequences caused by wave uprushes affecting coast and bottom constructions.

The research is based on state task No 2014/64, state project "Scientific researches organization".

References

[1] Connor J and Brebbia K 1979 *Finite Element Techniques for Fluid Flow* (London and Boston: Newnes–Butterworth)

[2] Patankar S V 1980 *Numerical Heat Transfer and Fluid Flow* (New York: Hemisphere Publ. Co.)

[3] Liu G R 2003 *Mesh free methods: moving beyond the finite element method* (London: CRC Press)

[4] Monaghan J. Smoothed particle hydrodynamics. *Ann. Rev. Astronand Astrophysics.* 1992. **No ~30,** 543–574.

[5] Frank A M 2001 *Discrete models of incompressible fluid (*Moscow: Fizmatlit)

[6] Facundo P. 2003 The meshless finite element method applied to a lagrangian particle formulation of fluid flows *Partial Fulfillment of the Requirements for the Degree Doctor of Philosophy. Instituto de Desarrollo tecnologico para la industria quimica (INTEC) universidad nacional del litoral noviembre* , 157 p.

[7] Onate E and Idelsohn S.R and Zienkiewicz O C 1996 A finite point method in computational mechanics. Applications to convective transport and fluid flow *International Journal for Numerical Methods in Engineering* 39 3839

[8] Traveroni L. Natural neighbor finite elements 1994 *In International Conference on Hydraulic Enginnering Software. Hydrosoft Proceedings 2.Computational Mechanics* Publications, pp. 291–297.

[9] Afanasiev K E, Karabtsev S.N., Rein, T.S., Stukolov, S.V. 2011 Numerical simulation of free boundary fluid flows by using mesh–free methods *Trudy X vserossiyskogo sezda po fundamentalnym problemam teoreticheskoy I prikladnoy mekhaniki*

[10] Skvortsov A V 2002 *Triangulyatsia Delone and its application* (Tomsk: Tomsk University)

[11] Sibson R.. Barnett In V. (ed.) A brief description a natural neighbor interpolation *Interpret multivariate data.* – Chichester: John Wiley, (1981)pp. 21–36.

[12] Belikov, V.V., Ivanov, V.D., Kontorovich, V.K., Korytnik, S.A., Semenov, A.U. 1997 *Nonsibson's interpolation– new method of inetrpolation of function's values on undefined point system* 37 11

[13] Seabra-Santos, F J and Renouard D P and Temperville A M 1987 *J. Fluid Mech.* 176 117.

[14] Liu P. L.-F. and Y. Cheng 2001 *Physics of fluids* **13** 1660.

Comparing thermal stimulation techniques in infrared thermographic inspection of corrosion in steel

A O Chulkov, V P Vavilov
National Research Tomsk Polytechnic University, Russia 634050, Tomsk, Lenin Av., 30
chulkovao@tpu.ru

Abstract
Remote detection of corrosion in metals is a developing application area of active thermal nondestructive testing. In this study, emphasis is made on the optimization of heating techniques that is of a major interest in practical surveys. Some popular data processing techniques, such as Fourier transform, correlation and principal component analysis, are also quantitatively compared in application to corrosion detection in 1-2 mm thick steel by applying a criterion of signal-to-noise ratio. The best inspection results have been obtained by using powerful halogen lamps and air blowers. Material loss of about 25 % with lateral dimensions greater than 10×10 mm can be reliably identified in practical tests. The use of Xenon flash tubes is inefficient because of significant steel thickness. LED panels have not provided expected results due to low absorption of LED quasi-monochromatic radiation.

Keywords: active thermal nondestructive testing, corrosion detection, thermal stimulation, signal-to-noise ratio, image processing

1. Introduction

Remote detection of corrosion in metals is one of quickly developing application areas of active thermal nondestructive testing (TNDT). The results of earlier publications can be reduced to the following statements [1–14]: 1) the most successful applications are related to thin (1-2 mm) aluminum panels used in aviation and thicker (1-6 mm) steel components used in power production and petrochemical industry (in some cases, material loss can be detected in 10-15 mm-thick steel), 2) aluminum specimens are to be "black-painted" to enhance optical absorptivity/emissivity, even if the coating may be not really black in the visual wavelength band, 3) in most cases, optical thermal stimulation, in particular, by applying powerful Xenon tubes and halogen lamps, is considered as most appropriate, 4) acquisition frequency should be quite high in the case of aluminum (30–200 Hz) and "medium" if one deals with steel (10-30 Hz), 5) the theory of TNDT is well-developed to reveal basic relationships between surface temperature and defect parameters [6, 7, 10–12]; if defect lateral size is 6-10 times greater than specimen thickness (1D defects), a rather simple inversion formula allows fairly accurate estimation of material loss: $\Delta L/L = 1 - T_{nd}/T_d$, where T_{nd}, T_d are the specimen excess temperatures in defect and non-defect areas respectively; 3D defects can be identified by introducing some correction coefficients [1] or general inversion formulas [12], 6) the detection limit of few percent of material loss has been reported in the inspection of aluminum [2–5], while, in the case of thick steel, the corresponding estimate is about 20 % [1, 8–11].

In this study, emphasis is made on the optimization of heating techniques that is of a major interest in practical surveys. Some popular data processing techniques are quantitatively compared in application to corrosion detection by using a criterion of signal-to-noise ratio (following the inspection philosophy described in [15, 16]).

2. TNDT of corrosion: recall to the basics

A 'classical' scheme of active one-sided TNDT is reminded in Figure 1. A sample with hidden corrosion is heated on the front surface (Figure 1*a*). Typically, a powerful pulse of optical radiation should be applied to quickly enhance surface temperature in non-defect and defect areas and then monitor these temperatures thermographically at the cooling stage; note that excess temperatures T_{nd} and T_d are counted from the initial temperature T_{in} that is typically the ambient temperature. Both defect detection and characterization are related to the analysis of differential temperature signals $\Delta T(\tau) = T_d(\tau) - T_{nd}(\tau)$ which evolve in time τ (Figure 1*b*). A maximum signal ΔT_m occurs at a particular time τ_m with both latter values being dependent on metal conductivity, material loss $\Delta L / L$, defect size and intensity of heat exchange with the ambient.

Three graphs in Figure 1*b* were calculated for the 1D material loss of 1 mm in a 5 mm-thick AISI 1010 steel by using Layer-3 Analytic software from Innovation, Ltd., Russia. In all cases, the injected energy was the same (10^4 J/m^2) but the heat pulse duration was 0.1, 1 and 10 s. In the first 2 cases, ΔT_m=0.30 °C, and only for the long (10 s) pulse the signal slightly decayed up to 0.28 °C.

If heating time τ_h is relatively long, the time instant τ_m may occur within the heat pulse, i.e. $\tau_m < \tau_h$, or immediately after heating stopped as shown in the middle and bottom graphs of Figure 1*b*. This is an undesirable situation because temperature signals are often corrupted by strong reflection noise. Therefore, either a heat pulse should be short enough or stimulating energy should be out of the spectral range of a used infrared (IR) imager.

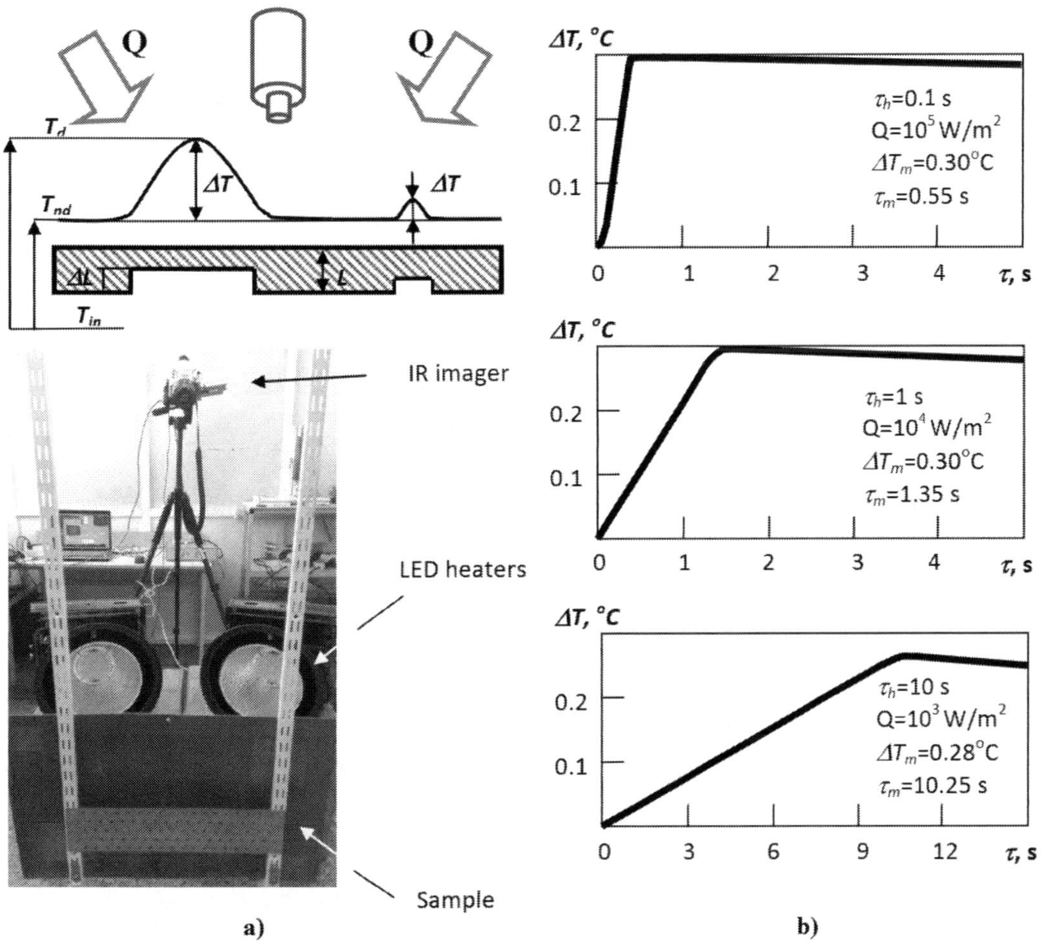

Figure 1. TNDT of hidden corrosion:
a – test scheme and experimental set-up,
b – evolution of temperature signals
(AISI 1010 steel, L= 5 mm, $\Delta L/L$=20 %)

3. Experimental setup and test samples

The photo of the experimental set-up used in corrosion detection is shown in Figure 1*a*. A test sample with rear-surface corrosion has been thermally stimulated on the front surface with its dynamic temperature distribution being recorded with a TH-9100 IR imager from NEC Avio (image format 320×240, temperature resolution up to 30 mK, maximum acquisition frequency 60 Hz). Thermal stimulation sources have been the following: 1) two Xenon flash tubes with the total energy of 3.2 kJ delivered for 5 ms, 2) two halogen lamp sources with the total power of 2 and 30 kW, 3) two LED panels with the total power of 1 kW, 4) two air blowers with the total power of 4 kW.

Two 1 and 2 mm-thick reference samples made of AISI 1010 steel have been used to evaluate efficiency of corrosion detection by applying the above-mentioned heat sources. Some square-shaped bottom "holes" were manufactured to simulate corrosion areas of varying size and material loss (Figure 2*a*). Both samples were painted with a yellow acrylic dye to simulate containers used in nuclear engineering for keeping radioactive wastes (perspective application area for active IR thermography).

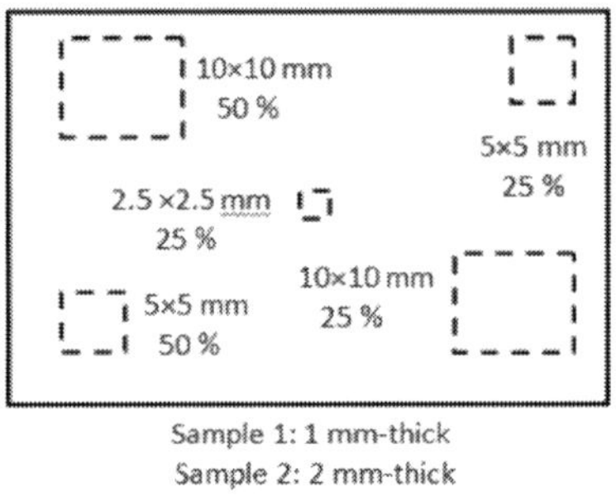

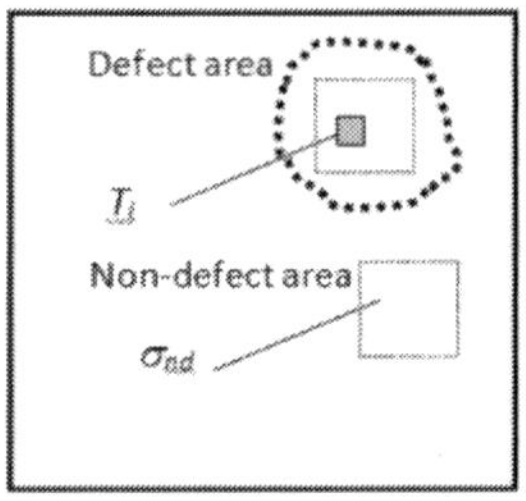

a) **b)**

Figure 2. Reference sample scheme (a) and signal-to-noise concept (b).

4. Results and discussion

In TNDT, optimization of means of thermal stimulation is typically done on reference samples by using the signal-to-noise ratio criterion; also, mass/size parameters may be involved in consideration. Some other factors can be also taken into account, such as safety requirements; for instance, heaters with open filament may be not allowed for use close to petrochemical tanks.

Signal-to-noise ratio S is a convenient optimization parameter to compare inspection procedures and processing algorithms. When calculating S, one must choose both defect and non-defect areas thus making S as quantitative but subjective parameter (Figure 2*b*). The simplest formula for S is:

$$S = (\overline{T}_d - \overline{T}_{nd}) / \sigma_{nd}, \qquad (1)$$

where $\overline{T}_d$, $\overline{T}_{nd}$ are the mean temperatures, and σ_{nd} is the standard deviation of temperature in a non-defect area. The disadvantage of Eq. (1) is that, in some special cases, separate sections of defect areas can provide signals both higher and lower than the background thus decreasing their 'arithmetic' contribution to S. A more correct expression would be as follows:

$$S = [\sqrt{\sum_{i=1}^{M} (T_i - \overline{T}_{nd})^2 / M}\,] / \sigma_{nd}, \qquad (2)$$

where T_i is the temperature (or any derivative signal) in the i-th pixel of a defect area, and M is the number of pixels in a chosen defect area.

Experimental results are illustrated with images in Figure 3. When applying flash heating, the acquisition frequency was 60 Hz (300 images in a sequence), in all other cases – 10 Hz (150 images in a sequence). All sequences have been processed by using the algorithms which are common in TNDT, namely: Fourier transform in time (pulse phase thermography), singular value decomposition (principle component analysis - PCA) and temporal correlation. To calculate S values, the corresponding square-shaped areas of interest were chosen on IR thermograms as shown in Figure 3.

The left column of images in Figure 3 is related to Sample 1 and the right one – to Sample 2. We remind that both sample contained the areas of simulated corrosion with the same material loss, so the only difference between results was in sample thickness. Visual perception of data in Figure 3 allows the following conclusions: 1) in accordance with the TNDT theory, the corrosion areas look warmer than the non-defective ones in raw images, while the processed images are characterized by 'defective' signals both higher and lower than the background, 2) all image processing techniques are based on the analysis of temperature evolution in time and lead to considerable reduction of uneven heating phenomena, 3) the visually best results appear when stimulating samples with the 30 kW halogen lamps and the 4 kW air blowers (in the latter case, absorptivity/emissivity phenomena are less significant in comparison to optical heating), 4) independently on sample thickness, material loss of 25 % with lateral dimensions greater than 10×10 mm can be reliably identified; smaller size defects can be also seen but probability of correct detection may be low.

Sample 1 (1 mm-thick) **Sample 2 (2 mm-thick)**

Heating with 2 halogen lamps 2×1 kW for 10 s

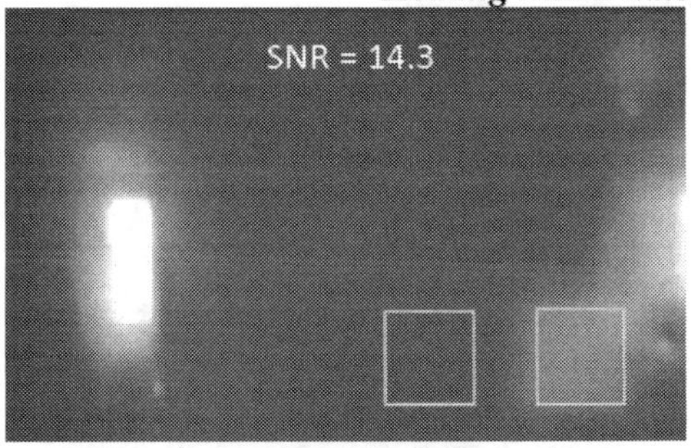

Raw image (sample deformation is seen on the left)

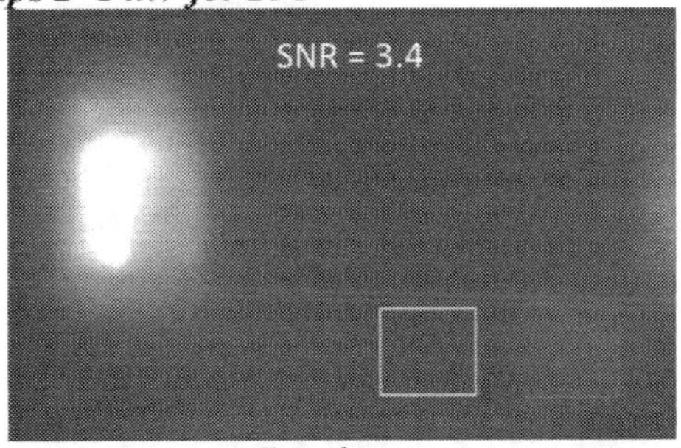

Raw image

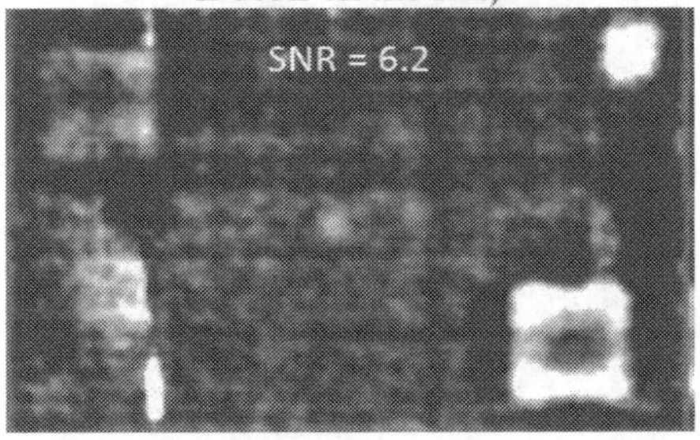

PCA image

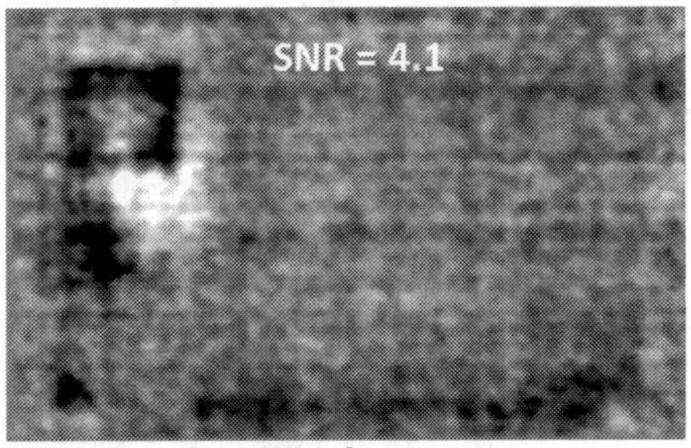

PCA image

Heating with 6 halogen lamps 6×5 kW for 3 s

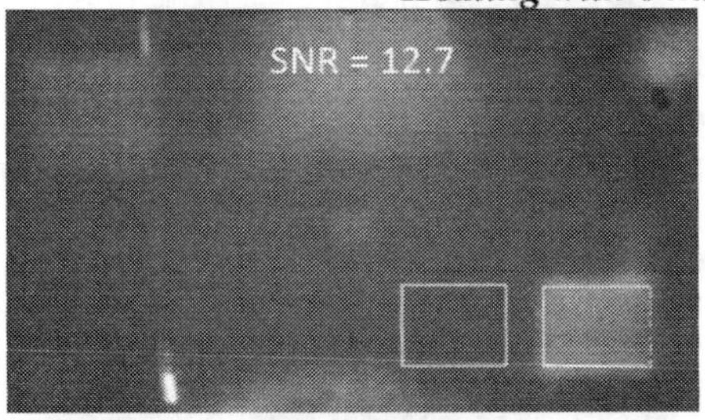

Raw image

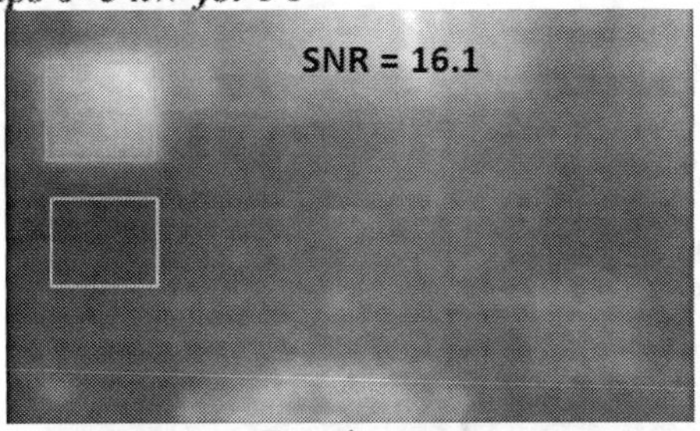

Raw image

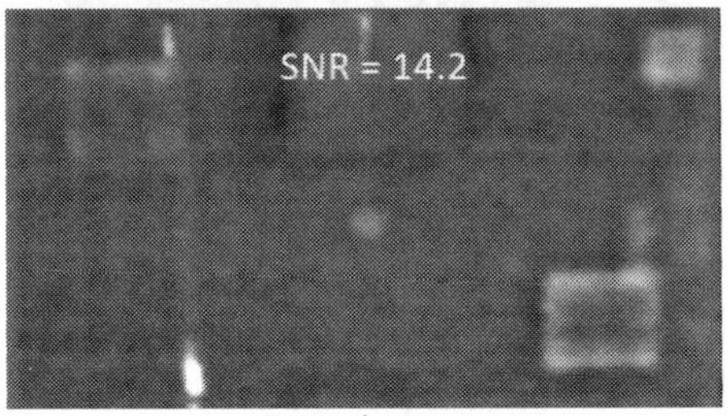

PCA image

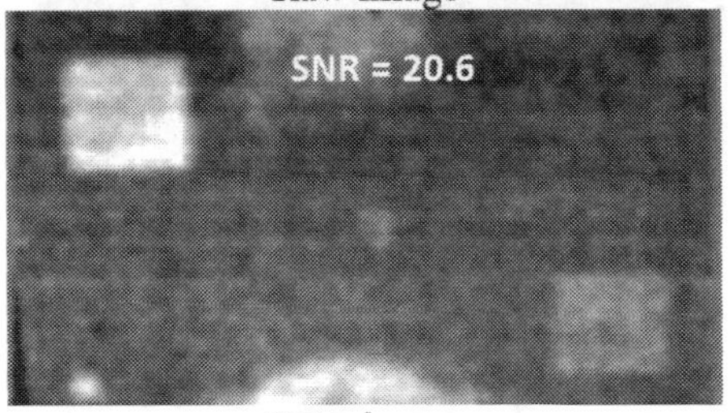

PCA image

Figure 3. Image processing in the inspection of steel reference samples (to continue)

IOP Conf. Series: Materials Science and Engineering **81** (2015) 012100 doi:10.1088/1757-899X/81/1/012100

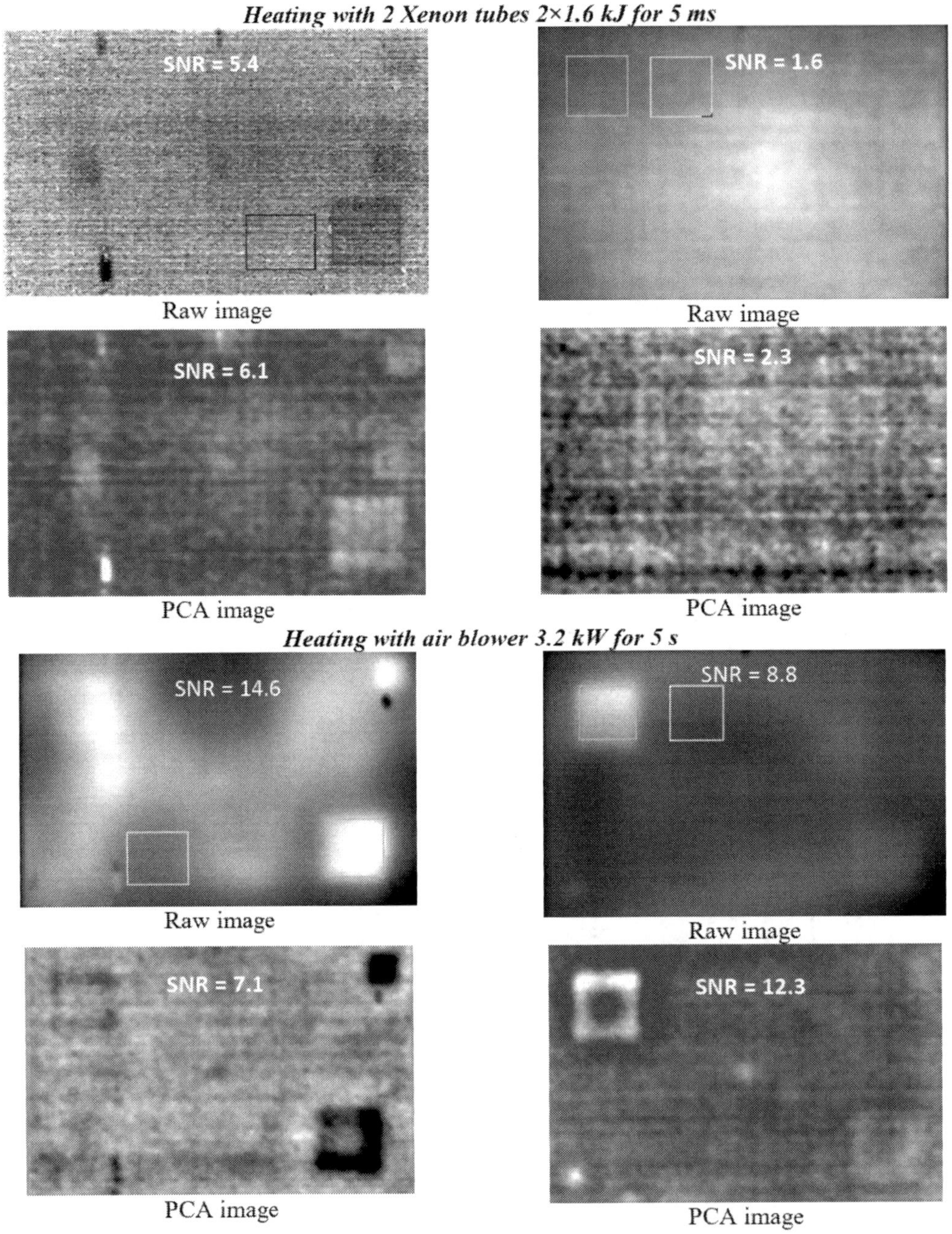

Figure 3. Image processing in the inspection of steel reference samples (to continue)

Heating with 2 LED panels 2×500 W for 5 ms

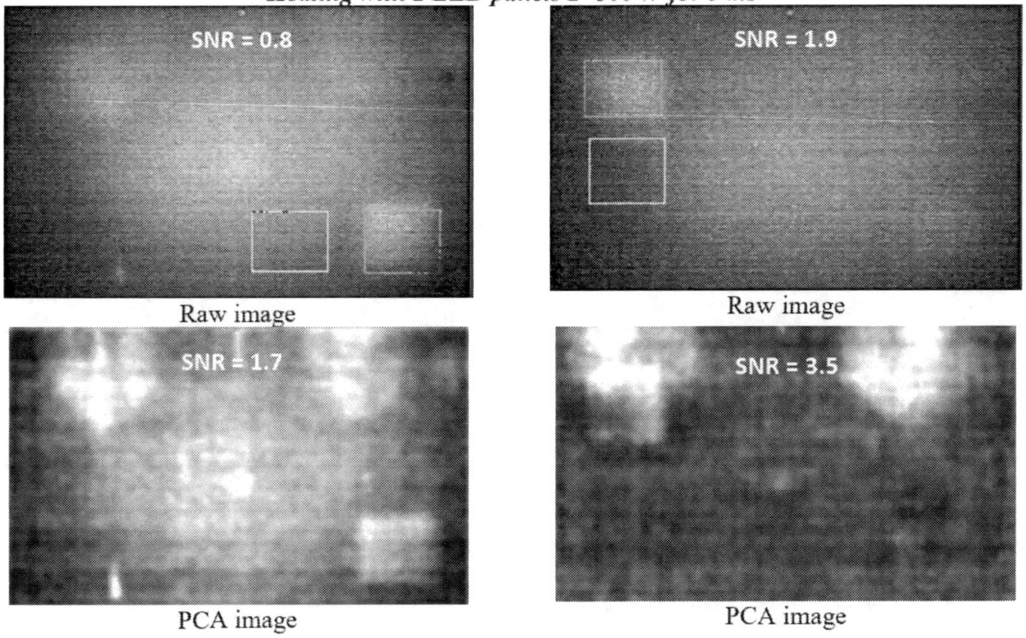

Figure 3. Image processing in the inspection of steel reference samples (continued)

The evaluation of experimental results by signal-to-noise ratio that is presented in Figure 4 leads to the same conclusions. The highest *S* values have been from 11 to 20,6 thus ensuring reliable detection of the corresponding defects. It is interesting that in some cases *S* values are higher in raw images than after processing but the visual perception of images is always in favor of image processing. This can be explained by a relatively small size of chosen defect and non-defect areas, while advantages of image processing appear in the processing of large format images. Besides, S values involve single pixel amplitudes without taking into account image texture heuristically evaluated by an operator.

The low efficiency of Xenon tubes can be explained by the significant thickness of the steel samples; this stimulation technique seems to be fairly good in the inspection of thin aluminum sheets. It is important noting the unsatisfactory results obtained with LED panels. This finding does not support the hopes of improving test results by using quasi-monochromatic LED radiation of which spectrum (about 0.7 μm) is far away from the spectral range of typical IR imagers (2–5 and 7–13 μm) thus producing no reflection noise. However, in our case, the absorption of the LED radiation by yellow-painted samples has been too low to build-up appropriate temperature signals.

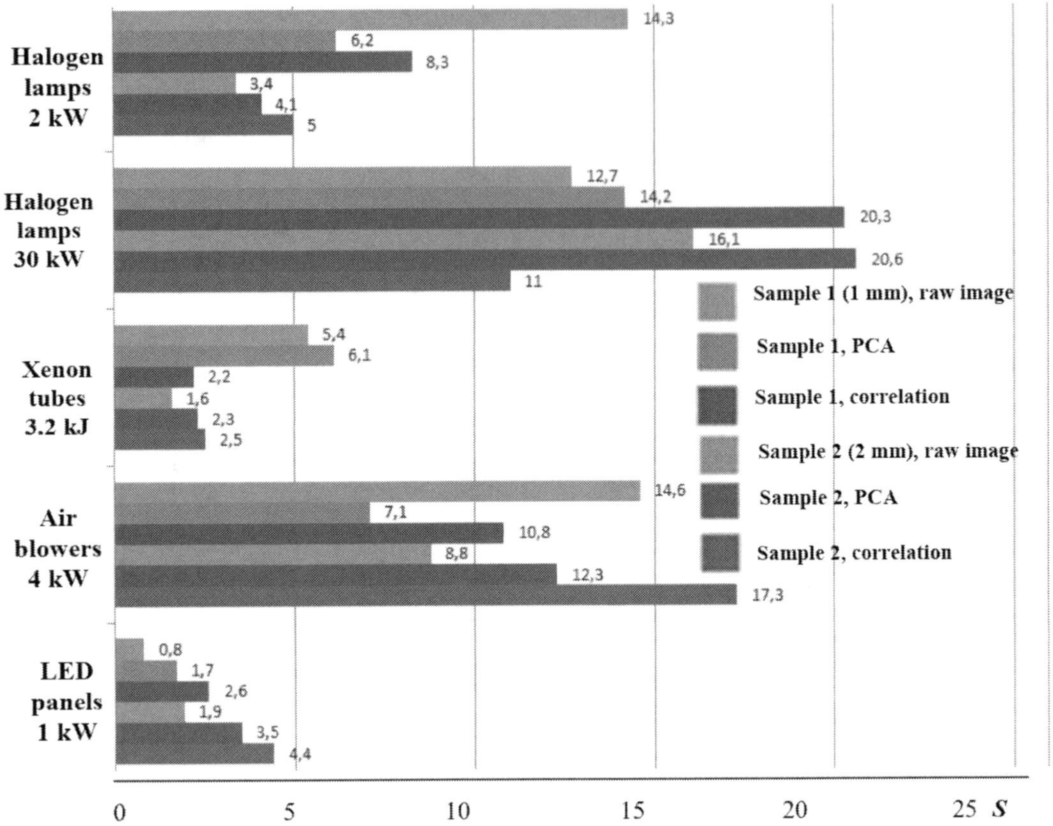

Figure 4. Evaluating thermal stimulation efficiency in corrosion detection.

5. Conclusion

The practical efficiency of several thermal stimulation techniques has been evaluated in the detection of hidden corrosion in 1–2 mm thick steel samples. Signal-to-noise ratio has been chosen as a comparison criterion to show that the best results are ensured when using powerful halogen lamps and air blowers (a required stimulation power is of some kW). Material loss of about 25 % with lateral dimensions greater than 10×10 mm can be reliably identified in practical surveys. The use of Xenon flash tubes with energy of few kJ is inefficient because of significant steel thickness. LED panels have not provided expected results due to low absorption of quasi-monochromatic LED radiation.

This research was supported by the NIR # 445 (ONG), State order of the Russian Ministry of Higher Education for 2014-2016.

References

[1] Vavilov V P, Chulkov A O, Burleigh D. 2012 *Proc. SPIE Thermosense-XXXIV* vol. 8354 pp 835401–835409
[2] Prabhu D R, Winfree W P 1993 *Rev. Progress in Quant. NDE* vol. 12 pp 1260–1265
[3] Syed HI, Winfree W P, Cramer K E, Howell P A 1993 *Rev. of Progress in Quant. NDE* vol.12 pp 724–729
[4] Alcott J 1994*Mater. Evaluation* **5** pp 64–73
[5] Prati J 2000 *Proc. SPIE* vol. 4020 pp 200–209

[6] Vavilov V, Grinzato E, Bison P G, Marinetti S and Bales M 1996 *Intern. J. Heat & Mass Transfer* vol. 39 pp 355–371
[7] Grinzato E and Vavilov V 1998 *Rev. Generale Termique* vol. 37 **8** pp 669–679
[8] Vavilov V, Shiryaev V and Grinzato E 1998 *Insight* vol. 40 **6** pp 408–410
[9] Vavilov V P, Nesteruk D A, Chulkov A O, Shiryaev V V 2013 *Rus. J. NDT* vol. 49 **11** pp 619–624
[10] Grinzato E, Bison P, Marinetti S and Vavilov V 2006 *Proc. SPIE* **6205** 620517
[11] Grinzato E, Vavilov V, Bison P G and Marinetti S 2007 *Infrared Physics & Technology* 49 pp 234–238
[12] Marinetti S and Vavilov V 2010 *Corrosion science* vol. 52 **3** pp 865-872
[13] Lugin S and Netzelmann U 2007 *NDT & E Intern* vol. 40 pp 220–228
[14] Woolard D F and Cramer K E 2004 *Proc. SPIE Thermosense-XXVI* vol. 5405 pp 366–373
[15] Vavilov V P 2011 *Proc. of the Fall 2011 ASNT Conf.* 8 p.
[16] Albendea P, Madruga F J, Cobo A, López-Higuera J M 2010 *Proc. 10th Intern. Conf. on Quant. Infr. Therm.* 8 p

Application of reconfigurable computing environments for image processing in X-ray tomography of materials

D V Shashev[1], S V Shidlovskiy[1,2], V I Syriamkin[1,3], A V Yurchenko [4]

[1] Department of Innovative Technologies, National Research Tomsk State University, Lenina Avenue, 36, Tomsk, RU
[2] Institute of Power Engineering, National Research Tomsk Polytechnic University, Lenina Avenue, 30, Tomsk, RU
[3] Institute of Cybernetics, National Research Tomsk Polytechnic University, Lenina Avenue, 30, Tomsk, Russia
[4]Institute of Non-Destructive Testing, National Research Tomsk Polytechnic University, Lenina Avenue, 30, Tomsk, Russia
E-mail: dshashev@mail.ru

Abstract. The article investigates the application of reconfigurable computing environments for image processing purposes, specifically for X-ray images.

1. Introduction

The methods of non-destructive examination of internal structure of objects are directly related to a variety of both scientific and applied domains (robotics, medicine, construction, aircraft and ship industry, etc.). The use of X-ray remains the most convenient and in some cases the only approach to such examinations.

One of the main objectives so far is still the making of X-ray microscopes having higher resolution than the visual range, but unlike electronic microscopes not demanding pre-treatment of samples [1].

Contemporary methods of X-ray tomography are based on the application of high-performance computing data processing methods to tomography data (figure 1).

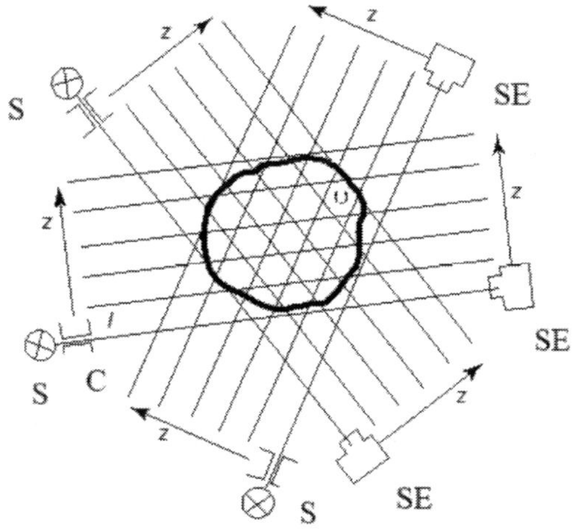

Figure 1. Tomographic scanner schematic.

A narrow beam of X-ray from the source 'S', formed by collimator 'C', passes through the object 'O', after which it is registered by the sensor element 'SE'. Upon simultaneous movement of the source and the detector along the given direction 'z', a sequential scanning of all the areas of the object is performed with the connection between radiation intensity '*I*', registered by SE, and linear absorption factor of the object's medium μ has the form of integral equation [2]:

$$I(z) = I_0 \exp\left[-\int_l \mu(z,l)dl \right],$$

where I_0 – incident radiation intensity, dl – absorption path element along beam '*l*' which corresponds to the scanning direction. The measurements are repeated for several scanning directions with respect to the object. In order to reconstruct the distribution of μ and accordingly the density and material composition in the volume of the object, special algorithms of computerized data processing are used.

After the scanning itself, at the output of a microtomograph we get a bulk flow of data in the form of X-ray images of the object, which should be treated as quickly and as thoroughly as possible to enable the analysis of the object with respect to various defects. This problem is efficiently solved by the application of *reconfigurable computing environments*.

2. Reconfigurable computing environment for binary image processing

The concept of reconfigurable computing environments (RCE) enables the solution of a vast number of scientific and industrial problems. In a special case this concept may be efficiently used to solve binary image processing problems. It should be noted that nowadays the digital image processing is one of innovative and rapidly developing scientific domains, essential for a vast variety of scientific disciplines such as robotics, medicine, aerospace and geophysical surveys, etc.

A reconfigurable computing environment is a discreet mathematical model of a high-performance system made of an array of identical computers with identical connection to one another, capable of dynamic adjustment to perform any preset logical functions [3–5].

The peculiarity of the reconfigurable computing environment in question is that every computer operates a given basic set of logical functions, specifically AND, OR, NOT.

Figure 2(a, b) shows an abstract 3D-model of an RCE setup for specific operation, while figure 2 (c, d) shows a generic scheme of an RCE built of these computers.

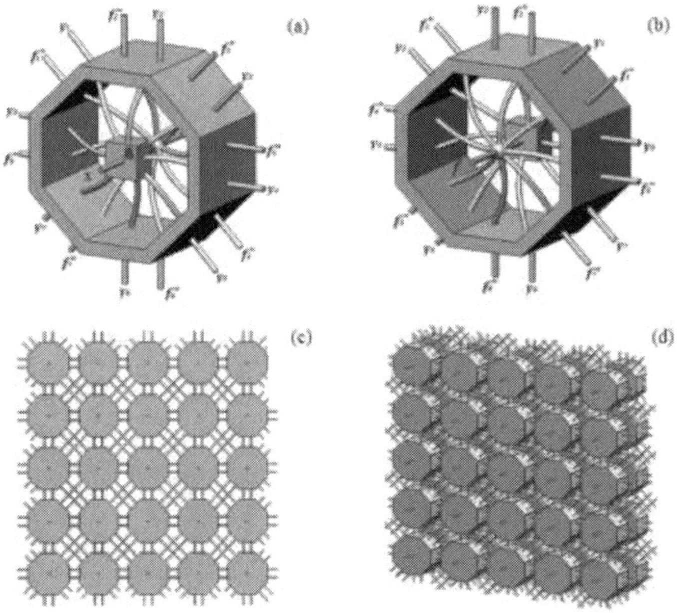

Figure 2. Abstract 3D-model of an RCE and a generic RCE scheme.

It should be noted that as seen in figure 2(d), the RCE may have a multi-layered structure. Such structure is conditioned by the complexity of performed operations.

2.1. Morphological binary image processing methods

In the frameworks of this paper, of interest are the morphological methods used for image processing, e.g. morphological filtering, refinement, contouring of a given object.

Mathematical morphology enables the view of a binary image (i.e. the one containing pixels with brightness values of 0 and 1 for background pixels and object pixels respectively) as a multitude of foreground pixels (having brightness values of 1), contained in the space of Z^2 [6]. Accordingly, the operations with multitudes can be performed with images.

Morphological operations are performed with two images: the one being processed and the special, which depends on the kind of operation and the given problem. Such special image in mathematical morphology is called a structure element or a primitive. The size of the structure element is usually 3×3, 4×4, and 5×5 pixels. This is conditioned by the very idea of morphological processing wherein typical details of an image are searched for. The sought-for detail is described by a primitive and morphological processing allows for highlighting or deleting such details in the whole of the processed image.

Set-theoretic operations form the main morphological image processing operations: erosion, dilation, open, close [6–8].

The operations of *dilation* and *erosion* are fundamental for morphological image treatment.

2.1.1. Erosion. Let us assume that A and B are multitudes from the space of Z^2. Erosion of multitude A with respect to B is denoted $A\,!\,B$ and is determined as

$$A\,!\,B=\left\{z\,|\,(B)_z\cap A^c=\varnothing\right\},$$

where A^c – negation of A; $(B)_z$ – parallel transfer of multitude B into point $z=(z_1,z_2)$ (shift to point z), i.e. the narrowing of multitude A with respect to primitive B is the multitude of all such z points, that primitive B would entirely be contained in A if shifted into them. In practice the operation of erosion 'narrows' or 'details' binary image objects. The method and the degree of narrowing is controlled by primitive B.

2.1.2. Dilation. Let us assume that A and B are multitudes from the space of Z^2. The dilation of multitude A with respect to multitude B is denoted and determined as follows:

$$A \oplus B = \left\{ z \,|\, \left(\hat{B}\right)_z \cap A \neq \varnothing \right\},$$

where $\hat{B}_z$ – central reflection of B (with respect to its origin of coordinates) with subsequent parallel transfer to point $z = (z_1, z_2)$ (shift into point z), i.e. the dilation of multitude A with respect to primitive B is the multitude of all such shifts of z, when the multitudes of $\hat{B}$ and A coincide at least in one element. In practice the operation of dilation 'builds-up' or 'thickens' binary image objects.

It should be noted that more often than not the operations of erosion and dilation are used together when processing the image. This is why morphological operations are also joined by the operations of *open* and *close*.

2.1.3. Open. The open of multitude A with respect to primitive B is denoted as $A \circ B$ and determined by the equation

$$A \circ B = \left(A \,!\;\; B\right) \oplus B$$

or by the equivalent formula

$$A \circ B = \bigcap \left\{ (B)_z \,|\, (B)_z \subseteq A \right\}.$$

In practice the operation of open smoothes the outlines of an object, quenches straits, and eliminates thin peaks.

2.1.4. Close. The close of multitude A with respect to primitive B is denoted $A \cdot B$ and determined by equation

$$A \bullet B = \left(A \oplus B\right) ! \;\; B.$$

In practice the operation of close smoothes the outlines of an object and in general 'fills' narrow gaps and long and thin cavities, it also eliminates small apertures and fills the gaps in the outline.

2.2. Modeling results

The reconfigurable computing environment for morphological binary image processing purposes was built and modeled in MATLAB Simulink software.

The general arrangement of an RCE computer and its structure in MATLAB Simulink are shown in figure 3.

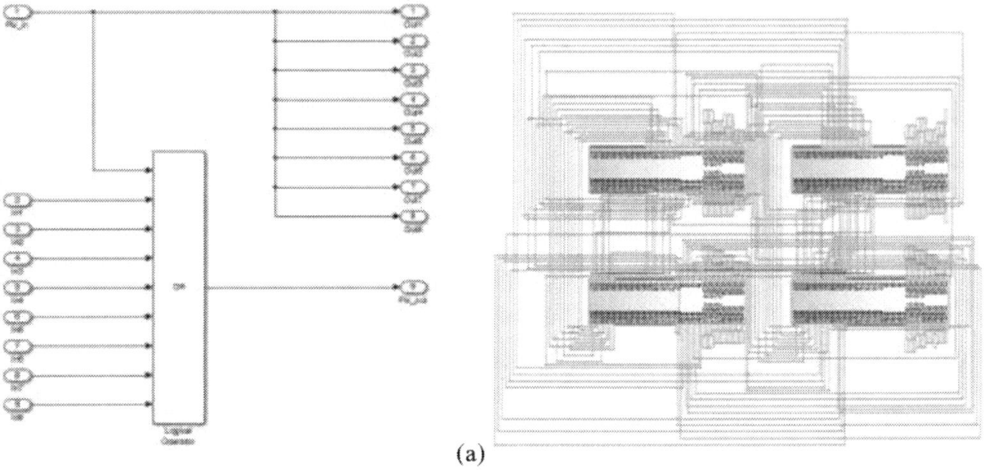

(a) (b)

Figure 3. RCE general arrangement (a) and RCE model structure (b) in MATLAB

As seen in figure 3(a) the model of an RCE computer for image processing purposes is a structure where the central component is a given logical function – in the case depicted in the figure it is OR – but in general the computer may be given have a series of logical functions or complex structures made of logical functions. The structure of the computer also contains a series of input and output tie-lines the number of which may vary in a wide range, depending on the given problem. Based on the specific case shown in the figure, we have 1 input and 1 output (Pix_in and Pix_out respectively) for a pixel of processed image and 8 inputs and outputs each for the tie-lines interconnecting the computers of the RCE.

Thus the structure of the end RCE model adjusted for specific image processing problem looks as shown in figure 3(b). It includes a large number of computers, interconnected by tie-lines, and each of the computers processes 1 pixel of the source image during one operating cycle of the computer.

The resulting RCE model for morphological binary image processing was tested using an X-ray image of a defective welding seam (figure 4).

More often than not, the end goal of digital image processing is a high quality identification of the object of interest from the source image for its subsequent analysis and decision making in case of an automated system. As seen in figure 4 after binarization of a gray scale image (figure 4(b)) we got a nondistinct image of the seam itself having many additional 'pixel' objects. Using the model RCE we got a homogeneous and distinct from the background object of interest which is suitable for subsequent analysis.

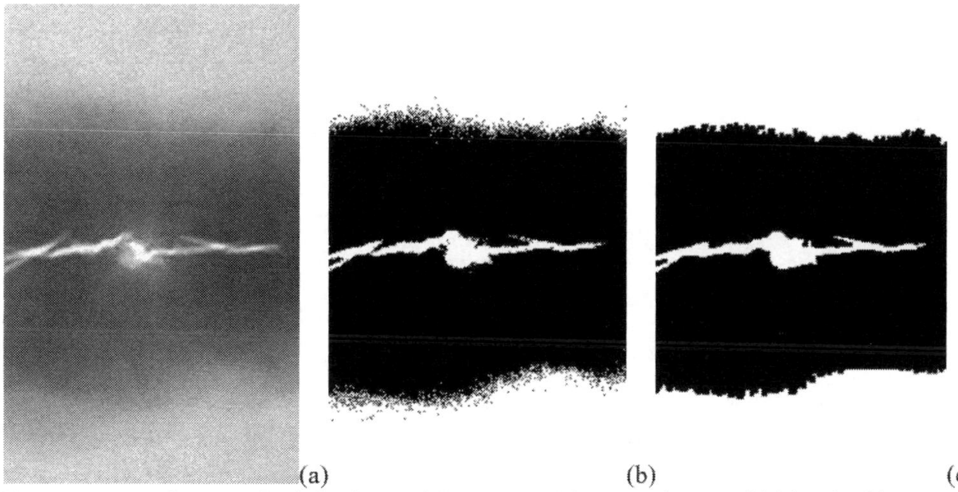
(a) (b) (c)

Figure 4. X-ray image of a defective welding seam: (a) source image; (b) binarized image; (c) image treated with morphological methods

3. Conclusion

Nowadays when solving a vast number of diverse problems one confronts a series of challenges such as: the use of high-performance computing systems with a parallel functioning architecture, the real-time processing of vast and voluminous flows of date.

A quality solution of these problems is achieved by the use of the concept of reconfigurable computing environments due to the advantages in the build-up of such systems. The article has shown an efficient use of RCE for binary image processing purposes with X-ray images.

This work was supported by the grant from the President of the Russian Federation (MD-411.2014.9) and was made on the program improving the competitiveness of National Research Tomsk State University.

References

[1] Webb S 1988 *The Physics of Medical Imaging* (Bristol, Philadelphia: Adam Hilger) p 103
[2] Troitskiy I N 1989 *Statisticheskaya teoriya tomografii* (Moskva: Radio i svyaz) pp 45–49
[3] Shidlovskiy S V 2006 *Avtomaticheskoye upravleniye. Perestraivayemyye struktury* (Tomsk: Tomskiy gosudarstvennyy universitet) p 22
[4] Shidlovskii S V 2006 Logical control in automatic systems *JCSSI* **45** pp 282–286
[5] Rusyn B, Kuzio M, Shmoylov V 2000 Reconfigurable high-performance computing systems based on homogeneous media *Auto. Cont. Comp. Scien.* **34** pp 58–65
[6] Gonzalez R C and Woods R E 2008 *Digital Image Processing* (New Jersey: Prentice-Hall) pp 519–532
[7] Shapiro L, Stockman G 2001 *Computer Vision* (New Jersey: Prentice-Hall) pp 58–102
[8] Seibold P, Gerke M, Masar I, Jelenciak F, Bahnik P, Bahr L 2014 Detection of power lines by image processing *Computational Vision and Medical Image Processing IV (Funchal, POR, 14–16 October 2013* (Florida: CRC Press) pp 203–209

Dynamic simulation of electrons in the injector of sealed vacuum chamber of industrial betatrons

A P Shestak, E L Malikov, K V Sukharnikov, M M Rychkov, M V Kuimova
National research Tomsk polytechnic university, Russia, Tomsk

Abstract. The paper researches some difficulties that emerge in the production of series small-sized betatrons. In particular, disadvantages of electron injector construction and attributed problems are described. Analysis of the results of mathematical simulation of the dynamics of electrons in the injector of the sealed acceleration vacuum chamber of serial betatrons was carried out. The problem areas were highlighted in the injector of electrons requiring to be improved.

Keywords: betatron, injector of electrons, acceleration vacuum chamber.

1. Introduction

In the tool base of industrial crack detection, betatron (as a source of continuous X-ray radiation) plays a significant role, especially when it comes to inspection at steel thicknesses exceeding 70 mm. The current state of unit record equipment allows using sources of low intensity, which is the betatron, for non-destructive testing of large-sized objects, since betatron can produce a sufficiently high energy of bremsstrahlung in comparison with an X-ray generator (there are industrial patterns of betatron with finite energy of electrons up to 25 MeV) [1]. In addition, in the last ten to fifteen years, betatron, as a type of bremsstrahlung source, found its application in a large number of other spheres. As examples of betatron application, one can mention the production of high-energy industrial tomographs [2], as well as the inspection system of vehicle transport [3]. Besides applications listed above, there are a few prototype models of medical betatrons for oncology diseases therapy [4]. Therefore, the task of enhancement of betatrons and improving their dose characteristics continues to be of current interest.

2. Theoretical part

Betatron is an accelerator with weak-focusing, whereby the final dose rate of both bremsstrahlung and outputted electron radiation (in the case of medical betatron) strongly depends on the conditions of injection and electron capture in the acceleration. Therefore, increasing the stability and efficiency of the system of betatron injection and electron injector is a very important task.

The sealed vacuum acceleration chamber of modern betatron is a glass bulb of toroidal-shape with two bends (Figure 1); the getter pump is installed in one bend, and high voltage electron injector is installed in the other.

Figure 1. Sealed vacuum acceleration chamber of 6 MeV betatron.

The acceleration vacuum chamber is installed into the working pole gap of a betatron so that the output slit of electron injector is within the radius of injection, i.e. in the position of maximum electron capture in acceleration. The capture efficiency is mainly determined by the electron optics of the injector. This dependence imposes strict requirements on the dimensional tolerances in the manufacture of injection unit of the betatron acceleration chamber. The production of the injector unit is relevant to "adaptor" assembly; i.e., the main splits between the anode and the focusing electrode are given by special technological mountings. However, the existing construction has a parameter whose value is visually monitored. It is referred to the depth of the cathode installation in the focusing electrode (parameter «α» Figure 2). Also, there is another problem associated with the cathode. Injector cathode of electron betatron relates to a filamentary type and because of design features is subjected to thermal expansion, which results in distortion of electron optics of the injector as a whole.

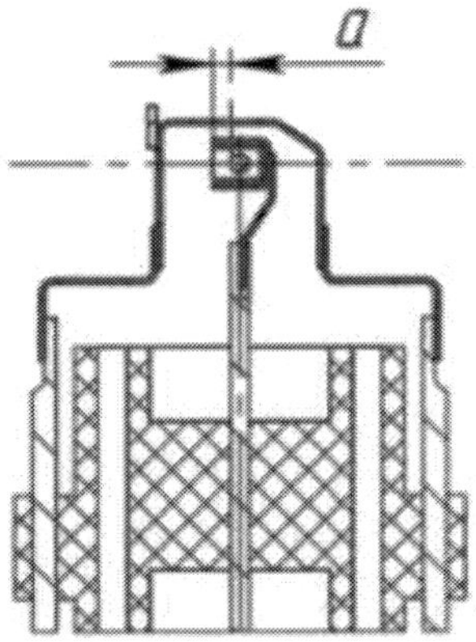

Figure 2. General form and cut of electron injector of betatron.

3. Simulation

Mathematical simulation was carried out with the finite-difference method of the dependence of the coefficient of injector's current transmission on the depth of the cathode installation into the focusing electrode for the impact assessment of this parameter on the optics of the injector. In this case the current transmission coefficient is the ratio of the total charge of the electrons emerging from the injector to the total charge of the electrons that overcome the accelerating gap of the injector, some of which face the anode.

Basic design elements that determine the electron optics were used in the geometric model of the injector. To simplify calculations the spiral cathode was replaced with the cylindrical of the same dimensions. Another simplification relates to the absence of the magnetic field. On the one hand, such simplification is justified by the fact that during the injection, the value of the magnetic field induction corresponds to the radius of electron injection, which is, for example, 78 mm for a betatron of 7 MeV, while the accelerating split in the injector is 3.4 mm. Thus, the electron beam, accounting for the magnetic field, will deviate from the axis of the focusing inside the injector at a distance equal to 0.074 mm (the size of the anode slit trench is 1.5 mm). On the other hand, the magnetic field is not provided on the specially designed vacuum stand to check the results of modeling (including and due to the above-described reason).

The following electrical parameters were based on the mathematical model: injection current is 1 A; accelerating difference of electric potential is 40 kV. It is worth mentioning that in the model as well as in a virtual injector, the anode and the focusing electrode are at the same electric potential. This fact negatively affects the focusing properties of the injector and as practice shows [5] considerably reduces the variation limits of deviation of the cathode's size landing inside the focusing electrode.

As a result of simulation we obtained the dependence of the coefficient of the current transmission Ki of injector on α is installation depth of the cathode into the focusing electrode. The reading of distance α was conducted from the exit plane of the focusing electrode to the parallel plane of the nearest surface of the cylindrical cathode (Figure 2). In simulation, the value of the parameter α varied in the range from 0 to 2 mm and at a pitch of 0.1 mm. The result of simulation is graphically presented in Figure 3.

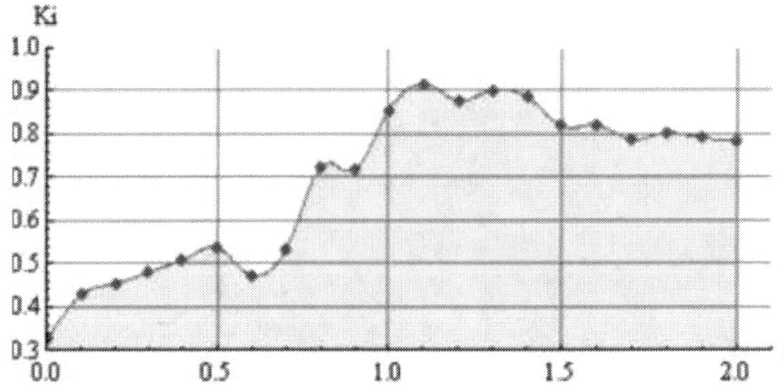

Figure 3. Graphical view of dependence **Ki(a)**

It is clearly seen that the maximum of the coefficient of current transmission is 0.92 and occurs at the installation depth of cathode of 1.1 mm, implying that about 8% of the electrons are lost on the anode. It is important to emphasize that the current between the focusing electrode and anode $\mathbf{I_{fe}}$ is not equal to the emission current from the cathode, as a portion of the electrons emitted from the cathode may undergo collision both with the cathode and the focusing electrode. Also, the number of these electrons increases with increased depth of cathode installation (Figure 4), and the current value $\mathbf{I_{fe}}$ decreases.

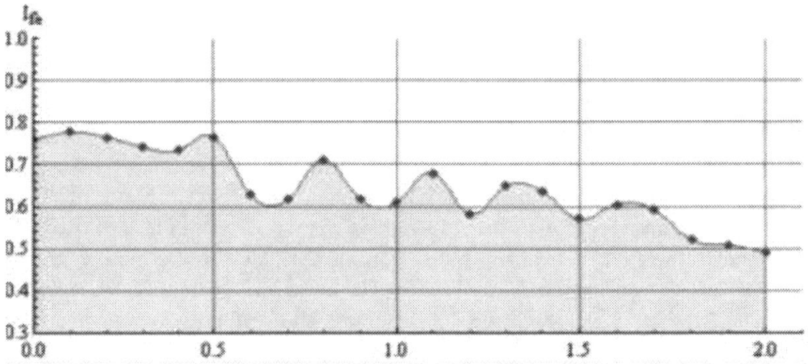

Figure 4. Graphical view of dependence $\mathbf{I_{fe}(\alpha)}$

As a result of the described processes, the current of the injector $\mathbf{I_{inj}}$ has the dependence $\mathbf{I_{inj}}(\alpha)$. Figure 5 shows that the maximum current of the injector occurs at the depth of cathode installation, 1.1 mm.

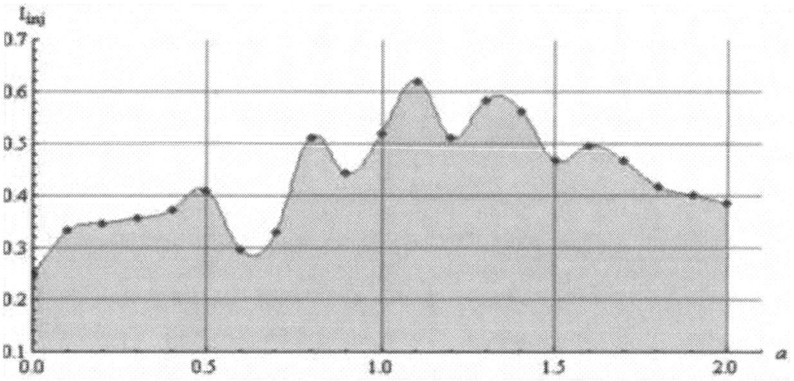

Figure 5. Graphical view of dependence $\mathbf{I_{inj}}(\alpha)$

The low level of injection current at cathode installation depths less than 1.1 mm is explained by the steradian predominance of collisional process of electrons with the anode. Figure 6 shows the distribution of the current density of the injector in the horizontal plane of the interpole space of a betatron, where $\alpha = 0$. And fall of current density for α more than 1.1 mm is due to the increasing number of collisions of electrons with the cathode and the focusing electrode (Figure 7).

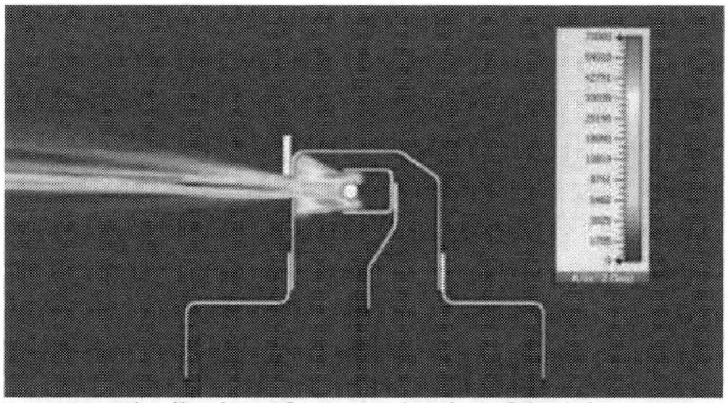

Figure 6. Distribution of current density of the injector at $\alpha = 0$.

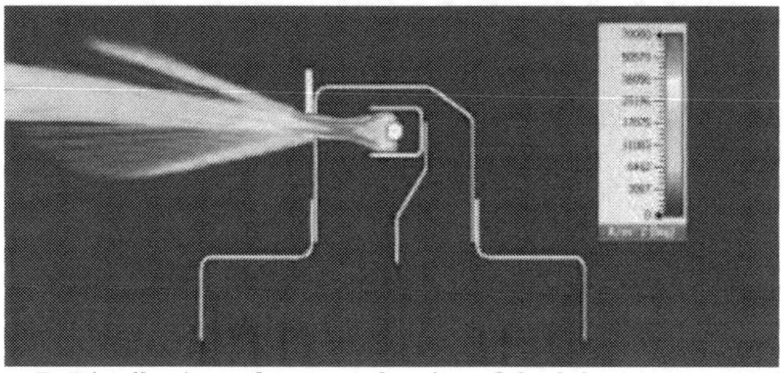

Figure 7. Distribution of current density of the injector at $\alpha = 1.2$ mm.

4. Conclusion

As the result of mathematical simulation, we obtained the dependence of the coefficient of the current transmission of the injector on the depth of the cathode installation into the focusing electrode. This dependence of the coefficient is compatible with experimental data in [5]. In addition, the graphical view of this relationship (Figure 3) allows setting the acceptable limits of the installation of the cathode into the focusing electrode. With **Ki(α)** and $\mathbf{I}_{inj}(\alpha)$ for the independent parameter **α,** one can set the following value with deviation limits from: $\alpha = 1,1^{+0,2}_{-0,1}$ **mm**. Within the range of values of parameter **α**, the coefficient of current transmission varies by no more than 6% from optimum. If parameter **α** deviates above specified limits, the value of the coefficient of current transmission falls below 85%.

We should not forget about another problem of currently available electron injectors. Deformation of the filamentary cathode due to thermal expansion can introduce significant distortions in the optics of the injector. The method by which the cathode of the injector for the betatron is produced, includes the process of carbonization of the cathode blank from thoriated tungsten wire. Thus the degree of carbonization should not be less than the value specified in the technical specifications (i.e., it may differ from the values of certain specifications upward). As a result of this manufacturing process, the structural stiffness of the cathode varies, and the corresponding deformation due to thermal expansion of a single cathode is not predictable.

Described in this article are drawbacks of currently available betatron injector technology, which lead directly or indirectly to highly-substantial variation of dose rate of bremsstrahlung, which is the main parameter of acceleration chambers. For example, according to experimental data, dispersion of dose rate for chambers BUK-6 ranged from 2 to 10 R/min, in the lot of 90 pieces (data was taken from the report for internal use of INT "TPU" in 2008).

Summarizing all above mentioned, it is necessary to change the design/construction of the injection unit of the acceleration chamber of betatron to obtain the required repeatability the dose rate of bremsstrahlung in a narrower range which is the main characteristics of the chambers. At the moment, this is achieved by screening of acceleration cameras at the stage of their installation in the betatron. In particular cases, the percentage of rejections reaches the value of 50% (at an average, rejections is 25%). Among possible solutions to the problem of repeatability of the output characteristics of betatron chambers, one can offer the transition to new construction materials, such as ceramics (now the injector is mounted on a glass pedestal), and the change in the actual design of the injector.

References

[1] JME – ADVANCED INSPECTION SYSTEMS [electronic resource]/BETATRON PORTABLE X-RAY –Access: http://www.jme.co.uk/JME-Products-Betatron.aspx

[2] E. I. Vaynberg, V. A. Kasyanov, V. L. Chakhlov, M. M. Stein Experience of using small-size betatron mib-5 in the structure of industrial computed tomograph bt-500xa http://www.ndt.net/article/wcndt2004/pdf/radiography/102_stein.pdf

[3] Bjorkholm P., Johnson J. Cargo Examination using the X-ray Radiation of High Energy. *Cargo Security Intelligence*. October 2004, Pages 141–147.

[4] David J. Lochman. Therapy in cancer of the head and neck with the betatron electron beam *The American Journal of Surgery*, Volume 98, Issue 6, **Pages 847–850**

[5] G. Baciu The electron injection process in the betatron *.Nuclear Instruments and Methods*, Volume 26, **Pages 173–178**

The increase of ultrasound measurements accuracy with the use of two-frequency sounding

Yu V Shulgina, A I Soldatov, Ya V Rozanova, A A Soldatov, E M Shulgin [1]

Tomsk Polytechnic University, Tomsk, Russia
[1]Joint-Stock Company «Scientific & Industrial Center «Polyus» Tomsk, Russia

shulgina@tpu.rul

Abstract In the article the new method for detection of the temporary position of the received echo signal is considered. The method consists in successive emission of sounded impulses on two frequencies and also the current study is concerned with the analysis of ultrasound fluctuation propagation time to and from the deflector on every frequency. The detailed description of the mathematical tool is presented in the article. The math tool used allows the authors to decrease the measurement error with help of calculations needed.

Introduction

As a matter of fact, blasting operations are the main step in mining preparation process of coal, iron ore and other mineral resources. What is more, expenses to carry out the blasting operations comprise a significant part of mined rock production cost. Hence the errors in hole depth measurement lead to increase of the mined rock production cost, and also it might result in the risk of a top horizon downfall where mining occurs.

A basic error in ultrasonic depth-gauge measurement is conditioned by inaccuracy in determination of ultrasonic impulse arrival. Generally the moment of ultrasonic impulse arrival is defined with a comparator but with help of complex shape of ultrasonic impulse, and the comparator operating time doesn't concur with the impulse initiation.

Nowadays existing ultrasonic hole depth-gauges have measurement errors and a high value of uncontrolled area. However the major issue concerning the hole formation depth measuring lies in a low accuracy of measurement connected with the hole shape fault, which takes place due to earth movement. Therefore, the development of universal method of acoustic impulse analysis which allows receiving the accurate information about the depth is the research objective of the current study.

As regards to the use of comparatively low frequencies in acoustic position, it leads to the situation when the time of propagation and the period of sounded fluctuation become commensurable quantities, thus as a rule, the received radio impulse but not its envelope is supplied directly to the comparator inlet. In case of square-wave modulation use or similar to it in shape, the maximum error will be equal to carrier frequency quarter-period.

For this reason the regulating system AGC to increase accuracy of acoustic measurements is used, however there is a range of problems solving which it is impossible to get a pulse with linearly-growing amplitude of the front edge of an echo signal envelope. Among them there are acoustic hole depth-gauges where unpredictable and imperfect hole shapes (fracture, roughness) often occur, and also the influence of a wave guided propagation pulse results in significant change of the shape of the received impulse, besides, the side reflection occurrence leads to infeasibility of the AGC system application. The shape model of the beginning of the echo signal of wave guided acoustic path is demonstrated in figure 1.

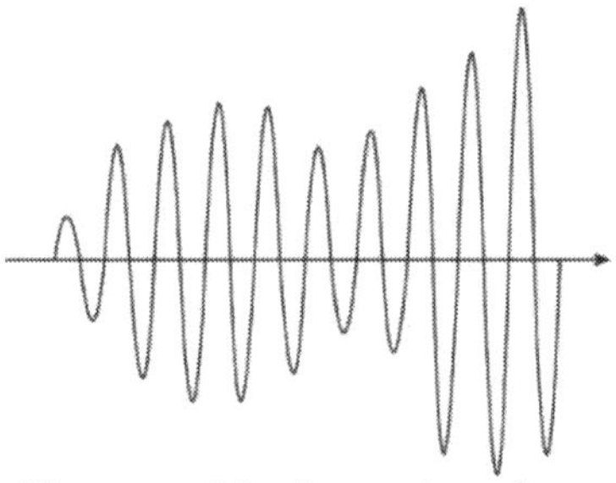

Figure 1. Oscillogram of the front edge of a received signal.

The method used to detect the accuracy of pulse initiation moment consists in ultrasonic signal supply to the comparator inlet and the moment of occurrence of a high logic level on the outlet of the device is considered to be the moment of echo impulse arrival. This method error makes up to tens of periods of the carrier frequency and it is impossible to take into account the error because of the unpredictable envelope change of the front edge of an echo signal at the wave guided propagation. Moreover the use of Hilbert transform for receiving the echo signal envelope and later detection of its initial point requires high computer performance and resources which portable apparatus don't have [1].

The research objective to determine the temporary position of the echo impulse is applicable and actual for acoustic instruments, the acoustic path of which contributes to the impulse envelope shape change. One of the innovative methods for the time delay detection of ultra sound propagation is the zero-crossing detecting [2–4].

1. Two-frequency sounding method description.

The authors of the present study offer a method which uses the signals of two different frequencies to measure the distance and as the result, it will allow excluding the major part of measurement mistakes connected with the wave guided propagation of an ultra sound and also, to increase the measurement accuracy.

The method principle consists in the emission of two signals at different frequencies and in measurement of two time intervals between the impulse emitted and the impulse received when the comparator comes into action (figure 2).

For two received signals of different frequencies the comparator response occurs at different time relative to signal stimulation – point t_1 and t_2, and the calculation of a temporary position of an echo impulse is estimated in relation to these points.

After the time intervals measurement between the emitted and received signals, the comparison of these intervals is carried out and the correction occurs according to the formula:

$$(\Delta t_1 - i \cdot T_1) - (\Delta t_2 - i \cdot T_2) = \min , \tag{1}$$

Where T_1 – the fluctuation period of the first ultra sound wave, T_2 – the fluctuation period of the second ultra sound wave, i– correction number, Δt_1 – the first measured time interval, Δt_2 – the second measured time interval. The formula $(\Delta t_1 - i \cdot T_1)$ is used to determine the reflection surface distance.

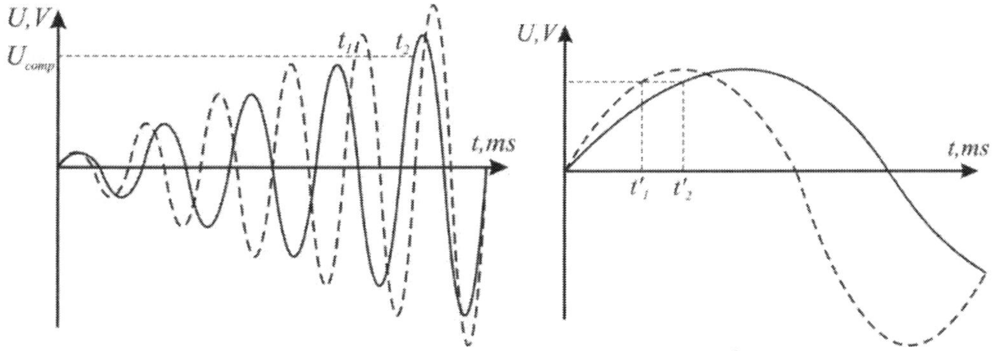

Figure 2. Oscillograms demonstrating the beginning of two echo signals (firm line – the first echo signal with the recurrence interval T2, dotted line – the second echo signal with the recurrence interval T1), a – the operation time of a comparator, b – correction works result, where $U_п$ – threshold comparator voltage; t'_1, t'_2– time intervals after correction.

2. Math modeling of errors

The time calculation of comparators shifting for different response levels has been carried out with the use of MathCad program and using its incorporated functions we solve the equation of the form:

$$k \cdot n \cdot \frac{2t}{T} \sin(n\omega t) = U_{comp}$$

Where n – correlation of frequencies $\frac{f_1}{f_2}$, t – the first period of comparator response, U_{comp} – comparator operation threshold, k – coefficient of the front edge signal growth.

Correction is carried out with the received time intervals.

To determine the limits of method applicability the math modeling has been used. As the point in question, coordinate plane zero has been chosen herewith the differential interval depending on iteration number is shown in figure 3.

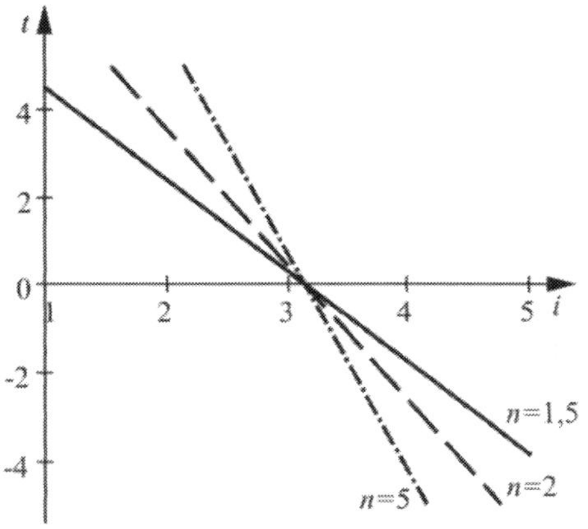

Figure 3. Dependance of differential interval on the iteration number for different ratio of emitted frequencies.

Figure 3 depicts iteration straight lines for different frequencies ratio. It is assumed from the graph that with the increase of the second operational frequency, the incidence angle of the graph to the axis increases as well. The cross-point with the axis indicates the number of iterations conducted to minimal discrepancy. In this case the comparator response occurred in the 4th period of received signal, thereafter, three time iterations lead to the minimal positive value of correction expression (1).

The accuracy of the impulse arrival moment detection will be defined with help of the signal phase at the comparator operation time, otherwise, the closer the comparator level to the change of a period, the more significant measurement error can be. Consequently, to increase the measurement accuracy it is necessary to increase the second emitted frequency; however the maximum possible measured distance decreases.

For the reason of error detection of the current method it was conducted the error measurement modeling. Changing the operation threshold of the comparator, the graph with the error dependence on the threshold voltage was received (Figure 4). Increasing the operation threshold when approaching to sinusoid peak leads to the gradual non linear increase of error. When the period changes the error reduces.

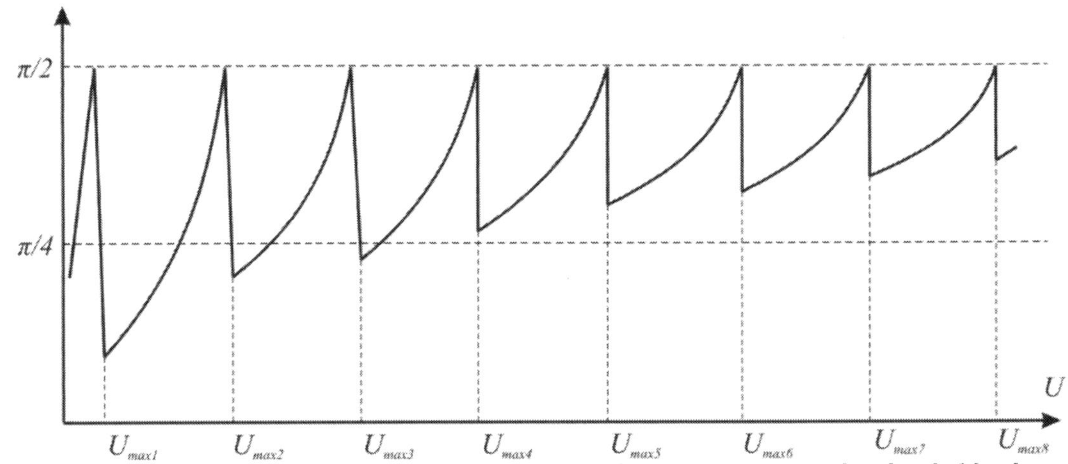

Figure 4. The measurement error dependence graph on the comparator operation threshold, where U_{max} – maximal amplitude of the signal, Δ – an error.

According to the graph data, the measurement error depends directly on the operation threshold chosen, but it doesn't exceed the quarter-period which can be taken as an advantage of the method. Besides it can be inferred from the graph that the measurement error at any operation threshold won't be equal to zero.

For further improvement of accuracy of the current method, the average error must be found for every period when the comparator response might occur. In this case, the number of conducted iterations to reach minimal expression value (1) and awareness about the average value of the error for the period can increase the measurement accuracy.

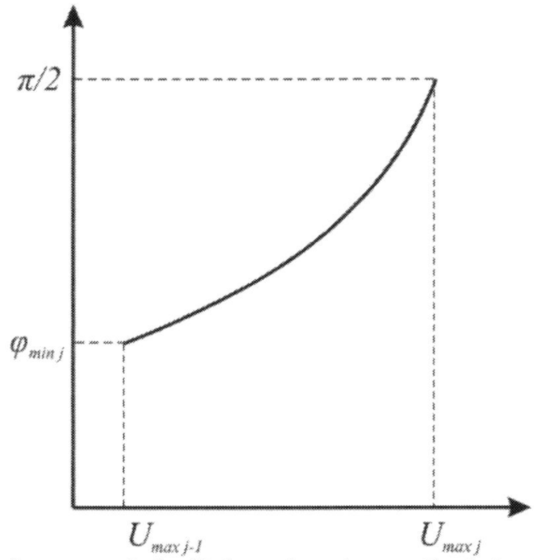

Figure 5. Minimal phase for any period will depend on the amplitude level of the previous period.

Thus, the following equitation is obtained:

$$\varphi_{\min_i} = \arcsin\left(\frac{U_{i-1}T}{t}\right)$$

where $U_{\max_{j-1}}$ – amplitude of the previous period of the received signal, $U_{\max_j}$ – amplitude j-the period when the comparator response occurs.

The average error value for j-period is defined with the formula:

$$\Delta_j = \frac{1}{\left(\varphi_{\min_j} - \frac{\pi}{2}\right)} \int_{\varphi_{\min_j}}^{\pi/2} k \cdot n \cdot \frac{2t}{T} \sin(n\omega t)dt,$$

Where n – frequency ratio $\frac{f_1}{f_2}$, k – coefficient showing the envelope slope of the front edge of a signal, $\varphi_{\min_j}$ – minimal phase when the comparator might come into action in j-period.

3. Conclusion

The obtained value Δ_j can be applied to increase the detection accuracy of the temporary position of the received signal, thus the expression for the moment determination of echo impulse arrival will be the following:

$$t_0 = (\Delta t_1 - i \cdot T_1) - \Delta_j,$$

where j=i-1, i – iteration number, j – the period number of echo signal arrival when the comparator operation occurred, T_1 – the fluctuation period of the first ultrasound wave of higher frequency, Δt_1 – time interval between the emitted impulse and the moment when the comparator comes into action.

IOP Conf. Series: Materials Science and Engineering **81** (2015) 012103 doi:10.1088/1757-899X/81/1/012103

References

[1] Huageng L, Xingjie F, Bugra E 2009 *AIAA Journal* **4** pp.923–932

[2] Basu B, Basu M 2011 *Proceedings of the 2011 14th European Conference on Power Electronics and Applications* pp 1–7

[3] Mažeika L, Draudvilienė L 2009 *Ultrasound* **64(4)** pp 18–21

[4] Molinaro A, Sergeyev Y D 2001 *Measurement: Journal of the International Measurement Confederation* **30(3)** pp 187–196

[5] Soldatov I, Chiglintseva J, 2009 *Proceedings* of *International Siberian Conference on Control and Communications* pp 313–317

Investigation of heating rate effect on solid-phase interaction in $Li_2CO_3 - Fe_2O_3$ reaction mixture

E N Lysenko[1], E V Nikolaev[1], E A Vasendina[1]

[1]National Research Tomsk Polytechnic University, 30 Lenin Ave, 634050, Tomsk, Russia

E-mail: lysenkoen@tpu.ru

Abstract. The influence of heating rate on solid-phase interaction in Li_2CO_3-Fe_2O_3 reaction mixture was investigated by thermal analysis method. The powder mixture components were in the ratio corresponding to $LiFe_5O_8$ ferrite. The ferrite synthesis was performed by thermal heating of mixture reagents in thermal analyzer up to 800 °C in air at various heating rates in the ranges (5-50) °C/min. The results showed that the heating rate affects the solid-phase interaction in $Li_2CO_3 - Fe_2O_3$ reaction mixture. The reaction phase formation is accompanied by heat endothermic effect, which was observed in the DSC curve in the form of a complex broad peak. For all samples, this complex peaks were decomposed into simpler peaks, and thereby, the enthalpies of the individual phase transitions were determined. It was shown that the heating rate affects the values of enthalpy and temperatures of heat endothermic effects, so that the high heating rate shifts the proceeding of reaction to higher temperatures.

1. Introduction

Lithium ferrite ($LiFe_5O_8$) is very important magnetic material for practical application in microwave techniques [1, 2], as a cathode material for chemical current sources [3], and in chemical sensors [4]. Also, $LiFe_5O_8$ is a convenient model material for the study of new phenomena in complex-oxide systems, such as the phenomenon of radiation-thermal intensification of solid-phase processes in ferrites [5]. In the early works [6, 7], we reported that the isothermal annealing of lithium carbonate and iron oxide mixture in condition of electron beam heating leads to increase the synthesis rate of lithium ferrite in comparison with the synthesis by thermal annealing. This result indicates the presence of radiation effect in stimulation of lithium ferrite solid-phase synthesis. However, the main features in solid-phase interaction during lithium ferrite synthesis under different heating rate have not been studied. Such investigation will help to correctly interpret the results from radiation-thermal synthesis of lithium ferrites by high-energy electron beam, where mainly large heating and cooling rates are used. This problem can be solved by using TG/DSC methods widely used in lithium ferrite analysis [8-10].

Thus, in this paper, we present the investigation results of heating rate effect on solid-phase interaction in $Li_2CO_3 - Fe_2O_3$ reaction mixture.

2. Experimental

The commercial Fe_2O_3 and Li_2CO_3 powders were used as the initial reagents of reaction mixture for lithium ferrite synthesis. The powders were dried for 7 hours at the temperature of about 200 °C in a laboratory furnace. After that, powders were weighed in the proportions with according to the equation

$Li_2CO_3+5Fe_2O_3 \rightarrow 2LiFe_5O_8 + CO2\uparrow$ and mixed by threefold rubbing through a sieve with a mesh of 110 microns. Samples were freely-placed in the crucibles, and the bulk density of powder was 0.95 g/cm^3. Calorimetric investigations were carried out by STA 449C Jupiter thermal analyzer (Netzsch-Geratebau GmbH, Germany). The powders were heated up to 800 °C in air at various heating rates in the ranges (5-50) °C/min and then were cooled with rate of 10 °C/min. Rates range was limited by instrumental capabilities of the analyzer.

3. Results and discussion

Figure 1 shows the DSC curves for reaction mixture under heating at 5 °C/min and 50 °C/min. Areas of the observed DSC peaks are enthalpy of reaction, normalized to the mass of the sample. The software package «Peak Separation» (Netzsch-Geratebau GmbH, Germany) was used for the separation of complex peaks into components. The DSC data (enthalpy S and temperatures T of peak position) are summarized in Table 1.

From calculation, the five endothermic peaks were observed on DSC curves for all samples under heating. The first three peaks S_1 (~545 °C), S_2 (~612 °C) and S_3 (~ 690 °C) are included in broad peak on DSC curve under heating. The fourth peak S_4 (~730 °C) is connected with lithium carbonate melting and decomposition; fifth $S_{\alpha\text{-}\beta}$ (~755 °C) is phase transition "order-disorder" in $LiFe_5O_8$ lithium ferrite [10].

The values of enthalpy S_1, S_2, S_3 decrease with increasing heating rate, and the enthalpy value of sharp S_4 peak increases. At the same time, the low-intensity peak S_3 is not possible to determine when high heating rates. Probably, the first three peaks are caused by the ion exchange between the reagent particles in the solid state, and the fourth peak - exchanges between the solid (α-Fe_2O_3) and liquid (Li_2CO_3) phases. Obviously, the increasing of heating rate reduces the time of interaction between the solid phase, and a large amount of unreacted lithium carbonate is present in mixture up to the temperature of Li_2CO_3 melting. The melting film instantly spreads over the surface of the Fe_2O_3 particles, increasing the area of interaction between lithium carbonate and iron oxide. Also, considering that this process is carried out at higher temperatures, it is easy to understand the reason a significantly higher efficiency of the second stage of lithium carbonate decomposition. Due to the increasing amounts of unreacted Li_2CO_3 (with increasing heating rate), the intensity of S_4 peak increases and overlaps adjacent S_3 and $S_{\alpha\text{-}\beta}$ peaks. It can be seen from Table 1, these peaks are confidently calculated at a heating rate of 20 °C/min inclusive, and so this rate can be taken as limiting in a detailed analysis of the DSC curves.

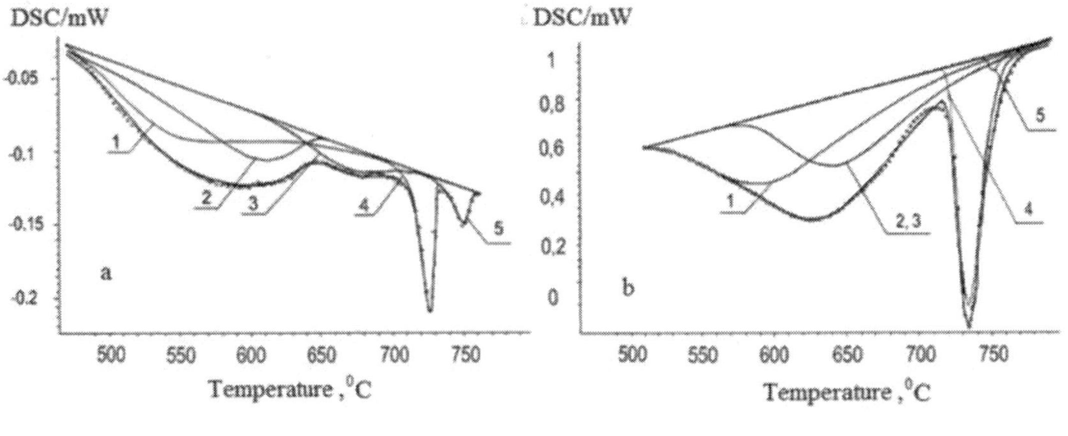

Figure 1. DSC curves for $Li_2CO_3 - Fe_2O_3$ mixture under heating at 5 °C/min (a) and 50°C/min (b). 1 - S_1, 2 - S_2, 3 - S_3, 4 - S_4, 5 - $S_{\alpha\text{-}\beta}$.

Table 1. DSC data for $Li_2CO_3 - Fe_2O_3$ mixture under heating.

Heating rate	S_1	T_1	S_2	T_2	S_3	T_3	S_4	T_4	$S_{\alpha\text{-}\beta}$	$T_{\alpha\text{-}\beta}$
°C/min	J/g	°C	J/g	°C	J/g	°C	J/g	°C	J/g	°C
5	46.2	548	38.3	600	10.6	670	13.5	727	4.7	750
10	43.6	562	30.9	605	8.9	683	16.2	728	4.5	751
20	42.6	575	34.7	622	5.8	693	20.7	729.5	2.9	750
30	41.7	583	34.6	628	-	-	23.8	730.4	1.9	750
40	40.7	587	34.1	635	-	-	24.7	733	1.5	751
50	39.8	597	33.2	646	-	-	24.7	733.9	0.9	751

As above, the peak $S_{\alpha\text{-}\beta}$ is proportional to the enthalpy of $\alpha \rightarrow \beta$ transition, and therefore indicates the amount of α-$LiFe_5O_8$ phase accumulated during the heating of the reaction mixture up to a temperature of 750-755 °C. Since with the increasing of heating rate, the total time of reaction mixture annealing is reduced, and so the decrease in peak intensity is observed (see Table 1). From the fact that a tenfold increasing in the heating rate leads to only a five-fold decreasing in the intensity of the $S_{\alpha\text{-}\beta}$ peak, we can conclude about the nonlinear dependence between the duration of annealing time and quantity synthesized $LiFe_5O_8$. According to result from DSC curves obtained in cooling regime, the content of lithium ferrite phase continues to increase. From this, the both a lithium ferrite and some intermediate phases are formed in the process of lithium carbonate decomposition. Most likely, lithium oxide and non-stoichiometric lithium ferrite phases ($LiFe_5O_8$, $LiFeO_2$) are as the intermediate phases after the Li_2CO_3 decomposition.

4. Conclusion

Paper presented the investigation of heating rate effect on solid-phase interaction in $Li_2CO_3 - Fe_2O_3$ reaction mixture using thermal analysis. The powders for reaction mixture were used to obtain $LiFe_5O_8$ lithium ferrite.

The results show that the heating rate affects the solid-phase interaction in $Li_2CO_3 - Fe_2O_3$ reaction mixture. The complex peaks are observed on the DSC curves, which were decomposed into simpler peaks, and thereby, the enthalpies of the individual phase transitions were determined.

5. Acknowledgements

This work was financially supported by The Ministry of Education and Science of the Russian Federation in part of the science activity program.

References:

[1] Baba P D, Argentina G M, Courtney W E, Dionne G F, Temme D H 1972 Fabrication and properties of microwave lithium ferrites *IEEE Trans. Magn.* **8** 83.

[2] Smit J, Wijn H P J 1959 *Ferrites. Physical properties of ferrimagnetic oxides in relation to their technical applications* (Eindhoven: Philips Technical Library) p 299.

[3] Wang X, Gao L, Li L, Zheng H, Zhang Z, Yu W, Qian Y 2005 Low temperature synthesis of metastable lithium ferrite: magnetic and electrochemical properties *Nanotechnology* **16** 2677.

[4] Rezlescu N, Doroftei C, Rezlescu E, Popa P D 2008 Lithium ferrite for gas sensing applications *Sensors and Actuators* B **133** 420.
[5] Surzhikov A P, Pritulov A M, Lysenko E N, Sokolovskiy A N, Vlasov V A, Vasendina E A 2010 Calorimetric investigation of radiation-thermal synthesised lithium pentaferrite. *J. Therm. Anal. Calorim.* **101** 11.
[6] Surzhikov A P, Pritulov A M, Lysenko E N, Sokolovskii A N, Vlasov V A, Vasendina E A 2012 Dependence of lithium–zinc ferrospinel phase composition on the duration of synthesis in an accelerated electron beam *J. Therm. Anal. Calorim.* **110** 733.
[7] Surzhikov A P, Lysenko E N, Vlasov V A, Malyshev A V, Nikolaev E V 2013 Investigation of the process of ferrite formation in the Li_2CO_3–ZnO–Fe_2O_3 system under high-energy actions *Russian Physics Journal* **56** 681.
[8] Berbenni V, Marini A, Matteazzi P, Ricceri R, Welham N J 2003 Solid-state formation of lithium ferrites from mechanically activated Li_2CO_3–Fe_2O_3 mixtures *J. Eur. Ceram. Soc.* **23** 527.
[9] Surzhikov A P, Frangulyan T S, Ghyngazov S A, Lysenko E N 2010 Investigation of oxidation processes in non-stoichiometric lithium–titanium ferrites using TG analysis *J. Therm. Anal. Calorim.* **102** 883.
[10] Surzhikov A P, Frangulyan T S, Ghyngazov S A, Lysenko E N 2012 Investigation of structural states and oxidation processes in $Li_{0.5}Fe_{2.5}O_{4-\delta}$ using TG analysis *J. Therm. Anal. Calorim.* **108** 1207.

SPECTRUM MONITORING OF ELECTROMAGNETIC SIGNALS FROM ROCKS TO CONTROL GEODYNAMIC PROCESSES UNDER WORKING MINE CONDITIONS

A A Bombizov,[1] A A Bespalko[2]

[1] Tomsk State University of Control Systems and Radioelectronics,
e-mail: unclelab@gmail.com
[2] National Research Tomsk Polytechnic University,
e-mail: besko48@tpu.ru

Abstract. The paper considers the techniques for continuous monitoring of samples and rocks destruction using the parameters of the measured electromagnetic signals which are generated in the destructive processes. The block diagram of the recorder is presented and the methods for processing of the measured electromagnetic signals to monitor and predict the geodynamic processes in rocks or the destruction of dielectric materials under electromagnetic interference. A phase modeling of the method to extract a useful signal from the background noise and electromagnetic interference has been implemented.

Keywords: electromagnetic signals, rocks, geodynamics, monitoring, spectrum analysis, event identification.

1. Introduction

Monitoring of the geomechanical state of a natural-technical system is essential for the safety of people and engineering facilities located in tectonically unstable areas. To efficiently determine the current state of an area and its geodynamic hazards, and to choose the proper way to control the massif state, it is promising to study the geodynamic and technology and nature related processes using the methods of mathematical and physical modeling, field geological and geophysical studies, deformation observations, electromagnetic and seismic monitoring [1–3].

In practice, a spatial-time analysis of microseismic area monitoring is primarily used to predict and prevent the formation of the environmentally fragile and geodynamically hazardous zones in the mine development areas. Two parameters that reflect time and space changes in the dynamic phenomena distribution are considered in [4, 5]: the time interval between seismic events and sampling fractal parameter for seismic events. Decrease of the time interval between the events of a certain energy class is considered to be the precursor of a powerful dynamic phenomenon. In addition, from the physical point of view, the value of the fractal distribution of seismic events of low power must correspond to the concentration of the seismic events in a small neighborhood of the future centre of a powerful dynamic phenomenon [4]. Unfortunately, the research in this field is complicated due to the specific conditions for observations and the presence of costly seismic stations that are not available in most working mines.

Another promising method to monitor changes in the stress-deformed state (SDS) of the rock mass and to predict geodynamic events, including those in a mine field, is the method based on dynamoelectric transformations in rocks and other dielectric structures, including structures that contain ferroelectrics and radioactive substances [1, 2, 6]. The development of the information system to monitor geodynamic events and other methods to control the geodynamic situation based on the research results on dynamoelectric transformations in rocks will significantly improve the accuracy of time and place predictions of these destructive phenomena.

The research aims at the development of electromagnetic monitoring of changes in the rock SDS and control of geodynamic processes in working mines, as well as the processes of dielectric materials destruction.

2. Monitoring and control techniques

To develop the information system to monitor and to predict geodynamic events, the investigation of the EMS parameters and electromagnetic emission characteristics of the rocks in the Tashtagolsky mine (Kemerovo region) are being carried out at the depths up to 800 meters before and during the process of technological explosion with masses up to 300 tons, and during the process of rock mass relaxation after the explosion [6, 7]. Determination of the EMS parameters obtained in the changes of the rock SDS is of special interest [8–10]. It is possible to estimate the SDS of the controlled rock mass while examining the spatial structure of the electromagnetic field and the complex amplitude of its spectral components. It allows obtaining an additional indicator of a short-term earthquake prediction [2]. It should be noted that work in the mine is not performed during the technological explosion, and the level of electromagnetic interference is almost zero.

The scientists of Tomsk Polytechnic University (TPU) found out that the EMS frequency range of the rock mass from 1 to 100 kHz is the most informative [2, 6]. The need to ensure a long-term recording in digital signal spectrograms with high frequency resolution and in real time causes the problem in the low and medium frequency regions in EMS monitoring. This requires large amount of storage and energy resources that limit the portable device possibilities for actual measurements, particularly, when used in mining installations.

The scientists from TPU and the pilot plant "Smena" of Tomsk State University of Control Systems and Radioelectronics are involved in the development of the diagnostic facilities to monitor the state of rocks and to predict the geodynamic events in natural conditions. An autonomous frequency recorder of electromagnetic and acoustic signals of rocks was developed as a result of cooperative efforts [11, 12]. The recorder is developed on the basis of the modular principle. A digital signal processor (DSP) is the main element of the device to control the peripheral units and to perform mathematical calculations. The individual controllers are responsible for work with the user interface and supply system to logically divide the tasks. The purpose of each peripheral module is shown in Figure 1.

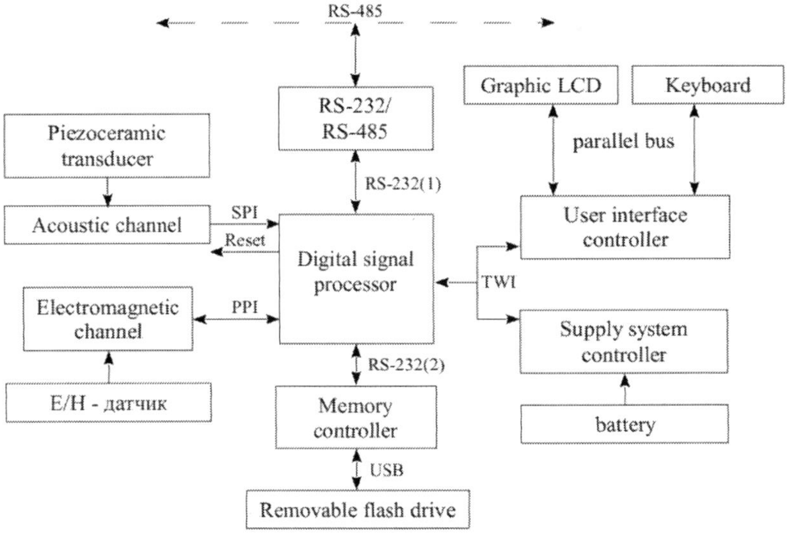

Figure 1. Block diagram of the recorder.

A keyboard and a graphic LCD are used for the information input and output in the initial processing. The device is powered by the battery to provide autonomous functioning of the complex for 3 days. RS-485 interface enables connection of several devices to the distributed network to analyze and provide connection during spatial processing of electromagnetic and acoustic signals.

The recorder is designed as a basic unit for receiving, mathematically processing and transmitting of data on electromagnetic and acoustic signals monitoring to the network. The acoustic signal recording provides the timing of the geodynamic events in the mine field. The main technical characteristics of the device are as follows: the frequency range of the signal analysis is from 1 to 100 kHz; the electromagnetic channel sensitivity is no less than 2 mV, and the acoustic channel sensitivity is no less than 5 mV; the dynamic range is no less than 60 dB; filtering is provided by Fast Fourier Transform (FFT); the refresh period of spectral characteristics is no less than16 ms; frequency step is no less than 1 kHz; 3 dB filter bands correspond to 3 kHz; rectangularity coefficient of 3 dB and 30 dB filters is 0.3; battery capacity is up to 16 A-hours; the autonomous work time of the unit is no less than 3 days; the mass of the unit with the battery is no more than 8 kg.

3. Simulation and mathematical processing

The device provides continuous recording of the electromagnetic emission with a sampling frequency of 1 MHz. In this regard, storage of the total amount of signals data with subsequent analysis in a laboratory or by a remote observer requires a portable autonomous device to possess a flash drive of an increased volume, a high data transmission rate to maintain communication in mines, and, in addition, it requires time consuming data processing. Therefore, to reduce the amount of information the device performs signal recording, real-time mathematical treatment (the window function, FFT, estimation of mathematical expectation and dispersion of a specified time interval) and stores the results in a removable flash drive. The operating algorithm for the mathematical block of the device is shown in Figure 2.

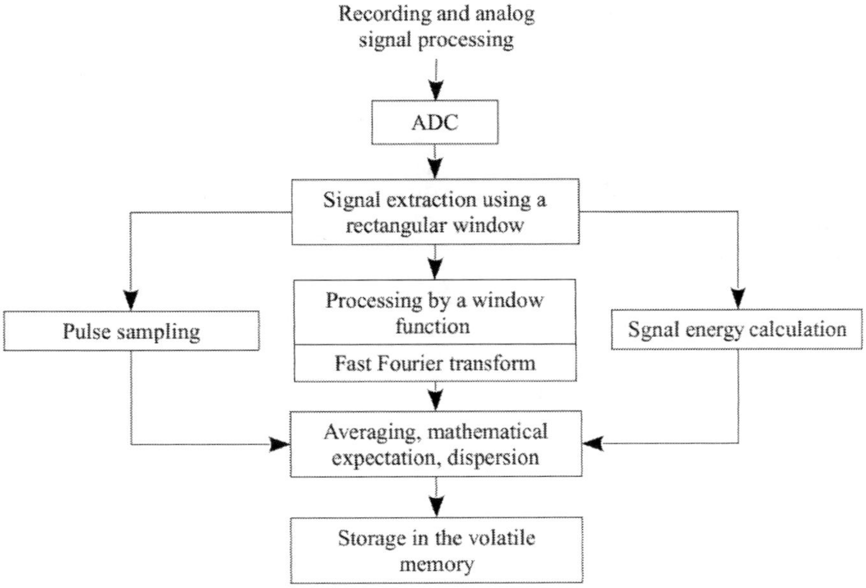

Figure 2. Algorithm for the mathematical block of the recorder.

To determine the system sensitivity at the stage of mathematical treatment of the signal is sufficiently relevant. Model this procedure. Assume that a useful signal with background noise caused by both imperfect device characteristics and continuous electromagnetic emission of the environment has been recorded. Therefore, identification is to be performed in the electromagnetic field of the event that in further analysis can appear to be the precursor of a geodynamic event.

For mathematical calculations, a useful sinusoidal signal emitted during rock crack propagation with an exponential rise and fall is chosen [13]. The dependence obtained by the authors in [13] was modified as where A_0 is the pulse amplitude; ω is pulse angular frequency ($\omega=2\pi f$); t_0 is entry of signal; τ_1, τ_2 are constants for rise time and fall time; T is the time for the pulse to reach its maximum amplitude. Here, τ_1 and τ_2 are substituted for the total time constant τ to independently control the rise time and fall time.

$$A = \begin{cases} A_0 \sin[\omega(t-t_0)]\left[1-\exp(\dfrac{-(t-t_0)}{\tau_1}\right], & t < T \\ A_0 \sin[\omega(t-t_0)]\left[1-\exp(\dfrac{-(T-t_0)}{\tau_1}\right]\exp(\dfrac{-(t-T)}{\tau_2}), & t \geq T \end{cases} , \qquad (1)$$

Figure 3a shows a simulated waveform for the two signals assumed to be ideal upon geodynamic events. The signal parameters are as follows: the amplitude for both signals is of 2.5 relative units; the frequency is of 20 and 80 kHz (the frequencies are chosen to check the technique for upper and lower frequencies of the range under study); time constants τ_1 = 2 μs, τ_2 = 10 μs for the first signal and τ1 = 4 μs, τ2 = 30 μs for the second one; T – t_0 = 12 μs for the first signal, T – t_0 = 34 μs for the second one. This signal waveform is characteristic of the formation of the destructive zones in the mine fields.

Next, since the system and environment are not ideal, additive noise in the form of stationary white noise with amplitude smaller than that of the 10 dB original signal is set (Figure 3b). This white noise can be observed in all measurements of the EMS in the Tashtagolsky mine, including those performed in the down-time period. This is due to the natural redistribution of the rock mass SDS that occurs in the mine field after the technological explosion and mining of the ore body [6, 7]. In addition, noise in the form of a stationary sinusoidal EMS with an amplitude modulation and a frequency of 40 kHz with a carrier frequency of 3 kHz and amplitude equal to the original signal are set (Figure 3c). Superposition of the sinusoidal EMSs is caused by the electromechanical equipment in the mine that emits this type of noise (fans, drills, generators, motors, transformers, etc.). As a result of the complication performed for the recorded EMS (Fig. 3c), it is difficult to define what is the useful component of the signal providing information on the emergence of the destructive zone in the rock mass and its development over time, as well as on the geodynamic event which has already occurred. Therefore, for a remote observer, it is necessary to extract the original EMS to make the decision. For this purpose, the algorithm of the mathematical block of the recorder was used (Figure 2).

Pulse identification can be performed with different techniques, for example, signal level detection using the wavelet analysis corresponding to the model in [13]. As can be seen from Figure 3, the level detection will not identify the first pulse signal.

A more complicated approach using wavelet transform does not provide a uniform frequency resolution and is closely related to the signal waveform in the search model. That is, if the parameters of pulse signals (shape, length, rise time and fall time) may be different,
the approach based on convolution with the known model of the signal cannot be sufficiently universal. Other techniques to identify the signals against the background noise can be found in [14, 15], where signal identification against the background noise is reduced to the statistical analysis of stationary processes in terms of both the time and the spectral-time representations. This approach is most appropriate for the analysis of electromagnetic radiation of rocks and detection of deviations from stationarity in the form of pulse signal.

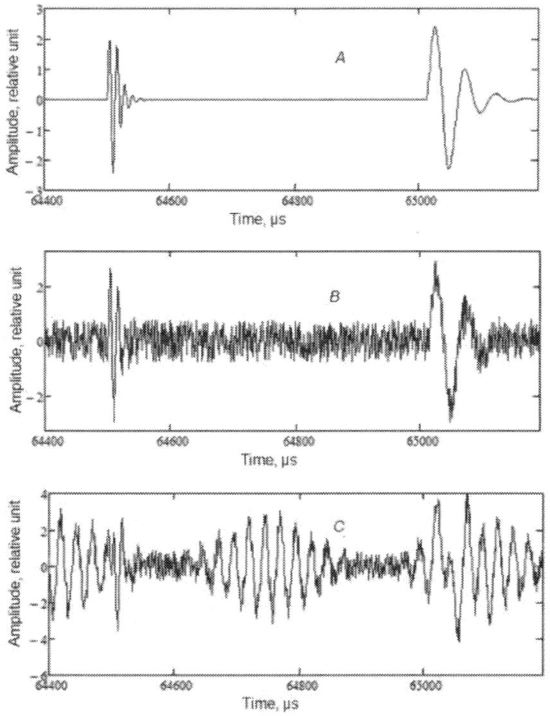

Figure 3. Informative electromagnetic signals of 20 kHz and 80 kHz (a); the same signals with white noise (b) white noise and sinusoidal electromagnetic interference (c).

As a result, the statistical analysis in the study was conducted using a fast Fourier transform (FFT) spectral-time decomposition of the total model signal shown in Figure 3. For the optimal relation between the analog filtering to identify the band with the frequency of 1–100 kHz and the compute capacity used for subsequent digital processing, the sampling rate was chosen equal to ~ 1 MHz. The width of the time window for FFT was ~ 1 ms (1024 samples), the frequency increment was 1 kHz. This relationship between frequency and time resolution is the most appropriate for spectral-time decomposition of the signal. Fast Fourier transform within a limited time interval may cause Gibbs effect, which can be minimized by processing the original signal with Kaiser Window with the coefficient of 9. Application of the window function, in contrast to the rectangle one, causes original signal losses, as processing reduces the amplitude of the signal near the edges of the time window. To eliminate these losses while calculating the spectral-time representation by means of the window function, fast Fourier transform is performed with the superposition of 50%, that is, 512 sample increment.

When considering the frequency-time representation of the signal, it is difficult to identify pulse signals against noise. The spectrum of the monophonic signal (Figure 3b) is illustrated in Figure 4, where the continuous line corresponds to the spectrum of the pulse signal with a frequency of 80 kHz over a time interval of 64–65 ms, and the dotted line is 20 kHz in the interval of 64.5 – 65.5 ms. Short duration of the pulse signals to be determined does not cause distinct maxima in the spectrum. The spectrum is strongly affected by the center frequency of 40 kHz with side harmonics, 3 kHz shifted from the center and by the bursts of white noise spectral components. Irregularity of the white noise spectrum is due to a small analysis time window.

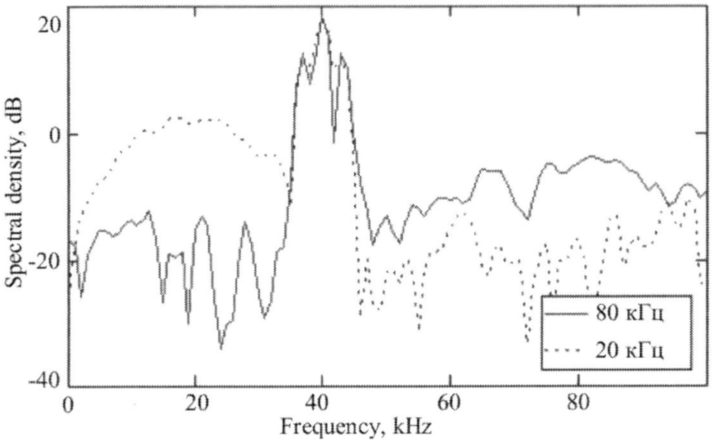

Figure 4. Amplitude spectra of the resultant signals to be determined.

To store the entire volume of the original data is difficult, and in addition, it may be not desirable, therefore, the recorder performs the spectral analysis and the calculated dependences on mathematical expectation (M_x), and the squared standard deviation (D_x) of the spectral amplitudes for each frequency component are stored in a removable media. The sample range to calculate M_x and D_x is set when starting the device. It can vary from 32 (accepted as the minimum calculation for D_x) to 1024 time windows being analyzed.

At the next simulation stage, statistical processing of spectral-time representation of the signal was performed. The time dependence for the squared standard deviation for spectral densities of the signal amplitudes was calculated by the expression [15]:

$$D(f,t) = \frac{\sum_{i=0}^{N-1}(x(f,t+\Delta t\cdot i)-\bar{x}(f,t..(t+\Delta t\cdot N)))^2}{N}, \tag{2}$$

where N is the value of the sequence of spectral densities to calculate the squared standard deviation n;
Δt is the minimum increment in the spectral-time representation;
$x(f,t+\Delta t\cdot i)$ is the spectral density at f frequency, at each increment in the time interval being calculated;
$\bar{x}(f,t..(t+\Delta t\cdot N))$ is mathematical expectation of the spectral density at f frequency, in the time interval from t to $(t+\Delta t\cdot N)$.

To perform calculation the simulated interval was divided into subintervals which consist of 32 samples ($N = 32$). The squared standard deviation of the spectral amplitude was calculated for each sub-interval by expression (2). Thus, the resulting volume of data on spectral-time representation was reduced by a factor of 32 times.

Figure 5 shows the dependence graph for the squared standard deviation (SD) of the signal amplitude spectral densities on frequency and time *D(f, t)*. Dark areas indicate large values of the square of the SD. The diagram in Figure 5 confirms that the event occurs after 64 ms from the beginning of the simulated interval.

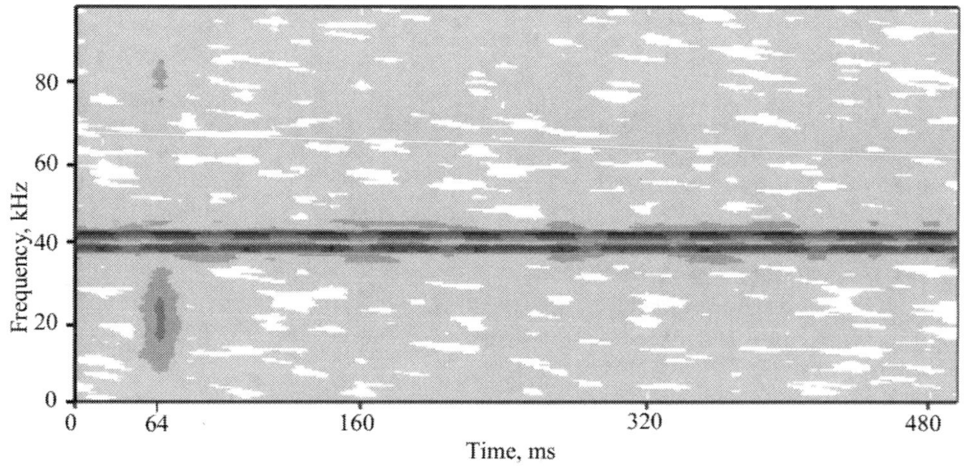

Figure 5. Time-and-frequency diagram of the squared standard deviation (SD) of the signal amplitude spectral densities *D(f, t)*.

Two distinct noises can be observed at the frequencies of 37 kHz and 43 kHz. To filter the pulse signal from stationary noise, the degree of dispersion *D(f, t)* was evaluated for all spectral components by the squared SD in the investigated time interval of 500 ms. The values below a specific threshold proportional to SD are eliminated from the initial analyzed *D(f, t)* by the following expression:

$$D_1(f,t) = \begin{cases} D(f,t), & D(f,t) \geq k \cdot \sigma(f) \\ 0, & D(f,t) < k \cdot \sigma(f) \end{cases} \qquad (3)$$

where *D(f, t)* is the initial squared standard deviation of the signal calculated by (2);
k is weight coefficient;
σ(f) is the standard deviation of the squared standard deviation for the spectral signal amplitudes calculated over the entire simulated time interval for each frequency component.

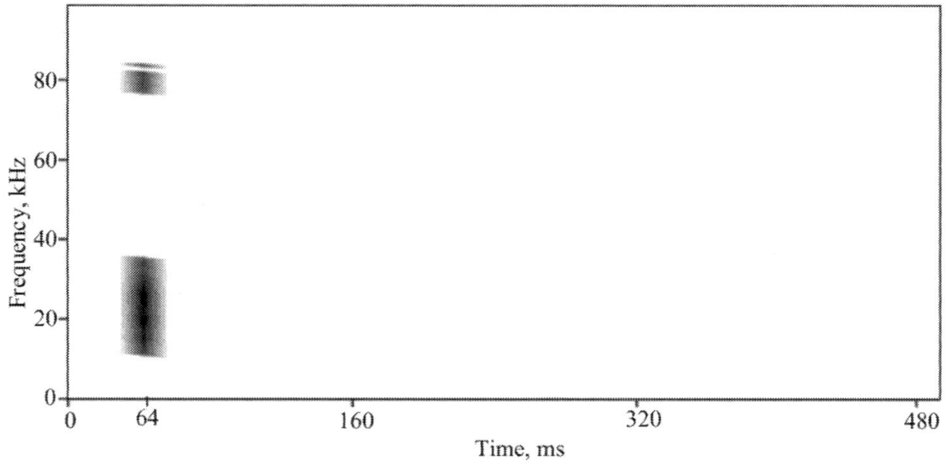

Figure 6. Identified parts of the pulse signal.

The data processed by expression (3) is shown in Figure 6. The weight coefficient k is 4.2 and it was chosen to ensure that the threshold value $k \cdot \sigma(f)$ is greater than the squared noise SD to obtain a "pure" characteristic.

The simulated SD was calculated over the entire time interval. But in processing of the measurement results within 2–3 days, this approach is not advisable as the background radiation intensity may vary with time. Therefore, in real time recording, the time interval for standard deviation estimation should be limited and chosen after preliminary evaluation of the environmental background radiation and in-situ calibration.

The weight factor to calculate the SD is chosen in the calibration process, in the time region free of signals caused by geodynamic events. Thus, the threshold $k \cdot \sigma(f)$ is expected to be of the value sufficient to exclude the stationary and quasi-stationary noise, and simultaneously to be most approximated to prevent filtering of the desired signal.

In simulation, the amplitude of the desired signal relative to the noise was chosen for the maximum probability of signal detection. Hence, for signals comparable to those in simulation with time parameters, the sensitivity of this processing method is about 10 dB.

Emission of pulse signals indicates deviation of the rock mass from the normal state. The analysis of the degree of deviation can be directly related to the prediction of a geodynamical event.

4. Conclusions

Thus, the paper presents a new type of a recorder to provide continuous recording of electromagnetic signals from rocks and dielectric materials in geodynamic processes, including those in the mine fields. This technique for processing the recorded electromagnetic signals may be used both to reduce the volume of the output data in continuous monitoring, and to obtain spectral-time images of electromagnetic pulses. The analysis of the data obtained in total processing of electromagnetic signals and noise showed that the most effective is standard deviation filtering of pulse signal to eliminate stationary noise. The analysis provides information on the development of destructive processes in the ore array or in the extended dielectric material destruction. The results obtained in processing the recorded electromagnetic signals from rocks can be used to transmit information in the network for remote monitoring. This sequence of data processing can be used successfully in the monitoring information network and for short-term prediction of geodynamic processes in underground and open pit mines.

The research was performed by the state task "SCIENCE" Ministry of Education and Science of the Russian Federation

References:

[1] Vorobyov A.A. Balance and conversion of energy in situ. *Tomsk: Publishing house TPU,* 1980. 210 p.

[2] Bespalko A.A., Yavorovich L.V. Variations of electromagnetic emission under changing stress-strain state of the rock mass after mass explosion. *Proceedings of the International Conference "Solar-Earth relationships and electromagnetic precursors of earthquakes." Petropavlovsk-Kamchatsky*: IKIR FEB RAS. 2004. 456 p.

[3] Eremenko A.A., Eremenko V.A., Gaydin L.P. Improvement of geotechnology development of rock-bump hazardous iron ore deposits under natural and man-made factors. *Collection of scientific articles: Publishing House Science*. 2008. 312p.

[4] Golitsyn B.B. Microseismic movement. *Selectas. M: Publishing House, AS USSR*, V. 2. 1960. Pp.126–416.

[5] Apparatus and method of seismic monitoring in the USSR. Ed. Aranovich R.I., D.P. Kirnos *M: Publishing House Science.* 1974. 364 p.

[6] Bespalko A.A., Yavorovich L.V., Fedotov P.I., Viitman E.V.. Mechanoelectrical transformation in rocks of the Tashtagolsky iron ore deposit. *Geodynamics.* 2008. **1 (7).** Pp. 54–60.
[7] Bespalko A.A., Surzhikov A.P., Yavorovich L.V., Khorsov N.N., Klimko V.K., Shtirts V.A., Shipeev O.V. Monitoring of changes in the stress state of the rock mass after mass explosion by electromagnetic emission parameters. *Physical Mesomechanics.* 2004. V. 7. **2** Pp. 253–256.
[8] Vostretsov A.G., Kulakov G.I., Timonenkov Yu.A., Yakovitskaya G.E. Prediction of rock destruction by the spectral characteristics of electromagnetic radiation signals. *Journal of Mining Science.* 1998. **4**. Pp. 21–25
[9] Bespalko A.A., Boukreev V.G., Mertvetsov A.N., Yavorovich L.V. Algorithm to search the precursors of rock destruction by time series of electromagnetic signals measurements. *Control. Diagnostics.* 2012.. **13.** Pp. 3–8.
[10] Bespalko A.A., Yavorovich L.V., Fedotov P.I., Popelyev A.I. Characteristics of the electromagnetic signal spectrum of rock samples under uniaxial compression. *Control. Diagnostics.* 2013. **9.** Pp. 25–33.
[11] Bombizov A.A., Bespalko A.A., Loshchilov A.G., Filatov A.V. Development of the device to monitor stress-deformed state of rocks using indirect indicators of electromagnetic and acoustic emission. *Selectas. TUSUR.* 2012. No. **2(26).** P.1. Pp. 24–27.
[12] Bombizov A.A., Bespalko A.A., Loshchilov A.G. Autonomous recorder of electromagnetic and acoustic signals to monitor mine installations. *Mining rules.* 2013. **1.** Pp.141–143.
[13] Rabinovitch. A., Frid V., Bahat D. Parameterization of electromagnetic radiation pulses obtained by triaxial fracture in granite samples. *Philos. Mag. Lett.* 1998. 77 (5). Pp. 289–293.
[14] Levin B.R.. Theoretical fundamentals of statistical radio engineering. 3rd ed., *M.: Radio i Svyaz'*, 1989. 656 p. **ISBN 5-256-00264-3.**
[15] Julius S. Bendat. Random Data: Analysis and Measurement Procedures: transl. from English. *Julius S. Bendat, Allan G. Piersol. M: Mir*, 1989. 540 p.

Neural geological-genetic and radiogeochemical forecast model of oil-bearing fields

S V Gorbachev[1], I K Kurkan[2]
[1]Department of Innovative Technologies National Research Tomsk State University, 634050 Russia, Tomsk, Lenina Avenue,36
[2]Head of laboratory Tomsk Polytechnic University, Lenina Avenue, 30, Tomsk, Russia

E-mail: spp03@sibmail.com

Abstract. In recent years, oil and gas exploration are increasingly turning to direct methods to identify accumulations of hydrocarbons (magnetometry, radiometry, geochemical methods, etc.). Similar works are tested high in the Tomsk region, near the Ob basin. In this paper we present some results of testing of geological and genetic models and radiogeochemical occurrence of hydrocarbons in relation to various oil and gas complexes, with the development of neural network methods of analysis and forecasting, formulated proposals for their integrated use.

1. Introduction

Systematic approach to the problems of recognition and classification of oil and gas complex objects as characterized by ill-conditioned, noisy background information requires high reliability, accuracy and reliability of the research results. Traditional leading geophysical methods used in prospecting and exploration of oil and gas fields, is seismic. It solves a number of problems of prediction of the geological section, but remains one of the most expensive methods of exploration. Despite progress in the development of seismic methods, the success of the forecast of deposits based on only this method is not sufficiently high, particularly in the case of complicated traps, which include traps, concentrated in the Cretaceous interval of the section of the sedimentary cover of West Siberia.

The way out of this situation can be widely used non-traditional, progressive methods of geochemistry and geophysics, their aggregation with the main, prevailing types of oil and gazopoiskovyh works. In modern practice petrosearch considered appropriate use of the cheapest methods of exploration on the largest possible area, followed by isolation of promising areas and focusing on them more detailed and expensive methods. In particular, in the United States on poorly known areas recommended to consistently apply: small-scale radiometric and magnetometric data obtained under the program NURE (National Assessment Program uranium reserves), geomorphological analysis Landsat, high sensitivity aeromagnetic and airborne gamma-spectrometric survey of large-scale, ground geochemical survey [1].

The search for oil and gas fields radiometric survey was first applied Bogoyavlinskim L.N. and Lomakin A.A. in 1926 in Maikop oil-rich region. From the measurements it was found a systematic decrease in the total radioactivity of the backoff oil deposits. Start of intensive research to identify opportunities for the use of radioactive methods for forecasting and search for oil and gas accounts for

50-60-ies of XX century. During this period, the method being tested radiometric surveys, and sometimes used in the US, Canada, France, the Soviet Union. Subsequently, in spite of some positive results, methods have not been widely used. Primarily, this was due to insufficient sensitivity of the apparatus is not clearly captures minor variations radioactive derivative of the oil and gas fields, the lack of an integrated approach and multi-variant interpretation of the radio geochemical anomalies.

In the last 30 years abroad (USA, China, France, Mexico) due to improved measuring techniques, teaching methods of work and interpretation of results using computer data processing systems there is increasing interest in radio geochemical methods, as one of the ways to search for hydrocarbon accumulations [1–3]. For a more complete understanding of these methods based on secondary changes in the surrounding rocks, in our opinion, it is necessary to include a fourth coordinate - time, and geologic time, and consider the appearance of deposits as part of an integrated system of natural geological forms of motion (the origin, development and destruction). Productive deposits that are in the integrity of the natural system as their component, in turn, are composed of many individual elements in the historical and genetic relationship with each other. These elements experienced qualitative and quantitative changes in the background continuously, intermittent motion and recorded in the pages of geological history in the form of geochemical lithological complexes caused by certain genetic spatio-temporal structural links with the surrounding geological environment [4].

2. Geological-genetic model of oil and gas complex

To create a geological-genetic model is important for the correct differentiation of the considered column on components. Their selection must be in a dialectical relationship with the natural system as a whole and productive thickness in particular. Using the known series is a chemical element, mineral, rock, geological body, you need to add the geological complex, characterized by its own peculiarities, however, similar to other, mainly in relation to geochemical formations.

For hydrocarbons in this complex include the presence of source rocks, migration in reservoirs, tanks, tires, favorable conditions for generation and preservation of these traps for quite a long geological period of time.

Now consider two geological stage sedimentation of organic material and enclosing strata and structural-lithological complex after the formation of deposits, i.e., its state at the moment. Including consideration of the mechanism of formation of oil, it is necessary to consider the structural factor at the time of burial of the sediments and their immersion depth of 1 km to 2 km under the action of heat begins the process of transformation of organic carbon compounds and polymers. This process occurs in the same structural buildings that exist at the moment. Thus, we can trace the migration routes of micro-oil and hydrothermal solutions as their imprint created in the process of formation of hydrocarbons and buried together with the already formed and sealed by a deposit after the onset of the stage of stabilization.

In the formation of a simple oil hydrocarbon spectrum which is characteristic of living organisms, is transformed into a complex spectrum of oil as a result of the formation during diagenesis of a wide range of hydrocarbon - derived primary biological structures, as well as the arrival of huge quantities of hydrocarbons that occur during thermal conversion of deeply buried organic matter. The greatest number of petroleum hydrocarbons formed at the expense of the latter in the subsurface is heated to temperatures from about 60 to 150 degrees [5].

Start warming up, the presence of catalysts (clayey sediments) leads to the formation lithology retort, converting organic matter into hydrocarbons (gas, oil and others), and compaction of sediments that occurs under the action of gravity and increasing as the burial of the sediments, leading to the expulsion of fluids from the parent species in the reservoir. The compressed fluids migrate through compactible precipitation from area of higher pressure usually up in the early stages of compaction, and later towards the neighboring less compressible to reservoir beds.

The peculiarity of the structural-tectonic position of oil and gas is confined to zones of high-temperature fluid flows. In the field of fluid flow are formed not only hydrocarbon deposits, but also complexes of mineral associations with certain regularities in the spatial distribution of the mineral component. In the oil and gas section are formed in the zone of secondary mineral formation, differing from each other by the nature and associations of mineral growths. In the loop petroleum, "halo blight" area of the reservoir, in the underlying reservoir rocks and in the far out of loop areas are formed of contrasting mineral paragenetic association tumors.

Hydrocarbon deposits are located in the vaults of the elevations of different orders, or in their slope parts. Oil and gas zones often bizreport on the most deep basins and valleys [4]. Conservation of oil and gas accompanies heat flux (it is the most intense in the deep hollows) and flow of fluids. Deep fluids include water, carbon dioxide, methane and other low molecular weight hydrocarbons, contribute to the transformation of organic matter and secondary mineral formation, having certain characteristics and patterns of mineral distribution in the sequence of mineral tumors and their zonal distribution. Epigenetic mineralization in the area hydrocarbon plume, on the sites of occurrence of oil and gas in the halo of the invasion of hydrocarbons and in the peripheral part of the sub-vertical hydrocarbon column has its own unique features and patterns that occur in a particular assemblage of tumors, the sequence of deposition and forms associations of minerals.

The main phase of the formation of oil fell on the depth interval of 2-3 km The maximum decrease in porosity when submerged rocks 3 km (depth of 1 km, corresponding to the beginning of intensive hydrocarbon formation to depths of more than 3 km, corresponding to the end) is about 20 %. Consequently, the layer of sediments 1 m, with an area of 100 square km and accordingly volume 10 * 109 cubic meters, during the formation of oil separated from 0.5 to 2*109 cubic meters of water.

Literature data and the results of scientific research carried out on the core material Vahsky, South Vakhsky, Koshinsky, Yuryev, Ininsky fields of Aleksandrovsky Bank, Nizhnetabagansky, Kalinovy, Ostankinsky fields of Pudinsky Bank of Western Siberia has shown that the pathways of hydrocarbon flow and areas of concentration, including in the emerging hydrocarbon formed mineral assemblages of non-contiguous areas and, in turn, carrying the information about the field of education [4,5].

Authigenous mineralization in zones of hydrocarbon streams and areas of their concentration is related to changing physico-chemical conditions at the contact of movable hydrocarbons and water is due to dissolved gases in the water, variations in pH and Eh environment, the activity of anaerobic bacteria, etc. In the process of secondary mineral formation in the zone of the hydrocarbon stream and the loop petroleum minerals mostly dissolved. In the area of concentration of hydrocarbons and halo blight in part formed the zone of disintegration. A large part of the dissolved substance paratragedy below the boundary of the water-oil contact, where a zone of cementation. Lots of cementation and decompression in the case of a stage of formation of the deposits and the formation of the intermediate water-oil contacts form a parallel-banded structure.

As a result of interaction of hydrocarbons with enclosing rocks over oil and gas deposits are sub-vertical zone-annular geochemical, geophysical and biogeochemical fields. This part of the geologic section above the hydrocarbon is the most informative for assessing hydrocarbon potential and is based on the vertical migration of hydrocarbons from hydrocarbons.

The decay products of hydrocarbons, carbon dioxide, water, hydrogen sulfide and other migrating by diffusion and filtration of the deposits of gases and water stimulate epigenetic conversion processes leading to the change of physico-chemical environmental parameters that appear in the transformation of species over the productive complex, the occurrence of specific mineral assemblages, changes in redox conditions and the reallocation of some chemical elements.

3. Radiogeochemical oil and gas field model

Radiogeochemical oil and gas field model is based on a common genetic trait of oil and gas complexes - the presence of post secondary altered rocks under the influence of migrating hydrocarbons and hydrothermal fluids during the formation of deposits and further his life up to the present time [6].

These methods are based on the understanding that the system is rock - water - oil is a complex geochemical environment where processes of transformation of the original constants are so strong that geological objects are changed beyond recognition. The breed is sometimes reaches the level of mono - mineralization [6].

Problems in the study of the formation of geochemical field of oil and gas fields, secondary mineral deposits in the halos of many papers. Such studies in the region in different years were engaged A.E. Kantorovich, O.L. Kuznetsov, R.S. Sahibgareev, E.N. Ilyasova, V.A. Ezhova, N.F. Stolbovaya, S.L. Shvartsev, G.I. Tishchenko, V.A. Bazhenov.

With the use of this research and the results of geochemical work in other oil and gas regions, as well as with independent mineralogical and petrographic studies (microscopic mineralogical, petrochemical, microstructural analysis, Resonant nuclear, chemical analysis) built geochemical model hydrocarbon deposits. Particular attention was paid to the construction of the model of secondary mineral. The model constructed has convincingly shown that the oil and gas potential in the circuit, in the underlying reservoir rocks and distant aquifer areas are formed of mineral paragenetic associations tumors other than the host reservoir rocks and is characteristic for the oil and gas saturation zone. Such associations for different minerals paleosrezov petroleum cut different. In the sub-surface zone characterized by a spectrum of newly formed minerals, including preference in setting mineralogical subsurface (ground surface) research should be given quite easily determined microscopically secondary minerals. Among them: skutterudite, fosforuranit, hydrogoethite, magnetite, Colombia and others. For geochemical halos characteristic abnormal distribution of U, Nb, Hg, Co, Ni, V, and others.

According to the model of formation of radio geochemical anomalies (Figure 1), the proposed S.J. Pirson [6] on the oil and gas fields there is a peculiar geochemical post, the origin of which is due to selective mineralization repressed during compaction water, which led to the migration and accumulation of oil and gas in structural and stratigraphic traps.

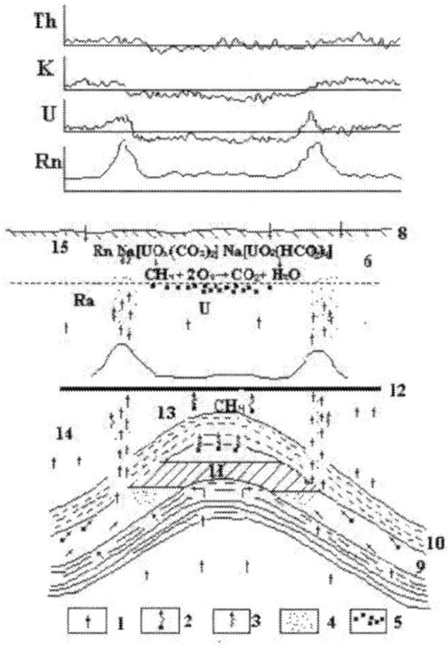

Figure 1. A model of radio geochemical anomalies over hydrocarbon deposit: 1 – water; 2 – oil; 3 – accumulation of Ra; 10 – limiting layer; 11 – trap; 12 – radiogeochemical anomaly according to drilling; 13 – a zone of increased migration, 14 – zone of reduced migration; 15 – area of penetration of atmospheric precipitation.

During electrophoresis under the influence of currents electrotelluric redistribution of radioactive elements. Thus, under the influence of epigenetic processes caused by the influence of hydrocarbon deposits, over oil and gas fields over a long geological time, is the formation of a kind of radiogeochemical field, characterized by fields of distribution of the total radioactivity, the levels of accumulation of radioactive elements and the nature of their relationship

Concentration field over oil and gas fields are characterized by a high degree of differentiation in the distribution of R, Th, U, and has a much more complex structure than abroad.

More clearly heterogeneous structure radiogeochemical field caused by the influence of hydrocarbon deposits, reviewed the analysis of Th/U, the intensity of redistribution of natural radionuclides and intensity of thermoluminescence.

At midday Square West in the selection radiogeochemical abnormalities were 91.7% of exploration wells penetrated the hydrocarbon deposits, 45.4% of them are in the loop area with high oil and gas potential.

In the North-Vasyugan area almost all exploratory wells, which gave commercial flow of hydrocarbons (with the exception of one well) are inside the contour petroleum potential.

All exploration wells in the Myldzhinskoye area, gave inflows of hydrocarbons found within the selected radio geochemical anomalies. Moreover, 69.2% of them fall into the contour plot with high petroleum potential [6].

The above abnormal patterns of radiochemical largely determines patterns of change in the physical properties of rocks in the oxidation zone of the deposits of various minerals and redistribution under the influence of migration of light hydrocarbons from the reservoir and atmospheric oxygen to form a powerful redox barrier in the upper part of the section (in the depth range of 300-500 m). X-ray radiometric analysis of samples of sludge blast holes confirms this phenomenon, showing abnormally high concentrations of thorium, iron, titanium and calcium on the background of a low manganese content in the area of fallow, consistent with a change in the magnetic properties. Hurricane calcium and anomalously high uranium recorded in the marginal zones of projection hydrocarbon deposits on the surface.

Comparison of anomalous geochemical effects with a series of geophysical methods for prediction of hydrocarbon accumulations and results of deep drilling shows high convergence manifestations of anomalies and forecast areas and the location of the identified oil and gas fields.

4. Neural network computer methods of processing and analyzing radio geochemical fields of oil and gas deposits

Radiogeochemical model oil and gas field allows you to create a multi-dimensional feature space, which can be used as a factual basis for the prediction of oil and gas potential circuit. In addition to traditional statistical methods for analyzing the mathematical apparatus of artificial neural networks, which allows you to enter a new level of interpretation of geological data related to intellectual analysis of the material [7,8].

Neural networks - this analytical technologies developed and verified by the nature millions of years of its existence. They allow us to solve the problem of forecasting, classification, pattern recognition [9], the search for optimal variants and absolutely indispensable in cases where under normal conditions, the solution of the problem is based on intuition and experience, rather than the strict (in the mathematical sense) its description. Relevance of research in this area is confirmed a mass of different applications of neural networks [10].

To improve the efficiency of post-processing of geochemical data the author developed and tested software for calculating the integrated predictive parameter based on neural network analysis methods, means training artificial neural network exploration well data [8]. In general terms, the problem reduces to the problem of non-formalized classification, and in accordance with the logging instruments and digitized test chart exploration wells in the simplest case can be used two classes (I - producing wells, II - "blank"), and more complex problems the number of classes may be increased on any grounds.

The input is calculated by digital maps of various radio geochemical parameters, and for each class is formed by the standard, assuming that all the points that fall in this space near the well, bear some information load. On the basis of the reference neural network is trained to classify the membership of each point of the map a particular class. For training on the basis of the standards of a two-layer neural network (Figure 2) with sigmoid activation function of neurons:

f(A)=A/(c+|A|),

where A - adder output neuron, the constant c - steepness parameter sigmoid.

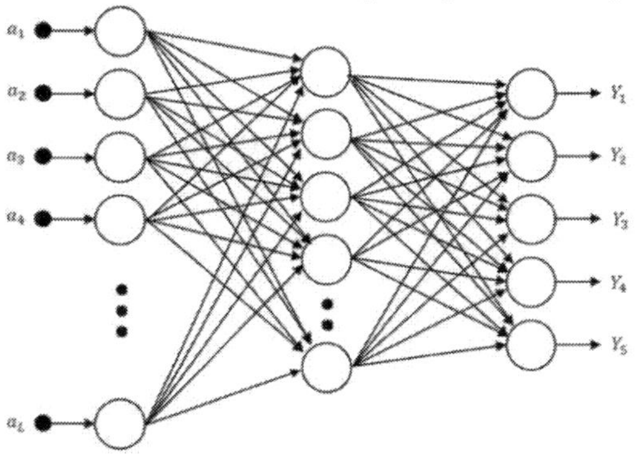

Figure 2. The two-layer neural network.

Neural network algorithms developed marked medal and diploma of Ministry of Industry "For the introduction of neural network methods of complex analysis and interpretation of geophysical and geochemical data" at the international exhibition of high technologies in St. Petersburg, and diploma at the "Siberian Athens" (Tomsk).

5. Prognostic criteria

Theoretical background and analysis of geological materials treated neural network programs in the North-Kalinovoye, North Luginetskoye, West Ostaninskaya, South Cheremshanskaya, pole and Krapivinskoe fields Tomsk region revealed the following predictive criteria.

Formation radiogeochemical anomalies over oil and gas fields based on the idea of the existence of deposits peculiar geochemical column, formed as a result of the election mineralization repressed during compaction of rock water movement which resulted in the migration and accumulation of oil traps [11]. Redistribution of stresses in areas forming anticlinal structures leads to elastic and plastic deformation of the sedimentary cover, creating a permeable zones favorable for the filtration of groundwater and hydrocarbon migration. In turn, the enrichment section of the geological hydrocarbon fraction having strong reducing properties and penetration into the formation oxidant - atmospheric oxygen - leads to a substantial change in the redox environment of the mine, the formation of geochemical barriers. Thus, under the influence of epigenetic processes caused by the influence of hydrocarbon deposits, the oil fields and gas complex formed radiogeochemical field characterized by a peculiar distribution of total radioactivity levels of accumulation of radioactive elements and features of their relationship.

To evaluate the effectiveness of geological neural network techniques were tested in a test mode for a single profile by identifying hydrocarbon deposits, where the opportunity to track changes informative features radiogeochemical fields associated with the identification of hydrocarbons, depending on their phase state (oil, gas, gas condensate) and stratigraphic position of the traps .

The above-mentioned group of fields is all the major oil and gas complexes: the Cretaceous, Jurassic and the contact zone of rocks of Paleozoic and Mesozoic. In addition, in order to improve the methodology of interpretation was conducted areal processing and neural network analysis integrated with other geological and geophysical methods in the South Cheremshanskaya oilfield.

In conjunction with the methods of forecasting seismic data convergence results are obtained using the analysis of gravitational and magnetic fields, which also recorded locations of anomalies in oil and gas accumulations.

The use of this group of methods is based on the laws of changes in the physical properties of rocks in the oxidation zone of the deposits of various minerals and redistribution under the influence of migration of light hydrocarbons from deposits and atmospheric oxygen to form a powerful redox barrier in the upper part of the section (in the depth range 300-500 m).

6. Conclusion

The experience of many years of geophysical studies for oil and gas south-eastern part of Western Siberia shows that the structure of the observed gravity, magnetic, radio geochemical fields natural way reflects the geological structure of the section and the spatial location of hydrocarbon deposits.
Introduction staging, common throughout the civilized world, before directing involves carrying out seismic surveys modern "direct" searching for oil and gas (magnetometry, radiometry, geochemical methods, etc.). At the heart of many of them is the "breath" of deposits - diffusion and filtration of hydrocarbon gases in the section to the surface of the Earth and those geochemical changes that occur under the influence of hydrocarbons in the context of rocks, soil, as well as contact between the accumulations of hydrocarbons and water or rocks.

To improve the efficiency of geological prediction methods developed complex neural network processing and analysis of multidimensional information. The use of this software provides an interactive interpretive processing seismic data, as well as a comprehensive interpretation of it with other geophysical methods on the PC, the formulation of the reporting cartographic material in accordance with the national standards.

Studies carried out in the framework of improving the competitiveness of the National Research Tomsk State University.

References

[1] Sanders D.F., Tompson C.K.Integrated exploration improves widcat success (part I). *World Oil*, September, 1987, pp 36-45

[2] Siegel F.R., Hu Decheng, Vaz J.E., Wang Zaiming, Viterito A. Areal thermoluminescence radiometric survey of Shengping oil using buried dosimeters. *Oil and Gas Journal*, july 3, 1989, pp. 53-57

[3] Wang Z., Qin D., Zhuang G., Zha Z., Wang S., Shen W. and Cai G. Application of thermoluminescence dozimetry in the exploration for oil and gas using Chinese GR-200 LiF (Mg,Cu,P) TLD. *Radiation Protection Dozimetry,* vol. 47, 1993, pp. 323-326

[4] Rostovtsev V.N. Geological and genetic model of productive horizon - the basis of the search difficult constructed oil and gas deposits *Collection of scientific papers SNIIGGiMS. - Novosibirsk*, 1997, pp.35-43

[5] Alekseev F.N., Rostovtsev V.N., Parovinchak M.S. New opportunities to improve the efficiency of exploration for oil and gaz.- *Tomsk: TSU,* 1997, pp.56-79

[6] Sobolev I.S., Rikhvanov L.P. et al. Prediction and prospecting of oil and gas radiogeochemical methods // *Journal Oil and Gas Geology*, 1999, **No 3**, pp.18-27

[7] Gorbachev S.V., Rudakov I.B. Neural network technology mineral exploration // *Journal "Proceedings of the higher educational institutions. Physics", Tomsk,* 2013, **No 10/2,** pp.16-17

[8] Syryamkin V.I., Gorbachev S.V. The program neural network predictive model for calculating oil and gas field on geophysical and geochemical data "Neuro-S". *ROSPATENT. Certificate* **No 2010613255** от 17.05.2010

[9] Bishop C.M. Neural networks and pattern recognition. *Oxford: Oxford Press*, 1995, pp.169-193

[10] Gorbachev S.V., Syryamkin V.I., Rudakov I.B. "Recognition of complex deposits of oil, gas, based on neuro-fuzzy portraits". *LAMBERT Academic Publishing, Saarbrucken,* 2013, pp.24-59

[11] J. Hunt. Geochemistry and geology of oil and gaza.- *Moscow: "The World"*, 1982, pp.16-58

IOP Conf. Series: Materials Science and Engineering **81** (2015) 012107 doi:10.1088/1757-899X/81/1/012107

Event-driven management algorithm of an Engineering documents circulation system

V Kuzenkov[1], A Zebzeev[2], E Gromakov[3]

[1] Graduate student of Tomsk Polytechnic University, CEO of JSC "TomskNIPIneft", Tomsk, 634050, Russia
[2] Graduate student of Tomsk Polytechnic University, lead engineer of JSC "TomskNIPIneft", Tomsk, 634050, Russia
[3] PhD, Tomsk Polytechnic University, Tomsk, 634050, Russia

Gromakov@tpu.ru

Abstract. Development methodology of an engineering documents circulation system in the design company is reviewed. Discrete event-driven automatic models using description algorithms of project management is offered. Petri net use for dynamic design of projects is offered.

1. Introduction

Project management in the oil-and-gas survey design companies (SDC) is an actual task. Various practices, in particularly PMI [1,2], are used for this. A complex flow of engineering documents is formed during the project activities execution process. The main disadvantages of the existing engineering documents circulation and design workflow management include the following:

- non-transparent of the project execution processes due to defect procedures of documentation process, distribution and keeping of documents, accounting documents;
- impossibility of modification process tracking of engineering document of design management in a multi-user environment SDC, complexity of detection and identification of insertion errors and difficulties to troubleshooting, respectively;
- impossibility of effective monitoring of the design process and objective assessment of the degree of the readiness of the project or its parts at a given time, including tracking their status within the life cycle and the location in the work process, as there is no centralized engineering project data management.

So, all SDC are constantly working on improving the engineering documents circulation system (EDCS): adaptation of packaged commercial products or development their own portal systems of design document management [3].

The aim of this work is evolution of EDCS development methodology in the design company independently. The following procedure for development of such EDCS is offered: generate detailed static (circuit) models describing the flow of project activities and information flows (engineering documents) accompanying their; develop a model of supervisory core which providing project activities management; convert these models into dynamic (coupled with the attributive settings of documents circulation in the company in ideal); use the results of the design for EDCS software

IOP Conf. Series: Materials Science and Engineering **81** (2015) 012107 doi:10.1088/1757-899X/81/1/012107

coding. It can be expected that such a sequence of design will provide the necessary system project manageability.

2. Description of design process engineering documents circulation system in the design company

The suitable option of the such development beginning is eEPC-notation of ARIS CASE-package, which is well suited for building the static models used at all levels of project management. Such description is used in the quality management system of many subsidiaries oil-and-gas companies, in particular, of "TomskNIPIneft". Quite easy eEPC model can be converted to a dynamic model, in particular to the Petri net (PN), because it represents a strict model concept of interleaving of eventfulness and workflow.

Visual format of eERS-notation of projects design is a best practice, which providing understability and consistency of automaton model semantics of the SDC. In addition to the formal verification of discrete event systems (DES) control It provides semantic verification of the model description of the project activities through a collective agreement and refinement of the structural features and specifications context. As well as, such descriptions have an essential advantages over operational monitoring of requirement changes of specifications at execution of projects.

Model of project activities in the eEPC-notation can be presented by a finite set of events E of a finite set of functions F, a finite set of connectors C, a connector function T, which shows each connector to the appropriate type, and a finite set of directed arcs A [4]:

$$T \in C \to (or, xor, and), \tag{1}$$

$$A \subseteq (E \cup F) \cup (F \times E) \cup (E \times C) \cup (C \times E) \cup (F \times C)(C \cup F) \cup (C \times C). \tag{2}$$

The resulting EPC diagrams consist of three types of event units E of functions F and connectors C. Type of each connector is set by the function T. The ratio A is specified by the directed arcs, which connect event functions and connectors. This ratios are mutually disjoint (two functions and two events don't get connected to each other). Also, only events are the beginning and end of the diagram. Event description of the project activities allows to classify project management in the DES class [5]. Figure 1 shows eEPC model of works organization of preparation and issuance of tasks for project parts working out (TPPW).

As is seen from the model this notation describes not only operating sequence full but also document contents used in the EDCS operations. DES of this type can be described by the triple (G, K, S), where G – is the projects, K – requirements (specifications), establishing procedures for their implementation, and S – supervisor which must have the manageability property of the design process of the SDC under given specification constraints of projects performance. The functioning of such a DES is characterized by a set of events E in the project life cycle and management sequence e_c based on the events from E. A set of management sequences S form an attribute content languages of DES: the language $L(G)$ generated by status of project G, the language $L(S)$ generated by the supervisor S.

Description of the project activity in this notation allows for EDCS to set the event sequence path of project performance and by this means to formalize the DES specification. This allows to set a conceptual description of the engineering documents circulation in the SDC using language $L(G)$ and $L(S)$. This formalization allows to determine sequence of control commands in the EDCS. This control commands are presented in accordance with the regulations of the DSC engineering documents. Engineering documents should be worked out by the project executors.

The EDCS structuring can start with the separation of all projects in the SDC. Let it be $G = (G^1, G^2, ..., G^n)$. For each G^i defined: a set of events that are supervisory commands E^i_c, a set of the expected events $E^i w$ that characterize the routine response of project activities (a set of expected reactions E^i_c to control of the project execution) and finally a set of events E^i_u is determined as external relative to G^i. Common alphabet for G^i is $E^i = (E^i_w, E^i_c, E^i_u)$.

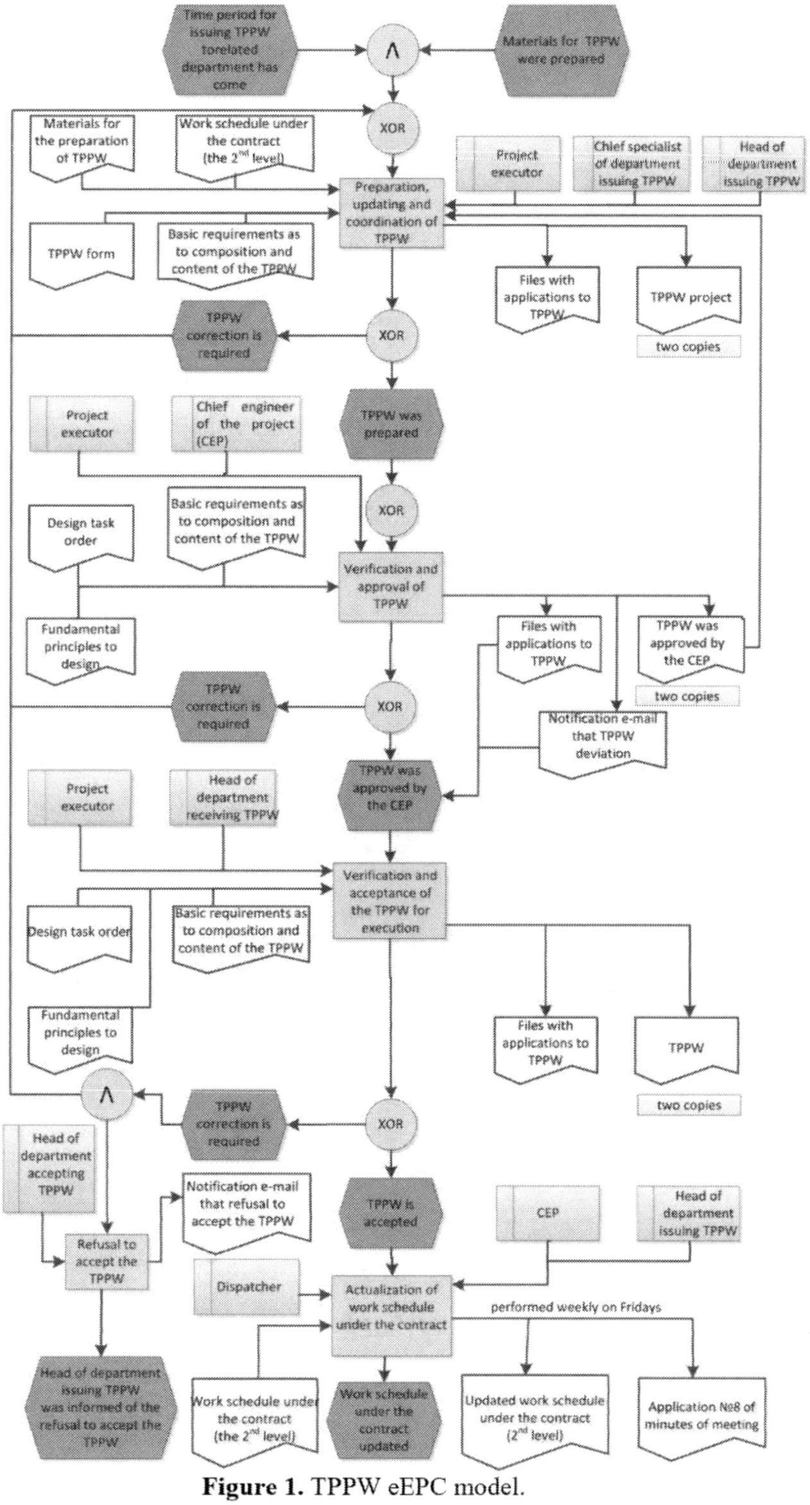

Figure 1. TPPW eEPC model.

Relation of the commands (E^i_c) of expected reactions and external conditions are usually presented in the form of diagrams in regulatory documents. It can be formalized by a finite state machine G^i [6]:

$$G^i = (Q^i, E^i, \delta, \Gamma^i, Q^i_m, q^i_0), \quad (3)$$

where:

Q^i – is a finite set of project states; E^i – is a set of project events; δ – is the transition function; Γ^i — is a set of acceptable events; $Q_m{}^i$ – is a set of states of Q (i.e. $\Gamma \subseteq Q$) called accept states; $q_0{}^i$ – is the start state, that is, the state of the automaton before any input has been processed, where $q_0 \in Q$.

As a result, at an early stage are determined: the composition of projects $G = (G^1, G^2 ... G^n)$ and a set of events E^i, each of which is structured on the sets E^i_w, E^i_c, E^i_u of routine, managed and external events. Let's consider procedural supervisor module which should form the procedure of preparing and description TPPW, detection modification history of engineering document of design management in a multiuser environment of SDC, provide the necessary management by coordination and approval of projects.

This procedure can be formalized machine G^5 which shown in Figure 2.

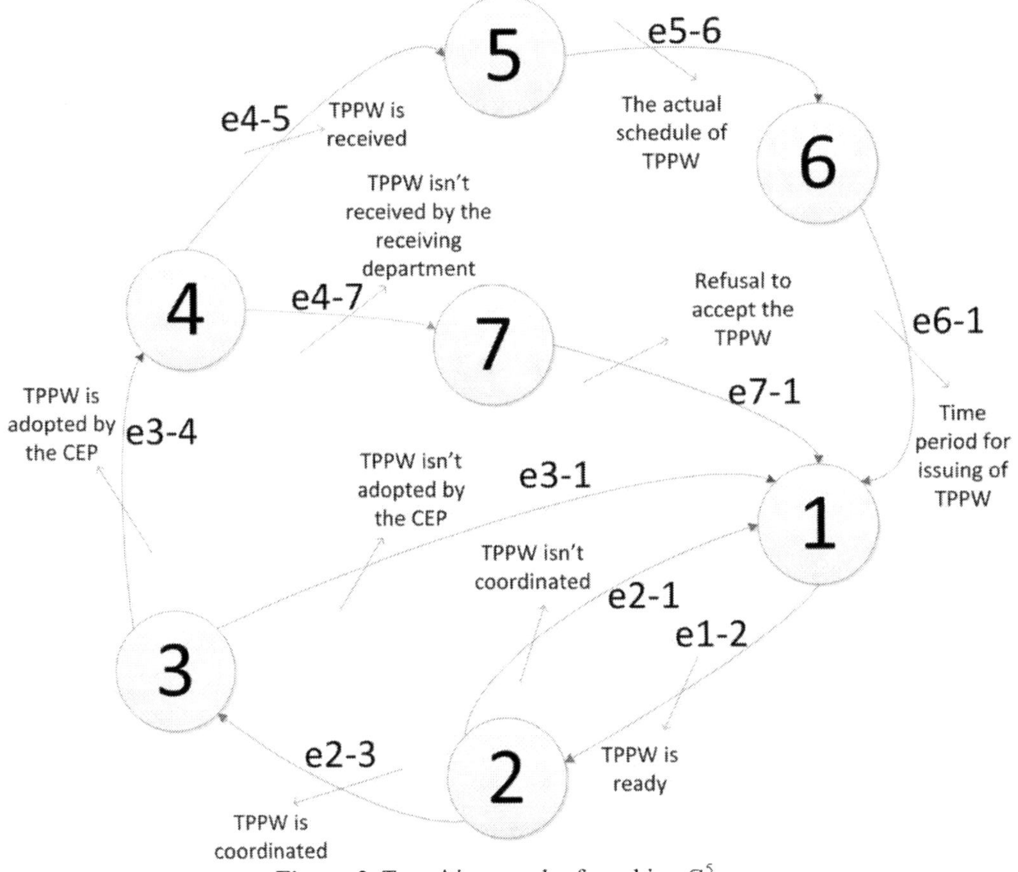

Figure 2. Transition graph of machine G^5.

The following states (from 1 to 7, respectively) generate in the machine: "TPPW preparation", "allows of TPPW in the department", "TPPW adoption by the chief engineer of the project (CEP)", "TPPW consideration by the receiving department", "updated work schedule under contract "," monitoring of TPPW dates", "TPPW deviation". As the graph shows, *e3-1, e3-4, e5-6* of a set of events E^i are related to the supervisory commands of the CEP. *e1-2, e2-1, e2-3, e4-1, e6-1, e7-1* are the expected events on the control commands. *e4-7* event does not related to the expected, because it usually appears after changing of external requirements to the project, unrecorded in preparation of TPPW. Automate description of project activities in the graph of projects eventness G^5 (Figure 2) allows to conclude that the engineering project documentation can be divided into several paths. One of them related to the conceptual type events (for example, changes in the project execution conditions, such as "*1-2-3-1*", "*1-2-3-4-7-1*"), the other - to the operating sequence of the project execution conditions ("*1-2-1*", "*1-2-3-4-5-6-1*"). From the *G*-automaton model of project activities of SDC follows that possible dead states of projects execution, related to stopping the project, can be solved by the changes of supervisor data and the status of the project or create a new project, which is ready to start.

According to our proposed methodology for the subsequent verification of the conceptual EDCS projects, its structural model must be converted to the PN notation, which will allow to establish manageability and observability of the decision. In order to reduce the PN dimensions and to minimize the number of possible system (structural) errors is logical to use operations patterns (standard samples) of the project activity [7,8]. During EDCS software development PN patterns are converted into object-oriented software modules that identify and show relation between classes or objects, without determining what the final classes or application objects to be used. The pattern denominates a solution the custom name that making it easier for communication developers between the individual EDCS modules, allowing to refer to the known patterns. Thus, due to the patterns, software component standardization is executed and the number of possible errors is reduced. Properly formulated design pattern allows to use it again and again after it finds a good solution. A set of operation patterns of the project activity in the SDC simplifies the EDCS developer a decision of the most suitable variant of its construction.

3. Summary

Automaton-oriented technology of the EDCS management in any (oil-and-gas, construction, energy or other) design company provides an understability and predictable during projects processes. The proposed algorithm preparation of the EDCS management will have consistensy property and will accord to established specifications (documented procedures in the SDC, operating procedures) as in supervisory management of projects portfolio as a management of portfolio individual projects.

References

[1] A Guide to the Project Management Body of Knowledge (PMBOK Guide) – Fifth Edition, Project Management Institute 2013.
[2] Ron Basu, Managing quality in projects: An empirical study, International Journal of Project Management 32 2014, pp. 178–187.
[3] Kuzenkov V, Zebzeev A, GromakovE., Resource leveling in the project design process by Petri net using, Advanced Materials Research Vol. 905 (2014) pp.752-756.
[4] Repin V.V., Eliferov V.G. Process approach to management. Modeling business processes. - M. Mann, Ivanov and Ferber 2013, 544 p.
[5] Christos G. Cassandras, Introduction to Discrete Event Systems. Second Edition. Springer Science & Business Media LLC 2008, 769 p.
[6] Linz, Peter. An Introduction to Formal Languages and Automata (5th ed.). Sudbury, MA: Jones & Bartlett Learning 2011, 437 p.
[7] N. Russell, A.H.M. ter Hofstede, W.M.P. van der Aalst, and N. Mulyar. Workflow Control-Flow Patterns: A Revised View. BPM Center Report BPM-06-22 2006, 134 p.
[8] W.M.P van der Aalst, A.H.M. ter Hofstede, B. Kiepuszewski, and A.P. Barros. Workflow Patterns. Distributed and Parallel Databases, **14(3),** p. 5-51, July 2003

IOP Conf. Series: Materials Science and Engineering **81** (2015) 012108 doi:10.1088/1757-899X/81/1/012108

Fuzzy control of the production environment process parameters

V N Izvekov
Tomsk Polytechnic University Lenina Avenue, 30, 634050, Tomsk, Russia
E-mail: izvekovvn@tpu.ru

Abstract. The fuzzy control process for support of given microclimatic production environment process parameters with loss of one from values, regulating regime of process was shown. The structural schematic decisions with algorithm of functioning and oriented to existing apparatus (means of realization) was presented.

1. Introduction

In according with the set of the normative documents [1–3] microclimatic parameters of production environment (temperature –T, humidity – Φ, air velocity) have to are supported with the tolerance limits.

These limits propose certain spread of the parameters (interval).

There are the set of productions (for example, medicine, production of superfine substances, crystals and others) where microclimatic parameters of production environment have to are supported with high accuracy that allow achieve high quality of the final product. This task is solved with help of programming devises for conditioning of production environment parameters. They provide the given parameters and their control is realized either individual sensors or combined control devices with information transfer to regulating devices and support of production environment parameters constantly.

2. The structural scheme of settings

The structural scheme similar setting may be illustrated Figure1.

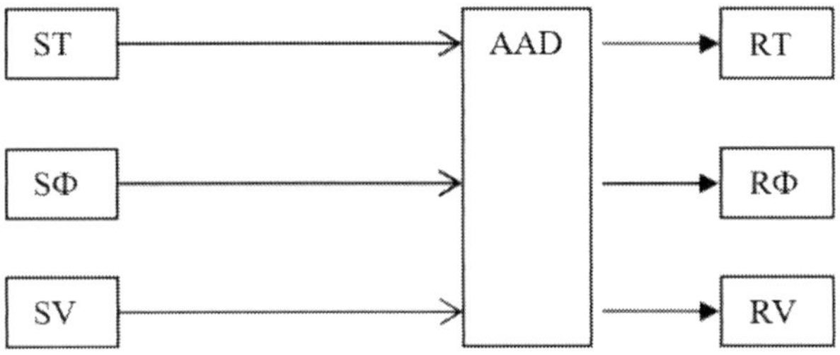

Figure 1. Structural scheme of setting: ST – sensor T; SΦ – sensor Φ; SV– sensor V; AAD – arrangement of assumption decision; RT – regulator T; RΦ – regulator Φ; RV – regulator V.

In common case the work regime of conditioner depends from informational parameters, that is

$$R_{\kappa} = f\,(T,\ \Phi,\ V).$$

Possibly we have situation, when one from sensors is put out of action, that lead to distortion of information about production environment parameters, violation of given state and disparity of final quality product with regard of established requirements.

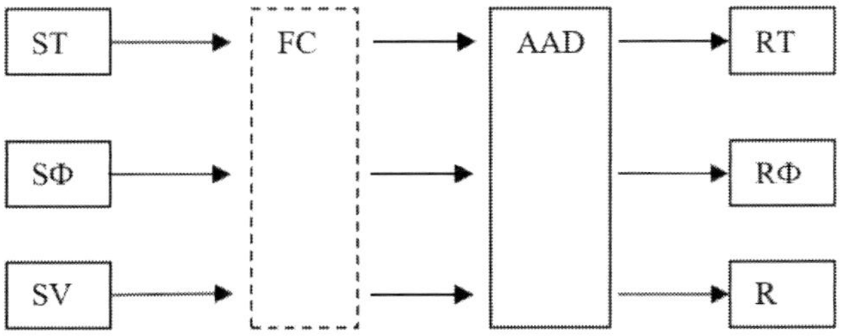

Figure 2. Structural scheme of setting: *ST* – sensor *T*; *SΦ* – sensor Ф; *SV*– sensor *V*; *FC* – fuzzy control; *AAD* – arrangement of assumption decision; *RT* – regulator *T*; *RΦ* – regulator *Φ*; *RV* – regulator *V*.

For registration loss of information from one of sensors (sensor *T*, sensor *Φ*, sensor *V*) that is support technical process framed given parameters, one may using in structural scheme by Figure 1 fuzzy controller (Figure 2).

3. Principal of fuzzy control action

Principal of action FC is founded on fuzzy logic conception [4,5]. There are basic fuzzy logic conceptions:
- linguistic variable (*LV*) (any physical value, which has more than only «YES» or «NO»);
- term (great number of *LV*);
- membership function (describe a degree of fulfilment *LV* by term);
- fuzzy base of knowledge (aggregate fuzzy rules determining interrelation between input and output values of researched objects).

The realization of fuzzy process consists of from following stages [5]:
- fuzzification (determination of confidence degree that output *LV* accepts – concrete term);

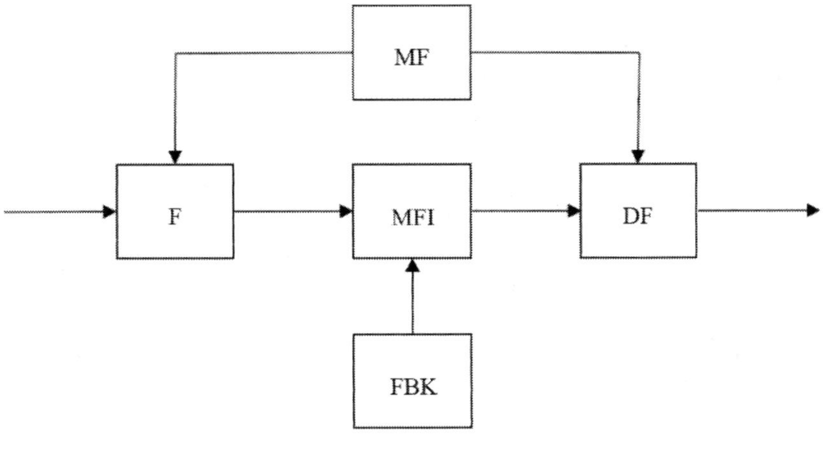

Figure 3. Fuzzy inference: *F* – fuzzificator; *MF* – membership functions; *MNI* – machine of fuzzy inference; *FBK* – fuzzy base of knowledge; *DF* – defuzzificator.

- fuzzy logic inference (calculation of true membership meaning *LV* by term on base of rules set, that is fuzzy base of knowledge);
- defuzzification (transformation of fuzzy meanings set *LV* to exact meanings).

The functional scheme of process is presented on Figure 3.

If we have given values LV and their rages with chosen membership functions, fuzzy base of knowledge FC with given parameters for case of loss information (when one from sensors either T, Φ or V is put out of action) may be realized.

Today there are many series FC microchips, realizing the scheme Figure 2. The proposed fuzzy control model of production environment parameters constancy allow compensate loss of information, that is provide supporting given parameters of production environment and adequate production quality.

For illustrating mentioned reasoning discuss the concrete example of device realizing by the scheme Figure 2.

We have meanings of production environment parameters with given accuracy of supporting

$$T = (22 \pm 1)\,^{o}C;\ \Phi = (60 \pm 2)\ \%;\ V = (0{,}2 \pm 0{,}05)\ m/s.$$

All rates measurable with well known multitude of meanings. From work experience with system the set rules of interrelation between parameters are known. Even if we have situation when one from sensors of parameters measuring is put out of action (for example sensor of V), that is information about V was loss. In this case we have task about appraisal of nearly meaning V if meanings E, Φ and their interrelations are known by following rules:

- IF *Temperature* (T) *low* END *Humidity* (Φ) *low* THAT *Air velocity* (V) *low*;
- IF *Temperature* (T) *middle* THAT *Air velocity* (V) *middle;*
- IF *Temperature* (T) *high* OR *Humidity* (Φ) *high* THAT *Air velocity* (V) *high.*

In this case T, Φ, V are linguistic variables. Describe theirs with help of universes (multitude of possibly meanings) and membership functions. For this we use the standard membership functions Z, P, L, S - types [4,5,7].

Temperature. Univers – piece [(21 - 23) °C].

Initial multitude of terms – {*High, Middle, Low*}.

Membership functions (Figure 4):

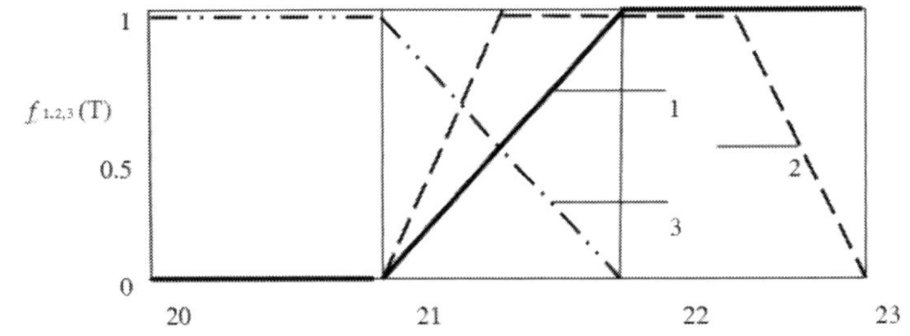

Figure 4. Membership functions of terms. 1 – T *high*; 2 – T *middle*; 3 – T *low*.

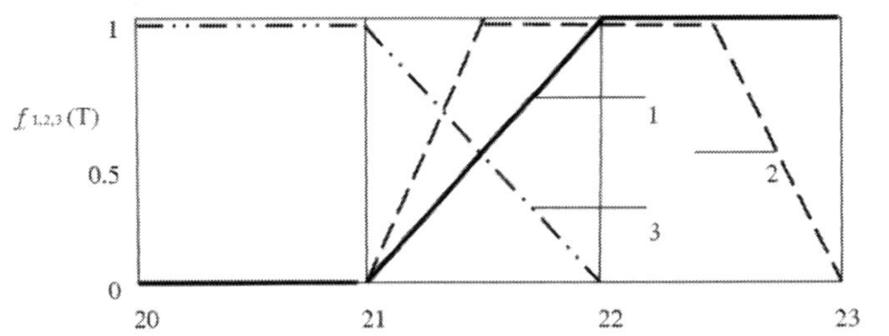

Figure 5. Membership functions of terms. 1 – Φ *high*; 2 – Φ *middle*; 3 – Φ *low*.

Humidity. Univers - piece [(58 - 62) %].
Initial multitude of terms - {*High, Middle, Low*}.
Membership functions: Figure 5.
Air velocity. Univers - piece [(0.15 – 0.25) m/s].
Initial multitude of terms - {*High, Middle, Low*}.
Membership functions (Figure 6):

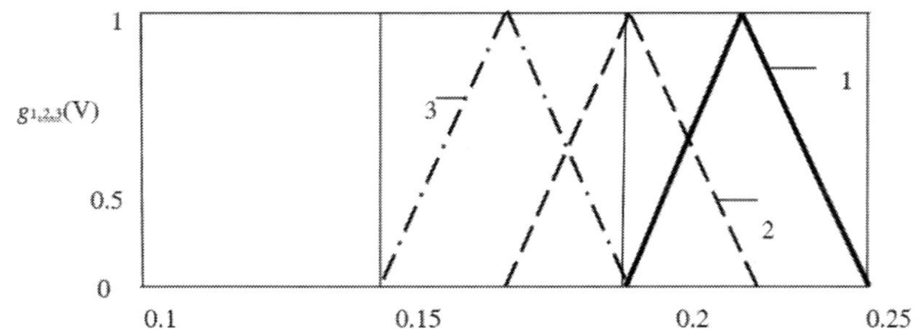

Figure 6. Membership functions of terms. 1 – V *high*; 2 – V *middle*; 3 – V *low*.

Fuzzification stage.
Measured meanings T and Φ are known (T = 21,5 °C; Φ = 60 %). Conduct the appraisal of necessary meaning V. According to input meanings T and Φ we find the degree of confidence affirmations by view " linguistic variable A is term of linguistic variable A" (conversion from given precise meanings to degree of confidences). In according to measured meanings and membership functions following degree of confidences may are founded.

- *Temperature* (T) *High* - 0.7;
- *Temperature* (T) *Middle* - 1.0;
- *Temperature* (T) *Low* - 0.3;
- *Humidity* (Φ) *High* - 0.7;
- *Humidity* (Φ) *Middle* - 1.0;
- *Humidity* (Φ) *Low* - 0.3.

Fuzzy inference stage.
We find degree of confidences of antecedents [4]:

- *Temperature* (T_l) Low END *Humidity* (Φ_l) Low,
  min (T_l, Φ_l) = min (0.3; 0.3) = 0.3;
- *Temperature* (T_m) *Middle* = 1.0;
- *Temperature* (T_h) *High* OR *Humidity* (Φ_h) High,
  max (T_h, Φ_h) = max (0.7; 0.7) = 0.7.

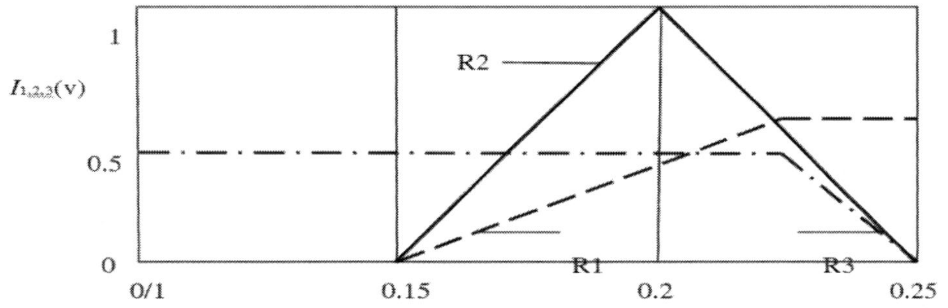

Figure 7. Functions of antecedent confidence: *R1* – rule 1; *R2* – rule 2; *R3* – rule 3.

Each from rules is a fuzzy implication [7]. We have degree of confidences of antecedents but degree of confidences of conclusion is given membership functions of adequate term [5]. Using one from means of fuzzy implication construction (minimum left and right parts) the new fuzzy variable may be obtained (in our case - implication to V) (Figure 7).

Composition (accumulation) stage.

On this stage it is necessary to combine all rules. There are several means of accumulation [5].

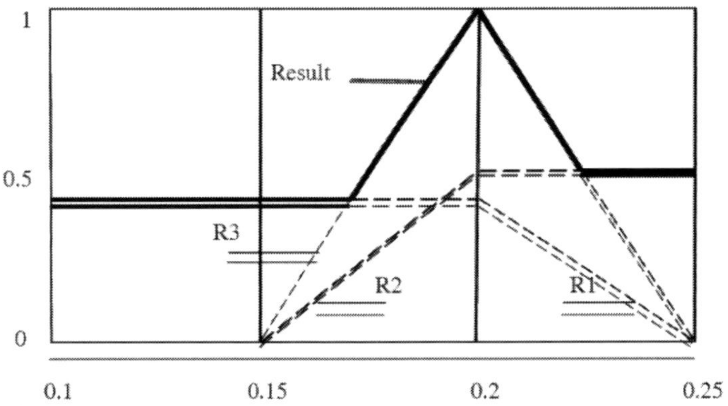

Figure 8. Membership function of output variable *V*.

One from the basic mean is construction of membership functions by rules numbers 1 – 3. In the result of construction following membership functions is obtained (Figure 8).

It is a new term of output variable V. Its membership function give degree of confidence for meanings input parameters and their rules of interrelations with output variable.

Defuzzification stage.

This stage is a stage of concrete meaning obtaining of from univers by given membership function. There are many methods of defuzzification but in our case sufficiently the method of the first maximum [8]. It's using to obtained membership function give the meaning of V = 0,2 m/s.

5. Conclusion

With the aid of obtained results fuzzy controller with given parameters for case of information loss (sensors are put out of action) about meaning either *T*,*Φ* or *V* may be realized.

Today there are many serial microchip complete sets fuzzy controllers for realization the scheme Figure 2 (for example, look at table 1). Their using in aggregate with the programme's resources of fuzzy systems projecting (for example, CAD - system fuzzy TECH 3.0) allow are designed the set of fuzzy control and regulation devises.

Table 1. Microchip complete sets fuzzy controllers

Platform	20 rules 2 in. and 1 out.	20 FAM-rules 2 in. and 1 out.	80 rules 3 in. and 1 out.
MCS-96, 16 bit, 80C196KD, integrated ROM, 20 MHz	0.28 ms 0.84 Kbyte ROM 63 byte RAM	0.29 ms 0.87 Kbyte ROM 63 byte RAM	0.43 ms 1.27 Kbyte ROM 69 byte RAM
MCS-51, 8 bit, 80C51, integrated ROM, 12 MHz	1.4 ms 0.54 Kbyte ROM 25 byte RAM	1.5 ms 0.58 Kbyte ROM 25 byte RAM	4.4 ms 1.0 Kbyte ROM 29 byte RAM

IOP Conf. Series: Materials Science and Engineering **81** (2015) 012108 doi:10.1088/1757-899X/81/1/012108

References

[1] GOST R ISO 14644-1-2002 "Clearn premises and controlled environments connected with them", part 1, Classification of air clean.
[2] GOST R 52249-2004 "The rules of the production and control quality medicine remedies ".
[3] OCT 42-510-98 " The rules of the production organization and control quality medicine remedies (GMP)".
[4] Zade L. A. The concept of a linguistic variable and its application to approximate reasoning. Part 1, 2, 3 // Information Sciences, n. 8 pp.199-249, pp.301-357; **n. 9** pp. 43-80.
[5] Applied fuzzy systems: Translation with japan./ K. Asai, Д. Vatada, C. Ivai and others.; under edit. T. Tarano, K. Asai, M. Sugeno. - M.: World, 1993.
[6] Efimov A.C., Morenov O.A. The basis of fuzzy logic, logic - . – Materials of seminar ITLab, NGU, Law Novgorod, 2004.
[7] Mamdani E. H. Applications of fuzzy algorithms for simple dynamic plant. Porc. IEE. vol. 121, n. **12**, pp. 1585-1588, 1974.
[8] Zimmermann H.-J. Fuzzy Sets, Decision Making and Expert Systems.- Kluwer:Dordrecht.-1987.- 335 p.
[9] Rotshtain A.П. "Intellectual technologies of identification: fuzzy logic, genetical algorithms, neuron nets." - Vinnica: UNIVERSUM- Vinnica, 1999. - 320 c.

IOP Conf. Series: Materials Science and Engineering **81** (2015) 012109 doi:10.1088/1757-899X/81/1/012109

Carrying out thermodynamic calculations and definition of the main reactions of decomposition of vapours of ethyl alcohol

A I Sechin, O S Kyrmakova, T A Ivanova
National Research Tomsk polytechnic university, 634050, Tomsk, Lenin Avenue, 30, Russia

E-mail: olia_917@mail.ru

Abstract. Thermodynamic opportunities of course of chemical reactions of decomposition of the vapors of ethyl alcohol necessary at development of devices where these reactions will take place are considered. The entalpiyny method of calculation of constants of balance of probable chemical reactions is given in the Excel editor. Independent reactions of process of oxidation are defined. By result of thermodynamic calculation of each reaction schedules of dependence of a constant of balance on environment temperature from which follows are constructed that one reactions proceed until the end of aside formation of the final products, and others are improbable or impossible. The analysis of the received results shows that reactions of oxidation will successfully proceed in the established directions, and for an intensification of process of decomposition it is necessary to provide a supply of some energy which quantity will be sufficient for oxidation of vapors of ethyl alcohol. Results of calculations showed good convergence with programs of thermodynamic calculations like "Aster – 4" and "TERRA".

1.Introduction

Problem of pollution of atmospheric air – one of the most serious global problems which the mankind faced. Danger of pollution of the atmosphere – that harmful substances get to clean air, pernicious for live organisms.

Gravity of the environmental problems connected with pollution of the atmosphere is illustrated by the following statistical data: only 15% of urban population of Russia breathe rather clean air. In 125 cities of the Russian Federation the exceeding maximum concentration limits of concentration of pollution of the atmosphere are annually fixed at 5-10 times [1]. Hit in it uncharacteristic physical, chemical and biological substances in the course of activity of the person is the main reason for air pollution.

In the course of production of medicinal substances, there is a need of cleaning of organized emissions in environment of vapors of various organic substances which are emitted as in synthesis processes, and processings of a number of chemicals. The main share of pollution of the atmosphere belongs to drying processes. The dividing systems representing two-level cleaning of dust fractions most often don't provide cleaning of a steam phase. There is a need of the solution of a question of cleaning of organized atmospheric emissions of vapors of organic chemistry.

For successful fight against similar pollution it is necessary to know as physical and chemical properties of the substances deleted from production volume, and the thermodynamic opportunities of course of chemical reactions necessary at development of devices where these reactions will take place.

This work is also devoted to the thermodynamic analysis of chemical catalytic reactions of decomposition of vapors of ethyl alcohol and definition of the main reactions.

2.Calculated part

The chemical thermodynamics considers interrelations between work and energy in relation to chemical transformations. As chemical transformation usually is followed by release or absorption of a certain amount of heat, it submits to the fundamental principles (beginnings) of thermodynamics. The chemical thermodynamics defines, first of all, conditions (such as temperature and pressure) courses of chemical reactions and equilibrium states which they reach [2].

For carrying out thermodynamic calculations it is necessary from possible reactions to define independent reactions of process of oxidation of ethyl alcohol.

For calculation of thermodynamic parameters of process of oxidation of ethyl alcohol balance reactions were made:

1) $C_2H_5OH + 3O_2 = 2CO_2 + 3H_2O$;
2) $C_2H_5OH \rightarrow C_2H_2 + H_2\uparrow$;
3) $C_2H_5OH \rightarrow CH_4 + H_2O + C$;
4) $C_2H_5OH \rightarrow CH_4 + CO + H_2\uparrow$;
5) $C_2H_5OH \rightarrow C_2H_4 + H_2O$;
6) $C_2H_5OH \rightarrow C_2H_2 + H_2O + H_2\uparrow$.

Number of independent reactions equally to number of substances (C_2H_5OH, O_2, CO_2, H_2O, C_2H_2, H_2, CH_4, C_2H_4, C) in system minus number of diverse atoms (to H_2, O_2, C) of which reagents are formed, i.e. the number of independent reactions is equal to six (9-3=6) [2]. Therefore, all presented reactions are independent.

Thermodynamic calculations were implemented on a method of calculation of the balance [3] considering dependence $\Delta H = f(T)$, by means of which $lgKp = f(T)$, and ΔH, $lgKp$ in the range of temperatures 0 ÷ 100 0C are counted.

Influence of temperature on a constant of balance is described by Vant Hoff's equation [4].

$$\frac{d\ln Kp}{dT} = \frac{\Delta H^0}{RT^2} \tag{2}$$

where, ΔH^0 – standard thermal effect of reaction; R – universal gas constant; T – temperature is ^{0}K; Kp – a balance constant.

In case of exact calculation and at big temperature intervals the equation (2) considers dependence of thermal effect of reaction on temperature. For this purpose we will substitute the equation

$$\Delta C_p = \Delta a + \Delta b \cdot T + \Delta c' \cdot T^2 \tag{3}$$

in the equation of the law of Kirchhoff [5].

Integrating, we will receive: in a final formula:

$$\left(\frac{d(\Delta H)}{dT}\right) \cdot p = \Delta C_p \tag{4}$$

$$\Delta H_T = \Delta H^0 + \Delta a \cdot T + \frac{\Delta b \cdot T^2}{2} + \frac{\Delta c \cdot T^3}{3} - \frac{\Delta c'}{T} \tag{5}$$

Combining the equation (2) with the equation (4) and integrating, we will receive the equation in technical system of coordinates:

$$\lg K_p = \frac{-\Delta H^0}{2.303 \cdot R \cdot T} + \frac{\Delta a}{R} \cdot \lg T + \frac{\Delta b \cdot T}{2.303 \cdot 2 \cdot R} + \frac{\Delta c \cdot T^2}{2.303 \cdot 6 \cdot R} + \frac{\Delta c' \cdot T^{-2}}{2.303 \cdot 2 \cdot R} + J''$$

According to the equation of an isobar of chemical reaction

$$\Delta G_T = -R \cdot T \cdot \ln K_p = -2.303 \cdot R \cdot T \cdot \lg K_p \tag{6}$$

$$\Delta G_T = \Delta H^0 - \Delta a \cdot T \cdot \ln T - \frac{\Delta b \cdot T^2}{2} - \frac{\Delta c \cdot T^3}{6} - \frac{\Delta c' \cdot T}{2} + J \cdot T \tag{7}$$

where, $J = -2.303 \cdot R \cdot J''$.

In the equations 3 – 6 unknown sizes are Δa, $\Delta в$, Δc, $\Delta c'$, ΔH_{298}, ΔG_{298}, ΔH^0, J, which are in compliance of the equations:

$$\Delta H_{298} = \sum \Delta H_{298(прод)} - \sum \Delta H_{298(исх)} \quad (8)$$

$$\Delta G_{298} = \sum \Delta G_{298(прод)} - \sum \Delta G_{298(исх)} \quad (9)$$

$$\Delta a = \sum \Delta a_{(прод)} - \sum \Delta a_{(исх)} \quad (10)$$

$$\Delta b = \sum \Delta b_{(прод)} - \sum \Delta b_{(исх)} \quad (11)$$

$$\Delta c = \sum \Delta c_{(прод)} - \sum \Delta c_{(исх)} \quad (12)$$

$$\Delta c' = \sum \Delta c'_{(прод)} - \sum \Delta c'_{(исх)} \quad (13)$$

Coefficients *a, в, c, c'* and ΔH_{298}, ΔG_{298} undertake taking into account the stekhiometricheskikh of coefficients of the equation of the counted reactions. For stay ΔH^0, which are an extrapolation constant we will substitute in the equation (2) values Δa, $\Delta в$, Δc, $\Delta c'$ at T=298 ^{0}K and ΔH_{298} we will receive in case of high temperatures:

$$\Delta H^0 = \Delta H_{298} - \left(\Delta a \cdot 298 + \frac{\Delta b}{2} \cdot 298^2 + \frac{\Delta c}{3} \cdot 298^3 - \frac{\Delta c'}{298} \right) \quad (14)$$

For finding of *J* we will substitute values ΔH^0 (the equation 14), ΔG_{298} and T = 298 ^{0}K in the equation (7), we will receive

$$J = \frac{(\Delta G_{298} - \Delta H^0)}{T} + \Delta a \cdot \ln T + \frac{\Delta b \cdot T}{2} + \frac{\Delta c \cdot T^2}{6} + \frac{\Delta c' \cdot T^{-2}}{2} = \frac{(\Delta G_{298} - \Delta H^0)}{298} + \quad (15)$$

$$+\Delta a \cdot 5.69 + \Delta b \cdot 149 + \Delta c \cdot 14801 + \Delta c' \cdot 5.63 \cdot 10^{-6}$$

We will substitute values 8 – 15 in 5 and we will receive:

$$\lg K_p = \frac{-\Delta H^0 / T + \Delta a \cdot 2.3 \cdot \lg T + 0.5 \cdot \Delta b \cdot T + 0.5 \cdot \Delta c' \cdot T^{-2} + J''}{2.303 \cdot R} \quad (16)$$

It is known that thermodynamic parameters share on extensive and intensive. The sizes proportional to weight (or to amount of substance) the considered thermodynamic system, carry to extensive, it is volume, internal energy, an enthalpy, etc., intensive sizes don't depend on the mass of thermodynamic system. It, for example, temperature and pressure.

3 Results

By result of thermodynamic calculation of each reaction schedules of dependence of a constant of balance of *lgKp* on environment temperature which are represented in figure 1 were constructed.

Reactions of decomposition can go in several directions. Everything depends on energy which is applied to these processes.

From figure 1 it is visible that reactions:

$C_2H_5OH \rightarrow C_2H_2 + H_2\uparrow$ и $C_2H_5OH \rightarrow C_2H_2 + H_2O + H_2\uparrow$

aren't feasible as have negative value of a constant of balance of lgKp and course of reaction towards formation of products it is impossible.

Following reactions:

$C_2H_5OH \rightarrow CH_4 + H_2O + C$;
$C_2H_5OH \rightarrow CH_4 + CO + H_2\uparrow$;
$C_2H_5OH \rightarrow C_2H_4 + H_2O$

have low probability of course towards formation of products.

Reaction with high probability of course until the end of aside formation of the final products the following:

$C_2H_5OH + 3O_2 = 2CO_2 + 3H_2O$.

After the analysis of schedules of dependence of a constant of balance on temperature of the environment the directions of decomposition of reactions were defined.

From schedules it is visible that one reactions proceed until the end of aside formation of the final products, and others are improbable or impossible.

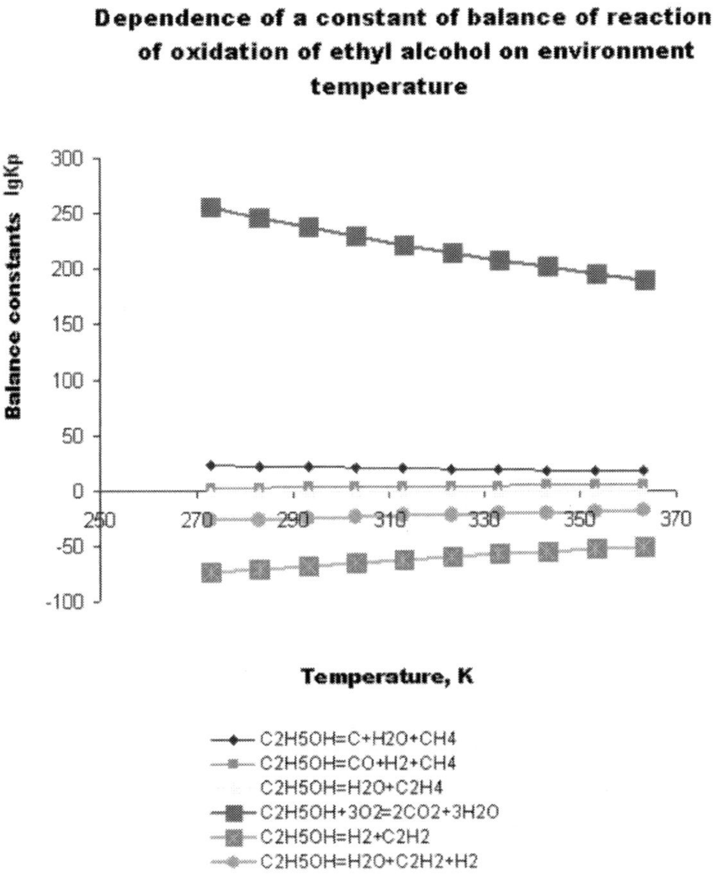

Figure 1. Dependence of a constant of balance of reaction of oxidation of ethyl lgKp alcohol on environment temperature.

The analysis of the received results allows to draw a conclusion that reactions of oxidation will successfully proceed in the established directions, and for an intensification of process of decomposition it is necessary to provide a supply of some energy which quantity will be sufficient for oxidation of vapors of alcohol, thereby minimizing technological emission on a steam component in environment.

4. Conclusion

As a result of the conducted research thermodynamic calculation of possible reactions of oxidation of alcohol was considered. Independent reactions of process of oxidation are defined. Thermodynamic calculations of process of oxidation of ethyl alcohol are carried out (balance constants, Gibbs's energy). The entalpiyny method of calculation of constants of balance of chemical reactions is given in the Excel editor. Results of calculations showed good convergence with programs of thermodynamic calculations like "Aster – 4" and "TERRA" [6, 7].

References:

[1] Smurov A., Snakin V., Komarova N. Current state of atmospheric air . *Ecology of Russia, Educational edition.* 2012. P.12–33
[2] Kosintsev V., Kulikova M., Sechin A., Bordunov S., Prokudin I. Physical and chemical bases of chemical processes of receiving inorganic salts: manual. Tomsk , Tomsk University Publ., 2008. 58
[3] Bezdenezhnyh A. Engineering methods of drawing up equations of speeds of reactions and calculation of kinetic constants. Chemistry , 1973
[4] Shilyaev M, Dorokhov A Criteria of a choice and comparison of devices of gas purification. Izv. higher education institutions. Construction, 1998, **6**. P.81–84
[5] Vladimirov L . The accelerated methods of thermodynamic calculations of chemical reactions. LSU Publ. , 1956
[6] Dyachenko A., Ftorammoniynaya technology of processing of mineral raw materials: the abstract of the thesis on competition of an academic degree the Dr.Sci.Tech. Tomsk, 2008
[7] Pushkarev A. The gaseous plasmochemical processes initiated by a pulse electron beam: the abstract of the thesis on competition of an academic degree of doctor of physical and mathematical sciences. The automated system of thermodynamic calculations of "TERRA" in the range of temperatures 300-10000 0K and pressure of initial mix of 0,01-0,1 MPas. Tomsk ,2007

To study the influence of snow cover on the power generated by photovoltaic modules

R V Meyta[1] , F V Savrasov[2]

[1] Postgraduate student, Tomsk Polytechnic University, Tomsk, Russia
[2] Senior tutor, PhD, Tomsk Polytechnic University, Tomsk, Russia

E-mail: savrasov@tpu.ru

Abstract. The possibility of increasing the power generation of PV modules due to radiation reflected from the snow is investigated. An experiment has been held to determine the value of the current generated by PV-modules on account of the underlying snow surface. Also the experiment has been held to determine the influence of the PV-module's location due the snow surface on the performance of solar cells.

1. Introduction

The area of using micro-devices for different purposes is expanding rapidly. However, many electronic devices are not integral parts of a hardware system, but autonomous systems that operate without direct human intervention. Accordingly, there is the problem of providing a continuous and stable power supply of these devices. Currently becoming more common for energy autonomous systems to get power supply from renewable energy sources, among which the most promising (because of its availability) is a photovoltaic modules [1, 2].

Despite the popularity of this type of energy in other countries, in Russia it is still a very limited kind of distribution. At the same time, using of this in some Russian regions is highly advisable due to climatic features a number of subjects of the Russian Federation. This feature is associated with the periodic occurrence of snow cover, which, under certain conditions, can be a natural concentrator of the solar energy that can be used for photo-conversion (for comparison – in countries with hot climates concentrators are made artificially; this fact causes a significant increase of the photovoltaic installation's cost).

Snow cover – a layer of snow on the surface of the earth, formed as a result of snowfall and blizzards. Snow cover has a low density, increasing over time, especially in the spring. Reflectivity (albedo) of fresh snow up to 90% of the value of incident light flux depending on the angle of incidence (for comparison, the albedo of dark soil is about 10%). The reflectivity of the melting snow is not more than 40% [3]. Most parts of the Russian are covered by snow during 4-5 months in the year, and in the northern regions – 8-9 months. The average duration of snow cover in Tomsk, where research was provided, is more than 6 months. It appears mainly in the beginning of October, and begins to defrost in the middle of March.

Accordingly, using PV-converters in the places of possible snow occurrence, it seems interesting to measure the influence of snow cover to the performance of a single photovoltaic module, and as a result, the performance of the photovoltaic system. Studies devoted to the investigation of this effect were considered, particularly, in [4, 5]; in this cases, the results of these studies provide information over a long period, and for a set of combined photovoltaic modules (PV-power stations).

As is known, the energy generated by PV modules, depends on the intensity of solar radiation and the angle of the receiving surface's module orientation relatively to the sun. In turn, the efficiency of the module is simultaneously influenced by both the direct rays, and the light reflected from the underlying surface and diffused in the surrounding atmosphere [3].

In addition to transparency and clouds, a great influence on the diffused radiation has the nature of the underlying surface. As the reflectivity of the underlying surface is increasing, the flux of scattered radiation is greatly increasing too. In the presence of snow cover reflection of direct solar radiation increases, and its secondary scattering in the atmosphere leads to an increase of the diffused radiation.

2. Practical experiments

2.1. Basic study of the effects of radiation reflected from the snow

To determine the quality influence of the snow cover's reflectivity on the power value generated by the PV module, the special mechanical system of sun tracking IGS-1 was developed and used (figure 1). It allows to place on a special rod small-sized photovoltaic module as the sensing element. The measure element of the standard pyranometer M-80 used as standard model.

Figure 1. Experimental system for tracking the sun.

Mechanical system allows rotating module in a horizontal plane from 0° to 360° and vertical plane from 0° to 90°, with the possibility of placing the receiving surface of the PV converters in parallel down the underlying surface.

The experiment consisted of 2 parts. During the first, IGS-1 was set to open unshaded horizontal surface with a clean and uniform snow on the black cloth to remove the effects of reflected sunlight on the measuring module (figure 1). During the second, IGS-1 was located directly on the snow.

The measuring of generated current was provided February 8, 2013, from 10:00 to 16:00 (on the Tomsk time) with a digital multi-meter M-832P. The reception surface of the module was located at the angle to the underlying surface equal to the latitude of the area (about 56°), and was oriented to the south.

As the experimental unit, photovoltaic module FSM-2-6 (rated power – 2 W, open circuit voltage 6 V, short-circuit current 0.5 A, dimensions 230×145×4 mm) made of monocrystalline silicon, standard produced by JSC 'NIIPP', was selected.

The venue of the experiment was chosen flat area close to the riverside Tom. Over the entire period of observation the sun was not eclipsed by clouds and near buildings and trees. The average ambient temperature during the period of observation was -18 °C.

During the measurements, it was observed that the PV module's voltage varies slightly depending on the sunlight. The results of measuring the current produced by the modules shown in figure 2.

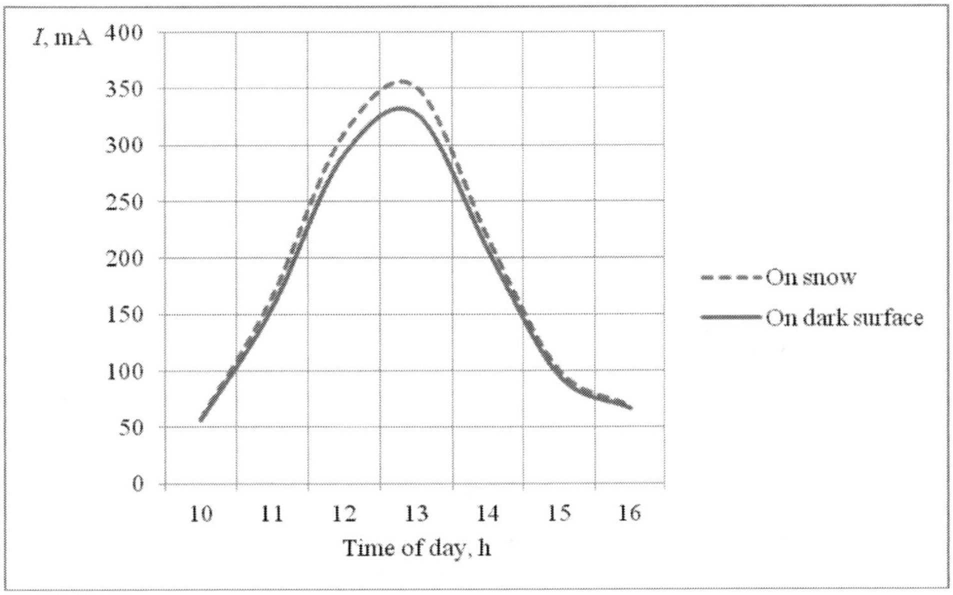

Figure 2. The measurement results of IGS-1 current depending on the type of the underlying surface.

From these data it is follows that photovoltaic module, disposed directly onto the snow, produces more energy than onto black cloth (to 7.3%). It should be noted the fact that at different illumination of photovoltaic module depending on the time of day (and, therefore, the position of the sun in the sky), there is a difference in the generated current values. Thus, the lighting the module disposed at angle in which the reflected insolation has less effect on PV modules, there is a decrease in the effectiveness of the current generation.

Total (all day) value of the current generated by the arrangement of photovoltaic module on the snow, exceeds the value of the module generated on a black canvas is greater than 6%. Accordingly, it is assumed that with the presence of stable snow cover with high albedo in a shorter duration of daylight of the year (for example, in December and January), the resulting value will be less, and in the longer (March) – more.

Thus, the experiment shows that the reflected radiation from the snow affects the current produced by the PV converter in the direction of its increase. Therefore, it makes sense to install PV modules, which are an integral part of the power supply devices, in the places with a high albedo (for example, on the roof, painted in bright colors, the roof of galvanized iron, etc.), or on the surface where snow cover is located within significant part of the year.

2.2. Determination of the current's quantify generated due to the different reflective surfaces

In order to identify the potential value of the amount of current generated by radiation reflected from the snow, the corresponding experiment was also provided. During this experiment, the solar module located in a plane parallel to the sun rays (PV converter's surface was pointing down), towards the sun, with a 2 sides the construction was protected with plywood plates, painted black, to minimize the impact of peripheral reflected and diffused solar radiation, as well as to remove the effects of direct radiation (figure 3). Generated current was measured for various types of underlying surface (snow or a fragment of black cloth, I_S and I_C, respectively).

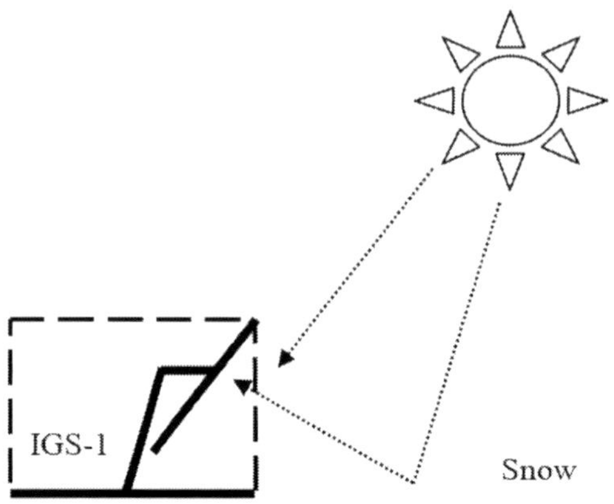

Figure 3. Experiment to determine the magnitude of the current generated due to the reflected radiation.

Also simultaneously current I_H measured; it produced by module located on the horizontal top surface (PV module's surface was pointing up – for a given location the exposure of reflected from the underlying surface radiation does not occur to the converters). There were measured two current's values: I_{H1} – for the case when the surface of the module was not covered with snow, and I_{H2} – for the case when the surface of the module was artificially caused snow layer thickness of 2 cm. The latter value was measured to determine the possibility of developing electricity by PV unit in the event of its bloom of snow cover (for example, in the case of snowfall or snowstorm).

For this experiment the photovoltaic module FSM-2-6 also was used. Readings were taken with a multi-meter M-832P. The experiment had been occurred for 3 days, at one and the same time (13:00). Selection of observation's dates was made in such a way as to make measurements at various cloudiness and atmospheric aerosol component saturation (low clouds, haze or fog, heavy cloudiness). The measurement results are presented in table 1.

Table 1. Determination of the amount of current generated due to the reflected radiation

Date of measurements	Air temperature (°C)	Cloudiness	I_{H1} (mA)	I_{H2} (mA)	I_S (mA)	I_C (mA)
07.03.13	-10	overcast	124	75	35	15
12.03.13	-8	clear	311	272	199	52
13.03.13	-5	haze	146	97	56	23

From the results for a horizontal-oriented module it allows that in times when the value of the direct solar radiation is small (foggy or cloudy day), the generated current's value less by 2-2.5 times than in the cloudless sky. The difference between produced values depending on the saturation of the atmosphere aerosols and clouds are not so significant (18% of the difference between the haze and strong cloudiness); that fact indicates a minor influence of the diffused radiation's nature on the silicon type, which the converters made of.

It also shows that only due the reflected from the surface radiation, PV panels are able to generate current; the difference depending on the intensity of insolation can be substantial (64% of the value of the current generated by the direct solar radiation, in clear weather, and 28-38% – in cloudy weather and fog, respectively). Depending on the degree of radiation's reflection, the current value also changes: when the reflected radiation is formed by a straight line, the presence of snow increases the value of 3.83 times in comparison with a dark surface; when the reflected radiation is formed by diffused, there is an increase of 2.3-2.4 times. The difference between the currents produced by the various types of surface in the case of a diffused light is 53-60%, depending on the nature of the diffuse solar radiation. It should be noted that, according to [6], the albedo of the snow surface in the overcast exceeds albedo in clear weather about 3-5%.

The presence of a thin snow layer on the surface of monocrystalline solar cells does not lead to a intensive decline in their performance. According to experiments, in clear weather performance module is 87% of the cases, when the surface is clean. In the presence in the air aerosols and cloudiness this value is 66 and 60%, respectively. These values are close enough to correspond to the results described in [7].

3. Conclusion

Summarizing the results, the following conclusions may be done:

- snow cover increases current generated from monocrystalline silicon PV modules up to 7.3%; in case of using automatic solar guidance systems (trackers), this value probably will be larger;
- the radiation reflected from the snow cover and its impact on the performance of the PV module is largely dependent on the presence of clouds and aerosols in the atmosphere; simultaneously, the difference in the production of current in cases of haze and strong cloudiness is not huge;
- the presence of a shallow snow layer on the solar cell's surface can reduce performance by 13% on a clear day, and 40% – in cloudy.

4. Acknowledgments

This work was supported as a part of state mission 'Nauka'.

References

[1] Alberola J, Pelegri J, Lajara R and Perez J J 2008 Solar inexhaustible power source for wireless sensor node *Proc. IEEE Int. Instrumentation and Measurement Technology Conf. (Canada, 12 May 2008)* pp 657-662

[2] Raghunathan V, Kansal A, Hsu J, Friedman J and Srivastava M Design Considerations for Solar Energy Harvesting Wireless Embedded Systems 2005 *Information Processing in Sensor Networks (Fourth Int. Symp. on 15 April 2005)* pp 457-462

[3] McEvoy A, Markvart T and Castaner L 2011 *Practical Handbook of Photovoltaics: Fundamentals and Applications* (Exeter: Imprint Academic Press) p 1268

[4] Andrews R W and Pearce J M 2012 Prediction of Energy Effects on Photovoltaic Systems due to Snowfall Events *Proc. 38th IEEE Photovoltaic Specialists Conf. (Austin, US, 3-8 June 2012)* pp 3386-3391

[5] Becker G, Schiebelsberger B, Weber W, Vodermayer C, Zehner M and Kummerle G 2006 An Approach to the Impact of Snow on the Yield of Grid Connected PV Systems Proc. 21th European Photovoltaic Solar Energy Conf. (Dresden, Germany, 4-8 Sept 2006) pp 2732-2735

[6] Wang X and Zender C S 2011 Arctic and Antarctic diurnal and seasonal variations of snow albedo from multiyear Baseline Surface Radiation Network measurements *Journal of Geophysical Research* **116** F03008

[7] Andrews R W, Pollard A and Pearce J M 2013 The effects of snowfall on solar photovoltaic performance *Solar Energy* **92** pp 84-97

The pipeline oil pumping engineering based on the Plant Wide Control technology

D P Starikov[1], E A Rybakov[1], E I Gromakov[1,2]

[1] Institute of Cybernetics, National Research Tomsk Polytechnic University, Tomsk, Russian Federation
[2] National Research Tomsk State University, Tomsk, Russian Federation

E-mail: dstarikov@me.com

Abstract. This article provides recommendations for the use technology Plant Wide Control to control the pumping of oil through the pipeline. The proposed engineering using pipeline management in general (Pipe Wide Control) will reduce the loss of electric power at the expense of the balance of pumping stations located along the pipeline route.

1. Introduction

The pumping oil through the pipeline main task is providing required operational modes: maintaining regulatory requirements set by the pressure in the main oil pipeline and performance targets in terms of the transfer. In many companies, the main oil pipelines transition are occurs on the frequency controlled pump units now. Energy consumption that can reach tens of megawatts of electricity per year, depending on the volume of transported oil and the quality of the dynamic characteristics of all automatic pressure regulators on the pipeline.

Oil pumping system is a set of regulating purposes, pumps and tanks controlled by automation system. In modern operating conditions the system requirements process automation oil pump are follows:

- Continuous maintenance of the specified setpoints (SP) pressure / flow in the pipeline;
- Flexible adjusting the set values;
- Operative stabilization of change (fall / increase) the pressure in the pipeline;
- Optimization of the production process transfer (in this paper establishes the requirement to control pumping stations pipeline provides energy-efficient mode of operation);
- Possibility of operative business process management transfer (increase / decrease the volume of pumping) ERP (MES) system.

The problem pumping oil efficiency improving is presented in papers [1–4]. An example the technological process of the "reprocessing pipe" is a refinery where the pumped oil is processed to the commercial product. Refineries optimize the entire cycle of processing efficiently solved using automated technology perspective Plant Wide Control (PWC) [5], thereby minimizing the loss of various resources due to the energy balance of all process units. However, the controlling and optimizing pumping oil problem with automatic optimization of energy consumption of pumps pumping oil with help of PWC technology was not solved.

2. The problem statement

The aim of this paper is the substantiation of innovative process management framework pumping oil on the main oil pipeline (Pipe Wide Control) at the expense-balanced operation of pumping stations on the pipeline.

Consider a production process diagram pumping oil main oil pipeline (Figure 1), which is includes the following elements:

- Tank (with automatic level control).
- Booster pump.
- Bulk pump (or pump system).
- The pipeline with consumers.

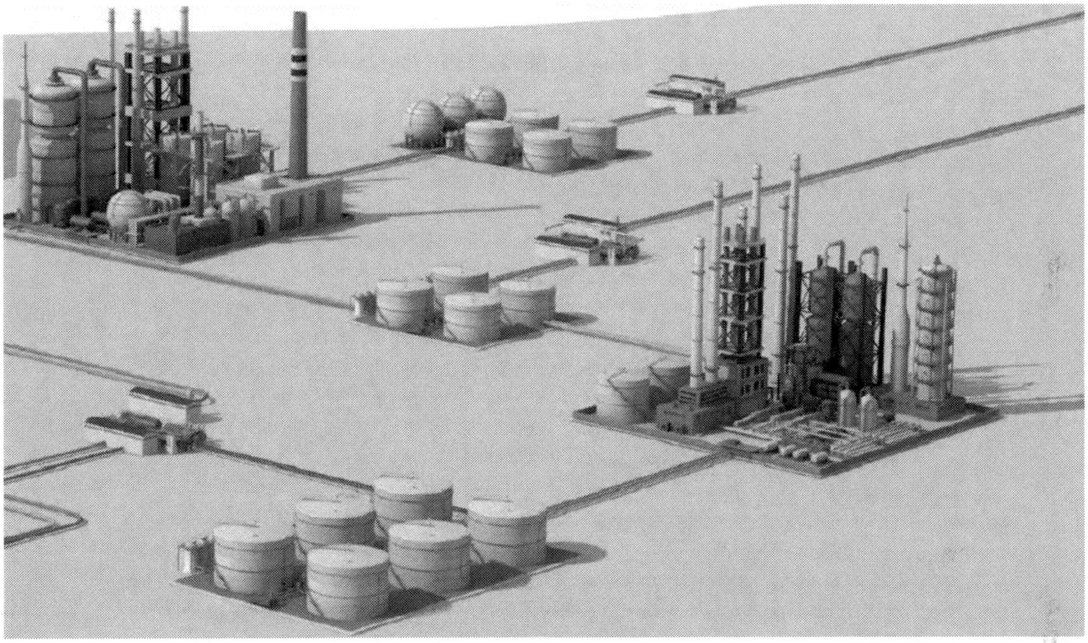

Figure 1. The typical scheme of pumping.

In [7,8] describes the energy-saving algorithms for controlling the pressure in the pipeline by the valve and the main pump. Such algorithms provide optimization of individual pumping stations, but do not provide the balance of its overall pipeline.

Oil transportation in the considered model structure is as follows: crude oil is pumped from the tank farm booster pump by the main pump station and then used for transfer of oil through the pipeline. Control it maintaining the pressure equal to the target value (SP) is carried out electronically frequency drive control and electric valves in the pipeline (pipeline). During operation of the pipeline may change the planned volume pump, and pressure in the pipeline (Figure 2) and the operating point (PT) in the feed-pressure coordinates (Q, H) continuously pump's drift (Figure 3).

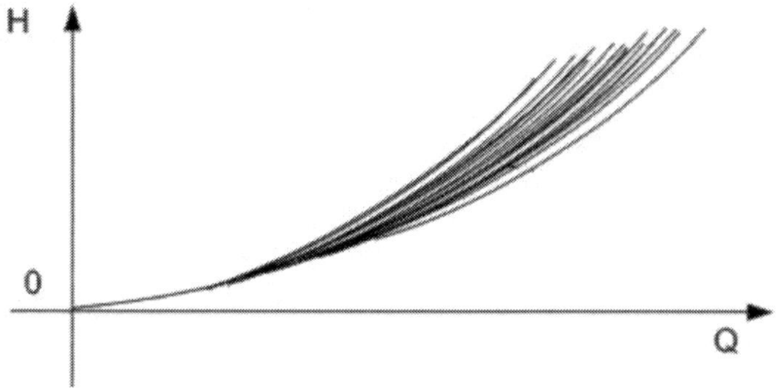

Figure 2. Changes in the characteristics of the pressure in the pipeline parameters.

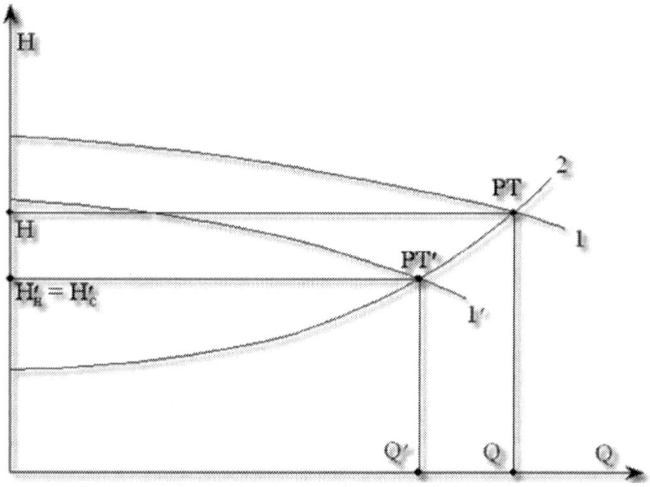

Figure 3. The pump operating point changing.

The reasons for changing the position of the operating point of the pump set may be different values of backwater pump due to differential pressures at the pump, pressure changes the pressure side of the pump caused by the change in cross-section pipes (the appearance of sediments, ice formation), throttling valves, oil consumption, and others. As a result of the pipeline route with the portions formed of different feed oil. This leads to the changes in the performance of the main pipeline and, often, to business losses oil-company.

Therefore there is a necessity for a coordinated automatic control of the whole complex of pumping oil, which are ensure the minimization of consumption of electric energy consumption of pumping units, and improve the efficiency of main oil pipeline as a whole.

3. Plant Wide Control application

PWC is a concept of automatic control of production in general with accent on hierarchy optimization problem. It may be represented as multilayered architecture of control actions. Here interaction between upper and lower layers is determined by control settings (set point, SP). SP is a degree of freedom for automatic regulators of a process (Figure 4). Values of those set points are being calculated to balance operation of technological units.

Set points of hierarchy located loops of pump control are used for maximize efficiency of a process on a bottom level of PWC structure. During PWC design the problems such operator control, start of control object, its shutdown, failures diagnostic, SIS design and ERP-locking are being solved.

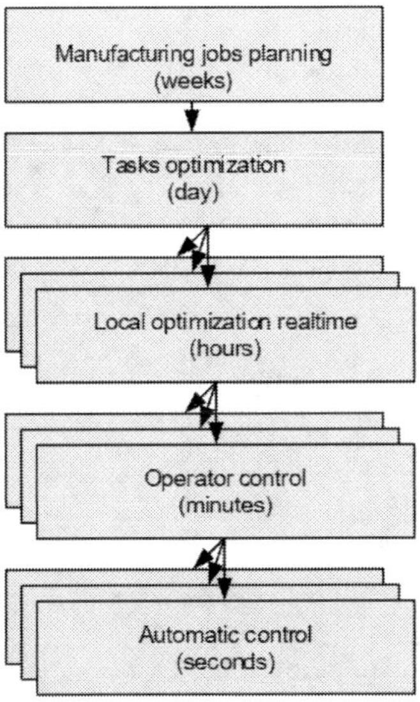

Figure 4. Production Automatic control layers (Plant Wide Control).

Controlling object in proposed PWC technology is process of an oil tranportatin and main algorithms on top level. It provides minimization of power consumption in general when limited pumping volumes.

Choosing optimal control strategy, its efficiency and attainability depend on "good choice" of free controlled variables.

For this degree of freedom is assigned to that independent controlled process variable which will effect on one or more crucial indicator, which characterizes value of a process. According to the task in the article, that parameter will be energy consumption of a pump unit.

Mathematically optimization task of oil transportation in general may be formalized as three components. It is minimization of energy cost (J), reachable of set points and technical specifications (s.t.) and operation limits (g):

$$\min J(x, u_{ss}, d),$$
$$s.t., f(x, u_{ss}, d),$$
$$g(x, u_{ss}, d) \leq 0,$$

where u_{ss} – controlled variables, which characterize efficiency of transportation (pump rotating speed, valve's opening degree, set points of automatic pressure regulators); d – disturbances; x –variables that characterize process of pumping condition (pump energy consumption, pressure and pressure in pipeline).

Then function of process cost J may be written as:

$$J = \sum_i k_{P_i} P_i + \sum_i k_{F_i} F_i + \sum_i k_{Q_i} Q_i,$$

where P_i – pressure in pipeline; F_i –quantity of pumping; Q_i –power consumption of pumps stations.

Now Model Predictive Control (MPC) algorithms are being used as optimization algorithms of operating control of process in oil productions. The idea of it is being implemented in automatic regulators and follows from modern problem of optimal control: determination of the best influence on certain time interval T in a future. Implementation of this requires experience of object dynamic characteristics, robust and high-speed computer-assisted calculation. It is because for each sampling step it is necessary to solve system of differential equations with initial conditionals.

But in case of oil transportation MPC solution is difficult because of complexity of hydrodynamic processes of oil pumping. Nonetheless the problem of pressure effective control may be solved. But better way is application of PID regulators. It should be kept in mind that suitable option of choosing controlled variables according Plant Wide Control is ISA-PID structure with calculated by optimization algorithms parameters [9]. Feature of this algorithm is possibility of simultaneous control of process two degrees of freedom: setting SP and measured variable x (pressure in pipeline) as follows on Figure 5.

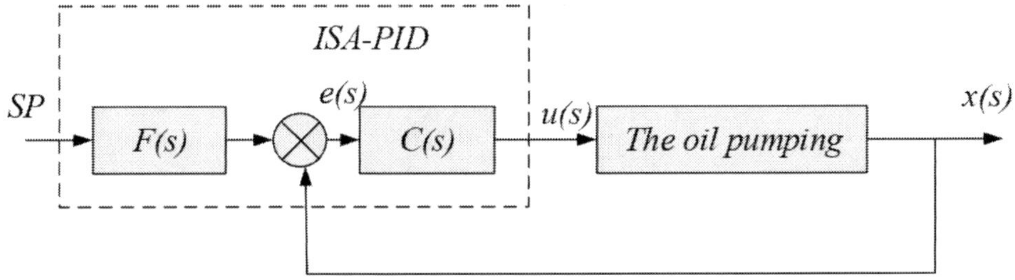

Figure 5. ISA-PID structure of oil transportation.

$$C(s) = PID(K_p, K_i) = K_p + \frac{K_i}{s},$$

$$F(s) = \frac{bK_p s + K_i}{K_p s + K_i},$$

$$u(s) = K_p\left(1 + \frac{1}{T_i s}\right)e(s) - K_p\left(\frac{T_d s}{\left(1 + T_d/N\right)s}\right).$$

Here transfer functions of PID-control of both degrees of freedoms; u(s) is control actions on process of oil transportation. Parameters of regulator are SP (setting point), x(s) (measured variable), e(s) (difference between set and actual values of a pressure). Advantage of this regulator is possibility to provide required smooth changing of pressure in pipeline and as a result decrease pump's power consuming during transients.

4. Conclusion

Recommendations of Plant Wide Control technology application for oil transportation in pipeline were proposed in the article.

Proposed engineering with pipe control in general application (Pipe Wide Control) will allow decreasing energy losses due to balancing operation of pumps stations located along the pipeline.

PWC algorithms will allow to control of pump units operating point and provide maximum throughput opportunity with minimization of energy consumption.

References

[1] Kutukov S.E. Infromation and analitical system of main pipelines *SIP RIA,* Moscow, 2002, 324 p.

[2] Shtukaturov K.Y., Katanov R.S., Abdullina A.D. Calculation of operation modes of pipeline with automatic regulations system. *Oil Business*, **2**: Ufa, 2007. pp.113-119.

[3] Bykov, K.V. Modernization of crude oil pipelines control system. *CINAREM'11: proceedings of the 6th IC on use of mineral resources*, 2011, p. 23-24.

[4] Jiang D J, Design And Application Of Automatic Monitoring System For Crude *Oil Pipeline Transportation*, Master Thesis, China, 2012, pp.128.

[5] Elvira Marie B. Aske Design of plantwide control systems with focus on maximizing Throughput, Thesis for the degree of philosophiae doctor, Trondheim, 2009, pp.126.

[6] Kadam, J.V., W. Marquardt, M. Schlegel, T. Backx, O.H. Bosgra, P.-J. Brouwer G. D¨ unnebier, D. van Hessem, A. Tiagounov and S. de Wolf Towards integrated dynamic real-time optimization and control of industrial processes. In: Proceedings Foundations of Computer-Aided Process Operations (FOCAPO2003). Coral Springs, Florida, 2003, pp. 593–596.

[7] Starikov D. P. , Rybakov E. A. , Gromakov E. I. Application of a Bypass Pipeline during Pressure Control *Communications in Computer and Information* Science 2014 **487**, p. 406-414.

[8] Starikov D. P. , Rybakov E. A. , Gromakov E. I. SMART Pressure Regulation In Main Oil Pipeline *2nd International Conference on Advanced ICT (Information and Communication Technology) for Education (ICAICTE2014): proceedings,* Dalian, August 16-17, 2014. Paris: Atlantis Press, 2014, p. 53-56.

[9] King, Myke, Process Control: A Practical Approach. Chichester, UK: John Wiley & Sons Ltd, 2011, pp.396.

Power supply of autonomous systems using solar modules

A V Yurchenko [1], L G Zotov[2], A D Mekhtiev[3], V V Yugai[4], G G Tatkeeva[5]

[1]Professor, Head of the Research Laboratory of Solar Elements, Tomsk, Russia
[2]Assistant professor of "Theoretical foundations of radio" Department in Novosibirsk State Technical University, Novosibirsk, Russia
[3] Head of "The technology and communication systems" Department in Karaganda State Technical University, Karaganda, Kazakhstan
[4] Senior lecturer of "The technology and communication systems" Department in Karaganda State Technical University, Karaganda,phD of KazNTU, Kazakhstan
[5] Dean of the "Power Engineering and Telecommunications" faculty in Karaganda State Technical University, Karaganda, Kazakhstan

E-mail: slawa_v@mail.ru

Abstract. The article shows the methods of constructing autonomous decentralized energy systems from solar modules. It shows the operation of up DC inverter. It demonstrates the effectiveness of DC inverters with varying structure. The system has high efficiency and low level of conductive impulse noise and at the same time the system is practically feasible. Electrical processes have been analyzed to determine the characteristics of operating modes of the main circuit elements. Recommendations on using the converters have been given.

1.Introduction

Basic requirements for the development of autonomous power supply (APS) systems are compact size, high efficiency and low level of impulse noise. Efficiency of modern electrical energy converting devices such as classic high frequency pulse-width inverters and DC-DC converters for autonomous energy supply systems is close to reaching its maximum limit. A further increase in the efficiency by a few percentages can be achieved by reducing the switching power losses in power keys by using methods of resonance, providing them with a mode of soft switching, as well as with the reduction in dynamic voltage drops when switching from cut off to saturation. Reduction of impulse noise level is achieved by a reasonable decentralization combined with multi-stroke mode of operation of used converters.

2. Studying of push-pull-push resonant down DC-DC converter with $cf = \frac{1}{2}$

Schematic diagram of the three-stroke resonant down DC-DC converter power circuit used in the APS for the lighting system is shown in Figure 1 and Figure 2 where timing diagrams explaining its operation [1].

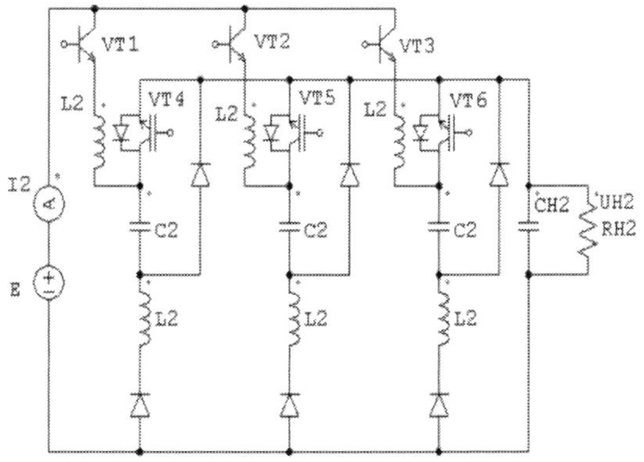

Figure 1. Three-stroke resonant down DC-DC converter with $cf = \frac{1}{2}$.

Converter is composed of k_2 down ECVs and each of them is composed of n_2 CDC. The operating principle of down ECV is periodic recharging of series-connected capacitors in its circuit from AB with voltage E by means of charging keys $VT1 - VT3$ which then have parallel discharge to the load though charging keys $VT4 - VT6$. As the source E recharges capacitors through series-connected load the ECV power circuit is simplified by the decrease of the number of CDCs by one. As a result the convertor output voltage is equal to $U_{H2} = \frac{E}{(n2+1)}$. Increasing of the number of ECV - k_2 also leads to a proportional increase of the maximum load current and a reduce of conductive impulse noise level.

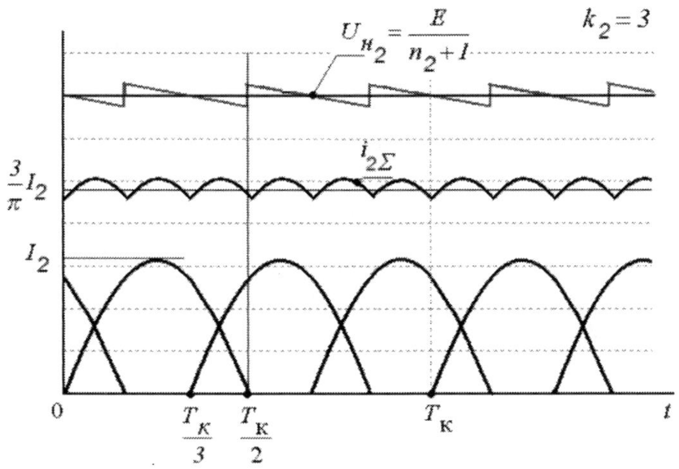

Figure 2. Timing diagrams explaining the operation of push-pull-push resonant down CP.

Charging current of the down CP is determined by the following expression (1)

$$i_2(t, \tau_2) = I_2 \cdot \sum_{i=0}^{k_2-1} [sin\omega_K \cdot (t - i \cdot \frac{T_K}{k_2} - \tau_2)] \cdot 1[i \cdot \frac{T_K}{k_2} + \tau_2, \frac{T_K}{2} + i\frac{T_K}{k_2} + \tau_2] \qquad (1)$$

where τ_2 - time offset of charging current impulse sequence related to the central point, $I_2 = \frac{\pi}{k_2} \cdot \frac{I_{H2}}{(n_2+1)}$ – the amplitude of the down CP charging current, $I_{H2} \approx \frac{E}{(n_2+1) \cdot R_{H2}}$ – average load current, [2–4]

The number of ECVs in k_2 groups (3), capacitance C and inductance L of modules (4) are determined from maximum required current $I_{нjcp}$, voltage ripple on capacitors ΔU_c and switching frequency of keys ω_K.

$$k_2 \cdot C_2 = \frac{2 \cdot \pi \cdot I_{H2}}{\omega_K \cdot (n_2+1) \cdot \Delta U_C} \quad (2)$$

$$L_2 = \frac{1}{\omega_K^2 \cdot C_2} \quad (3)$$

$\omega_K = (2\pi / T_K) = (1 / \sqrt{L_2 \cdot C_2})$ – resonant frequency of charging circuit, L_2, C_2 – CDC inductance and capacitance of down CP. A further decrease of the impulse noise level on AB clamps during simultaneous operation of up and down CP in APS is achieved by optimal time offset choice - $\tau_{1,}\tau_2$ of charging currents $i_1(t,\tau_1)$ and $i_2(t,\tau_2)$. Generalized optimization criterion is formulated as follows: at the same frequency of commutation of keys $\omega_K = (2\pi / T_K)$, operating with a duty ratio equal to two, to determine time offsets $\tau_1, \tau_2, ..., \tau_K$, delivering minimum to the square of the real value of the variable component of the total charging current, consumed by K converters from AB (4)

$$i^2_{\partial\Sigma\sim}(\tau_j) = \frac{k_{min}}{T_K} \cdot \int_0^{\frac{T_K}{k_{min}}} [\sum_{j=1}^{K} i_j(t,\tau_j) - i_{0\Sigma}]^2 \cdot dt - min \quad (4)$$

Here k_{min} is minimum CP strokes, which consumes current from AB, $i_{0\Sigma} = \frac{1}{T_K} \cdot \int_0^{T_K} i_\Sigma(t,\tau_j) \cdot dt = \sum_{j=1}^{K} I_j \cdot \frac{k_j}{\pi}$ - constant component of the total charge current which is independent from τ_j. If k_{max} - is maximum CP strokes, then for APS made of two capacitors ($K = 2$) analysis gives the following expression for the optimal time offsets $\tau_1 = 0, \tau_2 = \frac{T_K}{2 \cdot k_{max}}$, $\tau_2 = \frac{T_K}{4 \cdot k_{max}}$ – respectively for even and odd values of k_{max}. If APS is made of 2 push-pull-push capacitors, i.e. $k_1 = k_2 =$, then the optimal offset on $\tau_2 = (T_K / 12)$, and respectively (5):

$$i^2_{min\partial\Sigma\sim}(\tau_2 = \frac{T_K}{12}) = \frac{1}{2} \cdot (1 + \frac{3\sqrt{3}}{2\pi}) \cdot (I_1^2 + I_2^2) + (\frac{\sqrt{3}}{2} + \frac{3}{\pi}) \cdot I_1 \cdot I_2 - [\frac{3}{\pi} \cdot (I_1 + I_2)]^2 \quad (5)$$

Its maximum (6) is achieved when $\tau_1 = \tau_2 = 0$:

$$i^2_{max\partial\Sigma\sim}(\tau_2 = 0) = \frac{1}{2}\cdot(1 + \frac{3\sqrt{3}}{2\pi})\cdot(I_1^2 + I_2^2) - [\frac{3}{\pi}\cdot(I_1 + I_2)]^2 \qquad (6)$$

It is important to note that for typical values $k_j = 2,3$ effectiveness of conducted. It is sufficient to compare the maximum and minimum values of $i^2_{\partial\Sigma\sim}$ to estimate it. If current impulse amplitudes are equal then the value of winning (7) is 15.46 times:

$$B = (\tau_2 = \frac{T_K}{4\cdot k_{max}}) = \frac{i^2_{max\partial\Sigma\sim}}{i^2_{min\partial\Sigma\sim}} = 15.46 \qquad (7)$$

Calculations based on the formulas given in the article show that for the APS with an output power in the AC circuit Pout~=10 kW and in the DC circuit (lighting) Pout=1 kW t is required to used 2 capacitors with capacity C = 6 mfd and voltage U=130 V and 6 capacitors with capacity C = 1 mfd and voltage U=260 V. For the lighting system 2 capacitors with capacity C = 13 mfd and voltage U=60 V are required.
The basic layout of a single-phase bridge voltage inverter based on up CP with varying structure is shown in Figure 3, and time diagrams of voltages and currents for explaining its operation are shown in Figure 4.

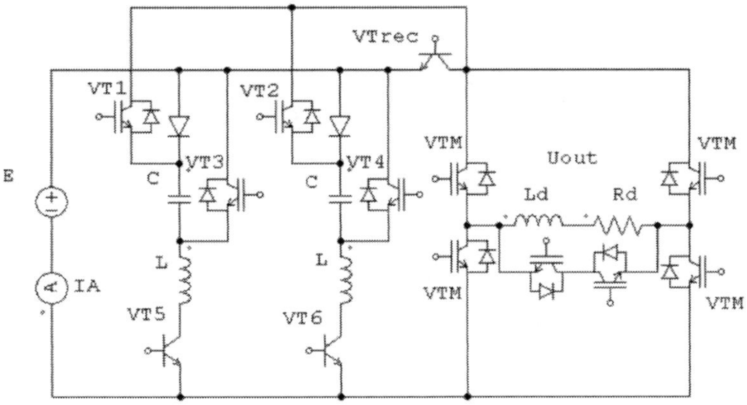

Figure 3. The basic layout of a single-phase bridge voltage inverter based on the up CP with varying structure.

The operating principle of the inverter is based on the change in the structure of the up DC-DC capacitor converter by alternate switching the transistor switches VT3,VT4 with frequency of 50 hz. Thus there is an abrupt change in conversion coefficient $K_n = 0,1,2$, which lead to the change of output voltage U_{out} , as shown in Figure 4. By the influence of harmonic PWM on VTM bridge transistors one

IOP Conf. Series: Materials Science and Engineering **81** (2015) 012112 doi:10.1088/1757-899X/81/1/012112

can obtain a sinusoidal current in the load I(Ld). An important advantage of the inverter under consideration is the reduced dynamic voltage drop on bridge transistors which does not exceed 224 V in the considered APS.

As shown in the Bibliography [4] this can further increase the efficiency of the inverter by 1.5-2% due to the reduction of power loss on switching losses on VTM bridge transistors. The disadvantage of the inverter under consideration is the pulsating current, consumed by AB.

The Figure 4 shows that the value of current ripple I A is twice the current ripple on the load which has a negative effect on AB operation.

A significant reduction in consumed current ripple is achieved by the construction of the APS according to a decentralized scheme Figure 5.

The effect is achieved by a uniform distribution of current consumed by decentralized load I (Ld) for the half or the whole period of their frequency. Timing diagrams of input and output currents of decentralized APS, confirming a sharp decline in the value of the ripple current AB are shown in Figure 6.

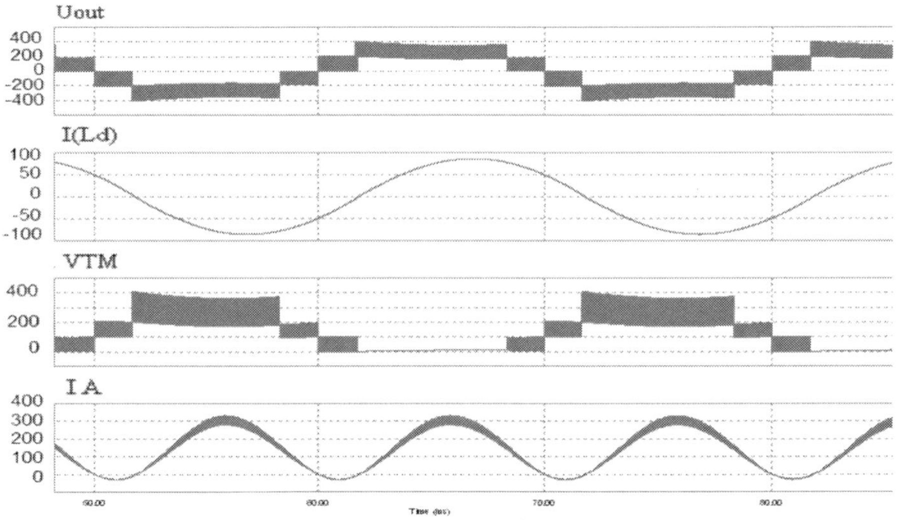

Figure 4. Timing diagrams explaining the operation of single-phase bridge voltage inverter based on up CP.

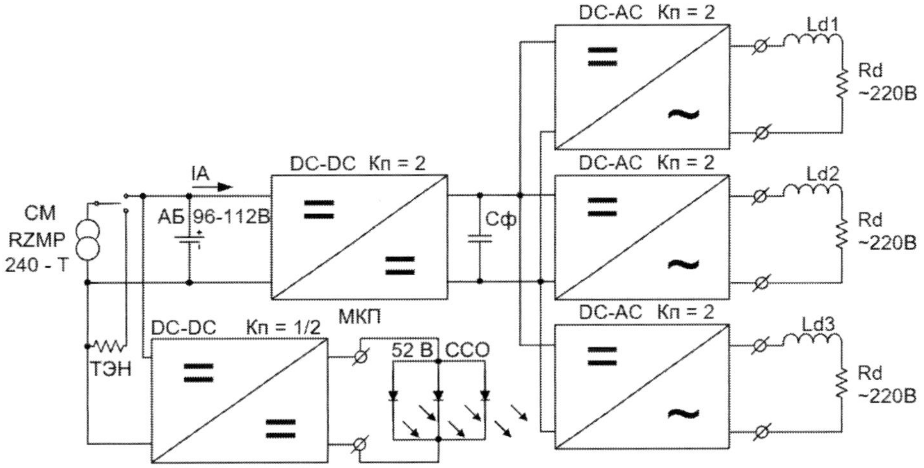

Figure 5. Structural diagram of decentralized APS.

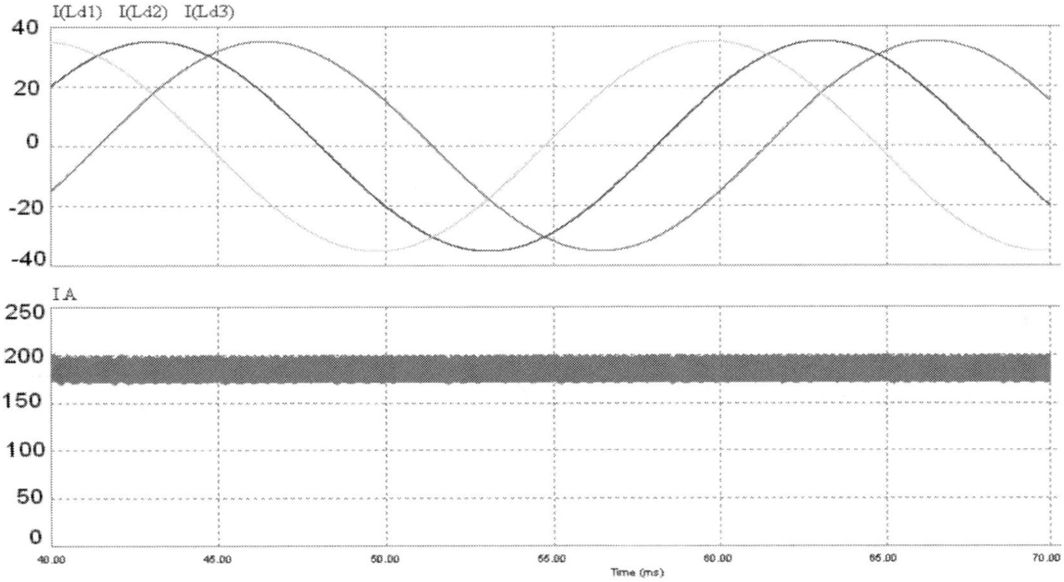

Figure 6. Timing diagrams explaining the operation of a single-phase bridge voltage inverter based on the control panel.

3. Studying of stroke capacitor resonant up DC/DC with $cp = 2$.

The importance of APS is that its power supply is performed from groups of 4 consecutive solar modules (SM) RZMP 240-T, which are parallel-connected (Figure 7). Such a solution allows applying the input battery (AB) with a sufficiently high output voltage 96-112 V and thereby reducing the cross section and the power losses in leading-in cables. Later on this voltage is doubled twice – firstly by DC-DC, and then by DC-AC converters. For the purpose of energy saving LED lighting system (LLS) is applied combined with double down DC-DC converter powering it [5].

Another distinctive feature of the APS is that in case of an increase in charge voltage above 112 volts SM disconnect from the battery and switch to heat the heating element, performing, for example, water heating.

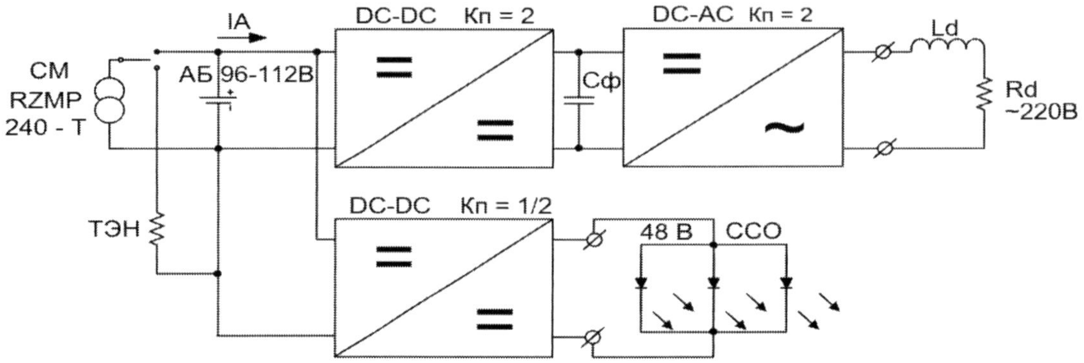

Figure 7. Structural layout of APS.

Schematic diagram of the power circuit 3-stroke capacitor resonant up DC-DC converters used in the APS is shown in Figure 8, and Figure 9 shows timing diagrams explaining its operation.

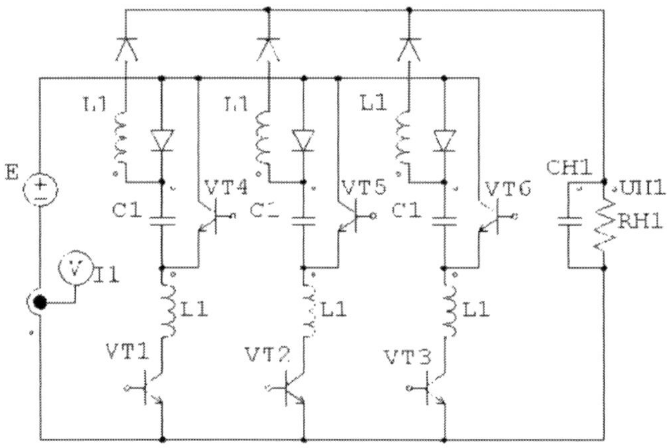

Figure 8. 3-stroke capacitor resonant up DC/DC cp = 2.

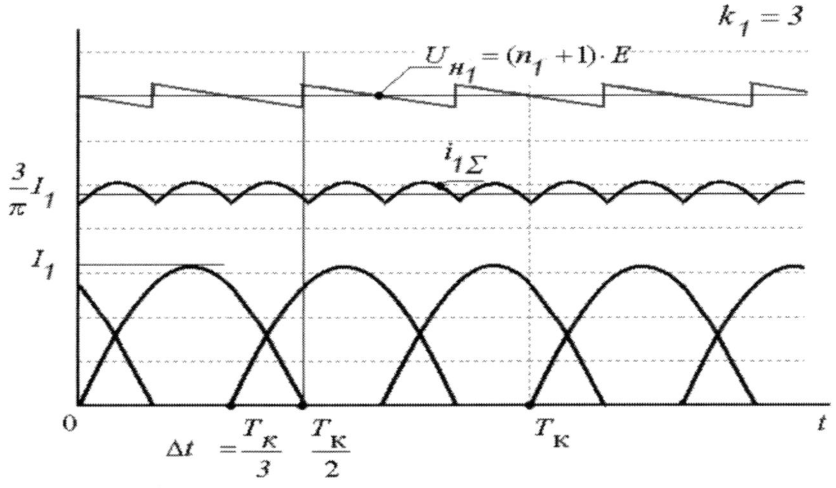

Figure 9. Timing diagrams explaining the operation of 3-stroke capacitor resonant up CP.

Downsizing and increase of efficiency is achieved correspondingly due to the increased frequency of conversion up to the magnitude of about 500 kHz or higher and due to a reduction in switching losses in high-frequency charging keys $VT1 - VT6$ of the power circuit of series reactors L1, providing them with a soft switching mode.

Initially, the converter consists of k1 (strokes) similar up converters (PVC), each containing n1 capacitor-diode circuit (CDC), operating on the load.

The operating principle of the up OCP consists in periodic parallel recharge of capacitors its circuits from constant voltage E on input source through charge keys $(VT1 - VT3)$ followed by their successive discharges to the load through the discharging keys $(VT4 - VT6)$ corresponding to this condition. Since the discharge of the capacitor to the load goes through a serial connection AB with voltage E, then the power circuit OKP simplifies by decrease in the number of CDC by one. As a result, the converter output voltage is equal to $U_{H1} = (n1 + 1) \cdot E$. Increase of strokes converter - k1 results in a proportional increase in the maximum load current and reduce impulse noise on the conductive terminals AB [6]. In order to minimize it recharging of capacitors ECV is made by positive impulses sinusoidal current with duration of $0{,}5 \cdot T_K$ which are universally distributed along the period T_k of switching frequency of charging keys with time offset related to each other equal to $(\Delta t)_1 = (T_K / k_1)$. Charging current ECV is determined as follows (8):

$$i_1(t, \tau_1) = I_1 \cdot \sum_{i=0}^{k_1 - 1} [\sin\omega_K \cdot (t - i\frac{T_K}{k_1} - \tau_1)] \cdot 1[i\frac{T_K}{k_1} + \tau_1, \frac{T_K}{2} + i\frac{T_K}{k_1} + \tau_1] \quad (8)$$

Where τ_1 is time offset of sequence of charging current impulses from the central point, $I_1 = \frac{\Delta E}{\rho} = \frac{\pi}{k_1} \cdot n_1 \cdot I_{H1}$ – is the amplitude of ECV charging current, $\Delta E -$ is voltage ripple on ECV capacitors, $\rho = \frac{1}{n_1} \cdot \sqrt{\frac{L1}{C1}}$ – characteristic independence of ECV charging circuit, $I_{H1} = (U_{H1}/R_{H1})$ - average load current

$$i_{1\partial\sim} = \sqrt{(i_{1\partial}^2 - i_{10}^2)} = \sqrt{\frac{1}{T_K} \cdot \int_0^{T_K} i_1^2(t) \cdot dt - [\frac{1}{T_K} \int_0^{T_K} i_1(t) \cdot dt]^2} \quad (9)$$

where $i_{10} = \frac{1}{T_K} \cdot \int_0^{T_K} i_1(t) \cdot dt = I_1 \cdot \frac{k_1}{\pi}$ – average charging current $i_1(t)$. By increasing the number of converter strokes k_1 there is a decrease of $i_{1\partial\sim}(k_1)$.

The active power of load is determined by average value. Thus comparing values $i_{1\partial\sim}$ for same average values i_{10}, one comes to conclusion that the decrease of real value of interference current $i_{1\partial\sim}$ for odd k_1 happens on a sufficiently lowered level [7].

For instance, level of interference for k_1 equal to three and six correspond. Therefore one can make a conclusion about appropriateness of using only odd number of ECVs in groups. From the law of conservation of charge, it follows that the average currents of all ECV keys are same and equal to $I_{\text{кср}} = \frac{1}{k_1} \cdot I_{H1}$. The amplitude of current going through charging and discharging ECV keys VT1-VT6 is determined by following expression (10):

$$I_{K1} = \frac{1}{n_1} \cdot I_1 = \frac{\pi}{k_1 \cdot (n_1 + 1)} \cdot I_{H1}. \quad (10)$$

The number of ECVs – k_1, capacity C1 and inductance L1 of circuits are determined from the maximum required current I_{Hmax}, ripple ratio on the first harmonic load voltage $K_{n1} = \frac{U_{H1}}{U_{H0}}$ and the frequency of key switching ω_K (11):

$$k_1 \cdot C1 = \frac{2 \cdot n_1 \cdot I_{H1max}}{K_{n1} \cdot \omega_K \cdot (n_1 + 1) \cdot E}, \quad L1 = \frac{1}{\omega_K^2 \cdot C1}. \quad (11)$$

4. Conclusions

- It is shown that by reducing the switching losses in the power keys of DC-DC converters and capacitor inverters with a variable structure and by application of multi-stroked operation mode APS has a higher efficiency and significantly lower level of impulse noise.
- The analysis of the electrical processes was conducted and it allows calculating the characteristics of operation modes of the main circuit elements and making recommendations on the use of converters only an odd number of ECV.
- The principle of construction of autonomous decentralized energy systems from solar modules was suggested. It is based on the capacitor, up and down, multi-stroked, resonant DC-DC converters, and up capacitor inverters with varying structure as well.

IOP Conf. Series: Materials Science and Engineering **81** (2015) 012112 doi:10.1088/1757-899X/81/1/012112

- It is shown that due to reducing the switching losses in the power keys of DC-DC converters and capacitor inverters with varying structure and due to application of multi-stroked mode of operation APS has a higher efficiency and significantly lower level of impulse noise.

Page layout (headers, footers, page numbers and margins)

If you don't wish to use the Word template provided, please set the margins of your Word document as follows.

References

[1] Weiss H H, Ince K and Zinovyev G S *Multi-Input Small-Power Renewable Energy Supply System Realized By Special Power Electronics* p.617-622

[2] Zotov L G 2006 *Adjustable DC down converter* (Patent RU №2284633: БИ №27)

[3] Zubritsky A 2002 *ScriptaPeriodica* **5(1)**, p. 46-52

[4] Nunez C and Visairo J N 2010 *EPE Journal* **20** p. 5-11

[5] Zotov L G 2005 *Scientific Vestnik* **1(19)** p. 83-88

[6] Smirnov D 2008 *Proceedings of the Institute of Engineering Physics* **2(8)** p. 10-13

[7] Nunez C, Lira J and Visario N 2010 *EPE journal* **4** p. 5-11

Studying mathematical model of mine and quarry pneumatic lifting equipment in "skip - guidance devices" systems

M V Kitaeva [1], Yu ANikolaev[2], A V Taranov[3], A D Mehtiev [4]
[1] National Research Tomsk Polytechnic University,Tomsk, Russia
[2] Professor of the "Energetics" Department in Karaganda State Technical University, Karaganda, Kazakhstan
[3] Head of "Energetics" Department in Karaganda State Technical University, Karaganda, Kazakhstan
[4] Head of "The technology and communication systems" Department in Karaganda State Technical University, Karaganda, Kazakhstan

E-mail: barton.kz@mail.ru

Abstract. The article is devoted to determining the maximum loads acting on the guide roller devices of skips in the trunk of mine and quarry overpass lifting equipment necessary to develop methods for calculating the guiding devices.

1.Introduction

There are two basic ways of mining: open pit and underground in the mining industry: open pit and underground. Underground mining less profitable, however, it is more environmentally friendly. In addition, this method is essentially inevitable due to the fact that the major mineral reserves are at greater depths and in the future will need to solve the problems of development of mineral deposits deep mines. The development of the mining sector and the deepening of mining operations require a new type of transport, justified and developed in the works of Bogatyrev M D, Kiselev V I, Kartaviy N G, Saginov A S, Nechyporenko S N, Piskunov O M, Y.A. Nikolaev Yu A – skip pneumatic lifting equipment designed to reduce the cost of transportation of the rock mass.

2.Analysis of mine SPLI work

The cycle of operation of pneumatic lifting equipment consists in the following (Figures 1 and 2). Skips 4, 5 are loaded at the lower end of the mine from dispensers 5, 7 jet of rock mass falling from a height of more than ten meters in a skip and has a significant dynamic loads on the bottom and sides of the latter. After loading the skip is included working body (turbo blower) 1 and compressed air begins to flow under the bottom of the skip. When a certain overpressure skip is accelerated and begins to accelerate to the steady speed, determines the performance blowers.

The main part of the way to the top mark of the mine skips moves at a constant speed, and when approaching the upper stop device 6, 9 is inhibited due to discharge of the compressed air to the atmosphere through the air valves 9 and 10. After entering the mine skip 4 in abutment device 6 it is unloaded through the side walls into the hopper 7, and 5 of career skip installation, after fixing in the upper rotary device 9, bends and unloaded into the hopper 8.

All the time of unloading the skip 4 mine pneumatic lifting installation it is supported by positive pressure in the cavity of the barrel 3. After unloading the skip 5 last of career installation returns to its original position, stops are removed and skip 5 under the action of gravity starts to fall down by squeezing under a cushion of air.

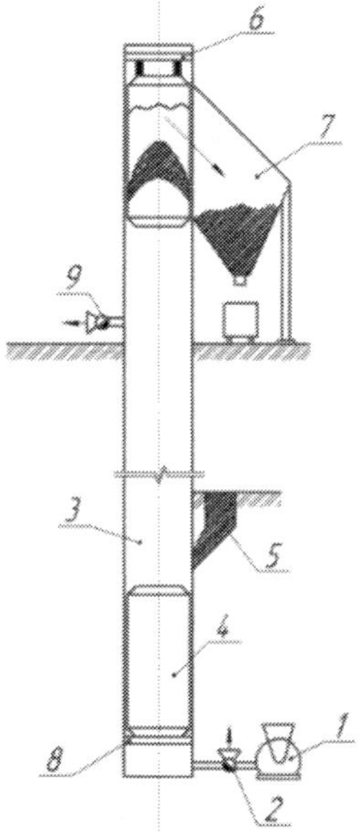

Figure 1. The structural layout of mine SPLI: 1 – working body (turbo blower); 2, 9 – air lock; 3 – overpass with the guide and sealing devices; 4 – skip; 5 – dispenser; 6 –the upper stop; 7 – unloading hopper; 8 – the lower landing device.

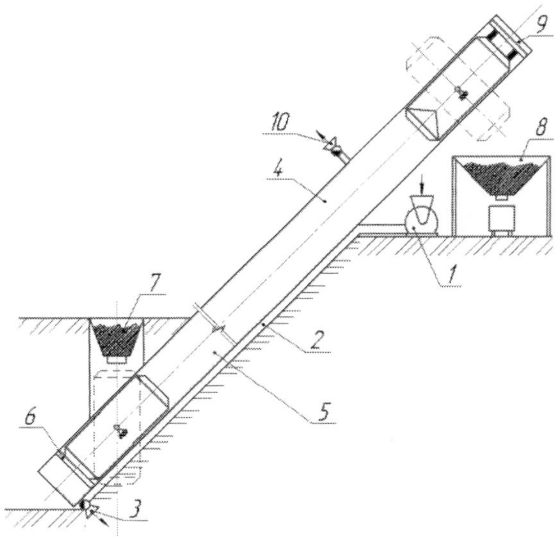

Figure 2. The structural layout of SPLI open pit: 1 – working body (turbo blower); 2 – air pipe; 3, 10 – air lock; 4 – overpass with the guide and sealing devices; 5 – skip; 6 – lower rotating part of the overpass; 7 – dispenser; 8 – unloading hopper; 9 – the upper rotating part of the overpass with a focus.

The lowering speed of skips 4 and 5 and their landing on the lower landing gear 6, 8 is regulated by air locks 2 and 3. The main and most crucial element of mine and career pneumatic lift systems are rails movement the skip device performance depends on the reliability and durability of these plants.

However, the task of creating the guide devices of skips for mines and quarries paid insufficient attention has hindered the implementation of these machines in production. Therefore there is a need to study and selection of the parameters guiding devices of skips of mine and career pneumatic lifting equipment. Large problem of determining the parameters of guiding devices, one of the main elements of these systems, is not sufficiently developed and needs to be resolved, as the reliable operation of skip pneumatic lift systems without solving this problem is impossible. To do this, you must develop a methodology for calculating the guiding devices of mine and career pneumatic lifting equipment, as one of the main elements of these installations.

In this regard, the choice of parameters and the design procedure of guiding devices of skips of mine and career pneumatic lifting equipment, is an actual problem.

According to the kinematics of the pneumatic lifting equipment, the following basic operating modes guiding devices:
- loading the skip;
- the skip entry in the guide device with maximum acceleration;
- motion of the skip in the guide device with maximum acceleration;
- the skip entry in the guide device with a maximum speed;
- uniform motion of the skip in guiding devices with a maximum speed;
- unloading the skip.

Under each mode guidance devices are subjected to static and dynamic loads (Figures 3 and 4), representing different kinds of danger.

Reaction of rollers depend on how far from the center axis of the vessel its mass is. Therefore, when designing pneumatic lifting equipment is necessary to ensure such conditions that its center of mass is as close as possible to the axis of symmetry. But in reality, this condition is very difficult to provide because of the uneven loading.

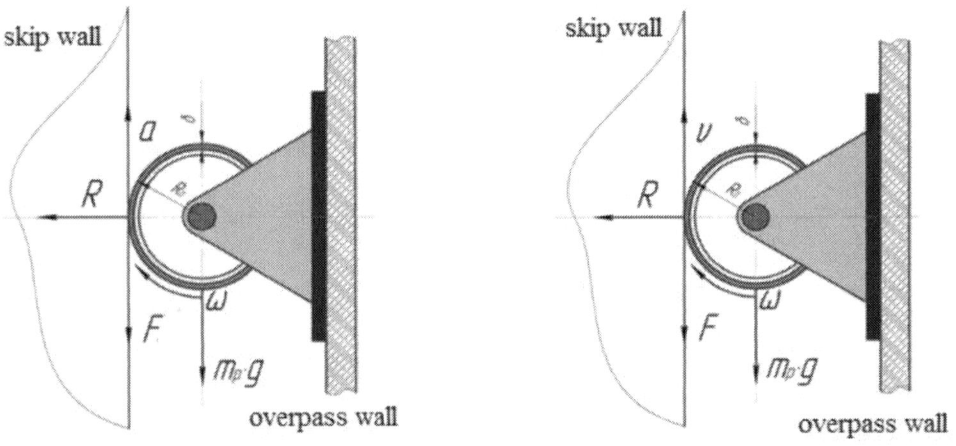

a) b)

Figure 3. Scheme of loads guiding devices pneumatic lifting equipment when rolling the skip on rollers: a) the movement of the skip to guide devices with maximum acceleration; b) uniform motion of the skip in guiding devices with a maximum speed.

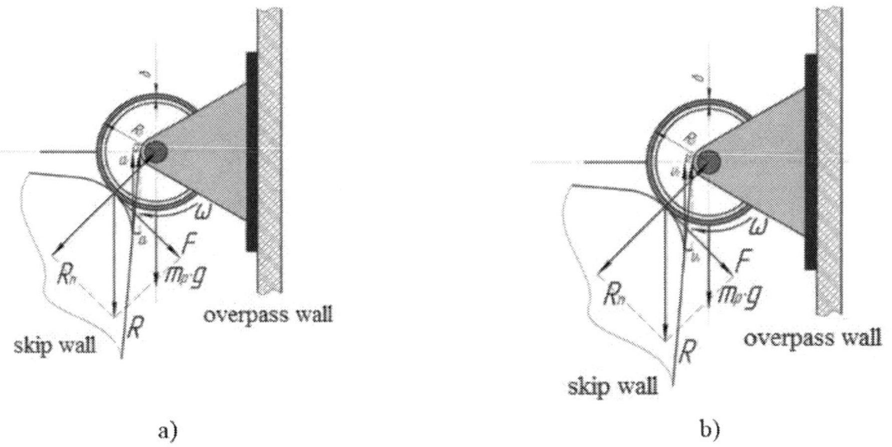

Figure 4. Scheme of loads guiding devices: a) the skip entry in the guide device with maximum acceleration; b) the skip entry in the guide device with a maximum speed.

In addition, it is necessary to take into account the dynamic processes when adjusting the speed of the vessel and the conditions for its entry in the guide roller devices.

The maximum values of reaction of rollers at a coal mine pneumatic lift installation depends only on the diameter (the skip-duty) and practically do not depend on the thickness of the wall of the skip. For ore mine pneumatic lifting installation with thicker walls the skip observed a significant reduction reaction of rollers by increasing the wall thickness of the skip. By increasing the diameter of the intensity of the reduction reaction the skip of rollers weakened.

The maximum values of reaction of rollers and coal and ore of career pneumatic lift installation depend not only on the diameter (the skip load capacity) and the wall thickness of the skip. By increasing the diameter of the intensity of the reduction reaction the skip of rollers weakened.

The analysis of the results of a numerical calculation using licensed software shows that determines the load in determining the design parameters of the skip, as a rule, local external forces generated by reactions of rollers guiding devices [1, 2].

To solve the problem of determining the dynamic interaction with the walls of the guide devices the skip (career overpass) developed the design scheme and a mathematical model of the motion of skips of mine and career pneumatic lift systems (Figures 5 and 6). A system of differential equations of motion the skip in the guide device.

Unlike the previously known mathematical models [3], the resistance in the guide devices are recorded according to the deformation of the roller over time. The study parameters and guiding devices pneumatic mine lifting installation during periods of unsteady motion the skip, as well as the regulation of its speed accomplished using a mathematical model, which is a system of seven differential equations that describe the motion the skip along the axis of the shaft. This mathematical model is based on the calculation scheme shown in Figure 5.

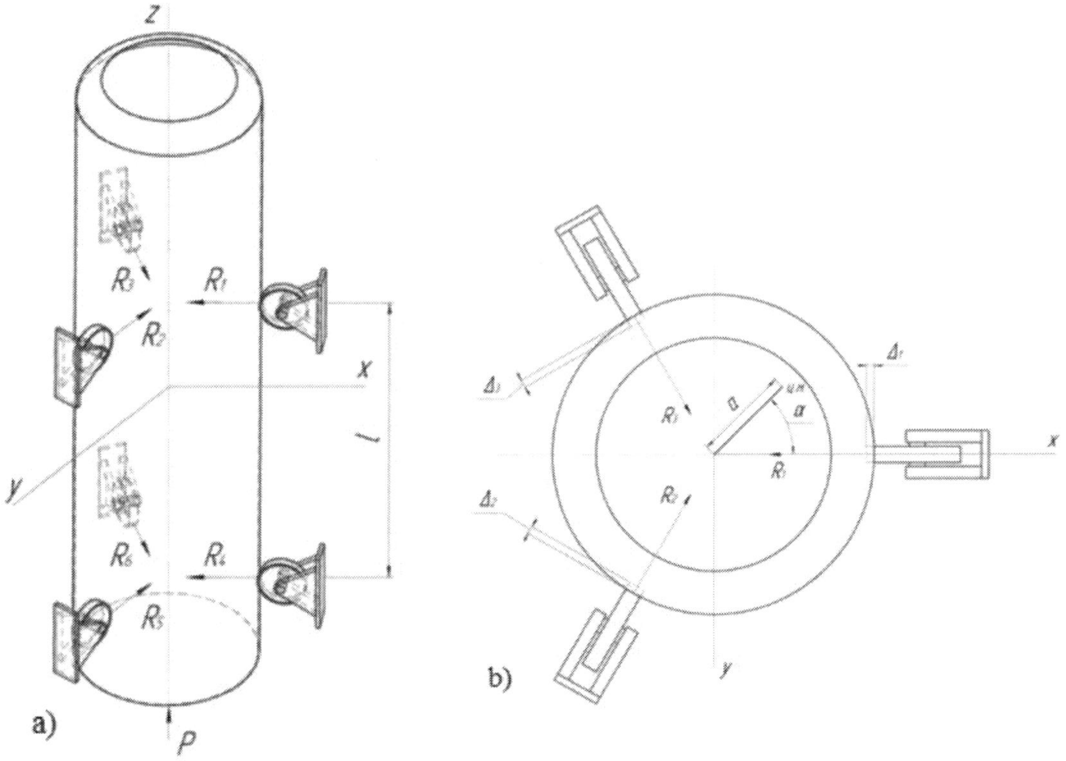

Figure 5. Design scheme of mine pneumatic lifting installation: a) general view; b) cross-section.

At constant pressure vessels under $P = const$ system of differential equations goes to the following form (1):

$$\begin{cases} \dfrac{d^2z}{dt^2} = \dfrac{1}{M}\left[(P-P_a)\cdot S_c - F - F_{fr.yy} - \dfrac{k_{fr}\cdot C}{R_0}\cdot(\Delta_1+\Delta_2+\Delta_3+\Delta_4+\Delta_5+\Delta_6)\right] \\ \dfrac{d^2\Delta_1}{dt^2} = -\dfrac{1}{M}\left(C\cdot\Delta_1+\beta\dfrac{d\Delta_1}{dt}\right);\ \dfrac{d^2\Delta_2}{dt^2} = -\dfrac{1}{M}\left(C\cdot\Delta_2+\beta\dfrac{d\Delta_2}{dt}\right); \\ \dfrac{d^2\Delta_3}{dt^2} = -\dfrac{1}{M}\left(C\cdot\Delta_3+\beta\dfrac{d\Delta_3}{dt}\right);\ \dfrac{d^2\Delta_4}{dt^2} = -\dfrac{1}{M}\left(C\cdot\Delta_4+\beta\dfrac{d\Delta_4}{dt}\right); \\ \dfrac{d^2\Delta_5}{dt^2} = -\dfrac{1}{M}\left(C\cdot\Delta_5+\beta\dfrac{d\Delta_5}{dt}\right);\ \dfrac{d^2\Delta_6}{dt^2} = -\dfrac{1}{M}\left(C\cdot\Delta_6+\beta\dfrac{d\Delta_6}{dt}\right); \end{cases} \quad (1)$$

where Z is current coordinate of movement the skip measured from the mark of the barrel bottom , m; M is the total mass of the payload and the body of the skip, kg; P is current absolute air pressure in the cavity beneath the vessel barrel, Pa; P_a is the current value of the atmospheric pressure, Pa;

S_c is sectional area of the skip, $m2$; $F = M \cdot g$ is useful resistance force, N; $F_{fr.yy}$ is resistance force the skip in stem sealing devices, N; $R_i = \frac{k_{fr} \cdot C}{R_0} \cdot \Delta_i$ is reaction of rollers of guiding devices,N; k_{fr} is coefficient of rolling friction, N; C is rigidity of guiding devices;

R_0 is radius of the roller of guide device, m; $\Delta_1, \Delta_2, \Delta_3, \Delta_4, \Delta_5, \Delta_6$ is deformation of rollers of guiding devices; a is the displacement of the center of mass of the laden the skip; α is the angle between the X axis and the direction of displacement of the center of mass; β is viscous friction coefficient.

Under $\alpha = 0$ and $P = const$ (2):

$$\begin{cases} \frac{d^2 z}{dt^2} = \frac{1}{M} \cdot \left[(P - P_a) \cdot S_c - F - F_{fr.yy} - 2 \cdot \frac{k_{fr} \cdot C \cdot x}{R_0} \right] \\ \frac{d^2 x}{dt^2} = -2 \cdot \frac{1}{M} \cdot \left(C \cdot x + \beta \frac{dx}{dt} \right) \end{cases} \quad (2)$$

The resultant of the research design scheme of mine pneumatic lifting installation and bears the highest possible operating forces is the basis for design choice guiding devices, calculation of stresses acting in their elements by using the mathematical apparatus of the finite element program ANSYS et al., the selection structural materials and the disc rim roller, testing the strength and stability.

The above mathematical model and its numerical solution results allow procedures to develop an algorithm for calculating the parameters guiding devices mine pneumatic lifting equipment.

To solve the problem of determining the dynamic interaction with the walls of the guide devices career skip developed analytical model and mathematical model of motion skips career pneumatic lifting equipment. A system of differential equations of motion skip in the guide device.

Unlike the previously known mathematical models resistance guiding devices recorded in the roller strain depending on time. The study parameters and guiding devices career pneumatic lifting mechanism during periods of unsteady motion skip, as well as the regulation of its speed accomplished using a mathematical model, which is a system of seven differential equations that describe the motion skip along the career of an overpass. This mathematical model is based on the calculation scheme shown in Figure 6.

At constant pressure vessels under $P = const$ (3):

$$\begin{cases} \frac{d^2 z}{dt^2} = \frac{1}{M} \left[(P - P_a) \cdot S_c - F \cdot \sin \beta_c - F_{fryr} - \frac{k \cdot C}{R_0} \cdot \sum_{i=1}^{n} \Delta_i \right] \\ \frac{d^2 \Delta_1}{dt^2} = -\frac{1}{M} \left(C \cdot \Delta_1 + \beta \frac{d\Delta_1}{dt} \right); \ \frac{d^2 \Delta_2}{dt^2} = -\frac{1}{M} \left(C \cdot \Delta_2 + \beta \frac{d\Delta_2}{dt} \right); \ldots \\ \ldots; \ \frac{d^2 \Delta_n}{dt^2} = -\frac{1}{M} \left(C \cdot \Delta_n + \beta \frac{d\Delta_n}{dt} \right) \end{cases} \quad (3)$$

where β_c is the angle of the career of an overpass to the horizon; n is the number of guiding devices are simultaneously in contact with loader.

The above mathematical model and the results of its numerical solution allowed us to develop an algorithm for calculating the parameters of procedures guiding devices career pneumatic lifting equipment.

Numerical study of the dynamics of motion skip pneumatic lifting equipment was carried out by a licensing system MatLab (Matrix Laboratory), which is an interactive system for performing engineering and scientific calculations, and is focused on work with arrays of data [4,5].

The system makes it easy to exchange information with the word processor Microsoft Word, in particular to carry any text and images to the clipboard or read text strings from the clipboard as executable instructions. The system MatLab are special means of solving the Cauchy problem for systems of ordinary differential equations given in explicit $\frac{dx}{dt}=F(t,x)$, and in the implicit $M\frac{dx}{dt}=F(t,x)$. Here M-matrix - the so-called ODE solver (solver ODE), allows the user to select the method, the initial conditions, etc.

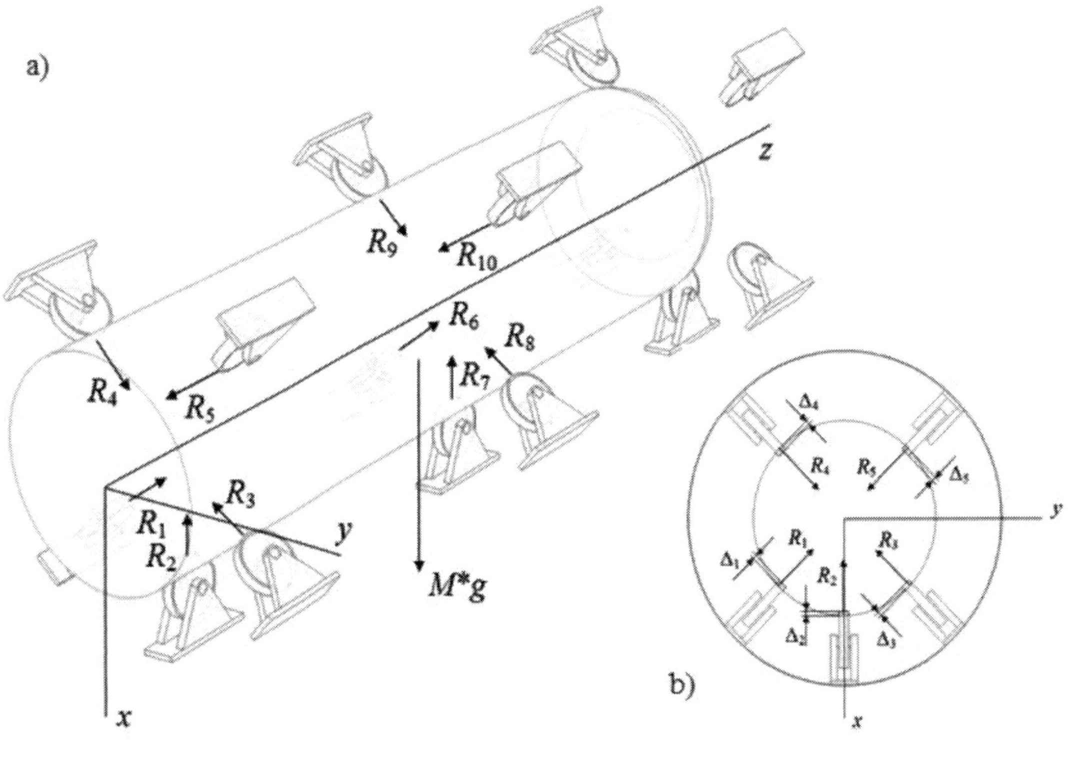

Figure 6. Open pit pneumatic lifting equipment scheme: a) general view; b) cross-section.

To solve ODE systems in MatLAB various methods implemented. Their implementation called ODE solvers. Solvers implement the following methods for solving systems of differential equations:

- *Ode* 45 is one-step explicit Runge-Kutta 4th and 5th order. This is a classic method recommended for the initial sample solutions. In many cases, it gives good results;
- *Ode* 23 is one-step explicit Runge-Kutta 2nd and 4th order. In moderate rigidity of the system of ODE and low accuracy requirements, this method can provide a gain in speed solutions;
- *Ode* 133 is multi-step method of Adams-Moulton-Bashvorta variable order. The adaptive method that can provide high accuracy solutions;

- $Ode\ 15s$ is multi-step method of variable order (from the first to the fifth, the default is 5), using the formula of numerical differentiation. This adaptive method, it should be applied, if the solver $Ode\ 45$ not provide solutions;
- $Ode\ 23\ s$ is one-step method, which uses a modified Rosenbrock formula of order 2. Can provide a high computational speed with low accuracy;
- $Ode\ 23t$ is trapezium method with interpolation. This method gives good results in solving problems describing oscillators with nearly harmonic output signal;
- $Ode\ 23tb$ is implicit Runge-Kutta method at the beginning of the solution and the method using the inverse of differentiation formula of order 2 in the future. At low precision, this method may be more efficient than $Ode\ 15s$.

All solvers can solve systems of equations of the explicit form $y' = F(t, y)$. Solvers $Ode\ 15s$, $Ode\ 23s$, $Ode\ 23t$, $Ode\ 23tb$ can solve the equations of implicit form $My' = F(t, y)$.

All of these methods are incorporated in a means of solving differential equations Simulink. Solution of the mathematical model in the application program Matlab Simulink is shown in Figure 7.

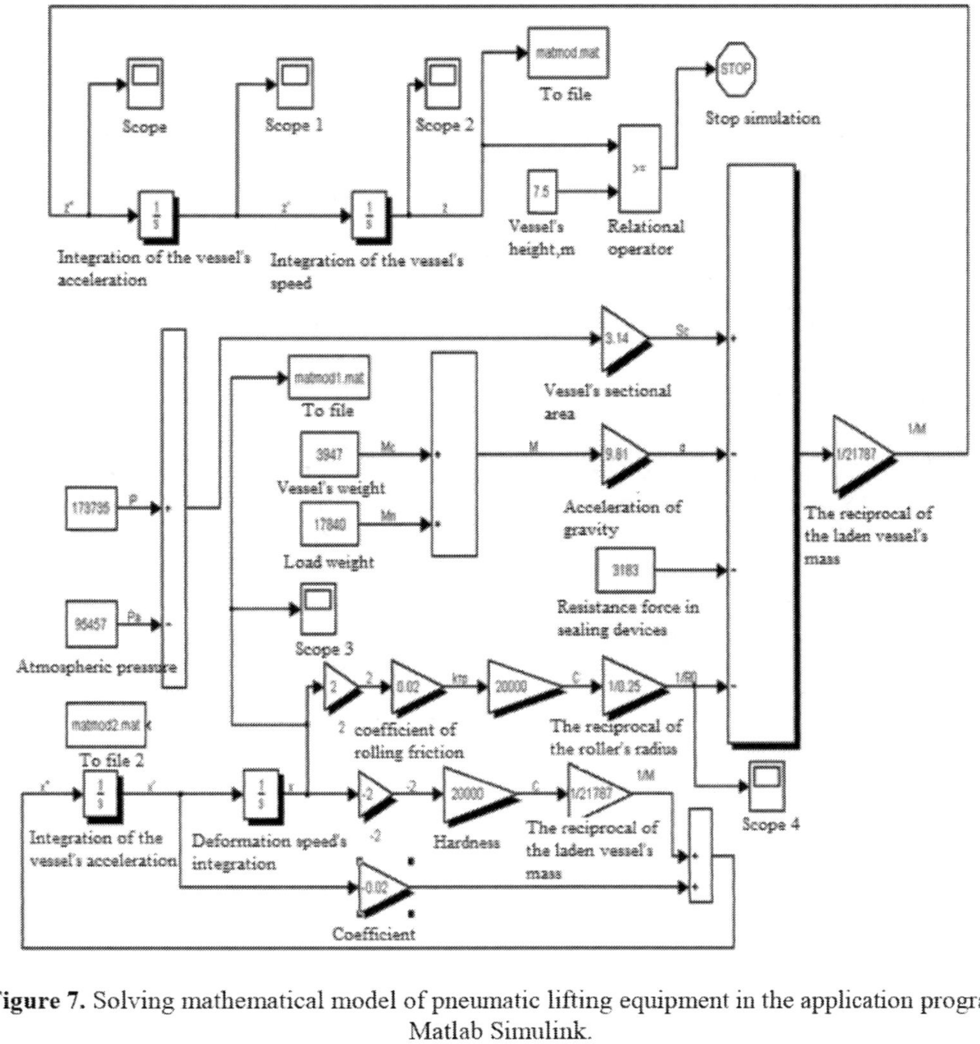

Figure 7. Solving mathematical model of pneumatic lifting equipment in the application program Matlab Simulink.

As described in the following functions for the solution of systems of differential equations, the following notation and rules:
F is the name of the ODE-file, that is a function of t and y, which returns a column vector;
tspan is the vector determining the integration interval $[t0\ tfinal]$. To obtain solutions at specific times $t0, t1...,tfinal$ (arranged in order of decreasing or increasing), you must use $tspan = [t0\ t1\ ...\ tfinal]$;
$y0$ is the vector of initial conditions;
options is argument to create functions odeset;
$p1, p2$ ... is Arbitrary parameters passed to the function F;
T, Y is solution matrix Y, where each line corresponds to the time returned in the column vector T.
We now turn to the description of functions for solving systems of differential equations (we denote the concept solver one of the possible numerical methods for solving ODE: $ode\ 45$, $ode\ 23$, $ode\ 113$, $ode\ 15s$, $ode\ 23s$, $ode\ 23t$, $ode\ 23tb$):
$[T, Y] = solver\ ('F', tspan, y0)$ integrates the system of differential equations of the form $y' = F(t, y)$ the interval tspan with the initial conditions $y0$. 'F' – a string containing the name of the ODE-file. The function F (t, y) must return a column vector. Each row in the array of solutions Y corresponds to the time returned in the column vector T.
$[T, Y] = solver\ ('F', tspan, y0, options)$ gives a solution similar to that described above, but with the parameters determined by the values of the argument options, created with odeset. Commonly used options include permissible value of the relative error RelTol (default $1e-3$) and the vector of allowable values of the absolute error AbsTol (all the components default to $1e-6$;
$[T, Y] = solver\ ('F', tspan, y0, options, p1, p2, ...)$ provides a solution similar to that described above, by passing additional parameters $p1, p2$ ... Use options = [] , if no options are specified.
ODE solver enables the production of solutions of systems of n equations. ODE system can be either homogeneous or heterogeneous.
From the graphs, we can draw conclusions about the nature of dynamical fluctuations skip to the transverse axis of the shaft direction (Figure 8,9). Periodic fluctuations are damped, where damping is exponentially damping factor for mechanical parameters of the dynamic system "skip - guiding device" has different meanings, which can be calculated from the equation.
The experimental model velocities and accelerations in the transverse direction are used to determine the loads on the rollers and the guide devices powers are compared with the values obtained from the mathematical model. Revealed that the maximum dynamic loads on the guide device by pneumatic skip lifting equipment as it moves are determined depending on the weight of the loaded skip, the shift of its center of mass, the angle of the overpass to the horizon and the coefficient of dynamic (3,4):

$$R_i = (1 + k_o \cdot V) \cdot M \cdot g \cdot \frac{a}{l}, \tag{3}$$

$$R_i = \frac{(1 + k_o \cdot V) \cdot M \cdot g \cdot \cos\beta_c}{n}, \tag{4}$$

Where V – the velocity of the skip, m/s; k_o is stiffness coefficient overpass; a is the displacement of the center of mass of the laden skip; l is the distance between adjacent tiers of guiding devices;
$k_d = 1 + k_c \cdot V$ is dynamic factor.

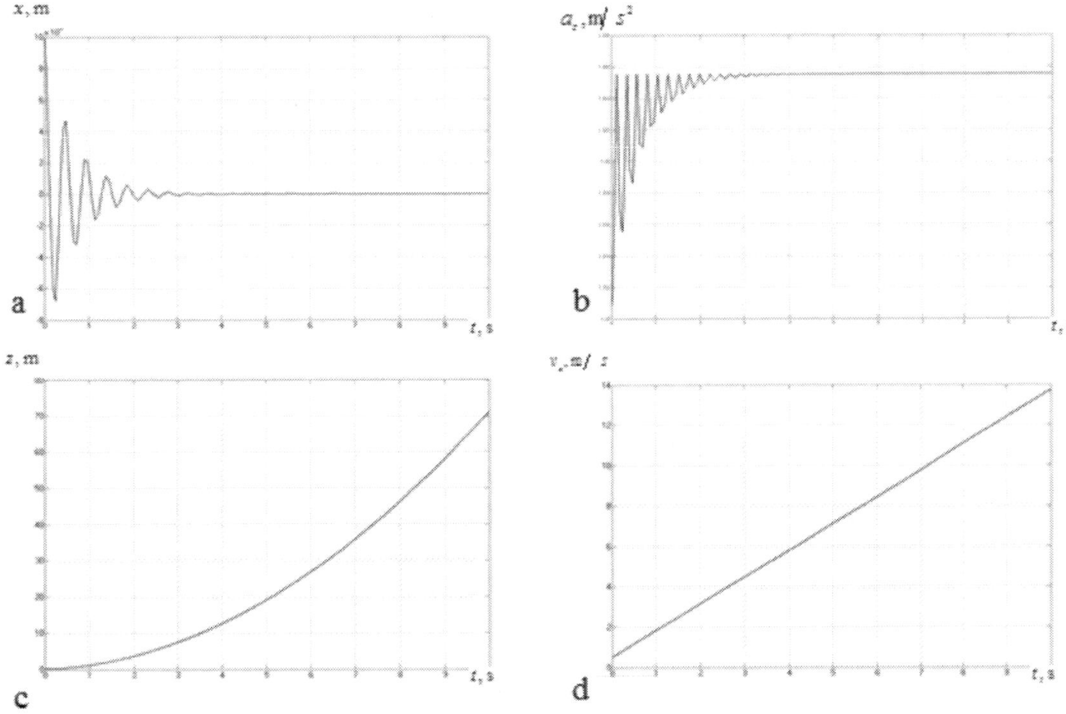

Figure 8. Results of the solution of the mathematical model of pneumatic lifting equipment in the application program Matlab Simulink:
a) dependence of the deformation roller from time to time; b) dependence of the acceleration skip from time to time; c) dependence on the time of travel Skip; d) dependence of the rate on the time skip.

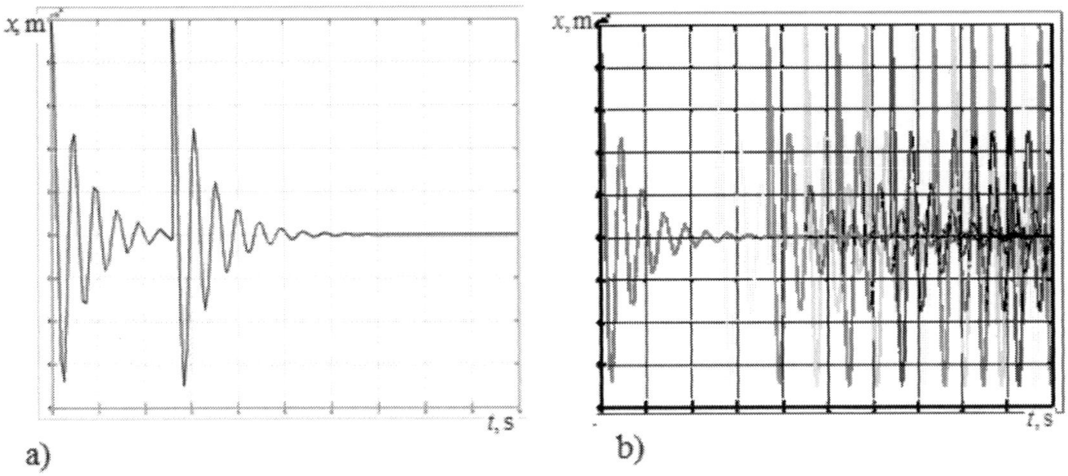

Figure 9. Walkthrough skip several tiers guiding devices: a) passage of two tiers; b) passage of fifteen tiers.

3.Conclusions

Developed and researched mathematical model of the motion skip in systems of quarry pneumatic lifting equipment and mine have established patterns of interaction "skip - guiding devices". It was found that the system of "skip - guiding devices" most loaded at the time of entry unevenly loaded skip stage vanes. Revealed that the maximum dynamic loads on the guide device by pneumatic skip lifting equipment as it moves are determined depending on the weight of the loaded skip, the shift of its center of mass, the angle of the overpass to the horizon and the coefficient of dynamic. The adequacy of the emerging dynamic loads on the roller guide device is confirmed by experimental research models skip pneumatic lifting equipment.

References

[1] Zhautikov B. A. Analysis of efforts on housing skip open pit pneumatic lifting equipment *Bulletin of the National Engineering Academy of the Republic of Kazakhstan* 2006 **3** P. 38–42.

[2] Nikolaev Yu A and Rudaya R P 1981 A mathematical model of pneumatic shaft skip lifting equipment *Proceedings of the universities Mining Journal* **10** P. 88.

[3] Chen K, Dzhiblin P and Irving A MATLAB's mathematical research Moscow: Mir, 2001 P. 346

[4] Chernih I.V. SIMULINK: creation environment engineering applications 2003 Moscow: DIALOG-MIFI, 2003 P. 496

[5] Porshnev S.V. Using Simulink package to describe the dynamics of oscillating systems with several degrees of freedom *Electronic scientific journal «Bulletin of Omsk State Pedagogical University»* 2006 P.1–5

Using of science technologies for mining machinery constructions' strength improvement

E V Yurchenko [1], A D Mehtiev [2], V V Yugai [3], F N Bulatbayev [4]

1 National Research Tomsk Polytechnic University,Tomsk, Russia
[2] Head of "The technology and communication systems" Department in Karaganda State Technical University, Karaganda, Kazakhstan
[3] Senior lecturer of "The technology and communication systems" Department in Karaganda State Technical University, Karaganda, Kazakhstan, phD in KazNTU.
[4] Deputy Director of "Correspondence and Distance learning faculty"

E-mail: slawa_v@mail.ru

Abstract. Recommendations for strengthening the brake construction in accident dangerous areas of fatigue destruction were developed. Computer modeling was made using the ANSYS program that helps to visualize stained condition of the construction for further practical testing of the strength and reliability improving technology of mining elevating machines' constructions, which are being in a long-term use, with a help of the strengthening elements. A way of construction strengthening, which eliminates the possibility of further fatigue destruction of the brake system elements, because of the load cycle in exploitation process.

1. Introduction

The trend of modern development of production technology of mining equipment is that the main task should not be placed on maintaining technology to achieve a high level, but the need for prompt changes. The most significant changes are not on the path of modernization, but due to a fundamental transformation of existing technologies in high-tech.

Improving the reliability of the mining elevating machine's (MEM) brake system (BS) requires, including scientific research of a construction's rational variant which can withstand long-term fatigue and continue to remain MEM operational for a long service life.

The ANSYS program, allows you to simulate the stress-strain state of the BS elements specifications and replace bulky full-scale tests of experimental samples by modelling find the best option reinforced construction, with higher strength parameters and reliability, also to develop practical recommendations to restore strength through the BS MEM repair works. With the gain elements it is possible to locate and stop the growth of fatigue cracks in the BS elements and to prolong its service life.

Ensuring the reliability of the Karaganda basin-ray's mining elevating machines' brake system is becoming increasingly important, as the load sharply increases due to increased mining in recent years. The load increasing directly affects the wear and the brake element failures. In the process of raising or lowering the load must be considered rapidly changing nature of the load on the brake system. Significant peak loads on the brakes occur during sudden stops of the body coiling rope in time of emergencies. Equally important is the number of cycles of the brake system's working during the day, which is from 500 to 1500 "ascending and descending" [1]. These factors contribute to the development of fatigue cracks in BS which can develop during operation and lead to structural failure. It is hard to predict the development and forming of such damages, only the most affordable and reliable method is destructive testing. According to the data presented in [2], fatigue cracks are formed in place of the lugs which fasten the brake linkage sleeve. After identifying the damages in the brake beam elements it is necessary to make its repair or replacement. The beam replacing associated with considerable economic and labour costs, as it requires disassembly of indigenous parts of hoisting machine, which in turn causes a length-tive simple lifting machine. Repair usually boils down to cutting and welding-arc welding cracks. However, this method does not provide a reliable operation of the construction, since the operation may continue to develop cracks cracks formed.

Considering these circumstances, it is appropriate to increase the strength of brake beams by strengthening its structure in the presence of stress concentrators places where eventually formed fatigue cracks. Strengthening MEM BS can significantly improve their durability.

2. The studying of stress-strain state of the mining elevating machine's brake beam construction

Based on the requirements of the studying object, is formed the ultimate goal of research, outlined the rational way to achieve it and formed a research program [1]. The last one should be regarded pre-loading process optimization and modes of operation of mining elevating machine's brake system, implementing this knowledge-based technology, which should provide the best technical and economic parameters of the studying object. An important component of this technology is to obtain an adequate model, which can be used to identify "critical points of failure" and "accident-prone zones of deformation." It is known that the details of the brake system are carry-out its functions only in the initial period of operation [1]. They often can be destroyed by fatigue reasons related to exposure to repeated cyclic loads. Mechanical stresses concentrators reduce stress durability during cyclic loading, and form the "accident-prone zones of deformation" in the BS constructions. Figure 1 shows a graphical representation of a mechanical stresses modeled in the ANSYS program. Marked decrease in fatigue strength of the beam structure in accident-prone zones 1 and 2, where the greatest concentration of stress in the process of using the pro-beam fatigue cracks is formed. Zone 1 and 2 are located in an area-weld the top and the side face of the beam (Figure 1). This is also related to metal structural changes around the suture zone and the concentration of residual stresses after welding.

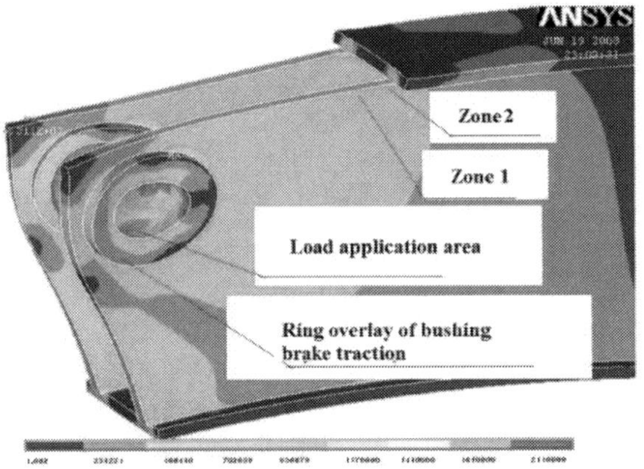

Figure 1. Graphical display of typical brake beam model with the "accident-prone zones" fatigue fracture 1 and 2.

Analysis of the fatigue fracture of metal brake beam with the cyclic load factor showed that the main factors in the development of fatigue cracks is:

- The amplitude of the stresses and deformations;
- Duration and number of cycles.

A transition from a hidden (fine) to ext-term (broad) fracture is observed in the process of destruction due to metal fatigue which accompanied by the accumulation of damages in the grain boundaries of the metal from cycle to cycle. Results modeling of structure cracks development using the ANSYS presented on Figure 2. The computer model allows you to visualize the development of fatigue cracks in the accident-prone zones 1 and 2, that will set the parameters of structural elements in the future [2].

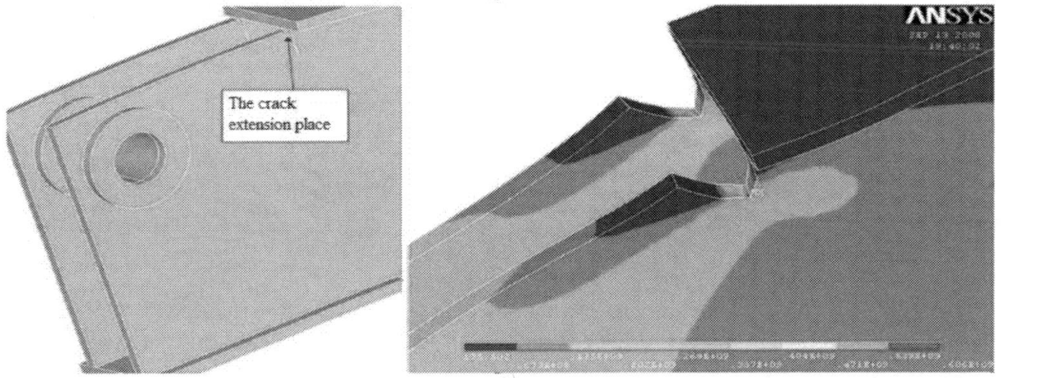

Figure 2. Graphical representation of a fatigue crack development process in the "accident-prone area."

A method of the construction enhancing based on the use of overhead elements, which can eliminate the possibility of further fatigue failure associated with multiple cyclic loads application BS elements during its operation. Overlay gain can be disc pad, edge, ring pad or combined options. Overhead elements mount on the side faces of the BS beam using a welded or adhesive bondings. These methods have been practically tested on existing MEM. The disc pad attaching by welding method requires measures for liquidation of the con-effects of welding deformation. The perspective direction is complement-installation work on the establishment of the use of adhesive pad. The present level of adhesive joints provide mechanical properties equal to connections made using arc welding. The advantage of adhesive joints is the lack of stress and deformations, also the formation of hardening structures of metal in the weld seam area because of the heat-affected influences occurring during welding process. The last are more efficient, related to the localization of "emergency - danger zones" and fatigue crack growth resistance. The technology of manufacture and installation of overhead components is simple and does not require sophisticated welding and mounting hardware. Figure 3 shows the elements of the proposed options for reinforcement of BS type "disk pad" and "edge" that are able to influence the nature of the development of cracks, significantly reducing stress concentrations and thus increase the durability and reliability. ANSYS software opportunities allow you to simulate the load on the beam elements associated with cyclical load and the asymmetry of the cycle.

During the simulation was able to establish the nature of changes in the rate of fatigue crack growth, which depends on the cycle asymmetry coefficient of stress intensity. Using the structural reinforcement of BS reduce the scope of cycle asymmetry coefficient of stress intensity, and therefore significantly reduce the fracture closure predisposition to change state on the verge of exhaustion and stress cycle asymmetry, which ultimately reduces the probability of the fatigue crack growth.

Terms and conditions of the process of formation and growth of fatigue cracks are directly depend on the cycle stress parameters (amplitude, asymmetry, frequency) and environmental impact (its reactivity, humidity, temperature), as well as the nature of the stress-strain state near the crack tip in the structural element which is defined by their geometry and dimensions.

The task of increasing the work durability and reliability of the top beam of brake system reduces to the definition of "dangerous places" in the design and measures for decreasing the stress and deformation. This can be done using the method of simulation of the development of cracks in the structure using the ANSYS program [3].

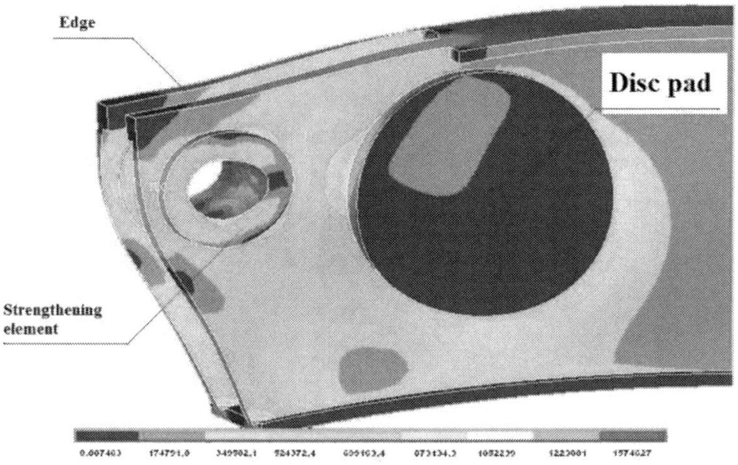

Figure 3. The design of beam elements with structural reinforcements.

The results of studies on the establishment of dependence reduction factor for the pass-β structural strength of the length of the growing crack in it; and reducing the dependence of values permissible for the construction of mechanical stress on the σk stresses growing crack in it, leading to the destruction of the structure presented in Figure 4*a* and Figure 4*b* respectively.

Computer modelling confirmed that the use of strengthening elements provide values reduction of mechanical stress and deformation at the crack tip, and significantly weaken the influence of stress concentrators (Figure 5).

The basis for calculation of the fatigue life (endurance) determining the ability of a material to resist the design cycle fatigue, which leads to the destruction of life and reduce the braking device. Calculation of endurance runs using three methods: calculation of the deformation, stresses calculation and fracture mechanics. All the listed methods are available in the module calculation of durable, available in ANSYS Fatigue Module. Computer simulation in the ANSYS Fatigue Module environment allowed to determine the geometric parameters of the overhead elements and rational coordinates of their location on the side faces of the beam structure [4].

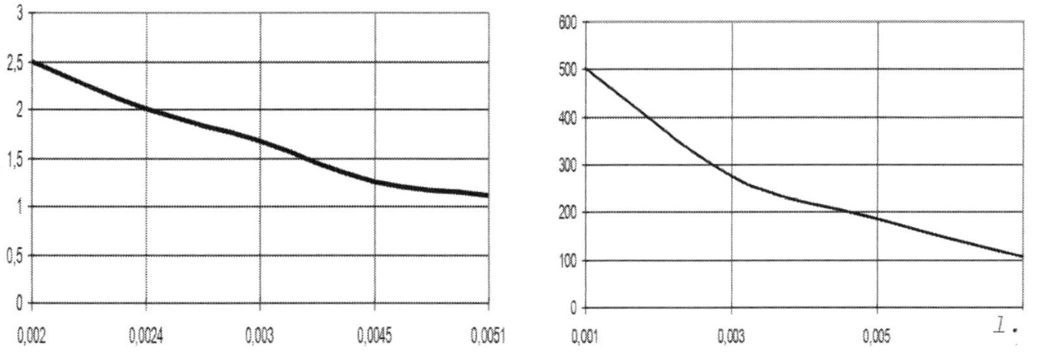

a) b)

Figure 4. The nature of the development of cracks in the MEM BS structure: a) the dependence of reducing the construction safety of the length of the growing crack in it; b) the dependence of reducing allowable stress values, leading to the destruction of the structure, increasing the length of the crack in it.

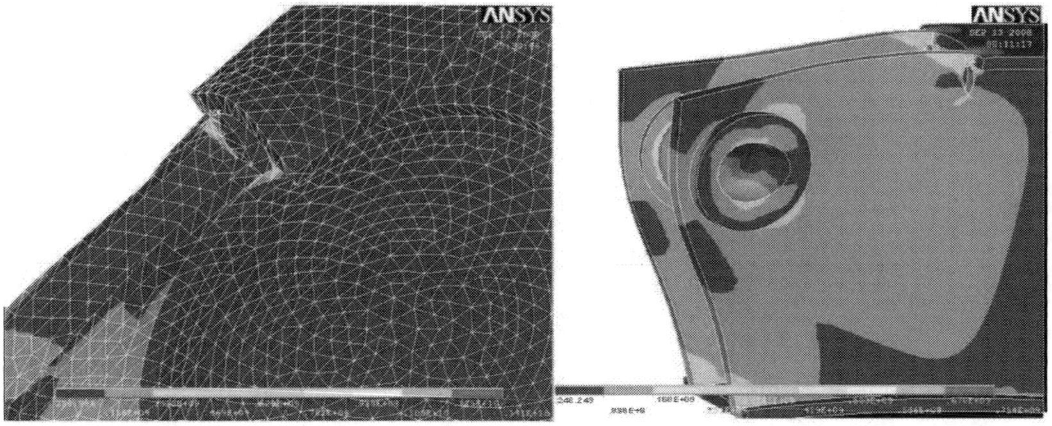

Figure 5. Graphic display of the localization of the cracks development in the MEM BS structure using reinforcements.

3. Practical implementation of knowledge-based technology to enhance emergency and hazardous areas of fatigue failure

Advantages of the practical implementation of knowledge-based technology is the use of ultrasonic destructive testing, which allows detecting at an early stage fatigue cracks and modeling. The modeling allows determining the design parameters of the gain elements to restore the parameters of reliability and durability of steel structures, without a considerable amount of field experiments. It will significantly reduce material costs for repairs and losses from downtime and to create high-optimal shape design with low metal content, without the use of expensive high alloys in the future.

Three-dimensional models, developed an automated way of solving the problem of the finite element method, are created based on the data of field surveys, analysis, technical documentation and design study of the stress-strain state of typical beams brake system the results are presented in the form of numerical values, graphs. The developed mathematical schemes have flexible program code and can be used as templates for algorithmization of solving such problems. During the operational load, analyzed the stress-strain state of the model and the major factors that lead to premature-time violations of brake system beams. As a result, a BS beam lifting machine's simulation-model developed, which allows you to set the static and cyclic loads in this construction depending on the type of use the momentum and technical conditions [5]. The use of simulation techniques makes it possible to eliminate the full-scale tests completely, to optimize the shape and construction parameters of the structural reinforcement elements, to reduce the values of maximum stress and achieve a leveling effect on the harmful effects of stress concentrators in the "critical points" at the operational design loads. Mounting technology is unique in its simplicity and does not require expensive equipment [6].

A number of field experiments conducted at existing MEM (diagram in Figure 6), showed that the brake system beam having a wall thickness of 20 mm without power in the "critical points" (model 14) and construction with a wall thickness of 20 mm, but with the combined option amplification "disk-pad edge" (Model 26) are capable to operate at the exploitation loads and retain its strength as construction with a wall thickness of 40 mm (Model 44).

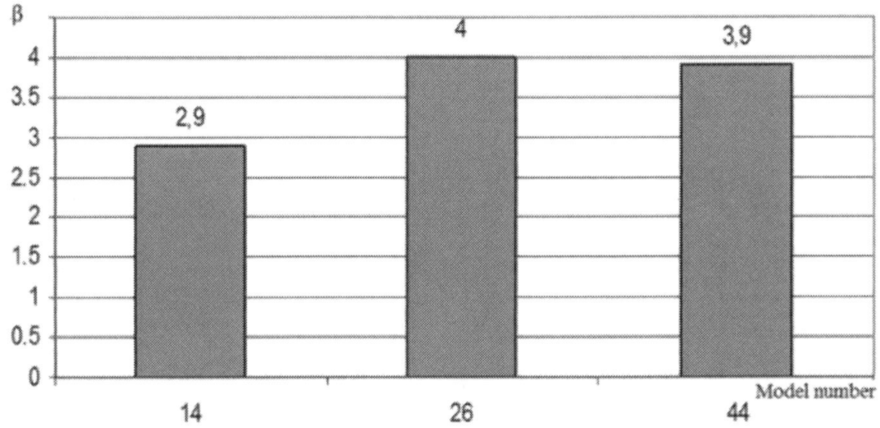

Figure 6. The safety coefficient for the brake system beam construction.

The installing disc pad allows to eliminate the hotbed of fatigue cracks occurrence, and the edge works to increase the structural strength of the beam. The technique was pro-operational testing in making repairs on existing MEM BS (Figure 7).

IOP Conf. Series: Materials Science and Engineering **81** (2015) 012114 doi:10.1088/1757-899X/81/1/012114

Figure 7. Disc pads' practical application for Novokramatorsky Engineering Factory mining elevating machine's brake system repair

4. Conclusions.

The results of computer modeling using ANSYS program allowed us to develop techniques for strength structural parts of mining machines improving that are using elements of the gain in continuous operation. It is established that at the top of the crack formed during loading the relative concentration of stress occurs, and then there is a redistribution of the latter, which causes further cracks development. It is necessary to have the distribution of stresses and deformations near the crack tip for quantitative characteristics of a material's ability to resist brake system beam construction violations.

BS construction's reinforcement elements allow eliminating the formation and growth of cracks in the field of bushings' mount; providing the necessary strength and durability; increase the service life of the brake system; significantly reduce the impact of stress concentrators, lower values of stress and deformation in the construction. Disc pad installing allows to eliminate the hotbed of fatigue cracks occurrence and the edge works to increase the structural strength of the beam and eliminates the possibility of construction's further deterioration associated with multiple cycles of load application during operation.

References

[1] Mehtiev A. D., Likhachev V. V. and Bulatbaev F. N. 2008 *The mining elevating machine's brake beam improving of work reliability methodics* - Karaganda: Proceedings of the university p. 72-74

[2] Mehtiev A.D. and Yugai V.V. 2011 Current status and problems of operation mine hoists Karaganda coal basin *Mining Machinery and Electrical Engineering* **6** p. 26–29

[3] Basov K. A. 2002 *ANSYS examples and problems* - Moscow: Computerpress p. 342

[4] Hemumov R. A., Kepler J. and Prokopenko V. I. 2004 *Application of the finite element method to the calculation of the structure* – Moscow:URSS p. 272

[5] Chigarev A.V. 2004 *ANSYS to engineers* – Moscow: Engineering p. 510

[6] Ogo O. M. 2008 *ANSYS Wokbench 10 and 11* - Moscow: Serviceprime p. 231

Improvement of optical systems for detection of smokes

V F Panin, A G Dashkovskii
National Research Tomsk Polytechnic University

E-mail: agd53@mail.ru

Abstract. The theory of electromagnetic radiation dispersion by polydisperse particles is analyzed. Methods of reliable optical indication of smokes to identify Fire Danger are considered. The conventional method of optical smoke detection implies measuring optical characteristics of the environment under control. After that the results obtained are converted into microphysical parameters which can be compared to the known microphysical properties of smokes.The calculated optical portrait of smokes is offered. The portrait of smokes is the field of representation points in coordinates of the degree of diffusion radiation polarization for two diffusion angles. Each of the spots indicates one of the numerous realizations of smokes. The direct match of the representation spots in the optical increases the probability of smoke A different way to protect optical system is to use the device with mutually orthogonal polarizers of the light source and detector. If hindrance is nonspherical aerosol, the signal from the device is used to correct the signals from smoke detectors.

1.Introduction

Smoke detection to identify fire condition in many cases is based on the registration of intensity of electromagnetic radiation (EMR), scattered the smoke particles at an angle Θ to the probing beam. It contributed to the smoke particles and particles such as dust, noise. To enhance the reliability of the identification of the fire needs to minimize the contribution of scattering dust particles in the total signal.

We shall consider the elements of the theory of dispersion of electromagnetic radiation by aerosol particles. Let flat linearly-polarized wave of electromagnetic radiation $\varepsilon_0 = \mathrm{Im}\ \{E_0\ e^{iwt-ikz}\}$ extend along the axis Z. At its interaction with a particle in a point of M the spherical scattered wave is formed:

$$\mathcal{E}(t, r, \Theta, \varphi) = \mathrm{Im}\{\dot{S}(\Theta, \varphi) \frac{e^{iwt-ikz}}{ikr}\}$$

here, $\dot{S}(\Theta, \varphi)$=$Se^{i\sigma}$ – the complex amplitude function defining the module and a phase of a wave, disseminated in the direction characterized by Θ, φ; r – distance; $k = \frac{2\pi}{\lambda}$ wave number.

Having chosen the plane of dispersion and having spread out the electric vector E_0 on mutually orthogonal components $\dot{E}_{0l}$ and $\dot{E}_{0r}$, it is possible to express $\dot{E}_l$ and $\dot{E}_r$ of a scattered wave through Maxwell-Jones's matrix:

$$\begin{bmatrix} \dot{E}_l(\Theta, \varphi) \\ \dot{E}_r(\Theta, \varphi) \end{bmatrix} = \begin{bmatrix} \dot{S}_2(\Theta, \varphi) & \dot{S}_3(\Theta, \varphi) \\ \dot{S}_4(\Theta, \varphi) & \dot{S}_1(\Theta, \varphi) \end{bmatrix} \mathrm{x} \begin{bmatrix} \dot{E}_0 \cos\varphi \\ \dot{E}_0 \sin\varphi \end{bmatrix} \frac{e^{-iwt+ikz}}{ikr} \quad (1)$$

Here, $\dot{S}_1 \div \dot{S}_4$ – components $\dot{S}(\Theta, \varphi)$. For smoke particles – spheres: $\dot{S}_1(\Theta, \varphi) = \dot{S}_1(\Theta)$, $\dot{S}_2(\Theta, \varphi) = \dot{S}_2(\Theta)$, $\dot{S}_3 = \dot{S}_4 = 0$, according to [1], thus:

$$\dot{E}_l = \dot{S}_2(\Theta)\dot{E}_{ol}\,\frac{e^{-ikr+ikz}}{ikr},\quad \dot{E}r = \dot{S}_1(\Theta)\dot{E}_{or}\,\frac{e^{-ikr+ikz}}{ikr}$$

The solutions for $\dot{S}_1$ and $\dot{S}_2$ according to [2] in [1] are presented in the form of:

$$\dot{S}_1(\Theta) = \sum_{n=1}^{\infty} \frac{2n+1}{n(n+1)}\left[\dot{a}_n \pi_n(\cos\Theta) + \dot{b}_n \tau_n(\cos\Theta)\right],$$

$$\dot{S}_2(\Theta) = \sum_{n=1}^{\infty} \frac{2n+1}{n(n+1)}\left[\dot{b}_n \pi_n(\cos\Theta) + \dot{a}_n\, \tau_n(\cos\Theta)\right].$$

$$\dot{a}_n = \frac{A_n(y)\psi_n(\rho) - m\psi_n'(\rho)}{A_n(y)\xi_n(\rho) - m\zeta(\rho)},\quad \dot{b}_n = \frac{mA_n(y)\psi_n(\rho) - \psi_n'(\rho)}{mA_n(y)\zeta(\rho) - \zeta_n'(\rho)}, \tag{2}$$

- $\pi_n(\cos\Theta), \tau_n(\cos\Theta)$ – angular functions, $= x = m_a \frac{2\pi a}{\lambda}$, $y = \frac{m_i}{m_a}\frac{2\pi a}{\lambda} = m\rho$, diffraction parameters, m_a, m_i – indices of refraction of environment and substances of a particle, $\psi_n(\rho), \zeta(\rho)$ – Bessel and Hankel's functions, $\psi_n(\rho), \zeta(\rho)$. Deyrmendzhan presented (2) in the form of [3]

$$\alpha_0 = \pi a^2 K(\rho, \dot{m}) = \pi a^2 \frac{2}{\rho^2} \sum_{n=1}^{\infty} (2n+1) Re\left(\dot{a}_n + \dot{b}_n\right)$$

$$\sigma_0 = \pi a^2 K_p(\rho, \dot{m}) = \pi a^2 \frac{2}{\rho^2} \sum_{n=1}^{\infty} (2n+1)\,\left[|\dot{a}_n|^2 + |\dot{b}_n|^2\right]$$

where K and K_p –are factors of the efficiency of weakening and diffusion.

The real stream of dispersed electromagnetic radiation is the mixture of simple waves, and the definition of the total result of their transformation is a statistical task. For this reason, the probing bunches of electromagnetic radiation (of optical range) are described with the help of statistical parameters that consider the characteristics of simple waves and additivity of incoherent components of intermixture: four square and bilinear forms regarding to $\dot{E}_l$ and $\dot{E}$. They determine completely the condition of polarization of electromagnetic radiation (Stokes parameters) and are registered by square devices:

$$\left.\begin{aligned} I &= \dot{E}_l\widehat{E_l} + \dot{E}_r\widehat{E_r} \qquad Q = \dot{E}_l\widehat{E_l} - \dot{E}_r\widehat{E_r} \\ U &= \dot{E}_l\widehat{E_r} + \dot{E}_r\widehat{E_l} \quad V = -i\left(\dot{E}_l\widehat{E_r} - \dot{E}_r\widehat{E_l}\right) \end{aligned}\right\}, \tag{3}$$

where I - intensity, интенсивность, energy stream through unit of area in unit of time, dimension is the same of Q, U, V. From (1) and (3) it is possible to formulate the equation of transformation of probing electromagnetic radiation.

$$\begin{bmatrix} I \\ Q \\ U \\ V \end{bmatrix} = \frac{1}{k^2r^2}\begin{bmatrix} 0{,}5(M_2+M_3+M_4+M_1) & 0{,}5(M_2+M_3-M_4-M_1) & (A_{23}+A_{41}) & -(D_{23}+D_{41}) \\ 0{,}5(M_2+M_3-M_4-M_1) & 0{,}5(M_2-M_3-M_4+M_1) & (A_{23}-A_{41}) & -(D_{23}-D_{41}) \\ (A_{24}+A_{31}) & (A_{24}+A_{31}) & (A_{21}+A_{34}) & -(D_{21}+D_{34}) \\ (D_{24}+D_{31}) & (D_{24}-D_{31}) & (A_{21}+A_{34}) & (A_{21}-A_{34}) \end{bmatrix}\begin{bmatrix} I_0 \\ Q_0 \\ U_0 \\ V_0 \end{bmatrix},$$

$$or\ [S_i] = [F_{ik}][S_{0k}],$$

where $[S_i]$, $[S_{0k}]$ – Stokes column vector, dispersed and probing bunches, $[F_{ik}]$ – matrix of light dispersion (MLD) or Muller matrix [5],

$M_k = S_k^2$, $A_{kj} = A_{jk} = \frac{1}{2}(\dot{S}_j\widehat{S_k} + \dot{S}_k\widehat{S_j})$, $D_{jk} =- D_{kj} = \frac{1}{2}(\dot{S}_j\widehat{S_k} - \dot{S}_k\widehat{S_j})$ so i-th Stokes parameter of dispersed electromagnetic radiation:

$$S_i = \sum_{k=1}^{4} F_{ik}\, S_{0k}$$

For a smoke particle (sphere):

$$\mathrm{F_{ik}} = \frac{1}{\mathrm{k^2r^2}}\begin{bmatrix} 1/2(\mathrm{M_2+M_1}) & 1/2(\mathrm{M_2\text{-}M_1}) & 0 & 0 \\ 1/2(\mathrm{M_2\text{-}M_1}) & 1/2(\mathrm{M_2+M_1}) & 0 & 0 \\ 0 & 0 & (\mathrm{A_{21}}) & \text{-}(\mathrm{D_{21}}) \\ 0 & 0 & (\mathrm{D_{21}}) & (\mathrm{A_{21}}) \end{bmatrix}$$

$$M_1 + M_2 = i_1 + i_2 = \alpha_1;\ M_1 - M_2 = i_1 - i_2 = \alpha_2;$$
$$2A_{12} = 2Re\{\dot{S}_1\widehat{S_2}\} = 2i_3 = \alpha_3;\ 2D_{21} = -2Im\{\dot{S}_1\widehat{S_2}\} = 2i_4 = \alpha_4.$$

During the transition from the matrix of light dispersion of a smoke particle to the matrix of light dispersion of particles in unit of volume, N, the function of their distribution by the sizes is considered

$$F(\rho) = Nf(\rho)$$

where (ρ)– spectral density, meeting the condition:

$$\int_{\rho_{min}}^{\rho_{max}} f(\rho)d\rho = 1$$

ρ_{min}, ρ_{max} – limits of the interval ρ. Therefore, the matrix of light dispersion of N particles system is:

$$[D_{ik}] = \begin{bmatrix} (D_{11}) & (D_{12}) & 0 & 0 \\ (D_{21}) & (D_{22}) & 0 & 0 \\ 0 & 0 & (D_{33}) & (D_{34}) \\ 0 & 0 & (D_{43}) & (D_{44}) \end{bmatrix} \qquad (4)$$

$$D_{ik} = k^{-3}\int_{\rho_{min}}^{\rho_{max}} F(\rho)\frac{ai}{2}d\rho$$

For optically active [6] smoke particles:

$$F(a) = NAa^{-\nu}$$

$$F(\rho) = k^{\nu}NA\rho^{\nu}$$

$$D_{ik} = \pi NAk^{\nu-3}\int_{\rho_{min}}^{\rho_{max}} \rho^{-\nu}\frac{ai}{2}d\rho$$

$$\alpha = \pi NAk^{\nu-3} \int_{\rho_{min}}^{\rho_{max}} \rho^{-(\nu-2)} K(\rho, \dot{m}) d\rho$$

$$\sigma = \pi NAk^{\nu-3} \int_{\rho_{min}}^{\rho_{max}} \rho^{-(\nu-2)} K(\rho, \dot{m}) d\rho$$

Using the matrices $[f_{ik}]$, $[f'_{ik}]$, $[\widetilde{f_{ik}}]$, received by rationing (4) on $\frac{\sigma}{4\pi}$, σ, D_{11}, as well as angle structure of dispersed electromagnetic radiation:

$$f'_{11}(\Theta) = \frac{D_{11}}{\sigma} = \varkappa$$

For calculation of $D_{ik}$the form of the light dispersion matrix from [3] is used:

$$[D'] = \begin{bmatrix} (D'_{11}) & 0 & 0 & 0 \\ 0 & (D'_{22}) & 0 & 0 \\ 0 & 0 & (D'_{33}) & (D'_{34}) \\ 0 & 0 & (D'_{43}) & (D'_{44}) \end{bmatrix}$$

$$D'_{11} = D_{11} + D_{21};\ D'_{22} = D_{11} - D_{21};\ D'_{33} = D_{33};\ D'_{44} = D_{44}; D'_{34} = D_{34};\ D'_{43} = D_{43}$$

Elements f'_{11}, $\widetilde{f_{11}}$, are received through D'_{ik}, by rationing on $\frac{\sigma}{4\pi}$, D_{11} [3]:

$$P_i(\Theta) = \frac{4\pi D'_{ik}}{\sigma} = \frac{4\pi}{\sigma} k^{-3} \int_{\rho_{min}}^{\rho_{max}} F(\rho) i_k d\rho$$

$$f'_{11} = \varkappa = \frac{D_{11}}{\sigma} = \frac{(D'_{11} + D'_{22})}{2\sigma} = \frac{(P_1 + P_2)}{8\pi}$$

$$\widetilde{f_{21}} = \frac{D_{21}}{D_{11}} = \frac{(P_1 - P_2)}{(P_1 + P_2)}$$

$$\widetilde{f_{33}} = \frac{D_{33}}{D_{11}} = \frac{2P_3}{(P_1 + P_2)}$$

$$\widetilde{f_{43}} = \frac{D_{43}}{D_{11}} = \frac{2P_4}{(P_1 + P_2)}$$

2. About smokes identification

Authors with employees are engaged in researches in the field of optical and micro structural characteristics of the smoke aerosols received at low-temperature pyrolysis (decay) of wooden, electro – heat-insulating, decorative, constructional and other materials and surfaces [7–9] since the beginning of the 1970th.

These researches were conducted with the Tomsk Scientific Center of Siberian Branch of the Russian Academy of Science, in particular, with the Doctors of Physico-mathematical Sciences: B.V. Kaul, I.V. Samokhvalov, V. F. Belov, O. A. Krasnov, V. S. Kozlov, to whom authors express their deep gratitude.

At the same time the authors of the research develop, test and implement highly sensitive to smoke, few dimensional sensors of smoke concentration and systems of fire detection [10–14] in cooperation with NPO Meridian, NPO Energiya, Institute of medicobiological problems of the Russian Academy of Sciences, nowadays the All-Russian Scientific Research Institute of Fire-prevention Defence, and with many other organizations.

Some results of the researches in the field of optical-polarizing protection of optical systems of early detection of the fire-dangerous state that, nowadays, are based, mainly, on registration of the radiation, dispersed by smoke particles, are given below. That means it is based on the registration of the first parameter of a Stokes column vector I – intensity (brightness). Therefore, the basis of optical-polarizing protection concept is the idea of using the Muller matrix and Stokes column vector or derivatives of them in order to identify the smokes of other parameters, for example, the degree of polarization of the radiation dispersed by the environment.

Two approaches to identification of smokes are formulated. According to the first one, the classical approach, optical characteristics of the environment are measured and transferred into microphysical parameters which are compared to known microphysics of smokes. Its basis is in the solution of the integrated equation:

$$\int_a^b K(x,y)f(x)\,dx = \varphi(y),\ a \le x \le b;\ \ c \le y \le d$$

or in the operator form:

$$Kf = \varphi \tag{5}$$

where, K – equation kernel, f – function of dispersion of particles according to the sizes, φ – result of optical experiment. So, the solution (5) is the solution of the system:

$$\sum_{i=1}^{n} K_{ij} f_i = \varphi_i,\ j = 1,2, \ldots n \tag{6}$$

The problem of the solution (6) – is to find out the closest to true solution from a set of F solutions. Trial and error method [15] is used under a known subclass of solutions M for correct tasks according to Tikhonov: the function f_δ, meeting the following requirement is selected:

$$\|Kf_\delta - \varphi_\delta\| \le \delta \tag{7}$$

where, $\delta = \Delta^2$, $\Delta -$ measurement error φ. If φ_δ does not belong to the set KM, the reverse operator $K^{-1}\varphi_\delta$ cannot make any sense.

In the method of regularization [15] there is a solution f_δ^α, minimizing the functional with the discrepancy (7) and with the functional of smoothness of solution with a regulating multiplier α. In [16] for definition of $f(x)$ and n multifrequency measurements $D_{11}(\pi)$ and α, are used. According to zero approach n^0 the method of regularization is used to find out f^1, that is further used by the following iteration to specify n with the discrepancy $\|D_{11}(\pi) - K_\pi(n)f^1\|$ etc.

Measuring the frequency ranges $D_{11}(\pi)$, α , $D_{11}(\Theta)$ it is possible to calculate $f(x),\ n,\ \chi$.

The complexity of multifrequency or multiangular [16] measurements of the elements of the light dispersion matrix, the necessity to use the computer programmes in order to convert the results of optical measurements and implementation of the criterion of "recognition" of a smoke etc. make the application of the "microphysical" approach of smokes identification very problematic.

Within the second approach the identification is organized by measuring of known parameters of light dispersion matrix of the environment and of their comparison with an optical portrait of smokes, to similar types of optical weather [17].

According to the theory of recognition of images an optical portrait the can be defined by a field of representation spots in n – measured space of the characteristic parameters of the light dispersion matrix, where each spot represents a certain degree of smoke (or hindrances). Characteristic parameters have to provide simplicity of measurements, compactness and discrepancy of fields of smokes and hindrances, diminutiveness of fire situation system. The simplest case of a field – in coordinates of two parameters, for example, the degree of polarization at two angles of dispersion, P_1 and P_2. At radiation of smoke particles by the monochromatic non- polarized radiation, the Stokes's vector parameter of dispersed electromagnetic radiation is:

$$[S_i] = [D_{ik}][S_{0K}] = \begin{bmatrix} (D_{11}) & (D_{12}) & 0 & 0 \\ (D_{21}) & (D_{22}) & 0 & 0 \\ 0 & 0 & (D_{33}) & (D_{34}) \\ 0 & 0 & (D_{43}) & (D_{44}) \end{bmatrix} \begin{bmatrix} I_0 \\ 0 \\ 0 \\ 0 \end{bmatrix} = I_0 \begin{bmatrix} (D_{11}) \\ (D_{21}) \\ 0 \\ 0 \end{bmatrix}$$

$$P =-\frac{Q}{I} =-\frac{D_{21}}{D_{11}} = \widetilde{f_{21}}$$

When installing a polarizar П with an azimuth $\psi = 0$ at a receiver with the angle Θ

$$[S_1(\Theta_1)] = [M_k][S_K(\Theta_1)] = K \begin{bmatrix} 1 & 1 & 0 & 0 \\ 1 & 1 & 0 & 0 \\ 0 & 0 & 0 & 0 \\ 0 & 0 & 0 & 0 \end{bmatrix} I_0 \begin{bmatrix} D_{11}(\Theta_1) \\ D_{21}(\Theta_1) \\ 0 \\ 0 \end{bmatrix} = KI_0 \begin{bmatrix} D_{11}(\Theta_1) + D_{21}(\Theta_1) \\ D_{11}(\Theta_1) + D_{21}(\Theta_1) \\ 0 \\ 0 \end{bmatrix}$$

Here $[M_k]$ and K – matrices of conversion and transmission coefficient П that are registered by the brightness receiver $\psi = 0$ и $\pi/2$, $S_{1\parallel}$ и $S_{1\perp}$, $S_{1\parallel} = KI_0(D_{11}(\Theta_1) + D_{21}(\Theta_1))$;

$S_{1\perp} = KI_0(D_{11}(\Theta_1) - D_{21}(\Theta_1))$, therefore $\frac{(S_{1\parallel} - S_{1\perp})}{(S_{1\parallel} + S_{1\perp})} = \frac{D_{21}(\Theta_1)}{D_{11}(\Theta_1)} =- P_1 = \widetilde{f_{21}}(\Theta_1)$

Measurements of $S_{1\parallel}$ end $S_{1\perp}$ are possible by cyclic turning of П on the angle $\pi/2$ and vice versa or by locating another identical receiver with П on the other side from the axis of the lighter at the angle θ_1. The same process is used to calculate P_2. A number of researches [13], of optical schemes for measuring of other elements of light dispersion matrix confirm the measurement scheme $P(\Theta_1)$ end $P(\Theta_2)$.

Realization of smokes identifier on the basis of control of environment-representing spot in coordinates P_1 end P_2.represents rather complex challenge. Therefore, the other principle of optical system protection from aerosol hindrances is investigated. It is based on the device with mutually orthogonal surfaces of polarization of light source and detector polarizers, for example, light source: $\psi = 0$, and detector: $\psi = \pi/2$. Such device is a cross-polarizing indicator of aerosol hindrances (dust) – CIH.

Stokes column vector of polarizing radiation of a light source:

$$[S_i] = K \begin{bmatrix} 1 & 1 & 0 & 0 \\ 1 & 1 & 0 & 0 \\ 0 & 0 & 0 & 0 \\ 0 & 0 & 0 & 0 \end{bmatrix} \begin{bmatrix} I_0 \\ 0 \\ 0 \\ 0 \end{bmatrix} = K \begin{bmatrix} I_0 \\ 0 \\ 0 \\ 0 \end{bmatrix}$$

Stokes column vector of radiation by dispersed smokes:

$$[S_i] = \begin{bmatrix} D_{11} & D_{12} & 0 & 0 \\ D_{21} & D_{22} & 0 & 0 \\ 0 & 0 & D_{33} & D_{34} \\ 0 & 0 & D_{43} & D_{44} \end{bmatrix} K \begin{bmatrix} I_0 \\ 0 \\ 0 \\ 0 \end{bmatrix} = KI_0 \begin{bmatrix} D_{11} + D_{12} \\ D_{21} + D_{22} \\ 0 \\ 0 \end{bmatrix}$$

At the polarizer output in front of the receiver:

$$[S_i]_\Phi = K \begin{bmatrix} 1 & -1 & 0 & 0 \\ -1 & 1 & 0 & 0 \\ 0 & 0 & 0 & 0 \\ 0 & 0 & 0 & 0 \end{bmatrix} KI_0 \begin{bmatrix} D_{11} + D_{12} \\ D_{21} + D_{22} \\ 0 \\ 0 \end{bmatrix} = K^2 I_0 \begin{bmatrix} (D_{11} + D_{12}) - (D_{11} + D_{22}) \\ -(D_{11} + D_{12}) + (D_{21} + D_{22}) \\ 0 \\ 0 \end{bmatrix} = \begin{bmatrix} 0 \\ 0 \\ 0 \\ 0 \end{bmatrix}$$

If for sphere particles $D_{11} = D_{12}$, $D_{21} = D_{12}$, so in a smoked environment an output optical signal of cross-polarizing indicator of aerosol hindrances is equal to zero, and in sensors it is formed "purely smoke" signals proportional to the size $I_0 D_{11}$.

In case of an aerosol hindrance with light dispersion matrix $[D^n_{ik}]$ sensors produce additional signals $I_0 D^{\,n}_{11}$, and CIH form dispersed radiation with Stokes column vector:

$$[S_i^n] = \begin{bmatrix} D_{11}^n & D_{12}^n D_{13}^n D_{14}^n \\ D_{21}^n & D_{22}^n D_{23}^n D_{24}^n \\ D_{31}^n & D_{32}^n D_{33}^n D_{34}^n \\ D_{41}^n & D_{42}^n D_{43}^n D_{44}^n \end{bmatrix} K \begin{bmatrix} I_0 \\ I_0 \\ 0 \\ 0 \end{bmatrix} = K I_0 \begin{bmatrix} D_{11}^n + D_{12}^n \\ D_{21}^n + D_{22}^n \\ D_{31}^n + D_{32}^n \\ D_{41}^n + D_{42}^n \end{bmatrix}$$

The sensor receives the radiation with parameters:

$$[S_i^n] = K \begin{bmatrix} 1 -1\,0\,0 \\ -1\;1\,0\,0 \\ 0\;\;0\,0\,0 \\ 0\;\;0\,0\,0 \end{bmatrix} K I_0 \begin{bmatrix} D_{11}^n + D_{12}^n \\ D_{21}^n + D_{22}^n \\ D_{31}^n + D_{32}^n \\ D_{41}^n + D_{42}^n \end{bmatrix} = K^2 I_0 \begin{bmatrix} D_{11}^n + D_{12}^n - D_{21}^n - D_{22}^n \\ -D_{11}^n - D_{12}^n + D_{21}^n + D_{22}^n \\ 0 \\ 0 \end{bmatrix}$$

with intensity $K^2 I_0 (D_{11}^n + D_{12}^n - D_{21}^n - D_{22}^n)$. The idea of system protection from aerosol hindrances is based on the selection of transformation coefficient $K_{К\text{-}П}$ of CIH signal, providing

$$K_Д I_0 D_{11}^n = K_{К\text{-}П} K^2 I_0 (D_{11}^n + D_{12}^n - D_{21}^n - D_{22}^n),$$

where, $K_Д$– transformation coefficient of optical signal of smoke detector and a component $K_Д I_0 D_{11}^n$–smoke detector signals before their processing in the main system unit.

3. Conclusions

1. Classical way of smoke detection in a controlled space is the measurements of optical characteristics of environment and their transfer into microphysical parameters with further comparison and known microphysics of smokes. The complexity of conversion procedures of optical characteristics into microphysical parameters makes the process of implementation of "microphysical" identification rather problematic.
2. More preferable way is the method of measuring of certain parameters of the light dispersion matrix of the controlled environment, for example, degree of radiation polarization by dispersed smokes at two diffusion angles, and their comparisons with an optical portrait of smokes. Optical smoke portrait is a set of diffusion spots for the maximum number of realization of smokes calculated using the known parameters of microphysics of real smokes: distribution of particles by the sizes, complex index of substance refraction of smoke particles. The fact of entrance of the current representing spot into the area of an optical portrait increases the probability of fire detection.
3. The other method is the use of cross-polarizing indicator of aerosol hindrances of CIH, when the polarization surfaces of light source and receiver polarizers are mutually orthogonic. Then, at output of the polarizer in front of the receiver all Stokes vector column components are equal to zero. It means that in a controlled space there are only smoke, i.e. spherical particles, and smoke sensors produce "purely smoke" signals, proportional to size $I_0 D_{11}$. In case of an aerosol hindrance (dust) with light dispersion matrix $[D_{ik}^n]$ smoke sensors produce additional signals $I_0 D_{11}^n$, and the CIH detector (receiver) produces a signal which is proportional to $K^2 I_0 (D_{11}^n + D_{12}^n - D_{21}^n - D_{22}^n)$. The idea of optical system protection from aerosol hindrances consists in selection of the transfer coefficient $K_{\text{-}П}$ of a CIH signal that provides the equality of $K_Д I_0 D_{11}^n = K_{К\text{-}П} K^2 I_0 (D_{11}^n + D_{12}^n - D_{21}^n - D_{22}^n)$ and the compensation of $K_Д I_0 D_{11}^n$component of CIH signals produced by smoke detectors before their processing in the main system unit.

References

[1] Van de Hulst 1981 *Light scattering by Small Particles* (New York: Dover Publications) 470 p
[2] Mie G 1908 *Annalen der Physik* 25 **3** pp 377 – 445
[3] Deirmendjian D 1969 *Electromagnetic radiation scattering by spherical polydisperse particles* (New York: Elsevier) 290 p

[4] Shifrin K S, Zelmanovich I L 1968 *Tablitsyi po svetorasseyaniyu v 2-h t. T.2. Tablitsyi matrits rasseyaniya i sostavlyayuschih rasseyannogo polya* (L.: Gidrometeoizd) 471 p.
[5] Gerrard A, Burch J M 1975 *Introduction to Matrix Methods in Optics* (London, New York: wiley) 355 p
[6] Zuev V E 1970 *Rasprostranenie vidimyih i infrakrasnyih voln v atmosphere* [in Russian] (M.: Sov. Radio) 496 p
[7] Kozlov V S, Panin V F, Rapoport G A 1976 Rasseyanie i refraktsiya optich. voln v atmosphere [in Russian] (Tomsk) p 78 – 95
[8] Grozdov G M, Kopyilov N P, Panin V F *Massovyie pozharyi* **5** pp 61 – 74
[9] Dashkovskiy A G, Kozlov V S, Panin V F 1986 *Materialyi VIII Vsesoyuzn. simpoz. po rasprost. lazern. izluch. v atmosfere* [in Russian] (Tomsk) pp 45 – 49
[10] Panin V F, Trampiltsev V N 1978 *Elektr. sistemyi i upr. imi.* [in Russian] **4** pp 52 – 55
[11] Panin V F, Trampiltsev V N 1980 *Vsesoyuzn. sovesch. po atmosfernoy optike. Tez. dokl.* ch. I. pp 26 – 29
[12] Shmoylov A V, Panin V F, Gusev A S 1984 A.S. 1065864 (SSSR) Ustroystvo dlya kontrolya avariynoy opasnosti *Opubl. v B. I.* **1**
[13] Dashkovskiy A G, Panin V F, Trampiltsev V N 1985 *Koord.-chuvstvit. fotoprYomn. i optiko-elektronnyie ustroystva na ih osnove* pp 131 – 132
[14] Balashov E V, Dashkovskiy A G, Panin V F A.S. 262081 (SSSR) Dyimovoy datchik Prioritet izobr. 28.04.86., zaregistr. v Gos. reestre izobr. SSSR 01.10.87.
[15] Tihonov A N, Arsenin V Ya 1986 *Metodyi resheniya nekorrektnyih zadach* [in Russian] (M: Nauka) 286 p.
[16] Naats I E 1980 Teoriya i metodyi lazernogo zondirovaniya aerozolnyih polidispersnyih sistem: Dis… dokt.f.–m. nauk Tomsk 437 p.
[17] Rozenberg G V 1981 *Nekotoryie probl. sovr. fiz. atmosferyi* [in Russian] (M.: Nauka) pp134 – 157

IOP Conf. Series: Materials Science and Engineering **81** (2015) 012116 doi:10.1088/1757-899X/81/1/012116

Radial distribution of a single-pass amplified radiation in the active elements of CuBr lasers

F A Gubarev[1,2], **M S Klenovskii**[1,3] **and Lin Li**[1]

[1] National Research Tomsk Polytechnic University, 30 Lenin Avenue, Tomsk 634050, Russia
[2] V.E. Zuev Institute of Atmospheric Optics, Russian Academy of Sciences, Siberian Branch, 1 Academician Zuev square, Tomsk 634021, Russia
[3] Institute of Electron Physics, National Academy of Sciences of Ukraine, 21 Universiteskaya St., Uzhgorod 88017, Ukraine

E-mail: gubarevfa@tpu.ru

Abstract. The paper presents the results of study of single-pass amplified radiation distribution of copper bromide vapor laser active elements used in high-speed laser monitors. The possibility of modifying the profile of a single-pass amplified light beam by changing the copper bromide vapor concentration is demonstrated. This means of influence on the radiation profile seems to be easiest due to implementation by varying only one parameter of operation. Gaussian, ring-shaped or flat profiles can be achieved depending on the temperature of the containers with copper bromide. The diameter of the beam becomes narrower when increasing the concentration of copper bromide vapor. This feature is characteristic of the discharge tubes as small (diameter 2.5, length 5 cm) and large (diameter 5 cm, length 90 cm) active volume.

1. Introduction

Laser monitor based on active media of copper vapor laser or copper bromide vapor laser is one of the effective instruments for the diagnostics of processes accompanied by intensive shielding background light [1–12]. Laser monitors allow observing processes such as surface modification for operational characteristics improvement, production of novel materials including nanostructures, processes of thermonuclear synthesis, plasma-induced processes, etc. There are a number of papers devoted to the application of such active media for imaging various objects but researching towards the development of practical laser monitors is carried out insufficiently.

In the work [13] it is reported about CuBr vapor lasers for different applications. In the work [12] one of these lasers was applied for objects visualization under conditions of intensive background lighting. In the paper [11] a construction of laser monitor based on an active medium of CuBr laser is represented and visualization results obtained by the laser monitor for some objects are also given. An optical part of the laser monitor construction is shown in figure 1. The main element of the laser monitor scheme is an active element of brightness amplifier.

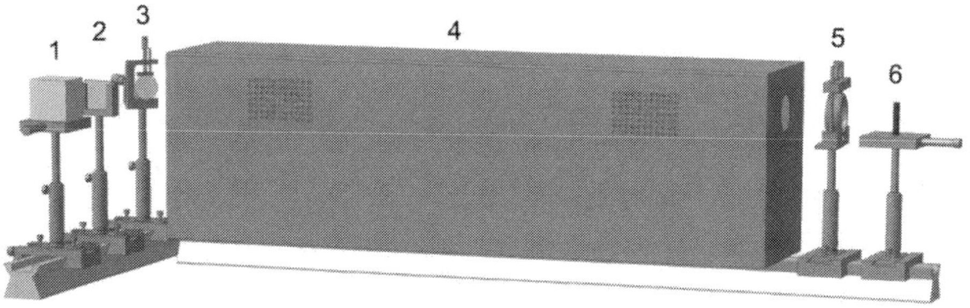

Figure 1. Laser monitor: 1 – CCD-camera, 2 – optical filters, 3 – deflecting mirror, 4 – brightness amplifier, 5 – lens, 6 – object under observation.

The applied laser monitor scheme implies that an active medium of CuBr laser generates light illuminating an object of visualization and simultaneously amplifies light reflected from the object. However, the presence of radial inhomogeneity of population inversion in the active medium accordingly leads firstly to non-uniform illuminating the visualization object and then to non-uniform amplifying light reflected from the visualization object. As a result the visualization object image is obtained with distortions. It should be noted that such distortions are only connected with incorrect brightness transfer of some parts of obtained picture but not with the picture shape deformations. Image shape deformations depend mainly on quality of passive optical part of the laser monitor scheme.

The radial inhomogeneity of population inversion in the active medium can be estimated from corresponding inhomogeneity of the gain profile. The gain profile as well as gain coefficient depend on the pumping condition of the active medium in general and on CuBr vapor concentration in the working volume of an active element [14] in particular. The aim of the current research was study the influence of CuBr vapor concentration on the radial distribution of radiation in CuBr vapor brightness amplifiers.

2. Experimental technique

Figure 2 shows the constructions of active elements used in the present work and applied earlier as brightness amplifier for laser monitors in the works [11, 12]. The active element of the brightness amplifier [11] that is shown in figure 2*a* consists of a gas-discharge tube (GDT) made of a fused quartz with a length of 50 cm and inner diameter of 2.5 cm. The maximum average lasing power of this active element enclosed in a parallel-mirror cavity was equal to 3.2 W at a pulse repetition frequency of 20 kHz and a pumping power of 600 W (the power consumed by a high-voltage power supply from the mains). The active element of the brightness amplifier [12] that is shown in figure 2*b* consists of GDT made of a fused quartz with a length of 90 cm and inner diameter of 5 cm. The maximum average lasing power of this active element was equal to 5 W at a pumping power of 1450 W. Performance of this GDT was studied in details in papers [15, 16]. Pumping of the active medium is realized by a circuit of pulse charging storage capacitor and then it direct discharging on the GDT by a thyratron [17]. Active admixture of HBr that is often used for lasing power increasing of CuBr laser [2] was not used in the current research.

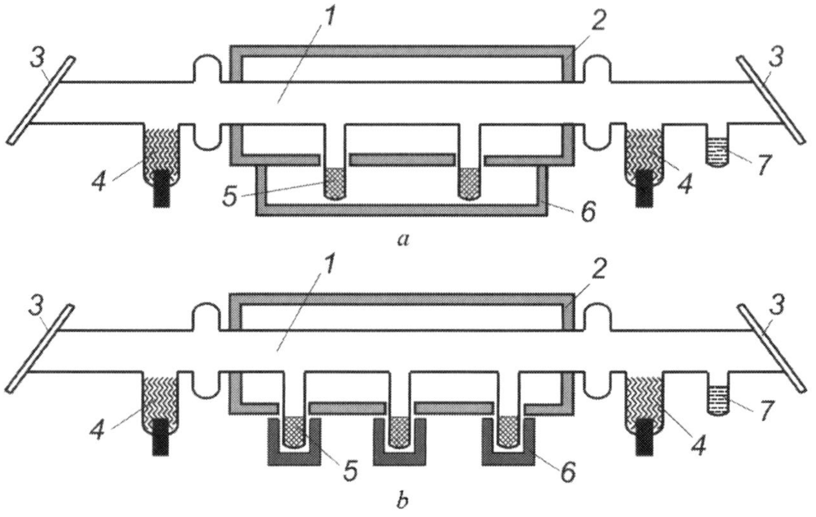

Figure 2. Active elements: 1 – quartz gas-discharge tube, 2 – main heater (heater of working area of GDT), 3 – output windows, 4 – copper electrodes, 5 – container with CuBr powder, 6 – auxiliary heater for containers with CuBr powder, 7 – container with adsorbent.

The both constructions of active elements have separated main heater for working area of GDT and auxiliary heater for containers with CuBr powder. That allows to influence on the amplifying characteristics of active element by varying CuBr vapor concentration. A feature of the first construction (figure 2*a*) is a common heater for all containers with CuBr powder. The second construction of the active element (figure 2*b*) has an individual heater for each container with CuBr powder. The advantages of the first configuration over the second are ease of design, manufacturing, simplicity of assembly, and less of electrical connections and wires. The disadvantage of this construction is the higher influence of the main heater and the energy dissipated by the discharge on the heating temperature of CuBr powder. As a result the heating temperature of CuBr powder is slightly changes every time when pumping conditions are varied. The second construction of the active element with individual heaters is characterized by minimal influence on the heating temperature of CuBr powder and accordingly on CuBr vapor concentration in the active medium.

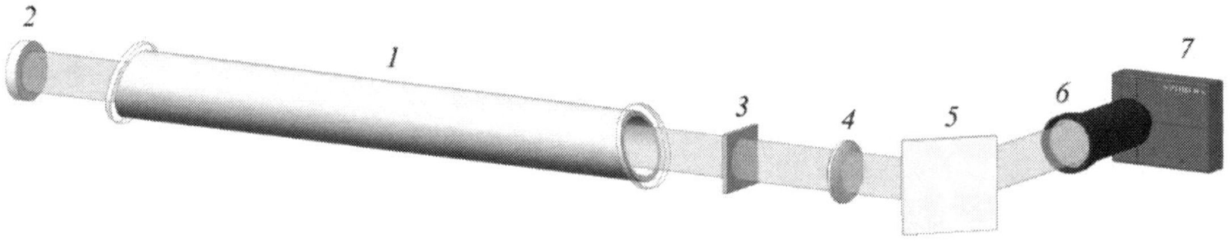

Figure 3. Experimental setup scheme: 1 – brightness amplifier, 2 – plane mirror, 3 – optical filters, 4 and 6 – lenses, 5 – beam-splitting plate with 20% reflectance, 7 – CCD-camera.

Study of single-pass amplification profiles of the active element is carried out by an experimental setup scheme shown in figure 3. Pumping active medium leads to population inversion of working laser levels in the active element and causes generating an amplified spontaneous emission (ASE). ASE reflected from the mirror is returned to the active medium again where it amplifies by intensity. The plane mirror simulates a visualization object with uniform reflectance that allows eliminating influence of reflected properties of the object on the analysis of amplifying characteristics of the active element. When beam profiles of ASE were studied the mirror was took away.

As a light analyzer a CCD-camera (beam profiler) Spiricon SP503U with BeamGage Standard 5.4 software was used. The size of the CCD-camera matrix is 6.4×4.8 mm^2. To match a cross section of the laser tube output beam with the matrix of the CCD-camera a fourfold zoom lens SPZ17024 mounted on the CCD-camera and Industar-51 lens were used. To eliminate the overexposure of the CCD-camera the output light was pre-attenuated by gray optical filters. To define an average radiation power a measuring head of power meter (L30C-SH from Ophir Optronics Solutions Ltd.) was set at the output of the laser tube.

3. Results and discussion

Figure 4 shows the profiles of single-pass amplified light beam corresponding to different average radiation power for the GDT of 2.5 cm bore. For this series of experiments the pumping power was 600 W and the CuBr heater temperature range was varied from 315 to 410 °C. These temperature values are comparatively low owing to closely spaced main heater of GDT with heater of containers with CuBr powder. The geometry of the optical scheme remained unchanged for all represented regimes. As one can see in figure 4 Gaussian, ring-shaped or flat profiles are achieved depending on the temperature of the containers with CuBr powder. Hence, increasing average radiation power that is caused by growing CuBr vapor concentration leads to significant contraction of light beam profiles. When average single-pass amplified radiation power is 0.2 W the light beam diameter is close to GDT diameter and reaches ~2 cm. When average single-pass amplified radiation power growths to 2.4 W the light beam diameter decreases twice and equals to ~1 cm. At reduction of CuBr vapor concentration in the active medium a contraction of lasing profile for the same active element was observed in the work [18].

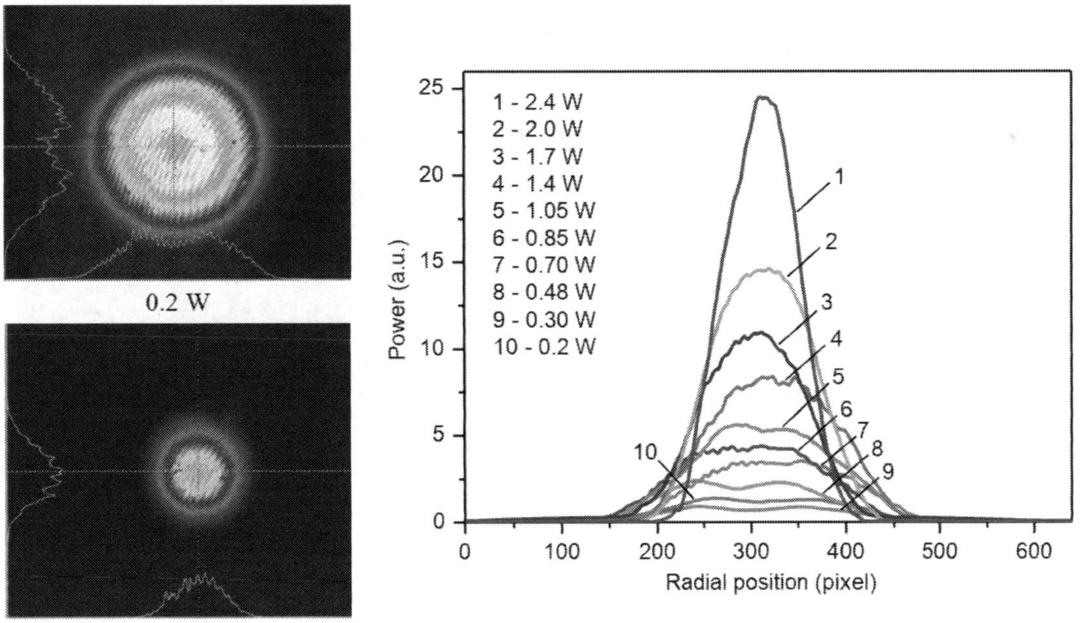

Figure 4. The profiles of a single-pass amplified light beam at various average radiation powers for the GDT of 2.5 cm bore.

A considerable contraction of profile light beam occurs also in the active element of big diameter. Moreover, in the GDT with diameter of 5 cm it is observed a regime when radiation virtually disappears on the axis (figure 5). These measurements were carried out at unchanged pumping power value of 1350 W. The temperature of the containers with CuBr powder was varied in the range from 440 to 570 °C. The considerable intensity dip on the beam axis for the GDT with diameter of 5 cm comparing to the GDT with diameter of 2.5 cm is probably caused by overheating on the axis of the gas-discharge tube owing to high pumping power.

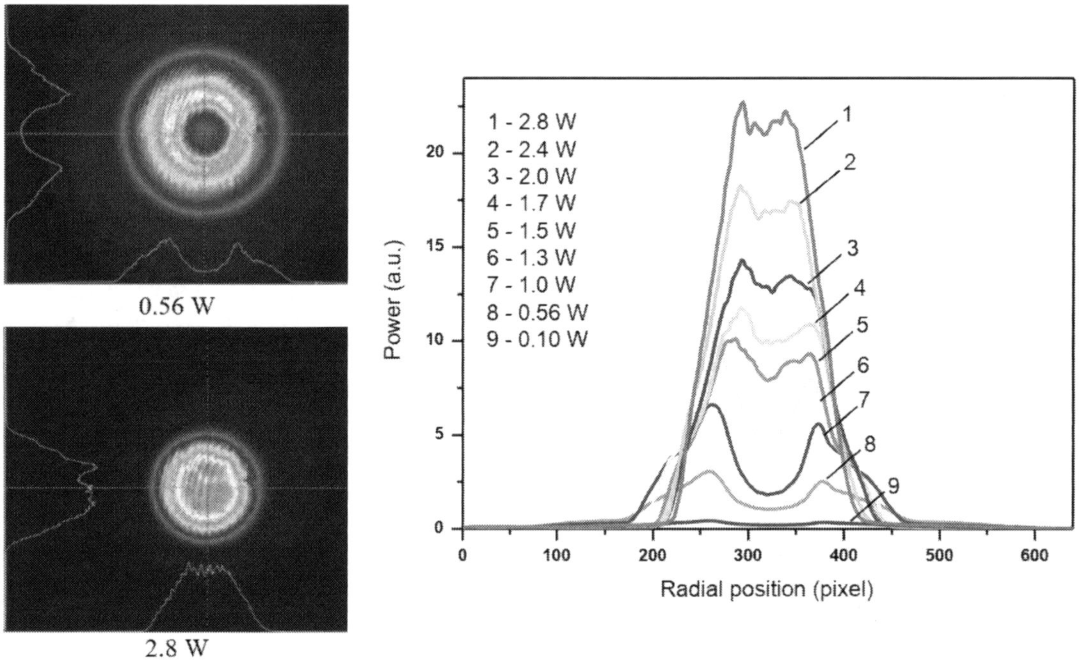

Figure 5. The profiles of a single-pass amplified light beam at various average radiation powers for the GDT of 5 cm bore.

The contraction of light beam profile that is caused by increasing CuBr vapor concentration was noted earlier in [14] where an active element of CuBr-laser with a length of 90 cm and diameter of 8 cm was used. The all containers with CuBr powder were heated by a common heater. In the work [14] the measurement of light beam profile was carried out by a master-oscillator power-amplifier scheme with a master-oscillator laser beam of 1.5 mm in diameter.

In laser projection microscopes and laser monitors the illumination of a visualization object and amplification of reflected light occur during the same pulse of radiation. To obtain an image with correct brightness a uniform illumination of a visualization object is preferable. Figure 6 shows beam profiles of ASE at various average radiation powers. Accordingly to presented data more uniform illuminating a visualization object occurs at a smaller amount of CuBr vapor concentration. More uniform illumination of a visualization object in combination with more uniform amplification profile subsequently provides higher quality of obtained images. This fact was noted in [19] which states that the working conditions of CuBr vapor laser active elements in the brightness amplification mode for obtaining high-contrast images differ from the conditions providing maximum average output power in the lasing mode. The difference of working conditions is in less optimal CuBr concentration.

Not only increasing CuBr vapor concentration leads to contraction of light beam profile. Such conditions as pulse repetition frequency and pumping power should be taken into consideration at optimizing performance of laser monitor. Significant contraction of light beam profile at increasing pulse repetition frequency was observed in [20].

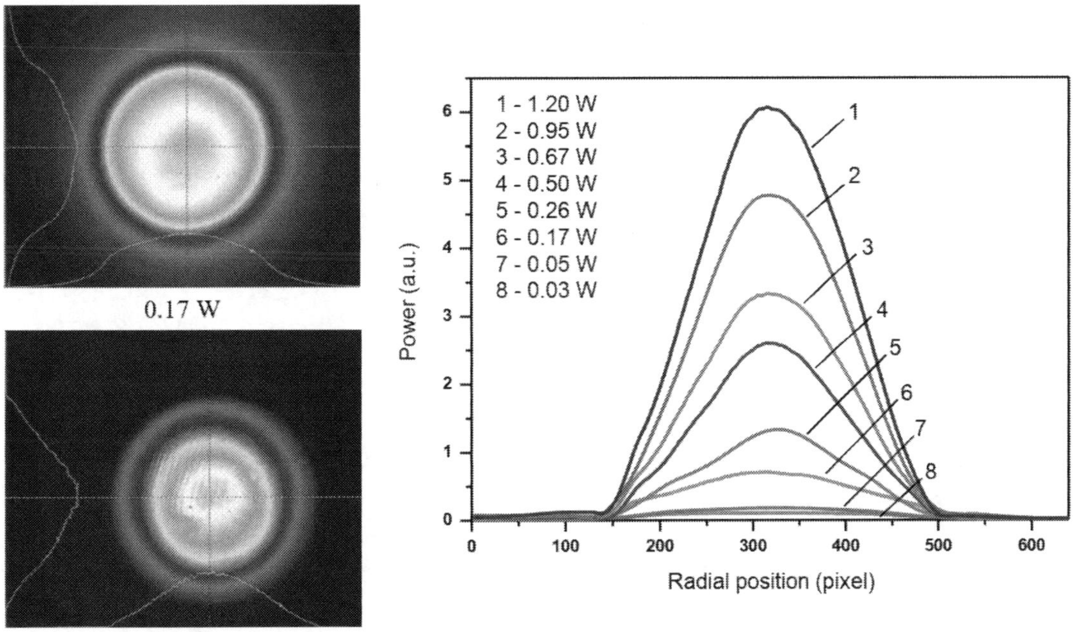

Figure 6. The beam profiles of ASE at various average radiation powers for the GDT of 5 cm bore.

4. Conclusions

The study showed that increasing temperature of containers with CuBr powder, i.e. growing CuBr concentration in the active medium, leads to considerable contraction of the output beam profile of CuBr brightness amplifiers. The profile contraction occurs for ASE, for a single-pass amplified light beam and even for a laser beam. It is observed for both gas-discharge tube with diameters of 2.5 and 5 cm and is accompanied by increasing output lasing power.

The uniform (flat top) of beam profile of ASE and gain defines image quality obtained with laser monitor. The most flat top of a single-pass amplified light beam profile for the GDT of 2.5 cm bore is observed within average output powers from 480 to 850 mW that is considerably less than maximum average output power (2.4 W) of a single-pass amplified radiation for this GDT. The optimal mode for the GDT of 5 cm bore lies within average output powers of a single-pass amplified radiation from 1 to 1.8 W that is also less than maximum value. The mentioned modes correspond to 10-15% temperature reduction of the containers with CuBr powder.

Acknowledgments

Author M.S. Klenovskii express thanks to Tomsk Polytechnic University for the opportunity of conducting research as a postdoc.

The authors are grateful to M.E. Levitsky for the technical support.

The work is supported by the Russian Science Foundation, project No 14-19-00175.

References

[1] Petrash G G 1991 *Optical Systems with Brightness Amplifiers* (Moscow: Nauka)
[2] Evtushenko G S, Shiyanov D V and Gubarev F A 2010 *Metal Vapor Lasers with High Pulse*

Repetition Rates (Tomsk: Tomsk Polytechnic University)
[3] Zemskov K I, Kazaryan M A, Matveev V M, Petrash G G, Samsonova M P and Skripnichenko A S 1984 *Sov. J. Quantum Electron.* **14** 288
[4] Batenin V M, Klimovsky I I and Selezneva L A 1988 *Soviet Phys. Dokl.* **33** 949
[5] Abrosimov G V, Pol'skii M M and Saenko V B 1988 *Sov. J. Quantum Electron.* **18** 544
[6] Abramov D V, Arakelyan S M, Galkin A F, Klimovskii I I, Kucherik A O and Prokoshev V G 2006 *J. Quantum Electron.* **36** 569
[7] Buzhinsky R O, Savransky V V, Zemskov K I, Isaev A A and Buzhinsky O I 2010 *Plasma Phys. Rep.* **36** 1269
[8] Yermachenko V M, Kuznetsov A P, Petrovskiy V N, Prokopova N M, Streltsov A P and Uspenskiy S A 2011 *Laser Phys.* **21** 1530
[9] Dashinimaeva E Z and Trigub M V 2014 *Proc. Int. Conf. of Young Specialists on Micro/Nanotechnologies and Electron Devices (EDM)* pp 257–261
[10] Evtushenko G S, Trigub M V, Gubarev F A and Torgaev S N 2011 *Bul. Tomsk Polytechnic Univ.* **319** 154
[11] Evtushenko G S, Trigub M V, Gubarev F A, Torgaev S N and Shiyanov D V 2013 *Testing. Diagnostics* **13** 42
[12] Evtushenko G S, Trigub M V, Gubarev F A, Evtushenko T G, Torgaev S N and Shiyanov D V 2014 *Rev. Sci. Instrum.* **85** 033111
[13] Andrienko O S, Gubarev F A, Dimaki V A, Ivanov A I, Levitsky M E, Sukhanov V B, Troitskii V O, Fedorov V F, Filonov A G and Shiyanov D V 2009 *Atmos. Ocean. Opt.* **22** 999
[14] Gubarev F A, Troitskiy V O, Trigub M V and Sukhanov V B 2011 *Opt. Commun.* **284** 2565
[15] Gubarev F A, Trigub M V and Fedorov K V 2014 *Proc. Int. Conf. of Young Specialists on Micro/Nanotechnologies and Electron Devices (EDM)* pp 309–313.
[16] Gubarev F A, Evtushenko G S, Vuchkov N K, Sukhanov V B and Shiyanov D V 2012 *Rev. Sci. Instrum.* **83** 055111
[17] Dimaki V A, Sukhanov V B, Troitskii V O and Filonov A G 2012 *Instrum. Exp. Techniq.* **55** 696
[18] Gubarev F A, Evtushenko G S and Li Lin 2014 *Testing. Diagnostics* **13** 160
[19] Trigub M V 2013 Laser monitor for non-destructive testing of goods and visualization of fast processes shielded by background lighting *Dissertation for a PhD degree in technical sciences*
[20] Trigub M V, Shiyanov D V and Vlasov V V 2014 *Proc. Int. Conf. of Young Specialists on Micro/Nanotechnologies and Electron Devices (EDM)* pp 301–304

Measurement error reducing in the ultrasound time-pulse systems

A I Soldatov, O A Kozhemyak, A A Soldatov and Yu V Shulgina

National Research Tomsk Polytechnic University
634050 Russia, Tomsk, 30 Lenin av.

E-mail: asoldatof@mail.ru

Abstract. Ultrasound wave propagation in bounded media results in change of the echo-pulse shape, therefore the use of traditional method of determining of the propagation time of the ultrasound pulse based on use of threshold device (comparator) gives a large error of measurement. To reduce this error the new methods of the determining of echo-pulse propagation time are proposed in this paper. The first method is the method based on the use of two comparators, the second one is the method based on second-order polynomial approximation of the echo-pulse envelope. Analysis showed that the steeper the rising edge of the echo-pulse envelope, the less the error of determining of the pulse start time. The more the difference between comparator thresholds the less the error of determining of the pulse start time. Application of the method of two comparators gives two times systematic error reducing. Minimal sampling of input signal is determined for the method based on echo-pulse envelope. Dependence of the error at start echo-pulse determining on adjacent samples amplitude ratio is obtained. Application of echo-pulse envelope gives three times systematic error reducing.

1. Introduction

Ultrasound devices based on time-pulse method are widely used nowadays. Basic error of measurement of such devices is due to inaccuracy of determining of the sound pulse start time. Usually sound pulse start time is determined by the comparator, but due to the complex shape of the sound pulse, the response time of the comparator does not coincide with the pulse start.

Another widely used method to determine echo-pulse start time is zero-crossing method [1, 2]. But it is impossible to determine the number of period in echo-pulse when the algorithm of calculation of zero-crossing instant was applied. Therefore it is impossible to determine the start time of echo-pulse with high accuracy.

One more method uses Hilbert transform. Disadvantage of this method is the complexity of the algorithm that can be implemented only with use of personal computer [3]. Implementation of Gilbert transform is impossible in portable devices, based on microcontrollers.

If the amplitude and shape of a signal does no change, then the error of determining of the time of signal propagation in the medium also remains constant and can be taken into account. However, in practice, signal amplitude is decreases at the signal propagation. This decreasing is due to signal divergence and losses in the medium. Automatic gain control maintains a constant amplitude signal. But if the signal shape changes in process of its propagation, the application of automatic gain control or a comparator with "tracing" threshold in this case does not solve the problem of accuracy in determining of the pulse start time. The fundamental basis of this phenomenon is the inequality of the phase velocities of different modes (Figure 1).

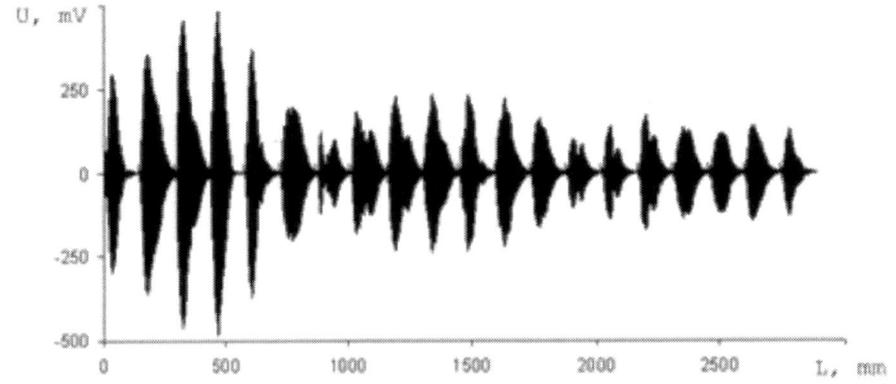

Figure 1. Change of the signal shape when it propagates through the waveguide.

Figure 2 shows two signals with the same amplitudes, but with different shape of the rising edge. The change of the rising edge of the pulse envelope at 10-15 % results in the error of determining of the pulse position at 2–3 periods of carrier frequency. This error is no longer constant, but varies depending on changes in the duration of the rising edge, and it is impossible take it into account in traditional ways.

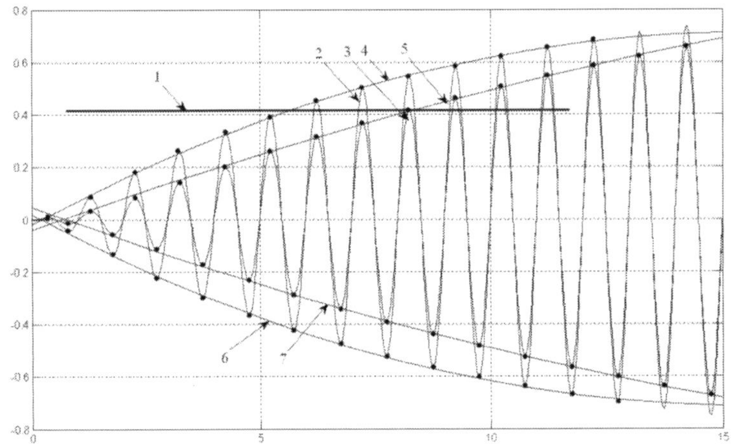

Figure 2. Determining of the coordinates of the pulse start time using the method of approximation of the signal envelope to 2nd power of a polynomial:
1) comparator threshold level; 2) the 1st signal with higher rising edge;
3) the 2nd signal with lower rising edge; 4) envelope of the positive extrema of the 1st signal; 5) envelope of the positive extrema of the 2nd signal; 6) envelope of the negative extrema of the 1st signal; 7) envelope of the negative extrema of the 2nd signal.

2. Description of the first method

It is possible to increase the accuracy of measurement, if we apply modern methods of the echo signals processing. One of such methods is the method of two comparators with different thresholds [4, 5] (Figure 3).

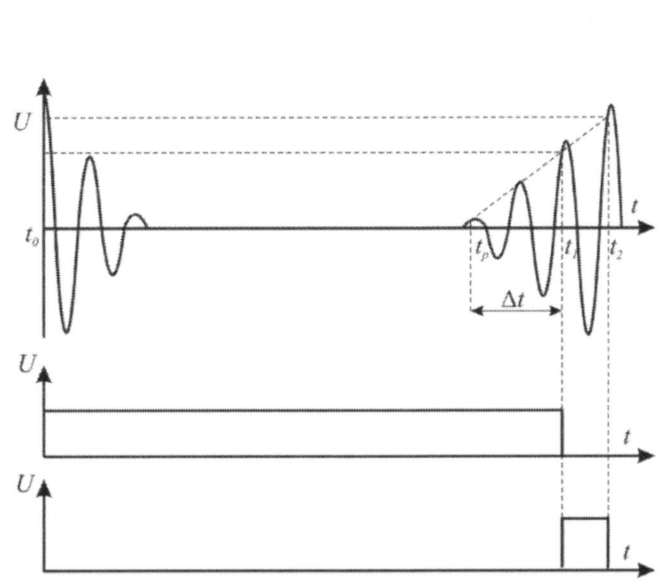

Figure 3. Determining of the echo pulse position using the method of two comparators.

At the instant t1 the 1st comparator triggers at the level of *U*1. At the instant t2 the 2nd comparator triggers at the level of *U*2. Constructing the line at the coordinates of these points allows determining the time coordinate *t* of the echo pulse start time by the expression:

$$t_p = t_1 - \frac{U_1}{U_2 - U_1} \cdot (t_2 - t_1) \tag{1}$$

Define the maximum error obtained by this method for the linear part of the envelope of the echo pulse rising edge:

$$U_{i+1}^{\max} = U_i^{\max} + k \cdot U_1, \tag{2}$$

where $i = 0, 1, \ldots,$ $k < 1$, U_1 is threshold voltage of the first comparator.

2.1Modeling

The maximum error is obtained, when the levels U_1 and U_2 reach the maxima values of the previous peaks (Figure 4).

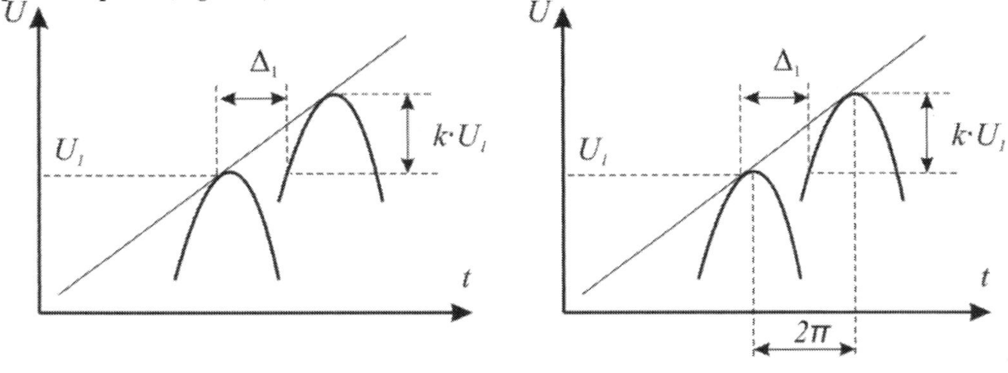

a) b)

Figure 4. Variants of error occurrence: a) due to the 1st comparator, b) due to the 2nd comparator.

Maximum amplitude of i-th period after comparator triggering is obtained using the expression:

$$U_i^{\max} = \frac{U_1}{\cos\frac{\Delta\varphi_1}{2}} \tag{3}$$

Using the expression (2), we determine the angle:

$$\cos\frac{\Delta\varphi_1}{2} = \frac{U_1}{U_1 + k \cdot U_1}$$

Hence

$$\frac{\Delta\varphi_1}{2} = \arccos\frac{1}{1+k} \tag{4}$$

Thus, the maximum error in determining of the angle by the 1st comparator is equal to:

$$2\cdot\pi - \frac{\Delta\varphi_1}{2} = 2\cdot\pi - \arccos\left(\frac{1}{1+k}\right)$$

Therefore, it is possible to define the error when we determine the time of the first sample (Figure4*a*):

$$\Delta_1 = T\left(1 - \frac{1}{2\cdot\pi}\cdot\arccos\frac{1}{1+k}\right) \tag{5}$$

Similarly, for the second sample we obtain:

$$U_2 + k\cdot U_1 = \frac{U_2}{\cos\frac{\Delta\varphi_2}{2}}$$

Denote

$$q = \frac{U_1}{U_2}$$

Using the expression (2), we determine the angle:

$$\frac{\Delta\varphi_2}{2} = 2\cdot\pi - \arccos\left(\frac{1}{1+q\cdot k}\right)$$

Therefore, it is possible to find the error when we determine the time of the second sample (Figure 4*a*):

$$\Delta_2 = T\left(1 - \frac{1}{2\cdot\pi}\cdot\arccos\frac{1}{1+q\cdot k}\right) \tag{6}$$

Figure 5 shows two cases: in one case the first sample has a maximum error and the second sample has a minimum error; while in the other case the first sample has a minimum error and the second sample has a maximum error. Points A and B are for the ideal envelope, and points C and D are when the errors in samples occur.

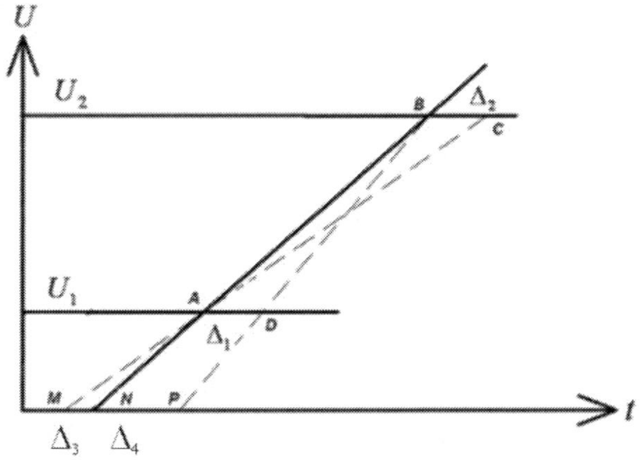

Figure 5. The envelope of the echo pulse rising edge for two samples:
NB is an ideal line,
MC is the line with the error introduced by the 2nd comparator,
BP is the line with the error introduced by the 1st comparator.

Using the similarity of triangles ΔADB and ΔNBP, it is possible to write the expression:

$$\frac{\Delta_4}{\Delta_1} = \frac{U_2}{U_2 - U_1}$$

Hence:

$$\Delta_4 = \Delta_1 \cdot \frac{1}{1-q}$$

Using the similarity of triangles ΔABC and ΔMAN, it follows:

$$\frac{\Delta_3}{\Delta_2} = \frac{U_1}{U_2 - U_1}$$

Thus, the total error is determined as a sum of errors:

$$\Delta_3 + \Delta_4 = \Delta_2 \cdot \frac{q}{1-q} + \Delta_1 \cdot \frac{q}{1-q} \qquad (7)$$

From equations (5), (6) and (7):

$$\Delta_3 + \Delta_4 = \frac{T}{1-q}\left[\begin{array}{l}(1+q) - \frac{1}{2\cdot\pi}(\arccos\frac{1}{1+k} + \\ +q\cdot\arccos\frac{1}{1+q\cdot k})\end{array}\right]$$

If the result of the experiment is affected by various factors, then their total effect is in accordance with the law of normal distribution (according to the theorem of Lyapunov). Therefore, we can assume that a random variable is normally distributed, and all values of t_0 are in the range:

$$t_0 \in \left(t_0 - \Delta_3, t_0 + \Delta_4\right)$$

According to the law of three sigma:

$$6\cdot\sigma = \Delta_3 + \Delta_4$$

That is we obtain σ:

$$\sigma = \frac{1}{6}(\Delta_3 + \Delta_4)$$

Error confidence interval of 95% in fractions of T:

$$\frac{\Delta}{T} = \frac{1}{1-q} \cdot \left[\begin{array}{l} 1 + q - \frac{1}{2 \cdot \pi} \cdot (\arccos \frac{1}{1+k} + \\ + q \cdot \arccos \frac{1}{1+k}) \end{array} \right] \qquad (8)$$

Figure 6 shows calculation results of expression (8) for different values of q.

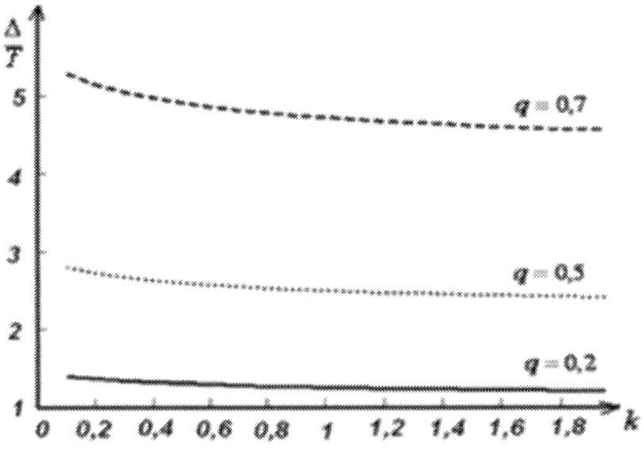

Figure 6. Value of time determining error vs. steepness of the envelope rising edge.

The steeper the rising edge of envelope (the larger k) the smaller the error in determining of the start time of the echo pulse. The error in determining of the echo pulse start time decreases as the difference in the thresholds of comparators U_1 and U_2 increases (the smaller q). Effect of the difference of thresholds is stronger than the effect of the steepness of the rising edge of the envelope. To estimate the accuracy of the theoretical results obtained using given method, laboratory tests were carried out. The tests showed that the accuracy depends on the threshold values. When performing the experiments we used a waveguide with a diameter of 52 mm. The frequency of ultrasonic vibrations was 40 kHz and step size was 20 ns.

2.2Experimental study

Figure 7 shows the dependence between the error of measurement and distance for the ratios of comparator thresholds of $U_1 = 0.21 \cdot U_m$; $U_2 = 0.85 \cdot U_m$. At other ratios of comparator thresholds the error of measurement increases and at similar thresholds the errors closes to the error of measurement for the method of one comparator.

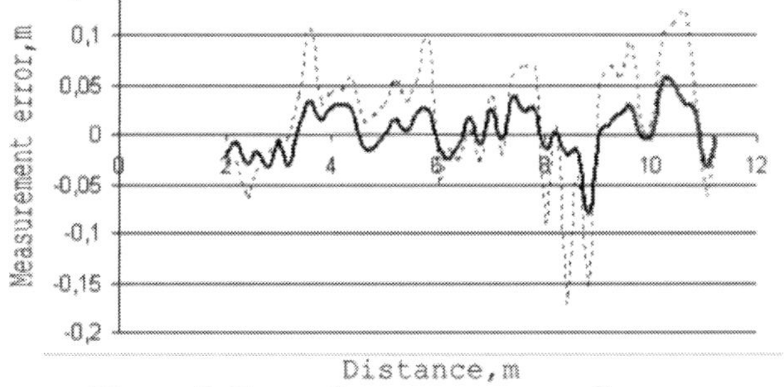

Figure 7. Error of measurement vs. distance:
dotted line – single comparator method,
full line – method of two comparators ($U_1 = 0.21 \cdot U_m$; $U_2 = 0.85 \cdot U_m$)

3. Description of the second method

Application of the method of two comparators is limited only by the case of linear rise of the echo pulse rising edge of the envelope.

$$s = a \cdot t^2 + b \cdot t + c \tag{9}$$

where s is the amplitude of the envelope, t is a time, a, b, c are the coefficients of the polynomial.

Since the shape of the echo pulse is asymmetric with respect to the t axis, two curves are used to approximate the rising edge of the pulse. One curve is the envelope for positive pulses, and the other one is the envelope for negative pulses (Figure 8). These curves have two common points, and one of them is taken as the start time of the echo pulse [6, 7].

3.1 Modeling

To define *a, b, c* coefficients the amplitude values of the ultrasonic pulse in three points were considered. These three points correspond to the peaks of a sinusoidal signal (the extrema) in three neighboring periods. Using these data, a system of equations is considered:

$$\begin{cases} s_1 = a \cdot t_1^2 + b \cdot t_1 + c \\ s_2 = a \cdot t_2^2 + b \cdot t_2 + c, \\ s_3 = a \cdot t_3^2 + b \cdot t_3 + c \end{cases} \tag{10}$$

where *s*1, *s*2, *s*3 are the amplitudes of ultrasound pulse in peaks at *t*1, *t*2, *t*3 instances. Using this system of equations (10) coefficients *a, b, c* can be calculated.

Similarly, we approximate the shape of rising edge of the pulse:

$$s = a \cdot t^3 + b \cdot t^2 + c \cdot t + d\,.$$

In some cases, the 3rd power of a polynomial approximates rising edge of the envelope of the echo pulse with better accuracy, which gives the best result. To obtain the values of *a, b, c, d* coefficients, a system of four equations is considered:

$$\begin{cases} s_1 = a \cdot t_1^3 + b \cdot t_1^2 + c \cdot t_1 + d \\ s_2 = a \cdot t_2^3 + b \cdot t_2^2 + c \cdot t_2 + d \\ s_3 = a \cdot t_3^3 + b \cdot t_3^2 + c \cdot t_3 + d \\ s_4 = a \cdot t_4^3 + b \cdot t_4^2 + c \cdot t_4 + d \end{cases} \tag{11}$$

When solving system of equation 9 and 10, the Cramer method is of a big help [8]. Since it can be no intersection of the envelopes, the point of shortest distance between the envelopes is taken as the echo pulse start time.

Contrary to the method of the comparator with fixed threshold, the error of method of the 2nd and the 3rd power of a polynomial does not depend on the amplitude of the echo pulse, while it depends on its shape.

In practice, the shape of the echo-pulse is affected by the noises, what results in errors at the calculation of envelope equations. To reduce the noise influence the least-square method has been applied for determining of the equation coefficients. When using this method we can consider the number of points that exceeds the number of unknown variables, and we construct an average envelope by these points. In general case, the least-squares method is used to estimate the unknown values according to measurements, having random errors [9].

The coefficients of approximation are calculated by:

$$\begin{cases} \sum_{i=1}^{n} t_i^2 \cdot s_i = a \cdot \sum_{i=1}^{n} t_i^4 + b \cdot \sum_{i=1}^{n} t_i^3 + c \cdot \sum_{i=1}^{n} t_i^2 \\ \sum_{i=1}^{n} t \cdot s_i = a \cdot \sum_{i=1}^{n} t_i^3 + b \cdot \sum_{i=1}^{n} t_i^2 + c \cdot \sum_{i=1}^{n} t_i , \\ \sum_{i=1}^{n} s_i = a \cdot \sum_{i=1}^{n} t_i^2 + b \cdot \sum_{i=1}^{n} t_i + c \cdot n \end{cases}$$

where n is the number of bend points taken out for polynomial approximation and the current number of an experimental point.

3.2 Experimental study

The proposed methods have been tested by constructing the envelopes of the 2nd and the 3rd order for the echo-pulse (Figure 8).

Figure 8 shows the ultrasound echo-pulse, whose shape was approximated by the envelopes of the 2nd order. The vertical line stands for the estimated time of ultrasound pulse start time.

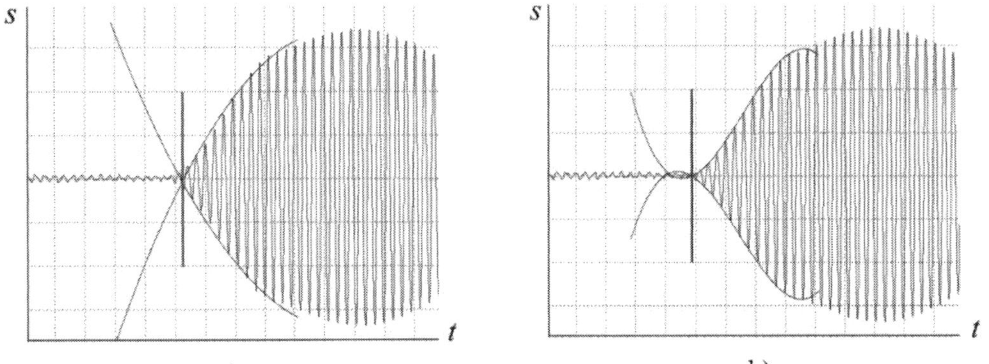

a) b)

Figure 8. Approximation of the ultrasound echo-pulse by:
a) the envelope of the 2nd order; b) the envelopes of the 3rd order.

To estimate the accuracy of the proposed method we derived the dependence of the measurement error of the level by the comparator method envelopes of the 2nd and 3r order and the distance L between a transmitter and a receiver. A round metal wave guide of 50mm was used in the experimental study. Ultrasound converters MA40V of 16 mm diameter were taken as a transmitter and receiver. They were placed in the center of the waveguide. The frequency of ultrasound oscillations was 40 kHz.

From Figure 9 it can be seen that systematic inaccuracy was reduced more than two times, and random error was 1.5 times lower. When using the methods of the envelopes of the 2nd and 3rd order the measurement errors do not sufficiently differ. Although the calculation of the echo pulse position takes much more time when using the method of envelope of the 3rd order.

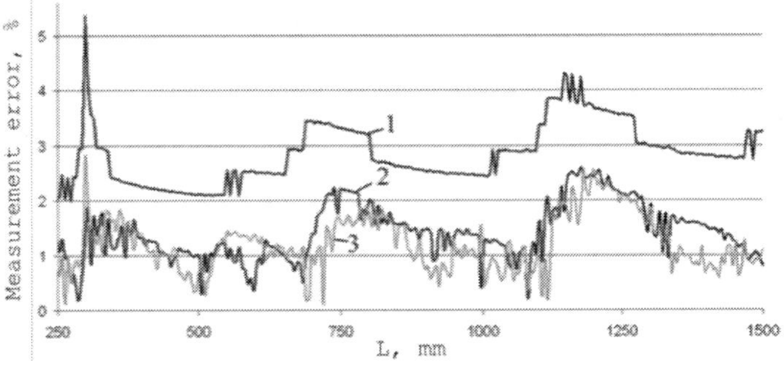

Figure 9. Measurement error vs. distance *L* between transmitter and receiver when using: 1) method of comparator; 2) method of envelope of the 2nd order; 3) method of envelope of the 3rd order.

From Figure 9 it can be seen that systematic inaccuracy was reduced more than two times, and random error was 1.5 times lower. When using the methods of the envelopes of the 2nd and 3rd order the measurement errors do not sufficiently differ. Although the calculation of the echo pulse position takes much more time when using the method of envelope of the 3rd order.

4. Conclusion

The use of new methods of echo-pulse processing allows sufficiently increasing of the accuracy of ultrasound devises based on pulse-time methods. Application of new methods of processing of echo-pulses can significantly reduce the measurement error of ultrasonic devices using time-pulse method.

Using a first-degree polynomial approximation for the envelope of the rising edge of the echo-signal allows a twofold increase of the measurement accuracy of ultrasonic devices. With the increase in the steepness of the rising edge of envelope of the echo-signal the error in determining of the start time of the pulse-echo decreases. Increase of the difference of comparators thresholds U_1 and U_2 also reduces an error in determining of the start time of the echo-pulse. However, the choice of optimum thresholds requires a detailed analysis of the acoustic channel measuring instrument. In addition, the calculation result will affect the accuracy of measuring the time interval between the insertion of the magnitude comparator and the dc component in the echo-signal.

Using a second-degree polynomial approximation for the envelope of the rising edge of the echo-signal is more preferable. However, to obtain high metrological characteristics it is necessary to provide at least 10 samples per period. Therefore, the application area of this method is limited to frequencies up to 10 MHz.

REFERENCES

[1] Mažeika L, Samaitis V, Burnham K, Makaya K 2011 *ULTRAGARSAS (ULTRASOUND)* **66** No 3 pp. 7–12
[2] Mažeika L, Draudvilienė L 2010*ULTRAGARSAS (ULTRASOUND)* **65** No.2 pp 7–12
[3] Soldatov A I , Tsekhanovskiy S A A method of compensation of the measurement error of the ultrasound level gauge. Russian Federation Patent # 2358243 filed 10.06.2009

IOP Conf. Series: Materials Science and Engineering **81** (2015) 012117 doi:10.1088/1757-899X/81/1/012117

[4] Soldatov A I, Tsekhanovskiy S A, Sorokin P V, Makarov V S Device for compensation of the measurement error of the ultrasound level gauge. Russian Federation Patent # 2396521, filed 10.08.2010
[5] Soldatov A I, Tsekhanovskiy S A, Sorokin P V, Makarov V S A method of compensation of the measurement error of the ultrasound level gauge. Russian Federation Patent №2389981, filed 20.05.2009
[6] Soldatov A I, Tsekhanovskiy S A, Sorokin P V, Makarov V S Device for compensation of the measurement error of the ultrasound level gauge. Russian Federation Patent # 2406979, filed 20.12.2010
[7] Korn G, Korn T 1974 *Handbook on mathematics for scientists and engineers* (Moscow) 832 pp.
[8] Andre Angot 1965*Mathematics for electrical engineers* (Moscow: Nauka) 799 pp.

Research of the photovoltaic properties of anodized films of Sn

D A Afanasyev[1], N Kh Ibrayev[1], G S Omarova[1], Zh K Smagulov[1]

[1]Karaganda, 100028, Kazakhstan, Institute of Molecular Nanophotonics, E.A. Buketov Karaganda State University

E-mail: a_d_afanasyev@mail.ru

Abstract. The results of studies of photovoltaic properties of solar cells based on porous tin oxide films, sensitized with an organic dye are presented. Porous films were prepared by electrochemical anodization of tin in alkaline electrolytes based on aqueous solution of NaOH and aqueous ammonia NH_4OH. It was found that the time of anodizing of the Sn films affects on conversion efficiency of light energy into electrical energy. Increasing of the sorption time leads to an increase of the number of molecules on the surface of the porous film. For the solar cell based on tin oxide there is a strong dark current, which significantly reduces the efficiency of conversion of light energy into electrical energy.

1. Introduction

Today metal oxides such as titanium oxide, tin oxide, zinc oxide, etc., still remain in the field of view of researchers. With the rapid development of nanotechnologies and their applications are discovered new properties of these substances on the micro- and nanoscale [1-3]. Electrochemical anodization is one of the promising methods of obtaining of nanostructures. Electrochemical oxidation and electrolytic formation of the oxide film on the surface of metals, alloys and semiconductors [4, 5] occurs during anodization. The oxide film protects the product against corrosion, has insulating properties, could be used as a good base for the paint coatings used for decorative purposes. For a number of metals, under certain conditions of anodizing it`s possible to obtain a porous oxide films with predetermined dimensions of the nanopore, film thickness and porosity. Such metals include aluminum, niobium, tantalum, titanium, zirconium [6]. For each of these metals its own conditions for the anodizing process exsist. Properties of the porous films strongly depend on the properties of metal.

An integral part of modern material science is the analysis of the microstructure, which acquires special value in relation to porous structures, where the state of the object (pore size, orderliness, and others) significantly affects the complex of the physical and chemical properties [7, 8]. It is therefore important to determine the impact of technical conditions of nanomaterials on their microstructural and geometrical properties. In this paper we present the results of the investigation of microstructural properties of the tin oxide films obtained by electrochemical oxidation.

2. Methods and Results

Granular tin with 99.99 % purity was used as a starting material. Tin plates were prepared from granules. The thicknesses of plates were equal to 0.5 mm and the dimensions were 2.5×2.5 cm. Chemical polishing of tin plates was carried out before anodizing. Before anodizing the surface of the polished plate was degreased with ethanol and then was cleaned in an ultrasonic bath.

Synthesis of porous SnO_2 was performed in a two-electrode electrochemical cell using a constant current source MPS-7081. The auxiliary electrode was a titanium plate, and a working electrode was polished plate of tin. As electrolyte an aqueous solution of NaOH and aqueous ammonia solution NH_4OH was used. The electrolyte was obtained on the basis of dual water filtration and deonization with system AquaMax Basic 360. The resistivity of the water was equal to 18.2 MΩ/cm. Anodizing of the porous film of tin oxide was carried out at room temperature (25 °C).

Voltage was kept constant (potentiostatic anodization mode) during the process of anodizing. After anodizing film was washed with distilled water and dried in an oven at a temperature of T = 50 °C.

An important step in controlling of the anodizing process is the measurement of the current density in the anodizing process. The current value allows estimating the reaction rate and the thickness of the oxide layer. Nature of the change of the current shows the step of forming of oxide film.

Figure 1 shows the current characteristics of the anodization process in an aqueous solution of NaOH. Gradual decrease in the current density with increasing time of anodizing is observed for the anodizing process. No increase of the current density anodizing and no expressed transitions between stages of formation of barrier and porous oxide layers [9].

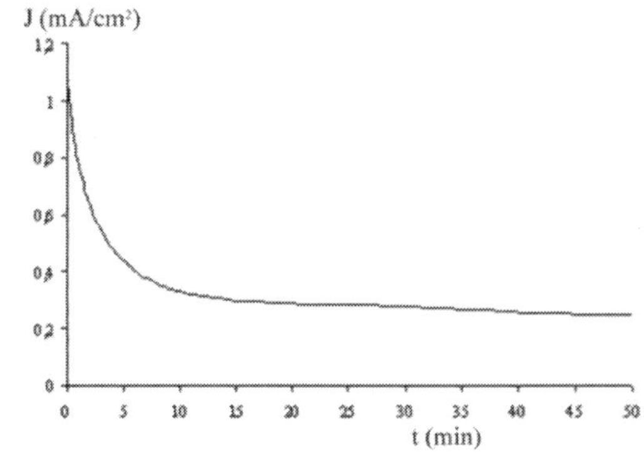

Figure. 1. Current characteristics tin anodizing process in an alkaline solution NaOH, U = 12 V.

When anodizing occurs in aqueous solution of NH_4OH the stage of increase of the current density (Figure 2) is observed. The growth of the barrier layer of tin oxide occurs at this stage. When the current density reach the maximal value it begin gradually decrease. At this stage the porous structure of the oxide film forms. However, analysis of the current characteristics and a comparison with the standard model representations, allows only predicting the probability of occurrence of an oxide film of porous type. The formation of pores in anodising tin could be confirmed with electron microscopy and probe measurements of obtained samples.

For the microstructural study of the films, anodized under various conditions of anodizing, the scanning electron microscope (SEM) JEOL JSM-5910, operating in the mode of secondary electron emission (SEI) with a permissive ability in a high vacuum of 3 nm was used. The electron energy of the probe was 20 keV, the probe current was 0,1nA, and the measurement time was 50 s. Working distance was 10 mm. All films were subjected to the control of the elemental composition. Microanalysis of quantitative elemental composition of the films was made by energy dispersive method with X-ray spectrometer where detector JXA-8200 is embedded in the SEM. Quantitative microanalysis of chemical elemental composition of the investigated films was carried out by a special program PHI-RHO-Z by mathematical processing of the energy-dispersive spectra using a computer that came with SEM. The study of the elemental composition of the film obtained in aqueous solution of NaOH, showed the presence of the tin atoms and oxygen, indicating the formation on the surface of an tin oxide film (Figure 3). Also the energy-dispersive spectrum contains sodium atoms.

The presence of sodium atoms is a negative factor that may affect on the further application of oxide films as a porous semiconductor layer in photovoltaic solar cells. Therefore, further anodizing of tin was held in NH_4OH.

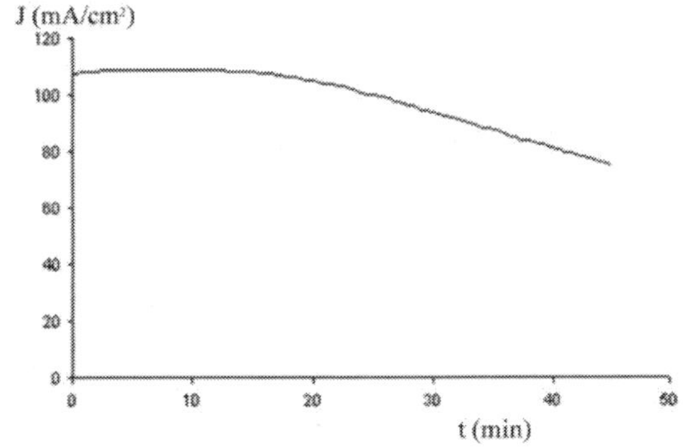

Figure 2. Current characteristics of tin anodizing process in an NH_4OH alkaline solution.

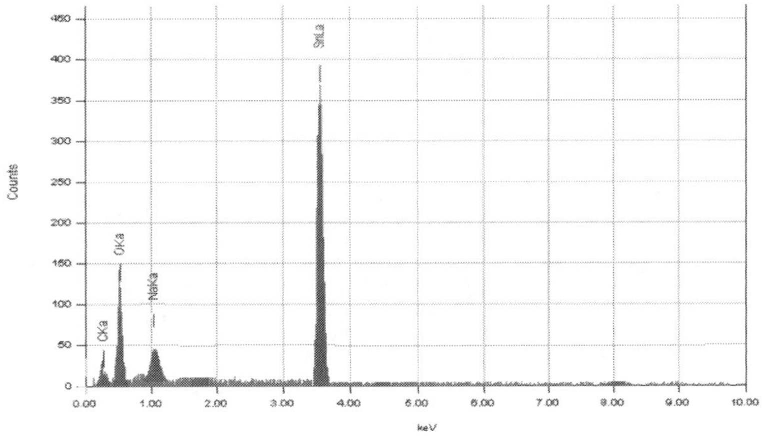

Figure 3. X-ray spectrum of the anodized surface of tin at the anodization in NaOH with the distribution of elements on the surface.

The microstructure of anodized tin oxide films in an electrolyte based on water and NH_4OH was carried out. The measurements were performed with a scanning probe microscope JSPM-5400 (JEOL) using the methods of atomic force microscopy (AFM). Results of anodizing in solution NH_4OH at room temperature are shown in Figure 4.

Reduction of temperature at the anodizing process leads to an increase of inhomogeneity of film relief. Against the background of the rough surface of the film the porous structure of the film is formed (Figure 4).

The microstructure of the surface of the anodized film of tin oxide was also investigated with a scanning electron microscope (SEM) JEOL JSM-5910 (Figure 5). Microscopic measurement of the film structure of the anodized tin was identical to AFM images of the surfaces (Figure 4). Similarly, the AFM images, images taken with the SEM testify the formation of a developed mesoporous surface. Inside the mesopores the pores of smaller size are formed. At the same time the structure obtained by electron microscopy measurements, corresponds to images obtained by the probe microscope (Figure 5).

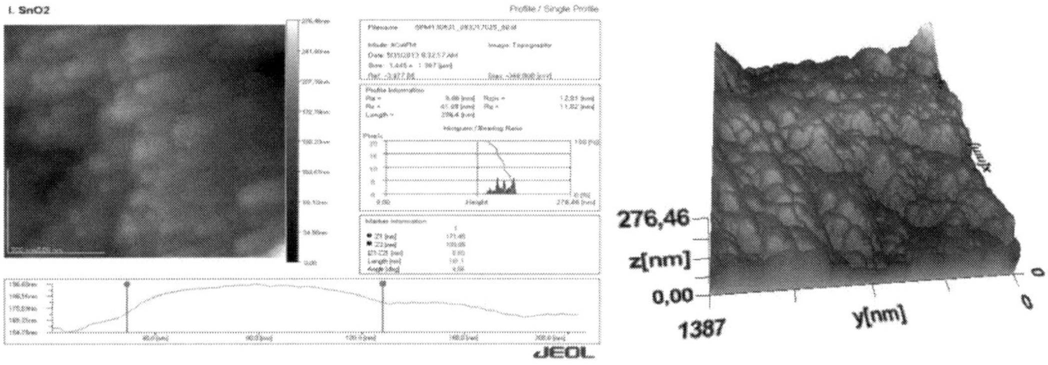

Figure 4. AFM images of a surface of a porous tin oxide obtained at room temperature in NH_4OH.

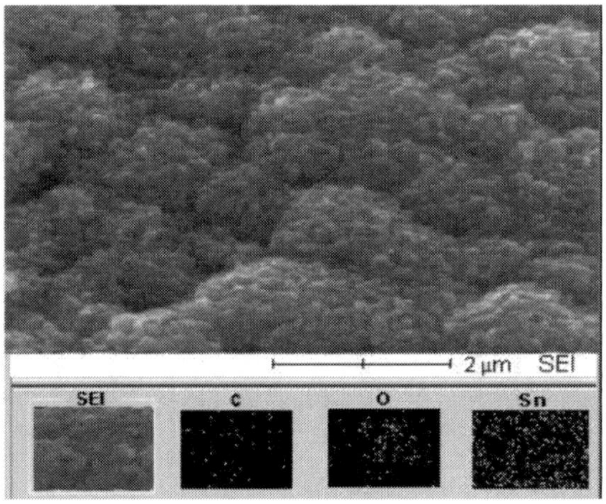

Figure 5. The microstructure of the surface of the tin oxide film.

Minor presence of carbon on the image (Figure 5) indicates the presence of organic compounds on the surface of the tin. Against the background of inhomogeneous relief of the film there is a uniformly distributed porous structure, which is evidence of the formation of the oxide layer of porous type for SnO_2. The pore size of the nanostructured tin oxide depends on the conditions of anodization. Typical pore sizes were determined from the AFM images of the films (Figure 4). The dimensions were equal to 100 nm for films obtained at room temperature.

The sorption capacity of porous films of anodized tin was studied. For this purpose sorption of dye molecules was performed from a solution. Dye D149 was chosen as adsorbate.

The choice of this dye due to the fact that its energy states are best suited to energy levels of tin oxide. Therefore, the solar cells on the basis of anodized oxide films of Sn and D149 should have high rates of conversion of light energy into electrical energy.

The electronic absorption spectra were measured with spectrophotometer Specol 1500. Absorption spectra of the dye D149 before adsorption and after adsorption of molecules on the surface of the nanoporous film of Sn oxide are shown on the figure 6.

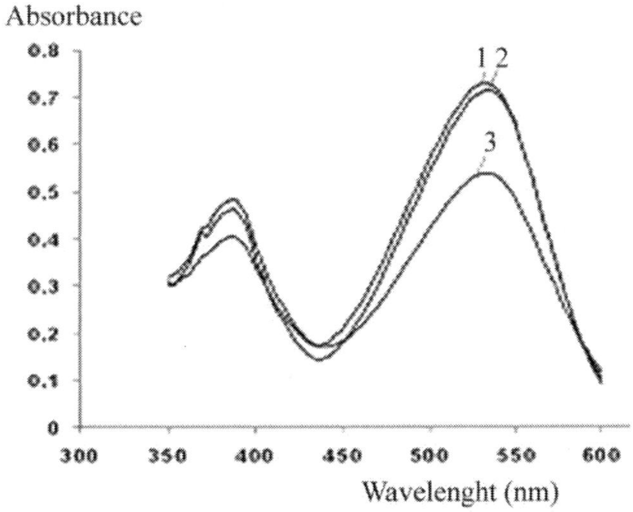

Figure 6. The absorption spectra of the D149 dye before and after sorption: 1 – before sorption 2 - 30 min sorption 3 - 40 min.

It is evidence that the process of sorption of molecules does not lead to a distortion of the absorption spectra. After sorption the optical density of the solution is reduced, which shows the decrease in the concentration of molecules in solution and the flow of physical adsorption of molecules on the surface of the anodized film tin oxide.

Increasing of the time of sorption leads to an increase in the number of molecules on the surface of the porous film (Figure 7). The concentration of molecules adsorbed on the 1 surface of the tin oxide during the 45 minutes was determined and it was $8{,}4 \cdot 10^{16}$ 1/cm^2.

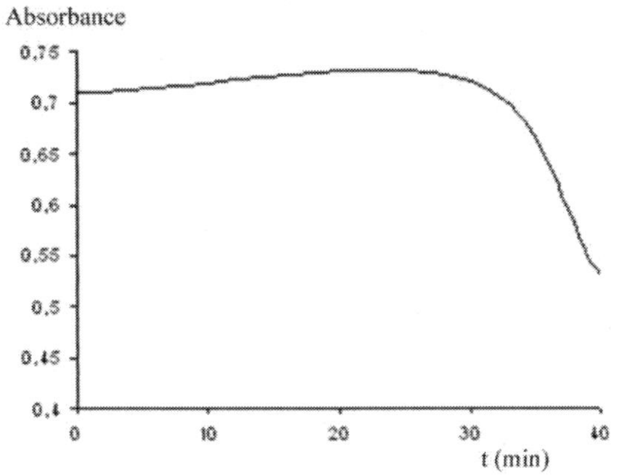

Figure 7. The dependence of the optical density of D149 dye versus sorption time into anodized Sn.

To test photovoltaic properties of the obtained films of nanoporous tin oxide solar was assembled cell according to the method closest to the methodology used by Grätzel.

The measurements were carried out in the photovoltaic mode at room temperature. Lighting of the cell surface was performed from the transparent ITO electrode, on the surface of which by method of irrigation from solution and annealing at a temperature of T=450 ^{0}C transparent thin platinum film was obtained. The power of xenon lamp was equal to 50 mW/cm^2. Photocurrent and photovoltage was not registered for films that were anodized in the electrolyte on the basis of NaOH. This can be explained by the presence of Na on the surface, which adversely affects on the photoconductive properties of tin oxide.

For films anodized in the electrolyte on the basis of NH4OH measuring of the dark and light current-voltage characteristics was performed. Figure 8 shows the dark and light I-V characteristics of the Gretzel cell sensitized with dye D149.

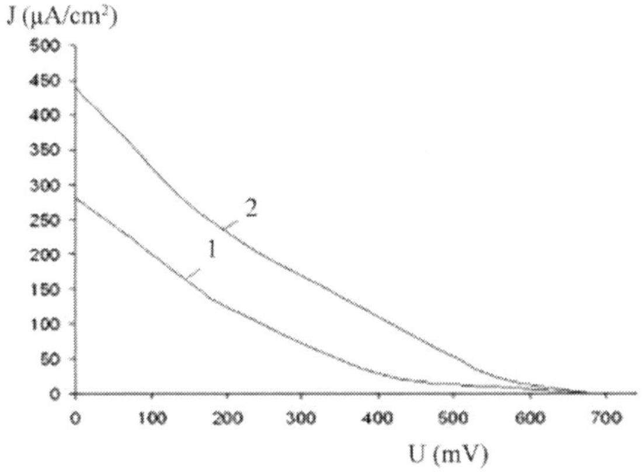

Figure 8. The current-voltage characteristics (CVC) of solar cell based on tin oxide sensitized D149 dye: 1 - the dark CVC, 2 - light CVC.

As can be seen from figure 8, curve 1, even in the absence of light exposure on the sample there is some voltage and current of the cell. Measurement of voltage and current specified by solar cell was carried out under changes of load resistence. When the light source is turned on, the voltage and current in the circuit, which includes Graetzel cell, significantly increased (curve 2). It shows that the light leads to the generation of free charge carriers in the cell and forming of photovoltage.

Analysis of obtained results showed the efficiency of solar cells based on porous films of tin oxide doped with dye molecules. The value of photovoltage and photocurrent shows a perspective of using tin oxide as a semiconductor film in dye-sensitized solar cells. However, for solar cells based on tin oxide exhibits a strong dark current, which significantly reduces the efficiency of conversion of light energy into electrical energy. To explain and eliminate this disadvantage it is necessary to perform of further study of the solar cells based on the films of tin oxide.

The results of the research can be done as number of conclusions:

1. The methodology of synthesis of porous films of Sn oxide by electrochemical anodization was elaborated.

2. The study of the elemental composition of the anodized film showed the presence of tin atoms and oxygen, which indicates the formation of tin oxide film on the surface. It was conducted the studies of the microstructure of the films of tin oxide anodized in the electrolyte on the basis of water and NaOH. Due to the presence of sodium atoms it was decided to abandon the use of this electrolyte for the preparation of porous tin oxide films.

3. The microstructure of anodized tin films obtained at different temperatures of the anodizing in NH_4OH solution was studied. For films obtained at different temperatures of anodizing different

microstructure of the porous films of tin oxide was observed. Microstructural data obtained for oxide films by scanning probe microscopy and electron microscopy correspond and complement each other.

4. Sorption capacity of tin oxide films was studied. It was found the effect of the chemical nature of the dye on the efficiency of sorption of dye in the film of anodized tin.

5. Photovoltaic properties of solar cells based on porous anodized tin oxide films, sensitized organic dye molecules D149 were studied.

References

[1] Kolmakov A, Zhang Y, Cheng G, Moskovits M 2003 *Adv. Materials.* **15**, 997
[2] Hu J, Bando Y, Liu Q, Golberg T D 2003 *Adv. Funct. Mater.* **13** 493
[3] Djurisic A B, Leung Y H. 2006 *Small.* **2** 944
[4] Belenky M A, Ivanov A F 1985 *Electrodeposition of metallic coatings, the reference* (Moscow;Metallurgy)
[5] Averianov E E 1988 *Handbook of anodizing* (Moscow: Mashinostroenie)
[6] Su Z, Zhou W 2008 *Adv. Mat.* **20** 3663
[7] Gusev A I 1991 *Physical chemistry of non-stoichiometric refractory compound* (Moscow: Nauka)
[8] Zainulin Y G, Alyamovskii SI, Gusev A I, Shveikin G P 1992 *The effect of high pressures and temperatures on the defective phase implementation* (Ekaterinburg: RISO, Ural branch of the Russian Academy of Sciences)
[9] Aitkenova G T, Afanasyev D A, Ibrayev N Kh, Khuanbai E 2014 *Bulletin of the Karaganda state University. Series physics.* **73** 4

IOP Conf. Series: Materials Science and Engineering **81** (2015) 012119 doi:10.1088/1757-899X/81/1/012119

Simulation of heat conductivity and charging processes in polar dielectrics induced by electron beam exposure

A Maslovskaya[1] and A Pavelchuk[1]

[1] Physics Engineering Department, Amur State University, 21, Ignatievskoe shosse, Blagoveshchensk, 675027, Russia
e-mail: maslovskayaag@mail.ru

Abstract. The article presents results of simulation of thermal and charging processes arising in polar dielectric materials analysed with scanning electron microscope techniques. Monte-Carlo simulation of electron transport in irradiated target was evaluated to realize field effect models. Simulation of thermal load and charging characteristics was performed using net-point methods. The analysis of injection, thermal and charging processes induced by electron irradiation in ferroelectrics subjected to experimental parameters are also discussed.

1. Introduction

Recently, one of the priority areas of condensed matter physics has focused on the problem of studying the basic laws and mechanisms of electron irradiation effects on dielectrics. This fact is due to wide applications of the analytical capabilities of scanning electron microscopy (SEM) for investigation and modification of various functional materials. Specific features of polar dielectrics and ferroelectrics, in particular, consist in sensitivity with respect to electrical and thermal exposure of the electron beam. This allows creating not only a potential contrast research methods, but also using an electron beam to study electrical properties of the samples as well as modify their polar structure. The electron beam injection activates the accumulation of absorbed electrons on the surface and in the bulk of sample, which, acting as an internal source, give rise to heating [1] and charging effects [2]. The previous studies based on experimental data has led to the point of view that effect of ferroelectric charging in the SEM has been considered as negative. To reduce the role of charging-up effects the low accelerating voltages (2-5 kV) have been employed [3]. However, the charging effect of ferroelectric material underlies a number of experimental techniques of direct polarization switching under electron beam exposure [4-6]. In addition the possibility of registration of the pyroelectric response due to thermal exposure of electron beam has been used to visualize ferroelectric domain structures with SEM pyroelectric mode [7].

Despite significant progress in the development of SEM experimental techniques the study of many effects is a difficult experimental problem. So in some cases, information on the behavior of the object can be obtained with the use of theoretical methods. The following main classes of models are usually considered to analyze the processes of electron beam interaction with solids. Firstly discrete stochastic models [8] based on Monte-Carlo method describe the electron transport in a sample. Secondly continuous-deterministic models enable field effects of injected charges [9-10] to be calculated.

However, information related to the specific study of polar dielectrics using different SEM modes as well as the study of corporate electron irradiation effects remains incomplete and requires further consideration. The present study was undertaken to simulate and analyze the dynamics of injection, thermal and charging processes occurring in the ferroelectrics irradiated by electron beam.

2. Simulation of field effects resulting from electron injection in irradiated polar dielectrics

2.1. Specification of electron injection process in irradiated sample

The first step both of thermal and charging processes simulation involves the estimation of initial electron distribution. The parameters of the electron irradiation doze and source function were estimated by a 3D-Monte-Carlo simulation of electron trajectories in the sample. We assumed that electrons with start beam energy E_0 fall perpendicularly to sample surface at point P_0. The electron position is defined by values of scattering angles: ω_i – azimuth angle and deviation angle $-\varphi_i$. The values of scattering angles and interaction mode are specified using random-number generator. An electron walks the distance s_i with energy E_i between stochastic scattering events:

$$s = -\lambda \ln(\xi), \quad \lambda = \frac{1}{N_0 \cdot \rho} \sum_{k=1}^{M} \frac{\omega_k \cdot A_k}{\sigma_k}, \tag{1}$$

where $\xi \in (0,1)$ is uniform distributed random quantity, λ is mean free path in m, A_k is atomic weight in g/mole, ω_k is weight fraction of element k, M is number of elements composing substance, σ_k is cross-section for each element k, in cm^2, N_0 is Avogadro's number in mole^{-1}, ρ is density in g/cm^3.
Mott cross-section was used to evaluate the total cross-section σ_k [11]. The computation of energy changes for inelastic electron scattering was based on modified form of Bethe law specified for multicomponent materials [8]:

$$\frac{dE}{ds} = -75800\frac{\rho}{E} \cdot \sum_{k=1}^{M} \omega_k \frac{Z_k}{A_k} \ln\left(\frac{1.166(E + 0.8J_k)}{J_k}\right), \text{ keV/cm}, \quad Jk = \left[9.76Z_k + 58.5Z_k^{-0.19}\right] \cdot 10^{-3} \tag{2}$$

where Z_k is atomic number of element k, J_k is mean ionization potential of element k in keV.
The equation (2) for each electron position was solved numerically taking into account the initial condition specified as energy value at previous electron position. The changes in electron trajectories as well as energy loss were calculated for each electron till corresponding energy value decreases to threshold energy E_{th}~0.5 keV. According to Monte-Carlo method one needs performing simulation for rather significant electron histories $N\approx$1000-10000 to achieve a statistical confidence.

2.2. Thermal process simulation of the electron beam interaction with ferroelectrics

In order to estimate the thermal effect of electron beam the some issuers should be considered: conditions of scanning modes, the geometry of model sample and internal source, thermalphysic characteristics of the object, conditions of experimental observation. The problem definition lies in the following. Focused electron beam has thermal effect on the sample. The geometry of model sample is shown in figure 1, where axis Z related to the symmetry axis of electron probe, the axes Y and X locate in sample surface plane. In addition «switching-on» of internal source occurs instantly in the sample and continuous heat source effects on the plane z=0 during the observation time.

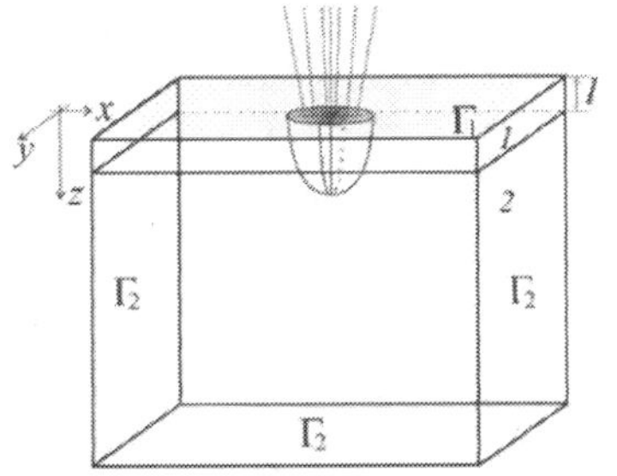

Figure 1. The geometry of model sample and approximation of internal heat source (for modes of high-voltage SEM).

The simulation can be performed for the sample covered by thin metal electrodes with thickness l. The half-spherical approximation of source was specified for the high-voltage SEM modes and the cylindrical approximation of the source was used for the low-voltage SEM modes. Contact resistance

is absent on the border surface. To approximate the internal source we employed Monte-Carlo simulation of electron trajectories in irradiated target.
The problem of nonstationary heat conductivity process simulation allows us to calculate the thermal distribution in polar material irradiated by focused electron beam particularly for the modes which use the thin metal electrodes covering the sample surface. The problem statement also includes initial, boundary conditions and conditions specified for metal-dielectric interface:

$$K_{xx}^{i}\frac{\partial^2 T_i}{\partial x^2}+K_{yy}^{i}\frac{\partial^2 T_i}{\partial y^2}+K_{zz}^{i}\frac{\partial^2 T_i}{\partial z^2}+f(x,y,z)=\rho_i c_i\frac{\partial T_i}{\partial t},\quad T_i(x,y,t_0)=T_0,$$

$$T_i|_{\Gamma_2}=0,\quad \left.\frac{\partial T_1(r,z,t)}{\partial z}\right|_{\Gamma_1}=0,\quad i=\overline{1,2},\quad k_1\left.\frac{\partial T_1(M,t)}{\partial z}\right|_{z=l}=k_2\left.\frac{\partial T_2(M,t)}{\partial z}\right|_{z=l},\quad T_1|_{z=l}=T_2|_{z=l},\tag{3}$$

where T_i is temperature in K, K_{xx}^{i}, K_{yy}^{i}, K_{zz}^{i} are components of heat conductivity coefficient in W/(m·K), ρ_i is density in kg/m^3, c_i is specific heat capacity in J/(kg·K) in the layer *i*, *l* is thickness of metal electrode in m, $f=W/V$ is volume power density of heat source in W/m^3.
The general computational scheme for the problem (3) is based on combined net-point method, which permits expressing the temperature distribution on each temporal layer. Model (3) can be used for thermal load computation in the following practical cases: 1) for application of low-voltage as well as high-voltage SEM modes, 2) for experimental modes analyzing samples covered by thin metal electrodes, 3) for nonlinear temperature regime, 4) for object scanned by electron beam moving along one of the coordinate on the surface crystal, 5) for object scanned by pulsing electron probe.

2.3. Simulation of charging processes in polar dielectrics exposed to electron beam

In practice electric field potential is one of the constitutive quantity which characterizes charging effect in dielectrics. The classical general model describing dielectric charging process is based on combined solution of continuity equation and local-instantaneous Poisson equation:

$$\begin{cases}\dfrac{\partial\rho}{\partial t}=G-\operatorname{div}\mathbf{j}\\[2mm] \operatorname{div}\mathbf{E}=\dfrac{\rho}{\varepsilon\varepsilon_0}\end{cases}\tag{4}$$

where ρ – charge density in C/m^3, $\mathbf{j}=\sigma\mathbf{E}$ is current density in A/m^2, σ – specific electric conduction in 1/(Ω·m), E is field intensity in V/m, ε is dielectric permittivity of sample, ε_0 is dielectric constant in C^2/(N ·m^2), G is generation component in C/(m^3·s).
Furthermore we used of relation between quantities which are associated with dimensions of distance and time: $\Delta t_{sc}=(\delta_{sc})^2/D$, where δ_{sc} is characterized by distance scale in m and D is defined as diffusion coefficient of electrons in m/s^2. This allows us to conclude that provided the irradiation occurs rather long time $t_{ir}>>\Delta t_{sc}$ then the stationary mode of charging process in some consideration is realized. The solving Poisson equation permits the potential distribution through crystal to be calculated depending on positions of injected charges. The problem statement of stationary charging process simulation is given by following form taking into account cylindrical symmetry of the problem:

$$\frac{\partial^2\varphi}{\partial r^2}+\frac{1}{r}\frac{\partial\varphi}{\partial r}+\frac{\partial^2\varphi}{\partial z^2}=-\frac{\rho(r,z)}{\varepsilon\varepsilon_0},\quad \mathbf{E}=-\operatorname{grad}\varphi,\quad 0\le r\le R,\quad 0\le z\le Z.\tag{5}$$

The boundary conditions are also specified to close the problem (5) as follows $\partial\varphi/\partial n=0$ at $r=0$ and at $z=0$; $\varphi=0$ at $r=R$ and at $z=Z$. Finite element method was used to calculate the potential as well as field intensity distribution created by volume charge distribution $\rho(r,z)$. In consideration of local character of charging process we also used modified computational scheme with minor size of elements covered injection area. Moreover the vector and absolute value of electron beam-induced polarization component can be estimated using the formula $\mathbf{P}=(\varepsilon-1)\varepsilon_0\mathbf{E}$.

3. Numerical experiment and result discussion

3.1. Estimation of electron beam injection characteristics

Figure 2 demonstrates the results of Monte-Carlo simulation of electron trajectories obtained for $LiNbO_3$ crystal. The analysis of computational results for typical ferroelectrics irradiated by electron beam with averages energies E_0=1-40 keV indicates that geometry of the area of electron beam interaction with samples can be characterized by semispherical or semi-ellipsoidal form (mostly for inorganic crystals such as $LiNbO_3$, $BaTiO_3$, $LiTaO_3$) and "pear-shaped" form (for organic type crystals such as TGS, DTGS, CH_4N_2S) depending on elemental composition.

The simulation results allow us to conclude that when start beam energy increases the form of the curve enveloping area of electron interaction with samples extends lengthwise coordinate axis *OZ*. Absolute value of electron injection depth does not depend on size of irradiated area on the sample surface. In addition the presence of metal electrode does not prevent electron injection into the sample and type of metal electrode (Ag, Au, Cu) results in approximately similar electron injection depth.

To initialize the internal source function let us consider function of electron density distribution, which can be specified using Gauss distribution for local irradiated area as follows:

$$I(r,z) = I_0 \cdot \exp\left(-\left(\sqrt{r^2+z^2} - \delta_1 \right)^2 / \left(2 \cdot (\delta_2)^2 \right) \right), \quad r = \sqrt{x^2+y^2} \qquad (6)$$

where I_0 is normalization constant; δ_1 and δ_2 are parameters numerically estimated by means of least square procedure.

The figure 3 illustrates computational result of the profile of normalized energy loss distribution obtained with use of equation (6).

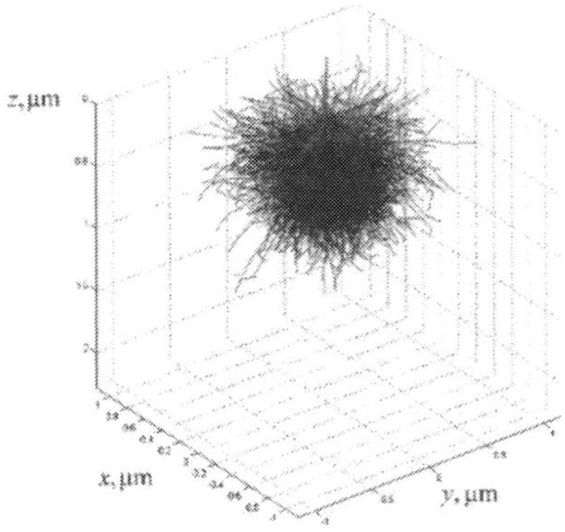

Figure 2. The result of Monte-Carlo simulation of the area of electron beam interaction with $LiNbO_3$ sample at start beam energy E_0=10 keV, beam diameter d=1 µm, N=10000 electron histories.

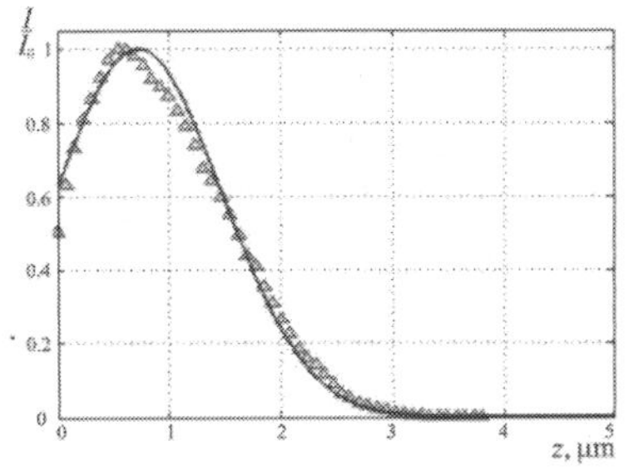

Figure 3. The approximation depth profile of electron energy loss distribution obtained by Monte-Carlo simulation of electron transport in $LiNbO_3$ at start beam energy E_0=25 keV, beam diameter d=2 µm, N=10000 electron histories.

3.2. Computation of electron beam-induced thermal loading effects in ferroelectrics

In order to perform simulation of temperature distributions the characteristic parameters need to be estimated. In general one should determine geometrical size of sample and electrode thickness, thermal characteristics of sample and electrode, the time of source action, initial temperature of the sample as well as accelerating voltages and probe current. The equation (6) is used to express volume power density of heat source taking into account the approximation of area of electron beam interaction with material and also the value of source power.

The simulation was performed taking into account various parameters of the samples and experimental conditions. Specifically uncoated sample surface as well as ferroelectric coated by with thin metal electrodes were considered. The curves in figures 4 and 5 demonstrate the calculated dynamic temperature profiles attributed to typical ferroelectric crystals.

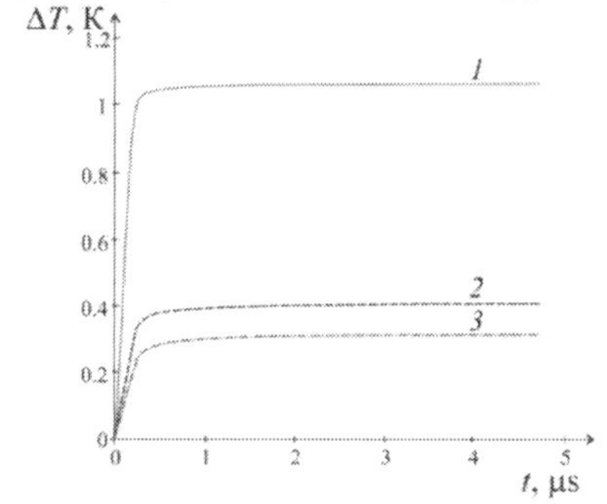

Figure 4. Dynamic dependence of overheat level $\Delta T = T - T_0$ ΔT for $LiNbO_3$ at electron beam position with energies: 1 – E_0=15 keV, 2 – E_0=25 keV, 3 – E_0=40 keV.

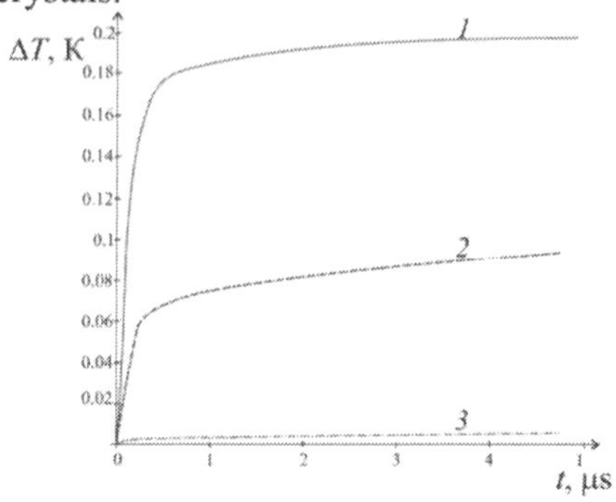

Figure 5. Temperature dynamics ΔT at electron beam position for $BaTiO_3$: 1 – observation on the uncoated sample, 2 – on the sample coated by silver electrode with thickness l=1 μm and 3 – l=10 μm.

The overheat level of typical ferroelectric crystals (TGS, DTGS, $LiNbO_3$, $LiTaO_3$, $BaTiO_3$, $NaNO_2$) in high-voltage SEM mode are estimated to be magnitudes of order of several Kelvins. These values can not cause heat damage of the samples although changing in temperature gradient can activate pyroelectric effect which impacts on videosignal. This aspect is essential point of ferroelectric domain structure observation with pyroelectric mode of SEM [7].

The figure 4 demonstrates plotted dynamic temperature profiles for $LiNbO_3$ at different values of accelerating voltages. The increasing in source power leads to decreasing in temperature. These results suggest that temperature gradient can affect videosignal and controlled heating level in SEM modes requires increasing source power due to increasing current of the probe. The figure 5 displays the plotted dynamic temperature profiles for $BaTiO_3$ depending on presence and thickness of metal electrode. According to the results the use of coated samples considerably reduces the thermal load. The use of modes with samples coated by metal electrode is essential for thermal estimations. The increasing of electrode thickness leads to decreasing of overheating level. Experimental methods (using thermocouple) result in rough results due to influence of thermocouple on temperature distribution in the sample volume. The latter has previously also been noted in [1].

3.3. Computing characteristics of charging process

The geometrical size of the sample, dielectric permittivity as well as initial charge distribution were determined to calculate following charging characteristics: potential, field intensity distribution and polarization component electron beam induced. The computation of ρ_0 requires initialization of accelerating voltages U, electron beam diameter d, surface density of injected charge σ_{surf} at given probe current I. The function $\rho_0(r,z)$ was defined using equation (6) based on Monte-Carlo simulation of electron trajectories in a sample. The figure 6 shows model potential distribution and depth profile of field intensity absolute value calculated for $LiNbO_3$. The magnitude of the field on the border of injection area is rather more than coercive field for $LiNbO_3$ ($2 \cdot 10^7$ V/m). The evidence from this study confirm that electron injection into the ferroelectric sample can result in charge accumulation. The field of accumulated charges is able to stimulate intensive repolarization processes in domains "tail-to-beam" orientated. This effect has been observed experimentally in [5]. The simulation results indicate that increase in start beam energy E_0 at fixed value of surface density of injected charges σ_{surf} leads to reducing levels of maximum values of φ, E and P due to increase in volume of internal source.

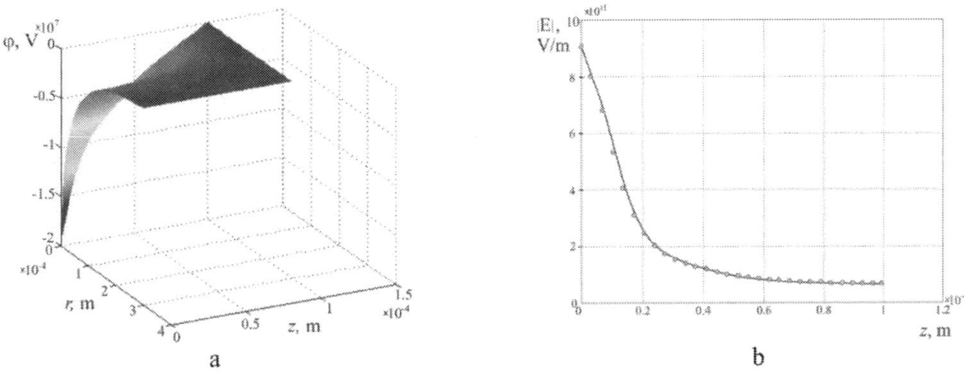

Figure 6. The simulation results of potential distribution – a, the depth profile of field intensity absolute value at point r=0 for $LiNbO_3$. The irradiation dose correspond to surface density of injected charges σ_{surf}=400 C/m^2 (irradiated area S=10^{-10} m^2, probe current I=200 pA), E_0=10 keV, d=14 μm.

Control of charging characteristic process needs to adjust the following parameters. The electron beam current I determines injected charge value ΔQ and therefore surface density of injected charges σ_{surf}. Moreover accelerating voltages U specify electron beam energy E_0 and volume of internal source ΔV.

4. Conclusion

Thus this study presents the simulation results of injection, thermal and charging processes of electron beam interaction with polar dielectrics. The description of geometrical form of electron beam interaction area was demonstrated using the Monte-Carlo simulation of electron transport. The criteria limiting the thickness of metal electrode were determined to analyze ferroelectrics with injective SEM modes. Also the thermal field simulation enables heat load of electron beam to be evaluated at various given SEM experimental parameters. The estimations of electrical characteristic due to charging effects at increased accelerating voltages indicate that controlled initial data of probe current as well as start beam energy can lead to arising significant field intensity values and as a consequence this can create conditions for domain structure reorientation under electron beam exposure.

Acknowledgments
The work has been supported by a scientific program of the Ministry of Education and Science of Russia (project No.1158.2014/9).

References

[1] Filippov M N 1993 *Izvestiya Akademii Nauk Seriya Fizicheskaya* **8** 163
[2] Cazaux J 1986 *J. Appl. Phys.* **59** 1418
[3] Le Bihan R 1989 *Ferroelectrics* **97** 19
[4] He J, Tang S H, Qin Y Q et al. 2003 *J. Appl. Phys.* **93** 9943
[5] Molina P, Ramírez M O, García-Sole J, Bausá L E 2009 *Optical Materials* **31** 1777.
[6] Maslovskaya A G, Kopylova I B 2009 *JETP* **108** 1144
[7] Latham R V 1976 *J. Phys. D: Appl. Phys.* **9** 2295
[8] Joy D C 1995 *Monte-Carlo modeling for electron microscopy and microanalysis* (New York: Oxford University Press)
[9] Bakaleinikov L A, Galaktionov E V, Tret'yakov V V, Tropp E A 2001 *Physics of the Solid State* **43** 779
[10] Suga H, Tadokoro H, Kotera M 1998 *Electron microscopy* **1** 177.
[11] Czyzewski Z, MacCallum D O, Roming A, Joy D C 1990 *J. Appl. Phys.* **68** 3066.

Hydrostatic pressure effects on structural and electronic properties of pentacene from first principles calculations

I A Fedorov, E S Rubtsova, N S Khaydukova and Y N Zhuravlev

Physics Faculty, Kemerovo State University, Krasnaya 6, 650043, Kemerovo, Russia

E-mail: ifedorov@kemsu.ru

Abstract. The structural and electronic properties of crystalline pentacene has been investigated within the framework of density functional theory including van der Waals interactions. The computed lattice parameters have good agreement with experimental data. We studied on the structural and electronic properties of the pentacene under the hydrostatic pressure of 0–10 GPa. The isothermal equations of state calculated from the results show good agreement with experiment in the pressure intervals studied. Using the Murnaghan equation of state it has been established that compression bulk modulus is B_0=11.42 GPa and it has good agreement with the experimental value. We have also calculated the quasiparticle band structure of pentacene with the G_0W_0 approximation.

Keywords: pentacene, density functional calculations, dispersion interaction, equation of state, quasiparticle band structure.

1. Introduction

Currently, oligoacenes are widely used for practical applications and due to their optical properties still have prospects for other practical applications. Pentacene is a typical representative of these compounds, besides, the first pentacene-based organic processor has recently been made. One of the most perspective trends of organic crystal applicability is the possibility to create new and simple semiconductor devices. Notable achievements have been made in this field and there are sufficient conditions for creating stable organic semiconductor transistor. The main advantage of organic semiconductors is that they are flexible and easy-to-manufacture, therefore, that makes it possible to create new pentacene –based electronic devices. Flexibility is one of the main advantages of organic semiconductors. For this reason it is interesting to investigate the effects of mechanical deformations on electronic properties. The perylene unit cell has the symmetry group $P\bar{1}$ under normal conditions [1]. The projection of the unit cell on the *cb* plane is shown in Fig. 1. The unit cell contains two molecules involved in van der Waals interactions. The pentacene molecule in the crystal is presented in figure 1.

2. Computional details

First-principles total energy calculations were carried out within density-functional theory with a plane-wave pseudopotential approach. The computations were performed with the Quantum ESPRESSO (QE) [2, 3] suite of electronic structure, using the Perdew–Burke–Ernzerhof (PBE) parametrization [4]. The cutoff energy of plane waves was set to 55 Ry. Brillouin zone sampling was performed by using the Monkhorst–Pack scheme [5] with a k-point grid of 3x2x2. The values of the kinetic energy cutoff and the **k**-point grid were determined to ensure the convergence of total energies. We used in our study the fully non-local exchange–correlation rev-vdW-DF2 [6, 7, 8]. Also, we used a London-type empirical correction for dispersion interactions DFT-D3(BJ) [9]. Structural data [1] were used as the initial geometry. Also, the calculations were performed with the CRYSTAL09 [10], using hybrid functional PBE0. In this case structural parameters, calculated using QE, were used as the input data. We have used the implementation of the G_0W_0 approximation [11, 12] provided by the code YAMBO [13]. The cutoff energy of plane waves was set to 65 Ry. The dielectric function is calculated using the plasmon-pole approximation. The GW dielectric matrix cutoff is 8 Ry. To obtain convergence, we used 400 unoccupied states.

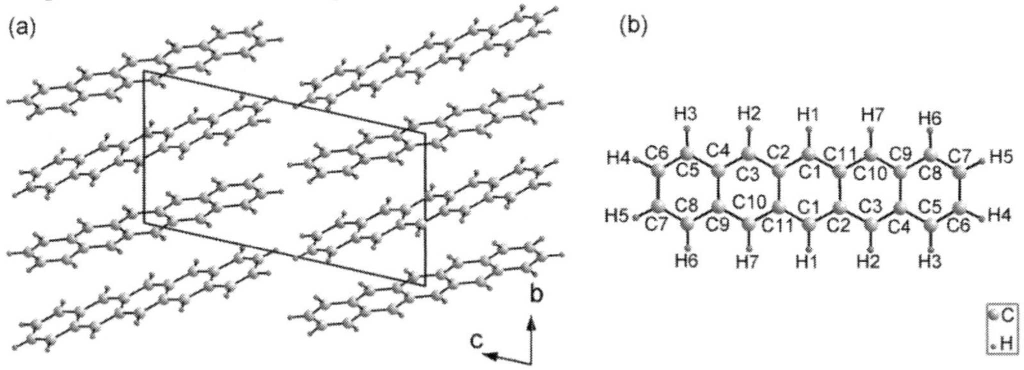

Figure 1. Crystal structure of pentacene viewed along a axis (a).The pentacene molecule in the crystal (b).

The main idea of the DFT-D algorithm consists in adding the empirical potential to the exchange–correlation potential. The total energy is given by

$$E_{\text{DFT-D}} = E_{\text{KS-DFT}} + E_{\text{disp}} \tag{1}$$

where $E_{\text{KS-DFT}}$ is the usual self-consistent Kohn–Sham energy as obtained from the chosen DF and E_{disp} is an empirical dispersion correction. The DFT-D2 scheme is included in the QE [2]. We carried out computations within the DFT-D3(BJ) [9].

3. Results and discussion

3.1. Equation of state

Our computed lattice parameters for the pentacene are given in table 1. Experimental values are also listed. The modern versions of the DFT-D and vdW-DF predict the lattice parameters that have good agreement with the experimental data.

Figure 2 shows the variation of the lattice parameters and unit cell volume for pentacene with pressure as predicted using DFT-D3(BJ). The obtained results were comparable to the experimental values [14]. There is good agreement between the computed and experimental values. As the computation was performed at T = 0 K, the V(P) dependence should be below experimental data (T=100 K). Generally, DFT-D3(BJ) leads to good agreement with the experimental data. There is the difference between the experimental and theoretical data at high pressure. Similar results were observed for the values calculated using rev-vdW-DF2 and other schemes [15].

Pressure dependence on volume is usually written in the form of isothermal equation of state for solids. The parameters of this equation are V_0 - equilibrium volume, B_0 - isothermal bulk modulus and B_0' - its pressure-derivative, defined at zero pressure. Two semi-empirical isothermal equation of state

(EOS) – Vinet, and Murnaghan – were used to analyze the compression curves for pentacene. These equations are as follows:

$$P(V) = (\frac{B_0}{B_0'})((\frac{V}{V_0})^{-B_0'} - 1), \quad (2)$$

$$P(V) = 3B_0(\frac{V}{V_0})^{-\frac{2}{3}}(1-(\frac{V}{V_0})^{\frac{1}{3}})\exp[\frac{3}{2}(B_0'-1)(1-(\frac{V}{V_0})^{\frac{1}{3}})]. \quad (3)$$

Table 1. Computed and experimental lattice parameters for pentacene at ambient pressure.

	DFT-D3(BJ)	Rev-vdW-DF2	Exp. [1]
a, Å	6.243	6.228	6.266
b, Å	7.769	7.572	7.775
c, Å	14.432	14.331	14.53
α, °	76.269	77.384	76.475
β, °	88.414	88.285	87.682
γ, °	84.573	84.446	84.684
V_0, Å^3	676.955	656.491	685.155

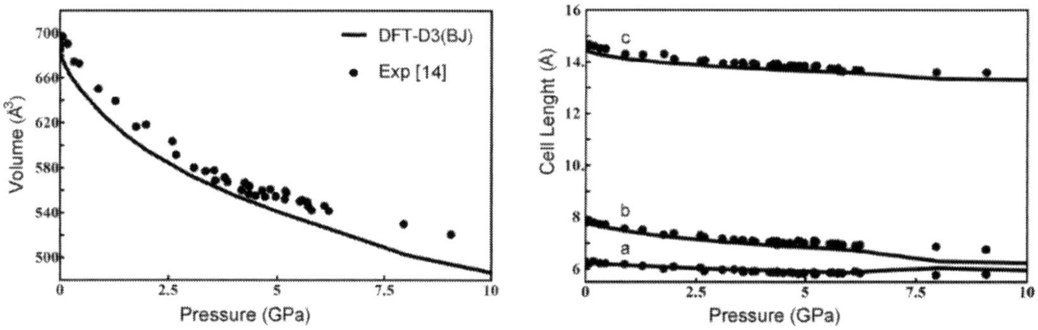

Figure 2. Unit cell dimensions as a function of pressure from DFT-D3(BJ) and experiment [14].

Table 2 shows parameter values which were fitted for the both isothermal equations of state. The computed bulk modulus and its pressure derivative have good agreement with the experimental values and computations of other authors. The values calculated using DFT-D3(BJ) have better agreement with the experimental data. Depending on the conditions of crystal growth, pentacene under normal conditions may be in two stable forms. There is only one phase at pressure 0.2 GPa and higher, which has been studied in present work. Thus, there are some problems in experimental definition of crystal volume under normal conditions, and this leads to scattering of parameter values of equations of state.

3.2. Band structures

The calculated band structure for crystalline pentacene is presented in figure 3. The reference point is the top of the valence band. The designation of points and choice of lines are in accordance with [16]. The values of band gaps (E_{gap}) are listed in table 3. At the ambient pressure the band gap E_{gap} = 1.72 eV (PBE0). We have also calculated the band structure of pentacene with the G_0W_0 approximation (E_{gap} = 2.03 eV). The top of the valence band and the minimum value of the conduction band is realized at the L point. Thus, the fundamental band gap is a direct gap.

Table 2. Comparison of bulk modulus (B_0) and its pressure-derivative (B_0') determined from the EOS analyses of pentacene.

Method	V_0(Å^3)	Murnaghan		Vinet	
		B_0(GPa)	B_0'	B_0(GPa)	B_0'
DFT-D3(BJ)	676.955	11.42	5.21	10.54	6.70
Rev-vdW-DF2	655.679	14.11	5.20	13.22	6.54
Exp. [14]	696.0±3.7	9.6±1.0	6.4±0.5	-	-

The band structure presents weakly dispersive energy levels originated in molecular orbitals. Figure 3 shows the band structures under hydrostatic pressure of 10 GPa. The pressure increases band dispersion and decreases band gap. Lower unoccupied bands are particularly strongly split. The calculated quasiparticle band gap at 10 GPa equal 1.05 eV. This behaviour of the band gap is confirmed by both experimental [17] and theoretical data [15].

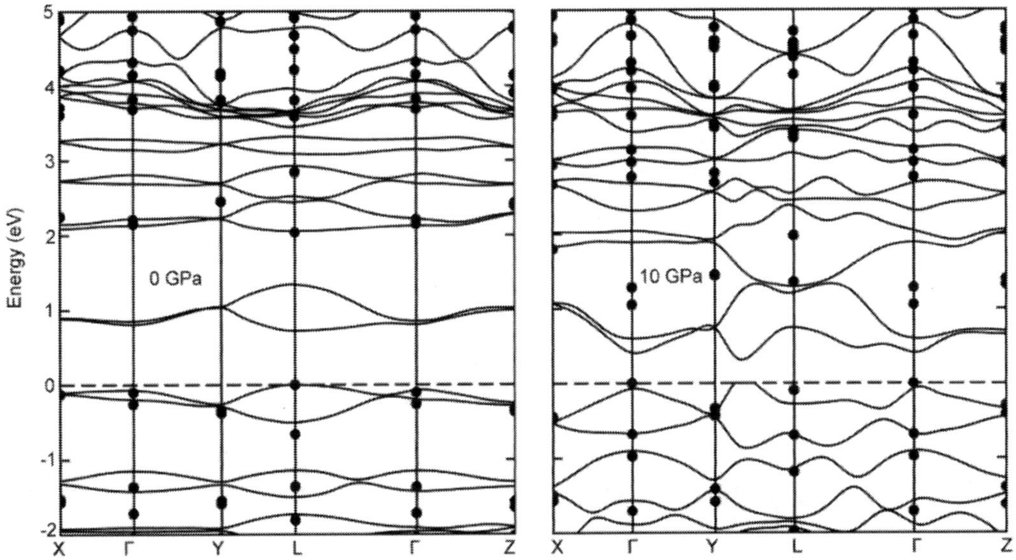

Figure 3. Band structure $E(\mathbf{k})$ of crystalline pentacene within GGA/PBE (solid lines) and G_0W_0 approximation (points) at ambient and hydrostatic pressure of 10 GPa . The top of the valence band indicated by the Fermi level (dashed line).

Table 3. The band gaps (eV) of pentacene obtained using various of the exchange–correlation functionals.

G_0W_0	PBE	PBE0	PBE[15]	HSE03[15]	PBE0[15]
2.03	0.54	1.72	0.858	1.431	2.031

One of the important characteristics of chemical bond is the distance between the atoms. The effect of pressure is shown in reduction of the distance between the molecules, however, the changes in bond lengths within the molecule are much smaller. Mulliken population analysis has allowed us to determine the crystal atomic charges (table 4). Though in the presented schemes they do not have a clear physical sense, they allow establishing the atom charge change.

3.3. *Electron densities*

The electron densities were calculated with PBE0 hybrid density functional. The distribution of total and deformation electron densities (difference between total and superposition atomic electron densities) on the plane is presented in figure 4. Pentacene molecule in the crystal is not planar molecule. The positive values are marked by solid line, negative by dashed line, while the null contour is marked by a dotted line. The interval between the isolines equals 0.05 e Å^{-3}. It has been established that when a chemical bond in the molecule is formed, the charge leakage occurs in the C–C bond line as well as in the C–H antibonding regions. The distribution of deformation density is typical for sp^2 hybridization. The peaks of electron density between carbon atoms are stretched perpendicular to the C–C bond lines. The nonhybridized p-orbitals are perpendicular to the molecule plane and responsible for the socalled p-stacking. The pressure of 10 GPa causes a slight change for the distribution of deformation density, atomic charges and interatomic distance. Thus, the pressure of 10 GPa has significant impact on the energy bands, which is caused by the convergence of molecules, but the molecules remain virtually undeformed.

Table 4. Atomic charges calculated according to Mulliken scheme for the pentacene (|e|).

Atom	Pressure, GPa		
	0	5	10
C(1)	-0.171	-0.175	-0.184
C(2)	-0.011	-0.021	-0.009
C(3)	-0.169	-0.175	-0.186
C(4)	-0.012	-0.025	-0.022
C(5)	-0.163	-0.164	-0.167
C(6)	-0.188	-0.204	-0.206
C(7)	-0.187	-0.197	-0.193
C(8)	-0.145	-0.139	-0.148
C(9)	-0.019	-0.028	-0.014
C(10)	-0.168	-0.171	-0.176
C(11)	-0.012	-0.019	-0.011
H(1)	0.179	0.191	0.192
H(2)	0.179	0.190	0.191
H(3)	0.177	0.182	0.181
H(4)	0.171	0.177	0.178
H(5)	0.171	0.179	0.179
H(6)	0.186	0.198	0.197
H(7)	0.179	0.190	0.192

4. Conclusions

According to the *ab initio* calculations, crystalline pentacene has been studied. The relaxed lattice parameters, band structure, distribution of electronic densities, bulk modulus and its pressure-derivative of the pentacene were computed in the framework of DFT. The determined dependencies of the values of lattice parameters on pressure are in good agreement with experimental data. It is established that within DFT-D3(BJ) bulk modulus is B_0=11.42 GPa, that has good agreement with experimental data. The quasiparticle band gap at ambient pressure and 10 GPa is computed with the G_0W_0 approximation and equals 2.03 and 1.05 eV, respectively. The pressure leads to the decrease in the distance between the crystal molecules and increases overlapping of their molecular orbitals.

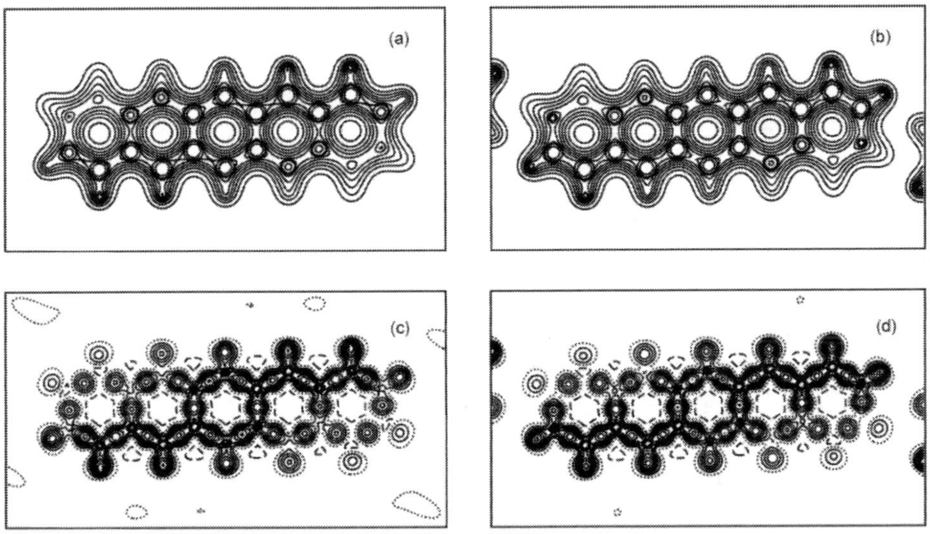

Figure 4. Total electron density map of crystalline pentacene at the ambient (a) and hydrostatic pressure of 10 GPa (b) (in logarithmic scale). Deformation electron density map at the ambient (c) and hydrostatic pressure of 10 GPa (d).

Acknowledgement
Computational resources were provided by the Center for collective use "High Performance Parallel Computing" of the Kemerovo State University.
This work was supported by the Ministry of Education and Science of the Russia Federation (Project №3.1235.2014K).

References
[1] Mattheus C C *et al* 2001 *Acta Cryst.* C **57** 939
[2] Barone V *et al* 2009 *Comput. Chem.* **30** 934
[3] Giannozzi P *et al* 2009 *J. Phys.: Condens. Matter* **21** 395502
[4] Perdew J P, Burke K and Ernzerhof M 1996 *Phys. Rev. Lett.* **77** 3865
[5] Monkhorst H J and Pack J D 1976 *Phys. Rev. B* **13** 5188
[6] Dion M *et al* 2004 *Phys. Rev. Lett.* **92** 246401
[7] Lee J-W *et al* 2007 *J. Appl. Phys.* **102** 124104
[8] Hamada I 2014 *Phys. Rev. B* **89** 121103
[9] Grimme S, Antony J and Krieg H 2010 *J. Chem. Phys.* **132** 154104
[10] R. Dovesi *et al* 2009 *CRYSTAL09 User's Manual* (Torino: University of Torino) p 306
[11] Onida G, Reining L and Rubio A 2002 *Rev. Mod. Phys.* **74** 601
[12] Hedin L and Lundqvist S 1970 *Solid State Phys.* **23** 1
[13] Marini A, Hogan C, Grüning M and Varsano D 2009 *Comput. Phys. Commun.* **180** 1392
[14] Oehzelt M, Aichholzer A and Resel R 2006 *Phys. Rev.* B **74** 104103
[15] Schatschneider B, Monaco S, Tkatchenko A and Liang J-J 2013 *J. Phys. Chem. A* **117** 8323
[16] Setyawan W and Curtarolo S 2010 *Comp. Mat. Sci.* **49** 299
[17] Farina L *et al* 2003 *High Press. Res.* **3** 349

Synthesis and investigation of the geometric characteristics of titanium dioxide nanotubes

T M Serikov[1], N Kh Ibrayev[1], V K Gladkova[1]
[1] Karaganda, 100028, Republic of Kazakhstan, Institute of Molecular Nanophotonics, E.A. Buketov Karaganda State University

E-mail: serikov-timur@mail.ru

Abstract. The nanotubes of titanium dioxide were produced by electrochemical anodization of metallic titanium in a fluorine-containing electrolyte. The influence of the electrolyte composition, voltage and time of anodization on the geometric characteristics of the synthesized nanotubes was investigated.

Introduction

Recently one of the researchers' attention objects is titanium dioxide nanotubes. Due to its adsorption, optical, electrical and catalytic properties the titanium dioxide is widely used in molecular electronics [1], solar energy [2-5] and photocatalysis [6].

The materials based on titanium dioxide nanotubes exhibit the necessary physical and chemical properties, which depend primarily on its obtaining conditions. There are several methods for producing titanium dioxide nanotubes. One of them belongs to a group of researchers led by Professor Latika Menon [7] from Northwestern University (USA). This method consists in carrying out the electrolysis of the chloride salts solution in the presence of a titanium foil. As a result of the electrolytic reaction (with the release of chlorine in an aqueous medium), the surface of the titanium foil was oxidized, and under certain conditions it was transformed into a highly oriented structure. Another method for the preparation of ordered titanium dioxide nanotubes is electrochemical anodization of metallic titanium in the electrolyte, which includes hydrofluoric acid or concentrated hydrogen peroxide [8]. However, these methods except the positive characteristics, such as controlled geometric characteristics of nanotubes, high surface area, high light absorption ability, also have negative characteristics: toxicity of anodization, the necessity of strict control of the temperature of environment and electrolyte, which requires additional equipment and cost. In addition, synthesized nanotubes have amorphous structure.

One of the methods for the preparation of titanium dioxide nanotubes is the anodic oxidation of metallic titanium in the fluorine-containing electrolyte. Parameters of nanotubes can be controlled by varying the oxidation conditions (voltage, time, composition of the electrolyte). Analysis of publications devoted to the synthesis of TiO_2 nanotubes in the fluorine-containing electrolyte, has shown that many of the works contradict each other and most of their claims are not supported by experimental data.

Though, one of the methods for obtaining the titanium dioxide nanotubes is the anodic oxidation of metallic titanium in the fluorine-containing electrolyte. Parameters of nanotubes can be controlled by varying the oxidation conditions.

This paper presents the results of highly ordered TiO_2 nanotubes synthesis, obtained by a two-step electrochemical anodization of metallic titanium in the fluorine-containing electrolyte while changing the composition of the electrolyte, voltage and anodization time.

Experimental part

The titanium foil with 120 microns thickness was subjected to chemical polishing in 100 ml of a solution which comprises: sulfuric acid - 60%, hydrofluoric acid - 25%, glycerol - 15%. The polishing the surface the titanium foil is mounted on the cathode, the anode was titanium platen. The polishing process occurs at a voltage of 20 V with vigorous solution stirring for 10 minutes. After polishing the sample was washed with distilled water, dried and then it was divided into the 2x1 cm plate. Before anodization the sample was sonicated into a 2 M HCl solution and acetone for 10 minutes to remove the surface dirt and degreasing.

As the electrolyte bases the ethylene glycol with NH_4F content and H_2O was used. Anodic oxidation of titanium was carried out in potentiostatic mode, in an electrochemical cell at temperature of 5-70 C. As the cathode platinum foil was used, as the anode - anodized material. The distance between the anode and cathode was 3 cm.

Electrochemical oxidation process consists of two stages, the first: anodization for 2 hours. The separation of formed film at the first stage was carried out in an ultrasonic bath with 1 M hydrochloric acid solution. The sample was then washed with much deionized water and dried at a temperature of 800^0C. The second stage was the same as the first one.

Results and discussion

The surface morphology of the porous titanium oxide obtained by a scanning electron microscope MIRA 3LMU (Tescan, Czech Republic) is shown in Figure 1. The film was prepared in a mixed solution of ethylene glycol, NH_4F (0,5%) and H_2O (3%) at anodization voltage of 50 V.

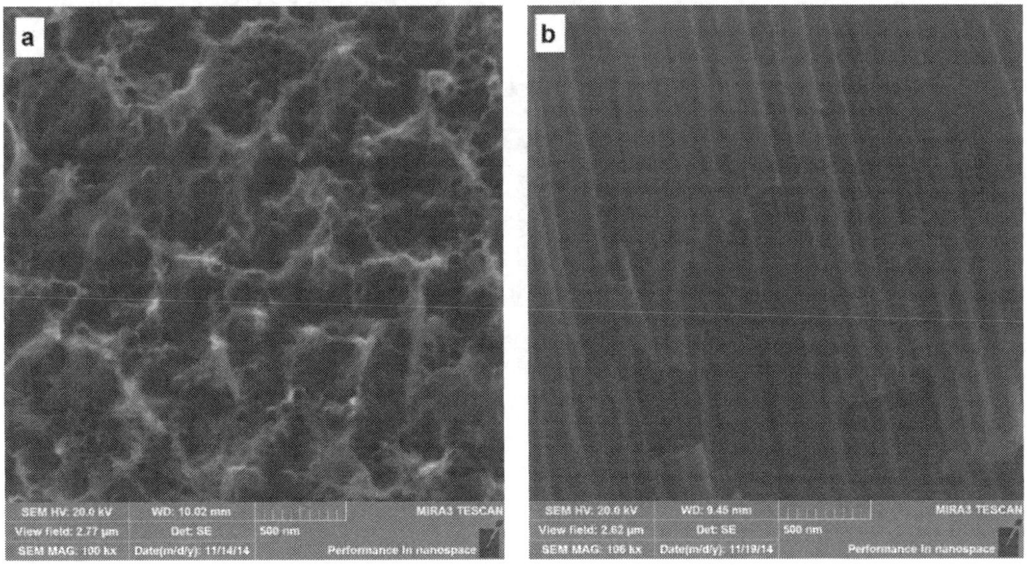

Figure 1. Micrographs of Ti foil after the first stage of anodization:a) the surface; б) cross-section.

The picture shows (Figure 1, a) that there are titanium dioxide nanotubes were formed on the surface of the titanium foil. They have a disordered structure and placed perpendicular to the base. On the surface there is a large amount of dirt and cracks. According to the literature [9-10], the source of pollution are by-products of hydrolysis $[TiF_6]^{2-}$. Their presence leads to low adhesion of the films, which further limits its use. The repeated anodization of titanium foil, after removing the first anodization film, is one of the ways of structuring the surface, removing contaminants and increasing the adhesion of the synthesized nanotubes.

After the second stage of the anodization the obtained TiO_2 nanotubes were sonicated to remove the byproducts and the surface oxide layer, washed with much distilled water and dried in a stream of nitrogen. Crystallization of the amorphous phase of TiO_2 was performed by the thermal treatment of samples in an oven at 450 °C for 3 hours while increasing the temperature at 50 °C per minute.

SEM images of porous titanium oxide are shown in Figure 2.

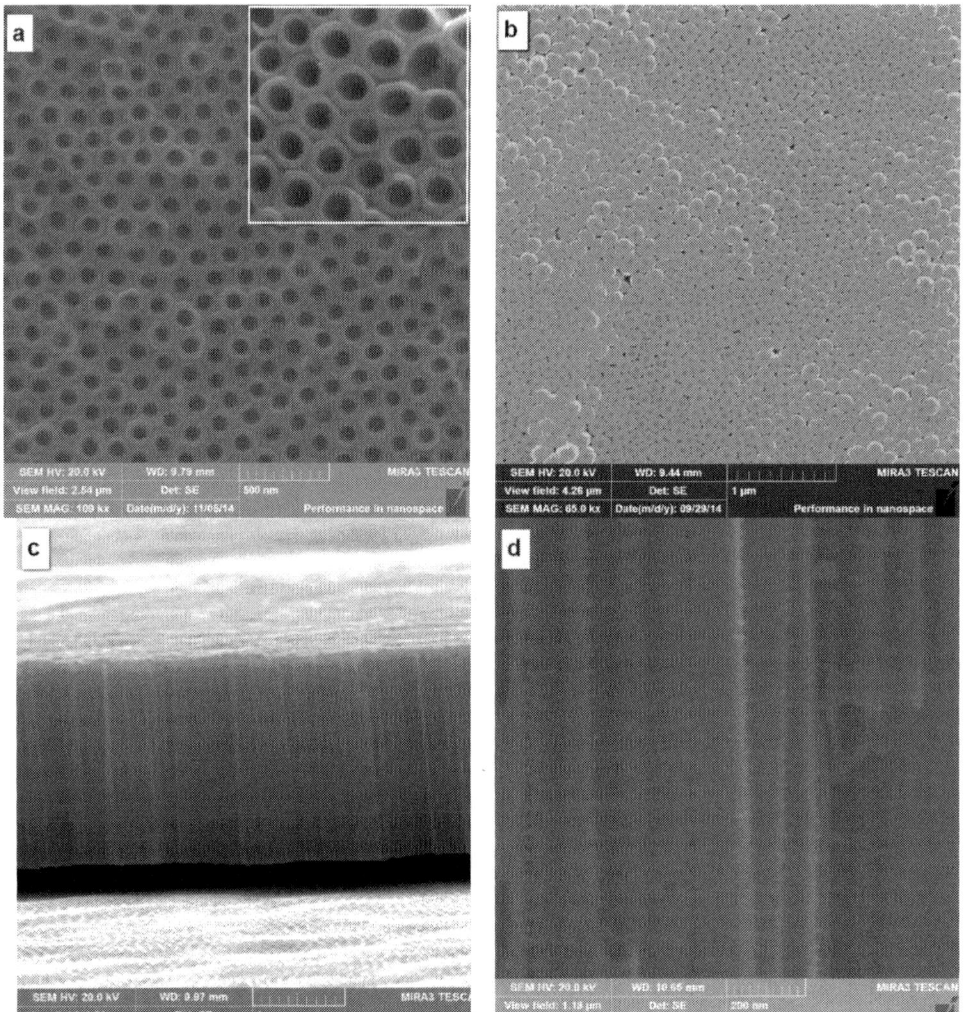

a – surface; b – barrier layer; c and d – cross-section.

Figure 2. Microstructure of titanium dioxide nanotubes.

Investigation of the ammonium fluoride (NH_4F) quantity effect in the electrolyte on the geometrical characteristics of obtained TiO_2 nanotubes was performed on samples obtained at 50 V and at temperature 5-7 °C at the duration of the second anodization stage of 1 hour. H_2O and NH_4F content in the electrolyte was varied from 0.1 to 0.5 (wt.%) and 2-6 (wt.%), respectively.

Table 1. NH_4F content effect on the lenth increasing of titanium dioxide nanotubes

The lenth of titanium dioxide nanotubes, μm					
Content of NH_4F, wt. %	Content of H_2O, wt. %				
	2	3	4	5	6
0.1	10	10	2.5	2	1.2
0.2	11	11	6	3	2.2
0.3	12	9	7	6.6	4.5
0.4	3	7	9	8	5.3
0.5	8.5	10	11	9	6

The experimental data presented in Table 1 shows that in the electrolyte consisting of ethylene glycol, 0.3 wt.% NH_4F and 2 wt.% H_2O the titanium dioxide nanotube growth rate is the highest. However, the coatings obtained under these conditions, exhibit poor adhesion to the surface of the metallic titanium, and the drying occurs at a partial TiO_2 peeling from the titanium foil.

When the water content in the electrolyte is 2– 6 wt.% the adhesion of the TiO_2 film to the surface of the titanium foil increases. TiO_2 nanotubes, obtained at the H_2O content of 4 wt. %, have a high bonding strength with titanium substrate. This conclusion is based on the results of measurements of films microhardness by Vickers and by the absence of delamination of the oxide layer from the substrate during drying.

The establishment of voltage influence on the structure of synthesized titanium dioxide nanotubes was carried out at temperature of 5-7 ° C in the range from 10 to 80 V in the electrolyte with following composition, 0,5 wt.% NH_4F, 3 wt.% H_2O and 97.7 wt.% ethylene glycol. The duration of the second stage of the anodizing was 1 hour.

Table 2. Influence of voltage on the inner diameter and the density of titanium dioxide nanotubes

№	Voltage, V	Inner diameter of nanotubes, nm	Density of nanotubes, $10^9/cm^2$
1	10	10	26
2	20	40	25
3	30	60	22
4	40	70	11
5	50	95	6
6	60	105	2.5
7	70	110	0.5
8	80	110	0.5

The formation of highly ordered and open nanotubes occurs at a voltage of 20–60 V. With increasing anodization voltage the inner diameter of nanotubes increases, but the number of nanotubes per unit area falls. The packing density of the titanium dioxide nanotubes was determined from the micrographs obtained by scanning electron microscope.

At voltages < 20 V the closing of nanotubes occurs, due to formation of the byproducts dense layer on the TiO_2 surface, which is formed by the hydrolysis of complex anions $[TiF_6]^{2-}$. Hydrolysis occurs due to the fact that the nanotube has pH gradient from the lower value at the bottom, to a larger near the mouth of the nanotubes [9]. With increasing the voltage above 60 V the tubes distribution over the surface becomes disordered.

From electron microscopic data it follows that the distance between the centers of the nanotubes depend on the voltage. This makes it possible to obtain arrays of nanotubes with certain packing density of nanotubes per unit area. From the micrographs it was also found that the wall thickness of the nanotubes doesn't depend on the voltage and it is 8-10 nm.

Investigation on the anodization duration influence on the titanium dioxide nanotube structure was carried out on samples synthesized at voltages of 50 V and a temperature of 5-7 ° C in an electrolyte with following composition: NH_4F (0,5%) and H_2O (3%). The duration of the second stage of the anodization varied from 1 to 30 hours.

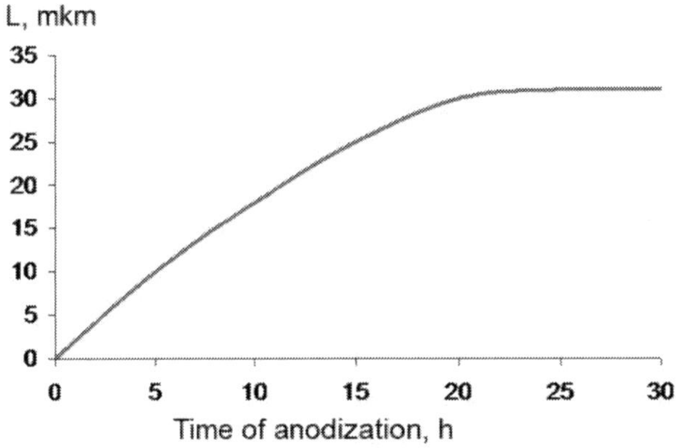

Figure 3. The dependence of the nanotubes length (L) on the duration of the second stage of anodization.

Figure 3 shows that in the early hours of anodization the nanotubes length increases linearly with anodization time and then, reaching 25-30 μm, the rate of nanotubes formation decreases.

The obtained data revealed that with increasing length of the titanium dioxide nanotubes the electrochemical oxidation rate falls. As a result of this the activity of fluoride ions is redirected from the bottom of nanotube to its sidewalls. This leads to the fact that with increasing anodization duration the wall thickness reduces and the inner diameter of the nanotubes is growing.

Conclusion

Thus, by the two stages electrochemical anodizing of metallic titanium in the fluorochemical electrolyte the highly ordered arrays of titanium dioxide nanotubes were synthesized, they were directed perpendicularly to the substrate. It was found that the electrolyte consisting of ethylene glycol, NH_4F (0,3 wt.%) and H_2O (2 wt.%) allows to receive the highest growth rate of titanium dioxide nanotubes. However, the array of nanotubes was characterized by a low adhesion with the metal substrate. When the water content in the electrolyte was 2 – 6 wt. % the adhesion of the TiO_2 film to the foil surface increased. With the growth of the anodization voltage the inner diameter of the nanotubes increased, but the number of nanotubes per unit area decreased. Found that the wall thickness of the titanium dioxide nanotubes depends on the duration of the anodization.

References

[1] Yuxin T, Yanyan Zh, Jiyang D, Jiaqi W 2014 *Advanced materials.* **26** 6111.
[2] Wu H P, Li L L, Chen C C, Diau E 2012 *Ceramics International.* **38** 6253.
[3] Chang H, Chen C, Kao M, Chien S, Chou C 2013 *Applied Surface Science.* **275** 252.
[4] Xi M, Zhang Y, Long L, Li X 2014 *Journal of Solid State Chemistry.* **219** 118.
[5] Li S, Liu Y, Zhang G, Zhao X, Yin J 2011 *Thin Solid Films.* **520** 689.
[6] Kandiel T A, Robben L, Alkaimad A, Bahnemann D 2013 *Photochemical and Photobiological Sciences.* **12** 602.
[7] Latika M 2012 *Nanmomaterials* **1** 152.

[8] Petukhov D I, Kolesnik I V, Eliseev E E, Lukashin A V, Tretyakov Yu D 2007 *Int. J. Alternative Energy Ecol.* **1** 65.
[9] Morozov A N 2014 Synthesis and catalytic properties of nanostructured coatings of titanium dioxide (Moscow).
[10] Mor G K,Varghese O K, Paulose M, Mukherjee N, Grimes C A 2003 *Journal of Materials Research.* **18** 2588.

Synthesis and Staging of the Phase Formation for Strontium Ferrites in Thermal and Radiation-Thermal Reactions

U V Ancharova[1], M A Mikhailenko[1], B P Tolochko[1], N Z Lyakhov[1], M V Korobeinikov[2], A A Bryazgin[2], V V Bezuglov[2], E A Shtarklev[2], A Yu Vlasov[2] and Z S Vinokurov[3]

[1] Institute of Solid State Chemistry and Mechanochemistry SB RAS, Novosibirsk, Russia
[2] Budker Institute of Nuclear Physics SB RAS, Novosibirsk, Russia
[3] Boreskov Institute of Catalysis SB RAS, Novosibirsk, Russia

E-mail: ancharova@gmail.com

Abstract. The strontium ferrite radiation-thermal synthesis process due to heating by intensive electron beam is described. The radiation-thermal and thermal strontium ferrite synthesis processes kinetics were studied and compared.

1. Introduction

The heating of the reaction mixture by the intensive electron beam greatly changes the chemical synthesis reaction kinetics as compared with the thermal processes. Investigations of the irradiation influence on the high-temperature synthesis kinetics have shown [1-4], that the electron irradiation significantly speeds up the synthesis process at relatively low temperatures for a number of different oxide systems ($NaFeO_2$, $BaTiO_3$, $NiFe_2O_4$, $LiFe_5O_8$, $MnFe_2O_4$, $ZnFe_2O_4$ and many others). It was found that the electron beam is a powerful factor in speeding up the solid state reactions, at least when the reaction rate is determined by the diffusion of reactants through the product layer.

This work is devoted to the study and description of radiation-thermal process in the synthesis of ferrite ceramic materials during irradiation by intensive electron beam with energy of 2.4 MeV.

2. Experimental

Reaction mixtures were prepared by a mechanochemical method in a planetary ball mill AGO-2 (acceleration of milling bodies 20 g) of the stoichiometric composition of iron oxide and the strontium carbonate of reagent grade or analytical grade purity.

Investigations of the reaction products and reaction kinetics in the in situ thermal and ex-situ radiation-thermal strontium ferrite synthesis were held by X-ray diffraction method using synchrotron radiation. Measurements were carried out at the 6th beamline "Precision diffractometry" of VEPP-3 storage ring (shared research center SSTRC [5], located in BINP SB RAS). X-ray beam of synchrotron radiation after collimator slits was monochromatized by reflection from Si (200) crystal planes and had the wavelength λ~1Å. The shorter wavelength (compared to λ~1.54Å from Si (111) crystal planes more intense reflection) have been chosen for decrease in fluorescence from iron atoms in the reaction mixtures on the one hand. In addition, on another - for increase in the range of reciprocal space registered vectors by the linear position-sensitive detector OD-3 [6] having a fixed angular range of 30 degrees. The detector was set to an angular range of 20-50 degrees, where the most significant diffraction reflections of the reaction products got.

The thermal synthesis of strontium ferrite had performed in air - by heating in high temperature chamber Anton Paar HTK 2000N, for *in situ* investigations of chemical reaction.

The radiation-thermal synthesis had carried out by irradiation with intensive electron beam. The radiation-thermal treatment performed using accelerator ILU-6, located in Budker Institute of Nuclear Physics SB RAS. The electron energy was set 2.4 MeV, the pulse beam current was 320 mA, the pulse duration was 0.5 ms. Figure 1 shows the reactor with the sample inside installed under the accelerator's beam extraction device.

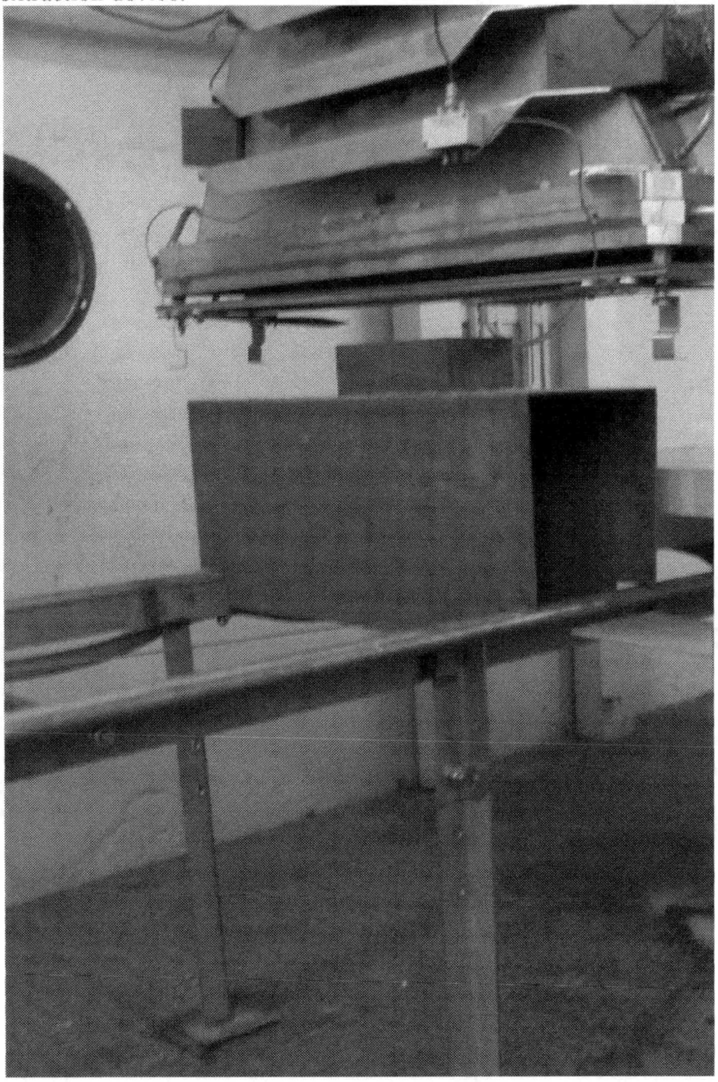

Figure 1. The reactor placed under the ILU-6 beam window

Treated mixtures were placed in a corundum crucible. The surface density of the layer was 0.2 - 0.25 g/sm^2, thus ensuring uniform electron beam heating inside the samples. Crucibles with reaction mixture or other samples were placed into a heat-insulated reactor having the side walls made of firebrick shown in Figure 2 and covered with a plate made of foam-quartz (density 0.15 g/sm^3) having thickness of 0.2 - 0.3 cm. The reactor was under the accelerator beam window as shown in Figure 1.

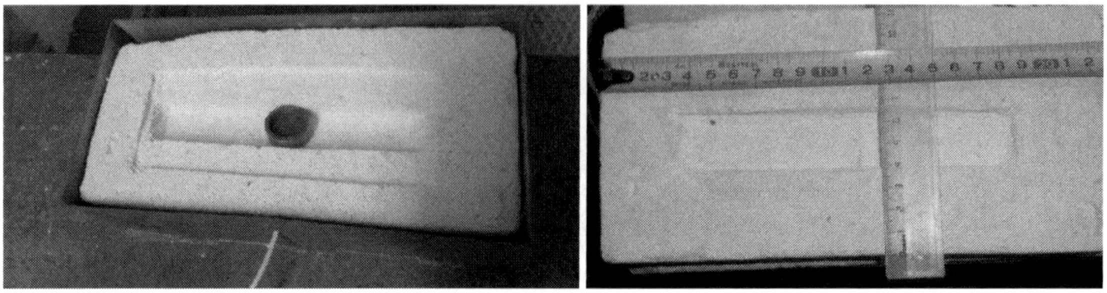

Figure 2. The inner working area of the reactor for radiation-thermal synthesis

The electron beam parameters (electron energy, pulse beam current and pulse repetition frequency) were set by the accelerator control program. The accelerator control program has the option to maintain the treated sample temperature according the temperature chart set in special accelerator control program window shown in Figure 3. Normally the program varies the pulse repetition rate to maintain the required temperature, next regulation level – pulse beam current variation.

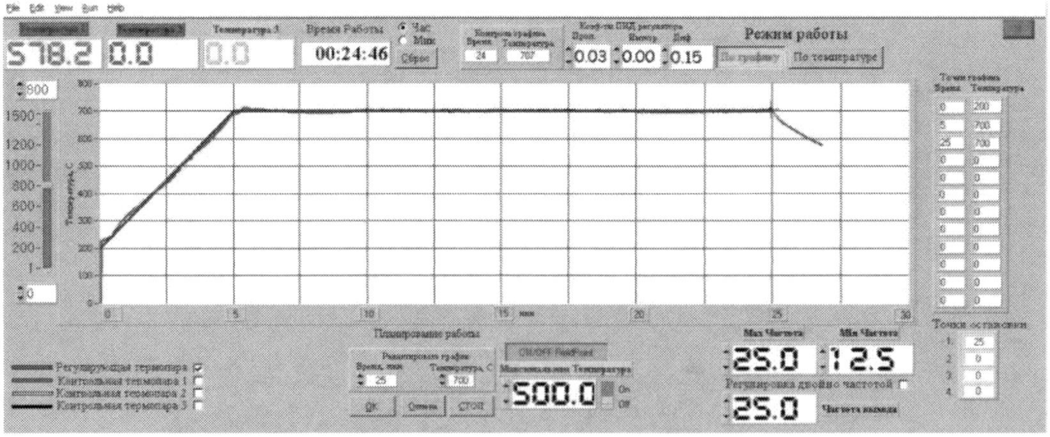

Figure 3. The temperature chart (blue line) and process temperature (red line) in the accelerator control program window

Usually the temperature chart for ferrites synthesis consisted of two parts: the linear heating and the temperature maintenance at the plateau. The cooling of reaction mixtures was carried out in a natural way (in some experiments the cooling was retarded due to controlled electron beam heating). The heating rate in all experiments was 100°C/min. The temperature plateau at set temperature of 700-1000°C was achieved in 10 minutes. The average beam current in the heating stage does not exceed 4.5 mA, and was 1.5-2.5 mA at the plateau.

3. Results

In the *in situ* X-Ray diffraction experiments with strontium ferrites thermal synthesis at temperatures of 700, 800 and 900°C it observed, that at the first synthesis stage during the mixtures heating with rate of 100°C/min the defects annealing goes. The X-ray diffraction patterns were analysed by Rietveld method. The observed line broadening caused by the size of coherent scattering regions and microstrains gradually decreases with heating. Then the reaction of the strontium ferrites high-temperature synthesis starts. Its ways differ depending on the synthesis temperature.

The strontium ferrites synthesis processes were studied at the same temperatures in different ways – the thermal (T) was studied *in situ* and radiation-thermal (RT) was studied *ex situ* ways.

The exposure durations have been chosen from 10 to 70 minutes for the RT processes reaction rate characterization. The conversion level value to the final product (strontium ferrite) over a characteristic time of 30 minutes was selected as a quantitative indicator (Table 1).

Table 1. Strontium ferrites synthesis over a characteristic time of 30 minutes.

Method	Temperature, °C	Phases, %						
		Initial		Intermediate and product				
		$SrCO_3$	Fe_2O_3	SrO	$SrFe_{12}O_{19}$	$Sr_3Fe_2O_7$	$Sr_4Fe_3O_{10}$	$SrFeO_3$
Thermal	700	57	32	0	0	0	0	11
	800	44	30	0	0	0	0	26
	900	4	12	24	10	17	0	33
Radiation-thermal	700	66	26	0	0	0	0	8
	800	55	18	0	0	0	0	27
	850	34	2	0	15	0	13	36
	900	21	1	0	13	0	18	47
	950	7	0	0	12	0	22	59
	1000	0	0	0	13	20	0	67
The trend of the time		decrease	decrease	increase-decrease	no	increase	increase	increase

At temperatures below 850°C there are no significant differences between reaction rates and ways depending on synthesis method – thermal or radiation-thermal. The reaction goes without the intermediate phases' formation: strontium carbonate directly reacted with iron oxide. Thus the irradiation with intensive electron beam does not speeds up the reaction of strontium carbonate phase with iron oxide, despite the well-known effect [1-4] of reactants diffusion speeding up due to irradiation by intensive electron beam. So, in this case, the reaction rate is not limited by mutual diffusion of the reactants. Apparently, the limiting factor at temperatures below the decomposition of carbonate is the carbon dioxide elimination from the system.

However, at temperatures above 850°C there is a number of differences in chemical reactions for the thermal and radiation-thermal strontium ferrite synthesis.

In situ diffraction experiments during the thermal heating of the reaction mixture showed that at 900°C an intensive decomposition of strontium carbonate $SrCO_3 \rightarrow SrO + CO_2$ goes over the time scales smaller than strontium ferrite formation rate. Just for ~ 10 minutes the strontium carbonate concentration reduced in ~ 20 times, while the strontium oxide concentration sharply increased from zero, reaching its maximum value. Thus the high temperature synthesis of strontium ferrite without irradiation at this temperature goes mainly by the iron oxide interaction directly with strontium oxide, which is the product of carbonate decomposition. However, there are small concentrations of the other intermediate phases in the reaction product in addition to the reactants, final product and strontium oxide. They are hexa- strontium ferrite $SrFe_{12}O_{19}$ and Ruddlesden-Popper phase (RP2) $Sr_3Fe_2O_7$.

Ex situ phase analysis of high-temperature radiation-thermal synthesis reaction products showed that the radiation-thermal heating of the reaction mixture does not cause the carbonate decomposition over the temperature range up to 1000°C, that is confirmed by experimental radiation-thermal treatment of the pure strontium carbonate. This phenomenon will be discussed in more details in our later works.

Because the strontium carbonate enters directly to the chemical reaction without stage of decomposition to the oxide, a set of intermediate phases in the reaction differs in case of radiation-thermal synthesis: instead of phase RP2 ($Sr_3Fe_2O_7$), phase RP3 ($Sr_4Fe_3O_{10}$) was formed. Despite the fact that the reaction goes by a different mechanism then in the thermal process, the reaction product phase (strontium ferrite) concentration grows faster in the radiation-thermal process.

4. Conclusions

The strontium ferrite synthesis speeding up was observed at temperatures above 850°C due to radiation-thermal treatment. In the temperature range up to 1000°C the strontium carbonate decomposition was not observed in the radiation-thermal process (unlike in the thermal process). This effect confirms that the chemical reactions kinetics in radiation-thermal and thermal synthesis of strontium ferrite differ thus forming the different sets of the intermediate phases.

5. Acknowledgments

This work was carried out on the equipment belonging to the shared research center "SSTRC" that is supported by the Ministry Education and Science of Russian Federation.

This work was supported by the Russian Foundation for Basic Research (RFBR) (Grant 12-03-01109-a).

References

[1] Lyakhov N Z, Boldyrev V V, Voronin A P, Gribkov O S, Bochkarev L G, Rusakov S V and Auslender V L 1995 *J. of Thermal Analyais* **43** 21

[2] Boldyrev V V, Voronin A P, Gribkov O S, Tkachenko E V, Karagedov G R, Yakobson B I and Auslender V L 1989 *J. Solid State Ionics* **36** 1

[3] Voronin A P, Neronov V A, Boldyrev V V, Melekhova T F, Auslender V L, Aleksandrov V V, Lyakhov N Z, Polyakov V A and Savinkina M A 1982 *J. Soviet Powder Metallurgy and Metal Ceramics* **21** 632

[4] Auslender V L, Bochkarev I G, Boldyrev V V, Lyakhov N Z and Voronin A P 1997 *J. Solid State Ionics* **101-103** 489-493

[5] Kuzin M V Shared research center "Siberian Synchrotron and Terahertz Radiation Centre" (Electronic Material) http://ssrc.inp.nsk.su/CKP/

[6] Aulchenko V M Detector OD-3M (Electronic Material) http://ssrc.inp.nsk.su/CKP/prod/detectors/OD3/

Radiation-Thermal Paraffin Cracking

M A Mikhailenko[1], U V Ancharova[1], B P Tolochko[1], M V Korobeinikov[2], A A Bryazgin[2], V V Bezuglov[2], E A Shtarklev[2], A Yu Vlasov[2], L M Pokrovsky[3] and K G Korolev[4]

[1] Institute of Solid State Chemistry and Mechanochemistry SB RAS, Novosibirsk, Russia
[2] Budker Institute of Nuclear Physics SB RAS, Novosibirsk, Russia
[3] Vorozhtsov Institute of Organic Chemistry SB RAS, Novosibirsk, Russia
[4] Boreskov Institute of Catalysis SB RAS, Novosibirsk, Russia

E-mail: mkor@inp.nsk.su

Abstract. The radiation-thermal paraffin cracking process was studied aiming to model heavy hydrocarbon radiation-thermal cracking. The skim yield dependences on the irradiation parameters are given.

1. Introduction

It happened that main patterns of the radiation influence on the chemical processes were studied in the 1950-s and 1960-s using the radiation sources having low dose rate. These experiments were carried out with continuous irradiation using Co^{60} sources and electron accelerators generating continuous electron beam having energy of 1 MeV and beam power up to 1 kW [1, 2]. It was observed that reaction rates depend on the dose rate (W) as $W^{0.5}$ thus determining the decrease in irradiation efficiency with the dose rate increase. So in the 1960-s it was postulated that radiation power usage efficiency in the radiation-chemical processes decreases with the dose rate increase.

In the 1970-s the great progress was achieved in the accelerator development and the dose rates considerably increased – the powerful industrial electron accelerators reached the energy up to 5 MeV and beam power up to 50-100 kW (compare with 1 MeV and 1 kW in the 60-s), maximum parameters now are 10 MeV and 700 kW.

The pulse radio frequency (RF) electron accelerators type ILU developed in the Budker Institute of Nuclear Physics can generate the beam with average power up to 20-50 kW.

The electron beam pulses in all RF accelerators have the thin structure – the beam pulse is formed by the current peaks, the frequency spectrum consists of the accelerating radio frequency (116-118 MHz for ILU machines) and its higher harmonics. These beam pulse RF components are decaying with the distance from the beam window due to electron beam scattering. So the products subjected to electron beam treatment are in point of fact irradiated by the continuous pulse without the thin RF structure. So in the absolute majority of the experiments one can neglect the thin beam pulse structure in the dose rate estimations and measurements.

The thin structure of the electron beam generated by ILU-6 accelerator with the main accelerating frequency harmonics (117 MHz) was registered at the distance of 5 cm from the beam window, but it was not easy to arrange this experiment.

The electrons in the beam are scattered in the atmosphere and in the treated products. The scattering angle is between 5-15 degrees for most of the scattering acts. Approximately half of the electron's energy is lost in portions of less than 100 eV, and the average energy spent in the scattering act for ionisation area formation is 40 eV [3] and 60 eV [2].

The more is the distance from the beam window the more homogeneous is the electron beam in distance and time due to scattering, its energy spectrum also widens and shifts to lower energy.

At the distance of more than 20 cm from the beam window of the ILU accelerators electron beam acts as a time continuum with pulse duration of 0.5 ms and the thin RF structure is not manifested. Even if the treated product is places immediately under the beam window the beam's thin RF structure erodes after passing the thin surface layer (not more than 10-50 mcm for the material density of 1 g/cm^3).

The conventional classification of radiation-induced hydrocarbons conversion is based on the temperature ranges typical for certain characteristic reactions:

- radiolysis with low G-values (radiative yield) at temperatures below 350°C;
- radiation-thermal cracking in the temperature range of 350-450°C;
- thermal cracking prevails at temperatures above 450°C [4-6].

The radiation-thermal cracking temperature range (350-450°C) is characterized by thermal initiation and radiation-induced hydrocarbons cracking chain reactions with prevailing contribution of radiation into the reaction rate.

It is accepted that thermal cracking prevails at temperatures above 450°C due to short lifetime of the excited radical states generated by ionizing radiation and low dose rates.

In recent years some new results were published [7] correcting the established theories and questioning short lifetime of the radiation-induced excited states. The heavy oils electron beam treatment by powerful accelerators gave the experimental basis for the corrections as it caused the sufficient increase in fractional yield with boiling temperature below 360°C comparing with previously observed results [8].

Basing on the data in [7] one can consider that the radiation-thermal cracking (RTC) rate is proportional to $W^{0,5} + W^{1,5}$ (where W is the dose rate). It means that at the same absorbed dose the product yield will first decrease and then increase with the increase in dose rate. The second component according to the authors [9] is caused by the molecules in the special excited state.

Summarizing all we can assume that the dose rate increase can result in higher efficiency of cracking process – not in its diminishing as it was considered in old works.

The aim of the work was to study the cracking products yield on the irradiation dose rate for the hydrocarbons in the thermal-radiation process. The paraffin was chosen as the model compound to simplify the work and the results interpretation.

2. Preliminary estimations and works

The radiation-thermal cracking process rates were calculated for the paraffin at various temperatures and some real dose rates. The experimental studies were done using powerful pulse electron accelerator ILU-6. Its standard beam pulse duration is 0.5 ms, usual working energy is 2.4-2.5 MeV, and the dose rates can be varied by controlling the pulse beam current from 80 to 420 mA and pulse repetition rate from 25 Hz (40 ms between the pulses) to 2 Hz (0.5 s between the pulses). The beam pulse duration of 0.5 ms is greater than the excited states lifetime, but nevertheless the pulse dose rate can influence on the cracking process when the average dose rate remains the same. So we decided to study the thermal-radiation cracking varying all parameters including pulse beam current variation with the matching pulse repetition rate variation to maintain the stable average beam current.

The dose rates were independently measured using the film dosimetry. The films were placed under the ILU-6 beam widow, the energy was set 2.4 MeV, pulse beam current was set 320 mA and number of pulses was set 500. The dose per pulse observed to be equal 0.15 kGy.

The pulse repetition rate of 25 Hz in this regime results in dose rate of 25 * 0.15 = 3.75 kGy/s, the pulse beam current increase up to 420 mA gives dose rate of 4.375 kGy/s. This regime corresponds the beam power of 12.5 kW. The ILU-10 accelerator can generate the beam with power of up to 50 kW, so the dose rate of 16 kGy/s is the achievable value.

The paraffin cracking process rates were calculated for dose rates of 0.3 kGy/s ($1.88 * 10^{18}$ eV/cm^2/s), 2.92 kGy/s ($1.82 * 10^{19}$ eV/cm^2/s), 16.1 kGy/s ($1 * 10^{20}$ eV/cm^2/s) and for thermal process without irradiation.

The Figure 1 presents the calculated cracking rates dependences on the temperature for the dose rates mentioned above for $C_{15}H_{32}$ assuming that the heavy hydrocarbon molecules are dissociated into 2 light molecules.

The calculated curves show that the intensive electron beam decreases the process temperature in 150-200° at the same production rate value.

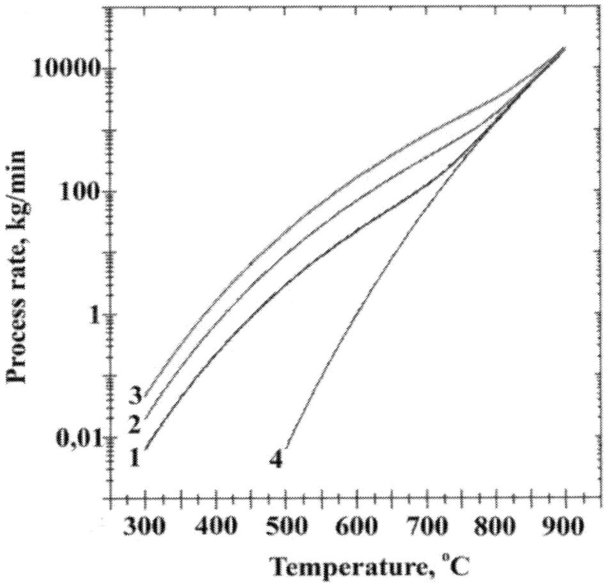

Figure 1. Calculated paraffin cracking rates at various dose rates

1- 0.3 kGy/s
2- 2.92 kGy/s
3- 16.1 kGy/s
4- Thermal cracking without radiation

3. Experimental studies

The paraffin produced by oil company Lukoil was chosen as a model hydrocarbon compound. The pulse RF electron accelerator ILU-6 with working electron energy or 2.4-2.5 MeV was used as the ionizing radiation source. The beam pulse duration was 0.5 ms, pulse repetition rate varied from 2 to 25 Hz, pulse beam power can be varied from 80 to 420 mA, the maximum beam power was 12.5 kW.

The paraffin was subjected to radiation-thermal cracking in the glass installation placed under the accelerator's beam window. The installation was supplied with backflow condenser preventing the removal of the cracking products with boiling temperature below 200°C from the reaction medium.

The reaction zone was heated by the electric heater via the sand bath. The chromel-alumel thermocouples were placed inside the installation in the reaction zone, in the backflow condenser and in the heater (in the sand bath) to measure the temperature in these points. The reaction zone temperature was stabilized using the heater control module.

The radiation-thermal cracking product yield was measured by run-down drum weight increase. The losses were determined as the difference between the initial paraffin weight and the sum of the distillate and leavings weights.

The resulting distillate fractions were assayed by gas chromatography-mass-spectrometry. The Hewlett Packard G1800A chromato-mass-spectrometer with nonpolar capillar chromatographic column HP5-MS having length of 30 m was used. Samples recording conditions were: probe volume – 1 μl, gas-carrier (helium) flow – 30 ml/min, column temperature during experiment raised from 50 to 300°C.

4. Results

The yields of the paraffin radiation-thermal cracking products were measured in the dose rate range 0.3-2.92 kGy/s (0.188-1.82*10^{19} eV/(cm^3*s). The experiments were organized so that only the fractions with boiling temperature below 200°C can come into run-down drum, heavier products returned into reaction zone.

The cracking products yields were measured at:

- pulse beam current variation, set dose of 710 kGy and dose rate of 0,3 kGy/s;
- dose rate variation, set dose of 750 kGy and pulse beam current of 320 mA.

In second experimental set the dose rates were controlled by pulse repetition frequency variation, the process time was set in inverse proportion to maintain the same total dose.

The calculated curves show that the intensive electron beam decreases the process temperature in 150-200° at the same production rate value.

IOP Conf. Series: Materials Science and Engineering **81** (2015) 012123 doi:10.1088/1757-899X/81/1/012123

The Figure 2 presents the calculated cracking rates dependences on the temperature for the dose rates mentioned above using the patterns suggested in [7] and some experimental results. We assumed that the heavy hydrocarbon molecules -- in our case paraffin $C_{15}H_{32}$ – are dissociated into 2 light molecules, and the secondary processes were not taken into account.

The paraffin cracking process rates were calculated for various dose rates, curve 1 corresponds dose rate of 0.3 kGy/s (1.88 * 10^{18} eV/cm^2/s), curve 1 corresponds dose rate of 2.92 kGy/s (1.82 * 10^{19} eV/cm^2/s), curve 3 corresponds dose rate of 16.1 kGy/s (1 * 10^{20} eV/cm^2/s) and curve 4 corresponds thermal process without irradiation. These curves witness that the powerful radiation source decreases in 150-200°C the process temperature for required cracking rate.

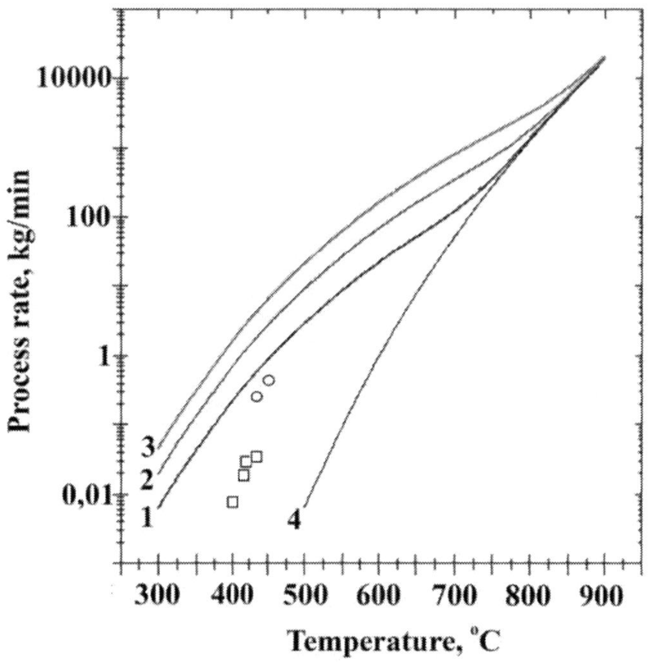

Figure 2. Calculated paraffin cracking rates at various dose rates and experimental data

1 - 0.3 kGy/s,

2 - 2.92 kGy/s

3 - 16.1 kGy/s

4 - Thermal cracking without radiation

Experimental data
□ – 0.3 kGy/s
○ – 2.92 kGy/s

The additional points on the chart show the experimental data get during our work in the temperature range 390-450°C and atmospheric pressure, the total dose range was 710-750 kGy.

The lower group of points marked □ was get at dose rate of 0.3 kGy/s, it corresponds the curve 1.

The upper group of points marked ○ was get at dose rate of 2.92 kGy/s, it corresponds the curve 2.

The dose rate was controlled by changing the pulse repetition frequency with the same pulse beam current.

The experimental process rate is lower than the calculated value – for the curve 1 in 50 times and for the curve 2 in 10 times, although the behavior is similar in both cases. The reasons of such great differences can be the small reactor dimensions, processes of radicals recombination and also the secondary processes.

The cracking product in the run-down drum were observed not in all experiments in the set at temperature of 350°C, dose 750 kGy and dose rate of 0.3 kGy/s. Nevertheless the reaction mixture (initial paraffin) weight loss was always observed. The dose rate increase to 2.92 kGy/s gave the same results.

The remaining paraffin assaying had shown the sufficient widening of the chain-length distribution thus witnessing the paraffin cracking due to radiation-thermal process.

The increase in process temperature gave more stable and reproducible results in distillate fraction mass, losses and conversion level. The temperature range of 410-415°C occurred to be optimal concerning the distillate fraction yield with relatively low losses. In this temperature range the boiling of the paraffin and reaction mixture remainder (after thermal-radiation treatment) is not intensive and at the same time the cracking products are rather efficiently removed from the reaction zone.

Figure 3 represents the chromatograms of the initial paraffin and the distilled fraction (the radiation-thermal cracking products). Initial paraffin was the mixture of n-alkanes starting from eicosane and finishing with tetracontane (C_{20}-C_{40}). The maximum is situated in the region of pentacosane (C_{25}). The 2-metylalkanes content is about 5% and the 3-metylalkanes content is about 4%. The alkenes were not found.

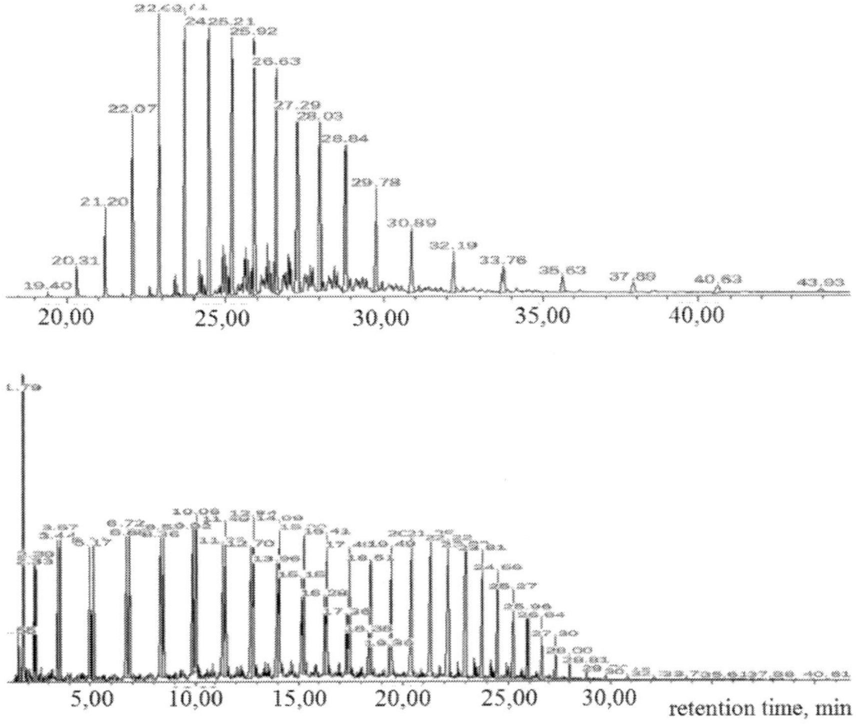

Figure 3. The chromatograms of the initial paraffin (above) and the distillate (below)

The distillate had given quite different picture – it shows n-alkanes from heptane to nonatriacontane (C_7-C_{39}) and the 1-alkenes from butane to tricosene (C_4-C_{23}) with maximums in the regions of n-heptane and n-tetradecane (C_7-C_{39}) and near 1-tridecene (C_{11}). The main components were n-alkanes – 60-62%, and 1-alkenes – 25-31%. The small quantities (3.2-6.7%) of the cyclic compounds and 1-alkenes isomers were found. There were no noticeable differences between the probes treated with various beam pulse current values.

The accelerator ILU-6 permitted to vary the pulse beam current from 80 to 420 mA at electron energy of 2.4-2.5 MeV. We studied the radiation-thermal cracking process at dose rate of 0.3 kGy/s and variable pulse beam current within these limits. The pulse repetition frequency was varied to maintain the constant dose rate. The distillate yield dependence on the pulse beam current value was not observed – the data spread was within the measurement error. The distillate fraction yield was 45-50% of the initial paraffin, the distillate contents were practically identical – although according the generally accepted theory the distillate fraction yield ought to decrease in 2 times due to process rate decrease.

Figure 4 shows the light fractions (distillate and losses) yield versus dose rate for radiation-thermal cracking at temperature of 410°C.

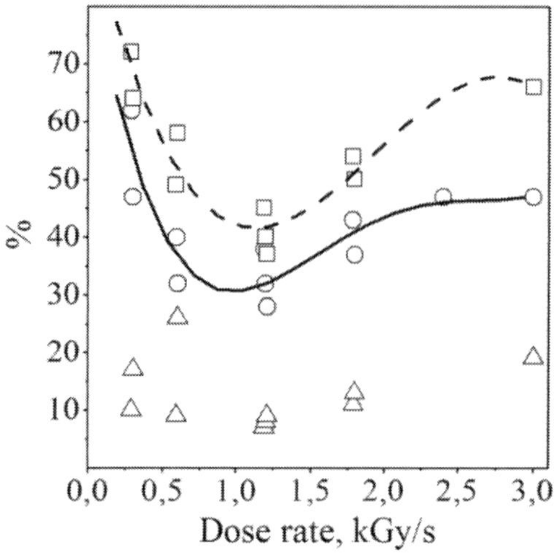

Figure 4. Light fraction yield and losses versus dose rate

○ — distillate yield
Δ — losses,
□ — sum losses + distillate yield

The total process doses were in the range of 710-750 kGy – practically equal in this experimental set (considering measurement errors and possible inaccuracies), and the process duration times were respectively varied. The distillate content noticeably differed on the dose rate. Surely this effect at higher dose rates could be caused by additional irradiation of the cracking products due to their higher concentration. The removal of the cracking products is more difficult for higher production rate.

Despite all the mentioned above drawbacks and difficulties in experimental results interpretations it is clearly seen that the decrease in light fraction yield and paraffin conversion level versus the dose rate increase was not observed in our experiments.

5. Conclusions

The high speed of the paraffin radiation-thermal cracking process was observed at temperature of 350°C under intensive electron beam. This temperature is less than the threshold of the thermal cracking process.

The effect of the cracking products yield decrease with dose rate increase was not observed in the dose rates up to 2.92 kGy/s.

Despite all the mentioned above drawbacks and difficulties in experimental results interpretations it is clearly seen that the decrease in light fraction yield and paraffin conversion level versus the dose rate increase was not observed in our experiments.

References

[1] Topchiev A V and Polak L S (Eds.) 1962 Hydrocarbon Radiolysis (Moscow: Publ. Acad. Sci. USSR) p 208

[2] Saraeva V V 1986 Radioliz uglevodorodov v zhidkoy faze. Sovremennoe sostoyaniye voprosa [in Russian] (Moscow: Moscow State University Publ.) pp. 8-9

[3] Pikaev A K 1985 Sovremennaya radiatsionnaya himiya. Osnovnye polozheniya. Eksperimentalnaya tehnika i metody [in Russian] (Moscow: Nauka) pp. 86-87

[4] Brodskii A M, Zvonov N V, Lavrovskii K P and Titov V B 1962 *Petroleum Chemistry USSR* **1** Issue 2 pp 256–271

[5] Lavrovsky K P 1976 Catalytic, Thermal, and Radiation-Chemical Conversion in Hydrocarbons (Moscow: Nauka) pp 312–373, 255–263.

[6] Lavrovsky K P [Ed. Nametkin N S, Compilers Kalinenko R A, Kolbanovsky Yu A, Lavrovskaya G K, et al.] 1978 Izbrannye trudy: Kataliticheskie, termicheskie i radiatsionno-himicheskie prevrascheniya uglevodorodov [in Russian] (Moscow: Nauka)

[7] Zaikin Yu A 2008 *Radiat. Phys. Chem.* **77**, Issue 9, pp 1069-1073

[8] Zaykina R F, Zaykin Yu A, Mamonova T B and Nadirov N K 2001 *Radiat. Phys. Chem.* **60**, Issue 3 pp 211-221

[9] Zaykin Yu A, Zaykina R F and Mirkin G 2003 *Radiat. Phys. Chem.* **67** Issues 3-4 pp 305-309

IOP Conf. Series: Materials Science and Engineering **81** (2015) 012124 doi:10.1088/1757-899X/81/1/012124

Radiation-Thermal Treatment in Ore Dressing

M V Korobeinikov[1], A A Bryazgin[1], V V Bezuglov[1], E A Shtarklev[1], A Yu Vlasov[1], S A Kondratyev[2] and V I Rostovtsev[2]

[1] Budker Institute of Nuclear Physics SB RAS, Novosibirsk, Russia
[2] Vorozhtsov Institute of Mining SB RAS, Novosibirsk, Russia

E-mail: mkor@inp.nsk.su

Abstract. The radiation-thermal electron beam treatment of ore concentrates and its influence on the useful fractions yield are described.

1. Introduction

During the past 25 years the mankind did not discover a single new considerable metalliferous deposit containing iron, lead-zinc, copper-nickel, tin or complex ores. Of late years the metals are obtained from the ores having complex material constitution and characterized by low dressability, low metal content, fine dissemination and similar surface properties of the minerals and barren rocks.

The future mankind welfare will depend on the increase in efficiency and all-inclusiveness of the ores treatment.

The electric energy consumption for ore dressing and pretreatment is about 10% of the total energy generated in the world. The Russian mining industry share in the gross domestic product is about 30%.

These are the basic facts for estimation of the potential benefits if the new more efficient technologies in ore dressing will be developed. The prominent Russian scientist I.N. Plaksin considered that one of the main directions in the ores treatment is the search for such energy depositions that will sufficiently increase the complexity and completeness of the useful components recovery. He was one of the pioneers who substantiated external actions efficiency in the ore dressing processes [1].

The studies [2-7] shows that preliminary (before grinding) electron beam treatment of the ores or electron beam treatment of the concentrates is an efficient instrument for mechanical and physical-chemical properties modification.

The electron beam treatment activates the surface and volumetric physical-chemical processes in minerals and ores resulting in their weakening and floatation properties changing. This treatment can also augment the magnetic properties of ferriferous sulphides [5].

The processes caused by electron beam greatly depend on the surface current density and total absorbed dose. At low current densities and low dose (less than 1 kGy) the main effect is radiation defects formation without noticeable changes in physical properties of the treated matters.

Absorbed dose range 1-10 kGy is characterized by electric charge accumulation and following electric discharges. The electric charge is not accumulated in conductive minerals (pyrite, galenite, etc.). In the non-conductors (quartz, sphalerite, etc.) the electric charge is accumulated until the electric strength threshold will be reached, and then the electric discharge happens forming the breakdown paths. After discharge the process repeats, and the breakdown paths can grow. The charge pulsations and discharges are followed by mechanical tension pulsations. The tension pulsations and breakdown paths cause the microcracks formation and growing causing minerals weakening [6]. The pulse electron beam is more efficient for these processes than the continuous electron beam.

The ore particles usually consist of the heterogeneous materials, and the breakdown paths and microcracks are usually forming and developing in the crystallite coalescence regions. The result is the following selective disintegration and increase in grinding rate process.

The high absorbed doses (of more than 100 kGy) get by intensive electron beam without forced cooling simultaneously heat the treated matters during the irradiation. The combined high temperature and irradiation action (radiation-thermal process) causes the physical-chemical processes to go differently than in the purely thermal process.

We have studied the radiation-thermal processes in some ferriferous sulphide minerals and their influence on the following concentrations process.

2. Magnetic separation and radiation-thermal treatment

The magnetic separation is widely used for extraction of the magnetic and weak-magnetic minerals from the ferriferous and other ores. It bases on the differences in magnetic properties of the different grinded ore components. This technology is easy in realization, highly efficient and ecologically friendly because it does not pollute the environment.

But the differences in magnetic properties of the minerals are usually not enough large for their efficient separation, especially for weak-magnetic iron ores. To improve the ore concentration process it is necessary to increase the magnetic properties differences by physical, physical-chemical and even by chemical effects.

A calcining at high temperature and certain conditions (oxidizing, neutral or reducing medium) converts the weak-magnetic ferriferous minerals into magnetic compounds (magnetite Fe_3O_4, hematite Fe_2O_3 or γ-hematite Fe_2O_3). This is the main industrial process used for modification of the ores' magnetic properties. Process temperature is 600°C and higher, process duration – 4-6 hours or more. Main problems of the process – high power consumption, long duration, high cost and air pollution.

The phase transitions induced by electron beam treatment are well known for decades. Our studies have shown that the treatment by intensive electron beam can modify the magnetic properties of the weak-magnetic minerals [8]. It is important for ore dressing intensification and minerals separation.

We have experimentally studied the radiation-thermal modification of some ferriferous sulfide minerals. The treatment carried out in the Budker Institute of Nuclear Physics, Novosibirsk, by pulse electron accelerator ILU-6, electron energy was 2.4 MeV. The pulse beam current was usually set to 320 mA, pulse duration was 0.5 ms, pulse repetition frequency was varied to control the average beam current and respectively the dose rate. The maximum beam power of this ILU-6 machine is 12.5 kW, so the beam power is enough to heat the ore samples to the required temperatures up to 600°C and more. (The maximum radiation-thermal process temperature reached on this installation was 1500°C).

The ore samples temperature can be monitored by the chromel-alumel thermocouples in the chosen points – 4 channels are envisaged. The process temperature displayed in the special window of the accelerator control program. The special accelerator program unit permits to control and stabilize the temperature of the treated samples according the preset temperature chart.

Various ferriferous crushed ores were placed under the accelerator beam window and treated in the atmosphere, see Figure 1.

Figure 1. Ore samples under ILU-6 accelerator beam window

The minerals' magnetic properties were assayed by specific magnetic moment measured by vibrating-coil magnetometer LDJ9600 (LDJ Electronic Inc., USA).

The phase transitions after radiation-thermal treatment were studied by X-ray diffraction analysis.

The specific magnetic moment values observed after treatment depended on process temperature and mineral fineness.

3. Experimental studies

The pyrite (FeS_2), arsenopyrite (FeAsS), chalcopyrite ($CuFeS_2$) and other crushed ores of various finenesses were subjected to radiation-thermal treatment on the ILU-6 accelerator in the temperature range of 300-600°C.

Figure 2 shows the temperature control window of the accelerator control program for one experiment – pyrite treatment with maximum temperature of 400°C and total absorbed dose about 300 kGy. The blue line indicates the preset temperature chart, the red line shows the real temperature during the experiment, the horizontal axis shows the time in minutes. The heating up to 400°C carried out in 5 minutes, then the beam was switched off and sample was cooled due to free convection (without forced cooling).

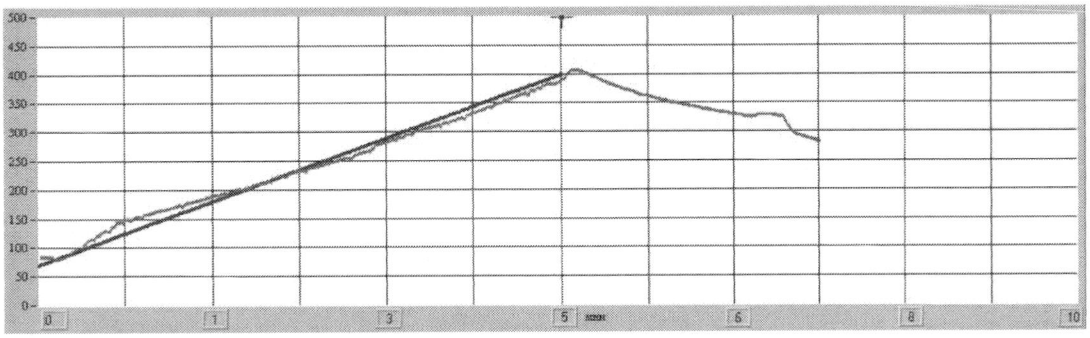

Figure 2. Pyrite radiation-thermal treatment temperature curve, time in minutes, dose 300 kGy

Blue line – preset temperature chart

Red line – pyrite temperature curve

The 5 minutes radiation-thermal treatment at temperature of 400°C greatly augmented the magnetic moments of the pyrite samples:

- Pyrite with fineness of 75-180 micrometers – specific magnetic moment have changed from $0.2*10^{-8}$ A·м^2/г to $10.5{\cdot}10^{-8}$, increase in $10.5/0.2 \approx 53$ times.
- Pyrite with fineness of 53-75 micrometers – specific magnetic moment have changed from $0.4*10^{-8}$ A·м^2/г to $18.1{\cdot}10^{-8}$, increase in $18.1/0.4 \approx 45$ times.

The difference for arsenopyrite is greater, see it below.

The radiation-thermal treatment increased the magnetic moments depending on the mineral type and fineness (size of the particles) at temperature of 400°C:

- in 53 times for pyrite with fineness of 75-180 micrometers
- in 45 times for pyrite with fineness of 53-75 micrometers
- in 291 times for arsenopyrite with fineness of 53-75 micrometers
- in 921 times for arsenopyrite with fineness of less than 53micrometers
- in 9.7 times for chalcopyrite with fineness of 75-180 micrometers.

The magnetic phases Fe_2O_3 and Fe_3O_4 in various combinations were observed after radiation-thermal treatment.

The phase Fe_2O_3 (as well as γ-Fe_2O_3) has the greater magnetic moment than Fe_3O_4 phase.

The magnetic properties of the ore samples subjected to radiation-thermal treatment depend on the magnetic phases content as well as on the samples fineness.

X-ray diffraction patterns in Figure 3 show pyrite phase structures before the treatment, after thermal treatment at temperature of 400°C and after radiation-thermal treatment at the same temperature.

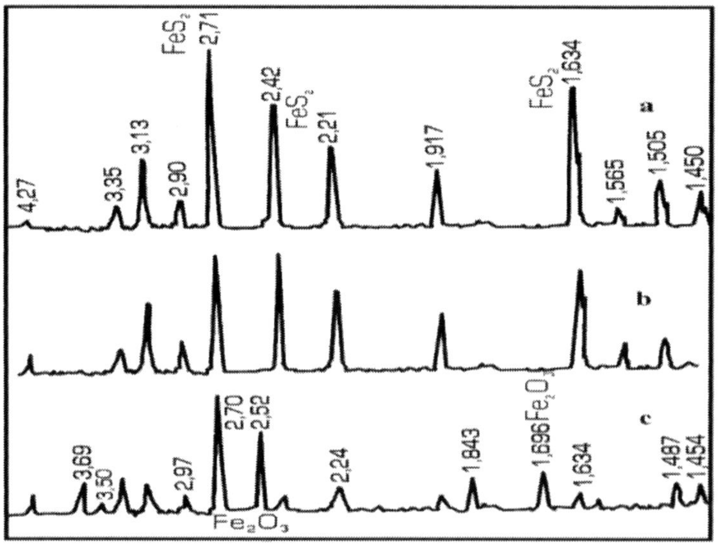

Figure 3. Pyrite (FeS_2) X-ray diffraction patterns

a – initial pyrite

b – pyrite after thermal heating up to 400°C

c – pyrite after radiation-thermal heating up to 400°C

The thermal treatment at temperature of 400°C did not result in magnetic phases formation.

The low temperature (400°C) nonmagnetic sulphides transition into magnetic compounds **was first time observed** in pyrite, chalcopyrite and arsenopyrite – it is the radiation-thermal treatment effect.

The temperature threshold for magnetic phase formation in thermal heating process is 600°C, it was confirmed by our experiments. The thermal treatment results in formation of the Fe_3O_4 phase that has the lower magnetic moment than Fe_2O_3 and γ-Fe_2O_3 phases.

The advantages of the radiation-thermal pyrite treatment – process temperature can be lower in 200°C, process time is shorter (5 minutes comparing with 4-6 hours), the magnetic moment is greater (due to Fe_2O_3 and γ-Fe_2O_3 phases formation) thus improving the magnetic separation process.

Figures 4 and 5 below show the chalcopyrite and marmatite X-ray diffraction patterns after radiation-thermal treatment at temperature of 600°C and dose of 1000 kGy.

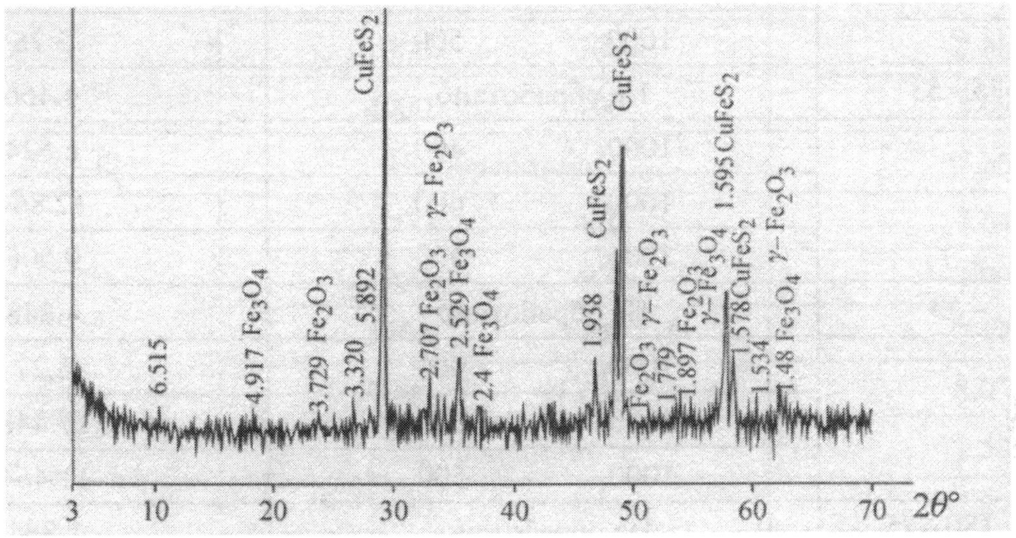

Figure 4. Chalcopyrite X-ray diffraction pattern after radiation-thermal treatment, temperature 600°C, dose 1000 kGy

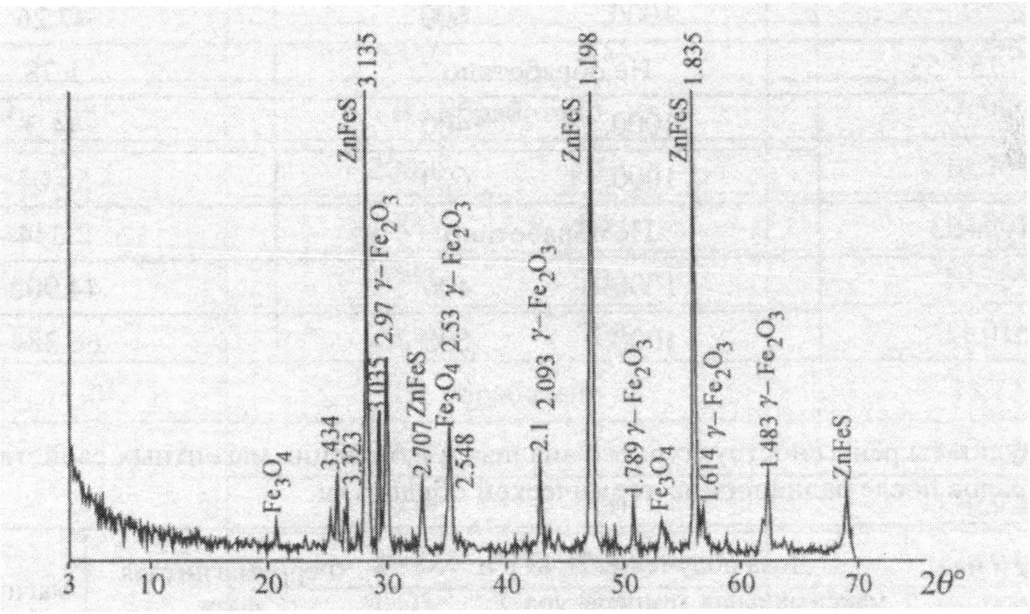

Figure 5. Marmatite X-ray diffraction pattern after radiation-thermal treatment, temperature 600°C, dose 1000 kGy

4. Conclusion

The radiation-thermal treatment gave good results – we observed low temperature (400°C) non-magnetic sulphides transition into magnetic compounds. Such low transient temperature was first time observed in pyrite, chalcopyrite and arsenopyrite. The thermal treatment at temperature of 400°C did not result in magnetic phase formation.

The radiation-thermal process completed in 5-10 minutes (for fineness of less than 180 micrometers). The traditional calcining process lasts 4-6 hours at temperature of 600°C or higher.

The magnetic moments after radiation-thermal treatment increased in 9-291 times. The phase content after radiation-thermal treatment get better comparing with the phase content after thermal treatment. Magnetic properties augmentation in ferriferous compounds owing to radiation-thermal treatment offer the challenge to develop the new more efficient and ecologically pure technologies for all-inclusive ore dressing.

Radiation-thermal treatment can increase efficiency and all-inclusiveness of the ores treatment, augment the useful products output and sufficiently decrease the energy consumption due to lower process temperature and shorter treatment time.

The radiation-thermal treatment permits to intensify and to better separate the compounds of tin ores concentrates containing arsenopyrite. About 90% of the ferriferous compounds and 70% of the arseno-compounds can be extracted into the magnetic fraction.

The radiation-thermal treatment can be built into the existing technological schemes with relatively low efforts – there is no need to change the dressing works equipment.

References

[1] Plaksin I N, Shafeev R Sh and Chanturiya V A 1970 O vliyanii ioniiruyuschih izlucheniy na flotatsionnye svoistva nekotoryh mineralov. In: Obogascheniye poleznyh iskopaemyh: selected studies [in Russian] (Moscow: Nauka)

[2] Chanturiya V A and Bunin I Zh 2007 Non-Traditional High-Energy Processes for Disintegration and Exposure of Finely Disseminated Mineral Complexes *J. of Mining Science* **43** no 3 pp 311–330

[3] Chanturiya V A and Bunin I Zh Netraditsionnye energeticheskie metody selektivnoy dezintegratsii tonkodispersnyh mineralnyh kompleksov blagorodnyh metallov (Non-Conventional Methods of Selective Disintegration of Finely Dispersed Mineral Complexes of Noble Metals). Novye tehnologii v nauke o zemle I gornom dele (New Technologies in the Earth Sciences and in Mining) [in Russian] 2012 Proceedings of the Vsesoyuznaya nauchno-prakticheskaya konferentsiya, Kabardino-Balkar State University (Nalchik, Russia)

[4] Chanturiya V A and Vigdergauz V E 1995 Nauchnye osnovy i perspektivy promyshlennogo ispolzovaniya energii uskorennyh electronov v obogatitelnyh processah (Scientific Basis and Prospects of Industrial Application of the Accelerated Electron Energy in Mineral Processing) [in Russian] *Gornyy zhurnal (Mining Journal)* **7**

[5] Bochkarev G R et al. 1997 21-26 September Prospects of electron accelerators used for realizing effective low-cost technologies of mineral processing Proceedinds of the XX International Mineral Processing Congress: [Aachen Germany Clausthal-Zellerfeld] GDMB **1**

[6] Bochkarev G R, Veigel't Yu P, Izotov A S and Rostovtsev V I 2001 Radiation Thermal Stresses in Minerals and Their Role in Magnetite Quartzite Beneficiation *J. of Mining Science* **37** no 3 pp 323-329

[7] Rostovtsev V I Technological and Economic Effect of Nonmechanical Energy use in Rebellious Mineral Processing 2013 *J. of Mining Science* **49** no 4

[8] Wang H, Bochkarev G R, Rostovtsev V I, Veigel't Yu P and Lu S.S 2004 Improvement of the magnetic properties of iron-bearing minerals during radiation-thermal treatment *J. of Mining Science* **40** no 4 pp 399-408

IOP Conf. Series: Materials Science and Engineering **81** (2015) 012125 doi:10.1088/1757-899X/81/1/012125

Study lasing properties of Rhodamine 4c dye in the pores of anodized aluminum

N Kh Ibrayev[1], A K Zeinidenov[1], A K Aimukhanov[1], Zh Zh Ahatova[1]
[1] Institute of Molecular Nanophotonics, E.A. Buketov Karaganda State University, *Karaganda, 100028, Kazakhstan*

E-mail: niazibrayev@mail.ru

Abstract. In this paper we study the spectral and luminescent properties of the Rhodamine 4C in porous anodic alumina matrix. It was the excitation of anodic aluminum oxide films doped with Rhodamine 4C molecules second harmonic of a Nd: YAG laser with a power of 0.8 mW/cm^2 on the background of the spectrum of spontaneous fluorescence appears band of stimulated emission with a peak of wavelength in 620 nm.

1. Introduction

Currently, active research in obtaining of lasing nanostructured materials doped with the laser dye [1-4] are performed. Dye molecules are deposited on the surface of the nanoparticles [5,6], or embedded in the nanoporous structure [7,8]. Cylindrical microcavities in such structures can maintain high quality mode due to the effect of total internal reflection of electromagnetic waves from the cavity walls [9]. Intensity of radiation processes are enhanced due to the effective concentration of the electromagnetic field inside the microcavity [10]. In the case of filling of microcavities with the laser dye, such system can be used as active medium for tunable laser. In this aspect, one of the promising materials, which can be used as the active medium is a porous anodic aluminum oxide doped with a laser dye. Structure of anodic aluminum oxide films is a system of ordered close-packed pores. The pores are perpendicular to the film surface, and the diameter and spacing can be varied by changing of the anodization conditions [11].

2. Experimental

This paper presents the results of a research of transformation of the spontaneous fluorescence to stimulated emission of Rhodamine 4C molecules embedded in porous anodic aluminum. Anodizing of aluminum was carried out in a two-electrode electrochemical cell using a DC power source MPS-7081. Auxiliary electrode was a platinum plate and polished aluminum foil was used as the working electrode. The electrolyte was obtained on the basis of deionized water that was purified with AquaMax 360 Basic system. The resistivity of the water was 18.2 M Ohms /cm.

The aluminum oxide matrixes are separated from unreacted aluminum by selectively dissolving of the latter in solution of $CuCl_2$ in HCl. The surface morphology and cross-cleaved samples, obtained with scanning electron microscope Leo Supra 50VP are presented on figure 1 The measurements were performed at an accelerating voltage of 7 kV in the high vacuum. The pores of ~ 50 nm diameter and the distance between the pores of about 105 nm are observed on the film surface (Figure 1a). On cross-cleaved sample parallel lines channels located perpendicular to the surface are observed (Figure 1b).

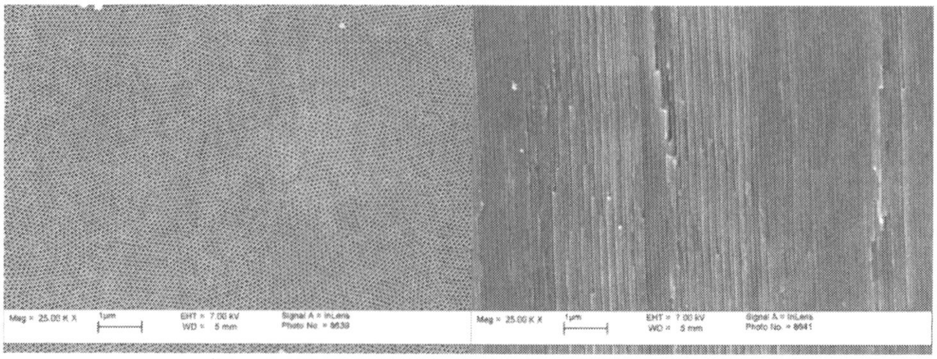

a) b)

Fugure 1. SEM images of porous anodic aluminum oxide film obtained by a two step anodizing: a) the surface of the oxide film after removal of the barrier layer, b) a cross-section.

Sorption of Rhodamine 4C molecules into the pores was carried out by keeping the porous aluminum in solution of dye with concentration C = 0.1 mMol within 3 hours, followed by drying the film in an oven at 100 °C for 1 hour. Measurement of absorption and fluorescence spectra of Rhodamine 4C in a film was carried on a spectrophotometer CM2203 (Solar).

3. Results and discussion

Figure 2 shows the absorption and fluorescence spectra of molecules of Rhodamine 4C embedded in a porous alumina. It can be seen that the absorption band of the dye in the matrix has a maximum at a wavelength of λ_{max} = 550 nm. Photoexcitation of the dye fluorescence was carried at a wavelength equal to λ = 530 nm. The fluorescence spectrum has a maximum at a wavelength of λ = 578 nm. The quantum yield of fluorescence (Φ_{fl}) Rhodamine 4C in the pores of the alumina was determined by using an absolute Reflectometric sphere AvaSphere-50-REFL (Avantes BV). The accuracy of determining the fluorescence quantum yield is not more than 5%.

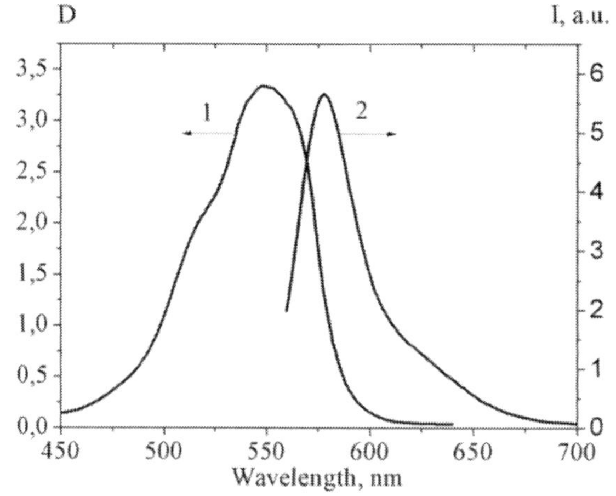

Figure 2. The absorption (1) and fluorescence (2) spectra of Rhodamine 4C in the porous of alumina.

Quantum yield of Rhodamine 4C fluorescence in pore alumina, defined by de Mello method for calculating of absolute quantum yield of mixed systems [12], was equal to Φ_{fl} = 0.3.

The measurement of the stimulated emission spectra of films was carried out on the special facility, its scheme is shown in Figure 3. The spectra were measured in the excitation of the samples by the second harmonic of Nd: YAG laser (1) (SOLAR LQ 215, λ_{gen} = 532 nm, E_{pulse} =90mJ, $\tau_{duration}$ = 10 ns) in the longitudinal version. The pumping radiation, after passing a diaphragm (3) with a lens (4) then focused on the surface of the film (5) in form of strip with area about 0.1 cm^2. Pump power density was varied using neutral density filter (2) and was 0.01-1 MW / cm^2. Stimulated emission using a converging lens (6) focused on the input fiber of spectrometer AvaSpec-2048. Stimulated emission spectra were measured on spectrometer AvaSpec-2048 (7), connected to a computer (8). The relative error in the determination of the spectra was 3%.

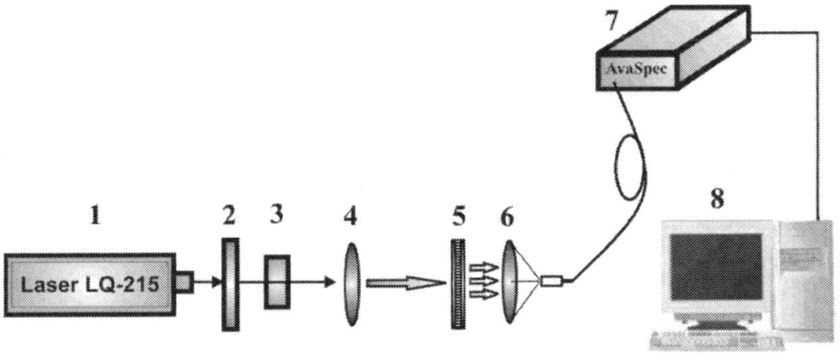

1– laser LQ 215; 2– neutral density filters; 3– diaphragm; 4, 6– converging lens; 5– test sample; 7– spectrometer AvaSpec-2048; 8– computer

Figure 3. The scheme of facility for investigation the properties of thin films stimulated emission.

Stimulated emission of the dye molecules in the anodic aluminum oxide film was obtained in the long-wavelength band of the fluorescence spectrum (Figure 4). With the power density of the pump source till 0.04 mW/cm^2 the only stationary fluorescence of dye was observed in the pores as seen in Figure 4, curve 1.

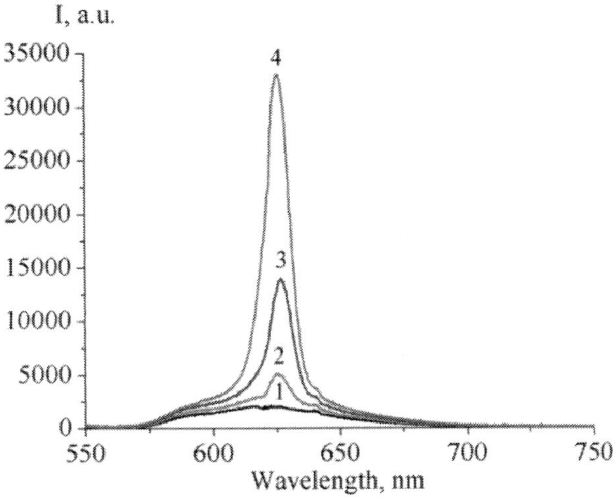

1 - P = 0,04 mW/cm^2; 2 - P = 0,2 mW/cm^2; 3 - P = 0,4 mW/cm^2; 4 - P = 0,8 mW/cm^2

Figure 4. The emission spectra of Rhodamine 4C molecules in anodic alumina at different pump power density.

When the power source of the pump reached 0.2 MW/cm^2 a narrow band of stimulated emission with a wavelength peak 625 nm appeared on the background of the fluorescence spectrum (curve 2). A further increase in the power density of the pump source increased the intensity and narrowing of the stimulated emission band (Figure 4, curves 3-4). When the density of the pump power was 0.8 mW/cm^2 the component of stimulated emission prevailed over spontaneous emission component (curve 4), however, completely eliminate the spontaneous emission component was not possible. The appearance of stimulated emission when using low quality resonator containing two glass plates due to the fact that the walls of cylindrical pores of the matrix acts as microresonators in addition to the Fresnel reflection from two glass plates in studied system.

Due to the effect of total internal reflection of electromagnetic waves from the walls of the pores the high energy density of the electromagnetic field concentrated in the microcavity, that greatly strengthens the emission intensity of the dye molecules in the pores.

Dependence of the half-width of the emission band ($\Delta\lambda_{1/2}$) and the luminescence intensity of the film versus the pump power density of stimulated emission spectra were obtained (Figure 5 a, b).

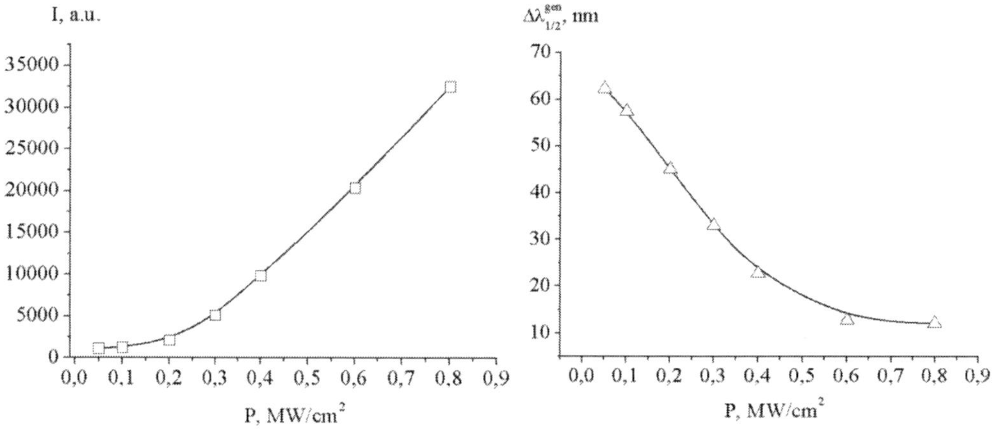

a) b)

Figure 5. Dependence of the stimulated emission intensity (a) and its half-width (b) via the pump power density.

When changing the pump power density from 0.1 to 0.8 MW/cm^2 the intensity at the maximum of emission spectrum is increased by almost of 35 times, and $\Delta\lambda_{1/2}$ decreases of 6.5 times. The narrowing of the emission band with increasing of excitation intensity indicates about the prevalence of stimulated emission over the spontaneous emission, and reveals that the system pass into generation mode. Further increase in the power density of the pumping source leads to a narrowing of the band of generation and increase of its intensity. Threshold of stimulated emission was determined on a tangent in Figure 5 (a) and equal to 0.2 MW/cm^2.

These results demonstrate the potential use of the porous alumina for developing of active elements of quantum electronics.

References:

[1] Bol'shchikov F A, Garibin E A and Gusev P E 2011 *Quantum electron* **41** 193 – 197.
[2] Shin H W, Cho S Y and Choi K H 2006 *Applied physics letters* **88** 263112.
[3] Costel A, Garcia-Moreno I and del Agua D 2007 *Journ. of Applied Physics* **101** 731 – 742.

[4] Maltseva E O and Plekhanov A I 2011 *NSU Bulletin* **6** 16 – 19.
[5] Alimov O K, Basiev T T, Orlovskii Yu V, Osiko V V and Samoilovich M I 2008 *Quantum electron* **38** 665 – 669.
[6] Donchenko V A, Geints Y E, Kharenkov V A and Zemlyanov A A 2013 *Optics and Photonics Journal* **3** 13 – 17.
[7] Deshpande A V and Kumar U 2009 *Journal of Non-Crystalline Solids* **355** 501 – 506.
[8] Beger V N 1998 *Technical Physics letters* **24** 369 – 370.
[9] Braginskii V B, Il'chenko V S and Gorodetskii M L 1990 *Physics-Uspekhi* **160** 157 – 159.
[10] Sherstnev V V, Krier A, Monakhov A M and Hil G 2003 *Electron. Lett.* **39** 916 – 917.
[11] Eliseyev A A, Lujashin A V and Tretyakov Yu D 2010 Functional nanomaterials, FIZMATLIT Moscow p. 456.
[12] De Mello J C, Wittmann H F and Friend R H 1997 *Advanced Materials* **9** 230 – 232.

IOP Conf. Series: Materials Science and Engineering **81** (2015) 012126 doi:10.1088/1757-899X/81/1/012126

Breaks of dose dependence of transient creep as result of competing influence of defects' fluxes on climb of dislocations

P Selyshchev
Department of Physics, University of Pretoria, South Africa

E-mail: *pavel.selyshchev@up.ac.za*

Abstract. In the framework of climb-glide model a theoretical approach is developed to describe transient creep under irradiation. It is obtained the explicit expression for creep rate which describes experimentally observed breaks of dose dependence of creep. It is shown that the breaks arise as result of competition of radiation and thermal fluxes of defects to dislocation. When interstitial and vacancy fluxes become equal, the dislocation cannot overcome the obstacle via climbing and cannot continue glide. Climb-glide mechanism does not contribute to the creep. The creep rate drops. Numbers of breaks depend on initial state of material and conditions of irradiation. Dose (time) of break appearance are obtained.

1. Introduction

Under loading and irradiation the resource of structural materials is largely determined by creep. New mechanisms of creep are developed under irradiation and lead to noticeable increase in creep. But it is more important there is a new behavior of creep under irradiation. Therefore often one distinguishes thermal and radiation creep [1,2].

If instantaneous deformation of the material is neglected, we can distinguish three stages of creep: transient (creep rate changes with time), stationary (creep rate is constant) and destroyed (accelerated growth of creep until destruction of material). The first two stages are the most important for problems of radiation materials science because to extend resource material we are interested in material before its destruction. In what follows we consider only the first two stages.

At high temperatures and low stresses the most part of deformation is accumulated during stationary stage of creep. When temperature decreases, the stationary stage shrinks. At not very high temperatures which correspond to operating temperatures of some nuclear power plants, the most part of deformation is accumulated during transient stage of creep [3]. Saturation of creep can be reached at large enough doses (more than hundreds dpa). In some cases self-oscillations of creep rate are observed [4,5] or the rate of creep can tend to different stationary values [6]. It means there are several different stationary regimes of creep. Sometimes the creep has only transient stage. Beside it is experimentally obtained the dose dependence of creep has complex behavior with several breaks [1,4,8-13].

In operating conditions of nuclear power plants the major contribution to the creep of irradiated materials gives "glide-climb" mechanism which is connected with dislocation motion. The motion of the dislocation combines glide and climb. The glide of the dislocation can be interrupted and blocked by an obstacle, such as clusters of impurity atoms, complexes, voids, loops, etc. In this case the dislocation can climb to a glide plane that does not intersect the obstacle and continue its glide in this plane.

Without irradiation climb of dislocations is provided by thermal vacancies. Under irradiation there is heating of the irradiated material due to relaxation of various radiation-induced excitations, for example excitations of electron and phonon subsystems. Consequently, the concentration of the thermally activated defects increases, and the creep rate increases.
However hallmark creep under irradiation is that the density of mobile radiation defects can be sufficiently high to provide climb of dislocations at virtually any temperature because vacancies and interstitials atoms are continuously generated by irradiation. Firstly, the irradiation generates along with the vacancies about the same amount of interstitial atoms. Interstitial atoms are considerably more mobile than vacancies. The activation energy of the interstitial atom is about 0.1 eV, which is several times smaller than the vacancy. Secondly, generation rate of the radiation defects do not depend on temperature. Thermal generation of vacancies increases exponentially with temperature and becomes noticeable at temperatures greater than about half the melting temperature. Interstitial atoms are not activated thermally. Third, the decreasing of the diffusion with decreasing temperature is compensated by increasing the density of defects, thus their fluxes on dislocation varies slightly. All these peculiarities influence on dynamics of point defects accumulation.
In this connection it is of interest to investigate peculiarities of the transient creep, which is connected with density of point defects and, therefore, is a function of irradiation conditions and the properties of the irradiated sample.
Here we use only one parameter of irradiation, namely defect generation rate, and don't investigate dependence of defect generation rate on the other parameters of irradiation because it is the defect generation rate that is determined creep in framework of climb-glide mechanism. But we take into account that the defect fluxes to dislocations, and so rate of defect accumulation are functions of the temperature. Thus the creep in framework of the climb-glide model can be described by only defect generation rate, temperature and parameters of the irradiated materials.

2. Physical model and principal equations

In this paper, we develop a theoretical approach [14], which was proposed to describe the stationary creep under irradiation by quantitative way.
Under irradiation due to radiation-induced generation of a large number of interstitial atoms a dislocation can climb an obstacle by absorbing both vacancies and interstitial atoms [1,2]. When there is excess of absorption of vacancies, the dislocation overcomes the obstacle by virtue in dissolution of the extra-plane and vacancy regime of creep is realized. When there is excess of absorption of interstitial atoms, the dislocation overcomes the obstacle by virtue in expanding of extra-plane. The interstitial regime of creep takes place. To continue glide it does not matter in which way the dislocation overcomes obstacle: due to preferential absorption of interstitial atoms or vacancies. The rate of the creep must be positive any way. Thus the rate of creep is proportional to the modulus of the difference of absorption of vacancies and interstitials.
Suppose the material contains no obstacles non-radiation origin. In this case propagation and immobilization of dislocations are precluded. Thus the dislocation density can be considered as constant, and the creep rate is given by [14]

$$\dot{\varepsilon} = \nu|\alpha_i C_i - \alpha_v C_v| \tag{1}$$

Here C_v and C_i are the concentrations of interstitial atoms and vacancies, $\alpha_i = z_d \rho_d D_i$, $\alpha_v = \rho_d D_v$, ρ_d is density of dislocations. z_d is their preference coefficients. $D_i = D_i^0 \exp(-E_m^i/kT)$ and $D_v = D_v^0 \exp(-E_m^v/kT)$ are the diffusion coefficients for interstitial atoms and vacancies respectively. E^i_m and E^v_m are the migration energies. T is the temperature of irradiated crystal, k is Boltzmann constant. $\nu=L/l$. Let us consider case of the constant stress when the height and concentration of the obstacles are constant parameters of model and dependence of equilibrium vacancy concentration on stress is negligible.
The system of equations for vacancies and interstitial atoms has typical form

$$\frac{dC_i}{dt} = K - \beta_i C_i - \gamma C_i C_v \tag{2}$$

$$\frac{dC_v}{dt} = K - \beta_v \left(C_v - C_v^e\right) - \gamma C_i C_v \tag{3}$$

where K is the point defect production rate. It is the main parameter of the irradiation for considered mechanisms. $\beta_i = (z_d \rho_d + z_k \rho_k) D_i$ and $\beta_v = (\rho_d + \rho_k) D_v$ are inverse life-times of vacancies and interstitial atoms which are determined by all (dislocation and non dislocation) sinks for defects, ρ_k is density of all sinks excluding sinks providing creep, z_k is their preference coefficients,$\gamma = \mu D_i$ is recombination coefficients, μ is constant. $C_v^e = C_v^{eo} \exp\left(-E_f^v / kT\right)$ is the thermodynamic equilibrium concentration of vacancies. E^v_f is energy of vacancy formation. C^{e0}_v and μ are material constants. Thermodynamic equilibrium concentration of interstitials is small and can be neglected.
C_i^0 and C_v^0 are initial concentrations of interstitial atoms and vacancies. Typically, the initial concentrations of defects are equal to their thermodynamic equilibrium concentrations. However, in a special way, such as pre-irradiation, the initial concentration of defects can be created different from the equilibrium. The thermal equilibrium concentration of vacancies and diffusion coefficients are exponential functions of temperature. In view of radiation heating the equation for the variation of crystal temperature has the form

$$c\rho \frac{dT}{dt} = Q + \theta_i \beta_i C_i + \theta_v \beta_v \left(C_v - C_v^e\right) + \theta_{iv} \gamma C_i C_v - h(T - T_e) \tag{4}$$

Here Q is rate of radiation heating, $c\rho$ is heat capacity of the sample per volume. Parameter h describes heat exchange between irradiated sample and environment. The environment temperature is constant and equal to T_e. Heat exchange depends on its nature and geometry of the sample. Without irradiation the temperature of the sample is equal to the temperature of environment. θ_i, θ_v and θ_{iv} are formation energies of one interstitial atom, one vacancy and the Frenkel pairs respectively. The rate of radiation heating is proportional to defect generation rate: $Q = \xi \theta_{iv} K$. Parameter ξ shows in how many times more irradiation energy goes into heating of the sample than for defect generation.

3. Results and discussion

Let consider change of creep rate from beginning of irradiation till stationary stage in qualitatively way.
During irradiation the concentrations of defects go to their stationary values. Creep tends to stationary stage, and the creep rate tends to a constant value. The stationary concentration of interstitial atoms is reached much faster because the characteristic lifetime of interstitial atoms is noticeably shorter than the lifetime of the vacancies.
In beginning of irradiation the concentrations of defects are equal to thermal equilibrium values, and glide-climb mechanism is provided by thermal vacancies. Thus this is a vacancy regime of creep.
The radiation-induced vacancies and interstitial atoms appear since beginning of irradiation. The concentration of interstitial atoms is growing rapidly, concentration of vacancies - slowly. Accordingly, the flux of interstitial atoms to the dislocation grows faster and the creep rate decreases. If the generation rate of defects is large enough, at some point in time flux of interstitial atoms to the dislocation exceeds the flux of vacancies. Thus interstitial regime of creep will replace the vacancy regime. At this time, the contribution of glide-climb mechanism into the creep is absent.
After that the interstitial regime of creep is developed and creep begins to increase. Increasing in the concentration of interstitial atoms is rapidly saturated, and the concentration of vacancies continues to increase. Further behavior of transient creep depends on conditions of irradiation. For example the flux

of vacancies to dislocations can again become more than flux of interstitial atoms and then exceed it. Rate of creep again becomes to equal zero and vacancy regime of creep again takes place. As the concentration of vacancies increases monotonically, and the concentration of interstitial atoms is almost constant, in further creep regime does not change. The creep will occur in the same regime as in the stationary stage. Thus, one or two dips of creep rate can take place depending on the regime of stationary creep.

Regimes of stationary creep without irradiation heating were investigated in [15] where it was shown that at low rate of the defect generation there is a vacancy regime of creep and at large - an interstitial regime. When the rate of defect generation is equal to critical value, the stationary creep rate equals zero. The critical value depends on the temperature of the sample. To account for the effect of radiation heating we take into account (4). Since the formation energy of the Frenkel pairs is approximately equal to the sum of the energies of vacancy and interstitial atom, we put $\theta_{iv} = \theta_i + \theta_v$. Then for stationary temperature of irradiated sample we obtain $T = T_e + (Q + \theta_{iv} K)/h$ and the critical value of the defect generation rate satisfy the equation

$$K_c = \frac{\beta_v(K_c)C_v^e(K_c)}{\sum(z_d - z_k)\rho_k}\left(\rho_i + \frac{\mu z_d \rho_v C_v^e(K_c)}{\sum(z_d - z_k)\rho_k}\right) \tag{5}$$

Where $\rho_i = z_d \rho_d + z_k \rho_k$ and $\rho_v = \rho_d + \rho_k$.

To obtain an expression for rate of a transient creep as a function of irradiation condition (rate of defect generation, radiation heating, temperature of environment and so on) we have to solve system (2) – (4) and to substitute obtained values as function of parameters into equation (1).

The explicit solution of equations (2) - (4) can be obtained for low intensity irradiation when recombination and radiation heating are neglected. For this approximation we find

$$T(t) = T_e \tag{6}$$

$$C_i(t) = \frac{K}{\beta_i} - \left(\frac{K}{\beta_i} - C_i^0\right)\exp(-\beta_i t) \tag{7}$$

$$C_v(t) = \frac{K}{\beta_v} + C_v^e - \left(\frac{K}{\beta_v} + C_v^e - C_v^0\right)\exp(-\beta_v t) \tag{8}$$

To take into account recombination the equations (2) - (4) are solved numerically. The numerical solution shows that the influence of recombination leads to a no monotonic change of concentration of interstitial atoms: in the beginning it sharply increases, reaches a maximum and then decreases to a stationary value. Vacancy concentration monotonically approaches to its stationary value. Under certain conditions, the positive feedback between concentration of defects and temperature of the sample can lead to the development of self-oscillations of temperature [4,5], but here this effect is not considered.

Substitute (6) - (8) into (1) we obtain

$$\dot{\varepsilon}(t) = \nu\left|\frac{\dot{\varepsilon}_\infty}{\nu} - \alpha_i\left(\frac{K}{\beta_i} - C_i^0\right)\exp(-\beta_i t) + \alpha_v\left(\frac{K}{\beta_v} + C_v^e - C_v^0\right)\exp(-\beta_v t)\right| \tag{9}$$

Where $\dot{\varepsilon}_\infty = \nu K\left(\frac{\alpha_i}{\beta_i} - \frac{\alpha_v}{\beta_v} - \frac{\alpha_v C_v^e}{K}\right)$ is the rate of stationary creep.

When initial concentrations of defects are thermodynamic equilibrium values, the expression (9) is transformed into

$$\dot{\varepsilon}(t) = \nu\left|\frac{\dot{\varepsilon}_\infty}{\nu} - \frac{z_d\rho_d}{\rho_i}K\exp(-\beta_i t) + \frac{\rho_d}{\rho_v}K\exp(-\beta_v t)\right| \tag{10}$$

Let rate of creep reach the first zero at time t_1. We can estimate its approximate value taking into account that vacancy concentration grows slower then concentration of interstitial atoms. Therefore we can put $\exp(-\beta_v t_1) \approx 1$ in (10) and find

$$t_1 \approx \frac{1}{\beta_i}\ln\left(\frac{\nu z_d\rho_v\rho_d K}{\rho_i(\rho_v\dot{\varepsilon}_\infty + \nu\rho_d K)}\right) \tag{11}$$

The first zero rate of creep takes place when the inequality $\alpha_i C_i^{st} - \alpha_v C_v^e > 0$ is satisfied.
Let us estimate approximate value of t_2 for which the second zero of creep rate is happen. Owing to the concentration of interstitial atoms reaches its stationary concentration much more quickly we can use stationary concentration of interstitial atoms at time t_2. Therefore we can put $C_i \approx C_i^{st}$ and $\exp(-\beta_i t_2) \approx 0$ in (10) and obtain

$$t_2 \approx \frac{1}{\beta_v}\ln\left(K\Big/\left(\beta_v C_v^e + \left(1 - \frac{\alpha_i\beta_v}{\alpha_v\beta_i}\right)K\right)\right) \tag{12}$$

The second zero rate of creep takes place when the next inequalities are satisfied:
1). $\alpha_i C_i^{st} - \alpha_v C_v^e > 0$ (the second minimum takes place only when the first minimum takes place);
2). $\alpha_i C_i^{st} - \alpha_v C_v^{st} < \alpha_v C_v^e$, this condition coincides with the condition for the implementation of the vacancy regime of stationary creep.
Time of appearance of the first and the second dip of dose dependence was estimated for model with next parameters: D_v^0= 3 10^{-5} $m^{-2}s^{-1}$, E_m^v=1.4 eV, C_v^0=1, E_f^v=1.8 eV, z_d=1.08, z_k=1, μ=8 10^{20} m^{-2}, ρ_d=$10^{11}m^{-2}$ and ρ_d+ρ_k=2 $10^{11}m^{-2}$. These parameters are corresponded to results of experiment [8] in which annealed pure nickel (99.96%) was irradiated inside nuclear reactor. Defect generation rate was 2 10^{-8} dpa/s, temperature of the sample was 720K. It was established experimentally that at stress σ=83 MPa the fracture of dose dependence of creep (dip of creep rate) is watched in 900 s after the start of irradiation. When the stress is σ=100 MPa the fracture is watched in 1000 s after beginning of irradiation.
Theoretical estimation of time of the first fracture is very small, about 10^{-13} s. The estimation for the second fracture gives about 1000 s. It well correlates with experimental result. Thus in this case the transient creep is developed due to both vacancy and interstitial mechanisms.
When temperature of irradiated sample grows, the time of the fracture appearance (it means change of creep mechanism) decreases.

4. Conclusions

The proposed model explains the experimentally observed behavior of the transient creep under irradiation. This non-monotonic behavior is connected with change of creep regime: in the beginning of the irradiation the thermal creep with vacancy regime is changed by transient creep with interstitial regime. After that the interstitial regime is again changed by vacancy regime and creep is developed through vacancy regime till stationary stage.
When creep regime changes, contribution of "glide-climb" mechanism is absent and dose dependence of transient creep has fractures or breaks. The analytical expression for rate of the transient creep allows obtaining dependence of peculiarities of creep on conditions of irradiation and properties of the irradiated material.
The measurement of the position of creep break as function of opportune parameter gives the precision method for determination of the material parameters under irradiation.

IOP Conf. Series: Materials Science and Engineering **81** (2015) 012126 doi:10.1088/1757-899X/81/1/012126

5. References

[1] Garner F and Gelles D 1988 *J.Nucl.Mater.* **159** 286

[2] Was G 2011 *Fundamentals of Radiation Materials Science* (Berlin: Springer)

[3] Karasev V 1981 *Voprosy Atomnoy Nauki i Tekhniki. Ser.: Fizika radiatsionnykh povrezhdeniy i radiatsionnoe materialovedenie* **19** 84

[4] Schule W and Hausen H 1994 *J. Nucl. Mater.* **212-215** 388

[5] Selishchev P, Karasev V. and Totskiy A. 1992 *Voprosy Atomnoy Nauki i Tekhniki. Ser.: Fizika radiatsionnykh povrezhdeniy i radiatsionnoe materialovedenie* **58-59** 18

[6] Selishchev P 2003 *Metally* **2** 89

[7] Selishchev P and Sugakov 2000 *Met.Phys.Adv.Tech..* **212-215** 604

[8] Karasev V, Totskiy A 1990 *Voprosy Atomnoy Nauki i Tekhniki. Ser.: Fizika radiatsionnykh povrezhdeniy i radiatsionnoe materialovedenie* **52** 37

[9] Karasev V, Totskiy A and Ozhigov L 1994 *Voprosy Atomnoy Nauki i Tekhniki. Ser.: Fizika radiatsionnykh povrezhdeniy i radiatsionnoe materialovedenie* **61** 3

[10] Toloczko M, Garner F and Eiholzer C 1994 *J.Nucl.Mater.* **212-215** 604

[11] Aitkhozhin E and Chumakov E 1996 *J.Nucl.Mater.* **233-237** 537

[12] Kohyama A, Kohno Y, Asakura K, Yoshino M, Namba C and Eiholzer C 1994 *J.Nucl.Mater.* **212-215** 751

[13] Kurato Y and Nakajima H 1996 *J.Nucl.Mater.* **228** 176

[14] Selishchev P and Sugakov 2000 *Met.Phys.Adv.Tech.* **22** 46

[15] Selishchev P and Sugakov 2013 Fatigue & *Fracture of Engineering Materials & Structures* **36** 1123

AUTHOR INDEX

AUTHOR INDEX

AUTHOR INDEX

AUTHOR INDEX